AF821028

CLYMER® MANUALS

OMC COBRA

STERN DRIVE SHOP MANUAL

1994- 2000

WHAT'S IN YOUR TOOLBOX?

YouTube

More information available at haynes.com

Phone: 805-498-6703

Haynes UK
Sparkford Nr Yeovil
Somerset BA22 7JJ England

Haynes North America, Inc
859 Lawrence Drive
Newbury Park
California 91320 USA

ISBN-10: 1-59969-107-8
ISBN-13: 978-1-59969-107-7
Library of Congress: 2006937773

Author: *Mark Rolling*
Technical Photography: *Mark Rolling*
Technical Illustrations: *Mitzi McCarthy*
Cover: *G. Randall Goss*

B739, 4AA1, 22-574

General Information 1

Troubleshooting 2

Lubrication, Maintenance and Tune-up 3

Lay-Up, Winterizing and Fitting Out 4

Adjustments 5

Engine 6

Fuel System 7

Cooling and Exhaust Systems 8

Electrical System 9

Drive Unit 10

Transom Bracket 11

Power Trim System 12

Steering System 13

Index 14

Wiring Diagrams 15

Common spark plug conditions

NORMAL

Symptoms: Brown to grayish-tan color and slight electrode wear. Correct heat range for engine and operating conditions.
Recommendation: When new spark plugs are installed, replace with plugs of the same heat range.

WORN

Symptoms: Rounded electrodes with a small amount of deposits on the firing end. Normal color. Causes hard starting in damp or cold weather and poor fuel economy.
Recommendation: Plugs have been left in the engine too long. Replace with new plugs of the same heat range. Follow the recommended maintenance schedule.

CARBON DEPOSITS

Symptoms: Dry sooty deposits indicate a rich mixture or weak ignition. Causes misfiring, hard starting and hesitation.
Recommendation: Make sure the plug has the correct heat range. Check for a clogged air filter or problem in the fuel system or engine management system. Also check for ignition system problems.

ASH DEPOSITS

Symptoms: Light brown deposits encrusted on the side or center electrodes or both. Derived from oil and/ or fuel additives. Excessive amounts may mask the spark, causing misfiring and hesitation during acceleration.
Recommendation: If excessive deposits accumulate over a short time or low mileage, install new valve guide seals to prevent seepage of oil into the combustion chambers. Also try changing gasoline brands.

OIL DEPOSITS

Symptoms: Oily coating caused by poor oil control. Oil is leaking past worn valve guides or piston rings into the combustion chamber. Causes hard starting, misfiring and hesitation.
Recommendation: Correct the mechanical condition with necessary repairs and install new plugs.

GAP BRIDGING

Symptoms: Combustion deposits lodge between the electrodes. Heavy deposits accumulate and bridge the electrode gap. The plug ceases to fire, resulting in a dead cylinder.
Recommendation: Locate the faulty plug and remove the deposits from between the electrodes.

TOO HOT

Symptoms: Blistered, white insulator, eroded electrode and absence of deposits. Results in shortened plug life.
Recommendation: Check for the correct plug heat range, over-advanced ignition timing, lean fuel mixture, intake manifold vacuum leaks, sticking valves and insufficient engine cooling.

PREIGNITION

Symptoms: Melted electrodes. Insulators are white, but may be dirty due to misfiring or flying debris in the combustion chamber. Can lead to engine damage.
Recommendation: Check for the correct plug heat range, over-advanced ignition timing, lean fuel mixture, insufficient engine cooling and lack of lubrication.

HIGH SPEED GLAZING

Symptoms: Insulator has yellowish, glazed appearance. Indicates that combustion chamber temperatures have risen suddenly during hard acceleration. Normal deposits melt to form a conductive coating. Causes misfiring at high speeds.
Recommendation: Install new plugs. Consider using a colder plug if driving habits warrant.

DETONATION

Symptoms: Insulators may be cracked or chipped. Improper gap setting techniques can also result in a fractured insulator tip. Can lead to piston damage.
Recommendation: Make sure the fuel anti-knock values meet engine requirements. Use care when setting the gaps on new plugs. Avoid lugging the engine.

MECHANICAL DAMAGE

Symptoms: May be caused by a foreign object in the combustion chamber or the piston striking an incorrect reach (too long) plug. Causes a dead cylinder and could result in piston damage.
Recommendation: Repair the mechanical damage. Remove the foreign object from the engine and/or install the correct reach plug.

Contents

QUICK REFERENCE DATA. IX

CHAPTER ONE
GENERAL INFORMATION 1

Manual organization
Warnings, cautions and notes
Safety
Model identification
Engine operation
Fasteners
Shop supplies
Galvanic corrosion
Propellers
Basic tools
Special tools
Precision measuring tools
Electrical system fundamentals
Basic mechanical skills
Specifications

CHAPTER TWO
TROUBLESHOOTING 37

Preliminary inspection
Electric starting system
Charging system
Ignition system
Basic fuel system troubleshooting
Electronic fuel injection (EFI) system
Diagnostic trouble codes
Power steering system
Power trim system
Cooling system troubleshooting
Warning system
Engine
Drive system
Specifications

CHAPTER THREE
LUBRICATION, MAINTENANCE AND TUNE-UP 137

Before each use
After each use
Lubrication
Fuel system maintenance
Engine coolant
Tune-up
Submersion
Specifications

CHAPTER FOUR
LAY-UP, WINTERIZING AND FITTING OUT. 165

Lay-up
Winterizing
Fitting out

CHAPTER FIVE
ADJUSTMENTS 174
Carburetor adjustment
Ignition timing adjustment
Shift cable adjustment
Throttle cable adjustment
Trim position sender and trim limit control module adjustment
Specifications

CHAPTER SIX
ENGINE 192
Engine operation
Special tools
Replacement parts
Special service precautions
Engine removal
Engine installation
Engine alignment
Rocker arm cover
Valve adjustment
Placing the No. 1 piston at TDC (firing position)
Lifter cover
Measuring camshaft lobe lift
Cylinder head
Valves and valve seats
Rocker arm and pushrod
Flywheel housing, drive coupler and flywheel
Oil pan
Oil pump
Harmonic balancer
Timing gear or chain cover and oil seal
Timing chain and sprockets
Balance shaft
Camshaft and lifters
Piston and connecting rod
Crankshaft and rear main seal
Break-in procedure
Specifications

CHAPTER SEVEN
FUEL SYSTEM 272
Servicing the fuel system
Mechanical low-pressure fuel pump
Low-pressure electric fuel pump
Flame arrestor and cover
Carburetor
Electronic fuel injection (EFI)
Throttle body
Intake plenum, fuel rail and injectors
Fuel pressure regulator
High-pressure electric fuel pump
Fuel reservoir
Intake manifold
Specifications

CHAPTER EIGHT
COOLING AND EXHAUST SYSTEMS 329
Thermostat
Hose replacement
Recirculation pump replacement
Seawater pump
Oil and power steering fluid coolers (replacement)
Exhaust elbow
Exhaust bellows, exhaust tube and flapper valve
Exhaust manifold
Specifications

CHAPTER NINE
ELECTRICAL SYSTEM 360
Battery
Alternator
Starter and starter solenoid
Ignition system
Starter slave solenoid
Starter relay
Fuel pump relay
Electronic fuel injection (EFI) system relay
Oil pressure sender
Oil pressure switch
Coolant temperature sender
Coolant temperature sensor
Engine overheat switch
Manifold air pressure sensor
Intake air temperature (IAT) sensor
Throttle position sensor (TPS)
Idle air control motor (IAC)
Idle speed control solenoid (ISC)
Knock sensor (KS)
Knock sensor module
Knock sensor amplifier
Engine control unit (ECU)
Engine control assembly (ECA)
Specifications

CHAPTER TEN
DRIVE UNIT. 415

Drive unit identification
Drive unit operation
Drive unit gear ratio
High altitude operation
Outdrive service precautions
Sacrificial anodes
Shifting and propeller rotation
Propeller
Drive unit
Universal joint service
Shift housing service
Lower gearcase
Drive unit pressure and vacuum tests
Specifications

CHAPTER ELEVEN
TRANSOM BRACKET 489

Pivot housing
Bellows and water hose
Gimbal bearing and seal
Gimbal ring
Inner transom plate and bracket housing
Specifications

CHAPTER TWELVE
POWER TRIM SYSTEM 505

Electrical trim system components
Hydraulic trim system components
Trim cylinder and hoses
Trim position sender
Filling and bleeding the trim system
Specifications

CHAPTER THIRTEEN
STEERING SYSTEM. 527

Safety precautions
Mechanical (manual) steering system
Power steering system
Specifications

INDEX 541

WIRING DIAGRAMS. 545

Quick Reference Data

MODEL:________________________________YEAR:________________
VIN NUMBER:____
ENGINE SERIAL NUMBER:______________________________________
CARBURETOR SERIAL NUMBER OR IDENTIFICATION MARK:_____________

MAINTENANCE TORQUE SPECIFICATIONS

Fastener	in.-lb.	ft.-lb.	N•m
Drive unit dipstick	48-72	–	5.4-8.1
Drive unit lubricant vent plug	48-72	–	5.4-8.1
Drive unit drain/fill plug	60-84	–	6.8-9.5
Spark plugs			
GM 3.0L models	–	20	27
GM 4.3L models			
Used cylinder head	–	15	20
New cylinder head	–	22	30
GM 5.0L and 5.7L models			
Used cylinder head	–	15	20
New cylinder head	–	22	30
Ford 5.0L and 5.8L models	120-180	–	14-20
GM 7.4L and 8.2L models	–	22	30
Rear drive cover (plastic) screws	108-132	–	12.2-14.9

RECOMMENDED LUBRICANTS

Lubricant	Application
SAE 30 NMMA-certified FC-W marine-use motor oil*	Engine crankcase–all models
SAE 30 motor oil: GM standard GM-6094-M*	Crankcase–GM engines
SAE 30 motor oil: Ford specification ESE-M2X153*	Crankcase–Ford engines
Mobil 1 SHC 75-90 fully synthetic gearcase lubricant	Drive unit
Volvo Penta Dura Plus synthetic gearcase lubricant	Drive unit
Ethylene glycol-based engine coolant	Closed cooling system
Dexron II automatic transmission fluid (ATF)	Power steering and power trim pumps
OMC power trim/tilt fluid	Power steering and power trim pumps
Volvo Penta power trim/tilt fluid (part No. 3851930)	Power steering and power trim pumps
Quicksilver power trim/tilt fluid (part No. 92-90199A21)	Power steering and power trim pumps
OMC EP/Wheel Bearing Grease	Gimbal bearing
OMC Moly Lube (part No. 175356)	Engine coupling splines
OMC Triple Guard Grease	Steering cable ram
Marine-grade water-resistant grease	Steering cable ram
Marine-grade fuel stabilizer	Fuel tank
OMC fuel conditioner and stabilizer (part No. 775614)	Fuel tank
Quicksilver fuel conditioner (part No. 92-858072KO1)	Fuel tank

(continued)

RECOMMENDED LUBRICANTS (continued)

Lubricant	Application
Marine fogging oil	Storage fogging agent
Quicksilver Storage Seal (part No. 92-858081KO1)	Storage fogging agent
*Recommended oil viscosity for operation in ambient air temperature exceeding 32° F (0° C). Refer to text for oil viscosity recommendation if operating engine at lower air temperatures. Refer to Chapter Three.	

ENGINE OIL AND DRIVE UNIT LUBRICANT CAPACITIES

Model	Capacity (approximate)
Engine oil capacity	
GM 3.0L models	
With oil filter change	4.0 qt. (3.8 L)
Without oil filter change	3.5 qt. (3.3 L)
GM 4.3L models	
With oil filter change	4.5 qt. (4.3 L)
Without oil filter change	4.0 qt. (3.8 L)
GM 5.0L models	
With oil filter change	6.0 qt. (5.7 L)
Without oil filter change	5.0 qt. (4.7 L)
Ford 5.0L models	
With oil filter change	6.0 qt. (5.7 L)
Without oil filter change	5.0 qt. (4.7 L)
GM 5.7L models	
57GL models	
With oil filter change	5.0 qt. (4.7 L)
Without oil filter change	4.0 qt. (3.8 L)
57GI and 57GSI models	
With oil filter change	6.0 qt. (5.7 L)
Without oil filter change	5.0 qt. (4.7 L)
Ford 5.8L models	
58FL (carburetor-equipped) models	
With oil filter change	6.0 qt. (5.7 L)
Without oil filter change	5.0 qt. (4.7 L)
58FI and 58FSI (EFI) models	
With oil filter change	6.0 qt. (5.7 L)
Without oil filter change	5.0 qt. (4.7 L)
GM 7.4L models	
74GL models	
With oil filter change	7.0 qt. (6.6 L)
Without oil filter change	6.0 qt. (5.7 L)
74GI and 74GSI models	
With oil filter change	9.0 qt. (8.5 L)
Without oil filter change	8.0 qt. (7.6 L)
GM 8.2L models	
With oil filter change	7.0 qt. (6.6 L)
Without oil filter change	6.0 qt. (5.7 L)
Drive unit lubricant capacity	
Cobra SX (single propeller)	71 oz. (2.1 L)
Cobra DP-S DuoProp (twin propeller)	81 oz. (2.4 L)

ENGINE OPERATING SPEED SPECIFICATIONS

Model	Idle speed (rpm)	Wide open throttle speed (rpm)
GM 3.0L models	650-750	4200-4600
GM 4.3L models		
43GL (carburetor-equipped) models	550-650	4200-4600

(continued)

ENGINE OPERATING SPEED SPECIFICATIONS (continued)

Model	Idle speed (rpm)	Wide open throttle speed (rpm)
GM 4.3L models (continued)		
43GS (carburetor-equipped) models	550-650	4200-4800
43GI (EFI) models	600*	4400-4800
GM 5.0L models		
50GL (carburetor-equipped) models	550-650	4400-4800
50GI (EFI) models	600*	4600-5000
Ford 5.0L models		
50FL (carburetor-equipped) models	550	4200-4600
50FI (EFI) models	600*	4200-4600
5.7L models		
57GL (EFI) models	600-650	4200-4600
57GS (EFI) models	550-650	4400-4800
57GI (EFI) models	600*	4200-4600
57GSI (EFI) models	600*	4600-5000
Ford 5.8L models		
58FL (carburetor equipped) models	550-650	4000-4400
58FI (EFI) models	600*	4200-4600
58FSI (EFI) models	600*	4600-5000
GM 7.4L models		
74GL (carburetor-equipped) models	550-650	4200-4600
74GI (EFI) models	600*	4200-4600
74GSI (EFI) models	600*	4800-5200
GM 8.2L models		
82GL (carburetor-equipped) models	550-650	4400-4800
82GSI (EFI) models	600*	4600-5000

*The idle speed is nonadjustable on this model.

SPARK PLUG SPECIFICATIONS

Model	Spark plug No.	Gap specification
GM 3.0L models		
AC plug	MR43LTS	
Champion plug	RS12YC	
NGK plug	TR5	
Plug gap	–	0.045 in. (1.14 mm)
GM 4.3L models		
1994-1996 (with 12-bolt intake)		
43GL and 43GS models		
AC plug	MR43T	
NGK plug	BR6FS	
Plug gap	–	0.045 in. (1.14 mm)
43GI models		
AC plug	R43TS	
NGK plug	BPR6FS	
Plug gap	–	0.045 in. (1.14 mm)
1997-2000 (with 8-bolt intake)		
AC plug	MR43LTS	
Champion plug	RS12YC	
NGK plug	BPR6EFS	
Plug gap	–	0.045 in. (1.14 mm)
GM 5.0L models		
1994-1997 (with 12-bolt intake)		
AC plug	MR43T	
NGK plug	BR6FS	
Plug gap	–	0.045 in. (1.14 mm)
1998-2000 (with 8-bolt intake)		
50GL and 50GS (carburetor-equipped)		
AC plug	MR43LTS	

(continued)

SPARK PLUG SPECIFICATIONS (continued)

Model	Spark plug No.	Gap specification
GM 5.0L models		
1998-2000 (with 8-bolt intake) (continued)		
50GL and 50GS (carburetor-equipped)		
Champion plug	RS12YC	
NGK plug	BPR6EFS	
Plug gap	–	0.035 in. (0.89 mm)
50GI (EFI)		
Champion plug	RS12YC	
NGK plug	BPR6EFS	
Plug gap	–	0.045 in. (1.14 mm)
Ford 5.0L models		
50FL (carburetor-equipped)		
AC plug	MR43T	
Motorcraft plug	ASF32C	
NGK plug	UR4	
Plug gap	–	0.035 in. (0.89 mm)
50FI (EFI)		
AC plug	MR43T	
Motorcraft plug	AWSF32C	
NGK plug	TR5	
Plug gap	–	0.045 in. (1.14 mm)
GM 5.7L models		
1994-1997 (with 12-bolt intake)		
57GL (carburetor-equipped)		
AC plug	MR43T	
NGK plug	BR6FS	
Plug gap	–	0.045 in. (1.14 mm)
57GI (EFI)		
AC plug	R43TS	
NGK plug	BPR6FS	
Plug gap	–	0.045 in. (1.14 mm)
1998-2000 (with 8-bolt intake)		
57GS (carburetor-equipped) models		
AC plug	MR43LTS	
Champion plug	RS12YC	
NGK plug	BPR6EFS	
Plug gap	–	0.035 in. (0.89 mm)
57GSI (EFI) models		
AC plug	MR43LTS	
Champion plug	RS12YC	
NGK plug	BPR6EFS	
Plug gap	–	0.045 in. (1.14 mm)
Ford 5.8L models		
58FL (carburetor-equipped) models		
AC plug	MR43T	
Motorcraft plug	ASF32C	
NGK plug	UR4	
Plug gap	–	0.035 in. (0.89 mm)
58FI (EFI) models		
AC plug	MR43T	
Motorcraft plug	AWSF32C	
NGK plug	BPR6EFS	
Plug gap	–	0.045 in. (1.14 mm)
GM 7.4L models		
1994-1997		
74GL (carburetor-equipped) models		
AC plug	MR43T	
NGK plug	BR6FS	
Plug gap	–	0.045 in. (1.14 mm)
74GI (EFI) models		
AC plug	R43TS	
NGK plug	BPR6FS	
Plug gap	–	0.045 in. (1.14 mm)

(continued)

SPARK PLUG SPECIFICATIONS (continued)

Model	Spark plug No.	Gap specification
GM 7.4L models (continued)		
1998-2000		
74GI (EFI) models		
AC plug	MR43LTS	
Champion plug	RS12YC	
NGK plug	BPR6EFS	
Plug gap	–	0.045 in. (1.14 mm)
74GSI (EFI) HP models		
AC plug	MR43T	
NGK plug	BR6FS	
Plug gap	–	0.045 in. (1.14 mm)
GM 8.2L models		
AC plug	MR43T	
NGK plug	BR6FS	
Plug gap	–	0.045 in. (1.14 mm)

Chapter One

General Information

This detailed and comprehensive manual covers OMC stern drive engines from 1994-2000.

This manual can be used by anyone from first-time do-it-yourselfers to professional mechanics. The text provides step-by-step information on maintenance, tune-up, repair and overhaul. Hundreds of photos and drawings guide the reader through every job.

A shop manual is a reference that should be used to find information quickly. Clymer manuals are designed with this in mind. All chapters are thumb-tabbed, and important items are indexed at the end of the manual. All procedures, tables, photos and instructions in this manual are designed for the reader working on the machine or using the manual for the first time.

Keep the manual in a handy place such as a toolbox or in the boat. It will increase understanding of how the boat runs, lower repair and maintenance cost and generally increase enjoyment of the boat.

Frequently used specifications and capacities from individual chapters are summarized in the *Quick Reference Data* at the front of the manual. Specifications concerning a specific system are included in the tables at the end of each chapter. **Tables 1-6** at the end of this chapter list technical abbreviations, conversion formulas, U.S. standard tap and drill sizes, metric tap and drill sizes, general torque recommendations and engine identification codes.

MANUAL ORGANIZATION

All dimensions and capacities are expressed in U.S. standard and metric units of measure.

This chapter provides general information on shop safety, tool use, service fundamentals and shop supplies.

Chapter Two provides methods and suggestions for quick and accurate diagnosis of engine trouble. The troubleshooting procedures describe typical symptoms and provide logical troubleshooting methods.

Chapter Three explains all periodic lubrication and routine maintenance necessary to keep the stern drive engine operating at peak efficiency. Chapter Three also includes all recommended tune-up procedures, which eliminate the need to refer to other chapters on the various systems.

Chapter Four describes the procedures for preparing the stern drive engine for storage. Preparation for operation after the lay-up period is also described in this chapter.

Chapter Five describes adjustments for the fuel, shifting and power trim systems.

Subsequent chapters describe specific systems and provide disassembly, repair and assembly procedures in simple step-by-step form.

Some of the procedures in this manual specify special tools. When possible, the tool is illustrated in use. Well-equipped mechanics may substitute similar tools or fabricate a suitable replacement. However, in some cases, the specialized equipment or expertise might make it impractical for the home mechanic to attempt the procedure. When necessary, such operations are identified in the text with the recommendation to have a dealership or specialist perform the task. It may be less expensive to have a professional perform these jobs, especially when considering the cost of the equipment. This is true regarding machine work for engine rebuilds because machinists spend years perfecting their trade, and even professional mechanics often rely upon their services.

WARNINGS, CAUTIONS AND NOTES

The terms WARNING, CAUTION, and NOTE have specific meanings in this manual.

A WARNING emphasizes areas where personal injury or even death could result from negligence. Mechanical damage may also occur. WARNINGS *are to be taken seriously.*

A CAUTION emphasizes areas where equipment damage could occur. Disregarding a CAUTION could cause permanent mechanical damage; however, personal injury is unlikely.

A NOTE provides additional information to make a step or procedure easier or clearer. Disregarding a NOTE could cause inconvenience or misdiagnosis but would not cause damage or injury.

SAFETY

Professional mechanics can work for years and never sustain a serious injury. Follow these guidelines and practice common sense to safely service the stern drive engine.

1. Do not operate the stern drive engine in an enclosed area. The exhaust gasses contain carbon monoxide, an odorless, colorless, tasteless and poisonous gas. Carbon monoxide levels build quickly in small enclosed areas and can cause unconsciousness and death in a short time. Make sure to properly ventilate the work area or operate the stern drive engine outside.
2. *Never* use gasoline or any extremely flammable liquid to clean parts. Refer to *Handling Gasoline Safely* and *Cleaning Parts* in this chapter.
3. *Never* smoke or use a torch in the vicinity of flammable liquids such as gasoline or cleaning solvent.
4. After removing the engine cover, allow the engine compartment to air out before performing any service work.
5. Use the correct type and size of tools to avoid damaging fasteners.

6. Keep tools clean and in good condition. Replace or repair worn or damaged equipment.
7. When loosening a tight or stuck fastener, always consider what would happen if the wrench should slip. In most cases, it is safer to pull on a wrench or ratchet than it would be to push on it. Be careful and protect yourself accordingly.
8. When replacing a fastener, make sure to use one with the same measurements, material and strength as the old one. Refer to *Fasteners* in this chapter for additional information.
9. Keep the work area clean and uncluttered. Keep all hand and power tools in good condition. Clean grease or oil from tools after using them. Unkept tools are difficult to hold and can cause injury. Replace or repair worn or damaged tools. Do not leave tools, shop rags or anything that does not belong in the hull.
10. Wear safety goggles during all operations involving drilling, grinding or the use of a cold chisel or *anytime* the eye safety is in question (when debris may spray or scatter). *Always* wear safety goggles when using solvent or compressed air.
11. Do not carry sharp tools in clothing pockets.
12. Always have an approved fire extinguisher available. Make sure it is rated for gasoline (Class B) and electric (Class C) fires. Read and fully understand the operating instructions for the fire extinguisher before beginning the work.
13. Do not use compressed air to clean clothes, the boat/engine or the work area. Debris might blow into eyes or skin. *Never* direct compressed air at anyone. Do not allow children to use or play with any compressed air equipment.
14. When using compressed air to dry rotating parts, hold the park so that it cannot rotate. The air jet is capable or rotating the parts at extremely high speed. The part may become damaged or disintegrate and cause serious injury.

Handling Gasoline Safely

Gasoline is a volatile flammable liquid and is one of the most dangerous items in the shop.

2

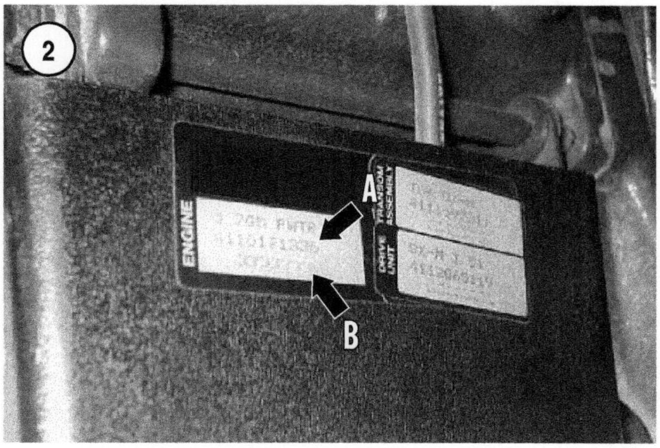

Because gasoline is used so often, many people forget that it is hazardous. Only use gasoline as fuel for gasoline internal combustion engines. Do not use it as a cleaner or degreaser. When working, keep in mind gasoline is always present in the fuel tank, fuel lines and carburetor or fuel rail. To avoid a disastrous accident when working around the fuel system, carefully observe the following precautions.

1. *Never* use gasoline to clean parts. Refer to *Cleaning Parts* in this chapter.
2. When working on the fuel system, work outside or in a well-ventilated area.
3. Do not add fuel to the tank or service the fuel system while near open flames, sparks or where someone is smoking. Gasoline vapor is heavier than air. It collects in low areas and is more easily ignitable than liquid gasoline.
4. Allow the engine to cool completely before working on any fuel system component.
5. When draining the carburetor, fuel rail or fuel lines, catch the fuel in a suitable plastic container. Then pour it into an approved gasoline storage device.
6. Do not store gasoline in a glass container. If the glass breaks, a serious explosion or fire can occur.
7. Immediately wipe up spilled gasoline with suitable shop towels. Store the towels in a metal container with a lid until they can be properly disposed of, or place them outside in a safe place so the fuel can evaporate.
8. Do not pour water onto a gasoline fire. Water spreads the fire and makes it more difficult to extinguish. Use a class B, BC or ABC fire extinguisher to extinguish the fire.
9. Always turn off the engine before refueling. Do not spill fuel onto the engine components. Do not overfill the fuel tank. Leave an air space at the top of the tank to allow the fuel to expand.

Cleaning Parts

Cleaning parts is one the more tedious and difficult service jobs performed in the home garage. There are many types of chemical cleaners and solvents available for shop use. Most are poisonous and extremely flammable. To prevent chemical exposure, vapor buildup, fire and serious injury, observe each product warning label and note the following:

1. Read the entire product label before using any chemical. Always know what type of chemical is being used and whether it is poisonous and/or flammable.
2. Do not use more than one type of cleaning solvent at a time. If mixing chemicals is necessary, measure the proper amounts according to the manufacturer.
3. Work in a well-ventilated area.
4. Wear chemical-resistant gloves.
5. Wear safety glasses.
6. Wear a vapor respirator if the instructions call for it.
7. Wash hands and arms thoroughly after cleaning parts.
8. Keep chemical products away from children and pets.
9. Thoroughly clean all oil, grease and cleaner residue from any parts that must be heated.
10. Use a nylon brush when cleaning parts. Metal brushes can cause a spark.
11. When using a parts washer, use only the solvent recommended by the equipment manufacturer. Make sure the parts washer is equipped with a metal lid that lowers if a fire occurs.

MODEL IDENTIFICATION

The Cobra SX drive system consists of three major assemblies: the engine, transom bracket and drive unit. Before servicing, troubleshooting or ordering parts for the engine, always identify the model number and serial number of these assemblies.

Engine Identification

Engine model name decal

A decal on the flame arrestor cover (**Figure 1**) or the plastic throttle body/intake cover on 5.0L and 5.8L Ford EFI models identifies the engine model name. The numbers on the decal (A, **Figure 1**) indicate the engine displacement in liters and tenths of a liter. The first letter following the numbers (B, **Figure 1**) indicates the source of the base engine. The letter *G* is used for General Motors (GM) base engines, and the letter *F* is used for Ford base engines. The letter following the *G* or *F* (C, **Figure 1**) identifies the version of the model. An *I* indicates the engine uses an electronic fuel injection (EFI) system. An *L* indicates the engine is the lower output version of the indicated displacement. An *S* indicates the standard or higher output version of the indicated displacement.

Engine identification tag

The engine identification tag (**Figure 2**, typical) is located on the rocker arm cover, fuse and circuit breaker housing, or ignition system mounting bracket (Ford EFI models). This tag includes the model identification number and engine serial number.

The model identification number (A, **Figure 2**) consists of nine characters. The first two numbers indicate the engine displacement in liters and tenths of a liter. For example, the number 57 would indicate the engine displaces 5.7 liters.

The character following the displacement identification numbers indicates the type of fuel system used on the engine. The fourth character indicates the engine version and is used to identify models produced with lower or higher power output for the given displacement. The fifth character is a letter that indicates whether the engine is produced as an OMC Cobra SX engine or Volvo Penta SX engine. The sixth character indicates the type of steering system on the engine. The seventh and eighth characters are letters that indicate the model year of the engine. The final character is a letter that indicates the production version for the given model year. The production version always starts with an A for each model year. The letter changes to B, C and so on as changes occur during the production year. The model year version letter is very important when ordering parts and servicing the engine.

Before servicing the engine or ordering replacement parts, compare the model identification number with the information in **Table 6** to correctly identify the engine.

The serial number on the identification tag (B, **Figure 2**) is unique to the engine and identifies the engine from other engines of the same model. Some states require this number for registration and title of the vessel. The serial number might also be required when ordering parts and servicing the engine.

Transom bracket identification tag

The transom bracket identification tag (**Figure 3**, typical) is affixed to the inner transom plate and above the steering cable connection point. The tag lists the seven-digit transom bracket model number (A, **Figure 3**) and the serial number (B) unique to the transom bracket. Some states require the number for registration and title of the vessel. The serial number might also be required when ordering parts and servicing the transom bracket. A decal listing the information on the transom bracket identification tag may be located next to the engine identification tag. The decal is supplied with the transom plate and is applied next to the engine tag during the original engine installation. This tag is useful when boat structure limits access to the inner transom plate.

Drive unit identification tag

The drive unit identification tag is riveted to the upper port side of the drive unit (**Figure 4**). The tag lists the drive unit model number (A, **Figure 4**), overall gear ratio (B) and serial number (C) unique to the drive unit. Some states re-

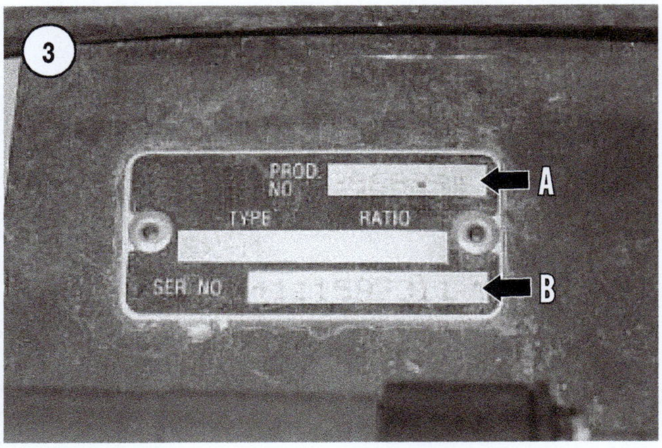

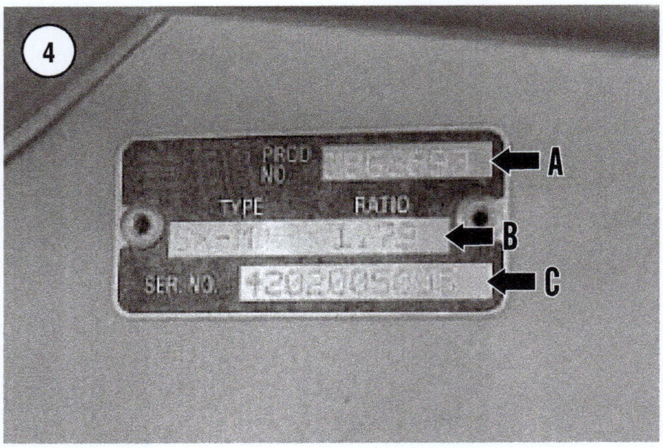

quire this number for registration and title of the vessel. All of the information on the tag must be supplied when ordering replacement parts for the drive unit.

NOTE
The manufacturer might make changes during a model year. When ordering parts, always order by model and serial number. For reference, record the model and serial number in the Quick Reference Data (QRD) section at the front of the manual. Compare new parts with old before purchasing them. If the parts are not alike, have the parts manager explain the difference and verify compatibility before installing the part.

ENGINE OPERATION

All marine engines, whether two- or four-stroke, gasoline or diesel, operate on the Otto cycle of intake, compression, power and exhaust phases. The models covered in this manual are a four-stroke design. **Figure 5** shows typical four-stroke gasoline engine operation.

FASTENERS

Proper fastener selection and installation is important to make sure that the stern drive engine operates as designed

(5) **TYPICAL FOUR-STROKE ENGINE OPERATION**

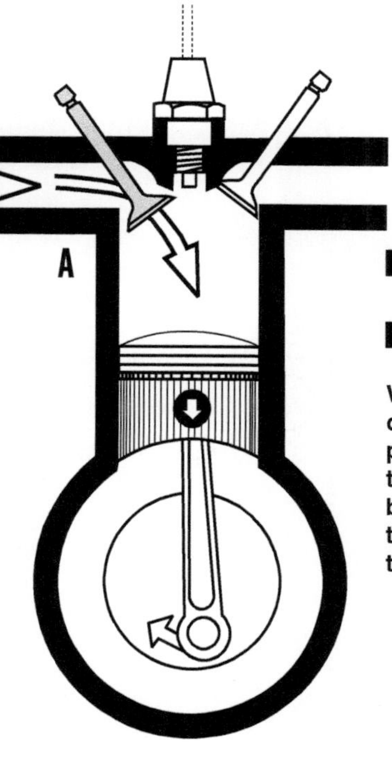

A

As the piston travels downward on the intake stroke, the exhaust valve is closed, and the intake valve opens and allows the fresh air/fuel mixture to be drawn into the cylinder. When the piston reaches the bottom of its travel (BDC), the intake valve closes and remains closed for the next one and one-half revolutions of the crankshaft.

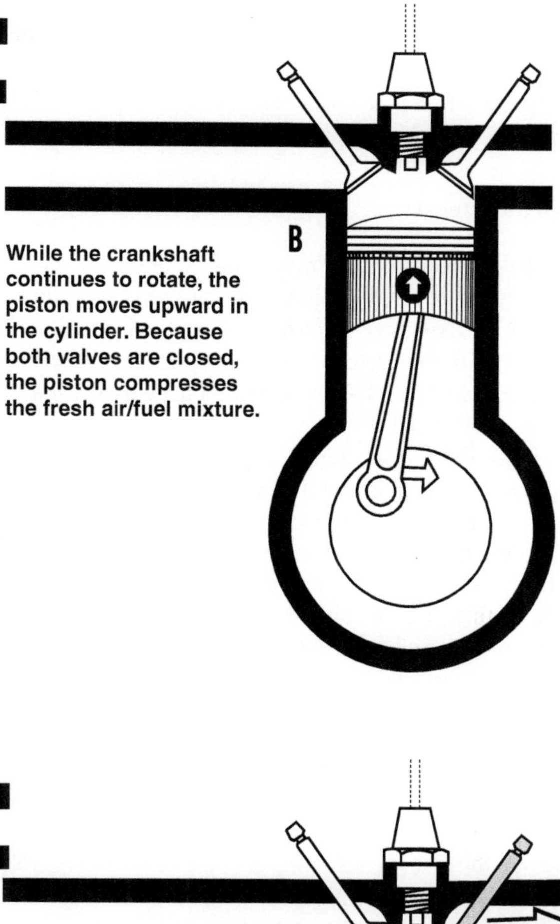

B

While the crankshaft continues to rotate, the piston moves upward in the cylinder. Because both valves are closed, the piston compresses the fresh air/fuel mixture.

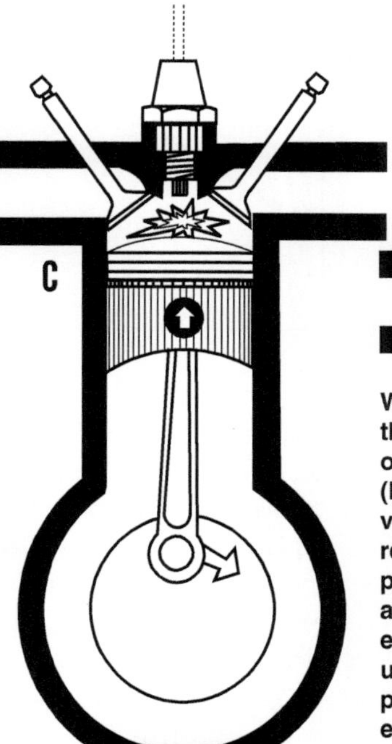

C

When the piston nears the top of its travel, the spark plug fires and ignites the compressed air/fuel mixture to start the power phase. Due to crankshaft rotation, the piston continues movement to top dead center (TDC) and is then pushed downward by expanding gasses formed by the burning fuel.

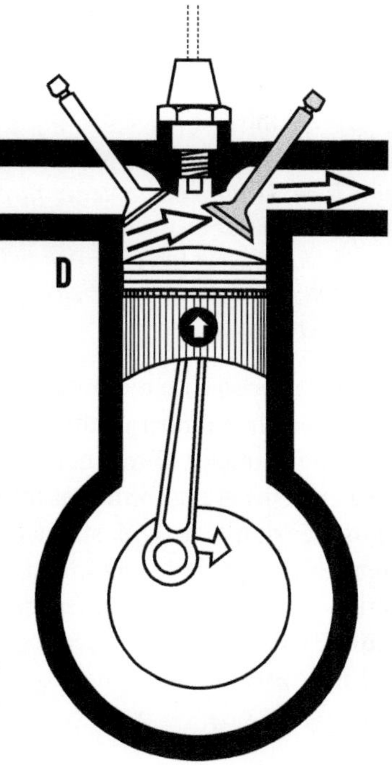

D

When the piston nears the bottom of its travel on the power stroke (BDC), the exhaust valve opens and remains open until the piston moves upward and nears the top of its exhaust stroke. The upward travel of the piston forces the exhaust gasses out of the cylinder. The exhaust valve closes to end the exhaust phase, and the cycle starts all over again.

and can be serviced efficiently. The choice of original equipment fasteners is not arrived at by chance. Make sure replacement fasteners meet all of the same requirements as the originals.

CAUTION
Stern drive engines commonly use fasteners made of stainless steel or other corrosion-resistant material in water-exposed locations. Never install a common steel fastener in a water-exposed location. The fastener will corrode, fail and cause damage or loss of engine components.

Threaded Fasteners

Threaded fasteners secure most of the components on the boat and engine. Most are tightened by turning them clockwise (meaning they have right-hand threads). If the clockwise rotation of the fastener loosens it, the fastener has left-hand threads. If a left-hand threaded fastener is used, it is noted in the manual.

Nuts, bolts and screws are manufactured in a wide range of thread patterns. To join a nut and bolt, the diameter of the bolt and the diameter of the hole in the nut must be the same, and the threads must match properly.

The best way to tell whether the threads on two fasteners match is to turn the nut on the bolt (or the bolt into the threaded hole in a piece of equipment) with fingers only. Make sure both pieces are clean; remove Loctite or other sealer residue from the threads. If force is required, check the thread conditions on each fastener. If the thread condition is good but the fasteners jam, the threads are not compatible. A thread pitch gauge (**Figure 6**) can also determine pitch.

NOTE
To make sure the fastener threads are not mismatched or cross-threaded, start all fasteners by hand. If a fastener is hard to start or turn, determine the cause before tightening it with a wrench.

Two dimensions are required to match the thread size of the fastener: the number of threads in a given distance and the outside diameter of the threads.

Two systems are currently used to specify threaded fastener dimensions: the U.S. standard system and the metric system. Although fasteners might appear similar, close inspection shows that the thread designs are not the same (**Figure 7**). Pay particular attention when working with unidentified fasteners; mismatching thread types can damage threads.

NOTE
OMC Cobra engines and stern drive units use both U.S. standard and International Organization for Standardization (ISO) metric fasteners.

6

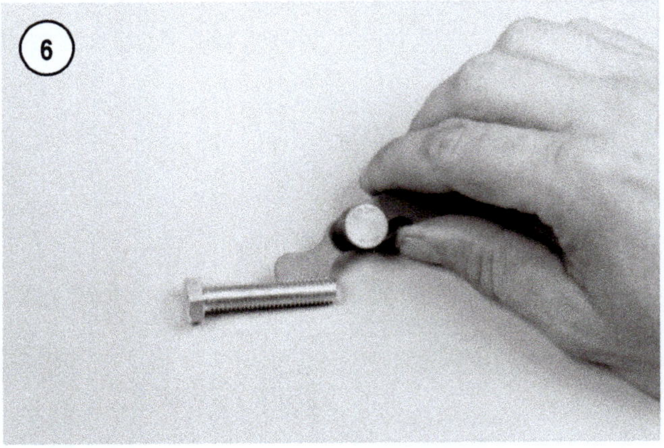

7

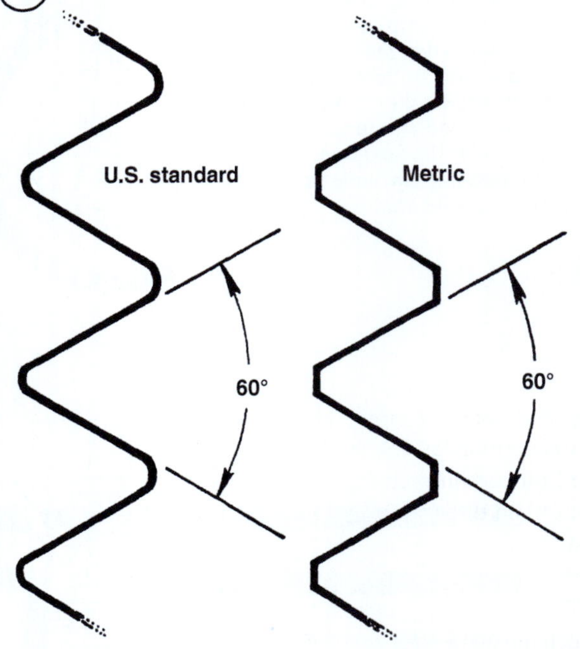

U.S. standard fasteners are sorted by grades (hardness/strength). Bolt heads are marked to represent different grades. No marks means the bolt is grade zero, two marks equal grade two, three marks equal grade five, four marks equal grade six, five marks equal grade seven, and six marks equal grade eight. It is important when replacing fasteners to make sure the replacements are of equal or greater strength than the original.

U.S. standard fasteners generally come in two pitches: coarse and fine. The coarse fasteners have fewer threads per inch than the fine. They are normally referred to by size, such as 1/2-16 or 3/8-24. In these names, the first number, 1/2 or 3/8 in the example, represents the measurement of the bolt diameter from the top of the threads to the top of the other side. The second number represents the number of threads per inch (16 or 24 in the case of the example).

International Organization for Standardization (ISO) metric threads come in three standard thread sizes: coarse,

8

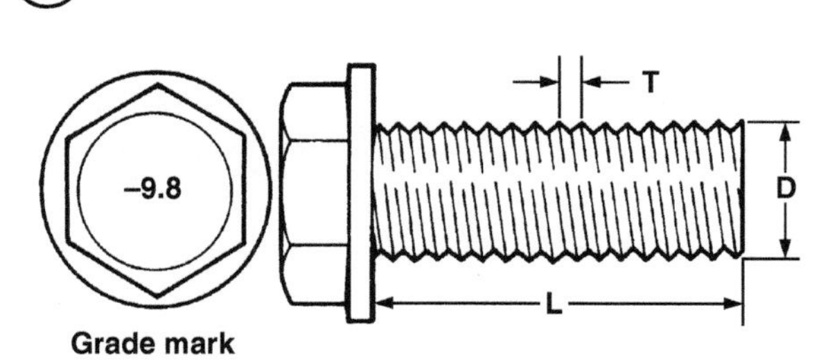

9

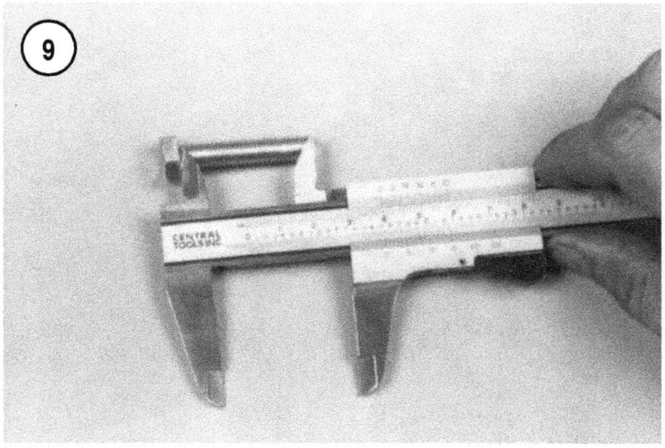

fine and constant pitch. The ISO coarse pitch is used for most common fastener applications. The fine pitch thread is used on certain precision tools and instruments. The constant pitch thread is used mainly on machine parts and not fasteners. The constant pitch thread, however, is used on all metric thread spark plugs.

The length (L, **Figure 8**), diameter (D) and distance between thread pitch crest (T) classify metric screws and bolts. The numbers 8-1.25 × 130 identify a typical bolt. The numbers indicate the bolt has a diameter of 8 mm. The distance between thread crests is 1.25 mm, and the length is 130 mm.

NOTE

*When purchasing a bolt, it is important to know how to specify bolt length. The correct way to measure bolt length is to measure the length, starting from underneath the bolt head to the end of the bolt (**Figure 9**). Always measure bolt length in this manner to avoid purchasing or installing bolts that are too long.*

The grade mark located on the top of the fastener (**Figure 8**) indicates the strength of metric screws and bolts. The higher the number, the stronger the fastener. Unnumbered fasteners are the weakest.

Many screws, bolts and studs are combined with nuts to secure particular components. To indicate the size of a nut, manufacturers specify the internal diameter and the thread pitch.

The measurement across two flats on a nut or bolt indicates the wrench size.

CAUTION

Do not install fasteners with a strength classification lower than what was originally installed by the manufacturer. Doing so may cause equipment failure and/or damage.

Torque Specification

The material used during the manufacturing of the engine might be subjected to uneven stresses if the fasteners of the various subassemblies are not installed and tightened correctly. Fasteners that are improperly installed or work loose can cause extensive damage. It is essential to use an accurate torque wrench, described in this chapter, with the torque specifications in this manual. If a torque is not listed, use the general torque recommendation in **Table 5** of this chapter.

Specifications for torque are provided in foot-pounds (ft.-lb.), inch-pounds (in.-lb.) and Newton-meters (N•m). Torque specifications for specific components (including all critical torque figures) are at the ends of the appropriate chapters. Torque wrenches are covered in the *Basic Tools* section of this chapter.

Self-Locking Fasteners

Several types of bolts, screws and nuts incorporate a system that creates interference between the two fasteners. Interference is achieved in various ways. The most common type is the nylon insert nut and a dry adhesive coating on the threads of a bolt.

Compared to standard fasteners, self-locking fasteners offer greater holding strength, which improves their resistance to vibration. Most self-locking fasteners cannot be reused. The material used to form the lock becomes distorted after the initial installation and removal. Discard and replace self-locking fasteners after removal. Do not replace self-locking fasteners with standard fasteners.

Washers

There are two basic types of washers: flat washers and lockwashers. Flat washers are simple discs with a hole to fit a screw or bolt. Lockwashers prevent fasteners from working loose. Washers may be used as spacers and seals or to help distribute fastener load and to prevent the fastener from damaging the component.

As with fasteners, when replacing washers, make sure the replacement washers are of the same design and quality as those originally installed by the manufacturer.

NOTE

Give as much care to the selection and purchase of washers as given to bolts, nuts and other fasteners. Avoid washers that are made of thin and weak material. These will deform and crush the first time they are used in a high-torque application and allow the nut or bolt to loosen.

Cotter Pins

A cotter pin is a split metal pin inserted into a hole or slot to prevent a fastener from loosening. In certain applications, the fastener must be secured in this way. For these applications, a cotter pin and castellated (slotted) nut are used.

To use a cotter pin, first make sure the diameter is correct for the hole in the fastener. After correctly tightening the fastener and aligning the holes, insert the cotter pin through the hole and bend the ends over the fastener (**Figure 10**). Cut the ends to a suitable length to prevent them from snagging on clothing or skin; remember that exposed ends of the pin cut flesh easily. When the cotter pin is bent and the ends are cut to length, the pin must be tight. If it can be wiggled, it is improperly installed.

Unless instructed to do so, never loosen a fastener to align the holes. If the holes do not align, tighten the fastener just enough to achieve alignment.

Cotter pins are available in various diameters and lengths. Measure length from the bottom of the head to the tip of the shorter end.

Do not reuse cotter pins because the ends can break and cause the pin to fall out and the fastener to loosen.

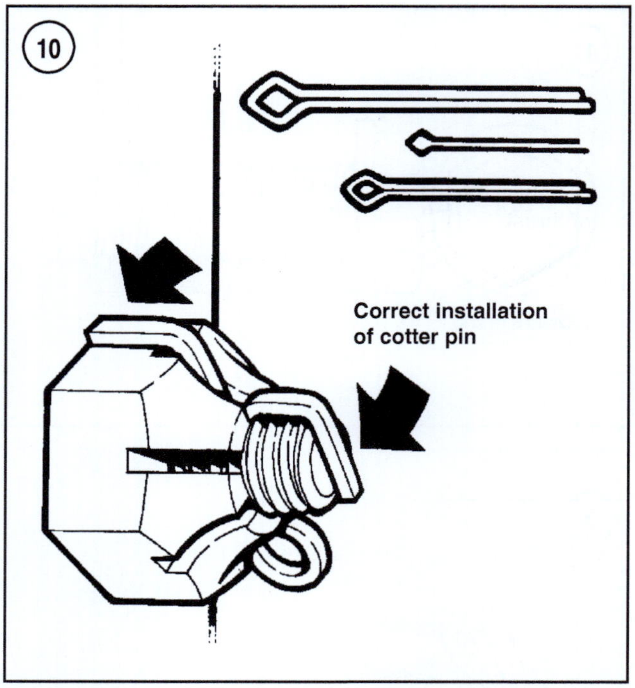

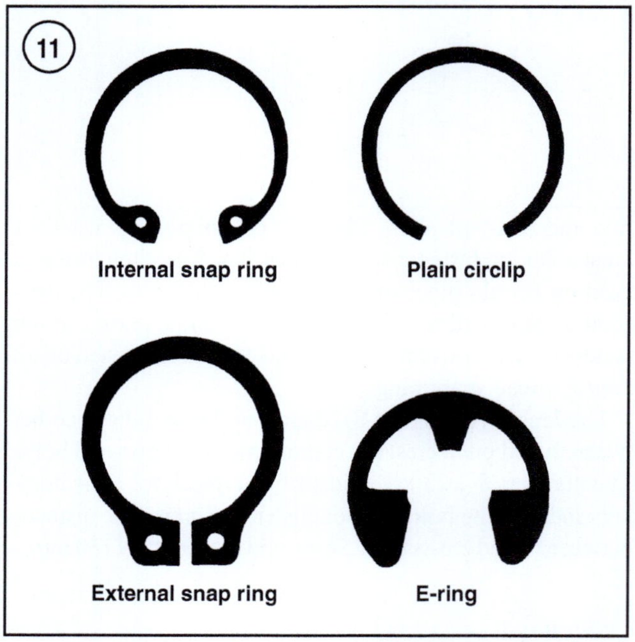

Snap Rings

Snap rings (**Figure 11**) are circular metal retaining clips. They help secure parts such as shafts, pins, gears or rods. External snap rings secure parts within housing bores. In some applications, in addition to securing components, snap rings of varying thicknesses also determine end play. These are usually called selective snap rings.

There are two basic types of snap rings: machined and stamped snap rings. Machined snap rings (**Figure 12**) can be installed in either direction because both faces have sharp edges. Stamped snap rings (**Figure 13**) have a sharp and a round edge. When installing a stamped snap ring in a thrust application, install the sharp edge facing away from the part producing the thrust.

Observe the following when installing snap rings:

1. Remove and install snap rings with snap ring pliers. Refer to *Basic Tools* in this chapter.
2. In some applications, it might be necessary to replace snap rings after removing them.
3. Compress or expand snap rings only enough to install them. If overly expanded or compressed, they lose retaining ability.
4. After installing a snap ring, make sure it seats completely.
5. Wear eye protection when removing or installing snap rings.

E-rings and circlips (**Figure 11**) are used when using a snap ring is not practical. Remove E-rings with a flat-blade screwdriver by prying between the shaft and E-ring. To install an E-ring, center it over the shaft groove and push or tap it into place.

SHOP SUPPLIES

Lubricants and Fluids

Periodic lubrication helps ensure long life for any type of equipment. The type of lubricant used is just as important as the lubrication service itself, though in an emergency, the

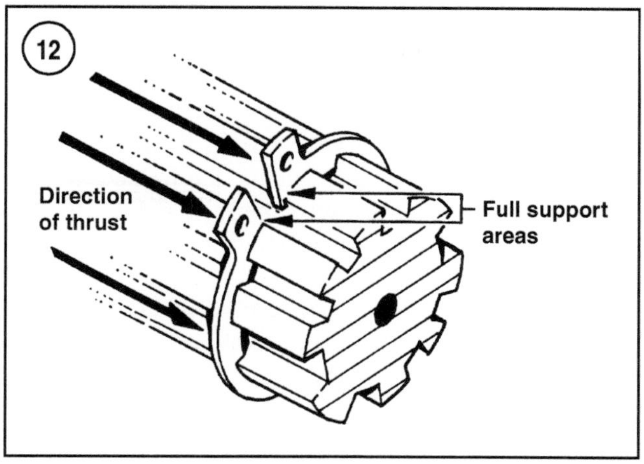

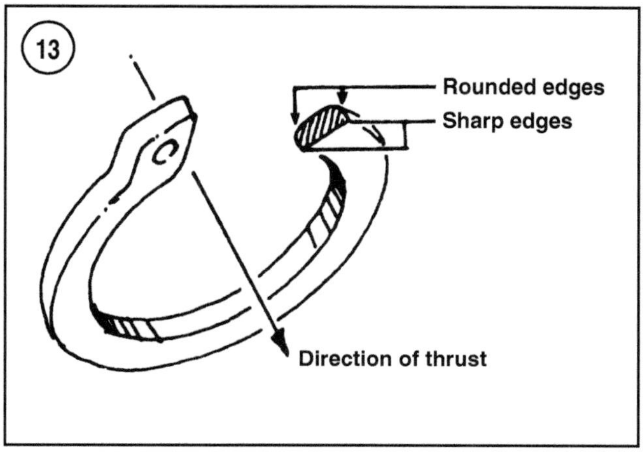

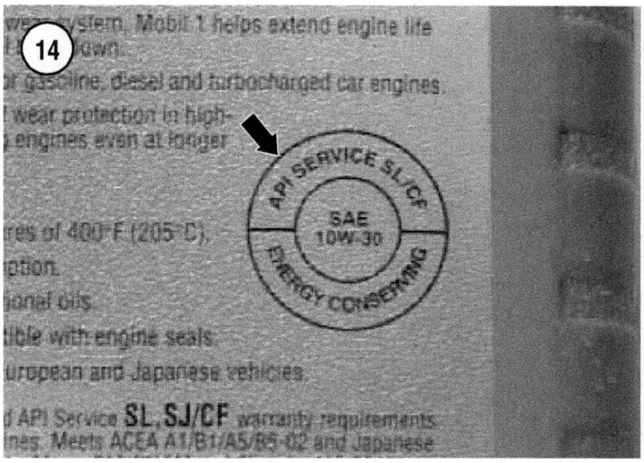

wrong type of lubricant is usually better than no lubricant at all. The following information describes the types of lubricants most often used on marine equipment. Follow the manufacturer's recommendation for lubricant types.

Generally, all liquid lubricants are called *oil.* They may be mineral-based (including petroleum bases), natural-based (vegetable and animal bases), synthetic-based or emulsions (mixtures). *Grease* is oil to which a thickening base was added so that the end product is semi-solid. Grease is often classified by the type of thickener added; lithium soap is commonly used.

Four-Stroke Engine Oil

Four-stroke oil for motorcycle, marine and automotive engines is classified by the American Petroleum Institute (API) and the Society of Automotive Engineers (SAE) in several categories. Oil containers display these classifications on the container cap or label (**Figure 14**).

The API oil classification is indicated by letters; oils for gasoline engines are identified by an *S*.

Viscosity is an indication of the oil thickness. The SAE uses numbers to indicate viscosity. Thin oils have low numbers, and thick oils have high numbers. A *W* after the number indicates that the viscosity testing was done at low temperature to simulate cold-weather operation. Four-stroke engine oils typically fall into the SAE 5-50 viscosity range.

Multigrade oils (for example 10W-40) are less viscous (thinner) and flow well at low temperatures. This allows the oil to perform efficiently across at wide range of engine operating conditions. The lower the number, the better the engine will start in cold climates. Higher numbers are usually recommended for engine operation in hot weather conditions. To help ensure the best performance and durability, use only the recommended oil grade and viscosity as described in Chapter Three.

Gearcase Lubricant

Gearcase lubricants are assigned SAE viscosity numbers under the same system as four-stroke engine oil. Gearcase lubricant falls in the SAE 75-250 range. Some gearcase lubricants, such as SAE 85-90, are multigrade. High-viscosity gearcase lubricant is a heavier oil designed to withstand the shock loading of high-performance models or models subjected to severe use. Always use the type of gearcase lubricant specified by the manufacturer.

Grease

Grease is lubricating oil with thickening agents added to it. The National Lubricating Grease Institute (NLGI) grades grease. Grades range from No. 000 to No. 6, with No. 6 being the thickest. Typical multipurpose grease is NLGI No. 2. For specific applications, manufacturers may recommend water-resistant grease or one with an additive such as molybdenum disulfide (MoS_2).

Cleaners, Degreasers and Solvents

Many chemicals are available to remove oil, grease and other residue.

Before using cleaning solvents, consider how they will be used and disposed of, particularly if they are not water-soluble. Local ordinances may require special procedures for disposal of many types of cleaning chemicals. Refer to *Safety* in this chapter for more information on solvent use.

Use electrical contact cleaner to clean wiring connections and components without leaving residue. Carburetor cleaner is a powerful solvent used to remove fuel deposits and varnish from fuel system components. Use this cleaner carefully because it can damage finishes.

Generally, degreasers are strong cleaners used to remove heavy accumulations of grease from engine and drive components.

Most solvents are used in a parts-washing cabinet for individual component cleaning. For safety, use only nonflammable or high-flash point solvents.

Gasket Sealant

Sealants are used with a gasket or seal and are occasionally used alone. Follow the manufacturer's recommendation when using sealants. Use extreme care when choosing a sealant different from the type originally recommended. Choose sealants based on resistance to heat and various fluids and sealing capabilities.

One of the most common sealants is RTV, or room-temperature vulcanizing sealant. This sealant cures at room temperature over a specific time period and allows the repositioning of components without damaging gaskets.

Moisture in the air causes RTV sealant to cure. Always install the tube cap as soon as possible after applying RTV sealant. RTV sealant has a limited shelf life and does not cure properly if the shelf life has expired. Keep partial tubes sealed and discard them if they have surpassed the expiration date.

Applying RTV sealant

Clean all old gasket residue from the mating surfaces. Remove all gasket material from blind threaded holes; it can cause inaccurate bolt torque. Spray the mating surfaces with aerosol parts cleaner. Then wipe dry with a lint-free cloth. The area must be clean and dry for the sealant to adhere.

Apply RTV sealant in a continuous bead 0.08-0.12 in. (2-3 mm) thick. Circle all the fastener holes unless otherwise specified. Do not allow any sealant to enter these holes. Assemble and tighten the fasteners to the specified torque within the time frame recommended by the RTV sealant manufacturer (usually within 10-15 minutes).

Gasket Remover

Aerosol gasket remover can help remove stubborn gasket material and prevent damage to the mating surfaces caused by using a scraping tool. Most of these products are very caustic. Follow the manufacturer's instructions.

Threadlocking Compound

A threadlocking compound is a fluid applied to the threads of fasteners. After tightening the fastener, the fluid dries and hardens to become a solid filler between the threads. This makes it difficult for the fastener to work loose from vibration or heat expansion and contraction. Some threadlocking compounds also provide a seal against fluid leaks.

Before applying threadlocking compound, remove any old compound from both thread areas and clean them with aerosol parts cleaner. Use the compound sparingly. Excess fluid can run into adjoining parts.

Threadlocking compounds come in different strengths. Follow the particular manufacturer's recommendations regarding compound selection. Two manufacturers of threadlocking compound are ThreeBond and Loctite, which offer a wide range of compounds for various strength, temperature and repair applications.

Applying threadlocking compound

Make sure all surfaces are clean. If a threadlocking compound was previously applied to the component, remove this residue.

Shake the container thoroughly and apply compound to both parts. Then assemble the parts and/or tighten the fasteners.

GALVANIC CORROSION

A chemical reaction occurs whenever two different types of metal are joined by an electrical conductor and immersed in an electrolytic solution such as water. Electrons transfer from one metal to the other through the electrolyte and return to the conductor.

The hardware on a boat is made of many different types of metal. The boat hull acts as a conductor between the metals. Even if the hull is wooden or fiberglass, the slightest film of water on the hull provides conductivity by acting as the electrolyte. This combination creates a good environment for electron flow. Unfortunately, electron flow results in galvanic corrosion of the metal involved and causes one of the metals to corrode or erode away. The amount of electron flow, and therefore the amount of corrosion, depends on several factors:

1. The types of metal involved.
2. The efficiency of the conductor.
3. The strength of the electrolyte.

Metals

The chemical composition of the metal used in marine equipment has a significant effect on the amount and speed of galvanic corrosion. Certain metals are more resistant to corrosion than others. These electrically negative metals

are commonly called *noble*; they act as the cathode in any reaction. Metals that are more subject to corrosion are electrically positive; they act as the anode in a reaction. The more noble metals include titanium, 18-8 stainless steel and nickel. Less noble metals include zinc, aluminum and magnesium. Galvanic corrosion becomes more excessive as the difference in electrical potential between the two metals increases.

In some cases, galvanic corrosion can occur within a single piece of metal. For example, brass is a mixture of zinc and copper, and when immersed in an electrolyte, the zinc portion of the mixture will corrode away as a galvanic reaction occurs between the zinc and copper particles.

Conductors

The hull of the boat often acts as the conductor between different types of metal. Marine equipment, such as the drive unit and transom shield can act as the conductor. Large masses of metal, firmly connected together, are more efficient conductors than water. Rubber mountings and vinyl-based paint can act as insulators between pieces of metal.

Electrolyte

The water in which a boat operates acts as the electrolyte for the corrosion process. The more efficient an electrolyte is, the more excessive and rapid the corrosion will be.

Cold, clean freshwater is the poorest electrolyte. Pollutants increase conductivity; therefore, brackish or saltwater is a more efficient electrolyte. This is one of the reasons that most manufacturers recommend a freshwater flush after operating in polluted, brackish or saltwater.

Protection from Galvanic Corrosion

Because of the environment in which marine equipment must operate, it is practically impossible to totally prevent galvanic corrosion. However, there are several ways in which the process can be slowed. After taking the precautions, the next step is to force the process into occurring only in certain places. This is the role of sacrificial anodes and impressed current systems.

Slowing corrosion

Some simple precautions can help reduce the amount of corrosion taking place outside the hull. These precautions are not substitutes for the corrosion protection methods discussed in *Sacrificial anodes* and *Impressed current systems* in this section, but they can help these methods reduce corrosion.

Use fasteners made of a material more noble than the parts they secure. If corrosion occurs, the parts they secure may suffer but the fasteners are protected. The larger secured parts are better able to withstand the loss of material. Also, major problems could arise if the fasteners corrode to the point of failure.

Keep all painted surfaces in good condition. If paint is scraped off and bare metal is exposed, corrosion rapidly increases. Use vinyl- or plastic-based paint, which acts as an electrical insulator.

Be careful when applying metal-based antifouling paint to the boat. Do not apply antifouling paint to metal parts of the boat, drive unit or transom shield. If applied to metal surfaces, this type of paint reacts with the metal and results in corrosion between the metal and the layer of paint. Maintain a minimum border of 1.0 in. (25.0 mm) border between the painted surfaces and any metal parts. Organic-based paints are available for use on metal surfaces.

When a corrosion protection device is used, remember that it must be immersed in the electrolyte along with the boat to provide any protection. If the drive unit is raised out of the water when the boat is docked, any anodes on the engine will be removed from the corrosion process and be ineffective. (Of course, the engine requires less protection when raised out of the water/electrolyte.) Never paint or apply any coating to anodes or other protection devices. Paint or other coatings insulate them from the corrosion process.

Any change in boat equipment, such as the installation of a new stainless steel propeller, affects the electrical potential and can cause increased corrosion. Always consider this fact when adding equipment or changing exposed materials. Install additional anodes or other protection equipment as required to ensure the corrosion protection system is up to the task. The expense to repair corrosion damage usually greatly exceeds that of additional corrosion protection.

Sacrificial anodes

Sacrificial anodes (**Figure 15**) are specially designed to do nothing but corrode. Properly fastening such pieces to the boat causes them to act as the anode in any galvanic reaction that occurs; any other metal in the reaction acts as the cathode and is not damaged.

Anodes are usually made of zinc, a less noble material. Some anodes are made of an aluminum and indium alloy.

This alloy is less noble than the aluminum alloy in the drive system components and provides the desired sacrificial properties. The aluminum and indium alloy is more resistant to oxide coating than zinc anodes. Oxide coating occurs as the anode material reacts with oxygen in the water. An oxide coating insulates the anode, dramatically reducing corrosion protection.

Anodes must be used properly to be effective. Simply fastening anodes to the boat in random locations does not do the job.

First, determine how much anode surface is required to adequately protect the equipment surface area. A good starting point is provided by the military specification MIL-A-818001, which states that 1 sq. in. of new anode protects either:

1. 800 sq. in. of freshly painted stainless steel.
2. 250 sq. in. of bare steel or bare aluminum alloy.
3. 100 sq. in. of copper or copper alloy.

This rule is valid for a boat at rest. When underway, additional anode area is required to protect the same surface area.

The anode must make good electrical contact with the metal that it protects. If possible, attach an anode to metal surfaces requiring protection.

Quality anodes have inserts around fastener holes that are made of a more noble material. Otherwise, the anode could erode around the fastener hole and loosen or possibly fall off, thereby losing needed protection.

Impressed current systems

An impressed current system can be added to any boat. The system generally consists of the anode, controller and reference electrode. The anode in this system is coated with a very noble material, such as platinum, so that it is almost corrosion-free and can last almost indefinitely. The reference electrode, under the boat waterline, allows the controller to monitor the current in the water for any indication of galvanic corrosion. If the controller senses current associated with galvanic corrosion, it applies positive battery voltage to the anode. Current then flows from the anode to all other metal components, regardless of how noble or non-noble these components may be. Essentially, the electrical current from the battery counteracts the galvanic reaction to dramatically reduce corrosion damage.

Only a small amount of current is needed to counteract corrosion. Using input from the reference electrode, the controller provides only the amount of current needed to suppress galvanic corrosion. Most systems consume a maximum of 0.2 A.h. at full demand. Under normal conditions, these systems can provide protection for 8-12 weeks without recharging the battery. Remember that this system must have a constant connection to the battery. Often, the battery supply to the system is connected to a battery switching device causing the operator to inadvertently shut off the system while docked.

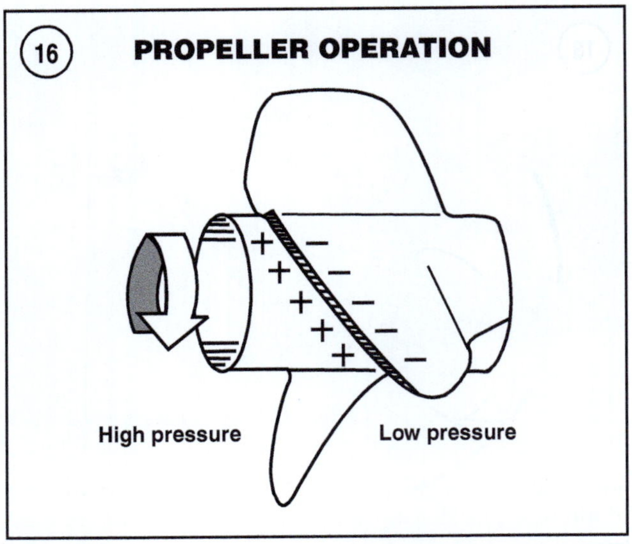

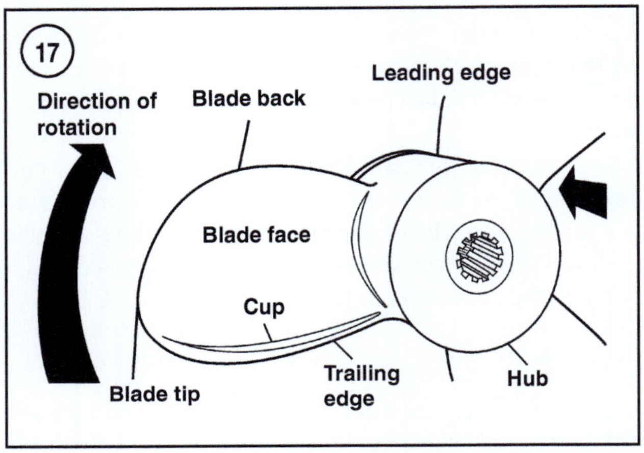

An impressed current system is more expensive to install than sacrificial anodes, but considering the low maintenance requirements and the superior protection it provides, the long-term cost might be lower.

PROPELLERS

The propeller is the final link between the boat drive system and the water. A perfectly maintained engine and hull are useless if the propeller is the wrong type, worn or damaged. Although propeller selection for a specific application is beyond the scope of this manual, the following provides the basic information needed to make an informed decision. A professional at a reputable marine dealership is the best source for a propeller recommendation.

How a Propeller Works

As the curved blades of a propeller rotate through the water, a high-pressure area forms on one side of the blade and a low-pressure area forms on the other side of the blade (**Figure 16**). The propeller moves toward the low-pressure area and carries the boat with it.

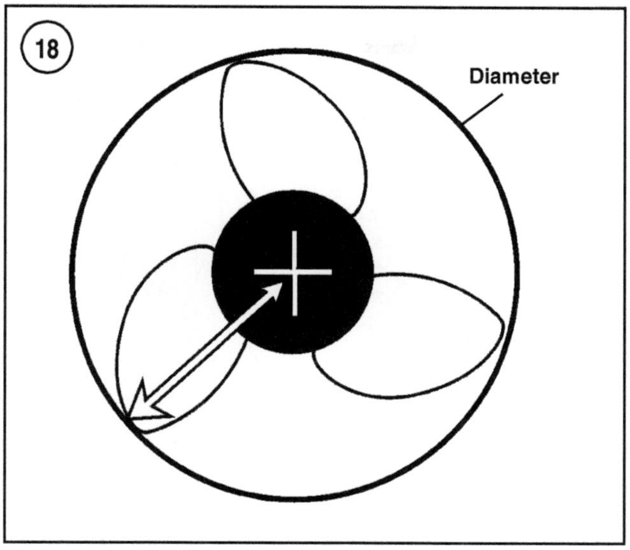

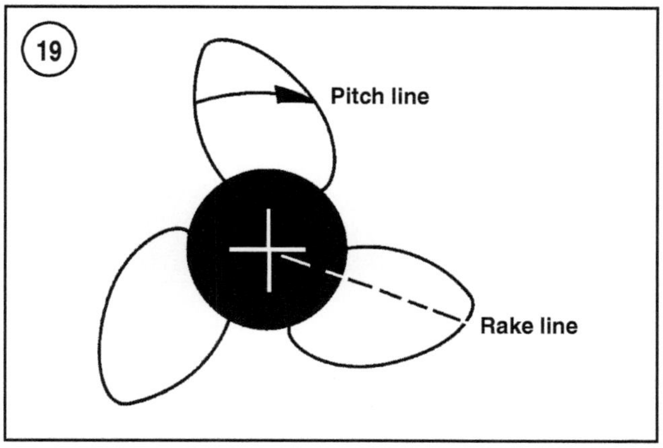

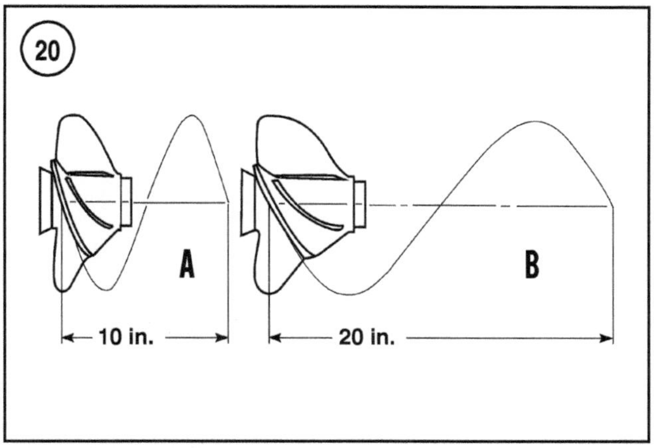

Propeller Parts

Although a propeller is usually a one-piece unit, it is made of several different parts (**Figure 17**). Variations in the design of these parts make different propellers suitable for different applications.

The blade tip is the point of the blade farthest from the center of the propeller hub or propeller shaft bore. The blade tip separates the leading edge from the trailing edge.

The leading edge is the edge of the blade nearest the boat. During forward operation, this is the area of the blade that first cuts through the water.

The trailing edge is the edge of the blade farthest from the boat. During reverse operation, this is the area of the blade that first cuts through the water.

The blade face is the surface of the blade that faces away from the boat. During forward operation, high pressure forms on this side of the blade.

The blade back is the surface of the blade that faces the boat. During forward operation, low pressure forms on this side of the blade.

The cup is a small curve or lip on the trailing edge of the blade. Cupped propeller blades generally perform better than noncupped propeller blades.

The hub is the center portion of the propeller. It connects the blades to the propeller shaft. On most drive systems, engine exhaust is routed out through the hub; in this case, the hub is made up of an outer and inner portion connected by ribs.

A diffuser ring is used on through-hub exhaust models to prevent exhaust gasses from entering the blade area.

Propeller Design

Changes in length, angle, thickness, material or propeller parts make different propellers suitable for different applications.

Diameter

Propeller diameter is the distance from the center of the hub or propeller shaft to the blade tip, multiplied by two. Essentially it is the diameter of a circle formed by the blade tips during propeller rotation (**Figure 18**).

Pitch and rake

Propeller pitch and rake describe the placement of the blades in relation to the propeller hub (**Figure 19**).

Pitch describes the theoretical distance the propeller would travel in one revolution. In A, **Figure 20**, the propeller would travel 10 in. in one revolution. In B, **Figure 20**, the propeller would travel 20 in. in one revolution. This distance is only theoretical. During typical operation, the propeller achieves only 75-85 percent of pitch. Slip rate describes the difference in actual travel relative to the pitch. Lighter, faster boats typically achieve a lower slip rate than heavier, slower boats.

Propeller blades can be constructed with constant pitch (**Figure 21**) or progressive pitch (**Figure 22**). On a progressive pitch propeller, the pitch starts low at the leading edge and increases toward the trailing edge. The propeller pitch specification is the average of the pitch across the entire blade. Propellers with progressive pitch usually provide better overall performance than constant pitch propellers.

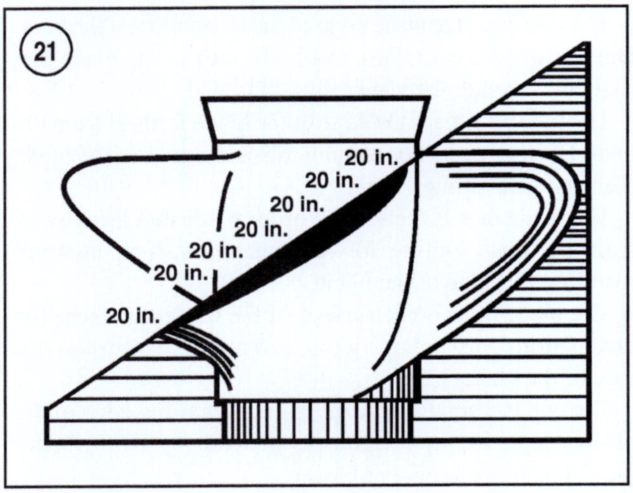

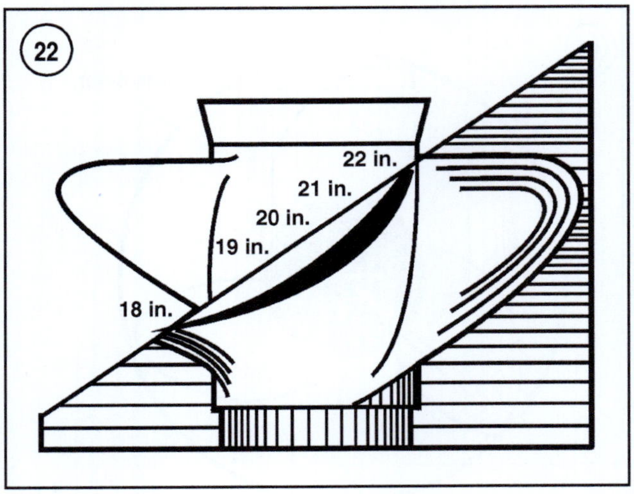

Constant pitch propellers are generally better suited for heavier, slower boats.

Blade rake is specified in degrees and is measured along a line from the center of the hub to the blade tip. A blade that is perpendicular to the hub (A, **Figure 23**) has 0° rake. A blade that is angled from perpendicular (B, **Figure 23**) has a rake expressed by its difference from perpendicular. Most propellers have rakes ranging from 0-20°. Lighter, faster boats generally perform better using a propeller with a greater amount of rake. Heavier, slower boats generally perform better using a propeller with less rake.

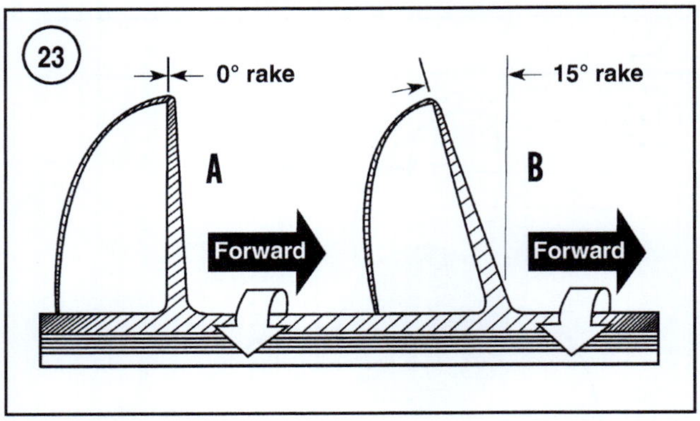

Blade thickness

Blade thickness is not uniform at all points along the blade. For efficiency, blades are as thin as possible at all points while retaining enough strength to move the boat. Blades are thicker where they meet the hub and thinner at the blade tips. This construction is necessary to support the heavier loads at the hub section of the blade. Overall blade thickness depends on the strength of the material used. When cut along a line from the leading edge to the trailing edge in the central portion of the blade, the propeller blade resembles an airplane wing. The blade face, where high pressure exists during forward rotation, is almost flat. The blade back, where low pressure exists during forward rotation, is curved with the thinner portions at the edges and the thickest portion at the center. Propellers that run only partially submerged, as in racing applications, may have a wedge-shaped cross-section (**Figure 24**). The leading edge is very thin, and the blade thickness increases toward the trailing edge, where it is thickest. If this type of propeller runs totally submerged, it is very inefficient.

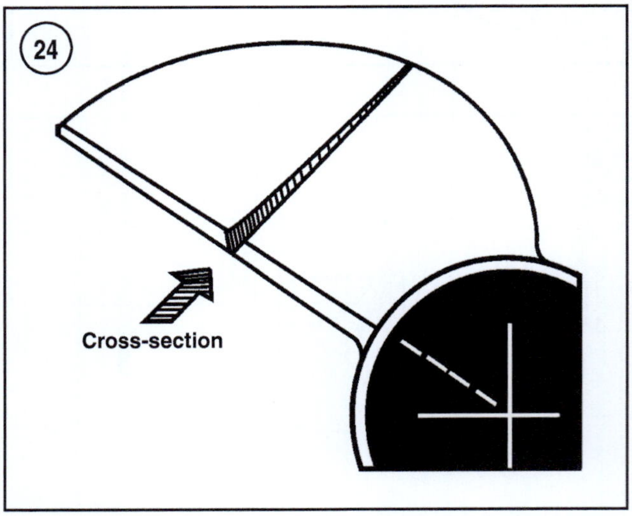

Number of blades

The number of blades on a propeller is a compromise between efficiency and vibration. A one-bladed propeller would be the most efficient, but it would create an unacceptable amount of vibration. As blades are added, efficiency decreases, but so does vibration. Most propellers have three or four blades, representing the most practical trade-off between efficiency and vibration.

Material

Propeller materials are chosen for strength, corrosion resistance and economy. Stainless steel, aluminum, plastic

25
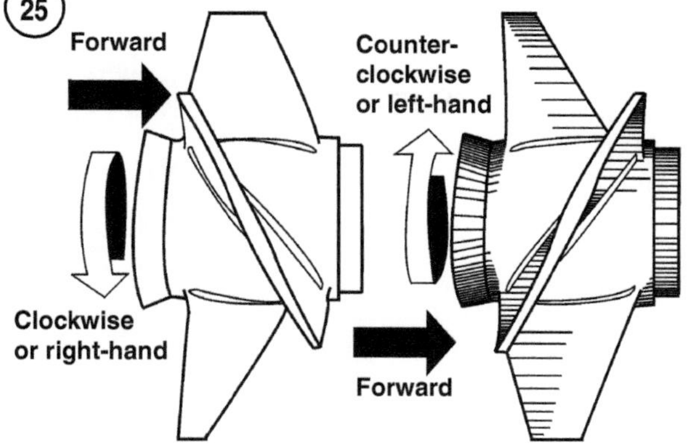

26
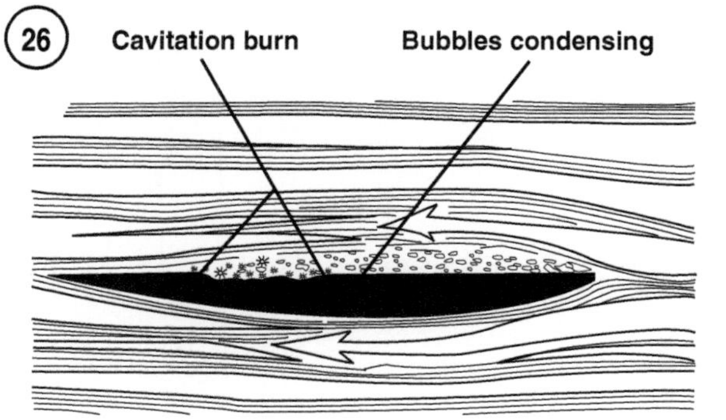

27
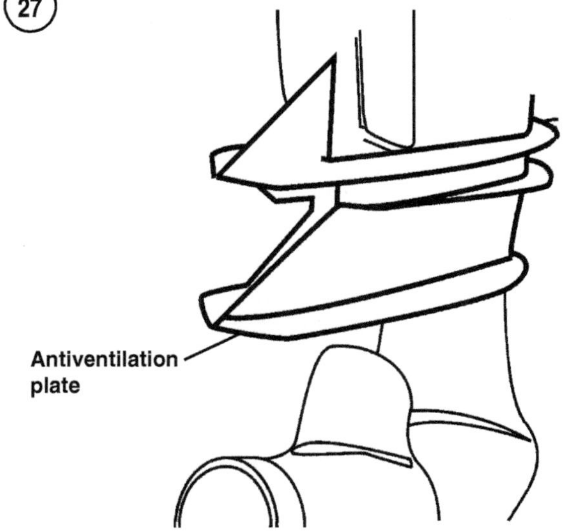

and bronze are the most commonly used materials. Bronze is quite strong but rather expensive. Stainless steel is more common than bronze because of the combination of strength at a lower cost. Aluminum alloy and plastic materials are the least expensive but lack the strength of stainless steel. Plastic propellers are more suited for lower horsepower applications.

Direction of rotation

Propellers are made for both right-hand and left-hand rotations, though right-hand is the most commonly used. As viewed from the rear of the boat while in forward, a right-hand propeller turns clockwise, and a left-hand propeller turns counterclockwise. On the boat, the direction of rotation is determined by observing the angle of the blades (**Figure 25**). Right-handed propeller blades slant from the upper left to the lower right; left-handed propeller blades are opposite.

Cavitation and Ventilation

Cavitation and ventilation are *not* interchangeable terms; they refer to two distinct problems encountered during propeller operation.

To help understand cavitation, consider the relationship between pressure and the boiling point of water. At sea level, water boils at 212° F (100° C). As pressure increases, such as within an engine cooling system, the boiling point of the water increases. It boils at a temperature higher than 212° F (100° C). The opposite is also true. As pressure decreases, water boils at a temperature lower than 212° F (100° C). If the pressure drops low enough, water will boil at room temperature.

During normal propeller operation, low pressure forms on the blade back. Normally, the pressure does not drop low enough for boiling to occur. However, poor propeller design, damaged blades or using the wrong propeller can cause unusually low pressure on the blade surface (**Figure 26**). If the pressure drops low enough, boiling occurs, and bubbles form on the blade surfaces. As the boiling water moves to a higher-pressure area of the blade, the boiling ceases and the bubbles collapse. The collapsing bubbles release energy that erodes the surface of the propeller blade.

Corroded surfaces, physical damage or even marine growth combined with high-speed operation can cause low pressure and cavitation on drive unit gearcase surfaces. In such cases, low pressure forms as water flows over a protrusion or rough surface. The boiling water forms bubbles that collapse as they move to a higher-pressure area toward the rear of the surface imperfection.

This entire process of pressure drop, boiling and bubble collapse is called *cavitation*. The ensuing damage is called *cavitation burn*. Cavitation is caused by a decrease in pressure, not an increase in temperature.

Ventilation is not as complex as cavitation. Ventilation refers to air entering the blade area, either from above the water surface or from a through-hub exhaust system. As the blades meet the air, the propeller momentarily loses contact with the water and subsequently loses most of its thrust. During ventilation, cavitation can also occur as the engine overspeeds and creates very low pressure on the back of the propeller blade.

Most marine drive systems have a plate (**Figure 27**) above the propeller designed to prevent surface air from entering the blade area. This plate is called the *antiventilation plate*, though it is often incorrectly called an *anticavitation plate*.

Most propellers have a flared section at the rear of the propeller called a diffuser ring. This feature forms a barrier and extends the exhaust passage far enough aft to prevent the exhaust gasses from ventilating the propeller.

A close fit of the propeller to the gearcase is necessary to keep exhaust gasses from exiting and ventilating the propeller. Using the wrong propeller-attaching hardware can position the propeller too far aft and prevent a close fit. The wrong hardware can also allow the propeller to rub heavily against the gearcase and cause rapid wear to both components. Wear or damage to these surfaces allows the propeller to ventilate.

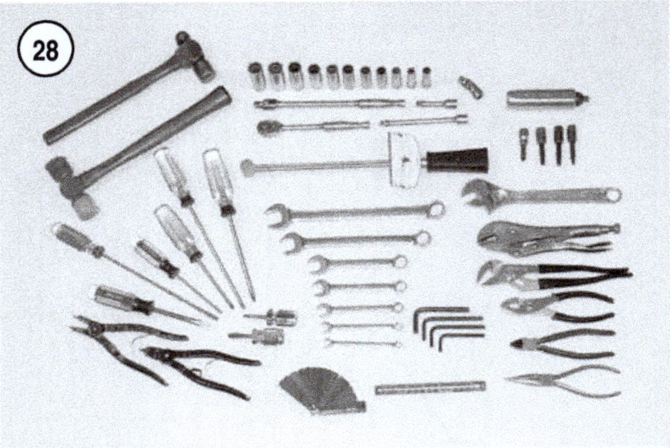

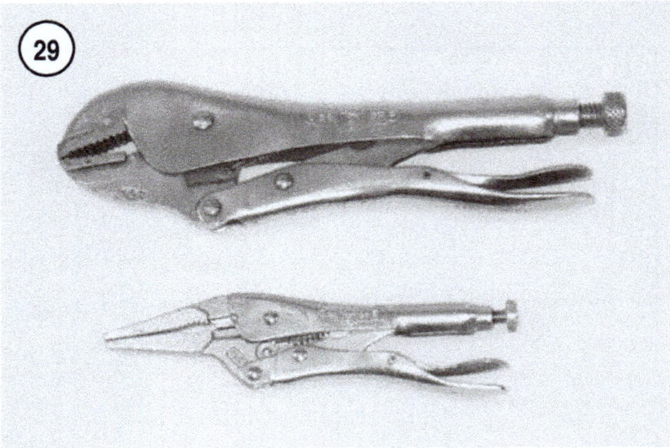

BASIC TOOLS

Most of the procedures in this manual can be carried out with simple hand tools and test equipment familiar to the home mechanic. Always use the correct tools for the job at hand. Keep tools organized and clean. Store them in a tool chest with related tools organized together.

After using a tool, wipe off dirt and grease with a clean cloth. Wiping off tools is especially important when servicing the craft in areas where they can come in contact with sand. Sand is very abrasive and causes premature wear to engine parts.

Quality tools are essential. The best are constructed of high-strength alloy steel. These tools are light, easy to use and resistant to wear. Working surfaces lack sharp edges, and the tools are carefully polished. They have an easy-to-clean finish and are comfortable to use. Quality tools are a good investment.

When purchasing tools to perform the procedures covered in this manual, consider the potential frequency of use. If starting a tool kit, consider purchasing a basic tool set (**Figure 28**) from a large tool supplier. These sets are available in many combinations and offer a substantial savings when compared to individually purchased tools. As work experience grows and tasks become more complicated, specialized tools can be added.

Screwdrivers

Screwdrivers of various lengths and types are mandatory for the simplest tool kit. The two basic types are the slotted tip (flat blade) and Phillips tip. These are available in sets that often include an assortment of tip sizes and shaft lengths.

As with all tools, use a screwdriver designed for the job. Make sure the size of the tip conforms to the size and shape of the fastener. Use them only for driving screws. Never use a screwdriver for prying or chiseling metal. Repair or replace worn or damaged screwdrivers. A worn tip can damage the fastener and make it difficult to remove.

Phillips screwdrivers are sized according to point size. They are numbered one, two, three, four. The degree of taper determines the point size; the No. 1 Phillips screwdriver is the most pointed. The points are more blunt as the number increases.

Pliers

Pliers come in a wide range of types and sizes. Though pliers are useful for holding, cutting, bending and crimping, they should never be used to turn bolts and nuts.

Each design has a specialized function. Slip-joint pliers are general-purpose pliers used for gripping and bending. Diagonal cutting pliers are needed to cut wire and can be used to remove cotter pins. Needlenose pliers are used to hold or bend small objects. Locking pliers (**Figure 29**) sometimes called vise-grips, are used to hold objects very tightly. They have many uses ranging from holding two parts together to gripping the end of a broken stud. Use caution when using locking pliers. The sharp jaws can damage the objects they hold.

Snap Ring Pliers

Snap ring pliers (**Figure 30**) are specialized pliers with tips that fit into the ends of snap rings to remove and install them.

30

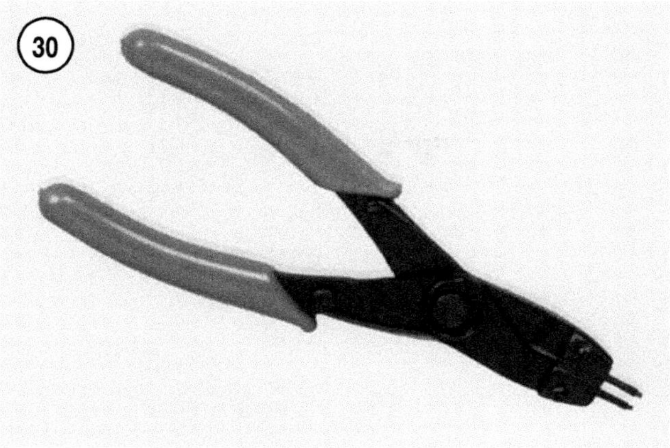

31

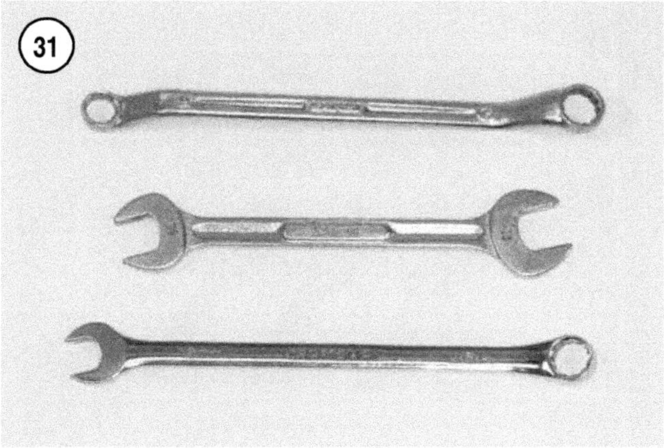

Snap ring pliers are available with fixed action (either internal or external) or with convertible action (one tool works both internal and external snap rings). They may have fixed tips or interchangeable ones of various sizes and angles. For general use, select convertible pliers with interchangeable tips.

WARNING
Snap rings can slip and fly off when removing and installing them. Also, the snap ring plier tips may break. Always wear eye protection when using snap ring pliers.

Hammers

Various types of hammers are available to fit a number of applications. A ball-peen hammer is used to strike another tool, such as a punch or chisel. Soft-faced hammers are required when a metal object must be struck without damaging it. *Never* use a metal-faced hammer on engine components because damage does occur in most cases.

Always wear eye protection when using hammers. Make sure the hammer face is in good condition and the handle is not cracked. Select the correct hammer for the job and make sure to strike the object squarely. Do not use the handle or side of the hammer to strike an object.

When striking a hammer against a punch, cold chisel or similar tool, the face of the hammer should be at least 1/2 in. larger than the head of the tool. When it is necessary to strike a steel part hard without damaging it, use a brass hammer. Brass will give when used on a harder object.

Wrenches

Box-end, open-end and combination wrenches (**Figure 31**) come in a variety of types and sizes.

The number stamped on the wrench refers to the distance between the work areas. This size must match the size of the fastener head.

The box-end wrench is an excellent tool because it grips the fastener on all sides. This factor reduces the chance of the tool slipping. The box-end is designed with either a six or 12-point opening. For stubborn or damaged fasteners, the six-point provides superior holding ability by contacting the fastener across a wider area at all six edges. For general use, the 12-point works well. It allows the wrench to be removed and reinstalled without moving the handle over such a wide arc.

An open-end wrench is fast and works best in areas with limited overhead access. It contacts the fastener at only two points and can slip under heavy force or if the tool or fastener is worn. A box-end wrench is preferred in most instances, especially when breaking loose and applying the final tightening force to a fastener.

The combination wrench has a box end on one end, and an open end on the other. This combination makes it a very convenient tool.

Adjustable Wrenches

An adjustable wrench or Crescent wrench can fit nearly any nut or bolt head that has clear access around the entire perimeter. Adjustable wrenches are best used as a backup wrench to keep a large nut or bolt from turning while the other end is being loosened or tightened with a box-end or socket wrench.

Adjustable wrenches contact the fastener at only two points, which makes them more subject to slipping off the fastener. The fact that one jaw is adjustable and might loosen only aggravates this shortcoming. Make certain the solid jaw is the one transmitting the force.

Socket Wrenches, Ratchets and Handles

Sockets that attach to a ratchet handle (**Figure 32**) are available in six-point or 12-point openings and different drive sizes. The drive size (1/4, 3/8, 1/2 and 3/4 in.) indicates the size of the square opening that accepts the ratchet handle. The number stamped on the socket is the size of the work area and must match the fastener head.

As with wrenches, a six-point socket provides superior-holding ability, and a 12-point socket needs to be moved only half as far to reposition it on the fastener.

Sockets are designated for either hand or impact use. Impact sockets are made of thicker material for more durability.

WARNING
Do not use hand sockets with air or impact tools because they may shatter and cause injury. Always wear eye protection when using impact or air tools.

Various handles are available for sockets. The speed handle is used for fast operation. Flexible ratchet heads in various lengths allow the socket to be turned with varying force and at odd angles. Extension bars allow socket setup to reach difficult areas. The ratchet is the most versatile. It allows the user to install or remove the nut without removing the socket.

Sockets combined with any number of drivers make undoubtedly the fastest, safest and most convenient tools for fastener removal and installation.

Impact Driver

An impact driver provides extra force for removing fasteners, by converting the impact of a hammer into a turning motion. This makes it possible to remove stubborn fasteners without damaging them. Impact drivers and interchangeable bits (**Figure 33**) are available from most tool suppliers. When using a socket with an impact driver, make sure the socket is designed for impact use. Refer to *Socket Wrenches, Ratchets and Handles* in this section.

WARNING
Do not use hand sockets with air or impact tools because they may shatter and cause injury. Always wear eye protection when using impact or air tools.

Impact drivers are great for the home mechanic because they offer many of the advantages of air tools without the need for a costly air compressor to run them.

Allen Wrenches

Allen or setscrew wrenches (**Figure 34**) are used on fasteners with hexagonal recesses in the fastener head. These wrenches come in L-shaped bar, socket and T-handle types. Allen bolts are sometimes called socket bolts.

Torque Wrenches

A torque wrench is used with a socket, torque adapter or similar extension to measure torque while tightening a fastener. Torque wrenches come in several drive sizes (1/ 4, 3/8, 1/2 and 3/4 in.) and have various methods of reading the torque value. The drive size is the size of the square drive that accepts the socket, adapter or extension. Common types of torque wrenches include the deflecting beam (A, **Figure 35**), the dial indicator (B) and audible click (C).

32

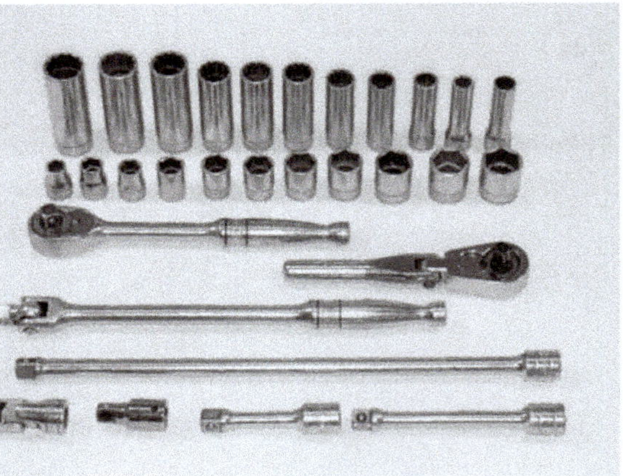

33

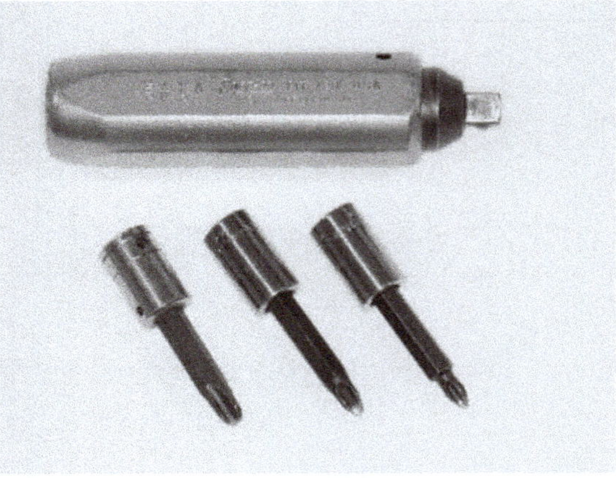

34

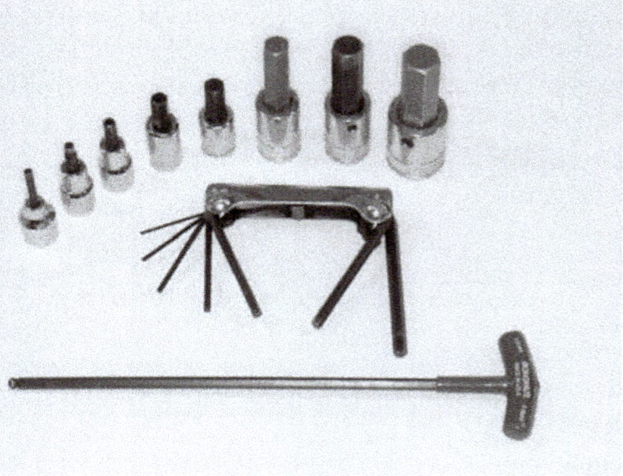

When choosing a torque wrench, consider the torque range, drive size and accuracy. The torque specifications in this manual provide an indication of the range required.

A torque wrench is a precision tool that must be properly cared for to remain accurate. Store torque wrenches in cases or separate padded drawers within a toolbox. Follow the manufacturer's care and calibration instructions.

Torque Adapters

Torque adapters or extensions extend or reduce the reach of a torque wrench. The torque adapter shown on the top of **Figure 36** is used to tighten a fastener that cannot be

35

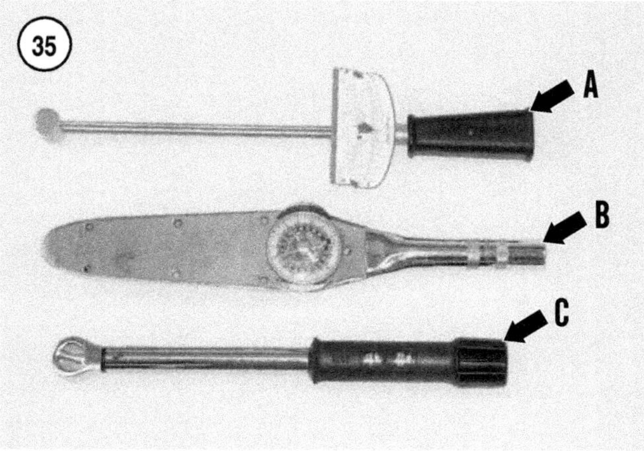

36

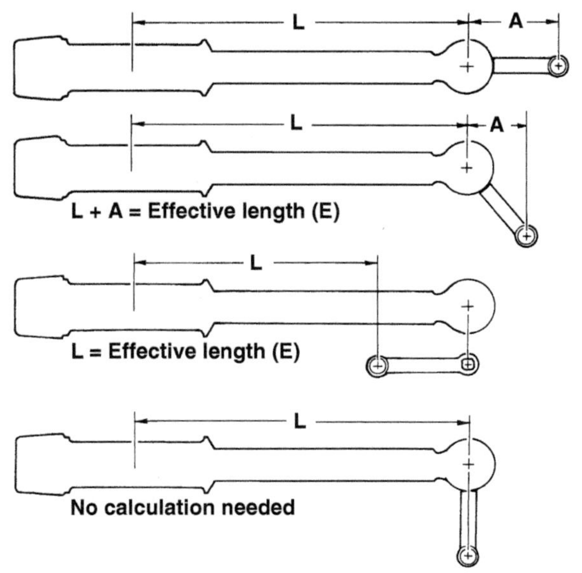

reached because of the size of the torque wrench head, drive and socket. If a torque adapter changes the effective lever length, the torque reading on the wrench will not equal the actual torque applied to the fastener. It is necessary to recalibrate the torque setting on the wrench to compensate for the change of lever length. When using a torque adapter at a right angle to the drive head, calibration is not required because the effective length has not changed.

To recalibrate a torque reading when using a torque adapter, use the following formula, and refer to **Figure 36**.

$$TW = \frac{TA \times L}{L + A}$$

TW is the torque setting or dial reading on the wrench.

TA is the torque specification (the actual amount of torque that should be applied to the fastener).

A is the amount that the adapter increases (or in some cases reduces) the effective lever length as measured along the centerline of the torque wrench (**Figure 36**).

L is the lever length of the wrench as measured from the center of the drive to the center of the grip.

The effective length of the torque wrench measured along the centerline of the torque wrench is the sum of L and A (**Figure 36**).

Example:

TA = 20 ft.-lb.

A = 3 in.

L = 14 in.

$$TW = \frac{20 \times 14}{14 + 3} = \frac{280}{17} = 16.5 \text{ ft.-lb.}$$

In this example, the torque wrench would be set to the recalculated torque value (TW = 16.5 ft.-lb.). When using a beam wrench, tighten the fastener until the pointer aligns with 16.5 ft.-lb. In this example, although the torque wrench is preset to 16.5 ft.-lb., the actual torque applied is 20 ft.-lb.

SPECIAL TOOLS

Some of the procedures in this manual require special tools. These are described in the appropriate chapter and are available from either the engine manufacturer or a tool supplier.

In many cases, an acceptable substitute may be found in an existing tool kit. Another alternative is to make the tool. Many schools with a machine shop curriculum welcome outside work that can be used as practical shop applications for students.

PRECISION MEASURING TOOLS

The ability to accurately measure components is essential to successfully rebuilding an engine. Equipment is manufactured to close tolerances, and obtaining consistently accurate measurements is essential to determining which components require replacement or further service.

Each type of measuring instrument is designed to measure a dimension with a certain degree of accuracy and within a certain range. When selecting the measuring tool, make sure it is applicable to the task.

As with all tools, measuring tools provide the best results if cared for properly. Improper use can damage the tool and result in inaccurate results. If any measurement is questionable, verify the measurement using another tool. A standard gauge is usually provided with measuring tools to check accuracy and calibrate the tool if necessary.

Precision measurements can vary according to the experience of the person performing the procedure. Accurate results are only possible if the mechanic possesses a feel for using the tool. Heavy-handed use of measuring tools produces less accurate measurements than if the fingertips grasp the tool gently so the point at which the tool contacts the object is easily felt. This feel for the equipment produces more accurate measurements and reduces the risk of damaging the tool or component. Refer to the following sections for specific measuring tools.

Feeler Gauges

The feeler or thickness gauge (**Figure 37**) is used for measuring the distance between two surfaces.

A feeler gauge set consists of an assortment of steel strips of graduated thickness. Blades can be of various lengths and angles for different procedures.

A common use for a feeler gauge is to measure valve clearance. Wire (round) gauges are used to measure spark plug gap.

To obtain a proper measurement using a feeler gauge, make sure the proper-sized blade passes through the gap with a slight drag. The blade should not need to be forced through, and should not have any play between the surfaces being measured.

Calipers

Calipers (**Figure 38**) are excellent tools for obtaining inside, outside and depth measurements. Although not as precise as a micrometer, they allow reasonable precision, typically to within 0.001 in. (0.05 mm). Most calipers have a range up to 6 in. (150 mm).

Calipers are available in dial, vernier or digital versions. Dial calipers have a dial gauge readout that provides convenient reading. Vernier calipers have marked scales that are compared to determine the measurement. Most convenient of all, the digital caliper uses an LCD to show the measurement.

To help ensure accurate readings, properly maintain the measuring surfaces of the caliper. There must not be any dirt or burrs between the tool and the object being measured. Never force the caliper closed around an object. Carefully close the caliper around the highest point so it can be removed with a slight drag. Some calipers require calibration. Always refer to the manufacturer's instructions when using a new or unfamiliar caliper.

To read the vernier caliper, refer to **Figure 39**. The fixed scale is marked in 1.0 mm increments. Ten individual lines on the fixed scale equal 1.0 cm. The movable scale is marked in 0.05 mm (five-hundredth) increments. To obtain a reading, establish the first number by location of the 0 line on the movable scale in relation to the first line to the left on the fixed scale. In this example, the number is 10.0 mm. To determine the next number, note which of the lines on the movable scale align with a mark on the fixed scale. A number of lines will seem close, but only one aligns exactly. In this case, 0.50 mm is the reading to add to the first number. The result of adding 10.0 mm and 0.50 mm is a measurement of 10.50 mm.

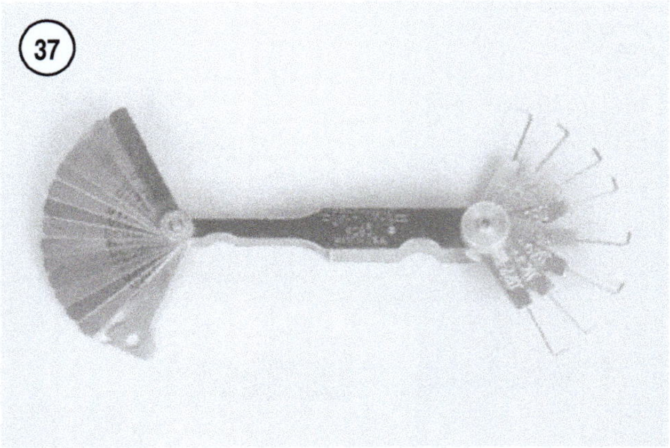

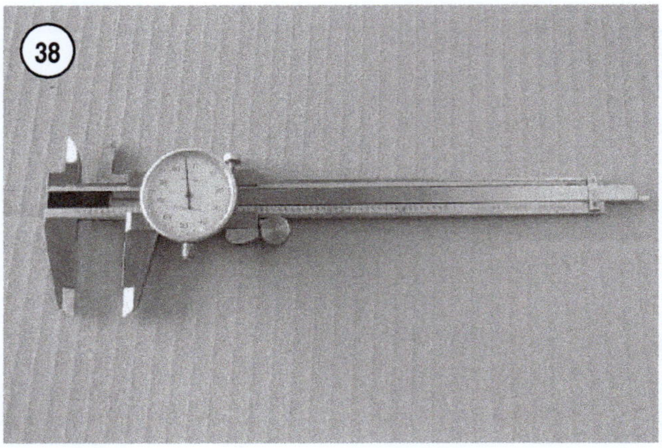

39 READING A VERNIER CALIPER

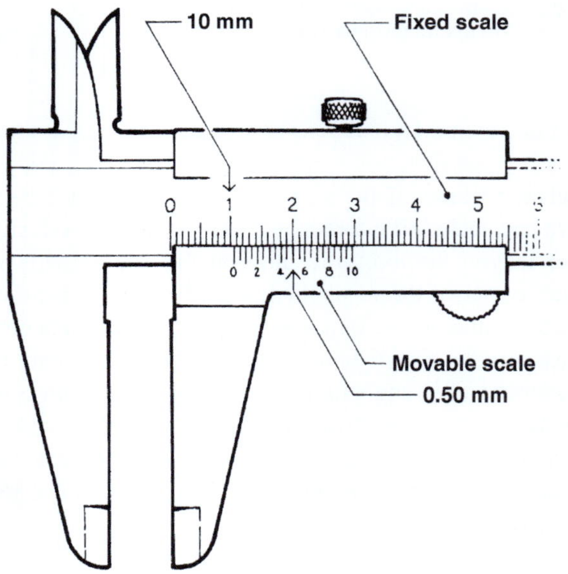

1.	Fixed scale reading:	10.00 mm
2.	Movable scale reading:	0.50 mm
3.	Total reading:	10.50 mm

40

DECIMAL PLACE VALUES*

0.1	Indicates 1/10 (one tenth of an inch or millimeter)
0.01	Indicates 1/100 (one hundredth of an inch or millimeter)
0.001	Indicates 1/1000 (one thousandth of an inch or millimeter)

*This chart represents the values of figures placed to the right of the decimal point. Use it when reading decimals from one tenth to one thousandth of an inch or millimeter. It is not a conversion chart (for example: 0.001 in. is not equal to 0.001 mm).

41

STANDARD INCH MICROMETER

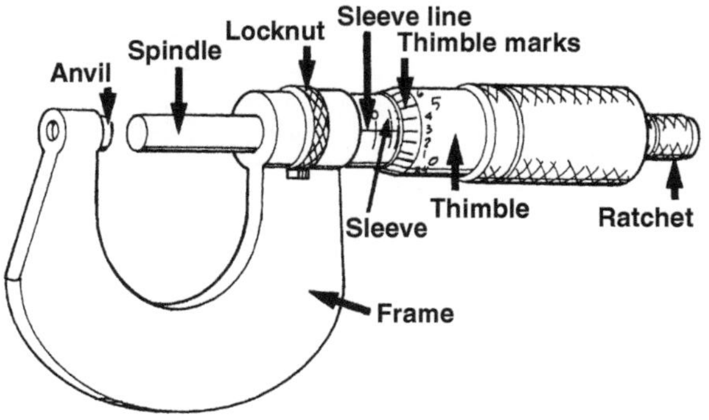

Micrometers

A micrometer is an instrument designed for linear measurement using the decimal division of the inch or meter (**Figure 40**). Although there are many types and styles of micrometers, most of the procedures in this manual call for an outside micrometer. The outside micrometer is used to measure the outside diameter of cylindrical forms and the thickness of materials.

Micrometer size indicates the minimum and maximum size of a part that it can measure. The usual sizes are 0-1 in. (0-25 mm), 1-2 in. (25-50 mm), 2-3 in. (50-75 mm) and 3-4 in. (75-100 mm).

Micrometers covering a wider range of measurement are available and use a large frame with interchangeable anvils of various lengths. This type of micrometer offers a cost savings; however, the overall size make it less convenient.

Reading a micrometer

When reading a micrometer, numbers are taken from different scales and added together. The following sections describe how to take measurements with various types of outside micrometers.

For accurate results, properly maintain the measuring surfaces of the micrometer. There cannot be any dirt or burrs between the tool and the measured object. Never force the micrometer closed around an object. Close the micrometer around the highest point so it can be removed with a slight drag. **Figure 41** shows the markings and parts of a standard inch micrometer. Be familiar with these terms before using a micrometer in the following sections.

Reading a standard inch micrometer

The standard inch micrometer is accurate to one thousandth of an inch or 0.001 in. The sleeve is marked in 0.025 in. increments. Every fourth sleeve mark is numbered 1, 2, 3, 4, 5, 6, 7, 8 and 9. These numbers indicate 0.100 in., 0.200 in., 0.300 in. and so on.

The tapered end of the thimble has twenty-five lines marked around it. Each mark equals 0.001 in. One complete turn of the thimble will align the zero mark with the first mark on the sleeve or 0.025 in.

When reading a standard inch micrometer, perform the following steps while referring to **Figure 42**.

1. Read the sleeve and find the largest number visible. Each sleeve number equals 0.100 in.
2. Count the number of lines between the numbered sleeve mark and the edge of the thimble. Each sleeve mark equals 0.025 in.
3. Read the thimble mark that aligns with the sleeve line. Each thimble mark equals 0.001 in.

NOTE
If a thimble mark does not align exactly with the sleeve line, estimate the amount between the lines. For accurate readings in ten-thousandths of an inch (0.0001 in.), use a vernier inch micrometer.

4. Add the readings from Steps 1-3.

Reading a vernier inch micrometer

A vernier inch micrometer is accurate to one ten-thousandth of an inch or 0.0001 in. It has the same marking as a

42

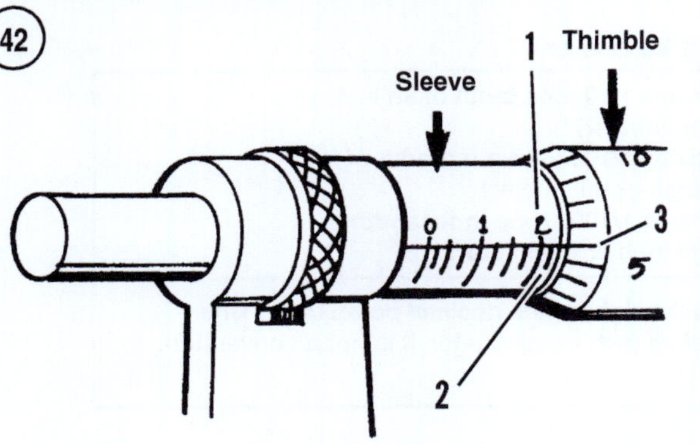

1. Largest number visible on the sleeve line	0.200 in.
2. Number on sleeve marks visible between the numbered sleeve mark and thimble edge	0.025 in.
3. Thimble mark that aligns with sleeve line	0.006 in.
Total reading	0.231 in.

43

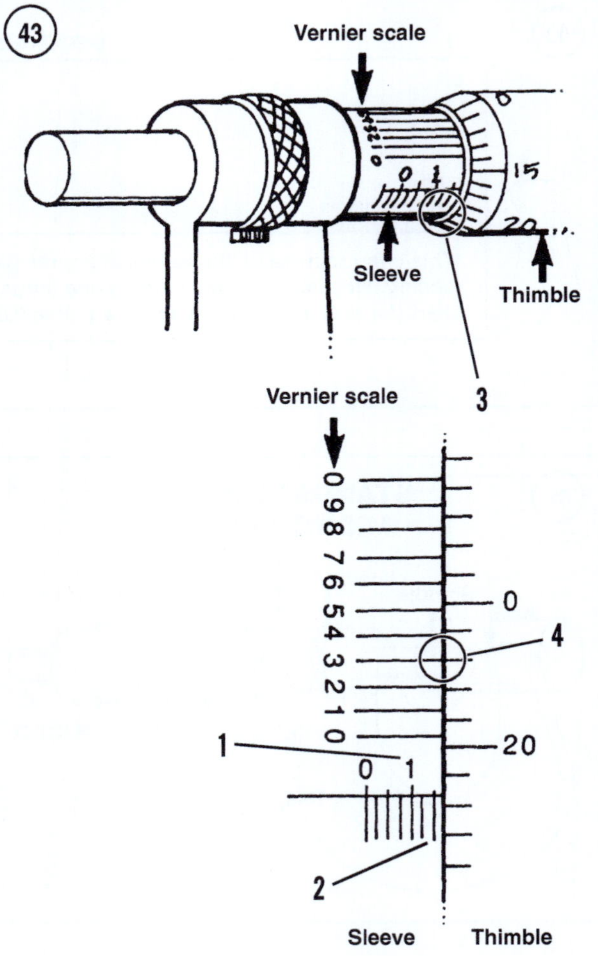

1. Largest number visible on sleeve line	0.1000 in.
2. Number of sleeve marks visible between the numbered sleeve mark and the thimble edge	2 × 0.025 in. = 0.0500 in.
3. Thimble is between 0.018 and 0.019 in. on the sleeve line	0.0180 in.
4. Vernier line coinciding with thimble line	0.0003 in.
Total reading	0.1683 in.

standard inch micrometer with an additional vernier scale on the sleeve (**Figure 43**).

The vernier scale consists of 11 lines marked 1-9 with a 0 on each end. These lines run parallel to the thimble lines and represent 0.0001 in. increments.

When reading a vernier inch micrometer, perform the following steps while referring to **Figure 43**.

1. Read the micrometer in the same way as a standard inch micrometer. This is the initial reading.
2. If a thimble mark aligns exactly with the sleeve line, reading the vernier scale is not necessary. If they do not align, read the vernier scale as described in Step 3.
3. Determine which vernier scale mark aligns with one thimble mark. The vernier scale number is the amount in ten-thousandths of an inch to add to the initial reading from Step 1.

Reading a standard metric micrometer

The standard metric micrometer (**Figure 44**) is accurate to one hundredth of a millimeter (0.01 mm). The sleeve line is graduated in millimeter and half-millimeter increments. The marks on the upper half of the sleeve line equal 1.00 mm. Every fifth mark above the sleeve line is identified with a number. The number sequence depends on the size of the micrometer. A 0-25 mm micrometer, for example, has sleeve marks numbered 0 through 25 in 5 mm increments. This numbering sequence continues with larger micrometers. On all metric micrometers, each mark on the lower half of the sleeve line equals 0.50 mm.

The tapered end of the thimble has fifty lines marked around it. Each mark equals 0.01 mm.

One complete turn of the thimble aligns the 0 mark with the first line on the lower half of the sleeve line or 0.50 mm.

When reading a standard metric micrometer, add the number of millimeters and half millimeters on the sleeve line to the number of hundredths of millimeters on the thimble. Perform the following steps while referring to **Figure 45**.

1. Read the upper half of the sleeve line and count the number of visible lines. Each upper line equals 1.00 mm.
2. Check whether the half-millimeter line is visible on the lower sleeve line. If so, add 0.50 mm to the reading in Step 1.
3. Read the thimble mark that aligns with the sleeve line. Each thimble mark equals 0.01 mm.

NOTE

If a thimble mark does not align exactly with the sleeve line, estimate the amount between

(44)

STANDARD METRIC MICROMETER

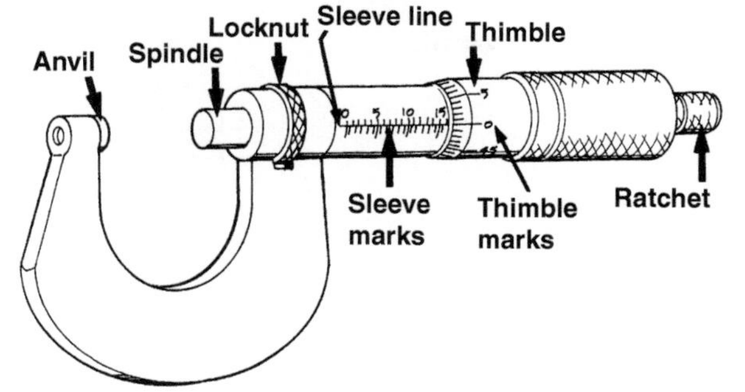

(45)

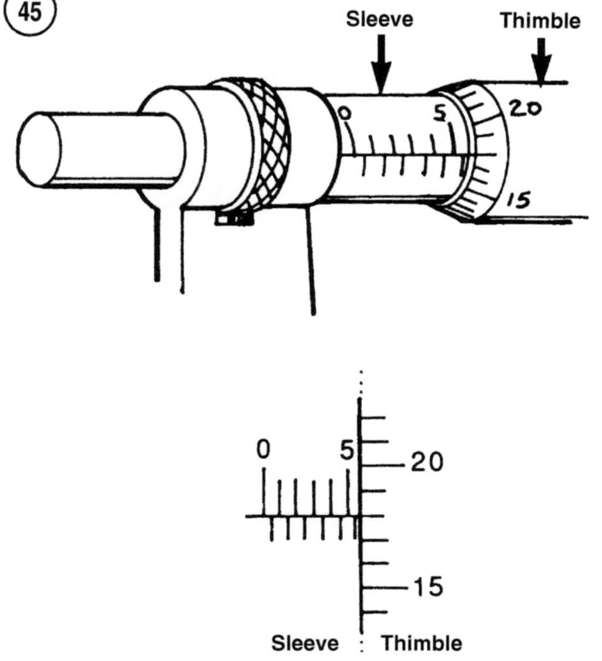

1. Upper sleeve line reading	5.00 mm
2. Lower sleeve line reading	0.50 mm
3. Thimble line coinciding with sleeve line	0.18 mm
Total reading	5.68 mm

(46)

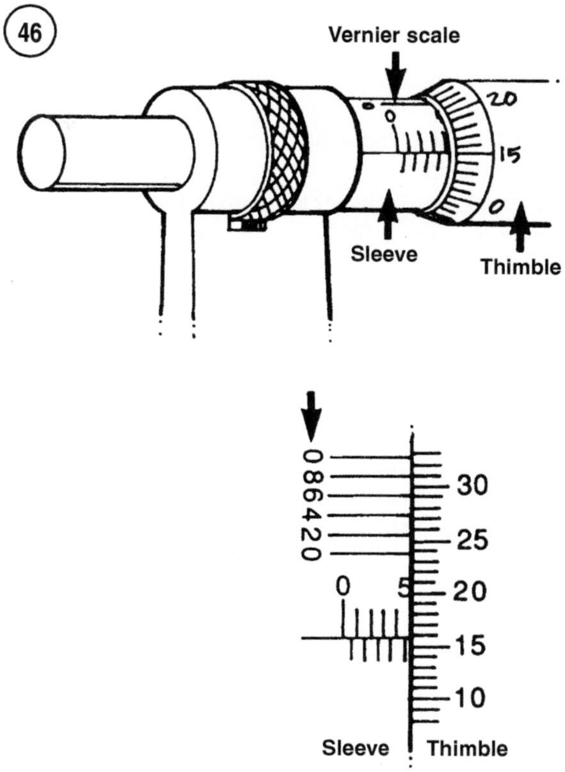

1. Upper sleeve line reading	4.000 mm
2. Lower sleeve line reading	0.500 mm
3. Thimble is between 0.15 and 0.16 mm lines on the sleeve line	0.150 mm
4. Vernier line coinciding with thimble line	0.008 mm
Total reading	4.658 mm

the lines. For accurate readings in two-thousandths of a millimeter (0.002 mm), use a metric vernier micrometer.

4. Add the readings from Steps 1-3.

Reading a metric vernier micrometer

A metric vernier micrometer (**Figure 46**) is accurate to two-thousandths of a millimeter (0.002 mm). It has the same markings as a standard metric micrometer with the addition of a vernier scale on the sleeve. The vernier scale consists of five lines marked 0, 2, 4, 6 and 8. These lines run parallel to the thimble lines and represent 0.002 mm increments.

When reading a metric vernier micrometer, refer to **Figure 46** and perform the following steps.

1. Read the micrometer in the same way as a standard metric micrometer. This is the initial reading.
2. If a thimble mark aligns exactly with the sleeve line, reading the vernier scale is not necessary. If they do not align, read the vernier scale as described in Step 3.
3. Determine which vernier scale mark aligns exactly with one thimble mark. The vernier scale number is the amount in two-thousandths of a millimeter to add to the initial reading from Step 1.

Micrometer Calibration

Before using a micrometer, check the calibration as follows.

1. Clean the anvil and spindle faces.

2A. To check a 0-1 in. or 0-25 mm micrometer:

a. Turn the thimble until the spindle contacts the anvil. If the micrometer has a ratchet stop, use it to ensure the proper amount of pressure is applied.

b. If the adjustment is correct, the 0 mark on the thimble will align exactly with the 0 mark on the sleeve line. If the marks do not align, the micrometer is out of adjustment.
c. Follow the manufacturer's instructions to adjust the micrometer.

2B. To check a micrometer larger than 1 in. or 25 mm, use the standard gauge supplied by the manufacturer. A standard gauge is a steel block, disc or rod that is machined to an exact size.

a. Place the standard gauge between the spindle and anvil and measure the outside diameter or length. If the micrometer has a ratchet stop, use it to ensure the proper amount of pressure is applied.
b. If the adjustment is correct, the 0 mark on the thimble will align exactly with the sleeve line. If the marks do not align, the micrometer is out of adjustment.
c. Follow the manufacturer's instructions to adjust the micrometer.

Micrometer care

Micrometers are precision instruments. Use and maintain them with great care.

Note the following:

1. Store micrometers in protective cases or separate padded drawers in a toolbox.
2. When in storage, make sure the spindle and anvil faces do not contact each other or another object. If they do, temperature changes and corrosion may damage the contact faces.
3. Do not clean a micrometer with compressed air. Dirt forced into the tool causes wear.
4. Lubricate micrometers with light oil to prevent corrosion.

Telescoping and Small-Bore Gauges

Use telescoping gauges (**Figure 47**) and small-bore gauges to measure bores. Neither gauge has a scale for direct reading. Use an outside micrometer to determine the reading.

To use a telescoping gauge, select the correct size gauge for the bore. Loosen the knurled end of the gauge to allow the post to extend. Compress the movable post and carefully insert the gauge into the bore. Carefully move the gauge in the bore to make sure it is centered. Tighten the knurled end of the gauge to hold the movable post in position. Carefully remove the gauge and measure the length of the post. Telescoping gauges are typically used to measure cylinder bores.

To use a small-bore gauge, select the correct size gauge for the bore. Carefully insert the gauge into the bore. Tighten the knurled end of the gauge to carefully expand the gauge fingers to the limit within the bore. Do not overtighten the gauge; there is no built-in release. Excessive tightening can damage the bore surface and the tool. Remove the gauge and measure the outside dimension. Small-bore gauges are typically used to measure valve guides.

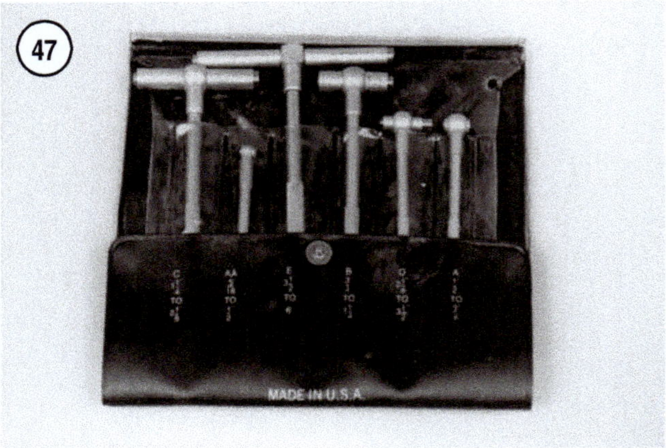

47

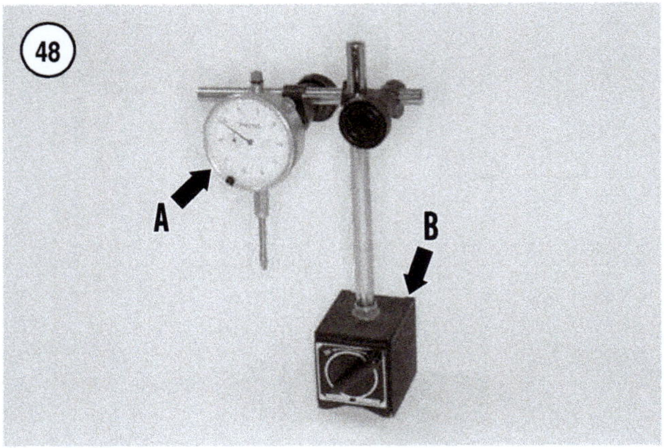

48

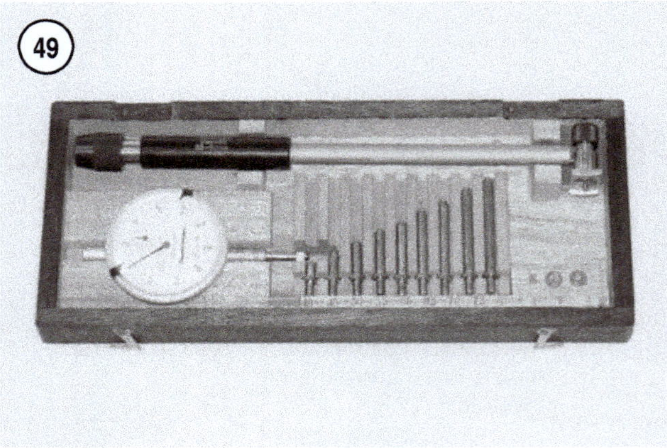
49

Dial Indicator

A dial indicator (**Figure 48**) is a gauge with a dial face and needle used to measure variations in dimensions and movements, such as crankshaft and gear shaft runout limits.

Dial indicators are available in various ranges and graduations and with three basic types of mounting bases: magnetic, clamp or screw-in stud. When purchasing a dial

50

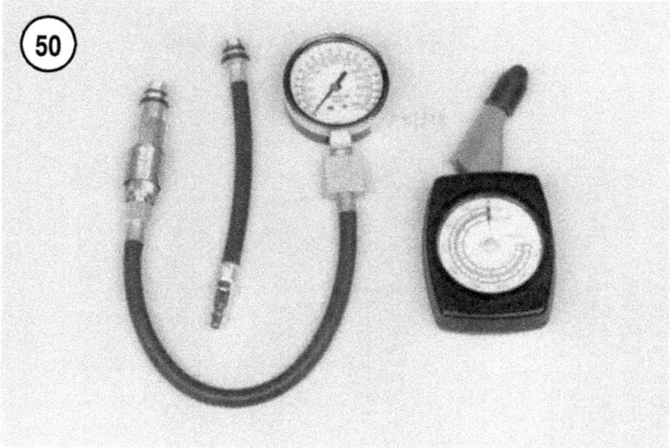

51

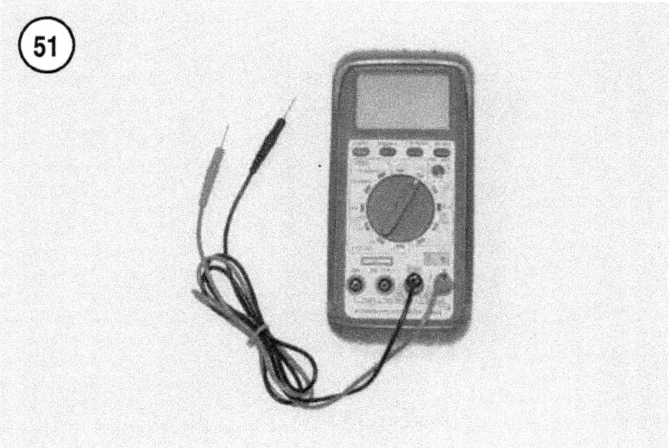

indicator, select the magnetic stand type (B, **Figure 48**) with a continuous dial (A).

Cylinder Bore Gauge

The cylinder bore gauge is a very specialized precision tool that is only needed for major engine repairs or rebuilds. The gauge set shown in **Figure 49** is comprised of a dial indicator, handle and a number of different length adapters (anvils) used to fit the gauge to various bore sizes. The bore gauge can be used to measure bore size, taper and out-of-round. When using a bore gauge, follow the manufacturer's instructions.

Compression Gauge

A compression gauge (**Figure 50**) measures the combustion chamber (cylinder) pressure, usually in psi or kg/cm^2. An engine is capable of mechanically generating pressure on the compression stroke. The gauge adapter is either inserted or screwed into the spark plug hole to obtain the reading. Disable the ignition or fuel control system so the engine does not start and hold the throttle in the wide-open position when performing a compression test. An engine that does not have adequate compression cannot be properly tuned.

Multimeter

A multimeter (**Figure 51**) is an essential tool for electrical system diagnosis. The voltage function indicates the voltage applied or available to various electrical components. The ohmmeter function test circuits for continuity, or lack of continuity, and measures the resistance of a circuit.

Some less expensive models contain a needle gauge and are known as analog meters. Most high-quality (but not necessarily expensive) meters available today contain digital readout screens. Digital multimeters are often know as DVOMs. When using an analog ohmmeter, the needle must be zeroed or calibrated according to the meter manufacturer's instructions. Some analog and almost all digital meters are self-zeroing, and no manual adjustment is necessary.

Some manufacturers' specifications for electrical components are based on results using a specific test meter. Results may vary if using a meter not recommended by the manufacturer.

ELECTRICAL SYSTEM FUNDAMENTALS

A thorough study of the many types of electrical systems used in engines today is beyond the scope of this manual. However, a basic understanding of electrical basics is necessary to perform simple diagnostic tests.

Voltage

Voltage is the electrical potential or pressure in an electrical circuit and is expressed in volts. The more pressure (voltage) in a circuit, the more work that can be performed.

Direct current (DC) voltage means the electricity flows in one direction. All circuits powered by a battery are DC circuits.

Alternating current (AC) means the electricity flows in one direction momentarily then switches to the opposite direction. Alternator output is an example of AC voltage. The voltage must be changed or rectified to direct current to operate in a 12-volt battery powered system.

Measuring voltage

Unless otherwise specified, perform all voltage tests with the electrical connectors attached.

When measuring voltage, select a meter range one scale higher than the expected voltage of the circuit to prevent damage to the meter. To determine actual voltage in a circuit, use a voltmeter. To simply check whether voltage is present, use a test lamp.

NOTE

When using a test lamp, either lead can be attached to ground.

1. Attach the negative meter test lead to a good ground (bare metal). Make sure the ground is not insulated with a rubber gasket or grommet.
2. Attach the positive meter test lead to the point being checked for voltage (**Figure 52**).
3. If necessary for the circuit being tested, turn on the ignition switch. This will be necessary if the point being tested only has power applied when the ignition switch is turned on, but the example in **Figure 52** shows a measurement at the positive battery terminal, which should always have voltage if the battery is charged. The test lamp should light, or the meter should display a reading. The reading should be within 1 volt of battery voltage. If the voltage is less, there is likely a problem in the circuit.

52 **MEASURING VOLTAGE**

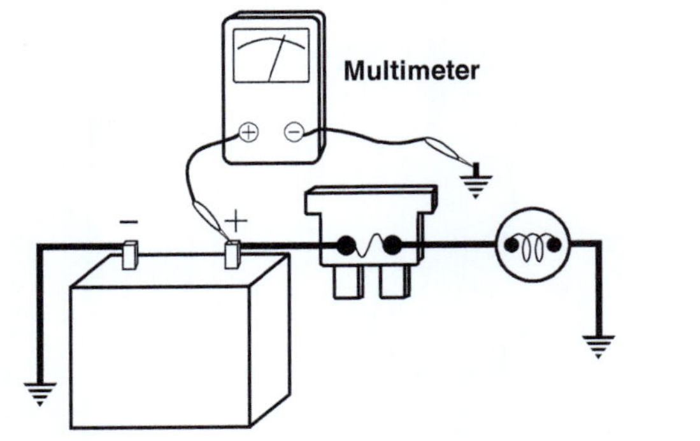

Voltage drop test

Resistance causes voltage to drop. This resistance can be measured in an active circuit by using a voltmeter to perform a voltage drop test. A voltage drop test compares the difference between the voltage available at the start of a circuit to the voltage at the end of the circuit, but it does so while the circuit is operational. If the circuit has no resistance, there will be no voltage drop. The greater the resistance, the greater the voltage drop will be. A voltage drop of 1 volt or more usually indicates excessive resistance in the circuit.
1. Connect the positive meter test lead to the electrical source (where the electricity is coming from).
2. Connect the negative meter test lead to the electrical load (where the electricity is going. See **Figure 53**.
3. If necessary, activate the component(s) in the circuit.
4. A voltage reading of 1 volt or more indicates excessive resistance in the circuit. A reading equal to battery voltage indicates an open circuit.

53 **VOLTAGE DROP TEST**

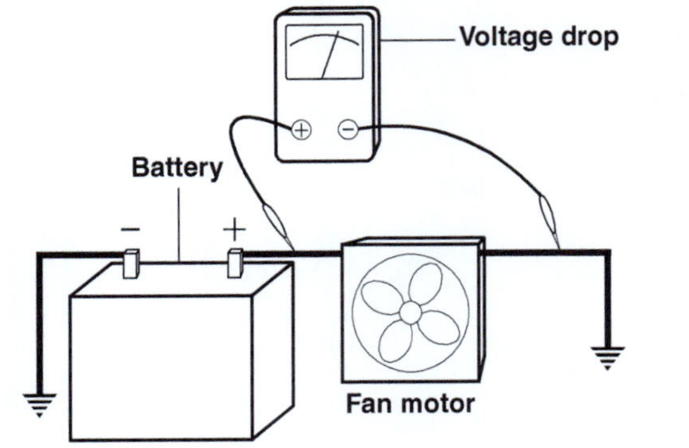

Resistance

Resistance is the opposition to flow of electricity within a circuit or component and is measured in ohms. Resistance causes a reduction in available current and voltage.

Resistance is measured in an inactive circuit with an ohmmeter. The ohmmeter sends a small amount of current into the circuit and measures how difficult it is to push the current through the circuit.

An ohmmeter, though useful, is not always a good indicator of the actual ability of the circuit under operating conditions because the meter uses low voltage (6-9 volts) to test the circuit. The voltage in an ignition coil secondary winding can be several thousand volts. Such high voltage can cause the coil to malfunction, even though it tests acceptably during a resistance test.

Resistance generally increases with temperature. Unless specified otherwise, perform all testing with the component or circuit at room temperature. Resistance tests performed at high temperatures might indicate high resistance readings and result in unnecessary replacement of a component.

Resistance and continuity test

CAUTION
Only use an ohmmeter on a circuit that has no voltage present. The meter will be damaged if it is connected to a live circuit. Remember, if using an analog meter, it must be calibrated each time it is used or if the scale is changed.

A continuity test can determine whether the circuit is complete. This type of test is performed with an ohmmeter or a self-powered test lamp.

1. Disconnect the negative battery cable.
2. Attach one test lead (ohmmeter or test lamp) to one end of the component or circuit.
3. Attach the other test lead to the opposite end of the component or circuit. See **Figure 54**.
4. A self-powered test lamp will come on if the circuit has continuity or is complete. An ohmmeter will indicate either low or no resistance if the circuit has continuity. An open circuit is indicated if the meter displays infinite resistance.

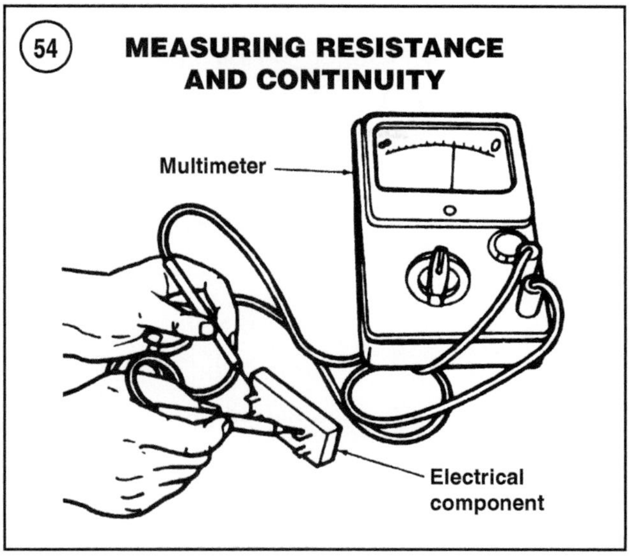

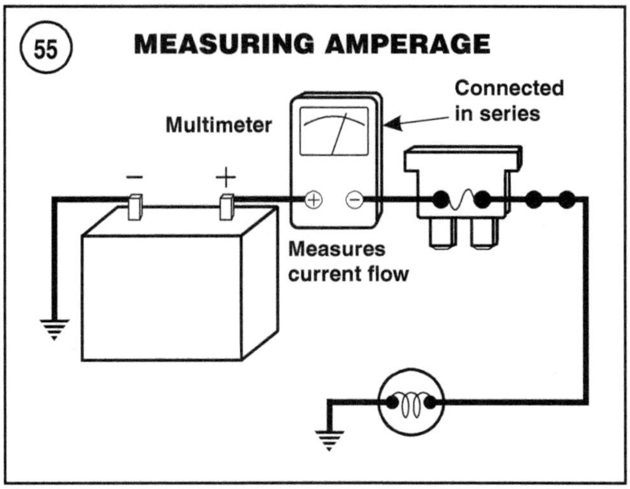

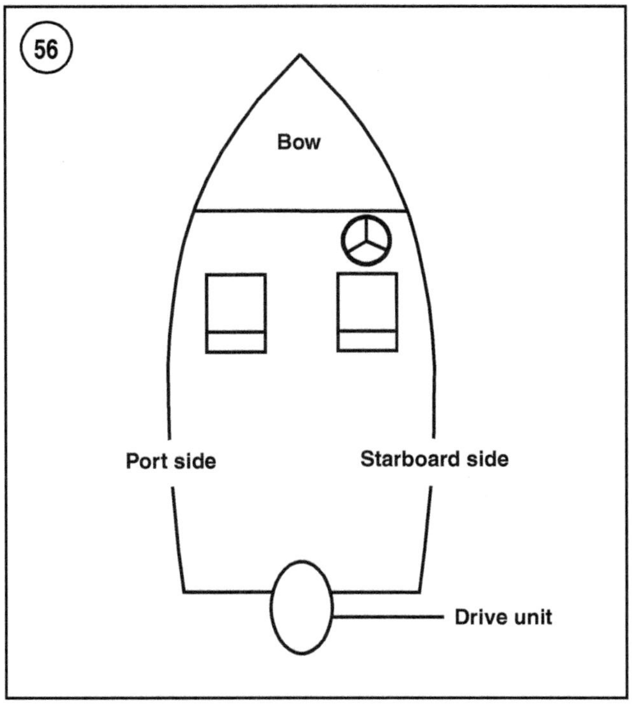

Amperage

Amperage is the unit of measure for the amount of current within a circuit. Current is the actual flow of electricity. The higher the current, the more work that can be performed up to a given point. If the current flow exceeds the circuit or component capacity, the system will be damaged.

Measuring amperage

An ammeter measures the current flow in a circuit (**Figure 55**). Amperage measurement requires that the circuit be disconnected and the ammeter be connected in series in the circuit. To prevent damage to the meter, always us an ammeter that can read higher than the anticipated current flow. Connect the red test lead to the electrical source and the black test lead to the electrical load.

BASIC MECHANICAL SKILLS

Most of the service procedures covered in this manual are straightforward and can be performed by anyone reasonable handy with tools. It is suggested, however, to consider your own capabilities carefully before attempting any operation involving major disassembly.

1. *Front*, as used in this manual, refers to the front of the engine or the side of the engine facing the front or *bow* of the boat (see **Figure 56**). The *left* side is referred to as the *port* side. Likewise, the *right* side is referred to as the *starboard* side. These terms are simple, but confusion can cause a major inconvenience during service.
2. When disassembling engine or drive components, mark the parts for location and mark all parts that mate together. Placing them in plastic sandwich bags can identify small parts, such as bolts. Seal the bags and label them with masking tape and a marking pen. Because many types of ink fade when applied to tape, use a permanent ink pen. If reassembly will occur immediately, place nuts and bolts in a cupcake tin or egg carton in the order of disassembly.
3. Protect finished surfaces from physical damage or corrosion. Keep gasoline off painted surfaces.
4. Apply penetrating oil to frozen or tight bolts. Then strike the bolt head a few times with a hammer and punch. (Use a screwdriver on screws.) Avoid the use of heat where possible, as it can warp, melt or affect the temper of parts. Heat also ruins finishes, especially paint, decals and plastics.
5. Unless otherwise noted, no parts removed or installed (other than bushings and bearings) in the procedures given in this manual should require unusual force during disassembly or assembly. If a part is difficult to remove or install, find out why before proceeding.
6. Cover all openings after removing parts or components to prevent objects such as dirt or small tools from falling in.
7. Read each procedure completely while looking at the actual parts before starting a job. Make sure to thoroughly

understand what is to be done and then carefully follow the procedure, step by step.

8. For the do-it-yourselfer, recommendations are occasionally made to refer service or maintenance to the dealership or specialist in a particular field. In these cases, the work will be done more quickly and economically than performing the job alone.

9. In procedural steps, the term *replace* means to discard a defective part and replace it with a new or exchange unit. *Overhaul* means to remove, disassemble, inspect, measure, repair or replace defective parts, reassemble and install major systems or parts.

10. Some operations require the use of a hydraulic press. If a suitable press is not available, it is wiser to have these operations performed by a shop equipped for such work rather than to try to do the job using makeshift equipment that can damage the engine.

11. Repairs go much faster and easier if the machine is clean before beginning work.

12. If special tools are required, make arrangements to get them before starting. It is frustrating and time-consuming to start a job and then be unable to complete it.

13. Make diagrams or take a picture wherever similar-appearing parts are found. For instance, intake manifold bolts often are not the same length. Do not try to remember where everything came from because mistakes are costly. It is also possible to become sidetracked and not return to work for days or even weeks, in which time, carefully laid out parts might have been disturbed.

14. When assembling parts, be sure all shims and washers are reinstalled in the original locations.

15. Whenever a rotating part butts against a stationary part, look for a shim or washer. Use new gaskets if there is any doubt about the condition of the used ones. A thin coating of silicone sealant on nonpressure gaskets may help them seal more effectively.

16. If it becomes necessary to purchase gasket material to make a gasket for the engine, measure the thickness of the used gasket (at an uncompressed point) and purchase the same type of gasket material with the same approximate thickness.

17. Heavy grease can be used to hold small parts in place if they fall out during assembly. However, keep grease and oil away from electrical components unless otherwise directed.

18. Never use wire to clean out fuel jets and air passages. They are easily damaged.

19. Take the time to do the job right. Do not forget that a newly rebuilt engine must be broken in just like a new one.

Removing Frozen Fasteners

If a fastener cannot be removed, several methods may be used to loosen it. First, apply penetrating oil such as Liquid Wrench, WD-40 or PB Blaster. Apply it liberally and let it penetrate for 10-15 minutes. Rap the fastener several times with a small hammer. Do not hit it hard enough to cause damage. Reapply the penetrating oil if necessary.

57

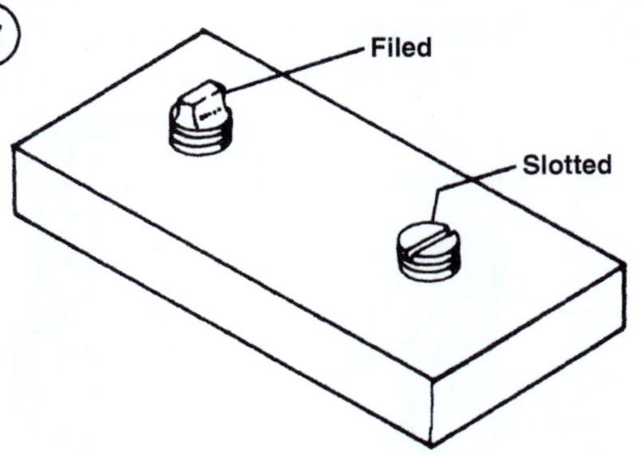

58

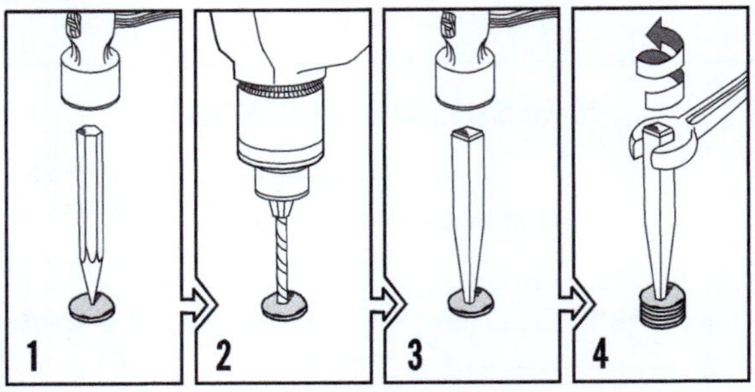

1. Center punch broken stud
2. Drill hole in stud
3. Tap in extractor
4. Remove broken stud

For frozen screws, apply penetrating oil as described previously. Then insert a screwdriver into the slot and rap the top of the screwdriver with a hammer. This technique loosens the corrosion so the screw can be removed in the normal fashion. If the screw head is too damaged to use this method, grip the head with locking pliers and twist the screw out.

Avoid applying heat unless specifically instructed because it can melt, warp or remove the temper from parts.

Removing Broken Fasteners

If the head breaks off a screw or bolt, several methods are available for removing the remaining portion. If a large portion of the remainder projects out, try gripping it with locking pliers. If the projecting portion is too small or a sufficient grip cannot be obtained on the protruding piece, file it to fit a wrench or cut a slot in it to fit a screwdriver (**Figure 57**).

If the head breaks off flush, use an extractor. To do this, center punch the exact center of the remaining portion of the screw or bolt. Drill a small hole in the top and tap the extractor into the hole. Back the fastener out with a wrench on the extractor (**Figure 58**).

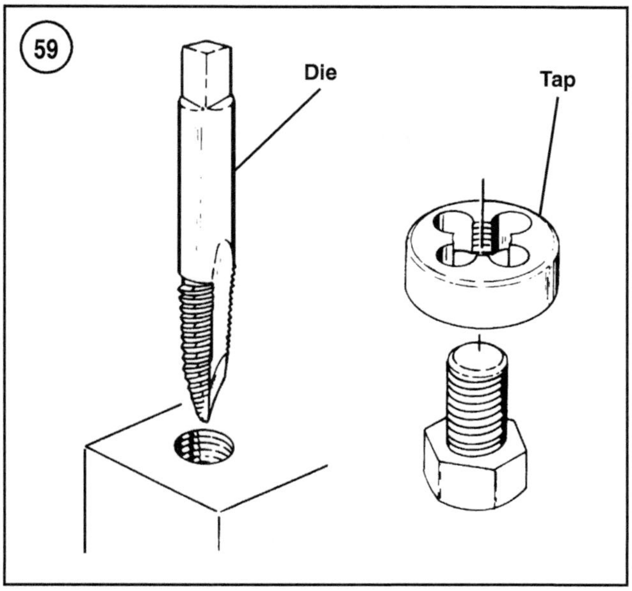

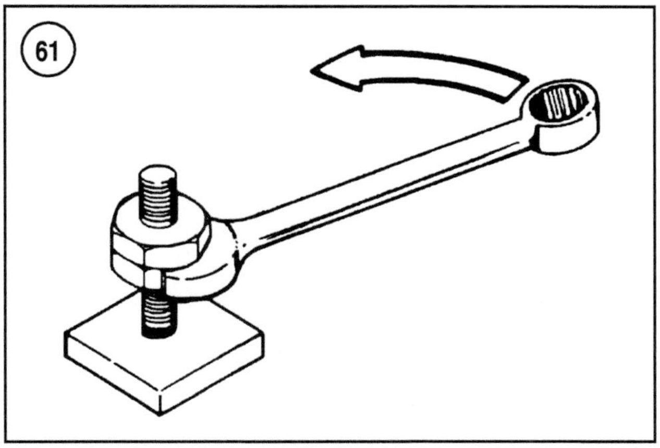

NOTE
Broken screw extraction sometimes fails to remove the fastener from the bore. If this occurs, or if the screw is drilled off-center and the threads are damaged, a threaded insert will be necessary to repair the bore. Check for one at a local dealership or supply store and follow the manufacturer's instructions for installation.

Repairing Damaged Threads

Occasionally, threads are stripped through carelessness or impact damage. Often, the threads can be repaired by running a tap (for internal threads or nuts) or die (for external threads on bolts) through the threads (**Figure 59**). Use only a specially designed spark plug tap to clean or repair spark plug threads.

If an internal thread is damaged, it may be necessary to install a thread insert. Follow the manufacturer's instructions when installing the insert.

If it is necessary to drill and tap a hole, refer to **Table 3** or **Table 4** for tap drill sizes.

Stud Removal/Installation

A stud removal tool (**Figure 60**, typical) that makes the removal and installation of studs easier is available from most tool suppliers. If one is not available, thread two nuts onto the stud and tighten them against each other to lock them in place, then remove the stud by turning the lower nut (**Figure 61**). Turn the top nut to install the stud using the same method.

NOTE
If the threads on the damaged stud do not allow installation of the two nuts, remove the stud with a pair of locking pliers or a stud remover.

Removing Hoses

When removing stubborn hoses, do not exert excessive force on the hose or fitting. Remove the hose clamp and carefully insert a small screwdriver or pick tool between the fitting and hose. Apply a spray lubricant under the hose and carefully twist the hose off the fitting. Clean the fitting of any corrosion or rubber hose material with a wire brush. Clean the inside of the hose thoroughly. Do not use any lubricant when installing the hose (new or used). The lubricant might allow the hose to come off the fitting even with the clamp secure.

Bearings

Bearings are used in the engine and drive unit assembly to reduce power loss, excessive heat and noise resulting from friction. Because bearings are precision parts, it is necessary to maintain them with proper lubrication and maintenance. If a bearing is damaged, replace it immediately. Bearing replacement procedures are included in the applicable chapter; however, use the following sections as a guideline.

NOTE
Unless otherwise specified, install bearings with the manufacturer's marking, symbol or number facing outward.

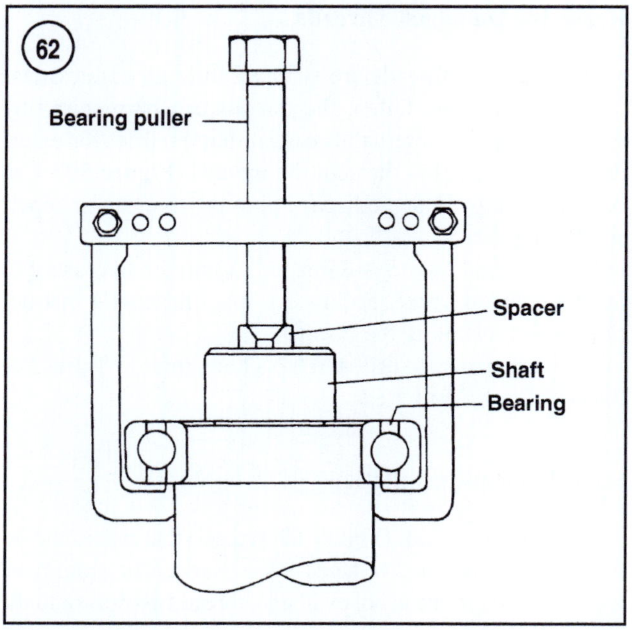

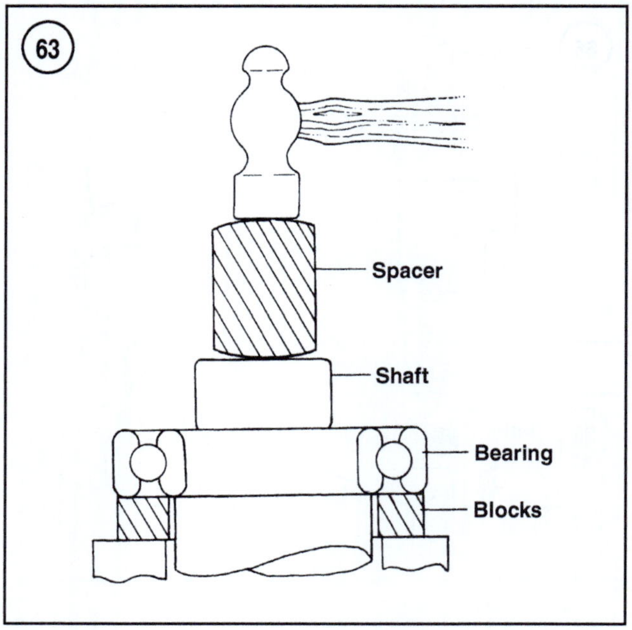

Removal

While bearings are normally removed only when damaged, there might be times when it is necessary to remove and reuse a bearing that is in good condition. However, improper removal will damage the bearing, shaft, and/or case half. Note the following when removing bearings.

1. When using a puller (**Figure 62**) to remove a bearing from a shaft, take care not to damage the shaft. Always place a piece of metal between the end of the shaft and the puller bolt. In addition, place the puller arms next to the inner bearing race.
2. When using a hammer to remove a bearing from a shaft, do not strike the hammer directly against the shaft. Instead, use a brass or aluminum spacer between the hammer and shaft (**Figure 63**) and make sure to support both bearing races with wooden blocks as shown.
3. The ideal method for bearing removal is with a hydraulic press. In order to prevent damage to the bearing and shaft or case, note the following when using a press:
   a. Always support the inner and outer bearing races with a suitably sized wooden or aluminum spacer ring (**Figure 64**). If only the outer race is supported, pressure applied against the balls or rollers and/or the inner race will damage them.
   b. Always make sure the press ram (**Figure 64**) aligns with the center of the shaft. If the ram is not centered, it might damage the bearing and/or shaft.
   c. The moment the shaft is free from the bearing, it will drop to the floor. Secure the shaft to prevent it from falling.

Installation

1. When installing a bearing into a housing, apply pressure to the *outer* bearing race (**Figure 65**). When installing a bearing onto a shaft, apply pressure to the *inner* bearing race (**Figure 66**).

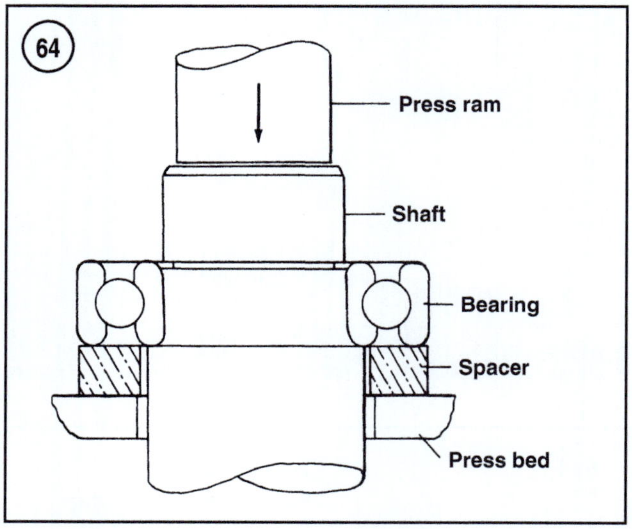

2. When installing a bearing as described in Step 1, use some type of driver. Never strike the bearing directly with a hammer, or the bearing will be damaged. Use a section of pipe or socket with a diameter that matches the bearing race. **Figure 67** shows the correct way to use a socket and hammer to install a bearing onto a shaft.

3. Step 1 describes how to install a bearing into a case half or over a shaft. When installing a bearing over a shaft *and* into a housing at the same time, however, a tight fit is required for both outer and inner bearing races. In this situation, install a spacer underneath the driver tool so that pressure is applied evenly across both races. See **Figure 68**. If the outer race is not supported as shown, the balls or rollers will push against the outer bearing race and damage it.

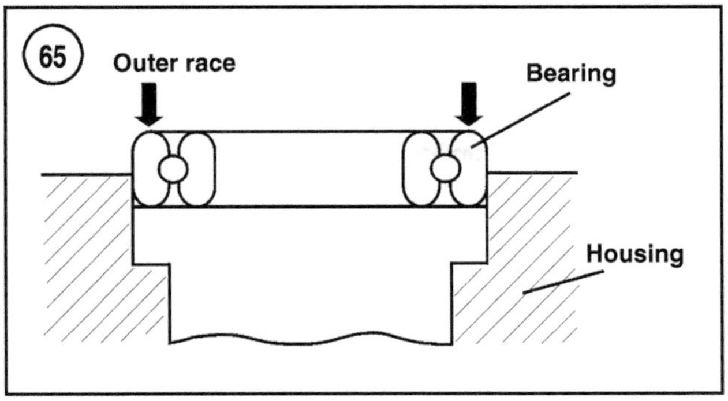

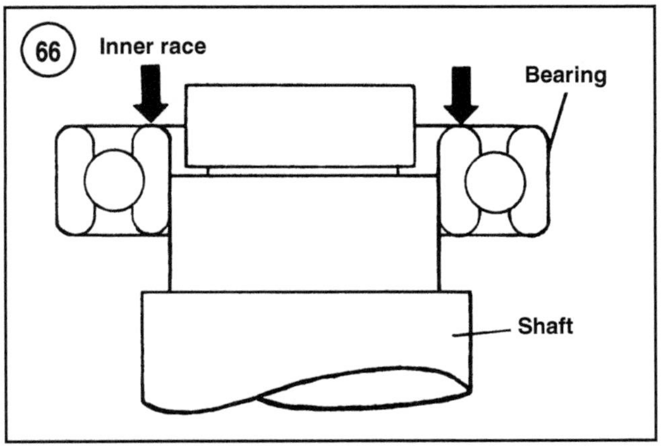

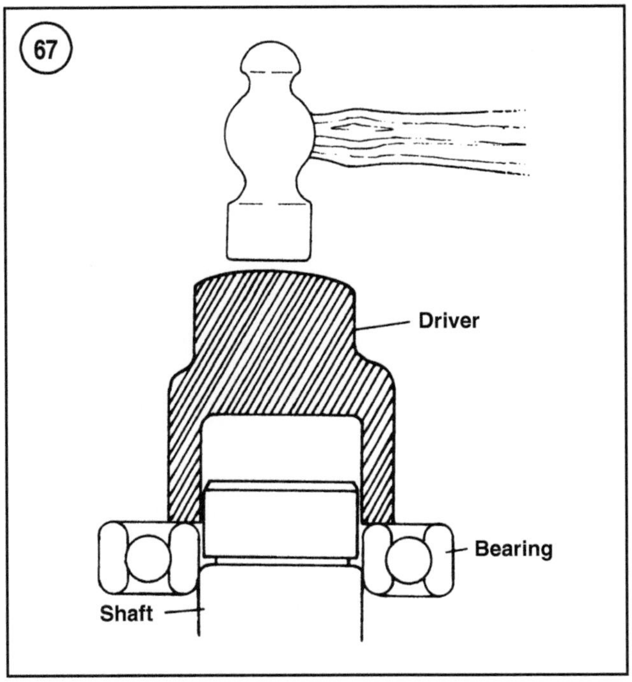

Installing an interference fit bearing over a shaft

When a tight fit is required, the bearing inside diameter will be smaller than the shaft. In this case, driving the bearing onto the shaft might cause bearing damage. Instead, heat the bearing before installation. Note the following:

1. Secure the shaft so that it is ready for bearing installation. While the parts are still cold, determine the proper size and gather all necessary spacers and drivers for installation.

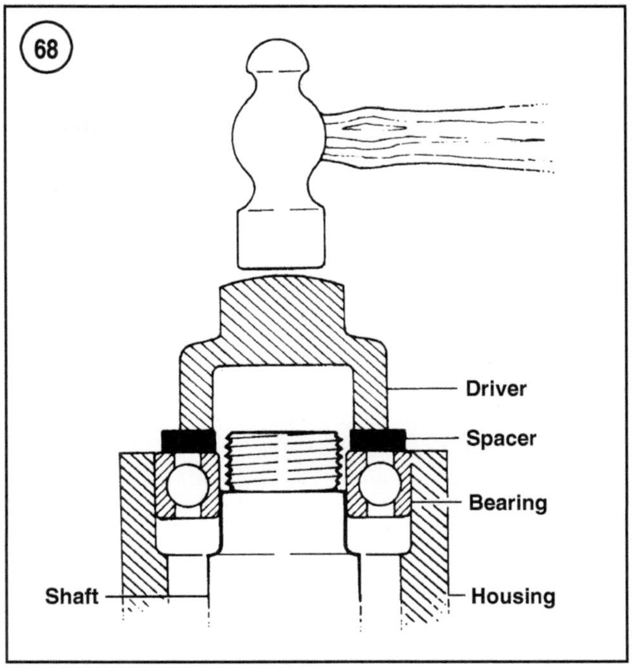

2. Clean the bearing surface on the shaft of all residues. Remove burrs with a file or sandpaper.
3. Fill a suitable pot or beaker with clean mineral oil. Place a thermometer rated higher than 248° F (120° C) into the oil. Support the thermometer so it does not rest on the bottom or side of the container.
4. Secure the bearing with a piece of heavy wire bent to hold it in the container. Hang the bearing so it does not touch the bottom or sides of the container.
5. Turn the heat ON and monitor the thermometer. When the oil temperature rises to approximately 248° F (120° C), remove the bearing from the container and quickly install it. If necessary, place a driver against the inner bearing race and tap the bearing into place. As the bearing cools, it will tighten on the shaft, so work quickly when installing it. Make sure the bearing is installed completely.

Replacing an interference fit bearing in a housing

Bearings are generally installed into a housing with a slight interference fit. Driving the bearing into the housing might damage the housing or bearing. Instead, heat the housing (to make the inner diameter of the bore larger) and chill the bearing (in order to make the outer diameter slightly smaller) before installation. This makes bearing installation much easier.

CAUTION

Before heating the housing in this procedure to remove the bearing(s), wash the housing thoroughly with detergent and clean water. In order to prevent a possible fire hazard, rinse and rewash the housing as required to remove all traces of oil and other chemical deposits.

1. While the parts are still cool, determine the proper size and gather all necessary spacers and drivers for installation.
2. Place the new bearing into a freezer to chill it and slightly reduce the outside diameter.
3. While the bearing is chilling, heat the housing to a temperature of about 212° F (100° C) in an oven or on a hot plate. An easy way to check whether the housing is hot enough is to drop tiny droplets of water onto the case. If they sizzle and evaporate immediately, the temperature is correct. Heat only one housing at a time.

CAUTION
Do not heat the housing with a propane or acetylene torch. Never bring a flame into contact with the bearing or housing. The direct heat will destroy the case hardening of the bearing and will likely warp the housing.

4. Remove the housing from the oven or hot plate using a thick kitchen potholder, heavy protective gloves or heavy shop cloths.

NOTE
A suitably sized socket and extension work well for removing and installing bearings.

5. Hold the housing with the bearing side down and tap out the used bearing. Repeat the process for all bearings in the housing.

NOTE
Always install bearings with the manufacturer's marking, symbol or number facing outward.

6. While the housing is still hot, install the chilled new bearing(s) into the housing. Install the bearing(s) by hand, if possible. If necessary, lightly tap the bearing(s) into the bore(s) with a socket placed against the outer bearing race. *Do not* install the bearings by driving against the inner bearing race. Drive each bearing into the bore until it seats completely.

Seal Replacement

Seals (**Figure 69**) are used to contain oil, water, grease or combustible gasses in a housing or shaft. Improper removal of a seal can damage components. Improper installation can damage the seal and/or other components. Note the following:

1. Prying is generally the easiest and most effective method of removing a seal from a housing. However, always place a shop towel under the pry tool to prevent damage to the housing.
2. Before installing the seal, pack waterproof grease into the seal lips.
3. In most cases, install seals with the manufacturer's marks or numbers facing outward.
4. Install seals with a driver placed against the outside of the seal as shown in **Figure 70**. Drive the seal squarely into the housing. Never install a seal by hitting directly against the top of the seal with a hammer.

69

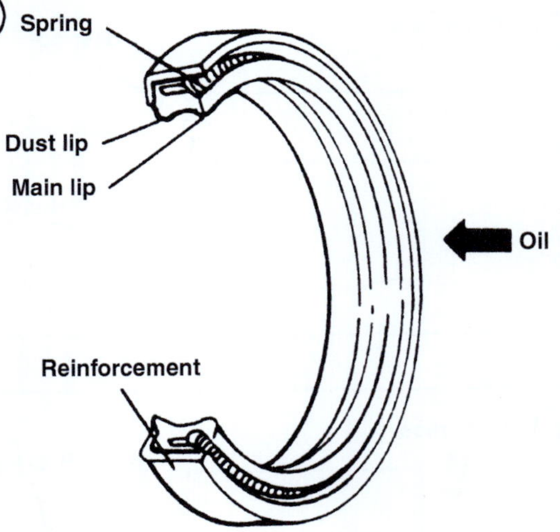

70

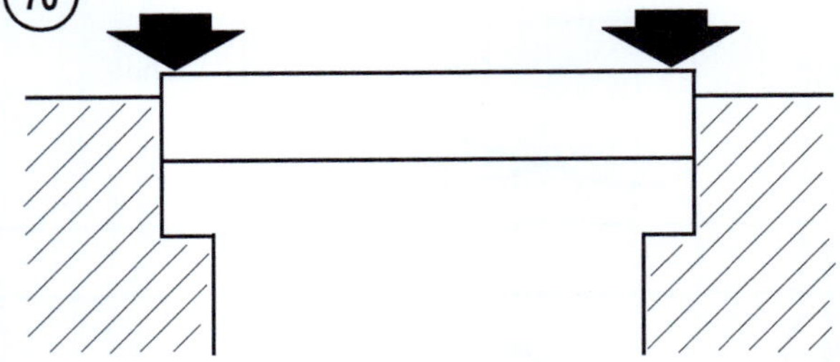

Table 1 TECHNICAL ABBREVIATIONS

ABDC	After bottom dead center
A.h.	Ampere hour
ATDC	After top dead center
BBDC	Before bottom dead center
BDC	Bottom dead center
BTDC	Before top dead center
C	Celsius (centigrade)
cc	Cubic centimeters

(continued)

Table 1 TECHNICAL ABBREVIATIONS (continued)

CCA	Cold cranking amps
CDI	Capacitor discharge ignition
CKP	Crankshaft position
CMP	Camshaft position
CT	Cylinder temperature
CTP	Closed throttle position
CTS	Coolant temperature sensor
cu.-in.	Cubic inch
DOHC	Dual-overhead camshafts
ECA	Engine control assembly
ECT	Engine coolant temperature sensor
ECU	Engine control unit
EFI	Electronic fuel injection
EM	Exhaust manifold
ETS	Engine temperature sensor
F	Fahrenheit
ft.-lb.	Foot-pounds
g	Gram
gal.	Gallons
hp	Horsepower
HPDI	High pressure direct injection
IAC	Idle air control motor
IAT	Intake air temperature sensor
in.	Inches
in. Hg	Inches of mercury
in.-lb.	Inch-pounds
ISC	Idle speed control solenoid
kg	Kilogram
kg/cm^2	Kilograms per square centimeter
kgm	Kilogram meters
km	Kilometer
KS	Knock sensor
L	Liter
m	Meter
MAG	Magneto
MAP	Manifold absolute pressure sensor
MCA	Marine cranking amps
mm	Millimeter
MPEFI	Multiport electronic fuel injection
MPI	Multiport injection
N.A.	Not available
N•m	Newton-meters
oz.	Ounce
OHC	Overhead camshaft
psi	Pounds per square inch
pto	Power take off
pts.	Pints
qt.	Quarts
RFI	Radio frequency interference
rpm	Revolutions per minute
TBI	Throttle body injection
TFI	Thick film integrated ignition
TPS	Throttle position sensor
WOT	Wide open throttle

Table 2 CONVERSION FORMULAS

Multiply	By	To get equivalent of
Length		
Inches	25.4	Millimeter
Inches	2.54	Centimeter
Miles	1.609	Kilometer

(continued)

Table 2 CONVERSION FORMULAS (continued)

Multiply	By	To get equivalent of
Length (continued)		
Feet	0.3048	Meter
Millimeter	0.03937	Inches
Centimeter	0.3937	Inches
Kilometer	0.6214	Mile
Meter	3.281	Foot
Fluid volume		
U.S. quarts	0.9463	Liters
U.S. gallons	3.785	Liters
U.S. ounces	29.573529	Milliliters
Liters	0.2641721	U.S. gallons
Liters	1.0566882	U.S. quarts
Liters	33.814023	U.S. ounces
Milliliters	0.033814	U.S. ounces
Milliliters	1.0	Cubic centimeters
Milliliters	0.001	Liters
Torque		
Foot-pounds	1.3558	Newton-meters
Foot-pounds	0.138255	Meter-kilograms
Inch-pounds	0.11299	Newton-meters
Newton-meters	0.7375622	Foot-pounds
Newton-meters	8.8507	Inch-pounds
Meter-kilograms	7.2330139	Foot-pounds
Volume		
Cubic inches	16.387064	Cubic centimeters
Cubic centimeters	0.0610237	Cubic inches
Temperature		
Fahrenheit	(°F −32) × 0556	Centigrade
Centigrade	(°C × 1.8) + 32	Fahrenheit
Weight		
Ounces	28.3495	Grams
Pounds	0.4535924	Kilograms
Grams	0.035274	Ounces
Kilograms	2.2046224	Pounds
Pressure		
Pounds per square inch	0.070307	Kilograms per square centimeter
Kilograms per square centimeter	14.223343	Pounds per square inch
Kilopascals	0.1450	Pounds per square inch
Pounds per square inch	6.895	Kilopascals
Speed		
Miles per hour	1.609344	Kilometers per hour
Kilometers per hour	0.6213712	Miles per hour

Table 3 U.S. STANDARD TAP AND DRILL SIZES

Tap size	Drill size	Decimal equivalent
#0-80	3/64	0.047
#1-64	No. 53	0.059
#1-72	No. 53	0.059
#2-56	No. 50	0.070
#2-64	No. 50	0.070
#3-48	5/64	0.078
#3-56	No. 46	0.081
#4-40	No. 43	0.890
#4-48	No. 42	0.935
#5-40	No. 39	0.938
#5-44	No. 37	0.104
#6-32	No. 36	0.107
#6-40	No. 33	0.113
#8-32	No. 29	0.136
#8-36	No. 29	0.136

(continued)

Table 3 U.S. STANDARD TAP AND DRILL SIZES (continued)

Tap size	Drill size	Decimal equivalent
#10-24	No. 25	0.150
#10-32	No. 21	0.159
#12-24	No. 16	0.177
#12-28	No. 14	0.182
1/ 4-20	No. 7	0.201
1/ 4-28	No. 3	0.213
5/16-18	F	0.257
5/16-24	I	0.272
3/8-16	5/16	0.3125
3/8-24	Q	0.332
7/16-14	U	0.368
7/16-20	25/64	0.390
1/ 2-13	27/64	0.422
1/ 2-20	29-64	0.453
9/16-12	31-64	0.484
9/16-18	33/64	0.516
5/8-11	17/32	0.531
5/8-18	37/64	0.578
3/ 4-10	21/32	0.656
3/ 4-16	11/16	0.688
7/8-9	49/64	0.766
7/8-14	13/16	0.813
1-8	7/8	0.875
1-14	15/16	0.938

Table 4 METRIC TAP AND DRILL SIZES

Metric tap (mm)	Drill size	Decimal equivalent	Nearest fraction (in.)
3 × 0.50	No. 39	0.0995	3/32
3 × 0.60	3/32	0.0937	3/32
4 × 0.70	No. 30	0.1285	1/8
4 × 0.75	1/8	0.125	1/8
5 × 0.80	No. 19	0.166	11/64
5 × 0.90	No. 20	0.161	5/32
6 × 1.00	No. 9	0.196	13/64
7 × 1.00	16/64	0.234	15/64
8 × 1.00	J	0.277	9/32
8 × 1.25	17/64	0.265	17/64
9 × 1.00	5/16	0.3125	5/16
9 × 1.25	5/16	0.3125	5/16
10 × 1.25	11/32	0.3437	11/32
10 × 1.50	R	0.339	11/32
11 × 1.50	3/8	0.375	3/8
12 × 1.50	13/32	0.406	13/32
12 × 1.75	13/32	0.406	13/32

Table 5 TORQUE RECOMMENDATIONS

Screw or nut size	in.-lb.	ft.-lb.	N•m
U.S. Standard			
6-32	9	–	1.0
8-32	20	–	2.3
10-24	30	–	3.4
10-32	35	–	4.0
12-24	45	–	5.1
1/4-20	70	–	7.9
1/4-28	84	–	9.5

(continued)

Table 5 TORQUE RECOMMENDATIONS (continued)

Screw or nut size	in.-lb.	ft.-lb.	N•m
U.S. Standard (continued)			
5/16-18	160	13	18
5/16-24	168	14	19
3/8-16	–	23	31
3/8-24	–	25	34
7/16-14	–	36	49
7/16-20	–	40	54
1/2-13	–	50	68
1/2-20	–	60	81
Metric			
M5	36	–	4.0
M6	70	–	8.0
M8	156	13	18
M10	–	26	35
M12	–	35	48
M14	–	60	81

Table 6 ENGINE IDENTIFICATION

Characters	Characteristics
Engine	
First and second characters	
30	3.0 Liter inline four-cylinder
43	4.3 Liter V configuration six-cylinder
50	5.0 Liter V configuration eight-cylinder
57	5.7 Liter V configuration eight-cylinder
58	5.8 Liter V configuration eight-cylinder
74	7.4 Liter V configuration eight-cylinder
82	8.2 Liter V configuration eight-cylinder
Third character	
2	Two-barrel carburetor fuel system
4	Four-barrel carburetor fuel system
F	Electronic fuel injection system
Fourth character	Letter indicating engine version
Fifth character	
C	Engine produced as an OMC Cobra SX unit
G	Engine produced as a Volvo Penta SX unit
Sixth character	
M	Manual steering system
P	Power assist steering system
Seventh and eighth characters	
RG	1991 model year
AM	1992 model year
JV	1993 model year
MD	1994 model year
HU	1995 model year
NC	1996 model year
KD	1997 model year
BY	1998 model year
WT	1999 model year
FS	2000 model year
Ninth character	Production version

Chapter Two

Troubleshooting

This chapter describes troubleshooting and testing procedures for all engine systems. An OMC Cobra SX stern drive engine comprises of three basic assemblies: the engine, transom bracket and the drive unit. Engine systems include the starting system, charging system, fuel system and ignition system. If done logically, troubleshooting is a relatively simple matter. Taking a haphazard approach might eventually solve the problem, but it can be very costly in terms of wasted time and unnecessary parts replacement.

The first step is to determine which system is malfunctioning. Subsequent steps help determine which part of the system is malfunctioning. Further testing determines which components of the system need adjustment or repair.

Most problems are very simple in nature and are easily corrected. Become familiar with the engine compartment. Note wire and hose connection points. Note the normal full throttle speed, idle speed, sound of the engine and any peculiarities. A rough idle, reduced power output, unusual sounds or strange odors usually indicate trouble. Perform a quick visual inspection at the first sign of trouble. Look for fluid leakage and loose wires, hoses or belts. If a quick inspection does not locate the problem, refer to *Preliminary Inspection* in this chapter for a list of additional items to check. **Tables 1-3** list starting, ignition and fuel system troubleshooting steps. **Tables 4-8** list recommendations for troubleshooting the charging system, power steering system, power trim system, cooling system and warning system. **Tables 9-12** list ignition, fuel and warning system specifications. **Tables 1-12** are located at the end of this chapter.

Proper lubrication, maintenance and periodic tune-ups will reduce the need to troubleshoot the engine. Even with the best of care, however, problems can and eventually do occur. Lubrication, maintenance and tune-up are described in Chapter Three.

PRELIMINARY INSPECTION

Every internal combustion engine requires three things to run (**Figure 1**): an uninterrupted supply of fuel and air in the proper proportion, adequate compression and a source of ignition at the proper time. If any are inadequate, the engine will not run properly. If any are missing, the engine will not run.

Perform the following steps if the engine fails to start, runs roughly, does not reach the maximum operating speed range or hesitates during rapid acceleration. The maximum

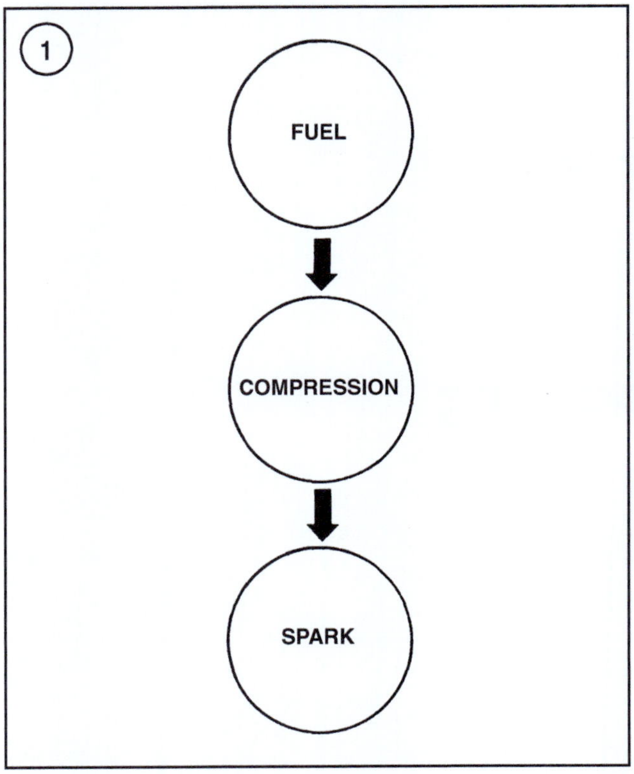

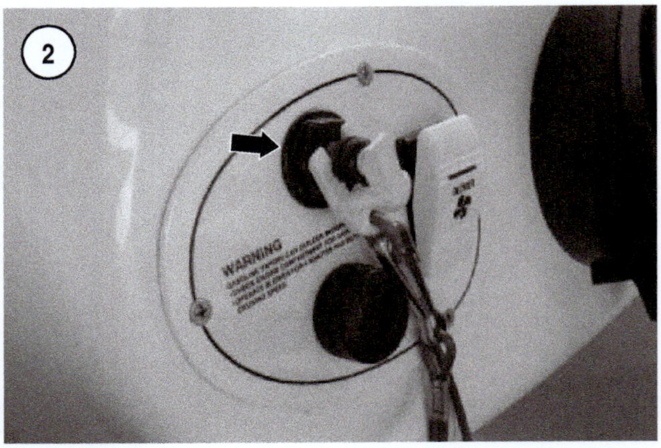

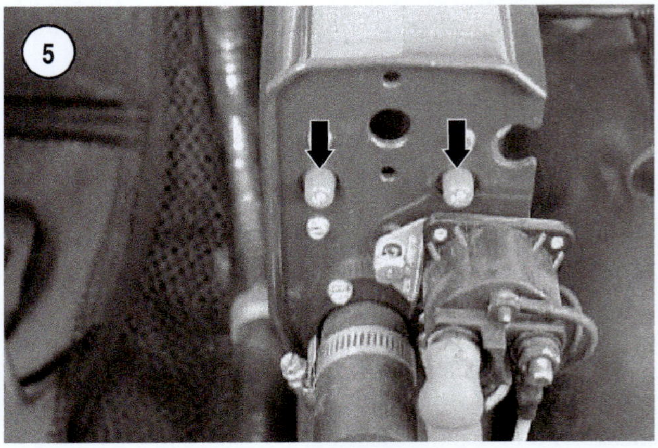

engine operating speed range is listed in the owner's manual.

1. If so equipped, check the position of the emergency stop (lanyard) switch (**Figure 2**).

2. Check for loose, dirty or damaged wire connections (**Figure 3**).

3. Check the fuses (**Figure 4**) and circuit breakers (**Figure 5**).

4. Supply the engine with fresh fuel.

5. Check and fully charge the battery as described in Chapter Nine.

6. Clean the flame arrestor. Refer to *Cleaning the Flame Arrestor* in Chapter Three.

7. Check for spark at the spark plugs as described in this chapter.

8. Clean, inspect or replace the spark plugs as described in Chapter Three.

9. Check for damaged or improperly adjusted throttle or shift cables.

10. If checking or correcting all of the previous items fails to identify or correct the malfunction, refer to **Tables 1-3** for additional items to check. Refer to the causes listed next to the symptom that best describes the malfunction. Perform the check, test or adjustment suggested as a corrective action.

11. It can be difficult to determine whether a starting problem is related to fuel, ignition or other causes. It is usually easiest to first verify that the ignition system is operating.

12. Check the fuel and fuel system if the ignition system is operating but the engine will not start or starting is difficult.

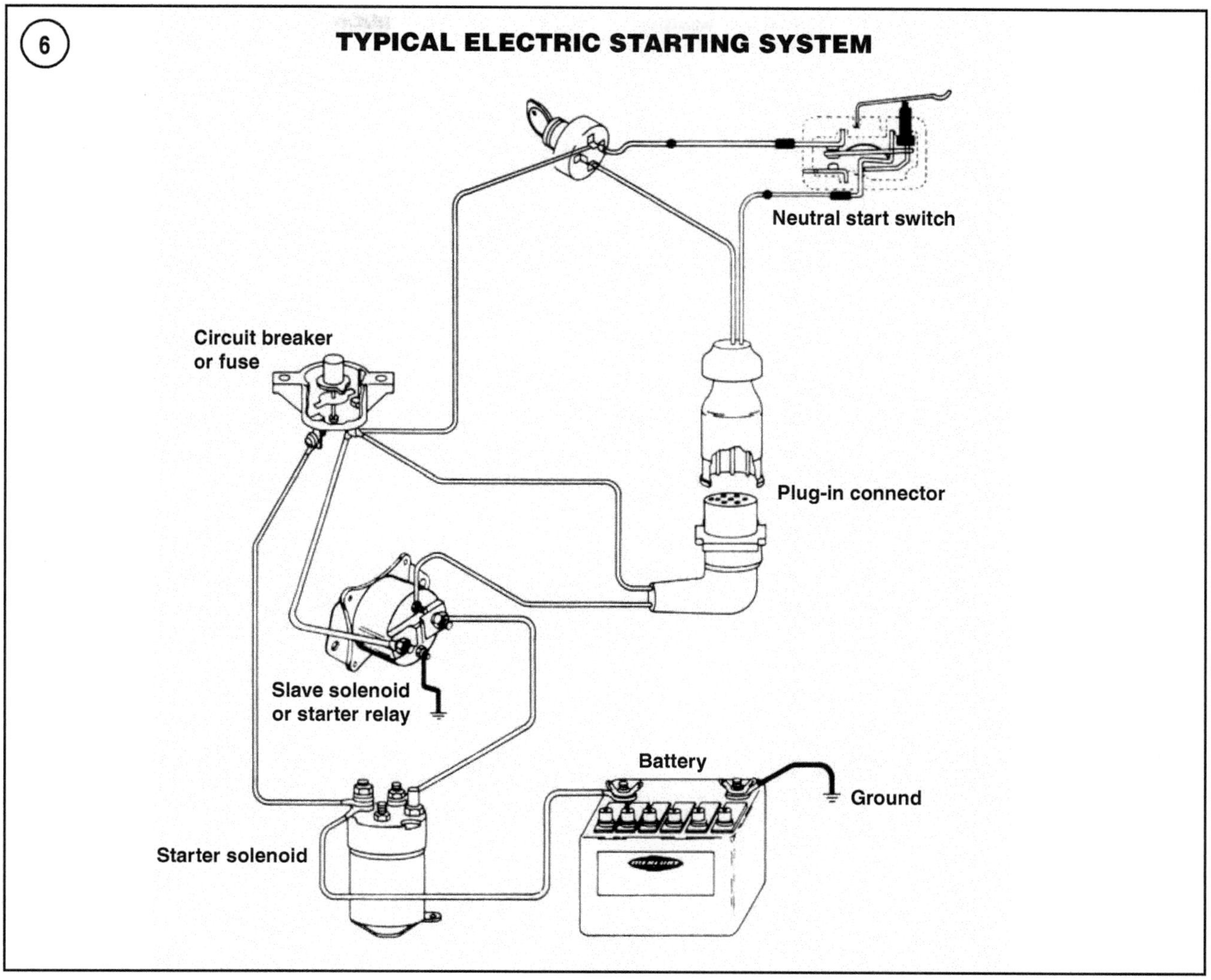

ELECTRIC STARTING SYSTEM

This section briefly describes how the electric starting system operates and provides testing instructions for the individual components.

The starting system consists of the battery, starter motor, starter relay or slave solenoid, ignition switch, neutral start switch and wiring.

Electric Starting System Operation

Refer to **Figure 6** for a typical starting system circuit. Battery current is supplied to the ignition switch from a circuit breaker or fuse. When the ignition switch is turned to START, battery current flows to the neutral start switch in the remote control assembly. Current flows through the switch to energize the slave solenoid or starter relay, provided the control is in NEUTRAL. When energized, the slave solenoid or starter relay supplies current to the starter solenoid. The solenoid then mechanically engages the starter pinion gear with the flywheel ring gear and supplies the current to operate the starter motor.

Due to the voltage drop associated with the long wires connecting the ignition switch to the engine, OMC Cobra stern drive engines must use either a slave solenoid or starter relay to activate the starter solenoid. These components perform the same function but are not interchangeable. The slave solenoid mounts onto the engine and uses eyelet or push-on connectors to connect to the engine wire harness. The starter relay plugs directly into the engine harness connector and is secured to the mounting bracket with a hook and loop band or a slide-on tab.

Once the engine is started and the ignition switch is released, the slave solenoid or starter relay de-energizes, thereby deactivating the starter solenoid. At this point, the solenoid disengages the starter pinion gear from the flywheel and interrupts the current to the starter motor.

Neutral Start Switch Test

The neutral start switch is incorporated into the remote control. The switch must open and interrupt the circuit connecting the ignition switch to the engine-mounted slave solenoid or starter relay anytime the control is in FORWARD or REVERSE. The switch must close and complete the circuit with the control in NEUTRAL. A faulty switch can prevent the starter from operating with the control in NEUTRAL or allow the starter to operate with the remote control in gear.

1. Disconnect the negative battery cable.

2. Locate the pair of yellow/red wires leading to the remote control housing (**Figure 7**). The wires are usually clamped to the remote control cables or instrument wiring harness. One of the wires connects to the ignition switch. The other wire connects to the instrument harness yellow/red wire that leads to the engine. Mark the wires to identify the connection points. Then disconnect the wires. In most cases, the wires are connected with crimp connectors. If necessary, cut the wires and install new connectors after testing.

3. Connect an ohmmeter between the disconnected yellow/red wires leading to the remote control.

4. Observe the meter while repeatedly shifting the remote control into FORWARD, NEUTRAL and REVERSE. Repeating the shifting cycles will often identify an intermittent switch malfunction.

5. The meter must indicate full continuity each time the remote control reaches NEUTRAL and no continuity each time the remote control reaches FORWARD or REVERSE. Disassemble the remote control and replace the neutral start switch if it fails to perform as specified.

6. Reconnect one of the yellow/red neutral start switch wires to the wire leading to the ignition switch. Reconnect the other yellow/red switch wire to the yellow/red instrument harness wire leading to the engine. Do not inadvertently connect the switch wires together or connect the ignition switch directly to the instrument harness wiring. Improper wire connections will prevent either the engine from starting in any gear or allow the engine to start in any gear.

WARNING
The engine might unexpectedly start or turn over during starting system testing. Keep all persons, clothing, tools or loose objects away from the propeller, propeller shaft, belts and pulleys during starting system testing.

CAUTION
If testing the starter motor with the boat out of the water, use a flush device to supply the engine with cooling water. Operation of the starter motor results in operation of the seawater pump. The seawater pump is quickly damaged if operated without adequate cooling water.

7. Connect the battery cables.

8. Pull the cord to activate the emergency stop (lanyard) switch (**Figure 2**). This prevents the ignition system from operating. The lanyard switch does not prevent starter motor operation. If the engine is not equipped with a lanyard switch, disconnect and ground the ignition coil secondary wire to prevent accidental starting.

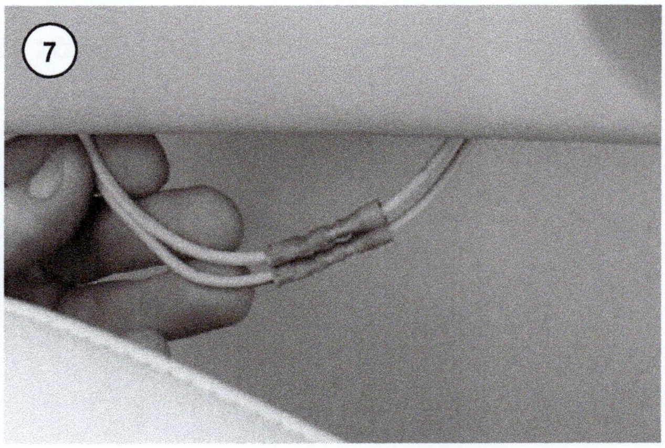

9. Verify proper neutral start switch operation as follows:
   a. Shift the remote control into FORWARD. Then attempt to start the engine. Repeat this step with the remote control in REVERSE. The starter must not operate with the remote control in FORWARD and REVERSE. Check for faulty wiring or replace the switch if the switch fails to perform as specified.
   b. Shift the remote control into NEUTRAL. Then attempt to start the engine. Check for faulty wiring, a faulty neutral start switch or other faulty starting system components if the starter fails to operate as specified.

10. Reattach the lanyard switch or reconnect the ignition coil secondary lead.

Ignition Switch

The ignition switch is secured to the dash with a decorative nut on the exposed side and a larger nut on the underside of the dash. On most switches, the wires are connected to the terminals with small screws. Other switches use push-on terminals. With either type, wire identification marks are molded into the housing next to each terminal. The test procedures in this section cover the standard three-terminal switch. Some switches have fourth and fifth terminals that allow use on inboard engines and some brands of outboard motors. If the switch has more than three terminals, perform the test using only the three indicated terminals.

The first test sequence measures the voltage at the switch terminals. The second sequence tests the internal switch circuits.

Voltage test

1. Pull the cord to activate the emergency stop (lanyard) switch (**Figure 2**). This prevents the ignition system from operating. The lanyard switch does not prevent starter motor operation. If the engine is not equipped with a lanyard

switch, disconnect and ground the ignition coil secondary wire to prevent accidental starting.

2. Disconnect the purple and yellow/red wires from the switch terminals. Do not disconnect the red or red/purple wire from the terminal.

3. Calibrate a voltmeter to the 20 or 40 VDC scale. Connect the negative meter test lead to the black ground wire at the tachometer or another suitable ground wire under the dash.

4. Touch the positive meter test lead to the red or red/purple wire terminal on the switch. The meter must show within 0.5 volt of battery voltage. If not, refer to the following:

   a. If the meter shows 0 volt, check for a blown fuse or tripped circuit breaker on the engine. Faulty wiring or connection is indicated if *all* of the fuses and circuit breakers check correctly.

   b. If the meter shows low voltage, check for loose or corroded wire terminals on the engine-mounted starter, starter solenoid or starter relay and at the dash-mounted instruments and switches. Make sure that no accessories (radios, pumps, fans) are wired into the ignition switch. The accessory load can cause a starting system or ignition system malfunction.

5. Place the ignition switch in OFF or STOP. Observe the meter while touching the positive test lead to the *S* and *I* terminals on the switch. The meter must indicate 0 volt at each connection point. If any voltage is indicated, the switch is shorted internally and must be replaced.

6. Connect the positive test lead to the *I* terminal of the ignition switch. Observe the meter while repeatedly turning the switch to ON or RUN and START. The switch must show within 0.5 volt of battery voltage with the ignition switch in ON or RUN and START. Replace the switch if it fails to perform as specified.

7. Connect the positive test lead to the *S* terminal of the ignition switch. Observe the meter while repeatedly turning the switch to ON or RUN and START. The switch must show within 0.5 volt of battery voltage with the ignition switch in START and 0 volt with the switch in ON or RUN. Replace the switch if it fails to perform as specified.

8. Connect the purple wire to the *I* terminal on the ignition switch. Connect the yellow/red wire to the *S* terminal on the switch. Route the wiring to prevent interference with moving components. Secure the wiring with clamps as needed.

9. Reattach the lanyard cord to the emergency stop (lanyard) switch to allow engine starting.

Internal circuit test

1. Disconnect the battery cables.

2. Disconnect the red or red/purple, purple and yellow/red wires from the ignition switch. If other wires are attached to the switch, note the wire routing and connection points. Then disconnect the wires.

3. Remove the ignition key. Then remove the decorative outer nut. Pull the ignition switch out of the dash. Insert the ignition key into the switch. Remove the lockwasher from the switch.

4. Connect an ohmmeter between the *B* or *Batt* terminal on the ignition switch and the *I* terminal on the switch. Note the meter reading while cycling the switch to OFF or STOP, ON or RUN and START. The meter must indicate no continuity with the ignition switch in OFF or STOP and full continuity with the ignition switch in ON or RUN and START. Replace the ignition switch if it fails to perform as specified.

5. Connect the ohmmeter between the *B* or *Batt* terminal and the *S* terminal on the switch. Note the meter reading while cycling the switch to OFF or STOP, ON or RUN and START. The meter must indicate no continuity with the ignition switch in OFF or STOP and ON or RUN and full continuity with the ignition switch in START. Replace the ignition key switch if it fails to perform as specified.

6. Install the lockwasher onto the switch and seat against the internal nut.

7. Remove the ignition key from the switch. Install the ignition switch into the dash opening and secure with the external decorative nut. If necessary, tighten the internal nut to secure the switch in the opening. Insert the ignition key into the switch.

8. Connect the red or red/purple wire to the *B* or *Batt* switch terminal. Connect the purple wire to the *I* switch terminal. Connect the yellow/red wire to the *S* switch terminal. Route the wiring to prevent interference with moving components. Secure the wiring with clamps as needed.

9. Connect the battery cables.

Starter Motor Cranking Voltage

This test measures the voltage at the starter battery cable during cranking. The battery must be fully charged, and the spark plugs must be installed for accurate testing.

WARNING

The engine might unexpectedly start or motor over during starting system testing. Keep all persons, clothing, tools or loose objects away from the propeller, propeller shaft, drive belts, pulleys and other moving components on the engine.

CAUTION

Iif testing the starter motor with the boat out of the water, use a flush device to supply the engine with cooling water. Operation of the starter motor results in operation of the seawater pump. The seawater pump is quickly damaged if operated without adequate cooling water.

1. Place the ignition switch in OFF or STOP and the remote control in NEUTRAL.

2. Pull the cord to activate the emergency stop (lanyard) switch (**Figure 2**). This prevents the ignition system from

operating. The lanyard switch does not prevent starter motor operation. If the engine is not equipped with a lanyard switch, disconnect and ground the ignition coil secondary wire to prevent accidental starting.

3. Carefully move the insulator boot away from the battery cable terminal on the starter solenoid.

4. Set a voltmeter to the 20 or 40 VDC scale. Touch the negative meter lead to an engine ground. Touch the positive meter lead to the battery cable to the starter solenoid terminal (A, **Figure 8**, typical). The meter must show battery voltage. If otherwise, test the battery as described in Chapter Nine and check for dirty terminals or a faulty battery cable.

5. Touch the positive meter test lead to the starter solenoid cable terminal (A, **Figure 8**). Touch the negative meter lead to an engine ground. Observe the meter while an assistant cycles the ignition switch to START. The meter must indicate 9.5-11.5 volts while the starter cranks the engine with adequate speed for starting. Refer to the following:

   a. If the voltage is less than 9.5 volts, and the starter rotates slowly, fully charge and test the battery as described in Chapter Nine. Test the starter solenoid if the battery tests satisfactorily.
   b. If the voltage exceeds 11.5 volts, but the starter does not operate or turns slowly, test the starter solenoid as described in this chapter. Repair the starter motor as described in Chapter Nine if the solenoid tests satisfactorily.

6. Touch the positive meter test lead to the braided wire terminal (B, **Figure 8**) on the starter solenoid. If the starter does not have a braided wire, touch the test lead to the large solenoid terminal opposite the battery terminal. Touch the negative meter lead to an engine ground. Observe the meter while an assistant cycles the ignition switch to START. The meter must indicate 9.5-11.5 volts while the starter cranks the engine with adequate speed for starting. Refer to the following:

   a. If the voltage is less than 9.5 volts, and the starter rotates slowly, fully charge and test the battery as described in Chapter Nine.
   b. If the voltage exceeds 11.5 volts, but the starter does not operate or operates slowly, repair the starter motor as described in Chapter Nine.
   c. If the voltage is 9.5-11.5 volts, but the starter rotates slowly or does not operate, check for a seized engine drive unit. Repair the starter as described in Chapter Nine if no fault is found with the engine or drive.

7. Make sure all terminal nuts are securely tightened and the wire terminals are not contacting other terminals. Reposition the wiring as necessary. To prevent terminal corrosion and starting system malfunction, thoroughly coat the terminals with Black Neoprene Dip. Purchase the neoprene dip (OMC part No. 909570) from an OMC dealership or automotive parts store.

8. Reposition the insulator over the battery cable terminal on the starter solenoid.

9. Reattach the lanyard cord to the emergency stop (lanyard) switch to allow engine starting. If the engine is not equipped with a lanyard switch, reconnect the ignition coil secondary wire.

Starter Solenoid

The starter solenoid mounts to the starter motor. The first test sequence measures the voltage supplied to the starter solenoid and checks for proper switching while mounted on the engine. The second test sequence checks for proper operation of the solenoid. The functional test is useful to test the solenoid if it is removed from the starter motor.

Voltage test

WARNING
The engine might unexpectedly start or turn over during starting system testing. Keep all persons, clothing, tools or loose objects away from the propeller, propeller shaft, drive belts, pulleys and other moving components on the engine.

CAUTION
If testing the starter solenoid with the boat out of the water, use a flush device to supply the engine with cooling water. Operation of the starter motor results in operation of the seawater pump. The seawater pump is quickly damaged if operated without adequate cooling water.

This test requires a voltmeter and clear access to the starter solenoid terminals. The starter and solenoid are mounted on the lower rear starboard side of the engine. Correct test results are only possible if the test leads make good contact to the correct terminals.

The starter motor might operate during testing. Avoid entrapment in belts, pulleys or other moving components on the engine. This procedure also tests the bypass circuit function of the solenoid. The bypass circuit uses the *R* terminal on the solenoid to supply voltage while the starter motor is operating to the ignition coil on some Ford 5.0L and 5.8L (carburetor-equipped) models. The circuit is also used on some GM (carburetor-equipped) models with an electric low-pressure fuel pump.

1. Place the ignition switch in OFF or STOP and the remote control in NEUTRAL.
2. Pull the cord to activate the lanyard switch (**Figure 2**). This prevents the ignition system from operating. The lanyard switch does not prevent starter motor operation. If the engine is not equipped with a lanyard switch, disconnect and ground the ignition coil secondary wire to prevent accidental starting.
3. Set the voltmeter to the 20 or 40 VDC scale.
4. Check for battery voltage at the solenoid as follows:
   a. Carefully pull the insulator boot away from the battery cable terminal (A, **Figure 8**).
   b. Touch the positive meter test lead to the battery cable terminal.
   c. Observe the meter while touching the negative meter test lead to an engine ground. The meter must indicate battery voltage. If not, test the battery as described in Chapter Nine and check for loose or dirty connections or a faulty battery switch.
5. Disconnect the negative battery cable.
6. Remove the brass nut (B, **Figure 8**). Then disconnect the large braided wire that connects the starter solenoid to the starter end cap or brush plate. Completely cover the wire and terminal with a thick wrapping of electrical tape and position it away from all starter solenoid terminals.
7. Reconnect the battery cable.
8. Touch the positive meter lead to the yellow/red wire terminal on the starter solenoid (C, **Figure 8**). Touch the negative meter lead to an engine ground. Maintain the test lead connection points and test the voltage from the slave solenoid or starter relay as follows:
   a. The meter must indicate 0 volt. If any voltage is indicated, disconnect the yellow/red wire from the terminal and repeat the test. If any voltage is indicated with the wire disconnected, the solenoid is shorted internally and must be replaced. If the meter indicates 0 volt *only* with the yellow/red wire disconnected, test the slave solenoid or starter relay as described in this section. Reconnect the yellow/red wire.
   b. Observe the meter while an assistant cycles the ignition switch to ON or RUN. The meter must indicate 0 volt. If any voltage is indicated, test the slave solenoid or starter relay as described in this section.
   c. Observe the meter while an assistant cycles the ignition switch to START. The meter must indicate battery voltage each time the switch is placed in START. If the meter indicates low or no voltage, perform the *Starter solenoid functional test* as described in this section.
9. Touch the positive meter lead to the large solenoid terminal (B, **Figure 8**) for the braided wire or wire opposite the battery cable wire terminal. The braided or large-gauge wire must be disconnected from the terminal as described in Step 6. Touch the negative meter lead to an engine ground. Maintain the test lead connections and test the solenoid switching function as follows:
   a. Place the ignition switch in OFF or STOP. The meter must indicate 0 volt. If any voltage is indicated, the solenoid is shorted internally and must be replaced.
   b. Observe the meter while an assistant cycles the ignition switch to START. The meter must indicate battery voltage, and the solenoid must make a clicking sound each time the switch is placed in START. Perform the *Starter solenoid functional test* as described in this section if the solenoid fails to perform as specified.
10. Test the bypass circuit operation as follows:
    a. Touch the positive meter lead to the purple wire terminal (D, **Figure 8**) on the solenoid. Touch the negative meter lead to an engine ground.
    b. Observe the meter while an assistant cycles the ignition switch to OFF or STOP, ON or RUN and START. The meter must indicate 0 volt each time the ignition switch is placed in OFF or STOP and ON or RUN. The meter must indicate battery voltage each time the switch is placed in START. Replace the starter solenoid if the switching circuit fails to perform as specified.
11. Disconnect the negative battery cable.
12. Remove the protective tape. Connect the braided cable or large-gauge wire to the starter solenoid terminal (B, **Figure 8**) and secure with the brass nut.
13. Make sure all terminal nuts are securely tightened and the wire terminals are not contacting other terminals. Reposition the wiring as necessary. To prevent terminal corrosion and starting system malfunction thoroughly coat the terminals with Black Neoprene Dip. Purchase the neoprene dip (OMC part No. 909570) from an OMC dealership or automotive parts store.
14. Reposition the insulator over the battery cable terminal on the starter solenoid.
15. Connect the battery cable.
16. Reattach the lanyard cord to the emergency stop (lanyard) switch to allow engine starting. If the engine is not equipped with a lanyard switch, reconnect the ignition coil secondary wire.

Starter solenoid functional test

This test requires a fully charged battery, ohmmeter and jumper wires.

1. Disconnect the battery cables.
2. Remove the starter motor then starter solenoid as described in Chapter Nine.
3. Connect an ohmmeter between the large terminals on the solenoid. The meter must indicate no continuity. Otherwise, the solenoid is shorted internally and must be replaced.

4. Test the solenoid plunger operation and starter switching circuits as follows:
   a. Connect a jumper wire to the positive battery terminal and the *S* terminal on the starter solenoid. Do not inadvertently connect the jumper to the *R* terminal.
   b. Connect a second jumper lead to the negative battery terminal.
   c. Observe the meter while repeatedly touching the second jumper lead to a clean metal surface on the solenoid housing. The meter must indicate full continuity, and the solenoid must make a clicking sound as the plunger retracts each time the jumper touches the housing. Replace the solenoid if it fails to perform as specified.
5. Test the bypass circuit in the solenoid as follows:
   a. Move the positive meter lead to the *R* terminal on the starter solenoid. Do not disconnect the meter lead from the other large terminal. Do not disconnect the jumper lead from the battery and solenoid.
   b. Observe the meter while repeatedly touching the second jumper lead to a clean metal surface on the solenoid housing. The meter must indicate full continuity each time the jumper touches the housing. Replace the solenoid if it fails to perform as specified.
6. Remove the jumper and meter leads.
7. Install the starter solenoid and then the starter motor as described in Chapter Nine.
8. Connect the battery cables.

Slave Solenoid Test

All Ford 5.0L and 5.8L V-8 (carburetor-equipped) models covered in this manual use a switch solenoid. This solenoid supplies battery current to the ignition coil while the starter operates. This bypass feature supplements the ignition voltage supplied by the ignition switch and wiring to provide higher ignition coil voltage while starting. On some later models, the bypass wire attaches to the starter solenoid bypass terminal in place of the slave solenoid bypass terminal. If the circuit is not used, an insulator cap is installed on the slave solenoid terminal. The starter solenoid bypass terminal provides the same function as the slave solenoid bypass circuit.

NOTE

The switch and standard solenoids are very similar in appearance and usually use the same number and size of terminals. The switch solenoid grounds through the mounting plate, and this can be verified by checking for continuity between the S terminal and the mounting plate. With a standard solenoid, the unit grounds through one of the smaller terminals. This can be verified by checking for continuity between the smaller terminals.

Use a test lamp for this procedure.

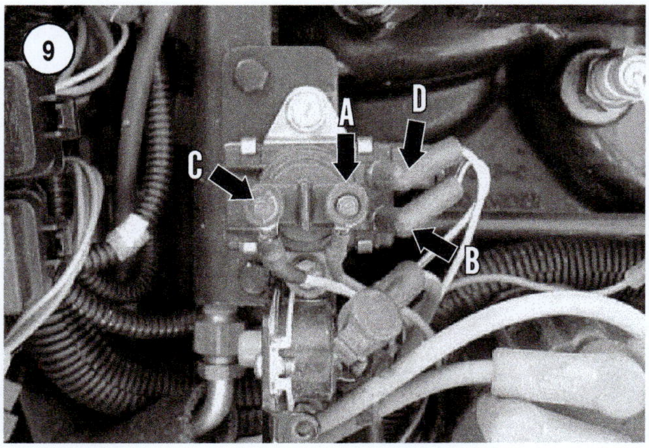

1. Place the ignition switch in OFF or STOP and the remote control in NEUTRAL.
2. Pull the cord to activate the emergency stop (lanyard) switch (**Figure 2**). This prevents the ignition system from operating. The lanyard switch does not prevent starter motor operation. If the engine is not equipped with a lanyard switch, disconnect and ground the ignition coil secondary wire to prevent accidental starting.
3. Connect the cord of the test lamp to an engine ground. Touch the test lamp probe to a positive battery voltage source to verify the ground and test lamp operation. Locate a proven ground or repair the test lamp as needed.
4. Touch the test lamp probe to the red or red/purple wire terminal on the slave solenoid (A, **Figure 9**, typical). The lamp must illuminate. If not, check for a blown fuse along the wire connecting the solenoid to the starter motor terminal. Repair the faulty wiring or connections if no fault is found with the fuse(s).
5. Test the voltage from the ignition switch and neutral start switch to the solenoid as follows:
   a. Disconnect the yellow/red wire from the smaller solenoid terminal (B, **Figure 9**, typical).
   b. Connect the test lamp probe to an engine ground as described in Step 3.
   c. Touch the test lamp probe to the disconnected yellow/red wire terminal.

11

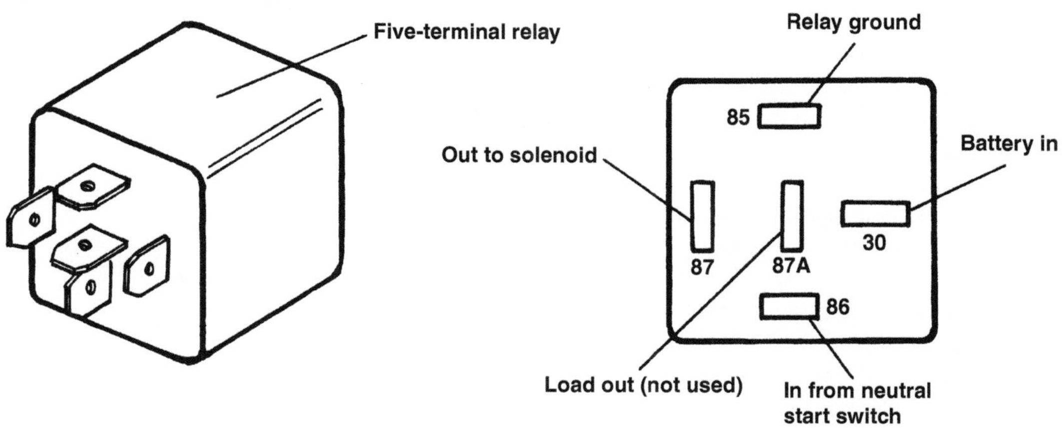

d. Observe the lamp while an assistant *repeatedly* turns the ignition switch to OFF or STOP, ON or RUN and START. The lamp must illuminate each time the ignition switch is in START and must not illuminate with the switch in OFF or STOP and ON or RUN. Test the ignition switch and neutral start switch as described in this section if the circuit fails to perform as specified. Repair the fault along the yellow/red wire leading to the instruments and switches if no fault is found with the ignition switch or neutral start switch.

e. Reconnect the yellow/red wire to the smaller solenoid terminal.

6. Test the solenoid ground as follows:

a. Connect the test lamp cord to a positive battery source. Touch the test lamp probe to an engine ground to verify the voltage source. If the lamp fails to illuminate, check for faulty engine ground cables, wiring or connections.

b. Touch the test lamp probe to the metallic solenoid mounting plate. The test lamp must illuminate. If not, correct the loose fasteners, dirty mating surfaces or damaged ground wires to restore the ground.

7. Test the solenoid operation as follows:

a. Connect the test lamp cord to an engine ground. Touch the test lamp probe to a positive battery source to verify the ground and lamp operation.

b. Touch the test lamp probe to the larger yellow/red terminal (C, **Figure 9**, typical) on the solenoid. Observe the lamp while an assistant *repeatedly* turns the ignition switch to OFF or STOP, ON or RUN and START. The lamp must illuminate, and the solenoid must make a clicking sound each time the ignition switch is in START. The lamp must not illuminate with the ignition switch in OFF or STOP and ON or RUN. Replace the slave solenoid if it fails to perform as specified.

8. On models with a switch solenoid, test the bypass switch circuit as follows:

a. Disconnect the purple wire from the smaller solenoid terminal.

b. Connect the test lamp cord to an engine ground. Touch the test lamp probe to a positive battery source to verify the ground and lamp operation.

c. Touch the test lamp probe to the small terminal on the solenoid for the purple wire (D, **Figure 9**, typical). Observe the lamp while an assistant *repeatedly* turns the ignition switch to OFF or STOP, ON or RUN and START. The lamp must illuminate each time the ignition switch is in START and must not illuminate with the ignition switch in OFF or STOP and ON or RUN. Replace the slave solenoid if it fails to perform as specified.

d. Reconnect the purple wire to the smaller solenoid terminal.

Starter Relay Test

The starter relay (**Figure 10**, typical) has a common five-terminal design. This procedure tests for proper current delivery to the relay terminals and checks for proper relay switching. Relays are relatively inexpensive. Replace the relay if in question.

Use a 12-volt test lamp for this procedure.

1. Place the ignition switch in OFF or STOP and the remote control in NEUTRAL.

2. Pull the cord to activate the emergency stop (lanyard) switch (**Figure 2**). This prevents the ignition system from operating. The lanyard switch does not prevent starter motor operation. If the engine is not equipped with a lanyard switch, disconnect and ground the ignition coil secondary wire to prevent accidental starting.

3. Disconnect the battery cables. Remove the starter relay as described in Chapter Nine. Reconnect the battery cables.

4. Refer to **Figure 11** to identify the relay terminals and the terminals in the wire harness or relay mounting bracket. The relay will only fit one way into the plug or bracket. Compare the orientation of the No. 30 and No. 87 terminal blades for reference.

5. Test the relay ground circuit as follows:

12 ALTERNATOR OUTPUT TEST

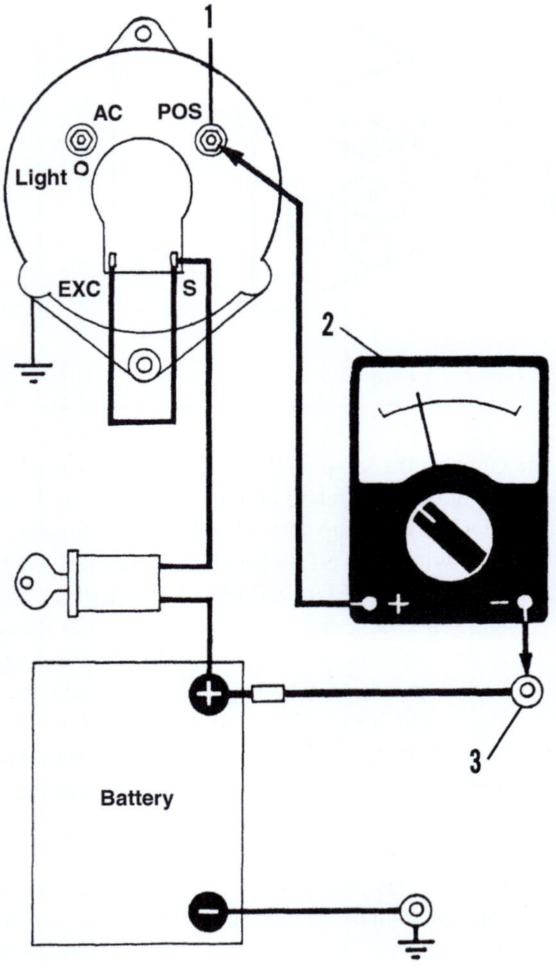

1. Alternator output terminal
2. Ammeter
3. Orange output wire terminal

a. Connect the cord of the test lamp to the positive battery terminal. Touch the test lamp probe to a good engine ground to verify the battery positive source and lamp operation.

b. Touch the test lamp probe to the No. 85 terminal in the harness connector that aligns with the black engine wire leading to the connector. If the No. 85 terminal does not align with the black wire, touch the test lamp probe to the No. 86 terminal in the harness connector. The lamp must illuminate. If not, check for a faulty terminal connection or fault in the black relay ground wire.

6. Test the battery voltage supply to the relay as follows:

a. Connect the cord of the test lamp to the negative battery terminal or good engine ground. Touch the test lamp probe to the positive battery terminal to verify the ground connection and lamp operation.

b. Touch the test lamp probe to the No. 30 terminal in the harness connector that aligns with the red or red/purple wire leading to the connector. If the No. 30 terminal does not align with the red or red/purple wire, touch the test lamp probe to the No. 87 terminal in the connector. The lamp must illuminate. If not, check for a blown fuse or circuit breaker or a faulty terminal or battery supply wire to the terminal.

7. Test the start circuit as follows:

a. Touch the test lamp probe to the No. 86 terminal in the harness connector that aligns with the yellow/red wire leading to the connector. If the No. 86 terminal does not align with the yellow/red wire, touch the test lamp probe to the No. 85 terminal in the connector. The lamp must not illuminate with the ignition switch at OFF or STOP. If the lamp illuminates, check the ignition switch as described in this section. If no fault is found with the ignition switch, check for a shorted instrument harness or engine wire harness.

b. Observe the lamp while an assistant cycles the ignition switch to ON or RUN and START. The lamp must not illuminate with the ignition switch in ON or RUN, and the lamp must illuminate with the switch in START. If otherwise, check for a faulty ignition switch, faulty instrument or engine wire harness.

8. Replace the relay if no faults are found with the wiring and connecting circuits and if the relay fails to supply current to the starter solenoid.

9. Install the starter relay as described in Chapter Nine.

10. Reattach the lanyard switch cord (**Figure 2**).

CHARGING SYSTEM

The charging system consists of the alternator, voltage regulator, battery, ignition switch, voltmeter, the necessary connecting wiring and circuit breaker or fuse. The alternator pulley is turned by the drive belt via the engine crankshaft pulley. As engine speed varies, the output from the alternator can vary, so a voltage regulator is used to control output within a safe range. The voltmeter indicates whether charging or discharging is occurring.

All models covered in this manual are equipped with either a 51- or 65-amp alternator with a transistorized regulator attached to the rear of the alternator housing. The output rating is stamped on the alternator frame.

A charging system malfunction is generally caused by a defective alternator, voltage regulator, battery, fuse or circuit breaker. The malfunction can also be caused by something as simple as incorrect drive belt tension. Always check these items before beginning the troubleshooting procedure. Refer to **Table 4** for troubleshooting recommendations if the battery loses charge at rest or if the dash-mounted voltmeter indicates discharging (fewer than 12.3 volts) or overcharging (more than 13.8 volts) with the engine running at normal cruising speeds. The recommended test procedures are described in this section.

(13) **ALTERNATOR EXCITATION CIRCUIT TEST**

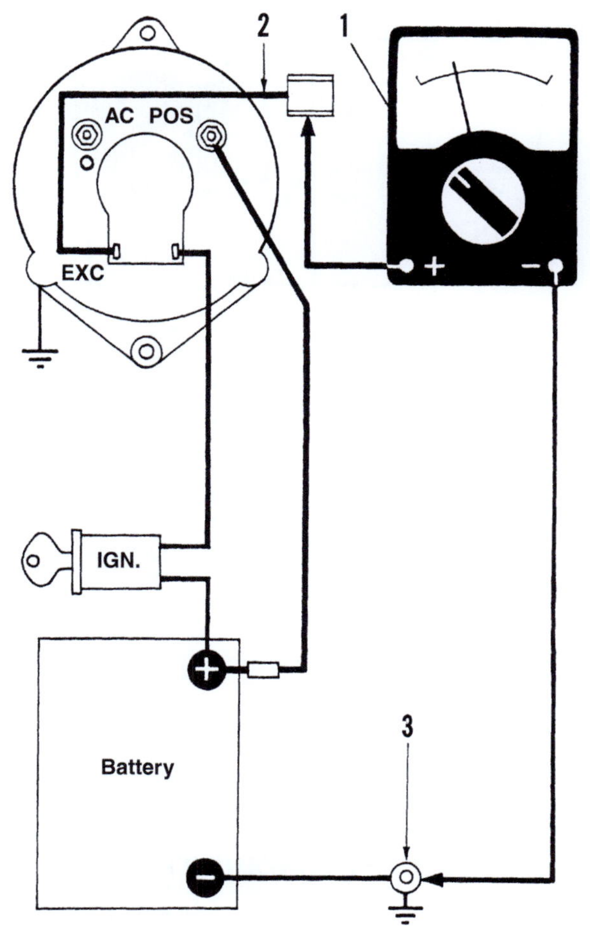

1. Voltmeter
2. Red/purple alternator wire
3. Engine ground terminal

Alternator Output Test

This test requires an ammeter and a load tester (such as Stevens LB-85 Load Bank) or other suitable carbon pile load set. This test measures the alternator output at various engine speeds to determine the alternator efficiency.

CAUTION
Never operate the engine without first providing cooling water. Use a flush test device or test tank, or operate the engine with the boat in the water. Operating the engine without providing adequate cooling water will quickly damage the seawater pump and can result in serious engine damage.

1. Disconnect the negative battery cable.
2. Disconnect the orange lead from the *POS* terminal stud (1, **Figure 12**) on the rear of the alternator.
3. Connect the ammeter (2, **Figure 12**) in series between the *POS* terminal stud (1) and the orange output wire terminal (3). Make sure the positive terminal on the ammeter connects to the *POS* and the negative terminal connects to the orange lead. Otherwise, the ammeter will read opposite the actual output.
4. Connect the battery cable.
5. Attach the load tester to the battery terminals.
6. Start the engine and operate at 1200-1500 rpm until it reaches normal operating temperature.
7. Return the engine to idle speed and switch off all accessories. Switch on the load tester and note the alternator output on the ammeter. The ammeter must indicate a minimum of 20 amps for either the 51-amp or 65-amp alternator. Switch off the load tester.
8. Raise the engine speed to 1500 rpm. Switch on the load tester and note the alternator output on the ammeter. The ammeter must indicate a minimum of 47 amps for the 51-amp alternator and 53 amps for the 65-amp alternator. Switch off the load tester.
9. Raise the engine speed to 2000 rpm. Switch on the load tester and note the alternator output on the ammeter. The ammeter must indicate a minimum of 51 amps for the 51-amp alternator and 56 amps for the 65-amp alternator. Switch off the load tester.
10. Stop the engine.
11. Disconnect the negative battery cable.
12. Disconnect the ammeter and reconnect the orange wire to the *POS* terminal stud on the rear of the alternator.
13. Connect the battery cable.
14. Test the excitation circuit, sensing circuit and diode trio if no output or low output was measured in Step 7, 8 or 9. Replace the alternator with a new or rebuilt unit if the excitation circuit, sensing circuit and diode trio test correctly.

Excitation Circuit Test

The transistorized regulator, attached to the rear of the alternator, contains excitation and sensing circuits. The excitation circuit, which is connected to the ignition switch circuit, sends a small amount of current to the alternator rotor field winding to build a magnetic field and initiate charging output. This allows the alternator to generate output more quickly after startup than relying solely on residual magnetism in the rotor.

This test requires an accurate voltmeter.

1. Place the ignition switch at OFF or STOP.
2. Set the voltmeter to the 20 or 40 VDC scale.
3. Unplug the red/purple alternator wire (2, **Figure 13**) from the engine harness connector. Connect the positive meter lead to the red/purple alternator wire terminal. Connect the negative meter lead to an engine ground (3, **Figure 13**). The meter must indicate 0 volt. Observe the meter while an assistant turns the ignition switch to ON or RUN. The meter must indicate 0 volt with the ignition switch in OFF or STOP and within 1 volt of battery voltage with the

ignition switch in ON or RUN. If otherwise, test the ignition switch as described in this chapter. If no fault is found with the ignition key switch, check all of the wiring for corroded or dirty terminals and damaged wiring.
4. Disconnect the meter leads. Reconnect the red/purple alternator wire to the engine harness connector.

Sensing Circuit Test

The transistorized regulator, attached to the rear of the alternator, contains excitation and sensing circuits. The sensing circuit allows the regulator to sense resistance inside the alternator and outside the alternator by comparing the voltage on the sensing circuit and output leads.

This test requires an accurate voltmeter.

1. Place the ignition switch in OFF or STOP.
2. Set the voltmeter to the 20 or 40 VDC scale.
3. Unplug the red/purple alternator wire (2, **Figure 13**) from the engine harness connector. Connect the positive meter lead to the red/purple alternator wire terminal. Connect the negative meter lead to the *S* terminal on the rear of the alternator. See **Figure 14**. Observe the meter while an assistant turns the ignition switch to ON or RUN. The meter should indicate within 1 volt of battery voltage with the ignition switch in ON or RUN. Voltage indicated with the ignition switch in OFF is not important for this test. If the voltage is less than specified, test the excitation circuit as described in this section. Replace the voltage regulator on the rear of the alternator if the excitation circuit tests correctly.
4. Disconnect the meter leads. Reconnect the red/purple alternator wire to the engine harness connector.
5. Remove the protective cap from the stud marked *LIGHT* on the rear of the alternator. Connect the positive meter lead to the *LIGHT* terminal. Connect the negative meter lead to an engine ground. See **Figure 15**. Observe the meter while an assistant turns the ignition switch to ON or RUN. The meter must indicate 1.5-3.0 volts with the ignition switch in ON or RUN. Voltage indicated with the ignition switch in OFF is not important for this test. Replace the voltage regulator on the rear of the alternator if the voltage indicated is not within the specification.
6. Disconnect the meter leads. Install the protective cap onto the *LIGHT* terminal.

Diode Trio Test

The diode trio is located within the alternator. After starting, this component supplies stator output to the voltage regulator to energize the rotor field. This prevents the regulator from relying solely on the excitation circuit for current to energize the rotor field.

This test requires an accurate voltmeter and a jumper wire.

1. Place the ignition switch in OFF or STOP.
2. Set the voltmeter to the 20 or 40 VDC scale.

14

SENSING CIRCUIT TEST

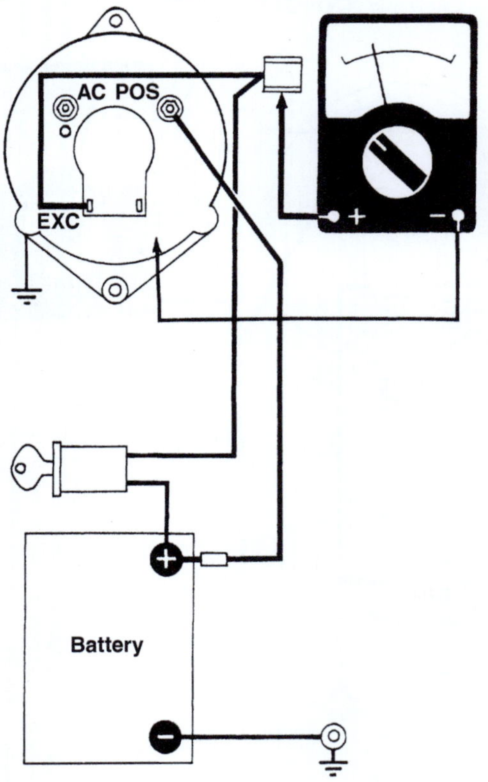

15

SENSING CIRCUIT TEST

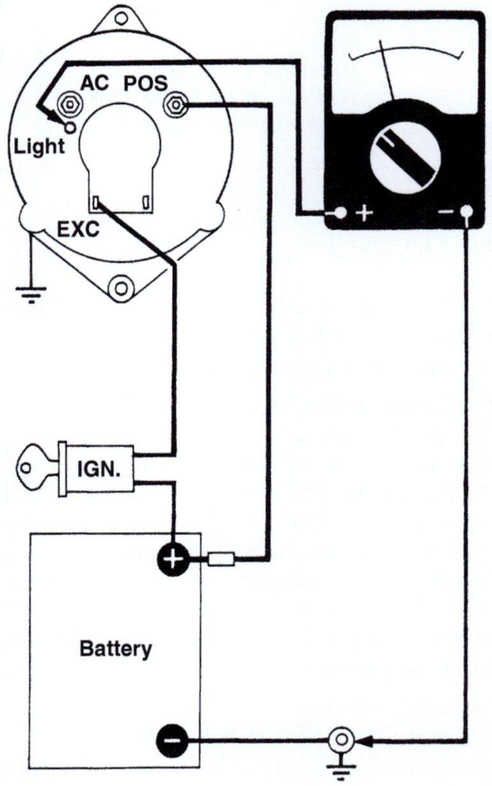

16

DIODE TRIO TEST

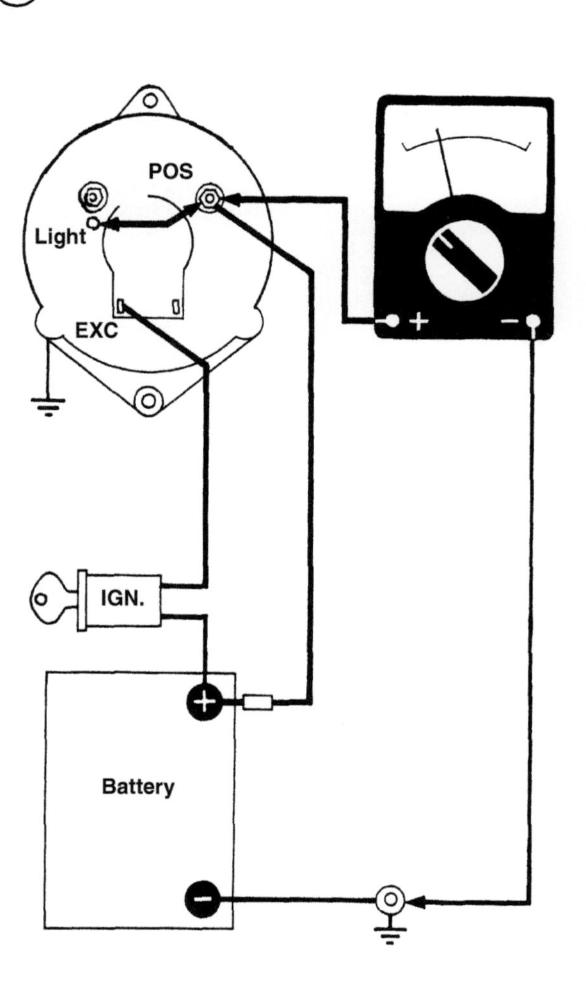

3. Remove the protective cap from the stud marked *LIGHT* on the rear of the alternator.
4. Connect a jumper lead to the *POS* terminal stud and the *LIGHT* terminal. See **Figure 16**.
5. Connect the positive meter lead to the *POS* terminal stud on the alternator. Connect the negative meter lead to an engine ground. See **Figure 16**.

CAUTION
Never operate the engine without first providing cooling water. Use a flush test device or test tank, or operate the engine with the boat in the water. Operating the engine without providing adequate cooling water will quickly damage the seawater pump and can result in serious engine damage.

6. Start the engine and operate at idle speed (approximately 600 rpm). Observe the meter. The diode trio is faulty if any voltage is indicated. If the diode trio is faulty, replace the alternator with a new or rebuilt unit.
7. Stop the engine. Remove the jumper lead. Reconnect the cap to the *LIGHT* terminal.

Alternator Ground Test

The alternator must have a good ground connection to operate at full efficiency. A faulty ground can cause no output, low output and repeat alternator failure. A digital multimeter is required for the procedure.

1. Disconnect the battery cables.
2. Connect the meter test leads. Select the ohms setting on the meter and note the internal resistance reading.
3. Touch the negative test lead to the negative battery terminal. Touch the positive test lead to an unpainted surface of the alternator housing and note the meter reading.
4. Subtract the internal resistance reading noted in Step 2 from the reading noted in Step 3. This is the actual resistance in the alternator ground circuit. A typical reading is 1-5 ohms. The resistance reading is affected by battery cable length and the length of the wires in the engine wire harness. If the resistance reading is significantly higher than 5 ohms, check for faulty wiring and loose or dirty wire terminals along the entire ground circuit.
5. Connect the battery cables.

Output Circuit Test

This test checks for an open or short circuit along the output circuit. A fault in this circuit can cause low output with the engine running or allow the battery to discharge with the engine off. A voltmeter and 12-volt test lamp are required for this procedure.

1. Disconnect the battery cables.
2. Disconnect the larger-gauge orange lead from the *POS* terminal stud (1, **Figure 12**, typical) on the rear of the alternator. Then disconnect the smaller-gauge red wire from the alternator.
3. Select the ohms setting on the meter. Touch the positive meter test lead to the terminal on the disconnected orange wire. Touch the negative test lead to an unpainted surface of the alternator case or other engine ground surface. The meter must indicate no continuity or very high resistance.
   a. If the meter indicates continuity or very low resistance (fewer than 5 ohms), inspect the orange wire for shorted wiring or faulty terminals.
   b. If no fault is found with the wiring or terminals, disconnect the instrument harness plug from the engine wire harness plug and note the meter reading. The instrument harness is shorted, or one or more of the electrical accessories is shorted if the reading changes to no continuity or the resistance increases significantly with the plug disconnected. Inspect the harness. If no fault is found with the harness, reconnect the harness to the engine. Then, to isolate the fault, note the meter reading while disconnecting each accessory connected to the red instrument harness wiring.
   c. If no fault is found with the instrument harness or connected accessories, note the meter reading each

time while disconnecting the engine control unit or engine control assembly (EFI models), circuit breaker, slave solenoid, starter relay and starter solenoid. A short is indicated in the component if the meter switches to no continuity or the resistance reading increases markedly with that component disconnected.

4. Reconnect the smaller-gauge red wire to the alternator. If disconnected, reconnect the instrument harness-to-engine harness plug.
5. Route the disconnected larger-gauge red or orange alternator output wire so the terminal cannot contact an engine ground or other component. Reconnect the battery cables.
6. Connect the cord of the test lamp to the terminal of the disconnected red or orange output lead. Touch the test lamp probe to an engine ground. The lamp must illuminate. If not, connect the test lamp cord to the positive battery terminal to verify the lamp is operational. If the lamp illuminates only when touching the battery terminal, check for a blown fuse, tripped circuit breaker or faulty terminal along the red or orange wire.
7. With the test lamp cord connected onto the orange output wire, touch the test lamp probe to the *POS* terminal stud (1, **Figure 12**, typical) on the rear of the alternator. Refer to the following:
   a. If the lamp illuminates brightly, the rectifier bridge has one or more failed diodes, the output terminal is shorted to ground, the output terminal is corroded, or the output terminal insulator is damaged. Replace the alternator with a new or rebuilt unit.
   b. If the lamp does not illuminate, is very dim or glows orange, any current draw on the battery with the engine off is from a different source. If in question, have the alternator tested by a reputable starter and alternator repair shop. Depending on the type of test lamp used, normal operation of the rectifier bridge may allow enough current flow to dimly illuminate the test lamp.
8. Disconnect the battery cables.
9. Reconnect the red or orange output lead to the alternator output terminal. Make sure to position the rubber insulating boot (**Figure 17**) fully over the terminal.
10. Reconnect the battery cables.

IGNITION SYSTEM

Three different types of ignition systems are used on the models covered in this manual. This section provides a brief description of the each system followed by troubleshooting procedures for the various components and systems. Refer to **Table 2** for a list of items to check if the engine will not start, or if the ignition system is suspected of causing an engine malfunction. Refer to the causes listed next to the symptom that best describes the malfunction. Perform the check, test or adjustment suggested under corrective action. Other useful information and tips are described with the specific test instructions.

17

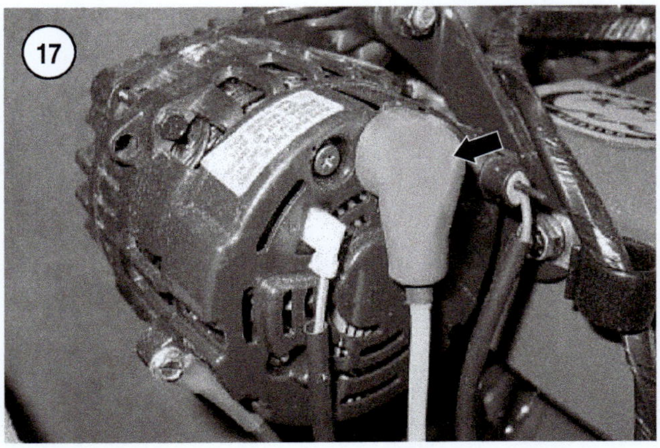

Basic Ignition System Troubleshooting

All models covered in this manual use a fully transistorized ignition system that offers higher spark output and greater reliability and requires less maintenance than the standard breaker-point systems used on earlier models.

Failure of the ignition system is rare and is usually caused by corroded, loose or dirty wire connections. This section describes basic troubleshooting procedures that apply to all of the ignition systems covered in this manual. Procedures that apply only to an individual ignition system are described after the system description.

Ignition system trouble may be roughly divided between trouble that affects a single cylinder and trouble that affects all cylinders. Trouble affecting a single cylinder is almost always caused by a faulty spark plug, spark plug wire or distributor cap. Trouble affecting all cylinders is almost always caused by a faulty ignition module, pickup coil, Hall effect sensor, rotor, shorted tachometer or low operating voltage (battery supplied). Refer to *Checking for spark at the ignition coil and spark plugs* and *Checking the spark plugs* in this section to determine whether an ignition system fault is affecting one or all cylinders.

Checking the spark plugs

No single component affects the engine operation more than the spark plug. Spark plugs operate in a harsh environment that includes high heat, high pressure and the corrosive effects of the fuel. Because of this environment, spark plug failure is inevitable. Marine engines operate for extended periods of time under high throttle loads, and the spark plugs are subjected to much harsher conditions than the plugs used in most internal combustion engines.

Remove the spark plugs as described in Chapter Three. Arrange the plugs to identify the cylinder in which they were installed. Compare the plugs to those shown in **Figure 18**. Spark plug condition is an important indicator of engine

(18)

SPARK PLUG CONDITION

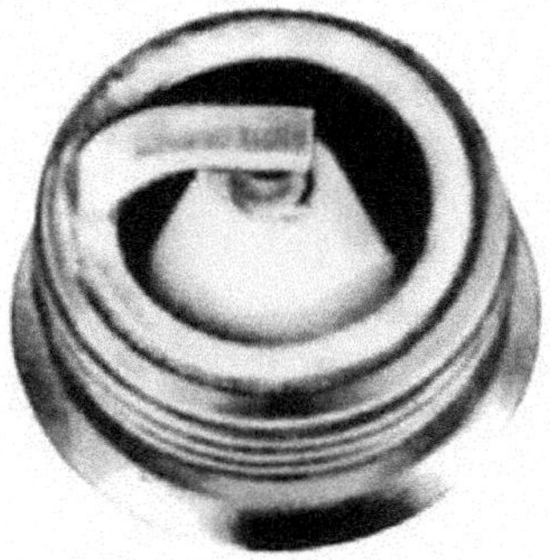

NORMAL
- Identified by light tan or gray deposits on the firing tip.
- Can be cleaned.

GAP BRIDGED
- Identified by deposit buildup closing gap between electrodes.

OIL FOULED
- Identified by wet black deposits on the insulator shell bore and electrodes.
- Caused by too much oil entering combustion chamber through worn rings and pistons, excessive clearance between valve guides and stems, or worn or loose bearings. Can be cleaned. If engine is not repaired, use a hotter plug.

CARBON FOULED
- Identified by dry, fluffy black carbon deposits on insulator tips, exposed shell surfaces and electrodes.
- Caused by a too-cold plug, weak ignition, dirty air cleaner, too-rich fuel mixture or excessive idling. Can be cleaned.

LEAD FOULED
- Identified by dark gray, yellow, tan or black deposits or a fused glazed coating on the insulator tip.
- Caused by highly leaded gasoline.

WORN
- Identified by severely eroded or worn electrodes.
- Caused by normal wear. Should be replaced.

FUSED SPOT DEPOSIT
- Identified by melted or spotty deposits resembling bubbles or blisters.
- Caused by sudden acceleration. Can be cleaned.

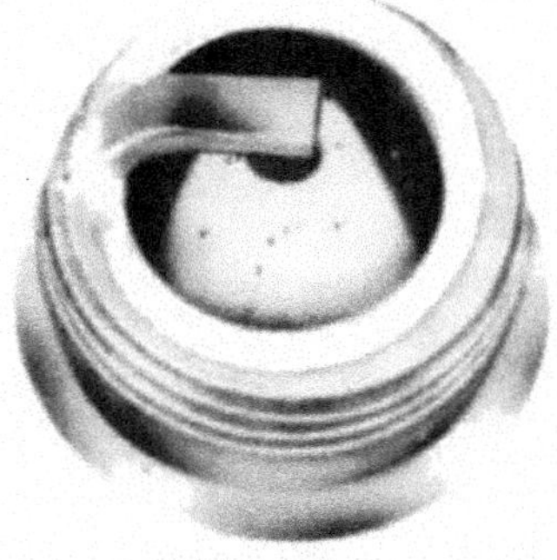

OVERHEATING
- Identified by a white or light gray insulator with small black or gray-brown spots with bluish-burnt appearance of electrodes.
- Caused by engine overheating, wrong type of fuel, loose spark plugs, a too-hot plug or incorrect ignition timing. Replace the plug.

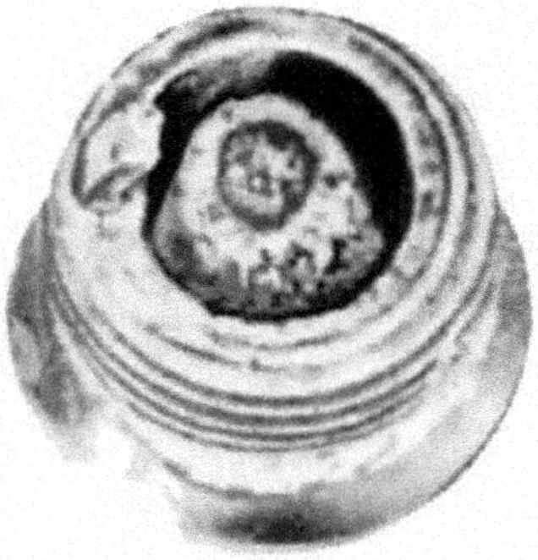

PREIGNITION
- Identified by melted electrodes and possibly blistered insulator. Metallic deposits on insulator indicate engine damage.
- Caused by wrong type of fuel, incorrect ignition timing or advance, a too-hot plug, burned valves or engine overheating. Replace the plug.

performance and condition. Spark plugs in a properly operating engine will have slightly pitted electrodes and a light tan insulator tip. **Figure 18** shows a normal plug and a number of others indicating trouble in their respective cylinders. All of the spark plugs should be nearly alike. If one or more of the spark plugs appear different from the rest, the cylinder is somehow different from the others. The difference could be a faulty plug, spark plug wire, distributor cap, ignition coil or fuel injector (multiport injection [MPI] models only) or internal engine damage that affects only that cylinder.

If the spark plugs reveal a problem that affects all of the cylinders, then the trouble is most likely caused by a faulty ignition module, pickup coil, Hall effect sensor, rotor, shorted tachometer or low operating voltage supplied by the battery. Do not forget to check all wiring connectors, fuses and circuit breakers. Refer to the wiring diagrams at the end of the manual to identify fuses, circuit breakers and wire connection points. After the inspection, clean, regap and install the spark plugs as described in Chapter Three. For future reference, note which cylinders are suspect.

Checking for spark at the ignition coil and spark plugs

The most basic ignition system test is checking for spark at the plugs and ignition coil(s). If a strong blue spark is present at the plugs and coil, the voltage supply is good, and the ignition system components are probably working correctly. A faulty spark plug wire or distributor rotor is indicated if good spark is present at the coil but not at the plugs.

If the engine will not start or runs improperly, and if good spark is present at the plugs, check or test the fuel system. This test requires a spark gap tester (**Figure 19**, typical). Purchase the spark gap tester from a tool supplier or an automotive parts store.

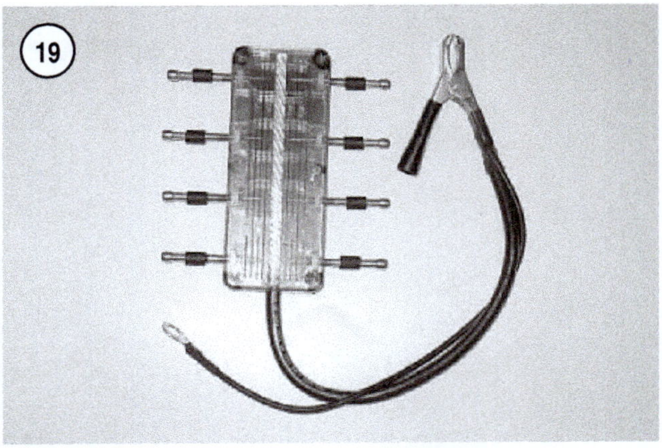

WARNING

High voltage is present in the ignition system. Never touch any wires or electrical components while running the engine or operating the electric starter. Never perform ignition system tests in wet conditions.

WARNING

The engine might unexpectedly start during testing. Keep all persons, clothing, tools or loose objects away from the drive belts and other moving components on the engine.

CAUTION

If testing the ignition system with the boat out of the water, use a flush device to supply the engine with cooling water. Operation of the starter motor results in operation of the seawater pump. Also, the engine might start during the troubleshooting procedure. The seawater pump is quickly damaged if operated without adequate cooling water.

1. Shift the engine into NEUTRAL.
2. On electronic fuel injection (EFI) models and carburetor-equipped models using a low-pressure electric fuel pump, remove the fuel pump relay as described in Chapter Nine. This will prevent the electric fuel pump(s) from operating during cranking.
3. Disconnect the high-tension ignition coil lead from the coil tower (**Figure 20**, typical). Refer to *Ignition Coil* in Chapter Nine to identify the ignition coil.
4. Connect the ground lead of the spark gap tester to a suitable engine ground.
5. Connect the spark gap tester to the coil tower. Adjust the gap to approximately 1/4 in. (6.35 mm).
6. Observe the spark gap tester while an assistant cranks the engine. A strong blue spark indicates the ignition system is producing output at the coil. If the spark is weak or absent, test the ignition system as described in this section.
7. Remove the spark gap tester and reconnect the ignition coil secondary wire to the coil tower (**Figure 20**).
8. Connect the ground lead of the spark gap tester to a suitable engine ground.
9. Note the cylinder number and then disconnect the spark plug wire from one of the spark plugs. Connect the spark plug wire to the spark gap tester. Adjust the gap to approximately 1/4 in. (6.35 mm).

10. Observe the spark gap tester while an assistant cranks the engine. A strong blue spark at the tester indicates the ignition system is producing spark to that cylinder. Record the test results for each cylinder.
11. Remove the spark gap tester and reconnect the spark plug wire.
12. Repeat Steps 9-11 for the remaining cylinders.
13. Note the test results and refer to the following:
    a. If spark is missing on all cylinders and the spark is good at the ignition coil, the fault is with the distributor rotor or coil wire. Replace the rotor and retest. Replace the coil wire if spark is not restored with the replacement rotor.
    b. If spark is good on some cylinders and weak or missing on others, the fault is with the spark plug wire(s) or distributor cap. Replace the distributor cap and retest. Replace the suspect spark plug wires if spark is not restored with the replacement cap.
14. Remove the spark gap tester and reconnect the spark plug lead.
15. If removed, install the fuel pump relay as described in Chapter Nine.

Checking for a shorted tachometer

The gray tachometer wire connects to the primary coil winding. The pulsating signal developed in the winding, which occurs during normal ignition system operation, provides the input to operate the tachometer. Any short along the wire from the coil to the tachometer terminal or an internally shorted tachometer will cause no spark or weak spark at the coil. A common complaint is misfiring at higher throttle settings.

WARNING
High voltage is present in the ignition system. Never touch any wires or electrical components while running the engine or performing a test. Never perform ignition system tests in wet conditions.

1. Check for spark at the coil as described in this chapter.
2. If spark is weak or missing on all cylinders, first inspect the gray wire for damaged insulation and the wire connection to the tachometer for shorting to ground or other terminals. Correct any faults and recheck for spark at the coil.
3. If no faults are found with the wiring, disconnect the gray wire from the tachometer. Tape the terminals to prevent short circuits during testing. Check for spark as described in this chapter. If good spark is restored or the misfire is absent with the gray tachometer wire disconnected, replace the tachometer.
4. If spark is not restored with the gray tachometer lead disconnected, disconnect the purple wire from the tachometer. Tape the terminals to prevent short circuits during testing. Check for spark as described in this chapter or operate the engine at the throttle setting where the misfire is occurring. If good spark is restored or the misfire is absent with the purple tachometer wire disconnected, replace the tachometer.
5. Reconnect the gray and purple tachometer wires. Orient the terminals and route the wiring to prevent shorting at the terminals and wire interference with moving components (such as steering components and control cables). Secure the wiring with plastic locking clamps as needed.

Checking for distributor rotation

The rotor on the distributor shaft directs the coil output to the individual cylinders, and a shaft-mounted device initiates spark at the coil. The distributor shaft must rotate to operate the ignition system.

WARNING
The engine might unexpectedly start during testing. Keep all persons, clothing, tools or loose objects away from the drive belts and other moving components on the engine.

CAUTION
If testing the ignition system with the boat out of the water, use a flush device to supply the engine with cooling water. Operation of the starter motor results in operation of the seawater pump. The seawater pump is quickly damaged if operated without adequate cooling water.

1. Remove the distributor cap as described in Chapter Nine. Do not remove the rotor for this test.
2. Pull the cord to activate the emergency stop (lanyard) switch (**Figure 2**). This prevents the ignition system from operating. The lanyard switch does not prevent starter motor operation. If the control does not incorporate an emergency stop (lanyard) switch, disconnect the coil wire from the distributor cap (**Figure 20**, typical) and connect to an engine ground to disable the ignition system.
3. Observe the distributor rotor (**Figure 21**) while an assistant cranks the engine. The rotor must rotate while cranking the engine. If it does not, the distributor gear is stripped and

22

PRESTOLITE BREAKERLESS INTEGRAL DISTRIBUTOR (BID) IGNITION SYSTEM

1. Distributor cap
2. Rotor
3. E-clip
4. Ignition module
5. Distributor shaft assembly
6. Bracket
7. Bolt
8. Retainer
9. High-tension lead
10. Ignition coil
11. Lockwasher
12. Nut
13. Bracket
14. Sleeve
15. Intake manifold
16. Gasket
17. Distributor gear
18. Washer
19. Bushing
20. Distributor housing
21. Hold-down clamp
22. Bolt

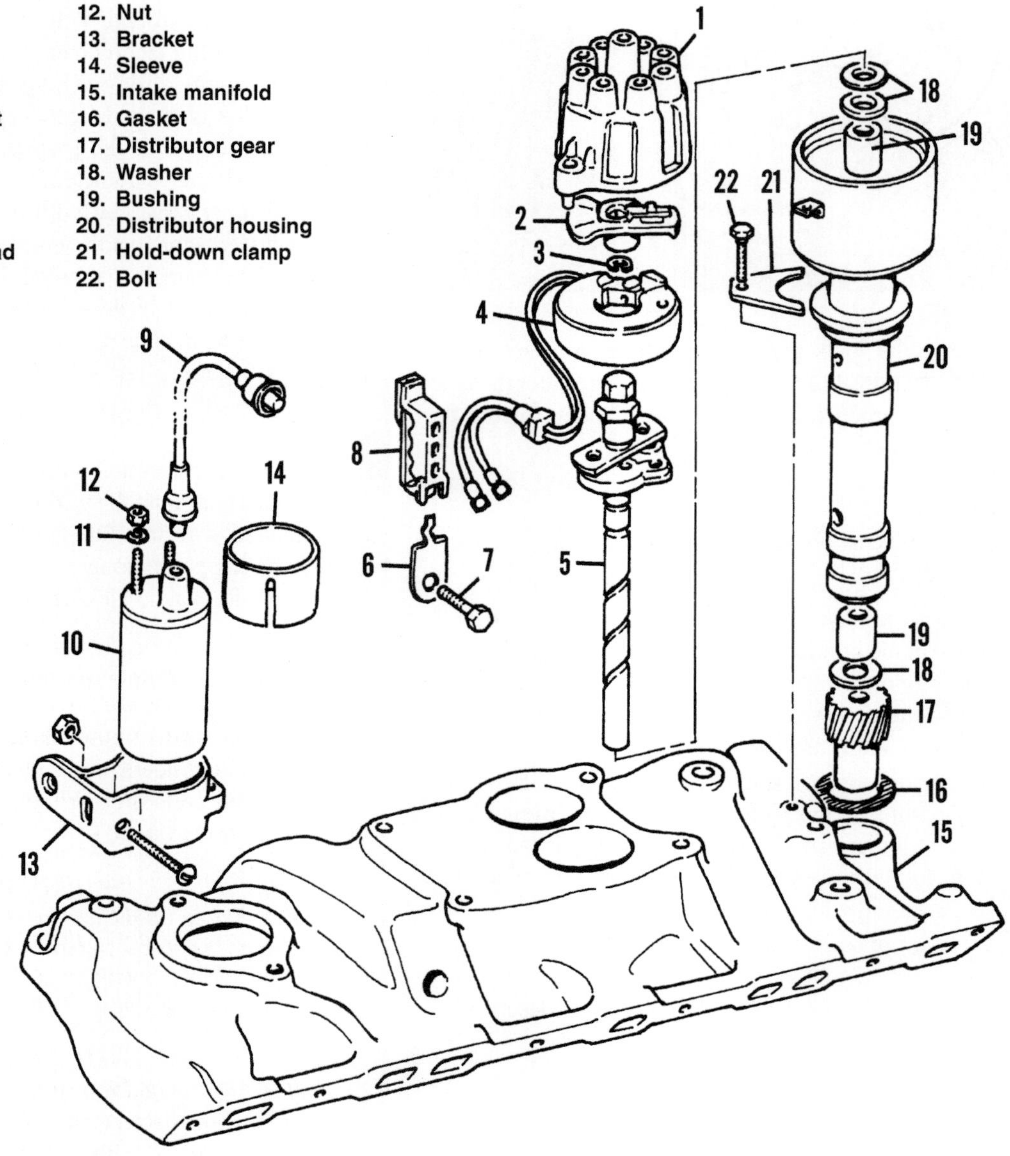

must be replaced. Refer to *Distributor Removal* in Chapter Nine. If no fault is found with the distributor gear, the camshaft, timing gear or timing chain, or related sprocket is faulty. Refer to Chapter Six for camshaft, timing gear, timing chain or sprocket removal, inspection and installation procedures.

4. Install the distributor cap as described in Chapter Nine.
5. Reattach the lanyard cord to the emergency stop (lanyard) switch or reconnect the coil wire to the distributor cap.

Testing the emergency stop (lanyard) switch

The emergency stop (lanyard) switch (**Figure 2**) is wired in series with the purple lead that connects the ignition switch *I* terminal to the instrument harness. The purple wire supplies current to operate the ignition system (carburetor-equipped models) or the system relay (fuel injection models). A fault with the switch can cause no spark or weak spark at the coil on carburetor-equipped models or no activation of the system relay on fuel injection models.

An ohmmeter is required for this procedure.

1. Disconnect the negative battery cable.
2. Locate the purple wires leading to the emergency stop (lanyard) switch. One of the wires connects to the ignition switch. The other wire connects to the purple instrument harness wire that leads to the engine. Mark the wires to identify the connection points and then disconnect the wires. In most cases, the wires are connected with crimp

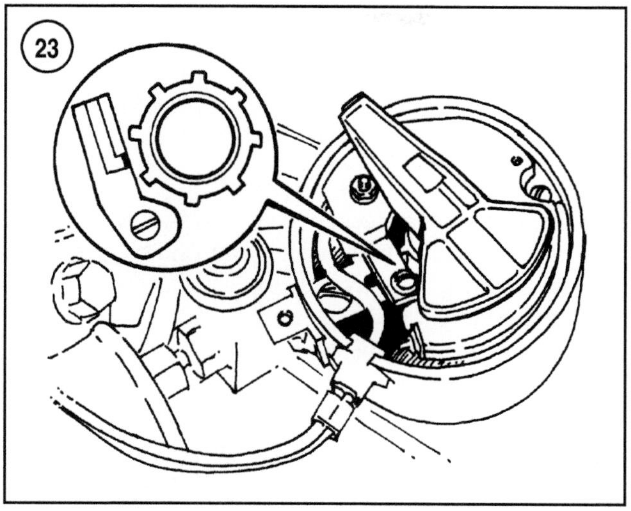

connectors. If necessary, cut the wiring and install new connectors after testing.

3. Connect the ohmmeter test leads to the disconnected purple wires leading to the switch. With the lanyard cord connected to the switch and the plunger depressed, the meter must indicate continuity. If not, the switch has failed open or has corroded internal contacts and must be replaced.

4. With the meter leads attached as described in Step 3, observe the meter while pulling the cord to dislodge the connector and allow the plunger to extend. The meter must now indicate no continuity. If not, the switch is shorted internally and must be replaced.

5. Repeat Step 3 and Step 4 several times to identify a possible intermittent switch failure.

6. Reconnect one of the purple switch wires to the wire leading to the *I* terminal on the ignition switch. Reconnect the other purple switch wire to the purple instrument harness wire leading to the engine. Do not inadvertently connect the switch wires together or connect the ignition switch directly to the instrument harness wiring.

CAUTION

If testing the ignition system with the boat out of the water, use a flush device to supply the engine with cooling water. Operation of the starter motor results in operation of the seawater pump. The seawater pump is quickly damaged if operated without adequate cooling water.

7. Connect the battery cable.

8. Start the engine. Pull the cord to dislodge the connector and allow the plunger to extend. The engine must stop immediately. If it does not, the switch, wiring or system relay (EFI models) is faulty. Repair the wiring or replace faulty components as needed. Do not operate the engine until proper switch operation is restored.

WARNING

High voltage is present in the ignition system. Never touch any wires or electrical components while running the engine or operating the electric starter. Never perform ignition system tests in wet conditions.

WARNING

The engine might unexpectedly start during testing. Keep all persons, clothing, tools or loose objects away from the drive belts and other moving components on the engine.

CAUTION

If testing the ignition system with the boat out of the water, use a flush device to supply the engine with cooling water. Operation of the starter motor results in operation of the seawater pump. Also, the engine might start during the troubleshooting procedure. The seawater pump is quickly damaged if operated without adequate cooling water.

Prestolite Breakerless Integral Distributor (BID) Ignition System

NOTE

With the manifolds, fuel system cooling system and electrical components installed, the General Motors (GM) 5.0L and 5.7L V-8 engines are very similar in appearance to the Ford 5.0L and 5.8L V-8 engines. If the model identification decal on the flame arrestor is missing, the engine can be quickly identified as a GM or Ford engine by noting the distributor location. On GM engines, the distributor is located at the rear (flywheel end) of the engine. On Ford engines, the distributor is located on the front of the engine.

The Prestolite breakerless integral distributor (BID) ignition system is used on all carburetor-equipped Ford 5.0L and 5.8L V-8 models. This system is also used on some GM 5.0L, 5.7L, 7.4L and 8.2L models. The Prestolite BID ignition system uses the type of distributor and ignition coil shown in **Figure 22**. The distributor houses the ignition module with an integrated impulse sensor, the rotor and the distributor shaft with an integrated impulse sender.

BID operation

When the ignition is switched ON, battery voltage is applied to the purple wire connected to the positive ignition coil terminal to supply the current needed to operate the ignition module. A black wire connects the module to the negative coil terminal.

Current supplied by the module flows through a winding in the integrated impulse sensor where it creates a small magnetic field. The camshaft-driven rotor in the distributor rotates at one-half crankshaft speed. As it rotates, the fingers of the impulse sender (**Figure 23**) pass by the impulse sensor (integrated into the module).

When one of the fingers aligns with the sensor, it disturbs the magnetic field and causes the sensor to switch on the module. Using the connecting black wire, the module then completes the ground circuit for the ignition coil. This allows battery current to flow through the primary winding and creates a strong magnetic field.

With more rotation, the finger moves out of alignment with the sensor, the magnetic field is restored, and the sensor switches off the module.

When switched off, the module interrupts the ground, which prevents battery current from flowing and causes the magnetic field to collapse. The collapsing field passes through the ignition coil secondary winding and induces very high voltage in this winding. The voltage flows from the coil tower (coil wire terminal) to the distributor rotor. The rotor directs the voltage to the distributor cap terminal for the individual spark plug. High-tension wires connect the cap terminals to the spark plugs.

The fuel/air mixture in the cylinder must ignite and burn at a precise time to push the piston down on the power stroke. Because fuel burns at a relatively constant rate, spark must occur earlier with increased engine speed. Ideally, the pressure from the burning fuel should peak as the piston starts moving downward on the power stroke. This provides the maximum push on the piston and maximum power output. If the spark timing is not advanced, the pressure peaks well after the piston starts moving downward and power output decreases.

Advance weights in the distributor automatically advance the ignition timing in proportion to engine speed. The ignition system does not provide engine overspeed protection. The system may be equipped with a tachometer amplifier to provide the stronger ignition system pulse required with some types of tachometers. The amplifier is wired in series in the gray wire connecting the negative coil terminal to the engine wire harness. The amplifier receives operating current from the primary coil winding via the gray wire connection. A separate black wire provides a ground circuit for the amplifier.

CAUTION
The air gap between the impulse sensor and impulse sender fingers must be correctly adjusted for proper ignition system operation. An improper gap can cause no spark, weak spark or incorrect ignition timing. Incorrect ignition timing can cause poor performance or serious engine damage.

Voltage test

Perform this procedure if the ignition system fails to deliver spark at the coil, the spark is weak, or the ignition system is misfiring. This test checks for battery voltage at the ignition coil and ignition module terminals. A voltmeter is required for this procedure.

WARNING
High voltage is present in the ignition system. Never touch any wires or electrical components while running the engine or performing a test. Never perform ignition system tests in wet conditions.

1. Make sure the emergency stop (lanyard) switch cord (**Figure 2**) is attached to the switch and is fully depressing the plunger. If not, the activated switch will open the circuit that supplies current to the ignition system.
2. Set the voltmeter to the 20 or 40 VDC scale.
3. Touch the positive meter lead to the positive battery terminal and the negative lead to the negative battery terminal. Charge the battery if the voltage is not above 12.0 volts. Record the battery voltage for reference during the test procedure.
4. Turn the ignition switch to ON or RUN. The dash gauges must turn on. If not, check for a blown fuse, tripped circuit breaker or loose wire connection. Correct faulty wiring, replace the fuse, or reset the circuit breaker to restore gauge operation. If this does not restore gauge operation, test the ignition switch as described in this chapter.
5. Connect the negative meter lead to an engine ground. Touch the positive meter lead to the positive ignition coil terminal. The purple wire connects to the positive coil terminal. The meter must indicate within 1 volt of the battery voltage recorded in Step 3. If not, check for corroded, dirty or loose wire terminals and damaged wiring.
6. Connect the negative meter lead to an engine ground. Touch the positive meter lead to the negative ignition coil terminal. The black wire connects to the negative coil terminal. With the ignition switch in ON or RUN, the meter must indicate 4.0-8.0 volts. If otherwise, refer to the following:
   a. If the voltage exceeds 8.0 volts, the distributor is not properly grounded to the engine, or the ignition module is not properly grounded to the distributor. Clean dirty mating surfaces and tighten loose fasteners to correct the ground and repeat the test.
   b. If the voltage is less than 4.0 volts, disconnect the black wire from the negative ignition coil terminal. Touch the positive meter lead to the coil terminal. Connect the negative meter lead to an engine ground. Turn the ignition switch to ON or RUN and note the voltage indicated on the meter. If the voltage is now near the battery voltage recorded in Step 3, the ignition module is faulty and must be replaced. If the voltage indicated remains less than 4.0 volts, the ignition coil primary winding has failed open, and the coil must be replaced.
7. Disconnect the negative battery cable.
8. Remove the distributor cap and rotor as described in Chapter Nine. Adjust the air gap between the impulse sender fingers and the impulse sensor (**Figure 24**) as described in the module installation instructions. Refer to *Ignition Module* in Chapter Nine. Do not disconnect the

24

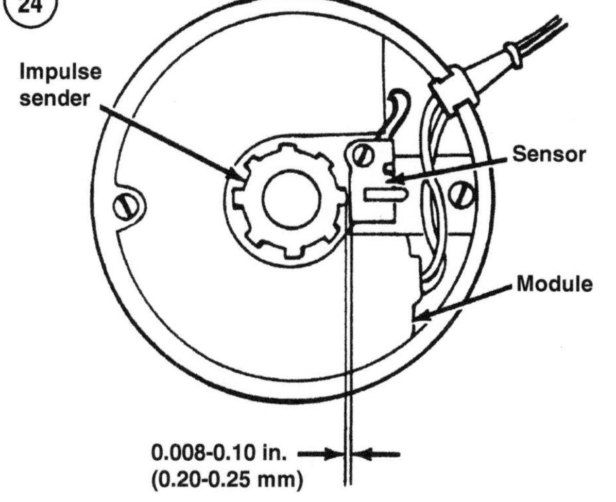

module wiring or install the rotor or distributor cap at this time.

9. Connect the battery cable. Turn the ignition switch to ON or RUN.
10. Connect the meter leads and measure the voltage as described in Step 5. Note the meter reading while inserting a feeler gauge between the impulse sender finger and the impulse sensor at the point indicated in **Figure 24**. If the voltage changes from 4.0-8.0 volts to near the battery voltage recorded in Step 3 when the feeler gauge passes through the gap, measure the ignition coil primary and secondary resistance as described in this section. If the voltage does not change and the system does not produce a spark, replace the ignition module as described in Chapter Nine.
11. Turn the ignition switch to OFF or STOP. Disconnect the test leads.
12. Disconnect the negative battery cable.
13. Install the rotor and distributor cap as described in Chapter Nine.
14. Connect the battery cable.
15. Adjust the ignition timing as described in Chapter Five.

Ignition coil

This section describes the resistance test for the primary and secondary ignition coil windings. Although the test might reveal an open or shorted circuit, it should not be considered a benchmark test for the coil. Due to the high voltage generated in the secondary winding, internal arcing can occur in partially shorted circuits or near windings with damaged insulation and cause an ignition misfire at any engine speed. This is usually not detected by a resistance test. However, when internal arcing occurs, a clicking noise is often heard emanating from the coil. Replace the coil if it fails either resistance test; internal arcing is suspected, and no spark, weak spark or an ignition misfire is present; and all other ignition system components of the system operate correctly.

NOTE

For accurate testing, perform the resistance test on the ignition coil with the coil and air temperature in the work area at approximately 68° F (20° C). The windings will have a higher resistance measurement at higher temperatures and a lower resistance measurement at lower temperatures.

An accurate digital ohmmeter is required for this procedure.

1. Remove the ignition coil as described in Chapter Nine.
2. Touch one ohmmeter lead to the positive coil terminal. Touch the other meter lead to the negative coil terminal. The meter must indicate 1.25-1.40 ohms. If not, the primary winding is shorted or open, and the coil must be replaced.
3. Touch one of the meter leads to the coil tower (high tension) terminal. Touch the other lead to either the positive or negative coil terminal. The meter must indicate 9400-11,700 ohms. If not, the secondary winding is shorted or open, and the coil must be replaced.
4. Install the ignition coil as described in Chapter Nine.

Testing the tachometer amplifier (BID)

A faulty tachometer can result in no or improper tachometer reading, a no-start condition or ignition misfiring at any engine speed.

1. Check the gray wire connections at the tachometer amplifier for corroded, dirty or loose terminals and damaged wiring. Correct faulty terminals or wiring as needed. Make sure the black ground wire for the amplifier has a secure, clean connection to the engine ground.

CAUTION

If testing the ignition system with the boat out of the water, use a flush device to supply the engine with cooling water. Operation of the starter motor results in operation of the seawater pump. Also, the engine might start during the troubleshooting procedure. The seawater pump is quickly damaged if operated without adequate cooling water.

2A. If the engine runs properly, but the tachometer is not operating, test the amplifier as follows:

a. Note the connection points and routing. Then disconnect all of the wires from the tachometer.
b. Temporarily connect a good compatible tachometer to the tachometer leads. It is not necessary to mount the tachometer in the dash for this test.
c. Start the engine. If the substitute tachometer operates, replace the original tachometer. If the substitute tachometer does not operate, replace the tachometer amplifier.

(25)

DELCO ELECTRONIC SPARK TIMING (EST) IGNITION SYSTEM (GM CARBURETOR AND EFI MODELS)

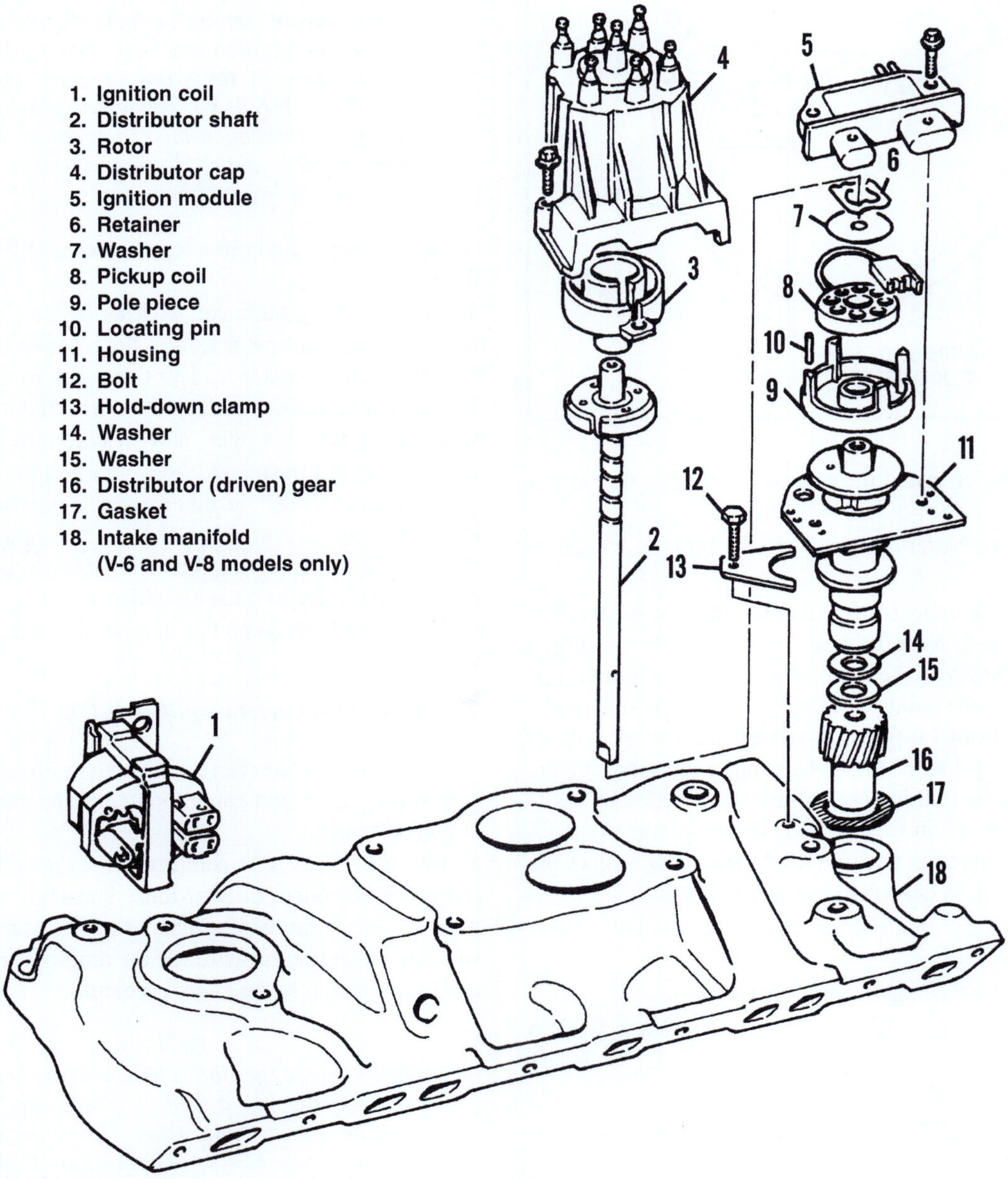

1. Ignition coil
2. Distributor shaft
3. Rotor
4. Distributor cap
5. Ignition module
6. Retainer
7. Washer
8. Pickup coil
9. Pole piece
10. Locating pin
11. Housing
12. Bolt
13. Hold-down clamp
14. Washer
15. Washer
16. Distributor (driven) gear
17. Gasket
18. Intake manifold (V-6 and V-8 models only)

d. Reconnect the tachometer wiring. Make sure the wire terminals are oriented to prevent contact with other tachometer terminals or wiring. Securely tighten the wire terminal nuts. They tend to work loose during boat operation.

2B. If the engine will not start or misfires, check for a shorted tachometer amplifier as follows:

a. Disconnect the two gray wires from the amplifier. Plug the gray wire leading to the ignition coil directly into the gray engine harness wire connectors to bypass the amplifier.
b. Start the engine and operate at the engine speed and under the conditions when the misfire is usually evident. Replace the amplifier if the engine operates properly only with the amplifier bypassed. Check for a shorted tachometer as described in this chapter if the engine still misfires with the amplifier bypassed.
c. Reconnect the gray coil wire and gray engine harness wire to the amplifier.

Delco Electronic Spark Timing (EST) Ignition System

NOTE

With the manifolds, fuel system cooling system and electrical components installed, the

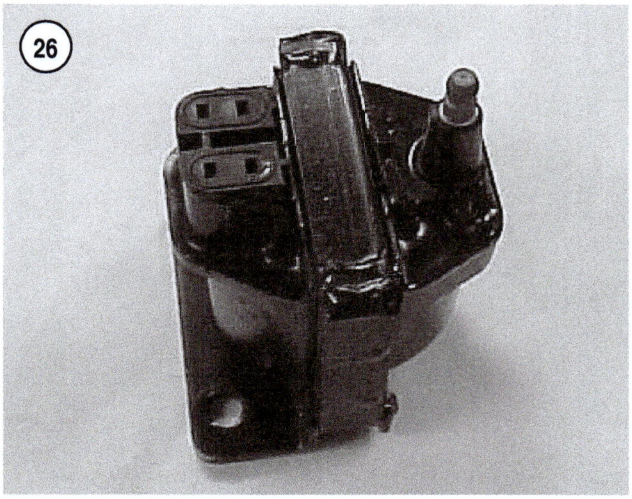

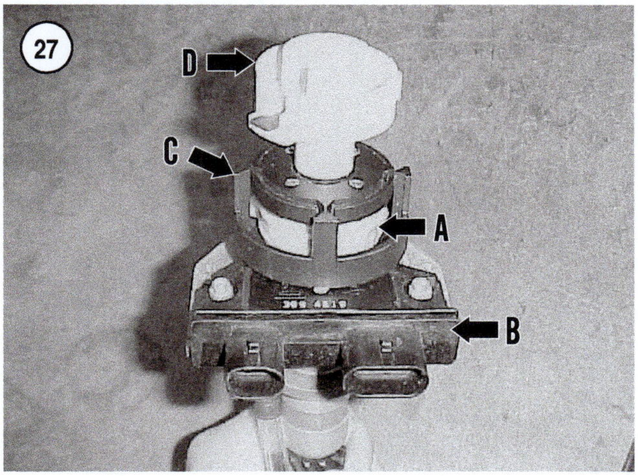

General Motors (GM) 5.0L and 5.7L V-8 engines are very similar in appearance to the Ford 5.0L and 5.8L V-8 engines. If the model identification decal on the flame arrestor is missing, the engine can be quickly identified as a GM or Ford engine by noting the distributor location. On GM engines, the distributor is located at the rear (flywheel end) of the engine. On Ford engines, the distributor is located on the front of the engine.

The Delco EST (electronic spark timing) ignition system is used on all GM 3.0L (four-cylinder) models, GM 4.3L (V-6) models, later GM (carburetor-equipped) models and all GM electronic fuel injection (EFI) models.

Major components of the system include the distributor, ignition coil, spark plugs and wires. The EST ignition system uses the distributor shown in **Figure 25**, typical. The EST ignition systems use the epoxy-coated ignition coil shown in **Figure 26**. The ignition coil has a very robust design, and failure is extremely rare. Coil failure is almost always caused by corroded or dirty terminals.

The EST distributor houses the pickup coil (A, **Figure 27**), ignition module (B), poles (C), rotor (D) and distributor shaft.

EST operation

When the ignition on carburetor-equipped models is switched ON, battery voltage is applied to the purple wire connected to the ignition coil. When the ignition on electronic fuel injection (EFI) models is switched ON, battery voltage is applied to the purple wire connected to the EFI system relay. The system relay switches on when it receives the voltage and supplies voltage to the purple wire connected to the ignition coil. A circuit inside the coil provides battery voltage to the pink wire leading to the ignition module. The pink wire supplies voltage to operate the ignition module.

The camshaft-driven rotor in the distributor rotates at one-half crankshaft speed. As it rotates, magnetic poles on the rotor shaft and pickup coil align and misalign. When they are aligned, a magnetic field passes through the pickup coil winding and creates an electric pulse. This pulse is directed to the timing circuits inside the ignition module.

The timing circuit modifies the pulse and directs it to a transistor in the module. The transistor functions as a switch to control current flow through the ignition coil primary winding.

The transistor switches ON when the pulse flows from the timing circuit and switches OFF when the pulse ceases. When switched on, the transistor provides a connection to ground for the ignition coil primary winding; this allows battery current to flow through the primary winding and creates a strong magnetic field.

When switched OFF, the transistor interrupts the ground, which prevents battery current from flowing and causes the magnetic field to collapse. The collapsing field passes through the ignition coil secondary winding inducing very high voltage current in this winding. The current flows from the coil tower (coil wire terminal) to the distributor rotor. The rotor directs the current to the distributor cap terminal for the individual spark plug. High-tension wires connect the cap terminals to the spark plugs.

The fuel/air mixture in the cylinder must ignite and burn at a precise time to push the piston down on the power stroke. Because fuel burns at a relatively constant rate, spark must occur earlier with increased engine speed. Ideally, the pressure from the burning fuel should peak as the piston starts moving downward on the power stroke. This provides the maximum push on the piston and maximum power output. If the spark timing is not advanced, the pressure peaks well after the piston starts moving downward and power output decreases.

The timing circuits in the module automatically advance the ignition timing in proportion to engine speed. The ignition system does not provide engine overspeed protection; however, the engine control unit (ECU) used on EFI models limits engine speed by shutting off the fuel injectors at a predetermined limit.

On EFI models, the ECU computes the optimum amount of spark advance for the given operating conditions and adjusts the timing accordingly. The ECU timing circuit essentially overrides the timing circuit in the ignition module. The timing circuit in the module assumes control of the ignition timing if a fault in the ECU circuitry or wiring should occur.

Voltage test

Perform this procedure if the ignition system fails to generate spark at the coil, the spark is weak, or the ignition system is misfiring. This test checks for battery voltage at the ignition coil and ignition module terminals. A 12-volt test lamp is required for this procedure.

WARNING
High voltage is present in the ignition system. Never touch any wires or electrical components while running the engine or performing a test. Never perform ignition system tests in wet conditions.

WARNING
The engine might unexpectedly start during testing. Keep all persons, clothing, tools or loose objects away from the drive belts and other moving components on the engine.

CAUTION
If testing the ignition system with the boat out of the water, use a flush device to supply the engine with cooling water. Operation of the starter motor results in operation of the seawater pump. The seawater pump is quickly damaged if operated without adequate cooling water.

1. Make sure the emergency stop (lanyard) switch cord (**Figure 2**) is attached to the switch and is fully depressing the plunger. If not, the activated switch will open the circuit that supplies current to the ignition system.
2. Turn the ignition switch to ON or RUN. The dash gauges must turn on. If not, check for a blown fuse, tripped circuit breaker or loose wire connection. Correct faulty wiring, replace the fuse or reset the circuit breaker to restore gauge operation. If this does not restore gauge operation, test the ignition switch as described in this chapter.
3. Disconnect the wire harness (A and B, **Figure 28**) from the ignition coil. Connect the ground lead of a test lamp to an engine ground.
4A. On carburetor-equipped models, touch the test lamp probe to the purple wire terminal in the gray connector. With the ignition switch in ON or RUN, the light must illuminate. If not, test the emergency stop (lanyard) switch as described in this chapter. Faulty instrument harness or engine harness wiring is indicated if no fault is found with the emergency stop (lanyard) switch.
4B. On electronic fuel injection (EFI) models, touch the test lamp probe to the purple wire terminal in the gray connector. With the ignition switch in ON or RUN, the lamp must illuminate. If not, test the EFI system relay as described in this chapter. Refer to *Electronic Fuel Injection (EFI) System.*

5. Reconnect the gray harness connector to the ignition coil. Do not connect the black harness connector to the coil for this test. Touch the test lamp probe to each of the two open terminals in the ignition coil. With the ignition switch in ON or RUN, the lamp must illuminate when touched to each terminal. Otherwise, the coil has failed open and must be replaced.
6. Reconnect the gray and black harness connectors to the ignition coil. Disconnect the two-wire terminal from the distributor-mounted ignition module (**Figure 29**). The harness contains a pink and brown wire. Touch the test lamp probe to each of the two open terminals in the harness connector. With the ignition switch in ON or RUN, the lamp must illuminate when touched to each terminal. Otherwise, the two-wire harness connecting the coil to the module has failed open and must be replaced.
7. Turn the ignition switch to OFF or STOP. Reconnect the two-wire harness connector onto the ignition module.

Pickup coil

The pickup coil (A, **Figure 27**) is located inside the distributor housing. The coil can be tested without removing the coil from the distributor.

An ohmmeter is required for this procedure.

30

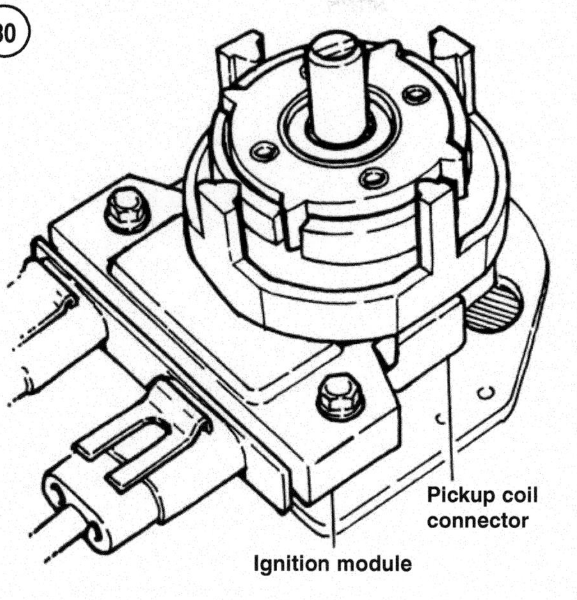

1. Disconnect the negative battery cable.
2. Remove the distributor cap as described in Chapter Nine.
3. Pull up on the locking tab and then disconnect the pickup coil connector (**Figure 30**) from the module.
4. Touch one of the meter leads to the white coil wire terminal. Touch the other meter lead to the green coil wire terminal. The meter must indicate 500-1500 ohms. If not, the pickup coil wire is shorted or open and must be replaced as described in Chapter Nine.
5. Plug the pickup coil connector (**Figure 30**) onto the module. The locking tab must engage the tab on the module.
6. Install the rotor and distributor cap as described in Chapter Nine.
7. Connect the battery cable.

Ignition module

Failure of the ignition module is relatively rare, and most faults are caused by corroded, dirty or defective wiring or terminals. Although test equipment is available, the cost far exceeds the cost of the module. Unless the equipment is available, test the module by a process of elimination.

Perform the following procedures.

1. Check for spark at the coil and spark plugs as described in this section.
2. Check for a shorted tachometer as described in this section.
3. Check the battery voltage to the ignition system as described in this section.
4. Test the pickup coil as described in this section.
5. Check for distributor rotation as described in this section.

CAUTION

If testing the ignition system with the boat out of the water, use a flush device to supply the engine with cooling water. Operation of the starter motor results in operation of the seawater pump. The seawater pump is quickly damaged if operated without adequate cooling water.

6. Observe the tachometer while cranking the engine. A reading of 200 rpm or more indicates the pickup coil and ignition module are operating correctly. Weak or missing spark is probably due to a faulty ignition coil or connecting wiring. Repair the wire or test the ignition coil resistance as described in this section. Then check for spark at the coil and spark plugs as described in this section.
7. Replace the ignition module as described in Chapter Nine and check for spark at the coil and spark plugs. If no spark is present with the replacement module, replace the ignition coil to restore ignition system operation.

Ignition coil

This section describes the resistance test for the primary and secondary ignition coil windings. Although the test might reveal an open or shorted circuit, it should not be considered a benchmark test for the coil. Due to the high voltage generated in the secondary winding, internal arcing can occur in partially shorted circuits or near windings with damaged insulation and cause an ignition misfire at any engine speed. This is usually not detected with a resistance test. However, when internal arcing occurs, a clicking noise is often heard emanating from the coil. Replace the coil if it fails either resistance test; internal arcing is suspected, and no spark, weak spark or an ignition misfire is present; and all other ignition system components of the system test correctly.

NOTE

For accurate testing, perform a resistance test on the ignition coil with the coil and air temperature in the work area at approximately 68° F (20° C). The windings will have a higher resistance measurement at higher temperatures and a lower resistance measurement at lower temperatures.

An accurate digital ohmmeter is required for this procedure. The test leads must have probes with diameters small enough to fit fully into the terminal openings in the coil, yet the probes must be long enough to reach the contacts. If in doubt, insert straightened paper clips into the openings and connect the leads to the ends.

1. Remove the ignition coil as described in Chapter Nine.
2. Carefully scrape the paint or other material to reveal bare metal on a small surface of the mounting bracket (A, **Figure 31**).

2

3. Touch one of the test leads to the cleaned surface of the mounting bracket (A, **Figure 31**) and the other test lead to the terminal in the plug opening (B). The meter must indicate no continuity. If not, replace the ignition coil as described in Chapter Nine.
4. Touch one of the test leads to the terminal in the plug opening (B, **Figure 31**) and the other test lead to the terminal in the other plug opening (C). The meter must indicate 1.35-1.45 ohms. If not, the primary winding is shorted or open, and the ignition coil must be replaced as described in Chapter Nine.
5. Touch one of the test leads to the terminal in the plug opening (D, **Figure 31**) and the other test lead to the coil wire terminal (E). The meter must indicate 6700-7500 ohms. If not, the secondary winding is shorted or open and must be replaced as described in Chapter Nine.
6. Apply paint to the bare spot on the mounting bracket to prevent corrosion.
7. Install the ignition coil as described in Chapter Nine.

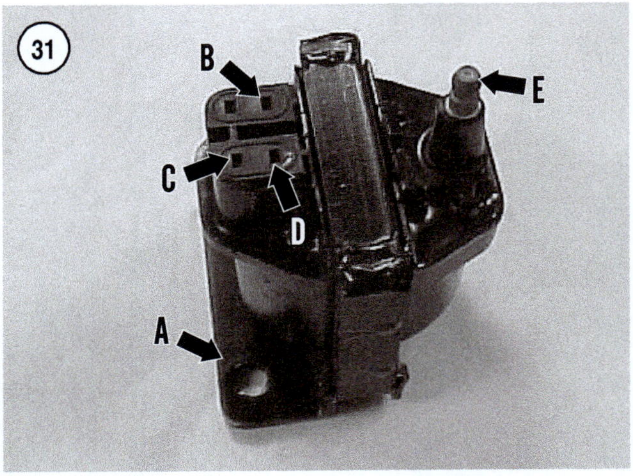

Thick Film Integrated (TFI) Ignition System

NOTE
With the manifolds, fuel system cooling system and electrical components installed, the General Motors (GM) 5.0L and 5.7L V-8 engines are very similar in appearance to the Ford 5.0L and 5.8L V-8 engines. If the model identification decal on the flame arrestor is missing, the engine can be quickly identified as a GM or Ford engine by noting the distributor location. On GM engines, the distributor is located at the rear (flywheel end) of the engine. On Ford engines, the distributor is located on the front of the engine.

The Thick Film Integrated (TFI) ignition system is used on Ford 5.0L and 5.8L electronic fuel injection (EFI) models.

The TFI distributor contains a Hall effect switch (7, **Figure 32**), distributor shaft and switch wheel. The ignition coil is mounted on a bracket near the distributor along with the ignition module. All TFI ignition components used for marine engines are robust, and failure is rare. Ignition system failure is almost always caused by corroded or dirty terminals.

Operation

The camshaft driven distributor shaft and switch wheel in the distributor rotate at one-half crankshaft speed.

When the ignition is switched ON, battery voltage is applied to the purple wire connected to the EFI system relay. The system relay switches on when it receives the voltage and supplies voltage to the purple/black wire connected to the ignition coil, ignition module and numerous other fuel system sensors. One of purple/black wires connects to a red wire that supplies voltage to operate the Hall effect sensor within the distributor.

The Hall effect switch serves as a trigger device by making and breaking a circuit when subjected to a magnetic field. This occurs as windows in the switch wheel on the distributor shaft (6, **Figure 32**) pass through a slot in the switch. The pulsating signal generated by the switch is referred to as the profile ignition pickup (PIP) signal and is used to time the ignition spark and fuel injector activation.

The PIP signal is directed to the ignition module (17, **Figure 32**) where it is used to control current flow through the primary winding in the ignition coil (18, **Figure 32**). When the ignition module switches on, it completes the ground circuit for the primary ignition coil winding. This allows battery current to flow through the primary winding and creates a strong magnetic field. When the ignition module switches off, it interrupts the ground, which prevents battery current from flowing and causes the magnetic field to collapse. The collapsing field passes through the ignition coil secondary winding and induces current with very high voltage in this winding. The current flows from the coil tower (coil wire terminal) to the distributor rotor. The rotor directs the current to the distributor cap terminal for the individual spark plug. High-tension wires connect the cap terminals to the spark plugs.

The fuel/air mixture in the cylinder must ignite and burn at a precise time to push the piston down on the power stroke. Because fuel burns at a relatively constant rate, spark must occur earlier with increased engine speed. Ideally, the pressure from the burning fuel should peak as the piston starts moving downward on the power stroke. This provides the maximum push on the piston and maximum power output. If the spark timing is not advanced, the pressure peaks well after the piston starts moving downward, and power output decreases. The timing circuit in the module automatically advances the ignition timing in proportion to engine speed.

During normal operation, the ignition module controls the ignition coil and timing advance only while starting. After the engine reaches normal idle speed, the engine control

(32)

THICK FILM INTEGRATED (TFI) IGNITION SYSTEM (FORD 5.0L AND 5.8L ELECTRONIC FUEL INJECTION [EFI] MODELS)

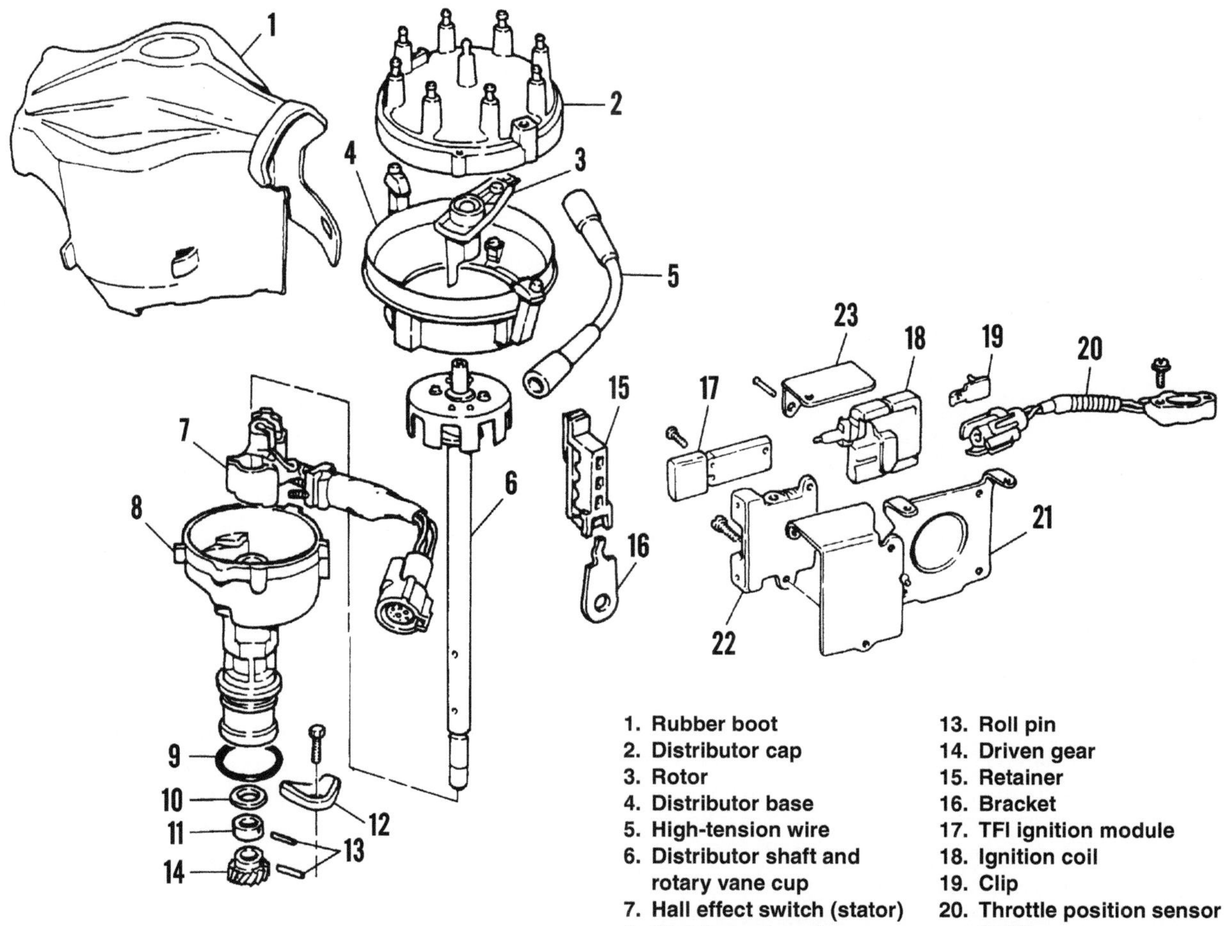

1. Rubber boot
2. Distributor cap
3. Rotor
4. Distributor base
5. High-tension wire
6. Distributor shaft and rotary vane cup
7. Hall effect switch (stator)
8. Distributor housing
9. O-ring
10. Spacer
11. Spacer
12. Hold-down clamp
13. Roll pin
14. Driven gear
15. Retainer
16. Bracket
17. TFI ignition module
18. Ignition coil
19. Clip
20. Throttle position sensor (TPS)
21. Bracket
22. Heat sink
23. Bracket

assembly (ECA) computes the optimum amount of spark advance for the given operating conditions and overrides the ignition timing circuit in the module. The timing circuit in the module assumes control of the ignition timing if a fault in the ECU circuitry or connecting wiring should occur and during ignition timing adjustment.

The ignition system does not provide engine overspeed protection, however, the ECA limits engine speed by shutting off the fuel injectors at a predetermined limit.

Voltage test

Perform this procedure if the ignition system fails to generate spark at the coil, the spark is weak, or the ignition system is misfiring. This test checks for battery voltage at the Hall effect switch, ignition coil and ignition module terminals. A 12-volt test lamp is required for this procedure.

WARNING

High voltage is present in the ignition system. Never touch any wires or electrical components while running the engine or performing a test. Never perform ignition system tests in wet conditions.

WARNING

The engine might unexpectedly start during testing. Keep all persons, clothing, tools or loose objects away from the drive belts and other moving components on the engine.

CAUTION

If testing the ignition system with the boat out of the water, use a flush device to supply the engine with cooling water. Operation of the starter motor results in operation of the seawater pump. The seawater pump is quickly damaged if operated without adequate cooling water.

1. Make sure the emergency stop (lanyard) switch cord (**Figure 2**) is attached to the switch and is fully depressing the plunger. If not, the activated switch will open the circuit that supplies current to the ignition system.
2. Turn the ignition switch to ON or RUN. The dash gauges must turn on. If not, check for a blown fuse, tripped circuit breaker or loose wire connection. Correct faulty wiring, replace the fuse, or reset the circuit breaker to restore gauge operation. If this does not restore gauge operation, test the ignition key switch as described in this chapter.
3. Check the voltage supplied to the Hall effect switch as follows:
   a. Connect the cord of the test lamp to an engine ground terminal. Touch the test probe to the battery positive terminal to verify the lamp and ground connection.
   b. Disconnect the engine wire harness connector from the Hall effect switch harness connector. Refer to *Hall Effect Switch* in Chapter Nine to identify the connectors.
   c. Turn the ignition switch to ON or RUN.
   d. Touch the test lamp probe to the terminal for the red wire in the disconnected engine harness connector. Do not inadvertently select the switch harness. The lamp must illuminate. If not, test the EFI system relay as described in this chapter. Refer to *Electronic Fuel Injection (EFI) System*. A fault in the purple/black and red wires connecting the power relay to the sensor is indicated if the relay tests correctly.
   e. Touch the test lamp probe to the terminal for the black wire in the disconnected engine harness connector. Do not inadvertently select the switch harness. The lamp must illuminate. If not, the black wire is open, or the wire terminal connection to ground is faulty.
4. Check the voltage supplied to the ignition coil as follows:
   a. Connect the test lamp cord to a good engine ground.
   b. Disconnect the engine wire harness connector from the ignition coil. Refer to *Ignition Coil* in Chapter Nine to identify the connector.
   c. Turn the ignition switch to ON or RUN.
   d. Touch the test lamp probe to the terminal for the purple/black wire in the disconnected engine harness connector. The lamp must illuminate. If not, test the EFI system relay as described in this chapter. Refer to *Electronic Fuel Injection (EFI) System*. A fault in the purple/black connecting the power relay to the ignition coil is indicated if the relay tests correctly.
5. Check the voltage supplied to the ignition module as follows:
   a. Disconnect the engine wire harness connector from the ignition module. Refer to *Ignition Module* in Chapter Nine to identify the connector.
   b. Turn the ignition switch to ON or RUN.
   c. Touch the test lamp probe to the terminal for the purple/black wire in the disconnected engine harness connector. The lamp must illuminate. If not, test the EFI system relay as described in this chapter. Refer to *Electronic Fuel Injection (EFI) System*. A fault in the purple/black wire connecting the power relay to the ignition module is indicated if the relay tests correctly.
6. Turn the ignition switch to OFF or STOP.
7. Reconnect the engine wire harness connectors to the Hall effect harness connector, ignition coil and ignition module.

Ignition coil

CAUTION
If testing the ignition system with the boat out of the water, use a flush device to supply the engine with cooling water. Operation of the starter motor results in operation of the seawater pump. The seawater pump is quickly damaged if operated without adequate cooling water.

Due to the high voltage generated in the secondary winding, internal arcing can occur in partially shorted circuits or near windings with damaged insulation and cause an ignition misfire at any engine speed. This is usually not detected with a resistance test. However, when internal arcing is occurring, a clicking noise is often heard emanating from the coil while attempting to start the engine.

If the engine will not start and the coil is suspect, test the voltage supplied to the ignition coil as described in this section. If the voltage is correct, attempt to start the engine while observing the tachometer. If the tachometer is registering cranking speed (approximately 200-300 rpm), the Hall effect switch and ignition module are probably working correctly. Check for spark at the coil as described in this section. If the spark is weak or absent, replace the ignition coil as described in Chapter Nine and repeat the spark test. Test the ignition module and Hall effect switch if spark is not restored with the replacement ignition coil.

Ignition module

CAUTION
If testing the ignition system with the boat out of the water, use a flush device to supply the engine with cooling water. Operation of the starter motor results in operation of the seawater pump. The seawater pump is quickly damaged if operated without adequate cooling water.

A stand-alone test for the ignition module requires expensive test equipment that makes it impractical for the home mechanic. The cost of the test equipment far exceeds the cost of the module. If the engine will not start and the ignition module is suspect, perform the following:

1. Test the voltage as described in this section. If the voltage is correct, attempt to start the engine while observing the tachometer.

2A. If the tachometer registers cranking speed (approximately 200-300 rpm), the ignition module is probably working correctly. Refer to the following:

   a. Check for spark at the coil as described in this chapter. If the spark is weak or absent, replace the ignition coil as described in Chapter Nine.
   b. Repeat the spark test. If spark is not restored with the replacement coil, replace the Hall effect switch as described in Chapter Nine.
   c. Repeat the spark test. If spark is not restored with the replacement coil and sensor, replace the ignition module as described in Chapter Nine to restore ignition system operation.

2B. If the tachometer is not registering cranking speed (approximately 200-300 rpm), the Hall effect switch and ignition module are suspect. Refer to the following:

   a. Check for spark at the coil as described in this chapter. If the spark is weak or absent, replace the Hall effect switch as described in Chapter Nine.
   b. Repeat the spark test. If spark is not restored with the replacement switch, replace the ignition module, as described in Chapter Nine, to restore ignition system operation.

Hall effect switch

A stand-alone test for the Hall effect switch requires expensive test equipment that makes it impractical for the home mechanic. The cost of the test equipment far exceeds the cost of the switch. If the engine will not start and the Hall effect switch is suspect, perform the following:

CAUTION
If testing the ignition system with the boat out of the water, use a flush device to supply the engine with cooling water. Operation of the starter motor results in operation of the seawater pump. The seawater pump is quickly damaged if operated without adequate cooling water.

1. Test the voltage as described in this section. If the voltage is correct, attempt to start the engine while observing the tachometer.

2A. If the tachometer registers cranking speed (approximately 200-300 rpm), the Hall effect switch is probably working correctly. Refer to the following:

   a. Check for spark at the coil as described in this chapter. If the spark is weak or absent, replace the ignition coil as described in Chapter Nine.
   b. Repeat the spark test. If spark is not restored with the replacement coil, replace the Hall effect switch as described in Chapter Nine.
   c. Repeat the spark test. If spark is not restored with the replacement coil and switch, replace the ignition module as described in Chapter Nine to restore ignition system operation.

2B. If the tachometer does not register cranking speed (approximately 200-300 rpm), the Hall effect switch and ignition module are suspect. Refer to the following:

   a. Check for spark at the coil as described in this chapter. If the spark is weak or absent, replace the Hall effect switch as described in Chapter Nine.
   b. Repeat the spark test. If spark is not restored with the replacement switch, replace the ignition module as described in Chapter Nine to restore ignition system operation.

BASIC FUEL SYSTEM TROUBLESHOOTING

Refer to **Table 3** for a list of items to check if the engine will not start, or if the fuel system is suspected of causing an engine malfunction. Other useful information and tips are described with the specific test instructions. Refer to *Carburetor Fuel System* or *Electronic Fuel Injection (EFI) System* in this section for specific troubleshooting procedures.

Most fuel system malfunctions (both carburetor and electronic fuel injection) are caused by the fuel used. Using the proper type of fresh, quality fuel and storing it properly, together with regular maintenance of the fuel filters will reduce fuel system problems. Fuel storage recommendations and filter maintenance are described in Chapter Three.

Fuel/Air Mixture

A gasoline engine must receive fuel and air mixed in precise proportions to operate efficiently at various loads and engine speeds. Determining whether a fuel system malfunction is causing an excessively rich or excessively lean condition can help you determine the cause of the malfunction and identify the corrective action.

An engine that is receiving too much fuel for the amount of air flowing into the engine is operating too rich. Symptoms of an excessively rich condition include rough operation (particularly at idle speed), excessive exhaust smoke, excessive sooting of the transom near the exhaust outlet, hesitation during rapid acceleration, dark black fluffy deposits on the spark plugs and stalling during extended periods or idling (commonly called loading up). Operating an engine that is too rich will result in increased fuel usage, increased exhaust emissions, premature fouling of the spark plugs, increased combustion chamber deposits and rapid blackening and dilution of the crankcase oil. To positively identify an excessively rich condition, remove and inspect the spark plugs as described in Chapter Three. Compare the spark plugs to those shown in **Figure 18**.

On carburetor-equipped models, an excessively rich condition is usually caused by one or more of the following:

1. An improperly adjusted carburetor.
2. A dirty or damaged flame arrestor.
3. Improper choke valve operation.
4. A fuel-saturated or improperly adjusted float.
5. A worn or leaking fuel inlet needle.
6. Damaged or stuck power valve.
7. Inadequate ventilation of the engine compartment.

On electronic fuel injection (EFI) models, an excessively rich condition is usually caused by one or more of the following:

1. A dirty or damaged flame arrestor.
2. Leaking fuel injector.
3. Excessively high fuel pressure.
4. Malfunction of a fuel injection system sensor.
5. Internal leak in fuel pressure regulator.
6. Inadequate ventilation of the engine compartment.

An engine receiving too little fuel for the amount of air flowing into the engine is operating too lean. Symptoms of an excessively lean mixture include hard starting, hesitation or backfiring during rapid acceleration, poor high speed performance, overheating (particularly at higher throttle settings) and overheating of the spark plug insulator tip. Operating an engine that is too lean will result in poor performance and promotes preignition and detonation that damages the pistons and valves. To positively identify an excessively lean condition, remove and inspect the spark plugs as described in Chapter Three. Compare the spark plugs to those shown in **Figure 18**.

On carburetor-equipped models, an excessively lean condition is usually caused by one or more of the following:

1. A restriction in the fuel supply line.
2. Air leaks in the fuel supply line.
3. Blockage or air leaks in the fuel filters.
4. Faulty fuel pump.
5. An improperly adjusted carburetor.
6. Plugged carburetor fuel jet or passage.
7. Improperly adjusted float.
8. A sticking fuel inlet needle.
9. Damaged or stuck power valve.

On EFI models, an excessively lean condition is usually caused by one of more of the following:

1. A restriction in the fuel supply line.
2. Air leaks in the fuel supply line.
3. Blockage or air leaks in the fuel filters.
4. Excessively low fuel pressure.
5. Malfunction of a fuel injection system sensor.
6. Faulty or blocked fuel injector(s).

Fuel inspection

Fuel-related problems are common with most marine engines. Gasoline has a relatively short shelf life and becomes stale within a few weeks under some conditions. This gasoline works fine in an automobile because the fuel is consumed in a week or so. Because a marine engine may sit idly for several weeks at a time, the gasoline often becomes stale.

33

As fuel evaporates, a gummy deposit usually forms in the carburetor or other fuel system components. These deposits can plug fuel filters, fuel lines, fuel pumps and small passages in the carburetor or fuel injection system.

Fuel stored in the fuel tank tends to absorb water vapor from the air. Over time, this water separates from the fuel and settles at the bottom of the fuel tank. Water in the fuel tank can lead to the formation of rust and other contaminants in the fuel tank. These contaminants block fuel filters and other fuel system passages. Inspect the fuel when the engine refuses to start and the ignition system is not at fault. An unpleasant odor usually indicates the fuel has exceeded the shelf life and should be replaced.

WARNING
Use extreme caution when working with the fuel system. Fuel is extremely flammable and if ignited can result in injury or death. Never smoke or allow sparks to occur around fuel or fuel vapor. Wipe up any spilled fuel at once with a shop towel and dispose of the shop towel in an appropriate manner. Check all fuel hoses, connections and fittings for leaks after any fuel system repair.

Most models covered in this manual are equipped with a spin-on water-separating fuel filter (**Figure 33**, typical) mounted on or near the engine. To inspect the fuel, place a suitable container under the filter and then remove the filter as described in Chapter Three. Refer to *Servicing the Water-Separating Fuel Filter*. Empty the fuel into the container. Then reinstall the filter as described in Chapter Three. Promptly clean up any spilled fuel. On models without a spin-on fuel filter, place the container under the mechanical or electric fuel pump. Then loosen the fuel inlet fitting at the pump (**Figure 34**, typical) to drain a sample of the fuel from the fuel supply hose or outlet line. Refer to Chapter Seven to identify the pump inlet fitting. Securely

tighten the fitting as described in Chapter Seven. Promptly wipe up any spilled fuel.

Inspect and carefully smell the fuel. An unusual odor, debris, cloudy appearance or the presence of water indicates a problem with the fuel. If any of these conditions are noted, dispose of all the fuel in an environmentally responsible manner. Contact a local marine dealership or automotive repair facility for information on the proper disposal of the fuel. Clean and inspect the entire fuel system if water or other contamination is in the fuel.

Checking for a restriction in the fuel supply hose

WARNING
Safely performing on-water testing or adjustments requires two people: one person to operate the boat and the other to monitor the gauges or test equipment and make necessary adjustments. All personnel must remain seated inside the boat at all times. Do not lean over the transom while the boat is underway. Use extensions to allow all gauges and test equipment to be located in normal seating areas.

Perform this check if the engine will not start or has excessively lean symptoms at higher engine speeds.

1. Remove the spin-on water-separating fuel filter (**Figure 33**, typical) as described in Chapter Three. Refer to *Servicing the Water-Separating Fuel Filter.* If the engine is not equipped with this filter, place the container under the mechanical or electric fuel pump and then loosen the fuel inlet fitting at the pump (**Figure 34**, typical) to drain a sample of the fuel from the fuel supply hose. Check for the presence of fuel. If no or very little fuel is found in the filter or fuel supply hose, perform the following:
   a. Check the fuel tank for fuel and check all hoses and connections for broken or loose clamps and loose or damaged fittings.
   b. Inspect the fuel line for cut, broken, cracked or weathered surfaces. If not found in excellent condition, replace the hose with a new U.S. Coast Guard-approved fuel hose.
   c. Locate the antisiphon valve threaded into the fuel supply hose fitting on the fuel tank. Remove and inspect the antisiphon valve for blockage, a loose fit to the fuel line fitting, cracking of the valve housing and damaged threads. Replace the valve with a new suitable replacement if any defects are evident.

WARNING
Never install a simple fitting in place of the antisiphon valve, modify the antisiphon valve or bypass the antisiphon valve. The valve provides an important safety feature by preventing fuel in the tank from siphoning into the engine compartment if a fuel hose, fuel line or other fuel system component develops a leak.

2. Disconnect the fuel supply hose from the inlet fitting on the spin-on fuel filter housing or the fuel pump inlet fitting. Drain residual fuel from the hose into a suitable container. If not contaminated, pour the fuel back into the boat fuel tank. Clean up any spilled fuel.
3. Temporarily connect a common outboard fuel supply hose with primer bulb to the fuel inlet fitting. The arrow on the primer bulb must face the hose end connected to the fitting.
4. Connect the other end of the fuel supply hose to a compatible outboard fuel tank filled with fresh fuel. Make sure the hose and portable tank fittings have at least a 1/4 in. (6.3 mm) inside diameter. Open the vent on the portable fuel tank.
5. Pump the primer bulb to prime the system and check for leakage at the fuel supply hose, hose fittings and fuel lines. Correct any leaks before proceeding.

WARNING
Safely performing on-water testing or adjustments requires two people: one person to operate the boat and the other to monitor the gauges or test equipment and make necessary adjustments. All personnel must remain seated inside the boat at all times. Do not lean over the transom while the boat is underway. Use extensions to allow all gauges and test equipment to be located in normal seating areas.

6. Operate the engine at the speed and under the conditions at which the malfunction occurs. Refer to the following.
   a. If the engine performs properly using a portable tank, the malfunction is caused by a blocked fuel tank pickup, damaged fittings, faulty antisiphon valve, loose or leaking fuel tank pickup, improper fuel tank venting, air leakage at fittings or faulty fuel hose. Inspect and repair or replace these components to restore proper engine operation. Proceed to Step 8.
   b. If the malfunction is still present, proceed to Step 7.
7. Have an assistant operate the engine at the speed and under the conditions at which the malfunction occurs. As soon as the symptoms occur, vigorously pump the primer bulb and note the reaction. Wait a few minutes for symptoms to reappear. Then repeat the pumping action. Refer to the following.

a. If the motor operates properly only after pumping the primer bulb, the fault is probably with the mechanical or electric low-pressure fuel pump. Replace the mechanical fuel pump as described in Chapter Seven. Test the low-pressure electric fuel pump as described in this chapter.
b. If the malfunction is still present (carburetor-equipped models), check the fuel pump fittings and all filters for blockage. If no blockage is found, disassemble, clean and assemble the carburetor as described in Chapter Nine.
c. If the malfunction is still present (electronic fuel injection [EFI] models), check the fuel pressure as described in this chapter. If the fuel pressure is correct, test the fuel injection system sensors as described in this chapter. Check the fuel injectors if no fault is found with the sensors.

8. Stop the engine. Disconnect the portable tank hose from the inlet fittings. Drain residual fuel from the hose. Reconnect the boat fuel tank supply hose. Securely tighten the clamp(s). Clean up any spilled fuel.
9. Operate the engine again to verify the fault. Often, the fault is corrected by securely tightening the supply hose clamp.

Checking the fuel tank vent

A faulty fuel tank vent can cause a no-start condition or restrict fuel flow at any engine speed. Typically, the symptom worsens the longer the engine operates. As fuel is drawn from the tank, improper venting will cause a vacuum to form in the area above the fuel. The more fuel withdrawn, the stronger the vacuum. In time, the vacuum will overcome the pumping ability of the fuel pump, and fuel starvation occurs. To verify a venting problem, simply open the fuel fill cap when the symptoms occur. Improper venting is evident if the symptoms quickly improve. Improper venting is usually caused by insects, insect nests or webs, or other debris in the vent fitting (on the boat) or in the vent hose connecting the vent fitting to the fuel tank.

Carburetor Fuel System

Refer to **Table 3** for a list of items to check if the engine will not start or if the fuel system is suspected of causing an engine malfunction.

Carburetor operation

The carburetor (**Figure 35**, typical) is designed to maintain the fuel/air proportion while providing for sudden acceleration and varying loads.

The choke valve (A, **Figure 35**) and electric choke heater (B) in the carburetor provide a richer mixture of fuel to air until the engine warms up. When cooled by surrounding air, the coil in the electric choke partially closes the choke valve to restrict air flow into the throttle openings. This de-

creases air pressure in the opening and causes the metering circuits to deliver more fuel. After starting the engine, current supplied by the alternator causes a heating element in the electric choke to heat the choke coil. This causes the coil to gradually open as the engine warms.

The accelerator pump provides the additional fuel required during rapid acceleration. The sudden manifold pressure change, which occurs during rapid acceleration, can momentarily disrupt fuel flow from the low-speed metering circuits at a time when more fuel is required. The accelerator pump provides an instantaneous spray of fuel into the carburetor opening to bridge the gap between the low-speed and high-speed fuel metering circuits.

The power valve provides additional fuel during heavy load conditions. Manifold pressure is applied to the diaphragm on the power valve. At idle and when the engine is lightly loaded, manifold pressure is low (high vacuum), and the power valve remains closed. While accelerating, when the engine is heavily loaded or when operating at higher throttle settings, the manifold pressure is higher (low vacuum), and the power valve opens. The open power valve allows additional fuel to bypass the normal metering circuits to provide the additional fuel required by heavy loading.

CAUTION
Various jets, passages and adjustment screws meter fuel and air to provide the precise amount of fuel for the engine model. Never use a carburetor or carburetor components from a different model on the engine. This can adversely affect the fuel metering and cause an excessively rich or excessively lean operating condition. Operating too rich or too lean can cause poor running quality and serious engine damage.

Checking for a flooding carburetor

A flooding carburetor can cause hard starting, stalling, rough idle, hesitation on acceleration, excessive exhaust smoke and fouled spark plugs.

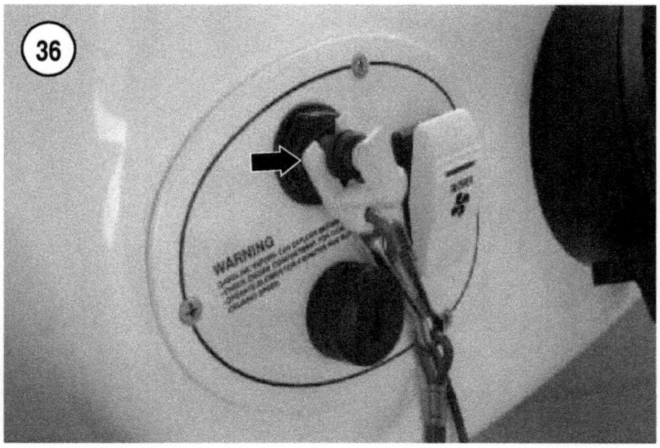

WARNING
Safely performing on-water testing or adjustments requires two people: one person to operate the boat and the other to monitor the gauges or test equipment and make necessary adjustments. All personnel must remain seated inside the boat at all times. Do not lean over the transom while the boat is underway. Use extensions to allow all gauges and test equipment to be located in normal seating areas.

WARNING
The engine might unexpectedly start during testing. Keep all persons, clothing, tools or loose objects away from the drive belts and other moving components on the engine.

CAUTION
If testing the fuel system with the boat out of the water, use a flush device to supply the engine with cooling water. Operation of the starter motor results in operation of the seawater pump. The seawater pump is quickly damaged if operated without adequate cooling water.

1. Remove the flame arrestor and cover as described in Chapter Seven.
2. Pull the cord to activate the emergency stop (lanyard) switch (**Figure 36**). This prevents the ignition system from operating. The lanyard switch does not prevent starter motor or fuel pump operation. If the control does not incorporate an emergency stop (lanyard) switch, disconnect the coil wire from the distributor cap and connect to an engine ground to disable the ignition system. The wire must make a secure connection to prevent arcing.
3. Observe the carburetor bowl overflow/vent tube(s) (C, **Figure 35**) for fuel flowing from the tubes while an assistant cranks the engine. A single tube is used on two-barrel carburetors, and two tubes are used on four-barrel carburetors. Stop cranking the engine *immediately* if fuel begins flowing from the tube(s). This indicates the carburetor is flooding due to a faulty needle and seat, improperly adjusted or damaged float, or excessively high fuel pressure. Disassemble, clean and assemble the carburetor as described in Chapter Seven. Replace the mechanical or electric low-pressure fuel pump if flooding is evident and no fault is found with the carburetor.
4. Stop the engine. Install the flame arrestor and cover as described in Chapter Seven.
5. Reattach the lanyard cord to the emergency stop (lanyard) switch or reconnect the coil wire to the distributor cap.

Checking the choke operation

Improper choke valve operation can cause hard starting, a rich running condition, rough idle, hesitation during rapid acceleration or poor overall performance.

WARNING
Safely performing on-water testing or adjustments requires two people: one person to operate the boat and the other to monitor the gauges or test equipment and make necessary adjustments. All personnel must remain seated inside the boat at all times. Do not lean over the transom while the boat is underway. Use extensions to allow all gauges and test equipment to be located in normal seating areas.

CAUTION
If testing the fuel system with the boat out of the water, use a flush device to supply the engine with cooling water. Operation of the starter motor results in operation of the seawater pump. The seawater pump is quickly damaged if operated without adequate cooling water.

1. Remove the cover and flame arrestor as described in Chapter Seven.
2. Note the position of the choke valve (A, **Figure 35**). The choke valve should be almost closing off the air horn opening if the engine and surrounding air is cooled to room temperature or cooler. If the valve is almost closed, proceed to Step 3. If not, refer to the carburetor disassembly procedures in Chapter Seven and remove the electric choke and coil. Do not remove the choke housing from the carburetor. Reinstall the choke as described in the assembly procedure. Replace the electric choke and coil if the choke does not perform as specified after proper installation.
3. Observe the choke valve while an assistant starts the engine. The choke valve should open a little more as soon as the engine starts. If the valve performs as specified, proceed to Step 4. If the valve does not open as specified, the choke piston and lever in the choke housing are sticking, dirty or damaged. Refer to the carburetor disassembly procedure in Chapter Seven. Then remove, clean and inspect the piston, lever and related components. Replace worn or damaged

components. Reinstall the piston and related components as described in the assembly procedures.

4. Observe the choke valve as the engine warms to normal operating temperature while running at idle speed. The choke should gradually open as the engine warms up and then should be completely open when the engine reaches normal operating temperature. If the choke valve opens as specified, proceed to Step 8. If not, proceed to Step 5.

5. Cautiously touch a finger to the side of the electric choke (B, **Figure 35**). The electric choke should feel warm. If not, proceed to Step 6. The choke valve, piston, or connecting lever or linkage is sticking if the electric choke is warm, yet the choke valve is not opening as specified. Refer to the carburetor disassembly procedures in Chapter Seven. Then remove, clean and inspect the piston, choke valve, lever and related linkage. Replace worn or damaged components. Reinstall the piston and related components as described in the assembly procedures.

6. Disconnect the purple/white (A, **Figure 37**) and black (B) wires from the electric choke terminals. Set a voltmeter to the 10 or 20 VDC scale.

7. Connect the positive meter lead to the purple/white wire (A, **Figure 37**) and the negative meter lead to the black wire (B). With the engine running at idle speed, the meter should indicate 6.5 volts or higher. Refer to the following:

   a. If the voltage is within the specification, but the choke did not become warm in Step 5, replace the electric choke as described in the carburetor disassembly and assembly instructions in Chapter Seven.
   b. If the voltage is below the specification, repeat the test with the negative test lead connected to a good engine ground. Repair or replace the ground wire if the voltage is now correct. If the voltage is below the specification using the alternate ground, repeat the test with the positive meter lead connected to the purple terminal leading into the alternator. If the voltage is now within the specification, repair the faulty terminal or open circuits along the purple/white wire. If the voltage is below the specification with the lead touching the alternator terminal, test the alternator excitation circuit as described in this chapter.

8. Stop the engine. Reconnect the purple/white and black wire to the choke. The terminals must fit securely. If necessary, disconnect the terminal(s) and pinch the wire or choke terminal to achieve a secure fit. Route the wiring to prevent interference with moving components or contact with hot engine surfaces. Secure the wiring with plastic locking clamps as needed.

9. Install the flame arrestor and cover as described in Chapter Seven.

Mechanical Fuel Pump Test

The mechanical fuel pump is used on carburetor-equipped GM 3.0L (four-cylinder) models, earlier (1994-1997) GM 5.0L and 5.7L (V-8) models and Ford 5.0L and 5.8L (V-8) models. A defective mechanical fuel pump can cause an inability to start the engine, carburetor flooding due to excessively high fuel pressure (rare event), stalling at idle, or misfiring and poor performance at higher throttle settings.

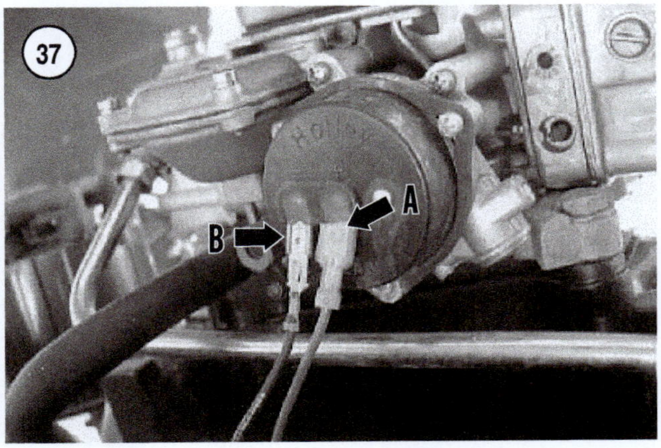

WARNING
The engine might unexpectedly start during testing. Keep all persons, clothing, tools or loose objects away from the drive belts and other moving components on the engine.

WARNING
Safely performing on-water testing or adjustments requires two people: one person to operate the boat and the other to monitor the gauges or test equipment and make necessary adjustments. All personnel must remain seated inside the boat at all times. Do not lean over the transom while the boat is underway. Use extensions to allow all gauges and test equipment to be located in normal seating areas.

CAUTION
If testing the ignition system with the boat out of the water, use a flush test device to supply the engine with cooling water. Operation of the starter motor results in operation of the seawater pump. The seawater pump is quickly damaged if operated without adequate cooling water.

This procedure requires a fuel pressure test gauge capable of accurately measuring 3-7 psi (20.7-48.3 kPa) and a suitable T-fitting for attaching the gauge to the fuel pump outlet and fuel pump-to-carburetor fuel line fittings.

1. To rule out other faults, inspect the fuel and check for a restriction in the fuel supply hose as described in this section.

2. Check for fuel in the sight tube (A, **Figure 38**, typical). Any fuel in the tube indicates the diaphragm in the pump is leaking, and the mechanical pump must be replaced as described in Chapter Seven.

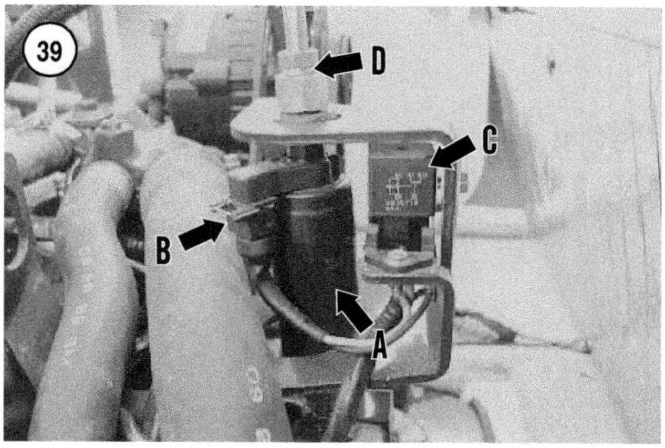

3. Place a suitable container under the fitting to capture spilled fuel. Disconnect the fuel line from the outlet fitting (B, **Figure 38**) on the fuel pump. Drain residual fuel from the hose. If the fuel is not contaminated, empty the fuel into the boat fuel tank.

4. Connect a suitable barbed T-fitting to the fuel pump outlet fitting and disconnected fuel line. Connect the hose of the fuel pressure gauge to the barbed fitting. Secure the pressure gauge hose with a suitable clamp. Securely tighten the fittings to prevent fuel leakage. Secure the fuel gauge and hose to prevent contact with moving or hot engine components. Clean up any spilled fuel before proceeding.

5. Pull the cord to activate the emergency stop (lanyard) switch (**Figure 36**). This prevents the ignition system from operating. The lanyard switch does not prevent starter motor or fuel pump operation. If the control does not incorporate an emergency stop (lanyard) switch, disconnect the coil wire from the distributor cap and connect to an engine ground to disable the ignition system. The wire must make a secure connection to prevent arcing.

6. Observe the fuel gauge and the fuel fittings for leakage while an assistant cranks the engine. Stop cranking the engine *immediately* if fuel leakage is detected. Correct fuel leaks before proceeding.

7. Observe the gauge while an assistant cranks the engine for approximately 20 seconds. The gauge must indicate 3-7 psi (20.7-48.3 kPa). Refer to the following:
   a. If the pressure on the gauge indicates 3-7 psi (20.7-48.3 kPa), the fuel pressure is correct at cranking speed. Proceed to Step 8.
   b. If the pressure on the gauge exceeds 7.0 psi (48.3 kPa), the fuel pressure is too high. Replace the fuel pump as described in Chapter Seven.
   c. If the pressure on the gauge is less than 3.0 psi (20.7 kPa), the fuel pump pressure is too low. Check for a flooding carburetor as described in this section. If the carburetor is not flooding, replace the fuel pump as described in Chapter Seven.

8. Reattach the lanyard cord to the emergency stop (lanyard) switch or reconnect the coil wire to the distributor cap to allow engine starting. Prepare the boat and engine for operation under normal operating conditions.

9. Observe the gauge while an assistant operates the engine under load at different speeds from idle to full throttle. The gauge must indicate 3-7 psi (20.7-48.3 kPa) at all engine speeds. If not, refer to the following.
   a. If the pressure on the gauge exceeds 7.0 psi (48.3 kPa) at any speed, replace the fuel pump as described in Chapter Seven.
   b. If the pressure on the gauge drops to less than 3.0 psi (20.7 kPa), with increased engine speed, check for a restriction in the fuel supply hose as described in this section. If no restriction is evident, replace the mechanical fuel pump as described in Chapter Seven.

10. Stop the engine. Remove the pressure gauge and T-fitting from the fuel pump and fuel line. Reconnect the fuel line, tighten the fittings and check for fuel leaks as described in Chapter Seven. Refer to *Mechanical fuel pump replacement.*

Low-Pressure Electric Fuel Pump Test (Carburetor-Equipped Models)

The low-pressure electric fuel pump (A, **Figure 39**) is used on carburetor-equipped GM 4.3L models and later (1998-2000) GM 5.0L and 5.7L models. A defective low-pressure electric fuel pump can cause an inability to start the engine, carburetor flooding due to excessively high fuel pressure, stalling at idle, or misfiring and poor performance at higher throttle settings.

Electrical circuit test

Voltage to operate the pump is provided by the battery and is controlled by the fuel pump relay. The relay receives battery voltage from the red/purple wire connected to the No. 30 relay terminal. The No. 86 relay terminal provides the relay ground. The starter relay applies battery voltage to the No. 85 relay terminal while cranking. This causes the relay to energize and apply battery voltage (supplied at the No. 30 relay terminal) to the No. 87 relay terminal. The No. 87 relay terminal directs the voltage through the yellow/red wire to the electric fuel pump and causes the electric pump

40

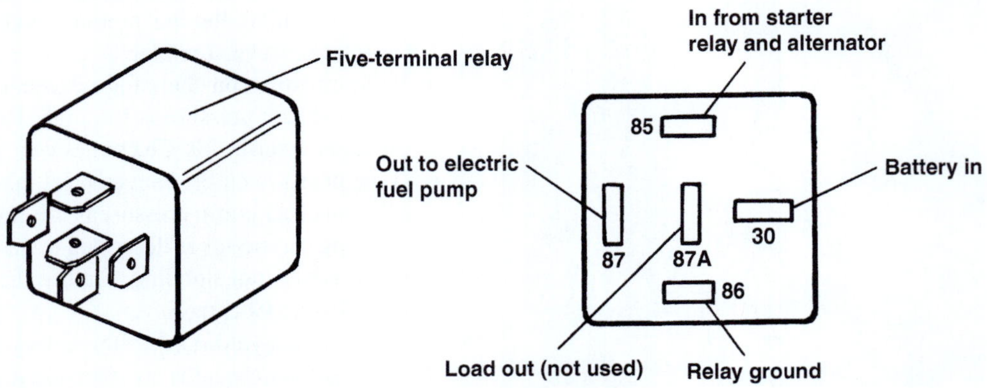

to operate and provide fuel to the carburetor. After starting, the alternator applies voltage to the No. 85 relay terminal to keep the relay energized and the fuel pump operating. If the engine should stall or is switched off with the ignition switch (or emergency stop [lanyard] switch), the alternator output ceases and stops applying voltage to the relay. The relay de-energizes and shuts off the fuel pump.

If the alternator output is missing or inadequate, the relay will not remain energized after starting. In such cases, the engine will start and run for a few minutes on the fuel pumped into the carburetor during the cranking mode and then stall when the fuel is burned. This cycle will repeat until the alternator output is restored.

Diodes are present in the green wire connecting the alternator-to-fuel pump relay wire and the yellow/red and then orange wire connecting the fuel pump relay to the starter relay. The diode on the yellow/red and then orange wire prevents the alternator voltage from feeding back into the starter relay circuit and energizing the starter relay. The diode in the green wire prevents voltage supplied by the starter relay from feeding back to and possibly damaging the alternator. If the starter engages while running, or if the alternator output at the No. 85 relay terminal is low or absent, test the diodes as described in this section.

WARNING
Safely performing on-water testing or adjustments requires two people: one person to operate the boat and the other to monitor the gauges or test equipment and make necessary adjustments. All personnel must remain seated inside the boat at all times. Do not lean over the transom while the boat is underway. Use extensions to allow all gauges and test equipment to be located in normal seating areas.

WARNING
The engine might unexpectedly start during testing. Keep all persons, clothing, tools or loose objects away from the drive belts and other moving components on the engine.

CAUTION
If testing the ignition system with the boat out of the water, use a flush test device to supply the engine with cooling water. Operation of the starter motor results in operation of the seawater pump. The seawater pump is quickly damaged if operated without adequate cooling water.

Use a 12-volt test lamp, mechanic's stethoscope and voltmeter for this procedure.

1. Pull the cord to activate the emergency stop (lanyard) switch (**Figure 36**). This prevents the ignition system from operating. The switch does not prevent the starter or electric fuel pump from operating in START mode. If the control does not incorporate an emergency stop (lanyard) switch, disconnect the coil wire from the distributor cap and connect it to an engine ground to disable the ignition system.
2. Push in the locking clip. Then carefully unplug the engine harness (B, **Figure 39**) from the fuel pump.
3. Connect the cord of the test lamp to the positive battery cable. Touch the test lamp probe to an engine ground to verify the lamp and a good engine ground. Touch the test lamp probe to the disconnected terminal for the black wire. The lamp must illuminate. If not, repair the faulty wiring or terminal along the black wire from the harness-to-relay mounting bracket and bracket-to-engine ground terminal.
4. Turn the ignition switch to OFF or STOP. Connect the cord of the test lamp to the negative battery cable. Touch the test lamp probe to the disconnected terminal for the black wire. The lamp must not illuminate. If the lamp illuminates, the fuel pump relay is shorted internally or stuck in the energized position and must be replaced as described in Chapter Nine.
5. Connect the test lamp as described in Step 4. Observe the lamp while an assistant turns the ignition switch to ON or RUN. The lamp must not illuminate. If the lamp illuminates, perform the following:
   a. Disconnect the green wire from the alternator. If the lamp extinguishes with the wire disconnected, the alternator is shorted internally. Replace the alternator

41

with a new or rebuilt unit as described in Chapter Nine.

b. Disconnect the yellow/red wire from the diode and orange wire leading into the fuel pump relay (C, **Figure 39**). If the lamp extinguishes with the wire disconnected, the yellow/red wire connecting to the starter relay is shorted to a positive battery source, or the starter relay is faulty. Inspect the wiring and test the starter relay as described in this chapter.
c. If the lamp remains illuminated in Steps 5a and 5b, the fuel pump relay is shorted internally or stuck in the energized position. Replace the relay as described in Chapter Nine.
d. Reconnect the yellow/red wire to the diode and connect the green wire to the alternator *L2* terminal.

6. Connect the test lamp as described in Step 4. Observe the lamp while an assistant turns the ignition switch to START. The lamp must illuminate. If the lamp illuminates with the switch in START, the fuel pump relay is operational. Note the test result and proceed to Step 15. If the lamp does not illuminate, proceed to Step 7.
7. Remove the fuel pump relay as described in Chapter Nine.
8. Refer to **Figure 40** to identify the relay terminals and the terminals in the relay mounting bracket. The relay will only fit one way into the bracket. Compare the orientation of the No. 30 and No. 87 terminal blades for reference.
9. Connect the cord of the test lamp to the positive battery terminal. Touch the test lamp probe to the No. 86 terminal in the mounting bracket. The lamp must illuminate. If not, check for a faulty connection or fault in the black relay ground wire.
10. Connect the cord of the test lamp to the negative battery terminal. Touch the test lamp probe to the No. 30 terminal in the mounting bracket. The lamp must illuminate. If not, check for a blown fuse or circuit breaker. Refer to the wiring diagrams at the end of the manual to identify any fuses or circuit breakers along the circuit. Check for a faulty connection or battery supply wire if no fault is found with the fuses or circuit breaker.
11. Connect the cord of the test lamp to the negative battery terminal. Touch the test lamp probe to the No. 85 terminal in the harness mounting bracket. Observe the lamp while an assistant places the ignition switch in OFF or STOP and ON or RUN. The lamp must not illuminate in either position. If the lamp illuminates, perform the following:

a. Disconnect the green wire from the alternator and repeat Step 11. If the lamp does not illuminate with the wire disconnected, the alternator is shorted internally. Replace the alternator with a new or rebuilt unit as described in Chapter Nine.
b. Disconnect the yellow/red wire from the diode and orange wire leading to the fuel pump relay (C, **Figure 39**) and repeat Step 11. If the lamp does not illuminate with the wire disconnected, the yellow/red wire connecting to the starter relay is shorted to a positive battery source. Inspect the wiring. If no fault is found with the wiring, test the starter relay as described in this chapter.

12. Connect the cord of the test lamp to the negative battery terminal. Touch the test lamp probe to the No. 85 terminal in the harness connector or mounting bracket. Observe the lamp while an assistant turns the ignition switch to START. The lamp must illuminate. If not, test the starter relay as described in this chapter.
13. Replace the fuel pump relay if the test lamp did not illuminate in Step 6, if there is no faulty wiring and if it tests correctly as described in Steps 7-12.
14. Install the fuel pump relay as described in Chapter Nine.
15. Reattach the lanyard cord to the emergency stop (lanyard) switch or reconnect the coil wire to the distributor cap.
16. Push in the locking clip and carefully plug the wire harness connector (B, **Figure 31**) onto the electric fuel pump.
17. Disconnect the negative battery cable. Disconnect the yellow/red wire from the starter solenoid (C, **Figure 41**). Completely cover the yellow/red wire terminal with electrical tape and route away from the starter solenoid terminals and any engine components to prevent arcing. Reconnect the battery cable.
18. Touch a mechanic's stethoscope to the body of the fuel pump (A, **Figure 39**). Listen carefully for fuel pump operation while an assistant turns the ignition switch to START position and back to ON or RUN. The fuel pump must operate, and the fuel pump and starter relays must make a clicking noise each time the switch is in the start position. Replace the electric fuel pump if the relays operate as specified, but the fuel pump does not operate. Repeat the test procedure if the relays do not operate as specified.
19. Place the ignition switch in OFF or STOP.
20. Disconnect the negative battery cable. Reconnect the yellow/red wire to the starter solenoid terminal (C, **Figure 41**). Make sure all terminal nuts are securely tightened and the wire terminals are not contacting other terminals. Reposition the wiring as necessary. To prevent corrosion and starting system malfunction, thoroughly coat the terminals

with OMC Black Neoprene Dip. Purchase the neoprene dip (OMC part No. 909570) from an OMC dealership.

21. Connect the battery cable.
22. Set the voltmeter to the 20 or 40 VDC scale.
23. Connect the negative meter lead to an engine ground. Touch the positive meter lead to the *L2* terminal on the back of the alternator. The green wire connects to this terminal. Do not disconnect the green wire for this procedure. Route the test leads to prevent interference with drive belts or other moving or hot engine components.
24. Observe the meter while an assistant starts the engine and operates at various engine speeds. The meter must indicate 13.9-14.7 volts at all speeds. If the meter indicates less than 13.9 volts or exceeds 14.7 volts, test the fuel pump harness diodes as described in this section. If the diodes are good, replace the alternator with a new or rebuilt unit as described in Chapter Nine.
25. Disconnect the test leads.

Harness diodes test

An ohmmeter is required for this procedure.

1. Disconnect the negative battery cable.
2. Remove the fuel pump relay as described in Chapter Nine.
3. Disconnect the green wire from the *L2* terminal on the back of the alternator.
4. Refer to **Figure 40** to identify the relay terminals and the terminals in the wire harness or relay mounting bracket. The relay will only fit one way into the bracket. Compare the orientation of the No. 30 and No. 87 terminal blades for reference.
5. Touch one of the ohmmeter leads to the No. 85 relay terminal in the mounting bracket. Touch the other lead to the green wire that was disconnected from the alternator. Note the meter reading. Reverse the test leads to change the polarity and again note the meter reading. The meter must show continuity when connected one way and no continuity when the polarity is reversed. Replace the green wire harness if the diode does not perform as specified.
6. Disconnect the yellow/red starter relay wire from the orange wire leading into the fuel pump relay mounting bracket.
7. Touch one of the ohmmeter leads to the No. 85 relay terminal in the mounting bracket. Touch the other lead to the orange wire that was disconnected from the yellow/red wire. Note the meter reading. Reverse the test leads to change the polarity and again note the meter reading. The meter must show continuity when connected one way and no continuity when the polarity is reversed. Replace the diode on the orange wire harness or relay mounting bracket if the diode does not perform as specified.
8. Reconnect the yellow/red wire to the diode and connect the green wire to the alternator *L2* terminal.
9. Install the fuel pump relay as described in Chapter Nine.
10. Connect the battery cable.

Pressure test

WARNING

Safely performing on-water testing or adjustments requires two people: one person to operate the boat and the other to monitor the gauges or test equipment and make necessary adjustments. All personnel must remain seated inside the boat at all times. Do not lean over the transom while the boat is underway. Use extensions to allow all gauges and test equipment to be located in normal seating areas.

WARNING

The engine might unexpectedly start during testing. Keep all persons, clothing, tools or loose objects away from the drive belts and other moving components on the engine.

CAUTION

If testing the ignition system with the boat out of the water, use a flush test device to supply the engine with cooling water. Operation of the starter motor results in operation of the seawater pump. The seawater pump is quickly damaged if operated without adequate cooling water.

This procedure requires a fuel pressure test gauge capable of accurately measuring 3.5-8.5 psi (24.1-58.6 kPa), a suitable T-fitting for attaching the gauge to the fuel pump, fuel lines and an accurate shop tachometer.

1. To rule out other faults, inspect the fuel and check for a restriction in the fuel supply hose as described in this section. Perform the following test if previous tests or checks indicate a no-start condition, lean operation or carburetor flooding, which might be due to a faulty low-pressure electric fuel pump.
2. Touch a mechanic's stethoscope to the body of the fuel pump (A, **Figure 39**). Listen carefully for fuel pump operation while an assistant turns the ignition switch to START and back to ON or RUN. The fuel pump must operate, and the fuel pump and starter relays must make a clicking noise each time the switch is in START. Refer to the following:
   a. If both relays click, and the fuel pump operates, the fuel pump relay and fuel pump are operating. Proceed to Step 3.
   b. If both relays click, but the fuel pump does not operate, test the electrical circuit for the low-pressure fuel pump as described in this section.
   c. If one or both relays do not click, and the fuel pump does not operate, test the electrical circuit for the low-pressure fuel pump as described in this section.
3. Place a suitable container under the fitting to capture spilled fuel and disconnect the fuel line from (D, **Figure 39**) the fuel pump. Drain residual fuel from the hose. If the fuel is not contaminated, empty the fuel into the boat fuel tank.

4. Connect a suitable T-fitting to the fuel pump and fuel line. Connect the fuel pressure gauge to the T-fitting. Secure the pressure gauge hose with a suitable clamp. Securely tighten the fittings to prevent fuel leakage. Secure the fuel gauge and hose to prevent contact with moving or hot engine components. Clean up any spilled fuel before proceeding.
5. Pull the cord to activate the emergency stop (lanyard) switch (**Figure 36**). This prevents the ignition system from operating. The lanyard switch does not prevent starter motor or fuel pump operation. If the control does not incorporate an emergency stop (lanyard) switch, disconnect the coil wire from the distributor cap and connect it to an engine ground to disable the ignition system. The wire must make a secure connection to prevent arcing.
6. Observe the fuel gauge and the fuel fittings for leakage while an assistant cranks the engine. Stop cranking the engine immediately if fuel leakage is detected. Correct fuel leaks before proceeding.
7. Observe the gauge while an assistant cranks the engine for approximately 20 seconds. The gauge must indicate 4.9-8.5 psi (33.8-58.6 kPa) while cranking. Refer to the following:
   a. If the gauge indicates 4.9-8.5 psi (33.8-58.6 kPa), the fuel pump pressure is correct at cranking speed. Proceed to Step 8.
   b. If the pressure exceeds 8.5 psi (58.6 kPa), the fuel pump pressure is too high. Replace the fuel pump as described in Chapter Seven.
   c. If the pressure is less than 4.9 psi (33.8 kPa), the fuel pump pressure is too low. Check for a flooding carburetor as described in this section. If the carburetor is not flooding, replace the fuel pump as described in Chapter Seven.
8. Reattach the lanyard cord to the emergency stop (lanyard) switch or reconnect the coil wire to the distributor cap.
9. Following the manufacturer's instructions, connect a shop tachometer to the engine.
10. Prepare the boat and engine for operation under normal conditions.
11. Observe the gauge and shop tachometer while an assistant operates the engine under normal load at the speeds listed in **Table 10**. Record the gauge reading at each speed.
12. Stop the engine. Compare the gauge reading at the given engine speeds with the fuel pressure specifications in **Table 10**. Refer to the following:
   a. If the pressure exceeds the fuel pressure specification at any speed, replace the low-pressure electric fuel pump as described in Chapter Seven.
   b. If the pressure is less than the specification at any speed (particularly at higher engine speeds), check for a restriction in the fuel supply hose as described in this section. If no restriction is evident, replace the low-pressure electric fuel pump as described in Chapter Seven.
13. Remove the pressure gauge and T-fitting from the fuel pump and fuel line. Reconnect the fuel line, tighten the fittings and check for fuel leaks as described in Chapter Seven. Refer to *Low-pressure electric fuel pump replacement.*
14. Disconnect the shop tachometer.

ELECTRONIC FUEL INJECTION (EFI) SYSTEM

Refer to **Table 3** for a list of items to check if the engine will not start or if the fuel system is suspected of causing an engine malfunction. Refer to the causes listed next to the symptom that best describes the malfunction. Perform the check, test or adjustment suggested under corrective action. Other useful information and tips are described with the specific test instructions.

NOTE

With the manifolds, fuel system cooling system and electrical components installed, the General Motors (GM) 5.0L and 5.7L V-8 engines are very similar in appearance to the Ford 5.0L and 5.8L V-8 engines. If the model identification decal on the flame arrestor is missing, the engine can be quickly identified as a GM or Ford engine by noting the distributor location. On GM engines, the distributor is located at the rear (flywheel end) of the engine. On Ford engines, the distributor is located on the front of the engine.

Fuel Delivery System

Two different types of fuel injection systems are used on the models covered in this manual, and the major differences are the mounting location of fuel injectors and location of the fuel pressure regulator.

The throttle body injection (TBI) system is used on GM 4.3L, 5.0L and 5.7L models. The multiport injection (MPI) system is used on Ford 5.0L and 5.8L and GM 7.4L and 8.2L models.

Throttle body injection (TBI)

On the throttle body injection (TBI) system, the throttle body is located on the intake manifold in the same location as the carburetor on models without electronic fuel injection. The throttle body houses the fuel injectors (**Figure 42**) along with the throttle position sensor (A, **Figure 43**) and idle air control motor (B, **Figure 43**). Pressurized fuel is delivered to the fuel injectors by internal passages in the throttle body assembly.

The high-pressure electric fuel pump (**Figure 44**) is mounted on the fuel reservoir at the rear of the engine. The low-pressure electric fuel pump mounts on a bracket at the front starboard side of the engine. Both the high-pressure and low-pressure electric fuel pumps are controlled by the

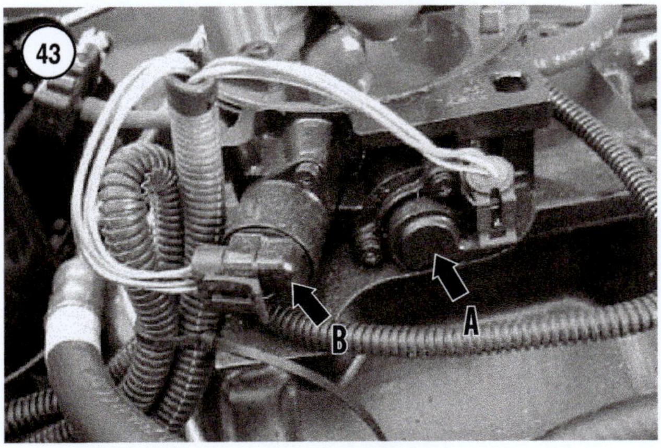

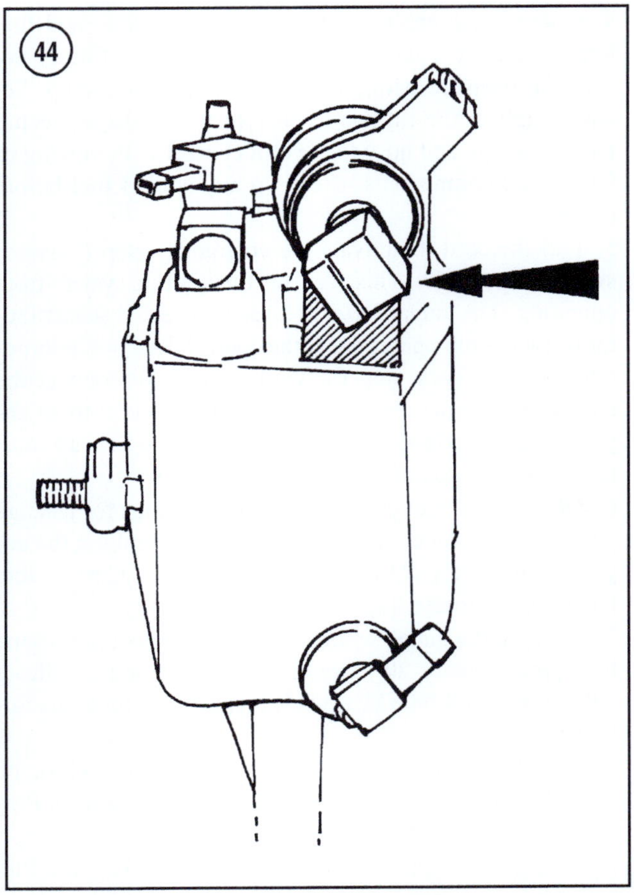

engine control unit (ECU) via the fuel pump relay. When operating, the low-pressure electric fuel pump draws fuel from the fuel tank and then through the water-separating fuel filter. The fuel passes through the fuel pump and then to the fuel reservoir. The fuel reservoir contains a float, inlet needle and seat arrangement much like the float system used on a carburetor. The fuel reservoir vents fuel vapor and provides a liquid fuel source for the high-pressure electric fuel pump. As fuel is used, the float drops with the fuel level and allows the inlet needle and seat to open. The low-pressure electric fuel pump replenishes the used fuel and maintains a consistent level.

Any fuel vapor escapes through the vent fitting and hose to the throttle body fitting. Water is drawn through dedicated passages in the fuel reservoir by hoses connected to the cooling system seawater pump. Water flowing through the passages cools the fuel in the reservoir to help prevent heat-related problems, such as vapor lock and fuel percolation.

The high-pressure electric fuel pump draws fuel from the fuel reservoir and directs it through the inlet line (A, **Figure 45**) to the fuel pocket in the throttle body assembly for use by the fuel injectors. The throttle body-mounted fuel pressure regulator opens when the fuel pressure reaches approximately 30 psi (207 kPa) and allows fuel to exit the fuel pocket and return to the fuel reservoir through the fuel return line (B, **Figure 45**). This arrangement allows fuel that

is not used by the fuel injectors to circulate back to the fuel reservoir for purging of air or fuel vapor, cooling and then the eventual return to the throttle body.

Multiport injection (MPI)

On multiport injection (MPI) systems, the throttle body (**Figure 46**, typical) is located on the front or side of the intake plenum. The throttle body used on MPI models houses the throttle position sensor (TPS) and idle air control motor (IAC) or idle speed control solenoid (ISC).

The plenum is located directly over the intake manifold. On MPI systems, the eight individual fuel injectors

46

47

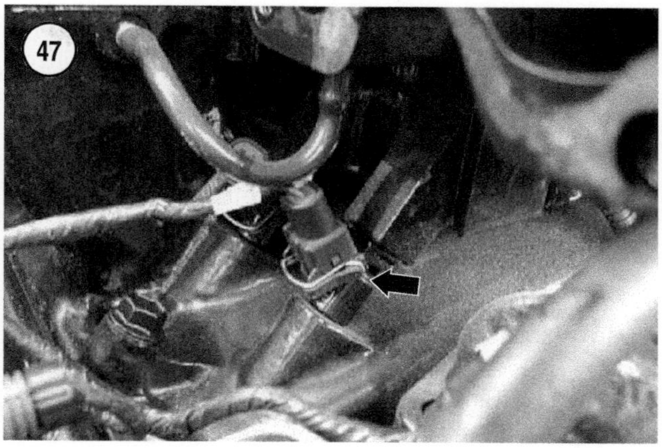

48

FUEL DELIVERY SYSTEM
(TYPICAL MULTIPOINT INJECTION [MPI] SYSTEM)

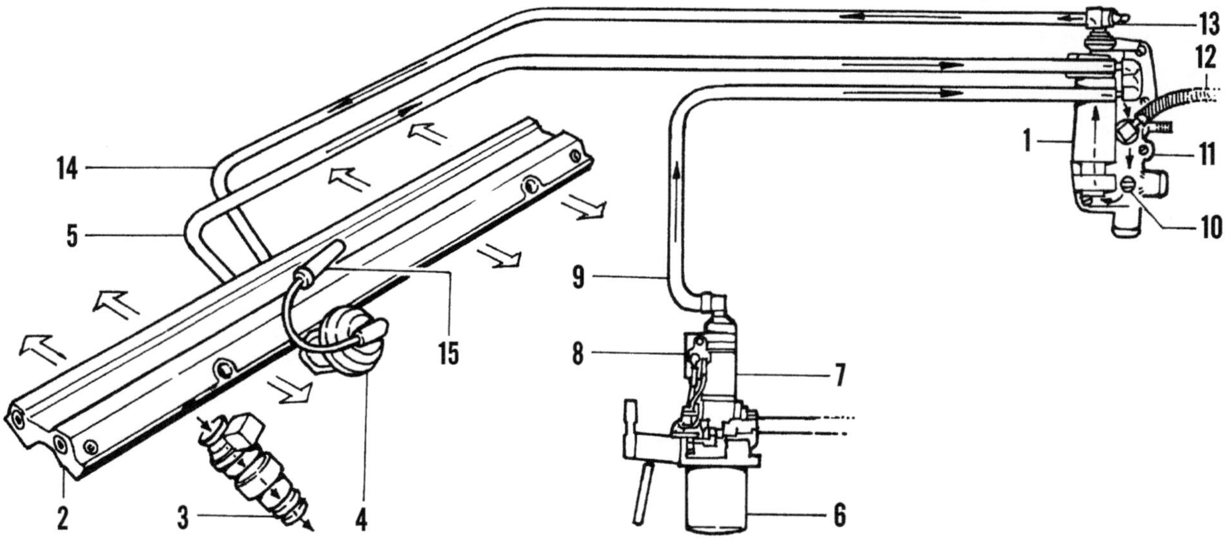

1. High-pressure electric fuel pump
2. Fuel rail
3. Fuel injector
4. Fuel pressure regulator
5. Fuel return line
6. Water-separating fuel filter
7. Low-pressure electric fuel pump
8. Circuit breaker
9. Fuel line (low-pressure pump to reservoir)
10. Low-pressure test port
11. Fuel reservoir
12. Vapor vent hose (to intake manifold or plenum)
13. High-pressure test port
14. Fuel delivery line (high-pressure)
15. Vacuum hose (pressure regulator to intake plenum or manifold)

(**Figure 47**, typical) fit into openings in the intake manifold that align with the intake passage for each cylinder. The fuel injectors spray fuel into the individual intake ports just above the intake valves.

The high-pressure electric fuel pump (1, **Figure 48**, typical) is mounted on the fuel reservoir at the rear of the engine. The low-pressure electric fuel pump (7, **Figure 48**, typical) mounts on a bracket at the front starboard side of the engine. Both the high-pressure and low-pressure electric fuel pumps are controlled by the engine control unit (ECU) via the fuel pump relay. When operating, the low-pressure electric fuel pump draws fuel from the fuel tank and then through the water-separating fuel filter (6, **Figure 48**, typical). The fuel passes through the fuel pump and to the fuel reservoir (11, **Figure 48**, typical). The fuel reservoir contains a float, inlet needle and seat arrangement much like the float system used on a carburetor. The fuel reservoir vents fuel vapor and provides a liquid fuel source for the high-pressure electric fuel pump. As fuel is used, the float drops with the fuel level allowing the inlet needle and seat to open. The low-pressure electric fuel pump replenishes the used fuel and maintains a consistent level.

Any fuel vapor escapes through the vent fitting and hose (12, **Figure 48**, typical) to the throttle body fitting. Water is drawn through dedicated passages in the fuel reservoir by hoses connected to the cooling system seawater pump. Water flowing through the passages cools the fuel in the reservoir to help prevent heat-related problems, such as vapor lock and fuel percolation.

The high-pressure electric fuel pump draws fuel from the fuel reservoir and directs it through the delivery line (14, **Figure 48**, typical) to the fuel rail (2) for use by the fuel injectors. The fuel rail-mounted fuel pressure regulator (4, **Figure 48**, typical) opens when the fuel pressure reaches 36-42 psi (248-290 kPa) and allows fuel to exit the fuel rail and return to the water-separating fuel filter housing through the fuel return line (5, **Figure 48**). This arrangement allows fuel that is not used by the fuel injectors to circulate back to the filter for purging of air or fuel vapor, cooling and then the eventual return to the fuel rail.

A dedicated hose applies manifold pressure (vacuum) to the diaphragm chamber in the fuel pressure regulator. This allows the regulator to change the fuel pressure to compensate for varying intake manifold pressure that occurs during normal operation. This feature provides a consistent pressure differential between fuel pressure delivered to the injectors and intake manifold pressure where the injector sprays fuel. This is needed to prevent variable manifold pressure from adversely affecting fuel delivery.

Fuel Injection Control System

The fuel injection control system consists of the engine control unit (ECU) or engine control assembly (ECA), manifold air pressure sensor (MAP), intake air temperature sensor (IAT), engine coolant temperature sensor (ECT), throttle position sensor (TPS), and idle air control motor (IAC) or idle speed control solenoid (ISC).

The ECU (**Figure 49**, typical) or ECA controls operation of the fuel injection system, ignition timing and warning system. All GM V-6 and V-8 EFI models use an ECU. All Ford 5.0L and 5.8L V-8 EFI models use an ECA. Though not interchangeable, the ECU and ECA are similar in appearance and serve the same function in the EFI system.

There have been two different versions of the ECU used on the GM V-6 and V-8 models.

All 1994 GM V-8 models use the marine electronic fuel injection (MEFI) 1 ECU. The MEFI 1 ECU can be identified by the harness connectors located on each end of the unit and the gray finish. The MEFI 2 ECU is used on 1995-2000 (V-6 and V-8) models, and like the MEFI 1 ECU, it has the harness connectors on the ends of the unit. The MEFI 2 unit is slightly shorter than the MEFI 1 unit and has a black finish. The internal circuitry of the MEFI 2 unit is very similar to the MEFI 1 circuitry, and on some models the MEFI 2 ECU may be used in place of the MEFI 1 version.

49

Operation of the ECU used on GM models is different than operation of the ECA on Ford models. Refer to the applicable section for a description of ECU or ECA operation.

ECU operation (GM V-6 and V-8 models)

Using a fuse-protected wire, battery voltage is continually supplied to the ECU. This wire provides the operating voltage when the ECU is switched on. Using a fuse-protected wire, battery voltage is also continuously supplied to the No. 30 terminal on the EFI system relay. A black wire connects the No. 85 relay terminal to an engine ground. When the ignition switch is turned to ON or RUN, battery voltage is supplied by the purple wire to the No. 86 terminal on the EFI system relay. This energizes the system relay. When energized, the relay supplies battery voltage to the fuel injectors, ignition coil and fuel pump relay. The relay also supplies the voltage to the No. 11 terminal on the J1 ECU connector.

The voltage from the relay powers up the ECU and causes it to supply a 5-volt power source to the TPS, MAP, IAT and ECT sensors. Each sensor in turn sends input back to the ECU where it is used to determine the throttle position, air pressure (elevation), air temperature and engine temperature. The IAC is not a sensor. The idle air control motor is simply an air valve used by the ECU to control the amount of air entering the intake manifold at lower engine speeds (particularly at idle speeds). The ECU sends pulsating voltage to the motor winding to open or close the air opening. The ECU uses input from the sensors to compute the IAC opening required to start the engine under the given conditions (primarily temperature and elevation).

Using a fuse-protected wire, battery current is supplied by the system relay to the No. 30 fuel pump relay terminal. A dark green/white wire connects the No. 85 fuel pump relay terminal to the No. 9 terminal on the J2 ECU connector. A black wire connects the No. 86 fuel pump relay terminal to an engine ground. Yellow/green wires connect the No. 87 fuel pump relay terminal to the low-pressure and high-pressure electric fuel pumps. When powered up, the ECU applies voltage to the No. 85 fuel pump relay. The voltage flows through a coil and then to ground using the

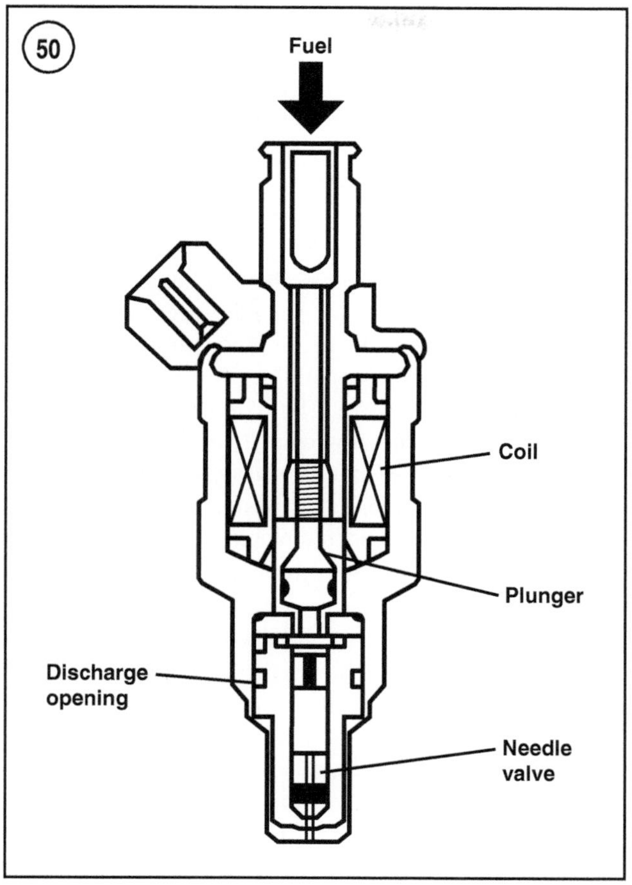

No. 86 relay terminal. This causes the relay to energize and complete the internal circuit that connects the No. 30 terminal to the No. 87 terminal and supplies battery voltage to both electric fuel pumps. The fuel pumps will operate for 2 seconds to prime the fuel system for starting and then shut off when the ECU removes the voltage from the No. 85 relay terminal.

The ECU controls fuel delivery by completing or interrupting the ground circuit for the fuel injectors. The injectors (**Figure 50**) are solenoid-activated fuel valves that, when activated, spray fuel in a very fine pattern. Because battery voltage is supplied to the injectors by the system relay, the injectors activate when the ECU provides the ground circuit using internal driver circuits. Input from the ignition system allows the ECU to determine engine speed and compute the number of times to open the injectors for the given engine speed. This is often referred to as the injector frequency. A higher frequency delivers more fuel, and a lower frequency delivers less fuel. Input from the sensors allows the ECU to determine the amount of fuel required for the given operating conditions. The ECU adjusts the fuel delivery by controlling the length of time the injectors are activated. This is often referred to as the duration or on-time. A longer duration or on-time delivers more fuel, and a shorter duration or on-time delivers less fuel.

While starting the engine, the ECU receives input from the ignition system and starts activating the injectors to provide fuel for starting.

After starting, the ECU maintains the fuel pump relay ground to keep the pumps operating as long as it receives input from the ignition system. The ECU continuously monitors input from the sensors to determine the precise amount of fuel needed and adjust the injector frequency and on-time accordingly. This allows the ECU to quickly change the fuel delivery to compensate for any change in throttle position, manifold vacuum, engine temperature and air temperature. Using input from the sensors, the ECU determines the optimum ignition timing for the given conditions and adjusts the timing accordingly.

The ECU controls idle speed by opening or closing the IAC and changing the ignition timing to provide the optimum idle speed and smoothest operation for the given operating conditions. This allows the fuel injection system to compensate for sudden loading and unloading that occurs when shifting into and out of gear. The IAC is opened more when the engine is cold, so the engine will idle considerably faster. The idle speed will drop gradually as the engine reaches normal operating temperature. The ECU uses input from the TPS to determine whether the throttle is in the idle position. The ECU will open the IAC with increased throttle opening to allow more air into the system. This increases power slightly and helps prevent stalling if the throttle quickly returns to idle speed. This will cause the engine to gradually return to normal idle speed if the throttle is closed abruptly.

All EFI models are equipped with a detonation (knock) sensor threaded into the side(s) of the cylinder block (**Figure 51**, typical). The detonation (knock) sensor creates a pulsating voltage signal when subjected to noise resonating in the cylinder block cooling passages. Some later models use two knock sensors (one on each side). This signal is directed to the knock sensor module. The knock sensor module amplifies the signal and determines whether the noise frequency is within the normal range for the engine. The knock sensor module applies 8 volts to the brown/red or brown/orange wire leading to the ECU, provided the signal indicates normal engine noise. If detonation (knock) develops, the knock sensor signal exceeds the normal range, and the knock sensor removes the 8-volt signal from the wire.

This causes the ECU to retard the ignition timing and increase fuel delivery until the detonation ceases.

The ECU continuously monitors the sensor and various operating system voltages. If any of the sensor inputs and some of the operating system voltages exceed a predetermined range, the ECU substitutes a limp-home value for the suspect sensor or operating system voltage. This allows continued operation; however, the engine might operate with reduced power or with less-than-desirable running quality. In this event, the ECU stores a diagnostic trouble code that is used to troubleshoot the fuel injection system. If the sensor input or operating system voltages return to the predetermined range, the ECU will return to using the input or voltage. However, the diagnostic trouble code remains stored in the ECU. Unlike some systems, disconnecting the battery does not erase the code from the ECU memory.

Operation of the warning horn is controlled by the ECU. The purple wire provides battery voltage to the warning horn. The tan/blue and then purple horn wire leads to the No. 12 terminal in the ECU J2 connector. The ECU controls operation of the horn by grounding the tan/blue wire. The oil pressure switch grounds the tan/blue wire if the oil pressure drops below a preset value. The ECU monitors the engine coolant temperature sensor to determine whether the engine is overheating.

The ECU will complete the ground circuit and turn on the warning horn if it determines that the engine is overheating from the engine temperature sensor input or if the oil pressure switch grounds the tan/blue wire. The horn will continue sounding until the engine temperature drops to an acceptable level, the oil pressure rises above the preset limit or the engine is shut off.

When powered up, the ECU sounds the horn to test the horn operation and indicate that the ECU is operational. This can be an important troubleshooting tool because it indicates that the EFI system relay is operating properly, the ECU is operational and the warning horn is operational. Keep this in mind if trouble should occur.

ECA operation (Ford 5.0L and 5.8L [V-8] models)

Battery voltage is continually supplied to the engine control assembly (ECA) by the circuit breaker-protected pink/blue wire. This wire provides the operating voltage when the ECA is switched on. Battery voltage is also continuously supplied to the red/purple terminal on the electronic fuel injection (EFI) system relay by the circuit breaker-protected red/purple wire. A black wire connects one relay terminal to an engine ground terminal. When the ignition switch is turned to ON or RUN, battery voltage is supplied to the purple wire on the EFI system relay via the ignition switch and instrument harness wiring. This energizes the system relay. When energized, the relay uses the purple/black wire to supply battery voltage to the ECA, fuel injectors, idle speed control solenoid (ISC), knock sensor amplifier, ignition coil, ignition module and fuel pump relay.

The voltage from the relay powers up the ECA and causes it to supply a 5-volt power source to the throttle position sensor (TPS), manifold air pressure sensor (MAP), intake air temperature sensor (IAT) and engine coolant temperature sensor (ECT). Each sensor sends input back to the ECA where it is used to determine the throttle position, air pressure (elevation), air temperature and engine temperature. The ISC is not a sensor. The idle speed control solenoid is an air valve used by the ECA to control the amount of air entering the intake manifold at lower engine speeds (particularly at idle speeds). The ECA sends variable pulsating voltage to the motor winding to open or close the air opening. The ECA uses input from the sensors to compute the ISC opening required to start the engine under the given conditions (primarily temperature and elevation).

Battery voltage is continuously supplied to the fuel pump relay by the circuit breaker-protected red/purple wire. Using the purple/black wire, the system relay supplies current to the fuel pump relay. A blue/orange wire connects one relay terminal to a terminal in the ECA connector. Green/yellow wires connect one of the relay terminals to the low-pressure and high-pressure electric fuel pumps. When powered up, the ECA uses the blue/orange wire to provide a ground for the fuel pump relay. The voltage flows through the purple/black wire, through the magnetic coil in the relay and then to the ground using the blue/orange wire and a ground circuit within the ECA. This causes the relay to energize and complete the internal circuit that essentially connects the red/purple and green/yellow relay terminals. The energized relay supplies battery voltage to both electric fuel pumps. The fuel pumps will operate for 2 seconds to prime the fuel system for starting and then shut off when the ECA removes the relay ground circuit.

The ECA controls fuel delivery by completing or interrupting the ground circuit for the fuel injectors. The injectors (**Figure 50**) are solenoid-activated fuel valves that, when activated, spray fuel in a very fine pattern. Because battery voltage is supplied to the injectors by the system relay, the injectors activate when the ECA provides the ground circuit using internal driver circuits. Input from the ignition system allows the ECA to determine engine speed and compute the number or times to open the injectors for the given engine speed. This is often referred to as the injector frequency. A higher frequency delivers more fuel, and a lower frequency delivers less fuel. Input from the sensors allows the ECA to determine the amount of fuel required for the given operating conditions. The ECA adjust the fuel delivery by controlling the length of time the injectors are activated. This is often referred to as the duration or on-time. A longer duration or on-time delivers more fuel, and a shorter duration delivers less fuel.

While starting the engine, the ECA receives input from the ignition system and starts activating the injectors to provide fuel for starting.

After starting, the ECA maintains the fuel pump relay ground to keep the pumps operating as long as it receives input from the ignition system. The ECA continuously monitors input from the sensors to determine the precise amount of fuel needed and adjusts the injector frequency and on-time accordingly. This allows the ECA to quickly change the fuel delivery to compensate for any change in throttle position, manifold vacuum, engine temperature and air temperature. Using input from the sensors, the ECA determines the optimum ignition timing for the given conditions and adjusts the timing accordingly.

The ECA controls idle speed by opening or closing the ISC and changing the ignition timing to provide the optimum idle speed and smoothest operation for the given operating conditions. This allows the fuel injection system to compensate for sudden loading and unloading that occurs when shifting into and out of gear. The ISC is opened more when the engine is cold, so the engine will idle considerably faster. The idle speed will drop gradually as the engine reaches normal operating temperature. The ECA uses input from the TPS to determine whether the throttle is in the closed idle position. The ECA will open the ISC with increased throttle opening to allow more air into the system. This increases power slightly and helps prevent stalling if the throttle quickly returns to idle speed. This will cause the engine to gradually return to normal idle speed if the throttle is closed abruptly.

All EFI models are equipped with a detonation (knock) sensor threaded into the side(s) of the cylinder block (**Figure 51**, typical). The detonation (knock) sensor creates a pulsating voltage signal when subjected to noise resonating in the cylinder block cooling passages. This signal is directed to the knock sensor amplifier, which amplifies the signal and then directs the signal to the ECA. The ECA uses the signal to determine whether the noise is associated with damaging detonation. If detonation (knock) develops, the ECA retards the ignition timing and increases fuel delivery until the detonation ceases.

The ECA continuously monitors the sensor and various operating system voltages. If any of the sensor inputs and some of the operating system voltages exceed a predetermined range, the ECA substitutes a limp-home value for the suspect sensor or operating system voltage. This allows continued operation; however, the engine might operate with reduced power or with less-than-desirable running quality. In this event, the ECA stores a diagnostic trouble code that is used to troubleshoot the fuel injection system. If the sensor input or operating system voltages return to the predetermined range, the ECA will return to using the input or voltage. However, the diagnostic trouble code remains stored in the ECA. Disconnecting the battery will erase the code from the ECA memory.

Operation of the warning horn is controlled by the engine overheat switch and oil pressure switch. The purple wire provides battery voltage to the warning horn via the ignition switch. The overheat switch grounds the tan/black wire if the engine temperature exceeds a preset value. The oil pressure switch grounds the tan/blue wire if the oil pressure drops below a preset value. The warning horn will sound if either switch completes the ground circuit. The horn will continue sounding until the engine temperature drops to an acceptable level, the oil pressure rises above the preset limit or the engine is shut off. The tan/black wire connecting to the overheat switch and oil pressure switch leads to a terminal on the ECA connector. The ECA uses this wire and an internal ground circuit to operate the warning horn when powering up to test the horn operation and indicate that the ECA is operational. This can be an important troubleshooting tool because it indicates that the EFI system relay is operating properly, the ECA is operational and the warning horn is operational. Keep this in mind if trouble should occur. A diode in the tan/black wire connecting the switches to the ECA prevents current from feeding back to the ignition circuit through either of the activated switches. Otherwise, under certain conditions, the ignition system would continue to operate with the ignition switch off.

2

EFI System Relay Test

GM V-6 and V-8 (EFI) models

The electronic fuel injection (EFI) system relay provides voltage to operate the fuel injectors and ignition coil(s) and powers up the engine control unit (ECU). A faulty EFI system relay will prevent the engine from starting or cause the engine to stall at any speed.

A 12-volt test lamp is required for this procedure.

1. Make sure the emergency stop (lanyard) switch is in RUN.
2. Turn the ignition to ON or RUN. The dash instruments must turn on, and the warning horn must sound for 2 seconds. Refer to the following:
   a. If the instruments do not turn on, check for a blown fuse, tripped circuit breaker or loose wire. Correct faulty wiring, replace the fuse or reset the circuit breaker to restore gauge operation. If this does not restore gauge operation, test the ignition switch as described in this chapter.
   b. If the instruments are on, but the warning horn does not sound, test the emergency stop (lanyard) switch as described in this chapter. Refer to *Basic Ignition System Troubleshooting*. If the switch tests correctly, proceed to Step 3.
   c. If the instruments are on, but the warning horn does sound, the system relay is operational. Proceed to Step 11 to check the injector wire circuit.
3. Remove the EFI system relay as described in Chapter Nine.
4. Refer to **Figure 52**, the relay terminals and the terminals in the relay mounting bracket to identify the relay terminals. The relay will only fit one way into the bracket. Com-

52

EFI SYSTEM RELAY (GM V-6 AND V-8 MODELS)

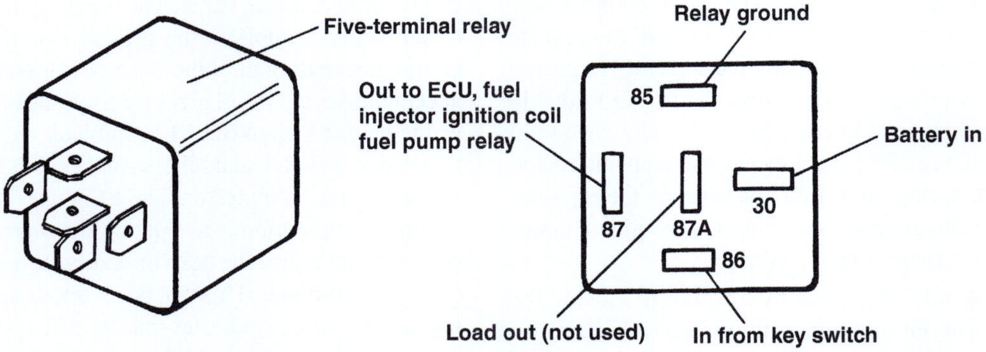

pare the orientation of the No. 30 and No. 87 terminal blades for reference.

5. Connect the cord of the test lamp to the positive battery terminal. Touch the test lamp probe to an engine ground to verify the lamp and ground. Touch the test lamp probe to the No. 85 terminal in the mounting bracket or harness plug. The lamp must illuminate. If not, check for a faulty connection or black relay ground wire.

6. Connect the test lamp cord to a good engine ground. Touch the test lamp probe to the No. 30 terminal in the mounting bracket or harness plug. The lamp must illuminate. If not, check for a blown fuse or circuit breaker. Refer to the wiring diagrams at the end of the manual to identify any fuses or circuit breakers along the circuit. Check for a faulty terminal or battery supply wire to the terminal if no fault is found with the fuses or circuit breaker.

7. Place the ignition switch in OFF or STOP. Connect the cord of the test lamp to the negative battery terminal. Touch the test lamp probe to the No. 86 terminal in the mounting bracket or harness plug. The lamp must not illuminate. If the lamp illuminates, test the ignition switch as described in this chapter. If the switch tests correctly, the purple wire leading into the relay mounting bracket is shorted to a positive battery source. Inspect the wiring and correct the fault before proceeding.

8. Connect the cord of the test lamp to the negative battery terminal. Touch the test lamp probe to the No. 86 terminal in the mounting bracket or harness plug. Observe the lamp while an assistant turns the ignition switch to ON or RUN. The lamp must illuminate. If not, test the ignition switch and lanyard switch as described in this chapter. If both switches test correctly, an open circuit or short to ground is present in the purple wire of the instrument or engine wire harness. Correct the wiring or connection before proceeding.

9. Install the system relay as described in Chapter Nine.

10. Turn the ignition switch to OFF or STOP. Disconnect the plug containing the red, pink or pink/white wire from one of the fuel injectors. Connect the cord of the test lamp to the negative battery terminal. Touch the test lamp probe to the pink/white, red or red/white wire terminal in the connector. The test lamp must not illuminate. If the lamp illuminates, the relay is shorted internally or stuck in the energized position and must be replaced.

11. Connect the cord of the test lamp to the negative battery terminal. Touch the test lamp probe to the pink/white or red/white wire terminal in the connector. Observe the lamp while an assistant turns the ignition switch to ON or RUN. The lamp must illuminate. If not, the relay has failed open or stuck in the de-energized position. Replace the relay and repeat the test. The pink/white or red/white wire is open or shorted to ground if the lamp does not illuminate with the replacement relay.

12. Turn the ignition switch to OFF or STOP.

Ford 5.0L and 5.8L (V-8) models

The electronic fuel injection (EFI) system relay provides current to operate the fuel injectors and ignition coil and powers up the engine control assembly (ECA). A faulty EFI system relay will prevent the engine from starting or cause the engine to stall at any speed.

A 12-volt test lamp is required for this procedure.

1. Make sure the emergency stop (lanyard) switch is in RUN.

2. Turn the ignition to ON or RUN. The dash instruments must turn on, and the warning horn must sound for 2 seconds. Refer to the following:

   a. If the instruments do not turn on, check for a blown fuse, tripped circuit breaker or loose wire connection. Correct faulty wiring, replace the fuse or reset the circuit breaker to restore gauge operation. If this does not restore gauge operation, test the ignition switch as described in this chapter.

   b. If the instruments are on, but the warning horn does not sound, test the emergency stop (lanyard) switch as described in this chapter. Refer to *Basic Ignition System Troubleshooting*. If the switch tests correctly, proceed to Step 3.

53

EFI SYSTEM RELAY (FORD 5.0L AND 5.8L V-8 MODELS)

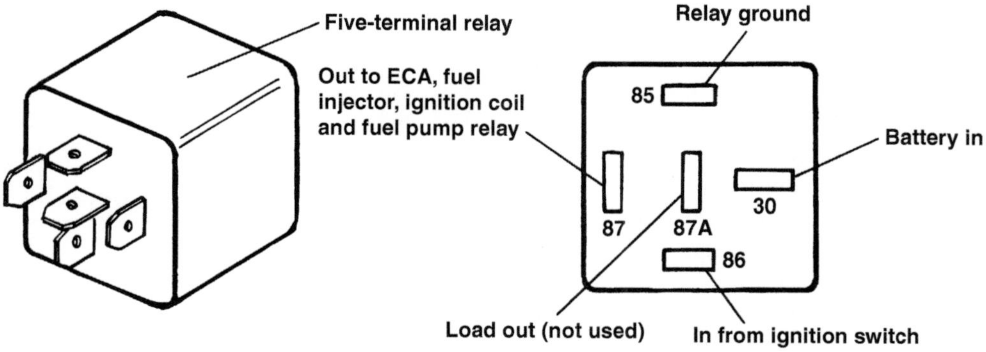

c. If the instruments are on, but the warning horn does sound, the system relay is operational. Proceed to Step 11 to check the injector wire circuit.

3. Remove the EFI system relay as described in Chapter Nine.

4. Refer to **Figure 53**, the relay terminals and the terminals in the relay mounting bracket to identify the relay terminals. The relay will only fit one way into the bracket. Compare the orientation of the No. 30 and No. 87 terminal blades for reference.

5. Connect the cord of the test lamp to the positive battery terminal. Touch the test lamp probe to an engine ground to verify the light and ground. Touch the test lamp probe to the terminal for the black wire in the mounting bracket or harness plug. The lamp must illuminate. If not, check for a faulty connection or black relay ground wire.

6. Connect the test lamp cord to a good engine ground. Touch the test lamp probe to the terminal for the red/purple wire in the mounting bracket or harness plug. The lamp must illuminate. If not, check for a blown fuse or circuit breaker. Refer to the wiring diagrams at the end of the manual to identify any fuses or circuit breakers along the circuit. Check for a faulty terminal or battery supply wire to the terminal if no fault is found with the fuses or circuit breaker.

7. Place the ignition switch in OFF or STOP. Connect the cord of the test lamp to the negative battery terminal. Touch the test lamp probe to the terminal for the purple wire in the mounting bracket or harness plug. The lamp must not illuminate. If the light illuminates, test the ignition switch as described in this chapter. If the switch tests correctly, the purple wire leading into the relay mounting bracket is shorted to a positive battery source. Inspect the wiring and correct the fault before proceeding.

8. Connect the cord of the test lamp to the negative battery terminal. Touch the test lamp probe to the terminal for the purple wire in the mounting bracket or harness plug. Observe the lamp while an assistant turns the ignition switch to ON or RUN. The lamp must illuminate. If not, test the ignition switch and lanyard switch as described in this chapter. If both switches test correctly, an open circuit or short to ground is present in the purple wire of the instrument or engine wire harness. Correct the wiring or terminal defects before proceeding.

9. Install the system relay as described in Chapter Nine.

10. Turn the ignition switch to OFF or STOP. Disconnect the plug containing the purple/black wire from one of the fuel injectors. Connect the cord of the test lamp to the negative battery terminal. Touch the test lamp probe to the purple/black wire terminal in the connector. The test lamp must not illuminate. If the lamp illuminates, the relay is shorted internally or stuck in the energized position and must be replaced.

11. Connect the cord of the test lamp to the negative battery terminal. Touch the test lamp probe to the purple/black wire terminal in the connector. Observe the lamp while an assistant turns the ignition switch to ON or RUN. The lamp must illuminate. If not, the relay has failed open or stuck in the de-energized position. Replace the relay and repeat the test. The purple/black wire is open or shorted to ground if the lamp does not illuminate with the replacement relay.

12. Turn the ignition switch to OFF or STOP.

Fuel Pump Relay Test (EFI Models)

WARNING

The engine might unexpectedly start during testing. Keep all persons, clothing, tools or loose objects away from the drive belts and other moving components on the engine.

CAUTION

If operating the starter with the boat out of the water, use a flush device to supply the engine with cooling water. Operation of the starter motor results in operation of the seawater pump. The seawater pump is quickly damaged if operated without adequate cooling water.

GM V-6 and V-8 (EFI) models

The fuel pump relay provides battery voltage to operate the low-pressure and high-pressure electric fuel pumps. The test procedure tests the complete fuel pump circuit. A test lamp and ohmmeter are required for this procedure.

1. Make sure the lanyard switch is in RUN.
2. Turn the ignition to ON or RUN. The dash instruments must turn on, the warning horn must sound for 2 seconds, and the electric fuel pumps must operate for 2 seconds. Touch the tip of a mechanic's stethoscope to the body to hear the pump operation as needed.
   a. If the instrumenst do not turn on, check for a blown fuse, tripped circuit breaker or loose wire connection. Correct faulty wiring, replace the fuse or reset the circuit breaker to restore gauge operation. If this does not restore gauge operation, test the ignition switch as described in this chapter.
   b. If the instruments are on, but the warning horn does not sound, test the emergency (lanyard) switch as described in this chapter. Refer to *Basic Ignition System Troubleshooting*. If the switch tests correctly, test the EFI system relay as described in this section.
   c. If the instruments are on, the warning horn did sound, but the fuel pumps do not operate, the system relay and ECU are operational. Proceed to Step 3.
   d. If the instruments are on, the warning horn did sound, and the fuel pumps operate, the fuel pump relay is operational. Remove the fuel pump relay as described in Chapter Nine. Then proceed to Step 16.
3. Lift the locking clip. Then disconnect the engine harness connector from the high-pressure electric fuel pump (**Figure 54**) and the low-pressure electric fuel pump (**Figure 55**). The high-pressure pump is mounted on the top of the fuel reservoir at the back of the engine. The low-pressure fuel pump is mounted on a bracket at the front starboard side of the engine.
4. Connect the cord of the test lamp to the positive battery cable. Touch the test lamp probe to an engine ground to verify the light and engine ground. Touch the test lamp probe to the terminal for the black wire in each of the fuel pump harness connectors. The lamp must illuminate for each connector. If not, repair the faulty wiring or terminal along the black wire from the suspect fuel pump connector to the engine ground.
5. Turn the ignition switch to OFF or STOP. Connect the cord of the test lamp to the negative battery cable. To verify the lamp, touch the test lamp probe to a positive battery terminal. Touch the test lamp probe to the terminal in each of the fuel pump harness connectors for the yellow/green wire. The test lamp must not illuminate in either connector. If the light illuminates, the fuel pump relay is shorted internally, the relay is stuck in the energized position, the yellow/green wire is shorted to ground, or the ECU is faulty. Proceed with testing to isolate the fault.
6. Connect the cord of the test lamp to the negative battery cable. Touch the test lamp probe to the terminal for the yellow/green wire in one of the fuel pump harness connectors. Observe the test lamp while an assistant turns the ignition switch to ON or RUN. Turn the ignition switch to OFF for a few seconds to reset the ECU. Then repeat the test for the other fuel pump harness connector. For each connector, the lamp must illuminate for 2 seconds and then extinguish.
   a. If the lamp illuminates as described, the fuel pump relay is operational, and wire connections to the pump test correctly. If the pump does not operate as noted in Step 2, check the pump pressure for the corresponding connector as described in this chapter.
   b. If the lamp does not illuminate for one or both connectors, the fuel pump relay, power supply to the relay, related wiring or ECU is faulty. Proceed with testing to isolate the fault.
7. Remove the fuel pump relay as described in Chapter Nine.
8. Refer to **Figure 56**, the relay terminals and the terminals in the relay mounting bracket or harness plug to identify the

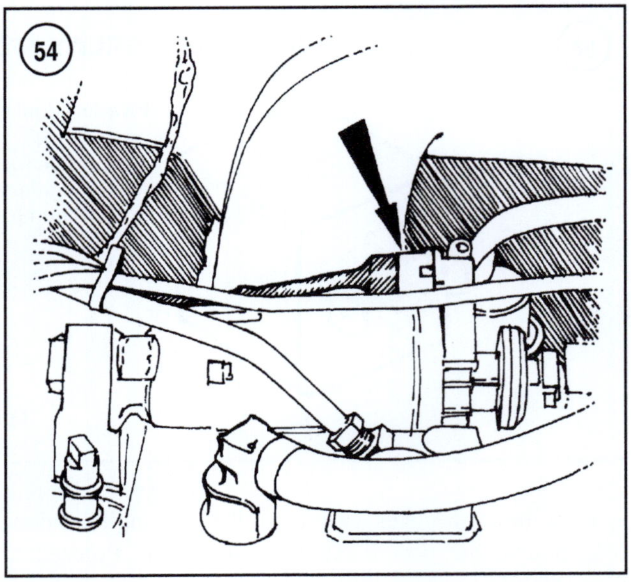
54

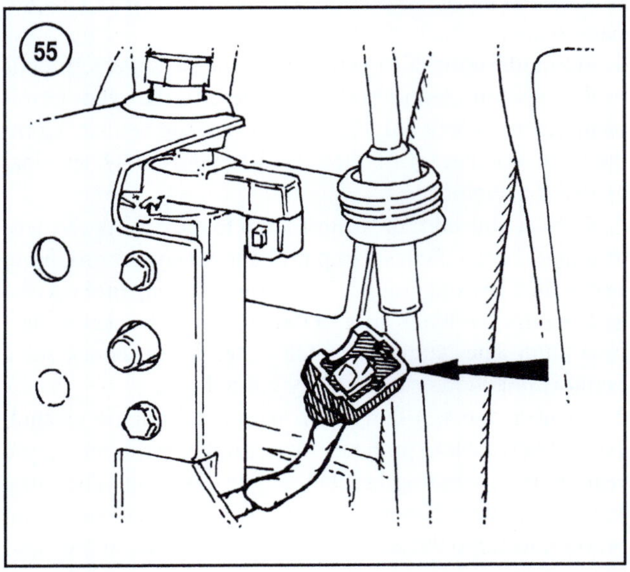
55

56

FUEL PUMP RELAY (GM V-6 AND V-8 [EFI] MODELS)

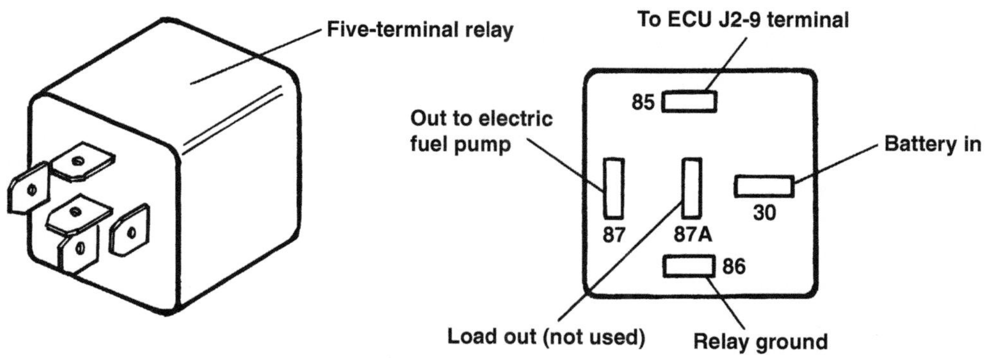

relay terminals. The relay will only fit one way into the bracket. Compare the orientation of the No. 30 and No. 87 terminal blades for reference.

9. Connect the test lamp cord to the negative battery terminal. Touch the test lamp probe to the No. 30 terminal in the mounting bracket or harness plug. The lamp must illuminate. If not, check for a blown fuse or circuit breaker. Refer to the wiring diagrams at the end of the manual to identify any fuses or circuit breakers along the circuit. Check for a faulty terminal or battery supply wire if no fault is found with the fuses or circuit breaker.

10. Connect the cord of the test lamp to the positive battery terminal. Touch the test lamp probe to the No. 86 terminal in the relay mounting bracket or harness plug. The lamp must illuminate. If not, repair the faulty wiring or terminal connecting the No. 86 terminal to the engine ground.

11. Touch one of the ohmmeter test leads to the No. 87 terminal in the relay bracket or harness plug. Note the meter reading while touching the other meter test lead to the yellow/green wire terminal in one of the disconnected fuel pump connectors. Repeat the test using the other pump connectors. The meter must indicate continuity at both connectors. If not, repair the open circuit along the yellow/green wire to restore fuel pump operation.

12. Touch one of the meter leads to the No. 87 terminal in the relay bracket or harness plug. Touch the other lead to an engine ground. The meter must indicate no continuity. If the meter shows continuity, the yellow/green wire in the fuel pump harness is shorted to ground, and the harness must be repaired or replaced.

13. Reconnect the harness connector to the low-pressure and high-pressure electric fuel pumps. Route the wiring to prevent interference with drive belts or other moving components. Secure the wiring with plastic locking clamps as needed.

14. Turn the ignition switch to OFF or STOP. Connect the cord of the test lamp to an engine ground. Touch the test lamp probe to the No. 85 terminal in the relay mounting bracket or harness plug. The lamp must not illuminate. If the lamp illuminates, the wire connecting the No. 85 relay terminal to the ECU is shorted to ground, or the ECU is faulty. Either condition will cause the fuel pump relay to remain in the energized position, and the pump will operate anytime the system relay is energized. Check the wiring and terminals for shorting. Replace the ECU only if all other components and wiring on the circuit test correctly.

15. Touch the test lamp probe to the No. 85 terminal in the relay mounting bracket or harness plug. Observe the test lamp while an assistant turns the ignition switch to ON or RUN. The lamp must illuminate for 2 seconds and then extinguish. If not, test the EFI system relay as described in this section. If the system relay tests correctly, the wire connecting the No. 85 relay terminal to the ECU is open, or the ECU is faulty. Check the wiring and terminals for faults and correct as needed. Replace the ECU only if all other components and wiring on the circuit test correctly. Failure of the ECU is extremely rare.

16. Disconnect and ground the spark plug leads to prevent the engine from starting during the following procedure.

17. Touch the test lamp probe to the No. 85 terminal in the relay mounting bracket or harness plug. Observe the test lamp while an assistant turns the ignition switch to START. The lamp must illuminate continuously while cranking the engine and extinguish a few seconds after stopping the engine. If not, test the ignition system as described in this section. If the ignition tests correctly, and spark is present at the plugs, the wiring connecting the ignition system to the ECU is open, or the ECU is faulty. Check the wiring and terminals and correct as needed. Replace the ECU only if all other components and wiring on the circuit test correctly. Failure of the ECU is extremely rare. Turn the ignition switch to OFF.

18. Replace the fuel pump relay if the fuel pump does not operate as described in Step 2, and all circuits test correctly in Steps 3-17.

19. Install the fuel pump relay as described in Chapter Nine.

20. Reconnect the spark plug leads.

Ford 5.0L and 5.8L (V-8) models

The fuel pump relay provides battery voltage to operate the low-pressure and high-pressure electric fuel pumps.

57

FUEL PUMP RELAY (FORD 5.0L AND 5.8L V-8 MODELS)

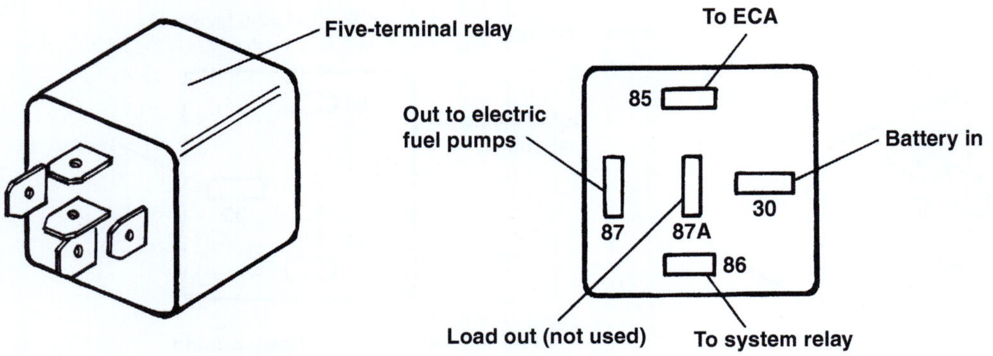

The test procedure tests the complete fuel pump circuit. A 12-volt test lamp and ohmmeter are required for this procedure.

1. Make sure the lanyard switch is in RUN.

2. Turn the ignition to ON or RUN. The dash instruments must turn on, the warning horn must sound for 2 seconds, and the low-pressure and high-pressure fuel pumps must operate for 2 seconds. Touch the tip of a mechanic's stethoscope to the body to hear the pump operation as needed.

   a. If the instruments do not turn on, check for a blown fuse, tripped circuit breaker or loose wire connection. Correct faulty wiring, replace the fuse or reset the circuit breaker to restore gauge operation. If this does not restore gauge operation, test the ignition switch as described in this chapter.

   b. If the instruments are on, but the warning horn does not sound, test the emergency (lanyard) switch as described in this chapter. Refer to *Basic Ignition System Troubleshooting*. If the switch tests correctly, test the EFI system relay as described in this section.

   c. If the instruments are on, and the warning horn sounds, but the fuel pumps do not operate, the system relay and ECA are operational. Proceed to Step 3.

   d. If the instruments are on, the warning horn sounds, and the fuel pumps operate, the fuel pump relay is operational. Remove the fuel pump relay as described in Chapter Nine. Then proceed to Step 17.

3. Lift the locking clip. Then disconnect the engine harness connectors from the high-pressure electric fuel pump (**Figure 54**) and the low-pressure electric fuel pump (**Figure 55**). The high-pressure pump is mounted on the top of the fuel reservoir at the back of the engine. The low-pressure fuel pump is mounted on a bracket at the front starboard side of the engine.

4. Connect the cord of the test lamp to the positive battery cable. Touch the test lamp probe to an engine ground to verify the lamp and engine ground. Touch the test lamp probe to the terminal for the black wire in each of the fuel pump harness connectors. The lamp must illuminate for each connector. If not, repair the faulty wiring or terminal along the black wire from the suspect fuel pump connector to the engine ground terminal.

5. Turn the ignition switch to OFF or STOP. Connect the test lamp cord to the positive battery cable. Touch the test lamp probe to the terminal in each of the fuel pump harness connectors for the green/yellow wire. The test lamp must not illuminate in either connector. If the lamp illuminates, the fuel pump relay is shorted internally, the relay is stuck in the energized position, the green/yellow wire is shorted to ground, or the ECU is faulty. Proceed with testing to isolate the fault.

6. Connect the test lamp cord to the negative battery cable. Touch the test lamp probe to the terminal for the green/yellow wire in one of the fuel pump harness connectors. Observe the test lamp while an assistant turns the ignition switch to ON or RUN. Turn the ignition switch to OFF for a few seconds to reset the ECU. Then repeat the test for the other fuel pump harness connector. For each connector, the lamp must illuminate for 2 seconds and then extinguish.

   a. If the lamp illuminates as described, the fuel pump relay is operational, and wire connections to the pump test correctly. If the pump does not operate as noted in Step 2, check the pump pressure for the corresponding connector as described in this chapter.

   b. If the lamp does not illuminate for one or both connectors, the fuel pump relay, power supply to the relay, related wiring or ECU is faulty. Proceed with testing to isolate the fault.

7. Remove the fuel pump relay as described in Chapter Nine.

8. Refer to **Figure 57**, the relay terminals and the terminals in the relay mounting bracket or harness plug to identify the relay terminals. The relay will only fit one way into the bracket. Compare the orientation of the No. 30 and No. 87 terminal blades for reference.

9. Connect the cord of the test lamp to the negative battery terminal. Touch the test lamp probe to the No. 30 terminal in the relay mounting bracket or harness plug for the red/purple wire. The lamp must illuminate. If not, check for a blown fuse or circuit breaker. Refer to the wiring diagrams at the end of the manual to identify any fuses or cir-

cuit breakers along the circuit. Check for a faulty terminal or battery supply wire to the terminal if no fault is found with the fuses or circuit breaker.

10. Place the ignition switch in OFF or STOP. Connect the cord of the test lamp to the negative battery terminal. Touch the test lamp probe to the No. 86 terminal in the relay mounting bracket or harness plug for the purple/black wire. The lamp must not illuminate. If the lamp illuminates, the EFI system relay is shorted or stuck in the energized position, or the purple/black wire is shorted to a positive battery source. Test the EFI system relay as described in this chapter. Repair the shorted purple/black wire if no fault is found with the system relay.

11. Connect the cord of the test lamp to the negative battery terminal. Touch the test lamp probe to the No. 86 terminal in the relay mounting bracket or harness plug for the purple/black wire. Observe the test lamp while an assistant turns the ignition switch to ON or RUN. The lamp must illuminate for 2 seconds and then extinguish. If not, test the EFI system relay as described in this section.

12. Place the ignition switch in OFF or STOP. Connect the cord of the test lamp to a positive battery terminal on the engine. Touch the test lamp probe to the No. 85 terminal in the relay mounting bracket or harness plug for the blue/orange wire. The lamp must not illuminate. If the lamp illuminates, the blue/orange wire is shorted to ground, or the ECA is faulty. Repair the shorted wire as needed. Either condition will cause the fuel pump relay to remain in the energized position, and the pumps will operate anytime the system relay is energized. Check the wiring and terminals for shorts. Replace the ECA only if all other components and wiring on the circuit test correctly. Failure of the ECA is extremely rare.

13. Connect the cord of the test lamp to a positive battery terminal on the engine. Touch the test lamp probe to the No. 85 terminal in the relay mounting bracket or harness plug for the blue/orange wire. Observe the test lamp while an assistant turns the ignition switch to ON or RUN position. The light must illuminate for 2 seconds and then extinguish. If not, test the EFI system relay as described in this section. If the relay tests correctly, replace the ECA only after testing all other related components and verifying the wiring. ECA failure is rare, and replacement seldom is necessary.

14. Touch one of the ohmmeter test leads to the No. 87 terminal for the green/yellow wire in the relay bracket or harness plug. Note the meter reading while touching the other meter lead to the green/yellow wire terminal in one of the disconnected fuel pump harness connectors. Repeat the test using the other pump harness connector. The meter must indicate continuity at both connectors. If not, repair the open circuit along the green/yellow wire to restore fuel pump operation.

15. Touch one of the meter leads to the No. 87 terminal for the green/yellow wire in the relay bracket or harness plug. Touch the other lead to an engine ground. The meter must indicate no continuity. If the meter shows continuity, the green/yellow wire in the fuel pump harness is shorted to ground, and the harness must be repaired or replaced.

16. Reconnect the harness connector to the low-pressure and high-pressure electric fuel pumps. Route the wiring to prevent interference with drive belts or other moving components. Secure the wiring with plastic locking clamps as needed.

17. Disconnect and ground the spark plug lead to prevent the engine from starting during the following procedure.

18. Connect the cord of the test lamp to a positive battery terminal on the engine. Touch the test lamp probe to the No. 85 terminal in the relay mounting bracket or harness plug for the blue/orange wire. Observe the test lamp while an assistant turns the ignition switch to START. The light must illuminate continuously while cranking the engine and extinguish a few seconds after stopping the engine. If otherwise, test the ignition system as described in this section. If the ignition tests correctly, and spark is present at the plugs, the wiring connecting the ignition system to the ECA is open, or the ECA is faulty. Check the wiring and terminals for faults and correct as needed. Replace the ECA only if all other components and wiring on the circuit test correctly. Failure of the ECA is extremely rare. Turn the ignition switch to OFF.

19. Replace the fuel pump relay if the fuel pump did not operate as described in Step 2, and all circuits test correctly in Steps 3-18.

20. Install the fuel pump relay as described in Chapter Nine.

21. Reconnect the spark plug leads.

Fuel Pressure Test (EFI Models)

WARNING
Safely performing on-water testing or adjustments requires two people: one person to operate the boat and the other to monitor the gauges or test equipment and make necessary adjustments. All personnel must remain seated inside the boat at all times. Do not lean over the transom while the boat is underway. Use extensions to allow all gauges and test equipment to be located in normal seating areas.

WARNING
The engine might unexpectedly start during testing. Keep all persons, clothing, tools or loose objects away from the drive belts and other moving components on the engine.

CAUTION
If testing the ignition system with the boat out of the water, use a flush test device to supply the engine with cooling water. Operation of the starter motor results in operation of the seawater pump. The seawater pump is

quickly damaged if operated without adequate cooling water.

Pressure test (low-pressure electric fuel pump [EFI models])

A suitable pressure gauge with a bleed-off hose (**Figure 58**), available from Kent-Moore (part No. J-34730-1A), is required for this procedure.

1. To rule out other faults, inspect the fuel and check for a restriction in the fuel supply hose as described in *Basic Fuel System Troubleshooting* in this chapter. Perform the following test if previous tests indicated a no-start condition, lean condition or fuel reservoir flooding, which might be due to a faulty low-pressure electric fuel pump.
2. Touch a mechanic's stethoscope to the body of the low-pressure electric fuel pump (A, **Figure 39**, typical). Listen carefully for fuel pump operation while an assistant turns the ignition switch to ON and back to OFF or STOP for a minimum of 30 seconds. The fuel pump must operate for approximately 2 seconds each time the ignition switch is in ON. If not, test the fuel pump relay as described in this section.
3. Turn the ignition switch to OFF.
4. Disconnect the negative battery cable.
5. Remove the cap from the low-pressure test port (10, **Figure 48**).
6. Wrap a suitable shop towel around the low-pressure test port to capture spilled fuel. Pull up the locking lever. Then quickly push the fuel gauge connector onto the test port and release the locking lever. If the test gauge threads onto the test port, hold the hose fitting in firm contact with the test port fitting and then quickly thread the hose fitting onto the test port.
7. Clean up any spilled fuel. Connect the battery cable.
8. Observe the pressure gauge while an assistant turns the ignition switch to ON or RUN. Note the pressure reading and then have the assistant turn the ignition switch to OFF or STOP. Wait approximately 30 seconds for the ECU to reset. Repeat this step several times to purge any air from the system. The pressure gauge must indicate 4.0-8.5 psi (27.6-58.6 kPa). Refer to the following:
   a. If the pressure is between 4.0-8.5 psi (27.6-58.6 kPa), the low-pressure electric fuel pump is operating correctly under key-on conditions. Proceed to Step 9.
   b. If the pressure is less than 4.0 psi (27.6), check for a flooded fuel reservoir as described in this section. If no flooding is evident, repeat the test with a portable tank and hose connected to the fuel inlet fitting in place of the boat fuel line and fuel supply hose. Refer to *Checking for a restriction in the fuel supply hose* in this chapter. If the pressure remains below the specification with the portable tank installed, the low-pressure electric fuel pump must be replaced as described in Chapter Seven.

58

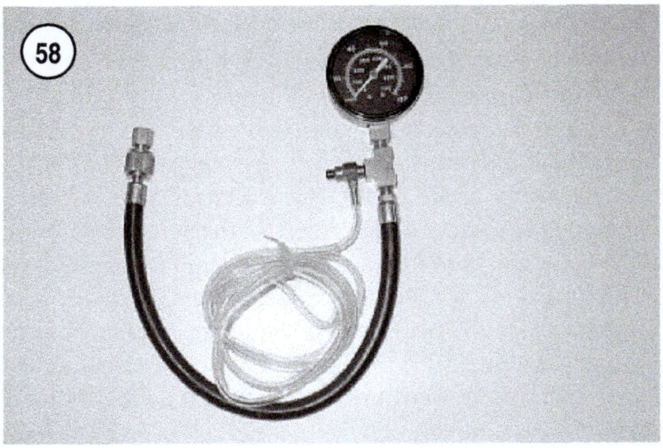

   c. If the pressure exceeds 8.5 psi (58.6 kPa), the fuel pump is producing excess pressure and may cause flooding of the fuel reservoir. Replace the low-pressure electric fuel pump as described in Chapter Seven.
9. Prepare the boat and engine for normal operation. Do not disconnect the fuel pressure gauge at this time; however, route the gauge hose to prevent contact with drive belts, pulleys and other moving components. Secure the hose with clamps as needed.
10. Observe the gauge while an assistant operates the engine under normal load at speeds from idle to wide open throttle. The pressure gauge must indicate 4.0-8.5 psi (27.6-58.6 kPa) at all engine speeds. Refer to the following:
   a. If the pressure is between 4.0-8.5 psi (27.6-58.6 kPa), the low-pressure electric fuel pump is operating correctly. Proceed to Step 11.
   b. If the pressure is less than 4.0 psi (27.6 kPa) at any engine speed, check for a flooded fuel reservoir as described in this section. If no flooding is evident, repeat the test with a portable tank and hose connected to the fuel inlet fitting in place of the boat fuel line and fuel supply hose. Refer to *Checking for a restriction in the fuel supply hose*. If the pressure remains below the specification with the portable tank installed, the low-pressure electric fuel pump must be replaced as described in Chapter Seven.
   c. If the pressure exceeds 8.5 psi (58.6 kPa), the low-pressure electric fuel pump is producing excess pressure and might cause flooding of the fuel reservoir. Replace the low-pressure electric fuel pump as described in Chapter Seven.
11. Return to port and stop the engine.
12. Disconnect the negative battery cable. Lift the locking lever and quickly pull the connector off the test port. If the test gauge threads onto the test port, hold the hose fitting in firm contact with the test port fitting. Then quickly unthread the hose fitting from the test port.
13. Connect the battery cable.
14. Clean up any spilled fuel and replace the cap.
15. After performing the fuel system service procedure, start the engine and immediately check for fuel leakage at

all fuel line or hose connections. Correct any fuel leakage before operating the engine.

Pressure test (high-pressure electric fuel pump [EFI models])

The fuel pressure test verifies that fuel is being supplied to the fuel injectors at the required pressure.

On throttle body injection (TBI) models, a suitable pressure gauge with a bleed-off hose (**Figure 58**) and adapters available from Kent-Moore (part No. J-34730-1A and J-29658-D) to connect to the pressure line and throttle body fitting are required for this procedure.

On multiport injection (MPI) models, a suitable pressure gauge with a bleed-off hose (**Figure 58**) available from Kent-Moore (part No. J-34730-1A) is required for this procedure.

WARNING
Safely performing on-water testing or adjustments requires two people: one person to operate the boat and the other to monitor the gauges or test equipment and make necessary adjustments. All personnel must remain seated inside the boat at all times. Do not lean over the transom while the boat is underway. Use extensions to allow all gauges and test equipment to be located in normal seating areas.

WARNING
The engine might unexpectedly start during testing. Keep all persons, clothing, tools or loose objects away from the drive belts and other moving components on the engine.

CAUTION
If testing the ignition system with the boat out of the water, use a flush device to supply the engine with cooling water. Operation of the starter motor results in operation of the seawater pump. The seawater pump is quickly damaged if operated without adequate cooling water.

1. To rule out other faults, inspect the fuel and check for a restriction in the fuel supply hose as described in this chapter. Refer to *Basic Fuel System Troubleshooting*. Perform the following test if previous tests indicated a no-start condition, lean condition or fuel reservoir flooding, which might be due to a faulty low-pressure electric fuel pump.
2. Touch a mechanic's stethoscope to the body of the high-pressure electric fuel pump (**Figure 54**). Listen carefully for fuel pump operation while an assistant turns the ignition switch to ON and back to OFF or STOP for a minimum of 30 seconds. The fuel pump must operate for approximately 2 seconds each time the switch is in ON. If not, test the fuel pump relay as described in this section.
3. Turn the ignition switch to OFF.
4. Disconnect the negative battery cable.

5A. On TBI models, connect the fuel pressure gauge as follows:

a. Remove the flame arrestor and cover as described in Chapter Seven.
b. Locate the fuel delivery line (A, **Figure 45**) at the rear of the throttle body assembly. Use compressed air to remove debris from the fittings and throttle body surfaces.
c. Wrap a suitable shop towel around the fitting to capture spilled fuel.
d. Hold the throttle body fitting with a suitable wrench to prevent rotation. Then use a flare nut wrench to loosen the fuel line fitting.
e. Slowly loosen the fuel line fitting by hand until fuel just starts dripping from the connection. Wait until fuel stops dripping. Then unthread the fitting.
f. Drain residual fuel in the line into a suitable container. Clean up any spilled fuel at once.
g. Remove the O-ring from the groove at the end of the fuel line fitting. Discard the O-ring.
h. Thread the gauge adapter onto the throttle body fitting and disconnected fuel delivery line fitting. Securely tighten the line and adapter.
i. Connect the pressure gauge to the adapter.

5B. On MPI models, connect the fuel pressure gauge as follows:

a. Remove the cap from the high-pressure test port (13, **Figure 48**).
b. Wrap a suitable shop towel around the high-pressure test port to capture spilled fuel. Pull up the locking lever. Then quickly push the fuel gauge connector onto the test port and release the locking lever. If the test gauge threads onto the test port, hold the hose fitting in firm contact with the test port fitting and then quickly thread the hose fitting onto the test port.
c. Clean up any spilled fuel.

6. Connect the battery cable.
7. Observe the pressure gauge while an assistant turns the ignition switch to ON or RUN. Note the pressure reading. Then have the assistant turn the ignition switch to OFF or STOP. Wait approximately 30 seconds for the ECU to reset. Repeat this step several times to purge any air from the system. The pressure gauge must indicate the specified pressure as follows:
   a. On TBI models, the gauge must indicate 27-31 psi (186-214 kPa).
   b. On MPI models, the gauge must indicate 36-42 psi (248-290 kPa).
8. If the pressure indicated in Step 7 is less than the minimum specification, replace the fuel pressure regulator as described in Chapter Seven and repeat the test. The high-pressure electric fuel pump is faulty, or blockage is present in a hose, line or filter if the pressure remains below the specification with the replacement regulator. Replace

the high-pressure electric fuel pump if no fault is found with the hoses, lines or filters.

9. If the pressure indicated in Step 7 exceeds the maximum specification, replace the fuel pressure regulator as described in Chapter Seven and repeat the test. The fuel return hose or related fitting is blocked if the pressure remains above the specification with the replacement regulator. Remove the blockage or replace the hose as needed to correct the fuel pressure.

10. Prepare the boat and engine for operation under actual running conditions. Do not disconnect the fuel pressure gauge at this time; however, route the gauge hose to prevent contact with drive belts, pulleys and other moving components. Secure the hose with clamps as needed.

11. Observe the gauge while an assistant operates the engine under normal load at speeds from idle to wide open throttle. The gauge must indicate the pressure range specified in Step 7 at all engine speeds. Refer to the following:

   a. If the fuel pressure is correct at all engine speeds, the fuel delivery system is operating properly. Any engine malfunction is related to old or contaminated fuel, faulty spark plugs or other ignition system components, the fuel injection control system or a mechanical problem with the engine. Check these items as described in the appropriate section of this chapter. Proceed to Step 12.
   b. If the fuel pressure exceeds the maximum specification at any engine speed, replace the fuel pressure regulator as described in Chapter Seven and repeat the test. The fuel return hose or related fitting is blocked if the pressure remains above the specification with the replacement regulator. Remove the blockage or replace the hose as needed.
   c. If the fuel pressure is less than the minimum specification at any engine speed (particularly at higher engine speed, test the electric boost pump (if so equipped) and check for a restriction in the fuel supply hose as described in this chapter. Refer to *Basic Fuel System Troubleshooting*. If no fault is found with the boost pump, and no restriction is evident, replace the fuel pressure regulator as described in Chapter Seven and repeat the test. Replace the high-pressure electric fuel pump as described in Chapter Seven if the pressure remains below the specification with the replacement regulator.

12. Return to port and stop the engine.

13A. On TBI models, disconnect the fuel pressure gauge as follows:

   a. Direct the bleed-off hose into the fuel tank fill opening.
   b. Open the valve and allow all fuel to drain from the hose.
   c. Wrap a suitable shop towel around the fitting to capture spilled fuel.
   d. Hold the throttle body fitting with a suitable wrench to prevent rotation. Then use a flare nut wrench to loosen the adapter and fuel line.
   e. Slowly loosen the fuel line fitting by hand until fuel just starts dripping from the connection. Wait until fuel stops dripping and then unthread the fitting.
   f. Drain residual fuel in the line into a suitable container. Clean up any spilled fuel at once.
   g. Install a new O-ring into the fuel delivery line fitting groove. Then thread the delivery line fitting into the throttle body fitting by hand.
   h. Hold the throttle body fitting with a suitable wrench to prevent rotation. Then tighten the fuel line fitting as described in Chapter Seven. Refer to *Throttle body (removal and installation)*. Do not allow the fuel line to twist during the tightening procedure. Twisting damages the line.
   i. Close the fuel tank fill cap.

13B. On MPI models, disconnect the fuel pressure gauge as follows:

   a. Direct the bleed-off hose into the fuel tank fill opening.
   b. Open the valve and allow all fuel to drain from the hose.
   c. Lift the locking lever and quickly pull the connector off the test port. If the test gauge threads onto the test port, hold the hose fitting in firm contact with the test port fitting. Then quickly unthread the hose fitting from the test port.
   d. Thread the cap onto the fuel pressure test port.
   e. Close the fuel tank fill cap.

14. Clean up any spilled fuel.

15. After performing the fuel system service procedure, start the engine and immediately check for fuel leakage at all fuel line or hose connections. Correct any fuel leakage before operating the engine.

16. On TBI models, install the flame arrestor and cover as described in Chapter Seven.

Checking for a flooded fuel reservoir

A flooded fuel reservoir will introduce fuel into the intake manifold via the vent hose. This will cause a rich running condition with symptoms that include rough or unstable idle, excessive exhaust smoke, decreased fuel economy, hesitation during rapid acceleration and spark plug fouling.

WARNING

Safely performing on-water testing or adjustments requires two people: one person to operate the boat and the other to monitor the gauges or test equipment and make necessary adjustments. All personnel must remain seated inside the boat at all times. Do not lean over the transom while the boat is underway. Use extensions to allow all gauges and test equipment to be located in normal seating areas.

CAUTION
If testing the ignition system with the boat out of the water, use a flush test device to supply the engine with cooling water. Operation of the starter motor results in operation of the seawater pump. The seawater pump is quickly damaged if operated without adequate cooling water.

1. Disconnect the vent hose (12, **Figure 48**) from the intake manifold, intake plenum or throttle body fitting. Cap the fitting to prevent a vacuum leak.
2. Direct the hose into a container suitable for holding fuel.
3. Observe the hose and have an assistant start the engine and operate at various engine speeds. Flooding is indicated if fuel spills from the hose at any engine speed. If flooding is evident, either the low-pressure electric fuel pump is developing too much pressure, or the components in the fuel reservoir that control fuel level are faulty. In either case, the fuel reservoir must be repaired or replaced as described in Chapter Seven.

Fuel Injector Test

A fault with one or more of the injectors can cause either an excessively rich or excessively lean condition on one or more cylinders. On models with throttle body injection (TBI), a faulty injector will tend to affect only one bank of cylinders. On models with multiport fuel injection (MPI), a faulty injector will affect only the cylinder in which it supplies fuel.

A leaking fuel injector introduces fuel into the intake manifold in excess of the amount determined by the ECU or ECA and will cause symptoms consistent with an excessively rich condition. Refer to *Basic Fuel System Troubleshooting* in this chapter. The engine might perform properly at higher engine speeds as the engine is able to burn the excess fuel.

A fuel injector with a poor spray pattern due to debris or deposits in the valve will usually cause a poor idle quality because the poor spray pattern can prevent the fuel from atomizing before it reaches the combustion chamber. Poor idle quality occurs because the fuel droplets will not burn efficiently at lower engine speeds. The engine will usually operate properly at mid-range and higher throttle settings as the higher velocity of the air entering the intake will promote better atomization. An excessively lean condition might occur at higher throttle settings because the condition causing the poor spray pattern usually restricts fuel flow to some degree.

Blockage in the fuel injector will cause an excessively lean operating condition because the injector will not spray the proper amount of fuel for the given operating conditions. The engine might perform properly at lower engine speeds if the blockage is not overly restrictive. The resulting lean condition will worsen as the engine speed and fuel demand increases.

Throttle body injection (TBI) models

The design of the throttle body injection (TBI) system allows easy visual inspection of the fuel injector spray pattern. Remove the flame arrestor and observe the fuel injectors (**Figure 42**) while an assistant operates the engine under normal operating condition at various engine speeds and throttle settings. Observe the injectors after stopping the engine to identify a leaking fuel injector.

Multiport injection (MPI) models

The design of the multiport injection (MPI) system prevents visual inspection of the fuel injector spray pattern. An easy way to identify injector trouble is to remove and inspect the spark plugs as described in Chapter Three. All spark plugs should appear the same. A suspect injector is one in which the spark plug from the corresponding cylinder appears markedly different from the others. Replace the suspect fuel injector and all spark plugs only after verifying that the injector is operating as described in this section, the ignition system is operating properly and a diagnostic code check does not indicate a faulty fuel injection control system. Run the engine for several hours. Then remove and inspect the spark plugs. An ignition system fault or mechanical problem affecting only the suspect cylinder is indicated if the spark plug for that cylinder still indicates an excessively rich or excessively lean condition.

Operational test

WARNING
The engine might unexpectedly start during testing. Keep all persons, clothing, tools or loose objects away from the drive belts and other moving components on the engine.

CAUTION
If testing the ignition system with the boat out of the water, use a flush device to supply the engine with cooling water. Operation of the starter motor results in operation of the seawater pump. The seawater pump is quickly damaged if operated without adequate cooling water.

CAUTION
Safely performing on-water testing or adjustments requires two people: one person to operate the boat and the other to monitor the gauges or test equipment and make necessary adjustments. All personnel must remain seated inside the boat at all times. Do not lean over the transom while the boat is underway.

Use extensions to allow all gauges and test equipment to be located in normal seating areas.

Perform this procedure if the engine will not start or if failure of one or more of the injectors is suspected of causing the engine to run improperly. Use a mechanic's stethoscope for this procedure.

1A. On models using a throttle body injection (TBI) system, remove the flame arrestor and cover as described in Chapter Seven.

1B. On models using a multiport fuel injection (MPI) system, remove the plastic engine cover to access the fuel injectors.

2. Start the engine. Touch the tip of the stethoscope to the body one of the fuel injectors (**Figure 42** or **Figure 47**). Listen for the distinct clicking noise. Touch the stethoscope to the body of each fuel injector on the engine and note the clicking noise. Stop the engine and refer to the following:

  a. If all injectors make the same clicking noise, the injectors are all operating. Low fuel pressure, an ignition malfunction or a mechanical problem in one or more cylinders is the likely cause of the engine malfunction.
  b. If none of the injectors are making a clicking noise, test the electronic fuel injection (EFI) system relay and check for voltage at the injectors as described in this section. An ignition system fault, faulty injector wiring or failure of all injectors is likely if no fault is found with the relay. Replace the ECU or ECA only if a fault is found with the injector voltage, and all other EFI components, ignition system components, wiring and terminals test correctly. Failure of the ECU or ECA is extremely rare.
  c. If one or more of the injectors sound markedly different from the others, check for voltage at the injector as described in this section. If the injector voltage is correct, remove and inspect the spark plugs for an excessively rich or excessively lean condition as described in this chapter. Refer to *Basic Fuel System Troubleshooting*. Replace the suspect injector if the spark plug in which it delivers fuel (MPI models) or the spark plugs on the cylinder bank in which it delivers fuel (TBI models) show an excessively rich or excessively lean condition.

3A. On models using a throttle body injection (TBI) system, install the flame arrestor and cover as described in Chapter Seven.

3B. On models using a multiport injection (MPI) system, install the plastic engine cover.

Injector voltage test

WARNING

The engine might unexpectedly start during testing. Keep all persons, clothing, tools or loose objects away from the drive belts and other moving components on the engine.

59

CAUTION

If testing the ignition system with the boat out of the water, use a flush device to supply the engine with cooling water. Operation of the starter motor results in operation of the seawater pump. The seawater pump is quickly damaged if operated without adequate cooling water.

CAUTION

Do not use a high-capacity test lamp to check the injector voltage. A high-capacity light might draw excessive amperage and damage the injector driver circuits in the engine control unit (ECU).

This procedure checks for voltage at the harness connectors to the injectors and checks for proper switching from the engine control unit (ECU).

Use a 12-volt test lamp for this procedure. Do not use a high-capacity test lamp. The test lamp must consume fewer than 0.3 amps when connected to a 12-volt source. To determine the amperage draw, connect the negative terminal of a digital ammeter (usually integrated into a digital multimeter) to the negative terminal of the battery. Touch the test lamp probe to the positive terminal of the battery. Connect the positive terminal of the ammeter to the cord of the test lamp and read the amperage draw. For accurate test results, the lamp must draw 0.1-0.3 amps.

1. Disconnect the high-tension coil lead from the distributor cap. Connect the coil lead securely to an engine ground to prevent engine starting.

2A. On models using a throttle body injection (TBI) system, remove the flame arrestor and cover as described in Chapter Seven.

2B. On models using a multiport injection (MPI) system, remove the plastic engine cover to access the fuel injectors.

3. Note the wire routing and connection points. Then disconnect the wire harness from each of the fuel injectors (**Figure 42** or **Figure 47**).

60

4. Connect the cord of the test lamp to the engine ground. Touch the test lamp probe to the positive battery cable to verify the test lamp and battery-to-engine ground connection.

5. With the ignition switch in OFF or STOP, touch the test lamp probe to the pink/white or red/white wire on GM models and the purple/black wire on Ford models. The lamp must not illuminate at any of the test ports. If the lamp illuminates, the EFI system relay or relay circuit is faulty. Test the relay as described in this section. If the relay and related circuits test correctly, the pink/white, red/white or purple/black wire in the suspect connector is shorted to a battery power source. Inspect the wiring and correct the fault before proceeding.

6. Turn the ignition switch to ON or RUN. Touch the test lamp probe to the pink/white, red/white or purple/black wire terminal in each of the injector harness connectors. The lamp must illuminate at each of the test ports. If not, the EFI system relay or relay circuit is faulty. Test the relay as described in this section. If the relay and related circuits test correctly, the pink/white, red/white or purple/black wire in the suspect connector is open or shorted to engine ground. Inspect the wiring and correct the fault before proceeding.

7. Touch the test lamp probe to the other wire terminal (not the pink/white, red/white or purple/black) in one of the injector harness connectors. Observe the test lamp while an assistant attempts to start the engine. The test lamp must flash repeatedly while cranking. Repeat this step for the remaining injector harness connector(s). Refer to the following.

   a. If the test lamp flashes at each connector, the injector voltage is correct, and the ECU or ECA is switching the injectors. Proceed to Step 8.
   b. If the test lamp does not flash at any connection point, test the ignition system as described in this chapter. If the ignition system tests correctly, the fault is due to faulty wiring between the ignition system and the ECU or ECA, faulty wiring between the ECU or ECA and the fuel injectors or a faulty ECU or ECA. Check the wiring and terminals for faults and correct as needed. Replace the ECU or ECA only if all other components and wiring on the circuit test correctly. Failure of the ECU or ECA is extremely rare.
   c. If the test lamp flashes at some connection points and not at others, the wiring connecting the suspect injector to the ECU or ECA to the injector is open or shorted to ground, or the ECU or ECA is faulty. A faulty injector driver circuit will affect one injector on TBI models and four injectors on MPI models. Check the wiring and terminals for faults and correct as needed. Replace the ECU or ECA only if all other components and wiring on the circuit test correctly. Failure of the ECU or ECA is extremely rare.

8. Turn the ignition switch to OFF or STOP.

9. Reconnect each of the connectors onto the injectors. Route the wiring to prevent contact with moving or hot engine components.

10A. On models using a TBI system, install the flame arrestor and cover as described in Chapter Seven.

10B. On models using an MPI system, install the plastic engine cover.

DIAGNOSTIC TROUBLE CODES

Ford 5.0L and 5.8L (V-8) Models

The engine control assembly (ECA) continuously monitors input voltage from various sensors. The ECA stores a diagnostic trouble code if voltage from one or more of the sensors exceeds the predetermined limit.

Reading and interpreting the trouble codes on Ford 5.0L and 5.8L models requires expensive tools and equipment backed by factory training. Specifications for the fuel system sensors can vary widely by model type, model year, elevation and production changes. If the cause of an engine malfunction cannot be identified by performing the tests described in this chapter, have the fuel injection system tested at an OMC dealership.

GM V-6 and V-8 Models

The engine control unit (ECU) continuously monitors input voltage from various sensors. The ECU stores a diagnostic trouble code if voltage from one or more of the sensors exceeds the predetermined limit. Unlike the ECA used on Ford 5.0L and 5.8L models, the ECU used on GM V-6 and V-8 models retains the trouble code if the battery is disconnected. The codes can be read and interpreted fairly easily with relatively inexpensive test equipment.

Two available tools for reading the stored codes are the diagnostic code tool (**Figure 59**) and the scan tool (**Figure 60**, typical).

Contact an OMC dealership for information about purchasing either tool.

Reading codes using a diagnostic code tool

Identify trouble codes by reading the flashing light sequence on the diagnostic code tool. Refer to the following example to understand how to identify codes.

A short-duration flash followed by a short-duration pause and then a short-duration flash indicate a 2 for the first digit of the trouble code. See **Figure 61**. The second digit of the trouble code is displayed after a slightly longer pause. In this example (**Figure 61**), the second digit is a 1. The trouble code displayed in this example is code 21. Each code flashes three times. A much longer pause separates the individual trouble codes. The flashing codes are displayed in numerical order. **Figure 61** displays the flashing light sequence for a repeating code 21. To make sure all codes are identified, always record them until they repeat.

1. Place the control in NEUTRAL idle. Turn the ignition switch to OFF or STOP. The ECU will not display codes unless the throttle position sensor reads a closed throttle value.
2. Connect the code tool as follows:
   a. Lift the locking tab and then remove the protective cap from the diagnostic link connector (DLC) plug (**Figure 62**).
   b. Lift the locking tab and carefully plug the tool onto the DLC plug.
   c. Move the tool switch (**Figure 63**) to OFF.
3. Turn the ignition switch to ON or RUN. The code light must illuminate steadily.
   a. If the light does not illuminate, check for a blown fuse, faulty wire connector or faulty throttle position sensor. Temporarily unplug the throttle position sensor connector to check for a fault with the sensor. The sensor is faulty if the light illuminates only with the sensor disconnected. Test the EFI system relay as described in this section if no fault is found with the wiring or throttle position sensor.
   b. If the light is flashing, the white/black wire connecting the DLC plug to the ECU is shorted to ground.
4. Move the tool switch (**Figure 63**) to ON. The light should display code 12. This code indicates the diagnostic system is functioning properly, If not, check for a fault along the wiring connecting the plug to the ECU.
5. Read the flashing light to identify trouble codes as described earlier in this procedure. Record all trouble codes.
6. Move the tool switch (**Figure 63**) to OFF. Turn the ignition switch to OFF or STOP.
7. Refer to **Table 11** to match the code(s) to the fault detected in the system. Refer to the information in this section for the stored trouble codes. Perform the suggested repair or test.
8. Clear the trouble code(s) as described in this section.
9. Remove the diagnostic code tool and install the cap onto the DLC plug (**Figure 62**).

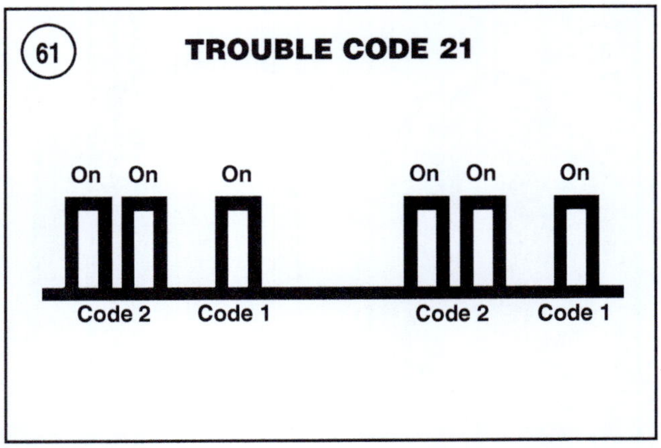

Clearing codes with a diagnostic code tool

The engine must be started to clear the codes.

CAUTION
If testing the ignition system with the boat out of the water, use a flush device to supply the engine with cooling water. Operation of the starter motor results in operation of the seawater pump. The seawater pump is quickly damaged if operated without adequate cooling water.

1. Place the control in NEUTRAL idle. Turn the ignition switch to OFF or STOP.
2. Connect the code tool as follows:
   a. Lift the locking tab. Then remove the protective cap from the diagnostic link connector (DLC) plug (**Figure 62**).
   b. Lift the locking tab and carefully plug the tool onto the DLC plug.
   c. Move the tool switch (**Figure 63**) to ON.
3. Turn the ignition switch to ON or RUN. Do not start the engine at this time.
4. Advance the throttle to the wide open position and then back to idle.
5. Move the tool switch (**Figure 63**) to OFF. Do not turn the ignition switch to OFF or STOP before starting the engine in the next step.

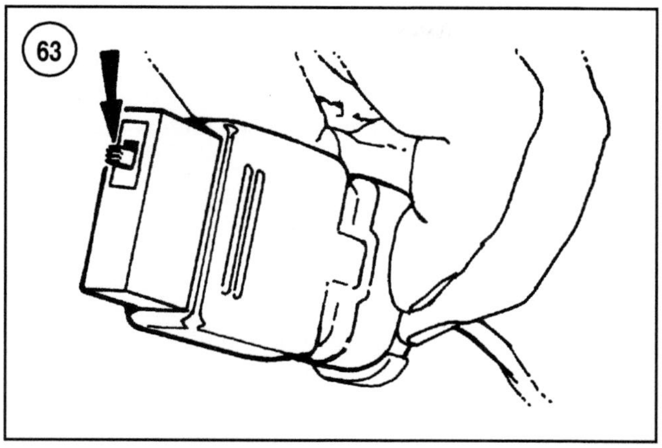

6. Start the engine and run for approximately 15 seconds. Turn the ignition switch to OFF or STOP.
7. Check for stored trouble codes as described in this section. Check for a discharged or faulty battery if the codes do not clear.
8. Remove the diagnostic code tool and install the cap onto the DLC plug (**Figure 62**).

Reading codes with a scan tool

1. Place the control in NEUTRAL idle. Turn the ignition switch to OFF or STOP.
2. Connect the scan tool as follows:
   a. Lift the locking tab. Then remove the protective cap from the diagnostic link connector (DLC) plug (**Figure 62**).
   b. Lift the locking tab and carefully plug the tool onto the DLC plug.
3. Turn the ignition switch to ON. Follow the instruction booklet and the on-screen instructions and retrieve the stored trouble codes.
4. Turn the ignition switch to OFF or STOP.
5. Refer to **Table 11** to match the code(s) to the fault detected in the system. Refer to the information in this section for the stored trouble codes. Perform the suggested repair or test.
6. Clear the trouble code(s) as described in this section.
7. Disconnect the scan tool and install the cap onto the DLC plug (**Figure 62**).

Clearing codes with a scan tool

CAUTION

If testing the ignition system with the boat out of the water, use a flush device to supply the engine with cooling water. Operation of the starter motor results in operation of the seawater pump. The seawater pump is quickly damaged if operated without adequate cooling water.

The codes must be cleared with the engine running.

1. Place the control in NEUTRAL idle. Turn the ignition switch to OFF or STOP.
2. Connect the scan tool as follows:
   a. Lift the locking tab. Then remove the protective cap from the diagnostic link connector (DLC) plug (**Figure 62**).
   b. Lift the locking tab and carefully plug the tool onto the DLC plug.
3. Start the engine and operate at idle speed. Follow the instruction booklet and the on-screen instructions to clear the trouble codes. The codes are cleared while running the engine.
4. Stop the engine. Check for trouble codes as described in this section. Repeat the code clearing procedure if the code(s) do not clear.
5. Disconnect the scan tool and install the cap onto the DLC plug (**Figure 62**).

Code 14 (GM models)

This code sets if the engine control (ECU) unit detects the engine coolant temperature (ECT) sensor voltage is not within the preset limit.

A sensor producing output voltage that is too low will cause the ECU to deliver too much fuel resulting in an excessively rich running condition.

A sensor producing output voltage that is too high will cause the ECU to deliver too little fuel resulting in an excessively lean running condition. Refer to *Basic Fuel System Troubleshooting* in this chapter. If a fault is detected, the ECU will calculate the fuel and timing for an engine operating at approximately 32° F (0° C). The substitute temperature will cause the ECU to increase the engine idle speed above the normal speed (600 rpm) with the engine at normal operating temperature.

If code 14 is stored, test the engine coolant temperature sensor (ECT) resistance as described in this section. If no fault is found with the sensor, check for an open or short circuit in the wiring connecting the sensor to the ECU. Refer to the wiring diagrams at the end of the manual to identify the sensor wiring and connection points. Clear the trouble code as described in this section.

Code 21 (GM models)

This code sets if the engine control unit (ECU) detects a throttle position sensor (TPS) voltage that is above or below the preset limit.

A sensor producing output voltage that is too high will cause the ECU to interpret it as a higher-than-actual throttle opening. Reacting to this, the ECU will deliver excess fuel and open the idle air control motor (IAC) as part of the programmed follower function. This will produce a very high idle speed and an excessively rich running condition. Refer to *Basic Fuel System Troubleshooting* in this chapter.

A sensor producing output voltage that is too low will cause the ECU to interpret it as a lower-than-actual throttle opening. Reacting to this, the ECU will deliver too little fuel and cause an excessively lean condition. Refer to *Basic Fuel System Troubleshooting* in this chapter. This fault will prevent the ECU from opening the IAC as part of the programmed follower function and cause stalling when the throttle returns to idle.

A faulty TPS can also prevent the ECU from detecting and reacting to rapid throttle opening and cause the engine to hesitate during acceleration.

If code 21 is stored, check for a short or open circuit in the wiring connecting the sensor to the ECU. Refer to the wiring diagrams at the end of the manual to identify the sensor wiring and connection points. If no fault is found with the wiring, replace the TPS as described in Chapter Nine. Clear the trouble code as described in this section.

Code 23 (GM models)

This code sets if the engine control unit (ECU) detects that the intake air temperature (IAT) sensor voltage is above or below the preset limit.

A sensor producing output voltage that is too low will cause the ECU to deliver too much fuel for the actual temperature of air entering the engine. The excess fuel will cause an excessively rich condition. Refer to *Basic Fuel System Troubleshooting* in this chapter.

A sensor producing output voltage that is too high will cause the ECU to deliver too little fuel for the actual temperature of air entering the engine. Running the engine with too little fuel will cause an excessively lean condition. See *Basic Fuel System Troubleshooting* in this chapter.

If code 23 is stored, test the intake air temperature sensor resistance as described in this section. If no fault is found with the sensor, check for an open or short circuit in the wiring connecting the sensor to the ECU. Refer to the wiring diagrams at the end of the manual to identify the sensor wiring and connection points. Clear the trouble code as described in this section.

Code 33 (GM models)

This code sets if the engine control unit (ECU) detects that the manifold air pressure (MAP) sensor voltage is above or below the preset limit.

A sensor producing output voltage that is too high will cause the ECU to deliver too much fuel for the actual air pressure in the intake manifold. The excess fuel will cause an excessively rich condition. Refer to *Basic Fuel System Troubleshooting* in this chapter.

A sensor producing output voltage that is too low will cause the ECU to deliver too little fuel for the actual air pressure in the intake manifold. Running the engine with too little fuel will cause an excessively lean condition. Refer to *Basic Fuel System Troubleshooting* in this chapter.

If code 33 is stored, check for an open or short circuit in the wiring connecting the sensor to the ECU. Refer to the wiring diagrams at the end of the manual to identify the sensor wiring and connection points. If no fault is found with the wiring, replace the sensor as described in Chapter Nine. Clear the trouble code as described in this section.

Code 42 (GM models)

This code sets if the engine control unit (ECU) detects a fault along the green/white wire connecting the ECU to the ignition module. The open or short circuit will prevent the ECU from controlling the ignition timing. The ECU will then relinquish control, and the timing advance is controlled by the ignition module. The engine will start and run with noticeably reduced power and efficiency, hard starting and noticeably rougher idle characteristics.

If code 42 is stored, check for an open or short circuit in the green/white wire or faulty terminals. Refer to the wiring diagrams at the end of the manual to identify the wire connections. If no fault is found with the wiring, replace the ignition module as described in Chapter Nine. Clear the trouble code as described in this section.

Code 43 (GM models)

This code sets if the engine control unit (ECU) fails to receive the signal from the knock control module. If code 43 is stored, check for an open or shorted wire connecting the knock sensor(s) to the knock control module and the knock control module to the ECU. If no fault is found with the wiring, replace the knock control module as described in Chapter Nine.

Code 51 (GM models)

The engine control unit (ECU) sets this code if it detects a failure of the calibration circuits. The calibration circuits control the fuel delivery, ignition, engine overspeed limits, idle speed value and other values unique to the engine in which the unit is used. This failure can cause a no-start condition, hard starting, poor performance or poor operating characteristics. If code 51 is stored, clear the trouble code as described in this section. Run or attempt to start the engine. Stop the engine and check for stored trouble codes as described in this section. If code 51 is stored again, replace the ECU as described in Chapter Nine.

Engine Coolant Temperature Sensor (ECT) Resistance Test

A fault with the engine coolant temperature sensor (ECT) can cause poor cold engine operating characteristics, excessive idle speed on a fully warmed engine or improper activation of the overheat warning system.

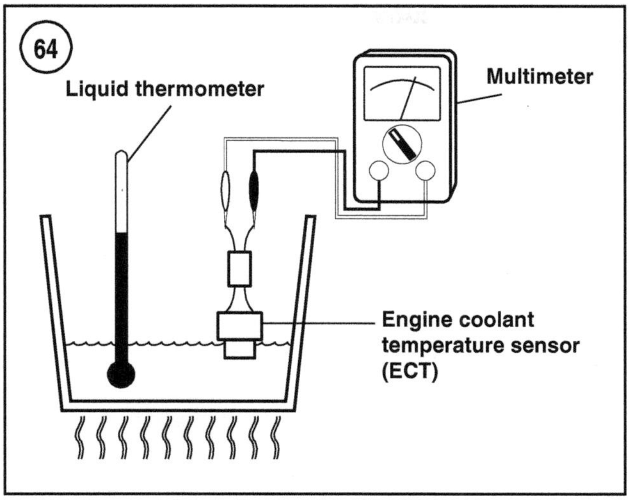

This test requires an accurate multimeter, a liquid thermometer and a container of water that can be heated.

1. Remove the ECT as described in Chapter Nine.
2. Calibrate the meter to the appropriate scale for the resistance specification in **Table 9**. Connect the meter leads between the two sensor terminals. Test lead polarity is not important for this test.
3. Suspend the sensor in a container of water that can be heated (**Figure 64**, typical). Make sure the sensor does not touch the bottom or sides of the container and that the tip of the sensor is completely below the water surface.
4. Add ice or heat the container until the water temperature reaches the temperatures specified in **Table 9**. Record the resistance at each temperature.
5. Compare the resistance readings with the specifications in **Table 9**. A slight variance in resistance (less than 10 percent) is common and does not indicate a faulty sensor. Replace the ECT if the resistance is significantly different from the specification at any of the specified temperatures.
6. Install the ECT as described in Chapter Nine.

Intake Air Temperature Sensor (IAT) Resistance Test

The intake air temperature sensor (IAT) is used only on multiport injection (MPI) models. A faulty intake air temperature sensor can cause poor cold engine operating characteristics and symptoms consistent with an excessively lean or excessively rich operating condition. Refer to *Basic Fuel System Troubleshooting* in this chapter.

This test requires an accurate ohmmeter and a thermometer suitable for accurately measuring air temperature.

1. Remove the IAT as described in Chapter Nine.
2. Calibrate the meter on the appropriate scale for the resistance specification in **Table 9**. Connect the meter leads between the two sensor terminals. Test lead polarity is not important for this test.
3. Wait a few minutes for the sensor to react to the temperature of the air in the work area. Compare the resistance reading on the meter and the surrounding air temperature with the specification in **Table 9**.
4. The resistance at the actual air temperature should be within the range for the nearest temperature listed in **Table 9**. Heat or cool the work area or move to a different area to test the sensor at various temperatures. A slight variance in resistance (less than 10 percent) is common and does not indicate a fault with the sensor. Replace the IAT if the resistance is significantly different from the specification at any of the specified temperatures.
5. Install the intake air temperature sensor as described in Chapter Nine.

Idle Air Control Motor (IAC) Test

CAUTION

If testing the ignition system with the boat out of the water, use a flush device to supply the engine with cooling water. Operation of the starter motor results in operation of the seawater pump. The seawater pump is quickly damaged if operated without adequate cooling water.

A faulty idle air control motor (IAC) can cause stalling at idle, excessive idle speed or stalling during deceleration. A faulty IAC will not cause operational problems at higher engine speeds. A diagnostic code tool (**Figure 59**) or a scan tool (**Figure 60**, typical) is required for this procedure.

1. Start the engine and run at idle speed until the engine reaches normal operating temperature.
2. Stop the engine. Lift the locking tab. Then remove the protective cap from the diagnostic link connector (DLC) plug (**Figure 62**). Lift the locking tab and carefully plug the tool onto the DLC plug. If using a diagnostic code tool, move the tool switch (**Figure 63**) to OFF.
3. Start the engine and raise the speed to approximately 1500 rpm in NEUTRAL. Then use the tool to place the engine in service mode. If using a diagnostic code tool, place the engine in service mode by moving the tool switch (**Figure 63**) to ON. If using a scan tool, follow the instruction booklet and the on-screen instructions to put the engine in service mode.
4. The engine speed must drop by 200 rpm or more while in service mode.
5. If the engine speed does not change as described in Step 4, perform the following:
   a. Stop the engine and disconnect the diagnostic code tool or scan tool.
   b. Remove the IAC as described in Chapter Nine. Clean and inspect the connecting wiring, IAC plunger and plunger opening in the throttle body.
   c. Install the IAC as described in Chapter Nine.
   d. Repeat the test to check for proper IAC operation. Replace the IAC if it fails to perform as specified.
6. Disconnect the code tool or scan tool and install the cap onto the DLC plug (**Figure 62**).

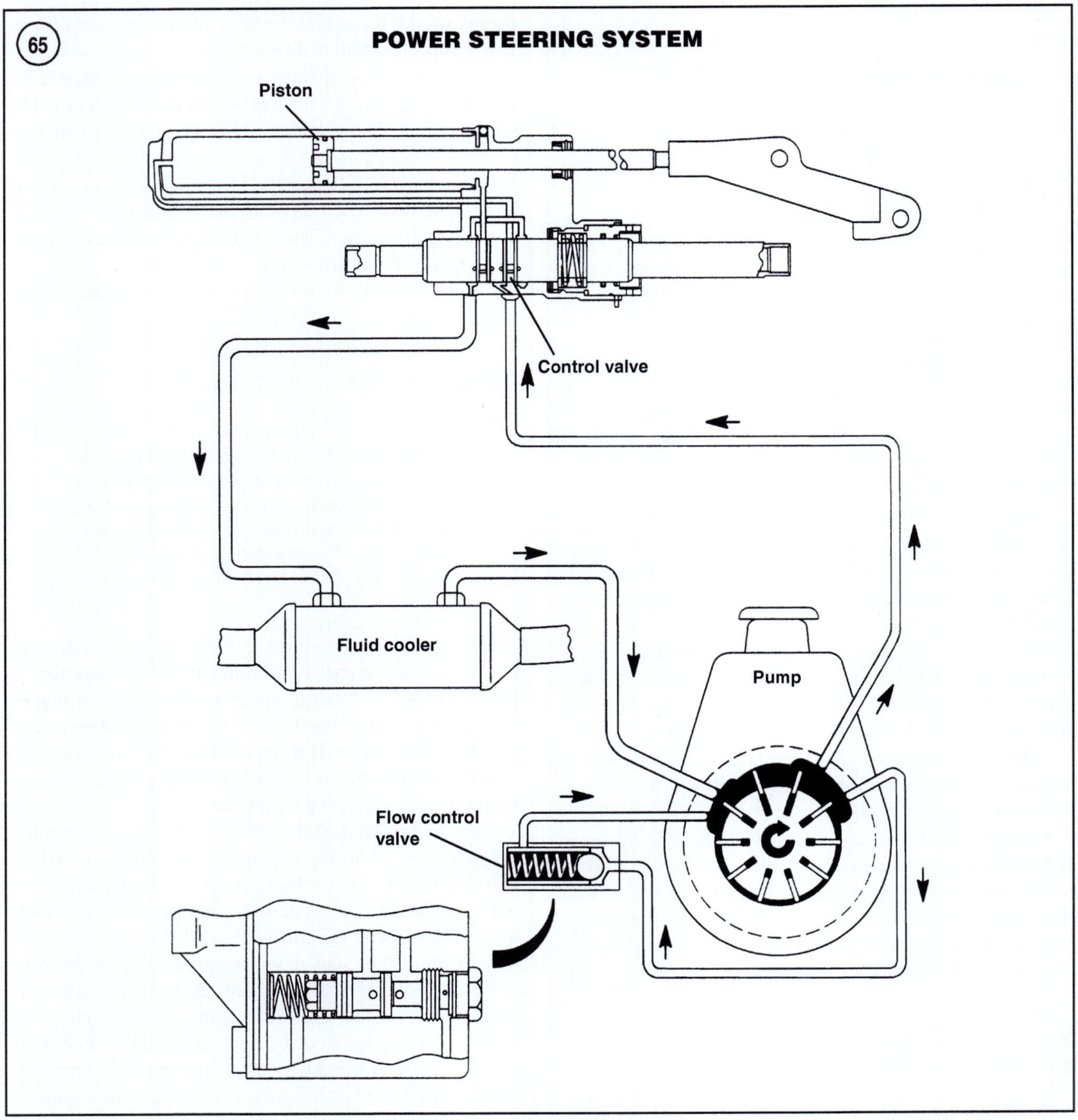

POWER STEERING SYSTEM

This section describes troubleshooting for the cable-actuated power steering system (**Figure 65**). The power steering system is very reliable, and malfunction is usually due to low fluid level or improper installation of the steering cable. At the first sign of trouble, check and correct the fluid level as described in Chapter Three. Check the cable installation as described in the steering cable installation instructions in Chapter Thirteen. Refer to *Steering Cable and Helm*. If no faults are found with the fluid level and cable installation, check the items listed in **Table 5**. Refer to the causes listed next to the symptom that best describes the malfunction. Check the items or perform the suggested repair listed as corrective actions.

Stiff or Binding Cable or Helm

A binding steering cable or helm can lead the operator to believe the power steering system is not operating. The cable must move the control valve for steering assist to occur. Always check for a faulty steering cable or helm before repair or replacement of other steering system components.

1. Temporarily disconnect the steering cable from the actuator and steering lever. Refer to *Steering Cable and Helm* in Chapter Thirteen.

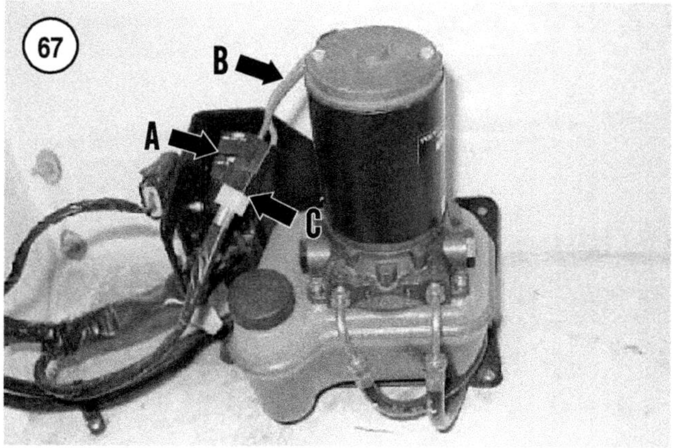

2. Rotate the steering wheel to the full port and starboard directions. The wheel must turn in both directions without excessive effort, binding or roughness. Otherwise, replace the steering cable and helm as described in Chapter Thirteen.
3. Reconnect the steering cable to the actuator and steering lever as described in Chapter Thirteen.

Power Steering Fluid Cooler

A faulty power steering fluid cooler can allow water into the power steering fluid, leak power steering fluid into the cooling water, restrict cooling water flow or restrict power steering fluid flow. If these symptoms occur, remove the fluid cooler as described in Chapter Eight. Refer to *Oil and Power Steering Fluid Cooler Replacement.*

Direct air through the passages to check for and remove restrictions. If suspected of leaking, have the cooler tested and repaired at a radiator repair shop.

POWER TRIM SYSTEM

This section describes troubleshooting procedures for the power trim system. The first section describes operation of the power trim electrical system followed by test procedures for the power trim electrical components. The second section describes the power trim hydraulic system followed by troubleshooting procedures. Refer to **Table 6** for a list of items to check if the trim system is malfunctioning. Refer to the causes listed next to the symptom that best describes the malfunction. Perform the test or adjustment suggested under corrective action. Other useful information and tips are described with the specific test instructions.

Electrical System Troubleshooting

A faulty power trim electrical system is almost always caused by a faulty trim switch or related wiring or failure of one or more of the relays.

First, refer to the appropriate system operation description in the following sections. Many times, the system is operating as designed, and the operator is unaware of the special features designed into the system. The system description will usually indicate by symptom which components are suspect. If faulty operation is evident, check the trim switch and trim limit control circuits. If the trim switch and trim limit control circuits test correctly, test the trim relay operation. If both subsystems test correctly, but the trim pump is not operating, the trim pump is faulty.

Electrical System Operation

Two different types of pumps are used on the models covered in this manual. Earlier models use the pump shown in **Figure 66**. This pump uses a top-mounted fluid reservoir and a circuit board attached to the side of the pump. The circuit board contains the up relay, down relay and circuit breaker. The board must be replaced as an assembly if any of the integral components fail.

Later models use a pump with a bottom-mounted fluid reservoir (**Figure 67**) with relays (A) and fuses attached to the pump mounting bracket. The individual relays can be replaced on the later-style pump. The electric control system and troubleshooting procedures for the earlier and later-style pumps are very similar.

SX single-propeller models (with earlier-style trim pump)

Refer to **Figure 68**.

The up and down relays integrated into the circuit board (**Figure 68**) provide the battery positive and ground circuits to operate the trim motor. The relays are controlled by the control or dash-mounted trim switch. A separate 50-amp circuit breaker provides battery voltage to the circuit board. A circuit breaker integrated into the circuit board provides battery power to the dash- or remote control-mounted trim switch. A dedicated circuit connects the trim pump and relay mounting bracket to the ground stud on the flywheel housing. The engine ground connects to the negative battery terminal.

When the trim system is not operating, both relays within the circuit board are deactivated and ground both the blue and green trim pump wires (**Figure 68**). The trim pump

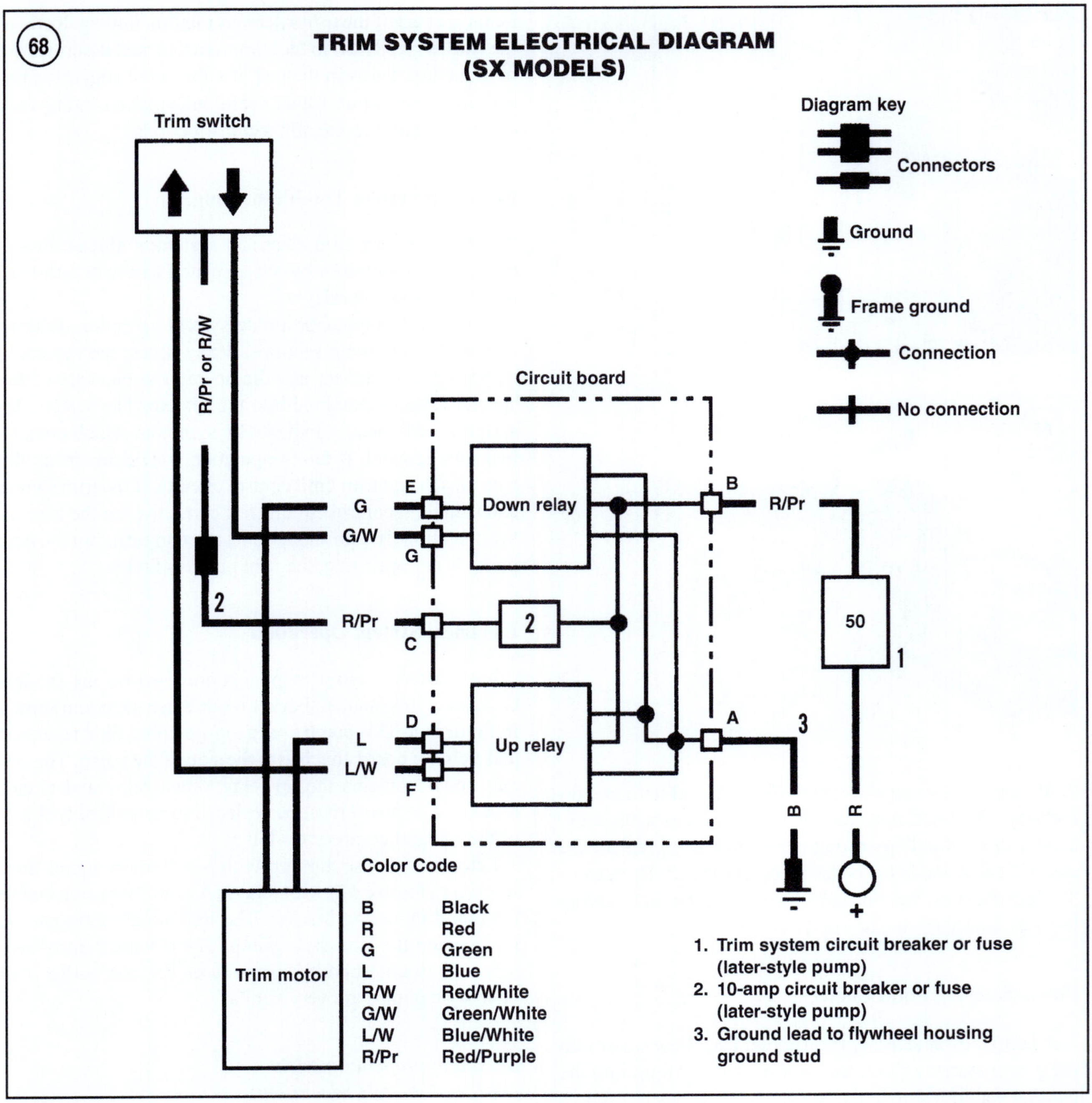

cannot operate because it has no connection to battery voltage.

When the trim switch toggles up, battery voltage is applied to the blue/white trim switch wire leading to the up trim terminal on the circuit board. This activates the up trim relay in the circuit board and causes it to apply battery voltage to the blue trim pump wire. Because the down relay is deactivated, it connects the green trim motor wire via the ground circuit to the battery ground. The trim motor turns the hydraulic pump in the direction needed to move the drive unit upward.

When the trim switch toggles down, battery voltage is applied to the green/white trim switch wire leading to the down trim terminal on the circuit board. This activates the down trim relay and causes it to apply battery voltage to the green trim pump wire. Because the up relay is deactivated, it connects the blue trim motor wire via the ground circuit to the battery ground. The trim motor turns the hydraulic pump in the direction needed to move the drive unit downward.

SX single-propeller models (with later-style pump)

The two trim pump-mounted relays (A, **Figure 67**) provide the battery positive and ground circuits to operate the trim motor. The relays are controlled by the control- or dash-mounted trim switch. Fused circuits provide battery power to the trim switch and both trim relays using the red/purple, red/white or blue/red wire. A dedicated circuit connects the trim pump and relay mounting bracket to engine ground. The engine ground connects to the negative battery terminal.

When the trim system is not operating, both trim relays are deactivated and ground both the blue and green trim pump wires (B, **Figure 67**). The trim pump cannot operate because it has no connection to battery voltage.

When the trim switch toggles up, battery voltage is applied to the blue/white trim switch wire leading to the up trim relay. This activates the up trim relay, causing it to apply battery voltage to the blue trim pump wire. Because the down relay is deactivated, it connects the green trim motor wire via the ground circuit to the battery ground. The trim motor turns the hydraulic pump in the direction needed to move the drive unit upward.

When the trim switch toggles down, battery voltage is applied to the green/white trim switch wire leading to the down trim relay. This activates the down trim relay and causes it to apply battery voltage to the green trim pump wire. Because the up relay is deactivated, it connects the blue trim motor wire via the ground circuit to the battery ground. The trim motor turns the hydraulic pump in the direction needed to move the drive unit downward.

DP-S twin-propeller models (with earlier-style pump)

Refer to **Figure 69**.

The up and down relays integrated into the circuit board (3, **Figure 69**) provide the battery positive and ground circuits to operate the trim motor. The relays are controlled by the control- or dash-mounted trim switch. A separate 50-amp circuit breaker (1, **Figure 69**) provides battery voltage to the circuit board. A circuit breaker integrated into the circuit board provides battery power to the dash- or remote control-mounted trim switch. A dedicated circuit connects the trim pump and relay mounting bracket to the ground stud on the flywheel housing. The engine ground connects to the negative battery terminal.

When the trim system is not operating, both relays within the circuit board are deactivated and ground both the blue and green trim pump wires. The trim pump cannot operate because it has no connection to battery voltage.

When the trim switch toggles up, battery voltage is applied to the blue/white trim switch wire leading to the trim limit control module (9, **Figure 69**) and then the up trim terminal on the circuit board. This activates the up trim relay in the circuit board and causes it to apply battery voltage to the blue trim pump wire. Because the down relay is deactivated, it connects the green trim motor wire via the ground circuit to the battery ground. The trim motor turns the hydraulic pump in the direction needed to move the drive unit upward.

When the trim switch toggles down, battery voltage is applied to the green/white trim switch wire leading to the trim limit control module and then the down trim terminal on the circuit board. This activates the down trim relay and causes it to apply battery voltage to the green trim pump wire. Because the up relay is deactivated, it connects the blue trim motor wire via the ground circuit to the battery ground. The trim motor turns the hydraulic pump in the direction needed to move the drive unit downward.

Using the resistance reading from the trim position sensor, the module limits the trim down to -2° and the trim up to 5° by internally opening the trim switch-to-relay circuits. To trim up or down beyond the limits, the operator must depress the override switch (16, **Figure 69**) on the control *and* the trim up or down switch. The override switch provides voltage to a module terminal that disables the limit function. This allows the additional trim up needed for shallow water operation or loading the boat onto a trailer or additional down trim as needed.

DP-S twin-propeller models (with later-style pump)

The two trim pump-mounted relays (A, **Figure 67**) provide the battery positive and ground circuits to operate the trim motor. The relays are controlled by the control- or dash-mounted trim switch. Fused circuits provide battery power to the trim switch and both trim relays using the red/purple, red/white or blue/red wire. A dedicated circuit connects the trim pump and relay mounting bracket to the engine ground. The engine ground connects to the negative battery terminal.

On DP-S models, the trim switch wiring and relay wiring connect to the transom-mounted trim limit control module (**Figure 70**).

When the trim system is not operating, both trim relays are deactivated and ground both the blue and green trim pump wires (B, **Figure 67**). The trim pump cannot operate because it has no connection to battery positive voltage.

When the trim switch toggles up, battery voltage is applied to the blue/white trim switch wire leading to the trim limit control module and then the up trim relay. This activates the up trim relay and causes it to apply battery voltage to the blue trim pump wire. Because the down relay is deactivated, it connects the green trim motor wire via the ground circuit to the battery ground. The trim motor turns the hydraulic pump in the direction needed to move the drive unit upward.

When the trim switch toggles down, battery voltage is applied to the green/white trim switch wire leading to the trim limit control module and then the down trim relay. This activates the down trim relay and causes it to apply battery voltage to the green trim pump wire. Because the up relay is deactivated, it connects the blue trim motor wire via the ground circuit to the battery ground. The trim motor turns

69

TRIM SYSTEM ELECTRICAL DIAGRAM (DP-S MODELS)

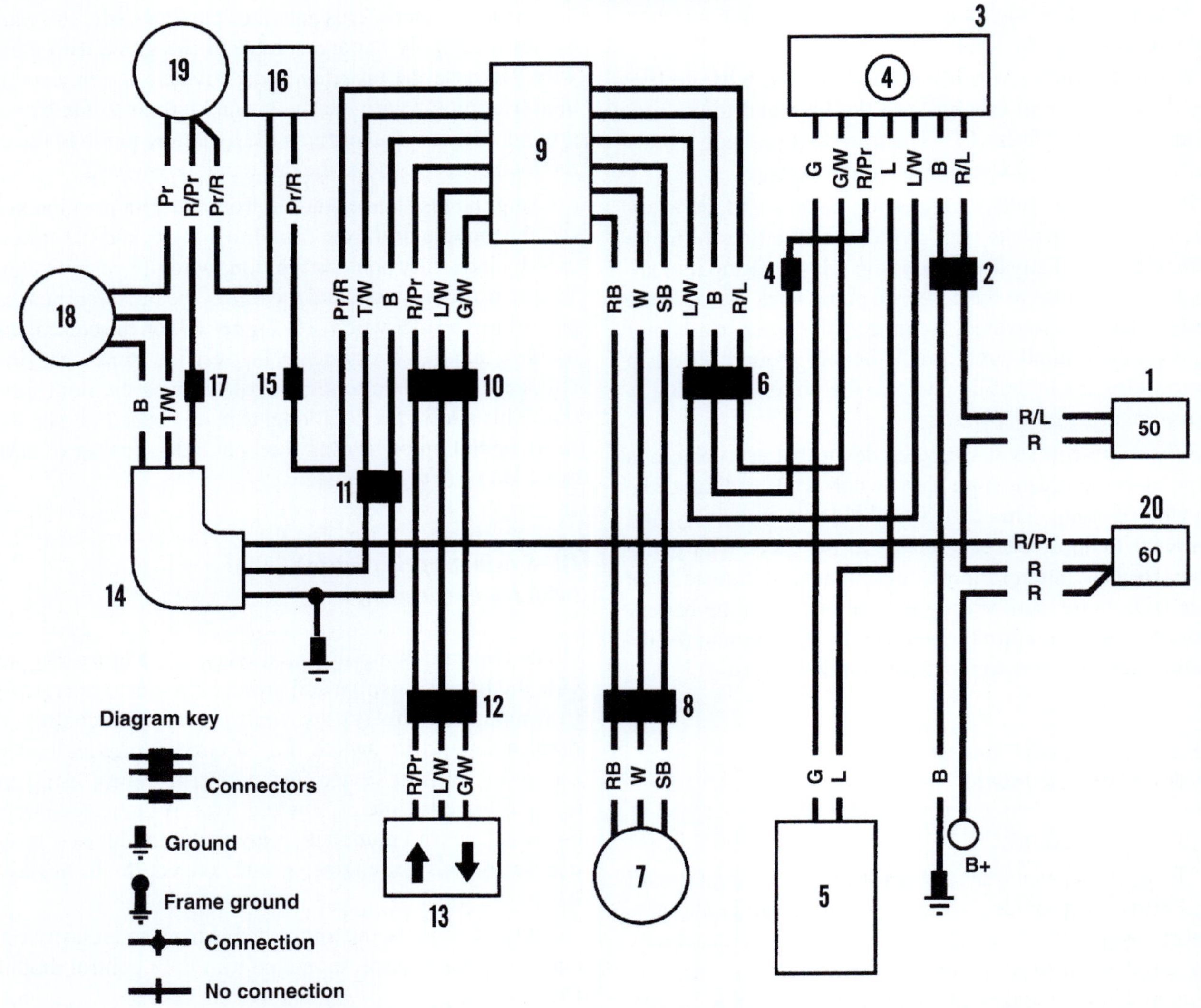

Color Code

B	Black
W	White
R	Red
G	Green
L	Blue
Pr	Purple
B/W	Black/White
R/W	Red/White
G/W	Green/White
L/W	Blue/White
T/W	Tan/White
R/L	Red/Blue
R/Pr	Red/Purple
Pr/R	Purple/Red
RB	Ribbed Black
SB	Smooth Black

1. 50-amp circuit breaker
2. Connector (battery power and ground leads)
3. Circuit board (integrated relays and circuit breaker)
4. 10-amp circuit breaker
5. Trim motor
6. Connector
7. Trim position sender
8. Connector
9. Trim limit control module
10. Connector
11. Connector
12. Connector
13. Trim switch
14. Main engine harness connector
15. Connector
16. Override switch
17. 20-amp fuse
18. Trim gauge
19. Ignition key switch
20. 60-amp circuit breaker

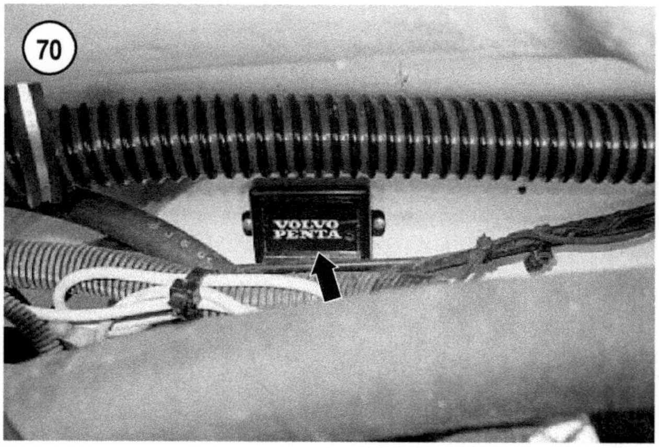

the hydraulic pump in the direction needed to move the drive unit downward.

Using the resistance reading from the trim position sensor, the module limits the trim down to -2° and the trim up to 5° by internally opening the trim switch-to-relay circuits. To trim up or down beyond the limits, the operator must depress the override switch on the control *and* the trim up or down switch. The override switch provides voltage to a module terminal that disables the limit function. This allows the additional trim up needed for shallow water operation or loading the boat onto a trailer or additional down trim as needed.

Trim Switch and Trim Limit Control Circuit Test

SX single-propeller models (with earlier-style pump)

These models do not use a trim limit control circuit. A 12-volt test lamp is required for this procedure.

Refer to **Figure 68**.

1. Locate the trim switch harness leading to the circuit board on the side of the trim pump (**Figure 66**).
2. Connect the cord of the test lamp to the positive terminal of the battery. Touch the test lamp probe to the black (ground) wire terminal on the circuit board (A, **Figure 68**) to verify the light and the ground connection for the trim pump and relays. The lamp must illuminate. If not, repair the ground connection or repair the faulty lamp.
3. Connect the cord of the test lamp to the black (ground) terminal on the circuit board. Touch the test lamp probe to the terminal on the circuit board for the red/purple wire (C, **Figure 68**) connecting to the trim switch harness. The lamp must illuminate. If not, check all fuses, circuit breakers and wiring providing battery voltage to the trim pump. Do not continue testing until battery voltage is restored to this wire.
4. Touch the test lamp probe to the blue/white wire terminal (F, **Figure 68**) on the circuit board. Observe the lamp while an assistant repeatedly toggles the trim switch to the UP trim position and back to OFF. The lamp must illuminate each time the switch reaches UP and must not illuminate with the switch in OFF. Replace the trim switch or repair faulty trim switch wiring if the switch does not perform as specified.
5. Touch the test lamp probe to the green/white wire terminal (G, **Figure 68**) on the circuit board. Observe the lamp while an assistant repeatedly toggles the trim switch into the DOWN trim position and back to OFF. The lamp must illuminate each time the switch reaches DOWN and must not illuminate with the switch in OFF. Replace the trim switch or repair faulty trim switch wiring if the switch does not perform as specified.
6. If no fault is found with the trim switch circuit, but the trim motor does not operate in one or both directions, the fault is with the circuit board or trim motor. Test the circuit board to determine whether the trim motor is faulty.

2

SX single-propeller models (with later-style pump)

These models do not use a trim limit control circuit. A 12-volt test lamp is required for this procedure.

Refer to **Figure 68**.

1. Locate the trim switch harness leading to the trim pump. Peel back the insulating tape to expose the green/white, blue/white and red/purple wires.
2. Connect the cord of the test lamp to the positive terminal of the battery. Touch the test lamp probe to the trim pump ground terminal (A, **Figure 68**) to verify the light and the ground connection for the trim pump and relays. The lamp must illuminate. If not, repair the ground connection or repair the faulty lamp.
3. Connect the cord of the test lamp to the trim pump ground terminal. Using the test lamp, probe the red/purple wire in the trim switch harness. The lamp must illuminate. If not, check all fuses, circuit breakers and wiring providing battery voltage to the trim pump. Do not continue with testing until battery voltage is restored to this wire.
4. Probe the blue/white wire in the harness. Observe the lamp while an assistant repeatedly toggles the trim switch to the UP trim position and back to OFF. The lamp must illuminate each time the switch reaches UP and must not illuminate with the switch in OFF. Replace the trim switch or repair faulty trim switch wiring if the switch does not perform as specified.
5. Probe the green/white wire in the harness. Observe the lamp while an assistant repeatedly toggles the trim switch into the DOWN trim position and back to OFF. The lamp must illuminate each time the switch reaches DOWN and must not illuminate with the switch in OFF. Replace the trim switch or repair faulty trim switch wiring if the switch does not perform as specified.
6. Repair the wire insulation damaged by the test lamp probe.
7. If no fault is found with the trim switch circuit, but the trim motor does not operate in one or both directions, the

fault is with the trim relays or trim motor. Test the relays to determine whether the trim motor is faulty.

DP-S twin-propeller models (with earlier-style pump)

These models use a trim limit control module to limit the trim system operation in both the up and down trim directions. A 12-volt test lamp is required for this procedure.

Refer to **Figure 69**.

1. Locate the trim limit control module (**Figure 70**) on the boat transom or other area in the engine compartment. Disconnect the trim switch and the wire harness leading to the trim pump relays from the module. Connect the trim switch harness to the relay harness to bypass the limit module.
2. Locate the trim switch harness leading to the circuit board on the side of the trim pump (**Figure 66**).
3. Connect the cord of the test lamp to the positive terminal of the battery. Touch the test lamp probe to the black (ground) wire terminal on the circuit board to verify the light and the ground connection for the trim pump and relays. The lamp must illuminate. If not, repair the ground connection or repair the faulty light.
4. Connect the cord of the test lamp to the black (ground) terminal on the circuit board. Touch the test lamp probe to the circuit board wire terminal for the red/purple wire connecting to the trim switch harness. The lamp must illuminate. If not, check all fuses, circuit breakers and wiring providing battery voltage to the trim pump. Do not continue with testing until battery voltage is restored to this wire.
5. Touch the test lamp probe to the blue/white wire terminal on the circuit board. Observe the lamp while an assistant repeatedly toggles the trim switch into the UP trim position and back to OFF. The lamp must illuminate each time the switch reaches UP and must not illuminate with the switch in OFF. Replace the trim switch or repair faulty trim switch wiring if the switch does not perform as specified.
6. Touch the test lamp probe to the green/white wire terminal on the circuit board. Observe the lamp while an assistant repeatedly toggles the trim switch into the DOWN trim position and back to OFF. The lamp must illuminate each time the switch reaches DOWN and must not illuminate with the switch in OFF. Replace the trim switch or repair faulty trim switch wiring if the switch does not perform as specified.
7. Reconnect the trim switch and trim relay harness connectors to the trim limit control module.
8. Place the drive unit in the full DOWN position. Turn the ignition switch to ON or RUN. Repeat Step 5 and refer to the following.
   a. If the lamp illuminates as specified, the switch and trim limit control circuit are providing current to the up trim relay. The up relay, down relay or trim motor is faulty if the trim motor does not operate.
   b. If the lamp does not illuminate, depress the override switch (16, **Figure 69**) and repeat Step 5. If the lamp illuminates as specified only with the override switch depressed, the trim limit control module is improperly adjusted, the override switch is faulty, the trim position sender is faulty, or the trim position sender is improperly adjusted. Test the override switch. If the override switch tests correctly, adjust the trim position sender and trim limit control module as described in Chapter Five. Repeat Step 8. If the lamp does not illuminate as specified, test the trim position sender. Replace the trim limit control module if no fault is found with the trim position sender.
9. Place the drive unit in the fully UP position. Turn the ignition switch to ON or RUN. Repeat Step 6 and refer to the following.
   a. If the lamp illuminates as specified, the switch and trim limit control circuit are providing current to the down trim relay. The up relay, down relay or trim motor is faulty if the trim motor does not operate.
   b. If the lamp does not illuminate, depress the override switch and repeat Step 6. If the lamp illuminates as specified only with the override switch depressed, the trim limit control module is improperly adjusted, the override switch is faulty, the trim position sender is faulty, or the trim position sender is improperly adjusted. Test the override switch. If the override switch tests correctly, adjust the trim position sender and trim limit control module as described in Chapter Five. Repeat Step 8. If the light does not illuminate as specified, test the trim position sender. Replace the trim limit control module if no fault is found with the trim position sender.
10. Turn the ignition switch to OFF or STOP.
11. If no fault is found with the trim switch and trim limit control circuit, but the trim motor does not operate in one or both direction, the fault is with the circuit board or trim motor. Test the circuit to determine whether the trim motor is faulty.

DP-S twin-propeller models (with later-style pump)

These models use a trim limit control module to limit the trim system operation in both the up and down trim operations. A 12-volt test lamp is required for this procedure.

Refer to **Figure 69**.

1. Locate the trim limit control module (**Figure 70**) on the boat transom or other area in the engine compartment. Disconnect the trim switch and the wire harness leading to the trim pump relays from the module. Connect the trim switch harness onto the relay harness to bypass the limit module.
2. Locate the trim switch harness leading to the trim pump. Peel back the insulating tape to expose the green/white, blue/white and red/purple wires.

3. Connect the cord of the test lamp to the positive terminal of the battery. Touch the test lamp probe to the trim pump ground terminal to verify the light and the ground connection for the trim pump and relays. The light must illuminate. If not, repair the ground connection or repair the faulty light.
4. Connect the cord of the test lamp to the trim pump ground terminal. Using the test lamp, probe the red/purple wire in the trim switch harness. The lamp must illuminate. If not, check all fuses, circuit breakers and wiring providing battery voltage to the trim pump. Do not continue with testing until battery voltage is restored to this wire.
5. Probe the blue/white wire in the harness. Observe the lamp while an assistant repeatedly toggles the trim switch into the UP trim position and back to OFF. The lamp must illuminate each time the switch reaches UP and must not illuminate with the switch in OFF. Replace the trim switch or repair faulty trim switch wiring if the switch does not perform as specified.
6. Probe the green/white wire in the harness. Observe the lamp while an assistant repeatedly toggles the trim switch into the DOWN trim position and back to OFF. The lamp must illuminate each time the switch reaches DOWN and must not illuminate with the switch in OFF. Replace the trim switch or repair faulty trim switch wiring if the switch does not perform as specified.
7. Reconnect the trim switch and trim relay harness connectors to the trim limit control module.
8. Place the drive unit in the full DOWN position. Turn the ignition switch to ON or RUN. Repeat Step 5 and refer to the following.
   a. If the lamp illuminates as specified, the switch and trim limit control circuit are providing voltage to the up trim relay. The up relay, down relay or trim motor is faulty if the trim motor does not operate.
   b. If the lamp does not illuminate, depress the override switch and repeat Step 5. If the lamp illuminates as specified only with the override switch depressed, the trim limit control module is improperly adjusted, the override switch is faulty, the trim position sender is faulty, or the trim position sender is improperly adjusted. Test the override switch. If the override switch tests correctly, adjust the trim position sender and trim limit control module as described in Chapter Five. Repeat Step 8. If the light does not illuminate as specified, test the trim position sender. Replace the trim limit control module if no fault is found with the trim position sender.
9. Place the drive unit in the fully UP position. Turn the ignition switch to ON or RUN. Repeat Step 6 and refer to the following.
   a. If the lamp illuminates as specified, the switch and trim limit control circuit is providing voltage to the down trim relay. The up relay, down relay or trim motor is faulty if the trim motor does not operate.
   b. If the lamp does not illuminate, depress the override switch and repeat Step 6. If the lamp illuminates as specified only with the override switch depressed, the trim limit control module is improperly adjusted, the override switch is faulty, the trim position sender is faulty, or the trim position sender is improperly adjusted. Test the override switch. If the override switch tests correctly, adjust the trim position sender and trim limit control module as described in Chapter Five. Repeat Step 8. If the light does not illuminate as specified, test the trim position sender. Replace the trim limit control module if no fault is found with the trim position sender.
10. Turn the ignition switch to OFF or STOP.
11. Repair the wire insulation damaged by the test lamp probe.
12. If no fault is found with the trim switch and trim limit control circuit, but the trim motor does not operate in one or both directions, the fault is with the trim relays or trim motor. Test the relays to determine whether the trim motor is faulty.

2

Trim switch test

A faulty trim switch can cause the trim pump not to operate in one or both directions. A multimeter is required for this procedure. Refer to **Figure 68** or **Figure 69**.

1. Disconnect the trim switch harness from the instrument harness. The connector should be located near the remote control- or dash-mounted trim switch.
2. Touch one of the ohmmeter test leads to the terminal for the red/purple or red/white trim switch wire. Test lead polarity is not important for this procedure.
3. Place the trim switch in OFF. Touch the other test lead to the terminals for the blue/white and then green/white trim switch wires. The meter must indicate no continuity at each connection point. If otherwise, the trim switch is shorted and must be replaced.
4. Toggle the trim switch to the UP trim position. Touch the other test lead to the terminals for the blue/white and then green/white trim switch wires. The meter must indicate continuity when touching the terminal for the blue/white wire and no continuity when touching the green/white wire. If otherwise, the trim switch has failed open or is shorted internally and must be replaced.
5. Toggle the trim switch to the DOWN trim position. Touch the other test lead to the terminals for the blue/white and then green/white trim switch wires. The meter must indicate no continuity when touching the terminal for the blue/white wire and continuity when touching the green/white wire. If otherwise, the trim switch has failed open or is shorted internally and must be replaced.
6. Connect the trim switch connector to the instrument harness connector.

Override switch test (DP-S [twin-propeller] models)

A faulty override switch can prevent the operator from tilting the drive unit up for shallow water operation or when loading the boat on the trailer. A fault can also prevent the limit system from operating. A 12-volt test lamp is required for this procedure. Refer to **Figure 69**.

1. Disconnect the purple/red wire connector (15, **Figure 69**) from the trim limit module harness.
2. Connect the cord of the test lamp to an engine ground or good ground terminal for one of the instruments. Touch the test lamp to the red wire or *Batt* terminal on the ignition switch to verify the light and battery voltage is applied to the ignition switch. Check for and correct faulty fuses, circuit breakers or wiring to restore battery voltage before proceeding.
3. Turn the ignition switch to ON or RUN. Touch the test lamp probe to the purple/red wire terminal (override switch) at the ignition switch. The lamp must illuminate. If not, test the ignition switch as described in this chapter. Restore correct ignition switch operation before proceeding.
4. Touch the test lamp probe to the terminal in the purple/red wire connector (15, **Figure 69**). Observe the test lamp while repeatedly depressing and releasing the tilt/override button. The test lamp must illuminate each time the button is depressed and extinguish each time the button is released. If otherwise, the switch is shorted or has failed open and must be replaced.
5. Turn the ignition switch to OFF. Reconnect the purple/red wire connector (15, **Figure 69**) to the trim limit module harness.

Testing the trim position sender

A fault with the trim position sender can prevent proper trim gauge or trim limit system operation. A digital ohmmeter is required for this procedure.

1. Disconnect the trim position sender wire connector from the instrument or trim system wiring (**Figure 71**).
2. Connect one of the test leads to the *A* terminal in the sender wire connector. Connect the other test lead to the *C* terminal.
3. Observe the meter reading while an assistant operates the trim to the full UP and full DOWN positions. The meter must show steadily increasing and decreasing resistance readings. Adjust the sender as described in Chapter Five if erratic or no change is evident. Repeat the test. Replace the sender if correct operation is not restored after adjustment.
4. Reconnect the trim position sender wire to the instrument or trim system wiring.

Trim pump circuit board test

Faulty circuit board operation can cause the trim pump not to operate in one or both directions (up or down). Re-

place the trim pump if the trim pump does not operate, but the circuit board tests correctly. A 12-volt test lamp is required for this test.

Refer to **Figure 68**.

1. The trim switch and limit control circuits must operate properly for this procedure. Before testing the circuit board operation, test the trim switch and trim limit control circuits as described in this section. The trim switch and limit control circuits must operate properly for this procedure.
2. Connect the cord of the test lamp to the positive terminal of the battery. Touch the test lamp probe to the black (ground) wire terminal on the circuit board (A, **Figure 68**) to verify the lamp and the ground connection for the circuit board. The lamp must illuminate. If not, repair the ground connection or repair the faulty lamp.
3. Disconnect the large-gauge blue and green trim pump wires from the circuit board terminals (D and E, **Figure 68**).
4. Connect the cord of the test lamp to the positive terminal of the battery. Touch the test lamp probe to the circuit board terminals for the blue (D, **Figure 68**) and then green (E) wire terminals. The lamp must illuminate when touching each terminal. If not, the relay within the circuit board connected to the suspect wire has failed open, is stuck in the energized position or has a faulty ground connection. Replace the circuit board to restore trim system operation.
5. Connect the cord of the test lamp the black circuit board ground terminal. Touch the test lamp probe to the circuit board terminal (B, **Figure 68**) for the red/purple wire. The lamp must illuminate. If not, check the fuses, circuit breakers and wiring along the circuit supplying battery voltage to the circuit board. Correct the fault to restore battery power to the circuit before proceeding.
6. Touch the test lamp probe to the circuit board terminal for the blue trim pump wire. Observe the lamp while an assistant repeatedly toggles the trim switch to the UP trim position and back to OFF. If the unit is a DP-S (twin-propeller) model, activate the tilt/override switch to prevent the trim limit system from interrupting circuit board operation. The lamp must illuminate each time the switch reaches the UP trim position and must not illuminate

with the switch in OFF. If otherwise, replace the circuit board.

7. Touch the test lamp probe to the circuit board terminal for the green trim pump wire. Observe the lamp while an assistant repeatedly toggles the trim switch into the DOWN trim position and back to OFF. If the unit is a DP-S (twin-propeller) model, activate the tilt/override switch to prevent the trim limit system from interrupting circuit board operation. The lamp must illuminate each time the switch reaches the DOWN trim position and must not illuminate with the switch in OFF. If otherwise, replace the circuit board.

8. Replace the trim motor if it fails to operate but the circuit board tests as specified. If the trim pump operates in one direction, it is capable of operating in the other.

9. Connect the large-gauge blue and green trim pump wires to the circuit board terminals (D and E, **Figure 68**).

Trim pump relay test

Faulty trim pump relay operation can cause the trim pump not to operate in one or both directions (up or down). Replace the trim pump if the trim pump does not operate, but the trim relays test correctly. A 12-volt test lamp is required for this test.

1. The trim switch and limit control circuits must operate properly for this procedure. Before testing the relay operation, test the trim switch and trim limit control circuits as described in this section. The trim switch and limit control circuits must operate properly for this procedure.

2. Connect the cord of the test lamp to the battery positive terminal. Touch the test probe to the relay ground terminal to verify the lamp and verify the ground connection for the relays and trim pump.

3. Disconnect the trim motor wire connector (C, **Figure 67**) from the relay harness. Touch the test lamp probe to the connector terminals for the blue and then green wire in the relay harness connector. Do not inadvertently touch the probe to the trim motor wire terminals. The lamp must illuminate when touching each terminal. If not, the relay connecting onto the suspect wire has failed open, is stuck in the energized position or has a faulty ground connection. Replace the suspect relay if no fault is found with the ground connection.

4. Connect the cord of the test lamp to the trim pump and relay ground terminal. Probe the heavy-gauge red/purple wire leading into each relay harness connector. The lamp must illuminate. If not, check the fuses, circuit breakers and wiring along the circuit supplying battery voltage to the relays. Correct the fault before proceeding.

5. Touch the test lamp probe to the relay harness terminal for the blue wire. Observe the lamp while an assistant repeatedly toggles the trim switch into the UP trim position and back to OFF. If the unit is a DP-S (twin-propeller) model, activate the tilt/override switch to prevent the trim limit system from interrupting relay operation. The lamp must illuminate each time the switch reaches the UP trim position and must not illuminate with the switch in OFF. If otherwise, replace the up trim relay.

6. Touch the test lamp probe to the relay harness terminal for the green wire. Observe the lamp while an assistant repeatedly toggles the trim switch into the DOWN trim position and back to OFF. If the unit is a DP-S (twin-propeller) model, activate the tilt/override switch to prevent the trim limit system from interrupting relay operation. The lamp must illuminate each time the switch reaches the DOWN position and must not illuminate with the switch in OFF. If otherwise, replace the down trim relay.

7. Replace the trim motor if it fails to operate, but the trim relays test as specified. If the trim pump operates in one direction, it is capable of operating in the other.

Power Trim Hydraulic System Troubleshooting

The power trim hydraulic system is very robust, and failure is relatively rare. A fault is almost always caused by low fluid level or air in the system due to leaking hoses, fitting or seals.

Hydraulic system testing using pressure gauges requires special training, equipment and experience working with hydraulic systems. The procedures described in this section will isolate a hydraulic system fault in almost all cases. If the fault is still present after performing all tests in this section, have the system pressure tested at an OMC dealership or other qualified repair facility. The hydraulic system is relatively simple; the major components include the trim pump, trim cylinder(s), and connecting hoses, lines and fittings. Read *Power trim hydraulic system operation* in this section before beginning the troubleshooting procedure. Many times, the system is operating as designed, and the operator is unaware of the operating features of the system. The description might indicate which components are suspect for the given symptom(s).

If a hydraulic fault is present, first check the trim fluid level as described in Chapter Three. If the fluid level is low, check for fluid leaks at all fittings and hoses. The trim fluid does not simply evaporate. If the fluid level is correct, refer to **Table 6** for a list of items to check. Refer to the causes listed next to the symptom that best describes the malfunction. Perform the check, test or adjustment suggested as a corrective action. Other useful information and tips are described with the specific test instructions.

Power trim hydraulic system operation

Refer to **Figure 72**.

The SX and DP-S drive units used on the models covered in this manual are equipped with two trim cylinders. The hydraulic pump (**Figure 72**) is driven by the trim motor and controlled by the trim switch and trim limit control system.

Operating the trim system in the up direction causes the trim motor to rotate the hydraulic pump gears (9, **Figure 72**) in a direction that moves the fluid in the down pressure

72

TRIM PUMP SCHEMATIC (TYPICAL)

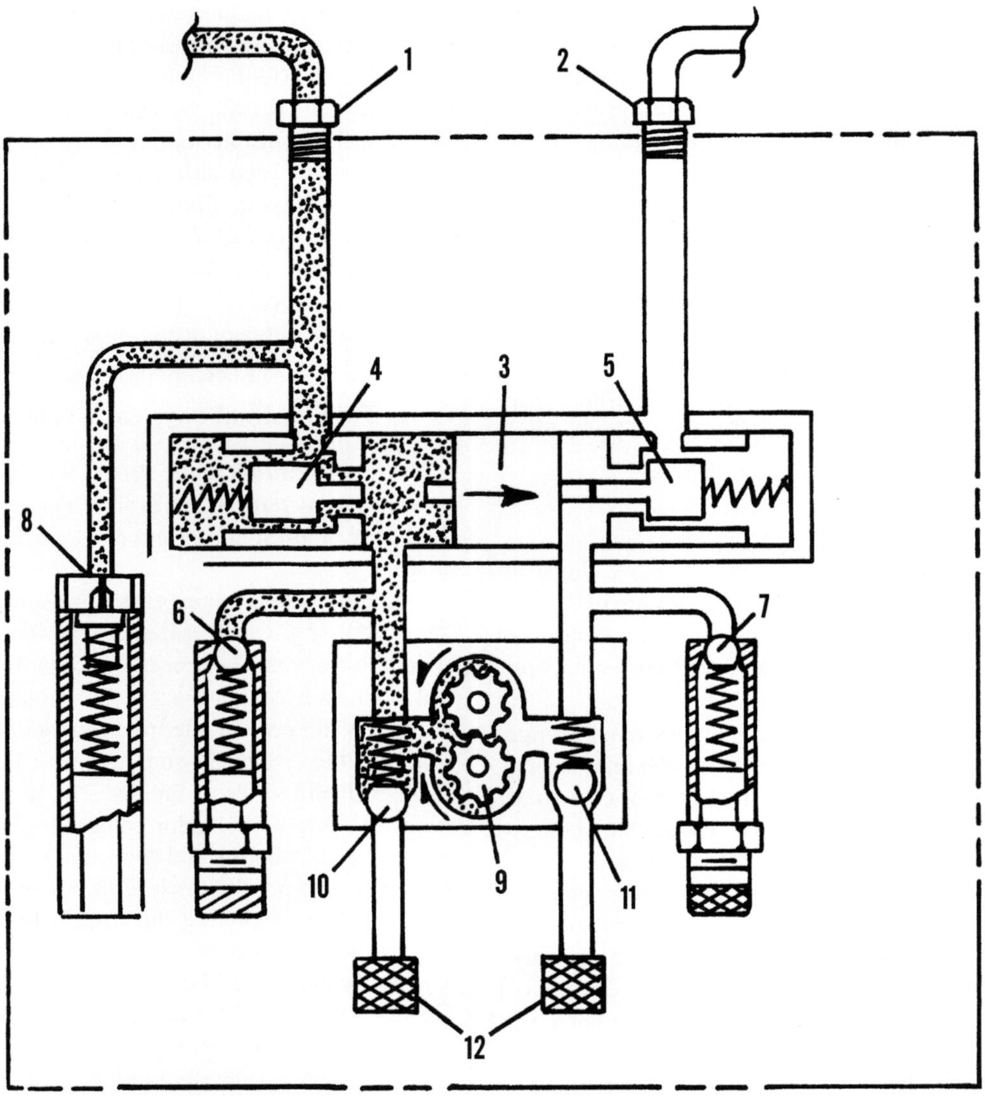

1. Up pressure line connection
2. Down pressure line connection
3. Shuttle piston
4. Up pressure check valve
5. Down pressure check valve
6. Up pressure relief valve
7. Down pressure relief valve
8. Thermal relief valve (not used on all models)
9. Pump gears
10. Pump inlet check ball
11. Pump outlet check ball
12. Inlet screen

side of the pump to the up pressure side of the pump. This causes the pump to draw fluid from the reservoir through the open check ball (11, **Figure 72**). Fluid moving to the up pressure side of the pump increases in pressure and closes the inlet check ball (10, **Figure 72**). This prevents fluid from simply returning to the reservoir. With further rotation, the pressure will rise and open the up pressure check valve (4, **Figure 72**). The pressure applied to the shuttle piston (3, **Figure 72**) causes it to slide in the bore and open the down pressure check valve (5). This allows the fluid to flow into the up side of the trim cylinder(s). The pressure developed in the cylinder is high enough to offset the weight of the drive and forward propeller thrust, and the cylinder rams extend and move the drive unit out or up. As the rams extend, the fluid in the down side of the cylinder returns through the down pressure line and enters the pump through the open down pressure check valve. If the rams fully extend, and the pump continues to rotate, the pressure will build to the point of opening the up pressure relief valve (6, **Figure 72**). This allows fluid to return to the reservoir and limits the up pressure to the predetermined value.

When the pump stops rotating, the pressure stops building in the pump, and the spring closes both check valves (4 and 5, **Figure 72**). This prevents fluid from returning to the pump from both the up and down sides of the cylinder(s). The closed up pressure check valve hydraulically locks the up side of the cylinders to prevent the ram from retracting from the weight of the drive unit and the load of the propeller thrust. The down pressure check valve hydraulically locks the down side of the cylinder(s) to prevent the drive from moving out or up when in REVERSE or during deceleration.

Operating the trim system in the down direction causes the trim motor to rotate the hydraulic pump gears (9, **Figure 72**) in a direction that moves the fluid in the up pressure side of the pump to the down pressure side of the pump. Fluid moving to the up pressure side of the pump increases in pressure and closes the check ball (11, **Figure 72**). This prevents fluid from simply returning to the reservoir. With further rotation, the pressure will raise and open the down pressure check valve (5, **Figure 72**). The pressure applied to the shuttle piston (3, **Figure 72**) causes it to slide in the bore and open the up pressure check valve (4).This allows the fluid to flow into the down side of the trim cylinder(s). The pressure developed in the cylinders is high enough to retract the cylinder ram and move the drive unit in or downward. As the rams retract, the fluid in the up side of the cylinder returns through the up pressure line and enters the pump through the open up pressure check valve. Because the cylinder volume on the up side of the cylinder(s) is greater than the down side, more fluid returns to the pump than is flowing out. Because of this, the fluid level in the reservoir rises while trimming down and lowers while trimming up. If the ram fully retracts, and the pump continues to rotate, the pressure will build to the point of opening the down pressure relief valve (7, **Figure 72**). This allows fluid to return to the reservoir and limits the down pressure to the predetermined value. Because the down side of the cylinder is not subjected to the load produced by forward propeller thrust, the down pressure relief valve opens at a much lower pressure.

If the drive unit is raised to the full up position, and the pump continues to operate, the up side of the cylinder is subjected to the maximum pressure determined by the up pressure relief valve setting. If the cylinders are then exposed to sunlight, the heating will dramatically increase the pressure in the cylinders and up pressure hoses/lines. The up pressure relief valve cannot relieve the excess pressure because of the closed up pressure check valve (4, **Figure 72**). This excess pressure is also applied to the up pressure check valve. Unless the pressure is reduced, the considerably lower down pressure is unable to open the up pressure check valve. This prevents fluid from returning to the up pressure line, and the drive cannot be trimmed down. To prevent this condition, the thermal relief valve (8, **Figure 72**) opens at a preset pressure and allows fluid in the cavity to return to the reservoir. This drops the pressure enough to allow the down pressure to open the up pressure check valve. Some models use a pump that does not incorporate a thermal relief valve.

If the drive unit should strike an underwater object, special valves incorporated into the cylinder pistons open and allow fluid to flow from the down side of the cylinder to the up side. This feature provides a shock-absorbing feature to help minimize damage to the hydraulic system and other components.

Some earlier models use a pump equipped with a manual release valve. The valve threads into the side of the pump housing and is retained into the bore with an internal snap ring. When opened, the valve essentially connects the up and down pressure cavities. This equalizes the pressure and allows manual tilting of the drive unit. Open the valve by fitting a screwdriver into the slot and rotating the valve counterclockwise. Close the valve by rotating clockwise until fully seated.

Check for an internal cylinder leak

An internal cylinder leak will cause the drive unit to drop down from the tilt position, tuck under or move out or up under propeller thrust or move up or out during deceleration. In extreme cases, the trim system may not be able to move the drive. This test checks for fluid flowing from a disconnected down pressure hose to check for fluid bypassing the cylinder piston check valves or O-rings.

1. Place the drive unit in the full UP position. If the system is unable to move the drive unit, use an overhead lift for assistance. On models so equipped, use the override or tilt switch to tilt the unit.
2. Support the drive with an overhead lift or other suitable means. For accurate test results, the drive must be in the full UP position and not move down.
3. Place a suitable container under the trim hose fittings to capture spilled fluid. Disconnect the down pressure hose from one of the trim cylinders. Refer to *Trim Cylinders* in Chapter Twelve to identify the down pressure hoses and corresponding cylinder fittings. It is not important which hose is disconnected. Direct the hose into the container to capture spilled fluid.
4. Observe the up pressure hoses, down pressure hose fitting and corresponding opening in the cylinder while an assistant operates the trim pump in the UP trim direction. On models so equipped, use the override or tilt switch to tilt the unit. The up pressure hoses should straighten slightly, indicating the pump is building pressure, and no fluid should exit the disconnected down pressure hose or corresponding opening in the cylinder. Refer to the following:
   a. If the up pressure hoses straighten as specified, and no fluid flows from hose or cylinder, the pump is pro-

73

STANDARD (OPEN) COOLING SYSTEM (TYPICAL FOUR-CYINDER MODEL)

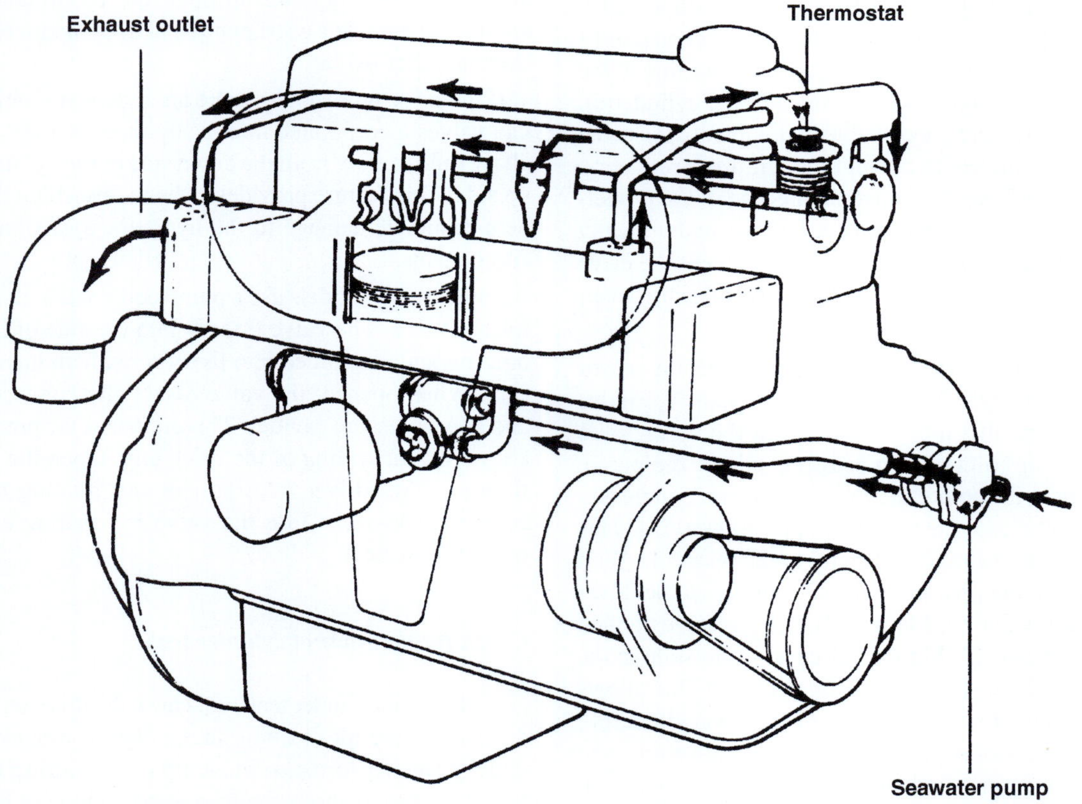

ducing up pressure, and the cylinders are not leaking internally.

b. If the up pressure hoses do not straighten as specified, the pump is probably not developing adequate up pressure for testing purposes. Replace or repair the trim pump as described in Chapter Twelve. Replace the up pressure relief valve during the repair. Repeat the test.
c. If fluid is flowing from the opening in cylinder, the trim cylinder is leaking internally. Repair or replace the cylinder as described in Chapter Twelve.
d. If fluid is flowing from the down pressure hose, the opposite trim cylinder is leaking internally. Repair or replace the cylinder as described in Chapter Twelve.

5. Reconnect the down pressure hose to the trim cylinder and tighten as described in Chapter Twelve. Refer to *Trim Cylinder.*

6. Remove the overhead support. Fill and bleed the system as described in Chapter Twelve.

COOLING SYSTEM TROUBLESHOOTING

Malfunction of the cooling system results in overheating or overcooling.

Overheating is almost always caused by insufficient water flowing through the engine cooling passages or heat exchanger (models with a closed cooling system), or the water fails to absorb heat from the engine. Overheating causes activation of the warning system, unusual odors or the presence of steam from the exhaust. Poor idle quality or stalling at idle might also occur with overheating.

If overheating is suspected, first refer to the cooling system operation description in the following section. Many times, the system is operating as designed, and the operator is unaware of the special features designed into the system. The system description will usually indicate by symptom which components are suspect.

Most overheating problems are directly related to excessive wear or failure of water pump components. Inspect and correct any faults found in the water pump prior to inspecting any other cooling system components. Before beginning the troubleshooting procedure, verify the engine temperature as described in this section. The dash-mounted gauge, engine temperature sender or wiring might be faulty and falsely indicating overheating. If overheating is verified, refer to Chapter Eight for water pump inspection and repair instructions.

If no fault is found with the water pump, refer to **Table 7** for a list of items to check. Refer to the causes listed next to

(74) **OPTIONAL CLOSED COOLING SYSTEM (TYPICAL V-8 MODEL)**

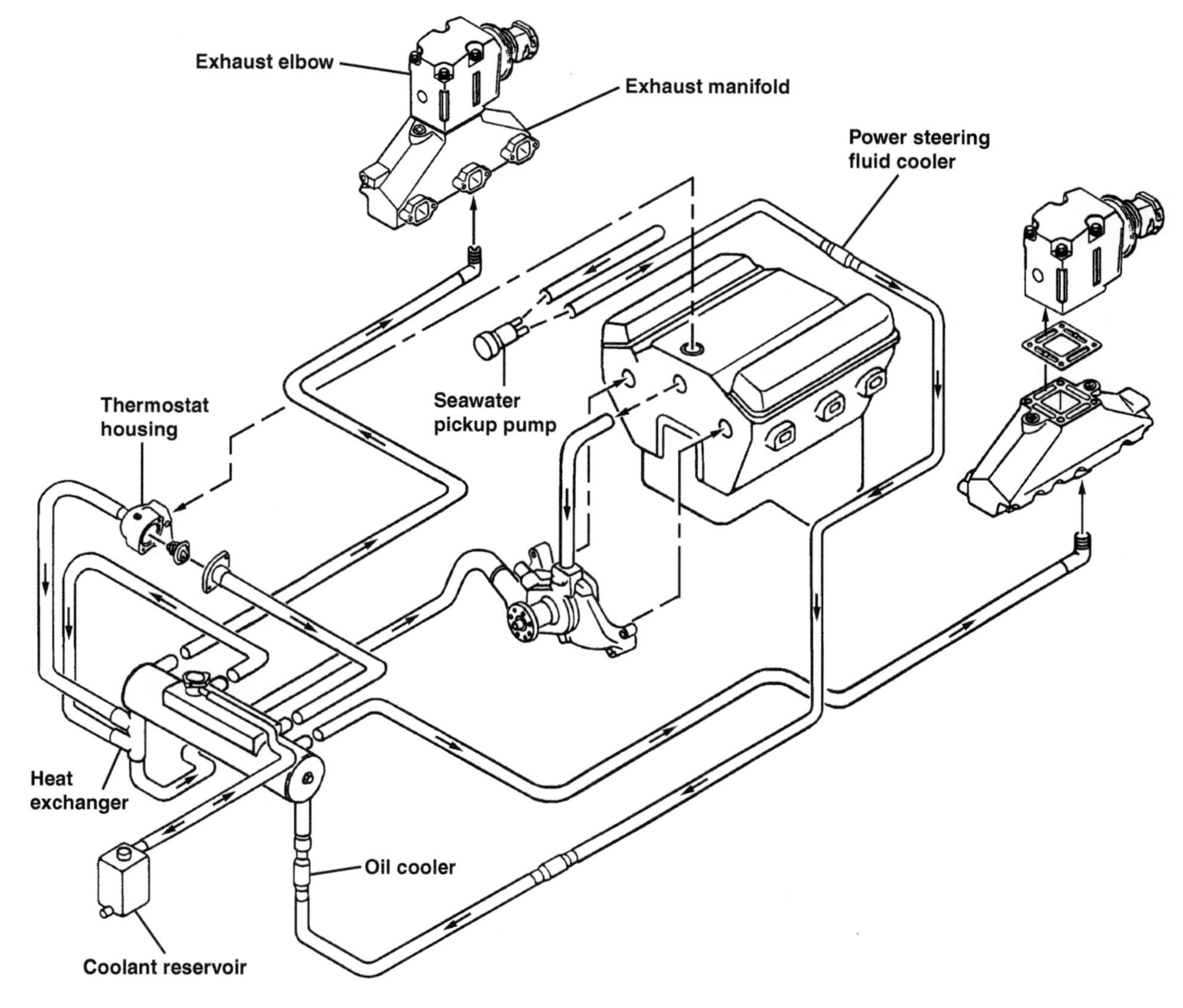

the symptom that best describes the malfunction. Perform the check, test or adjustment suggested under corrective action. Other useful information and tips are described with the specific test instructions.

Overcooling is almost always caused by failure of the thermostat. First, verify the engine temperature as described in this section. If the engine is operating at a temperature that is too low, test the thermostat as described in this section.

Cooling System Operation

On models equipped with a standard (open) cooling system, water in which the boat is being operated is picked up at the drive unit by the seawater pump and circulated throughout the engine to absorb heat. The heated water is transferred to the exhaust manifold(s) where it is expelled overboard with the exhaust gasses.

Some engines are equipped with a closed cooling system that is divided into two separate subsystems. One system is a closed system of coolant (antifreeze and water mixture) circulating in the engine cooling passages and absorbing the engine heat. Heat absorbed by the coolant is then transferred (exchanged) to the seawater in the heat exchanger. Water in which the boat is being operated is picked up at the drive unit by the seawater pump and transferred through the passages in the heat exchanger, where it absorbs heat from the coolant. The heated seawater from the exchanger is transferred to the exhaust manifold(s) where it is expelled overboard with the exhaust gasses. **Figure 73** shows a typical standard (open) cooling system. **Figure 74** shows a typical closed cooling system. The location of the pumps and path of flow may differ slightly from those shown in the illustrations.

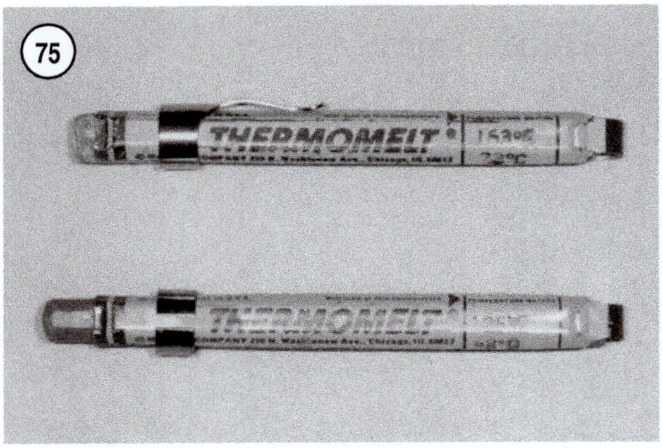

75

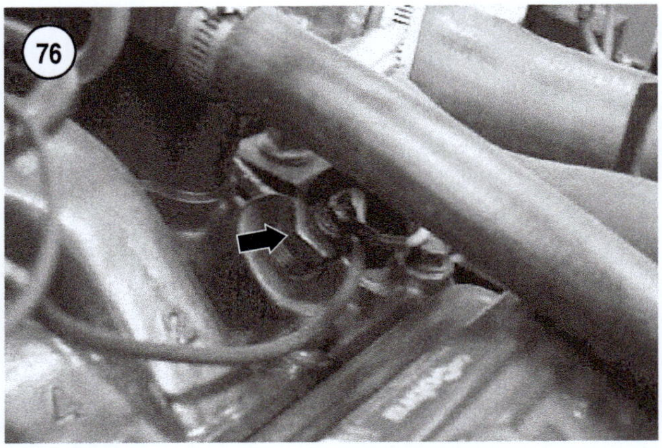
76

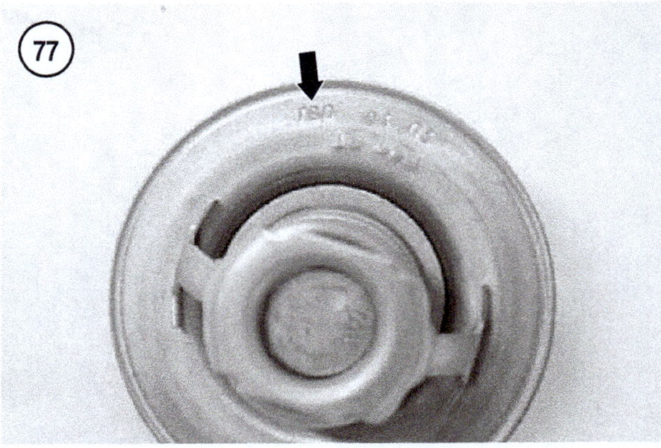
77

78

Verifying the Engine Temperature

Always verify the actual temperature of the engine using Thermomelt sticks or a digital pyrometer before testing other cooling system components. Thermomelt sticks (**Figure 75**) resemble crayons and are designed to melt at a specific temperature. The melting temperatures are listed on the sides of the sticks or the labels. Purchase the Thermomelt sticks or digital pyrometer from a tool supplier.

Troubleshooting an overheating or overcooling problem using a flush/test attachment is difficult if not impossible. Water supplied by the flushing adapter tends to mask problems with the cooling system. Perform this test with the boat in the water under actual operating conditions.

If using Thermomelt sticks, hold the sticks against the brass body of the temperature sender (**Figure 76**), overheat switch or engine temperature sensor (EFI models). Refer to Chapter Nine to locate these components. Try to check the temperature immediately after or during the suspected overheat condition. Hold different temperature sticks to the power head to determine the temperature range the engine is reaching.

If using a pyrometer, attach the sender to or aim the pyrometer against the brass body of the temperature sender, overheat switch or engine temperature sensor (EFI models). Refer to Chapter Nine to locate these components. Monitor the engine temperature as the engine warms up.

Stop the engine if the temperature exceeds 195° F (91° C). Normal operating temperature is 140-180° F (60-82° C). Overheating is evident if higher-than-normal temperature is noted. Test the temperature sender or switch if an alarm or gauge indicates overheating, but the Thermomelt sticks indicate normal operating temperature.

Compare the actual engine temperature with the temperature on the dash instruments. A faulty sender, gauge or wiring is evident if the readings are substantially different.

Overcooling is occurring if the engine fails to reach the normal operating temperature. Due to variations in the temperature of the water in which the engine is operating, it can take as little as 5 minutes or as long as 20 minutes of operating at idle speed to reach normal operating temperature. If operating at low speed only in very cold water, the engine might not reach normal temperature. This occurs because the very cool water flowing through the thermostat bypass is able to absorb more heat than the engine produces at lower operating temperature. A faulty gauge, temperature sender, temperature sensor or related wiring is evident if the gauge or fuel injection system indicates overheating or normal operating temperature, yet the engine is actually overcooling.

79

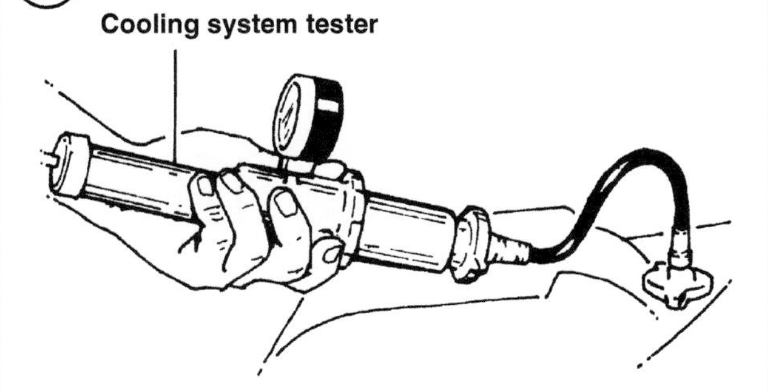

80

81

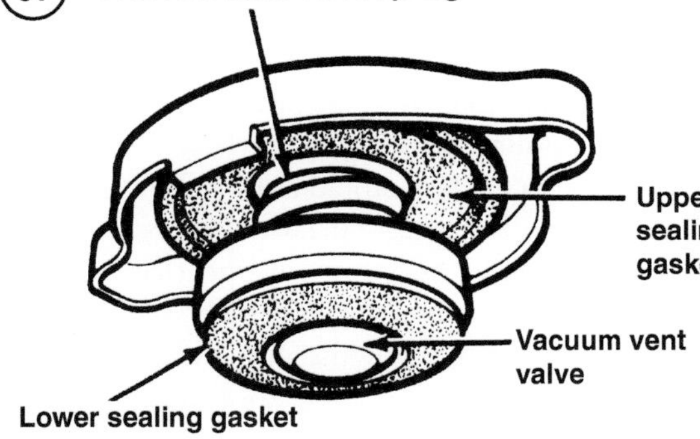

82

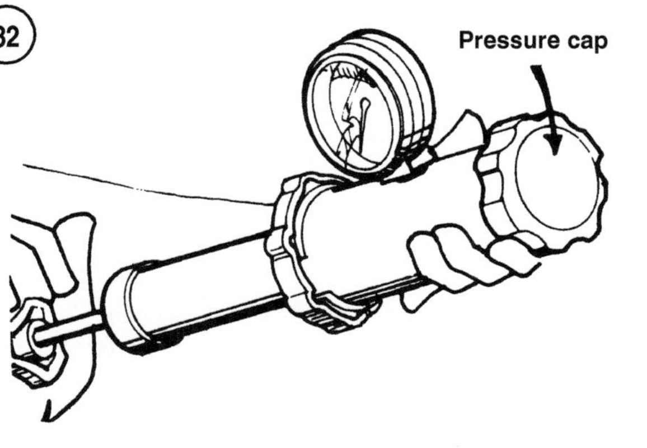

Thermostat Test

Test the thermostat if the engine overheats or is running too cool. A thermometer, piece of string and a container of water that can be heated are required.

1. Remove the thermostat as described in Chapter Eight. Note the temperature value stamped on the thermostat (**Figure 77**). Suspend the thermostat into the container of water with the string tied to the thermostat as shown in **Figure 78**.
2. Place the liquid thermometer in the container and begin heating the water. Observe the temperature of the water and the thermostat while heating the water.
3. The thermostat must start opening within a few degrees of the number stamped on the thermostat (**Figure 77**). If not, discontinue testing and replace the thermostat.
4. Heat the water to 25° F (14° C) above the number stamped on the thermostat. The thermostat must be fully opened at this temperature. If not, replace the thermostat.
5. Discontinue the testing and install the thermostat as described in Chapter Eight.

Closed Cooling System Pressure Test

If the coolant reservoir in a closed cooling system requires frequent filling, it probably has a leak. Small leaks might not be easy to locate, because the hot coolant evaporates as quickly as it leaks out and prevents the formation of telltale rusty or grayish-white staining. A pressure test of the coolant side of the closed cooling system will usually pinpoint the leak. This procedure tests the pressure cap for leaking and proper unseating pressure and tests the closed cooling system for leaking. A cooling system pressure tester (**Figure 79**) is required for this procedure. Purchase the tester from a tool supplier.

1. Allow the engine to cool completely. Never remove the pressure cap from a warm or hot engine.
2. Remove the pressure cap from the heat exchanger or integrated reservoir (**Figure 80**). Wash the cap in clean water to remove debris and deposits. Inspect the cap for cut, cracked or deteriorated seals and gaskets (**Figure 81**, typical). Make sure the locking tabs on the cap are not bent or damaged. Replace the cap if these or other defects are evident.
3. Attach the cap to the pressure tester (**Figure 82**). Dip the pressure cap into clean water.
4. Observe the cap and gauge while pumping the tester to apply pressure against the cap. Note the pressure in which bubbles indicate the pressure cap is unseating. Small bubbling does not indicate unseating. Large bubbles coupled with a rapid pressure drop indicates unseating. The cap should unseat at 13-15 psi (90-103 kPa). Replace the cap if unseating occurs at a higher or lower pressure.
5. Remove the cap from the water. Pump the gauge to apply 14 psi (97 kPa) of pressure against the cap. Wait approximately 30 seconds and check the gauge. Replace the cap if the pressure has dropped to less than 11 psi (76 kPa).

6. Remove the cap from the pressure tester.
7. Inspect the filler neck and sealing surfaces (**Figure 83**) for nicks, distortion, and dirty or contaminated surfaces. Wipe the surfaces clean. Make sure the locking cams are not bent or damaged.
8. Check and correct the coolant level as described in Chapter Three.
9. Connect the pressure tester to the filler neck (**Figure 79**). Pump the tester to apply 17 psi (116 kPa) to the system. Wait 2 minutes and then check the gauge. Any noticeable pressure drop indicates leaking. If leaking, check all hoses, drain plugs, core plugs, the thermostat housing, intake manifold and recirculation pump for leaking coolant. Listen for a hissing while the system is under pressure. If hissing is noted, apply soap suds to the pressure tester-to-filler neck connection to check for leakage.
10. If no external coolant leaks are evident, the leakage might be a leaking heat exchanger or a cooling passage leak in the engine. Remove the heat exchanger and have it pressure-tested at a reputable radiator shop. If the heat exchanger is not leaking, and the coolant requires frequent filling, the leak is evidently in the engine cooling system. Typically, this is due to a blown head gasket, cracked or porous cylinder head, leaking intake manifold gasket, cracked or porous intake manifold or cracked or porous cylinder block. Check the oil for a milky appearance and inspect the spark plugs for crystalline deposits to verify internal leakage. Internal leakage will almost always allow coolant into the oil pan or combustion chamber(s).
11. Remove the tester.
12. If removed, reinstall the heat exchanger.
13. Correct any external leaks or internal leaks.
14. Check and correct the coolant level as described in Chapter Three.

WARNING SYSTEM

CAUTION
The warning system alerts the operator if a cooling or lubrication system failure occurs. Continued operation with the warning system activated can lead to serious and expensive engine damage.

If the warning horn is sounding, first check the oil pressure and engine temperature displayed on the dash instruments. Then stop the engine. Refer to the warning system operation description in this section. Many times, the system is operating correctly, and the operator is unaware of the system features. The system description and gauge readings will usually indicate by symptom which system has failed. If the gauge indicates overheating, troubleshoot the cooling system as described in this chapter. If the gauge indicates low oil pressure, test the oil pressure as described in this chapter. Refer to *Engine*.

If the warning horn is sounding, and the cooling and lubrication systems are operating properly, or if the gauges indicate

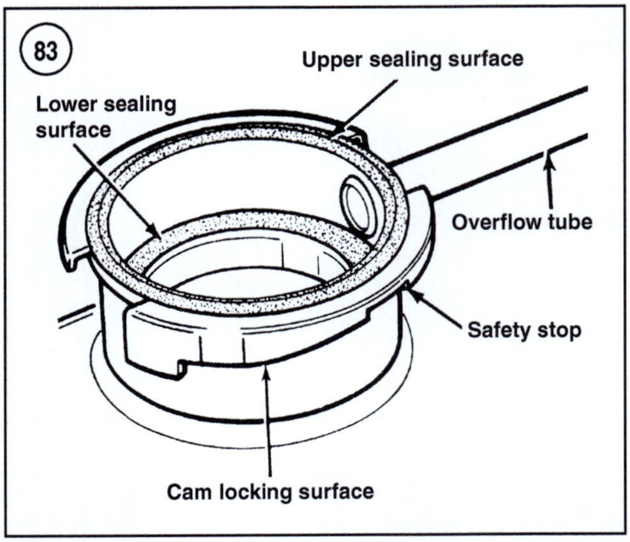

overheating or low oil pressure, and the warning horn is not sounding, refer to **Table 8** for a list of items to check. Refer to the causes listed next to the symptom that best describes the malfunction. Perform the check, test or adjustment suggested as a corrective action. Other useful information and tips are described with the specific test instructions.

Warning System Operation (Carburetor-Equipped Models)

The warning system components used on carburetor equipped models include the oil pressure switch, engine overheat switch and the warning horn.

Battery voltage is applied to the positive terminal of the dash-mounted warning horn via the purple wire connecting to the ignition circuit. The tan/blue warning horn wire connects to the negative terminal of the warning horn and to the oil pressure switch and engine overheat switch. The switches control operation of the warning horn by completing or removing the ground circuit for the warning horn.

The oil pressure switch (A, **Figure 84**) threads into the cylinder block and is exposed to oil pressure developed by the lubrication system. Do not confuse the oil pressure switch with the larger oil pressure sender (B, **Figure 84**). With the engine at rest, no oil pressure is developed, and the switch is closed. The closed switch grounds the tan/blue wire to complete the ground circuit for the warning horn. When the ignition switch is turned to ON or RUN, battery voltage is applied to the positive terminal on the warning horn, and it sounds to indicate the warning horn, tan/blue wire and oil pressure switch are operational. This is commonly referred to as the *self-test* feature. When the engine is started and the oil pressure exceeds 5 psi (34 kPa), the oil pressure switch removes the ground circuit for the horn. The horn stops sounding. If the oil pressure should drop below the specification, the switch will close, complete the ground circuit and sound the horn. The horn will continue

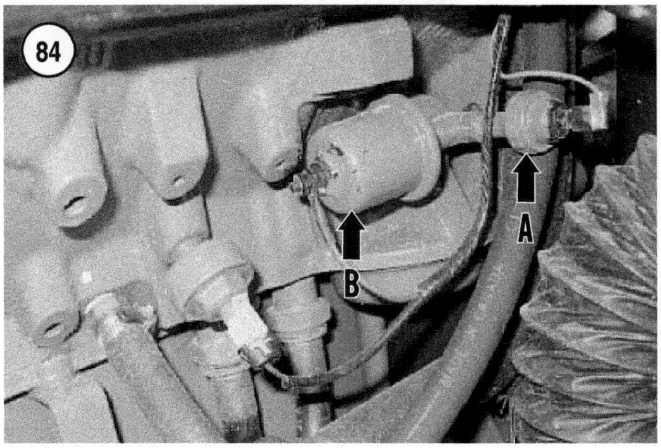

sounding unless the minimum oil pressure is restored, or the ignition switch is turned to OFF or STOP.

The overheat switch (**Figure 85**) threads into the thermostat housing, and the tip of the switch is exposed to the water or coolant exiting the engine. The engine overheat switch is normally open. If the switch is exposed to water or coolant temperature exceeding 190-200° F (88-93° C), the switch closes, completes the circuit and causes the warning horn to sound. The horn will continue sounding until the engine is shut off or the engine temperature drops below 150-170° F (66-77° C).

GM electronic fuel injection (EFI) models

The warning system components used on electronic fuel injection (EFI) models include the oil pressure switch, dash-mounted warning horn, engine temperature sensor and engine control unit (ECU).

Battery voltage is applied to the positive terminal of the warning horn via the purple wire connected to the ignition circuit. The tan/blue wire connects to the negative terminal of the warning horn and the ECU. An additional tan/blue wire connects the ECU to the oil pressure switch. The switch is used to activate the warning system in the event of low oil pressure.

Using internal circuitry, the ECU controls sounding of the warning horn by completing or removing the warning horn ground. The ECU uses input from the oil pressure switch and engine temperature sensor to determine if overheating or low oil pressure is evident.

When the ignition switch is turned to ON or RUN, the ECU sounds the warning horn for approximately 2 seconds to indicate the warning horn and ECU are operational. This is commonly referred to as the *self-test* feature.

The oil pressure switch (A, **Figure 84**) threads into the cylinder block and is exposed to oil pressure developed by the lubrication system. Do not confuse the oil pressure switch with the larger oil pressure sender (B, **Figure 84**). With the engine at rest, no oil pressure is developed, and the switch is closed. The closed switch grounds the tan/blue wire leading to the ECU. The ECU will not sound the warning horn unless input from the various EFI systems indicate the engine is running. When the engine is started, and the oil pressure exceeds 5 psi (34 kPa) the oil pressure switch opens the grounded tan/blue wire. If the oil pressure should drop below the specification with the engine running, the switch will close and ground the tan/blue wire leading to the ECU. The ECU will then sound the warning horn. The horn will continue sounding until the minimum oil pressure is restored, or the ignition switch is turned to OFF or STOP.

The ECU uses input from the engine temperature sensor to continuously monitor the temperature of the water or coolant in the engine. If the sensor input detects a water or coolant temperature exceeding 200° F (93° C), it completes the ground circuit to sound the horn. The horn will continue sounding unless the engine is shut off, or the engine temperature drops below approximately 180° F (82° C).

Ford electronic fuel injection (EFI) models

The warning system components used on electronic fuel injection (EFI) models include the oil pressure switch, dash-mounted warning horn, engine overheat switch and engine control assembly (ECA).

Battery voltage is applied to the positive terminal of the warning horn via the purple wire connected to the ignition circuit. The tan/black wire connects to the negative terminal of the warning horn, the oil pressure switch and the engine overheat switch. The tan/black wire also connects to the ECA. The switches are used to activate the warning system in the event of low oil pressure or engine overheating.

Using internal circuitry, the ECA controls the warning horn by completing or removing the horn ground. When the ignition switch is turned to ON or RUN, the ECA sounds the warning horn for approximately 2 seconds to indicate the warning horn and ECA are operational. This is commonly referred to as the *self-test* feature.

The oil pressure switch (A, **Figure 84**) threads into the cylinder block and is exposed to oil pressure developed by the lubrication system. Do not confuse the oil pressure switch with the larger oil pressure sender (B, **Figure 84**). When the engine is started, and the oil pressure exceeds 5 psi (34 kPa), the oil pressure switch removes the ground circuit for the horn. If the oil pressure should drop below the specification, the switch will close to complete the ground circuit and sound the horn. The horn will continue sounding until the minimum oil pressure is restored, or the ignition switch is turned to OFF or STOP.

The overheat switch (**Figure 85**) threads into the intake manifold, and the tip of the switch is exposed to the water or coolant exiting the engine. The engine overheat switch is normally open. If the water or coolant temperature exceeds 190-200° F (88-93° C), the switch closes and grounds the tan/blue wire, causing the warning horn to sound. The horn will continue sounding unless the engine is shut off or the engine temperature drops below 150-170° F (66-77° C).

86

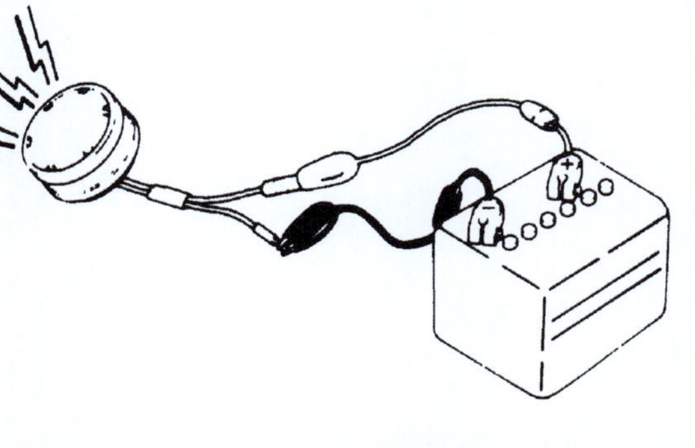

Warning Horn Test

1. Remove the warning horn as described in Chapter Nine.
2. Using jumper leads, connect the leads of the horn to the battery terminals (**Figure 86**). Polarity is not important for this test. Replace the warning horn if it fails to emit a loud tone.
3. Install the warning horn as described in Chapter Nine.
4. If the engine uses two warning horns, test the other horn using the same procedure.

Oil Pressure Switch Test

This test requires a multimeter and a vacuum/pressure pump. Purchase the vacuum/pressure pump from a local automotive parts or tool supply store. Test the switch as follows:

1. Remove the oil pressure switch as described in Chapter Nine.
2. Carefully slide an appropriately sized hose over the threaded fitting of the oil pressure switch (**Figure 87**). Fasten a clamp over the hose to ensure an airtight connection. Do not apply pressure at this time.
3. Connect the meter leads to the switch terminals and the body of the switch as shown in **Figure 87**. If necessary, use a clamp to secure the test lead to the switch body. The meter must indicate continuity. If otherwise, the switch must be replaced.
4. Observe the meter and the pressure gauge while *slowly* applying pressure to the switch. Note the pressure reading when the meter changes to no continuity.
5. Observe the meter and pressure gauge while *slowly* relieving the pressure. Note the pressure reading when the meter changes to continuity.
6. Repeat Step 5 and Step 6 several times to check for an intermittent switch failure. The meter must change to no continuity each time the pressure reaches approximately 5 psi (34 kPa) and continuity each time the pressure drops below the specification. Replace the oil pressure switch if it fails to perform as specified.

87

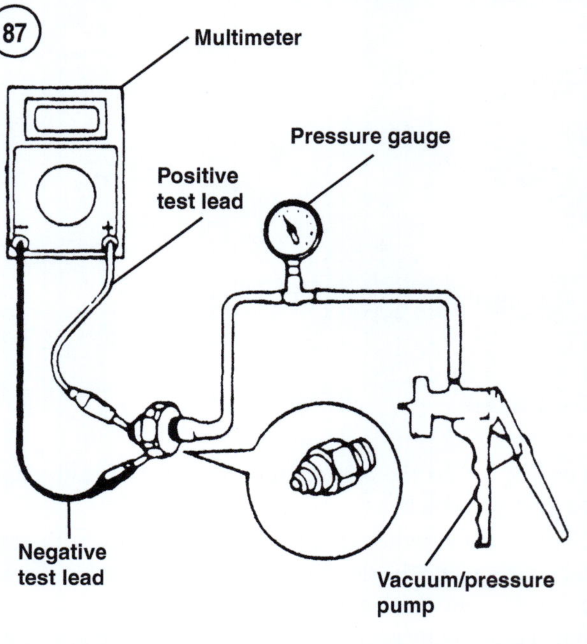

7. Remove the test leads and vacuum/pressure pump from the oil pressure switch. Clean all hose material from the threaded section of the switch.
8. Install the oil pressure switch as described in Chapter Nine.

Engine Overheat Switch Test

A faulty engine overheat switch can cause the warning horn with sound when the engine is not overheating or prevent the warning horn from sounding if the engine overheats.

This test requires an ohmmeter, liquid thermometer and a container of water that can be heated.

1. Remove the engine overheat switch as described in Chapter Nine.
2. Fill the container with cool water and suspend the tip of one overheat switch in the water (**Figure 88**). Make sure the switch does not contact the bottom or sides of the con-

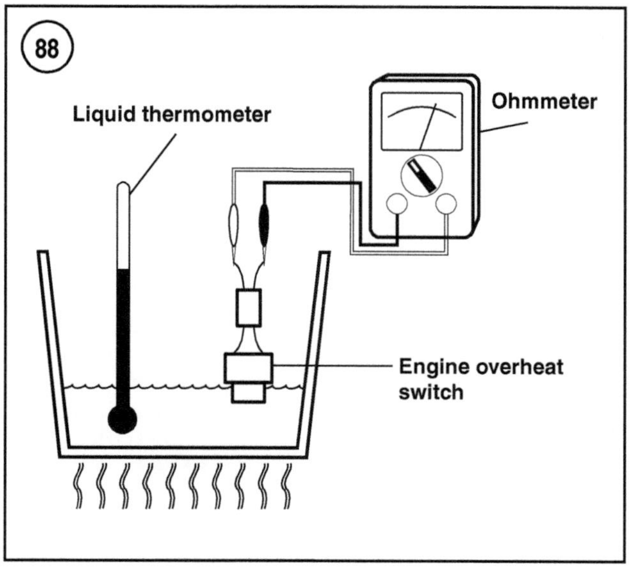

tainer. Place a thermometer in the container with the overheat switch shown in **Figure 88**.

3. Connect one of the ohmmeter leads to the brass body of the switch. Use a clamp to secure the wire as needed. Connect the other test lead to the switch terminal. If placed in cool water, the meter must indicate no continuity. If the meter indicates continuity, the switch is shorted and must be replaced. No further testing is required for a shorted switch.

4. Begin heating and gently stirring the water while observing the meter. Note the temperature when the meter switches from no continuity to continuity. This should occur at a temperature of approximately 190-200° F (88-93° C). Discontinue the test and replace the switch if the water boils before the meter reading changes.

5. Allow the water to gradually cool and note the temperature at which the meter changes from continuity to no continuity. This should occur at a temperature of 150-170° F (66-77° C).

6. Replace the engine overheat switch if it does not perform as specified.

7. Install the engine overheat switch as described in Chapter Nine.

Testing the Engine Coolant Temperature Sensor (ECT)

The engine coolant temperature sensor (ECT) provides input to the ECU to operate the fuel injection and warning systems. If suspected of causing a malfunction of the warning system, test the ECT as described under *Electronic Fuel Injection (EFI) System* in this chapter.

2

ENGINE

This section provides instructions to determine whether an engine malfunction is caused by mechanical failure. Cylinder compression and oil pressure testing are also described in this section.

Engine Noises

Some engine noise is generated during normal engine operation. A ticking noise or a heavy knocking noise intensifying when accelerating is a reason for concern.

If a worn or damaged component is suspected of causing an engine noise, consider having a professional technician listen to the engine. In many cases, only the trained ear of the technician can determine what component has failed.

Engine repairs can be costly and time consuming. Investigate all noises thoroughly before disassembling the engine.

A broken or loose bracket or fastener can cause a noise that is easily mistaken for an internal engine failure.

Ticking noise

The ticking noise is common when valve adjustment is required, a rocker arm fastener is loose or a valve train component has failed. Adjust the valves or tighten the rocker arm fasteners as described in Chapter Six. Then listen to the engine.

A ticking noise can also result from a damaged piston. Inspect the spark plugs for damage or aluminum deposits (**Figure 89**). If metal deposits are found on the spark plug, complete engine disassembly and repair is required. Perform a compression test as described in this chapter. Remove the cylinder head and inspect the valves, gaskets, pistons and cylinders if low compression is revealed. Refer to Chapter Six for the Cylinder head removal and installation procedures.

If the spark plugs and compression are good, but the ticking noise persists, refer to *Valve Train Failure* in this section for additional information.

CAUTION
Running the engine with an abnormal noise can result in increased damage that dramatically increases the cost of repair.

Whirring or squealing noise

A whirring or squealing noise is usually related to a problem with main or connecting rod bearings. Often, the noise becomes louder if the throttle is abruptly reduced to idle from a higher speed.

Sometimes the cylinder creating the noise can be identified with a mechanic's stethoscope. Touch the tip of the probe to the engine while listening. Compare the noise emanating from one area of the engine, cylinder head or crankcase with the noise from the same area but different cylinder. A noise common to one cylinder only indicates a problem with the respective connecting rod bearing. A noise common to all cylinders indicates a problem with crankshaft main bearings. Test the oil pressure as described in this section to determine whether the noise is caused by insufficient lubrication. Be aware that the accessory drive belts, timing chain or gear, and pulleys can generate a fair amount of noise. Use the stethoscope to make sure the noise is emanating from the engine before proceeding with an engine repair.

WARNING

Use extreme caution when working on or around a running engine. Never wear loose clothing. Make sure no one goes near the pulleys or drive belts. Never allow anyone near the propeller or propeller shaft while the engine is running.

Knocking noise

Use a mechanic's stethoscope to determine whether the noise is emanating from the engine or elsewhere. The noise will be more pronounced in the crankcase area if a problem exists in the crankshaft and connecting rods. Special insulated pliers are available to allow spark plug wire removal while running the engine. The noise might lessen when the spark plug wire is disconnected from the suspected cylinder. This procedure is difficult and can result in damage to the electrical system. To reduce the chance of damage to the ignition system, ground the spark plug promptly. Another method for isolating the cylinder is to remove one spark plug wire and attach it to an engine ground. Start the engine and listen to the noise. Install the spark lead and repeat the process for another cylinder. If the noise is quieter when one lead is grounded, that cylinder is suspect. The stethoscope method of isolating cylinders is more effective for an amateur technician.

If a knocking noise is detected, always perform an oil pressure test as described in this section. Knocking noise combined with low or unstable oil pressure generally indicates fault with the crankshaft main and/or connecting rod bearings.

Bearing failure causes metal particles to accumulate in the oil pan. These particles are picked up by the oil pump and deposited in the oil filter. Inspect the oil filter if bearing failure is suspected. Remove the filter as described in Chapter Three. Cut open the filter and inspect the filter element. Bearing failure is likely if a significant amount of metal debris is found.

90

Lubrication System Failure

Lubrication system failure can result in catastrophic engine failure. Lubrication system failure leads to scuffed pistons (**Figure 90**), excessive wear of valve train components (**Figure 91**) and eventual engine failure.

Failure of the lubrication system results from the following common causes:

1. Incorrect oil level.
2. Oil leakage or high oil consumption.
3. Using contaminated, diluted or the wrong type of oil.
4. Failure of the oil pump drive system.

CAUTION

If the lubrication system fails, damage can occur in a matter of seconds. To help prevent serious engine damage, slow down and stop the engine at once if low oil pressure is indicated by the warning system.

Incorrect oil level

Incorrect oil level can result from improperly filling the engine or improperly checking the engine oil level.

An oil level that is too high allows the crankshaft and other components to agitate the oil and cause the formation of bubbles or foam. This foamy oil can cause a drop in oil pressure and activate the low oil pressure warning system. In some cases, the oil pressure can remain above the low oil pressure activation point, yet the engine will still suffer from inadequate lubrication.

Oil level checking and oil filling instructions are provided in Chapter Three.

Oil leakage or high oil consumption

A rapid drop in oil level can result from oil leakage or from high oil consumption. Common leakage points are at the rocker arm/valve cover area, oil filter mounting surface, fuel pump mounting surface or oil pan-to-cylinder block mating surfaces. Finding the point of leakage can be difficult. Air flowing around the engine often distributes a thin film of oil on all external surfaces. To locate the point of leakage, carefully clean the engine using a shop towel and degreasing agent. Run the engine until it reaches operating temperature. Turn off the engine. Then wipe a white towel across all surfaces of the engine. Oil leakage is readily detected on the towel when the leakage point is contacted. Pinholes or casting flows can cause leakage to occur at other points on the engine. Many times, simply tightening fasteners corrects oil leakage. Replace gaskets, seals or other affected components if the leakage continues.

All engines consume some oil while running. Some of the oil lubricating the cylinders and valves is drawn into the combustion chamber and subsequently burned. Oil consumption rates vary by model, condition of the engine and how the engine is used. Engines with high operating hours or engines with worn internal components generally burn much more oil than new or low-hour engines. Damage to the piston and cylinders from detonation or preignition can cause increased oil consumption. New or recently rebuilt engines generally consume oil during the break-in period. After the break-in period, the oil consumption should drop to a normal level.

A typical symptom of excessive oil consumption is blue smoke coming from the exhaust during hard acceleration or high-speed operation. Inspection of the spark plugs usually reveals fouling or an oily film on the spark plug electrodes.

Perform a compression test if smoking or an oil-fouled plug is noted. Worn or damaged components will generally cause a low compression reading. Compression test instructions are provided in this section.

Contaminated, diluted or wrong type of oil

Contaminants enter the engine oil during normal operation. Dirt or dust enters the engine along with the air used during normal operation. Other particles form during the combustion process. Dirt, dust and other particles are captured by the lubricating oil and circulated throughout the engine. Most of the larger particles are captured in the oil filter. Smaller particles circulate through the engine. Frequent oil changes flush these particles from the oil before they reach high concentrations. High concentrations of these particles cause increased wear of internal engine components.

During normal operation, unburned fuel and water vapor accumulate in the lubricating oil. Oil also absorbs heat from the engine during operation. This heating causes the unburned fuel and water to evaporate from the oil and form crankcase vapor. The crankcase ventilation system returns the vapor to the combustion chamber where it burns or exits through the exhaust.

A faulty fuel system can dramatically increase the amount of unburned fuel in the oil. High levels of unburned fuel might not evaporate quickly enough to prevent oil dilution.

Failure to reach normal engine operating temperature prevents the fuel and water vapor from evaporating. Failure to reach normal operating temperature is generally caused by a faulty or improperly installed thermostat.

Wipe a sample of the oil from the dipstick between your finger and thumb. Compare the thickness of the oil with a sample of new oil. Smell the oil then check for a white or light color residue.

A very thin feel, fuel smell in the oil or white residue indicates oil dilution. If oil dilution is indicated, test the cooling and fuel system as described in this chapter.

Using the wrong grade, type or weight of oil can lead to increased engine wear and/or complete engine failure. Oil of poor quality might not provide the level of protection required by the engine. Using the wrong type of oil usually leads to serious engine damage. Never use two-stroke outboard oil in a four-stroke engine. Using the improper weight of oil can also prevent correct oil circulation during cold engine operation or allow excessive thinning of the oil at higher temperatures. Oil recommendations are provided in Chapter Three.

Oil pump failure

Failure of the lubrication system results in rapid wear of internal components and eventual engine seizure. Causes of oil pump failure include:

1. Worn, damaged or broken oil pump components.
2. Worn or damaged crankshaft or connecting rod bearings.
3. A blocked, damaged or loose oil pickup tube or screen.
4. A faulty or stuck oil pressure relief valve.

Failure of the lubrication system causes a loss of oil pressure and activation of the low oil pressure warning system. Continued operation results in rapid wear of internal components followed by eventual engine seizure. Check the oil pressure as described in this section if the warning system activates, and low oil level is ruled out as the cause.

Detonation

Detonation damage is the result of the heat and pressure in the combustion chamber becoming too great for the fuel that is used. Fuel normally burns at a controlled rate during combustion. The fuel explodes violently if the heat and pressure become too high. These violent explosions in the combustion chamber cause serious damage to internal engine components. The piston typically suffers the brunt of the damage. Detonation usually occurs only at higher engine speeds or during heavy acceleration. When detonation is occurring, a pinging noise from the engine accompanies a loss of power. Inspect the spark plugs if detonation damage is suspected. Aluminum deposits (**Figure 89**) or a melted electrode (**Figure 92**) indicate probable detonation damage. Perform a compression test to determine the extent of damage to the engine. Repair the engine if low compression is indicated. Correct the causes of detonation to prevent additional engine damage or repeat failures.

Conditions that promote detonation include:

1. Using a fuel with too low of an octane rating.
2. Excessive carbon deposits in the combustion chamber.
3. Overheating of the engine.
4. Using the incorrect propeller or overloading the engine.
5. Excessively lean fuel delivery.
6. Excessively advanced ignition timing.

Preignition

Preignition is caused by a glowing object inside the combustion chamber causing early ignition. The flame from the early ignition collides with the flame front initiated by the spark plug. A violent explosion occurs as these flame fronts collide. The piston suffers the brunt of the damage caused by this explosion. Inspect the spark plugs and perform a compression test if preignition is suspected. Aluminum deposits on the spark plug are likely when preignition has occurred. Repair the engine if low compression is revealed. Correct the causes of preignition to prevent additional engine damage or repeat failures.

Conditions that promote preignition include:

1. Excessive carbon deposits in the combustion chamber.
2. Using the wrong heat range of spark plug.
3. Overheating of the engine.

Engine Seizure

Engine seizure results from failure of internal engine components. This can occur at any engine speed. Should the failure occur at higher engine speeds, the engine might abruptly stop or gradually slow down and stall. Typically, the seizure is caused by a crankshaft, connecting rod or piston failure. Although this cause is not as common, engine seizure also occurs from a failure of the valve train components. Engine seizure prevents the electric starter from cranking the engine. Major repair is almost always required when engine seizure occurs.

92

Bear in mind that failure of the drive unit can also prevent engine rotation. Before suspecting the engine, remove the drive unit as described in Chapter Ten. Disassemble the engine if it remains seized with the drive unit removed. Water in the cylinders can also cause engine seizure. The water will cause a hydraulic lock of the piston as it comes up on the compression stroke. Remove the spark plugs to check for this condition.

Water Entering a Cylinder

Water can enter a cylinder for a number of reasons including:

1. Water entering the carburetor or throttle body openings.
2. Shutting the engine off at higher engine speeds.
3. Inadequate exhaust elbow height.
4. Engine dieseling.
5. Leaking cylinder head gaskets.
6. Cracked cylinder head.
7. Leaking water jacket in cylinder head.
8. Leaking exhaust manifold(s).
9. Leaking water jacket in cylinder block.
10. Leaking water jacket in intake manifold.

The typical symptom is rough running, especially at idle. The engine might run correctly at higher speed because a small amount of water cannot prevent normal combustion at higher speeds. If water enters the cylinders, remove the spark plugs and ground the spark plug leads. Crank the engine over to remove residual water from the cylinders. Change the engine oil. Then start the engine and operate for a minimum of 20 minutes to evaporate residual water from the oil passages.

Water entering the carburetor or throttle body opening

The presence of water in the throttle body opening indicates water is entering the cylinders through the throttle body opening. This occurs from submersion, a leaking en-

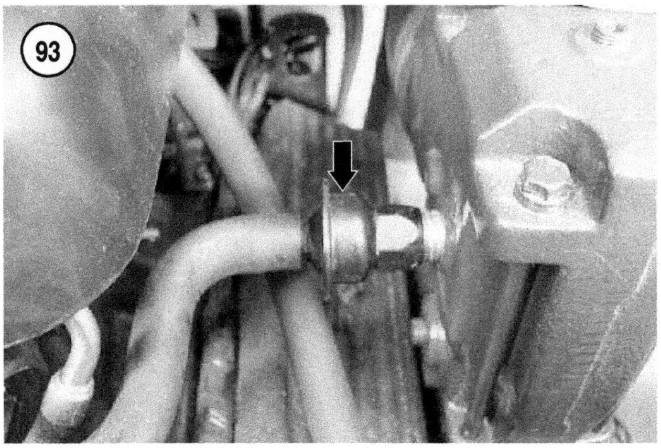

gine cover or subjecting the engine to a considerable amount of water spray. Water in the fuel allows water to enter the cylinders along with the fuel and incoming air. The heavier, less volatile water usually settles in the intake manifold or other passages. Inspect the fuel for water as described under *Inspecting the Fuel* in this chapter.

Shutting off the engine at higher engine speeds

Unless necessary to prevent an accident, never shut the engine off at speeds higher than idle. High vacuum forms in the intake manifold under this circumstance. This vacuum is applied to the exhaust passages due to camshaft overlap. High vacuum in the exhaust passages causes water in the exhaust system to flow into the cylinders. Serious engine damage can occur as a result. Later models are equipped with hydraulic lock prevention hoses and valves (**Figure 93**). The valve opens when subjected to high vacuum and allows air entry through the connecting hoses to reduce the vacuum. This can prevent or minimize the amount of water drawn in the cylinders.

Inadequate exhaust elbow height

The top of the exhaust elbow must extend a minimum of 15 in. (381 mm) above the water line. Otherwise, water can siphon into the engine through the exhaust system. Measure the distance with the boat in the water with normal loading and full fuel tank(s). Install exhaust riser kits if the water line-to-exhaust elbow measurement is less than the minimum specification.

Engine dieseling

Dieseling is a condition in which the engine runs after the ignition switch is turned OFF. Dieseling occurs if the fuel in the combustion chamber ignites from a source other than the spark plugs. It can be caused by excessive engine temperature, excessive combustion chamber deposits, poor fuel quality and improper carburetor adjustment. Excessively high idle speed is a primary cause for dieseling. The engine might run for several seconds after the ignition switch is turned off and might actually run backward briefly at the end of the event. The backward engine rotation will draw water into the engine from the exhaust system. If sufficient water enters the engine, hydraulic lock will occur. If dieseling occurs, check the cooling system, spark plug and carburetor adjustments.

Leaking cylinder head gasket or cracked/leaking cylinder head water jacket

A leaking cylinder head gasket results from a failure of the gasket that seals the cylinder head to the cylinder block. Damage to the gasket is likely if the engine has overheated. A cracked cylinder head is usually caused by expansion as water freezes in the cylinder head. A crack can also occur from rapid expansion and contraction that occurs from overheating. A leaking cylinder head water jacket is usually the result of corrosion damage. Symptoms of these faults include:

1. Water in the oil.
2. Water entering the cylinder(s).
3. Overheating (particularly at higher engine speeds).
4. Rough running (particularly at lower engine speeds).
5. External water or exhaust leakage at the cylinder head-to-cylinder block mating surfaces.

Perform a compression test if any of these symptoms are noted. Bear in mind that cylinder head gasket leakage is not always verified by a compression test. Typically, slightly lower compression is noted on two adjoining cylinders. Only removal and inspection of the gasket and mating surfaces verify a gasket failure.

If water is entering the cylinders, and external causes such as water in the fuel or silencer cover are ruled out, remove the cylinder head and inspect the cylinders and piston domes. Compare the appearance of the affected cylinder with other cylinders on the engine. Cylinders affected by water intrusion usually have significantly less carbon deposited on the piston and cylinders. Rusting or corrosion of the valves, valve seats, cylinders and piston dome is common with this condition. A complete engine repair is required if rusted cylinders or pitted pistons are evident. Refer to Chapter Six for cylinder head removal and installation instructions.

Leaking exhaust manifold water jacket

A leaking exhaust manifold allows water to enter the cylinders from the exhaust passages. High-performance four-stroke engines, such as marine engines, have an aggressive camshaft lift and duration creating a fair amount of valve overlap. Valve overlap allows reverse exhaust flow (reversion) under certain conditions. If water is present in the exhaust passages, reverse exhaust flow can increase the chance of water entering the cylinders through the exhaust valves. Leaking exhaust elbow gaskets allow water into the exhaust passages. Repair leaking gaskets if external leakage is evident at the mating surfaces. Removal and installation of the exhaust system components is described in Chapter Eight.

Leaking cylinder block water jacket

Leakage from the cylinder block water jacket can allow water to leak into the cylinders or oil. Water leaking from the cylinder block water jackets is almost always caused by cracks caused by freezing the water in the jacket. Porosity or excessive corrosion damage are other common causes. Leaking water jackets in the cylinder block can be difficult to find. Casting flaws, pinholes and cracks might or might not be visible. Replace the cylinder block if water is entering the cylinder, all other causes are ruled out, and no visible defects in the gaskets or mating surface can be found. Continued operation with water intrusion results in eventual engine failure.

Leaking intake manifold water jacket

Failure of the intake manifold water jacket can allow water to enter the cylinder(s) or oil. This failure almost always is due to gasket failure at the cylinder head-to-intake manifold mating surfaces for the water passage. Failure of the gasket is common after the engine has overheated. Remove the intake manifold as described in Chapter Seven. If water is entering the cylinder(s) or oil, and other causes are ruled out, inspect the gasket.

Water Entering the Oil

Normal condensation causes water to form in the oil. This is especially common if the engine is operated in cool and humid environments. Regular oil changes remove normal amounts of water before the engine suffers serious harm. Excessive amounts of water will dilute the oil and cause serious engine damage. Symptoms of excessive water in the oil include:

1. Water on the dipstick.
2. Water in oil drained from the crankcase.
3. A light color or milky oil appearance.

Excessive amounts of water in the crankcase are generally caused by water entering the cylinders; leaking water jackets in the cylinder block, cylinder heads or intake manifold; or internal leakage in the oil cooler.

If the described symptoms occur, check for water entering the cylinders as described in this section. If water is not entering the cylinders, remove the oil cooler as described in Chapter Eight. Have the oil cooler pressure tested at a radiator repair shop. If the cooler is not leaking, check for a leaking intake manifold. If the intake manifold is not leaking, and all other causes are ruled out, replace the cylinder block.

Valve Train Failure

Failure of the valve train can cause low compression, excessive engine noise or an inability to start the engine. Components constituting the valve train include the camshaft, lifters, timing gear, timing chain, valves, valve springs and cylinder head. This section describes common causes of valve train failure.

Timing chain, sprocket or gear failure

CAUTION
Attempts to start the engine with improper valve timing or a failed timing gear, sprocket or chain can lead to damaged valves, pistons and other engine components.

Failure of the gear, sprocket or chain will result in an inability to start the engine, rough running or poor performance.

An excessively worn timing chain or sprocket might allow the chain to jump over one or more of the teeth on the gear or sprocket. This results in improper valve timing causing rough operation and poor performance. Complete failure of the timing chain or sprocket prevents camshaft rotation. If failure of the sprocket or chain is suspected of causing a no-start condition, check for distributor rotation as described in this chapter. Refer to *Basic Ignition System Troubleshooting*.

Sticking, worn or damaged valves

Sticking, worn or damaged valves cause low compression, rough operation, poor performance and/or backfiring.

Corrosion or heavy carbon deposits can cause valves to stick in the open position. Valves can become bent or damaged from contacting the top of the piston or foreign objects in the combustion chamber.

Any of these conditions result in an inability to fully close the affected valve. Rough operation at lower engine speed is common with a sticking, worn or damaged valve. Backfiring occurs when a leaking valve allows the burning fuel to enter the intake manifold or exhaust passages. Backfiring or a popping noise coming from the intake usually is the result of a stuck, worn or damaged intake valve. Backfiring or a popping noise coming from the exhaust is usually the result of a stuck, worn or damaged exhaust valve.

Perform a compression test if any listed symptoms are noted.

Improper valve adjustments can cause the same symptoms as sticking, worn or damaged valves. Check the valve adjustment before disassembling the engine. Valve adjustment is described in Chapter Six.

If backfiring is noted, run the engine with one spark plug wire at a time grounded. The backfire will stop when the spark plug wire is grounded on the affected cylinder. Remove the cylinder head and inspect the valves when back-

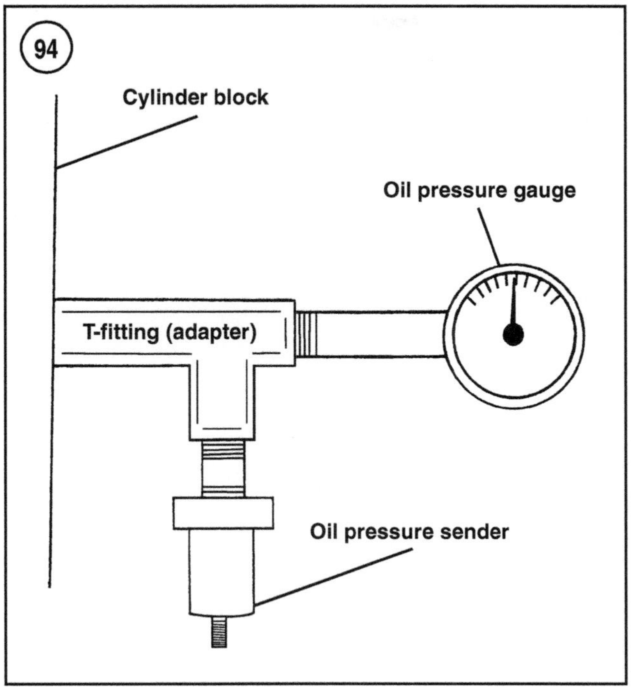

firing and low compression are evident on the same cylinder, and improper valve adjustment is ruled out.

Sticking valves are often the result of improper long-term storage, water entering the cylinders or submersion of the engine. Using the wrong type of oil, improper fuel system operation or lugging the engine all contribute to increased deposits or wear of the valve and seat area.

Oil recommendations are provided in Chapter Three. Prevent lugging by selecting the correct propeller for the given engine and boat combination. Refer to Chapter One for propeller selection information.

Worn valve guides

Worn valve guides prevent consistent closing of the valves. A compression test usually does not generally verify an excessively worn valve guide because the valve tends to seat normally at cranking speed. Continued operation with worn valve guides will cause the valve seats to wear unevenly and eventually cause low compression and poor performance.

High oil consumption, fouled spark plugs and blue exhaust smoke during deceleration are typical symptoms of excessively worn valve guides and valve stem seals. Only disassembly of the cylinder head and subsequent measurement can confirm excessively worn valve guides. Refer to Chapter Seven.

Camshaft failure

Failure of any of the camshaft lobes (**Figure 91**) can cause the same symptoms as a sticking or damaged valve. Excessively worn camshaft lobes are generally the result of improper valve adjustment; using the incorrect type, grade or weight of oil; a high number of operating hours; or oil dilution. Refer to *Lubrication System Failure* in this section for additional information on oil dilution. Check for excessively worn camshaft lobes by measuring the camshaft lobe lift as described in Chapter Six.

2

Oil Pressure Test

An oil pressure gauge, T-fitting adapter and a shop tachometer are required to perform an oil pressure test. This test must be performed under actual operating conditions.

Test the oil pressure at the threaded opening for the oil pressure sender (**Figure 94**). Refer to *Oil Pressure Sender Replacement* in Chapter Nine to locate the sender.

Use a T-fitting adapter with the same threads as the oil pressure sender on the male and female ends. One of the female ends must have the same thread size and pitch as the oil pressure gauge. Oil pressure gauges and threaded adapters are available from most automotive parts stores. Tighten the adapters securely before running the engine.

WARNING
Safely performing on-water testing or adjustments requires two people: one person to operate the boat and the other to monitor the gauges or test equipment and make necessary adjustments. All personnel must remain seated inside the boat at all times. Do not lean over the transom while the boat is underway. Use extensions to allow all gauges and test equipment to be located in normal seating areas.

Test the oil pressure as follows:

1. Run the engine until it reaches normal operating temperature.
2. Remove the oil pressure sender as described in Chapter Nine.
3. Install the male end of the adapter into the oil pressure sender opening. Securely tighten the adapter.
4. Install the oil pressure sender into one of the female openings of the adapter. Securely tighten the oil pressure switch.
5. Connect the engine wire harness connector to the oil pressure sender to allow verification of the oil pressure gauge during testing.
6. Install the threaded adapter of the oil pressure gauge (**Figure 94**) into the remaining female opening of the adapter. Securely tighten all fittings. Secure the oil pressure gauge hose to prevent it from contacting any moving components.
7. Following the manufacturer's instructions, attach the shop tachometer to the engine.
8. Start the engine and immediately check for and correct any oil leakage at the gauge and adapter fittings.

9. Shift the engine into FORWARD. Observe the oil pressure gauge while an assistant operates the engine in forward at idle speed. Stop the engine *immediately* if the gauge indicates no oil pressure. The gauge must indicate a minimum of 4 psi (28 kPa) at idle speed. Note the pressure while the assistant advances the throttle to RUN the engine at 2000 rpm. The gauge must indicate 30-60 psi (207-414 kPa) oil pressure with the engine at normal operating temperature. A slightly higher reading might be indicated if the engine is running too cool. Compare the actual engine oil pressure with the pressure displayed on the dash gauge. A faulty sender, gauge or wiring is evident if the readings are substantially different. Refer to the following:

a. If the oil pressure exceeds 60 psi (414 kPa) at idle speed, the engine is not reaching normal operating temperature, or the oil pressure relief valve is stuck. Verify the engine temperature as described in this chapter. Refer to *Cooling System Troubleshooting*. Replace the oil pump as described in Chapter Six if the oil pressure at idle exceeds the specification at normal operating temperature.

b. If the oil pressure is less than 4 psi (28 kPa) at idle speed, the engine is running too hot; the oil is too thin or diluted, the oil pump is faulty; or the engine has worn or failed bearing(s), crankshaft or camshaft. Verify the engine temperature as described in this chapter. Refer to *Cooling System Troubleshooting*. Inspect the oil for dilution as described in this section. Refer to *Lubrication System Failure*. Replace the oil pump as described in Chapter Six if the engine is operating at the correct temperature, and the oil is not diluted. Faulty bearings, crankshaft or camshaft is evident if the oil pressure is not corrected with the replacement oil pump. Replacement of these components is described in Chapter Six.

c. If the oil pressure exceeds 60 psi (414 kPa) at 2000 rpm, the engine is not reaching normal operating temperature, or the oil pressure relief valve is stuck. Verify the engine temperature as described in this chapter. Refer to *Cooling System Troubleshooting*. Replace the oil pump as described in Chapter Six if the oil pressure at 2000 rpm exceeds the specification at normal operating temperature.

d. If the oil pressure is less than 30 psi (207 kPa) at 2000 rpm, the engine is running too hot; the oil is too thin or diluted; the oil pump is faulty, or the engine has worn or failed bearing(s), crankshaft or camshaft. Verify the engine temperature as described in this chapter. Refer to *Cooling System Troubleshooting*. Inspect the oil for dilution as described in this section. Refer to *Lubrication System Failure*. Replace the oil pump as described in Chapter Six if the engine is operating at the correct temperature, and the oil is not diluted. Faulty bearings, crankshaft or camshaft is evident if the oil pressure is not corrected with the replacement oil pump. Replacement of these components is described in Chapter Six.

10. Return to port and stop the engine.
11. Remove the shop tachometer, oil pressure gauge and adapter. Install the oil pressure sender as described in Chapter Nine.

95

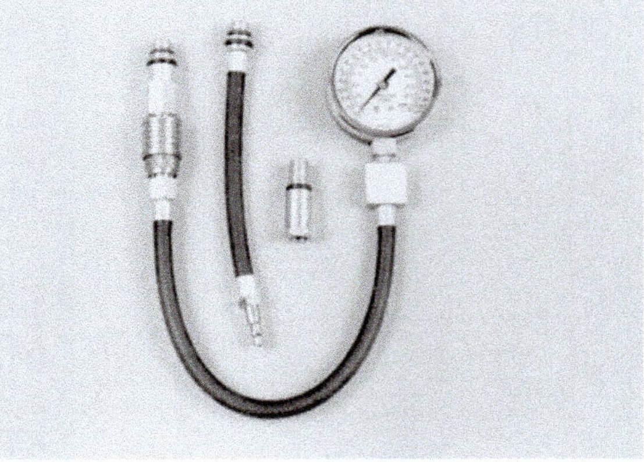

Compression Test

A quality compression gauge and adapters (**Figure 95**) are required for accurate compression testing. They are available at automotive parts stores and from tool suppliers. A small can of engine oil may also be required.

1. Remove all of the spark plugs as described in *Spark plug removal* Chapter Three. Connect the spark plug leads to a suitable engine ground such as a clean cylinder head bolt. Activate the lanyard safety switch to further disable the ignition system.
2. Install the compression gauge into the No. 1 spark plug hole. Refer to the following:
   a. On GM 3.0L models, the front cylinder is the No. 1 cylinder.
   b. On GM V-6 and V-8 models, the front cylinder on the port bank is the No. 1 cylinder. The front cylinder on the starboard side is the No. 2 cylinder. Odd-numbered cylinders are on the port cylinder bank, and even-numbered cylinders are on the starboard bank.
   c. On Ford 5.0L and 5.8L (V-8) models, the No. 1 cylinder is the front cylinder on the starboard side of the engine. Cylinders No. 2, 3 and 4 are arranged from front to back on the starboard side. Cylinder No. 5 is the front cylinder on the port side of the engine. Cylinders 6, 7 and 8 are arranged front to back on the port side of the engine.
3. Securely tighten the hose adapter. Position the throttle in the wide open position during testing.
4. Stay clear of the remaining spark plug openings during testing. Observe the compression gauge and operate the electric starter. Crank the engine a minimum of five revolutions at normal cranking speeds. Record the compression reading.
5. Repeat Steps 2-4 for the remaining cylinders. Record all compression readings.

96

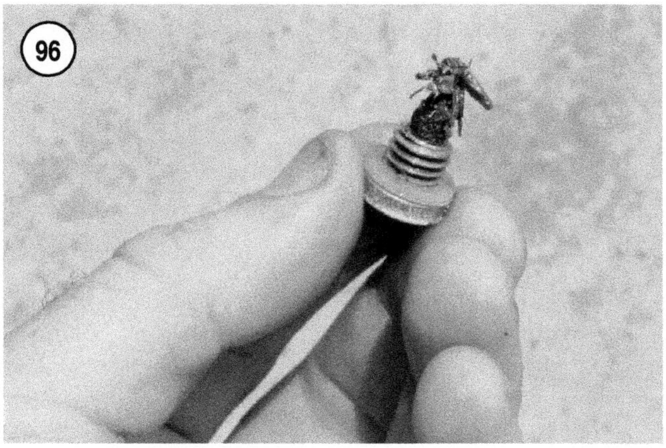

6. Low compression is evident if the compression pressure varies more than 10 percent between cylinders or if any cylinder has less than 100 psi (690 kPa) of pressure.
7. If low compression is noted, squirt approximately 1 teaspoon of clean engine oil into the suspect cylinder through the spark plug hole. Rotate the engine several revolutions to distribute the oil in the cylinder. Then repeat Steps 2-4.
   a. If the compression increases significantly, the piston rings and cylinder are excessively worn.
   b. If the compression does not increase, the compression leakage is the result of a worn valve face or seat.
8. Remove the compression gauge. Place the remote control in the idle position.
9. Install the spark plugs as described in Chapter Three. Refer to *Spark plug installation*. Place the lanyard safety switch in the normal RUN position.

DRIVE SYSTEM

Failure of the drive system will cause shifting difficulty, unusual noise or vibration. This section describes the causes of drive system failure and describes the troubleshooting procedures to pinpoint the cause.

Shifting Difficulty

Shifting difficulty can cause an inability to shift the engine into one or both gears, delayed shifting, shifting without moving the control lever or excessive shifting effort.

Inability to shift into gear or delayed shifting

Inability to shift into gear or delayed shifting is generally caused by improper shift cable adjustment, a faulty shift cable, faulty shift linkage, faulty or improperly assembled shift housing components, faulty cone clutch or failure of the internal drive unit. If these symptoms are evident, first inspect the drive unit lubricant as described in Chapter Three. If large metal particles are found on the drain/plug (**Figure 96**), disassemble and inspect the drive unit as described in Chapter Ten. If the lubricant inspection does not indicate an internal problem, adjust the shift cable as described in Chapter Five. If the adjustment does restore proper shifting, disassemble, inspect and reinstall the shift housing as described in Chapter Ten. Replace any faulty or questionable components. Make sure to adjust the shift shoe stop bolt as described in the assembly instructions. Replace the cone clutch as described in Chapter Ten if the fault persists.

2

Shifting into or out of gear without moving the control lever

Shifting into or our of gear without moving the control lever is usually caused by improper shift cable adjustment or worn or improperly assembled shift housing components. In rare cases, the failure is due to failure of gears, shafts or bearings in the drive unit. If this symptom is evident, first adjust the throttle cable as described in Chapter Five. If this does not restore proper shifting, disassemble, inspect and reinstall the shift housing as described in Chapter Ten. Replace any faulty or questionable components. Make sure to adjust the shift shoe stop bolt as described in the assembly instructions. If no fault is found with the adjustments, inspect the drive unit lubricant as described in Chapter Three. If large metal particles are found on the drive unit drain plug (**Figure 96**), disassemble and inspect the drive unit as described in Chapter Ten.

Excessive shifting effort

Excessive shifting effort is generally caused by a stiff or binding shift cable, improper shift cable adjustment, worn or improperly assembled shift housing components or failure of internal drive unit components. If excessive shift effort is evident, first adjust the shift cable as described in Chapter Five. Move the remote control while the shift cable is disconnected to check for a binding or stiff shift cable. Replace the cable as needed. If this does not restore proper shifting, disassemble, inspect and reinstall the shift housing as described in Chapter Ten. Replace any faulty or questionable components. Make sure to adjust the shift shoe stop bolt as described in the assembly instructions. If no fault is found with the adjustments, inspect the drive unit lubricant as described in Chapter Three. If large metal particles are found on the drain/plug (**Figure 96**), disassemble and inspect the drive unit as described in Chapter Ten.

Unusual Drive Unit Noise

Normal drive unit noise is barely noticeable over normal engine noise. A rough growling, grinding or knocking noise or a loud high-pitched whine is a reason for concern. These types of noises usually indicate failure of bearings, gears or other drive unit components. If any unusual noise is present, inspect the drive unit lubricant as described in Chapter Three. Failure of internal drive unit components is evident if large metal particles are found on the drain plug (**Figure 96**).

Knocking noise

A knocking noise is almost always caused by failure of the universal joints, especially if the noise is only present or worsens as the drive is steered to the port and starboard limits. Improper engine alignment can generate similar knocking noises. Though less likely, failure of internal drive unit components can also cause knocking noise; however, the noise does not worsen as the drive unit is steered to the port and starboard limits. If a knocking noise is evident, remove the drive unit and inspect the universal joint shaft as described in Chapter Ten. While the drive unit is off, check the engine alignment as described in Chapter Six. If the alignment is correct, and no fault is found with the universal joints, inspect the drive unit lubricant as described in Chapter Three. Failure of internal drive unit components is evident if large metal particles are found on the drain plug (**Figure 96**). Replace the universal joints if no fault is found with the engine alignment and gearcase lubricant. A faulty universal joint might only be evident after visual inspection of the cap, cross and rollers.

Grinding noise

A grinding noise indicates failure of gears, bearings or shafts in the drive unit. If a grinding noise is evident, inspect the drive unit lubricant as described in Chapter Three. Failure of internal drive unit components is evident if large metal particles are found on the drain plug (**Figure 96**). Drive unit repair is described in Chapter Ten.

High-pitched whine

A high-pitched whine normally indicates a faulty bearing, improper bearing preload or misaligned gears. Depending on the nature of the failure, metal material might not be found in the gearcase lubricant. Disassemble, inspect and reassemble the drive unit, as described in Chapter Ten, to restore normal drive unit operation.

Growling noise

A growling noise is almost always caused by failure of the gimbal bearing. Improper engine alignment can generate a similar noise. If this noise is present, remove the drive unit as described in Chapter Ten and inspect the gimbal bearing as described in Chapter Eleven. While the drive is off, check the engine alignment as described in Chapter Six. Correct the engine alignment or replace the bearing to restore normal drive unit operation.

Vibration

In almost all cases, an unusual vibration is due to a bent propeller shaft or damaged propeller. The propeller can appear to be in perfect condition but still be unbalanced. The best way to solve propeller vibration is to have the propeller(s) checked at a propeller repair shop. Another option is to try a different propeller or pair of propellers. Always check for a bent propeller shaft if vibration is present. Refer to *Before each use* in Chapter Three. A bent propeller shaft is usually caused by striking an underwater obstacle while underway. In many cases, other damage occurs to the drive unit.

Never operate the engine if excessive vibration is present. The vibration places excessive stress on gears, bearings, shafts and other engine components. Operating the engine with excessive vibration can seriously compromise the durability of the entire stern drive system.

Table 1 STARTING SYSTEM TROUBLESHOOTING

Symptom	Possible causes	Corrective action
Starter does not operate		
	Engine not in neutral	Shift into neutral
	Weak or discharged battery	Charge and test the battery (Chapter Nine)
	Loose battery cable at starter	Tighten battery cable terminal
	Dirty or corroded battery terminals	Thoroughly clean battery terminals
	Faulty neutral start switch	Test neutral start switch
	Faulty ignition switch	Test ignition switch
	Faulty starter solenoid	Test the starter solenoid
	Faulty starter relay	Test starter relay
	Loose or dirty wire connection	Clean and tighten starter wire connections
	Faulty electric starter	Test starter motor cranking voltage
	Faulty instrument harness wiring	Check for open or shorted wiring
	Faulty engine wiring harness	Check for open or shorted wiring

(continued)

Table 1 STARTING SYSTEM TROUBLESHOOTING (continued)

Symptom	Possible causes	Corrective action
Starter does not operate (continued)	Faulty instrument harness plug	Check for faulty terminals in plug
Starter turns slowly		
	Weak or discharged battery	Charge and test the battery (Chapter Nine)
	Dirty or corroded battery terminals	Thoroughly clean battery terminals
	Loose or dirty wire connections	Clean and tighten starter wire connections
	Engine is in gear	Adjust shift cable (Chapter Five)
	Faulty electric starter	Test starter motor cranking voltage
	Internal engine damage	Check engine compression
	Internal drive unit damage	Inspect gearcase lubricant for debris
Starter engages flywheel but does not turn	Weak or discharged battery	Charge and test the battery (Chapter Nine)
	Dirty or corroded battery terminals	Thoroughly clean battery terminals
	Loose or dirty wire connections	Clean and tighten starter wire connections
	Water in the cylinder	Inspect spark plug for water contamination
	Oil in the cylinder	Inspect spark plugs for oil contamination
	Damaged starter pinion gear	Inspect starter pinion gear
	Damaged flywheel gear teeth	Inspect flywheel gear teeth
	Faulty electric starter	Test starter motor cranking voltage
	Seized engine	Check engine for damage
	Seized drive unit	Inspect gearcase lubricant for debris
	Improper valve timing or adjustment	Check valve timing and adjustment
Noisy starter operation		
	Dirty or dry starter pinion gear	Repair starter motor (Chapter Nine)
	Damaged starter pinion gear teeth	Repair starter motor (Chapter Nine)
	Damaged or corroded flywheel gear	Inspect flywheel drive gear teeth
	Loose starter mounting bolt	Tighten starter mounting bolt
	Worn or dry starter bearings	Repair starter motor (Chapter Nine)
Starter operates forward or reverse gear	Faulty neutral start switch	Test neutral start switch
	Improper wiring	Check wiring to neutral start switch
Starter operates without turning key switch	Faulty ignition switch	Test ignition switch
	Faulty starter relay	Test starter relay
	Faulty starter solenoid	Test starter solenoid
	Loose wiring at starter terminals	Check and correct the wiring
	Faulty instrument harness wiring	Check for open or shorted wiring
	Faulty engine wiring harness	Check for open or shorted wiring
	Faulty instrument harness plug	Check for faulty terminals in plug
	Faulty fuel pump harness diode	Replace the harness or diode

Table 2 IGNITION SYSTEM TROUBLESHOOTING

Symptom	Possible causes	Corrective action
Engine will not start		
	Lanyard switch activated	Check lanyard switch position
	No spark at plugs	Check for spark at spark plug leads
	Faulty distributor rotor	Inspect or replace rotor (Chapter Nine)
	Faulty lanyard switch	Test lanyard switch
	Shorted tachometer	Check for shorted tachometer
	Distributor not rotating	Check for distributor rotation
	Faulty tachometer amplifier	Test the tachometer amplifier
	Faulty EFI system relay	Test EFI system relay
	Low voltage supply to system	Check battery voltage to system Perform ignition system voltage test Troubleshoot ignition system
	Faulty crankshaft position sensor	Troubleshoot ignition system
	Faulty ignition module	Test ignition module Perform ignition system voltage test Troubleshoot ignition system
	Faulty pickup coil	Test pickup coil

(continued)

Table 2 IGNITION SYSTEM TROUBLESHOOTING (continued)

Symptom	Possible causes	Corrective action
Engine will not start (continued)		
	Fouled spark plugs	Check or replace spark plug(s)
	Faulty ignition coil	Check battery voltage to system
		Test ignition coil
		Troubleshoot ignition system
	Faulty engine control unit (ECU)	Check for diagnostic trouble codes
	Faulty engine control assembly (ECA)	Test at local dealership
Stalls or runs rough at idle		
	Fouled spark plug(s)	Check or replace spark plug(s)
	Faulty distributor rotor	Inspect or replace rotor (Chapter Nine)
	Faulty distributor cap	Inspect or replace rotor (Chapter Nine)
	Faulty spark plug wires	Inspect spark plug wires
	Low voltage supply to system	Check battery voltage to system
		Perform ignition system voltage test
		Troubleshoot ignition system
	Faulty lanyard switch	Test lanyard switch
	Shorted tachometer	Check for shorted tachometer
	Faulty tachometer amplifier	Test the tachometer amplifier
	Faulty ignition coil	Check battery voltage to system
		Test ignition coil
		Troubleshoot ignition system
Idle speed too high		
	Incorrect ignition timing	Check and adjust ignition timing
	Incorrect sensor to sender air gap	See Ignition module replacement (Chapter Nine)
	Faulty engine temperature sender	Check for diagnostic trouble codes
	Faulty throttle position sensor	Check for diagnostic trouble codes
Idle speed too low		
	Incorrect ignition timing	Check and adjust ignition timing
	Incorrect sensor to sender air gap	See Ignition module replacement (Chapter Four)
Misfire or poor high-speed performance		
	Incorrect ignition timing	Check and adjust ignition timing
	Incorrect sensor to sender air gap	See Ignition module replacement (Chapter Four)
	Engine reaching rev limit	Check full speed engine rpm
	Fouled spark plug(s)	Inspect or replace spark plug(s)
	Faulty distributor rotor	Inspect or replace rotor (Chapter Nine)
	Faulty distributor cap	Inspect or replace rotor (Chapter Nine)
	Faulty spark plug wires	Inspect spark plug wires
	Low voltage supply to system	Check battery voltage to system
		Perform ignition system voltage test
		Troubleshoot ignition system
	Faulty lanyard switch	Test lanyard switch
	Shorted tachometer	Check for shorted tachometer
	Faulty tachometer amplifier	Test the tachometer amplifier
	Faulty ignition coil	Check battery voltage to system
		Test ignition coil
		Troubleshoot ignition system

Table 3 FUEL SYSTEM TROUBLESHOOTING

Symptom	Possible causes	Corrective action
Engine will not start		
	Plugged fuel tank vent	Check the fuel tank vent
	Old or contaminated fuel	Inspect the fuel
	Air or fuel leaks in hose fittings	Inspect the hoses and fitting
	Restricted fuel supply	Check for restricted fuel supply
	Blocked fuel filter	Inspect the fuel filters (Chapter Seven)
	Faulty mechanical fuel pump	Test the mechanical fuel pump
	Faulty low-pressure electric pump	Test the low-pressure electric pump
	Faulty high-pressure electric pump*	Test the fuel pressure
	Faulty EFI system relay*	Test the EFI system relay
	Flooding carburetor	Check for flooding carburetor

(continued)

Table 3 FUEL SYSTEM TROUBLESHOOTING (continued)

Symptom	Possible causes	Corrective action
Engine will not start (continued)		
	Flooding fuel cell*	Check for flooding fuel cell
	Choke valve malfunction	Check the choke valve operation
	Fuel injectors not operating*	Check for fuel injector operation
	Faulty engine temperature sensor*	Test the engine temperature sensor
	Stuck carburetor inlet needle	Repair the carburetor (Chapter Seven)
	Improper float level adjustment	Repair the carburetor (Chapter Seven)
	Blocked carburetor passages	Repair the carburetor (Chapter Seven)
	Faulty accelerator pump	Repair the carburetor (Chapter Seven)
	Faulty EFI sensor*	Check for diagnostic trouble codes
	Faulty idle air control motor (IAC)*	Test the idle air control motor (IAC)
Stalls or runs rough at idle		
	Improper idle speed adjustment	Adjust idle speed (Chapter Five)
	Old or contaminated fuel	Inspect the fuel
	Plugged fuel tank vent	Check the fuel tank vent
	Blocked carburetor passages	Repair the carburetor (Chapter Seven)
	Faulty idle air control motor (IAC)*	Test the idle air control motor (IAC)
	Air or fuel leaks in hose fittings	Inspect the hoses and fitting
	Restricted fuel supply	Check for restricted fuel supply
	Blocked fuel filter	Inspect the fuel filters (Chapter Seven)
	Faulty mechanical fuel pump	Test the mechanical fuel pump
	Faulty low-pressure electric pump	Test the low-pressure electric pump
	Faulty high-pressure electric pump*	Test the fuel pressure
	Faulty EFI system relay	Test the EFI system relay
	Flooding carburetor	Check for flooding carburetor
	Flooding fuel cell*	Check for flooding fuel cell
	Choke valve malfunction	Check the choke valve operation
	Faulty engine temperature sensor*	Test the engine temperature sensor
	Sticking carburetor inlet needle	Repair the carburetor (Chapter Seven)
	Improper float level adjustment	Repair the carburetor (Chapter Seven)
	Faulty EFI sensor*	Check for diagnostic trouble codes
	Leaking or blocked fuel injectors*	Check the fuel injectors
Idle speed too high		
	Improper throttle cable adjustment	Adjust the throttle cable
	Improper idle speed adjustment	Adjust idle speed (Chapter Five)
	Faulty idle air control motor (IAC)*	Test the idle air control motor (IAC)
	Faulty engine temperature sensor*	Test the engine temperature sensor
	Faulty EFI system sensor*	Check for diagnostic trouble codes
	Faulty throttle position sensor (TPS)	Check the throttle position sensor
Hesitation during acceleration		
	Improper carburetor adjustment	Adjust the carburetor (Chapter Five)
	Faulty accelerator pump	Repair the carburetor (Chapter Seven)
	Faulty throttle position sensor*	Check for diagnostic trouble code
	Blocked carburetor passages	Repair carburetor (Chapter Seven)
	Improper idle speed adjustment	Adjust idle speed (Chapter Five)
	Old or contaminated fuel	Inspect the fuel
	Plugged fuel tank vent	Check the fuel tank vent
	Air or fuel leaks in hose fittings	Inspect the hoses and fitting
	Restricted fuel supply	Check for restricted fuel supply
	Blocked fuel filter	Inspect the fuel filters (Chapter Seven)
	Faulty mechanical fuel pump	Test the mechanical fuel pump
	Faulty low-pressure electric pump	Test the low-pressure electric pump
	Faulty high-pressure electric pump*	Test the fuel pressure
	Flooding carburetor	Check for flooding carburetor
	Flooding fuel cell*	Check for flooding fuel cell
	Choke valve malfunction	Check the choke valve operation
	Sticking carburetor inlet needle	Repair the carburetor (Chapter Seven)
	Improper float level adjustment	Repair the carburetor (Chapter Seven)
	Faulty EFI system sensor*	Check for diagnostic trouble codes
	Leaking or blocked fuel injectors*	Check the fuel injectors
Misfire or poor high-speed performance		
	Plugged fuel tank vent	Check the fuel tank vent
	Old or contaminated fuel	Inspect the fuel
	Air or fuel leaks in hose fittings	Inspect the hoses and fitting

(continued)

Table 3 FUEL SYSTEM TROUBLESHOOTING (continued)

Symptom	Possible causes	Corrective action
Misfire or poor high-speed performance (continued)		
	Restricted fuel supply	Check for restricted fuel supply
	Blocked fuel filter	Inspect the fuel filters (Chapter Seven)
	Choke valve malfunction	Check the choke valve operation
	Faulty mechanical fuel pump	Test the mechanical fuel pump
	Faulty low-pressure electric pump	Test the low-pressure electric pump
	Faulty high-pressure electric pump*	Test the fuel pressure
	Blocked fuel injectors*	Check the fuel injectors
	Flooding carburetor	Check for flooding carburetor
	Sticking carburetor inlet needle	Repair the carburetor (Chapter Seven)
	Improper float level adjustment	Repair the carburetor (Chapter Seven)
	Blocked carburetor passages	Repair the carburetor (Chapter Seven)
	Faulty EFI sensor*	Check for diagnostic trouble codes
Excessive exhaust smoke		
	Improper carburetor adjustment	Adjust the carburetor (Chapter Five)
	Flooding carburetor	Check for flooding carburetor
	Flooding fuel cell*	Check for flooding fuel cell
	Choke valve malfunction	Check the choke valve operation
	Sticking carburetor inlet needle	Repair the carburetor (Chapter Seven)
	Improper float level adjustment	Repair the carburetor (Chapter Seven)
	Faulty engine temperature sensor*	Test the engine temperature sensor
	Faulty manifold air pressure sensor (MAP)	Check for diagnostic trouble codes.
	Leaking fuel injectors*	Check the fuel injectors
	Fault intake air temperature sensor (IAT)	Test the intake air temperature sensor
	Faulty mechanical fuel pump	Test the mechanical fuel pump
	Faulty low-pressure electric pump	Test the low-pressure electric pump
	Faulty high-pressure electric pump*	Test the fuel pressure
	Faulty EFI sensor*	Check for diagnostic trouble codes

*Applies only to electronic fuel injection (EFI) models.

Table 4 CHARGING SYSTEM TROUBLESHOOTING

Symptom	Possible causes	Corrective action
Low charging output		
	Loose drive belt	Adjust drive belt tension (Chapter Nine)
	Faulty battery	Charge and test the battery (Chapter Nine)
	Dirty or loose battery cables	Clean and check terminals (Chapter Nine)
	Blown fuse	Check fuses
	Tripped circuit breaker	Check circuit breaker
	Excitation circuit fault	Test the excitation circuit
	Faulty sensing circuit	Test the sensing circuit
	Faulty voltage regulator	Perform alternator output test
	Faulty diode trio	Perform diode trio test
	Faulty alternator	Perform alternator output test
Battery is overcharging		
	Dirty or loose battery cables	Clean and check terminals (Chapter Nine)
	Faulty sensing circuit	Test the sensing circuit
	Faulty voltage regulator	Perform alternator output test
	Faulty alternator	Perform alternator output test
Battery discharges at rest		
	Faulty battery	Charge and test the battery (Chapter Nine)
	Shorted boat accessory wiring	Check accessory wiring
	Shorted boat accessory	Check accessories
	Faulty voltage regulator	Perform alternator output test
	Faulty diode trio	Perform diode trio test
	Faulty alternator	Perform alternator output test
	Shorted instrument harness wiring	Check wiring
	Shorted engine harness wiring	Check wiring
	Excitation circuit fault	Test the excitation circuit
	Faulty sensing circuit	Test the sensing circuit

Table 5 POWER STEERING SYSTEM TROUBLESHOOTING

Symptom	Possible causes	Corrective action
Steers hard in both directions		
	Low fluid level	Check fluid level
	Loose drive belt	Adjust belt tension
	Improper cable installation	Check cable installation
	Binding cable or helm	Check cable and helm
	Air in system	Bleed the system
	Faulty pump	Repair pump
	Faulty actuator	Replace actuator
Steers hard in one direction only		
	Improper cable installation	Check cable installation
	Binding cable or helm	Check cable and helm
	Faulty actuator	Replace actuator
Noisy pump		
	Low fluid level	Check fluid level
	Loose drive belt	Adjust belt tension
	Air in system	Bleed the system
	Contaminated fluid	Inspect the fluid
	Loose pump	Check the pump mounting
	Sticking flow control valve	Repair pump
	Faulty pump	Repair pump
Noisy pump (only in full turn)		
	Low fluid level	Check fluid level (Chapter Three)
	Improper cable installation	Check cable installation
	Steering lever contacting transom	Check transom and lever clearance
Steering wheel jerks while turning	Low fluid level	Check fluid level
	Loose drive belt	Adjust belt tension
	Improper cable installation	Check cable installation
	Binding cable or helm	Check cable and helm
	Air in system	Bleed the system
	Faulty actuator	Replace actuator
Steers hard only when turning quickly	Low fluid level	Check fluid level
	Loose drive belt	Adjust belt tension
	Binding cable or helm	Check cable and helm
	Sticking flow control valve	Repair pump
	Faulty pump	Repair pump
	Faulty actuator	Replace actuator
Fluid appears milky		
	Water in system	Inspect fluid and check fluid cooler
	Air in system	Bleed the system
Losing fluid		
	External leak	Inspect hoses and fittings
	Leaking fluid cooler	Inspect fluid cooler
Rising fluid level		
	Leaking fluid cooler	Inspect fluid cooler
Steering wheel creeps or rotates without assistance	Improper cable installation	Check cable installation
	Faulty actuator	Replace actuator

Table 6 POWER TRIM SYSTEM TROUBLESHOOTING

Symptom	Possible causes	Corrective action
Trim motor does not operate (both directions)	Faulty fuse or circuit breaker	Check fuse and circuit breakers
	Faulty trim switch	Test the trim switch and limit circuits
	Faulty trim limit system	Test the trim switch and limit circuits
	Faulty wiring or terminals	Check wiring and terminals

(continued)

Table 6 POWER TRIM SYSTEM TROUBLESHOOTING (continued)

Symptom	Possible causes	Corrective action
Trim motor does not operate (both directions) (continued)		
	Faulty trim motor	Test the relays
	Faulty relay(s)*	Test the relays
	Faulty circuit board*	Test the circuit board
	Faulty trim position sender	Test the trim switch and limit circuits
Trim motor does not run in the up direction		
	Faulty trim switch	Test the trim switch and limit circuits
	Faulty trim limit system	Test the trim switch and limit circuits
	Faulty wiring or terminals	Check wiring and terminals
	Faulty circuit board*	Test the circuit board
	Faulty relay(s)	Test the relays
Trim motor does not run in the down direction		
	Faulty trim switch	Test the trim switch and limit circuits
	Faulty trim limit system	Test the trim switch and limit circuits
	Faulty wiring or terminals	Check wiring and terminals
	Faulty circuit board*	Test the circuit board
	Faulty relay(s)	Test the relays
Trim motor operates slowly		
	Low battery voltage	Charge and test the battery (Chapter Nine)
	Faulty trim motor wiring	Check wiring and terminals
	Faulty trim motor	Replace trim motor
Noisy trim motor		
	Low trim fluid level	Check fluid level
	Air in hydraulic system	Fill and bleed system
	Blocked pump pickup screen*	Replace pickup screen
	Faulty up pressure relief valve	Replace up pressure relief valve
	Faulty down pressure relief valve	Replace down pressure relief valve
	Faulty hydraulic pump	Replace hydraulic pump
	Faulty trim motor	Replace trim motor
Drive unit will not trim up or down		
	Low trim fluid level	Check fluid level
	Air in hydraulic system	Fill and bleed system
	Blocked pump pickup screen*	Replace pickup screen
	Damaged shaft coupler	Remove trim motor
	Faulty hydraulic pump	Replace hydraulic pump
Drive unit will not trim up (trims down)		
	Low trim fluid level	Check fluid level
	Air in hydraulic system	Fill and bleed system
	Blocked pump pickup screen*	Replace pickup screen
	Faulty up pressure relief valve	Replace up pressure relief valve
	Worn or damaged hydraulic pump	Replace hydraulic pump
Drive unit will not trim down (trims up)		
	Air in hydraulic system	Fill and bleed system
	Blocked pump pickup screen*	Replace pickup screen
	Faulty down pressure relief valve	Replace down pressure relief valve
Drive leaks down or tucks under in forward		
	Leaking hose, line or fitting	Inspect all hoses, lines and fittings
	Air in the hydraulic system	Fill and bleed the system
	Internal cylinder leak	Check for internal trim cylinder(s) leak
	Debris in piston check valve	Check for internal trim cylinder(s) leak
	Faulty thermal relief valve*	Replace thermal relief valve
	Faulty up pressure check valve	Repair hydraulic pump
Drive moves up or out in reverse		
	Leaking hose, line or fitting	Inspect all hoses, lines and fittings
	Air in the hydraulic system	Fill and bleed the system
	External cylinder leak	Inspect cylinder(s)
	Internal cylinder leak	Check for internal trim cylinder(s) leak
	Debris in piston check valve	Check for internal trim cylinder(s) leak
	Faulty down pressure check valve	Repair hydraulic pump
Drive moves up or out during deceleration		
	Leaking hose, line or fitting	Inspect all hoses, lines and fittings
	Air in the hydraulic system	Fill and bleed the system
	External cylinder leak	Inspect cylinder(s)
	Internal cylinder leak	Check for internal trim cylinder(s) leak

(continued)

Table 6 POWER TRIM SYSTEM TROUBLESHOOTING (continued)

Symptom	Possible causes	Corrective action
Drive moves up or out during deceleration (continued)	Debris in piston check valve	Check for internal trim cylinder(s) leak
	Faulty down pressure check valve	Repair hydraulic pump
Drive will not trim out against propeller thrust (does not leak down)	Air in the hydraulic system	Fill and bleed the system
	Blocked pump pickup screen*	Replace pickup screen
	Faulty up pressure relief valve	Replace up pressure relief valve
	Worn or damaged hydraulic pump	Replace hydraulic pump
Drive will not trim down from the full up position	Low fluid level	Check fluid level
	Blocked pump pickup screen*	Replace pickup screen
	Faulty down pressure relief valve	Replace down pressure relief valve
	Faulty thermal relief valve*	Replace thermal relief valve

*Not used on all trim pumps.

Table 7 COOLING SYSTEM TROUBLESHOOTING

Symptom	Possible causes	Corrective action
Engine is overheating (all engine speeds)	Faulty seawater pump	Repair the water pump (Chapter Eight)
	Loose seawater pump drive belt	Check the belt tension (Chapter Eight)
	Blocked water inlets in drive	Inspect the water inlets
	Faulty thermostat	Test the thermostat
	Blocked thermostat housing	Inspect thermostat housing (Chapter Eight)
	Blocked power steering fluid cooler	Remove and inspect cooler (Chapter Eight)
	Blocked or collapsed hoses	Inspect hoses (Chapter Eight)
	Blocked oil cooler	Remove and inspect cooler (Chapter Eight)
	Blocked heat exchanger	Inspect heat exchanger (Chapter Three)
	Blocked exhaust elbow	Remove and inspect elbow (Chapter Eight)
	Blocked exhaust manifold	Remove and inspect manifold (Chapter Eight)
	Blown cylinder head gasket	Perform compression test
	Cracked or porous cylinder head	Inspect the oil for milky appearance
		Inspect the spark plugs for water
	Cracked or porous cylinder block	Inspect the oil for milky appearance
		Inspect the spark plugs for water
	Cracked or porous intake manifold	Inspect the oil for milky appearance
		Inspect the spark plugs for water
Engine is overheating (lower speeds only)	Faulty thermostat	Test the thermostat
	Loose seawater pump drive belt	Check the belt tension (Chapter Eight)
	Blocked thermostat housing	Inspect thermostat housing (Chapter Eight)
	Faulty seawater pump	Repair the water pump (Chapter Eight)
	Damaged recirculation pump	Remove and inspect pump (Chapter Eight)
Engine is overheating (higher speeds only)	Faulty seawater pump	Repair the water pump (Chapter Eight)
	Loose seawater pump drive belt	Check the belt tension (Chapter Eight)
	Blocked water inlets in drive	Inspect the water inlets
	Faulty thermostat	Test the thermostat
	Blocked thermostat housing	Inspect thermostat housing (Chapter Eight)
	Blocked power steering fluid cooler	Remove and inspect cooler (Chapter Eight)
	Blocked or collapsed hoses	Inspect hoses (Chapter Eight)
	Blocked oil cooler	Remove and inspect cooler (Chapter Eight)
	Blocked heat exchanger	Inspect heat exchanger (Chapter Three)
	Blocked exhaust elbow	Remove and inspect elbow (Chapter Eight)
	Blocked exhaust manifold	Remove and inspect manifold (Chapter Eight)
	Blown cylinder head gasket	Perform compression test
	Cracked or porous cylinder head	Inspect the oil for milky appearance
		Inspect the spark plugs for water
	Cracked or porous cylinder block	Inspect the oil for milky appearance
		Inspect the spark plugs for water

(continued)

Table 7 COOLING SYSTEM TROUBLESHOOTING (continued)

Symptom	Possible causes	Corrective action
Engine is overheating (higher speeds only) (continued)	Cracked or porous intake manifold	Inspect the oil for milky appearance
		Inspect the spark plugs for water
Engine is running too cool (lower speeds only)	Normal cold water operation	Check operation in warmer water
	Faulty thermostat	Test the thermostat
	Faulty thermostat housing	Replace thermostat housing
Engine is running too cool (higher engine speed only)	Faulty thermostat	Test the thermostat
	Faulty thermostat housing	Replace thermostat housing
Engine is running too cool (all engine speeds)	Faulty thermostat	Test the thermostat
	Faulty thermostat housing	Replace thermostat housing
Temperature gauge does not match actual temperature	Faulty engine temperature sender	Replace engine temperature sender (Chapter Eight)
	Faulty sender or gauge wiring	Check the wiring
	Faulty temperature gauge	Replace the gauge
Oil pressure gauge does not match actual oil pressure	Faulty oil pressure sender	Replace oil pressure sender (Chapter Eight)
	Faulty sender or gauge wiring	Check the wiring
	Faulty oil pressure gauge	Replace the gauge

Table 8 WARNING SYSTEM TROUBLESHOOTING

Symptom	Possible causes	Corrective action
Warning horn sounding	Engine is overheating	Check temperature gauge
	Overheating exhaust elbow	Check exhaust elbow temperature
	Low oil pressure	Check oil pressure gauge
	Faulty engine overheat	Test engine overheat switch
	Faulty exhaust overheat switch	Test exhaust overheat switch
	Faulty engine temperature sensor (EFI)	Test engine temperature sensor
	Faulty oil pressure switch	Test oil pressure switch
	Shorted tan/blue switch wire	Check wiring
	Shorted tan/blue ECU wire (EFI)	Check wiring
	Shorted tan/blue warning horn wire	Check wiring
Warning horn does not self test	Faulty ignition switch	Test ignition switch
	Faulty oil pressure switch	Test oil pressure switch
	Open tan/blue switch wire	Check wiring
	Open tan/blue ECU wire (EFI)	Check wiring
	Open tan/blue warning horn wire	Check wiring
	Open purple warning horn wire	Check wiring
	Faulty ECU	Check for diagnostic trouble codes
Engine overheating—Warning horn not sounding	Faulty engine overheat switch	Test engine overheat switch
	Faulty exhaust overheat switch	Test exhaust overheat switch
	Faulty engine temperature sensor (EFI)	Test engine temperature sensor
	Open tan/blue ECU wire (EFI)	Check wiring
	Open tan/blue warning horn wire	Check wiring
	Open purple warning horn wire	
	Faulty ECU	Check for diagnostic trouble codes
Low oil pressure—Warning horn not sounding	Faulty oil pressure switch	Test oil pressure switch
	Open tan/blue ECU wire (EFI)	Check wiring
	Open tan/blue warning horn wire	Check wiring
	Open purple warning horn wire	Check wiring
	Faulty ECU	Check for diagnostic trouble codes

Table 9 IGNITION AND EFI SYSTEM RESISTANCE SPECIFICATIONS

Test	Specification (ohms)
Fuel injection system components	
Engine temperature sensor (ECT)	
Testing temperature	
212° F (100° C)	177
158° F (70° C)	467
104° F (40° C)	1800
68° F (20° C)	3520
41° F (5° C)	7500
Air temperature sensor (IAT)	
Testing temperature	
104° F (40° C)	1459
95° F (35° C)	1802
86° F (30° C)	2238
77° F (25° C)	2296
68° F (20° C)	3520
59° F (15° C)	4450
50° F (10° C)	5670
Ignition system components	
Ignition coil	
Delco EST	
Primary winding	1.35-1.45
Secondary winding	6700-7500
Prestolite BID	
Primary winding	1.25-1.40
Secondary winding	9400-11,700
Pickup coil	
Delco EST	500-1500

Table 10 FUEL PRESSURE SPECIFICATIONS

Model	Specification psi (kPa)
Mechanical fuel pump	3-7 (20.7-48.3)
Electric fuel pump	
Carburetor equipped models	
At cranking speed	4.9-8.5 (33.8-58.6)
At idle speed (600 rpm)	4.9-8.5 (33.8-58.6)
At 1000 rpm	4.8-8.4 (33.1-57.9)
At 1500 rpm	4.7-8.3 (32.4-57.2)
At 2000 rpm	4.6-8.2 (31.7-56.5)
At 2500 rpm	4.5-8.1 (31.0-55.8)
At 3000 rpm	4.4-8.0 (30.3-55.2)
At 3500 rpm	4.3-7.9 (29.7-54.5)
At 4000 rpm	4.0-7.7 (27.6-53.1)
At 4600 rpm	3.5-7.3 (24.1-50.3)
Electronic fuel injection (EFI) models	
Low-pressure electric fuel pump	4.0-8.5 (27.6-58.6)
High-pressure electric fuel pump	
Throttle body injection (TBI) models	27-31 (186-214)
Multiport injection (MPI) models	36-42 (248-290)

Table 11 DIAGNOSTIC TROUBLE CODES (ALL GM EFI MODELS)

Code	Fault
12	Normal condition (diagnostic system is operational)
14	Engine temperature sensor (ECT) out of range
21	Throttle position sensor (TPS) out of range
23	Intake air temperature sensor (IAT) reading out of range

(continued)

Table 11 DIAGNOSTIC TROUBLE CODES (ALL GM EFI MODELS) (continued)

Code	Fault
33	Manifold air pressure (MAP) sensor out of range
42	Shorted ignition control or bypass circuit
43	Incorrect knock sensor module voltage
51	Engine control unit (ECU) calibration failure

Table 12 ENGINE OPERATING SPEED SPECIFICATIONS

Model	Idle speed (rpm)	Wide open throttle speed (rpm)
GM 3.0L four-cylinder models	650-750	4200-2600
GM 4.3L V-6 models		
43GL (carburetor-equipped) models	550-650	4200-4600
43GS (carburetor-equipped) models	550-650	4200-4800
43GI (electronic fuel injection) models	600*	4400-4800
GM 5.0L V-8 models		
50GL (carburetor-equipped) models	550-650	4400-4800
50GI (electronic fuel injection) models	600*	4600-5000
Ford 5.0L V-8 models		
50FL (carburetor-equipped) models	550	4200-4600
50FI (electronic fuel injection) models	600*	4200-4600
GM 5.7L V-8 models		
57GL (carburetor-equipped) models	600-650	4200-4600
57GS (carburetor-equipped) models	550-650	4400-4800
57GI (electronic fuel injection) models	600*	4200-4600
57GSI (electronic fuel injection) models	600*	4600-5000
Ford 5.8L V-8 models		
58FL (carburetor-equipped) models	550-650	4000-4400
58FI (electronic fuel injection) models	600*	4200-4600
58FSI (electronic fuel injection) models	600*	4600-5000
GM 7.4L V-8 models		
74GL (carburetor-equipped) models	550-650	4200-4600
74GI (electronic fuel injection) models	600*	4200-4600
74GSI (electronic fuel injection) models	600*	4800-5200
GM 8.2L V-8 models		
82GL (carburetor-equipped) models	550-650	4400-4800
82GSI (electronic fuel injection) models	600*	4600-5000

*The idle speed is nonadjustable on this model.

Chapter Three

Lubrication, Maintenance and Tune-up

When operating properly, the OMC Cobra stern drive engine provides smooth operation, reliable starting and excellent performance. This chapter describes the procedures necessary to keep the engine and drive in peak condition.

Table 1 lists maintenance torque specifications for fasteners and spark plugs. Tighten fasteners or plugs not listed in **Table 1** to the general torque recommendations listed in Chapter One. **Table 2** lists maintenance intervals for all engine systems and components. If the boat is used for continuous heavy-duty, high-speed operation or under other severe operating conditions, all maintenance operations, including lubrication procedures, should be performed more frequently. **Table 3** lists engine oil and drive unit lubricant capacities. Refer to **Table 4** for recommended lubricants. **Table 5** lists spark plug specifications. **Tables 1-5** are located at the end of this chapter.

If the engine becomes partially or completely submerged, refer to *Submersion* in this chapter for service instructions. Service the engine as soon as possible to minimize the damage.

If the boat is not used regularly, moisture and dirt will collect in and on the engine and stern drive. The moisture and dirt will eventually lead to corroded surfaces and other damage. It is good to keep the engine and accessory units free of dirt and grease buildup. Frequent cleaning allows detection and correction of leaks before serious damage occurs.

BEFORE EACH USE

To avoid engine damage or equipment failure, certain items must be checked before use. This section describes inspection of the following fluids and components.

1. Fuel system leakage.
2. Engine oil level.
3. Drive unit lubricant level.
4. Power steering fluid level.
5. Power trim fluid level.
6. Propeller and propeller shaft.
7. Steering system.
8. Coolant level.
9. Emergency stop (lanyard) switch operation.

NOTE

If the boat is stored in a wet slip, it might be impractical to inspect drive system components such as the drive lubricant level and the propeller and shaft. In such instances, per-

form the inspections when the boat is removed from the water for routine maintenance.

Fuel Leakage Check

Before starting the engine for the first time each day, remove the engine cover or hatch and check for the presence of gasoline fumes or visual evidence of fuel leakage. Operate the bilge blower for a few minutes. If strong fumes can be smelled, determine the source and correct the leakage before operating or servicing the engine.

Engine Oil Level Check

CAUTION
Never operate the engine with the oil level above the full line or below the fill line on the dipstick. Overfilling can result in foaming of the oil. Operation with a low or high oil level can result in inadequate lubrication of internal components and certain engine damage.

To avoid overfilling the oil, wait until the engine has been switched off for 30 minutes or more. This allows the oil to drain back into the oil pan.

On GM 3.0L four-cylinder models, the dipstick (**Figure 1**) is located on the starboard side of the engine. On all other models, the dipstick (**Figure 2**, typical) is located on the port or starboard side of the engine and near the exhaust elbow or manifold. The yellow or red color makes it easy to identify.

On GM 3.0L four-cylinder models, the oil fill cap (**Figure 3**) attaches to the rocker arm cover. On V-6 and V-8 models, the oil fill cap (**Figure 4**, typical)is located on one of the rocker arm covers. On some models, a fill cap is present on both rocker arm covers.

Check the oil level as follows:

1. Pull the dipstick from the tube.
2. Wipe the dipstick with a clean shop towel.
3. Fully insert the dipstick into the tube.
4. *Slowly* pull the dipstick from the tube and note the oil level. It should fall within the *SAFE* area (**Figure 5**).
5. Inspect the oil for the presence of water, a milky appearance or significant fuel odor. Refer to *Engine* in Chapter Two if any of these conditions is noted. Also refer to *Water Entering the Oil* and *Lubrication System Failure.*
6. If necessary, add oil until the level is even with the upper *SAFE* marking. Do not overfill the engine. It is better for the oil level to be slightly low than overfilled. If the engine is overfilled, drain the excess oil as described under *Changing the Oil* in this chapter.
7. Insert the dipstick fully into the tube.

5

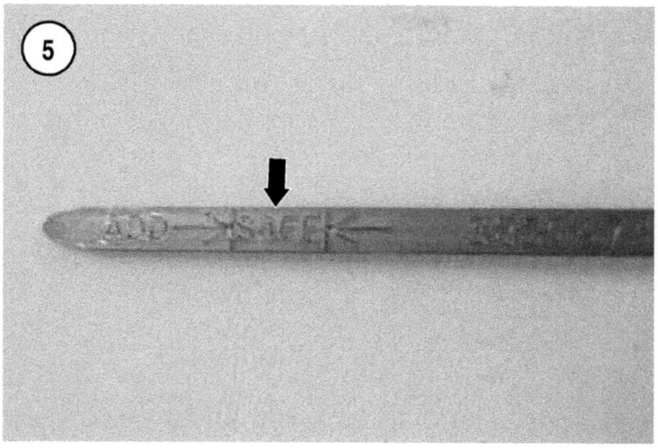

6

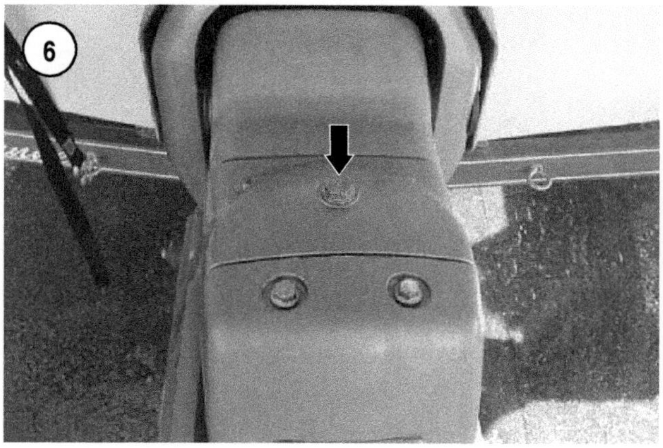

7

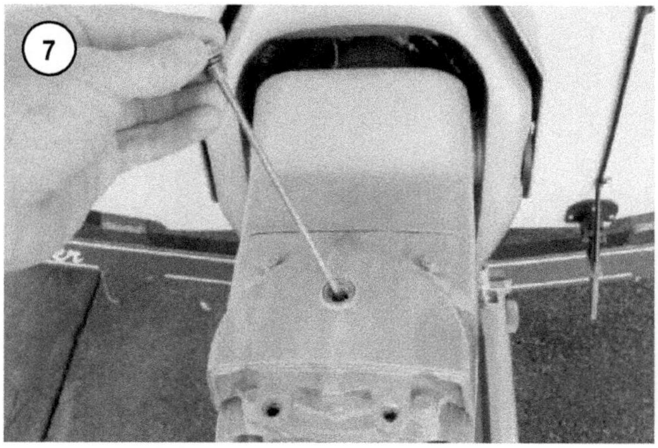

8

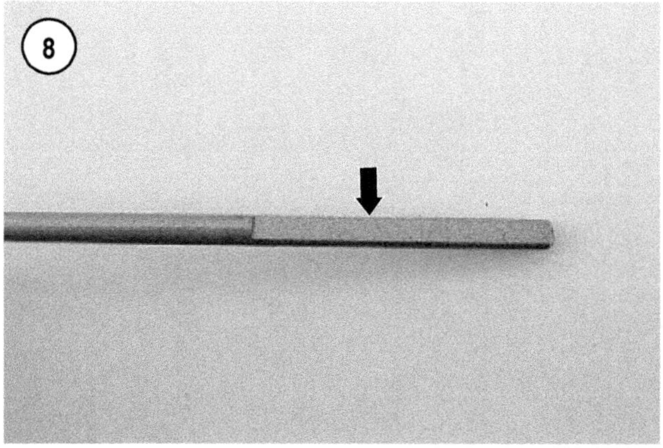

Drive Unit Lubricant Level Check

NOTE
To avoid overfilling, wait until the engine has been switched off for 30 minutes or more before checking the drive unit lubricant level. Because the lubricant expands when warmed, testing the lubricant level too soon after engine operation will give a false reading.

1. Place the drive in the vertical position. Locate the gearcase lubricant level dipstick (**Figure 6**) on the top of the drive unit.
2. Carefully loosen the dipstick. Use a wrench on the hex section to avoid damaging the screw slot. Allow any air pressure in the drive unit to dissipate. Unthread and remove the dipstick (**Figure 7**). Inspect the O-ring on the dipstick for torn, flattened or damaged surfaces. Replace the O-ring unless found in excellent condition.
3. Wipe the dipstick with a clean cloth or paper towel.
4. Reinstall the dipstick and screw fully into the opening. Wait a few minutes. Then remove the dipstick and note the lubricant level. The lubricant level must be within the flattened area of the dipstick (**Figure 8**). Refer to the following:
   a. If the lubricant level is low, add gearcase lubricant as described in *Drive Unit Lubricant Change* in this chapter. If the unit requires more than about 2 oz. (59 mL), a lubricant leak is probable. If necessary, pressure test and vacuum test the drive unit as described in Chapter Ten.
   b. If the lubricant level is too high, drain excess oil as described in *Drive Unit Lubricant Change*. Drain only a few ounces. Then quickly replace the fill plug. Allow a few minutes for the level to stabilize and recheck the level. Drain or fill the unit as necessary to correct the level.
5. If the lubricant level is correct, lubricate the O-ring with gearcase lubricant. Thread the dipstick fully into the opening. Use a wrench on the hex section to tighten the dipstick to 48-72 in.-lb. (5.4-8.1 N•m).
6. If a low lubricant level persists, check for loose drain/fill plugs. Then pressure test and vacuum test the drive unit as described in Chapter Ten.

Power Steering Fluid Level Check

Use Dexron III automatic transmission fluid or one of the common inboard engine manufacturer's recommended power trim/tilt and power steering fluid (such as Volvo Penta part No. 3851039 or Mercruiser part No. 92-90199A12) in the power steering system.

1. Locate the power steering pump (**Figure 9**, typical) on the front port or starboard side of the engine.
2. Place the drive unit in the straight-ahead position. Make sure the engine is stopped for a minimum of 5 minutes to allow the fluid level to stabilize.

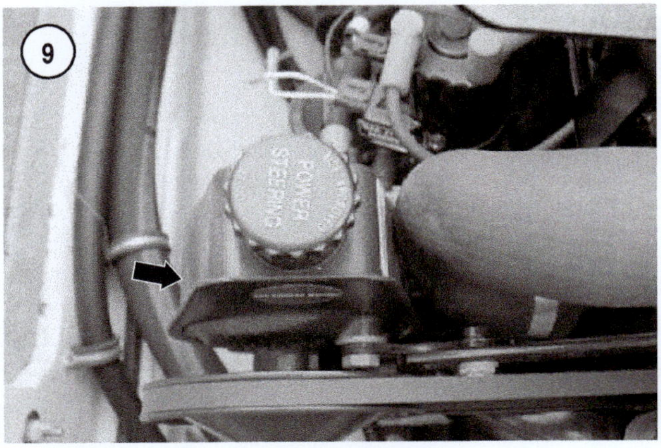

9

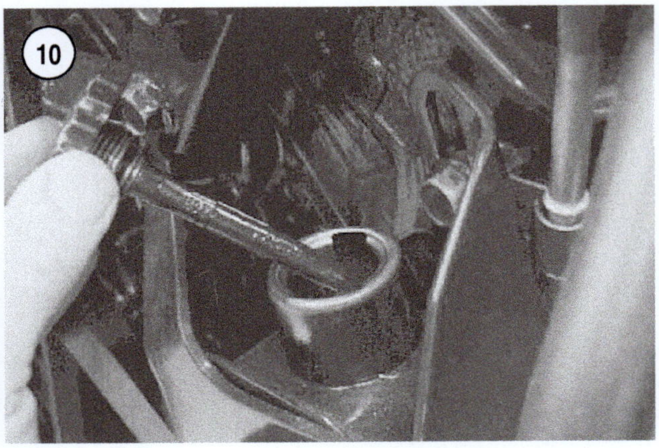

10

3. *Slowly* loosen and remove the fill cap (**Figure 10**). Wipe the dipstick with a clean shop towel and reinstall it into the pump reservoir.

4. Wait a few minutes for the fluid level to stabilize. Then remove the cap and note the fluid level on the dipstick (**Figure 11**). Correct the fluid level as follows:

   a. If the engine is warm, add fluid until the level just reaches the upper end of the range (**Figure 11**).

   b. If the engine is cold, add fluid until the level just reaches the groove below the *ADD* mark (**Figure 11**).

   c. If the reservoir is empty, fill and bleed the power steering system as described in Chapter Thirteen. Locate and correct the cause of the low fluid level before operating the engine.

   d. If the fluid level is too high (warm or cold engine), use a syringe or other means to remove excess fluid.

5. With the engine off, cycle the steering several times to the full port and starboard limits. Place the drive in the straight-ahead position and recheck the fluid level. Correct the level as necessary.

6. Install the fill cap.

Power Trim Fluid Level Check

The power trim pump mounts in or near the engine compartment. Two different types of pumps are used on the models covered in this manual, and the fluid level check procedures are different. Earlier models use a pump (**Figure 12**) with a level/fill plug on the side of the reservoir. Later models use a pump (**Figure 13**) with the fill cap and transparent fluid reservoir that allows visual inspection of the fluid level. Use Dexron III automatic transmission fluid or one of the common inboard engine manufacturer's recommended power trim/tilt and power steering fluids (such as Volvo Penta part No. 3851039 or Mercruiser part No. 92-90199A12) in the power trim system.

Earlier power trim pump

1. Place the drive in the full UP position.

11

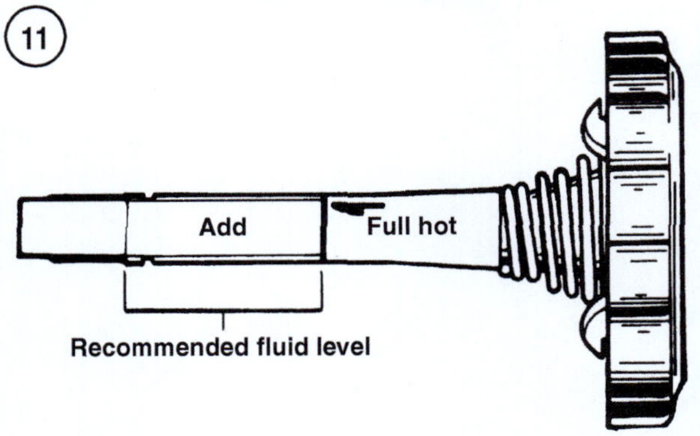

12

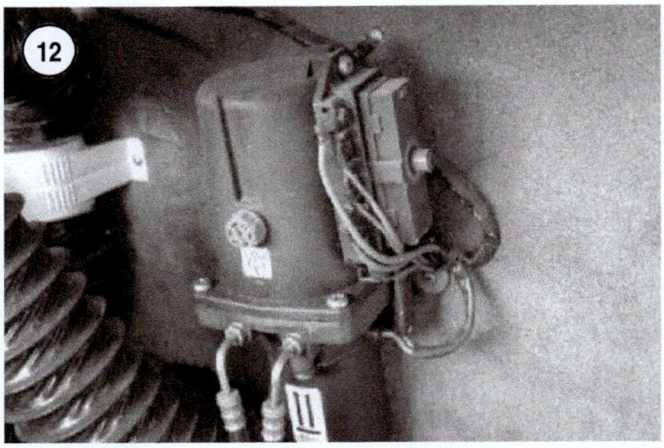

13

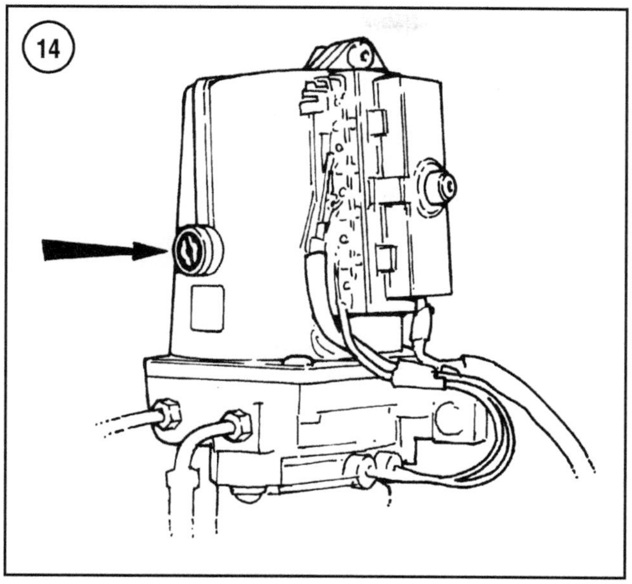

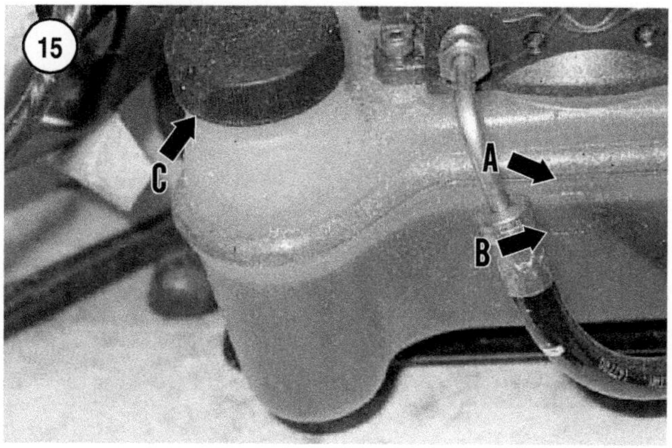

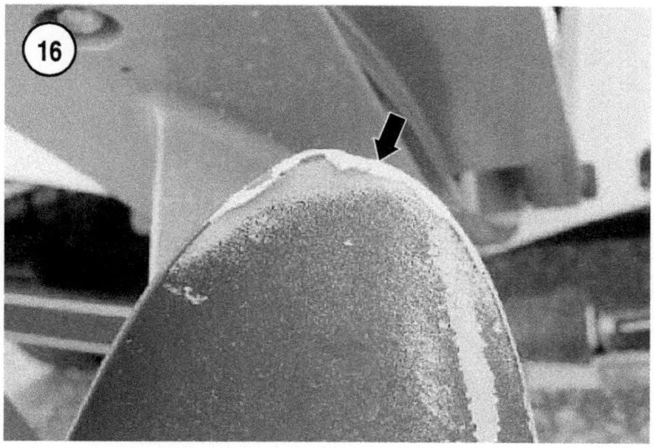

2. Slightly loosen the level/fill plug (**Figure 14**) and allow a few minutes for any internal pressure to dissipate. Remove the level/fill plug. Inspect the seal on the plug for tears, deterioration or other defects and replace as needed.

3. Note the fluid level. With the drive tilted fully up, the fluid must be even with the bottom of the plug opening.

   a. If the fluid has a milky appearance, check for oil leaks at the trim cylinders, hoses and fittings. Fill and bleed the trim system as described in Chapter Twelve.
   b. If the fluid level is low, add the recommended fluid. Check for external leaks and correct before operating the engine.

4. Install and securely tighten the level/fill plug.

5. Cycle the trim system to the full DOWN and UP limits several times to purge any air from the internal passages. Check for fluid leaks and correct as needed.

6. Trim the drive to the full UP position and again check the fluid level. Correct as needed.

3

Later power trim pump

1. Place the drive in the full DOWN position.
2. Note the fluid level in the translucent reservoir. With the drive down, the fluid must be clear, and the level must be between the *FULL* (A, **Figure 15**) and *ADD* (B) marks.
3. Remove the fill cap (C, **Figure 15**) and inspect the fluid.
   a. If the reservoir is overfilled, remove excess fluid using a syringe or other means.
   b. If the reservoir is empty, check for a leaking hose or fitting and cracks in the reservoir. Cracking of the reservoir is a common occurrence. Correct the oil leak. Then fill and bleed the power trim system as described in Chapter Twelve.
   c. If the reservoir is low, add fluid until the level just reaches the *FULL* mark. Cycle the trim to the full UP and DOWN limits several times to purge any air from internal passages. While operating the trim system, check for oil leaks at the hoses, fittings or trim cylinders. Correct any leaks before operating the engine.
   d. If the fluid has a milky appearance, check for evidence of a high water level in the vessel. Check for a leaking external trim hose or fitting. Fill and bleed the system as described in Chapter Twelve.
4. Install the cap. Do not overtighten the cap.

Propeller and Propeller Shaft Inspection

Inspect the propeller for cracked, damaged (**Figure 16**) or missing blades (**Figure 17**). Operating the engine with a damaged propeller results in decreased performance, excessive vibration and increased wear of the drive unit components. Straighten small bent areas with locking pliers. Repair small nicks with a metal file. To prevent an unbalanced condition, do not remove excessive amounts of material from the propeller. If significant damage is evident, have the propeller repaired at a propeller repair shop.

To check for a bent propeller shaft, disconnect the battery cables. Shift the drive unit into NEUTRAL. Spin the propeller while observing the end of the propeller shaft (**Figure 18**, typical). If the shaft wobbles noticeably, the propeller shaft is bent and must be replaced. Refer to *Lower Gearcase Disassembly or Assembly* in Chapter Ten.

Steering System Inspection

Before starting the engine, rotate the steering wheel to the clockwise and counterclockwise limits. Note any binding or excessive slack as the wheel changes direction. Inspect and repair worn or faulty components. Never operate the engine if the steering is binding or loose.

Coolant Level Check

Coolant level inspection applies only to engines equipped with a closed cooling system. The engine coolant section of the closed cooling system must be filled with a 50:50 solution of pure water and ethylene glycol-based antifreeze. The antifreeze mixture does not require draining during winter months unless the engine will be subjected to temperatures below the rating of the water and antifreeze mixture. The raw water section, however, must be drained *completely* before exposing the engine to freezing temperatures.

Allow the engine to cool completely before checking the coolant level. The coolant level drops as the engine cools.

Some closed cooling systems are equipped with a separate overflow reservoir (**Figure 19**, typical). The translucent reservoir allows visual inspection of the coolant level. Other systems commonly used on smaller engines do not use an overflow reservoir. The systems used on smaller engines have an expansion chamber integrated into the heat exchanger (**Figure 20**, typical). With this design, the fill cap must be removed to check the coolant level.

WARNING
Do not remove the pressure cap from the closed cooling system if the engine is warm. Hot coolant can spray out and cause serious injury.

NOTE
Closed cooling system kits are available from numerous sources. Cooling system design, hose routing and fluid capacity will vary by manufacturer. Refer to the information supplied by the cooling system manufacturer for specific information. This section describes maintenance for the typical closed cooling system.

With separate overflow reservoir

1. Allow the engine to cool completely before checking the coolant level. The coolant level drops as the engine cools.
2. Check the coolant level in the translucent reservoir. The level must be between the *ADD* and *FULL* lines. Correct the coolant level as follows:
   a. If the reservoir is empty, check for loose hose clamps, faulty hoses or a leaking recirculation pump. If the source of the leak is not visually apparent, pressure test the cooling system as described in Chapter Two.

17

18

19

20

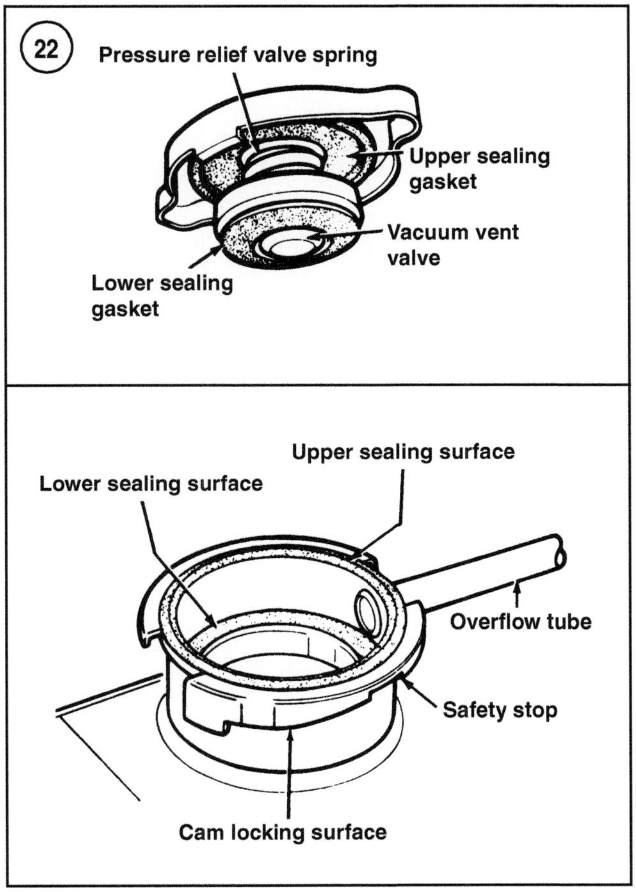

Refill the cooling system as described in this chapter (refer to *Changing the Engine Coolant*).

b. If the reservoir is low but not empty, carefully remove the cap and add the required amount of coolant-and-water mixture into the reservoir. Replace the cap. Check and correct the level after operating the engine long enough to reach full operating temperature. If the level has dropped significantly, check for loose hose clamps, faulty hoses or a leaking recirculation pump. If the source of the leak is not visually apparent, pressure test the cooling system as described in Chapter Two.

c. If the reservoir is overfilled, use a syringe or other means to remove excess coolant.

3. Start the engine. After the engine reaches normal operating temperature, check the level in the overflow reservoir. Cool the engine and correct the coolant level as needed.

Without separate overflow reservoir

1. Allow the engine to cool completely before removing the pressure cap.
2. Cover the cap (**Figure 21**, typical) with a heavy shop towel. Using gloves, push down on the pressure cap. Rotate the cap counterclockwise until the tabs are clear of the cam locking surfaces. Lift the cap from the reservoir.
3. Wipe debris and crystallized coolant from the cap and filler neck opening. Inspect the upper and lower gaskets on the cap (**Figure 22**) for cracked, missing or otherwise damaged surfaces. Make sure the cam locking surfaces on the filler neck (**Figure 22**) and cap are not damaged. Replace the cap or have the filler neck repaired if necessary. Damage to the seal area on the cap or filler neck will allow coolant leakage.
4. The coolant level should be within 1 in. (25 mm) of the filler neck opening. If necessary, *slowly* pour the coolant-and-water mixture into the opening until full. If more than a few ounces is required, check for loose hose clamps, faulty hoses or a leaking recirculation pump. If the source of the leak is not visually apparent, pressure test the cooling system as described in Chapter Two.

NOTE
To prevent the false indication of leakage, always wipe spilled coolant from the filler neck before installing the cap.

5. Wipe spilled coolant from the filler neck. Align the locking tabs on the cap with the notches on the filler neck. Push down on the cap to clear the cam locking surfaces. Then rotate the cap clockwise until it touches the stop (**Figure 22**). Release the cap.
6. If coolant was added, start the engine and allow it to reach operating temperature. Allow the engine to cool completely. Then check and correct the coolant level.
7. After starting the engine, check for coolant leakage around the cap and filler neck and correct leakage as needed.

Emergency Stop (Lanyard) Switch Operation

This switch is either incorporated in the remote control (**Figure 23**) or mounted on the dashboard (**Figure 24**). Check the operation of the switch before getting underway.

1. Make sure the lanyard cord is connected to the switch. If the boat is not in the water, prepare the engine for operation with a flush/test device or in a test tank. Start the engine. Do not shift into gear.
2. Pull the lanyard cord from the switch.
3. If the engine fails to stop, turn the ignition switch OFF. Repair or replace the switch before operating the engine.

AFTER EACH USE

Certain maintenance procedures must be performed after each outing of the craft. Observing these requirements can dramatically reduce corrosion of engine components and extend the life of the engine.

1. Flush the cooling system.
2. Clean debris or contaminants from the engine exterior surfaces.
3. Inspect the propeller and drive unit for damage.

Flushing the Cooling System

Flush the cooling system after each use to prevent deposit buildup and corrosion in the cooling passages. This is even more important if the engine was operated in salt, brackish or polluted water.

Engines equipped with a closed cooling system also benefit from regular flushing, because it reduces deposit buildup and corrosion in the heat exchanger, exhaust manifold(s) and fluid coolers.

Many later models are factory equipped with a freshwater flushing system (**Figure 25**, typical). This system allows connection of a garden hose to the cooling system to allow flushing of the cylinder block (standard cooling system only) and exhaust manifold(s) with the boat in the water. Never start the engine while using the freshwater flushing system because the seawater pump will introduce contaminated water into the freshwater.

For boats stored on a trailer or boat lift, flush the engine using a flush/test device (**Figure 26**, typical) connected to a garden hose. Available from marine dealerships or marine supply stores, the flush/test device flushes the entire seawater side of the cooling system, including the water passages in the drive unit. The engine can also be flushed by operating in a suitable test tank filled with clean water.

CAUTION
Never run the engine without first providing cooling water. Use either a test tank or a flush/test device.

CAUTION
Some models are equipped with a freshwater flush system that also serves as a means to drain the cylinder block. If the engine is equipped with a closed cooling system, this system is used only to drain the coolant from the cylinder block. If the engine has the flush system and is equipped with a closed cooling system, do not connect a garden hose to the drain hose and flush the system. The coolant will be diluted by the water, diminishing the freeze protection, corrosion protection and cooling system efficiency.

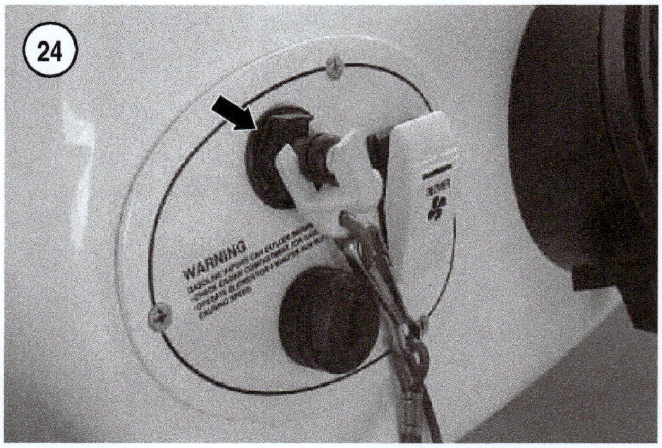

Using a flush/test device

WARNING
Stay clear of the propeller shaft while running the engine on a flush/test device. Remove the propeller(s) before running the engine to prevent injury or death. Always disconnect the battery cables before removing or installing the propeller.

1. Remove the propeller(s) as described in Chapter Ten.
2. Connect the garden hose onto the adapter.

NOTE
The arms on most flush/test devices are not long enough, when installed from the front of the drive unit, to position the cups over the water inlets. If necessary, install the device from the rear of the drive unit.

3. Beginning at the rear of the drive unit, slide the cups over each side of the gearcase. Position the cups over the water inlets.
4. Turn on the water. Make sure the flush adapter is positioned over the water inlets. Start the engine and run at fast idle (approximately 1500 rpm) in NEUTRAL until the engine reaches full operating temperature.
5. Continue to run the engine until the water exiting the engine is clear, and the engine has run for a minimum of 5

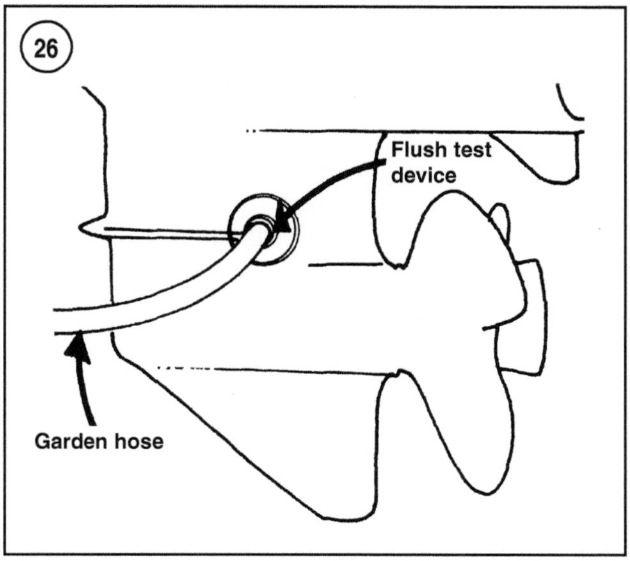

minutes. Monitor the engine temperature. Stop the engine if it begins to overheat or if water is not exiting the exhaust.

6. Throttle back to idle and stop the engine. Remove the flush/test adapter. Install the propeller(s) as described in Chapter Ten.

7. Allow the drive unit to remain in the vertical position for a few minutes to allow complete draining of the cooling passages.

8. Install the propeller(s) as described in Chapter Ten.

Using the freshwater flushing system

1. Remove the cap (A, **Figure 25**) from the flushing system hose. On some models, the adapter cap is brass, and on other models, the cap is blue plastic.

2. Detach the brass adapter (B, **Figure 25**) from the clamp and carefully thread it onto the hose fitting. Thread the garden hose onto the adapter.

3. Turn on the water. Immediately check for leakage from the adapter. Tighten the adapter fittings if leaking.

4. Leave the water on for a minimum of 5 minutes. Turn the water off and disconnect the water hose.

5. Remove the brass adapter from the flush hose. Connect the adapter to the clamp. Thread the cap onto the hose and securely tighten.

Using a test tank

To ensure adequate flushing of the cooling system, make sure the test tank is of an adequate size (minimum of 100 gallons). The contaminated water remaining in the engine after operation will mix into the freshwater in the tank. If a small test tank must be used, and the engine was operated in saltwater, heavily polluted or brackish water, repeat the flush operation two or more times with the tank drained and refilled with freshwater after each flushing operation.

1. Remove the propeller(s) as described in Chapter Ten.

2. Position the drive unit full UP. Position the boat with the drive unit directly over the test tank. Tilt the drive DOWN to position the water inlets well below the water level. Add water, if necessary, until the water level is a minimum of 6 in. (152 mm) above the water inlets.

3. Start the engine and run at fast idle (approximately 1500 rpm) in NEUTRAL until the engine reaches full operating temperature.

4. Continue to run the engine until the engine has run for a minimum of 5 minutes. Monitor the engine temperature. Stop the engine if it begins to overheat.

5. Throttle back to idle and stop the engine. Position the drive unit full UP and move the boat away from the test tank.

6. Position the drive full DOWN. Allow the drive unit to remain in the vertical position for a few minutes to allow complete draining of the cooling passages.

7. Drain the contaminated water from the test tank and refill with freshwater.

8. Install the propeller(s) as described in Chapter Ten.

Cleaning the Engine

Clean all external engine surfaces after each use. This step slows corrosion, reduces wear on gearcase and trim system seals and makes worn or damaged components easier to see.

Never use strong cleaning solutions or solvents to clean the engine. Mild dish soap and pressurized water adequately clean most debris from the engine. To prevent water intrusion, never direct water into the exhaust opening(s). Avoid directing high-pressure water spray at any openings, seals, plugs, wiring or wire grommets. The water might bypass seals and contaminate the trim fluid or enter the trim sender.

Rinse all external surfaces with clean water to remove soap residue. Wipe the engine with a soft cloth to prevent water spots.

Use a dry cloth to wipe oil or debris from components inside the engine compartment. Never direct water onto any

27

28

electrical or fuel system components on the engine. These components are not waterproof.

Propeller and Drive Unit Inspection

Inspect the drive unit after each outing. Smooth minor chips in the skeg with a file. Major damage to the skeg should be repaired at a propeller repair shop. Gearcase removal and installation is described in Chapter Ten. Touch up chipped or abraded paint with the recommended paint available at an OMC or Volvo Penta dealership.

Inspect the propeller(s) and propeller shaft(s) as described in this chapter.

LUBRICATION

Proper lubrication is absolutely critical to engine operation. Proper lubrication helps prevent wear to the engine, drive unit and other areas; guards against corrosion; and provides smooth operation of moving parts.

Stern drive engines operate in a corrosive environment and often require special types of lubricants. Using the wrong type of lubricant can cause serious engine damage or substantially shorten the life of the engine.

Special lubricant pumps, available from most automotive parts stores, are required for some of the procedures described in this section. The lubricants described in these procedures are available from most marine suppliers or marine dealerships.

Engine Oil and Filter Change

Change the engine oil and filter after the initial 20 hours of operation and then every 100 hours of operation or once a season, whichever comes first. Always replace the filter when changing the oil to prevent contaminants in the filter from mixing with the fresh oil.

Use only the recommended oil filter available from an OMC or Volvo Penta dealership. Oil filters designed for an automotive application might not meet the filtration requirements for the engine. Using the wrong filter can cause increased engine wear or serious engine damage.

OMC recommends using a high-quality FC-W motor oil certified for marine use by the NMMA (National Marine Manufacturers Association) in all models covered by this manual (refer to **Table 4** for additional lubricant recommendations). If this type of oil is not available, use a quality single viscosity SG/CD service classification oil that meets General Motors Standard GM-6094-M for GM engines and Ford Specification ESE-M2C153-E for Ford engines. The classification is printed on the container label.

Engine oil weight recommendations vary with the temperature of the air where the engine is operated. If the maximum anticipated air temperature will not exceed 0° F (-18° C), use an SAE 10W oil. If the minimum anticipated temperature exceeds 0° F (-18° C), and the maximum anticipated temperature will not exceed 32° F (0° C), use an SAE 20W-20 oil. If the maximum anticipated temperature exceeds 32° F (0° C), use an SAE 30 oil.

NOTE
In some instances, the oil fill cap on the engine is imprinted with engine oil weight and service classification specifications. The information on the cap might or might not reflect the actual engine oil requirements. Disregard the specification on the oil fill cap.

Oil filter change

Use an oil filter wrench to loosen and tighten the oil filter. Other tools can damage the filter. Never overtighten the replacement filter because future removal will be difficult, and overtightening can damage the filter. A damaged filter can rupture while underway and cause oil leakage and possible engine damage. Oil filter mounting locations vary by model.

On GM 3.0L four-cylinder models, the oil filter (**Figure 27**) mounts on the starboard side of the cylinder block. On GM V-6 and V-8 models, the oil filter mounts on the rear port side of the cylinder block. On Ford 5.0L and 5.8L V-8 models, the oil filter mounts on the front port side of the cylinder block. An optional remote oil filter mounting kit is available for these models and is factory installed on some

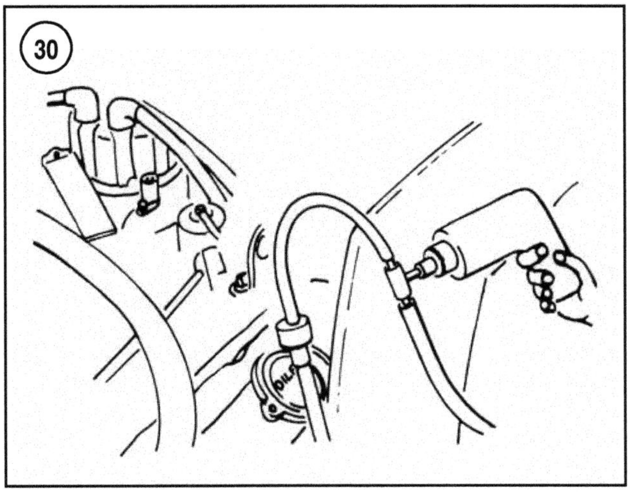

electronic fuel injection (EFI) models. The kit positions the oil filter (**Figure 28**) on the top port side of the engine just forward of the exhaust elbow. Consider adding this kit if boat structure prevents reasonable access to the filter.

1. Disconnect the battery or switch off the battery switch to prevent accidental arcing.
2. Place a suitable container or shop towel under the filter to capture spilled oil.
3. Loosen the filter until the filter just breaks contact with the mating surface. Allow a few minutes for the oil to drain from the filter. Turn the filter until free from the threaded adapter.
4. Remove the gasket (**Figure 29**) from the cylinder block or adapter if it does not come off with the filter.
5. Empty the filter into a suitable container. Dispose of the filter in an environmentally responsible manner. Contact a marine dealership, auto repair shop or automotive parts store for information on the proper disposal of the oil filter.
6. Thoroughly clean the filter mating surface on the cylinder block or remote filter adapter. Apply a light coat of oil to the gasket on the new filter.
7. Thread the filter onto the threaded adapter until the gasket just touches the mating surface. Use a filter wrench and tighten the filter exactly one turn. Do not overtighten the filter.
8. Change the engine oil as described in this chapter.
9. Connect the battery cables or turn on the battery switch.
10. Start the engine and immediately check for oil leakage. Monitor the oil pressure gauge. The oil pressure must rise within a few seconds of starting the engine. Correct oil leakage or low oil pressure before operating the engine.

Engine oil change

If easily accessible, remove the drain bolt from the oil pan fitting and drain the oil into a suitable container. This method allows complete draining of the oil and most of the contaminants that rest on the bottom of the pan. In most boats, it is difficult if not impossible to place a container under the pan. The most common method of removing the oil is to draw the oil from the pan through the dipstick tube. This requires an oil pumping device (**Figure 30**) that can be attached to the dipstick tube. This type of device is available from marine dealerships, marine supply stores and most automotive parts stores. Because the dipstick tube attaches to a fitting on the bottom of the oil pan, most of the oil and contaminants can be removed by this method.

1. Disconnect the battery cables or turn off the battery switch to prevent accidental arcing.
2. Remove the oil fill cap to allow air into the pan as the oil is removed.

3A. To drain oil directly from the pan:
   a. Place a suitable container under the oil drain bolt.
   b. Slowly loosen and remove the oil drain bolt, dipstick tube fitting and sealing washers as described in Chapter Six (refer to *Oil Pan*.
   c. Clean the bolt, sealing washers and dipstick tube fitting.
   d. After draining, install the drain bolt, sealing washers and dipstick tube fitting as described in Chapter Six (refer to *Oil Pan*).

3B. To pump oil using the dipstick tube:
   a. Remove the oil fill cap to allow air into the pan as the oil is removed.
   b. Remove the dipstick.
   c. Attach the oil pump hose to the dipstick tube.
   d. Pump the oil until the pump starts to ventilate. This indicates the oil has been sufficiently removed, and air is being drawn into the dipstick tube.
   e. Wipe clean and then install the dipstick.

NOTE

With the manifolds, fuel system, cooling system and electrical components installed, the General Motors (GM) 5.0L and 5.7L V-8 engines are very similar in appearance to the Ford 5.0L and 5.8L V-8 engines. If the model identification decal on the flame arrestor is missing, the engine can be quickly identified as a GM or Ford engine by noting the distributor location. On GM engines, the distributor is located at the rear (flywheel end) of the en-

gine. On Ford engines, the distributor is located on the front of the engine.

4. Refer to **Table 3** to determine the engine oil capacity. Slowly pour 90 percent of this amount into the oil fill opening. Install the oil fill cap.
5. Change the oil filter as described in this chapter.
6. Connect the battery cables or turn on the battery switch.
7. Start the engine and monitor the oil pressure gauge. The pressure must rise within a few seconds of starting the engine. Otherwise, check for an improperly installed oil filter.
8. Stop the engine and allow at least 30 minutes for the oil to drain back into the oil pan. Check and correct the oil level as described in this chapter (refer to *Before Each Use*).

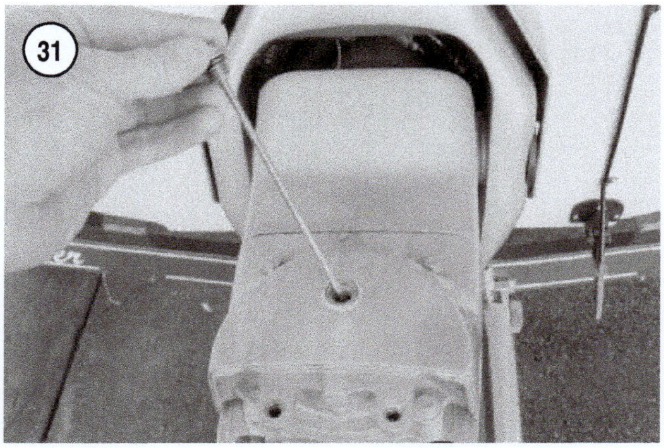

Drive Unit Lubricant Change

Change the drive unit lubricant after the initial 20 hours of operation and then after every 100 hours of operation or once a season, whichever comes first. OMC recommends using Volvo Penta Dura Plus GL5 Synthetic gear lubricant or Mobil One SHC SAE 75-90 Fully Synthetic gearcase lubricant in the Cobra SX and Cobra DP-S DuoProp drive units used on all models covered in this manual. Other types of gearcase lubricant might not meet the requirements of the drive unit. Using the wrong type of lubricant can result in increased wear, corrosion and/or failure of internal components. The wrong type of lubricant can also cause improper shifting into or out of gear. Refer to **Table 3** to determine the approximate drive unit capacity. A lubricant pump is required for this operation. A special adapter (OMC or Volvo Penta part No. 3855932) is required to connect the lubricant pump hose into the fill opening on models using a DP-S DuoProp drive unit. Purchase the lubricant, pump and adapter from an OMC or Volvo Penta dealership.

CAUTION

Never use automotive gearcase or transmission lubricant in the drive unit. Automotive lubricants are usually not suitable for marine applications. Using lubricants other than the recommended lubricants can lead to increased wear, corrosion, improper shifting and drive unit failure.

NOTE

A small amount of very fine particles are usually present on the drain plug. These fine particles form during normal operation. The presence does not necessarily indicate a problem. The presence of large particles, however, indicates a potential problem within the drive unit.

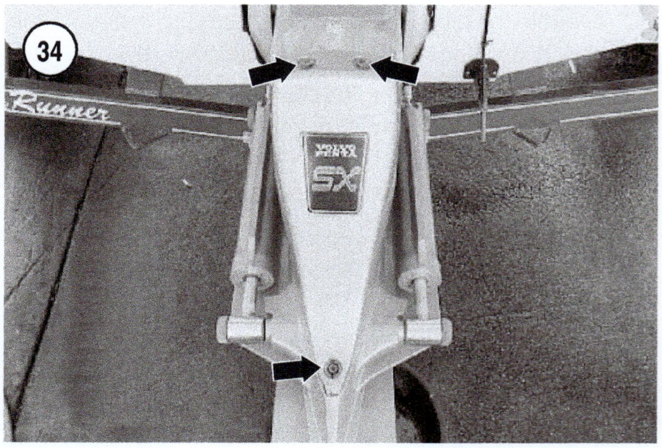

1. Disconnect the battery cables.
2. Lower the drive unit and place a suitable container under the drive to capture lubricant.
3. Remove the dipstick (**Figure 31**) to allow air to enter the lubricant cavities as the lubricant drains. Inspect the O-ring

35

36

37

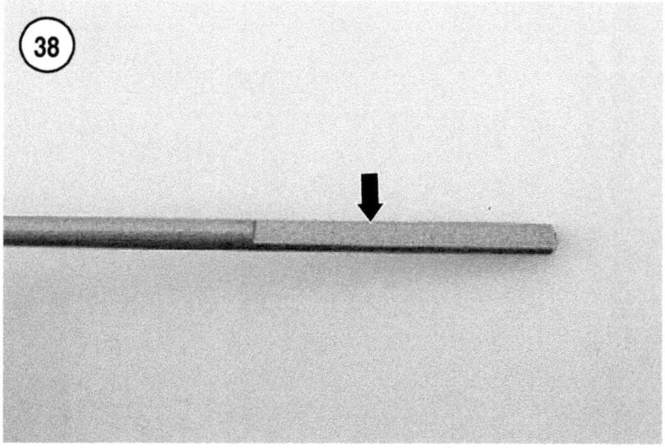
38

on the dipstick for torn, flattened or deteriorated surfaces. Replace the O-ring if these or other defects are evident.

4A. On models using a Cobra SX (single-propeller) drive unit, carefully remove the drain plug (**Figure 32**) on the lower port side of the gearcase.

4B. On models using a Cobra DP-S (twin-propeller) drive unit, remove the propellers as described in Chapter Ten. Use an 8 mm Allen adapter to remove the drain plug from the bearing carrier (**Figure 33**).

5. Allow the lubricant to fully drain. In cooler weather, this might take 30 minutes or longer.

6. Remove the three screws (**Figure 34**). Then pull the plastic cover from the drive unit (**Figure 35**). Remove the level/vent plug (**Figure 36**) from the rear of the drive unit.

7. Inspect the O-ring on the drain plug, level/vent plug and dipstick for torn, flattened or deteriorated surfaces. Replace the O-ring(s) if these or other defects are evident.

8A. On models using a Cobra SX (single-propeller) drive unit, fill the drive unit with lubricant as follows:

a. Thread the lubricant pump into the drain plug opening (**Figure 32**).
b. *Slowly* pump lubricant into the drive unit until it just starts flowing out of the vent plug opening (**Figure 37**).
c. Lubricate the O-ring on the vent plug with gearcase lubricant. Then thread the plug into the opening (**Figure 36**). Tighten the plug to (48-72 in.-lb. (5.4-8.1 N•m).
d. Wait 5 minutes without adding lubricant for the lubricant level to stabilize.
e. *Slowly* pump lubricant into the drive unit while frequently checking the level with the dipstick. Stop when the fluid level reaches the flat surface on the dipstick (**Figure 38**).
f. Lubricate the O-ring on the dipstick with gearcase lubricant. Then thread it into the opening. Tighten the dipstick to 48-72 in.-lb. (5.4-8.1 N•m).
g. Lubricate the O-ring on the drain plug with gearcase lubricant. Unthread the lubricant pump hose and then *quickly* thread the drain plug into the opening. Work quickly to minimize the amount of spilled lubricant. Tighten the drain plug to 60-84 in.-lb. (6.8-9.5 N•m).

8B. On models using a Cobra DP-S (twin-propeller) drive unit, fill the drive unit with lubricant as follows:

a. Thread the adapter (OMC or Volvo Penta part No. 3855932) into the drain plug opening (**Figure 39**) and tighten securely.
b. Thread the lubricant pump onto the adapter (**Figure 40**).
c. *Slowly* pump lubricant into the drive unit until it just starts flowing out of the vent plug opening (**Figure 37**).
d. Lubricate the O-ring on the vent plug with gearcase lubricant. Then thread the plug into the opening. Tighten the plug to 48-72 in.-lb. (5.4-8.1 N•m).

e. Wait 5 minutes without adding lubricant for the lubricant level to stabilize.

f. *Slowly* pump lubricant into the drive unit while frequently checking the level with the dipstick. Stop when the fluid level reaches the flat surface on the dipstick (**Figure 38**).

g. Lubricate the O-ring on the dipstick with gearcase lubricant. Then thread it into the opening. Tighten the dipstick to 48-72 in.-lb. (5.4-8.1 N•m).

h. Lubricate the O-ring on the drain plug with gearcase lubricant. Unthread the lubricant pump hose and adapter from the drain plug opening and then *quickly* thread the drain plug into the opening. Work quickly to minimize the amount of spilled lubricant. Tighten the drain plug to 60-84 in.-lb. (6.8-9.5 N•m).

9. Allow the fluid level to stabilize for about 30 minutes. Check and correct the drive unit lubricant level as described in this chapter (refer to *Before Each Use*).

10. Install the plastic cover onto the rear of the drive unit and secure with the three screws (**Figure 34**). Tighten the screws evenly to 108-132 in.-lb. (12.2-14.9 N•m).

WARNING

Stay clear of the propeller shaft(s) while running an engine on a flush/test adapter. Remove the propeller(s) before running the engine. Disconnect the battery cables before removing or installing the propeller(s).

CAUTION

Never run the engine without first providing cooling water. Use either a test tank or a flush/test adapter. Remove the propeller before running the engine.

11. Start the engine and allow it to reach operating temperature. Shift the engine into FORWARD and REVERSE to purge any trapped air and stabilize the lubricant level. Stop the engine and allow it to cool for approximately 30 minutes. Check and correct the drive unit lubricant level as described in this chapter (refer to *Before Each Use*).

12. On models using a Cobra DP-S (twin-propeller) drive unit, install the propellers as described in Chapter Ten.

Steering Cable Lubrication

Apply a quality water-resistant grease to the steering cable ram (**Figure 41**) after the first 20 hours of operation and then every 50 hours of operation or 60 days, whichever comes first.

1. Disconnect the negative battery cable.
2. Rotate the steering wheel clockwise to the full starboard limit.
3. Apply a light coat of grease to the exposed ram of the steering cable (**Figure 41**).
4. Rotate the steering wheel to the full port and starboard limits several times to distribute the grease.
5. Rotate the steering wheel to the straight-ahead position.

39

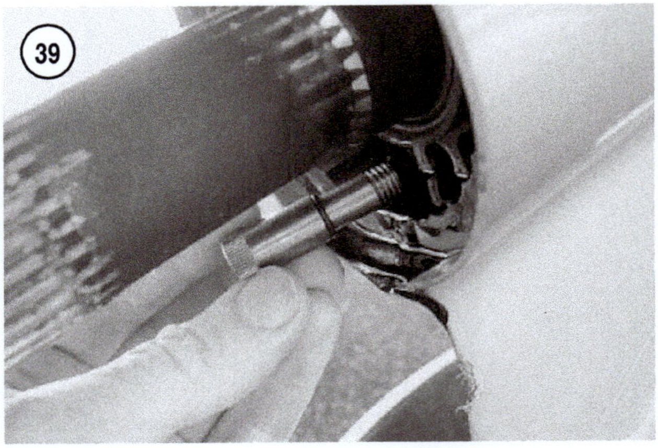

40

41

42

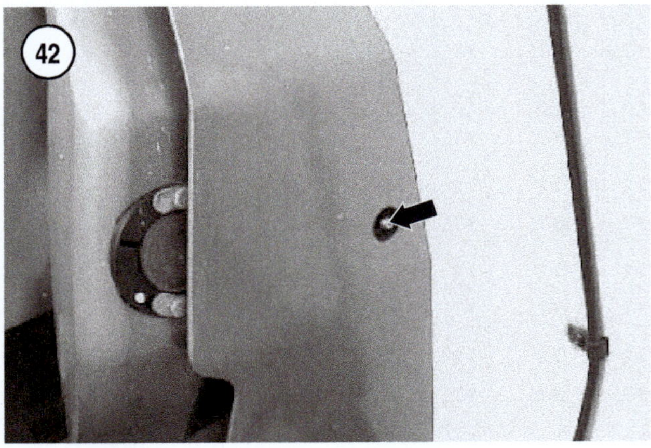

43

44

45

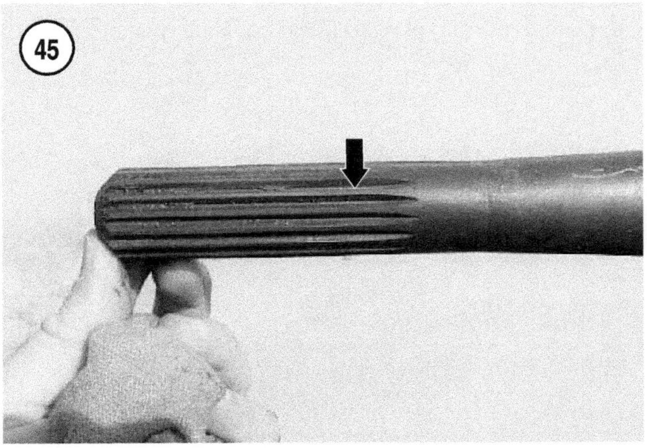

6. Connect the battery cable.

Gimbal Bearing Lubrication

Lubricate the gimbal bearing once each year or every 100 hours of operation, whichever comes first.

Use OMC EP/Wheel Bearing Grease or equivalent to lubricate the bearing. Although the bearing may be lubricated with the drive installed, the manufacturer recommends first removing the drive unit. This allows visual verification of proper lubrication.

1. Remove the drive unit as described in Chapter Ten.
2. Locate the gimbal bearing grease fitting (**Figure 42**) on the starboard side of the transom shield.

3A. If the drive unit is removed, pump grease into the fitting until the old grease exits the gimbal bearing (**Figure 43**) and new grease appears.

3B. If the drive unit is installed, pump approximately 4 oz. of grease into the fitting.

4. Install the drive unit as described in Chapter Ten.

Engine Coupling Lubrication

Lubricate the splines in the engine coupling (**Figure 44**) once a year or every 100 hours of operation, whichever comes first. Also lubricate the splines anytime the drive unit is removed. Use OMC Moly Lube (OMC part No. 175356) to lubricate the coupling splines. Purchase the Moly Lube from an OMC dealership.

1. Remove the drive unit as described in Chapter Ten.
2. Check and correct the engine alignment as described in Chapter Six.
3. Clean old grease and hard-packed material from the U-joint shaft yoke (**Figure 45**). Use a blunt instrument to remove aluminum material from the yoke splines. Clean the yoke with aerosol carburetor cleaner to remove residual grease residue.
4. Working through the gimbal bearing opening, use a long stick, socket extension or brush to apply OMC Moly Lube to the drive coupling splines. Apply enough grease just to coat the splines. Do not fill the opening with grease. Otherwise, the drive shaft will hydraulically lock in the coupler and prevent installation of the drive unit.
5. Apply a coating of OMC Moly Lube to the U-joint shaft yoke splines.
6. Install the drive unit as described in Chapter Ten.

Checking Engine Alignment

Check the engine alignment once a year or every 100 hours of use, whichever comes first, and anytime the drive unit is removed. Engine alignment is described in Chapter Six.

FUEL SYSTEM MAINTENANCE

CAUTION

Never run the engine on stale fuel. Varnish-like deposits form in the fuel system as the fuel deteriorates. These deposits block fuel passages and cause a lean condition in the combustion chamber. Damage to the pistons, valves and other engine components result from operating the engine with an excessively lean fuel mixture.

Fuel Requirements

Refer to *Lay-Up* in Chapter Four for information on fuel additives.

All models covered in this manual are designed to use regular unleaded fuel. Use fuel with an average octane rating of 89 or higher. Using fuel with a lower octane rating can result in power loss or serious engine damage. If a fuel with an average octane rating of 86-88 must be used on 4.3L, 5.0L or 5.7L models, the ignition timing must be readjusted to help prevent serious engine damage. Ignition timing adjustment and specifications are provided in Chapter Five.

Water-Separating Fuel Filter Maintenance

Debris can enter the fuel tank when adding fuel, from the fuel tank vent, during manufacture of the tank or when the tank was installed in the boat. Although the engine may have multiple fuel filters, the primary filter captures this material before it contaminates the carburetor, fuel pump or other fuel system components. Other filters on the engine are designed to capture debris or material that forms within the fuel system components.

All V-6 and V-8 models covered in this manual are equipped with a spin-on water-separating fuel filter (**Figure 46**) as the primary fuel filter in the system. This filter is readily accessible and easy to service. On GM 3.0L models, the primary filter is located within the mechanical fuel pump (**Figure 47**). Refer to Chapter Seven to replace the fuel pump filter on 3.0L models.

Inspection/replacement

Although the fuel filter is reusable, it is good to replace the filter after the first 20 hours of operation and then once a year or every 100 hours, whichever comes first.

1. Disconnect the negative battery cable.
2. Place a suitable container or shop towels under the water-separating fuel filter to capture spilled fuel.
3. Using an oil filter wrench or other suitable tool, carefully loosen the filter.
4. Remove the filter. Then promptly clean up any spilled fuel. Remove the seal from the filter housing if not found on the filter.
5. Empty the filter into a transparent container. Inspect the fuel sample for the presence of water, rust or other contaminants. Inspect the fuel tank if a significant amount of rust is present. Drain and refill the tank if a significant amount of water is present. Replace the filter if significant contamination is evident.
6. Inspect the filter for rust or damage. Replace the filter if any damage or evidence of leakage is present.
7. Apply a light coat of engine oil to the filter seal (**Figure 48**).

NOTE
Always fill the spin-on fuel filter with clean, fresh fuel prior to installation. Otherwise, the air introduced into the system can prevent proper priming of the fuel system upon starting.

8. Carefully pour clean, fresh fuel into the filter until the fuel level just reaches the threaded opening. Thread the filter onto the adapter until the seal just contacts the housing. Tighten the filter exactly one turn.
9. Clean up any spilled fuel. Wipe the filter dry to allow easy detection of leaking fuel.

10. Connect the battery cable.

WARNING
Stay clear of the propeller shaft(s) while running an engine on a flush/test adapter. Remove the propeller(s) before running the engine. Disconnect the battery cables before removing or installing the propeller(s).

CAUTION
Never run the engine without first providing cooling water. Use either a test tank or a flush/test adapter. Remove the propeller before running the engine.

11. Start the engine and immediately check for leakage at the filter. Stop the engine and correct the leakage if present. Allow the engine to reach operating temperature and then stop the engine. Wait a few minutes. Then check for fuel leakage at the filter. Correct any leakage before putting the engine into service.

Cleaning the Flame Arrestor

During normal engine operation, belt material, engine compartment insulation, oily deposits from the crankcase ventilation system and other material contaminates the flame arrestor screen (**Figure 49**, typical). Not only does this restrict air flow and decrease engine power, it also presents a fire hazard if subjected to an engine backfire. To help maintain maximum performance and minimize the fire hazard, clean the flame arrestor after every 50 hours of operation or if significant debris is noted on the screen.

1. Disconnect the negative battery cable.
2. Remove the flame arrestor cover and flame arrestor as described in Chapter Seven.

WARNING
Avoid serious bodily injury or death from fire or explosion. Replace the flame arrestor if any physical damage is evident. Operating the engine with a faulty or damaged flame arrestor can allow an engine backfire to ignite fuel vapor or other flammable material in the engine compartment.

3. If a large amount of black powder-like material is found in the flame arrestor, inspect the drive belts for excessive wear and replace as needed.
4. Use solvent to thoroughly clean the flame arrestor. Use a plastic bristle brush to remove fiber material from the fins or screen. Never use a wire brush to clean the arrestor. Use compressed air to remove loose debris and thoroughly dry the arrestor.
5. Inspect the flame arrestor for damage to the screen or mating surfaces. Replace the flame arrestor if any damage is evident.
6. Connect the battery cable.

PCV Valve Cleaning

Clean the positive crankcase ventilation (PCV) valve once a year or every 100 hours, whichever comes first. A PCV valve is not used on all models. If so equipped, the PCV valve is located on the breather hose at the opening into the rocker arm cover (**Figure 50**, typical).

1. Disconnect the negative battery cable.
2. Carefully pull the valve and hose from the rocker arm grommet.
3. Carefully pull the PCV valve from the hose fitting. Inspect the connecting hose for splitting, deterioration or other defects. Inspect the plastic hose fitting or cracks or a loose fit onto the valve. Replace the hose and/or fitting as needed.
4. Inspect the rubber grommet in the rocker arm cover for damage. A damaged grommet allows oil leakage and prevents proper operation of the crankcase ventilation system. Make sure the opening in the rocker arm cover fits into the slot in the grommet. For easier installation, apply a soap-and-water solution to the grommet prior to installation.
5. Direct aerosol carburetor cleaner into the PCV hose fitting and opening on the opposite end. Wait a few minutes. Then blow compressed air into the opening to remove the cleaner and contaminants. Repeat this step several times

until all deposits are removed and the cleaner exiting the valve is clear.

6. Shake the valve to check for a stuck check valve. The check valve must rattle while shaking the PCV valve. Replace the PCV valve if the check valve cannot be freed by cleaning.

7. Carefully insert the PCV valve into the grommet. The hose fitting end must face upward. Carefully seat the valve until the lip contacts the grommet.

8. Connect the plastic fitting and hose to the valve fitting.

9. Connect the battery cable.

10. After starting the engine, immediately check for oil leakage at the rocker arm cover grommet. Replace the grommet if any oil leakage is evident.

ENGINE COOLANT

Change the engine coolant every 2 years or 200 hours, whichever comes first, and anytime the coolant is drained to replace a hose or other engine components.

Over time, certain chemicals in the coolant react with minerals in the water mixed with the coolant or metals in the cylinder block, and crystalline deposits form to coat the passages in the heat exchanger and cylinder block. These crystal deposits insulate the coolant from the components and prevent efficient heat transfer. As they build, these deposits restrict coolant flow, increasingly compromise the cooling system and cause the engine to overheat.

Change the coolant on a regular basis to minimize these deposits. Heavy deposits can be removed using a commercially available cleaner. Always follow the manufacturer's instructions when using these products.

Most closed cooling system manufacturers recommend using an ethylene glycol-based, extended-life 5/100 coolant in the system. Read the container label carefully and make sure the coolant meets these requirements. If this type of coolant is not available, use a quality ethylene glycol-based coolant. Do not use propylene glycol-based coolant in the closed cooling system.

Dilute the coolant with an equal amount of clear, purified water. The cooling system must operate with a 50:50 mixture for efficient cooling and protection against freezing. Read the container carefully. Some brands and types of coolant are premixed and do not require dilution with water.

NOTE

Closed cooling system kits are available from numerous sources. Cooling system design, hose routing and fluid capacity vary by manufacturer. Refer to the information supplied by the cooling system manufacturer for specific information. This section describes maintenance on the typical closed cooling system.

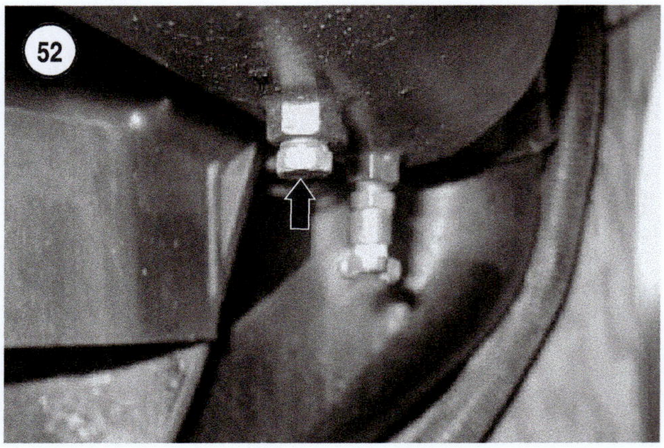

Draining

WARNING

Do not remove the pressure cap, drain plugs or hoses from the closed cooling system if the engine is warm. Hot coolant can spray out and cause serious injury.

NOTE

The old coolant must be drained from the heat exchanger, cylinder block and connecting hoses. Some closed cooling systems also flow coolant through the exhaust manifold(s), and they must also be drained. If in question, re-

54

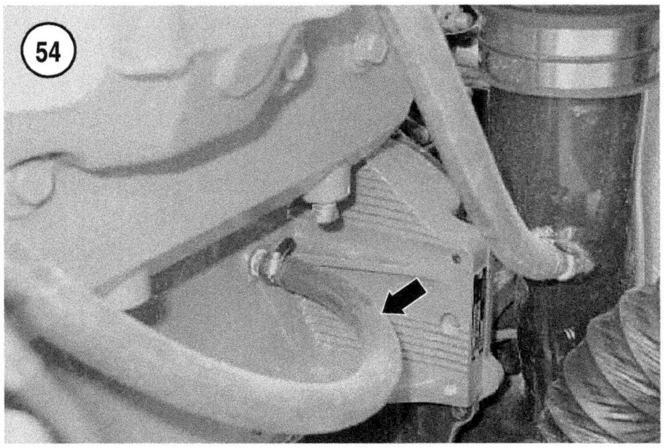

55

56

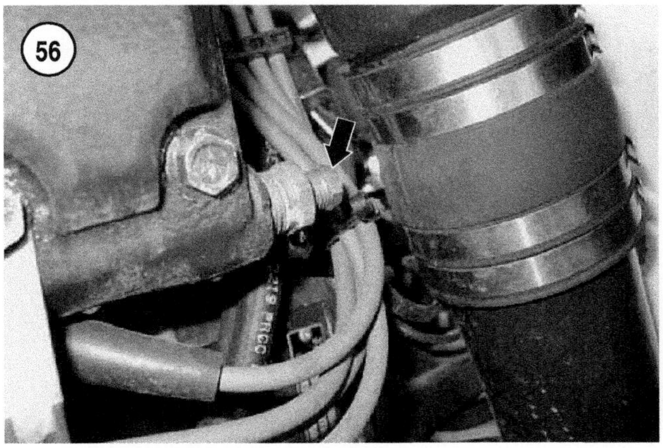

57

move the manifold drain plug or cap to check for coolant. Exhaust manifold draining procedures are described in Chapter Four (Winterizing).

1. Disconnect the negative battery cable.

2. Make sure the engine is off and allowed enough time to completely cool to normal room temperature.

3. Cover the cap (**Figure 51**, typical) with a heavy shop towel. Using gloves, push down on the pressure cap. Rotate the cap counterclockwise until the tabs are clear of the cam locking surfaces. Lift the cap from the reservoir.

4. Wipe debris and crystallized coolant from the cap and filler neck opening. Inspect the cap and filler neck as described in this chapter (refer to Coolant Level Check).

5. Locate the drain plug (**Figure 52**, typical) on the bottom of the heat exchanger. If two plugs are present, the plug nearest the end usually drains seawater, and the remaining plug drains coolant.

6. Place a suitable container under the fitting. Then remove the plug and allow the exchanger to fully drain.

CAUTION

Some models are equipped with a freshwater flush system that also serves as a means to drain the cylinder block. If the engine is equipped with a closed cooling system, this system is used only to drain the coolant from the cylinder block. If the engine has a flush system and is equipped with a closed cooling system, do not connect a garden hose to the drain hose and flush the system. The coolant will be diluted by the water, diminishing the freeze protection, corrosion protection and cooling system efficiency.

7. Refer to *Winterizing* in Chapter Four to determine the cylinder block drain locations. Most earlier models use plugs (**Figure 53**), and some later models use a single block drain hose (**Figure 54**). Place a suitable container under the hoses or fittings. Then remove the plug(s) from the block or cap from the drain hose.

8. If the closed cooling system flows coolant through the exhaust manifolds, refer to Chapter Four to determine the exhaust manifold drain locations. Place a suitable container under the cap or fitting. Then remove the clamp and cap (**Figure 55**, typical) or plug (**Figure 56**, typical) from the manifold(s).

9. Allow the coolant to fully drain. Make sure the boat is level to ensure full drainage.

10. If the system has a separate overflow reservoir, place a suitable container under the reservoir. Remove the clamp and disconnect the hose from the fitting on the bottom of the reservoir (**Figure 57**, typical). Drain the reservoir and hose. Reconnect the hose and secure with the clamp.

Filling

1. Reconnect hoses and install all plugs into the cylinder block, manifold and heat exchanger.

2. *Slowly* pour the coolant-and-water mixture into the fill neck until the level is within 1 in. (25 mm) of the top of the filler neck opening. Wait a few minutes for trapped air to purge and the level to drop. Then add additional coolant. Continue until the coolant level stabilizes.

3. If the system has a separate overflow reservoir, *slowly* pour the coolant-and-water mixture into the reservoir until the level reaches the FULL marking.

4. Wipe spilled coolant from the filler neck. Align the locking tabs on the cap with the notches on the filler neck. Push down on the cap to clear the cam locking surfaces. Then rotate the cap clockwise until it touches the stop. Release the cap.

WARNING

Stay clear of the propeller shaft(s) while running an engine on a flush/test adapter. Remove the propeller(s) before running the engine. Disconnect the battery cables before removing or installing the propeller(s).

CAUTION

Never run the engine without first providing cooling water. Use either a test tank or a flush/test adapter. Remove the propeller before running the engine.

5. Connect the battery cable.

6. Start the engine and immediately check for coolant leaking around the cap and filler neck. Correct any leakage before operating the engine.

7. Stop the engine and allow the engine to completely cool.

8. Check and correct the coolant level as described in this chapter.

TUNE-UP

A tune-up is a series of adjustments, tests and parts replacements performed to restore performance lost from wear, corrosion or deterioration of one or more engine components.

Compression Test

An engine with low compression on one or more cylinders cannot be properly tuned or expected to deliver satisfactory performance. Correct the cause of low compression before proceeding with the tune-up. Compression testing procedures are described in Chapter Two.

Spark Plugs

All of the OMC Cobra stern drive engines covered in this manual use a transistorized ignition system. These systems produce higher energy than conventional breaker-point systems used on earlier products. The transistorized systems provide longer spark plug life and less chance of spark plug fouling. Nevertheless, spark plugs operate in a harsh environment and eventually require replacement.

The spark plugs should be cleaned and regapped or replaced once each season or every 100 hours, when performing a complete tune-up, if the engine is misfiring or if a decrease in performance is evident.

However, an engine that is used primarily at mid-range cruising speed under normal operating conditions might not require spark plug replacement until after 200 hours or more of operation. On the other hand, an engine operated primarily at idle speed in very cold operating conditions might require spark plug replacement every 50 hours of operation or less. Generally, an engine operated under heavy load or in extreme climates will require more frequent spark plug replacement.

Spark plug condition can reveal much about the condition of the engine. Regular inspection allows problems to be identified and corrected before expensive engine damage occurs. Remove the spark plugs and compare them to the ones shown in **Figure 58**.

Correct any engine-related problems before installing new spark plugs. Operating the engine with an incorrectly adjusted carburetor, dirty flame arrestor, sticking choke valve, ignition system malfunction or fuel injection system malfunction will usually result in rapid fouling of the new spark plugs.

Removal

NOTE

With the manifolds, fuel system, cooling system and electrical components installed, the General Motors (GM) 5.0L and 5.7L V-8 engines are very similar in appearance to the Ford 5.0L and 5.8L V-8 engines. If the model identification decal on the flame arrestor is missing, the engine can be quickly identified as a GM or Ford engine by noting the distributor location. On GM engines, the distributor is located at the rear (flywheel end) of the engine. On Ford engines, the distributor is located on the front of the engine.

1. Mark the cylinder numbers on the spark plug leads before disconnecting them from the spark plugs. Refer to the following:
   a. On GM 3.0L (four-cylinder) models, the No. 1 cylinder is the front cylinder. No. 2 cylinder is the second from the front and so on.
   b. On GM V-6 or V-8 models, the No. 1 cylinder is the front cylinder on the port side, and the No. 2 cylinder

58

SPARK PLUG CONDITION

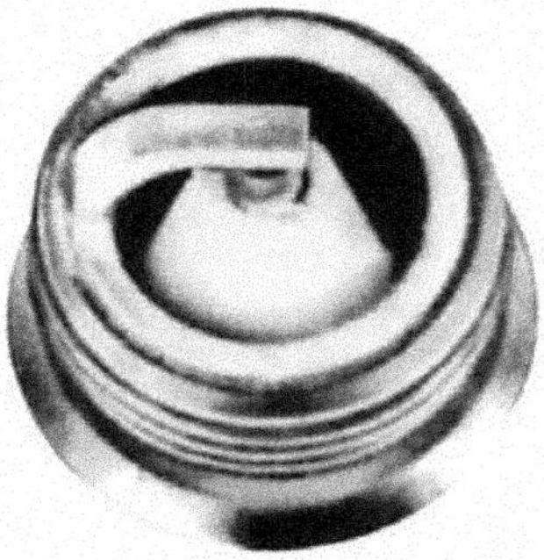

NORMAL
- Identified by light tan or gray deposits on the firing tip.
- Can be cleaned.

GAP BRIDGED
- Identified by deposit buildup closing gap between electrodes.

OIL FOULED
- Identified by wet black deposits on the insulator shell bore and electrodes.
- Caused by too much oil entering combustion chamber through worn rings and pistons, excessive clearance between valve guides and stems, or worn or loose bearings. Can be cleaned. If engine is not repaired, use a hotter plug.

CARBON FOULED
- Identified by dry, fluffy black carbon deposits on insulator tips, exposed shell surfaces and electrodes.
- Caused by a too-cold plug, weak ignition, dirty air cleaner, too-rich fuel mixture or excessive idling. Can be cleaned.

LEAD FOULED
- Identified by dark gray, yellow, tan or black deposits or a fused glazed coating on the insulator tip.
- Caused by highly leaded gasoline.

WORN
- Identified by severely eroded or worn electrodes.
- Caused by normal wear. Should be replaced.

FUSED SPOT DEPOSIT
- Identified by melted or spotty deposits resembling bubbles or blisters.
- Caused by sudden acceleration. Can be cleaned.

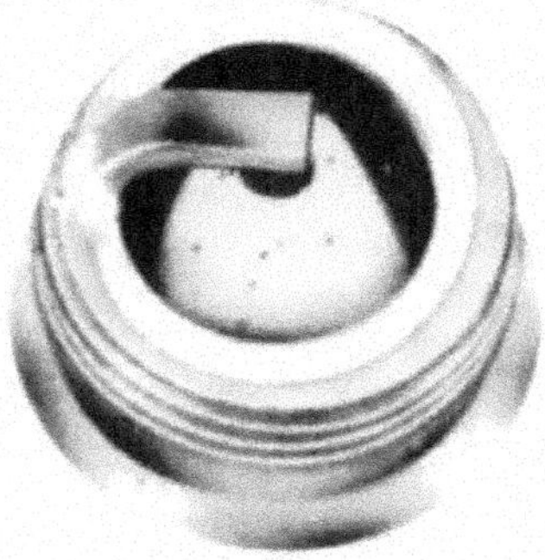

OVERHEATING
- Identified by a white or light gray insulator with small black or gray-brown spots with bluish-burnt appearance of electrodes.
- Caused by engine overheating, wrong type of fuel, loose spark plugs, a too-hot plug or incorrect ignition timing. Replace the plug.

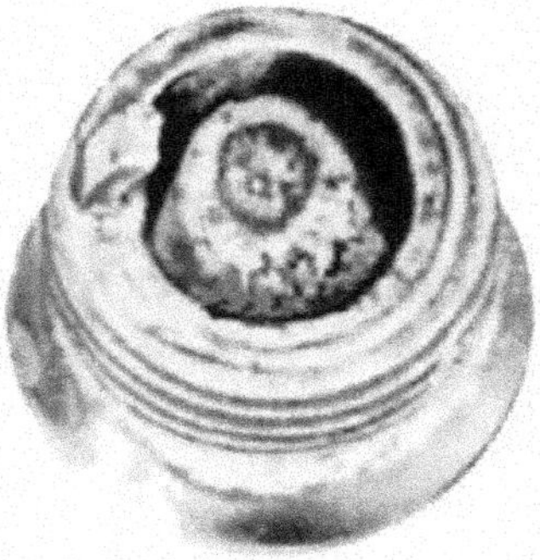

PREIGNITION
- Identified by melted electrodes and possibly blistered insulator. Metallic deposits on insulator indicate engine damage.
- Caused by wrong type of fuel, incorrect ignition timing or advance, a too-hot plug, burned valves or engine overheating. Replace the plug.

is the front cylinder on the starboard side. Odd-numbered cylinders are on the port side, and even-numbered cylinders are on the starboard side.

c. On Ford 5.0L and 5.8L V-8 models, the No. 1 cylinder is the front cylinder on the starboard side. The No. 5 cylinder if the front cylinder on the port side. The cylinders are numbered from front to rear (1, 2, 3 and 4 on the starboard side and 5, 6, 7 and 8 on the port side). The firing order is 1-3-7-2-6-5-4-8.

2. Disconnect the spark plug wires by twisting and pulling on the wire boot (**Figure 59**). If working on a hot engine or in cramped quarters, grip the boot with spark plug boot pliers. (**Figure 60**).

3. Use compressed air to blow debris from around the spark plug openings before removing the spark plugs. If the spark plug threads are corroded, apply a penetrating oil and allow it to soak in.

NOTE
On Ford 5.0L and 5.8L models, it can be difficult to secure a typical spark plug socket, extension and ratchet onto the No. 2, No. 3, No. 5 and No. 6 cylinders. If necessary, fit a thin-wall socket with a hex on the top over the spark plug. Engage a wrench onto the hex end of the socket to tighten and loosen the spark plug.

4. Remove each of the spark plugs from the cylinder head(s). Mark the cylinder number on the plugs or arrange them in a fashion that indicates the cylinder placement.

5. Sometimes the threads in the cylinder head are damaged during spark plug removal. This condition can be repaired without removing the cylinder head by installing a special threaded insert. Have the repair performed by a professional if you are unfamiliar with the repair or do not have access to the equipment.

6. Clean the spark plug opening with a thread chaser (**Figure 61**). Thread the chaser by hand into each spark plug opening. Several passes might be required to remove all carbon or corrosion deposits from the threads. Flush all debris from the openings with compressed air.

59

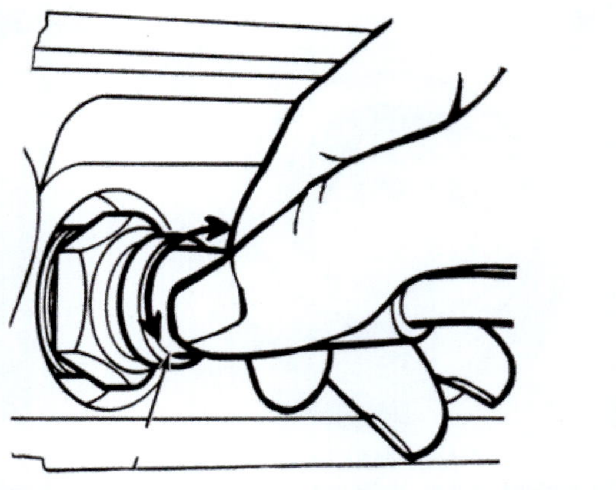

60

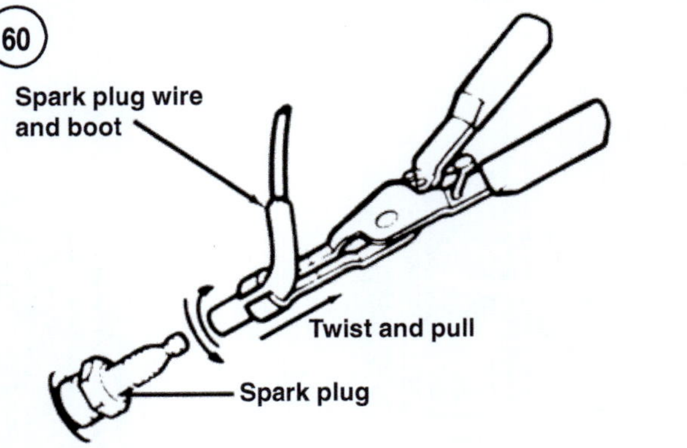

61

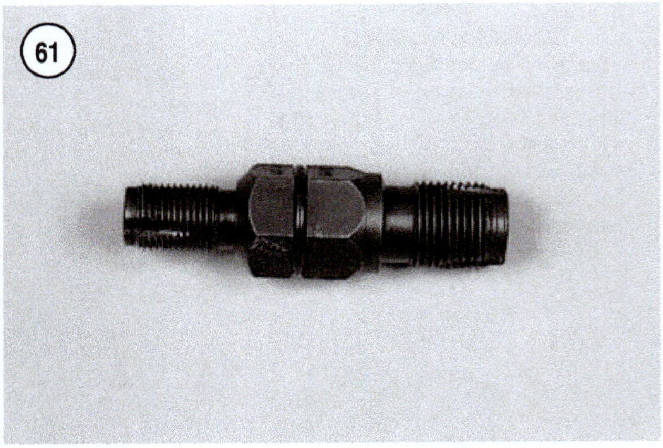

Inspection

Compare the spark plugs with those shown in **Figure 58**. Spark plugs can give a clear indication of problems in the engine sometimes before symptoms occur. Additional inspection and testing might be required if an abnormal condition is evident. Refer to Chapter Two for troubleshooting instructions.

Installation

Use a gap-adjusting tool (**Figure 62**) to adjust the spark plug gap (**Table 4**). Never tap the plug against a hard object to close the gap. The ceramic insulator can crack and break away. Gapping tools are available at most automotive parts stores.

1. Refer to **Table 4** to determine the spark plug gap.

2. Check the gap using a wire feeler gauge (**Figure 63**) of the same thickness as the recommended gap. The gauge should pass between the electrodes (**Figure 64**) with slight drag.

3. Open or close the gap as necessary.

4. Inspect the spark plug electrodes for parallel surfaces (**Figure 64**). Carefully bend the electrode until the surfaces are parallel and the gap is correct.

62

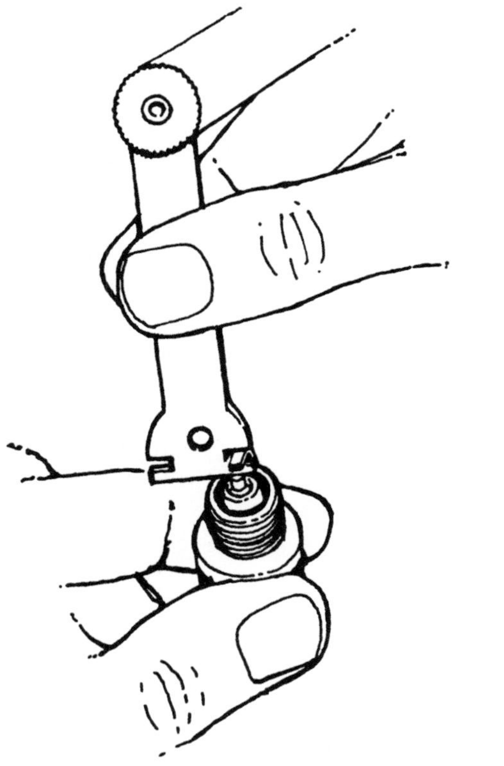

63

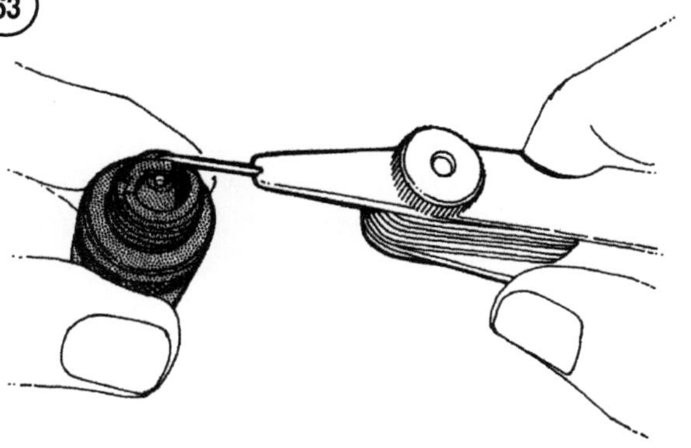

64

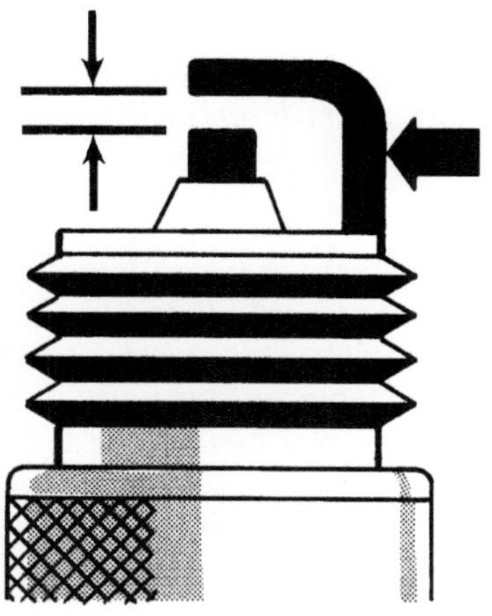

65

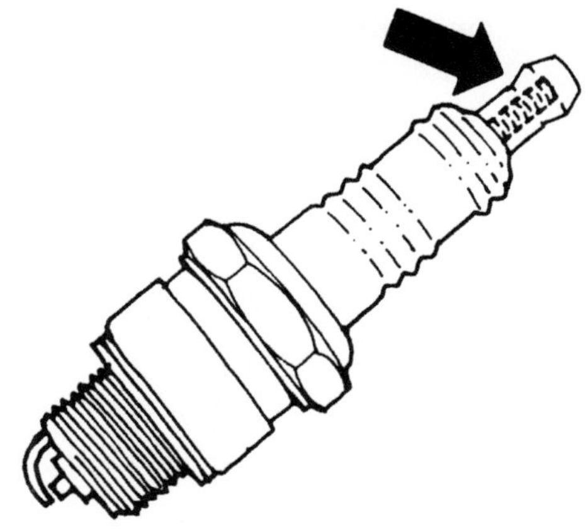

NOTE
On some brands of spark plugs, the terminal end must be installed before installing the spark plug. Thread the terminal onto the spark plug as shown in ***Figure 65****.*

5. Apply a very light coat of oil to the spark plug threads and thread them in by hand. Use a torque wrench to tighten the spark plug as follows:
   a. On 3.0L four-cylinder models, tighten the spark plugs to 20 ft.-lb. (27 N•m).
   b. On GM 4.3L, 5.0L and 5.7L models, tighten the spark plugs to 15 ft.-lb. (20 N•m).
   c. On Ford 5.0L and 5.8L models, tighten the spark plugs to 120-180 in.-lb. (14-20 N•m).
   d. On GM 7.4L and 8.2L models, tighten the spark plugs to 22 ft.-lb. (30 N•m).
6. Apply a light coat of silicone lubricant to the inner surface of the spark plug boots. Carefully slide the cap over the correct spark plug. Snap the connector in the boot fully onto the spark plug.

Valve Adjustment

The models covered in this manual use hydraulic lifters, and periodic valve adjustment is not required during a tune-up.

Excessive valve clearance causes valve system noise and can increase wear on some valve train components. Insufficient clearance can result in rapid wear of valve train components, reduced power and rough engine operation. Valve noise or insufficient clearance is generally the result of improper assembly or damaged components. If valve noise is evident, and improper valve adjustment is suspected as the cause, adjust the valves as described in Chapter Six.

Carburetor Adjustment

Proper carburetor adjustment is essential for smooth and efficient operation. Carburetor adjustment performed during a tune-up includes idle mixture and idle speed adjustment. Carburetor adjustment is described in Chapter Five.

Ignition Timing Adjustment

The initial timing must be adjusted on all models covered in this manual. On all electronic fuel injection (EFI) models and some carburetor-equipped models, a special procedure must be followed. Failure to follow this procedure can result in poor performance or serious engine damage. Ignition timing adjustment is described in Chapter Five.

Test-Running the Engine

NOTE
Perform this test with an average load in the boat and the drive unit properly trimmed to the correct angle for the best performance.

1. Operate the engine using a flush/test device or in a test tank to ensure correct starting and idling before testing the boat on the water.
2. Following the manufacturer's instructions, connect a shop tachometer to the engine.
3. Start the engine and run at fast idle (approximately 1500 rpm) in NEUTRAL until it reaches normal operating temperature.
4. Return the engine to idle speed. Note the idle speed while an assistant operates the boat. Adjust the idle speed as described in Chapter Five.
5. Accelerate to wide open throttle and note the engine speed. Compare the speed on the shop tachometer with the speed indicated on the dash tachometer. Check for improper adjustment of the dash tachometer if the speeds vary more than 200 rpm. Refer to Chapter Two (**Table 5**) to determine the correct engine operating range. Check the propeller for damage or incorrect pitch if the engine speed is not within the recommended range. Refer to *Preliminary Inspection* in Chapter Two if the correct propeller is installed, but the engine fails to reach the recommended speed range.
6. Try a rapid acceleration and run the engine at various speed ranges. Refer to *Preliminary Inspection* in Chapter Two if rough operation is noted at any speed range, or hesitation occurs during rapid acceleration.

SUBMERSION

If the engine has been subjected to complete submersion, three factors need to be considered. Was the engine running when the submersion occurred? Was the engine submerged in salt, brackish or polluted water? How long has the engine been retrieved from the water?

Complete disassembly and inspection of the engine is required if the engine was submerged when running. Internal damage to the engine, especially a bent connecting rod, is likely should this occur. Refer to Chapter Six for engine repair instructions.

Many components of the engine suffer the corrosive effects of submersion in salt, brackish or polluted water. The symptoms might not occur for some time after the event. Salt crystals form in many areas of the engine and promote intense corrosion in that area. The wire harness and connections are usually damaged very quickly. It is difficult to remove all of the salt crystals from the harness connectors. Replace the wire harness and clean all electrical connections to ensure a reliable repair. The starter motor, relays and any switch on the engine usually fail if they are not thoroughly cleaned of all salt residue.

Retrieve and service the engine as soon as possible. After retrieval, vigorously wash all debris from the engine with freshwater. Complete engine disassembly and inspection is required when sand, silt or other gritty material is noted inside the throttle opening. Refer to Chapter Six for engine repair instructions.

Service the engine quickly to make sure it is running within a few hours after retrieval. This is especially important if the engine was submerged in salt, brackish or polluted water.

Completely disassemble and inspect the internal engine components if the engine was not serviced in a timely manner.

Perform the following steps as soon as the engine is retrieved from the water.

1. Remove the engine cover and flame arrestor and vigorously wash all material from the engine with freshwater. Completely disassemble and inspect the engine if sand, silt or gritty material is present inside the throttle opening.
2. Dry the exterior of the engine with compressed air. Remove the spark plugs and ground all spark plug leads.
3. Remove the drive unit as described in Chapter Ten.
4. Drain all water and residual fuel from the fuel system. Replace all fuel filters on the engine. Refer to Chapter Seven.
5. Drain the engine oil. Disconnect the battery cables.
6. Manually rotate the harmonic balancer (clockwise) to force water from the cylinders. Rotate the flywheel several times and note whether the engine is turning freely. Completely disassemble and inspect the engine if interference is noted.
7. Pour approximately one teaspoon of engine oil into each spark plug opening. Repeat Step 6 to distribute the oil in the cylinders.
8. Disconnect all electrical connections and inspect the terminals. Dry all exterior surfaces and wire connectors with compressed air. Remove, disassemble and inspect the electric starter motor as outlined in Chapter Nine. Service and test the battery (refer to *Chapter Nine*). Have the alternator cleaned and inspected at a reputable starter and alternator repair shop.

9. Replace the oil filter and fill the engine with fresh oil. Clean and install the spark plugs. Reconnect all wire harnesses and battery terminals.

10. Install the drive unit as described in Chapter Ten.

11. Provide the engine with a fresh supply of fuel. Start the engine and run it at a low speed for a few minutes. If the engine cannot be started, refer to Chapter Two for troubleshooting instructions. If unusual noises are noted, stop the engine immediately and investigate. Allow the engine to run at low speed for a minimum of 30 minutes to dry any residual water from the engine. Promptly investigate any unusual noises or unusual running conditions.

12. Change the engine oil again and replace the oil filter. Perform all maintenance items listed in **Table 2**.

Table 1 MAINTENANCE TORQUE SPECIFICATIONS

Fastener	in.-lb.	ft.-lb.	N•m
Drive unit dipstick	48-72	–	5.4-8.1
Drive unit lubricant vent plug	48-72	–	5.4-8.1
Drive unit drain/fill plug	60-84	–	6.8-9.5
Spark plugs			
GM 3.0L models	–	20	27
GM 4.3L models			
Used cylinder head	–	15	20
New cylinder head	–	22	30
GM 5.0L and 5.7L models			
Used cylinder head	–	15	20
New cylinder head	–	22	30
Ford 5.0L and 5.8L models	120-180	–	14-20
GM 7.4L and 8.2L models	–	22	30
Rear drive cover (plastic) screws	108-132	–	12.2-14.9

Table 2 MAINTENANCE INTERVALS

Before each use	
	Check fuel system for leaks
	Check engine oil level
	Check drive unit lubricant level
	Check power steering fluid level*
	Check power trim fluid level
	Check propeller for damage
	Check propeller shaft for bending
	Inspect steering system
	Check coolant level (closed cooling system)*
	Check lanyard (safety stop) switch operation
After each use	
	Flush the cooling system
	Clean the engine
	Inspect the propeller(s) for damage
	Inspect the drive unit for damage
First 20 hours of operation	
	Change the engine oil
	Change the drive unit lubricant
	Check power steering fluid reservoir*
	Check the power trim fluid level
	Clean or replace primary fuel filter
	Adjust shift linkage (Chapter Five)
	Lubricate steering cable ram
	Check drive belt tension (Chapter Nine)
	Tighten hose clamp and engine mounts
Every 50 hours of operation	
	Check drive belt tension (Chapter Nine)
	Clean flame arrestor
	Check rubber hoses for deterioration
Every 60 days or 50 hours of operation	
	Lubricate the steering cable ram

(continued)

Table 2 MAINTENANCE INTERVALS (continued)

Every 60 days or 50 hours of operation (continued)	
	Inspect drive unit sacrificial anode (Chapter Ten)
	Inspect transom shield sacrificial anode (Chapter Eleven)
Once a year or every 100 hours of operation	
	Change spark plugs
	Change the engine oil
	Change the drive unit lubricant
	Clean or replace primary fuel filter
	Check shift linkage adjustment (Chapter Five)
	Lubricate the engine drive coupler splines
	Check engine alignment (Chapter Six)
	Lubricate gimbal bearing
	Clean PCV valve*
	Inspect seawater pump impeller (Chapter Eight)
Every two years or 200 hours of operation	
	Replace seawater pump impeller (Chapter Eight)
	Change closed cooling system coolant*

*This maintenance item does not apply to all models.

Table 3 ENGINE OIL AND DRIVE UNIT LUBRICANT CAPACITIES

Model	Capacity (approximate)
Engine oil capacity	
GM 3.0L models	
With oil filter change	4.0 qt. (3.8 L)
Without oil filter change	3.5 qt. (3.3 L)
GM 4.3L models	
With oil filter change	4.5 qt. (4.3 L)
Without oil filter change	4.0 qt. (3.8 L)
GM 5.0L models	
With oil filter change	6.0 qt. (5.7 L)
Without oil filter change	5.0 qt. (4.7 L)
Ford 5.0L models	
With oil filter change	6.0 qt. (5.7 L)
Without oil filter change	5.0 qt. (4.7 L)
GM 5.7L models	
57GL models	
With oil filter change	5.0 qt. (4.7 L)
Without oil filter change	4.0 qt. (3.8 L)
57GI and 57GSI models	
With oil filter change	6.0 qt. (5.7 L)
Without oil filter change	5.0 qt. (4.7 L)
Ford 5.8L models	
58FL (carburetor-equipped) models	
With oil filter change	6.0 qt. (5.7 L)
Without oil filter change	5.0 qt. (4.7 L)
58 FI and 58FSI (EFI) models	
With oil filter change	6.0 qt. (5.7 L)
Without oil filter change	5.0 qt. (4.7 L)
GM 7.4L models	
74GL models	
With oil filter change	7.0 qt. (6.6 L)
Without oil filter change	6.0 qt. (5.7 L)
74GI and 74GSI models	
With oil filter change	9.0 qt. (8.5 L)
Without oil filter change	8.0 qt. (7.6 L)
GM 8.2L models	
With oil filter change	7.0 qt. (6.6 L)
Without oil filter change	6.0 qt. (5.7 L)
Drive unit lubricant capacity	
Cobra SX (single-propeller)	71 oz. (2.1 L)
Cobra D-S DuoProp (twin-propeller)	81 oz. (2.4 L)

Table 4 RECOMMENDED LUBRICANTS

Lubricant	Application
SAE 30 NMMA-certified FC-W marine-use motor oil*	Engine crankcase–all models
SAE 30 motor oil: GM standard GM-6094-M*	Crankcase–GM engines
SAE 30 motor oil: Ford specification ESE-M2X153*	Crankcase–Ford engines
Mobil 1 SHC 75-90 fully synthetic gearcase lubricant	Drive unit
Volvo Penta Dura Plus synthetic gearcase lubricant	Drive unit
Ethylene glycol-based engine coolant	Closed cooling system
Dexron II automatic transmission fluid (ATF)	Power steering and power trim pumps
OMC power trim/tilt fluid	Power steering and power trim pumps
Volvo Penta power trim/tilt fluid (part No. 3851930)	Power steering and power trim pumps
Quicksilver power trim/tilt fluid (part No. 92-90199A21)	Power steering and power trim pumps
OMC EP/Wheel Bearing Grease	Gimbal bearing
OMC Moly Lube (part No. 175356)	Engine coupling splines
OMC Triple Guard Grease	Steering cable ram
Marine-grade water-resistant grease	Steering cable ram
Marine-grade fuel stabilizer	Fuel tank
OMC fuel conditioner and stabilizer (part No. 775614)	Fuel tank
Quicksilver fuel conditioner (part No. 92-858072KO1)	Fuel tank
Marine fogging oil	Storage fogging agent
Quicksilver Storage Seal (part No. 92-858081KO1)	Storage fogging agent

*Recommended oil viscosity for operation in ambient air temperature exceeding 32° F (0° C). Refer to text for oil viscosity recommendation if operating engine at lower air temperatures. Refer to Chapter Three.

Table 5 SPARK PLUG SPECIFICATIONS

Model	Spark plug No.	Gap specification
GM 3.0L models		
AC plug	MR43LTS	
Champion plug	RS12YC	
NGK plug	TR5	
Plug gap	–	0.045 in. (1.14 mm)
GM 4.3L models		
1994-1996 (with 12-bolt intake)		
43GL and 4.3GL models		
AC plug	MR43T	
NGK plug	BR6FS	
Plug gap	–	0.045 in. (1.14 mm)
43GI models		
AC plug	R43TS	
NGK plug	BPR6FS	
Plug gap	–	0.045 in. (1.14 mm)
1997-2000 (with 8-bolt intake)		
AC plug	MR43LTS	
Champion plug	RS12YC	
NGK plug	BPR6EFS	
Plug gap	–	0.045 in. (1.14 mm)
GM 5.0L models		
1994-1997 (with 12-bolt intake)		
AC plug	MR43T	
NGK plug	BR6FS	
Plug gap	–	0.045 in. (1.14 mm)
1998-2000 (with 8-bolt intake)		
50GL and 50GS (carburetor-equipped)		
AC plug	MR43LTS	
Champion plug	RS12YC	
NGK plug	BPR6EFS	
Plug gap	–	0.035 in. (0.89 mm)
50GI (EFI)		
Champion plug	RS12YC	
NGK plug	BPR6EFS	
Plug gap	–	0.045 in. (1.14 mm)

(continued)

Table 5 SPARK PLUG SPECIFICATIONS (continued)

Model	Spark plug No.	Gap specification
Ford 5.0L models		
50FL (carburetor-equipped)		
AC plug	MR43T	
Motorcraft plug	ASF32C	
NGK plug	UR4	
Plug gap	–	0.035 in. (0.89 mm)
50FI (EFI)		
AC plug	MR43T	
Motorcraft plug	AWSF32C	
NGK plug	TR5	
Plug gap	–	0.045 in. (1.14 mm)
GM 5.7L models		
1994-1997 (with 12-bolt intake)		
57GL (carburetor-equipped)		
AC plug	MR43T	
NGK plug	BR6FS	
Plug gap	–	0.045 in. (1.14 mm)
57GI (EFI)		
AC plug	R43TS	
NGK plug	BPR6FS	
Plug gap	–	0.045 in. (1.14 mm)
1998-2000 (with 8-bolt intake)		
57GS (carburetor-equipped) models		
AC plug	MR43LTS	
Champion plug	RS12YC	
NGK plug	BPR6EFS	
Plug gap	–	0.035 in. (0.89 mm)
57GSI (EFI) models		
AC plug	MR43LTS	
Champion plug	RS12YC	
NGK plug	BPR6EFS	
Plug gap	–	0.045 in. (1.14 mm)
Ford 5.8L models		
58FL (carburetor-equipped) models		
AC plug	MR43T	
Motorcraft plug	ASF32C	
NGK plug	UR4	
Plug gap	–	0.035 in. (0.89 mm)
58FI (EFI) models		
AC plug	MR43T	
Motorcraft plug	AWSF32C	
NGK plug	BPR6EFS	
Plug gap	–	0.045 in. (1.14 mm)
GM 7.4L models		
1994-1997		
74GL (carburetor-equipped) models		
AC plug	MR43T	
NGK plug	BR6FS	
Plug gap	–	0.045 in. (1.14 mm)
74GI (EFI) models		
AC plug	R43TS	
NGK plug	BPR6FS	
Plug gap	–	0.045 in. (1.14 mm)
1998-2000		
74GI (EFI) models		
AC plug	MR43LTS	
Champion plug	RS12YC	
NGK plug	BPR6EFS	
Plug gap	–	0.045 in. (1.14 mm)
74GSI (EFI) HP models		
AC plug	MR43T	
NGK plug	BR6FS	
Plug gap	–	0.045 in. (1.14 mm)
GM 8.2L models		
AC plug	MR43T	
NGK plug	BR6FS	
Plug gap	–	0.045 in. (1.14 mm)

Chapter Four

Lay-Up, Winterizing and Fitting Out

All major engine systems require some preparation before storage. Perform the procedures described in *Lay-Up* if the engine will not be operated for several weeks or longer and especially if the engine is used or stored in saltwater or heavily polluted water.

The engine must be winterized before being exposed to freezing temperatures. Water remaining in the cooling system can freeze, expand and cause serious damage to the cylinder block, manifolds, heat exchanger(s) and other components. Always winterize the engine before transporting or storing the boat in high elevation areas that can experience freezing temperatures even during the summer months.

Fitting out prepares the engine for operation. Perform these procedures before returning the engine to service.

LAY-UP

When preparing the engine for long-term storage, the objective is to prevent corrosion or deterioration during the storage period.

Perform any maintenance that will come due during the storage period. Maintenance requirements and procedures are described in Chapter Three. Lay-up maintenance requirements include the following:

1. Change the engine oil and filter.
2. Change the gearcase lubricant.
3. Check the fluid level in the power steering system.
4. Check the fluid level in the power trim system.
5. Remove and maintain the battery as described in Chapter Nine.
6. Check or change the engine coolant (in a closed cooling system).
7. Lubricate all steering, throttle and control linkages.
8. Inspect and clean all sacrificial anodes.
9. Inspect all belts and hoses.
10. Empty water or sediment from the water-separating fuel filter as described in Chapter Three.

Besides the above maintenance items, the procedures described in this chapter will prepare the drive unit for storage. To reduce corrosion and prevent freeze damage, drain all of the water from the cooling system as described in *Winterizing* in this chapter.

External Surfaces

Clean vegetation, dirt or deposit buildup from the exterior of the drive unit and transom shield. Scrub the engine components and apply a good corrosion prevention spray available from most marine dealerships and marine supply stores. Flush the cooling system as described in Chapter Three.

Fuel

CAUTION
Never run the engine without first providing cooling water. Use either a test tank or a flush/test adapter.

Serious problems can be avoided if the fuel system is properly prepared for storage. Drain as much fuel from the fuel tank as possible. Inspect or replace all fuel filters.

Fuel additive

NOTE
*Inspect the fuel fill hose for dips or low places (**Figure 1**) that can prevent all of the fuel additive from flowing into the fuel tank.*

Use a fuel stabilizer to help prevent the formation of gum or varnish during the storage period. Be aware that some additives adversely affect some fuel system components if mixed or used incorrectly. Deterioration of hoses, check valves and other nonmetallic components can occur. Never mix these additives at a rate greater than specified on the label.

Always inspect the fuel fill hose for dips or low spots (**Figure 1**) before pouring a fuel additive into the tank. Dips or low spots can prevent all of the additive from flowing into the tank. If necessary, lift the hose after pouring the additive into the hose or add fresh fuel to wash the additive into the tank.

Before storing the engine, run the engine for a minimum of 15 minutes to distribute the stabilizer throughout the fuel system. If the engine is equipped with a four-barrel carburetor (**Figure 2**), the engine must be operated for several minutes under actual operating conditions at a high throttle setting to transfer the treated fuel into the secondary fuel metering passages.

Fogging the Engine

If the engine will not be used for several weeks or longer, treat the cylinders with a fogging agent. This step can prevent corrosion inside the engine during the storage period. This is especially important if the engine is used or stored in saltwater or polluted water. The agent used and method for application depends on the type of fuel system used on the engine.

WARNING
Stay clear of the propeller shaft while running the engine on a flush/test adapter. Remove the propeller(s) before running the engine to prevent injury or death. Always disconnect the battery cables before removing or installing the propeller.

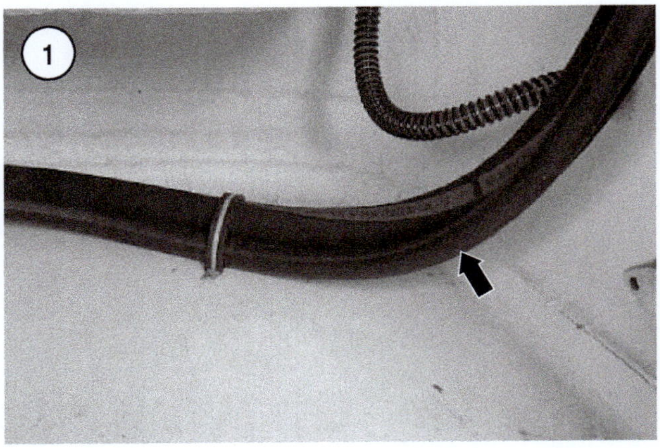

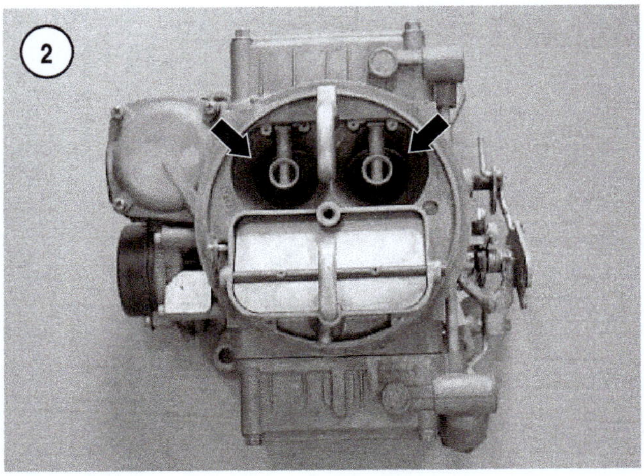

CAUTION
Never run the engine without first providing cooling water. Use either a test tank or a flush/test adapter.

Fogging carburetor-equipped engines

1. Add the required amount of fuel stabilizer to the fuel tank. Refer to *Fuel additive* in this section. Run the engine for a minimum of 10 minutes to distribute the treated fuel throughout the fuel system. If the engine is equipped with a four-barrel carburetor, the engine must be operated for sev-

4

5

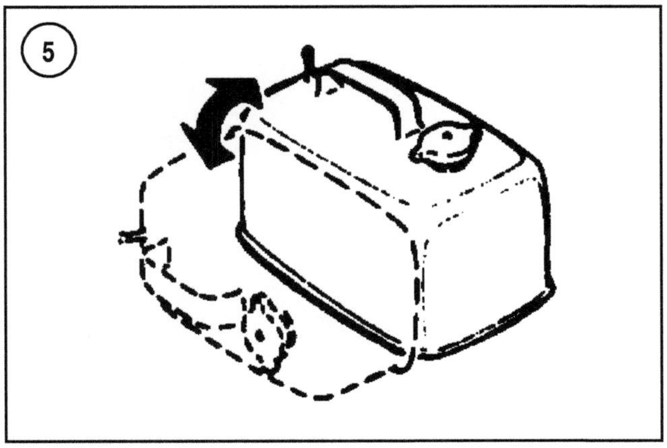

6

eral minutes under actual operating conditions to transfer the treated fuel through the secondary fuel metering passages.

2. Remove the cover and flame arrestor (**Figure 3**) as described in Chapter Seven.

3. Raise the engine speed to approximately 1500 rpm. Spray the fogging agent into the carburetor throat (**Figure 4**, typical) in 5-10 second intervals. Continue to spray the agent into the engine until heavy smoke comes from the exhaust. This indicates the agent has passed through the engine. Stop the engine at this point.

4. Spray more of the fogging agent into the carburetor. Activate the lanyard switch to prevent accidental starting. Operate the starter for a few seconds to distribute the agent over the cylinder walls. Reset the lanyard switch.

5. Install the flame arrestor and cover as described in Chapter Seven.

Fogging electronic fuel injection (EFI) engines

1. Add the required amount of fuel stabilizer to the fuel tank. Refer to *Fuel additive* in this section. Run the engine for a minimum of 10 minutes to distribute the treated fuel throughout the fuel system.

2. Turn the engine off and disconnect the negative battery cable.

3. Pour a mixture of 2 gal. (7.6 L) of fresh fuel, enough fuel stabilizer for 2 gal. (7.6 L) of fuel and 26 oz. (0.8 L) of OMC or Volvo Penta fogging oil or equivalent into a portable outboard fuel tank. Make sure the tank is equipped with a 3/8 in. (9.5 mm) inner diameter fuel supply hose. Rock the tank for several minutes to thoroughly mix the contents (**Figure 5**).

4. Trace the fuel supply hose from the fuel tank to the inlet fitting on the water-separating fuel filter (**Figure 6**). Loosen the clamp screw and carefully pull the supply hose off the fitting. Drain the hose into the portable fuel tank.

5. Connect the portable fuel tank hose to the fuel cell fitting and secure with the hose clamp. Route the hose to prevent interference with belts or other moving components on the engine. Squeeze the primer bulb to transfer treated fuel into the system in preparation for starting.

6. Connect the battery cable.

7. Start the engine and operate at 1500 rpm for 10 minutes to distribute the fogging mixture throughout the fuel system and into the cylinders (via the fuel injectors). A hint of smoke should be detected in the exhaust to indicate the mixture is distributed throughout the engine. If necessary, run the engine for a few more minutes or until smoking is evident.

8. Stop the engine and disconnect the portable fuel tank hose. Drain the hose into the portable tank.

9. Connect the boat fuel supply hose to the fuel filter and secure it with the hose clamp. Route the supply hose to prevent interference with belts or other moving components. Secure the hose with clamps as necessary.

Drive Unit Storage

Always place the drive unit in the full DOWN straight steering position during storage. This prevents the rubber bellows and other rubber components in the transom shield from developing a set and helps prevent water from collecting in the cooling and exhaust passages. Water remaining in these passages can freeze and cause serious damage to the drive unit.

Cover all exhaust openings to keep out insects or small mammals, which can enter the engine during the lay-up period.

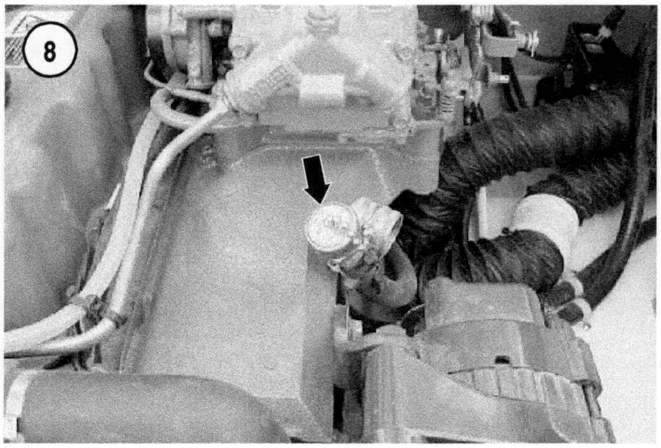

Remove the gearcase drain plug and check for water or other contaminants as described in Chapter Three. Drain the lubricant if contaminated. Disassemble and inspect the drive unit as soon as possible to minimize damage caused by corrosion.

NOTE
Some early production Cobra SX drives are equipped with a stainless steel drain plug on the port side of the pivot housing. If so equipped, remove the plug after performing all lay-up procedures. Insert a stiff wire into the opening to clear any corrosion or other debris. Continue until all water drains from the passage. Direct compressed air into the passage to remove residual water and debris. Then install and securely tighten the plug.

Battery Storage

After treating the fuel system, fogging the engine and placing the drive in the full DOWN position, remove the battery and recharge at 30-day intervals as described in Chapter Nine. A discharged battery can freeze and rupture.

WINTERIZING

Winterizing is a series of steps to drain all water from the cooling system. The exhaust manifolds and seawater pump must be drained on all models. The cylinder block must be drained on all models with a standard cooling system. The heat exchanger must be drained on all models equipped with a closed cooling system. On models with a closed cooling system, the cylinder block is filled with a coolant-and-water solution and does not require draining for storage provided that the coolant mixture can withstand the lowest anticipated temperature without freezing.

NOTE
Manufacturers often make changes in production that affect hose routing or water drain locations. Always refer to the owner's manual for specific drain locations.

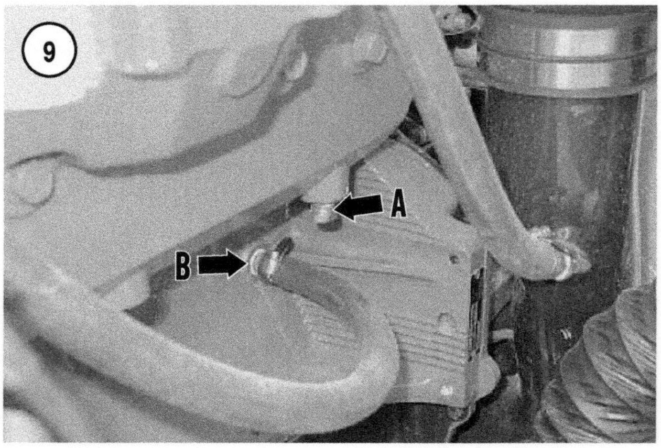

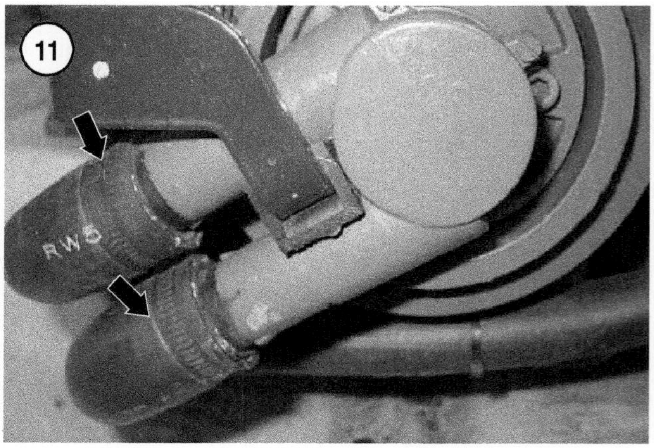

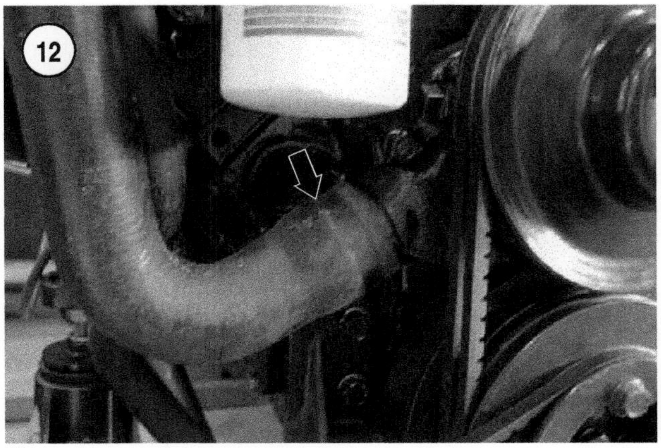

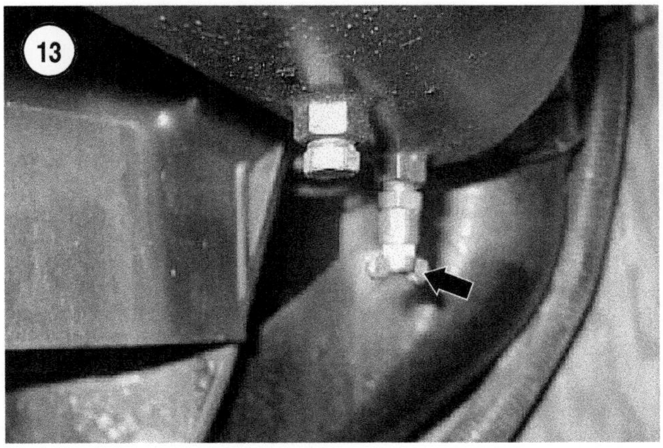

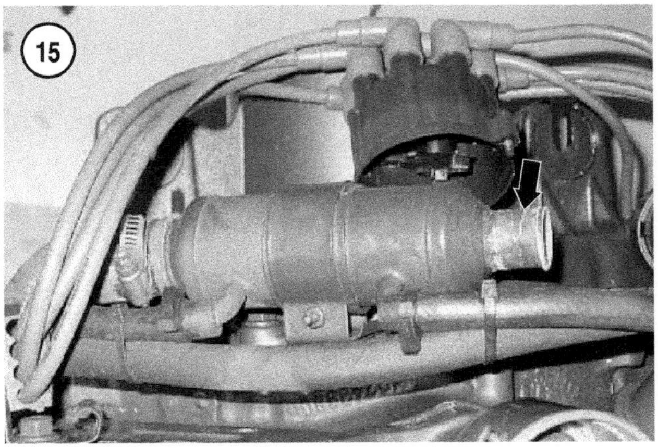

NOTE

Closed cooling system kits are available from numerous sources. Cooling system design, hose routing and fluid capacity will vary by manufacturer. Refer to the information supplied by the cooling system manufacturer for specific information. This section describes draining water from the typical closed cooling system.

GM 3.0L Models

Earlier 3.0L models have the cylinder block drain plug on the port side of the cylinder block (**Figure 7**, typical). Later 3.0L models are equipped with a freshwater flushing system that also is used to drain the cylinder block. The system consists of a brass fitting (**Figure 8**) and a hose that connects to a fitting on the port side of the cylinder block (B, **Figure 9**). The exhaust manifold drain plug on both earlier and later models is located on the rear and bottom side of the intake/exhaust manifold (A, **Figure 9**).

Earlier production models use a belt-driven seawater pump (**Figure 10**). Later production models use a crankshaft-driven seawater pump (**Figure 11**). The inlet and outlet hoses must be removed to drain the pump. The large recirculation pump hose (**Figure 12**, typical) must be removed to drain the recirculation pump.

On models with a closed cooling system, the seawater drain plug for the heat exchanger (**Figure 13**) is located on the bottom and nearest the end of the heat exchanger. The coolant drain plug is generally located farthest from the end of the exchanger. The heat exchanger end cap (**Figure 14**, typical) can also be removed to drain seawater from the heat exchanger. Removing the end cap will allow visual inspection for debris or excessive deposit buildup.

On models with a power steering cooler, the fluid cooler (**Figure 15**, typical) is located on the rear of the engine above the flywheel. To drain the cooler, remove the inlet hose that leads to the inner transom plate and the outlet hose that leads to the seawater pump and drains when removed from the seawater pump fitting.

1. Treat the fuel system and fog the engine as described in this chapter. Place the drive unit in the full DOWN position. Disconnect the negative battery cable. Make sure the engine is level. Tilting can prevent proper draining.

2A. On models with a standard cooling system, perform the following:

   a. On earlier models with block drain plugs, remove the brass drain plug (**Figure 7**, typical) from the port side of the cylinder block. Insert a stiff wire into the plug opening to clear sand and other debris. Continue clearing debris until the cylinder block is fully drained.
   b. On later models with a freshwater flush system, remove the brass cap (**Figure 8**) from the freshwater flush hose. Route the brass fitting to a location lower than the oil pan and allow the water to fully drain

from the block. If water does not drain, trace the hose to the fitting on the port side of the cylinder block (B, **Figure 9**). Remove the hose from the fitting and insert a stiff wire into the fitting to clear debris. Use compressed air to clear debris from the hose. Once fully drained, reconnect the drain hose to the fitting.

c. Remove the drain plug from the exhaust manifold (A, **Figure 9**). Insert a stiff wire into the plug opening to clear sand and other debris. Continue clearing debris until the manifold is fully drained.

d. Loosen the clamp. Then disconnect the recirculation pump hose (**Figure 12**, typical) from the recirculation pump. Tilt the hose to allow full drainage.

2B. On models with a closed cooling system, perform the following:

a. Remove the drain plug from the exhaust manifold (A, **Figure 9**). Insert a stiff wire into the plug opening to clear sand and other debris. Continue clearing debris until the manifold is fully drained.

b. Remove the drain plug from the heat exchanger (**Figure 13**, typical). The water drain fitting is generally located nearest the end of the heat exchanger. Insert a stiff wire into the plug opening to clear sand and other debris. Continue clearing debris until the exchanger is fully drained. If the drain plug is inaccessible, damaged or seized, remove the bolt and carefully pry the cap (**Figure 14**, typical) from the heat exchanger. Remove and replace the gasket from the cap upon removal.

CAUTION

The inlet and outlet hoses connected to the crankshaft-driven seawater pump are the same diameter. Always mark the hoses and pump prior to disconnecting them to make sure they are not switched when reconnected. Running the engine with the hoses switched will damage the impeller or other pump components.

3. Fully loosen the hose clamps. Then carefully disconnect the inlet and outlet hoses from the seawater pump (**Figure 10** or **Figure 11**). Carefully push the hoses off the seawater pump fittings with a blunt screwdriver. Work carefully to avoid damaging the hoses or fittings. Use compressed air to clear residual water from the hoses and fittings. If the boat is stored in the water, insert a plug into the end of the inlet hose to prevent water from siphoning into the bilge. Secure the plug into the hose with a clamp.

4. On models with power steering, loosen the hose clamp and then carefully remove the large-diameter inlet hose from the power steering fluid cooler (**Figure 15**, typical). The inlet hose leads to the inner transom. Carefully tilt the cooler down on the inlet hose end to fully drain it. If the boat is stored in the water, insert a plug into the end of the inlet hose to prevent water from siphoning into the bilge. Secure the plug into the hose with a clamp.

16

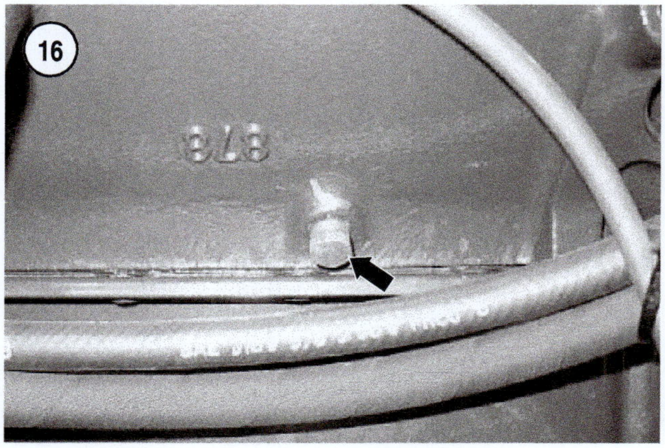

17

18

5. Direct compressed air into all disconnected hoses and open fittings or nipples to clear any residual water or debris.

NOTE

Some early production Cobra SX drives are equipped with a stainless steel drain plug on the port side of the pivot housing. If so equipped, remove the plug after performing the winterizing procedure. Insert a stiff wire into the opening to clear any corrosion or other debris. Continue until all water drains from the passage. Direct compressed air into the passage to remove residual water and debris. Then install and securely tighten the plug.

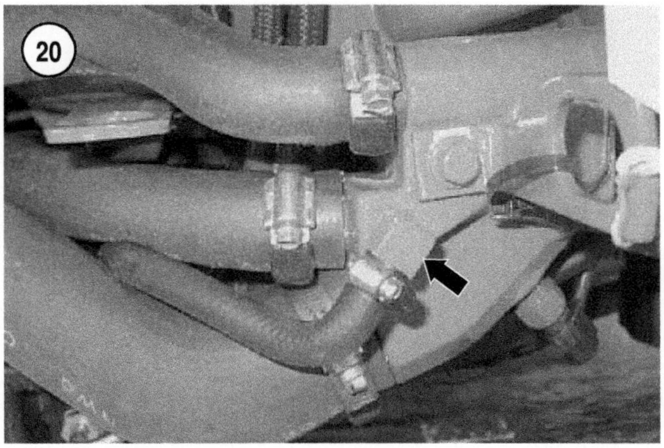

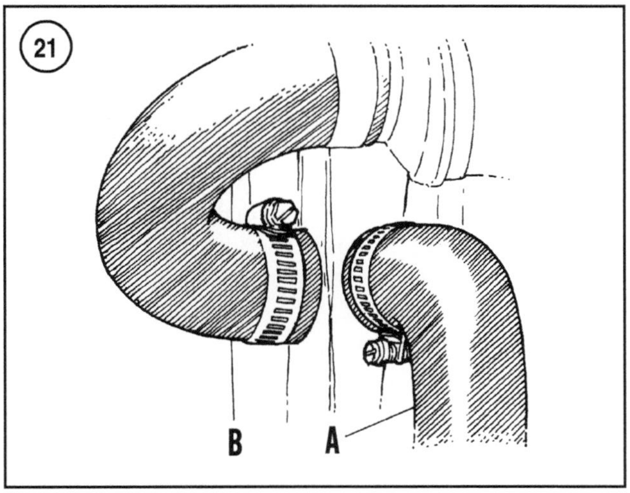

V-6 and V-8 Models

Earlier production GM models and all Ford models have cylinder block drain plugs (**Figure 16**, typical) on the port and starboard sides of the cylinder block. On models with electronic fuel injection, the drain plug on the starboard side of the cylinder block is located behind the knock sensor. If the engine is equipped with electronic fuel injection, and the drain plug cannot be found on the side of the block with the knock sensor, remove the knock sensor to drain seawater from that side of the block. Some later models are equipped with a freshwater flushing system that can also be used to drain the cylinder block. However, on some models, the freshwater flush hose connects to a hose leading to the seawater pump and cannot be used to drain the cylinder block. If in question, trace the flush hose to the connection on the cylinder block (**Figure 17**). The flush hose cannot be used to drain the block if it leads into the seawater pump hose.

Early GM 4.3L V-6 models have a seawater drain plug on the front of the intake manifold (**Figure 18**). The large recirculation pump hose (**Figure 12**, typical) must be removed to access the plug.

Early models are equipped with exhaust manifold drain caps (**Figure 19**, typical) located at the rear of each manifold. Later models are equipped with drain plugs located on the rear of each manifold (**Figure 20**).

Earlier production models use a belt-driven seawater pump (**Figure 10**). Later production models use a crankshaft-driven seawater pump (**Figure 11**). The inlet and outlet hoses and the large recirculation pump hose (**Figure 12**, typical) must be removed to drain the recirculation pump.

On models with electronic fuel injection, the cooling water hoses (A and B, **Figure 21**) must be disconnected from the fuel reservoir for proper drainage.

On models with a closed cooling system, the seawater drain plug for the heat exchanger (**Figure 13**) is located on the bottom near the end of the heat exchanger. The coolant drain plug is located farther from the end of the exchanger. The heat exchanger end cap (**Figure 14**, typical) can also be removed to drain seawater from the heat exchanger. Removing the end cap will allow visual inspection for debris or excessive deposit buildup.

The power steering fluid cooler (**Figure 15**, typical) is located on the rear of the engine above the flywheel. On models with electronic fuel injection, the power steering fluid cooler mounts on the fuel reservoir at the rear of the engine. The inlet hose leads to the inner transom plate. The outlet hose leads to the seawater pump.

NOTE
If the engine is equipped with electronic fuel injection, and the drain plug cannot be found on the side of the block with the knock sensor, remove the knock sensor to drain seawater from that side of the block. Someone might have threaded the knock sensor into the block instead of using a T-fitting that accommodates the knock sensor and drain plug.

1. Treat the fuel system and fog the engine as described in this chapter. Place the drive unit in the full DOWN position. Disconnect the negative battery cable. Make sure the engine is level. Tilting can prevent proper drainage.

2A. On models with a standard (open) cooling system, drain seawater from the cylinder block and exhaust manifolds as follows:

a. On early models, remove the brass drain plugs (**Figure 16**) from the port and starboard sides of the cylinder block. Insert a stiff wire into the plug openings to clear sand and other debris. Continue clearing debris until the cylinder block is fully drained.

b. On later models with a freshwater flush system that connects to the cylinder block fittings, remove the brass cap (**Figure 8**, typical) from the freshwater flush hose. Route the brass fitting to a location lower than the oil pan and allow the water to fully drain from the block. While draining the block, trace the water drain hoses to the fittings on the sides of the cylinder block (**Figure 17**, typical). Both hoses must be warm to the touch. If one of the hoses is cool, or water does not drain from the fitting, remove the hose(s) from the fitting(s) and insert a stiff wire into the fitting(s) to clear debris. Use compressed air to clear debris from the hoses. After fully drained, reconnect the drain hoses to the fitting(s).

c. On early models, loosen the clamp and carefully pry the cap (**Figure 19**) off the exhaust manifold nipple. Work carefully to avoid tearing the cap. Insert a stiff wire into the nipple opening to clear sand and other debris. Continue clearing debris until each manifold is fully drained.

d. On later models, remove the drain plug from the rear of each exhaust manifold (**Figure 20**). Insert a stiff wire into the plug opening to clear sand and other debris. Continue clearing debris until each manifold is fully drained.

e. Disconnect the hoses from the fittings on the opposite ends of the manifolds. Clear sand or debris from the hose fittings to ensure complete draining of the manifolds.

f. Loosen the clamp. Then disconnect the recirculation pump hose (**Figure 12**, typical) from the recirculation pump. Tilt the hose to allow full drainage.

g. On GM 4.3L V-6 models, remove the drain plug from the front of the intake manifold (**Figure 18**). Insert a stiff wire into the plug opening to clear sand or other debris. Continue clearing debris until the intake manifold is fully drained.

2B. On models with a closed cooling system, drain seawater from the exhaust manifold and heat exchanger as follows:

a. On early models, loosen the clamp and carefully pry the cap (**Figure 19**) off the exhaust manifold nipple. Work carefully to avoid tearing the cap. Insert a stiff wire into the nipple opening to clear sand and other debris. Continue clearing debris until each manifold is fully drained.

b. On later models, remove the drain plug from the rear of each exhaust manifold (**Figure 20**). Insert a stiff wire into the plug opening to clear sand and other debris. Continue clearing debris until each manifold is fully drained.

c. Disconnect the hoses from the fittings on the opposite ends of the manifolds. Clear sand or debris from the hose fittings to ensure complete draining of the manifolds.

d. Remove the drain plug from the heat exchanger (**Figure 13**, typical). The water drain plug is generally located near the end of the heat exchanger. Insert a stiff wire into the plug opening to clear sand and other debris. Continue clearing debris until the exchanger is fully drained. If the drain plug is inaccessible, damaged or seized, remove the bolt and carefully pry the cap (**Figure 14**, typical) from the heat exchanger. Remove and replace the gasket from the cap upon removal.

CAUTION

The inlet and outlet hoses connected to the crankshaft-driven seawater pump are the same diameter. Always mark the hoses and pump prior to disconnecting them to make sure they are not switched when reconnected. Running the engine with the hoses switched will damage the impeller or other pump components.

3. Fully loosen the hose clamps. Then carefully disconnect the inlet and outlet hoses from the seawater pump (**Figure 10** or **Figure 11**). Carefully push the hoses off the seawater pump fittings with a blunt screwdriver. Work carefully to avoid damaging the hoses or fittings. Use compressed air to clear residual water from the hoses and fittings. If the boat is stored in the water, insert a plug into the end of the inlet hose to prevent water from siphoning into the bilge. Secure the plug into the hose with a clamp.

4. Loosen the hose clamp. Then carefully remove the large-diameter inlet hose from the power steering fluid cooler (**Figure 15**, typical). The inlet hose leads to the inner transom. Carefully tilt the cooler down on the inlet hose end to fully drain. If the boat is stored in the water, insert a plug into the end of the inlet hose to prevent water from siphoning into the bilge. Secure the plug into the hose with a clamp.

5. On models with electronic fuel injection, drain residual water from the fuel reservoir as follows:

a. Loosen the clamps and then carefully remove the cooling water hoses (A and B, **Figure 21**) from the fuel reservoir nipples. Work carefully to avoid damaging the hoses or nipples.

b. Insert a stiff wire into the fuel reservoir nipples to clear sand and other debris. Continue clearing debris until the cell is fully drained.

6. Direct compressed air into all disconnected hoses and open fittings or nipples to clear any residual water or debris.

NOTE

Some early production Cobra SX drives are equipped with a stainless steel drain plug on the port side of the pivot housing. If so equipped, remove the plug after performing the winterizing procedure. Insert a stiff wire into the opening to clear any corrosion or other debris. Continue until all water drains from the passage. Direct compressed air into the passage to remove residual water and debris. Then install and securely tighten the plug.

FITTING OUT

Fitting out is a series of steps that prepare the engine for operation after storage or winterization.

1. Perform all required maintenance as described in Chapter Three.
2. Service the water pump as described in Chapter Eight.
3. Check and correct all fluid levels as described in Chapter Three.
4. Supply the engine with fresh fuel.
5. Remove any covers from the exhaust openings.
6. Service and install the battery as described in Chapter Nine.
7. If removed, fit the end cap (with new gasket) onto the heat exchanger. Secure the end cap with the bolt. Do not overtighten the bolt.
8. If the boat was winterized and then stored in the water, remove the plugs from the inlet water hoses to the seawater pump and power steering fluid cooler (if so equipped).
9. If removed, connect the water hoses to the seawater pump, power steering fluid cooler, exhaust manifolds and fuel reservoir (EFI models).
10. Install any drain plug or cap into the cylinder block, exhaust manifold(s) and intake manifold (GM V-6 models).
11. Install the cap onto the freshwater flush system hose (if so equipped) and securely tighten. Secure the freshwater flush hose to prevent interference with belts or other moving components.
12. Supply cooling water and then start the engine. Immediately check for leaking fuel, water, coolant, oil or other fluids. Stop the engine and correct any leakage before operating the engine.
13. Run the engine at low speed until the engine reaches normal operating temperature. Check for proper operation of the cooling, electrical and warning systems. Refer to Chapter Two for troubleshooting procedures if a malfunction is indicated.
14. Shift the engine into FORWARD and REVERSE and cycle the steering to the port and starboard limits. Advance the throttle slightly and then return to the idle setting. Do not operate the engine if a malfunction is evident in the shifting, steering or throttle control systems.

Chapter Five

Adjustments

This chapter describes adjustments for the carburetor, ignition timing, shift cable, throttle cable, trim position sender and trim limit control module. Perform the adjustments after servicing the related system or if improper adjustment is suspected of causing a malfunction. **Table 1** lists torque specifications for most applicable fasteners. Use the general torque recommendations listed in Chapter One for fasteners not listed in **Table 1**. **Tables 2-4** list service specifications for the procedures performed in this chapter. **Tables 1-4** are located at the end of this chapter.

CARBURETOR ADJUSTMENT

This section describes accelerator pump clearance, final float level (center-hinge carburetor only), idle speed and idle mixture screw adjustment.

Accelerator Pump Clearance Adjustment

Adjust the accelerator pump lever clearance if hesitation or backfiring during rapid acceleration occurs.

1. Disconnect the negative battery cable.
2. Remove the flame arrestor as described in Chapter Seven.
3. Place the throttle in the FORWARD wide open position.
4. Manually depress the accelerator pump lever on the port side of the carburetor.
5. Use feeler gauges to measure the clearance between the pump lever and the head of the screw on the pump arm (**Figure 1**). The clearance must be 0.010-0.015 in. (0.25-0.38 mm). If adjustment is needed, turn the screw to achieve the required clearance.
6. Return the throttle to idle.
7. Install the flame arrestor as described in Chapter Seven.
8. Connect the battery cable.

Final Float Level Adjustment (Early Models with a Center-Hinged Float)

Perform the final float level adjustment after a carburetor repair and if improper float level adjustment is suspected of causing the engine to malfunction. This adjustment applies only to models using a carburetor with a center-hinged float. Refer to *Carburetor* in Chapter Seven to identify carburetors using the center-hinged float.

WARNING
Stay clear of the propeller and propeller shaft while running an engine on a flush adapter.

CAUTION
Never operate the engine without first providing cooling water. Use a flush test device or

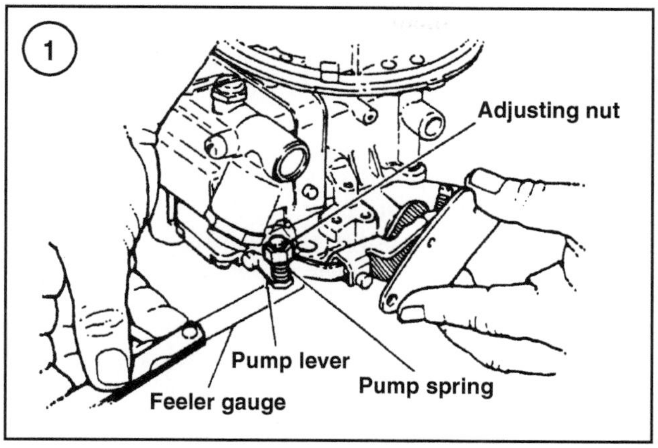

test tank or operate the engine under actual conditions (boat in the water). Operating the engine without providing adequate cooling water will quickly damage the seawater pump and can result in serious engine damage.

1. Remove the flame arrestor cover as described in Chapter Seven. Do not remove the flame arrestor for this procedure.
2. Following the manufacturer's instructions, connect a shop tachometer to the engine. Do not rely on the boat tachometer for carburetor adjustments.
3. Start the engine and operate at 1200-1500 rpm in NEUTRAL until it reaches normal operating temperature.
4. Return the engine to idle speed.
5. Place a shop towel under the float bowl to capture spilled fuel.
6. Carefully remove the sight plug (A, **Figure 2**) on the side of the float bowl.
7. The fuel must be even with the bottom of the threaded plug opening. If adjustment is required, loosen the screw (B, **Figure 2**) and *slowly* rotate the adjusting nut (C) until the fuel level is correct. Hold the nut to prevent rotation and then tighten the screw. Check the fuel level and correct as needed.
8. Install the sight plug and securely tighten.
9. On models with accessible idle mixture screws (**Figure 3**), adjust the idle mixture as described in this section.
10. Adjust the idle speed as described in this section.
11. Stop the engine and remove the shop tachometer.
12. Install the flame arrestor cover as described in Chapter Seven.

5

Idle Mixture Screw Adjustment

WARNING

Safely performing on-water adjustments requires two people: one to operate the boat and another to perform the adjustment(s). Always route hoses and harnesses from test equipment to prevent entanglement with drive belts, linkages and other moving components. Both persons must remain seated at all times. Add extension harnesses or hoses to test equipment to allow adjustment while remaining seated.

CAUTION

Never operate the engine without first providing cooling water. Use a flush test device or test tank or operate the engine under actual conditions (boat in the water). Operating the engine without providing adequate cooling water will quickly damage the seawater pump and can result in serious engine damage.

On early models, the idle mixture screws (**Figure 3**) are easily accessible on the sides of the float bowl. On later models, the idle mixture screws are factory-adjusted, and a plug is installed over the opening (**Figure 4**) to prevent tampering. On models with the tamper-proof plugs, adjust the mixture screws only if improper adjustment is suspected of causing the engine to malfunction. Always rule out improper timing adjustment; a faulty ignition system; old, stale or contaminated fuel; and low cylinder compression before adjusting the idle mixture screws.

1. Disconnect the negative battery cable.
2. Remove the flame arrestor cover as described in Chapter Seven. Do not remove the flame arrestor for this procedure.
3. On later models with plugs covering the idle mixture screws, carefully pry the plugs out of the idle mixture screw openings on the sides of the metering block (**Figure 4**).
4. Turn the idle mixture screws clockwise and record the number of turns from the original position until the screws are lightly seated. The screws should be turned out one-half to one turn out from lightly seated position.
5. If both screws are within the specification, turn the needles in until lightly seated then back the screws out the number of turns recorded in Step 4. If either of the screws are not within the adjustment range, adjust the screw(s) to three-quarters of a turn out from a lightly seated position.
6. Connect the battery cable.
7. Following the manufacturer's instructions, connect a shop tachometer to the engine. Do not rely on the boat tachometer for carburetor adjustments.
8. Prepare the boat and engine for operation. The idle mixture screws cannot be accurately adjusted while using a flush test device for cooling water and with the drive out of the water. Start the engine and operate at 1200-1500 rpm in NEUTRAL until it reaches normal operating temperature.
9. Return the engine to idle speed. Have the assistant shift the engine into FORWARD. Allow a few minutes for the idle to stabilize and then note the tachometer reading. The tachometer must indicate an idle speed within the specification in **Table 2**. If not, adjust the idle speed as described in this chapter.
10. Observe the shop tachometer while turning one of the idle mixture screws clockwise until the engine speed just begins to drop (mixture too lean). Repeat this step for the other idle mixture screw.
11. Observe the shop tachometer and record the number of turns the screw is moved while turning one of the idle mixture screws counterclockwise until the engine speed increases and then just begins to drop (mixture too rich). Repeat this step for the other idle mixture screw.
12. Turn the idle mixture screws clockwise to the midpoint between the setting in Step 10 (too lean) and the setting in Step 11 (too rich). This should provide the smoothest idle and highest idle speed.
13. On models with tamper-proof plugs, carefully push the plugs into the screw opening (**Figure 4**).
14. Adjust the idle speed as described in this section.
15. Remove the shop tachometer.
16. Install the flame arrestor cover as described in Chapter Seven.

Idle Speed Adjustment (Carburetor-Equipped Models Only)

WARNING

Safely performing on-water adjustments requires two people: one who is qualified to operate the boat and another to perform the adjustment(s). Always route hoses and harnesses from test equipment to prevent entanglement with drive belts, linkages and other moving components. Both persons must remain seated at all times. Add extension harnesses or hoses to test equipment to allow adjustment while remaining seated.

CAUTION

Never operate the engine without first providing cooling water. Use a flush test device or test tank or operate the engine under actual conditions (boat in the water). Operating the engine without providing adequate cooling water will quickly damage the seawater pump and can result in serious engine damage.

CAUTION

Do not attempt to adjust the idle speed on electronic fuel injection (EFI) models. On EFI models, the idle speed is controlled by the engine control unit (ECU) and the idle air control motor (IAC). Do not tamper with the factory-adjusted idle stop screw on the throttle body. Tampering with the screw might prevent the EFI system from properly controlling the idle speed and cause hard starting, excessive idle speed or stalling. The throttle body must be replaced if tampering is causing these or other symptoms. The manufacturer does not provide adjustment instructions for the idle stop screw.

NOTE

With the manifolds, fuel system, cooling system and electrical components installed, the General Motors (GM) 5.0L and 5.7L V-8 engines are very similar in appearance to the Ford 5.0L and 5.8L V-8 engines. If the model identification decal on the flame arrestor is missing, the engine can be quickly identified as a GM or Ford engine by noting the distributor location. On GM engines, the distributor is located at the rear (flywheel end) of the engine. On Ford engines, the distributor is located on the front of the engine.

1. Following the manufacturer's instructions, connect a shop tachometer to the engine. Do not rely on the boat tachometer for carburetor adjustments.
2. Prepare the boat and engine for operation. The idle speed cannot be accurately adjusted while using a flush test device for cooling water and with the drive out of the water.

Start the engine and operate at fast idle (1200-1500 rpm) in NEUTRAL until it reaches normal operating temperature.

3. Return the engine to idle speed. Have the assistant shift the engine into FORWARD.

NOTE
It might be necessary to disconnect the throttle cable from the linkage so the throttle lever on the carburetor remains in contact with the stop screw with the drive in FORWARD.

4. Allow a few minutes for the idle to stabilize and then note the tachometer reading. The idle speed must be within the specification in **Table 2**. If not, adjust as follows:
   a. Make sure the throttle stop is firmly against the idle speed adjusting screw. If not, temporarily disconnect the throttle cable from the carburetor linkage and hold the linkage *firmly* against the screw. The stop and screw (**Figure 5**) are located on the side of the carburetor.
   b. Observe the tachometer and turn the idle speed screw (**Figure 5**) to achieve the middle of the idle speed range specified in **Table 2**. Make sure the stop is in *firm* contact with the adjusting screw during the adjustment.
5. Stop the engine. If disconnected, connect the throttle cable to the carburetor linkage. Make sure the throttle stop is against the idle speed adjusting screw. If not, adjust the throttle cable as described in this chapter.
6. Start the engine and note the idle speed. Readjust the idle speed as necessary. Have the assistant shift the engine into FORWARD. Motor to a safe area. Then have the assistant advance the throttle and return to idle speed. Wait a few minutes and note the idle speed. Repeat this operation several times. Each time the throttle returns to idle, the idle speed must return to the desired range (**Table 2**) within a few minutes. If not, adjust the throttle cable as described in this chapter.
7. Remove the shop tachometer and return to port.

IGNITION TIMING ADJUSTMENT

Operating the engine with incorrect ignition timing will result in loss of power, poor fuel economy or serious engine damage. Frequently check and adjust the ignition timing to ensure peak performance. The ignition timing procedure for many models is determined by the type of ignition system used on the engine. Refer to Chapter Nine to identify the ignition system.

WARNING
Safely performing on-water adjustments requires two people: one to operate the boat and another to perform the adjustment(s). Always route hoses and harnesses from test equipment to prevent entanglement with drive belts, linkages and other moving components. Both persons must remain seated at all times. Add extension harnesses or hoses to test equipment to allow adjustment while remaining seated.

CAUTION
Never operate the engine without first providing cooling water. Use a flush test device or test tank or operate the engine under actual conditions (boat in the water). Operating the engine without providing adequate cooling water will quickly damage the seawater pump and can result in serious engine damage.

CAUTION
The ignition timing specification may vary by the antiknock index rating of the fuel being used. Operating the engine with the timing advanced too far for the fuel being used can damage the engine. If the engine might use a fuel with a lower octane rating at any time, adjust the timing to the specification for the fuel with the lower octane rating.

NOTE
With the manifolds, fuel system, cooling system and electrical components installed, the General Motors (GM) 5.0L and 5.7L V-8 engines are very similar in appearance to the Ford 5.0L and 5.8L V-8 engines. If the model identification decal on the flame arrestor is missing, the engine can be quickly identified as a GM or Ford engine by noting the distributor location. On GM engines, the distributor is located at the rear (flywheel end) of the engine. On Ford engines, the distributor is located on the front of the engine.

GM Carburetor-Equipped Models (Using a Delco EST Ignition System)

The timing adapter (OMC or Volvo Penta part No. 885163-6) is required to adjust the ignition timing on these models.

1. Disconnect the negative battery cable.
2. Remove the flame arrestor cover as described in Chapter Seven. Do not remove the flame arrestor for this procedure.

3. Following the manufacturer's instructions, connect a suitable timing light to the No. 1 cylinder spark plug lead. On 3.0L four-cylinder models, the No. 1 spark plug and lead are closest to the front of the engine. On V-6 or V-8 models, the No. 1 spark plug and lead are on the port side closest to the front of the engine.
4. Following the manufacturer's instructions, connect a shop tachometer to the engine. Do not rely on the boat tachometer for ignition timing adjustment.
5. Disable the ignition advance as follows:
   a. Remove the plastic cover to access the four-terminal harness connector on the distributor module (**Figure 6**).
   b. Carefully plug the adapter (Volvo Penta part No. 885163-6) onto the distributor module connector (**Figure 7**). The locking tab on the adapter must engage the tab on the module connector.
   c. Connect the wire end of the adapter to a battery positive power source. The large red wire terminal on the starter motor is a suitable connection point.
6. Locate the timing marks (**Figure 8**, typical) on the timing tab or harmonic balancer decal and the line on the harmonic balancer or notch in the timing chain cover (**Figure 9**). If necessary, clean the timing marks. Apply white paint or chalk to the marks and into the machined line to make them more visible.
7. Connect the battery cable.
8. Start the engine and operate at 1200-1500 rpm in NEUTRAL until it reaches normal operating temperature. Return the engine to idle speed.
9. Direct the timing light at the timing marks. The flash of the timing light will make the moving marks appear to stand still. The timing marks must align for the base ignition timing specified in **Table 3**.
10. If timing adjustment is required, proceed as follows:
   a. Support the distributor to prevent free rotation. Then use a special distributor wrench to loosen the distributor hold-down clamp bolt (**Figure 10**). Purchase the special wrench from a tool supplier or automotive parts store. Attempting to tighten the bolt after the timing adjustment or using makeshift tools will usually cause the distributor to rotate and disturb the timing adjustment.
   b. *Slowly* rotate the distributor (**Figure 11**) until the cover mark aligns with the balancer mark for the ignition timing specified in **Table 3**.
   c. Hold the distributor in position and tighten the hold-down clamp bolt to 18 ft.-lb. (24 N•m).
   d. Check the ignition timing. Readjust the timing if it changed while tightening the bolt. Continue adjusting until the ignition timing is correct and the hold-down clamp bolt is tight.
11. Stop the engine. Lift the locking tab and then unplug the adapter (**Figure 7**) from the ignition module. Disconnect the adapter wire from the battery power source. Slide the plastic cover onto the ignition module plug (**Figure 6**).

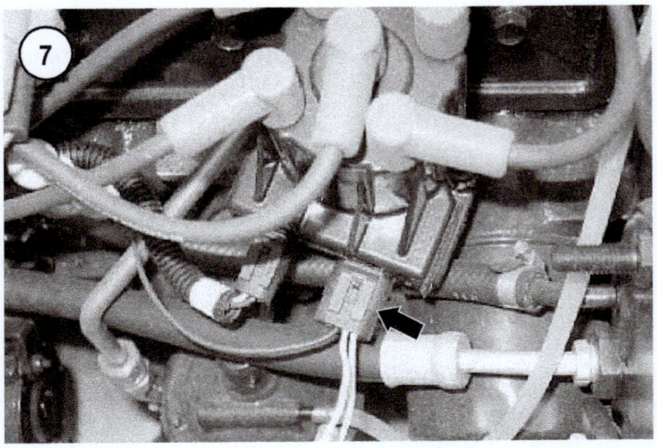

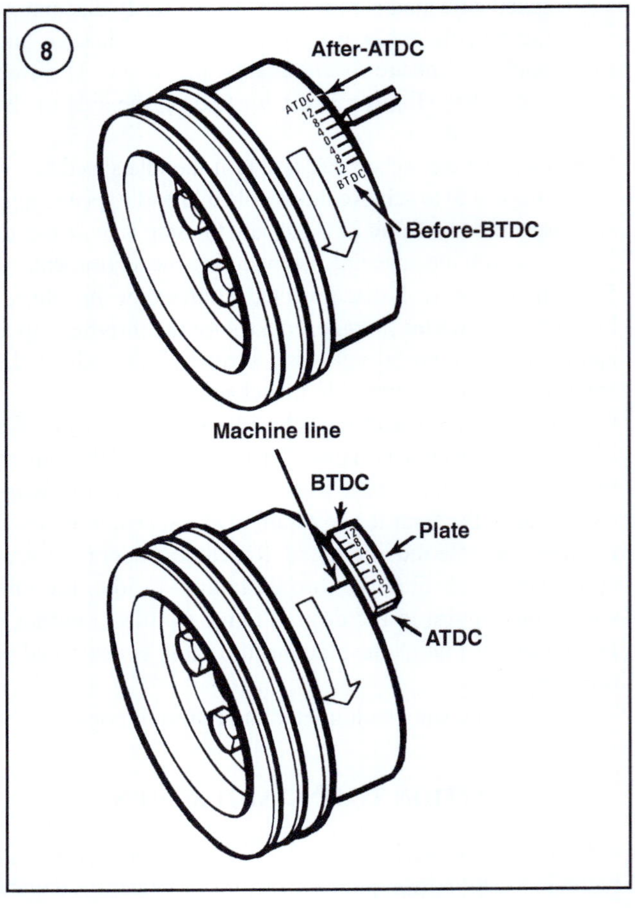

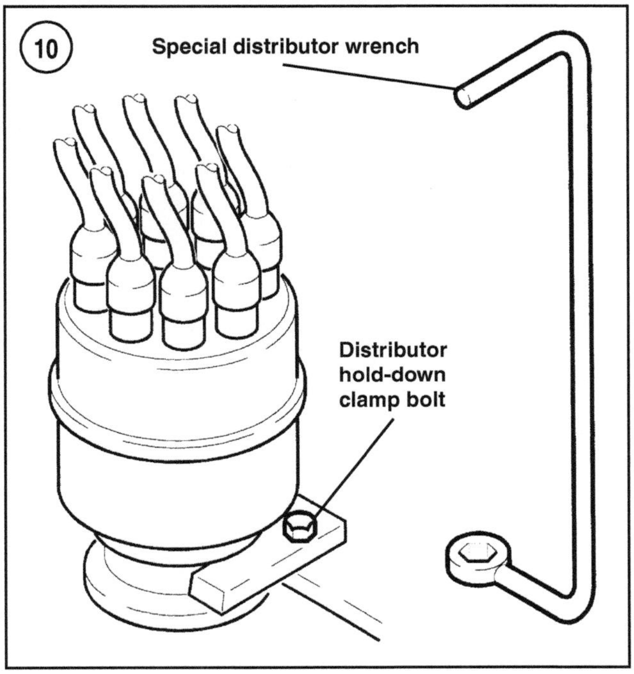

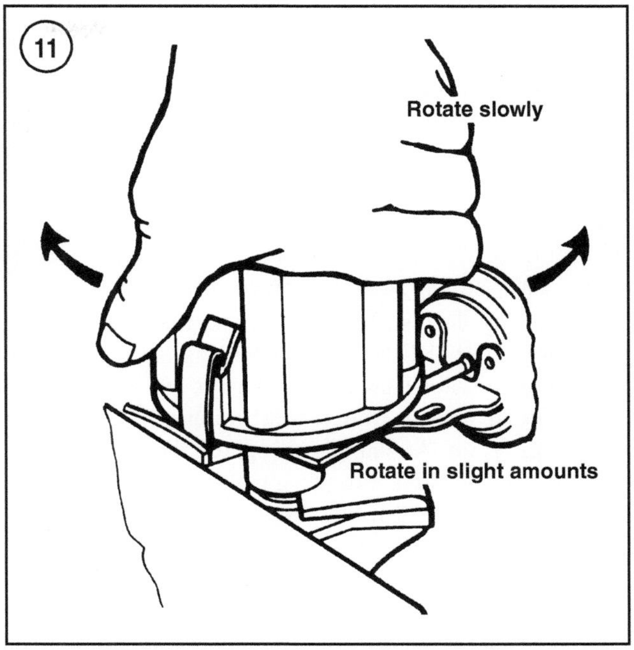

Do not disconnect the timing light or shop tachometer at this time.

12. Start the engine and then direct the timing light at the timing marks. The timing must now be advanced several degrees from base timing. If not, repeat the ignition timing adjustment procedure. Replace the ignition module as described in Chapter Nine if the timing fails to advance as specified with the adapter disconnected. Stop the engine.

13. Prepare the engine for operation.

14. Start the engine and operate in NEUTRAL at 1200-1500 rpm until the engine reaches normal operating temperature. Have the assistant shift the engine into FORWARD. Motor to a safe operating area. Observe the ignition timing and tachometer while an assistant advances the throttle to wide open. Note the timing at 4200-4600 rpm. Have the assistant return the engine to idle speed. Allow a few minutes for a cooling off period.

15. Compare the wide open throttle ignition timing with the specification in **Table 3**. Repeat the idle speed timing adjustment procedure if the timing is not within the specification. Replace the ignition module, as described in Chapter Nine, if the idle speed timing is correct, but the ignition timing at wide open throttle does not advance as specified.

16. Stop the engine. Disconnect the timing light and shop tachometer. Then return to port.

17. Install the flame arrestor cover as described in Chapter Seven.

Carburetor-Equipped Models (Using a Prestolite BID Ignition System)

1. Disconnect the negative battery cable.
2. Remove the flame arrestor cover as described in Chapter Seven. Do not remove the flame arrestor for this procedure.
3. Following the manufacturer's instructions, connect a suitable timing light to the No. 1 cylinder spark plug lead. On GM V-8 models, the No. 1 spark plug and lead are on the port side closest to the front of the engine. On Ford V-8 engines, the No. 1 spark plug and lead are on the starboard side closest to the front of the engine.
4. Following the manufacturer's instructions, connect a shop tachometer to the engine. Do not rely on the boat tachometer for ignition timing adjustment.
5. Locate the timing marks (**Figure 8**, typical) on the timing tab or harmonic balancer decal and the machined line on the harmonic balancer or notch in the timing chain cover (**Figure 9**). If necessary, clean the timing marks. Apply white paint or chalk to the marks and into the machined line to make them more visible.
6. Connect the battery cable.
7. Start the engine and operate at 1200-1500 rpm in NEUTRAL until it reaches normal operating temperature. Return the engine to idle speed.
8. Direct the timing light at the timing marks. The flash of the timing light will make the moving marks appear to stand

still. The timing marks must align with idle speed ignition timing specified in **Table 3**.

9. If the timing is not as specified, stop the engine. Remove the distributor cap and adjust the air gap between the sensor and impulse sender as described in Chapter Nine. Always correct the air gap before attempting to adjust the ignition timing.

10. If timing adjustment is required, and the sensor and impulse sender air gap is correct, proceed as follows:

   a. Support the distributor to prevent free rotation. Then use a special distributor wrench to loosen the distributor hold-down clamp bolt (**Figure 10**). Purchase the special wrench from a tool supplier or automotive parts store. Attempting to tighten the bolt after the timing adjustment or using makeshift tools will usually cause the distributor to rotate and disturb the timing adjustment.
   b. *Slowly* rotate the distributor (**Figure 11**) until the timing chain cover mark aligns with the ignition timing mark specified in **Table 3**.
   c. Hold the distributor in position and tighten the hold-down clamp bolt to 18 ft.-lb. (24 N•m).
   d. Check the ignition timing. Readjust the timing if it changed while tightening the bolt. Continue adjusting until the ignition timing is correct and the hold-down clamp bolt is tight.

11. Stop the engine. Do not remove the timing light or shop tachometer at this time.

12. Prepare the boat and engine for operation.

13. Start the engine and operate in NEUTRAL at 1200-1500 rpm until the engine reaches normal operating temperature. Have the assistant shift the engine into FORWARD. Motor to a safe operating area. Observe the ignition timing and tachometer while an assistant advances the throttle to wide open. Note the timing at 4400-4800 rpm. Have the assistant return the engine to idle speed. Allow a few minutes for a cooling off period.

14. Compare the wide open throttle ignition timing with the specification in **Table 3**. Repeat the idle speed timing adjustment procedure if the timing is not within the specification. Replace the ignition module as described in Chapter Nine if the idle speed timing is correct, but the ignition timing at wide open throttle does not advance as specified.

15. Stop the engine. Return to port. Disconnect the timing light and shop tachometer.

16. Install the flame arrestor cover as described in Chapter Seven.

GM V-6 and V-8 Electronic Fuel Injection (EFI) Models

Ignition timing adjustment for these models requires a marine diagnostic trouble code tool. If the tool is not available, a suitable tool can be fashioned with a short section of wire or a paper clip.

CAUTION

Purchase the marine diagnostic trouble code tool from a tool supplier or marine dealership that services OMC, Volvo Penta or Mercruiser engines. Make sure the tool is suitable for use on the engine. The diagnostic tools available for automobile applications might not be suitable for marine engines and can damage the EFI system.

NOTE

With the manifolds, fuel system, cooling system and electrical components installed, the General Motors (GM) 5.0L and 5.7L V-8 engines are very similar in appearance to the Ford 5.0L and 5.8L V-8 engines. If the model identification decal on the flame arrestor is missing, the engine can be quickly identified as a GM or Ford engine by noting the distributor location. On GM engines, the distributor is located at the rear (flywheel end) of the engine. On Ford engines, the distributor is located on the front of the engine.

1. Disconnect the negative battery cable.

2. Remove the flame arrestor cover as described in Chapter Seven. Do not remove the flame arrestor for this procedure.

3. Following the manufacturer's instructions, connect a suitable timing light to the No. 1 cylinder spark plug lead. The No. 1 spark plug and lead are on the port side closest to the front of the engine.

4. Following the manufacturer's instructions, connect a shop tachometer to the engine. Do not rely on the boat tachometer for ignition timing adjustment.

5. Locate the timing marks (**Figure 8**, typical) on the timing tab or harmonic balancer decal and the machined line on the harmonic balancer or notch in the timing chain cover (**Figure 9**). If necessary, clean the timing marks. Apply white paint or chalk to the marks and into the machined line to make them more visible.

6. Connect the battery cable.

7. Start the engine and operate at 1200-1500 rpm in NEUTRAL until it reaches normal operating temperature.

13

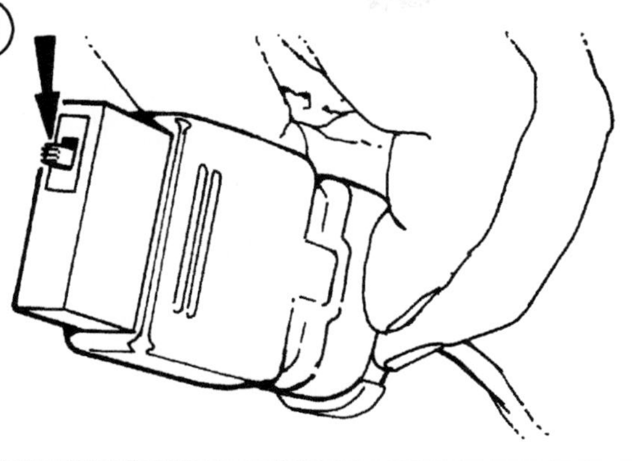

Return the engine to idle speed. Allow a few minutes for the idle to stabilize.

8. Using the fast idle feature of the remote control, advance the engine speed to 1500 rpm. Refer to *Electronic Fuel Injection* in Chapter Seven to identify the ECU.

9A. If using a marine diagnostic trouble code tool, place the engine in service mode as follows:

a. Lift the locking tab and then remove the protective cap from the diagnostic link connector (DLC) plug (**Figure 12**).
b. Lift the locking tab and carefully plug the code tool onto the DLC plug.
c. Move the tool switch (**Figure 13**) to ON.

9B. If not using a marine diagnostic trouble code tool, place the engine in service mode as follows:

a. Lift the locking tab and then remove the protective cap from the diagnostic link connector (DLC) plug (**Figure 12**).
b. Insert a small-diameter jumper wire or paper clip into the plug openings for the white/black and black wires. These should be the wire terminals labeled *A* and *B*. The terminal lettering is molded into the side of the plug.

10. The engine speed should decrease a few seconds after putting the engine into service mode. Check for proper tool or jumper wire connections if the engine speed does not decrease as specified.

11. Direct the timing light at the timing marks. The flash of the timing light will make the moving marks appear to stand still. The timing marks must align with the base ignition timing specified in **Table 3**.

12. If timing adjustment is required, proceed as follows:

a. Support the distributor to prevent free rotation. Then use a special distributor wrench to loosen the distributor hold-down clamp bolt (**Figure 10**). Purchase the special wrench from a tool supplier or automotive parts store. Attempting to tighten the bolt after the timing adjustment or using makeshift tools will usually cause the distributor to rotate and disturb the timing adjustment.
b. *Slowly* rotate the distributor (**Figure 11**) until the timing marks align for the ignition timing specified in **Table 3**.
c. Hold the distributor in position and tighten the hold-down clamp bolt to 18 ft.-lb. (24 N•m).
d. Check the ignition timing. Readjust the timing if it changed while tightening the bolt. Continue adjusting until the ignition timing is correct and the hold-down clamp bolt is tight.

13A. If using a marine diagnostic trouble code tool, move the tool switch (**Figure 13**) to OFF. The engine speed should increase. Return the engine to idle speed. Lift the locking tab and then unplug the code tool from the DLC connector.

13B. If not using a marine diagnostic trouble code tool, remove the jumper wire or paper clip from the DLC connector terminals. The engine speed should increase. Return the engine to the idle speed.

14. Lift the locking tab and then plug the protective cap onto the DLC connector (**Figure 12**). The locking tab must engage the tab on the connector.

15. Stop the engine. Do not remove the timing light or shop tachometer at this time.

16. Prepare the boat and engine for operation.

17. Start the engine and operate in NEUTRAL at 1200-1500 rpm until the engine reaches normal operating temperature. Have the assistant shift the engine into FORWARD. Motor to a safe operating area. Observe the ignition timing and tachometer while an assistant slowly advances the throttle to wide open. The timing must advance as the engine speed increases.

18. Observe the ignition timing as the assistant returns to idle speed and shifts into NEUTRAL. The timing advance must drop as the engine speed is reduced. The idle timing must vary several degrees with the engine in neutral at idle speed.

19. The timing advance is controlled by the engine control unit (ECU) and varies by temperature, barometric pressure and engine load, and the ignition timing will vary widely. The actual timing is not relevant for this test. It is only important that the timing does advance as the speed increases and varies under normal neutral idle speeds. If the timing does not advance as speed increases, and the engine performs poorly, replace the ignition module as described in Chapter Nine. If the ignition timing is not variable at neutral idle speed, recheck the ignition timing as described in this procedure.

20. Stop the engine. Remove the timing light and shop tachometer. Then return to port.

21. Install the flame arrestor cover as described in Chapter Seven.

5

Ford 5.0L and 5.8L V-8 Electronic Fuel Injection (EFI) Models

NOTE

With the manifolds, fuel system, cooling system and electrical components installed, the General Motors (GM) 5.0L and 5.7L V-8 engines are very similar in appearance to the Ford 5.0L and 5.8L V-8 engines. If the model identification decal on the flame arrestor is missing, the engine can be quickly identified as a GM or Ford engine by noting the distributor location. On GM engines, the distributor is located at the rear (flywheel end) of the engine. On Ford engines, the distributor is located on the front of the engine.

1. Disconnect the negative battery cable.
2. Following the manufacturer's instructions, connect a suitable timing light to the No. 1 cylinder spark plug lead. The No. 1 spark plug and lead are on the starboard side closest to the front of the engine.
3. Following the manufacturer's instructions, connect a shop tachometer to the engine. Do not rely on the boat tachometer for ignition timing adjustment.
4. Locate the timing marks (**Figure 8**, typical) on the harmonic balancer and timing chain cover. If necessary, clean the timing marks. Apply white paint or chalk to the marks and into the machined line to make them more visible.
5. Connect the battery cable.
6. Start the engine and operate at 1200-1500 rpm in NEUTRAL until it reaches normal operating temperature. Return the engine to idle speed.
7. Remove the shorting bar (**Figure 14**) from the double wire *SPOUT* connector. The connector is located near the oil fill cap, and the wiring leads into the engine wire harness.
8. Direct the timing light at the timing marks. The flash of the timing light will make the moving marks appear to stand still. The timing marks must align with the idle speed ignition timing specified in **Table 3**.
9. If timing adjustment is required, proceed as follows:
   a. Support the distributor to prevent free rotation. Then use a special distributor wrench to loosen the distributor hold-down clamp bolt (**Figure 10**). Purchase the special wrench from a tool supplier or automotive parts store. Attempting to tighten the bolt after the timing adjustment or using makeshift tools will usually cause the distributor to rotate and disturb the timing adjustment.
   b. *Slowly* rotate the distributor (**Figure 11**) until the timing marks align for the ignition timing specified in **Table 3**.
   c. Hold the distributor in position and tighten the hold-down clamp bolt to 18 ft.-lb. (24 N•m).
   d. Check the ignition timing. Readjust the timing if it changed while tightening the bolt. Continue adjusting until the ignition timing is correct and the hold-down clamp bolt is tight.

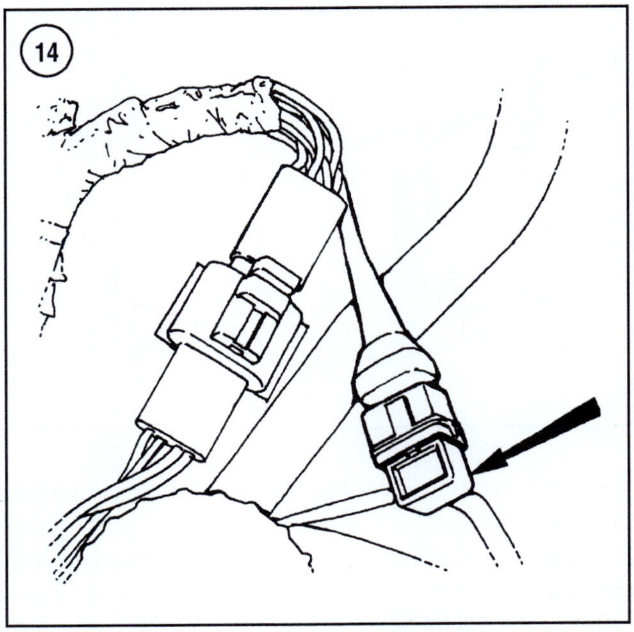

10. Stop the engine. Do not remove the timing light or shop tachometer at this time.
11. Plug the shorting bar into the *SPOUT* connector. Route the shorting bar and *SPOUT* harness wiring away from secondary wires.
12. Prepare the boat and engine for operation.
13. Start the engine and operate in NEUTRAL at 1200-1500 rpm until the engine reaches normal operating temperature.
14. With the engine at idle speed, observe the ignition timing. The timing with the shorting bar installed must be advanced from the ignition timing noted in Step 3 with the shorting bar removed. If the timing is not advanced, check for a loose fit of the shorting bar to the *SPOUT* connector or faulty terminals.
15. Have the assistant shift the engine into FORWARD. Motor to a safe operating area. Observe the ignition timing and tachometer while an assistant slowly advances the

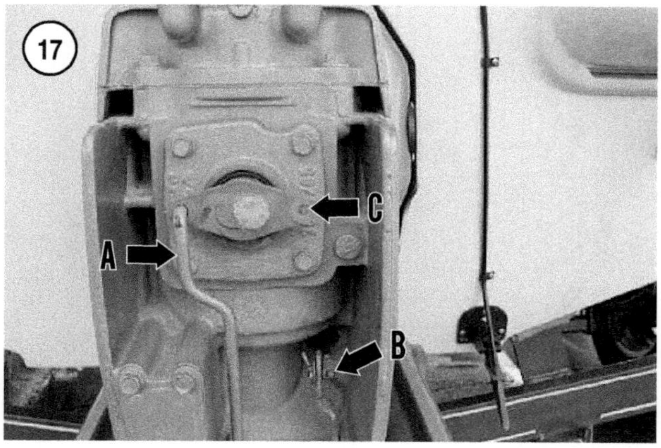

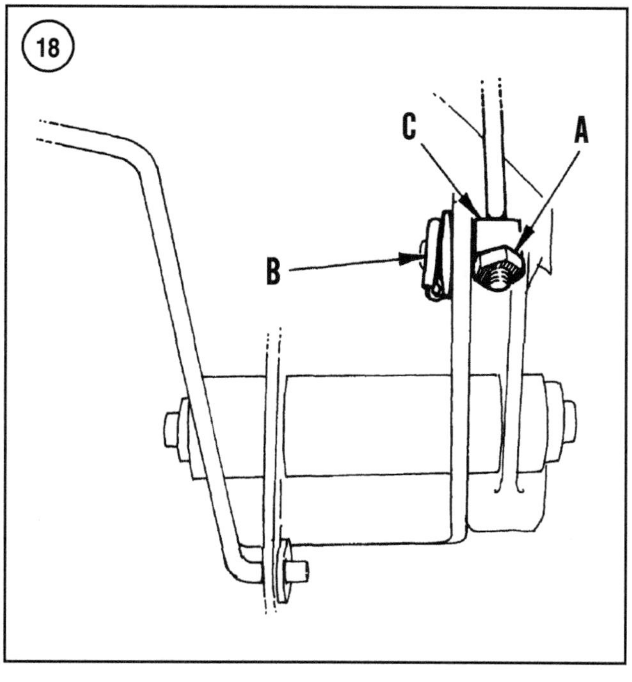

throttle to wide open. The timing must advance as the engine speed increases.

16. Observe the ignition timing as the assistant returns to idle speed and shifts into NEUTRAL. The timing advance must drop as the engine speed is reduced. The idle timing must vary several degrees with the engine in neutral gear at idle speed.

17. The timing advance is controlled by the engine control unit (ECU) and varies by temperature, barometric pressure and engine load, and the ignition timing will vary widely. The actual timing is not relevant for this test. It is only important that the timing does advance as the speed increases and varies under normal neutral idle speeds. Refer to *Ignition System* in Chapter Two if the timing does not advance as speed increases, and the engine performs poorly.

18. Stop the engine. Remove the timing light and shop tachometer. Then return to port.

19. Install the flame arrestor cover as described in Chapter Seven.

SHIFT CABLE ADJUSTMENT

Adjust the shift cable after replacing the cable, remote control cable, drive unit, upper drive shaft housing or pivot housing and if improper adjustment is suspected of causing a shifting system malfunction.

1. Disconnect the negative battery cable.
2. Shift the remote control into NEUTRAL.
3. Remove the three screws (**Figure 15**) that secure the plastic cover to the upper housing. Then pull the cover off the housing. (**Figure 16**).
4. Locate the shift cable (B, **Figure 17**) and linkage (A) at the rear of the upper housing. Remove the jam nut (A, **Figure 18**) from the shift cable.
5. Remove the cotter pin (B, **Figure 18**) and washer from the brass cube nut (C). Pull the cube nut out of the bell crank opening. Unthread the brass cube nut from the shift cable.
6. Measure the distance from the lower edge of the shift link to the center of the pivot on the clevis end (1, **Figure 19**). The distance must be approximately 5-15/16 in. (151 mm). If not, adjust the linkage as follows:
   a. Remove the cotter pin (6, **Figure 19**) and pull the link out of the bellcrank. Discard the cotter pin.
   b. Rotate the link into or out of the clevis end to achieve the desired distance.
   c. Fit the link into the bellcrank. Install the four washers (7, **Figure 19**) onto the link.
   d. Install a *new* cotter pin (6, **Figure 19**) into the link to secure the link and washers to the bellcrank. Bend over both prongs of the cotter pin.
7. Manually move the bellcrank as far forward as possible with hand pressure. Make a mark on the shift housing that aligns with the tip of the shift arm (C, **Figure 17**). Manually move the bellcrank as far rearward as possible. Again make a mark on the shift housing that aligns with the tip of the shift arm. The marks indicate the full engagement points for forward and reverse operation.
8. Manually move the bellcrank forward until you feel the neutral detent in the shift mechanism. This should position the shift arm midway between the full engagement mark. Move the arm up and down to find the midpoint in the free

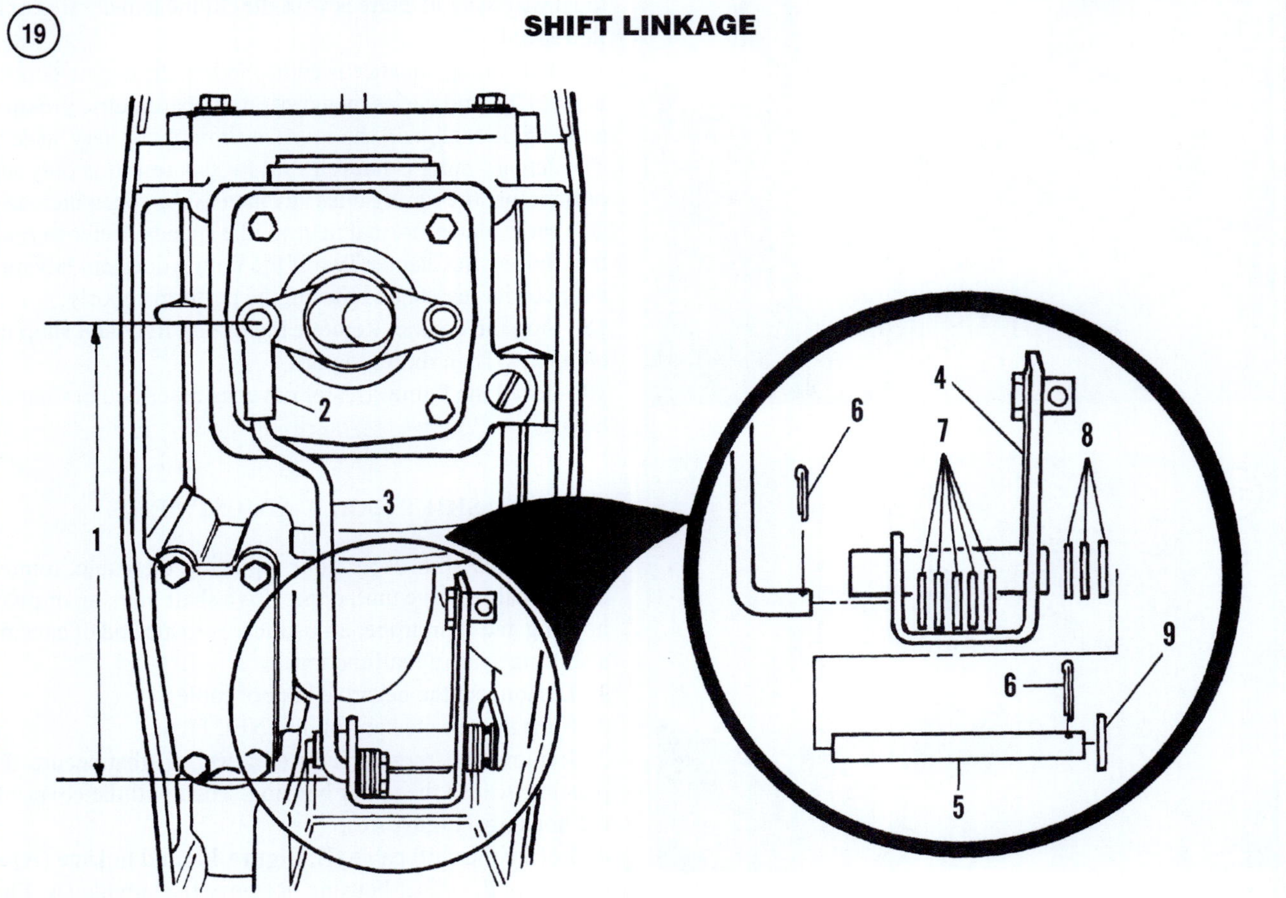

1. Shift link dimension (5-15/16 in. [151 mm])
2. Adjustable clevis end
3. Shift link
4. Bellcrank
5. Bellcrank shaft
6. Cotter pin
7. Flat washer (OMC or Volvo Penta part No. 3852193-6)
8. Flat washer (OMC or Volvo Penta part No. 3852083-9)
9. Flat washer (OMC or Volvo Penta part No. 3852083-9)

play in the neutral detent. Make a mark on the shift housing that aligns with the tip of the shift arm with the shift arm at the center of the neutral detent play. Maintain the shift arm in this position during the adjustment.

9. Place the remote control in NEUTRAL. Push and pull on the cable to determine the midpoint of the free play in the cable. Maintain the cable at the midpoint of play during the adjustment.

10. Thread the brass cube nut (C, **Figure 18**) onto the shift cable and position it so that when all play in both the remote control cable (B, **Figure 17**) and the shift linkage is centered, the brass cube nut will slide freely in the opening in the bellcrank arm.

11. Secure the brass cube nut into the bellcrank arm with a washer and a *new* stainless steel cotter pin (B, **Figure 18**). Bend both prongs of the cotter pin to secure the nut.

12. Thread the jam nut (A, **Figure 18**) over the shift cable until it contacts the brass cube nut. Hold the cube nut with pliers and tighten the jam nut against the cube nut.

13. Shift the remote control into FORWARD. Rotate the propeller (rear propeller on DP-S models) counterclockwise to verify full engagement. The propeller should lock when rotated counterclockwise and spin when rotated

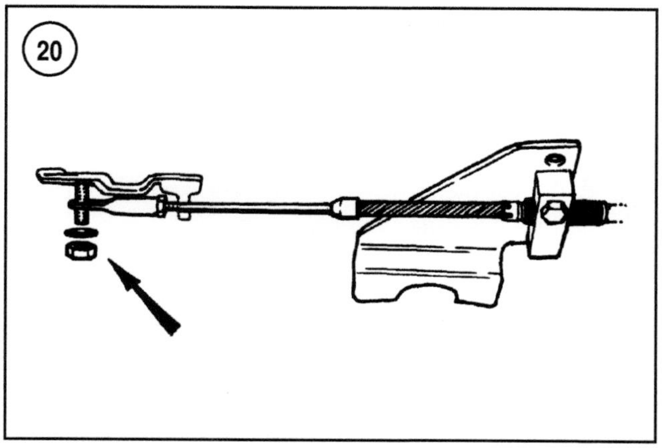

clockwise. Note the position of the shift arm relative to the mark made on the shift housing. The tip of the arm should be at or near the mark.

14. Shift the remote control into REVERSE. Rotate the propeller clockwise to verify full engagement. The propeller should lock when rotated clockwise and spin when rotated counterclockwise. Note the position of the shift arm relative to the mark made on the shift housing. The tip of the arm should be at or near the mark.

15. Adjustment is correct when the full engagement is verified in Step 13 and Step 14 and when the shift arm reaches or is near the marks on the shift housing. The shift cable is attached to the wrong attaching points in the remote control if full engagement is not realized, or the shift arm does not align with the marks as specified. Remove the remote control and attach the cable to a connection point that yields more travel to provide full engagement. If full engagement is realized in one gear but not the other, readjust the shift cables as described in this procedure. A faulty remote control cable, remote control or shift housing component is likely if adjustment does not correct the shift engagement.

16. Fit the plastic cover onto the upper drive shaft housing (**Figure 16**). Install the three screws (**Figure 15**) and tighten evenly to 108-132 in.-lb. (12.2-14.9 N•m).

17. Connect the battery cable.

18. Start the engine and allow it to reach operating temperature. Shift the engine repeatedly into FORWARD and REVERSE. Repeat the adjustment procedure if improper or delayed shifting is noted while shifting. *Do not* operate the engine if any shifting malfunction is evident.

THROTTLE CABLE ADJUSTMENT

1. Disconnect the negative battery cable.
2. Remove the flame arrestor cover as described in Chapter Seven.
3. Place the remote control into NEUTRAL idle position.
4. Remove the cotter pin, locking clip or elastic locknut and washer from the carburetor or throttle body linkage arm to release the cable end (**Figure 20**, typical).
5. Loosen the jam nut on the throttle cable to allow free rotation of the cable connector on the end of the arm.
6. Thread the connector off the end of the cable.
7. Hold the carburetor or throttle body linkage arm against the idle speed adjusting screw or throttle stop.
8. Push the threaded end of the cable *lightly* into the cable jacket to remove any slack.
9. Maintain light pressure on the cable to remove slack. Then thread the connector onto the cable end until the opening aligns perfectly with the post or stud on the carburetor or throttle body linkage arm. Do not install the connector over the post or stud at this time.
10. Mark the threaded end of the cable at the edge of the connector. Then thread the connector off the cable. Measure the distance from the threaded end to the mark to determine the amount of thread engagement into the connector. The connector must thread onto a cable a minimum of 0.373 in. (9.5 mm). If the thread engagement is less than the specification, move the cable anchor onto a different opening in the cable bracket to achieve at least the minimum engagement.
11. Thread the connector onto the cable and align with the post or stud as described in Step 9. Fit the connector over the post. Make sure the edge of the connector reaches or covers the mark made in Step 10. Secure the connector to the post or stud with a washer and *new* cotter pin, elastic locknut or locking clip.
12. Hold the connector with pliers to prevent rotation. Then securely tighten the jam nut against the connector.
13. Verify that the carburetor or throttle body lever is lightly contacting the idle speed adjusting screw or stop screw. Using light pressure, attempt to push the throttle linkage away from the adjusting screw or idle stop. The linkage must not move with light pressure. If the linkage moves, adjust the connector to provide more distance between the pivot points. Do not increase the distance enough to excessively preload the cable. Preloading the cable will cause remote control binding and excessive cable and linkage wear.
14. Move the throttle to the wide open and back to the idle positions several times while observing the throttle linkages. The throttle linkage arm on the carburetor or throttle body must be very close to or lightly contacting the wide open throttle stop at full throttle and contact the idle speed adjusting screw or idle stop at the idle position. If necessary, remove the remote control and connect the throttle cable to a different attaching point to achieve more or less travel from the remote control.
15. Install the flame arrestor cover as described in Chapter Seven.
16. Connect the battery cable.

TRIM POSITION SENDER AND TRIM LIMIT CONTROL MODULE ADJUSTMENT

On Cobra SX (single-propeller) drives, the trim position sender functions only to operate the trim gauge. On DP-S (twin-propeller) drives, the sender operates the gauge and provides input to the trim limit control module.

Trim Position Sender (Cobra SX [Single-Propeller] Drives])

> *NOTE*
> *If incorrectly wired during installation, the trim position sender will not work properly. Make sure a ribbed black wire is inserted into the A terminal opening on the connector plug and a smooth black lead is inserted into the C terminal opening on the connector plug.*

1. Place the drive unit in the full DOWN position. The cylinder rams must be fully retracted. Disconnect the negative battery cable.
2. Remove the two screws and washers (**Figure 21**). Then pull the trim position sender out of the hinge pin and gimbal ring opening.
3. Disconnect the three-pin sender harness connector from the engine harness (**Figure 22**).
4. Calibrate an ohmmeter on the 20- or 40-ohm scale.
5. Connect the meter leads to pin *A* and pin *C* of the sender harness connector plug.
6. *Slowly* rotate the sender shaft until the ohmmeter reads approximately 11 ohms.
7. Carefully guide the sender shaft into the hinge pin. Rotate the sender body *slightly* if necessary to align the sender shaft with the hinge pin opening. *Do not* rotate the sender shaft to align the shaft and hinge pin.
8. Hold the sender and pivot housing in position. Then thread the screws and washers (**Figure 21**) into the sender and gimbal ring openings. Tighten the screws to 18-24 in.-lb. (2.0-2.7 N•m).
9. Note the ohmmeter. Remove the sender and readjust the sender if the meter does not indicate 10-12 ohms.
10. Connect the three-pin sender harness connector to the engine harness. Route the wiring to prevent interference with steering system components or other moving components. Secure the wiring with clamps as needed.
11. Connect the battery cable.
12. Turn the ignition switch to ON or RUN position.

> *NOTE*
> *Battery voltage, charging system and other factors may affect the reading on the trim gauge. For this reason, the trim sender adjustment must be biased to the full DOWN position. This will allow the needle just to reach the bottom of the gauge when the drive just reaches the full down position. The needle should reach or be near the top of the gauge when the drive reaches the full up position.*

13. Observe the gauge while operating the trim system to the full up and down limits. The needle on the gauge must just reach the bottom of the gauge when the drive reaches the full down position, and the needle must reach near the top of the gauge when the drive reaches the full up position. If necessary, loosen the screws and rotate the sender in the gimbal ring opening to achieve the proper adjustment.

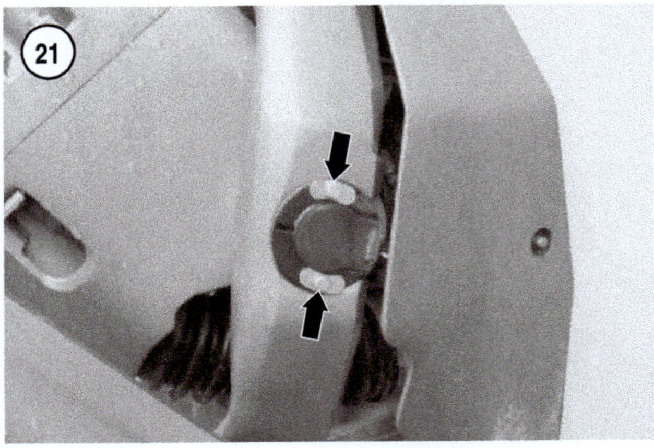

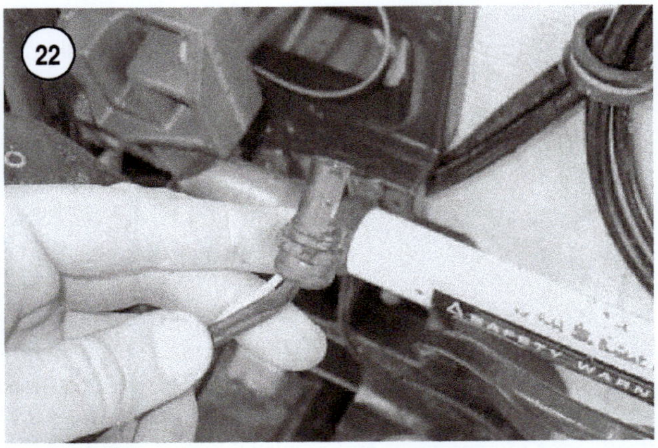

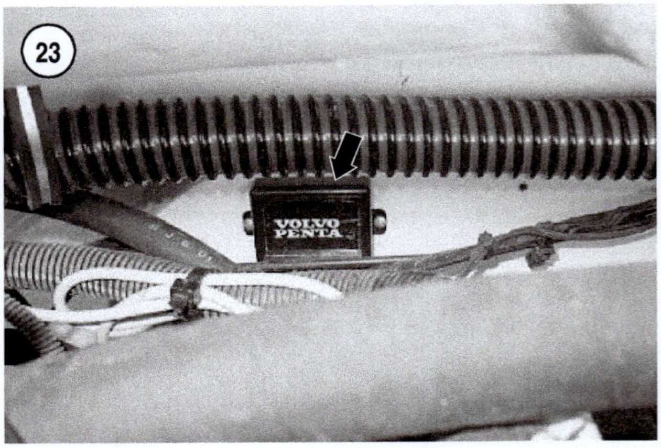

14. Turn the ignition switch to OFF or STOP.

Trim Position Sender and Limit Module Adjustment (Cobra DP-S [Twin-Propeller] Drives)

> *NOTE*
> *If incorrectly wired during installation, the trim position sender will not work properly. Make sure a ribbed black wire is inserted into the A terminal opening on the connector plug and a smooth black lead is inserted into the C terminal opening on the connector plug.*

24

TRIM LIMIT CONTROL MODULE PROGRAMMING (DP-S [TWIN-PROPELLER] DRIVES)

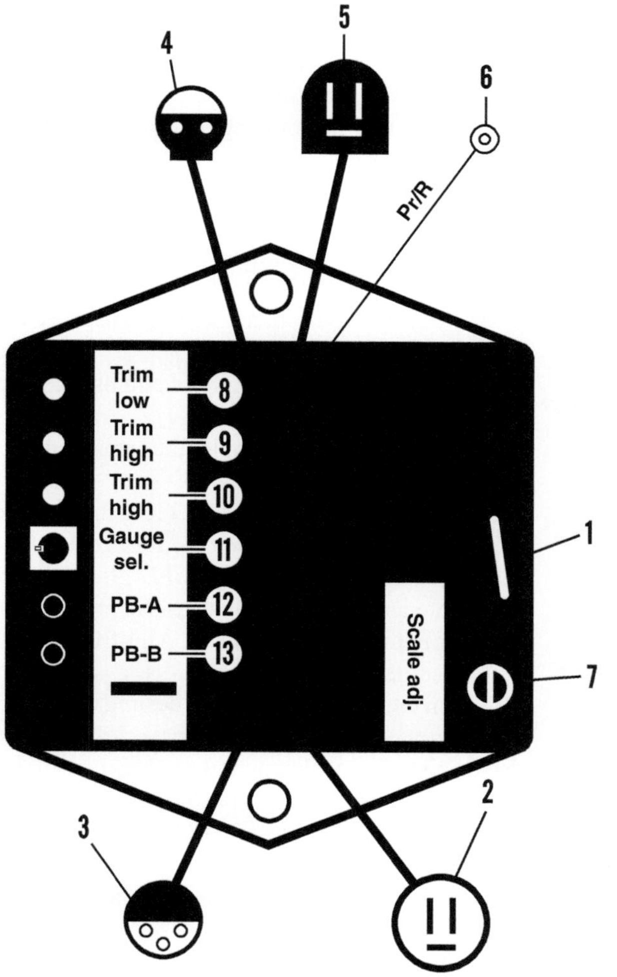

1. Jumper wire (do not cut)
2. Connect to trim relay harness
3. Connect to trim position sender harness
4. Connect to engine harness connector
5. Connect to trim switch harness
6. Connect to tilt/override switch
7. Scale adjustment potentiometer
8. Trim low LED (trim in circuit)
9. Trim high LED (trim out circuit)
10. Tilt high LED (override circuit)
11. Rotary gauge selection switch
12. Programming button – A
13. Programming button – B

1. Place the drive unit into the full DOWN position. The cylinder rams must be fully retracted. Disconnect the negative battery cable.

2. Remove the two screws and washers (**Figure 21**). Then pull the trim position sender out of the hinge pin and gimbal ring opening.

3. Disconnect the three-pin sender harness connector from the engine harness (**Figure 22**).

4. Calibrate an ohmmeter on the 20- or 40-ohm scale.

5. Connect the meter leads to pin *A* and pin *C* of the sender harness connector plug.

6. *Slowly* rotate the sender shaft until the ohmmeter reads approximately 11 ohms.

7. Carefully guide the sender shaft into the hinge pin. Rotate the sender body *slightly* if necessary to align the sender shaft with the hinge pin opening. *Do not* rotate the sender shaft to align the shaft and hinge pin.

8. Hold the sender and pivot housing in position. Then thread the screws and washers (**Figure 21**) into the sender and gimbal ring openings. Tighten the screws to 18-24 in.-lb. (2.0-2.7 N•m).

9. Note the ohmmeter. Remove the sender and readjust the sender if the meter does not indicate 10-12 ohms.

10. Connect the three-pin sender harness connector to the trim limit control module harness. Route the wiring to prevent interference with steering system components or other moving components. Secure the wiring with clamps as needed.

11. If the transom angle is unknown, measure the angle with a protractor, making sure to establish the level (angle) of the hull first. The transom angle must be 10-15° with 12-13° being the most common. Record the measured transom angle.

12. Locate the trim limit control module (**Figure 23**) on the boat transom or other surface in the engine compartment.

13. Connect the battery cable.

14. Turn the ignition switch to ON or RUN.

15. Turn the gauge selector switch (11, **Figure 24**) to the 0 position. All three LED should blink simultaneously.

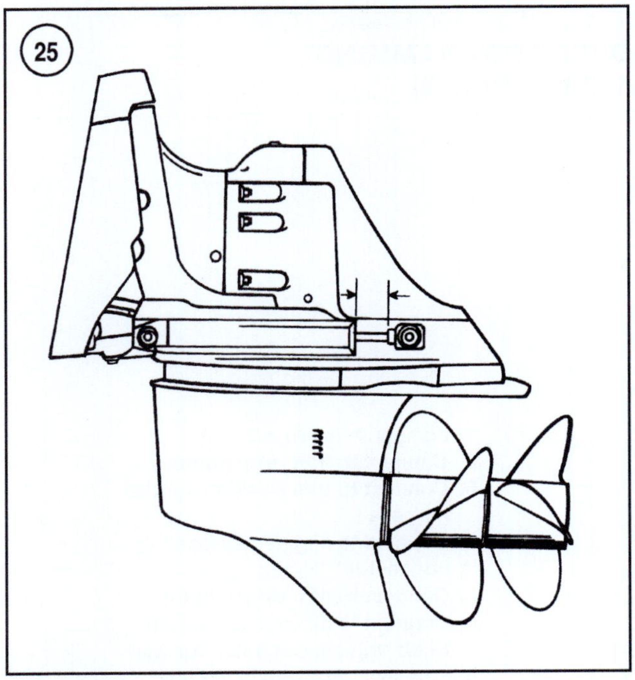

16. Operate the trim to position the drive unit in the -2° trim position by measuring the ram extension from the end of the trim cylinder cam to the eyelet boss as shown in **Figure 25**. Adjust the trim to achieve the specified distance for the recorded transom angle (Step 11).

a. For a 10° transom angle, the rams must extend 1.89 in. (48 mm).
b. For an 11° transom angle, the rams must extend 1.97 in. (50 mm).
c. For a 12° transom angle, the rams must extend 2.13 in. (54 mm).
d. For a 13° transom angle, the rams must extend 2.28 in. (58 mm).
e. For a 14° transom angle, the rams must extend 2.44 in. (62 mm).
f. For a 15° transom angle, the rams must extend 2.56 in. (65 mm).

17. After adjusting the trim to achieve the specified dimension, press the PB-A button (12, **Figure 24**) for at least 4 seconds. The green trim-low LED (8, **Figure 24**) should illuminate and then extinguish as the program is accepted by the trim limit control module. All LEDs will blink as the button is released.

18. Operate the trim to position the drive unit in the +5° trim position by measuring the ram extension from the end of the trim cylinder cam to the eyelet boss as shown in **Figure 25**. Adjust the trim to achieve the specified distance for the recorded transom angle (Step 11).

a. For a 10° transom angle, the rams must extend 2.87 in. (73 mm).
b. For an 11° transom angle, the rams must extend 2.99 in. (76 mm).
c. For a 12° transom angle, the rams must extend 3.19 in. (81 mm).
d. For a 13° transom angle, the rams must extend 3.35 in. (85 mm).
e. For a 14° transom angle, the rams must extend 3.50 in. (89 mm).
f. For a 15° transom angle, the rams must extend 3.66 in. (93 mm).

19. After adjusting the trim to achieve the specified dimension, press the PB-B button (13, **Figure 24**) for at least 4 seconds. The yellow trim-high LED (9, **Figure 24**) should illuminate and then extinguish as the program is accepted by the trim limit control module. All LEDs will blink as the button is released.

20. Trim the drive unit up to the desired maximum tilt position. Allow at least 1/2 in. (13 mm) clearance between the drive unit and any obstruction, such as a swim platform.

21. After adjusting the trim to the desired maximum tilt position, press the PB-A *and* PB-B buttons (12 and 13, **Figure 24**) simultaneously for at least 4 seconds. The red tilt-high LED should illuminate and then extinguish as the program is accepted into the trim limit control module. All LEDs will blink as the buttons are released.

22. Rotate the gauge selector switch (11, **Figure 24**) to the correct position as specified in **Table 4** for the brand and part number of the gauge used in the boat.

23. Trim the drive unit to the full DOWN position by depressing the override and trim in buttons simultaneously.

24. Turn the ignition switch to ON or RUN. Adjust the scale adjustment potentiometer (7, **Figure 24**) until the trim gauge reads full down.

25. Turn the ignition switch to OFF or STOP position to end the programming procedure.

Table 1 ADJUSTMENT TORQUE SPECIFICATIONS

Fastener location	in.-lb.	ft.-lb.	N•m
Distributor hold-down clamp bolt	–	18	24
Rear drive cover (plastic) screws	108-132	–	12.2-14.9
Trim position sender screws	18-24	–	2.0-2.7

Table 2 ENGINE OPERATING SPEED SPECIFICATIONS

Model	Idle speed (rpm)	Wide open throttle speed (rpm)
GM 3.0L models	650-750	4200-4600
GM 4.3L models		
43GL (carburetor-equipped) models	550-650	4200-4600
43GS (carburetor-equipped) models	550-650	4200-4800
43GI (EFI) models	600*	4400-4800
GM 5.0L models		
50GL (carburetor-equipped) models	550-650	4400-4800
50GI (EFI) models	600*	4600-5000
Ford 5.0L models		
50FL (carburetor-equipped) models	550	4200-4600
50FI (EFI) models	600*	4200-4600
GM 5.7L models		
57GL (carburetor-equipped) models	600-650	4200-4600
57GS (carburetor-equipped) models	550-650	4400-4800
57GI (EFI) models	600*	4200-4600
57GSI (EFI) models	600*	4600-5000
Ford 5.8L models		
58FL (carburetor-equipped) models	550-650	4000-4400
58FI (EFI) models	600*	4200-4600
58FSI (EFI) models	600*	4600-5000
GM 7.4L models		
74GL (carburetor-equipped) models	550-650	4200-4600
74GI (EFI) models	600*	4200-4600
74GSI (EFI) models	600*	4800-5200
GM 8.2L models		
82GL (carburetor-equipped) models	550-650	4400-4800
82GSI (EFI) models	600*	4600-5000

*The idle speed is nonadjustable on this model.

Table 3 IGNITION TIMING AND FUEL REQUIREMENT SPECIFICATIONS

Model	Timing Specification	Minimum AKI rating
GM 3.0L models		
1994-1996		
Base timing	0° TDC*	–
Timing at 600 rpm	10° BTDC	–
Timing at 2500 rpm	20° BTDC	–
Timing at 4600 rpm	23° BTDC	–
Minimum fuel AKI rating	–	87
1997-2000		
Base timing	2° ATDC*	–
Timing at 600 rpm	8° BTDC	–
Timing at 2500 rpm	18° BTDC	–
Timing at 4600 rpm	21° BTDC	–
Minimum fuel AKI rating	–	86
GM 4.3L models		
43GL and 43GS (carburetor-equipped) models		
1994-1996		
Base timing	5° ATDC*	–
Timing at 600 rpm	6° BTDC	–
Timing at 2500 rpm	15° BTDC	–
Timing at 4600 rpm	18° BTDC	–
Minimum fuel AKI rating	–	87
1997-2000		
Base timing	1° BTDC*	–
Timing at 600 rpm	18° BTDC	–
Timing at 2500 rpm	25° BTDC	–
Timing at 4600 rpm	25° BTDC	–
Minimum fuel AKI rating	–	86

(continued)

Table 3 IGNITION TIMING AND FUEL REQUIREMENT SPECIFICATIONS (continued)

Model	Timing Specification	Minimum AKI rating
GM 4.3L models (continued)		
43GI (electronic fuel injection) models		
Base timing	8° BTDC*	–
Timing at 1600 rpm	20° BTDC	–
Minimum fuel AKI rating	–	86
GM 5.0L models		
50GL (1997-2000) carburetor-equipped models		
Base timing	5° BTDC*	–
Timing at 600 rpm	3° BTDC	–
Timing at 2400 rpm	18° BTDC	–
Timing at 4600 rpm	26° BTDC	–
Minimum fuel AKI rating	–	86
50GI (electronic fuel injection) models		
Base timing	8° BTDC*	–
Timing at 1600 rpm	19° BTDC	–
Minimum fuel AKI rating	–	86
Ford 5.0L models		
50FL (carburetor-equipped) models		
Timing at 950 rpm	10° BTDC	–
Timing at 2050 rpm	20° BTDC	–
Timing at 4600 rpm	28° BTDC	–
Minimum fuel AKI rating	–	86
50FI (electronic fuel injection) models		
Base timing	5° BTDC*	–
Minimum fuel AKI rating	–	86
GM 5.7L models		
57GL (1994-1996) carburetor-equipped models		
Timing at 600 rpm	3° BTDC	–
Timing at 2800 rpm	22° BTDC	–
Timing at 4600 rpm	26° BTDC	–
Minimum fuel AKI rating	–	87
57GS (1996-2000) carburetor-equipped models		
Timing at 600 rpm	3° BTDC	–
Timing at 2800 rpm	22° BTDC	–
Timing at 4600 rpm	26° BTDC	–
Minimum fuel AKI rating	–	86
57GI (1994-1996) electronic fuel injection models		
Base timing	8° BTDC*	–
Timing at 1600 rpm	19° BTDC	–
Minimum fuel AKI rating	–	87
57GSI (1997-2000) electronic fuel injection models		
Base timing	8° BTDC*	–
Timing at 1600 rpm	19° BTDC	–
Minimum fuel AKI rating	–	86
Ford 5.8L models		
58FL (carburetor-equipped) models		
Timing at 950 rpm	10° BTDC	–
Timing at 2050 rpm	20° BTDC	–
Timing at 4400 rpm	28° BTDC	–
Minimum fuel AKI rating	–	86
58FI and 58FSI (electronic fuel injection) models		
Base timing	5° BTDC*	–
Minimum fuel AKI rating	–	86
GM 7.4L models		
74GL (carburetor-equipped) models		
Timing at 600 rpm	3° BTDC	–
Timing at 2800 rpm	22° BTDC	–
Timing at 4400 rpm	26° BTDC	–
Minimum fuel AKI rating	–	87
74GI (electronic fuel injection models)		
Base timing	10° BTDC*	–
Minimum fuel AKI rating		
1994-1996	–	87
1997-2000	–	86

(continued)

Table 3 IGNITION TIMING AND FUEL REQUIREMENT SPECIFICATIONS (continued)

Model	Timing Specification	Minimum AKI rating
GM 7.4L models (continued)		
74GSI (electronic fuel injection)		
Base timing	10° BTDC*	–
Minimum fuel AKI rating		
1994-1996	–	89
1997-2000	–	86
GM 8.2L models		
82GL (carburetor-equipped) models		
Base timing	3° BTDC*	–
Minimum fuel AKI rating	–	87
82GSI (electronic fuel injection) models		
Base timing	10° BTDC*	–
Minimum fuel AKI rating	–	86

*Idle timing with the ignition system in service mode.
During normal operation timing is variable at idle speed.

Table 4 GAUGE SETTINGS FOR TRIM LIMIT CONTROL MODULE

Gauge used	Selector switch position
Faria part No. GP9318B	3
Faria part No. GP9374D	5
Medallion part No. SIK-140-30W, Rev.A	4
Medallion part No. SIK-112-30W, Rev.A	4
OMC (all compatible part numbers)	3
VDO part No. 01-210-415	1
VDO part No. X-19-270-3123	2
VDO Vanguard part No. 1-275-810-136A	2
Volvo Penta part No. 3851788	1
Volvo Penta part No.857449	2
Teleflex part No. 20742 and 58031	4

Chapter Six

Engine

Torque specifications for most fasteners are listed in **Table 1**. Use the torque recommendations in Chapter One for fasteners not listed in **Table 1**. **Tables 2-6** list engine service specifications. **Tables 1-6** are at the end of this chapter.

ENGINE OPERATION

On General Motors (GM) 3.0L four-cylinder models, the cylinders are numbered 1-2-3-4 from front to rear, and the firing order is 1-3-4-2.

On General Motors (GM) 4.3L V-6 models, the cylinders are numbered from front to rear (1-3-5 on the port side and 2-4-6 on the starboard side). The firing order is 1-6-5-4-3-2.

On General Motors (GM) 5.0L, 5.7L, 7.4L and 8.2L V-8 models, the cylinders are numbered from front to rear (1-3-5-7 on the port side and 2-4-6-8 on the starboard side). The firing order is 1-8-4-3-6-5-7-2.

On Ford 5.0L and 5.8L V-8 models, the cylinders are numbered from front to rear (1-2-3-4 on the starboard side and 5-6-7-8 on the port side). The firing order is 1-3-7-2-6-5-4-8.

The intake and exhaust valves (one each per cylinder) are operated via pushrods and rocker arms by a camshaft located in the cylinder block. Camshaft motion is transferred through hydraulic lifters to the rocker arms by pushrods. The rocker arms move on ball pivots located on individual threaded shoulder studs or shoulder bolts.

On 3.0L models, the camshaft is gear- driven and is supported by four bearings. The oil pump, mounted at the bottom front of the cylinder block, is driven by the distributor shaft via a gear on the camshaft.

On all other models covered in this manual, the camshaft is chain-driven and is located between the two cylinder banks. The camshaft is supported by four bearings on 4.3L models and five bearings on 5.0L and 5.7L models. The oil pump, mounted at the bottom rear of the cylinder block, is driven by the distributor shaft via a gear on the camshaft.

A gear-driven balance shaft is located above the camshaft on 4.3L models. The balance shaft dampens much of the vibration in the 1700-2000 rpm range.

The cast-iron cylinder head(s) contain intake and exhaust valves with integral valve guides. Induction hardened valve seats are used on 3.0L models. This process hardens the exhaust valve contact areas on the cylinder head. Pressed-in exhaust valve seats are used on V-6 and V-8 models. This allows replacement of the seat if it cannot be resurfaced to the specifications. On 3.0L models, the cylinder head must be replaced if the seat cannot be resurfaced.

The crankshaft is supported by five main bearings on four-cylinder and V-8 models and four main bearings on V-6 models. One main bearing provides crankshaft thrust control. Crankshaft rotation is counterclockwise when viewed from the flywheel end of the engine. The crankshaft drives the camshaft via gears (3.0L models) or chain and sprockets (V-6 and V-8 models). The camshaft turns at one-half crankshaft speed.

NOTE

With the manifolds, fuel system, cooling system and electrical components installed, the General Motors (GM) 5.0L and 5.7L engines are very similar in appearance to the Ford 5.0L and 5.8L engines. If the model identification decal on the flame arrestor is missing, the engine can be quickly identified as a GM or Ford engine by noting the distributor location. On GM engines, the distributor is located at the rear (flywheel end) of the engine. On Ford engines, the distributor is located on the front of the engine.

SPECIAL TOOLS

Where special tools are required or recommended, the tool part numbers are provided in the procedure. Some of the tools are offered by Outboard Marine Corporation (OMC).

Many of the repair procedures for General Motors (GM) engines require tools with a J prefix. These tools are available from Kent-Moore Tools, 29784 Little Mack, Roseville, MI 48066, or a GM automobile dealership.

Many repair procedures for Ford engines require tools with T prefix. These tools are available from the OTC division of SPX Corporation, 655 Eisenhower Drive, Owatonna, Minnesota 55060.

REPLACEMENT PARTS

There are several differences between automotive engines and engines adapted for marine applications. For example, the cylinder head gasket must be corrosion-resistant. Marine engines use stainless steel or composite gaskets instead of the standard steel gasket used in automotive applications. Brass core plugs must be used instead of the steel plugs used in automotive blocks. Because marine engines run at or near maximum speed most of the time, they use special lifters, valve springs, pistons, bearings, camshafts and other heavy-duty components.

SPECIAL SERVICE PRECAUTIONS

When working on the engine, taking basic precautions will make your work easier, faster and more accurate:

1. Always make notes, drawings or photographs of all external engine components *before* beginning disassembly. An incorrectly routed hose or wire can interfere with linkage movement and result in a dangerous lack of throttle control. Hoses might abrade and leak if allowed to contact sharp, hot or moving parts. Check for components that can be installed in two or more positions and mark them accordingly.
2. Mark the up and forward directions of all components before removing them. If a cluster of components share common wires or hoses, try to remove the entire assembly intact. This reduces the time required to disassemble and assemble the engine and reduces the chance of improper connections during assembly.
3. Mount the engine to an engine stand if the repair involves the pistons, rods or crankshaft. Support the engine with an overhead lift or blocks as needed for other repairs.
4. Use muffin tins or egg cartons to organize fasteners as they are removed. Mark all fasteners to make sure they are reinstalled in the correct locations.
5. Use special tools where noted. In some cases, it might be possible to perform the procedure with makeshift tools, but this is not recommended. The use of makeshift tools can damage components or cause serious injury.
6. Use a vise with protective jaws to hold housings or components. If protective jaws are not available, insert wooden blocks on each side of the part before clamping it in the vise. Never clamp thin castings, such as the piston skirt, which can be damaged by the clamping force.
7. Remove and install pressed-on parts with the appropriate mandrel, support and hydraulic press. Do not try to pry or hammer them from the component.
8. Tighten all fasteners to the specification in **Table 1**. Use the general torque recommendation listed in the *Quick Reference Data* section at the front of the manual for fasteners not listed in **Table 1**.
9. Apply the recommended sealant or threadlocking compound to the outer surface of seals. Lubricate seal lips during assembly.
10. Prior to installation, apply engine oil or other recommended lubricant to all internal components.
11. Work in a clean area and where there is good lighting and sufficient space for storing components. Keep small containers on hand for storing small parts. Cover parts with clean shop towels when you are not working with them.
12. Replace all seals, gaskets and O-rings if removed. These parts are inexpensive compared to the damage they will cause if they fail.

WARNING

The engine is heavy, awkward to handle and has numerous sharp edges. It can shift or drop suddenly during removal. To prevent se-

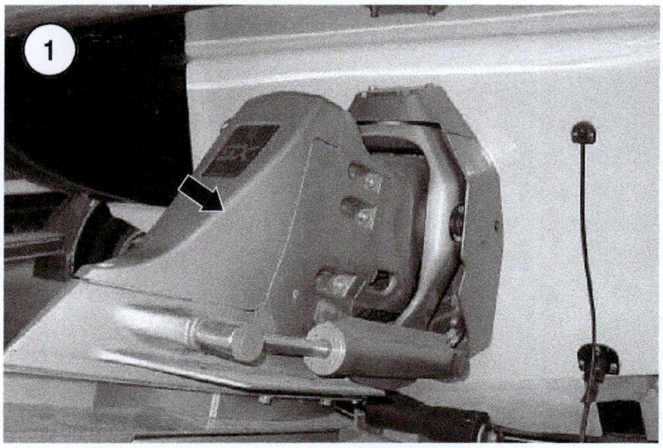

rious injury, never place the body where any sudden or unexpected engine movement can contact the body.

13. Never place any part of the body where a moving or falling engine can trap, cut or crush it.
14. If you must push the engine during removal or installation, use a board or similar tool to keep your hands and arms out of danger.
15. Make sure any lifting device, such as a hoist, is designed and has enough capacity to lift the engine.
16. Make sure the lifting device is securely attached to safe lifting points on the engine.
17. If difficulty is encountered while removing the engine, stop lifting, lower the engine back onto the mounts and make sure the engine has been completely separated from the vessel.

ENGINE REMOVAL

Some service procedures can be performed with the engine in the boat; others require removal. Boat design and the service procedure that must be performed determine whether the engine must be removed.

1. Remove the engine hatch, seating and any panels or structures that can interfere with engine removal or access to any of the engine mounts.

NOTE
In many instances, boat design or add-on structures can prevent removal of the engine as an assembly. In such cases, remove the exhaust manifold or other readily accessible external components to achieve the needed clearance. After installing the engine, follow the appropriate procedures to reinstall these components.

2. Disconnect the negative and then positive battery cables. As a precaution, remove the battery from the boat.
3. Remove the drive unit (**Figure 1**) as described in Chapter Ten.

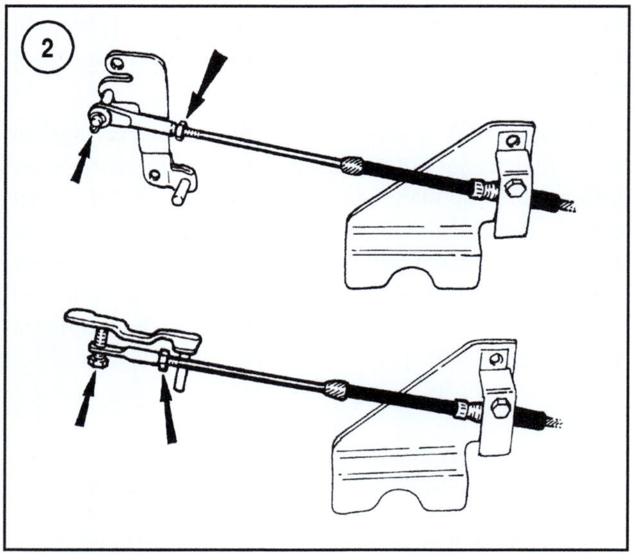

4. Disconnect the throttle cable from the carburetor or throttle body linkage (**Figure 2**, typical).
5. Detach the instrument harness (**Figure 3**, typical) and trim switch harness plugs (**Figure 4**, typical) from the engine harness.
6. Disconnect the fuel supply hose from the fuel pump (**Figure 5**, typical) or fuel filter inlet (**Figure 6**, typical). Plug the hose and fitting to prevent leakage and the entry of contamination.

5

6

7

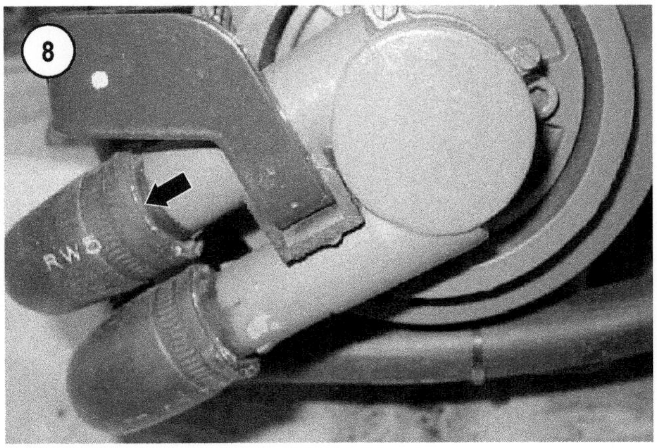
8

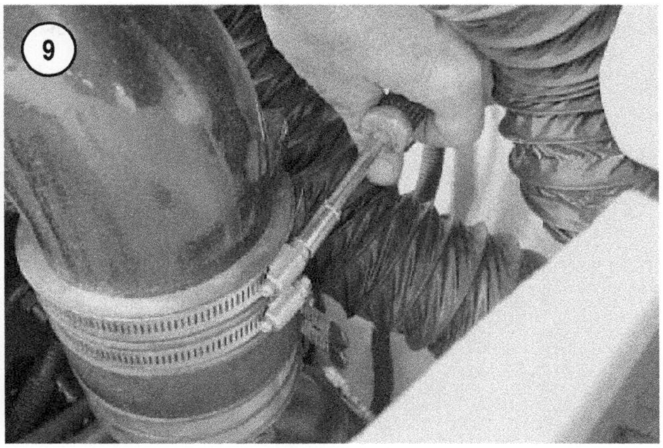
9

10

7A. On early models with a belt-driven seawater pump, loosen the hose clamp and detach the seawater inlet hose (**Figure 7**, typical) from the seawater pump. Drain all water from the hose.

7B. On later models with a crankshaft-driven seawater pump, loosen the hose clamp and detach the seawater inlet hose (**Figure 8**, typical) from the seawater pump. Drain all water from the hose.

8. Remove the engine drain plugs and drain all water and/or coolant (closed cooling system models) from the engine. Refer to *Winterizing* in Chapter Four for drain plug locations.

9A. On 3.0L models, loosen the clamps on the exhaust bellows (**Figure 9**) and then pull the bellows free from the manifold. If necessary, pry the bellows loose from the exhaust elbow with a blunt pry bar. Work carefully to avoid tearing the rubber bellows.

9B. On V-6 and V-8 models, loosen the upper clamps (two on each bank) that secure the exhaust bellows (**Figure 10**, typical) to the exhaust tubes.

10. On models equipped with power steering, place a suitable container or shop towel under the fittings to capture spilled fluid. Then use a flare nut wrench to loosen and disconnect both power steering hydraulic hoses from the control valve (**Figure 11**, typical). Use the special connector (OMC or Volvo Penta part No. 3853315) to connect the hoses together. Plug the control valve openings to prevent leakage and entry of contamination. If the special connector

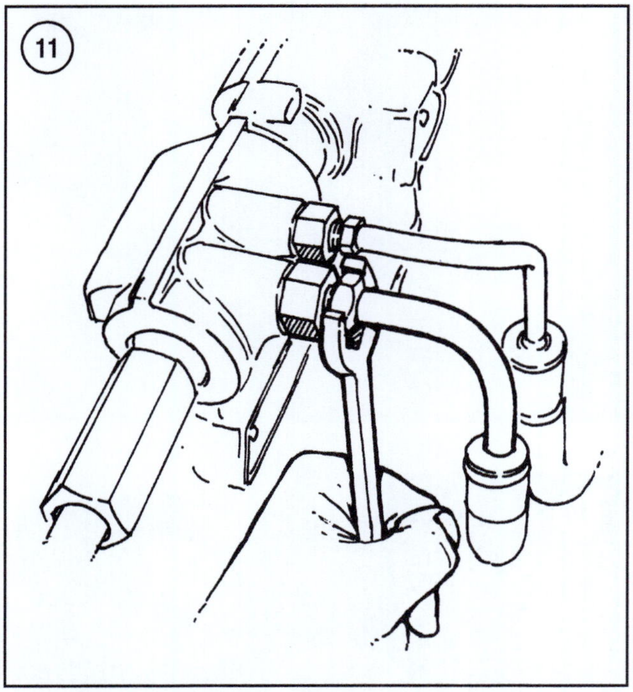
11

is not available, cover the hose fittings to prevent contamination and secure the hoses to a location well above the power steering pump to prevent fluid loss.

11. Disconnect the trim sender harness (**Figure 12**) from the engine harness connector.

12. Move the shift cable and cable hose away from the engine to prevent interference during engine removal and installation. Secure the cable and hose with suitable clamps if necessary.

13. Loosen the clamp and disconnect the seawater inlet hose from the transom shield water tube (**Figure 13**, typical).

14. Detach any wires, hoses or accessories that can interfere with engine removal. Accessory ground wires are often connected to the flywheel housing ground studs and are not usually apparent at a quick glance. Check thoroughly for overlooked wiring before lifting the engine.

NOTE

At this point, there should be no hoses, wires or linkages connecting the engine to the boat, drive unit or transom shield. Make sure nothing remains to interfere with engine removal.

15. Attach a suitable overhead hoist to the engine lifting brackets (**Figure 14**). The hoist must have a minimum lift capacity of 1500 lb. (680 kg). Raise the hoist just enough to remove slack from the chain. *Do not* lift the engine at this time.

16. Remove the lock nut (**Figure 15**) and then carefully lift the washer from both rear engine mounts. If necessary, carefully pry the washer from the mounting arm recess. If the bolt rotates while attempting to loosen the nut, fully tighten the nut. Then again attempt to loosen the nut. If the bolt again rotates, engage an appropriate thin profile wrench onto the bolt head on the lower side of the mount while loosening the nut.

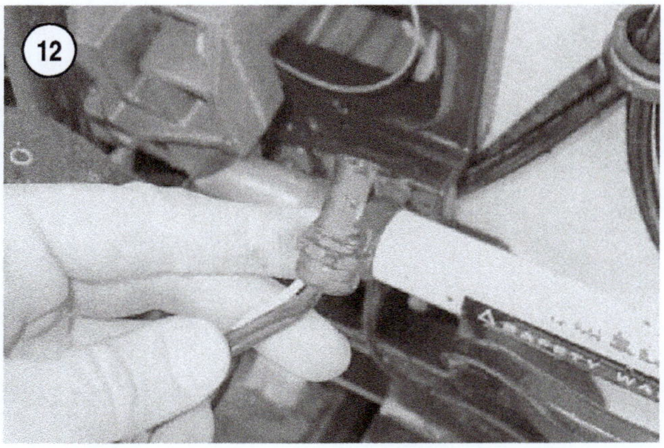
12

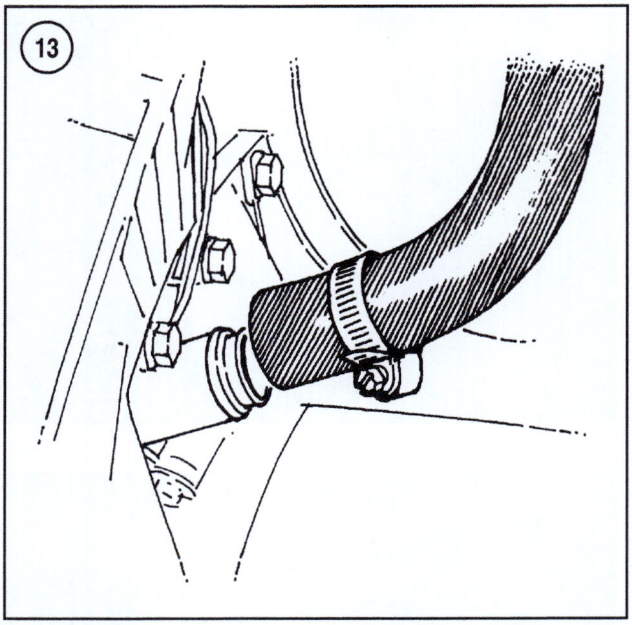
13

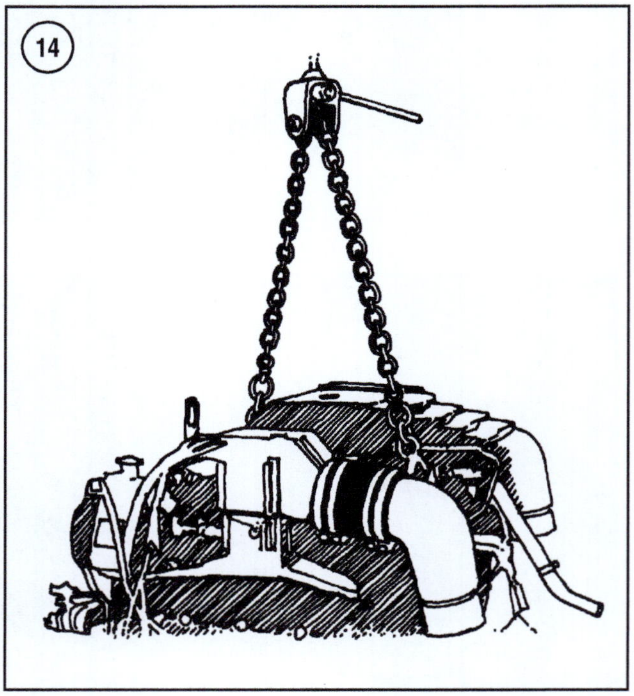
14

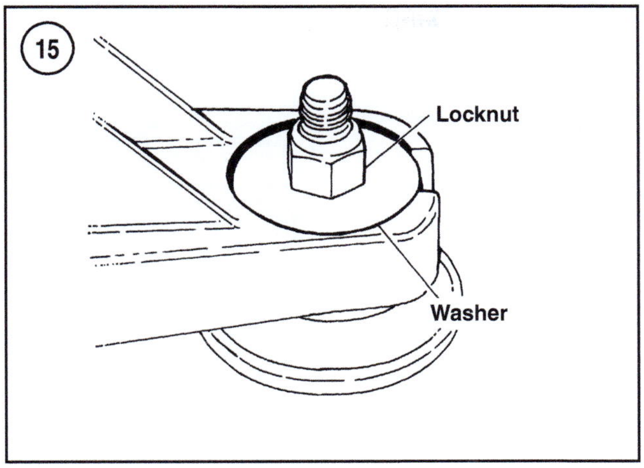

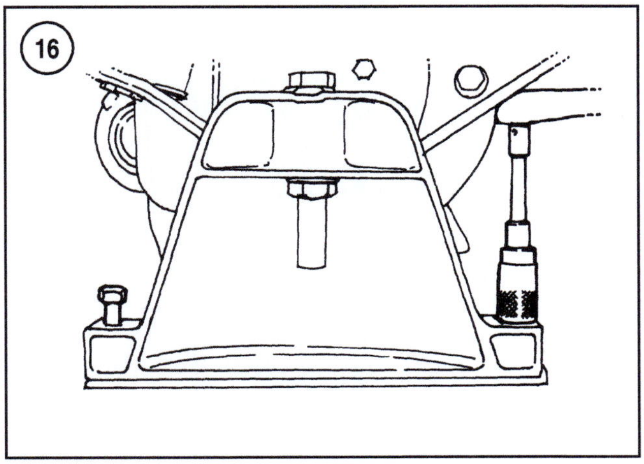

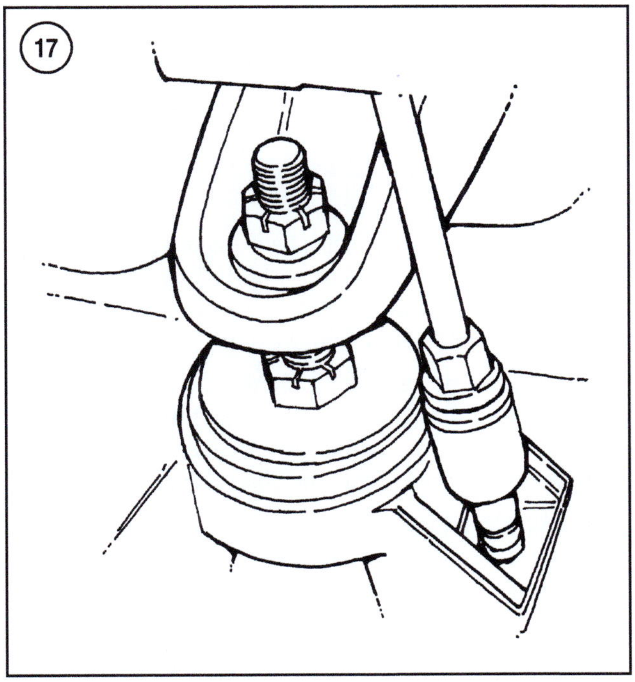

NOTE
Do not loosen or move the alignment adjusting nuts or complete realignment will be required when the engine is reinstalled. Engine alignment should not be disturbed if the engine mounts are detached as described in Step 17.

17A. On 3.0L models, remove the lag bolts (**Figure 16**) that secure the front engine mount to the mount pad.

17B. On V-6 and V-8 models, remove the lag bolts (**Figure 17**) that secure the side engine mounts (one on each side) to the mounting stringers.

18. Lift the engine just enough to relieve pressure from the engine mounts.

19. Again check for any wiring, cables or hoses that can interfere with engine removal and detach if necessary. Lift the engine from the boat.

20. Use a suitable detergent to remove oily deposits from the bilge area.

21. Wipe hardened grease and other debris from the engine coupler splines (**Figure 18**). Use a pick to remove stubborn deposits from the bottom of the splines. Then thoroughly clean the splines using an aerosol carburetor cleaner. Use compressed air to dry the splines after cleaning. Inspect the splines for excess wear. The splines should be square at the top. Replace the drive coupler if the splines are sharpened from excessive wear.

22. Use a wire brush to remove resin material, sealant or other contaminants from the mount lag bolts. Inspect the bolts for corrosion, damaged threads or other defects. Replace damaged or corroded lag bolts with suitable stainless steel units. Do not use steel or zinc-plated bolts. Only good quality stainless steel lag bolts will withstand the potential for corrosion in this area.

23. Inspect the lag bolt openings in the boat structure for stripped or oversized openings and deteriorated or questionable condition of the surrounding structure. Have questionable boat structure repaired by a qualified fiberglass repair technician before reinstalling the engine.

NOTE
Excessive engine coupler spline wear is the result of incorrect engine alignment, lack of sufficient lubrication, high operating hours, long period of operation at idle speed or a combination of these conditions. If excessive wear is evident, refer to Chapter Three for

coupler spline lubrication intervals and this chapter for the engine alignment procedure.

ENGINE INSTALLATION

1. Apply a coating of OMC Moly Lube (OMC part No. 175356) to the splines within the coupler or output shaft splines. Purchase the Moly Lube from an OMC, Johnson or Evinrude dealership.
2. Install any shims or spacers that were used with the engine mounts.
3. Carefully lower the engine into the vessel until the flywheel housing is aligned with the transom shield. Make sure the engine is positioned forward enough to prevent the housing from contacting the shield. Make sure to guide the exhaust elbow or exhaust tube into the exhaust bellows while lowering the engine. To ease installation and prevent damage to the rubber, apply a soap-and-water solution to the inner bellows surfaces.
4. Carefully move the engine rearward while guiding the rear mount studs into the slots in the mount arms. Continue until the stud is centered in the recess for the washer (**Figure 15**). When centered, install the washer over each stud and seat in the recess. Thread the locknut onto the stud to retain the washer. Do not tighten the locknut at this time.
5. Align the lag bolt opening in the engine mounts with the corresponding openings in the mount pad or stringers. Then slowly lower the engine until a slight amount of slack is present in the lifting chain.
6. Apply a light coating of marine-grade silicone sealant to the threads. Then thread the lag bolts into the mounts and boat structure. Securely tighten the lag bolts.
7. Tighten both rear-mounting locknuts (**Figure 15**) to 28-30 ft.-lb. (38-41 N•m).
8. Detach the lifting chain from the engine lifting brackets.
9. Check and correct the engine alignment as described in this chapter.
10. Fit the seawater inlet hose onto the transom shield water tube (**Figure 13**, typical). Secure the hose onto the tube with the clamp.
11. Connect the trim sender harness (**Figure 12**) onto the engine harness. Route the sender harness to prevent contact with the steering system or other moving components. Secure the harness with plastic locking clamps as needed.
12. On models equipped with power steering, thread the power steering hose fittings into the control valve openings (**Figure 11**, typical). Using a flare nut wrench, tighten the smaller fitting to 124-142 in.-lb. (14-16 N•m) and the larger fitting to 15-17 ft.-lb. (20-23 N•m). Make sure to route the hoses to prevent them from twisting, kinking or contacting moving components. Hold the hose fittings in position while tightening the fittings.
13. Securely tighten the clamps that secure the exhaust bellows to the exhaust elbow or exhaust tubes (**Figure 9**, typical or **Figure 10**, typical). Make sure the clamps are positioned over the exhaust tube or elbow before tightening.

19

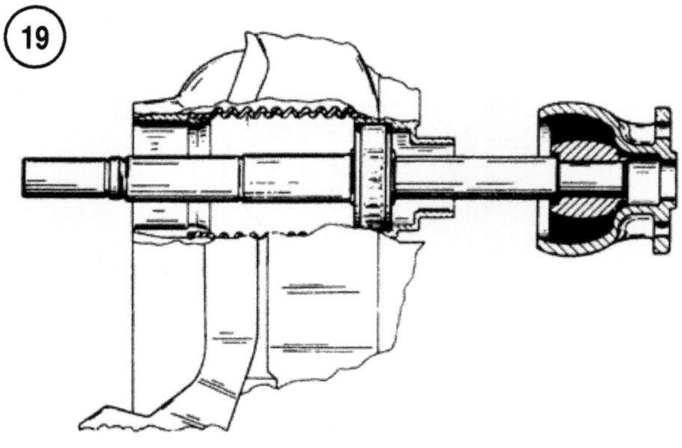

20

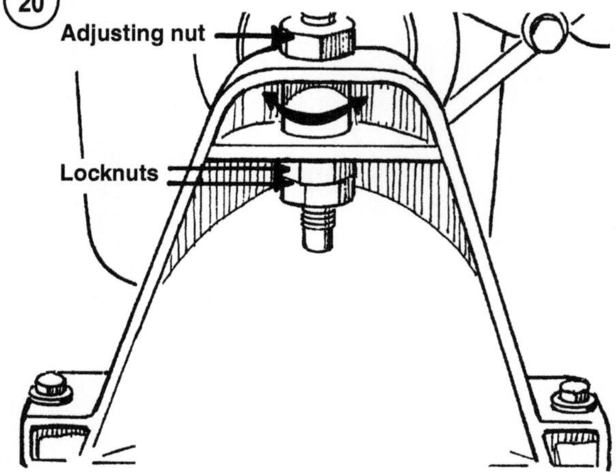

14. Install all of the engine drain plugs into the engine. Refer to *Winterizing* in Chapter Four for drain plug locations.

15A. On earlier models with a belt-driven seawater pump, connect the seawater inlet hose (**Figure 7**, typical) onto the seawater pump inlet fitting. Push the hose fully over the fitting, then secure the hose with a clamp.

15B. On later models with a crankshaft-driven seawater pump, connect the seawater inlet hose (**Figure 8**, typical) onto the seawater pump inlet fitting. Push the hose fully over the fitting. Then secure the hose with a clamp.

16. Remove any plugs or caps. Then connect the fuel supply hose to the fuel pump (**Figure 5**, typical) or fuel filter inlet (**Figure 6**, typical). Secure the hose to the fitting with a suitable clamp.

17. Connect the instrument harness (**Figure 3**, typical) and trim switch harness (**Figure 4**, typical) to the engine harness. Secure the instrument harness plug to the engine harness plug with a screw-type hose clamp. Do not overtighten the hose clamp. Overtightening damages the pins and sockets in the plugs.

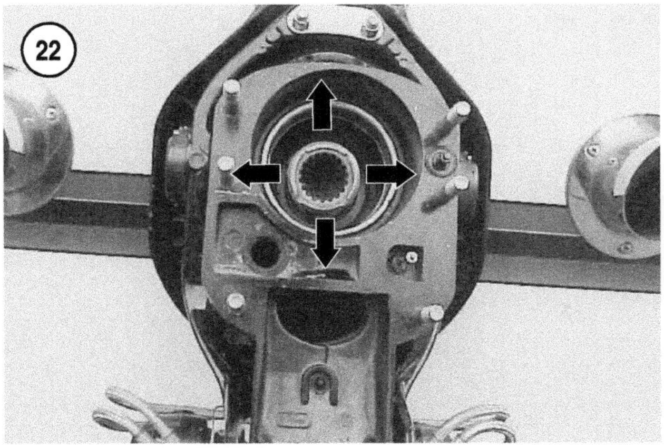

18. Connect the throttle cable to the carburetor or throttle body linkage (**Figure 2**, typical). Adjust the throttle cable as described in Chapter Five.

19. Install the drive unit (**Figure 1**) as described in Chapter Ten.

20. On models equipped with a closed cooling system, fill the engine and heat exchanger with the recommended coolant as described in Chapter Three.

21. On models equipped with power steering, fill and bleed the power steering system as described in Chapter Thirteen.

22. Install the battery into the boat and secure as described in Chapter Nine (refer to *Battery Maintenance*). Connect the positive and then negative battery cables.

23. Check and correct the engine oil level as described in Chapter Three.

24. Check for any disconnected wiring or hoses prior to attempting to start the engine. Make sure all accessory ground wires are properly connected to the flywheel housing ground stud.

25. Adjust the alternator drive belt as described in Chapter Nine. Adjust the seawater pump drive belt as described in Chapter Eight. Adjust the power steering pump drive belt as described in Chapter Thirteen.

26. Start the engine and immediately check for fuel, oil or water leakage. Stop the engine and correct any leakage before running the engine. Check for proper steering, shifting, throttle and cooling system operation. Correct any malfunction before putting the engine into service.

27. Install the engine hatch, seating and any panels or structures that were removed prior to engine removal.

ENGINE ALIGNMENT

Check the engine alignment at the intervals specified in Chapter Three (see **Table 1**) and anytime the drive, engine or transom bracket is removed from the boat. An alignment tool and thread-on handle (OMC or Volvo Penta part No. 3851083-0 and 38505009-31) are required to check the alignment. The tool is inserted into the gimbal bearing and splined section of the drive coupler (**Figure 19**). The alignment is correct when the tool passes freely into the drive coupler.

6

1. Remove the drive unit as described in Chapter Ten. Thread the handle onto the alignment tool.

2. Apply OMC Moly Lube (OMC part No. 175356) to the coupler end of the alignment tool. Purchase the Moly Lube from an OMC, Johnson or Evinrude dealership. Guide the tool through the gimbal bearing and into the drive coupler (**Figure 19**). Do not drive the tool into the drive coupler. Otherwise, removal is difficult, and the drive coupler can be damaged. The tool must slide freely into and out of the drive coupler. If the tool slides freely, refer to step 11.

3A. On 3.0L models, fully loosen the two lower locknuts (**Figure 20**). The nut on the top side of the mount bracket is the adjusting nut.

3B. On V-6 and V-8 models, fully loosen the upper locknut (**Figure 21**) on both the port and starboard side engine mounts. The lower nut is the adjusting nut.

4. Turn the adjusting nut one-half turn in the direction required to raise the engine. On V-6 and V-8 models with side engine mounts, the adjusting nuts must be raised and lowered the same amount on each side. The number of threads exposed on the mount stud must be the same on each side.

5. Attempt to insert the alignment tool. Push firmly on the end of the threaded handle while moving the end in all four directions (**Figure 22**). This is necessary to align the gimbal bearing with the drive coupler.

6. Repeat Step 4 and Step 5 until the alignment tool slips into the drive coupler. If the adjusting nuts reach the upper limit, and the tool cannot be inserted, turn the adjusting nuts in the direction required to lower the engine. Lower the engine one-half turn at a time and repeat Step 5 until the tool slips into the drive coupler.

7. Pull the alignment tool from the drive coupler. Coat the drive coupler end of the tool with OMC Moly Lube. Carefully insert the tool into the drive coupler. Do not rotate the tool.

8. Mark the top side of the tool. Without rotating, carefully pull the tool from the drive coupler. Inspect the spline wipe marks in the greased surface (**Figure 23**). The wipe marks

must be consistent around the entire surface. Otherwise, perform additional adjustments as follows:

a. If the marks are heavier on the top, the engine is too high in the front. Lower the engine one-quarter turn of the adjusting nut.
b. If the marks are heavier on the bottom, the engine is too low in the front. Raise the engine one-quarter turn of the adjusting nut.
c. Continue until the spline marks are consistent at all points and the tool slides freely into and out of the drive coupler. Align the gimbal bearing (refer to Step 5) after turning the adjusting nuts.

9A. On 3.0L models, securely tighten the upper locknut against the mount bracket. Tighten the lower locknut against the upper locknut to 115-140 ft.-lb. (156-190 N•m). Tighten the adjusting nut against the front engine mount to 115-140 ft.-lb. (156-190 N•m).

9B. On V-6 and V-8 models, hold the locknut (upper) with a wrench while tightening the adjusting nut against the port side mount bracket to 100-120 ft.-lb. (136-163 N•m). Repeat the tightening procedure for the starboard side.

10. Check for proper engine alignment as described in Step 2. Loosen the locking nuts and repeat the alignment procedure as needed.

11. Use a long stick, socket extension or brush and apply OMC Moly Lube to the drive coupler splines. Apply enough grease just to coat the splines. Do not fill the opening with grease. Otherwise, the drive shaft will hydraulically lock in the coupler and prevent installation of the drive unit.

12. Install the drive unit as described in Chapter Ten.

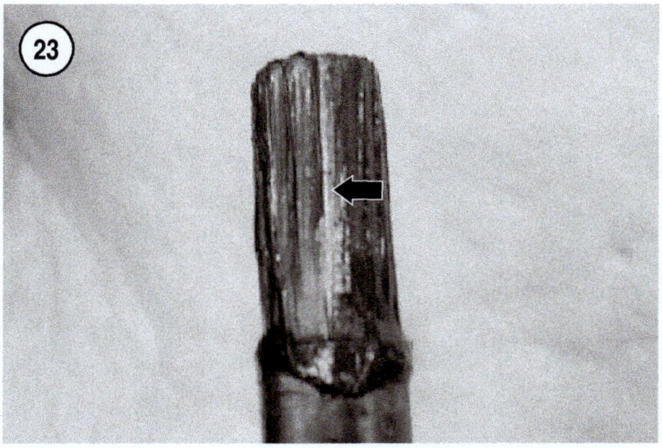

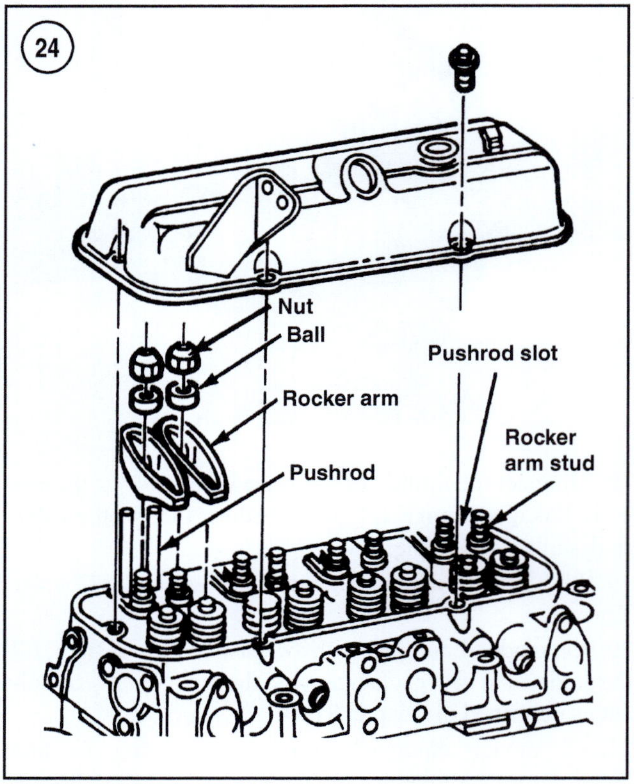

ROCKER ARM COVER

Removal/Installation

NOTE
With the manifolds, fuel system, cooling system and electrical components installed, the General Motors (GM) 5.0L and 5.7L V-8 engines are very similar in appearance to the Ford 5.0L and 5.8L V-8 engines. If the model identification decal on the flame arrestor is missing, the engine can be quickly identified as a GM or Ford engine by noting the distributor location. On GM engines, the distributor is located at the rear (flywheel end) of the engine. On Ford engines, the distributor is located on the front of the engine.

1. Disconnect the battery cables.
2. On 3.0L models, remove the flame arrestor cover and the fuel pump-to-carburetor fuel line. Disconnect the throttle cable from the carburetor linkage.
3. On Ford 5.0L and 5.8L models, remove the exhaust manifolds as described in Chapter Eight.
4. Disconnect the crankcase ventilation hose from the valve cover.

NOTE
Mark the rocker arm covers to identify the port and starboard sides. The oil fill cap orientation will be incorrect if the covers are installed in the wrong locations.

5. Disconnect or remove any components that can interfere with removal of the rocker arm cover.
6. Remove the fasteners and lift the rocker arm cover (**Figure 24**, typical) from the cylinder head.
7. Clean the rocker arm cover(s) with solvent and dry with compressed air. Inspect the cover gasket or O-ring seals for damage. Replace the gasket or O-ring seals as needed. Inspect the cover for distorted, cracked, rusted or pitted surfaces. Replace the cover if these or other defects are noted.
8. Carefully fit the rocker arm cover(s) onto the cylinder head(s).

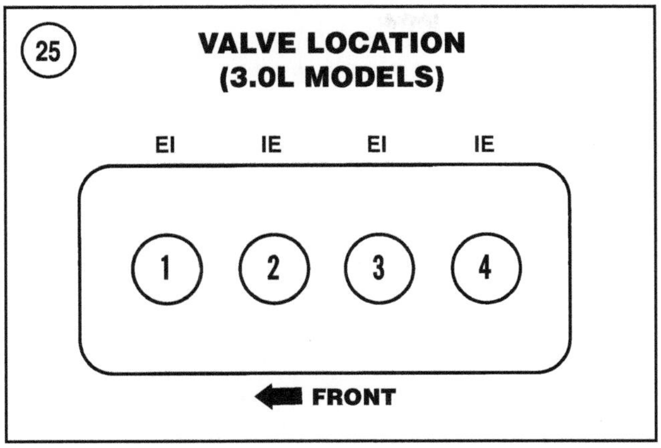

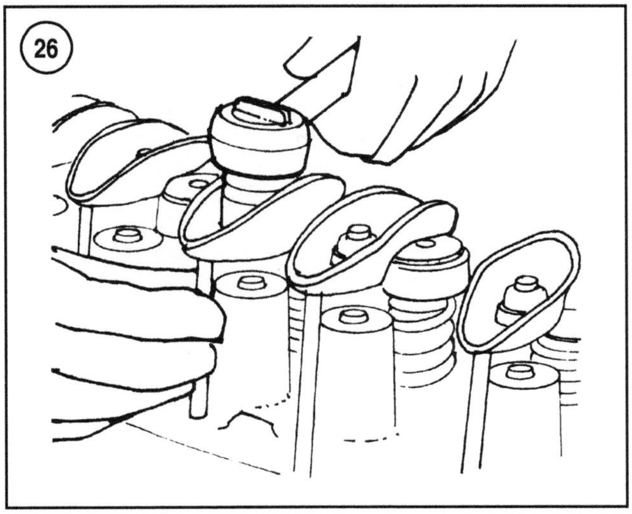

9. Thread the cover fasteners into the cover and cylinder head openings. Tighten the cover fasteners as follows:
   a. On 3.0L models, tighten the cover screws in a crossing pattern, starting from the center and working outward, to 106 in.-lb. (12 N•m).
   b. On GM 4.3L, 5.0L and 5.7L models, tighten the rocker arm cover screws, starting in the center and working outward, to 106 in.-lb. (12 N•m).
   c. On Ford 5.0L and 5.8L V-8 models, tighten the rocker arm cover screws, starting from the middle and working outward, to 35-53 in.-lb. (4-6 N•m).
   d. On GM 7.4L and 8.2L models, tighten the rocker arm cover screws, starting in the center and working outward, to 115 in.-lb. (13 N•m).
10. Install or reposition and components removed to access the rocker arm cover(s).
11. On 3.0L models, install the flame arrestor and the fuel pump-to-carburetor fuel line as described in Chapter Seven. Connect the throttle cable and adjust as described in Chapter Five.
12. On Ford 5.0L and 5.8L models, install the exhaust manifolds as described in Chapter Eight.
13. Connect the crankcase ventilation hose to the rocker arm cover fitting.
14. Connect the battery cables.

VALVE ADJUSTMENT

Adjust the valves if lifter noise is present and if replacing the cylinder head, camshaft, lifters, pushrods or other valve train components.

Valve adjustment is required only on GM 3.0L, earlier 4.3L V-6 (with rocker arm nuts), 5.0L and 5.7L models. Later 4.3L models and all other models covered in this manual use torque-to-specification rocker arm fasteners, and adjustment is not required. However, on such models, the rocker arm fasteners must be tightened in the proper sequence (refer to *Cylinder Head* in this chapter).

3.0L Models

6

1. Disconnect the negative battery cable and remove the spark plugs.
2. Position the No. 1 piston at TDC as described in this chapter.
3. Remove the rocker arm cover (**Figure 24**) as described in this chapter.
4. With the No. 1 piston at TDC, adjust the following valves (**Figure 25**).
   a. No. 1 cylinder intake.
   b. No. 2 cylinder intake.
   c. No. 1 cylinder exhaust.
   d. No. 3 cylinder exhaust
   e. No. 4 cylinder intake.

NOTE
Turn the pushrod to determine whether there is any lash between the rocker arm and the pushrod. The pushrod will turn freely if lash is present.

5. To adjust valve lash, loosen the adjusting nut until some lash is felt at the pushrod (**Figure 26**). While turning the pushrod, slowly tighten the nut just until all lash is removed. Then tighten the nut exactly one turn.
6. Rotate the crankshaft clockwise exactly one turn and align the notch in the harmonic balancer with the 0 mark on the timing tab (**Figure 27**). The No. 4 piston is now at TDC.

CAUTION
Do not turn the crankshaft counterclockwise, or the seawater pump will be damaged.

7. With the No. 4 piston at TDC, repeat Step 5 and adjust the following valves.
   a. No. 2 cylinder exhaust.
   b. No. 3 cylinder intake.
   c. No. 4 cylinder exhaust.
8. Install the rocker arm cover as described in this chapter.
9. Install the spark plug and tighten to 20 ft.-lb. (27 N•m). Connect the plug leads. Install or reposition any components removed to access the rocker arm covers.

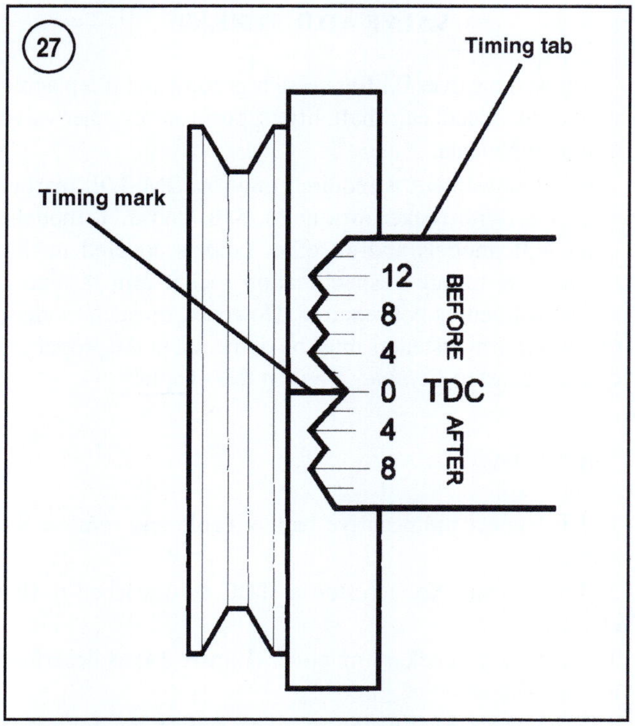

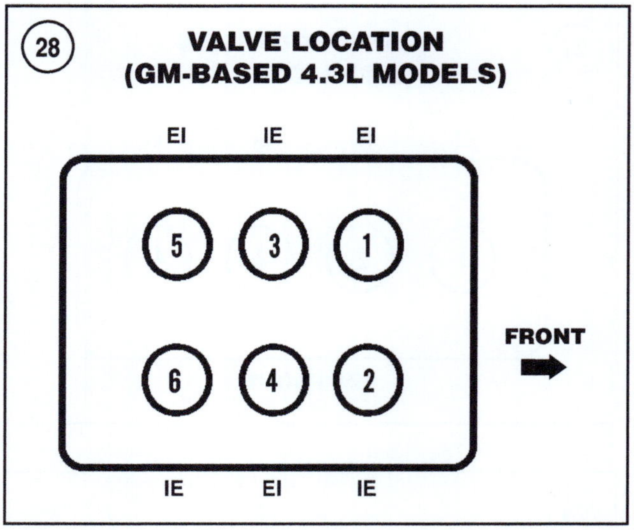

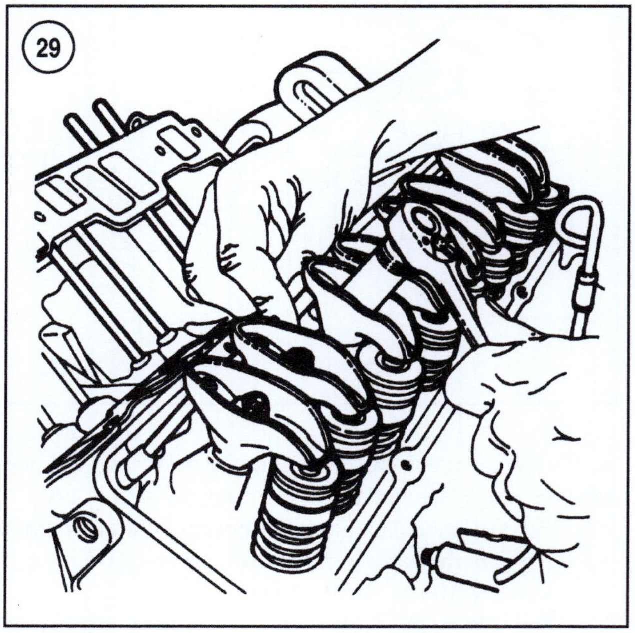

4.3L Models (with Rocker Arm Adjusting Nuts)

NOTE

Earlier 4.3L models (1994-1996) use a nut and pivot ball to secure the rocker arm onto the rocker arm stud. Later 4.3L models (1997-2000) use a shoulder bolt and pivot ball to secure the rocker arm on the cylinder head. The valves can be adjusted only on the earlier models with the rocker arm nut. The later models with the shoulder bolt cannot be adjusted. If in question, remove the rocker arm cover as described in this chapter and identify the rocker arm attaching hardware.

1. Disconnect the negative battery cable and remove the spark plugs.
2. Place the No. 1 piston at TDC as described in this chapter.
3. Remove the rocker arm cover (**Figure 24**) as described in this chapter.
4. With the No. 1 piston at TDC, adjust the following valves (**Figure 28**):
   a. No. 1 cylinder intake.
   b. No. 1 cylinder exhaust.
   c. No. 2 cylinder intake.
   d. No. 5 cylinder exhaust.
   e. No. 3 cylinder intake.
   f. No. 6 cylinder exhaust.

NOTE

Turn the pushrod to determine whether there is any lash between the rocker arm and pushrod. The pushrod will turn freely if lash is present.

5. To adjust the valve lash, loosen the adjusting nut until some lash is felt at the pushrod (**Figure 29**). While turning the pushrod, slowly tighten the adjusting nut just until all lash is removed. Then tighten the nut exactly one turn.
6. Rotate the crankshaft clockwise exactly one turn and align the machined line on the harmonic balancer with the 0 mark on the timing plate (**Figure 30**).

CAUTION

Do not rotate the crankshaft counterclockwise or the seawater pump will be damaged.

7. Repeat Step 5 and adjust the following valves:
   a. No. 4 cylinder intake.
   b. No. 5 cylinder intake.
   c. No. 2 cylinder exhaust.

30

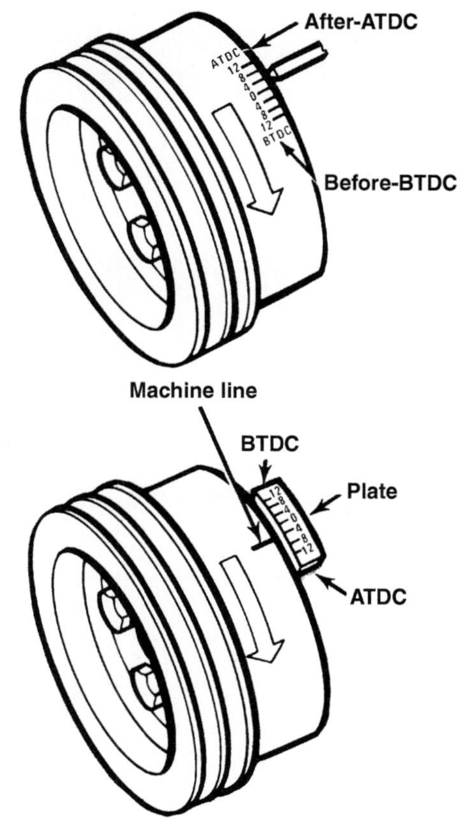

31

VALVE LOCATION (GM-BASED 5.0L AND 5.7L MODELS)

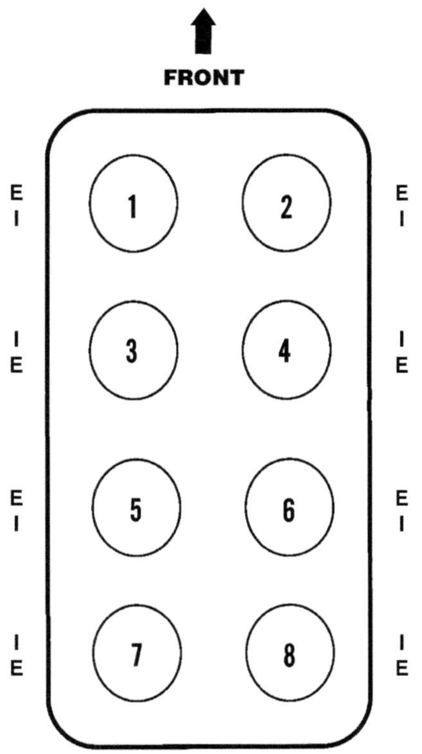

d. No. 6 cylinder intake.
e. No. 3 cylinder exhaust.
f. No. 4 cylinder exhaust.

8. Install the rocker arm covers as described in this chapter.
9. Install the spark plugs and tighten to 22 ft.-lb. (30 N•m) if the cylinder head is new and 15 ft.-lb. (20 N•m) if the cylinder head is used. Connect the plug leads and install or reposition any components removed to access the rocker arm cover.
10. Connect the battery cable.

GM 5.0L and 5.7L Models

1. Disconnect the negative battery cable and remove the spark plugs.
2. Place the No. 1 piston at TDC as described in this chapter.
3. Remove the rocker arm cover (**Figure 25**) as described in this chapter.
4. With the No. 1 piston at TDC, adjust the following valves (**Figure 31**):
   a. No. 1 cylinder intake.
   b. No. 1 cylinder exhaust.
   c. No. 2 cylinder intake.
   d. No. 3 cylinder exhaust.
   e. No. 4 cylinder exhaust.
   f. No. 5 cylinder intake.
   g. No. 7 cylinder intake.
   h. No. 8 cylinder exhaust.

NOTE
Turn the pushrod to determine whether there is any lash between the rocker arm and pushrod. The pushrod will turn freely if lash is present.

5. To adjust the valve lash, loosen the adjusting nut until some lash is felt at the pushrod (**Figure 29**). While turning the pushrod, slowly tighten the adjusting nut just until all lash is removed. Then tighten the nut exactly one turn.

6A. On earlier models with a 12-bolt intake manifold, rotate the crankshaft clockwise exactly one turn and align the machined line on the harmonic balancer with the 0 mark on the timing plate (**Figure 30**).

6B. On later models with an eight-bolt intake manifold, rotate the crankshaft clockwise exactly one turn and align the notch on the timing chain cover with the 0 mark on the harmonic balancer decal (**Figure 32**).

CAUTION
Do not rotate the crankshaft counterclockwise, or the seawater pump will be damaged.

7. Repeat Step 5 and adjust the following valves:
   a. No. 2 cylinder exhaust.
   b. No. 3 cylinder intake.
   c. No. 4 cylinder intake.
   d. No. 5 cylinder exhaust.
   e. No. 6 cylinder intake.

6

f. No. 6 cylinder exhaust.
g. No. 7 cylinder exhaust.
h. No. 8 cylinder intake.

8. Install the rocker arm covers as described in this chapter.
9. Install the spark plugs and tighten to 22 ft.-lb. (30 N•m) if the cylinder head is new and 15 ft.-lb. (20 N•m) if the cylinder head is used. Connect the plug leads and install or reposition any components removed to access the rocker arm cover.
10. Connect the battery cable.

PLACING THE NO. 1 PISTON AT TDC (FIRING POSITION)

NOTE
With the manifolds, fuel system, cooling system and electrical components installed, the General Motors (GM) 5.0L and 5.7L V-8 engines are very similar in appearance to the Ford 5.0L and 5.8L V-8 engines. If the model identification decal on the flame arrestor is missing, the engine can be quickly identified as a GM or Ford engine by noting the distributor location. On GM engines, the distributor is located at the rear (flywheel end) of the engine. On Ford engines, the distributor is located on the front of the engine.

Intake Manifold, Rocker Arm Cover or Lifter Cover Installed

1. Disconnect the negative battery cable and remove the spark plugs.
2. Place a thumb over the No. 1 cylinder spark plug opening to detect internal air pressure.
   a. On 3.0L models, the No. 1 cylinder is closest to the front of the engine (**Figure 25**).
   b. On GM V-6 and V-8 models, the No. 1 cylinder is on the port side closest to the front of the engine.
   c. On Ford 5.0L and 5.8L models, the No. 1 cylinder is on the starboard side closest to the front of the engine.
3. Observe the timing marks (**Figure 30** or **Figure 32**) and turn the harmonic balancer or pulley until air pressure forms in the No. 1 cylinder. This indicates the cylinder is on the compression stroke. Continue rotating the balancer or pulley until the timing mark aligns with the 0 mark (TDC).
4. Install spark plugs and tighten to specification in **Table 1**.
5. Connect the spark plug leads.
6. Connect the battery cable.

Intake Manifold, Rocker Arm Cover or Lifter Cover Removed

1. Remove the spark plugs.

2A. On 3.0L models, remove the lifter cover or rocker arm cover as described in this chapter.

32

33 VALVE LOCATION (FORD-BASED 5.0L AND 5.8L MODELS)

2B. On all other V-6 and V-8 models, remove the rocker arm cover as described in this chapter or intake manifold as described in Chapter Seven.

3. To identify the valve location for the No. 1 cylinder, refer to **Figure 25** (3.0L models), **Figure 28** (4.3L models), **Figure 31** (GM 5.0L and 5.7L models), **Figure 33** (Ford 5.0L and 5.8L models) or **Figure 34** (GM 7.4L and 8.2L models).

4. Attach a large breaker bar and socket to the center harmonic balancer attaching bolt. If a bolt is not used in this location, rotate the engine using a large pry bar between pulley mounting bolts.

5. Observe the No. 1 intake rocker arm or lifter while rotating the harmonic balancer clockwise. Stop when the No. 1 intake valve lifter or rocker arm starts moving downward. Observe the timing marks (**Figure 32**) and turn the harmonic balancer clockwise until the timing mark aligns with the 0 mark.

6A. On 3.0L models, install the lifter cover or rocker arm cover as described in this chapter.

6B. On V-6 and V-8 models, install the rocker arm cover as described in this chapter or intake manifold as described in Chapter Seven.

7. Install spark plugs and tighten to specification in **Table 1**.

34 **VALVE LOCATION (GM-BASED 7.4L AND 8.2L MODELS)**

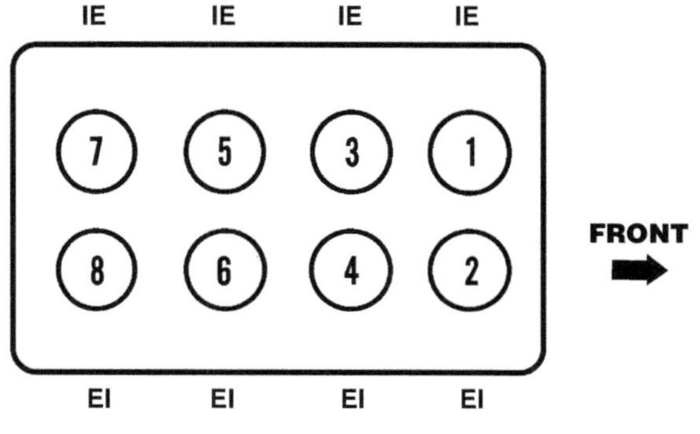

35

36

LIFTER COVER

3.0L Models

Removal/installation

1. Disconnect the negative battery cable.

2. Remove the retaining bolts and then carefully pry the lifter cover (**Figure 35**) from the starboard side of the cylinder block. Work carefully to avoid bending the cover while pulling the cover from the area behind the distributor.

3. Remove and discard the cover gasket. Clean the cover with solvent and dry with compressed air. Clean any remaining gasket or sealant material from the cylinder block-to-cover mating surface.

4. Apply a light coat of 3-M Weatherstrip Adhesive to the new gasket and then position it onto the lifter cover. Allow the adhesive to dry. Then carefully position the cover onto the cylinder block. Work carefully to avoid dislodging or damaging the gasket while fitting the cover behind the distributor.

5. Apply a light coat of RTV to the bottom of the retaining bolts. Then thread them into the cover and cylinder block openings.

6. Tighten the cover retaining bolts, starting in the center and working outward, to 65 in.-lb. (7 N•m).

7. Connect the battery cable.

MEASURING CAMSHAFT LOBE LIFT

This measurement checks for excessive camshaft wear. This procedure requires a dial indicator and a clamp for attaching the indicator to the cylinder head or block. A magnetic mount or mount that can be attached to the rocker arm stud will perform well for this measurement. Mark the original location of all valve train components before removing them. *All* parts must be reinstalled in their original locations.

NOTE

With the manifolds, fuel system, cooling system and electrical components installed, the General Motors (GM) 5.0L and 5.7L V-8 engines are very similar in appearance to the Ford 5.0L and 5.8L V-8 engines. If the model identification decal on the flame arrestor is missing, the engine can be quickly identified as a GM or Ford engine by noting the distributor location. On GM engines, the distributor is located at the rear (flywheel end) of the engine. On Ford engines, the distributor is located on the front of the engine.

3.0L Models

1. Disconnect the negative battery cable and remove the spark plugs.

2. Remove the rocker arm cover (**Figure 24**, typical) as described in this chapter.

3. Remove the lifter cover (**Figure 35**) as described in this chapter.

4. Remove the nut and pivot ball (**Figure 36**) from each rocker arm. Mark the original location and remove the rocker arms and pushrods from the cylinder head studs. Inspect all parts for deterioration, wear or damage and replace if necessary. Inspect the rocker arm studs for excessive wear on the sides. Replace the cylinder head if any of the

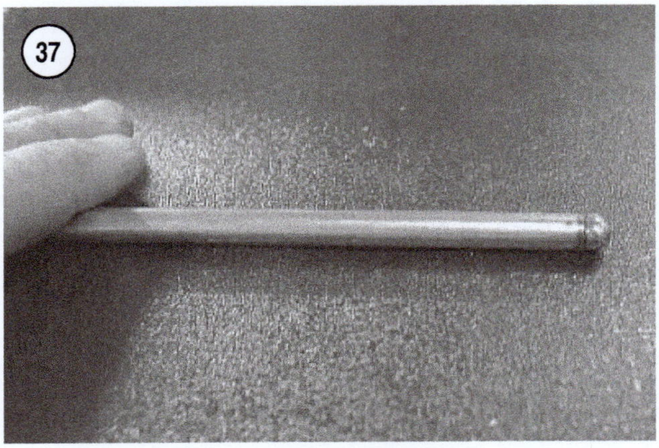

studs are heavily worn on one side. This condition usually affects only a single stud.

NOTE

Anytime the rocker arm cover is removed, inspect the rocker arm studs for grooving or excessive wear on the side of the studs that aligns with the rocker arms. Grooving or excessive wear in this area usually indicates the stud is improperly aligned in the stud boss. This defect is common on 3.0L models. The misalignment will cause the rocker arm to contact and wear heavily on the side of the stud. Eventual stud failure and damage to other engine components will result.

5. Roll the pushrods across a flat surface to check for damage (**Figure 37**). Replace bent pushrods.
6. Attach a large breaker bar and socket to the harmonic balancer bolt. If a bolt is not used, rotate the engine with a large pry bar between the pulley mounting bolts.
7. Observe the No. 1 exhaust valve lifter (see **Figure 25**) while rotating the engine clockwise. Stop when the lifter drops in the bore and the base of the cam lobe aligns with the bottom of the lifter (**Figure 38**).
8. Insert the pushrod into the corresponding slot in the cylinder head and seat it into the recess in the lifter.
9. Using a suitable mount (**Figure 39**), attach the dial indicator to the cylinder head. Align the plunger of the dial indicator with the tip of the pushrod. Rotate the face of the dial indicator to align the needle with the 0 mark.
10. Observe the dial indicator while rotating the harmonic balancer exactly one revolution clockwise. The amount of needle movement equals camshaft lobe lift. Record the lobe lift.
11. Repeat Steps 7-10 for the remaining camshaft lobes. Record each measurement.
12. The camshaft lobe lift must be 0.248-0.258 in. (6.299-6.553 mm). Replace the camshaft if any of the measurements are less than the minimum specification.
13. Install the pushrods, roller rocker arms and nuts as described in this chapter.
14. Adjust the valves as described in this chapter.

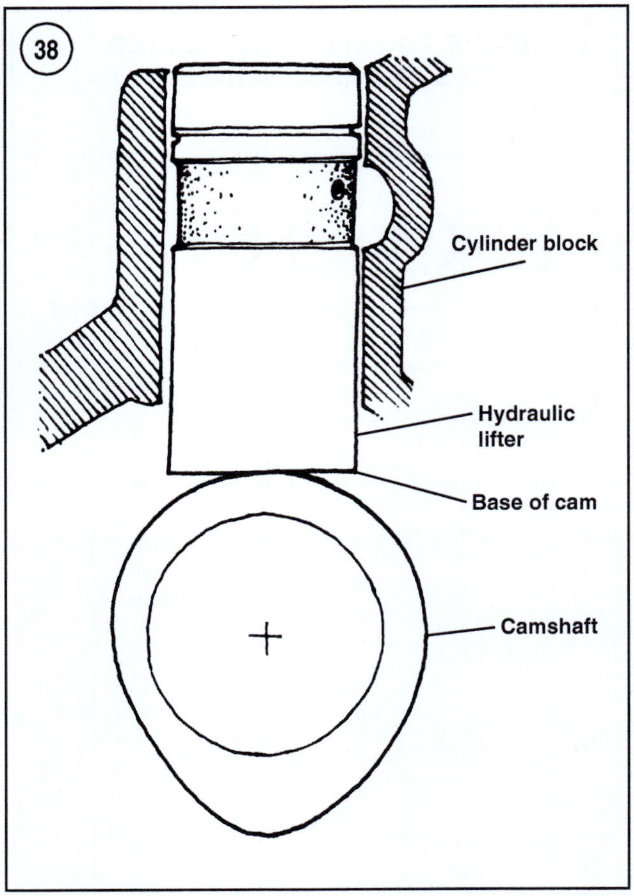

MEASURING CAMSHAFT LOBE LIFT (3.0L MODELS)

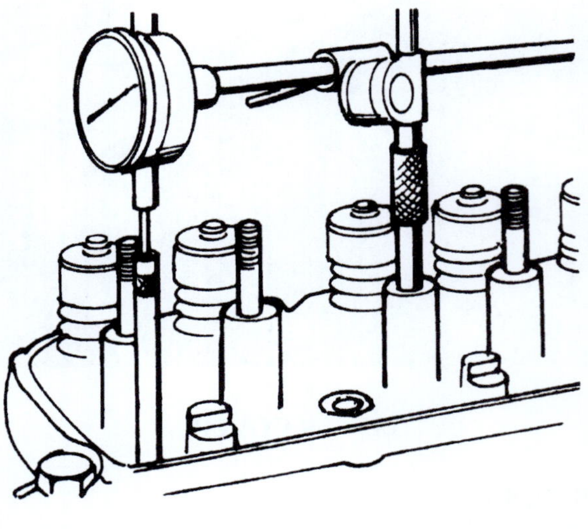

15. Install the rocker arm cover as described in this chapter.
16. Install the lifter cover as described in this chapter.
17. Install the spark plugs and tighten to 20 ft.-lb. (27 N•m). Connect the plug leads and install or reposition any components removed to access the rocker arm cover.
18. Connect the negative battery cable.

40 **VALVE SYSTEM COMPONENTS (4.3L [1997-2000], 7.4L AND 8.2L MODELS)**

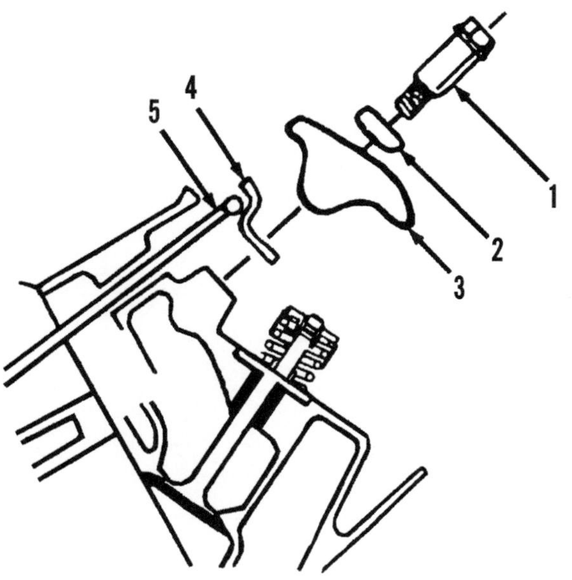

1. Shoulder bolt
2. Pivot ball
3. Rocker arm
4. Pushrod guide (7.4L and 8.2L only)
5. Pushrod

41

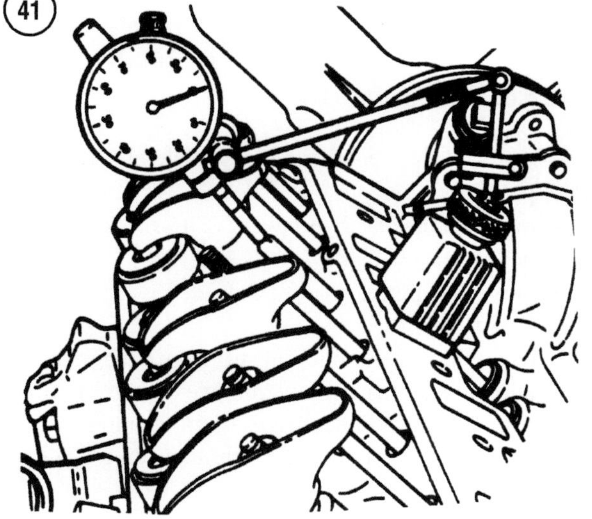

4.3L Models

NOTE
Earlier 4.3L models (1994-1996) use a nut and pivot ball to retain the rocker arm on the rocker arm stud. Later 4.3L models (1997-2000) use a shoulder bolt and pivot ball to retain the rocker arm on the cylinder head.

1. Disconnect the negative battery cable and remove the spark plugs.
2. Remove the rocker arm covers (**Figure 24**, typical) as described in this chapter.
3. Remove the intake manifold as described in Chapter Seven.
4A. On 1994-1996 models with a 12-bolt intake manifold, remove the nut and pivot ball (**Figure 36**) from each rocker arm.
4B. On 1997-2000 models with an eight-bolt intake manifold, remove the shoulder bolt (1, **Figure 40**) and pivot ball (2) from each rocker arm.
5. Mark the original location and remove the rocker arms and pushrods from the cylinder head studs. Inspect all parts for deterioration, wear or damage and replace if necessary.
6. Roll the pushrods across a flat surface (**Figure 37**) to check for damage. Replace bent pushrods.
7. Attach a breaker bar and socket to the harmonic balancer attaching bolt.
8. Observe the No. 1 exhaust valve lifter (see **Figure 28**) while turning the harmonic balancer clockwise. Stop when the lifter drops in the bore and the base of the cam lobe (**Figure 38**) aligns with the center of the roller on the bottom of the lifter.
9. Insert the pushrod into the corresponding slot in the cylinder head and seat it into the recess in the lifter.
10. Using a suitable mount (**Figure 41**), attach the dial indicator to the cylinder head. Align the plunger of the dial indicator with the tip of the pushrod. Rotate the face of the dial indicator to align the needle with the 0 mark.
11. Observe the dial indicator while rotating the harmonic balance exactly one revolution clockwise. The amount of needle movement equals camshaft lobe lift. Record the lobe lift.
12. Repeat Steps 7-11 for the remaining camshaft lobes. Record each measurement.
13A. On 1994-1996 models with a 12-bolt intake manifold, the camshaft lobe lift must be a minimum of 0.234 in. (5.944 mm) for the intake valve lobes and 0.258 in. (6.553 mm) for the exhaust valve lobes. Replace the camshaft if any of the measurements are less than the minimum specification.
13B. On 1997-2000 models with an eight-bolt intake manifold, the camshaft lobe lift must be a minimum of 0.276 in. (7.010 mm) for the intake valve lobes and 0.285 in. (7.239 mm) for the exhaust valve lobes. Replace the camshaft if any of the measurements are less than the minimum specification.
14A. On 1994-1996 models with a 12-bolt intake manifold, install the pushrods, rocker arms, pivot balls and nuts. Then adjust the valves as described in this chapter.
14B. On 1997-2000 models with an eight-bolt intake manifold, install the pushrods, rocker arms, pivot balls and shoulder bolts as described in this chapter.
15. Install the intake manifold as described in Chapter Seven.

6

16. Install the rocker arm covers as described in this chapter.
17. Install the spark plugs and tighten to 22 ft.-lb. (30 N•m) if the cylinder head is new and 15 ft.-lb. (20 N•m) if the cylinder head is used. Connect the plug leads and install or reposition any components removed to access the rocker arm covers.
18. Connect the negative battery cable.

GM 5.0L and 5.7L Models

1. Disconnect the negative battery cable and remove the spark plugs.
2. Remove the rocker arm covers (**Figure 24**, typical) as described in this chapter.
3. Remove the intake manifold as described in Chapter Seven.
4. Remove the adjusting nut and pivot ball (**Figure 36**) from each rocker arm. Mark the original locations and remove the rocker arms and pushrods from the cylinder heads.
5. Roll the pushrods across a flat surface (**Figure 37**) to check for damage. Replace bent pushrods.
6. Attach a breaker bar and socket to the harmonic balancer attaching bolt.
7. Observe the No. 1 exhaust valve lifter (**Figure 31**) while turning the harmonic balancer clockwise. Stop when the lifter drops in the bore and the base of the cam lobe aligns with the bottom of the lifter (**Figure 38**).
8. Insert the pushrod into the corresponding slot in the cylinder head and seat it into the recess in the lifter.
9. Using a suitable mount (**Figure 41**), attach the dial indicator to the cylinder head. Align the plunger of the dial indicator with the tip of the pushrod. Rotate the face of the dial indicator to align the needle with the 0 mark.
10. Observe the dial indicator while rotating the harmonic balancer exactly one revolution clockwise. The amount of needle movement equals camshaft lobe lift. Record the lobe lift.
11. Repeat Steps 7-10 for the remaining camshaft lobes. Record each measurement.
12A. On 1994-1997 models with a 12-bolt intake manifold, the camshaft lobe lift must be a minimum of 0.269 in. (6.832 mm) for the intake valve lobes and 0.276 in. (7.010 mm) for the exhaust valve lobes. Replace the camshaft if any of the measurements are less than the minimum specification.
12B. On 1998-2000 models with an eight-bolt intake manifold, the camshaft lobe lift must be 0.274-0.278 in. (6.960-7.061 mm) for the intake valve lobes and 0.278-0.283 in. (7.061-7.188 mm) for the exhaust valve lobes. Replace the camshaft if any of the measurements are less than the minimum specification.
13. Install the pushrods, rocker arms, pivot balls and nuts as described in this chapter.
14. Adjust the valves as described in this chapter.

42

15. Install the intake manifold as described in Chapter Seven.
16. Install the rocker arm covers as described in this chapter.
17. Install the spark plugs and tighten to 22 ft.-lb. (30 N•m) if the cylinder head is new and 15 ft.-lb. (20 N•m) if the cylinder head is used. Connect the plug leads and install or reposition any components removed to access the rocker arm covers.
18. Connect the negative battery cable.

Ford 5.0L and 5.8L Models

1. Disconnect the negative battery cable and remove the spark plugs.
2. Remove the rocker arm covers (**Figure 24**, typical) as described in this chapter.
3. Remove the intake manifold as described in Chapter Seven.
4. Remove the nut and fulcrum seat from each rocker arm. Mark the original locations and remove the rocker arms and pushrods from the cylinder heads. Mark the mounting location on all components to make sure they are installed into the original locations during assembly.
5. Roll the pushrods across a flat surface (**Figure 37**) to check for damage. Replace bent pushrods.
6. Attach a breaker bar and socket to the harmonic balancer attaching bolt.
7. Observe the No. 1 exhaust valve lifter (see **Figure 33**) while turning the harmonic balancer clockwise. Stop when the lifter drops in the bore and the base of the cam lobe aligns with the bottom of the lifter (**Figure 38**).
8. Insert the pushrod into the corresponding slot in the cylinder head and seat it into the recess in the lifter.

9. Using a suitable mount (**Figure 42**), attach the dial indicator to the cylinder head. Align the plunger of the dial indicator with the tip of the pushrod. Fit a section of suitable rubber tubing over the dial indicator plunger to hold it in place against the center of the pushrod. Rotate the face of the dial indicator to align the needle with the 0 mark.
10. Observe the dial indicator while rotating the harmonic balancer exactly one revolution clockwise. The amount of needle movement equals camshaft lobe lift. Record the lobe lift.
11. Repeat Steps 7-10 for the remaining camshaft lobes. Record each measurement.
12A. On 5.0L models, the camshaft lobe lift measurement must be 0.255-0.260 in. (6.477-6.604 mm) for the intake lobes and 0.273-0.278 in. (6.934-7.061 mm) for the exhaust lobes. Replace the camshaft if any of the measurements are less than the minimum specification.
12B. On 5.8L models, the camshaft lobe lift measurement must be 0.273-0.278 in. (6.934-7.061 mm) for the intake and exhaust lobes. Replace the camshaft if any of the measurements are less than the minimum specification.
13. Install the pushrods, rocker arms, fulcrum seat and bolts as described in this chapter.
14. Check the valve clearance as described in this chapter.
15. Install the intake manifold as described in Chapter Seven.
16. Install the rocker arm covers as described in this chapter.
17. Install the spark plugs and tighten to 120-180 in.-lb. (14-20 N•m). Connect the plug leads and install or reposition any components removed to access the rocker arm covers.
18. Connect the negative battery cable.

GM 7.4L and 8.2L Models

1. Disconnect the negative battery cable and remove the spark plugs.
2. Remove the rocker arm covers (**Figure 24**, typical) as described in this chapter.
3. Remove the intake manifold as described in Chapter Seven.
4. Remove the shoulder bolt (1, **Figure 40**) and pivot ball (2) from each rocker arm.
5. Mark the original location and remove the rocker arms and pushrods from the cylinder head studs. Inspect all parts for deterioration, wear or damage and replace if necessary.
6. Roll the pushrods across a flat surface (**Figure 37**) to check for damage. Replace bent pushrods.
7. Attach a breaker bar and socket to the harmonic balancer attaching bolt.
8. Observe the No. 1 exhaust valve lifter (see **Figure 34**) while turning the harmonic balancer clockwise. Stop when the lifter drops in the bore and the base of the cam lobe (**Figure 38**) aligns with the center of the roller on the bottom of the lifter.
9. Insert the pushrod into the corresponding slot in the cylinder head and seat it into the recess in the lifter.
10. Using a suitable mount (**Figure 41**), attach the dial indicator to the cylinder head. Align the plunger of the dial indicator with the tip of the pushrod. Rotate the face of the dial indicator to align the needle with the 0 mark.
11. Observe the dial indicator while rotating the harmonic balance exactly one revolution clockwise. The amount of needle movement equals camshaft lobe lift. Record the lobe lift.
12. Repeat Steps 7-10 for the remaining camshaft lobes. Record each measurement.
13A. On all 1994-1996 7.4L models, the camshaft lobe lift must be 0.232-0.236 in. (5.893-5.994 mm) for the intake lobes and 0.251-0.255 in. (6.375-6.477 mm) for the exhaust lobes. Replace the camshaft if any of the measurements are less than the minimum specification.
13B. On 1997-2000 7.4L GL and GI models, the camshaft lobe lift must be a minimum of 0.282 in. (7.163 mm) for the intake lobes and 0.284 in. (7.214 mm) for the exhaust lobes. Replace the camshaft if any of the measurements are less than the minimum specification.
13C. On 1997-2000 7.4L GS and GSI models, the camshaft lobe lift must be a minimum of 0.234 in. (5.944 mm) for the intake lobes and 0.252 in. (6.401 mm) for the exhaust lobes. Replace the camshaft if any of the measurements are less than the minimum specification.
13D. On 8.2L models, the camshaft lobe lift must be 0.249-0.253 in. (6.325-6.426 mm) for the intake and exhaust lobes. Replace the camshaft if any of the measurements are less than the minimum specification.
14. Install the pushrods, rocker arms, pivot balls and shoulder bolts as described in this chapter.
15. Install the intake manifold as described in Chapter Seven.
16. Install the rocker arm covers as described in this chapter.
17. Install the spark plugs and tighten to 22 ft.-lb. (30 N•m). Connect the plug leads and install or reposition any components removed to access the rocker arm covers.
18. Connect the negative battery cable.

CYLINDER HEAD

The cylinder head(s) can be removed and installed without removing the engine from the boat.

Take photographs or make sketches of bracket orientation, hose and wire routing before removing the cylinder head(s). This step saves a great deal of time during assembly and helps prevent incorrect assembly.

Mark the mounting location and orientation of the pushrods, rocker arm and mounting hardware prior to removing them.

The cylinder head uses bolts of different lengths. Label the bolts or make a diagram to show the location of each bolt as it is removed from the cylinder head.

43

CYLINDER HEAD TIGHTENING SEQUENCE (GM 3.0L MODELS)

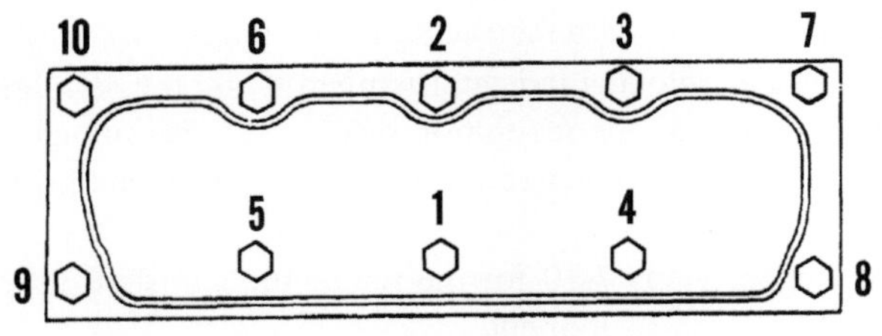

44

CYLINDER HEAD TIGHTENING SEQUENCE (GM 4.3L MODELS)

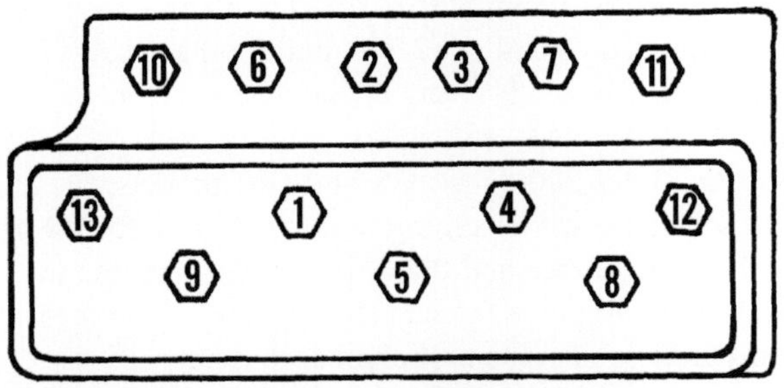

45

CYLINDER HEAD TIGHTENING SEQUENCE (GM 5.0L AND 5.7L MODELS)

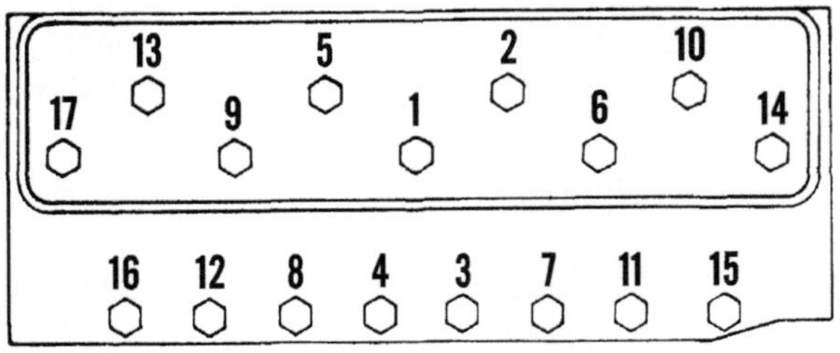

46

CYLINDER HEAD TIGHTENING SEQUENCE (FORD 5.0L AND 5.8L MODELS)

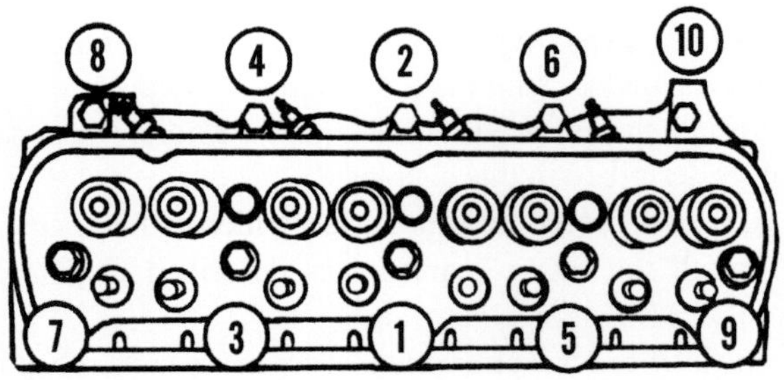

47

CYLINDER HEAD TIGHTENING SEQUENCE (GM 7.4L AND 8.2L MODELS

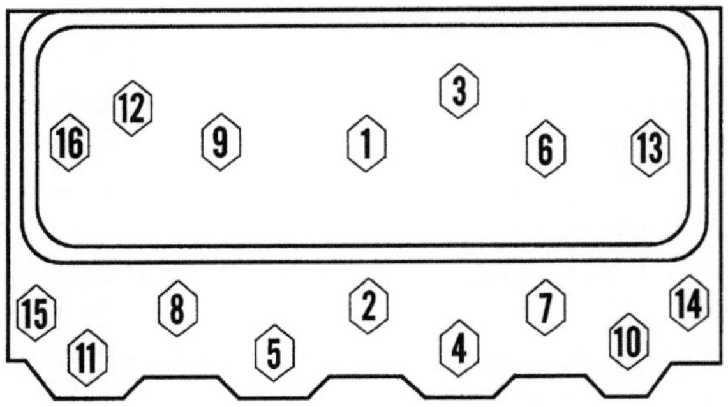

Use a wire brush to remove all sealant from the cylinder head bolts. Use a thread chaser to clean any sealant or corrosion from the bolt openings in the cylinder block. Apply a light coat of GM Sealant (GM part No. 12346004) or equivalent to the cylinder head bolt threads prior to installing them. Purchase the sealant from a GM automobile dealership.

Replace the head gasket each time the cylinder head bolts are loosened. Apply a light coat of OMC or Volvo Penta Gasket Sealing Compound (part No. 11411570-1) to both sides of the ribbed stainless steel gaskets. Do not apply sealant to composition gaskets.

Follow the specified head bolt tightening sequence and procedure. Many engines use a torque-and-turn tightening sequence. The bolts are tightened in sequence to a specified torque and then turned an additional amount. Use an angle torque gauge and tighten the bolts the specified degrees of rotation. Purchase the torque angle gauge from a tool supplier or automotive parts store.

NOTE

With the manifolds, fuel system, cooling system and electrical components installed, the General Motors (GM) 5.0L and 5.7L V-8 engines are very similar in appearance to the Ford 5.0L and 5.8L V-8 engines. If the model identification decal on the flame arrestor is missing, the engine can be quickly identified as a GM or Ford engine by noting the distributor location. On GM engines, the distributor is located at the rear (flywheel end) of the engine. On Ford engines, the distributor is located on the front of the engine.

Removal

1A. On models with a standard (open) cooling system, drain water from the cylinder block and manifold as described in Chapter Four.

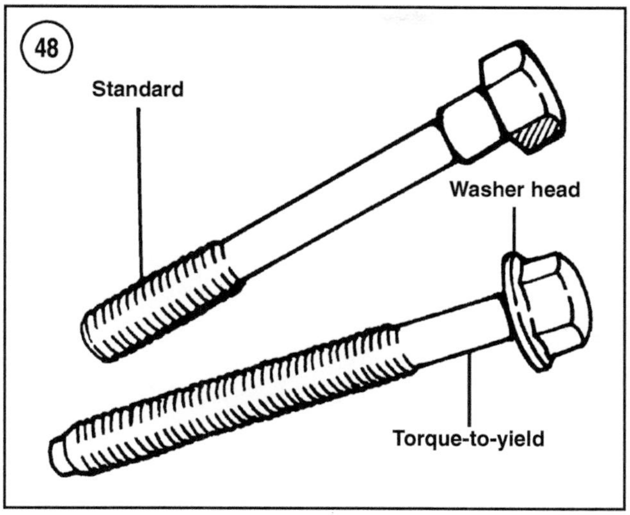

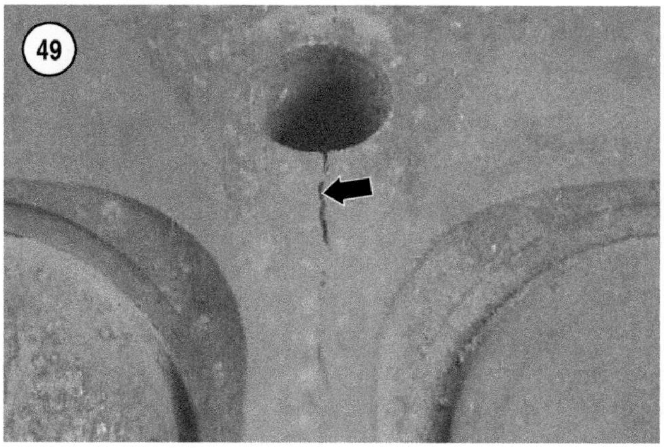

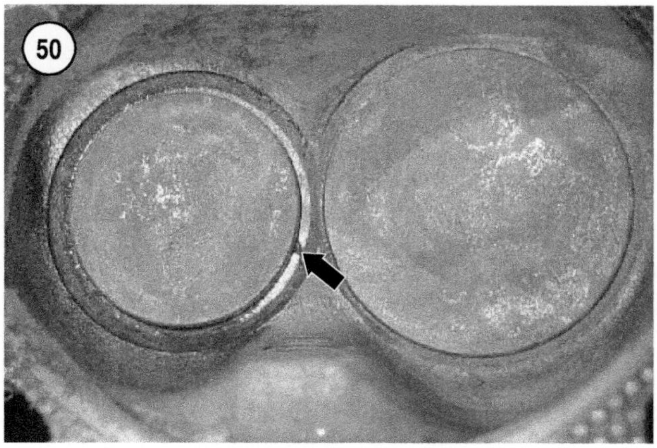

1B. On models with a closed cooling system, drain water from the exhaust manifolds and heat exchanger as described in Chapter Four. Drain coolant from the system as described in Chapter Three.

2. Remove the exhaust manifold as described in Chapter Eight.

3. On V-6 and V-8 models, remove the intake manifold as described in Chapter Seven.

4. Remove the rocker arm cover(s), rocker arms and pushrods as described in this chapter.

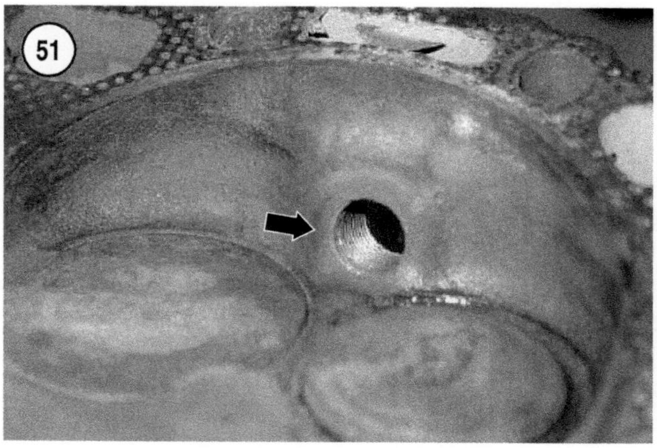

5. Relocate or remove any brackets, hoses, lines or other components that will interfere with cylinder head removal.

6. Remove the spark plugs.

7. Refer to the appropriate diagrams (**Figures 43-47**) and loosen the cylinder head bolts in the reverse order of the tightening sequence.

8. Remove the cylinder head bolts. Note each bolt length upon removal.

9. Carefully pry the cylinder head from the cylinder block. Use a blunt pry bar and work carefully to avoid damaging the mating surfaces.

10. Remove the gasket from the cylinder block or head. Carefully scrape all gasket material from the mating surfaces. Do not scratch or gouge the mating surfaces. Clean the surfaces with aerosol carburetor cleaner.

11. Inspect the piston crowns for damage or excessive carbon deposits. Remove carbon deposits from the pistons with a blunt scraper. If any of the pistons are damaged, disassemble the cylinder block and replace them as described in this chapter.

12. On Ford 5.0L and 5.8L models, compare the head bolts with those shown in **Figure 48**. Standard head bolts can be cleaned and reused. Discard torque-to-yield head bolts anytime they are loosened and replace with new ones because they cannot be reused. The torque-to-yield bolts can be identified by the washer head and longer threaded section.

Inspection

1. Scrape all carbon deposits from the combustion chamber with a blunt scraper. Do not nick or gouge the cylinder head.

2. Use solvent and thoroughly clean the cylinder head.

3. Inspect the cylinder head for surface cracks (**Figure 49**), damaged valve seats (**Figure 50**) and cracks near the spark plug openings (**Figure 51**). Replace the cylinder head if the casting is cracked. If cracked, have the seats replaced at a reputable machine shop.

4. Inspect the valves for broken and chipped or burned edges (**Figure 52**). Replace the valve and seat if this or other valve damage is evident.

5. Check for valve recession by comparing the valve depth in the valve seat with other valves in the cylinder head. The valve is recessed if it appears to be lower in the seat than the others. Have the cylinder head repaired and valves replaced if valve recession is evident.
6. Check for cylinder head warp by placing a straightedge at various positions across the cylinder head mating surface (**Figure 53**). Hold the straightedge firmly against the surface and check the gap at various points along the straightedge with a feeler gauge. The thickness of the feeler gauge that can pass under the straightedge indicates the amount of warp. Have the cylinder head resurfaced if surface damage is evident or if the warp exceeds 0.003 in. (0.076 mm) at any 6 in. (15 cm) span or 0.007 in. (0.178 mm) along the length of the cylinder head. Do not remove more than 0.010 in. (0.254 mm) of material from the cylinder head surface. Replace the cylinder head if a greater amount of material must be removed to true the surface.
7. Check the cylinder block for warp by placing a straightedge at various points on the cylinder head mating surface (**Figure 54**). Hold the straightedge firmly against the head and check the gap at various points along the straightedge with a feeler gauge. The thickness of the feeler gauge that can pass under the straightedge indicates the amount of warp. Have the cylinder block resurfaced if the warp exceeds 0.004 in. (0.102 mm) along the length of the straightedge or 0.002 in. (0.051 mm) within a 6 in. (15 cm) span. Do not remove more than 0.010 in. (0.254 mm) material from the cylinder block surface. Replace the cylinder block if a greater amount of material must be removed to true the surface.
8. Position the cylinder head so the ports are higher than the valves. Pour light solvent into the ports (**Figure 55**). Check for solvent leaking between the valves and seats. Have the cylinder head repaired if solvent leaks from the valves.
9. Inspect the dowel pins in the cylinder block. They must fit tightly in the openings. Replace damaged or loose pins.
10. Check the rocker arm studs, if so equipped, for wear or damage and replace if necessary. Studs with a hex head are threaded into the head and are removed by turning counterclockwise. Studs without a hex head are pressed into the head, and a suitable puller and installation tool must be used to replace them. Oversized studs are available to replace loose pressed-in studs. Have pressed-in studs installed at a reputable machine shop.

52

53

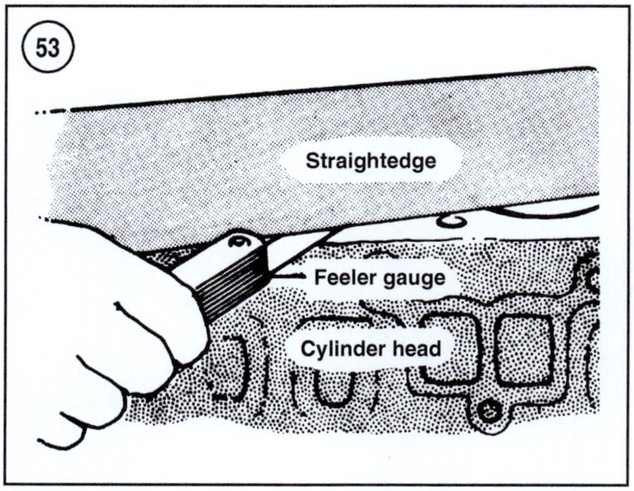

Installation

The cylinder head bolt tightening procedure for earlier 5.0L and 5.7L models is different from the procedure used for later models. Earlier models use 12 bolts to secure the intake manifold to the cylinder head. Later models use eight bolts to secure the intake manifold and are referred to as Gen+ models. The later GM 4.3L, 5.0L and 5.7L Gen+ engines use a torque-and-turn bolt tightening procedure that requires a torque angle gauge. Some Ford 5.0L and 5.8L models might use torque-to-yield head bolts that must not be reused and also require a torque angle gauge. To identify which tightening procedure to follow, compare the head bolts with those shown in **Figure 48**. Purchase the torque angle gauge from a tool supplier. Follow the tool manufacturer's instructions when using the tool.

1. Use an aerosol carburetor cleaner or other suitable solvent and thoroughly clean the cylinder head and cylinder block mating surfaces. Any residual gasket material or other contaminants can allow water, coolant or exhaust to leak.

2A. If a stainless steel gasket is used, apply a light coat of Permatex Gasket Sealing Compound or equivalent to both sides of the new gasket(s). Purchase the gasket sealing compound from an automotive parts store.

2B. If a composition gasket is used, do not apply sealant to the gasket or mating surfaces.

3. Thoroughly inspect the gasket for marks indicating which side of the gasket faces upward. The *up* or *this side up* inscription is sometimes difficult to locate. Align the cylinder bore and dowel pin openings and place the gasket(s) onto the cylinder block.

4. Align the openings with the dowel pins and carefully fit the cylinder head(s) onto the cylinder block.

54

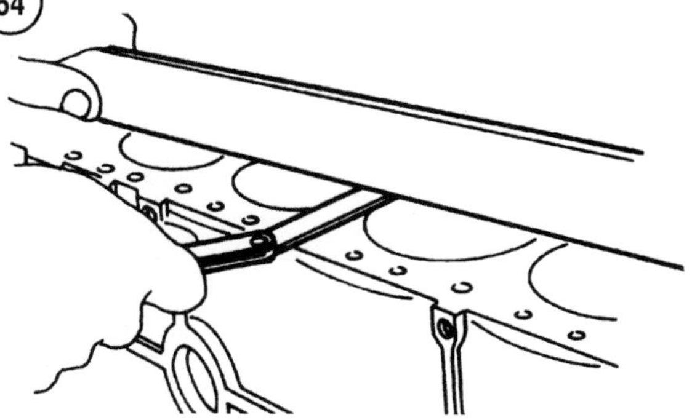

55

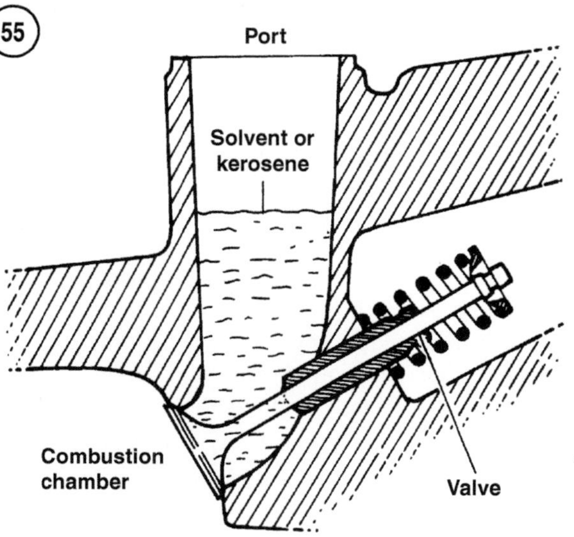

5. Apply a light coat of GM Sealant (GM part No. 12346004) or equivalent to the cylinder head bolt threads prior to installing them. Purchase the sealant from a GM automobile dealership. Install the bolts and thread them by hand into the cylinder head and cylinder block openings. If bolts of different lengths are used, make sure they are installed into the original locations.

6A. On GM 3.0L models, tighten the head bolts as follows:

a. Tighten the head bolts in sequence (**Figure 43**) to 35 ft.-lb. (47 N•m).
b. Tighten the head bolts in sequence a second time to 65 ft.-lb. (88 N•m).
c. On 1994-1996 models, tighten the head bolts a final time to 95 ft.-lb. (129 N•m).
d. On 1997-2000 models, tighten the head bolts a final time to 90 ft.-lb. (122 N•m).

6B. On 1994-1996 GM 4.3L models with a 12-bolt intake manifold, tighten the head bolts as follows:

a. Tighten the head bolts in sequence (**Figure 44**) to 24 ft.-lb. (33 N•m).
b. Tighten the head bolts in sequence a second time to 45 ft.-lb. (61 N•m).
c. Tighten the head bolts in sequence a final time to 65 ft.-lb. (88 N•m).

6C. On 1997-2000 GM 4.3L models with an eight-bolt intake manifold, tighten the head bolts as follows:

a. Tighten the head bolts in sequence (**Figure 44**) to 22 ft.-lb. (30 N•m).
b. Use a torque angle gauge and tighten the bolts in sequence a final time. Tighten the shorter bolts an additional 55°. Tighten the medium-length bolts an additional 65°. Tighten the longer bolts an additional 75°.

6D. On 1994-1997 GM 5.0L and 5.7L models with a 12-bolt intake manifold, tighten the head bolts as follows:

a. Tighten the head bolts in sequence (**Figure 45**) to 20 ft.-lb. (27 N•m).
b. Tighten the head bolts in sequence a second time to 40 ft.-lb. (54 N•m).
c. Tighten the head bolts in sequence a final time to 65 ft.-lb. (88 N•m).

6E. On 1998-2000 GM 5.0L and 5.7L models, with an eight-bolt intake manifold, tighten the head bolts as follows:

a. Tighten the head bolts in sequence (**Figure 45**) to 22 ft.-lb. (30 N•m).
b. Use a torque angle gauge and tighten the bolts in sequence (**Figure 45**) a final time. Tighten the shorter bolts an additional 55°. Tighten the medium-length bolts an additional 65°. Tighten the longer bolts an additional 75°.

6F. On Ford 5.0L and 5.8L models using standard head bolts (**Figure 48**), tighten the head bolts as follows:

a. Tighten the head bolts in sequence (**Figure 46**) to 95-105 ft.-lb. (129-142 N•m).
b. Tighten the head bolts in sequence a final time to 105-112 ft.-lb. (142-152 N•m).

6G. On Ford 5.0L and 5.8L models using torque-to-yield head bolts (**Figure 48**), tighten the head bolts as follows:

a. Tighten the head bolts in sequence (**Figure 46**) to 25-35 ft.-lb. (34-47 N•m).
b. Tighten the head bolts in sequence a second time to 45-55 ft.-lb. (61-75 N•m).
c. Use a torque angle gauge and tighten the bolts in sequence an additional 90°.

6H. On GM 7.4L and 8.2L models, tighten the head bolts as follows:

a. Tighten the head bolts in sequence (**Figure 47**) to 35 ft.-lb. (47 N•m).
b. Tighten the head bolts in sequence a second time to 65 ft.-lb. (88 N•m).
c. Tighten the head bolts in sequence a final time to 85 ft.-lb. (115 N•m).

7. Install the pushrods and rocker arms as described in this chapter. Do not install the rocker arm cover at this time.

8. Reinstall any brackets, hoses, lines or other components that were removed or repositioned for cylinder head removal.

9. On GM 3.0L, 1994-1996 4.3L V-6, 5.0L V-8 and 5.7L V-8 models, adjust the valves as described in this chapter.

10. On Ford 5.0L and 5.8L models, check the valve clearance as described in this chapter.

11. On V-6 and V-8 models, install the intake manifold as described in Chapter Seven.

12. Install the spark plugs and tighten to specifications in **Table 1**.

13. Install the rocker arm covers as described in this chapter.

14. Install the exhaust manifold as described in Chapter Eight.

15. On models with a closed cooling system, fill the cooling system with the proper coolant-and-water mixture as described in Chapter Three.

16. Connect the plug leads and install or reposition any components removed to access the rocker arm cover(s).

17. Connect the battery cables.

18. Set the ignition timing as described in Chapter Five.

VALVES AND VALVE SEATS

Valve and valve seat repairs require special knowledge and expensive machine tools. Have all valve and valve seat repairs performed by a dealership or reputable machine shop. Remove the cylinder head(s), complete disassembly procedures except for valve removal, and then take head(s) in for inspection and service. Valve and valve seat repair specifications are given in **Table 2** at the end of this chapter.

ROCKER ARM AND PUSHROD

NOTE
With the manifolds, fuel system, cooling system and electrical components installed, the General Motors (GM) 5.0L and 5.7L engines are very similar in appearance to the Ford 5.0L and 5.8L engines. If the model identification decal on the flame arrestor is missing, the engine can be quickly identified as a GM or Ford engine by noting the distributor location. On GM engines, the distributor is located at the rear (flywheel end) of the engine. On Ford engines, the distributor is located on the front of the engine.

3.0L Models

CAUTION
The rocker arm, pivot ball, pushrod and lifter establish a wear pattern to each other during use and into the valve that opens. To maintain the wear pattern and prevent premature wear, improper valve adjustment or premature failure, always reinstall these components in the original locations. Install the pushrods with the same ends facing up as when originally installed.

1. Remove the rocker arm and lifter covers as described in this chapter.

2. Place the No. 1 piston in the TDC firing position as described in this chapter.

3. Remove the nut and ball (**Figure 36**) from each rocker arm. Mark the original locations and remove the rocker arms and pushrods from the cylinder head studs.

NOTE
Anytime the rocker arm cover is removed, inspect the rocker arm studs for grooving or excessive wear on the side of the studs that aligns with the rocker arms. Grooving or excessive wear in this area usually indicates the stud is improperly aligned in the stud boss. This defect is common on 3.0L models. The misalignment will cause the rocker arm to contact and wear heavily on the side of the stud. Eventual stud failure and damage to other engine components will result.

4. Inspect all parts for deterioration, wear or damage and replace if necessary. Inspect the rocker arm studs for excessive wear on the sides. Replace the cylinder head if any of the studs are heavily worn on one side. This condition usually affects only a single stud.

5. Roll the pushrods across a flat surface (**Figure 37**) to check for damage. Replace bent pushrods.

6. Insert the pushrods (5, **Figure 56**) through the respective slots in the cylinder head. Used pushrods must be installed into the original locations in the cylinder head.

7. Fit the lower end of the pushrods into the recesses in the lifters (6, **Figure 56**). Make sure the lifters are seated against the camshaft lobes (7, **Figure 56**). Push down on each lifter to verify proper seating.

8. Guide the opening in each rocker arm (3, **Figure 56**) over the rocker arm studs. Rotate the rocker arms on the stud to align the recesses with the tops of the pushrods. The tips of the rocker arms must align with the tops of the valve stems.

9. Fit the pivot balls over the stud with the rounded sides facing the rocker arms. Set the balls into the recesses in the rocker arms. Thread the adjusting nuts by hand onto the studs. Do not tighten the nut at this time.

10. Tighten the adjusting nuts (**Figure 26**) for the following valves (**Figure 25**) until a slight amount of lash exists between the rocker arm and pushrod. *Do not* remove all the lash.

   a. No. 1 cylinder intake.
   b. No. 2 cylinder intake.
   c. No. 1 cylinder exhaust.
   d. No. 3 cylinder exhaust.
   e. No. 4 cylinder intake.

56

PUSHROD, ROCKER ARM AND LIFTER (TYPICAL)

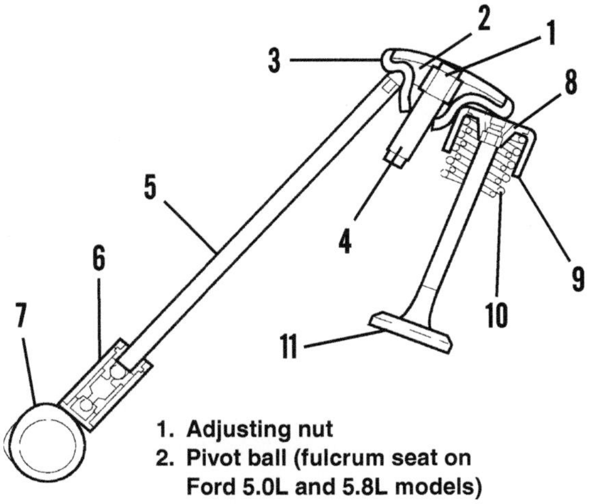

1. Adjusting nut
2. Pivot ball (fulcrum seat on Ford 5.0L and 5.8L models)
3. Rocker arm
4. Stud
5. Pushrod
6. Hydraulic lifter
7. Camshaft lobe
8. Valve spring retainer
9. Shield
10. Valve spring
11. Valve

11. Rotate the crankshaft clockwise exactly one turn and align the notch in the harmonic balancer with the 0 mark on the timing tab (**Figure 27**). The No. 4 piston is now at TDC.

CAUTION
Do not turn the crankshaft counterclockwise because the seawater pump will be damaged.

12. With the No. 4 piston at TDC, tighten the adjusting nuts (**Figure 26**) for the following valves (**Figure 25**) until a slight amount of lash exists between the rocker arm and pushrod. *Do not* remove all the lash.
    a. No. 2 cylinder exhaust.
    b. No. 3 cylinder intake.
    c. No. 4 cylinder exhaust.
13. Rotate the crankshaft clockwise exactly one turn and align the notch in the harmonic balancer with the 0 mark on the timing tab (**Figure 27**). The No. 1 piston is now at TDC.
14. Adjust the valves as described in this chapter.
15. To provide initial lubrication, coat the rocker arms, pushrods, pivot balls and lifters with engine oil.
16. Install the rocker arm and lifter covers as described in this chapter.
17. Install the spark plugs and tighten to 20 ft.-lb. (27 N•m). Connect the plug leads and install or reposition any components removed to access the rocker arm cover.
18. Connect the negative battery cable.

1994-1996 GM 4.3L Models with Adjusting Nut

CAUTION
The rocker arm, pivot ball, pushrod, lifter and valve establish a wear pattern to each other during use. To maintain the wear pattern and prevent premature wear, improper valve adjustment or premature failure, always reinstall these components in the original locations. Install the pushrods with the same ends facing up as when originally installed.

1. Remove the rocker arm covers as described in this chapter.
2. Place the No. 1 piston in the TDC firing position as described in this chapter.
3. Remove the nut and pivot ball (**Figure 36**) from each rocker arm. Mark the original location and remove the rocker arms and pushrods from the cylinder head studs.
4. Inspect all parts for deterioration, wear or damage and replace if necessary.
5. Roll the pushrods across a flat surface (**Figure 37**) to check for damage. Replace bent pushrods.
6. Insert the pushrods (5, **Figure 56**) through the respective slots in the cylinder head. Used pushrods must be installed into the original locations in the cylinder head.
7. Peer through the adjoining pushrod and oil drain openings in the cylinder head to view the lifter and lifter bore. Fit the lower ends of the pushrods into the recesses in the lifters (6, **Figure 56**). Make sure the lifters are seated against the camshaft lobes (7, **Figure 56**). Use the pushrods and push down on the lifters to verify proper seating.
8. Guide the opening in each rocker arm (3, **Figure 56**) over the rocker arm studs. Rotate the rocker arms on the studs to align the recesses with the tops of the pushrods. The slots at the tips of the rocker arms must align with the tops of the valve stems.
9. Fit the pivot balls over the studs with the rounded sides facing the rocker arms. Set the balls into the recesses in the rocker arms. Thread the adjusting nuts by hand onto the studs. Do not tighten the nuts at this time.
10. Tighten the adjusting nut (**Figure 29**) for the following valves (**Figure 28**) until a slight amount of lash exists between the rocker arm and pushrod. *Do not* remove all the lash.
    a. No. 1 cylinder intake.
    b. No. 1 cylinder exhaust.
    c. No. 2 cylinder intake.
    d. No. 3 cylinder intake.
    e. No. 5 cylinder exhaust.
    f. No. 6 cylinder exhaust.
11. Rotate the crankshaft clockwise exactly one turn and align the notch in the harmonic balancer with the 0 mark on the timing tab (**Figure 30**). The No. 4 piston is now at TDC.

CAUTION
Do not turn the crankshaft counterclockwise because the seawater pump will be damaged.

6

12. With the No. 4 piston at TDC, tighten the adjusting nut (**Figure 29**) for the following valves (**Figure 28**) until a slight amount of lash exists between the rocker arm and pushrod. *Do not* remove all the lash.
   a. No. 2 cylinder exhaust.
   b. No. 3 cylinder exhaust.
   c. No. 4 cylinder intake.
   d. No. 4 cylinder exhaust.
   e. No. 5 cylinder intake.
   f. No. 6 cylinder intake.
13. Rotate the crankshaft clockwise exactly one turn and align the notch in the harmonic balancer with the 0 mark on the timing tab (**Figure 30**). The No. 1 piston is now at TDC.
14. Adjust the valves as described in this chapter.
15. To provide initial lubrication, coat the rocker arms, pushrods and pivot balls with engine oil.
16. Install the rocker arm covers as described in this chapter.
17. Install the spark plugs and tighten to 22 ft.-lb. (30 N•m) if the cylinder head is new and 15 ft.-lb. (20 N•m) if the cylinder head is used. Connect the plug leads and install or reposition any components removed to access the rocker arm cover.
18. Connect the negative battery cable.

1997-2000 GM 4.3L Models with Shoulder Bolt

CAUTION
The rocker arm, pivot ball, pushrod, lifter and valve establish a wear pattern to each other during use. To maintain the wear pattern and prevent premature wear, improper valve adjustment or premature failure, always reinstall these components in the original locations. Install the pushrods with the same ends facing up as when originally installed.

1. Disconnect the negative battery cable and remove the spark plugs.
2. Place the No. 1 piston in the TDC firing position as described in this chapter.
3. Remove the rocker arm covers (**Figure 24**, typical) as described in this chapter.
4. Remove the shoulder bolts (1, **Figure 40**) and lift the rocker arms (3) from the cylinder heads. Mark the original locations and remove the pushrods (5, **Figure 40**) from the cylinder heads. Note the original location and remove the pivot ball from each rocker arm.
5. Roll the pushrods across a flat surface to check for damage (**Figure 37**). Replace bent pushrods.
6. Insert the pushrods (5, **Figure 40**) through the respective slots in the cylinder head. Used pushrods must be installed into the original locations in the cylinder head.
7. Peer through the adjoining pushrod and oil drain openings in the cylinder head to view the lifter and lifter bore. Fit the lower ends of the pushrods into the recesses in the lifters. Make sure the lifters are seated against the camshaft lobes. Use the pushrods and push down on the lifters to verify proper seating.
8. Align the openings in the roller rocker arm with the shoulder bolt openings in the cylinder heads. Fit the tops of the pushrods into the rounded recesses in the rocker arms. The slots near the tips of the rocker arms must align with the tops of the valve stems.
9. Insert the pivot balls (2, **Figure 40**) into the rocker arms (3) with the rounded sides facing the rocker arms.
10. Guide the shoulder bolts (1, **Figure 40**) through the pivot balls and rocker arms. Thread the shoulder bolts by hand into the respective cylinder head openings. *Do not* tighten the bolts at this time.
11. Tighten the rocker arm shoulder bolts to 22 ft.-lb. (30 N•m) for the following valves (**Figure 28**):
   a. No. 1 cylinder intake.
   b. No. 1 cylinder exhaust.
   c. No. 2 cylinder intake.
   d. No. 3 cylinder intake.
   e. No. 5 cylinder exhaust.
   f. No. 6 cylinder exhaust.

CAUTION
Do not turn the crankshaft counterclockwise because the seawater pump will be damaged.

12. Rotate the crankshaft clockwise exactly one turn and align the notch in the timing chain cover with the 0° TDC marking on the harmonic balancer decal (**Figure 32**). The No. 4 piston is now at TDC.
13. Tighten the rocker arm shoulder bolts to 22 ft.-lb. (30 N•m) for the following valves (see **Figure 28**):
   a. No. 2 cylinder exhaust.
   b. No. 3 cylinder exhaust.
   c. No. 4 cylinder intake.
   d. No. 4 cylinder exhaust.
   e. No. 5 cylinder intake.
   f. No. 6 cylinder intake.
14. Rotate the crankshaft clockwise exactly one turn and align the notch in the timing chain cover with the 0° TDC mark on the harmonic balancer (**Figure 32**). The No. 1 piston is now at TDC.
15. To provide initial lubrication, coat the rocker arms, pushrods and pivot balls with engine oil.
16. Install the rocker arm covers as described in this chapter.
17. Install the spark plugs and tighten to 22 ft.-lb. (30 N•m) if the cylinder head is new and 15 ft.-lb. (20 N•m) if the cylinder head is used. Connect the plug leads and install or reposition any components removed to access the rocker arm cover.
18. Connect the negative battery cable.

GM 5.0L and 5.7L Models

CAUTION
The rocker arm, pivot ball, pushrod, lifter and valve establish a wear pattern to each other during use. To maintain the wear pattern and prevent premature wear, improper valve adjustment or premature failure, always reinstall these components in the original locations. Install the pushrods with the same ends facing up as when originally installed.

1. Remove the rocker arm covers as described in this chapter.
2. Place the No. 1 piston in the TDC firing position as described in this chapter.
3. Remove the nut and pivot ball (**Figure 36**) from each rocker arm. Mark the original locations and remove the rocker arms and pushrods from the cylinder head studs.
4. Inspect all parts for deterioration, wear or damage and replace if necessary.
5. Roll the pushrods across a flat surface (**Figure 37**) to check for damage. Replace bent pushrods.
6. Insert the pushrods (5, **Figure 56**) through the respective slots in the cylinder head. Used pushrods must be installed into the original locations in the cylinder head.
7. Peer through the adjoining pushrod and oil drain openings in the cylinder head to view the lifter and lifter bore. Fit the lower ends of the pushrods into the recesses in the lifters (6, **Figure 56**). Make sure the lifters are seated against the camshaft lobes (7, **Figure 56**). Use the pushrods and push down on the lifters to verify proper seating.
8. Guide the opening in each rocker arm (3, **Figure 56**) over the rocker arm studs. Rotate the rocker arms on the studs to align the recesses with the tops of the pushrods. The slots at the tips of the rocker arms must align with the tops of the valve stems.
9. Fit the pivot balls over the studs with the rounded sides facing the rocker arms. Set the balls into the recesses in the rocker arms. Thread the adjusting nuts by hand onto the studs. Do not tighten the nuts at this time.
10. Tighten the adjusting nut (**Figure 29**) for the following valves (**Figure 31**) until a slight amount of lash exists between the rocker arm and pushrod. *Do not* remove all the lash.
   a. No. 1 cylinder intake.
   b. No. 1 cylinder exhaust.
   c. No. 2 cylinder intake.
   d. No. 3 cylinder exhaust.
   e. No. 4 cylinder exhaust.
   f. No. 5 cylinder intake.
   g. No. 7 cylinder intake.
   h. No. 8 cylinder exhaust.
11. Rotate the crankshaft clockwise exactly one turn and align the notch in the harmonic balancer with the 0 mark on the timing tab (**Figure 32**) or align the machined line on the harmonic balancer with the 0 mark on the timing tab (**Figure 30**). The No. 4 piston is now at TDC.

CAUTION
Do not turn the crankshaft counterclockwise because the seawater pump will be damaged.

12. With the No. 4 piston at TDC, tighten the adjusting nut (**Figure 29**) for the following valves (**Figure 31**) until a slight amount of lash exists between the rocker arm and pushrod. *Do not* remove all the lash.
   a. No. 2 cylinder exhaust.
   b. No. 3 cylinder intake.
   c. No. 4 cylinder intake.
   d. No. 5 cylinder exhaust.
   e. No. 6 cylinder intake.
   f. No. 6 cylinder exhaust.
   g. No. 7 cylinder exhaust.
   h. No. 8 cylinder intake.
13. Rotate the crankshaft clockwise exactly one turn and align the notch in the harmonic balancer with the 0 mark on the timing tab (**Figure 32**) or align the machined line on the harmonic balancer with the 0 mark on the timing tab (**Figure 30**). The No. 1 piston is now at TDC.
14. Adjust the valves as described in this chapter.
15. To provide initial lubrication, coat the rocker arms, pushrods and pivot balls with engine oil.
16. Install the rocker arm covers as described in this chapter.
17. Install the spark plugs and tighten to 22 ft.-lb. (30 N•m) if the cylinder head is new and 15 ft.-lb. (20 N•m) if the cylinder head is used. Connect the plug leads and install or reposition any components removed to access the rocker arm cover.
18. Connect the negative battery cable.

6

Ford 5.0L and 5.8L Models

CAUTION
The rocker arm, fulcrum seat, pushrod, lifter and valve establish a wear pattern to each other during use. To maintain the wear pattern and prevent premature wear, improper valve adjustment or premature failure, always reinstall these components in the original locations. Install the pushrods with the same ends facing up as when originally installed.

1. Remove the rocker arm covers as described in this chapter.
2. Remove the spark plugs.
3. Place the No. 1 piston in the TDC firing position as described in this chapter.
4. Mark the circumference of the harmonic balancer at 90° increments relative to the 0° BTDC mark. See **Figure 57**.
5. Remove the nuts that retain the rocker arms for the No. 1 cylinder intake and exhaust valves (**Figure 33**) onto the stud. Mark the mounting locations on the components. Then lift the rocker arms and fulcrum seats from the cylinder head studs.

6. Mark the mounting locations and top ends of the two pushrods. Then carefully pull them out of the lifter and cylinder head openings.

7. Observe the rocker arms and pushrods for the No. 3 cylinder intake and exhaust valves (**Figure 33**) while rotating the harmonic balancer exactly 90° in the clockwise direction. The No. 3 cylinder valves rocker arms should allow the valves to close completely, and the pushrods should drop along with the lifters. This indicates the No. 3 cylinder is in the TDC firing position.

8. Starting with the No. 3 cylinder, remove the remaining rocker arms and pushrods by repeating Steps 5-7 and following the engine firing order, 1-3-7-2-6-5-4-8. Remove the pair of rocker arms and pushrods only when the corresponding cylinder is in the TDC firing position.

9. After removing the rocker arms for the No. 8 cylinder, rotate the harmonic balancer 90° clockwise to again position the No. 1 piston in the 0° TDC firing position.

10. Inspect all parts for deterioration, wear or damage and replace if necessary.

11. Roll the pushrods across a flat surface (**Figure 37**) to check for damage. Replace bent pushrods.

12. Insert the pushrods for the No. 1 cylinder through the respective slots in the cylinder head. Used pushrods must be installed into the original locations and orientations in the cylinder head. Peer through the adjoining pushrod and oil drain openings in the cylinder head to view the lifter and lifter bore. Fit the lower ends of the pushrods into the recesses in the lifters. Make sure the lifters are seated against the camshaft lobes. Use the pushrods and push down on the lifters to verify proper seating.

13. Guide the openings in the No. 1 cylinder rocker arms over the rocker arm studs. Rotate the rocker arms on the studs to align the recess with the top of the pushrods. The tip of the rocker arms must align with the top of the valve stems.

14. Fit the fulcrum seats over the studs and seat into the rocker arms. If new rocker arms or fulcrum seats are installed, coat the contact surfaces with Ford Multipurpose Grease DOAZ-19584-AA or equivalent.

15. Thread the nuts by hand onto the No. 1 cylinder rocker arm studs. Verify proper alignment of the rocker arm with the pushrods and valve stems. Then tighten the rocker arm retaining nuts to 18-25 ft.-lb. (24-34 N•m).

16. Rotate the harmonic balancer exactly 90° in the clockwise direction to place the next cylinder in the firing order in the 0° TDC firing position.

17. Starting with the No. 3 cylinder, install the remaining rocker arms and pushrods by repeating Steps 12-16 and following the engine firing order, 1-3-7-2-6-5-4-8. Install the pair of pushrods and rocker arms only when the corresponding cylinder is in the TDC firing position.

18. After installing the rocker arms for the No. 8 cylinder, rotate the harmonic balancer 90° clockwise to place the No. 1 piston in the 0° TDC firing position again.

57

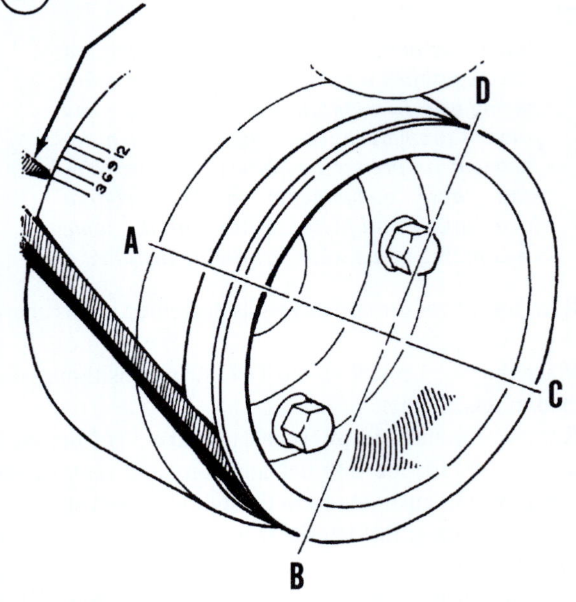

19. Install the rocker arm cover as described in this chapter.

20. Install the spark plugs and tighten to 120-180 in.-lb. (14-20 N•m).

GM 7.4L and 8.2L Models

CAUTION

Valve train components establish a wear pattern to each other during use. To maintain the wear pattern and prevent premature wear, improper valve adjustment or premature failure, always reinstall these components in the original locations. Install the pushrods with the same ends facing up as when originally installed.

1. Disconnect the negative battery cable and remove the spark plugs.

2. Place the No. 1 piston in the TDC firing position as described in this chapter.

3. Remove the rocker arm covers (**Figure 24**, typical) as described in this chapter.

4. Remove the shoulder bolts (1, **Figure 40**) and lift the rocker arms (3) from the cylinder heads. Mark the original locations and remove the pushrods (5, **Figure 40**) from the cylinder heads. Note the original location then remove the pivot ball from each rocker arm.

5. Roll the pushrods across a flat surface to check for damage (**Figure 37**). Replace bent pushrods.

6. Insert the pushrods (5, **Figure 40**) through the respective slots in the cylinder head. Used pushrods must be installed into the original locations in the cylinder head.

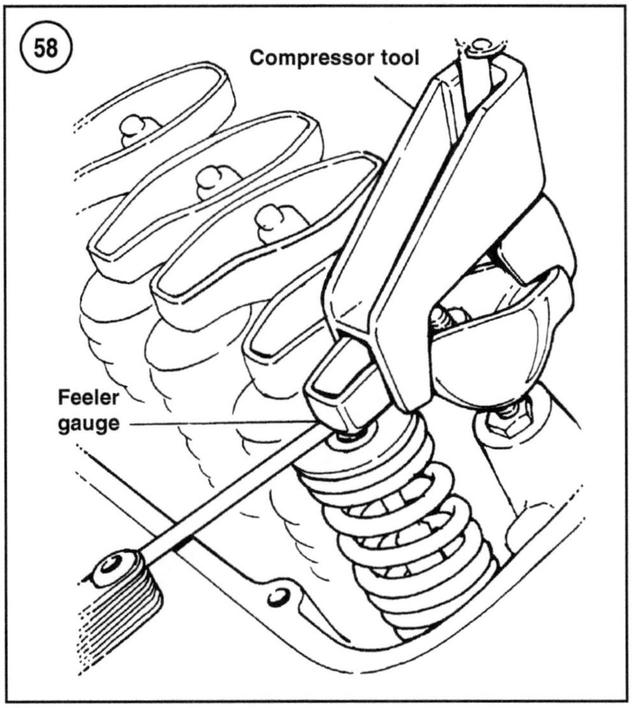

7. Peer through the adjoining pushrod and oil drain openings in the cylinder head to view the lifter and lifter bore. Fit the lower end of the pushrods into the recesses in the lifters. Make sure the lifters are seated against the camshaft lobes. Use the pushrods and push down on the lifters to verify proper seating.
8. Align the openings in the roller rocker arm with the shoulder bolt openings in the cylinder heads. Fit the top of the pushrods into the rounded recesses in the rocker arms. The tips of the rocker arms must align with the tops of the valve stems.
9. Insert the pivot balls (2, **Figure 40**) into the rocker arms (3) with the rounded sides facing the rocker arms.
10. Guide the shoulder bolts (1, **Figure 40**) through the pivot balls and rocker arms. Thread the shoulder bolts by hand into the respective cylinder head openings. *Do not* tighten the bolts at this time.
11. Tighten the rocker arm shoulder bolts to 40 ft.-lb. (54 N•m) for the following valves (**Figure 34**):
   a. No. 1 cylinder intake.
   b. No. 1 cylinder exhaust.
   c. No. 2 cylinder intake.
   d. No. 3 cylinder exhaust.
   e. No. 4 cylinder exhaust.
   f. No. 5 cylinder intake.
   g. No. 7 cylinder intake.
   h. No. 8 cylinder exhaust.

CAUTION
Do not turn the crankshaft counterclockwise because the seawater pump will be damaged.

12. Rotate the crankshaft clockwise exactly one turn and align the timing pointer with the 0° TDC mark on the harmonic balancer or timing chain cover (**Figure 30**). The No. 4 piston is now at TDC.
13. Tighten the rocker arm shoulder bolts to 40 ft.-lb. (54 N•m) for the following valves (**Figure 34**):
   a. No. 2 cylinder exhaust.
   b. No. 3 cylinder intake.
   c. No. 4 cylinder intake.
   d. No. 5 cylinder exhaust.
   e. No. 6 cylinder intake.
   f. No. 6 cylinder exhaust.
   g. No. 7 cylinder exhaust.
   h. No. 8 cylinder intake.
14. Rotate the crankshaft clockwise exactly one turn and align the timing pointer with the 0° TDC mark on the harmonic balancer or timing chain cover (**Figure 30**). The No. 1 piston is now at TDC.
15. To provide initial lubrication, coat the rocker arms, pushrods and pivot balls with engine oil.
16. Install the rocker arm covers as described in this chapter.
17. Install the spark plugs and tighten to 22 ft.-lb. (30 N•m). Connect the plug leads and install or reposition any components removed to access the rocker arm cover.
18. Connect the negative battery cable.

Valve Clearance Measurement (Ford 5.0L and 5.8L Models)

This procedure requires a lifter compressor tool (OMC part No. T71P-6513B) or equivalent and common feeler gauges. The lifter compressor tool can be purchased from tool suppliers, marine dealerships or automotive parts stores.

1. Remove the rocker arm covers as described in this chapter.
2. Remove the spark plugs.
3. Place the No. 1 piston in the TDC firing position as described in this chapter.
4. Mark the circumference of the harmonic balancer at 90° increments relative to the 0° BTDC mark. See **Figure 57**.
5. Attach the lifter compressor tool to the No. 1 cylinder intake valve rocker arm as shown in **Figure 58**. Refer to **Figure 33** to identify the valve. Slowly apply pressure against the tool handle to bleed the lifter down completely. Then hold the lifter compressor tool to keep the lifter bottomed. With the lifter bottomed, use feeler gauges to measure the clearance between the tip of the rocker arm and the valve stem (**Figure 58**).
6. The valve clearance must be 0.071-0.193 in. (1.803-4.902 mm) for 5.0L models and 0.098-0.198 in. (2.489-5.029 mm) for 5.8L models. Repeat this step for the No. 1 cylinder exhaust valve and then refer to the following:
   a. If the clearance is less than the specification, remove the corresponding pushrod as described in this chapter. Measure the length of the pushrod. Select an al-

59

FLYWHEEL HOUSING, DRIVE COUPER AND FLYWHEEL (TYPICAL)

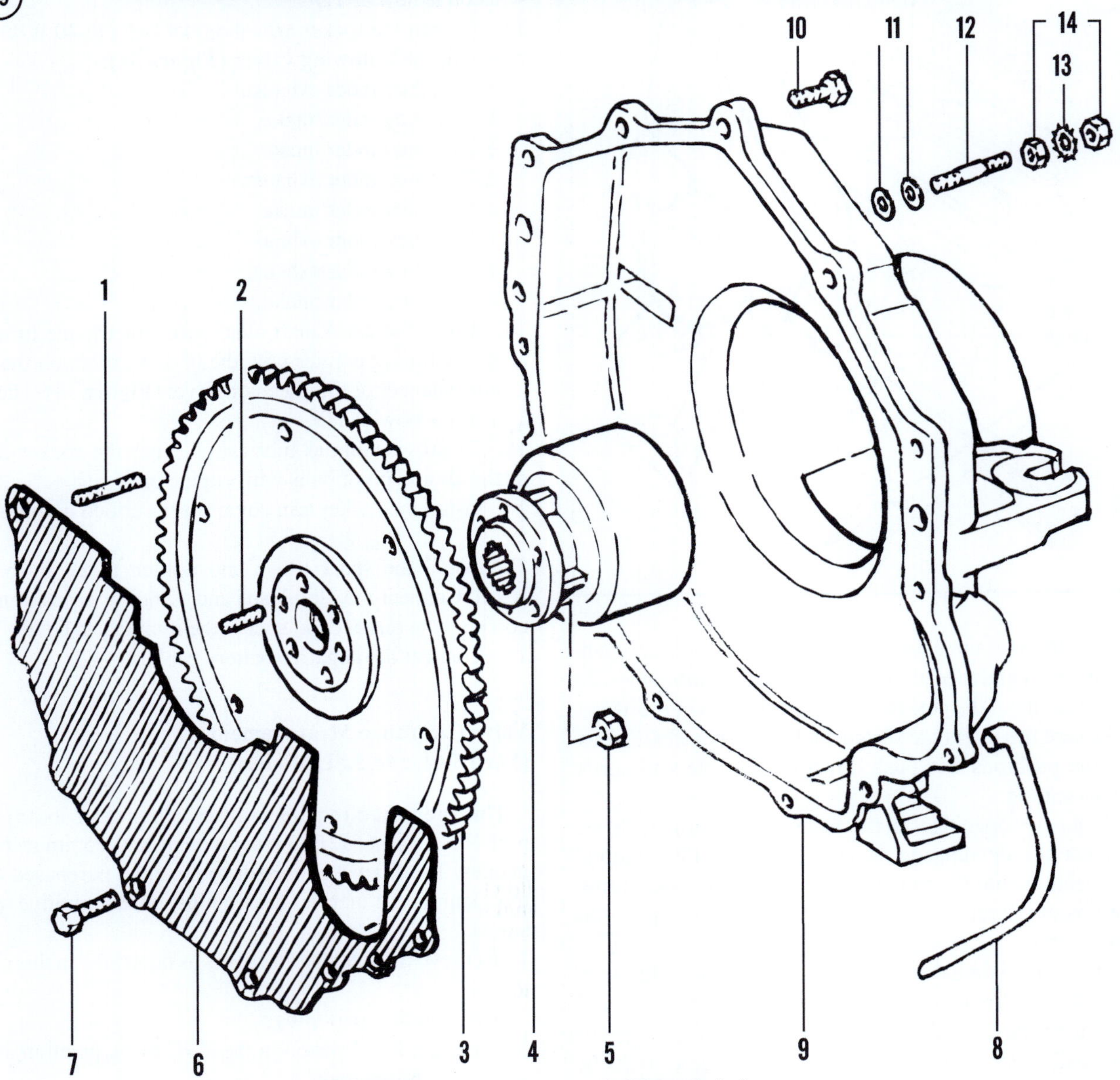

1. Stud
2. Alignment dowel
3. Flywheel
4. Drive coupler
5. Locknut
6. Shield
7. Screw
8. Water drain hose
9. Flywheel housing
10. Bolt
11. Washer
12. Stud
13. Washer
14. Nut

ternate pushrod that is shorter in length by the desired amount to provide the correct valve clearance. Install the pushrod and repeat the valve clearance measurement as described in this chapter. Alternate pushrods are available from an OMC or Volvo Penta dealership.

b. If the clearance is greater than the specification, remove the corresponding pushrod as described in this chapter. Measure the length of the pushrod. Select an alternate pushrod that is longer in length by the desired amount to provide the correct valve clearance. Install the pushrod and repeat the valve clearance measurement as described in this chapter. Alternate lengths of pushrods are available from an OMC or Volvo Penta dealership.

c. If the valve clearance cannot be corrected by installing the available alternate pushrods, check the camshaft lobe lift as described in this chapter. A faulty lifter, rocker arm, rocker arm stud valve or valve seat is evident if no fault is found with camshaft lobe lift.

60

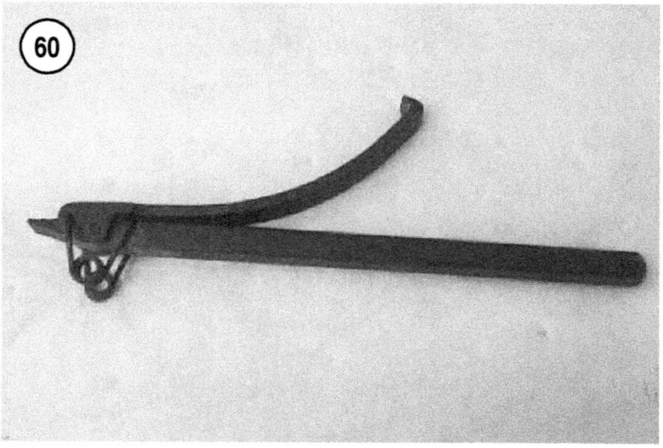

Inspect or replace these components, as described in this chapter to correct the valve clearance.

7. Observe the rocker arms and pushrods for the No. 3 cylinder intake and exhaust valves (**Figure 33**) while rotating the harmonic balancer exactly 90° in the clockwise direction. The No. 3 cylinder intake valve rocker arms should allow the valves to close completely, and the pushrods should drop along with the lifters. This indicates the No. 3 cylinder is in the TDC firing position.
8. Starting with the No. 3 cylinder, check the valve clearance for the remaining valves by repeating Steps 5-7 and following the engine firing order, 1-3-7-2-6-5-4-8. Check the valve clearance on the intake and exhaust valves only when the corresponding cylinder is in the TDC firing order.
9. After checking the valve clearance for the No. 8 cylinder, rotate the harmonic balancer 90° clockwise to place the No. 1 piston in the 0° TDC firing position again.
10. Install the rocker arm cover as described in this chapter.
11. Install the spark plugs and tighten to 120-180 in.-lb. (14-20 N•m).

FLYWHEEL HOUSING, DRIVE COUPLER AND FLYWHEEL

NOTE

With the manifolds, fuel system, cooling system and electrical components installed, the General Motors (GM) 5.0L and 5.7L engines are very similar in appearance to the Ford 5.0L and 5.8L engines. If the model identification decal on the flame arrestor is missing, the engine can be quickly identified as a GM or Ford engine by noting the distributor location. On GM engines, the distributor is located at the rear (flywheel end) of the engine. On Ford engines, the distributor is located on the front of the engine.

Flywheel Housing

1. Remove the engine as described in this chapter.
2. Remove the spark plugs.
3. Remove the starter motor as described in Chapter Nine.
4. To allow flywheel housing removal, suspend the rear of the engine with a suitable lift. It is acceptable to rest the engine on the side engine mounts, but *DO NOT* rest the engine on the oil pan. Doing so will probably damage the oil pan and does not provide enough stability.
5. On GM models, remove the screws (7, **Figure 59**, typical) and pull the shield (6) from the flywheel housing.
6. Use soap and water to remove grease and other contaminants from the shield.
7. Remove the nuts (14, **Figure 59**) and washers (13) from the studs (12). Support the flywheel housing (9, **Figure 59**) while removing the remaining bolts (10, **Figure 59**) and washers (11). Carefully pull the flywheel housing from the engine block. If difficulty is encountered, check for overlooked fasteners. If necessary, carefully pry the housing loose. Do not damage the mating surfaces. 6
8. Use a suitable solvent to remove oily deposits or corrosion from the flywheel housing. Dry the housing with compressed air.
9. Inspect the housing for cracks or other damage. Pay particular attention to the areas near the rear engine mount(s). Replace the housing if cracked or damaged.
10. Inspect the splines in the drive coupler for worn or damaged surfaces. Inspect the drive coupler for melted or loose rubber. Replace the drive coupler as described in this chapter if these or other defects are evident.
11. Fit the flywheel housing over the alignment dowels and fit the flywheel housing onto the cylinder block.
12. Support the flywheel housing while threading the washers (13, **Figure 59**) and nuts (14) onto the studs (12). Then thread the remaining washers (11, **Figure 59**) and bolts (10) into the flywheel housing and cylinder block. Tighten the fasteners in a crossing pattern to 40-45 ft.-lb. (54-61 N•m) for GM 3.0L-8.2L models and 28-36 ft.-lb. (38-49 N•m) for Ford 5.0L and 5.8L engines.
13. On GM models, position the shield (6, **Figure 59**) onto the flywheel housing. Make sure the starter motor slot in the shield is on the starboard side. Align the openings and then thread the screws (7, **Figure 59**) into the shield and flywheel housing openings. Tighten the screws evenly to 62-80 in.-lb. (7-9 N•m). Fit the water drain hose (8, **Figure 59**) into the opening in the flywheel housing. Direct the hose facing downward.
14. Install the starter motor as described in Chapter Nine.
15. Install the spark plugs and tighten to the specification in **Table 1**.
16. Install the engine as described in this chapter.

Drive Coupler

1. Remove the flywheel housing as described in this chapter.
2. Attach a flywheel holding tool (**Figure 60**) to the flywheel teeth to prevent rotation.

3. Remove the nuts (5, **Figure 59**) that secure the drive coupler and flywheel to the crankshaft. Support the flywheel while carefully pulling the drive coupler (4, **Figure 59**) from the flywheel. If necessary, carefully pry the coupler from the flywheel. Do not damage the mating surfaces. Thread one of the nuts onto a stud to secure the flywheel.

CAUTION
Make sure the mating surfaces of the crankshaft, flywheel and drive coupler are completely clean and the surfaces are true. Debris or uneven surfaces will prevent proper seating of the parts and allow the fasteners to loosen.

4. Use a suitable solvent to clean grease and other contaminants from the drive coupler. Dry the coupler with compressed air. Carefully scrape rust or contaminants from the mating surface of the flywheel. The surface must be completely cleaned prior to installing the coupler. Residual debris can prevent proper seating of the coupler and allow the fasteners to loosen.
5. Inspect the stud openings in the drive coupler for elongation or cracks. Inspect the mating surfaces of the coupler and flywheel for transferred material (rough or uneven mating surfaces) or other defects. Inspect the splined section of the drive coupler for wear or other defects. Inspect the coupler for melted or loose rubber. Replace the drive coupler if transferred material, damaged rubber, worn splines or other defects are evident. If transferred material or other defects are found on the flywheel, remove the flywheel as described in this chapter and have the mating surface machined. Do not remove more than 0.020 in. (0.5 mm) from the surface. Replace the flywheel if surface defects cannot be corrected by machining.
6. Support the flywheel and then remove the nut from the stud. Support the flywheel while fitting the drive coupler (4, **Figure 59**) onto the studs. Seat the coupler against the flywheel. Then thread the nuts (5, **Figure 59**) onto the studs. Engage a flywheel holding tool (**Figure 60**) onto the flywheel teeth to prevent rotation and then tighten the nuts in a crossing pattern to 40-45 ft.-lb. (54-61 N•m). Use a crowfoot or similar attachment to tighten the coupler nut.
7. Apply a coat of OMC Moly Lube (OMC part No. 175356) onto the coupler splines. Purchase the Moly Lube from an OMC, Johnson or Evinrude dealership.
8. Install the flywheel housing as described in this chapter.

Flywheel

1. Remove the flywheel housing as described in this chapter.
2. Remove the drive coupler as described in this chapter.
3. Pull the flywheel (3, **Figure 59**) from the crankshaft studs. If removal is difficult, rock the flywheel on the studs until free from the crankshaft. If necessary, pry between the cylinder block and flywheel. Work carefully to avoid damaging the flywheel teeth or cylinder block surfaces.

61 **OIL PAN AND GASKET (TYPICAL)**

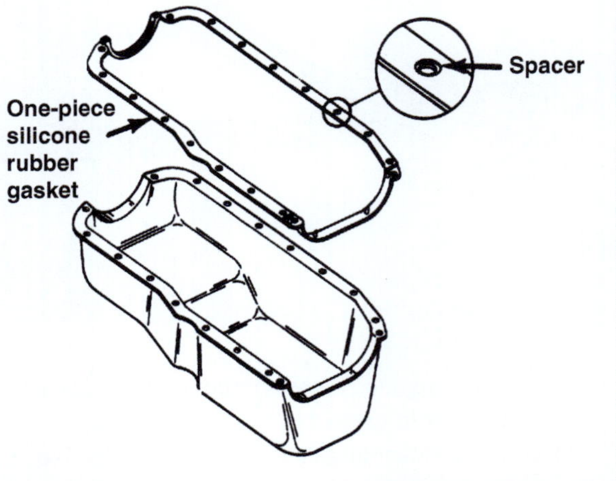

CAUTION
Make sure the mating surfaces for the crankshaft, flywheel and drive coupler are completely clean and the surfaces are true. Debris or uneven surfaces will prevent proper seating of the parts and allow the fasteners to loosen.

4. Use a suitable solvent to clean grease and other contaminants from the flywheel and crankshaft mating surfaces. Dry the surfaces with compressed air. Carefully scrape rust or contaminants from the mating surface of the flywheel. The surface must be completely cleaned prior to installing the coupler. Residual debris can prevent proper seating of the flywheel and allow the fasteners to loosen.

5. Inspect the flywheel-to-crankshaft mating surfaces for transferred material (rough or uneven mating surfaces) or other defects. If transferred material or other defects are found on the flywheel, have the mating surface machined. Do not remove more than 0.020 in. (0.5 mm) from the surface. Replace the flywheel if surface defects cannot be corrected by machining. If transferred material is found on the crankshaft, remove the crankshaft as described in this chapter and have it machined to remove the imperfection. Do not remove more than 0.020 in. (0.5 mm) from the surface.

6. Inspect the alignment dowel (2, **Figure 59**, typical) for wear, bending or other defects and replace as necessary. Using locking pliers, carefully twist and pull the dowel from the crankshaft. Carefully tap the replacement dowel into the opening. Drive gently to avoid damaging the replacement dowel.

7. Inspect the studs (1, **Figure 59**) for wear, cracks or damage and replace as needed.

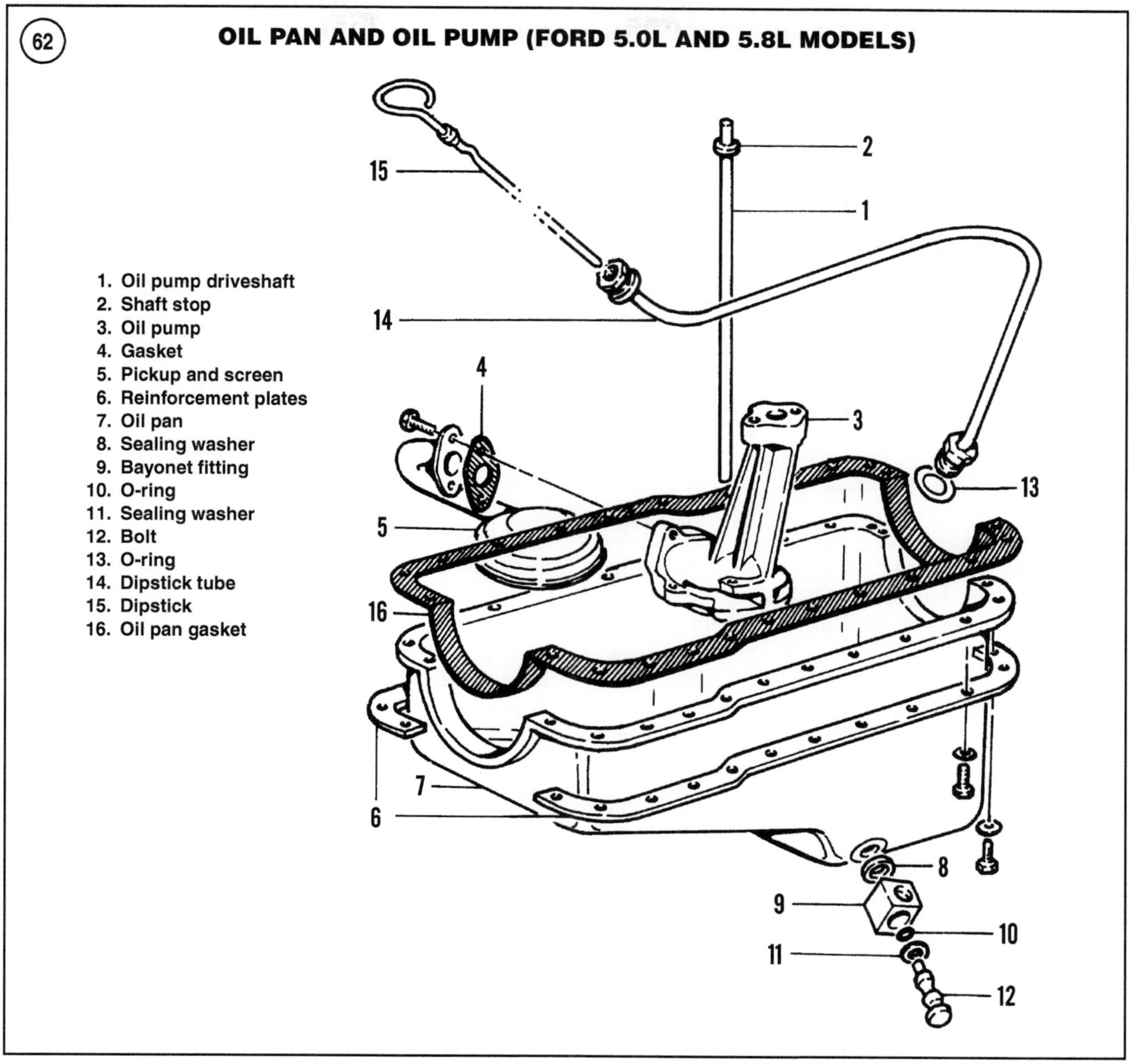

8. Align the dowel with the corresponding opening. Then fit the flywheel onto the crankshaft. Fully seat the flywheel onto the crankshaft.

9. Install the drive coupler as described in this chapter.

10. Attach a dial indicator mount to the cylinder block to position the stem in contact with the flat surface of the flywheel just inside the gear teeth. The stem must be perpendicular to the flywheel. Observe the dial indicator while rotating the flywheel. The flywheel runout indicated on the dial indicator must not exceed 0.008 in. (0.203 mm). If the runout exceeds the specification, the flywheel is warped, or debris or surface imperfections are preventing the flywheel from seating properly against the crankshaft flange. remove the flywheel and inspect the flywheel-to-crankshaft mating surfaces for burrs, debris or other defects. Clean debris or dress burrs as needed. Replace the flywheel if no defects are evident on the mating surfaces.

11. Remove the dial indicator and mount.

12. Install the flywheel housing as described in this chapter.

OIL PAN

All models covered in this manual use a one-piece oil pan gasket (**Figure 61**, typical). This type of gasket has molded-in metal spacers at the fastener openings, which prevent excessive gasket crushing and resulting oil leakage. Although the gasket is designed for reuse, it is good to replace the gasket anytime the oil pan is removed. Small tears might not be easy to see but will still allow oil to leak into the bilge. Unless the boat provides good access to the oil pan fasteners and clearance to remove the pan, remove the engine for oil pan removal.

1. Remove the engine as described in this chapter.

63

OIL PAN AND OIL PUMP (TYPICAL GM MODELS)

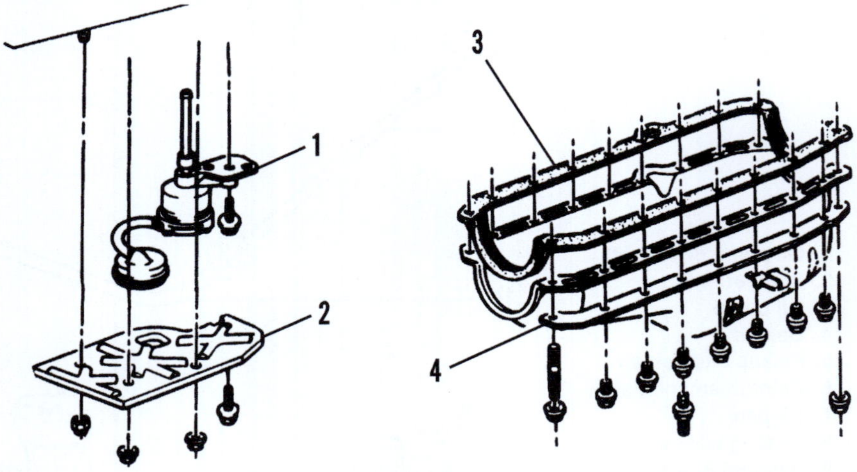

1. Oil pump
2. Splash plate
3. Oil pan gasket
4. Reinforcement

2. Place a suitable container under the oil pan and then loosen the bolt (12, **Figure 62**, typical) from the bayonet fitting (9) at the bottom of the pan. Allow the oil to drain completely from the pan. Carefully remove the bolt from the fitting and then pull the fitting away from the pan. Capture the sealing washers (8 and 11, **Figure 62**, typical) from both sides of the fitting. Inspect the O-ring (10, **Figure 62**) for tears, flatness or deterioration and replace as needed.

3. Remove the starter motor as described in Chapter Nine.

4A. On GM models, support the oil pan and remove the fasteners. Remove the reinforcement (4, **Figure 63**) if one is present. Carefully pull the oil pan from the cylinder block. If necessary, use a dull scraper and pry the pan from the cylinder block. Do not damage the gasket or mating surface.

4B. On Ford models, support the oil pan (7, **Figure 62**). then remove the screws and washers along the side rails of the pan. Remove the reinforcement plates (6, **Figure 63**) and then carefully remove the oil pan from the cylinder block.

5. If necessary, use a dull scraper and pry the pan from the cylinder block. Do not damage the gasket or mating surface.

6. Carefully remove the gasket from the oil pan or cylinder block. Work carefully to avoid damaging the gasket. Inspect the gasket for torn or damaged surfaces and replace if necessary.

7. Thoroughly clean the oil pan with solvent and dry with compressed air. Inspect the oil pan for cracks, dents or corrosion damage and replace if necessary.

8. Inspect the screen portion of the oil pump (**Figure 64**, typical) for debris or sludge deposits. Clean the screen with solvent if necessary.

9. Use aerosol carburetor cleaner and a clean shop towel to remove any remaining gasket sealant or oil film from the oil pan and cylinder block mating surfaces.

10A. On GM 3.0L models, insert the protrusions on the gasket into the small openings at the bottom of the timing gear cover.

10B. On GM V-6 and V-8 models, apply a bead of RTV silicone sealant on the corners where the oil pan gasket surface joins the timing chain cover and rear main seal carrier (**Figure 65**). Apply just enough sealant to fill in the gap.

11. Carefully position the gasket on the cylinder block. Align the openings in the metal spacers with the fastener openings.

12. On GM V-6 and V-8 models, align the front of the gasket with the groove in the timing chain cover. Align the rear of the gasket with the groove in the rear main seal carrier.

13. Carefully guide the oil pump screen into the oil pan cavity while fitting the oil pan onto the cylinder block. On GM models, the groove at the front and rear of the pan must fit over the gasket. Verify proper gasket position and then seat the oil pan against the cylinder block.

14. Align the screw openings. Then install the reinforcements (6, **Figure 62** or 4, **Figure 63**) onto the pan rails.

15. Thread the screws by hand into the opening or nuts onto the corner studs. Starting from the center and working outward, tighten the oil pan fasteners in a pattern crossing from one side to the other, as follows:

  a. On GM 3.0L models, tighten the screws that thread into the cylinder block to 80 in.-lb. (9 N•m). Tighten the screws that thread into the timing gear cover to 44 in.-lb. (5.0 N•m). Tighten the nuts at the rear of the pan to 165 in.-lb. (19 N•m).
  b. On GM 4.3L models with a stamped steel oil pan, tighten the rail screws to 80 in.-lb. (9 N•m). Tighten the corner nuts to 165 in.-lb. (19 N•m).
  c. On GM 4.3L models with a cast aluminum oil pan, tighten the rail screws to 221 in.-lb. (25 N•m). Tighten the corner nuts to 203 in.-lb. (23 N•m).
  d. On GM 5.0L and 5.7L models, tighten the screws or nuts at the corners of the mating surface to 177 in.-lb.

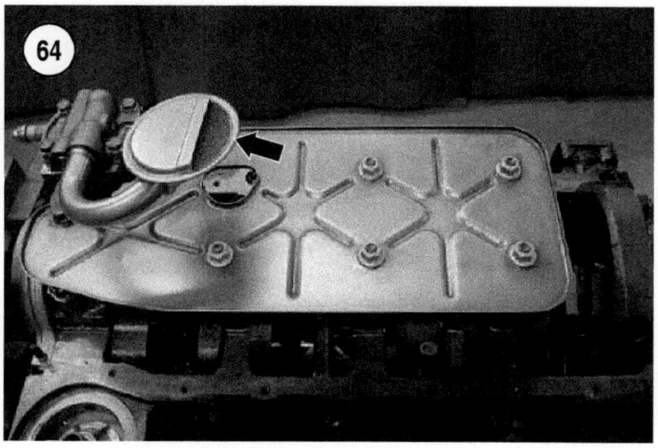

(20 N•m). Tighten the screws or nuts that run down the side rails to 106 in.-lb. (12 N•m).

e. On Ford 5.0L and 5.8L models, tighten the screws to 108-132 in.-lb. (12-15 N•m).
f. On GM 7.4L and 8.2L models, tighten the screws that thread into the cylinder block to 200 in.-lb. (23 N•m). Tighten the screws that thread into the timing chain cover to 70 in.-lb. (8 N•m).

16. Install the starter motor as described in Chapter Nine.
17. Align the fitting on the bottom of the dipstick tube with the drain opening in the oil pan. Fit the inner sealing washer (8, **Figure 62**) onto the surface between the bayonet fitting and the oil pan. Insert the outer sealing washer (11, **Figure 62**) onto the drain bolt. Make sure the sealing washers are in position and then thread the bolt into the fitting and oil pan opening. Do not tighten the bolt at this time.
18. Align the screw opening in the support near the top of the dipstick tube with its corresponding threaded. Opening. Thread the screw into the support and cylinder block. Tighten the support screw to 71 in.-lb. (8.0 N•m). Tighten the bayonet fitting drain bolt to 177-212 in.-lb. (20-24 N•m).
19. Install the starter motor as described in Chapter Nine.
20. Install the engine as described in this chapter.
21. Fill the crankcase with fresh oil as described in Chapter Three.
22. Start the engine and immediately check for oil leakage. Correct the leakage before operating the engine.

OIL PUMP

NOTE

With the manifolds, fuel system, cooling system and electrical components installed, the General Motors (GM) 5.0L and 5.7L engines are very similar in appearance to the Ford 5.0L and 5.8L engines. If the model identification decal on the flame arrestor is missing, the engine can be quickly identified as a GM or Ford engine by noting the distributor location. On GM engines, the distributor is located at the rear (flywheel end) of the engine. On Ford engines, the distributor is located on the front of the engine.

6

Removal and Installation

1. Attach the engine to a suitable engine stand and rotate the engine to position the oil pan side up. Remove the oil pan as described in this chapter.
2. On GM 5.0L and 5.7L models, remove the nuts from the stud bolts. Then remove the splash plate (2, **Figure 63**, typical).

3A. On GM 3.0L models, remove the fasteners that retain the oil pickup tube bracket on the cylinder block. Remove the two bolts that secure the oil pump to the cylinder block. Carefully pull the oil pump and pickup tube/screen assembly from the cylinder block.

3B. On GM V-6 and models, remove the mounting bolt(s) and carefully pull the oil pump and pickup tube/screen assembly (1, **Figure 63**) from the cylinder block.

3C. On Ford 5.0L and 5.8L models, remove the bolt securing the pickup tube support bracket to the cylinder block. Remove the two bolts securing the oil pump (3, **Figure 62**) to the cylinder block. Carefully lift the oil pump and pickup tube/screen assembly from the cylinder block. Remove the gasket from the pump-to-cylinder block mating surface. Discard the gasket.

4. On V-6 and V-8 models, pull the oil pump drive shaft (1, **Figure 62**, typical) from the cylinder block.
5. Inspect the oil pump for damage or excessive wear as described in *Disassembly, Inspection and Assembly* in this section.
6. If removed, install the distributor into the cylinder block opening.
7. On GM V-6 and V-8 models, if it is dislodged, fit the plastic sleeve over the lower end of the oil pump drive shaft. The sleeve should be flush with end of the shaft. Carefully insert the oil pump drive shaft into the opening in the cylinder block. If the distributor is installed, rotate the shaft to align the slot in the shaft with the pin in the distributor. The shaft will slide into the distributor gear when aligned.
8. On Ford 5.0L and 5.8L models, guide the oil pump drive shaft (1, **Figure 62**) into the distributor socket. Rotate the shaft to align the shaft with the distributor opening. Then

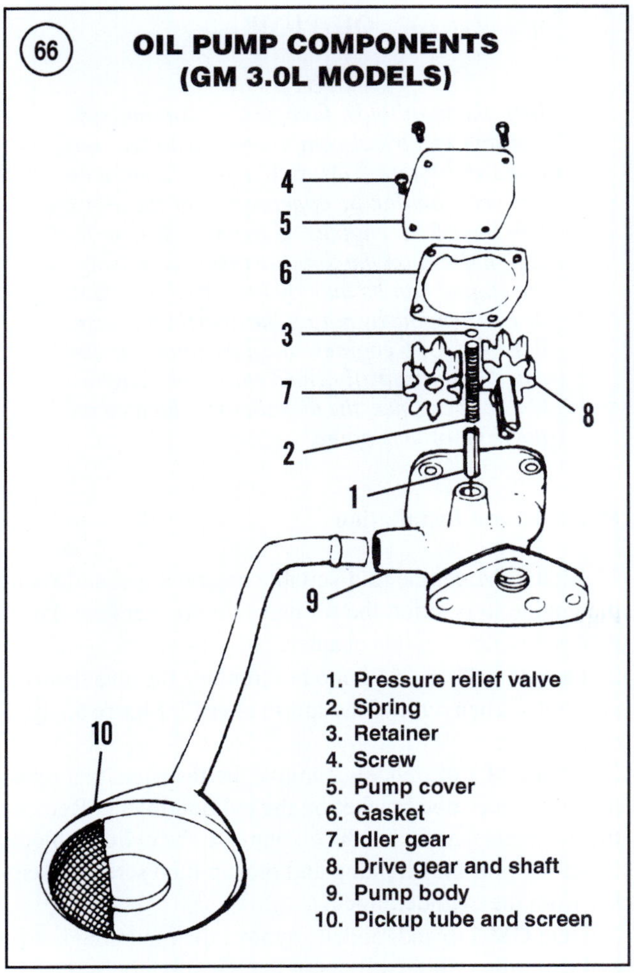

observe the stop (2, **Figure 62**) while seating the shaft in the distributor. The stop must contact the cylinder block at the same time the shaft seats into the distributor opening. If not, remove the shaft and move the stop up or down on the shaft as needed to provide correct contact. Install a new oil pump gasket onto the cylinder block. Align the gasket with the bolt and oil passage openings.

9. Rotate the oil pump body to align the pump with the extension shaft or distributor shaft (3.0L). The oil pump will seat against the cylinder block as the shafts align.

10. Rotate the oil pump on the mount to align the pickup tube/screen and mounting bolt openings.

11. On GM 5.0L and 5.7L models, fit the splash plate (2, **Figure 63**, typical) over the stud bolts. Make sure the bolt opening in the plate aligns with the corresponding mount bolt opening in the oil pump.

12. Thread the mount bolt(s) into the pump and cylinder block and tighten as follows:

a. On GM 3.0L models, tighten the mount bolts to 110-120 in.-lb. (12.4-14 N•m).

b. On GM 4.3L, 5.0L, 5.7L and 7.4L models, tighten the mount bolt to 65 ft.-lb. (88 N•m).

c. On Ford 5.0L and 5.8L models, tighten the pump mounting bolts to 22-32 ft.-lb. (30-43 N•m).

d. On GM 8.2L models, tighten the mount bolt to 70 ft.-lb. (95 N•m).

13. On GM 5.0L and 5.7L models, thread the nuts onto the studs and tighten to 30 ft.-lb. (41 N•m) to secure the splash plate.

14. On GM 3.0L models, move the pickup tube bracket to align the bolt opening with the corresponding opening in the cylinder block. Thread the bolt into the bracket and cylinder block and tighten to 180-216 in.-lb. (20-24 N•m).

15. On Ford 5.0L and 5.8L models, move the pickup tube bracket to align the bolt opening with the corresponding

70

71

72
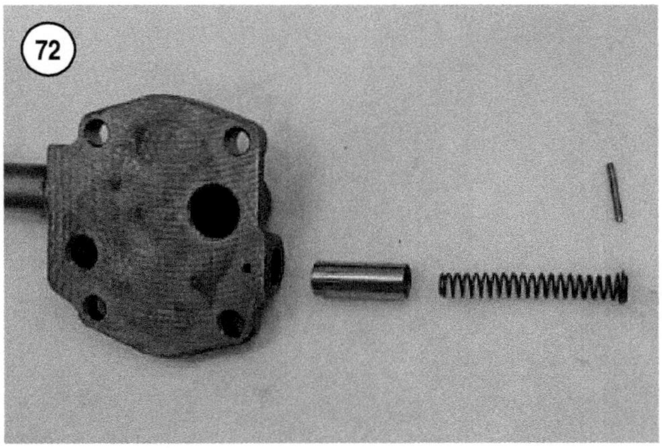

opening in the cylinder block. Install the bracket bolt and tighten to 22-32 ft.-lb. (30-43 N•m).

16. Install the oil pan as described in this chapter.

Disassembly, Inspection and Assembly

Internal oil pump components are not available separately. On GM models with a removable pickup tube and screen, use a pickup tube installer (Kent-Moore part No. J21882) to install the pickup tube and screen assembly into the pump cover. Otherwise, the tube might fit improperly and cause low oil pressure and engine damage.

GM 3.0L models

Refer to **Figure 66** for this procedure.

1. Remove the screws and carefully pull the cover from the pump body (**Figure 67**). Remove and discard the gasket (6, **Figure 66**).
2. Make match marks on the drive gear and idler gear teeth (**Figure 68**) prior to removing them. If reused, the teeth on one gear must align with the corresponding teeth on the other gear, or the gears will wear excessively or fail prematurely.
3. Pull the drive gear (**Figure 69**) and then the idler gear (**Figure 70**) out of the pump body.
4. Use a pin punch to carefully drive the retaining pin from the pump body (**Figure 71**). Remove the spring and pressure relief valve from the body (**Figure 72**).
5. Clean all components in solvent and dry with compressed air. Inspect the components as follows:
   a. Inspect the gears for wear, rough surfaces, excessive polishing or discoloration.
   b. Inspect the cover and pump body for wear, grooves, discoloration, cracks or roughness.
   c. Inspect the pressure relief valve for wear or roughness.
   d. Inspect the spring for corrosion or lost spring tension.
   e. Replace the oil pump assembly if these or other defects are evident.
6. Inspect the pickup tube and screen for corrosion or other damage. If either the pickup tube or screen is defective, replace the assembly as follows:
   a. Make match marks on the pickup tube and oil pump cover.
   b. Clamp the cover in a vise with protective jaws.
   c. Carefully twist the tube and screen assembly out of the cover opening.
   d. Transfer the match marks onto the replacement tube and screen assembly.
   e. Fit the pickup tube into the cover opening.
   f. Align the match marks. Then use Kent-Moore tool part No. J21882 (**Figure 73**) to drive the pickup tube into the cover until the bead on the tube contacts the cover.
7. Align the match marks and install the drive gear and idler gear into the pump body (**Figure 68**). The gears must spin freely in the pump body. If not, replace the oil pump assembly.
8. Insert the pressure regulator valve and then the spring into the pump body. Depress the spring until it is below the retaining pin bore. Then use a pin punch to carefully drive the retaining pin into the opening in the pump body.
9. Apply engine oil to the gears. Rotate the drive shaft to distribute the oil.
10. Install a new gasket (6, **Figure 66**) onto the pump body. Do not apply sealant to the gasket.
11. Fit the pump cover onto the pump body. Rotate the cover to properly position the pickup screen and screw openings. Thread the screws into the cover and pump body.

6

73

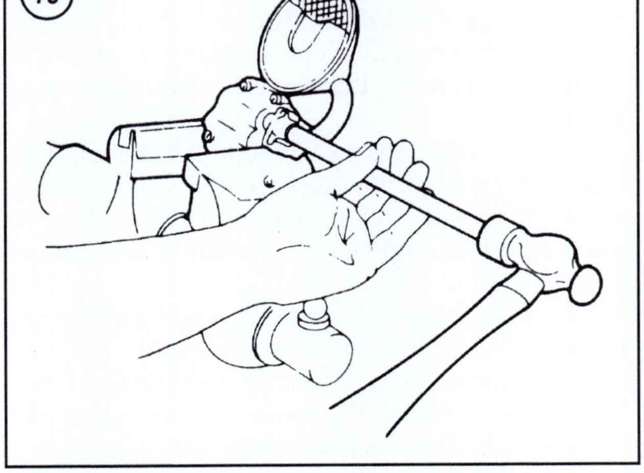

Tighten the screws in a crossing pattern to 72 in.-lb. (8 N•m).

GM 4.3L, 5.0L and 5.7L models

Refer to **Figure 74** for this procedure.

1. Carefully pull the drive extension shaft (1, **Figure 74**) and coupling (2) off the pump drive shaft.
2. Remove the screws and carefully pull the cover from the pump body (**Figure 67**).
3. Make match marks on the drive gear and idler gear teeth (**Figure 68**) prior to removing them. If reused, the teeth on one gear must align with the corresponding teeth on the other gear, or the gears will wear excessively or fail prematurely.
4. Pull the drive gear (**Figure 69**) and then the idler gear (**Figure 70**) out of the pump body.
5. Use a pin punch to carefully drive the retaining pin from the pump body (**Figure 71**). Remove the spring and pressure relief valve from the body (**Figure 72**).
6. Clean all components in solvent and dry with compressed air. Inspect the components as follows:
   a. Inspect the gears for wear, rough surfaces, excessive polishing or discoloration.
   b. Inspect the cover and pump body for wear, grooves, discoloration, cracks or rough surfaces.
   c. Inspect the pressure relief valve for wear or roughness.
   d. Inspect the spring for corrosion or lost spring tension.
   e. Replace the oil pump assembly if these or other defects are evident.
7. Inspect the pickup tube and screen (6, **Figure 74**) for corrosion or other damage. If either the pickup tube or screen is defective, replace the assembly as follows:
   a. Make match marks on the pickup tube and oil pump cover.
   b. Clamp the cover in a vise with protective jaws.
   c. Carefully twist the tube and screen assembly out of the cover opening.
   d. Transfer the match marks onto the replacement tube and screen assembly.
   e. Fit the pickup tube into the cover opening.
   f. Align the match marks. Then use Kent-Moore tool part No. J21882 to drive the pickup tube into the cover until the bead on the tube contacts the cover. See **Figure 73**.
8. Align the match marks and install the drive gear and idler gear into the pump body (**Figure 68**). The gears must spin freely in the pump body. If not, replace the oil pump assembly.
9. Insert the pressure relief valve and then the spring into the pump body. Depress the spring until it is below the retaining pin bore. Then use a pin punch to carefully drive the retaining pin into the opening in the pump body.

74

OIL PUMP COMPONENTS (GM 4.3L, 5.0L AND 5.7L MODELS)

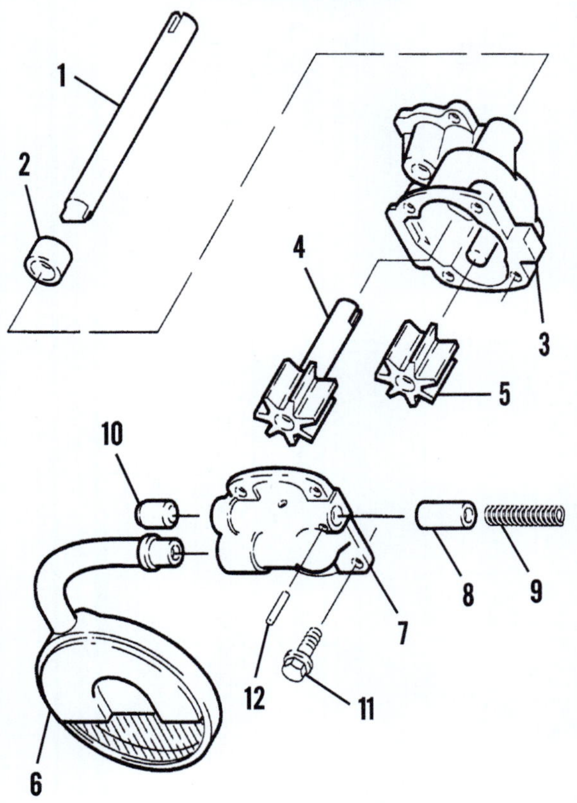

1. Drive extension shaft
2. Plastic coupling
3. Pump body
4. Drive gear and shaft
5. Idler gear
6. Pickup tube and screen
7. Pump cover
8. Pressure relief valve
9. Spring
10. Plug
11. Screw
12. Retaining pin

75

OIL PUMP COMPONENTS (FORD 5.0L AND 5.8L MODELS)

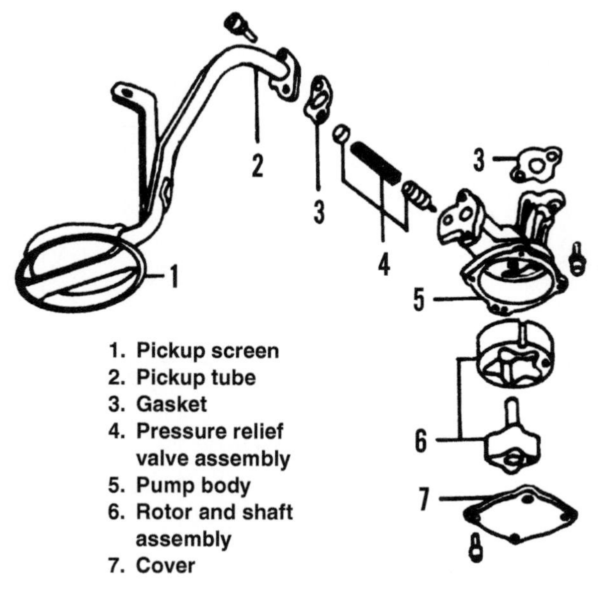

76

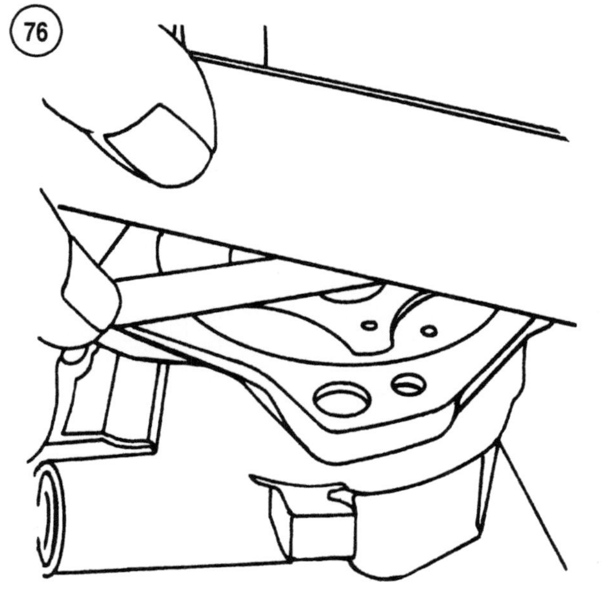

10. Apply engine oil to the gears. Rotate the drive shaft to distribute the oil.

11. Fit the pump cover onto the pump body. Rotate the cover to properly position the pickup screw and align the screw openings. Thread the screws into the cover and pump body. Tighten the screws in a crossing pattern to 106 in.-lb. (12 N•m).

Ford 5.0L and 5.8L models

Refer to **Figure 75** for this procedure.

77

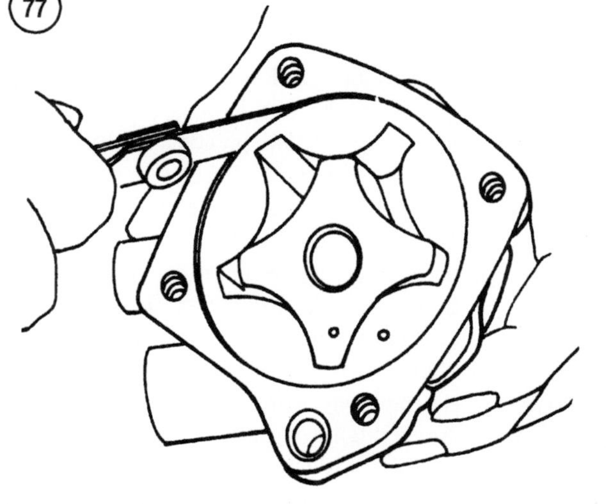

1. Remove the four screws and then carefully pull the cover (7, **Figure 75**) from the pump body.
2. Note the dimple mark on both the inner and outer rotor. These index marks designate the side of the components that contact the pump cover. If the outer rotor has operated with the dimple facing inward, it must be reinstalled in the original orientation. Otherwise, the component might wear excessively or fail prematurely.
3. Carefully push on the exposed shaft to remove the inner rotor from the pump housing. Then pull the outer rotor out of the pump housing.
4. Carefully remove the two screws and then pull the pickup tube (2, **Figure 75**) from the pump body.
5. Pull the plug, spring and relief valve (4, **Figure 75**) out of the pump body.
6. Clean all components in solvent and dry with compressed air.
7. Inspect the components as follows:
   a. Inspect the gears for wear, roughness, excessive polishing or discoloration.
   b. Inspect the cover and pump body for wear, grooves, discoloration, cracks or roughness.
   c. Inspect the pressure relief valve for wear or roughness.
   d. Inspect the spring for corrosion or lost spring tension.
   e. Replace the oil pump assembly if these or other defects are evident.
8. Install the inner and shaft assembly into the pump body. Place a straightedge across the rotor and pump body. Use a feeler gauge to measure the clearance between the rotor and the straightedge as shown in **Figure 76**. The inner rotor-to-pump cover clearance must be 0.001-0.004 in. (0.025-0.102 mm). Replace the pump assembly if the clearance exceeds the specification.
9. Note the orientation of the dimple marks described in Step 2. Then install the outer rotor into the housing along with the inner rotor and shaft. Use a feeler gauge to measure the clearance between the outer rotor and the pump housing as shown in **Figure 77**. The outer rotor-to-pump body

clearance must be 0.001-0.004 in. (0.025-0.102 mm). Replace the pump assembly if the clearance exceeds the specification.

10. Inspect the pickup tube (2, **Figure 75**) and screen (1) for corrosion cracks along the tube, damage to the screen or other defects. Replace the tube and pickup assembly unless found to be in excellent condition.

11. Apply engine oil to the gears. Rotate the drive shaft to distribute the oil.

12. Fit the pump cover onto the pump body. Rotate the cover to align the screw openings. Thread the screws into the cover and pump body. Tighten the screws in a crossing pattern to 72-120 in.-lb. (8-14 N•m).

13. Position the pump body with the relief valve opening facing upward. Apply engine oil into the opening. Then insert the relief valve into the pump body opening with the pointed end facing in. Install the spring and then plug into the opening.

14. Install a new gasket (3, **Figure 75**) onto the pump body. Align the screw openings and then fit the pickup tube (2, **Figure 75**) onto the pump body. Verify proper gasket alignment. Then thread the screws into the pickup tube and pump body openings. Tighten the screws evenly to 120-180 in.-lb. (14-20 N•m).

GM 7.4L and 8.2L models

Refer to **Figure 78** for this procedure.

1. Carefully pull the drive extension shaft (1, **Figure 78**) and coupling (2) off the pump drive shaft.
2. Remove the screws and carefully pull the cover from the pump body (**Figure 67**).
3. Make match marks on the drive gear and idler gear teeth (**Figure 68**) prior to removing them. If reused, the teeth on one gear must align with the corresponding teeth on the other gear, or the gears will wear excessively or fail prematurely.
4. Pull the drive gear (**Figure 69**) and then the idler gear (**Figure 70**) out of the pump body.
5. Use a pin punch to carefully drive the retaining pin from the pump body (**Figure 71**). Remove the spring and pressure relief valve from the body (**Figure 72**).
6. Clean all components in solvent and dry with compressed air. Inspect the components as follows:
   a. Inspect the gears for wear, roughness, excessive polishing or discoloration.
   b. Inspect the cover and pump body for wear, grooves, discoloration, cracks or roughness.
   c. Inspect the pressure relief valve for wear or roughness.
   d. Inspect the spring for corrosion or lost spring tension.
   e. Replace the oil pump assembly if these or other defects are evident.

CAUTION
On 7.4L and 8.2L models, the pickup tube is welded onto the oil pump body. The oil pump assembly must be replaced if the pickup tube or screen requires replacement.

(78) **OIL PUMP COMPONENTS (GM 7.4L AND 8.2L MODELS)**

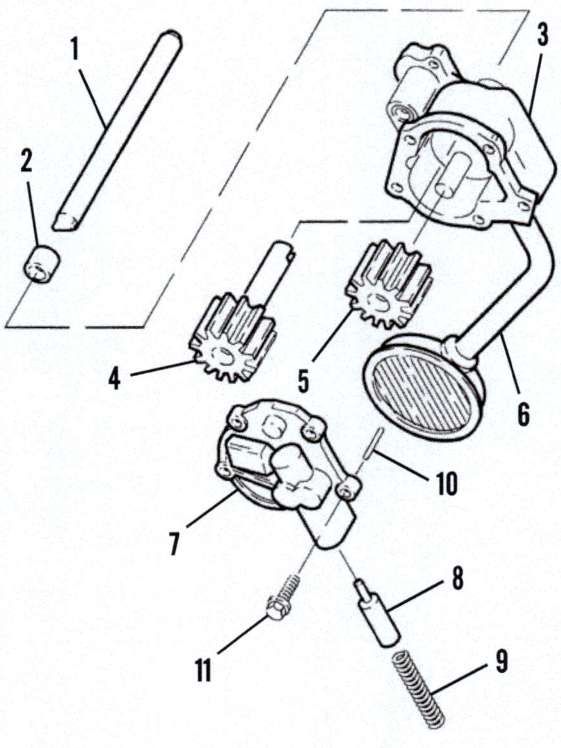

1. **Drive extension shaft**
2. **Coupling**
3. **Pump body**
4. **Drive gear and shaft**
5. **Idler gear**
6. **Pickup tube and screen**
7. **Pump cover**
8. **Pressure relief valve**
9. **Spring**
10. **Retaining spring**
11. **Screw**

7. Inspect the pickup tube and screen for corrosion or other damage. Replace the oil pump assembly if damage is evident.

8. Align the match marks and install the drive gear and idler gear into the pump body (**Figure 68**). The gears must spin freely in the pump body. If not, replace the oil pump assembly.

9. Insert the pressure relief valve and then the spring into the pump body. Depress the spring until it is below the retaining pin bore. Then use a pin punch to carefully drive the retaining pin into the opening in the pump body.

10. Apply engine oil to the gears. Rotate the drive shaft to distribute the oil.

79

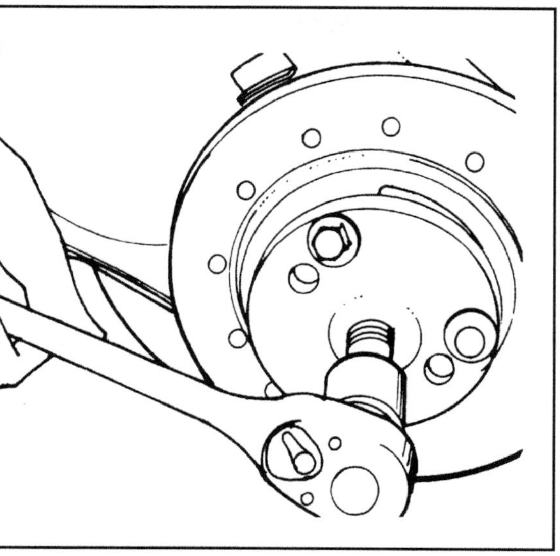

80

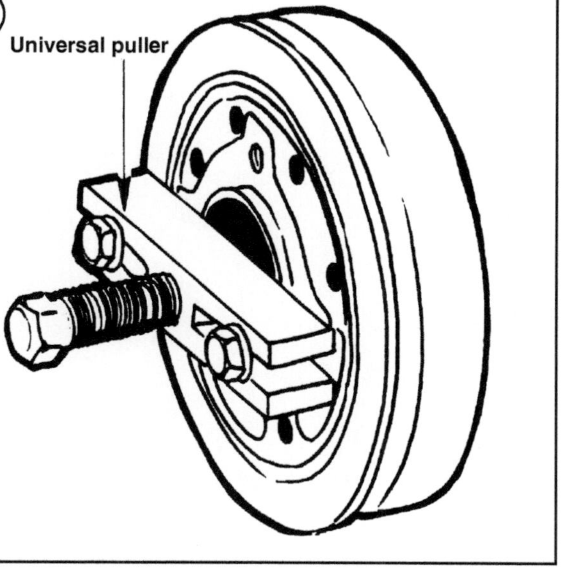

11. Fit the pump cover onto the pump body. Rotate the cover to properly position the pickup screw and align the screw openings. Thread the screws into the cover and pump body. Tighten the screws in a crossing pattern to 106 in.-lb. (12 N•m).

HARMONIC BALANCER

NOTE
With the manifolds, fuel system, cooling system and electrical components installed, the General Motors (GM) 5.0L and 5.7L engines are very similar in appearance to the Ford 5.0L and 5.8L engines. If the model identification decal on the flame arrestor is missing, the engine can be quickly identified as a GM or Ford engine by noting the distributor location. On GM engines, the distributor is located at the rear (flywheel end) of the engine. On Ford engines, the distributor is located on the front of the engine.

Removal and Installation

Use only the recommended tool (or equivalent) specified in the procedure to remove and install the harmonic balancer. Makeshift tools can damage the balancer or other engine components.

The balancer can be replaced without removing the engine. If boat structure prevents access to the balancer, remove the engine as described in this chapter.

1. Disconnect the battery cables.
2. On GM 3.0L models, hook the cable of an overhead hoist onto the front lifting eye of the engine. Operate the hoist enough to support the weight of the engine. Remove the four bolts and spacer block that attach the mount assembly to the front of the cylinder block. Remove the lag screw to free the mount from the engine compartment floor. Then remove the front engine mount.
3. Loosen the tension and remove the drive belts as described under *Alternator* in Chapter Nine.
4. Remove the pulley from the recirculation pump. Refer to *Recirculation Pump* in Chapter Eight.

5A. On 3.0L models, remove the three bolts and then carefully pull the pulley and harmonic balancer assembly off the pulley hub. If necessary, tap the pulley and balancer assembly loose with a rubber mallet.

5B. On earlier V-6 and V-8 models with a belt-driven seawater pump, remove the bolts and then carefully pull the drive belt pulley from the harmonic balancer. If necessary, tap the pulley loose with a rubber mallet.

5C. On later V-6 and V-8 models with a crankshaft-driven seawater pump, remove the seawater pump as described in Chapter Eight.

6. On V-6 and V-8 models, remove the retaining bolt and washer from the center of the balancer.

7A. On GM 4.3L, 5.0L and 5.7L models, attach the removal tool (Kent-Moore part No. J-39046) to the balancer as shown in **Figure 79**. Thread the appropriate bolts from the removal tool into the tool and balancer. Be careful not to cross thread the bolts. The openings use fine threads. Securely tighten the bolts.

7B. On Ford 5.0L and 5.8L models, attach the universal puller (OTC or SPX part No. T58P-6316-D) or equivalent to the balancer as shown in **Figure 80**. Thread the appropriate bolts into the puller and balancer openings. Make sure the bolts do not extend beyond the threaded openings enough to contact the front cover. If necessary, use shorter bolts. Securely tighten the bolts.

7C. On GM 7.4L and 8.2L models, attach the removal tool (Kent-Moore part No. J-23523-F) to the balancer as shown in **Figure 79**. Thread the appropriate bolts from the removal tool into the tool and balancer. Be careful not to cross thread the bolts. The openings use fine threads. Securely tighten the bolts.

6

8. On GM 3.0L models, attach the removal tool (Kent-Moore part No. J-6978-E) to the pulley hub as shown in **Figure 81**. Thread the appropriate bolts from the removal tool into the tool and hub. Be careful not to cross thread the bolts. The openings use fine threads. Securely tighten the bolts.
9. Thread the puller bolt into the removal tool until it contacts the crankshaft. Use the bolt with a tip that matches the crankshaft. Most of the models covered in this manual are tapered at the bolt opening.
10. Turn the puller bolt and remove the balancer or pulley hub (3.0L models) from the crankshaft.
11. Use solvent to clean the balancer, pulley hub (3.0L) and exposed end of the crankshaft.
12. Inspect the drive key in the crankshaft for corrosion or damage and replace as needed.
13. Inspect the balancer or pulley hub (3.0L) at the seal contact surface (**Figure 82**). Replace the balancer or hub if deeply grooved or pitted at this surface.
14. Inspect the balancer for protruding rubber material, cracks or other defects. Replace the balancer if these or other defects are evident.

NOTE
If the harmonic balancer is replaced, the seal in the timing gear or chain cover must also be replaced.

15. Inspect the seal (**Figure 83**) in the cover for wear or damage. Replace the seal if defective or if replacing the balancer. Refer to *Timing Gear or Chain Cover and Oil Seal* in this chapter.
16. On GM 3.0L models, fit the pulley and harmonic balancer assembly onto the pulley hub. One of the openings is for a 3/8 in. bolt, and the remaining two use 5/16 in. bolts. Rotate the pulley on the hub to align the openings. Thread the three bolts into the corresponding openings and securely tighten.
17. Apply a bead of RTV (**Figure 84**) sealant to the balancer or pulley hub key slot. Apply engine oil to the seal lip and the exposed end of the crankshaft.
18. Align the balancer or pulley hub (3.0L) key slot with the crankshaft drive key. Fit the balancer onto the crankshaft. Verify proper key-to-slot alignment. Then push the balancer or hub onto the crankshaft. Support the balancer to prevent it from falling while arranging the installation tool.
19A. On GM 3.0L models, fit the driver and pilot onto the clamp portion of the installation tool (Kent-Moore tool part No. J8792). Guide the arms of the clamp onto the balancer with the pilot facing toward the balancer as shown in **Figure 85**. Using a plastic mallet, drive against the driver portion of the tool until the balancer fully seats against the crankshaft gear. Remove the installation tool.
19B. On GM 4.3L, 5.0L and 5.7L, thread the bolt from the installation tool (Kent-Moore part No. J-39046) fully into the crankshaft threads. Thread the puller hub onto the bolt until it contacts the balancer. Turn the puller hub (**Figure 86**) until the balancer fully seats against the crankshaft. Do not allow the balancer or center bolt to rotate during balancer installation.
19C. On Ford 5.0L and 5.8L, thread the bolt from the installation tool (OTC or SPX T58P-6316-A) fully into the

81

CRANKSHAFT PULLEY HUB REMOVAL (GM 3.0L MODELS)

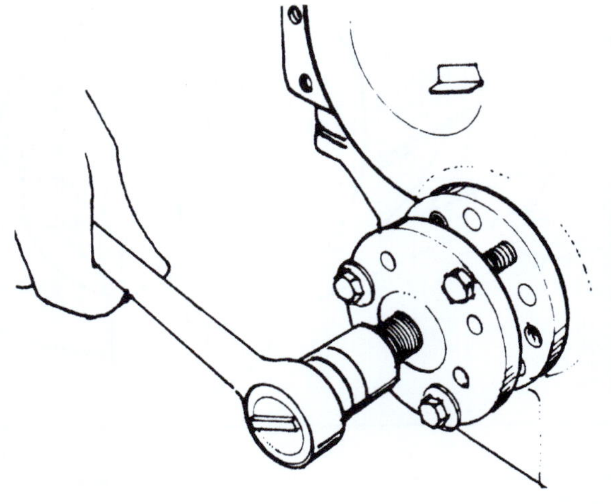

82

83

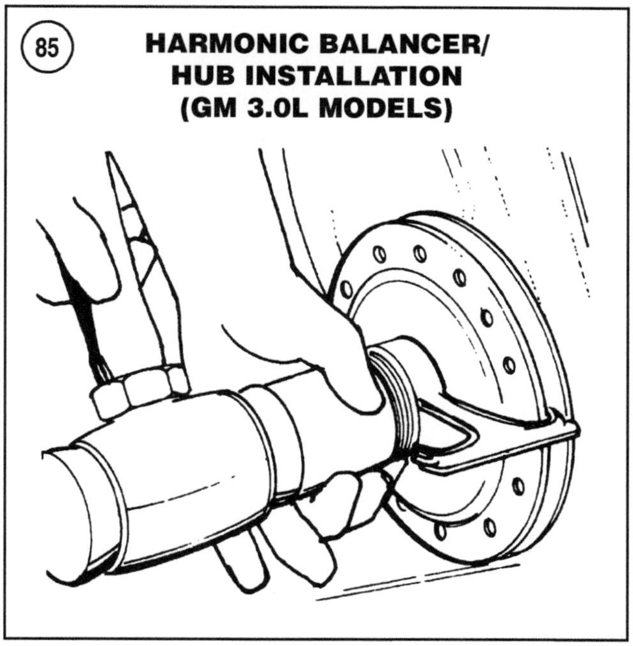

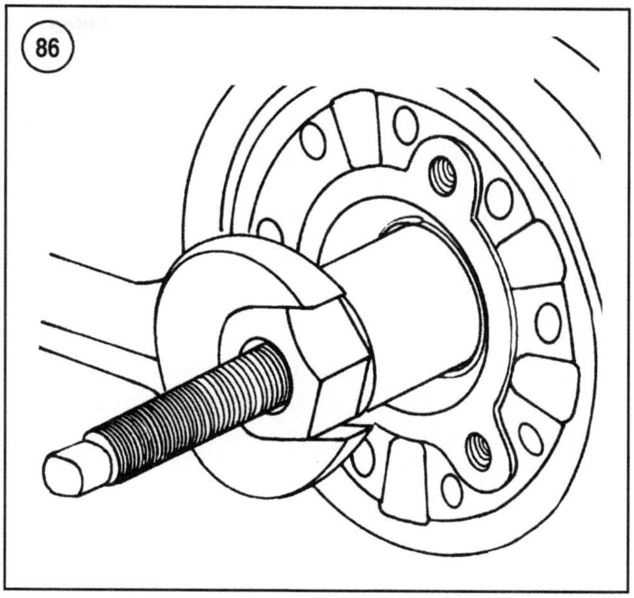

crankshaft threads. Thread the puller hub onto the bolt until it contacts the balancer. Turn the puller hub (**Figure 86**, typical) until the balancer fully seats against the crankshaft. Do not allow the balancer or center bolt to rotate during balancer installation.

19D. On GM 7.4L and 8.2L models, thread the bolt from the installation tool (Kent-Moore part No. J-23523-F) fully into the crankshaft threads. Thread the puller hub onto the bolt until it contacts the balancer. Turn the puller hub (**Figure 86**) until the balancer fully seats against the crankshaft. Do not allow the balancer or center bolt to rotate during balancer installation.

20. Remove the installation tool.

21. On V-6 and V-8 models, install the harmonic balancer retaining bolt and washer. Tighten as follows:

a. On GM 4.3L, 5.0L and 5.7L models, tighten the bolt to 74 ft.-lb. (100 N•m).

b. On Ford 5.0L and 5.8L models, tighten the bolt to 70-90 ft.-lb. (95-122 N•m).

c. On GM 7.4L models, tighten the bolt to 48 ft.-lb. (65 N•m) for 1994-1996 models and 85 ft.-lb. (115 N•m) for 1997-2000 models.

d. On GM 8.2L models, tighten the bolt to 73 ft.-lb. (99 N•m) for 1994-1996 models and 90 ft.-lb. (122 N•m) for 1997-2000 models.

22A. On earlier V-6 and V-8 models with a belt-driven seawater pump, install the crankshaft pulley onto the harmonic balancer. Tighten the pulley bolts to the general torque specification listed in Chapter One.

22B. On later V-6 and V-8 models with a crankshaft-driven seawater pump, install the seawater pump as described in Chapter Eight.

23. Install the pulley onto the recirculation pump. Refer to *Recirculation Pump* in Chapter Eight.

24. Install the drive belts and adjust the tension as described under *Alternator* in Chapter Nine.

25. On GM 3.0L models, fit the front engine mount and spacer block onto the cylinder block. Thread the four bolts by hand into the mount and cylinder block openings. Securely tighten the mount bolt. Align the lag bolt opening in the front mount with the corresponding openings in the engine compartment floor. Then slowly lower the engine until the mount contacts the boat. Apply a good quality marine sealant to the lag bolts. Thread the bolts into the mount and boat structure and securely tighten. Remove the overhead hoist cable.

26. Connect the battery cables.

27. Supply the engine with cooling water. Start the engine and immediately check for oil leakage from the front cover area. Correct any leakage before continuing operation.

TIMING GEAR OR CHAIN COVER AND OIL SEAL

Remove the cover for seal replacement or as necessary to access the timing sprockets and chain. On GM 4.3L models

(1997-2000), and GM 5.0 and 5.7L models (1998-2000) the timing chain cover is made of a composite material. It must be replaced if removed, or the cover might leak oil. On all other models, the cover is made of metal and may be reinstalled. The seal can be replaced without removing the cover from the engine. However, the preferred method is to remove the cover first. Seal removal with the cover on the engine requires special tools, and improper removal or installation can damage the cover, crankshaft and other engine components.

NOTE

With the manifolds, fuel system, cooling system and electrical components installed, the General Motors (GM) 5.0L and 5.7L engines are very similar in appearance to the Ford 5.0L and 5.8L engines. If the model identification decal on the flame arrestor is missing, the engine can be quickly identified as a GM or Ford engine by noting the distributor location. On GM engines, the distributor is located at the rear (flywheel end) of the engine. On Ford engines, the distributor is located on the front of the engine.

Timing Gear or Chain Cover

Removal and installation

1. Remove the oil pan as described in this chapter.

2A. On GM 3.0L models, remove the harmonic balancer and pulley hub as described in this chapter.

2B. On V-6 and V-8 models, remove the harmonic balancer as described in this chapter.

3. Remove the recirculation pump as described in Chapter Eight.

4. On Ford 5.0L and 5.8L models with a mechanical fuel pump, remove the mechanical fuel pump as described in Chapter Seven.

5. Remove the bolts from the timing gear or chain cover. Use a blunt scraper and carefully pry the cover from the cylinder block (**Figure 87**, typical).

6A. On GM 4.3L, 5.0L and 5.7L and V-8 models with a composite-material timing chain cover, discard the cover. The cover cannot be reused after removal.

6B. On all models using a metal timing chain or gear cover, carefully scrape gasket material from the cover mating surfaces. Thoroughly clean the cover with solvent and dry with compressed air. Inspect the cover for cracks, bent surfaces and corrosion damage. Corrosion can cause pin-sized holes that are difficult to detect yet allow significant oil leakage while running the engine.

7. Replace the front cover oil seal as described in this section.

8. On models using a metal timing chain cover, apply a light coating of gasket sealing compound to both surfaces of the new gasket. Fit the gasket onto the cover.

87

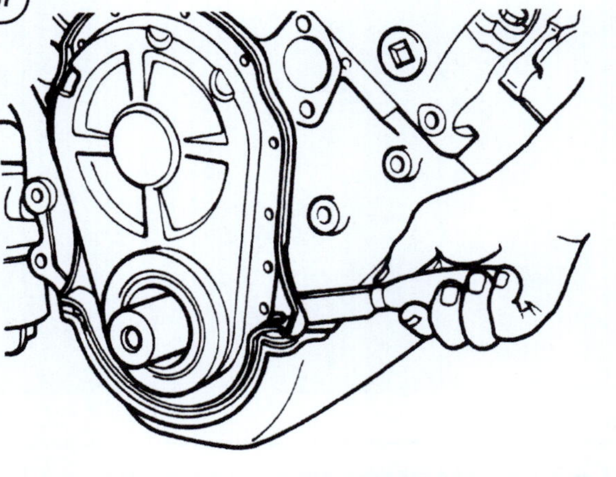

9. On GM V-8 models with a metal timing chain cover, apply a bead of RTV sealant to the two cylinder block surfaces where the cover, oil pan and cylinder block mate. This is necessary to prevent oil leakage in the corner where the oil pan gasket and cover gasket do not contact.

10A. On GM 3.0L models, install the timing gear cover as follows:

a. Slide the cover centering tool (Kent-Moore part No. J- 35468) over the end of the crankshaft.
b. Carefully guide the seal over the centering tool. Then seat the cover and gasket against the cylinder block.
c. Thread the screws by hand into the cover and cylinder block. Install the oil pan as described in this chapter. Do not tighten the oil pan screws as this time. Tighten the cover screws in a crossing pattern to 27-35 in.-lb. (3-4 N•m).
d. Carefully pull the centering tool off the crankshaft. Tighten the oil pan screws as described in this chapter.

10B. On GM 4.3L, 5.0L and 5.7L models with a metal timing chain cover, install the timing chain cover as follows:

a. Apply a coat of engine oil to the seal lip. Then carefully guide the seal over the crankshaft. To avoid wiping the RTV sealant from the cylinder block surfaces, position the cover toward the down side while guiding the cover in position. When the cover contacts the locating pins, carefully move the cover upward to align the pin openings and compress the RTV sealant. Then seat the cover against the cylinder block.
b. Verify proper gasket and screw opening alignment. Then thread the screws by hand into the cover and cylinder block. Tighten the cover screws in a crossing pattern to 80 in.-lb. (9 N•m).
c. Install the oil pan as described in this chapter.

10C. On GM 4.3L, 5.0L and 5.7L models with a composite timing chain cover, install the timing chain cover as follows:

a. Apply a coat of engine oil to the seal lip. Then carefully guide the seal over the crankshaft while positioning the cover on the cylinder block. Align the cylinder block aligning pins with the corresponding openings in the cover. Then seat the cover against the cylinder block.
b. Verify that the seal is present in each screw opening in the cover. Then thread the screws by hand into the cover and cylinder block. Starting from the center and working outward, tighten the cover screws in a crossing pattern to 106 in.-lb. (12 N•m).
c. Install the oil pan as described in this chapter.

10D. On Ford 5.0L and 5.8L models, install the timing chain cover as follows:

a. Apply a coat of engine oil to the seal lip. Then carefully guide the seal over the crankshaft while positioning the cover on the cylinder block. Align the cylinder block aligning pins with the corresponding openings in the cover. Then seat the cover against the cylinder block.
b. Carefully guide the cover centering tool (OTC or SPX part No. T61P-6019-B) through the seal opening and over the end of the crankshaft.
c. Verify proper gasket and screw opening alignment and then thread the screws by hand into the cover and cylinder block.
d. Install the oil pan as described in this chapter. Do not tighten the oil pan screws as this time.
e. Tighten the front cover screws to the general torque recommendation in Chapter One. Carefully pull the centering tool off the crankshaft.
f. Tighten the oil pan screws as described in this chapter.

10E. On GM 7.4L and 8.2L models, install the cover as follows:

a. Apply a coat of engine oil to the seal lip. Carefully guide the seal over the crankshaft. To avoid wiping the RTV sealant from the cylinder block surfaces, position the cover toward the down side while guiding the cover into position.
b. When the cover contacts the locating pins, carefully move the cover upward to align the pin openings and compress the RTV sealant. Then seat the cover against the cylinder block.
c. Verify proper gasket and screw opening and then thread the screws by hand into the cover and cylinder block. Tighten the screws in a crossing pattern to 60-84 in.-lb. (7-9 N•m).
d. Install the oil pan as described in this chapter.

11. On Ford 5.0L and 5.8L models with a mechanical fuel pump, install the mechanical fuel pump as described in Chapter Seven.

12. Install the recirculation pump as described in Chapter Eight.

13A. On GM 3.0L models, install the pulley hub and harmonic balancer as described in this chapter.

13B. On V-6 and V-8 models, install the harmonic balancer as described in this chapter.

14. Connect the battery cables.

15. Supply the engine with cooling water. Start the engine and immediately check for oil, water or coolant leakage from the front cover area. Correct any leakage before continuing operation.

Front Cover Oil Seal

The seal can be replaced without removing the timing chain or gear cover from the engine. However, the preferred method is to remove the cover first. Seal removal with the cover on the engine requires special tools, and improper removal or installation can damage the cover, crankshaft and other engine components.

With cover installed on engine

1. Remove the harmonic balancer and pulley hub (3.0L models) as described in this chapter.

2A. On GM models, carefully pry the used seal from the cover with a suitable pry bar. Work carefully to avoid damaging the front cover or seal bore. Discard the seal upon removal.

2B. On Ford 5.0L and 5.8L models, use OTC or SPX part No. T70P-6B070-B to pull the seal out of the front cover.

3. Use aerosol solvent to clean all residue from the seal bore. Dry the surfaces with compressed air.

4. Apply a light coat of Loctite 271 to the seal bore and outer diameter of the new seal.

5. Position the new seal into the seal bore with the seal lip facing inside the engine.

6. Carefully drive the seal into the bore using the recommended tool:

a. On GM 3.0L models, use Kent-Moore part No. J-23043.
b. On GM V-6 and V-8 models, use Kent-Moore part No. J-35468
c. On Ford 5.0L and 5.8L models, use OTC or SPX part No. T70P-6B070-B.

7. Wipe any residual Loctite from the seal and front cover surfaces. Apply a coat of engine oil to the seal lip.

8A. On GM 3.0L models, install the pulley hub and harmonic balancer as described in this chapter.

8B. On V-6 and V-8 models, install the harmonic balancer as described in this chapter.

9. Connect the battery cables.

10. Supply the engine with cooling water. Start the engine and immediately check for oil, water or coolant leakage from the front cover area. Correct any leakage before continuing operation.

With cover removed from engine

1. Place the cover on a clean work surface with the seal side facing upward. Place a block of wood under the seal bore for support.

2A. On GM models, carefully pry the used seal from the cover with a suitable pry bar. Work carefully to avoid damaging the front cover or seal bore. Discard the seal upon removal.

2B. On Ford 5.0L and 5.8L models, use OTC or SPX part No. T70P-6B070-B to pull the seal out of the front cover.

3. Use aerosol solvent to clean all residue from the seal bore. Dry the surfaces with compressed air.

4. Apply a light coat of Loctite 271 to the seal bore and outer diameter of the new seal.

5. Position the new seal into the seal bore with the seal lip facing inside the engine. Place a block of wood under the seal bore for support.

CAUTION

Do not use excessive force when driving the seal into the front cover. Excess force can damage the seal or front cover and cause the seal to tilt in the bore.

6. Carefully drive the seal fully into the bore (**Figure 88**) using the recommended tool:

   a. On GM 3.0L models, use Kent-Moore part No. J-23043.
   b. On GM V-6 and V-8 models, use Kent-Moore part No. J-35468
   c. On Ford 5.0L and 5.8L models, use OTC or SPX part No. T70P-6B070-B.

7. Wipe any residual Loctite from the seal and front cover surfaces. Apply a coat of engine oil to the seal lip.

TIMING CHAIN AND SPROCKETS

NOTE

With the manifolds, fuel system, cooling system and electrical components installed, the General Motors (GM) 5.0L and 5.7L engines are very similar in appearance to the Ford 5.0L and 5.8L engines. If the model identification decal on the flame arrestor is missing, the engine can be quickly identified as a GM or Ford engine by noting the distributor location. On GM engines, the distributor is located at the rear (flywheel end) of the engine. On Ford engines, the distributor is located on the front of the engine.

A timing chain and sprockets are used on all V-6 and V-8 engines. GM 3.0L models are equipped with timing gears in place of a chain and sprockets.

1. Disconnect the battery cables. Position the No. 1 piston at the TDC firing position as described in this chapter.

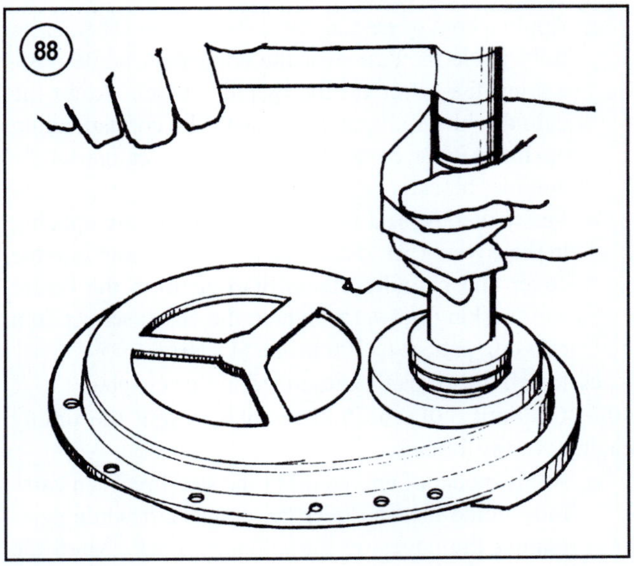

2. To prevent possible interference between the pistons and valves, remove the rocker arms and pushrods as described in this chapter.

3. Remove the timing chain cover as described in this chapter.

4. Measure the timing chain deflection as follows:

   a. Rotate the crankshaft counterclockwise to tighten the chain on the starboard side.
   b. Scribe a reference mark on the cylinder block that aligns with the tight side of the chain at a point midway between the sprockets.
   c. Rotate the crankshaft clockwise to tighten the chain on the port side.
   d. Pull the loose side of the chain (starboard side) toward the starboard side. Scribe a reference mark on the cylinder block that aligns with the chain. The distance between the two marks indicates the timing chain deflection. Replace the timing chain and both sprockets if the deflection exceeds 3/8 in. (10 mm) for GM V-6 and V-8 models and 1/2 in. (13 mm) for Ford V-8 models.

5. Rotate the crankshaft to align the marks on the camshaft and crankshaft sprockets (**Figure 89**).

6. On Ford 5.0L and 5.8L V-8 models, remove the bolt from the end of the camshaft sprocket. Then pull the washer and fuel pump drive eccentric off the sprocket.

NOTE

Earlier production 4.3L engines use stud bolts and nuts to retain the camshaft sprocket and balance shaft drive gear on the camshaft. Later production engines use bolts to retain the sprocket and balance shaft.

NOTE

The sprocket fits snugly on the camshaft, but should come off easily. If necessary, tap the lower edge of the sprocket with a rubber mallet to dislodge it from the camshaft.

89

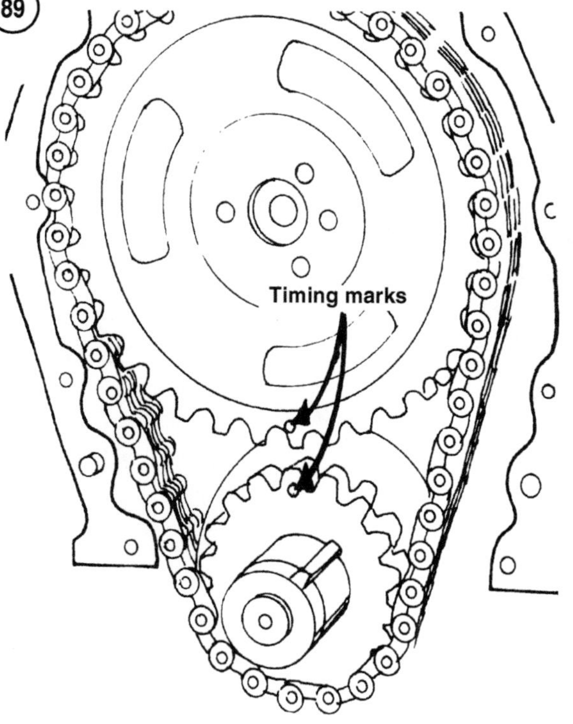

90

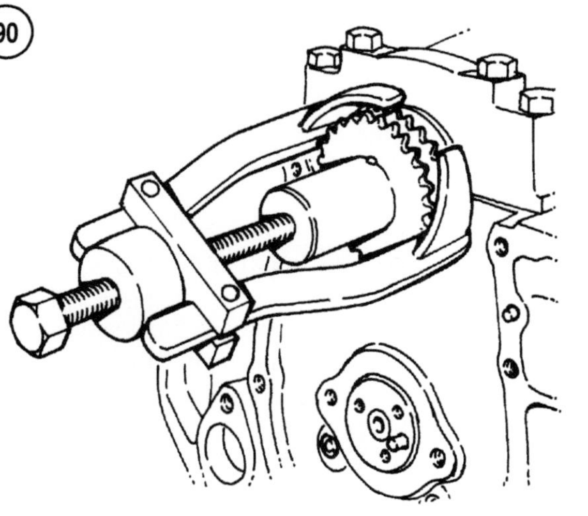

7. Remove the bolts or nuts. Then carefully pull the camshaft sprocket and timing chain from the engine.

8. Use a two-jaw puller to remove the crankshaft sprocket (**Figure 90**).

9. Clean the chain and sprockets with solvent and dry with compressed air.

10. Inspect the timing chain and both sprockets for wear, cracks or damage. Replace all three components if any defects are evident.

11. Inspect the drive key in the crankshaft for corrosion, bending or other damage and replace as needed.

12. Install the crankshaft sprocket as follows:

a. Align the key slot in the crankshaft sprocket with the crankshaft key.
b. Using a suitable driver, tap the crankshaft sprocket onto the crankshaft until it seats against the shoulder on the crankshaft.

13. Align the locating pin and fit the camshaft sprocket without the chain onto the camshaft. Do not install the bolts at this time. Rotate the camshaft and crankshaft sprockets as needed to align the timing marks (**Figure 89**). Remove the camshaft sprocket. Do not inadvertently rotate the camshaft or crankshaft.

14. Fit the timing chain onto the camshaft sprocket and crankshaft sprocket while installing the camshaft sprocket onto the camshaft. The timing marks must align as shown in **Figure 89**. Do not drive the camshaft sprocket onto the camshaft, which can dislodge the plug at the rear of the motor.

15A. On GM V-6 and V-8 models, thread the fasteners into the sprocket and camshaft. Tighten the bolts or nuts as follows:

a. On GM 4.3L models, tighten the bolts or nuts to 21 ft.-lb. (28 N•m).
b. On GM 5.0L and 5.7L models, tighten the bolts to 18 ft.-lb. (24 N•m).
c. On GM 7.4L models, tighten the bolts to 20 ft.-lb. (27 N•m).
d. On GM 8.2L models, tighten the bolts to 25 ft.-lb. (34 N•m).

15B. On Ford 5.0L and 5.8L models, fit the fuel pump drive eccentric and washer onto the camshaft sprocket. Thread the bolt into the washer, eccentric and sprocket. Tighten the eccentric bolt to 40-45 ft.-lb. (54-61 N•m).

16. Lubricate the timing chain and sprockets with engine oil.

17. Install the timing chain cover as described in this chapter.

18. Install the rocker arms and pushrods as described in this chapter.

19. Tighten the rocker arm nuts and/or adjust the valves as described in this chapter.

BALANCE SHAFT

A balance shaft is used on 4.3L models. The balance shaft can be replaced with complete disassembly of the engine.

Removal and Installation

1. Remove the timing chain and sprockets as described in this chapter.

2. Remove the intake manifold as described in Chapter Seven.

3. Wedge a piece of wood between the drive gear (16, **Figure 91**) and the driven gear (10). Remove the Torx bolt (11,

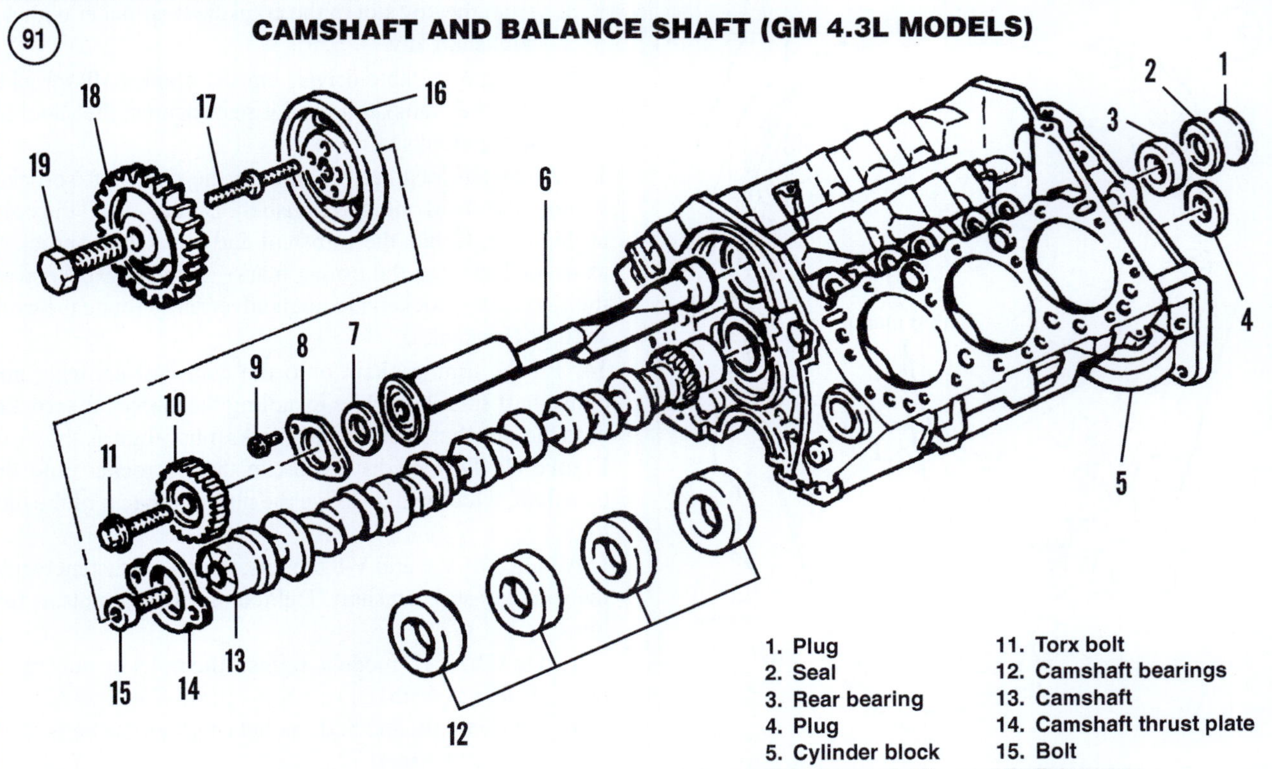

1. Plug
2. Seal
3. Rear bearing
4. Plug
5. Cylinder block
6. Balance shaft
7. Bearing
8. Thrust plate
9. Torx bolt
10. Driven gear
11. Torx bolt
12. Camshaft bearings
13. Camshaft
14. Camshaft thrust plate
15. Bolt
16. Drive gear
17. Stud bolt
18. Camshaft sprocket
19. Bolt

Figure 91) from the driven gear. Carefully pull the driven gear off the balance shaft.

NOTE

On earlier production engines, stud bolts are used to retain the balance shaft drive gear on the camshaft. Later production engines do not use stud bolts. On later engines, the bolts that retain the camshaft sprocket also retain the drive gear on the camshaft.

4. If so equipped, remove the stud bolt (17, **Figure 91**) from the drive gear. Later models do not use stud bolts.
5. Carefully pull the drive gear off the camshaft.
6. Remove the Torx bolts (9, **Figure 91**). Remove the thrust plate (8, **Figure 91**).
7. Insert a pry bar between the cylinder block and a solid section of the balance shaft (6, **Figure 91**). Carefully pry the balance shaft forward and remove it from the cylinder block.
8. Inspect the balance shaft bearings (3 and 7, **Figure 91**) for excessive wear, discoloration or roughness. If any defects are evident, have a reputable machine shop replace the bearings.
9. Clean the balance shaft with solvent and dry with compressed air. Inspect the rear bearing surface of the shaft for wear, discoloration or roughness. Replace the balance shaft and bearings if any defects are evident.
10. Inspect the gears for missing or damaged teeth, excessive wear and discoloration. Replace both gears if these or other defects are evident.
11. Lubricate the balance shaft and bearings with engine oil. Align the rear of the balance shaft with the bearing bore while guiding the balance shaft into the cylinder block. With a rubber mallet, carefully drive the balance shaft into the cylinder block until the front bearing (7, **Figure 91**) seats in the bore.
12. Fit the balance shaft thrust plate (8, **Figure 91**) over the balance shaft and seat against the cylinder block. Thread the two Torx bolts (9, **Figure 91**) into the thrust plate and cylinder block. Tighten the bolts evenly to 124 in.-lb. (14 N•m).
13. Fit the driven gear (10, **Figure 91**) onto the balance shaft. Rotate the gear as necessary to align the locating tang. When aligned, the gear seats against the shaft. Apply Loctite 271 to the threads and then thread the Torx bolt (11, **Figure 91**) into the gear and shaft. Do not tighten the bolt at this time.
14. Fit the drive gear (16, **Figure 91**) onto the camshaft. The timing marks must align as shown in **Figure 92**. Remove the drive gear and rotate the camshaft or balance shaft to achieve proper alignment.

92

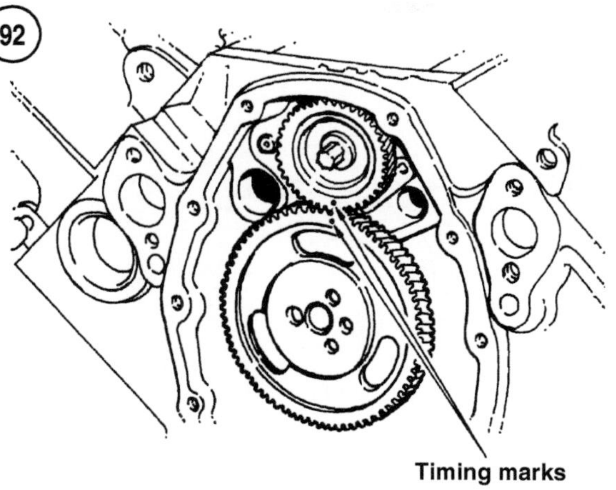

93 **MEASURING TIMING GEAR RUNOUT (GM 3.0L MODELS)**

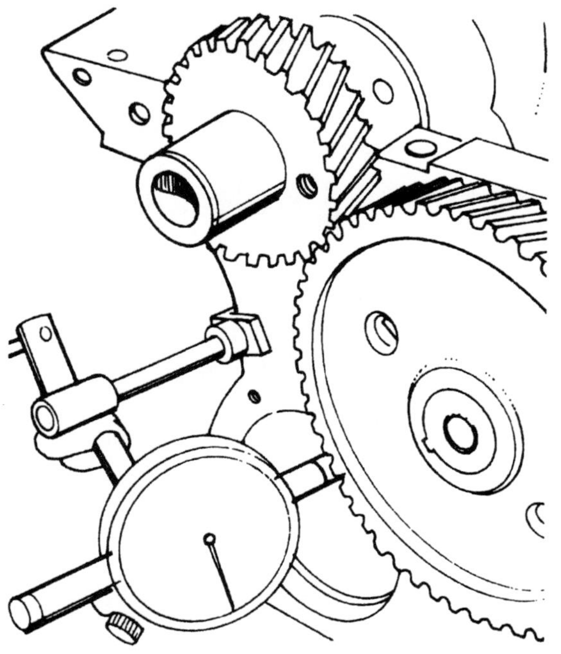

15. If so equipped, thread the stud bolts into the driven gear and camshaft. Wedge a piece of wood between the drive and driven gears. Tighten the bolts to 21 ft.-lb. (28 N•m).

16. Tighten the balance shaft Torx bolt (9, **Figure 91**) to 177 in.-lb. (20 N•m). Use an angle torque gauge to tighten the bolt an additional 35°.

17. Remove the wooden wedge. Clear any wood chips from the gears with compressed air. Lubricate the gear teeth with engine oil.

18. Install the timing chain, sprockets, timing chain cover and oil pan as described in this chapter.

19. Install the intake manifold as described in Chapter Seven.

CAMSHAFT AND LIFTERS

Always replace the lifters if replacing the camshaft. Operating the engine with used lifters will damage a new camshaft. If reusing lifters, always install them in the original locations. Otherwise, the camshaft and lifters will wear excessively and fail.

On GM models, apply General Motors Cam and Lifter Prelube to the camshaft lobes and lifters prior to installation. On Ford models, apply Ford Multipurpose Grease (part No. DOAZ-19584-AA) to the camshaft lobes and bottom of the lifters prior to installation.

Repair of defective camshaft bearings should only be attempted by a mechanic with the proper equipment and experience. The bearings and cylinder block are easily damaged without the proper equipment, and the installation equipment is more expensive than having the bearings professionally replaced. Improper bearing installation can also result in camshaft failure.

6

NOTE

With the manifolds, fuel system, cooling system and electrical components installed, the General Motors (GM) 5.0L and 5.7L engines are very similar in appearance to the Ford 5.0L and 5.8L engines. If the model identification decal on the flame arrestor is missing, the engine can be quickly identified as a GM or Ford engine by noting the distributor location. On GM engines, the distributor is located at the rear (flywheel end) of the engine. On Ford engines, the distributor is located on the front of the engine.

3.0L Models

Removal

1. Remove the oil pan and timing gear cover as described in this chapter. Remove the distributor as described in Chapter Nine.
2. Remove the lifter cover, pushrods and rocker arms as described in this chapter.
3. Remove the fuel pump as described in Chapter Seven.
4. Measure the timing gear runout prior to removing the camshaft. Instructions follow:
   a. Clamp a dial indicator onto the cylinder block.
   b. Position the plunger perpendicular to the front surface of the timing gear (**Figure 93**). The plunger must touch the gear as close to the teeth as possible. Zero the dial indicator.
   c. Observe the dial indicator while rotating the crankshaft. The needle movement indicates the timing gear runout. Replace the timing gear if the runout exceeds 0.004 in. (0.102 mm).
5. Measure the timing gear backlash prior to removing the camshaft. Instructions follow:
   a. Clamp a dial indicator onto the cylinder block.

94 **MEASURING TIMING GEAR BACKLASH (GM 3.0L MODELS)**

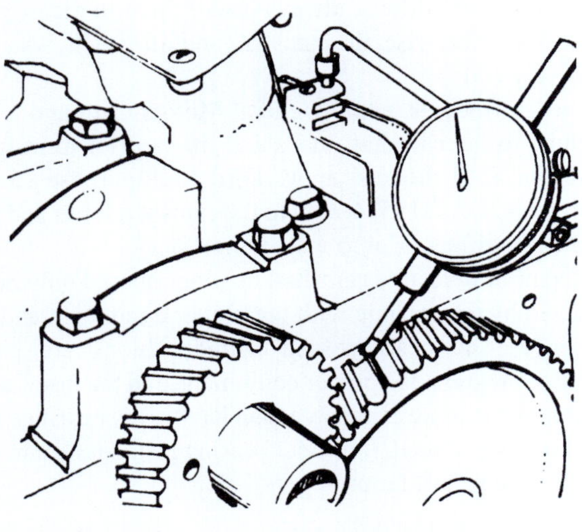

b. Position the plunger in contact with one of the timing gear teeth (**Figure 94**). Zero the dial indicator.

c. Observe the dial indicator while gently rotating the camshaft timing gear clockwise and counterclockwise. Do not drive the crankshaft gear. The needle movement indicates the timing gear backlash.

d. Replace the timing gear and crankshaft gear if the backlash is greater than 0.006 in. (0.152 mm) or less than 0.004 in. (0.102 mm).

6. Remove the lifters from the cylinder block and note the original locations. Used lifters must be installed in the original locations, or the lifters and camshaft lobes will wear excessively and fail.

7. Rotate the crankshaft gear until the timing marks align as shown in **Figure 95**. Remove the thrust plate bolts through the access holes in the timing gear (**Figure 95**).

8. Carefully pull the camshaft and gear assembly from the cylinder block. Support the camshaft during removal. Otherwise, the camshaft bearings might be damaged as the lobes pass through them.

Inspection

1. Inspect the camshaft thrust plate and corresponding surfaces on the camshaft for uneven wear, roughness or discoloration. Replace the thrust plate and/or camshaft and lifters if necessary.

2. Use feeler gauges and measure the gap between the thrust plate and the camshaft surface (**Figure 96**). The gap indicates the amount of camshaft end play necessary for maintaining a film of lubricant on the thrust surfaces. Replace the timing gear and thrust plate if the end play measurement is less than 0.003 in. (0.076 mm.) or greater than 0.008 in. (0.203 mm).

95

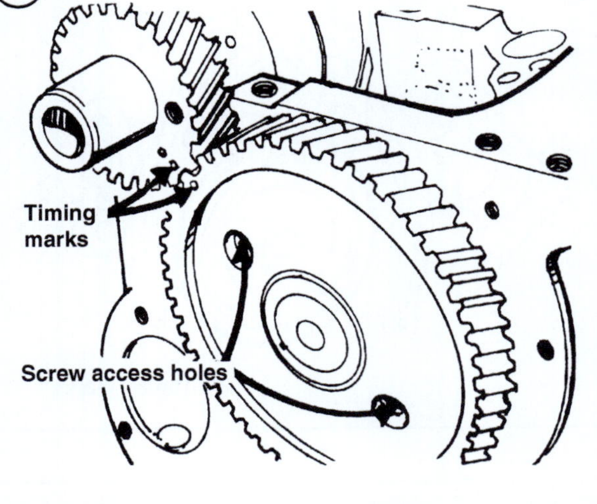

96 **MEASURING CAMSHAFT END PLAY (GM 3.0L MODELS)**

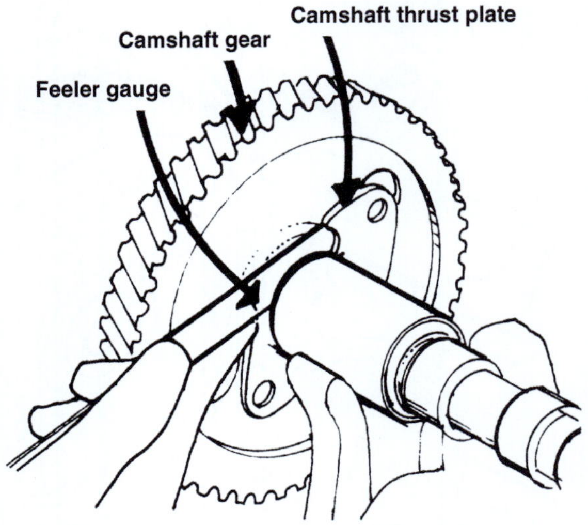

3. Inspect the timing gear for excess wear or roughness. Inspect the gear teeth for missing sections or uneven wear. Replace the timing gear and wear plate if any defects are evident. Replacement instructions follow:

a. Place the camshaft on a press with the gear resting on the table and the camshaft in the table opening.

b. Support the camshaft. Press the camshaft from the gear using a suitable mandrel. The mandrel must contact only the end of the camshaft.

c. Remove the thrust plate from the camshaft or timing gear.

d. Place the new gear on the table of the press with the timing mark facing downward.

e. Align the new thrust plate with the camshaft bore of the gear. Align the key slot in the gear with the key in

97

98

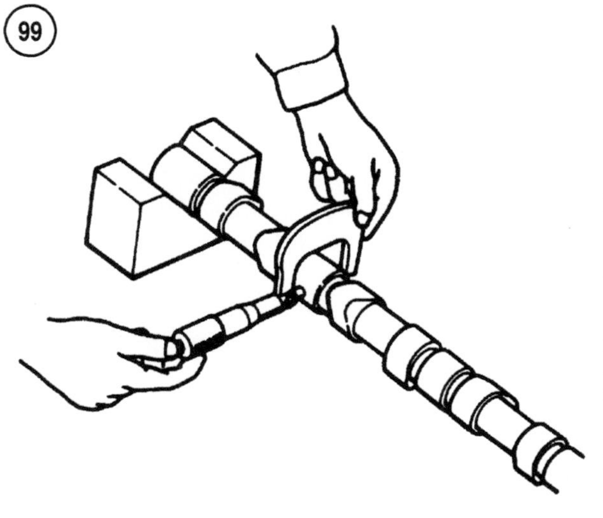
99

the camshaft. Apply engine oil to the gear bore. Insert the camshaft into the gear bore.

f. Press the camshaft into the gear until fully seated. Make sure the thrust plate remains centered over the gear bore during the procedure.

4. Inspect the camshaft bearings for rough surfaces, uneven wear or discoloration. Have the bearings replaced by a reputable machine shop if these or other defects are evident.

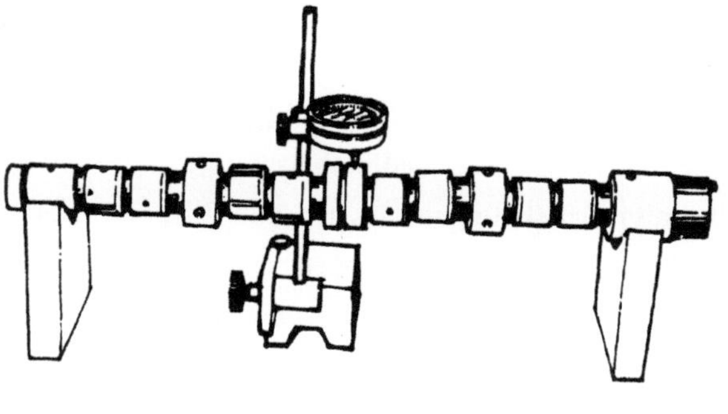
100 MEASURING CAMSHAFT RUNOUT

5. Inspect the lifters for wear on the lobe contact surfaces (**Figure 97**) and scuffing on the sides or other defects. Replace all lifters if any of them are defective.

6. Inspect the camshaft lobes for wear or rounded surfaces (**Figure 98**). Replace the camshaft, lifters and bearings if any of the lobes are worn excessively or damaged.

7. Measure each of the camshaft bearing journals at several locations around the diameter (**Figure 99**). Record all measurements. The difference between the largest and smallest measurement is the amount the journal is out of round. Replace the camshaft, lifters and bearings if any of the journals has a diameter less than 1.868 in. (47.440 mm) or if any journal is out of round more than 0.001 in. (0.025 mm).

8. Support the camshaft on V-blocks as shown in **Figure 100**. Set the plunger of a dial indicator so it touches one of the middle camshaft journals. Observe the dial indicator while slowly rotating the camshaft. The needle movement indicates the amount of camshaft journal runout. Replace the camshaft, lifters and bearings if the runout exceeds 0.002 in. (0.051 mm).

9. Inspect the distributor/oil pump drive gear on the camshaft for missing teeth, uneven wear, discoloration and excessive wear. Replace the camshaft if these or other defects are evident.

Installation

1. Apply engine oil to the camshaft bearings. Apply General Motors Cam and Lifter Prelube to the camshaft lobes and bottom of the lifters.

2. Carefully guide the camshaft into the cylinder block and bearings. Support the camshaft to prevent damaged bearings. Align the timing marks (**Figure 95**) and gear teeth. Then seat the camshaft into the cylinder block.

3. Align the thrust plate with the access holes in the gear (**Figure 95**). Install the thrust plate bolts and tighten evenly to 80 in.-lb. (9 N•m).

4. Apply General Motors Cam and Lifter Prelube to the wear surfaces. Then install the lifters into the cylinder block. If reused, the lifters must be installed in the original locations. Seat the lifters against the camshaft lobes.

6

5. Install the pushrods and rocker arms as described in this chapter.

6. Install a new gasket onto the lifter cover. Install the cover onto the cylinder block as described in this chapter.

7. Install the distributor as described in Chapter Nine.

8. Install the timing gear cover, oil pan, pulley hub and harmonic balancer as described in this chapter.

9. Install the fuel pump as described in Chapter Seven.

10. Adjust the valves as described in this chapter.

11. Install the rocker arm cover as described in this chapter.

12. Connect the battery cables. Adjust the ignition timing as described in Chapter Five.

GM 4.3L, 5.0L, 5.7L, 7.4L and 8.2L Models

Removal

1. Remove the oil pan and timing chain cover as described in this chapter.

2. Remove the intake manifold as described in Chapter Seven.

3. On models with a mechanical fuel pump, remove the pump and driver rod as described in Chapter Seven.

4. Remove the pushrods and rocker arms as described in this chapter.

5. Remove the timing chain and sprockets as described in this chapter.

6. On 4.3L models, remove the balance shaft as described in this chapter.

7A. On earlier models with flat lifters (**Figure 97**), use a lifter removal tool to remove the lifters from the cylinder block and note the original locations. Used lifters must be installed in the original locations, or the lifters and camshaft lobes will wear excessively and fail.

7B. On later models with roller lifters, scribe match marks on the roller lifters, guides and retainers (**Figure 101**) to make sure the roller turns in the original direction after installation. Remove the bolts and carefully lift the lifter guides and retainers (**Figure 102**, typical) from the cylinder block. Keep all components arranged in a manner consistent with the original mounting locations. Remove the lifters from the cylinder block. Note the original placement of each lifter in the cylinder block. Used lifters must be reinstalled in the original locations, or the lifters and camshaft will wear excessively and fail.

8. On models with roller lifters, remove the two bolts and then remove the thrust plate/retainer from the cylinder block.

9. Thread two 5/16 × 18 bolts into the camshaft bolt openings. Carefully pull the camshaft from the cylinder block. Support the camshaft during removal to prevent damage to the camshaft bearing surfaces.

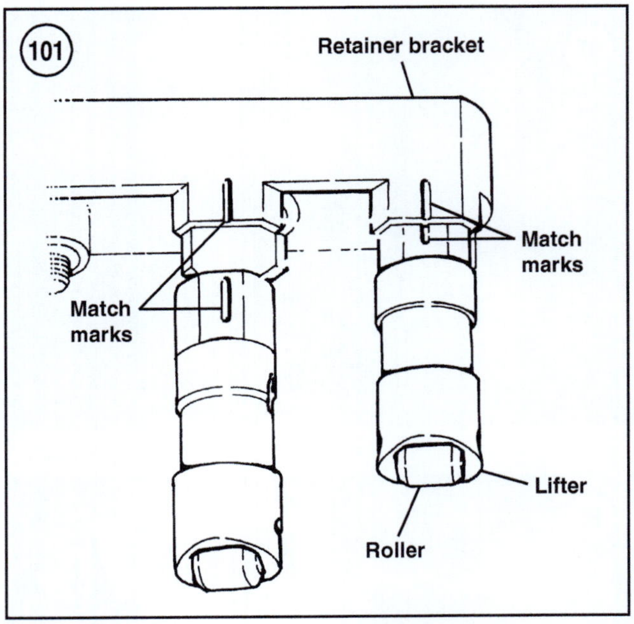

Inspection

1. Inspect the camshaft bearings in the cylinder block for rough surfaces, uneven wear or discoloration. Have the bearings replaced by a reputable machine shop if these or other defects are evident.

2A. On models with flat lifters, inspect the lifters for wear on the lobe contact surfaces and scuffing on the sides or other defects. Replace all lifters if any of them are defective.

2B. On models with roller lifters, inspect the lifters for discoloration, loose fitting rollers, scuffing on the sides, or other defects. Replace all of the lifters if any of them are defective.

3. Inspect the camshaft lobes for excess wear or rounded surfaces (**Figure 98**). Replace the camshaft, lifters and bearings if any of the lobes are worn excessively or damaged.

4. On models with roller lifters, inspect the lifter guides and retainers for excessive wear or damage. Replace any defective components.

5. Inspect the lifter bores in the cylinder block for excessive wear, scuffing or discoloration. Replace the cylinder block if any imperfections in the bore cannot be dressed by polishing with crocus cloth.

6. Measure each of the camshaft bearing journals at several locations around the diameter (**Figure 99**). Record all measurements. The difference between the largest and smallest measurement is the amount the journal is out of round. Replace the camshaft, lifters and bearings if any of the journals has a diameter less than the minimum specification in **Table 2** or if any journal is out of round more than 0.001 in. (0.025 mm).

7. Support the camshaft on V-blocks as shown in **Figure 100**. Set the plunger of a dial indicator so it touches one of the middle camshaft journals. Observe the dial indicator

102

while slowly rotating the camshaft. The needle movement indicates the amount of camshaft journal runout. Replace the camshaft, lifters and bearings if the runout exceeds 0.002 in. (0.051 mm).

8. Inspect the distributor/oil pump drive gear on the camshaft for missing sections, uneven wear, discoloration and excessive wear. Replace the camshaft if these or other defects are evident on the gear.

9. On models with roller lifters, inspect the thrust plate for uneven wear, rough surfaces or discoloration. Replace the thrust plate if these or other defects are evident.

Installation

1. Apply engine oil to the camshaft bearings. Apply General Motors Cam and Lifter Prelube to the camshaft lobes and bottom of the lifters (flat lifter) or lifter roller (roller lifters).

2. Carefully guide the camshaft into the bearings. Support the camshaft during installation to prevent damaged bearings.

3. On models with roller lifters, install the camshaft thrust plate and bolts. If used, make sure the thrust plate is installed with the same side facing the cylinder block as when removed. If necessary, inspect the thrust plate for wear polishing from the camshaft surfaces. Install the thrust plate bolts and tighten evenly to the following specification:

   a. On 4.3L models, tighten the thrust plate bolts to 124 in.-lb. (14 N•m).
   b. On 5.0L and 5.7L models, tighten the thrust plate bolts to 106 in.-lb. (12 N•m)
   c. On 7.4L and 8.2L models, tighten the thrust plate bolts to 120 in.-lb. (14 N•m).

4A. On earlier models with flat lifters, install the lifters into the original locations. Seat the lifters against the camshaft lobes.

4B. On later models with roller lifters, install the lifters into the respective locations in the cylinder block. Seat the lifter rollers against the camshaft lobes. Align the match marks and install the guides and lifter retainers. Tighten the retainer bolts to the following specification:

   a. On 4.3L models, tighten the lifter retainer bolts to 12 ft.-lb. (16 N•m).
   b. On 5.0L and 5.7L models, tighten the lifter retainer bolts to 18 ft.-lb. (24 N•m).
   c. On 7.4L and 8.2L models, tighten the lifter retainer bolts to 19 ft.-lb. (26 N•m).

5. On 4.3L models, install the balance shaft as described in this chapter.

6. Install the timing chain and sprockets as described in this chapter.

7. Install the pushrods and rocker arm as described in this chapter.

8. Install the timing chain cover, oil pan and harmonic balancer as described in this chapter.

9. Tighten the rocker arm nuts/bolts or adjust the valves as described in this chapter.

10. Install the intake manifold as described in Chapter Seven.

11. On models with a mechanical fuel pump, install the driver rod and fuel pump as described in Chapter Seven.

12. Install the rocker arm covers as described in this chapter.

13. Connect the battery cables.

14. Adjust the ignition timing as described in Chapter Five.

Ford 5.0L and 5.8L Models

Removal

1. On models with a mechanical fuel pump, remove the pump and driver rod as described in Chapter Seven.

2. Remove the oil pan and timing chain cover as described in this chapter.

3. Remove the intake manifold as described in Chapter Seven.

4. Remove the pushrods and rocker arms as described in this chapter.

5. Measure the camshaft end play as follows:

   a. Attach a suitable mount and dial indicator to the front of the engine as shown in **Figure 103**. Adjust the mount to position the stem of the dial indicator in contact with the eccentric bolt head.
   b. Push the camshaft sprocket toward the rear of the engine as far as possible with hand pressure. Hold the sprocket in this position and zero the dial indicator.
   c. Note the dial indicator movement while gently prying the camshaft sprocket forward using a suitable tool. *Do not* use excessive force. Use just enough force to register the end play. The end play displayed on the dial indicator must be 0.001-0.009 in. (0.025-0.229 mm). Replace the thrust plate behind the camshaft sprocket if the end play exceeds the specification.
   d. Remove the dial indicator and mount.

6. Remove the timing chain and sprockets as described in this chapter.

7. Use a lifter removal tool to remove the lifters from the cylinder block and note the original locations. Used lifters must be reinstalled in the original locations, or the lifters and camshaft lobes will wear excessively and fail.
8. Remove the two bolts and then remove the thrust plate/retainer from the cylinder block.
9. Carefully pull the camshaft from the cylinder block. Support the camshaft during removal to prevent damage to the camshaft bearing surfaces.

Inspection

1. Inspect the camshaft bearings in the cylinder block for roughness, uneven wear or discoloration. Have the bearings replaced at a reputable machine shop if these or other defects are evident.
2. Inspect the lifters for wear on the lobe contact surfaces, scuffing on the sides and other defects. Replace all lifters if any of them are defective.
3. Inspect the camshaft lobes for excessive wear or rounded surfaces (**Figure 98**). Replace the camshaft, lifters and bearings if any of the lobes are worn excessively or damaged.
4. Inspect the lifter bores in the cylinder block for excessive wear, scuffing or discoloration. Replace the cylinder block if any imperfections in the bore cannot be dressed by polishing with crocus cloth.
5. Measure each of the camshaft bearing journals at several locations around the diameter (**Figure 99**). Record all measurements. The difference between the largest and smallest measurement is the amount the journal is out of round. Replace the camshaft, lifters and bearings if any of the journals has a diameter less than the minimum specification in **Table 2** or if any journal is out of round more than 0.005 in. (0.127 mm).
6. Support the camshaft on V-blocks as shown in **Figure 100**. Set the plunger of a dial indicator so it touches one of the middle camshaft journals. Observe the dial indicator while slowly rotating the camshaft. The needle movement indicates the amount of camshaft journal runout. Replace the camshaft, lifters and bearings if the runout exceeds 0.002 in. (0.051 mm).
7. Inspect the distributor/oil pump drive gear on the camshaft for a missing section, uneven wear, discoloration and excessive wear. Replace the camshaft if these or other defects are evident on the gear.
8. Inspect the thrust plate for uneven wear, rough surfaces or discoloration. Replace the thrust plate if the camshaft end play exceeds the specification given in **Table 2** or if these or other defects are evident.

Installation

1. Apply engine oil to the camshaft bearings. Apply Ford Multipurpose Grease (part No. DOAZ-19584-AA) onto the camshaft lobes and bottom of the lifters (flat lifter) or lifter roller (roller lifters).
2. Carefully guide the camshaft into the bearings. Support the camshaft during installation to prevent damaged bearings.
3. Install the camshaft thrust plate and bolts. If used, make sure the thrust plate is installed with the same side facing the cylinder block as when removed. If necessary, inspect the thrust plate for wear polishing from the camshaft surfaces. Install the thrust plate bolts and tighten evenly to the general tightening torque specification listed in Chapter One.
4. Install the lifters into the cylinder block. If reused, the lifters must be installed in the original locations. Seat the lifters against the camshaft lobes.
5. Install the timing chain and sprockets as described in this chapter.
6. Install the pushrods and rocker arm as described in this chapter.
7. Install the timing chain cover, oil pan and harmonic balancer as described in this chapter.
8. Check the valve clearance as described in this chapter.
9. Install the intake manifold as described in Chapter Seven.
10. On models with a mechanical fuel pump, install the driver rod and fuel pump as described in Chapter Seven.
11. Install the rocker arm covers as described in this chapter.
12. Connect the battery cables.

103

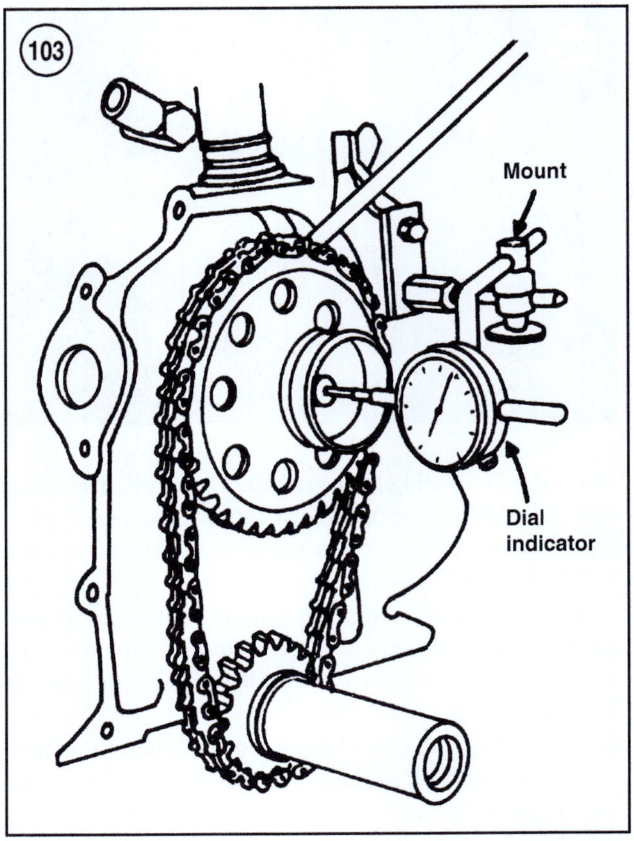

104

105

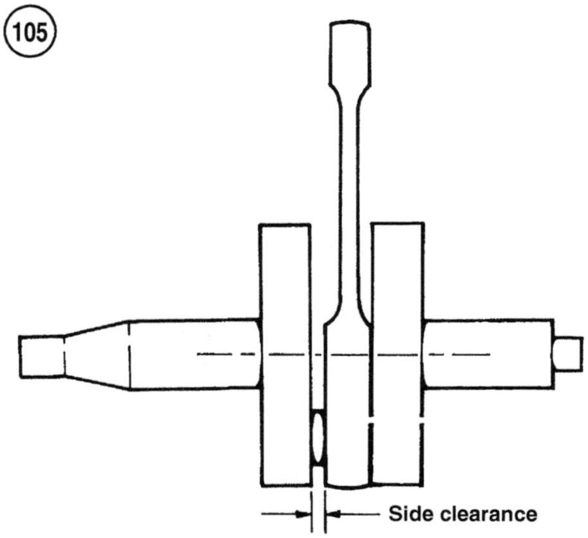

13. Adjust the ignition timing as described in Chapter Five.

PISTON AND CONNECTING ROD

Removal

CAUTION
To prevent engine damage caused by installing the piston and/or connecting rods into the wrong locations, always identify the cylinder number and mark all parts prior to complete disassembly. On GM 3.0L models, the cylinders are numbered 1-2-3-4 from front to rear. On GM 4.3L models, the cylinders are numbered from front to rear (1-3-5 on the port side and 2-4-6 on the starboard side. On GM 5.0L, 5.7L, 7.4L and 8.2L models, the cylinders are numbered from front to rear (1-3-5-7 on the port side and 2-4-6-8 on the starboard side. On Ford 5.0L and 5.8L models, the cylinders are numbered from front to rear (1-2-3-4 on the starboard side and 5-6-7-8 on the port side).

NOTE
With the manifolds, fuel system, cooling system and electrical components installed, the General Motors (GM) 5.0L and 5.7L engines are very similar in appearance to the Ford 5.0L and 5.8L engines. If the model identification decal on the flame arrestor is missing, the engine can be quickly identified as a GM or Ford engine by noting the distributor location. On GM engines, the distributor is located at the rear (flywheel end) of the engine. On Ford engines, the distributor is located on the front of the engine.

1. Remove the cylinder head(s), oil pan, oil pump and timing gear or chain cover as described in this chapter.
2. Mark the cylinder number on the side of the selected connecting rod and cap (**Figure 104**, typical) as follows:
   a. Rotate the crankshaft until one of the pistons is at the bottom of the stroke.
   b. On 3.0L models, mark the cylinder number on the port side of the connecting rod and cap. The port side is the side opposite the camshaft.
   c. On V-6 and V-8 models, mark the cylinder numbers on the port side of the rod for pistons on the port side of the engine and the starboard side for pistons on the starboard side of the engine. The numbers must face toward the same side of the block as the cylinder bore for the piston in which it is connected.
3. Measure the connecting rod side clearance as follows:
   a. On GM 3.0L models, move the connecting rod cap toward one side of the journal. Insert a feeler gauge between the opposite side of the connecting rod and the journal (**Figure 105**, typical). The clearance equals the thickness of the feeler gauge that passes between the rod and journal with slight drag.
   b. On GM 4.3L models, push the connecting rod toward the front of the engine. Insert a feeler gauge between the rear side of the connecting rod and the journal (**Figure 105**, typical). The clearance equals the thickness of the feeler gauge that passes between the rod and journal with slight drag.
   c. On GM and Ford V-8 models, push each connecting rod on the selected journal toward the crankshaft contact surfaces. Insert a feeler gauge between the two connecting rods (**Figure 106**). The clearance equals the thickness of feeler gauge that passes between the connecting rods with a slight drag.
   d. The connecting rod side clearance must be within the specification in **Table 4**. If the clearance exceeds the specification, replace the worn or damaged connecting rod(s) and/or crankshaft during engine assembly. A clearance less than the specification is generally caused by improper assembly that can be corrected during the assembly procedures.
   e. Repeat this step for the remaining connecting rods or pair of rods.

6

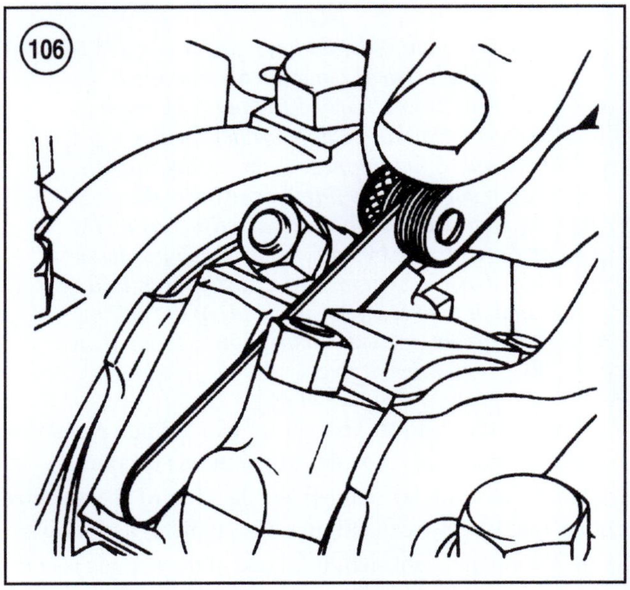

106

107

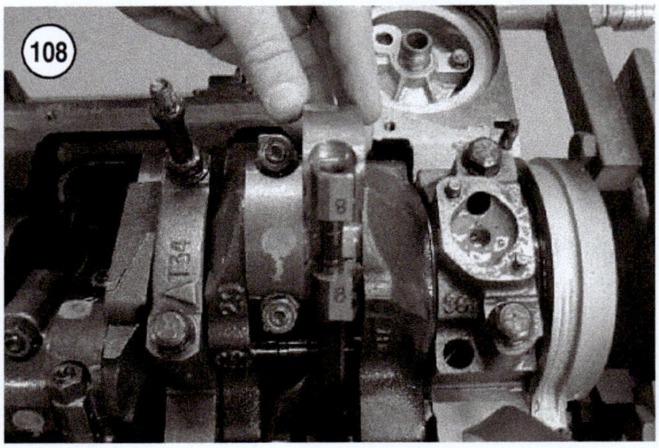

108

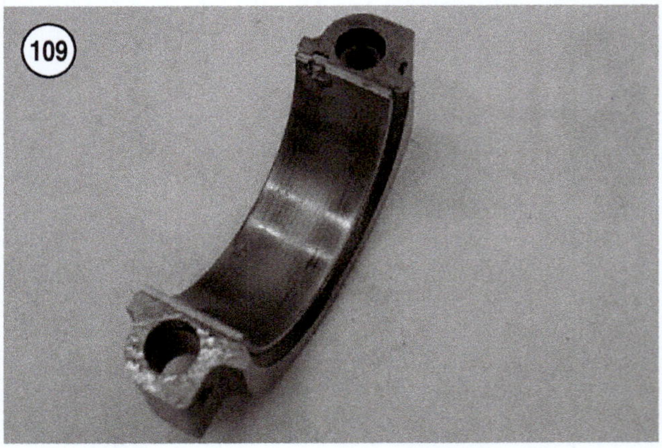

109

110

4. If present, remove the ridge from the top of the cylinder bore prior to removing the pistons. Even a small ridge due to cylinder bore wear and carbon buildup can make piston removal difficult and may damage the piston. Remove the ridge using a ridge reamer as follows:
   a. Rotate the crankshaft to position one of the pistons at the bottom of the stroke.
   b. Following the manufacturer's instructions, install the ridge reamer into the cylinder.
   c. Operate the tool to remove the ridge.
   d. Use a shop vacuum to remove all shavings from the cylinder.
   e. Repeat these steps for the remaining cylinders.
5. Evenly loosen and remove the connecting rod nuts (**Figure 107**, typical). Tap on the bottom of the connecting rod cap until it is free of the connecting rod (**Figure 108**). Reinstall the insert bearing (**Figure 109**) if it dislodges from the rod cap.
6. Using a wooden dowel, carefully push the piston and connecting rod from the cylinder block (**Figure 110**). Upon removal, install the rod cap onto the rod and secure with the nuts. The bearing tangs must align on the same side as shown in **Figure 111**.
7. Mark the cylinder number on the inner surface of the piston. Do not scratch the top or side of the piston. Remove and mark all pistons and connecting rods as described.
8. Remove the crankshaft and seal carrier as described in this chapter.
9. Use a piston ring expander (**Figure 112**) and remove the rings from the pistons. Use a broken ring to carefully scrape carbon from the ring grooves (**Figure 113**). Do not remove aluminum material from the ring groove.

Piston, Connecting Rod and Cylinder Bore Measurement and Inspection

1. Clean the pistons and rods with solvent. Use a brush with plastic bristles to remove carbon from the piston sur-

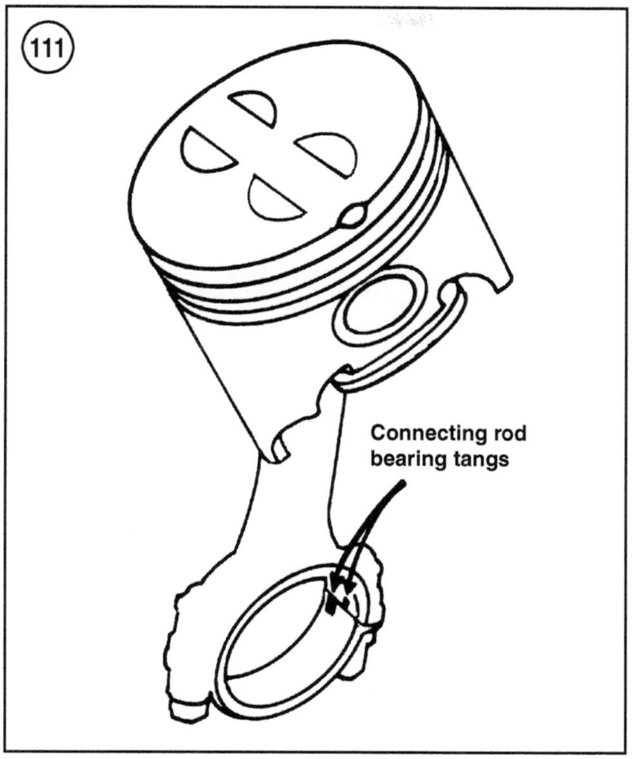

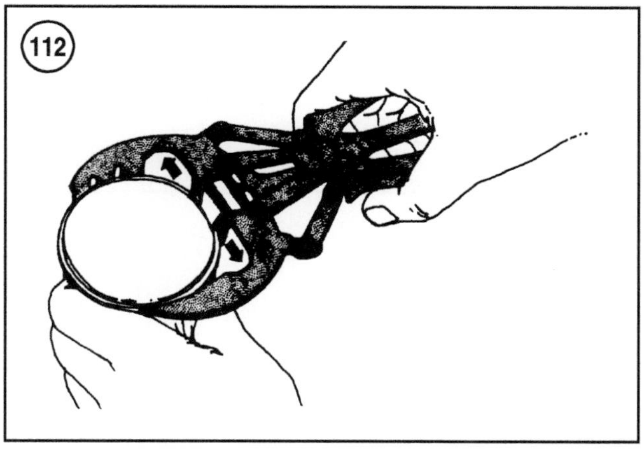

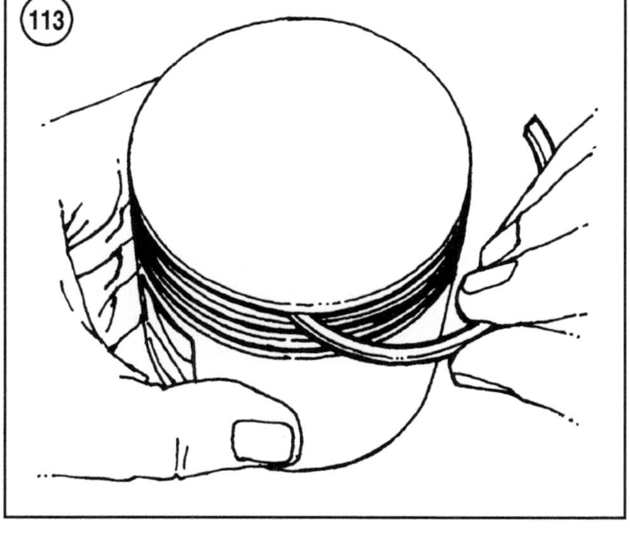

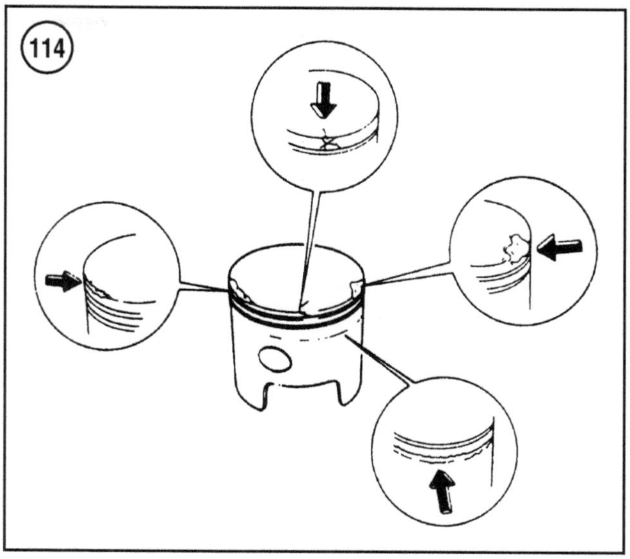

faces. Never use a wire brush because small pieces of wire can become embedded in the piston dome and cause preignition or detonation damage.

2. Inspect the edge of the piston crowns for cracks, erosion or missing sections (**Figure 114**). Inspect the sides of the pistons for scuffing (**Figure 115**), deep scratches or other damage.

3. Rock each piston on the connecting rod while checking for binding or a loose fit of the piston pin and piston. Replace the piston and pin if the pin fits loosely or if other defects are evident.

4. Inspect the cylinder bore for cracking, deep scratches, transferred piston material or corrosion pitting. Replace the cylinder block if cracks are visible. Have the cylinder bored to accept oversized pistons if it is deeply scratched or pitted. Remove transferred piston material with a blunt scraper.

NOTE

7.4L models are produced as both standard- and high-output models and must be identified before beginning the repair procedure. The 7.4 GL and 7.4GI models are lower-output models, and the 7.4 GS and 7.4GSI models are the high-output versions. If the

6

identification decal is missing from the engine, or the output specification is in question, inspect the intake manifold and cylinder head intake ports. The standard-output models use round or oval ports. The high-output models use rectangular ports.

5. Using an outside micrometer, measure the piston diameter at two points: aligned with the piston pin and at 90° to the pin (**Figure 116**). Record the measurement for each piston.
6. Insert the outer edge of a new upper compression ring into the upper ring groove as shown in **Figure 117**. Roll the ring around the circumference of the piston to check for binding. If binding is evident, the upper ring groove has collapsed, and the piston must be replaced. Repeat this step for the remaining pistons.
7. Check the piston and piston rings for proper fit as follows:
   a. Temporarily install new rings onto the pistons.
   b. Use feeler gauges and measure the clearance between the sides of the ring and the ring groove (**Figure 118**). The clearance equals the thickness of the feeler gauge that passes between the ring and groove with slight drag.
   c. Compare the clearance with the ring groove clearance specifications in **Table 5**. Replace the piston if the clearance is not within the specification.
   d. Remove the piston rings.
8. On GM 3.0L models, use feeler gauges to measure the width of the oil ring grooves. The groove width equals the thickness of the feeler gauge that can be inserted into the groove with slight drag. The oil ring groove width must be 0.188-0.189 in. (4.775-4.801 mm). If otherwise, replace the piston.
9. Measure the cylinder bore diameter, taper and out-of-round as follows:
   a. Use a cylinder bore gauge (**Figure 119**) to measure the cylinder bore diameter. Take the measurements at three or more evenly spaced depths in the bore and then again at 90° to the first measurement (**Figure 120**). Take the top measurement at a point even with the upper travel of the rings.
   b. Subtract the smallest diameter at the bottom of the bore from the largest diameter at the top of the bore. This calculation indicates the amount of cylinder bore taper. Record the cylinder bore taper calculation.
   c. Subtract the smallest diameter from the largest diameter measurement at each depth in the cylinder bore. This calculation indicates the amount the cylinder is out of round. Record the cylinder out-of-round calculation.
   d. Repeat this step for the remaining cylinders. Record all measurements.
10. Compare the cylinder diameter measurements, cylinder taper and out-of-round calculations to the specifications in **Table 6**. If any of the measurements exceed the specifications, have a reputable machine shop bore the cylinder to accept an oversized piston. It is not necessary to bore all of the cylinders because the oversized pistons weigh the same as standard pistons. Oversized piston availability varies by model. Always supply the oversized piston to the machine shop before boring the cylinder. This allows the machinist to precisely fit the piston to the cylinder bore.

11. Determine the piston-to-piston skirt clearance as follows:

116

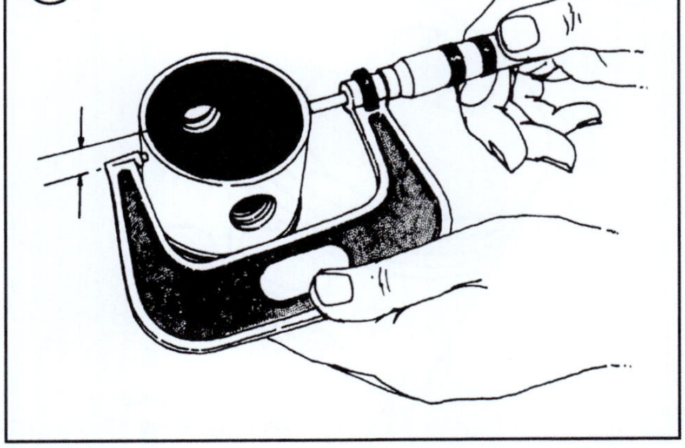

117

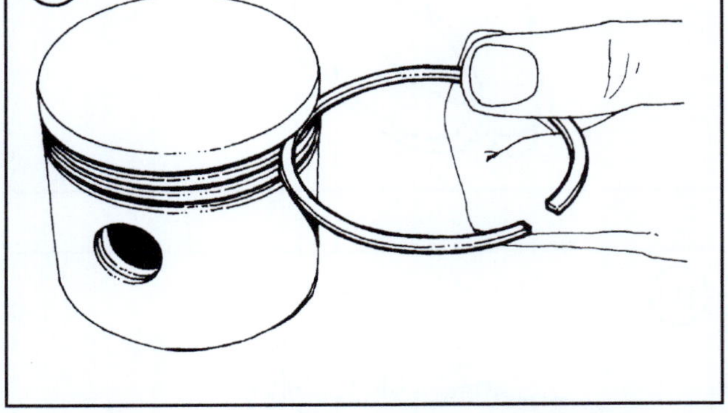

118

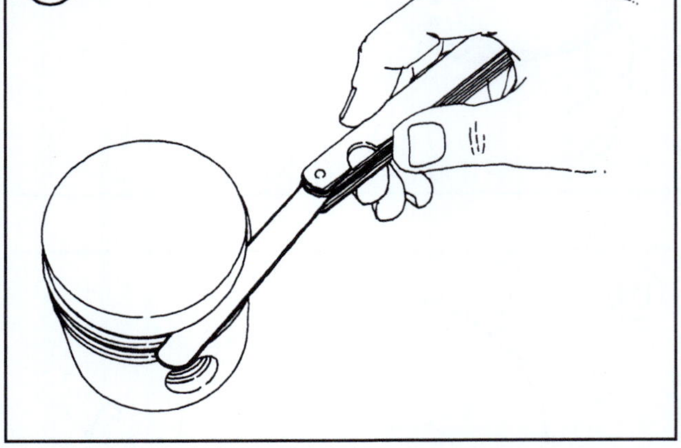

119

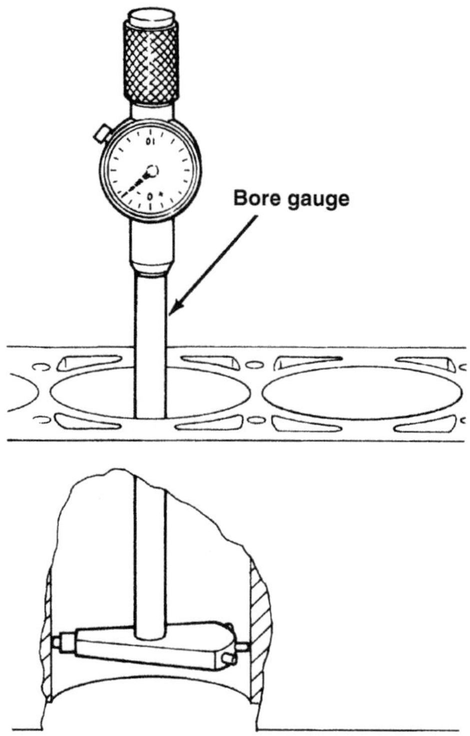

120

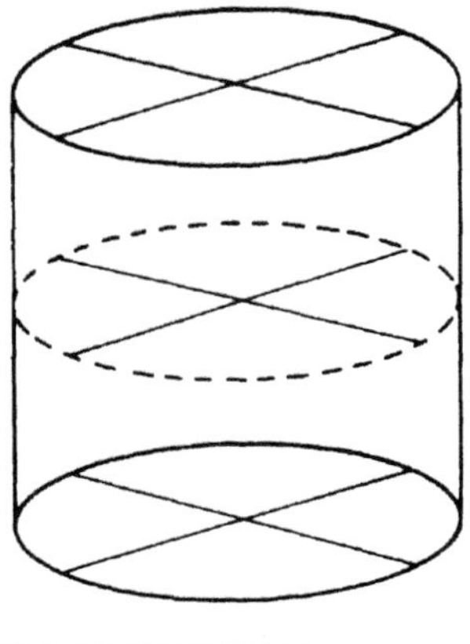

121

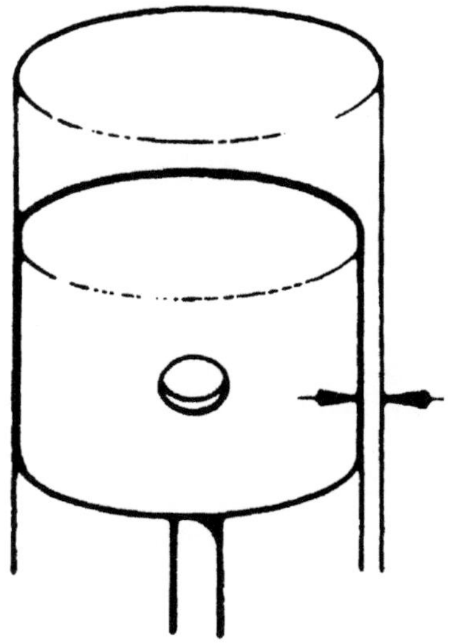

a. Subtract the piston skirt diameter, recorded in Step 5, from the smallest cylinder diameter to determine the minimum skirt clearance for the cylinder (**Figure 121**).

b. Subtract the piston diameter from the largest cylinder diameter to determine the maximum skirt clearance for the cylinder.

c. Compare the minimum and maximum clearances with the piston skirt clearance specification in **Table 5**. If the piston skirt clearance is below the specification, the piston diameter is too large for the bore, and/or the cylinder bore is at or below the minimum diameter. Repeat the measurements and calculation. Replace the piston or bore the cylinder to accept an oversized piston as necessary to achieve the required clearance. If the piston skirt clearance is above the specification, the piston diameter is too small for the cylinder bore, and/or the cylinder bore is at or above the maximum diameter. Repeat the measurements and calculations. Replace the piston or bore the cylinder to accept an oversized piston as necessary to achieve the required clearance.

d. Repeat this step for the remaining cylinders.

12. Hone the cylinders to achieve a 45-60° crosshatch pattern before installing the pistons. If you are unfamiliar with the honing operation or lack the proper equipment, have a reputable machine shop hone the cylinders. Thoroughly clean the cylinders with warm soapy water after the honing operation. Wipe a clean white shop towel through the bore to check for residual material. Any residual abrasive material from the honing process will cause rapid wear and improper seating of the piston rings. To prevent corrosion, coat the cylinder walls and other unpainted surfaces with engine oil.

13. Remove the insert bearings from the connecting rods. Inspect the rod surfaces for discoloration, corrosion pitting, roughness or other damage. Have a reputable machine shop remove the piston and replace the connecting rod if defective. Inspect the crankshaft for defects if discoloration, scratches, corrosion pitting or transferred material is found on the bearings.

14. Inspect the rod bolts and nuts for stretched or damaged threads. Replace the rod bolts and nuts unless they are in excellent condition.

15. Inspect the connecting rod for bending or twisting of the I-beam section. Have a reputable machine shop remove the piston and replace the rod if bent or twisted.

Piston and Connecting Rod Installation

1. Install the crankshaft, flywheel and drive plate as described in this chapter.

2. Measure the ring end gap as follows:

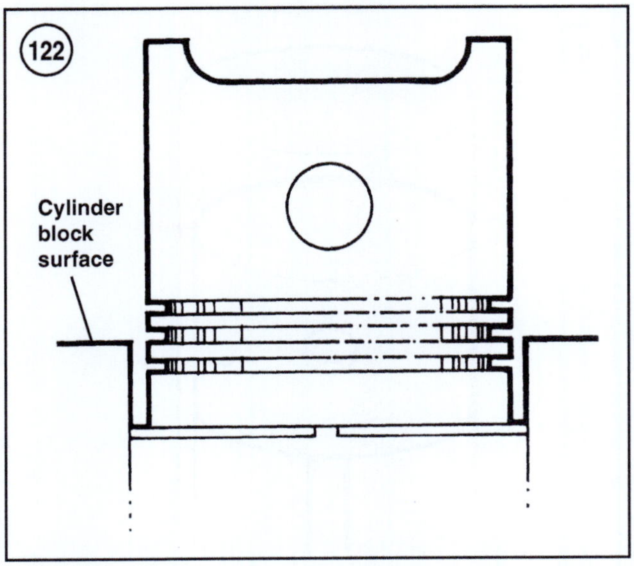

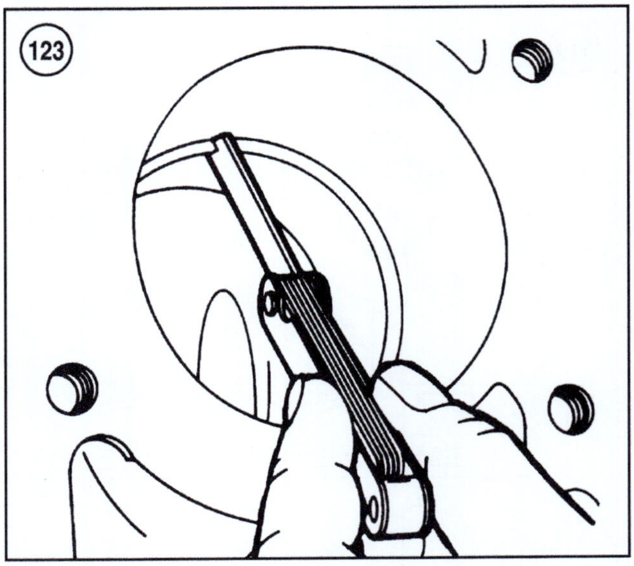

a. Insert one of the new compression (upper) rings into one of the cylinder bores. Use a piston without rings to push the ring into the bore to a depth near the lowest point of ring travel (**Figure 122**).
b. Measure the ring end gap (**Figure 123**) with a feeler gauge.
c. Compare the end gap with the specification in **Table 5**.
d. If the ring end gap is incorrect, remove the ring and try a different ring. Continue until a ring with the correct gap is located for the cylinder. If unable to find rings with the correct gap, check the cylinder bore diameter.
e. Measure and select the second compression ring and both oil ring rails as described. Attach a tag with the cylinder number to the selected rings.
f. Select rings for each cylinder.

3. Coat the cylinder bore with engine oil.
4. Use a ring expander (**Figure 112**) to install the rings into piston grooves as shown in **Figure 124**. The marks on the compression rings must face the top of the piston.
5. Make sure the piston is correctly attached to the connecting rod. If not, have a reputable machine shop remove the piston pin and reinstall the piston onto the connecting rod. Refer to the following:

a. On GM 3.0L models, the connecting rod bearing tangs (**Figure 111**) must face the port side of the engine or opposite the camshaft side when installed. The notch in the piston dome (**Figure 125**) must face toward the front of the engine when installed.
b. On GM 4.3L, 5.0L and 5.7L models, the connecting rod bearing tangs (**Figure 111**) must face the port side on odd-numbered cylinders and the starboard side on even-numbered cylinders. The notch in the piston dome (**Figure 125**) must face toward the front of the engine when installed.
c. On Ford 5.0L and 5.8L models, the side of the connecting rod with the cylinder number mark (**Figure 104**, typical) must face outward or toward the closest

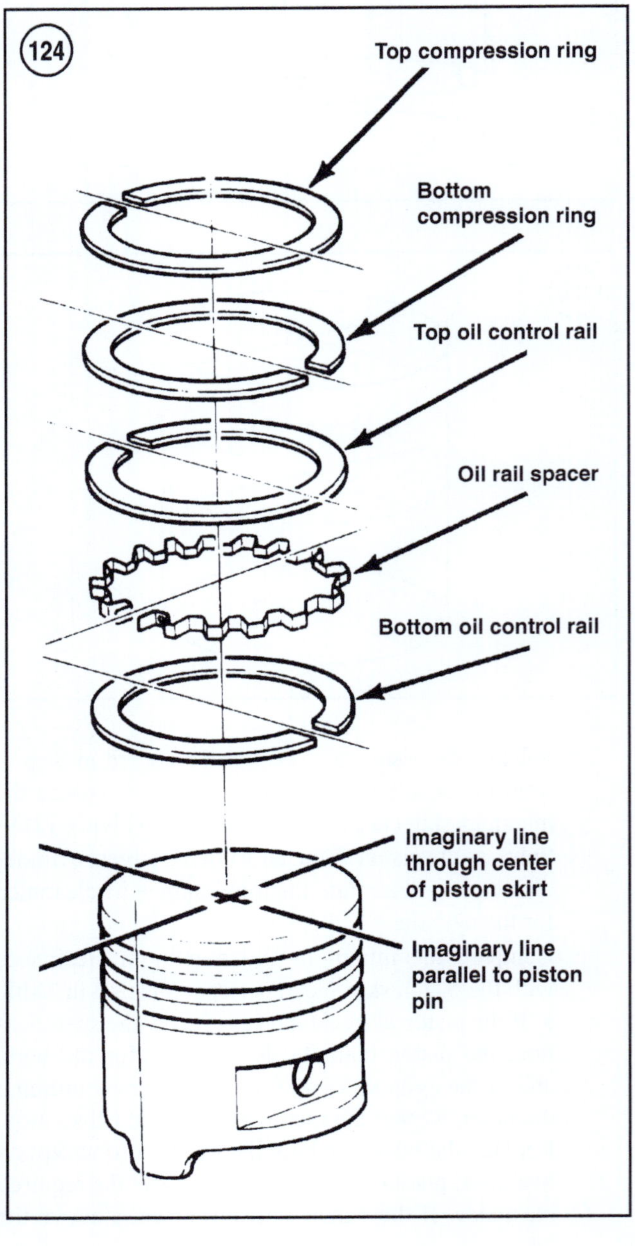

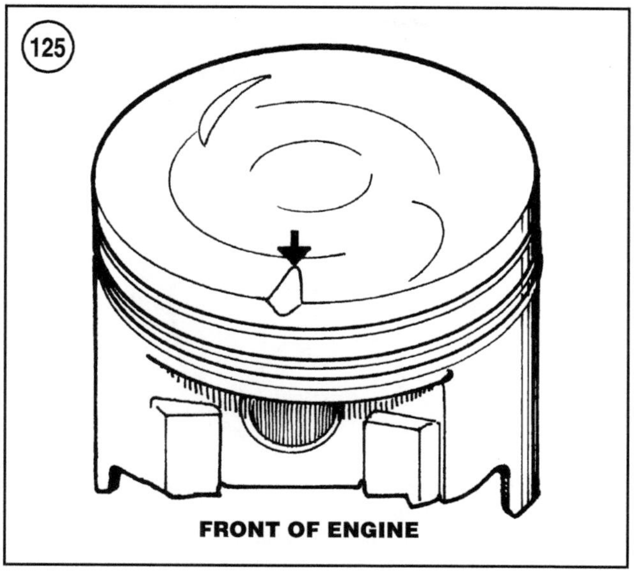

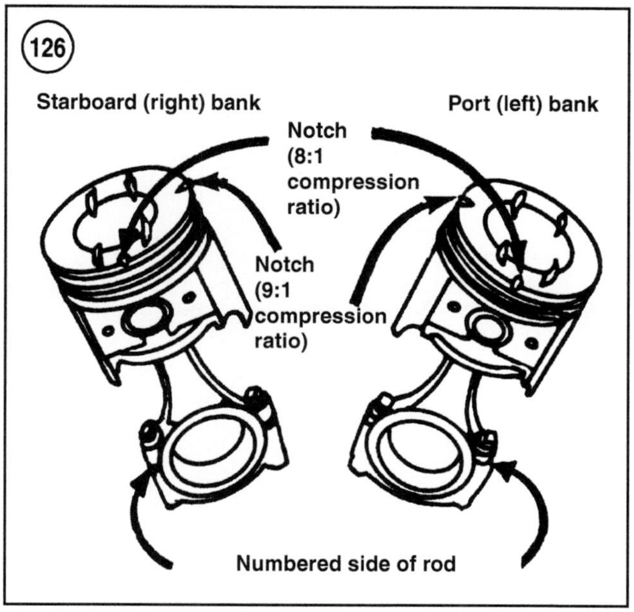

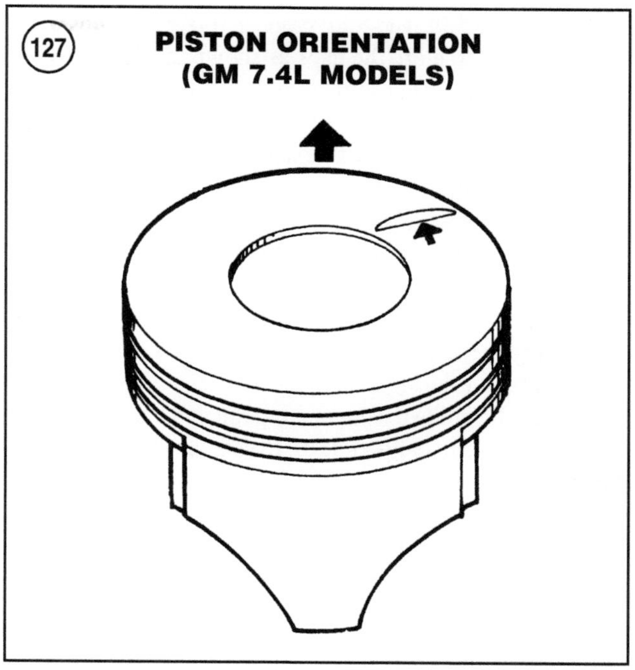

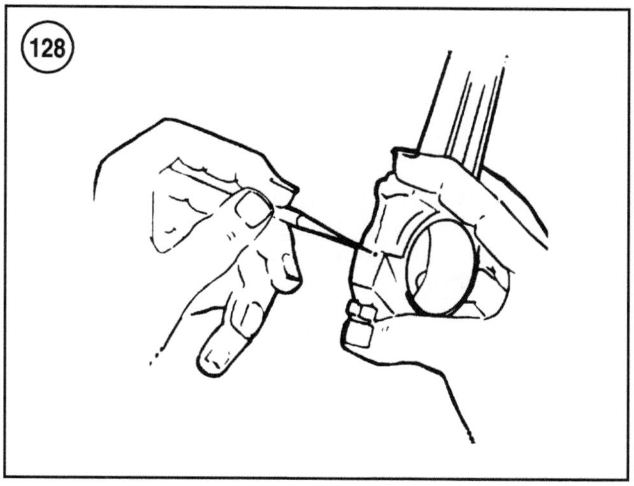

side of the cylinder block. On models with a notch cast into part of the piston dome that is aligned with the wrist pin bore (**Figure 126**), the notch must face toward the front of the engine. On models with a larger valve relief notch cast into the side of the piston dome perpendicular to the wrist pin bore (**Figure 126**), the notch must face toward the center of the engine when installed.

d. On GM 7.4L GL or GI models (standard output), the notch in the piston dome (**Figure 125**) must face toward the front of the engine. If the piston does not have a notch, the single valve relief depression in the piston dome (**Figure 127**) must face toward the center of the engine. If the piston does not have the notch or single valve relief depression, the small dimple in the piston dome must face toward the outside of the engine on cylinder Nos. 2, 4, 6 and 8. The dimple must face toward the center of the engine on cylinder Nos. 1, 3, 5 and 7. The hole in the connecting rod (**Figure 128**) must face toward the camshaft. The connecting rod bearing tangs (**Figure 111**) must face the port side on odd-numbered cylinders and the starboard side on even-numbered cylinders.

e. On 7.4L GS and GSI (high output) models and all 8.2L models, The larger valve relief depression in the piston dome must face toward the center of the engine. The hole in the connecting rod (**Figure 128**) must face toward the camshaft. The connecting rod bearing tangs (**Figure 111**) must face the port side on odd-numbered cylinders and the starboard side on even-numbered cylinders.

6. Align the piston ring gaps as follows:

   a. On all GM four-cylinder V-6 and V-8 models, rotate the rings in the grooves to align the gaps as shown in **Figure 129**.

b. On Ford 5.0L and 5.8L models, rotate the rings in the grooves to align the gaps as shown in **Figure 130**.

7. Remove the rod cap and bearings. Wipe all oil or debris from the connecting rod. To prevent the rod bolts from damaging the crankshaft, fit short pieces of rubber tubing over the rod bolts.

8. Rotate the crankshaft to position the selected cylinder crankpin at the bottom of the stroke.

9. Fit a ring compressor over the piston. Tighten the compressor until the rings fully compress into the grooves. Do not overtighten the compressor. The piston must slide from the compressor during installation.

10. Carefully guide the connecting rod and piston skirt into the cylinder block (**Figure 110**). Rotate the piston and position the notch, valve relief or reference hole as described in Step 5.

11. Hold the ring compressor tightly against the cylinder head mating surface. Use a wooden dowel and soft mallet (**Figure 131**) to carefully tap the piston into the bore. Do not use excessive force. Check for insufficient ring compression if the piston will not enter the bore with light tapping.

12. Carefully guide the connecting rod toward the crankshaft while pushing the piston down in the bore. Fit the new bearing into the connecting rod. The bearing tang must fit into the notch in the connecting rod (**Figure 111**, typical). Seat the rod and bearing against the crankpin journal.

13. Place a section of Plastigage (**Figure 132**) onto the crankpin journal. The Plastigage must be as long as the width of the bearing and parallel with the bottom of the journal (**Figure 133**).

CAUTION

Do not rotate the crankshaft with Plastigage installed onto the crankshaft connecting rod journal.

14. Fit the new bearing into the connecting rod cap (**Figure 109**). The bearing tang must fit into the notch in the cap. Install the rod cap onto the connecting rod (**Figure 108**). The bearing tangs must align on the same side of the rod. Carefully seat the cap onto the rod. To check for proper alignment, drag a sharpened pencil tip along the machined side of the cap and rod. The pencil tip must not catch on the split line. If the cap and rod do not align, make sure the correct cap is correctly installed. If installed correctly, the connecting rod cap might be distorted, and the rod and cap must be replaced.

15. Tighten the connecting rod fasteners as follows:

a. On GM 3.0L models, tighten the connecting rod nuts evenly to 203 in.-lb. (23 N•m). Check for proper rod cap alignment as described in Step 14. Tighten the connecting rod nuts a second time to 35 ft.-lb. (47 N•m). Tighten the nuts a final time to 45 ft.-lb. (61 N•m).

b. On 1994-1996 GM 4.3L models with a 12-bolt intake manifold, tighten the connecting rod nuts evenly to 203 in.-lb. (23 N•m). Check for proper rod cap alignment as described in Step 14. Tighten the connecting rod nuts a second time to 35 ft.-lb. (47 N•m). Tighten the nuts a final time to 45 ft.-lb. (61 N•m).

c. On 1997-2000 GM 4.3L models with an eight-bolt intake manifold, tighten the connecting rod nuts

(129) PISTON RING GAP ALIGNMENT (GM FOUR-CYLINDER, V-6 AND V-8 MODELS)

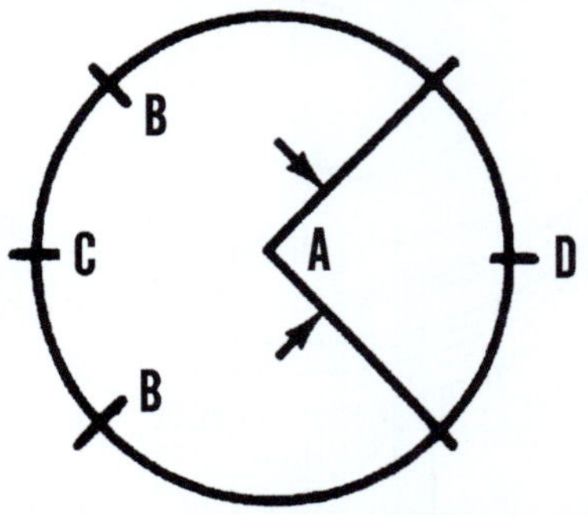

A. Oil ring spacer gap (tang in hole or slot within arc)
B. Oil ring rail gaps
C. Second compression ring (port side)
D. Top compression ring (starboard side)

(130) PISTON RING GAP ALIGNMENT (FORD 5.0L AND 5.8L MODELS)

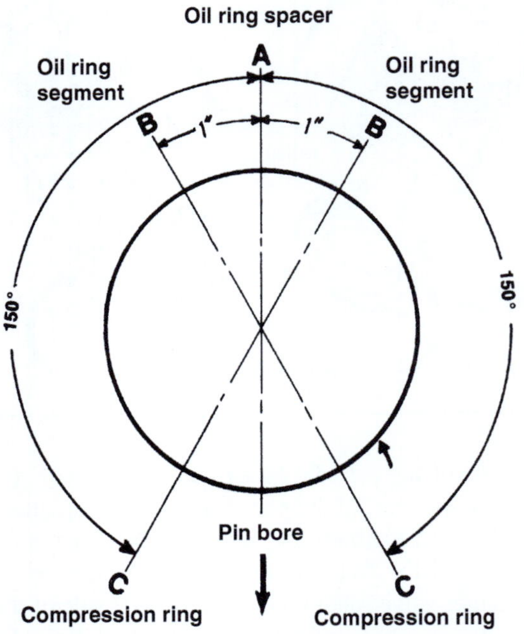

A. Position toward rear of engine
B. Position oil ring rails within 1 in. (25.4 mm) from oil ring spacer
C. Position compression ring 150° from oil ring spacer

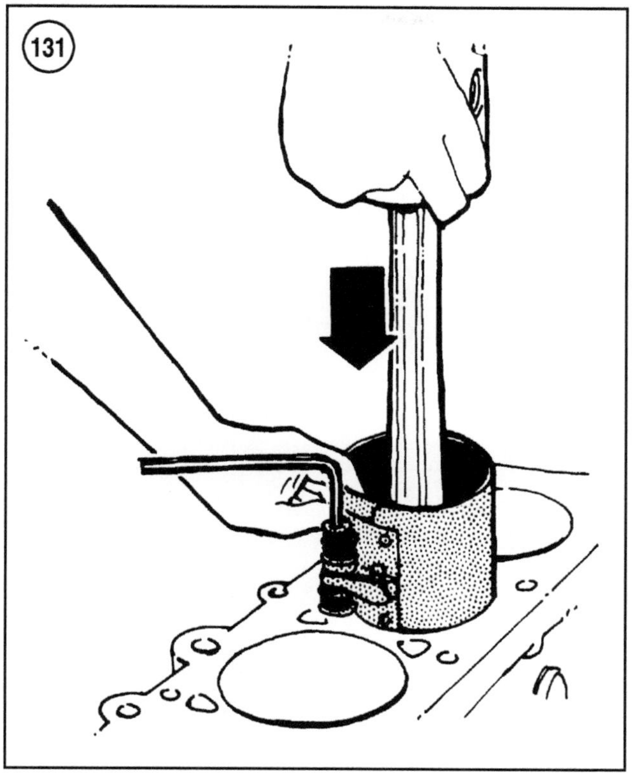

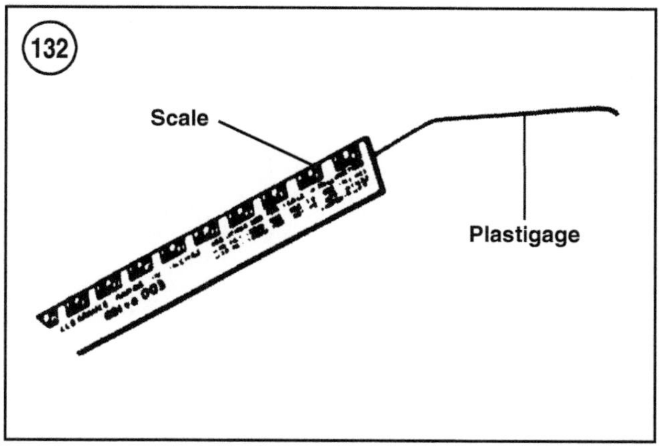

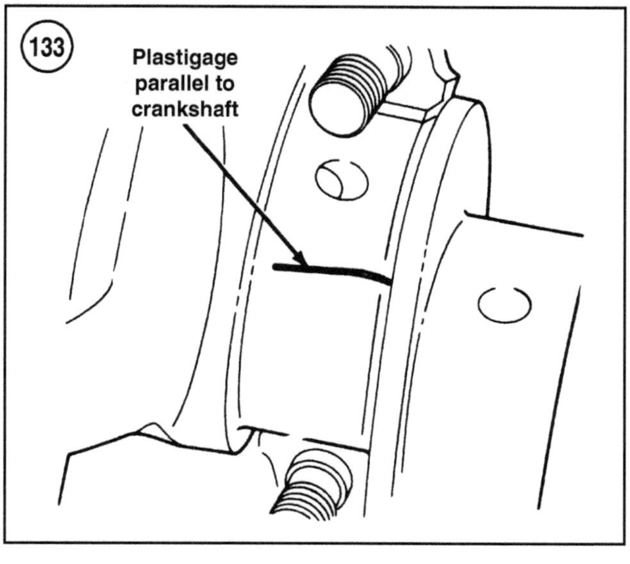

evenly to 20 ft.-lb. (27 N•m). Check for proper rod cap alignment as described in Step 14. Use an angle torque gauge and tighten the rod nuts an additional 70° of rotation.

d. On 1994-1997 GM 5.0L and 5.7L models with a 12-bolt intake manifold, tighten the connecting rod nuts evenly to 203 in.-lb. (23 N•m). Check for proper rod cap alignment as described in Step 14. Tighten the connecting rod nuts a second time to 35 ft.-lb. (47 N•m). Tighten the nuts a final time to 45 ft.-lb. (61 N•m).

e. On 1998-2000 GM 5.0L and 5.7L models with an eight-bolt intake manifold, tighten the connecting rod nuts evenly to 20 ft.-lb. (27 N•m). Check for proper rod cap alignment as described in Step 14. Use an angle torque gauge and tighten the rod nuts an additional 55° of rotation.

f. On Ford 5.0L models, tighten the connecting rod nuts evenly to 10 ft.-lb. (14 N•m). Check for proper rod cap alignment as described in Step 14. Tighten the nuts evenly a final time to 19-24 ft.-lb. (26-33 N•m).

g. On Ford 5.8L models, tighten the connecting rod nuts evenly to 10 ft.-lb. (14 N•m). Check for proper rod cap alignment as described in Step 14. Tighten the nuts evenly a second time to 20 ft.-lb. (27 N•m). Tighten the nuts evenly a final time to 40-45 ft.-lb. (54-61 N•m).

h. On GM 7.4L and 8.2L models with 3/8-24 nuts, tighten the connecting rod nuts evenly to 10 ft.-lb. (14 N•m). Check for proper rod cap alignment as described in Step 14. Tighten the nuts evenly a second time to 20 ft.-lb. (27 N•m). Tighten the connecting rod nuts evenly a final time to 48 ft.-lb. (65 N•m).

i. On GM 7.4L and 8.2L models with 7/16-20 nuts, tighten the connecting rod nuts evenly to 20 ft.-lb. (27 N•m). Check for proper rod cap alignment as described in Step 14. Tighten the nuts evenly a second time to 50 ft.-lb. (68 N•m). Tighten the connecting rod nuts evenly a final time to 73 ft.-lb. (99 N•m).

CAUTION
Do not rotate the crankshaft with Plastigage installed onto the crankshaft connecting rod journal.

16. Evenly loosen the nuts and remove the rod cap. Gauge the connecting rod bearing clearance by comparing the width of the flattened Plastigage with the scale on the envelope (**Figure 134**). Compare the clearance with the specifications in **Table 4**. Refer to the following:

a. If the bearing clearance exceeds the specification, remove the piston and connecting rod as described in this section. Then measure the crankshaft crankpin as described in *Crankshaft Removal and Inspection* in this chapter. If the measurement is within the specification given in **Table 4**, install either 0.001 in. or 0.002 in. undersized bearings into the rod and cap to correct the clearance. Never install bearings with an undersized amount that exceeds the out-of-round measurement of the journal. Otherwise, inadequate clearance could occur at the wider journal diameter. Repeat the bearing clearance measurement.

b. If the bearing clearance is less than the specification, remove the piston and connecting rod as described in this section. Then measure the crankshaft crankpin as described in *Crankshaft Removal and Inspection* in this chapter. If the measurement is within the specification, check for improper rod cap or bearing installation, debris or an oil film between the rod and/or rod cap and bearings. Install the piston and connecting rod as described in this section. Repeat the bearing clearance measurement.

c. If the bearing clearance is within the specification, use a fingernail to scrape the Plastigage from the journal and bearing. Apply engine oil to the crankshaft and bearing surfaces. Install the rod cap and tighten the nuts to the specification as described in Step 15.

17. Rotate the crankshaft several revolutions. If the crankshaft binds or turns roughly, check for improper piston or connecting rod installation.

18. Repeat Steps 4-17 for the remaining pistons and connecting rods.

19. Measure the connecting rod side clearance as follows:

a. On GM 3.0L models, move the connecting rod cap toward one side of the journal. Insert a feeler gauge between the opposite side of the connecting rod and the journal (**Figure 105**, typical). The clearance equals the thickness of the feeler gauge that passes between the rod and journal with slight drag.

b. On GM 4.3L models, push the connecting rod toward the front of the engine. Insert a feeler gauge between the rear side of the connecting rod and the journal (**Figure 105**, typical). The clearance equals the thickness of the feeler gauge that passes between the rod and journal with slight drag.

c. On GM and Ford V-8 models, push each connecting rod on the selected journal toward the crankshaft contact surfaces. Insert a feeler gauge between the two connecting rods (**Figure 106**). The clearance equals the thickness of the feeler gauge that passes between the connecting rods with slight drag.

d. The connecting rod side clearance must be within the specification in **Table 4**. If the clearance exceeds the specification, disassemble the engine and check for excessive wear on the corresponding connecting rod and crankshaft surfaces. Replace the suspect connecting rod or pair of connecting rods and recheck the clearance. If this does not correct the excessive clearance, replace the crankshaft. If the clearance is less than the specification, remove the suspect piston and connecting rod and check for improper installation of the connecting rod, connecting rod cap or piston.

e. Repeat this step for the remaining pairs of connecting rods.

(134)

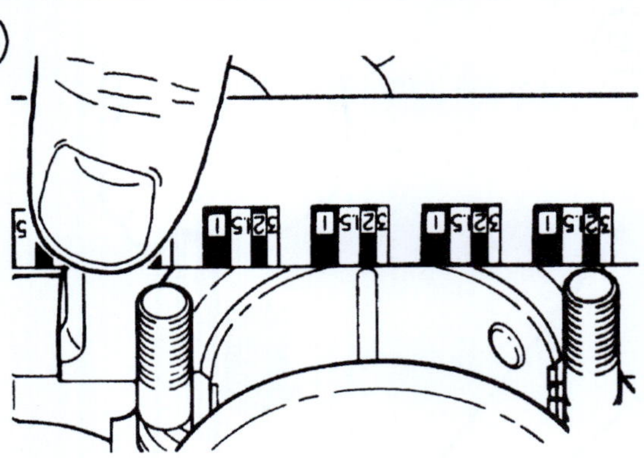

(135) **CRANKSHAFT AND CYLINDER BLOCK (TYPICAL GM MODEL)**

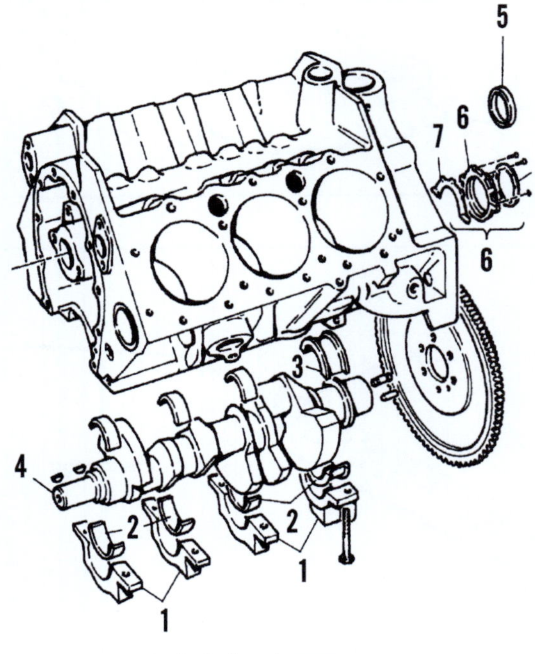

1. Main bearing caps
2. Main bearing inserts
3. Rear main bearing inserts
4. Crankshaft
5. Rear main seal
6. Seal carrier
7. Gasket

20. Install the oil pump, timing gear/chain cover, oil pan and cylinder head(s) as described in this chapter.

CRANKSHAFT AND REAR MAIN SEAL

Replace the rear main oil seal and crankshaft main bearings each time the crankshaft is removed. The rear main oil seal (5, **Figure 135** or 4, **Figure 136**) can be replaced without removing the crankshaft, pistons or connecting rods from the engine.

(136) **CRANKSHAFT AND CYLINDER BLOCK (FORD 5.0L AND 5.8L MODELS)**

1. Woodruff key
2. Crankshaft
3. Main bearing insert (upper halves)
4. Rear main oil seal
5. Flywheel
6. Sprocket
7. Main bearing insert (lower halves)
8. Bolts
9. Main bearing caps
10. Front seal
11. Flywheel cover
12. Engine coupler

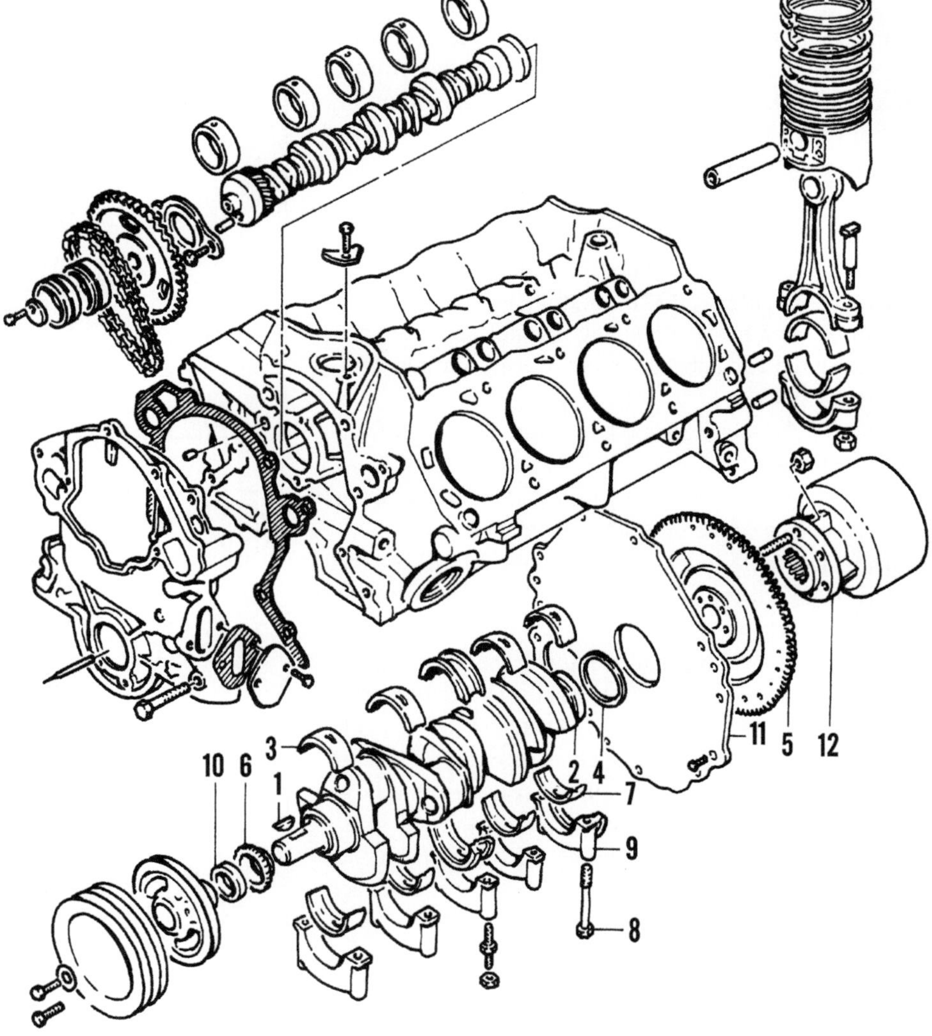

6

The main bearing caps (1, **Figure 135** or 9, **Figure 136**) are machined to match the cylinder block. The caps must be installed in the proper locations and with the arrows on the casting facing forward. Otherwise, the crankshaft will bind in the cylinder block, wear rapidly and fail. The cap from one cylinder block cannot be installed onto a different cylinder block.

To compensate for manufacturing variations, some GM engines are equipped with 0.009 in. undersized main bearing inserts. On such engines, the cylinder block is stamped with a 009 mark on the middle forward crankshaft flyweight. However, because 0.009 in. undersized bearings are not available separately, replace the bearings with 0.010 in. undersized main bearing inserts and check the main bearing clearance during assembly. On an engine with very low operating hours, the crankshaft might require machining to accept the 0.010 in. undersized bearings.

To fit precise factory tolerances, some GM engines are equipped with a 0.001 in. undersized and a standard bearing insert on the same journal. On such engines, use the standard or 0.001 in. undersized bearings for service replacements.

NOTE
With the manifolds, fuel system, cooling system and electrical components installed, the General Motors (GM) 5.0L and 5.7L engines are very similar in appearance to the Ford 5.0L and 5.8L engines. If the model identification decal on the flame arrestor is missing, the engine can be quickly identified as a GM or Ford engine by noting the distributor location. On GM engines, the distributor is located at the rear (flywheel end) of the engine.

On Ford engines, the distributor is located on the front of the engine.

Rear Main Seal Replacement

The rear main seal can be replaced without removing the oil pan or seal carrier. To avoid seal, carrier or cylinder block damage, use only the recommended tool listed in the instructions to install the rear main seal.

1. Remove the engine as described in this chapter.
2. Remove the drive coupler and flywheel as described in this chapter.

3A. On GM models, insert a pry bar into one of the three seal retainer slots (**Figure 137**). Carefully pry the seal from the carrier or main bearing housing. Do not allow the pry bar to contact the crankshaft or main bearing surface. Discard the seal.

3B. On Ford 5.0L and 5.8L models, remove the seal from the cylinder block as follows:

   a. Carefully punch two holes, one on each side, into the metal casing of the oil seal with a punch or awl.
   b. Thread suitable screws into the holes. Make sure the screws do not contact the cylinder block or rear main bearing surfaces.
   c. Fit the jaws of a slide hammer under the screw heads. Hold the jaws in position and use the slide hammer to pull the seal from the cylinder block. Discard the seal.

4. Use a quick-drying solvent, such as carburetor cleaner, to clean the seal contact surfaces in the seal carrier. Allow the solvent to dry completely before installing the new seal.
5. Lubricate the seal lip with engine oil. Then insert the new seal into the bore with the lip side facing the crankshaft. Make sure the seal is not cocked in the bore.
6. Use the recommended tool to push the oil seal into the bore. On Ford 5.0L models, push the seal in until the casing of the oil seal is within 0.005 in. (0.127 mm) of the rear face of the cylinder block. On all other models, the installation tool will position the seal in the bottom of the bore.
   a. For GM 3.0L, 4.3L 5.0L and 5.7L models, use Kent-Moore part No. J-65621.
   b. For Ford 5.0L models, use OTC or SPX part No. T-82-L-6701.
   c. For Ford 5.8L models, use SPX T-65-P-6701-A.
   d. For GM 7.4L and 8.2L models, use Kent-Moore part No. J-38841.
7. Install the flywheel, engine coupler and flywheel housing as described in this chapter.
8. Install the engine as described in this chapter.

Rear Main Seal Carrier Replacement

NOTE

A rear main seal carrier is used only on General Motors 3.0L, 4.3L, 5.0L and 5.7L models. On all other models, the rear main seal fits into a recess in the cylinder block and rear main bearing cap.

1. Remove the engine as described in this chapter.
2. Remove the drive coupler and flywheel as described in this chapter.
3. Remove the oil pan as described in this chapter.
4. Remove the rear main seal from the carrier as described in this chapter.
5. Remove the fasteners and carefully pry the carrier (6, **Figure 135**) from the rear main bearing. Remove and discard the gasket (7, **Figure 135**).
6. Clean any gasket material or sealant from the carrier and rear main mating surfaces.
7. Fit a new gasket onto the seal carrier. Do not apply sealant to the gasket. Align the locating pin and studs. Then fit the carrier onto the rear main bearing.

137

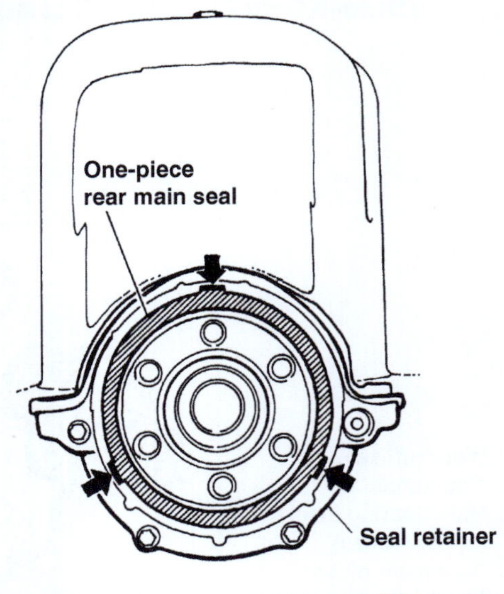

138

MEASURING CRANKSHAFT END PLAY (TYPICAL GM MODEL)

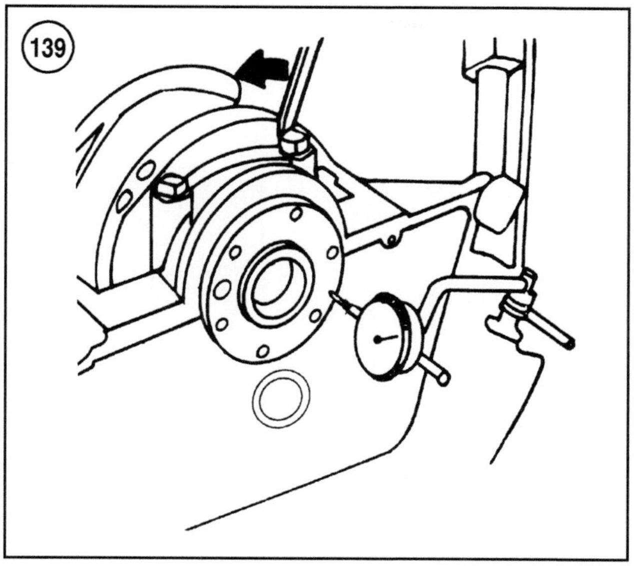

8. Apply a light coat of Loctite Pipe Sealant with Teflon to the threads of the carrier fasteners. Install the fasteners and tighten evenly to 133 in.-lb. (15 N•m) for 3.0L and 4.3L models and 106 in.-lb. (12 N•m) for 5.0L and 5.7L models.
9. Install the rear main seal into the carrier as described in this chapter.
10. Install the oil pan as described in this chapter.
11. Install the flywheel and drive coupler as described in this chapter.
12. Install the engine as described in this chapter.

Crankshaft Removal

Refer to **Figure 135** or **Figure 136** for this procedure.
1. Remove the engine as described in this chapter.
2. Remove the drive coupler and flywheel as described in this chapter.
3. Remove the oil pan, oil pump, timing chain/gear and cylinder head(s) as described in this chapter.
4. Remove the pistons and connecting rods as described in this chapter.
5. On GM 4.3L models, remove the balance shaft as described in this chapter.
6. Remove the lifters and camshaft as described in this chapter.
7. Position the cylinder block with the crankshaft facing upward.
8A. On GM 3.0L, 4.3L, 5.0L, 5.7L, 7.4L and 8.2L models, measure the crankshaft end play as follows:
   a. Use a pry bar to gently pry the crankshaft toward the front of the cylinder block (**Figure 138**).
   b. Maintain moderate pressure on the pry bar. Use a feeler gauge to measure the clearance between the rear main bearing thrust surface and the corresponding thrust surface on the crankshaft. The crankshaft end play equals the thickness of the feeler gauge that passes through the gap with slight drag.
   c. Compare the end play measurement with the specification in **Table 4**. If the end play exceeds the specification, remove the crankshaft and check for wear on the rear main bearing and crankshaft thrust surfaces. End play that is less than the specification is generally the result of improper rear main bearing thrust surface alignment that can be corrected during crankshaft installation.

8B. On Ford 5.0L and 5.8L models, measure the crankshaft end play as follows:
   a. Attach a dial indicator to the rear of the cylinder block to position the indicator stem in contact with the crankshaft flange as shown in **Figure 139**.
   b. Use a pry bar to gently pry the crankshaft toward the front of the engine.
   c. Set the dial indicator to zero. Observe the end play measurement on the dial indicator while carefully prying the crankshaft as far as possible toward the rear of the engine.
   d. Compare the end play measurement with the specification in **Table 4**. If the end play exceeds the specification, remove the crankshaft and check for wear on the center main bearing and crankshaft thrust surfaces. End play that is less than the specification is generally the result of improper center main bearing thrust surface alignment that can be corrected during crankshaft installation.

9. Starting at the front, number the main bearing caps to indicate the mounting locations on the cylinder block. Note the cast-in arrows or triangle marks on the bearing caps (A, **Figure 140**, typical). The arrows or triangles must face the front of the engine.
10A. On GM 3.0L, 4.3L, 5.0L and 5.7L models, remove the rear main seal carrier as described in this chapter.
10B. On Ford 5.0L and 5.8L models and GM 7.4L and 8.2L models, remove the rear main seal as described in this chapter.
11. Fully loosen the main bearing bolts (B, **Figure 140**, typical) and then remove the main bearing caps one at a time. Note the original mounting locations of each cap. Remove the lower bearing inserts (**Figure 141**, typical) from the main bearing caps.

6

12. Carefully lift the crankshaft from the cylinder block. Place the crankshaft on a sturdy work surface with good lighting. Remove the upper main bearing inserts from the cylinder block.

13. Remove any remaining fittings, brackets or other components from the cylinder block.

14. Thoroughly clean the cylinder block with pressurized soapy water. Direct the spray through all passages and into any crevices. Rinse the block with clean water and dry with compressed air. Direct air through all passages to remove residual water or debris. To prevent corrosion, promptly coat all unpainted surfaces with engine oil.

141

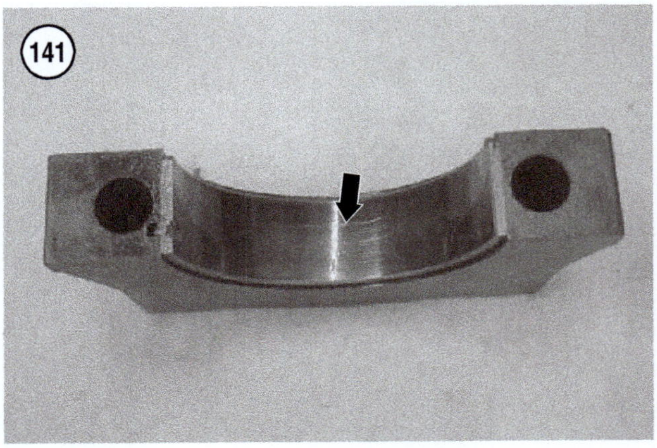

Crankshaft Inspection and Measurement

1. Inspect the main bearing inserts for discolored, scratched, corroded, pitted or rough surfaces. Inspect the crankshaft for damaged surfaces if these or other defects are found with the bearings.

2. Inspect the crankshaft main bearing and crankpin journals for cracks, discoloration, corrosion, pitting, scratches or transferred bearing material. If the surfaces are lightly discolored, scratched, corroded, rough or pitted, have a reputable machine shop machine the crankshaft to accept undersized bearings. Contact an OMC Cobra dealership for undersized bearing availability before machining the crankshaft. Replace the crankshaft if cracking or heavy discoloration is evident.

3. Use an accurate outside micrometer to measure the diameter of the main and crankpin journals (**Figure 142**). Take measurements at several points along the length of the journals and at several points around the circumference of the journals. Record all measurements for each journal. Refer to the following:

   a. Subtract the smallest diameter at one side of the journal from the largest diameter at the other side of the journal to determine the amount of journal taper. If the journal taper exceeds the specification in **Table 4**, have the crankshaft machined to accept the next undersized bearing inserts. Contact an OMC Cobra dealership for undersized bearing availability before machining the crankshaft.

   b. Subtract the smallest diameter at one point along the width of the journal from the largest diameter taken around the circumference of the journal (and at the same point along the length) to determine the amount of journal out-of-round. If the journal out-of-round exceeds the specification in **Table 4**, have the crankshaft machined to accept the next undersized bearing inserts. Contact an OMC dealership for undersized bearing availability before machining the crankshaft.

   c. Compare the journal measurement with the specification in **Table 4**. Have the crankshaft machined to accept the next undersized bearing inserts if any of the measurements is below the minimum specification. Contact an OMC Cobra dealership for undersized bearing availability before machining the crankshaft.

4. Support the crankshaft with V-blocks under the front and rear main bearing journals (**Figure 143**). Mount a dial indicator so the plunger touches one of the center main bearing journals. Zero the dial indicator. Observe the needle movement while slowly rotating the crankshaft. The needle movement indicates the amount of crankshaft runout. Replace the crankshaft if the runout exceeds the specification in **Table 4**.

5. On GM 3.0L models, inspect the crankshaft timing gear for worn, damaged or missing teeth. Replace the gear if these or other defects are evident. Replacement instructions follow:

   a. Pull the gear from the crankshaft using a three-jaw puller or other suitable tool.

   b. Inspect the drive key and key slot in the crankshaft for excessive wear or damage and replace as needed.

   c. Align the key slot and fit the new gear onto the crankshaft. Using a section of tubing or other suitable driver, tap the gear onto the crankshaft. Make sure the gear seats fully against the shoulder on the crankshaft.

Crankshaft Installation

Refer to **Figure 136** or **Figure 137** for this procedure.

1. Install the new main bearing inserts into the cylinder block. The inserts with the grooves fit into the cylinder block. The inserts without grooves fit into the main bearing caps. Install the insert with the side thrust surface into the rear of the engine on GM four-cylinder, V-6 and V-8 models. Install the insert with the side thrust surface into the middle location on Ford 5.0L and 5.8L models. Do not lubricate the inserts at this time.

2. Carefully lower the crankshaft into the cylinder block. Seat the crankshaft against the main bearings. Do not rotate the crankshaft.

3. Place a section of Plastigage (**Figure 132**) onto the main bearing journals. The Plastigage must span the length of the journal and align with the bottom of the journal (**Figure 133**).

142

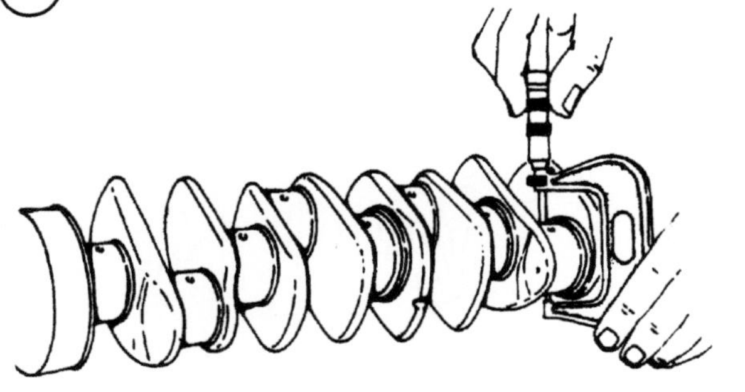

143

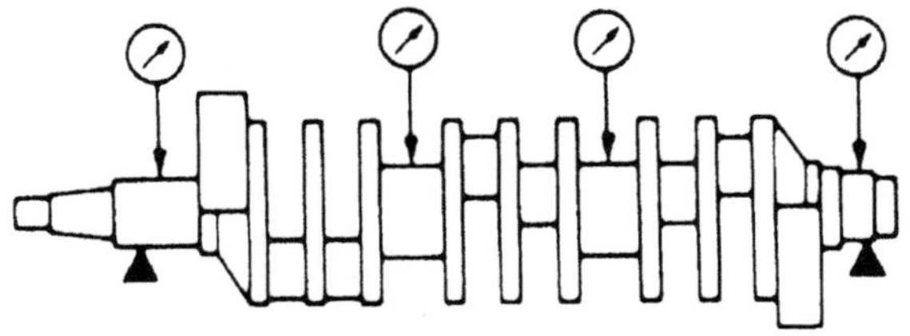

4. Fit the bearing inserts into the main bearing caps (**Figure 141**). The bearing tang must fit into the notch in the cap. Install the main bearing caps onto the corresponding locations on the cylinder block. The arrow or triangle mark (A, **Figure 140**, typical) must point toward the front of the engine.
5. Thread the bolts (B, **Figure 140**) by hand into the main bearing caps. Do not rotate the crankshaft.

CAUTION
Do not tighten the rear main bearing cap bolts (on GM models) or center main bearing (on Ford models) in Step 6. The main thrust surfaces must be aligned before tightening the main bearing cap bolts or center main bearing, or the thrust surfaces will wear excessively and fail the bearing. Tighten the front and center main bearings as described in Step 6. Then align the bearing and tighten the rear main bearing as described in Step 7.

6. Tighten the main bearing cap bolts as follows:
   a. On GM 3.0L models, tighten the front and center main bearing bolts evenly to 55-70 ft.-lb. (75-95 N•m). Tighten the rear main bearing bolts evenly to 110-144 in.-lb. (12-16 N•m).
   b. On 1994-1996 GM 4.3L models with a 12-bolt intake manifold, tighten the front and center main bearing bolts evenly to 80 ft.-lb. (108 N•m). Tighten the rear main bearing bolts evenly to 35 in.-lb. (4 N•m).
   c. On 1997-2000 GM 4.3L models with an eight-bolt intake manifold, tighten the front and center main bearing bolts evenly to 77 ft.-lb. (104 N•m). Tighten the rear main bearing bolts evenly to 124 in.-lb. (14 N•m).
   d. On 1994-1997 GM 5.0L and 5.7L models with a 12-bolt intake manifold and two-bolt main bearings, tighten the front and center main bearing bolts evenly to 80 ft.-lb. (108 N•m). Tighten the rear main bearing bolts evenly to 35 in.-lb. (4 N•m).
   e. On 1994-1997 GM 5.0L and 5.7L models with a 12-bolt intake manifold and four-bolt main bearings, tighten each pair of front and center main bearing inner bolts (closest to the crankshaft journal) evenly to 80 ft.-lb. (108 N•m). Then tighten each pair of front and center main bearing outer bolts (farthest from the crankshaft journal) evenly to 70 ft.-lb. (95 N•m). Tighten the rear main bearing bolts evenly to 35 in.-lb. (4 N•m).
   f. On 1998-2000 GM 5.0L and 5.7L models with an eight-bolt intake manifold, tighten each pair of front and center main bearing inner bolts (closest to the crankshaft journal) evenly to 77 ft.-lb. (104 N•m). Then tighten each pair of front and center main bearing outer bolts (farthest from the crankshaft journal) evenly to 67 ft.-lb. (91 N•m). Tighten the rear main bearing bolts evenly to 35 in.-lb. (4 N•m).
   g. On Ford 5.0L and 5.8L models, tighten each pair of the front, second, fourth and fifth main bearing bolts evenly to 60-70 ft.-lb. (81-95 N•m) for 5.0L models and 95-105 ft.-lb. (129-142 N•m) for 5.8L models. Tighten the center main bearing bolts to 35 in.-lb. (4 N•m).
   h. On GM 7.4L models, tighten each pair of front and center main bearing bolts evenly to 100 ft.-lb. (136 N•m). Tighten the rear main bearing bolts evenly to 35 in.-lb. (4 N•m).
   i. On GM 8.2L models, tighten each pair of front and center main bearing bolts evenly to 110 ft.-lb. (149 N•m). Tighten the rear main bearing bolts evenly to 35 in.-lb. (4 N•m).

7A. On GM four-cylinder, V-6 and V-8 models, insert a pry bar between one of the center main bearing caps and a crankshaft flyweight. Carefully pry the crankshaft toward the rear of the cylinder block. Move the pry bar to the other side of the bearing cap and carefully pry the crankshaft toward the front of the cylinder block (**Figure 144**). Maintain moderate pressure on the pry bar while tightening the rear main bearing cap (**Figure 145**) to the following specification.
   a. On GM 3.0L models, tighten the rear main bearing bolts evenly to 55-70 ft.-lb. (75-95 N•m).
   b. On 1994-1996 GM 4.3L models with a 12-bolt intake manifold, tighten the rear main bearing bolts evenly to 80 ft.-lb. (108 N•m).

c. On 1997-2000 GM 4.3L models with an eight-bolt intake manifold, tighten the rear main bearing bolts evenly to 77 ft.-lb. (105 N•m).

d. On 1994-1997 GM 5.0L and 5.7L models with a 12-bolt intake manifold and two-bolt main bearings, tighten the rear main bearing bolts evenly to 80 ft.-lb. (108 N•m).

e. On 1994-1997 GM 5.0L and 5.7L models with a 12-bolt intake manifold and four-bolt main bearings, tighten the rear main bearing inner bolts (closest to the crankshaft journal) evenly to 80 ft.-lb. (108 N•m). Then, tighten the rear main bearing outer bolts (furthest from the crankshaft journal) evenly to 70 ft.-lb. (95 N•m).

f. On 1998-2000 GM 5.0L and 5.7L models with an eight-bolt intake manifold, tighten the rear main bearing inner bolts (closest to the crankshaft journal) evenly to 77 ft.-lb. (104 N•m). Then, tighten rear main bearing outer bolts (furthest from the crankshaft journal) evenly to 67 ft.-lb. (91 N•m).

g. On Ford 5.0L and 5.8L models, tighten the rear main bearing bolts evenly to 60-70 ft.-lb. (81-95 N•m) for 5.0L models and 95-105 ft.-lb. (129-142 N•m) for 5.8L models.

h. On GM 7.4L models, tighten the rear main bearing bolts evenly to 100 ft.-lb. (136 N•m).

i. On GM 8.2L models, tighten the rear main bearing bolts evenly to 110 ft.-lb. (149 N•m).

7B. On Ford 5.0L and 5.8L models, align the thrust bearing and tighten the rear main bearing cap as follows:

a. Insert a suitable pry bar between the fourth main bearing cap and the crankshaft flyweight.

b. Pry the crankshaft forward against the thrust surface of the upper half of the center main bearing as shown in **Figure 144**.

c. To align both the upper and lower bearing thrust surfaces, maintain pressure to hold the crankshaft forward and use an additional pry bar to pry the center main bearing cap toward the rear of the engine (**Figure 146**)

d. Maintain pressure to hold the crankshaft forward (**Figure 145**). Then tighten the center main bearing bolts evenly to 60-70 ft.-lb. (81-95 N•m) for 5.0L models and 95-105 ft.-lb. (129-142 N•m) for 5.8L models.

8. Evenly loosen the main bearing bolts. Remove the main bearing caps. Determine the main bearing clearance by comparing the width of the flattened Plastigage (**Figure 134**) with the marks on the envelope. Compare the clearance to the specification in **Table 4**. Refer to the following:

a. If the clearance exceeds the specification, measure the crankshaft as described in *Crankshaft Inspection and Measurement* in this section. Install either 0.001 or 0.002 in. undersized bearings to correct excessive clearance. Never install bearings with an undersized amount that exceeds the out-of-round measurement of the journal. Otherwise, inadequate clearance could occur at the wider journal diameter.

b. If the clearance is less than the specification, measure the crankshaft as described in *Crankshaft Inspection and Measurement* in this section. Check for improper bearing cap or bearing insert installation, debris or an oil film between the cap and/or cylinder block and the bearings.

(144) **THRUST BEARING ALIGNMENT (TYPICAL)**

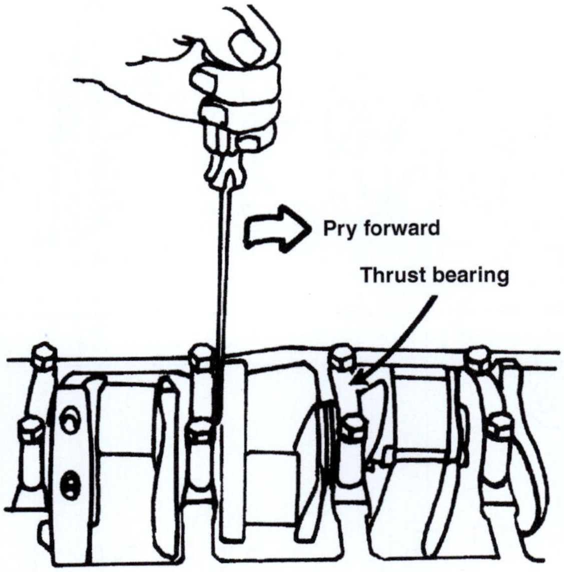

(145) **THRUST BEARING ALIGNMENT/TIGHTENING (TYPICAL)**

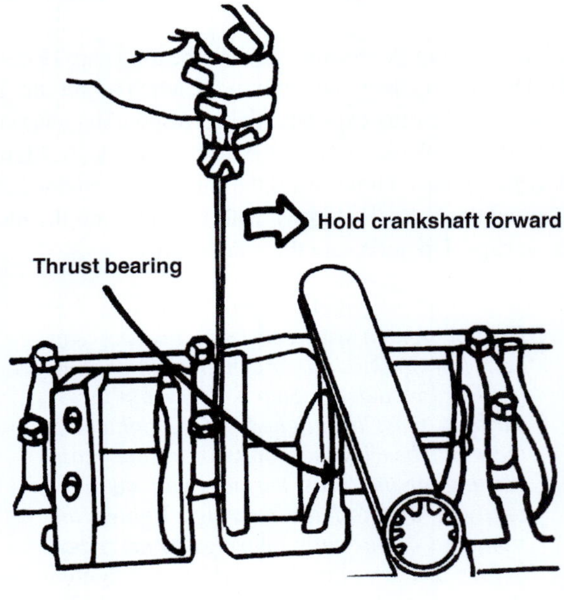

146

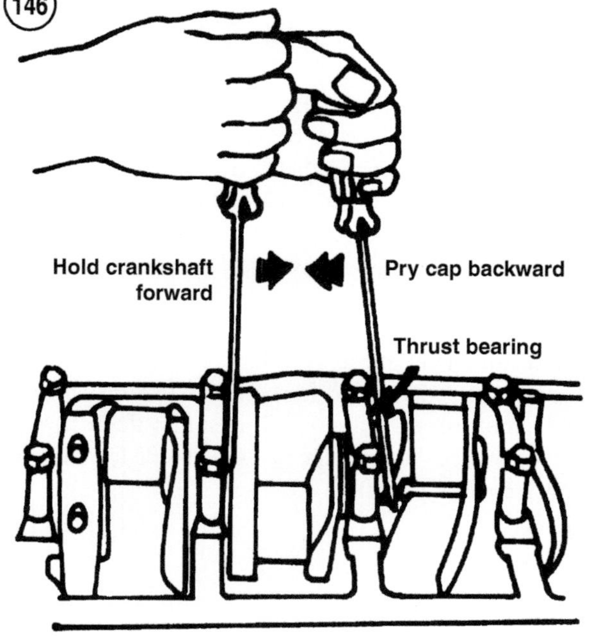

c. If the clearance is within the specification, use a fingernail to scrape the Plastigage from the journal and bearings. Apply engine oil to the crankshaft and bearing surfaces. Install the main bearing caps. Tighten the bearing caps and align the thrust bearing surfaces as described in Step 6 and Step 7.

9A. On GM four-cylinder, V-6 and V-8 models, measure the crankshaft end play as follows:

a. Use a pry bar to gently pry the crankshaft toward the front of the cylinder block (**Figure 145**).
b. Maintain moderate pressure on the pry bar. Use a feeler gauge to measure the clearance between the rear main bearing and the thrust surface on the crankshaft. The crankshaft end play equals the thickness of the feeler gauge that passes through the gap with slight drag.
c. Compare the measurement with the crankshaft end play specification in **Table 4**.
d. If the end play exceeds the specification, remove the crankshaft and check for wear on the rear main bearing and crankshaft thrust surfaces. Replace the rear main bearing and repeat the measurement. Replace the crankshaft if excess end play persists with the new bearing.
e. If the end play is less than the specification, loosen the rear main bearing cap bolts and align the thrust bearing as described in Step 7A. Tighten the rear main bolts as described and repeat the measurement. Replace the rear main bearing if inadequate clearance persists.

9B. On Ford 5.0L and 5.8L models, measure the crankshaft end play as follows:

a. Attach a dial indicator to the rear of the cylinder block to position the indicator stem in contact with the crankshaft flange as shown in **Figure 139**.
b. Use a pry bar to gently pry the crankshaft toward the front of the engine.
c. Set the dial indicator to zero. Observe the end play measurement on the dial indicator while carefully prying the crankshaft as far as possible toward the rear of the engine.
d. Compare the measurement with the crankshaft end play specification in **Table 4**.
e. If the end play exceeds the specification, remove the crankshaft and check for wear on the center main bearing and crankshaft thrust surfaces. Replace the center main bearing and repeat the measurement. Replace the crankshaft if excess end play persists with the new bearing.
f. If the end play is less than the specification, loosen the rear main bearing cap bolts and align the thrust bearing as described in Step 7B. Tighten the rear main bolts as described and repeat the measurement. Replace the rear main bearing if inadequate clearance persists.

10. Rotate the crankshaft several revolutions. Check for improper main bearing installation or inadequate clearance if binding or roughness is evident.

11A. On GM 3.0L, 4.3L, 5.0L and 5.7L models, install the rear main seal retainer and then the rear main seal as described in this chapter.

11B. On Ford 5.0L and 5.8L and GM 7.4L and 8.2L models, install the rear main seal as described in this chapter.

12. Install the pistons and connecting rods as described in this chapter.

13. Install the camshaft and lifters as described in this chapter.

14. On GM 4.3L models, install the balance shaft as described in this chapter.

15. On V-6 and V-8 models, install the timing chain and sprockets as described in this chapter.

16. Install the timing gear or chain cover as described in this chapter.

17. Install the harmonic balancer as described in this chapter.

18. Install the oil pump and oil pan as described in this chapter.

19. Install the cylinder head(s) as described in this chapter.

20. Install the flywheel, drive coupler and flywheel housing as described in this chapter.

21. Install any remaining fittings, brackets or other components that were removed from the cylinder block.

BREAK-IN PROCEDURE

Perform the break-in procedure anytime internal components are replaced. During the first few hours of operation, many of the engine components must avoid full load until fully seated. Failure to properly break in the engine can result in engine failure, decreased performance, shortened engine life and increased oil consumption.

An engine needs approximately 20 hours of running time to be fully broken in. Increased oil consumption may be expected during this period. Check the oil level frequently as described in Chapter Three during break-in. Also check the tightness of all external fasteners during the break-in period.

1. During the first 10 minutes of operation, start the engine and immediately advance the throttle to 1500 rpm in NEUTRAL. After the engine reaches operating temperature, reduce the throttle to idle speed for 1-2 minutes. Then return to 1500 rpm for 1-2 minutes.
2. After the first 10 minutes, return the engine to idle speed and shift into FORWARD. Advance the throttle quickly to plane the boat. Then immediately reduce the throttle just to maintain the boat on plane. Do not exceed 3500 rpm after planing the boat. During the next 2 hours of operation, repeatedly advance the throttle to three-quarters open for 3 minutes and then return to planing speed for 3 minutes. Continue this procedure during the 2-hour period.
3. During the next 8 hours of the break-in period, operate the engine at full open throttle for no more than 2 minutes at a time. Follow the 2 minutes of full open throttle operation with a few minutes of operation at idle speed to allow a cooling period. Otherwise, operate the engine at varying speeds of less than three-quarters full open.
4. During the next 10 hours of operation, operate the engine at full open throttle for no more than 10 minutes at a time and then operate at a lower speed. Otherwise, operate the engine at three-quarters full open throttle or less. Periodically reduce the engine to idle speed for a few minutes to allow a cooling period.

Table 1 ENGINE TORQUE SPECIFICATIONS

Fastener	in.-lb.	ft.-lb.	N•m
Balance shaft (4.3L models)			
Thrust plate	124	–	14
Driven gear stud bolt	–	21	28
Torx bolt*			
First step	177	–	20
Final step			
Camshaft sprocket/eccentric			
GM 4.3L models	–	21	28
GM 5.0L and 5.7L models	–	18	24
Ford 5.0L and 5.8L models	–	40-45	54-61
GM 7.4L models	–	20	27
GM 8.2L models	–	25	34
Camshaft thrust plate			
GM 3.0L models	80	–	9
GM 4.3L models	124	–	14
GM 5.0L and 5.7L models	106	–	12
GM 7.4L and 8.2L models	120	–	14
Connecting rod*			
GM 3.0L models			
First step	203	–	23
Second step	–	35	47
Final step	–	45	61
GM 4.3L models			
With 12-bolt intake (1994-1996)			
First step	203	–	23
Second step	–	35	47
Final step	–	45	61
With eight-bolt intake (1997-2000)			
First step	–	20	27
Final step			
GM 5.0L and 5.7L models			
With 12-bolt intake (1994-1997)			
First step	203	–	23
Second step	–	35	47
Final step	–	45	61
With eight-bolt intake (1998-2000)			
First step	–	20	27
Final step			

(continued)

Table 1 ENGINE TORQUE SPECIFICATIONS (continued)

Fastener	in.-lb.	ft.-lb.	N•m
Connecting rod* (continued)			
Ford 5.0L models			
First step	–	10	14
Final step	–	19-24	26-33
Ford 5.8L models			
First step	–	10	14
Second step	–	20	27
Final step	–	40-45	54-61
GM 7.4L and 8.2L models			
With 3/8-24 nuts			
First step	–	10	14
Second step	–	20	27
Final step	–	48	65
With 7/16-20 nuts			
First step	–	20	27
Second step	–	50	68
Final step	–	73	99
Crankshaft main bearing*			
GM 3.0L models	–	55-70	75-95
GM 4.3L models			
With a 12-bolt intake (1994-1996)	–	80	108
With an eight-bolt intake (1997-2000)	–	77	104
GM 5.0L and 5.7L models			
With a 12-bolt intake (1994-1997)			
Using two-bolt main bearings	–	80	108
Using four-bolt main bearings			
Inner bolts	–	80	108
Outer bolts	–	70	95
With an eight-bolt intake (1998-2000)			
Inner bolts	–	77	104
Outer bolts	–	67	91
Ford 5.0L models	–	60-70	81-95
Ford 5.8L models	–	95-105	129-142
GM 7.4L models	–	100	136
GM 8.2L models	–	110	149
Cylinder head*			
GM 3.0L models			
1994-1996			
First step	–	35	47
Second step	–	65	88
Final step	–	95	129
1997-2000			
First step	–	35	47
Second step	–	65	88
Final step	–	90	122
GM 4.3L models			
1994-1996 (with 12-bolt intake)			
First step	–	24	33
Second step	–	45	61
Final step	–	65	88
1997-2000 (with eight-bolt intake)			
First step	–	22	30
Final step			
GM 5.0L and 5.7L models			
1994-1997 (with 12-bolt intake)			
First step	–	20	27
Second step	–	40	54
Final step	–	65	88
1998-2000 (with eight-bolt intake)			
First step	–	22	30
Final step			
Ford 5.0L and 5.8L models			
With standard head bolts			
First step	–	95-105	129-142

(continued)

Table 1 ENGINE TORQUE SPECIFICATIONS (continued)

Fastener	in.-lb.	ft.-lb.	N•m
Cylinder head* (continued)			
Ford 5.0L and 5.8L models			
With standard head bolts			
Final step	–	105-112	142-152
With torque-to-yield head bolts			
First step	–	25-35	34-47
Second step	–	45-55	61-75
Final step			
GM 7.4L and 8.2L models			
First step	–	35	47
Second step	–	65	88
Final step	–	85	115
Dipstick tube support	71	–	8
Engine mount			
Rear locknut	–	28-30	38-41
Front/side engine mount adjusting nut			
3.0L models	–	115-140	156-190
V-6 and V-8 models	–	100-120	136-163
Flywheel/drive coupler nut	–	40-45	54-61
Flywheel housing fastener			
GM 3.0L-8.2L models	–	40-45	54-61
Ford 5.0L and 5.8L models	–	28-36	38-49
Flywheel housing shield screw	62-80	–	7-9
Harmonic balancer bolt			
GM 4.3L, 5.0L and 5.7L models	–	74	100
Ford 5.0L and 5.8L models	–	70-90	95-122
GM 7.4L models			
1994-1996 models	–	48	65
1997-2000 models	–	85	115
GM 8.2L models			
1994-1996 models	–	73	99
1997-2000 models	–	90	122
Lifter cover retaining bolt			
3.0L four-cylinder models	65	–	7
Lifter retainer (roller lifters) bolt			
4.3L V-6 models	–	12	16
5.0L and 5.7L models	–	18	24
7.4L and 8.2L models	–	19	26
Oil pan			
GM 3.0L models			
Oil pan-to-cylinder block screws	80	–	9
Oil pan-to-timing gear cover screw	44	–	5
Rear oil pan nuts	165	–	19
GM 4.3L models			
With stamped steel oil pan			
Oil pan rail screws	80	–	9
Corner nuts	165	–	19
With case aluminum oil pan			
Oil pan rail screws	221	–	25
Corner nuts	203	–	23
GM 5.0L and 5.7L models			
Oil pan rail screws	106	–	12
Corner nuts or screws	177	–	20
Ford 5.0L and 5.8L models			
Oil pan screws	108-132	–	12-15
GM 7.4L and 8.2L models			
Oil pan-to-cylinder block screws	200	–	23
Oil pan-to-timing chain cover	70	–	8
Oil pan fitting drain bolt	177-212	–	20-24
Oil pickup tube bracket bolt			
GM 3.0L four-cylinder models	180-216	–	20-24
Ford 5.0L and 5.8L models	–	22-32	30-43
Oil pickup tube-to-pump body screw			
Ford 5.0L and 5.8L models	120-180	–	14-20

(continued)

Table 1 ENGINE TORQUE SPECIFICATIONS (continued)

Fastener	in.-lb.	ft.-lb.	N•m
Oil pump mounting bolts			
GM 3.0L four-cylinder models	110-120	–	12-14
GM 4.3L, 5.0L, 5.7L and 7.4L models	–	65	88
Ford 5.0L and 5.8L models	–	22-32	30-43
GM 8.2L models	–	70	95
Oil pump cover screw			
GM 3.0L models	72	–	8
GM V-6 and V-8 models	106	–	12
Ford 5.0L and 5.8L models	72-120	–	8-14
Oil splash plate nut			
GM 5.0L and 5.7L models	–	30	41
Power steering hose fitting			
Larger fitting	–	15-17	20-23
Smaller fitting	124-142	–	14-16
Rear main seal carrier			
GM 3.0L models	133	–	15
GM 4.3L models	133	–	15
GM 5.0L and 5.7L models	106	–	12
Rocker arm cover screw			
GM 3.0L models	106	–	12
GM 4.3L, 5.0L and 5.7L models	106	–	12
Ford 5.0L and 5.8L models	35-53	–	4-6
GM 7.4L and 8.2L models	115	–	13
Rocker arm retaining nut			
Ford 5.0L and 5.8L models	–	18-25	24-34
Rocker arm shoulder bolt			
GM 4.3L models (1996-2000)	–	22	30
GM 7.4L and 8.2L models	–	40	54
Spark plugs			
GM 3.0L models	–	20	27
GM 4.3L, 5.0L and 5.7L models			
Used cylinder head	–	15	20
New cylinder head	–	22	30
Ford 5.0L and 5.8L models	120-180	–	14-20
GM 7.4L and 8.2L models	–	22	30
Timing chain cover screw			
GM 4.3L, 5.0L and 5.7L models			
With metal timing chain cover	80	–	9
With composite timing chain cover	106	–	12
GM 7.4L and 8.2L models	60-84	–	7-9
Timing gear cover screw			
GM 3.0L models	27-35	–	3-4

*Refer to the chapter for tightening procedures and torque specifications.

6

Table 2 CAMSHAFT SERVICE SPECIFICATIONS

Measurement	Specification
Camshaft end play	
GM 3.0L models	0.003-0.008 in. (0.076-0.203 mm)
Ford 5.0L and 5.8L models	0.001-0.009 in. (0.025-0.229 mm)
Journal diameter	
GM 3.0L models	1.868-1.870 in. (47.440-47.490 mm)
GM 4.3L models	1.8682-1.8692 in. (47.452-47.478 mm)

(continued)

Table 2 CAMSHAFT SERVICE SPECIFICATIONS (continued)

Measurement	Specification
Journal diameter (continued)	
GM 5.0L and 5.7L models	1.8677-1.8697 in. (47.440-47.490 mm)
Ford 5.0L and 5.8L models	
No. 1 journal (front)	2.0805-2.0815 in. (52.845-52.870 mm)
No. 2 journal	2.0655-2.0665 in. (52.464-52.489 mm)
No. 3 journal	2.0505-2.0515 in. (52.083-52.108 mm)
No. 4 journal	2.0355-2.0365 in. (51.702-51.727 mm)
No. 5 journal (rear)	2.0205-2.0215 in. (51.321-51.346 mm)
GM 7.4L models	
1994-1996	1.9482-1.9492 in. (49.484-49.510 mm)
1997-2000	
74GL and 74GI models	1.9477-1.9497 in. (49.472-49.522 mm)
74GS and 74GSI models	1.9482-1.9492 in. (49.484-49.510 mm)
GM 8.2L models	1.9482-1.9492 in. (49.484-49.510 mm)
Journal out-of-round (maximum)	
GM models	0.001 in. (0.0025 mm)
Journal runout (maximum)	
GM models	0.002 in. (0.051 mm)
Ford 5.0L and 5.8L models	0.005 in. (0.127 mm)
Lobe lift	
GM 3.0L models	
Intake and exhaust lobes	0.248-0.258 in. (6.299-6.553 mm)
GM 4.3L models	
1994-1996 (with 12-bolt intake manifold)	
Intake lobes	0.234 in. (5.944 mm)*
Exhaust lobes	0.258 in. (6.553 mm)*
1997-2000 (with eight-bolt intake manifold)	
Intake lobes	0.276 in. (7.010 mm)*
Exhaust lobes	0.285 in. (7.239 mm)*
GM 5.0L and 5.7L models	
1994-1997 models (with 12-bolt intake manifold)	
Intake lobes	0.269 in. (6.832 mm)*
Exhaust lobes	0.276 in. (7.010 mm)*
1998-2000 models (with eight-bolt intake manifold)	
Intake lobes	0.274-0.278 in. (6.960-7.061 mm)
Exhaust lobes	0.278-0.283 in. (7.061-7.188 mm)
Ford 5.0L models	
Intake lobes	0.255-0.260 in. (6.477-6.604 mm)
Exhaust lobes	0.273-0.278 in. (6.934-7.061 mm)
Ford 5.8L models	
Intake lobes	0.273-0.278 in. (6.934-7.061 mm)
Exhaust lobes	0.273-0.278 in. (6.934-7.061 mm)
GM 7.4L models	
1994-1996	
Intake lobes	0.232-0.236 in. (5.893-5.994 mm)
Exhaust lobes	0.251-0.255 in. (6.375-6.477 mm)
1997-2000	
74GL and 74GI models	
Intake lobes	0.282 in. (7.163 mm)*
Exhaust lobes	0.284 in (7.214 mm)*
74GS and 74GSI models	
Intake lobes	0.234 in. (5.944 mm)*
Exhaust lobes	0.252 in. (6.401 mm)*
GM 8.2L models	
Intake lobes	0.249-0.253 in. (6.325-6.426 mm)
Exhaust lobes	0.249-0.253 in. (6.325-6.426 mm)
Timing chain deflection (maximum)	
GM V-6 and V-8 models	3/8 in. (10 mm)
Ford 5.0L and 5.8L models	1/2 in. (13 mm)
Timing gear backlash	
GM 3.0L models	0.004-0.006 in. (0.102-0.152 mm)
Timing gear runout (maximum)	
GM 3.0L models	0.004 in. (0.102 mm)

*Minimum specification

Table 3 VALVE SERVICE SPECIFICATIONS

Measurement	Specification
Valves	
Face angle	
Ford 5.0L and 5.8L models	44°
GM four-cylinder, V-6 and V-8 models	45°
Valve stem-to-rocker arm clearance	
Ford 5.0L models	0.071-0.193 in. (1.803-4.902 mm)
Ford 5.8L models	0.098-0.198 in. (2.489-5.029 mm)
Stem-to-guide clearance	
GM 3.0L models	
Intake valve	0.0010-0.0037 in. (0.025-0.094 mm)
Exhaust valve	0.0010-0.0037 in. (0.025-0.094 mm)
GM 4.3L models	
Models with 12-bolt intake manifold	
Intake valve	0.0010-0.0037 in. (0.025-0.094 mm)
Exhaust valve	0.0010-0.0047 in. (0.025-0.119 mm)
Models with eight-bolt intake manifold	
Intake valve	0.0010-0.0037 in. (0.025-0.094 mm)
Exhaust valve	0.0010-0.0037 in. (0.025-0.094 mm)
Ford 5.0L and 5.8L models	
Intake valve	0.0010-0.0027 in. (0.025-0.069 mm)
Exhaust valve	0.0015-0.0032 in. (0.038-0.081 mm)
GM 5.0L and 5.7L models	
Models with 12-bolt intake manifold	
Intake valve	0.0010-0.0037 in. (0.025-0.094 mm)
Exhaust valve	0.0010-0.0047 in. (0.025-0.119 mm)
Models with eight-bolt intake manifold	
Intake valve	0.0010-0.0037 in. (0.025-0.094 mm)
Exhaust valve	0.0010-0.0037 in. (0.025-0.094 mm)
GM 7.4L V-8 models	
Intake valve	0.0010-0.0037 in. (0.025-0.094 mm)
Exhaust valve	0.0010-0.0049 in. (0.025-0.124 mm)
GM 8.2L V-8 models	
Intake valve	0.0010-0.0037 in. (0.025-0.094 mm)
Exhaust valve	0.0010-0.0049 in. (0.025-0.124 mm)
Valve seat	
Seat angle	
Ford 5.0L and 5.8L models	45°
GM four-cylinder, V-6 and V-8 models	46°
Seat runout (maximum)	0.002 in. (0.051 mm)
Seat width	
GM 3.0L models	
1994-1996 models	
Intake valve	0.0625-0.218 in. (1.588-5.537 mm)
Exhaust valve	0.0625-0.094 in. (1.588-2.388 mm)
1997-2000 models	
Intake valve	0.050-0.070 in. (1.270-1.778 mm)
Exhaust valve	0.060-0.080 in. (1.524-2.032 mm)
GM 4.3L models	
With 12-bolt intake manifold	
Intake valve	0.0625-0.218 in. (1.588-5.537 mm)
Exhaust valve	0.0625-0.094 in. (1.588-2.388 mm)
With eight-bolt intake manifold	
Intake valve	0.040-0.065 in. (1.016-1.651 mm)
Exhaust valve	0.0650-0.0980 in. (1.651-2.489 mm)
Ford 5.0L and 5.8L models	
Intake valve	0.060-0.080 in. (1.524-2.032 mm)
Exhaust valve	0.060-0.080 in. (1.524-2.032 mm)
GM 5.0L models	
With 12-bolt intake manifold	
Intake valve	0.0311-0.0625 in. (0.790-1.588 mm)
Exhaust valve	0.0625-0.0937 in. (1.588-2.380 mm)
With eight-bolt intake manifold	
Intake valve	0.0449-0.0701 in. (1.141-1.781 mm)
Exhaust valve	0.0650-0.0980 in. (1.651-2.489 mm)

(continued)

Table 3 VALVE SERVICE SPECIFICATIONS (continued)

Measurement	Specification
Seat width (continued)	
GM 5.7L models	
With 12-bolt intake manifold	
Intake valve	0.0311-0.0625 in. (0.790-1.588 mm)
Exhaust valve	0.0625-0.0937 in. (1.588-2.380 mm)
With eight-bolt intake manifold	
Intake valve	0.0402-0.0650 in. (1.021-1.651 mm)
Exhaust valve	0.0591-0.1008 in. (1.501-2.560 mm)
GM 7.4L models	
GL and GI models	
Intake valve	0.0300-0.0600 in. (0.762-1.524 mm)
Exhaust valve	0.0600-0.0950 in. (1.524-2.413 mm)
GSI models	
Intake valve	0.0311-0.0625 in. (0.790-1.588 mm)
Exhaust valve	0.0625-0.0937 in. (1.588-2.380 mm)
GM 8.2L models	
Intake valve	0.0300-0.0600 in. (0.762-1.524 mm)
Exhaust valve	0.0600-0.0950 in. (1.524-2.413 mm)

Table 4 CRANKSHAFT/OIL PUMP AND CONNECTING ROD SERVICE SPECIFICATIONS

Measurement	Specification
Connecting rod bearing clearance	
GM 3.0L models	0.0009-0.0027 in. (0.023-0.069 mm)
GM 4.3L models	
With a 12-bolt intake (1994-1996)	0.0013-0.0030 in. (0.033-0.076 mm)
With an eight-bolt intake (1997-2000)	0.0010-0.0030 in. (0.025-0.076 mm)
GM 5.0L and 5.7L models	
With a 12-bolt intake (1994-1997)	0.0013-0.0030 in. (0.033-0.076 mm)
With an eight-bolt intake (1998-2000)	0.0010-0.0030 in. (0.025-0.076 mm)
Ford 5.0L and 5.8L models	0.0008-0.0025 in. (0.020-0.064 mm)
GM 7.4L and 8.2L models	0.0011-0.0030 in. (0.028-0.076 mm)
Connecting rod side clearance	
GM 3.0L models	0.008-0.015 in. (0.203-0.381 mm)
GM 4.3L models	
With a 12-bolt intake (1994-1996)	0.008-0.014 in. (0.203-0.356 mm)
With an eight-bolt intake (1997-2000)	0.006-0.017 in. (0.152-0.432 mm)
GM 5.0L and 5.7L models	
With a 12-bolt intake (1994-1997)	0.008-0.014 in. (0.203-0.356 mm)
With an eight-bolt intake (1998-2000)	0.006-0.024 in. (0.152-0.610 mm)
Ford 5.0L and 5.8L models	0.010-0.023 in. (0.254-0.584 mm)
GM 7.4L and 8.2L models	0.0013-0.023 in. (0.033-0.584 mm)
Crankpin journal diameter	
GM 3.0L models	2.0989-2.0995 in. (53.312-53.327 mm)
GM 4.3L models	
With a 12-bolt intake (1994-1996)	2.0988-2.0998 in. (53.310-53.335 mm)
With a eight-bolt intake (1997-2000)	2.2487-2.2497 in. (57.117-57.142 mm)
GM 5.0L and 5.7L models	2.0978-2.0998 in. (53.284-53.335 mm)
Ford 5.0L models	2.1228-2.1236 in. (53.919-53.939 mm)
Ford 5.8L models	2.3103-2.3111 in. (58.682-58.702 mm)
GM 7.4L and 8.2L models	2.1990-2.1996 in. (55.855-55.870 mm)
Crankpin journal out-of-round (maximum)	
GM four-cylinder, V-6 and V-8 models	0.001 in. (0.025 mm)
Ford 5.0L and 5.8L models	0.0006 in. (0.015 mm)
Crankpin journal taper (maximum)	
GM four-cylinder, V-6 and V-8 models	0.001 in. (0.025 mm)
Ford 5.0L and 5.8L models	0.0006 in. (0.015 mm)
Crankshaft end play	
GM 3.0L models	0.002-0.006 in. (0.051-0.152 mm)
GM 4.3L models	
With a 12-bolt intake (1994-1996)	0.002-0.006 in. (0.051-0.152 mm)

(continued)

Table 4 CRANKSHAFT/OIL PUMP AND CONNECTING ROD SERVICE SPECIFICATIONS (continued)

Measurement	Specification
Crankshaft end play (continued)	
GM 4.3L models	
With an eight-bolt intake (1997-2000)	0.002-0.008 in. (0.051-0.203 mm)
GM 5.0L and 5.7L models	0.002-0.008 in. (0.051-0.203 mm)
Ford 5.0L and 5.8L models	0.002-0.008 in. (0.051-0.203 mm)
GM 7.4L models	
74GL and 74GI models	0.005-0.011 in. (0.127-0.279 mm)
74GS and 74GSI models (high output)	0.006-0.010 in. (0.152-0.254 mm)
GM 8.2L models	0.006-0.010 in. (0.152-0.254 mm)
Crankshaft runout (maximum)	
All models	0.002 in. (0.051 mm)
Flywheel runout (maximum)	0.008 in. (0.203 mm)
Main bearing journal diameter	
GM 3.0L models	2.2979-2.2994 in. (58.367-58.405 mm)
GM 4.3L models	
With a 12-bolt intake (1994-1996)	
Front main (No. 1) journal	2.4484-2.4493 in. (62.189-62.212 mm)
Center main (No. 2 and 3) journals	2.4481-2.4490 in. (62.182-62.205 mm)
Rear main (No. 4) journal	2.4479-2.4488 in. (62.177-62.199 mm)
With an eight-bolt intake (1997-2000)	
Front main (No. 1) journal	2.4488-2.4495 in. (62.199-62.217 mm)
Center main (No. 2 and 3) journals	2.4485-2.4494 in. (62.192-62.215 mm)
Rear main (No. 4) journal	2.4480-2.4489 in. (62.180-62.202 mm)
GM 5.0L and 5.7L models	
Front main (No. 1) journal	2.4484-2.4493 in. (62.189-62.212 mm)
Center main (No. 2-4) journals	2.4481-2.4490 in. (62.182-62.205 mm)
Rear main (No. 5) journal	2.4479-2.4491 in. (62.177-62.207 mm)
Ford 5.0L models	2.4482-2.4490 in. (62.184-62.205 mm)
Ford 5.8L models	2.9994-3.0002 in. (76.185-76.205 mm)
GM 7.4L and 8.2L models	
All journals	2.7482-2.7489 in. (69.804-69.822 mm)
Main bearing journal out-of-round	
All models	0.001 in. (0.025 mm)
Main bearing journal taper (maximum)	
All models	0.001 in. (0.025 mm)
Main bearing oil clearance	
GM 3.0L models	
All journals	0.0003-0.0029 in. (0.008-0.074 mm)
GM 4.3L V-6 models	
With a 12-bolt intake (1994-1996)	
Front main (No. 1) bearing	0.0010-0.0015 in. (0.025-0.038 mm)
Center main (No. 2 and No. 3) bearings	0.0010-0.0025 in. (0.025-0.064 mm)
Rear main (No. 4) bearing	0.0025-0.0035 in. (0.064-0.089 mm)
With an eight-bolt intake (1997-2000)	
Front main (No. 1) bearing	0.0008-0.0015 in. (0.020-0.038 mm)
Center main (No. 2 and No. 3) bearings	0.0011-0.0025 in. (0.028-0.064 mm)
Rear main (No. 4) bearing	0.0017-0.0035 in. (0.043-0.089 mm)
GM 5.0L and 5.7L V-8 models	
With a 12-bolt intake	
Front main (No. 1) bearing	0.0010-0.0025 in. (0.025-0.064 mm)
Center main (No. 2-4) bearings	0.0010-0.0025 in. (0.025-0.064 mm)
Rear main (No. 5) bearing	0.0015-0.0030 in. (0.038-0.076 mm)
With an eight-bolt intake (1998-2000)	
Front main (No. 1) bearing	0.0010-0.0020 in. (0.025-0.051 mm)
Center main (No. 2-4) bearings	0.0010-0.0025 in. (0.025-0.064 mm)
Rear main (No. 5) bearing	0.0010-0.0030 in. (0.025-0.076 mm)
Ford 5.0L models	
Front main (No. 1) bearing	0.0010-0.0020 in. (0.0254-0.051 mm)
Center and rear (No. 2-5) bearings	0.0005-0.0024 in. (0.013-0.061 mm)
Ford 5.8L models	
No. 1 (front) bearing	0.0008-0.0026 in. (0.020-0.066 mm)
No. 2-5 bearings	0.0005-0.0024 in. (0.013-0.061 mm)
GM 7.4L and 8.2L models	
Front main (No. 1) bearing	0.0010-0.0030 in. (0.025-0.076 mm)

(continued)

6

Table 4 CRANKSHAFT/OIL PUMP AND CONNECTING ROD SERVICE SPECIFICATIONS (continued)

Measurement	Specification
Main bearing oil clearance (continued)	
GM 7.4L and 8.2L models	
Center main (No. 2-4) bearings	0.0010-0.0030 in. (0.025-0.076 mm)
Rear main (No. 5) bearing	0.0025-0.0040 in. (0.064-0.102 mm)
Oil pump	
Ford 5.0L and 5.8L models	
Inner rotor-to-pump cover	0.001-0.004 in. (0.025-0.102 mm)
Outer rotor-to-pump body	0.001-0.004 in. (0.025-0.102 mm)

Table 5 PISTON SERVICE SPECIFICATIONS

Measurement	Specification
Cylinder bore-to-piston skirt clearance	
GM 3.0L models	0.0025-0.0035 in. (0.064-0.089 mm)
GM 4.3L models	
1994-1996	0.0007-0.0027 in. (0.018-0.069 mm)
1997-2000	0.0007-0.0024 in. (0.018-0.061 mm)
GM 5.0L and 5.7L models	
1994-1997	0.0007-0.0027 in. (0.018-0.069 mm)
1998-2000	0.0007-0.0026 in. (0.018-0.066 mm)
Ford 5.0L and 5.8L models	0.0018-0.0026 in. (0.046-0.066 mm)
GM 7.4L models	
74GL and 74 GI models	0.0018-0.0030 in. (0.046-0.076 mm)
74GS and 74GSI models	0.0025-0.0075 in. (0.064-0.191 mm)
GM 8.2L models	0.0040-0.0060 in. (0.102-0.152 mm)
Ring end gap	
GM 3.0L models	
Top compression ring	0.010-0.020 in. (0.254-0.508 mm)
Second compression ring	0.013-0.025 in. (0.330-0.635 mm)
Oil ring rail	0.015-0.055 in. (0.381-1.397 mm)
GM 4.3L models	
1994-1996 (with 12-bolt intake)	
Top compression ring	0.010-0.020 in. (0.254-0.508 mm)
Second compression ring	0.010-0.025 in. (0.254-0.635 mm)
Oil ring rail	0.015-0.055 in. (0.381-1.397 mm)
1997-2000 (with eight-bolt intake)	
Top compression ring	0.010-0.035 in. (0.254-0.889 mm)
Second compression ring	0.010-0.035 in. (0.254-0.889 mm)
Oil ring rail	0.015-0.065 in. (0.381-1.651 mm)
GM 5.0L and 5.7L models	
1994-1997 (with 12-bolt intake)	
Top compression ring	0.010-0.020 in. (0.254-0.508 mm)
Second compression ring	0.010-0.025 in. (0.254-0.635 mm)
Oil ring rail	0.015-0.055 in. (0.381-1.397 mm)
1998-2000 (with eight-bolt intake)	
Top compression ring	0.012-0.026 in. (0.305-0.660 mm)
Second compression ring	0.012-0.026 in. (0.305-0.660 mm)
Oil ring rail	0.010-0.030 in.(0.254-0.762 mm)
Ford 5.0L and 5.8L models	
Top compression ring	0.010-0.020 in. (0.254-0.508 mm)
Second compression ring	0.010-0.020 in. (0.254-0.508 mm)
Oil ring rail	0.015-0.055 in. (0.381-1.397 mm)
GM 7.4L models	
74GL or 74GI models	
Top compression ring	0.010-0.028 in. (0.254-0.711 mm)
Second compression ring	0.016-0.034 in. (0.406-0.864 mm)
Oil ring	0.010-0.040 in. (0.254-1.016 mm)
74GS or 74 GSI models	
Top compression ring	0.010-0.034 in. (0.254-0.864 mm)
Second compression ring	0.016-0.034 in. (0.406-0.864 mm)
Oil ring	0.010-0.040 in. (0.254-1.016 mm)

(continued)

Table 5 PISTON SERVICE SPECIFICATIONS (continued)

Measurement	Specification
Ring end gap (continued)	
GM 8.2L models	
Top compression ring	0.010-0.028 in. (0.254-0.711 mm)
Second compression ring	0.016-0.034 in. (0.406-0.864 mm)
Oil ring	0.010-0.040 in. (0.254-1.016 mm)
Ring groove clearance	
GM 3.0L four-cylinder models	
Top and second compression rings	0.0020-0.0035 in. (0.051-0.089 mm)
GM 4.3L models	
With 12-bolt intake manifold	
Top and second compression ring	0.0012-0.0032 in. (0.031-0.081 mm)
Oil ring	0.0020-0.0070 in. (0.051-0.178 mm)
With eight-bolt intake manifold	
Top and second compression ring	0.0012-0.0032 in. (0.031-0.081 mm)
Oil ring	0.0020-0.0078 in. (0.051-0.198 mm)
GM 5.0L and 5.7L models	
With 12-bolt intake manifold	
Top compression ring	0.0012-0.0032 in. (0.031-0.081 mm)
Oil ring	0.0020-0.0070 in. (0.051-0.178 mm)
With eight-bolt intake manifold	
Top and second compression ring	0.0012-0.0027 in. (0.031-0.069 mm)
Oil ring	0.0020-0.0060 in. (0.051-0.152 mm)
Ford 5.0L and 5.8L models	
Top and second compression rings	0.0020-0.0040 in. (0.051-0.120 mm)
GM 7.4L and 8.2L models	
Top and second compression rings	0.0012-0.0039 in. (0.031-0.099 mm)
Oil ring	0.0050-0.0075 in. (0.127-0.191 mm)
Ring groove width	
GM 3.0L models	
Oil ring groove	0.188-0.189 in. (4.775-4.801 mm)
Ring width	
GM 3.0L models	
Top and second compression rings	0.0775-0.0780 in. (1.969-1.981 mm)

Table 6 CYLINDER BORE SPECIFICATIONS

Measurement	Specification
Cylinder block warpage	0.004 in. (0.102 mm)
Cylinder bore diameter	
GM 3.0L models	4.0007-4.0017 in. (101.618-101.643 mm)
GM 4.3L models	4.0007-4.0017 in. (101.618-101.643 mm)
GM 5.0L models	3.7350-3.7384 in. (94.869-94.955 mm)
Ford 5.0L models	4.0007-4.0017 in. (101.618-101.643 mm)
GM 5.7L models	4.0007-4.0017 in. (101.618-101.643 mm)
Ford 5.8L models	
GM 7.4L models	
74GL and 74GI models	4.2500-4.2507 in. (107.950-107.968 mm)
74GS and 74GSI models	4.2451-4.2525 in. (107.826-108.014 mm)
GM 8.2L models	4.4655-4.4662 in. (113.424-113.442 mm)
Cylinder bore out-of-round (maximum)	
GM 3.0L models	0.002 in. (0.051 mm)
GM V-6 and V-8 models	0.002 in. (0.051 mm)
Ford 5.0L and 5.8L models	0.001 in. (0.025 mm)
Cylinder bore taper (maximum)	
GM 3.0L models	0.002 in. (0.051 mm)
GM V-6 and V-8 models	0.001 in. (0.025 mm)
Ford 5.0L and 5.8L models	0.001 in. (0.025 mm)
Cylinder head warpage	0.007 in. (0.178 mm)

Chapter Seven

Fuel System

Review *Safety* in Chapter One before performing any service procedures on the fuel system.

Refer to *Servicing the Fuel System* in this chapter prior to removal and disassembly of any fuel system component. This section provides information needed to perform a safe, reliable and effective repair.

Table 1 lists torque specifications for most fuel system fasteners. Use the torque recommendations in Chapter One for fasteners not listed in **Table 1**. **Table 1** is located at the end of this chapter.

WARNING

Use caution when working with the fuel system. Never smoke around fuel or fuel vapors. Make sure no flame or source of ignition is present in the work area. Flame or sparks can ignite fuel or vapor and cause a fire or explosion.

SERVICING THE FUEL SYSTEM

Always use gloves and eye protection when working with the fuel system. Take all necessary precautions against fire or explosion. To prevent arcing or accidental starting, always disconnect the battery cables *before* servicing any marine engine.

Pay close attention when removing and installing components (especially carburetor components) to avoid installing them in the wrong locations during assembly.

Capture fuel from disconnected hoses or fittings with a small container or clean shop towel. Try to use a clear container because it allows visual inspection of the fuel. Clean the fuel tank and all other fuel delivery components if water or other contamination is noted in the fuel.

On models equipped with electronic fuel injection (EFI), always relieve the fuel pressure as described in the removal or disassembly procedures before disconnecting any fuel system hose or fitting.

When servicing the fuel system, inspect all hoses for leakage or deterioration. Damaged fuel hoses pose a safety hazard. In addition, pieces of deteriorated or damaged hoses can break free and block fuel passages in the system.

Gaskets, Seals and O-Rings

To ensure a safe and reliable repair, use only factory-recommended replacement parts. Some commonly available seals or O-rings are not suitable for contact with fuel.

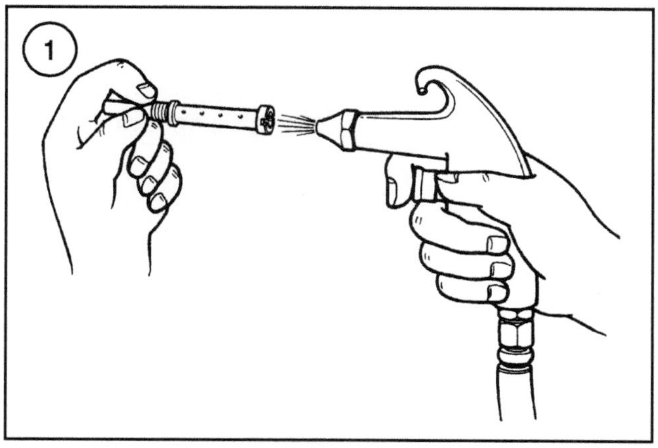

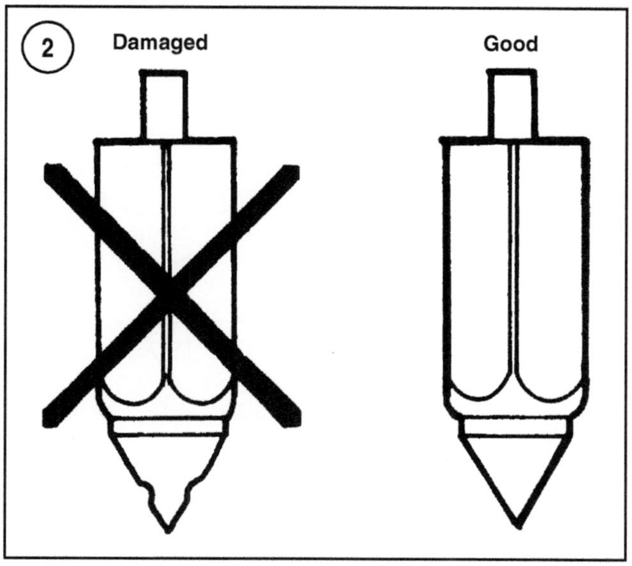

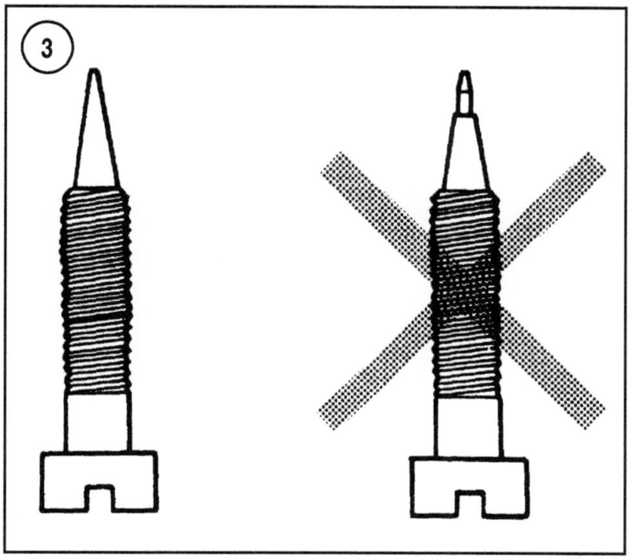

To avoid potential fuel or air leakage, replace all disturbed seals and O-rings anytime a fuel system component is removed from the engine. Consider reusing a seal or O-ring only if a replacement is not available, and the original is in excellent condition.

To help avoid the potential for contamination during lengthy downtime, have the required gaskets or repair kit on hand prior to removal and disassembly of the component(s).

Cleaning Fuel System Components

The most important step in carburetor, throttle body and fuel rail repair is the cleaning process. Use only solvents suitable for use on carburetors. Some cleaning agents can damage fuel system components. Aerosol carburetor cleaners are available at most automotive parts stores. They effectively remove most stubborn deposits. Avoid using any solvents not suitable for aluminum.

Remove all plastic or rubber components from the assembly before cleaning them with solvent. Gently scrape away gasket material with a scraper. Never scrape away any metal from the component. Use a stiff parts-cleaning brush and solvent to remove deposits from the carburetor bowl. Never use a wire brush because the sealing surfaces can quickly become damaged. Blow out all passages and orifices with compressed air (**Figure 1**). A piece of straw from a broom works well to clean small passages. Never use stiff wire for this purpose because the wire might enlarge the size of the passage and possibly alter the fuel system calibration. If the deposits are particularly difficult to remove, allow components to soak in the solvent for several hours.

Use great care and patience when removing fuel jets and other threaded or pressed-in components. Clean the passage without removing the jet if it cannot be removed without causing damage. Carburetor fuel jets are easily damaged.

One small particle in the carburetor or fuel injection system can compromise the cleaning process. Continue to clean until *all* deposits and debris are removed.

Inspecting Fuel System Components

Place all components on a clean surface as they are removed from the carburetor, throttle body or fuel rail. Arrange these components in a manner consistent with the provided illustrations to save time and make sure the parts are installed in the correct locations.

Inspect the inlet needle (carburetor-equipped models) for wear or deterioration (**Figure 2**). Replace the inlet needle unless the tip is perfectly cone-shaped (**Figure 2**). Inspect the inlet needle seat for grooved or damaged surfaces. Carburetor flooding is likely if a worn or faulty inlet needle or seat is used.

Inspect the tip of the idle mixture adjusting screw (carburetor-equipped models) for wear or damage (**Figure 3**). Damage to the tip usually occurs from improper seating of the screw during adjustment. In many instances, the seat for the tip is also damaged. Damage to the screw or seat will cause rough idle or improper off idle engine operation. Replace the screw or carburetor when worn or faulty components are noted.

Inspect the float (**Figure 4**, typical) for wear or damage. Some floats are made of a translucent material to allow visual detection of fuel within the float.

Push a thumbnail gently against the material on nontranslucent floats. A leaking or saturated float is indicated if fuel appears at the thumbnail contact area. Replace the float if visibly damaged, leaking or saturated. Check the float for free movement while moving on the float pin. Replace the float if it does not move freely.

Check and correct float level settings (**Figure 5**, typical) prior to assembling the carburetor. Use an accurate ruler or a caliper with depth-reading capability. Set the float *exactly* as specified to help ensure proper carburetor operation. Specific instructions are provided in the assembly instructions.

When servicing the carburetor or throttle body, move the throttle lever from closed to wide open throttle. Remove the throttle plate and repeat this step if binding or rough operation is noted. Continued binding indicates a bent throttle shaft. If free movement is noted with the plate removed, an improperly aligned or damaged throttle plate is evident. Replace the throttle shaft, carburetor or throttle body as necessary to eliminate binding. Apply a suitable threadlocking compound and stake all throttle plate retaining screws to prevent loosening.

Fuel jets (**Figure 6**) meter the fuel flow through various passages in the carburetor. Along with other components, they allow the carburetor to deliver the precise amount of fuel needed for the engine. Fuel jet sizes vary by model and carburetor location on the engine. Fuel jets normally have numbers stamped on the sides or ends. Note the fuel jet number and location in the carburetor prior to removal. Always reinstall the fuel jets and other carburetor components in the correct locations.

Purchase replacement jets from an OMC dealership or a carburetor specialty shop. For proper engine operation, replacement jets must have the same sizes and shapes of openings as the original fuel jets. Improper engine operation, increased exhaust emissions or potentially serious engine damage can result from using incorrect fuel jets.

Using the engine at elevations above 5000 ft. (1524 m) might require alternate fuel jets to achieve optimal engine operation. If necessary, contact an OMC dealership in the area with a similar elevation for recommended jet changes.

Never install a damaged jet in the carburetor. The fuel or air flow characteristics might be altered. Altering the fuel and air flow can cause an engine malfunction or potentially serious power head damage.

MECHANICAL LOW-PRESSURE FUEL PUMP

A mechanical low-pressure fuel pump (**Figure 7**, typical) is used on all GM 3.0L models, 1994-1997 GM 5.0L and 5.7L models and 1994-1998 Ford 5.0L and 5.8L carburetor-equipped models. An electric fuel pump is used on all other models covered in this manual.

4

5

FLOAT LEVEL ADJUSTMENT (TYPICAL)

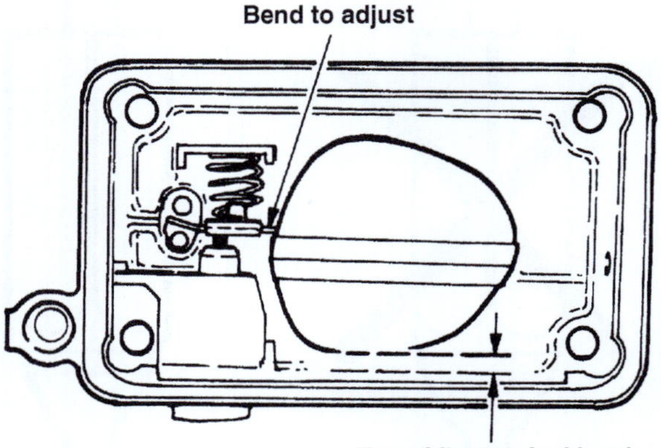

WARNING

Use only the manufacturer-recommended parts when replacing fuel hoses or tubing. Other types of hoses or tubing might not meet the requirements and fail prematurely, leak fuel and cause an extreme fire and explosion hazard.

WARNING

Rust deposits in the fuel filter might indicate rusting of the boat fuel tank and the need for immediate inspection or replacement. Rusting will eventually allow fuel to leak from the tank and cause a dangerous fire and explosion hazard.

CAUTION

Rust or other deterioration in the fuel tank will migrate to and block filters and other fuel system components. The blockage will restrict fuel flow and lead to a lean operating condition and possible engine damage.

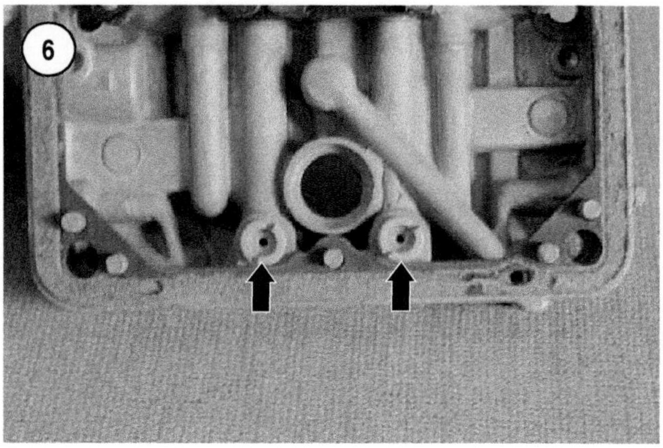

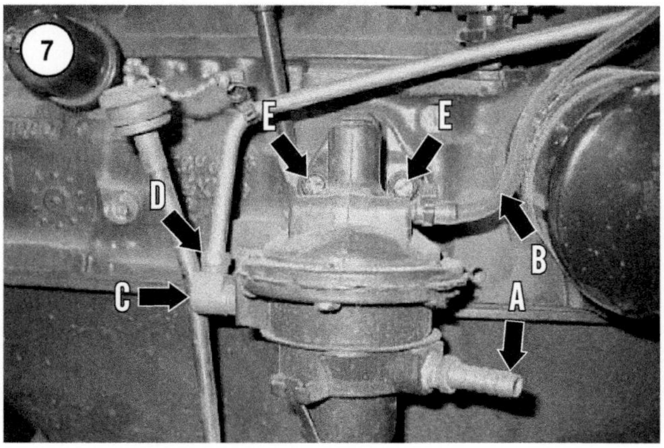

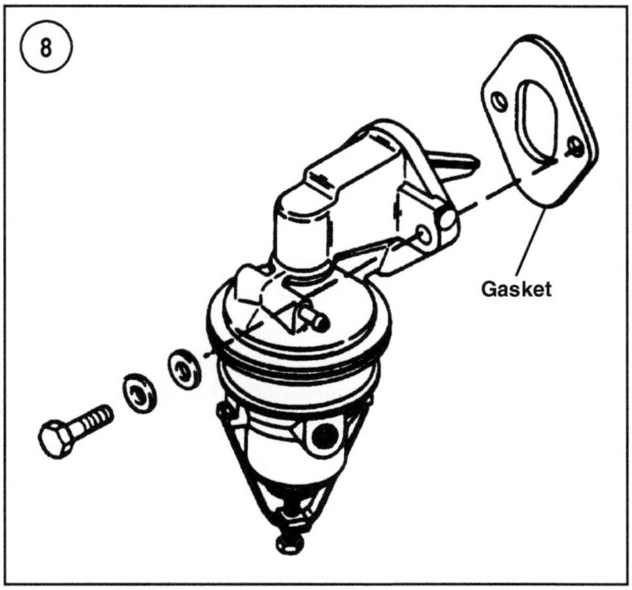

GM 3.0L Models

Fuel pump removal and installation

Two bolts secure the mechanical fuel pump to the starboard side of the cylinder block. Replace the pump if faulty. The manufacturer does not offer a repair kit. The replaceable fuel filter is located in the canister in the bottom of the fuel pump. Fuel filter replacement is described in this section.

1. Disconnect the negative battery cable. Use compressed air to blow debris and loose material off the mechanical fuel pump and cylinder block mating surfaces.
2. Place a suitable container or shop towel under the fuel pump to capture spilled fuel.

WARNING
Use only the manufacturer-recommended parts when replacing fuel hoses or tubing. Other types of hoses or tubing might not meet the requirements and fail prematurely, leak fuel, causing an extreme fire and explosion hazard.

3. Loosen the clamp and then push the fuel supply hose off the fuel pump inlet fitting (A, **Figure 7**). Drain residual fuel from the supply hose. Inspect the fuel supply hose for brittle, cracked, deteriorated or damaged surfaces. Replace the fuel supply hose with the manufacturer recommended marine grade and U.S. Coast Guard-approved hose.
4. Disconnect the sight tube (B, **Figure 7**) from the fuel pump fitting. Inspect the sight tube for brittle, deteriorated or cracked surfaces. Replace the sight tube with the manufacturer-recommended tube if these or other defects are evident. *Do not* replace the tube with any other type of hose or tubing.
5. Engage a wrench onto the elbow fitting (C, **Figure 7**) to prevent rotation. Then use another wrench to *slightly* loosen the fuel line fitting (D, **Figure 7**). Wrap a shop towel around the fitting. Then loosen the fitting by hand. Allow residual fuel to drain from the fuel line. When completely drained, unthread the fitting and disconnect the fuel line from the pump.
6. Support the pump and then remove the mounting bolts (E, **Figure 7**). Pull the pump off the cylinder block. Remove the gasket (**Figure 8**) from the mounting plate or fuel pump. Discard the gasket.
7. Carefully scrape all gasket material from the cylinder block and fuel pump mating surfaces. *Do not* damage the mating surfaces.
8. If the fuel pump will be replaced, remove the brass elbow fitting (C, **Figure 7**) from the pump. Clean all old sealant from the fitting.
9. Apply a coat of OMC Gasket Sealing Compound (OMC part No. 1141570-0) to both sides of the *new* gasket (**Figure 8**) and the threads of the fuel pump mounting bolts. Fit the gasket onto the pump. Insert the bolts and washers into the pump and gasket openings to hold the gasket in position.
10. Guide the pump arm into the cylinder block opening. The top of the pump arm must contact the camshaft lobe. Seat the fuel pump and gasket against the cylinder block. Thread the mounting bolts into the block. Then tighten the bolts evenly to 20-25 ft.-lb. (27-34 N•m).
11. If removed, apply a light coat of Loctite Pipe Sealant with Teflon to the threaded section of the brass elbow and

7

the fuel inlet fittings. Purchase the sealant from an OMC dealership (OMC part No. 1141570), automotive parts store or GM automobile dealership (GM part No. 12346004). *Do not* allow sealant into the fuel passage of the fittings. Thread the elbow fitting (C, **Figure 7**) into the fuel pump and tighten to 48-59 in.-lb. (5-7 N•m). Tighten *slightly* more if necessary to position the fuel line opening facing in the up direction. Thread the fuel inlet fitting (A, **Figure 7**) into the fuel pump by hand and then tighten to 48-59 in.-lb. (5-7 N•m).

12. Fit the fuel line fitting into the brass elbow fitting. Thread the line fitting into the elbow by hand. Engage a wrench onto the elbow fitting to prevent rotation. Then tighten the fuel line fitting to 177-212 in.-lb. (20-24 N•m).

13. Install the hose clamp over the end. Then push the fuel supply hose fully over the fuel inlet fitting (A, **Figure 7**). Align the hose clamp over the fitting and then securely tighten the clamp.

14. Push the sight tube (B, **Figure 7**) fully over the fuel pump fitting. Secure the tube to the fitting and fuel line with plastic locking clamps. Do not tighten the clamp to the point that the tube begins to collapse.

WARNING
Stay clear of the propeller and propeller shaft while running an engine on a flush adapter.

CAUTION
If running the engine with the boat out of the water, use a flush device or other means to supply the engine with cooling water. The seawater pump is quickly damaged if operated without adequate cooling water.

15. Connect the battery cable. Start the engine and immediately check for fuel leaking from the fitting and oil leaking from the pump-to-cylinder block mating surfaces. Correct any leaks before operating the engine.

Fuel filter replacement

The mechanical fuel pump fuel filter can be replaced without removing the fuel pump.

Refer to **Figure 9** for this procedure.

1. Disconnect the negative battery cable. Use compressed air to blow debris and loose material off the mechanical fuel pump.

2. Place a suitable container or shop towel under the fuel pump to capture spilled fuel.

WARNING
Use only the manufacturer-recommended parts when replacing fuel hoses or tubing. Other types of hoses or tubing might not meet the requirements and fail prematurely, leak fuel and cause an extreme fire and explosion hazard.

9 FUEL FILTER REPLACEMENT (GM 3.0L MODELS)

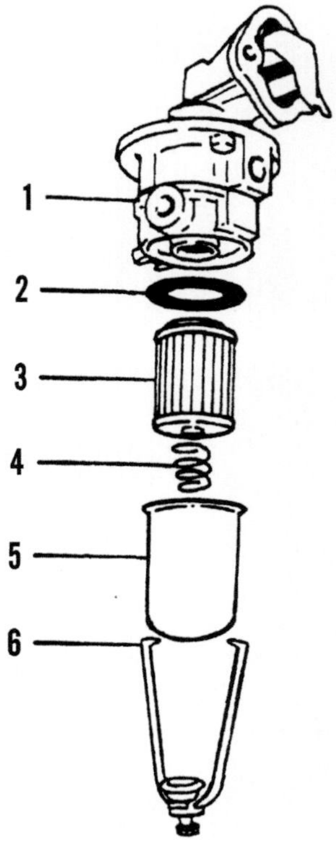

1. Mechanical fuel pump
2. Rubber gasket
3. Filter element
4. Spring
5. Filter canister
6. Retainer

3. Loosen the clamp and then push the fuel supply hose off the fuel pump inlet fitting (A, **Figure 7**). Drain residual fuel from the supply hose. Inspect the fuel supply hose for brittle, cracked, deteriorated or damaged surfaces. Replace the fuel supply hose with the manufacturer-recommended marine grade and U.S. Coast Guard-approved hose if these or other defects are evident.

4. Fully loosen the screw at the bottom of the retainer (6, **Figure 9**).

5. Support the canister (5, **Figure 9**) and then move the retainer to one side. Do not disconnect the retainer from the pump. Lower the canister, filter element (3, **Figure 9**) and spring (4) from the pump.

6. Remove the filter element and spring from the canister. Remove the rubber gasket (2, **Figure 9**) from the filter element or mating surface on the mechanical fuel pump. Discard the filter element and rubber seal.

7. Use aerosol carburetor cleaner to remove all contaminants from the canister, retainer and filter mating surfaces in the mechanical fuel pump.

WARNING
Rust deposits in the fuel filter might indicate rusting of the boat fuel tank and the need for immediate inspection or replacement. Rusting will eventually allow fuel to leak from the tank and cause a dangerous fire and explosion hazard.

CAUTION
Rust or other deterioration in the fuel tank will migrate to and block filters and other fuel system components. The blockage will restrict fuel flow and lead to a lean operating condition and possible engine damage.

8. If rust is found in the filter or canister, inspect the fuel tank for rust damage. Replace the fuel tank if rust or deterioration is evident.

9. Inspect the spring (4, **Figure 9**) for corrosion or lost spring tension and replace as needed. Inspect the canister for corrosion pitting or other damage and replace as needed.

10. Insert the spring (4, **Figure 9**) into the canister and rest over the raised boss in the bottom. Install the *new* filter element (3, **Figure 9**) into the canister (5) with the open side facing upward. Then raise the boss on the bottom centered over the spring.

11. Fit a *new* rubber gasket (2, **Figure 9**) onto the filter. Guide the canister, filter and gasket into the recessed fuel pump mating surface. Seat the canister against the pump. Maintain pressure to collapse the spring. Then swing the retainer (6, **Figure 9**) over the canister.

12. Make sure both retainer arms are positioned in the slots in the pump body. Center the screw and clamping pad over the depression in the bottom of the canister and then securely tighten the screw.

13. Install the hose clamp over the end and then push the fuel supply hose fully over the fuel inlet fitting (A, **Figure 7**). Align the hose clamp over the fitting. Then securely tighten the clamp.

14. Clean up any spilled fuel.

WARNING
Stay clear of the propeller and propeller shaft while running an engine on a flush adapter.

CAUTION
If running the engine with the boat out of the water, use a flush device or other means to supply the engine with cooling water. The seawater pump is quickly damaged if operated without adequate cooling water.

15. Connect the battery cable. Start the engine and immediately check for leaking fuel. Correct any leaks before operating the engine.

GM 5.0L and 5.7L (1994-1997) Models

The fuel pump mounts on the starboard side of the cylinder block directly below the No. 2 spark plug opening.

Removal and installation

1. Disconnect the battery cables.

2. Place a suitable container or shop towel under the fuel pump to capture spilled fuel.

WARNING
Use only the manufacturer-recommended parts when replacing fuel hoses or tubing. Other types of hoses or tubing might not meet the requirements and fail prematurely, leak fuel and cause an extreme fire and explosion hazard.

3. Loosen the clamp or fitting and then disconnect the supply hose or line from the fuel pump inlet fitting (A, **Figure 7**). Drain residual fuel from the supply hose.
   a. If a hose connects to the fuel inlet fitting, Inspect the hose for brittle, cracked, deteriorated or damaged surfaces. Replace the fuel supply hose with the manufacturer recommended marine grade and U.S. Coast Guard-approved hose.
   b. If a metal line and fitting connect to the inlet fitting, inspect the line for twisted, cracked, corroded, bent or kinked areas. Inspect the fitting for damaged threads, damaged seating surfaces or rounding of the hex section. Replace the line and fitting as an assembly if these or other defects are evident.

4. Disconnect the sight tube (B, **Figure 7**) from the fuel pump fitting. Inspect the sight tube for brittle, deteriorated or cracked surfaces. Replace the sight tube with the manufacturer-recommended tube if these or other defects are evident. *Do not* replace the tube with any other type of hose or tubing.

5. Engage a wrench onto the elbow (C, **Figure 7**) or outlet line fitting that threads into the pump body to prevent rotation. Then use a flare nut wrench to *slightly* loosen the outlet fuel line fitting (D, **Figure 7**). Wrap a shop towel around the fitting and then loosen the fitting by hand. Allow residual fuel to drain from the fuel line. When completely drained, unthread the fitting and disconnect the outlet fuel line from the pump. Cap or plug the hoses or lines to prevent leakage or contamination.

6. Loosen the two pump attaching bolts (E, **Figure 7**) and lockwashers. Do not remove the bolts at this time.

7. Hold the pump in position while an assistant rotates the crankshaft pulley in the clockwise direction (as viewed from the pulleys). Have the assistant stop rotating the engine when you feel a reduction in tension against the pump. This indicates the low point of the camshaft eccentric is aligned with the pump rod.

8. Remove the bolts and pull the pump off the mounting plate (**Figure 8**). Drain any residual fuel from the inlet and outlet fittings. Remove the gasket from the fuel pump or mounting pump mating surface. Discard the gasket.

9. Remove the two screws at the bottom of the mounting plate. Then carefully pull the plate off the cylinder block. Be prepared to capture the pump pushrod as the plate is removed. Note which end of the pushrod was driving against the fuel pump and mark accordingly. If reused, the pushrod must be installed with the same end driving against the camshaft eccentric. Remove the gasket from the mounting plate or cylinder block mating surface. Discard the gasket. Promptly clean up any spilled fuel.

10. Inspect the pushrod for excessive wear, discoloration or bending and replace as needed. Discoloration or excessive wear is caused insufficient lubrication. Insufficient lubrication is usually caused by fuel leaking from the fuel pump and diluting the lubricating oil. Replace the fuel pump if it is suspected of leaking.

11. Thoroughly clean the fuel pump, mounting plate and mounting surface on the cylinder block. Make sure to scrape all residual gasket material and sealant from the mating surfaces. *Do not* damage the mating surfaces.

12. If the fuel pump will be replaced, remove the pump inlet and outlet fittings. Clean all old sealant from the fitting.

13. Apply a heavy coat of clean grease to the end of the pushrod. Then insert the pushrod into the bore in the mount opening. If reusing the pushrod, make sure the end that was driving against the pump arm is facing outward. Push on the rod until it fully seats against the camshaft eccentric.

14. Apply a coat of OMC Gasket Sealing Compound (OMC part No. 1141570-0) to both sides of the *new* mounting plate-to-cylinder block gasket and the threads of the mounting plate bolts. Fit the gasket onto the mounting plate. Then fit the plate and gasket onto the cylinder block. Make sure the gasket is located between the mounting plate and block. Move the gasket to perfectly align the four bolt openings in the plate and gasket with the corresponding openings in the cylinder block. Thread the two screws into the lower openings and tighten by hand only.

15. Apply a coat of OMC Gasket Sealing Compound (OMC part No. 1141570-0) to both sides of the *new* fuel pump-to-mounting plate gasket and the mounting bolts. Fit the gasket onto the fuel pump mating surface.

16. Position the pump with the pump arm angled downward and toward the mounting plate. Carefully guide the pump onto the mounting plate. To avoid dislodging the pump rod, guide the pump arm into the lower end of the elongated opening in the mounting plate. Seat the pump against the mounting plate. Move the gasket and pump to align the mounting bolt opening. Then thread the bolts into the pump, mounting plate and cylinder block openings. Tighten the bolts evenly to 20 ft.-lb. (27 N•m). Then tighten the two lower mounting plate bolts to 15 ft.-lb. (20 N•m).

17. If removed, apply a light coat of Loctite Pipe Sealant with Teflon to the threaded section of the fuel inlet and outlet fittings. Purchase the sealant from an OMC dealership (OMC part No. 1141570), automotive parts store or GM automobile dealership (GM part No. 12346004). *Do not* allow sealant into the fuel passage of the fittings. Thread the elbow or outlet fitting (D, **Figure 7**) into the fuel pump and tighten to 48-59 in.-lb. (5-7 N•m). If necessary, tighten the elbow *slightly* to properly align with the fuel outlet line. Thread the fuel inlet barb fitting (A, **Figure 7**) or flare fitting into the fuel pump by hand. Then tighten to 48-59 in.-lb. (5-7 N•m). If the inlet line is a metal line, tighten the inlet fitting *slightly* to align the fitting opening with the inlet line.

18. Thread the fuel outlet line fitting into the elbow or outlet fitting on the fuel pump. Engage a wrench onto the fuel pump elbow or fitting to prevent rotation. Then tighten the fuel line fitting to 177-212 in.-lb. (20-24 N•m).

19A. If a rubber fuel inlet hose is used, install the hose clamp over the end and then push the fuel supply hose fully over the fuel inlet fitting (A, **Figure 7**). Align the hose clamp over the fitting and then securely tighten the clamp.

19B. If a metal fuel inlet line is used, thread the fuel inlet line fitting into the inlet fitting on the fuel pump. Engage a wrench onto the fuel pump inlet fitting to prevent rotation. Then tighten the fuel line fitting to 177-212 in.-lb. (20-24 N•m).

20. Push the sight tube (B, **Figure 7**) fully over the fuel pump fitting. Secure the tube onto the fitting and fuel line with plastic locking clamps. Do not tighten the clamp to the point that the tube begins to collapse.

21. Connect the battery cables.

WARNING

Stay clear of the propeller and propeller shaft while running an engine on a flush adapter.

CAUTION

If running the engine with the boat out of the water, use a flush device or other means to supply the engine with cooling water. The seawater pump is quickly damaged if operated without adequate cooling water.

22. Start the engine and immediately check for fuel leaking from the fitting and oil leaking from the pump-to-cylinder block mating surfaces. Stop the engine and repair any leaks before operating the engine.

Ford 5.0L and 5.8L Carburetor-Equipped Models

The fuel pump mounts on the port side of the timing chain cover.

Removal and installation

1. Disconnect the battery cables.

10

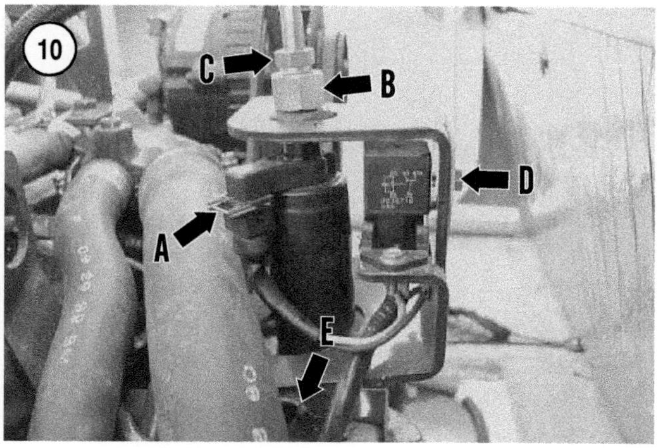

2. Place a suitable container or shop towel under the fuel pump to capture spilled fuel.

WARNING
Use only the manufacturer-recommended parts when replacing fuel hoses or tubing. Other types of hoses or tubing might not meet the requirements and fail prematurely, leak fuel and cause an extreme fire and explosion hazard.

3. Loosen the fitting that connects the fuel line to the fuel pump inlet fitting. The line connecting to the inlet fitting leads to the water-separating fuel filter and then the fuel tank. Drain residual fuel from the supply hose. Inspect the line for twisted, cracked, corroded, bent or kinked areas. Inspect the fitting for damaged threads, damaged seating surfaces or rounding of the hex section. Replace the line and fitting as an assembly if these or other defects are evident.

4. Disconnect the sight tube (B, **Figure 7**, typical) from the fuel pump fitting. Inspect the sight tube for brittle, deteriorated or cracked surfaces. Replace the sight tube with the manufacturer-recommended tube if these or other defects are evident. *Do not* replace the tube with any other type of hose or tubing.

5. To prevent rotation, engage a wrench onto the outlet line fitting that threads into the pump body. The outlet line leads to the carburetor. Then use a flare nut wrench to *slightly* loosen the outlet fuel line fitting. Wrap a shop towel around the fitting and then loosen the fitting by hand. Allow residual fuel to drain from the fuel line. When completely drained, unthread the fitting and disconnect the outlet fuel line from the pump. Cap or plug the hoses or lines to prevent leakage or contamination.

6. Loosen the two pump-attaching bolts (E, **Figure 7**, typical). Do not remove the bolts at this time.

7. Hold the pump in position while an assistant rotates the crankshaft pulley in the clockwise direction (as viewed from the pulleys). Have the assistant stop rotating the engine when you feel a reduction in tension against the pump. This indicates the low point of the camshaft eccentric is aligned with the pump arm.

8. Remove the bolts and pull the pump off the timing chain cover. Drain any residual fuel from the inlet and outlet fittings. Remove the gasket from the fuel pump or mounting pump mating surface. Discard the gasket.

9. Thoroughly clean the fuel pump and mounting surface on the front cover. Make sure to scrape all residual gasket material and sealant from the mating surfaces. *Do not* damage the mating surfaces.

10. Apply a coat of OMC Gasket Sealing Compound (OMC part No. 1141570-0) to both sides of the *new* fuel pump gasket (**Figure 8**) and the mounting bolts. Fit the gasket onto the fuel pump mating surface.

11. Position the pump with the pump arm angled slightly upward. Carefully guide the pump arm into the cover opening. Seat the pump against the front cover. Move the gasket and pump to align the mounting bolt opening. Then thread the bolts into the pump and front cover openings. Tighten the mounting bolts evenly to 18-20 ft.-lb. (24-27 N•m).

7

12. Thread the fuel outlet line fitting by hand onto the fuel pump outlet fitting. To prevent rotation, engage a wrench onto the fuel pump outlet fitting. Then tighten the fuel line outlet fitting to 120-144 in.-lb. (13-16 N•m).

13. Thread the fuel inlet line fitting by hand onto the fuel pump inlet fitting. Engage a wrench onto the fuel pump inlet fitting to prevent rotation. Then tighten the fuel line inlet fitting to 120-144 in.-lb. (13-16 N•m).

14. Push the sight tube (B, **Figure 7**, typical) fully over the fuel pump fitting. Secure the tube to the fitting and fuel line with plastic locking clamps. Do not tighten the clamp to the point that the tube begins to collapse.

15. Connect the battery cables.

WARNING
Stay clear of the propeller and propeller shaft while running an engine on a flush adapter.

CAUTION
If running the engine with the boat out of the water, use a flush device or other means to supply the engine with cooling water. The seawater pump is quickly damaged if operated without adequate cooling water.

16. Start the engine and immediately check for fuel leaking from the fitting and oil leaking from the pump-to-cylinder block mating surfaces. Stop the engine and repair any leaks before operating the engine.

LOW-PRESSURE ELECTRIC FUEL PUMP

Removal and Installation

The low-pressure electric fuel pump (**Figure 10**, typical) mounts onto a bracket at the front of the engine. The low-pressure electric fuel pump supplies fuel to the fuel reservoir (EFI models).

To prevent premature failure of the pump, replace the water-separating spin-on fuel filter (**Figure 11**, typical) when replacing the pump and at the specified maintenance intervals. Replacement of the water-separating fuel filter and replacement intervals are described in Chapter Three. Refer to *Fuel Filter Maintenance*.

1. Disconnect the negative battery cable. Use compressed air to blow debris and loose material off the electric fuel pump, fuel filter and mounting bracket.
2. Depress the locking clip (A, **Figure 10**). Then carefully unplug the harness connector from the pump.
3. Place a suitable container under the fuel pump and filter to capture spilled fuel.
4. Loosen the clamp and then *carefully* push the fuel supply hose off the filter bracket inlet fitting (**Figure 11**). Allow the fuel to drain fully from the inlet fitting and the supply hose. Inspect the fuel supply hose for brittle, cracked, deteriorated or damaged surfaces. Replace the fuel supply hose with the manufacturer-recommended marine grade and U.S. Coast Guard-approved hose if these or other defects are evident.
5. To prevent rotation, engage a wrench onto the brass fitting on the top of the pump (B, **Figure 10**). Use another wrench to *slightly* loosen the fuel line fitting (C, **Figure 10**). Wrap a shop towel around the fitting and then loosen the fitting by hand. Allow residual fuel to drain from the fuel line. When completely drained, unthread the fitting and disconnect the fuel line from the pump fitting.
6. Remove the screw (D, **Figure 10**). Then pivot the guard over the top of the fuel pump fitting and remove the guard from the bracket.
7. Support the fuel pump. Then remove the two bolts (E, **Figure 10**) that secure the fuel filter bracket to the cylinder head. Guide the fuel pump out of the upper bracket while removing the pump and filter assembly. Remove the grommet from the upper end of the fuel pump. Inspect the grommet for torn or deteriorated surfaces and replace as needed.
8. Drain residual fuel from the inlet fitting. Then place the filter and pump assembly onto a clean work surface.
9. Engage a wrench onto the brass fitting on the filter bracket. Engage another wrench onto the hex-shaped end at the bottom of the pump. Hold the wrench to prevent loosening of the brass fitting. Then unthread the pump using the other wrench. Remove the O-ring from the brass fitting or pump inlet opening. Discard the O-ring.
10. Engage a wrench onto the hex-shaped end at the bottom of the fuel pump. Hold the wrench and use another wrench to remove the brass fitting (B, **Figure 10**) at the top of the fuel pump. Remove the O-ring from the fitting or pump outlet opening. Discard the O-ring.
11. Lubricate. Then install *new* O-rings onto the fuel pump outlet fitting (B, **Figure 10**) and the brass fitting in the fuel filter bracket. Work carefully to avoid tearing the O-rings.

11

12

FLAME ARRESTOR AND COVER (TYPICAL)

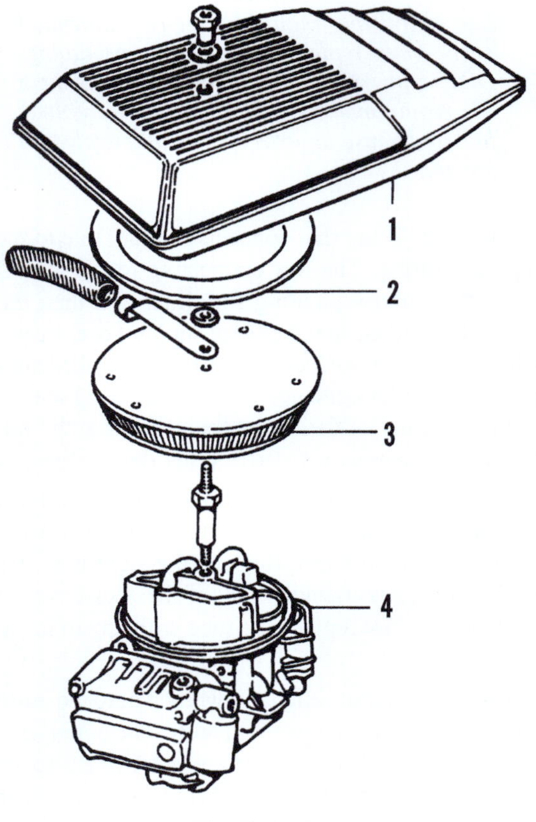

1. Plastic cover
2. Foam cushion
3. Flame arrestor
4. Carburetor

12. Thread the fuel pump inlet (lower end) by hand onto the filter bracket fitting. Engage a wrench onto the filter bracket fitting. Hold the wrench to prevent the fuel filter bracket fitting from rotating. Then tighten the pump to 97-124 in.-lb. (11-14 N•m).

13. Thread the fitting (B, **Figure 10**) by hand into the outlet opening in the fuel pump. Engage a wrench onto the

hex-shaped end at the bottom of the fuel pump. Hold the wrench to prevent the fuel filter bracket fitting or fuel pump from rotating. Then tighten the pump outlet fitting to 60-84 in.-lb. (7-10 N•m).

14. Fit the grommet onto the upper end of the fuel pump. Guide the upper end of the pump through the bracket opening while installing the filter and pump assembly. Align the openings and then thread the mounting bolts into the bracket and cylinder head openings. Tighten the bolts evenly to 20-25 ft.-lb. (27-34 N•m).

15. Fit the tip at the lower end of the guard into the opening in the fuel pump bracket. Then pivot the guard over the fuel pump outlet fitting. Align the openings and then thread the screw (D, **Figure 10**) into the bracket. Securely tighten the screw.

16. Thread the fuel line fitting (C, **Figure 10**) by hand into the fuel pump outlet fitting (B). Engage a wrench onto the outlet fitting. Then tighten the pump outlet fitting to 60-84 in.-lb. (7-10 N•m).

17. Depress the locking clip (A, **Figure 10**). Then carefully plug the engine harness connector onto the fuel pump.

18. Install the hose clamp over the end. Then push the fuel supply hose fully over the fuel inlet fitting (**Figure 11**) on the fuel filter bracket. Align the hose clamp over the fitting and then securely tighten the clamp.

19. Clean up any spilled fuel.

WARNING
Stay clear of the propeller and propeller shaft while running an engine on a flush adapter.

CAUTION
If running the engine with the boat out of the water, use a flush device or other means to supply the engine with cooling water. The seawater pump is quickly damaged if operated without adequate cooling water.

20. Connect the battery cable. Start the engine and immediately check for leaking fuel. Correct any leaks before operating the engine.

FLAME ARRESTOR AND COVER

Carburetor-Equipped or Throttle Body Injection (TBI) Models

Removal and installation

1. Disconnect the negative battery cable.

2. Remove the nut and carefully lift the plastic cover (1, **Figure 12**, typical) from the carburetor or throttle body. Remove the thin aluminum washer from the plastic cover (or the step on the nut).

3. Remove the second nut and carefully remove the breather hose and, on later models, the hydrolock prevention hoses from the flame arrestor and mounting stud (**Figure 13**).

4. Carefully lift the flame arrestor off the carburetor or throttle body unit. On EFI models, work carefully to avoid damaging the fuel injector wiring or connectors. 7

WARNING
Replace the flame arrestor if any physical damage is evident. Operating the engine with a faulty or damaged flame arrestor can allow an engine backfire to ignite fuel vapor or other flammable material in the engine compartment.

5. Clean and inspect the flame arrestor as described in Chapter Three.

6A. On carburetor-equipped models, align the stud with the opening in the flame arrestor and then fit the arrestor onto the carburetor. Rotate the flame arrestor to align the mounting bosses. When aligned, the flame arrestor will fully seat on the carburetor body.

6B. On throttle body injection (TBI) models, carefully fit the flame arrestor onto the throttle body unit without damaging the fuel injector wiring or connectors. Rotate the arrestor to align the stud with the opening. Then seat the arrestor onto the throttle body.

7. Fit the breather and hydrolock prevention system hoses (later models) onto the stud or openings in the flame arrestor. Thread the thinner nut onto the stud and tighten to 24-35 in.-lb. (3-4 N•m).

8. Fit the plastic cover (1, **Figure 12**) onto the flame arrestor. The longer end must fit over the back of the engine.

9. Install the thin aluminum washer over the stud and center in the depression in the plastic cover. Thread the second nut, stepped side down, onto the stud. Tighten the nut to 24-35 in.-lb. (3-4 N•m).

10. Connect the battery cable.

Ford 5.0L and 5.8L Electronic Fuel Injection (EFI) Models

Removal and installation

1. Disconnect the negative battery cable.
2. Remove the four screws and then lift the plastic cover off the engine.
3. Support the lower bracket. Then remove the two screws (**Figure 14**) from the flame arrestor and throttle body.
4. Cut the plastic locking clamp. Then pull the breather hose off the rubber gasket fitting.
5. Remove the flame arrestor and rubber gasket from the throttle body.
6. Inspect the rubber gasket for torn, pinched or deteriorated surfaces and replace as needed.
7. Clean and inspect the flame arrestor as described in Chapter Three.
8. Install the rubber gasket onto the throttle body. The hose fitting on the gasket must face down.
9. Fit the flame arrestor onto the throttle body. Work carefully to avoid pushing the gasket out of position or pinching the gasket.
10. Move the flame arrestor to align the screw openings. Then thread the two screws (**Figure 14**) by hand into the flame arrestor and throttle body openings.
11. Verify proper gasket alignment and push the flame arrestor toward the throttle body. Maintain pressure on the flame arrestor while tightening the two screws (**Figure 14**) to 24-48 in.-lb. (3-5 N•m).
12. Push the breather hose fully onto the rubber gasket fitting. Secure the hose in position with a plastic locking clamp.
13. Install the plastic cover onto the top of the engine. Install the four cover screws and tighten to 24-36 in.-lb. (3-4 N•m).
14. Connect the battery cable.

GM 7.4L and 8.2L Electronic Fuel Injection (1998 74GI and 82GI) Models

The flame arrestor is located on the front and starboard side of the intake plenum. A worm-screw hose clamp secures the flame arrestor to the throttle body.

Removal and installation

1. Disconnect the negative battery cable.
2. Loosen the clamp screw (**Figure 15**) and then carefully pull the flame arrestor off the throttle body flange.
3. Clean and inspect the flame arrestor as described in Chapter Three.
4. Guide the flame arrestor onto the throttle body flange. Rotate the flame arrestor to properly position the flame arrestor. The arrestor must not contact belts, wiring or other engine components.

14

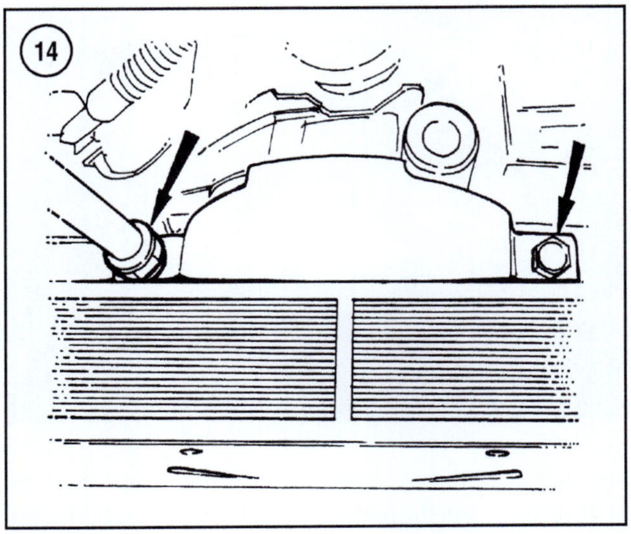

15

5. Make sure the arrestor is fully seated onto the throttle body flange. Center the clamp over the clamping surface, then securely tighten the clamp screw (**Figure 15**).
6. Connect the battery cable.

GM 7.4L and 8.2L Electronic Fuel Injection (74GSI and 82GSI) Models

On 74GSI and 82GSI models, the flame arrestor is located on the front of the intake plenum. A nut and washer secure the flame arrestor to the throttle body.

Removal and installation

1. Disconnect the negative battery cable.
2. Disconnect the breather hoses from the flame arrestor fittings.
3. Remove the nut and washer and then carefully pull the flame arrestor off the throttle body.
4. Remove the gasket from the flame arrestor or throttle body. Inspect the gasket for torn, deteriorated or folded surfaces. Replace the gasket unless it is in excellent condition.
5. Clean and inspect the flame arrestor as described in Chapter Three.

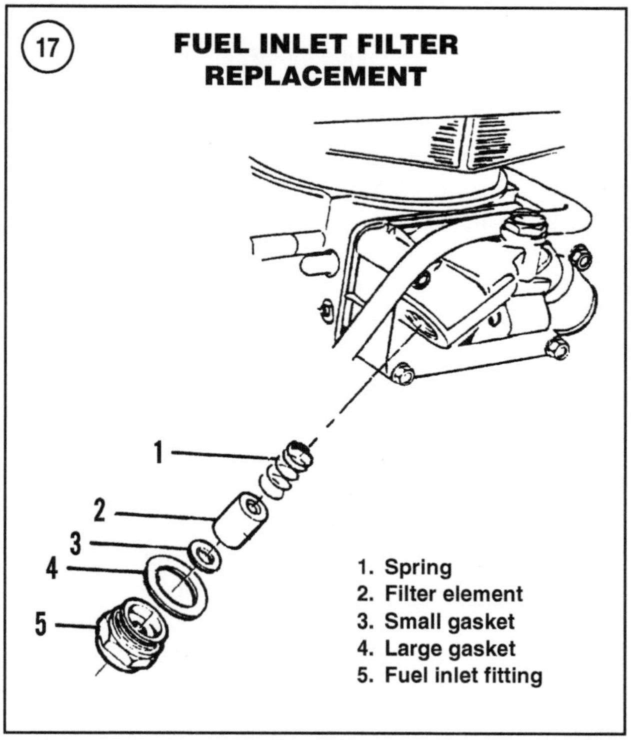

6. Install the gasket onto the flame arrestor. Make sure the gasket is properly seated.
7. Guide the stud into the flame arrestor opening while fitting the flame arrestor onto the throttle body.
8. Seat the throttle body on the flame arrestor. Make sure the gasket is properly positioned. Then install the washer and thread the nut onto the mounting stud. Securely tighten the nut.
9. Connect the breather hoses to the flame arrestor fittings. Secure the hoses with plastic locking clamps.
10. Connect the battery cable.

CARBURETOR

This section describes removal, repair and installation of the two-barrel and four-barrel carburetors and replacement of the carburetor fuel inlet filter.

Fuel Inlet Filter Replacement

Replace the fuel inlet filter if filter blockage is suspected or if contaminants are found in other fuel system components.

1. Disconnect the battery cables.
2. Remove the flame arrestor as described in this chapter.
3. Use compressed air to blow debris and loose material off the fuel line and carburetor fitting.
4. Disconnect the fuel inlet line as follows:
   a. Hold the inlet fitting (A, **Figure 16**, typical) with a wrench. Then use a flare nut wrench to *slightly* loosen the fuel line fitting (B, **Figure 16**).
   b. Wrap a shop towel around the fitting and then loosen the fitting by hand. Allow residual fuel to drain from the fuel line.
   c. When completely drained, unthread the fitting and disconnect the fuel line from the inlet fitting.

7

5. Use a socket to *slightly* loosen the inlet fitting (5, **Figure 17**). Wrap a shop towel around the fitting then loosen by hand. Allow residual fuel to drain from the inlet fitting cavity.
6. Unthread and remove the inlet fitting. Pull the spring (1, **Figure 17**) out of the inlet fitting cavity in the carburetor. Inspect the spring for corrosion damage or lost spring tension. Replace the spring if these defects are evident. If corroded, disassemble the carburetor and clean the passages as described in this section.
7. Pull the filter element (2, **Figure 17**) out of the fitting. Remove the gaskets (3 and 4, **Figure 17**) from the inlet fitting. Discard the gaskets. Clean the inlet fitting with solvent and dry with compressed air.

WARNING
Rust deposits in the fuel filter might indicate rusting of the boat fuel tank and the need for immediate inspection or replacement. Rusting will eventually allow fuel to leak from the tank and cause in a dangerous fire and explosion hazard.

8. Inspect the filter for sediment or rust deposits. If rust is found in the filter, inspect the fuel tank for rust damage. Replace the fuel tank if rust or deterioration is evident. White sediment usually indicates water has contaminated the fuel system. Corrosion of the fuel lines, fuel pump and fuel tank creates the white sediment. Inspect a fuel sample for water. If water is found, drain the tank and refill with fresh fuel before operating the engine. Discard the filter. If the filter must be reused, thoroughly clean the filter with solvent and dry with compressed air. *Do not* reuse the filter if it is damaged or if it cannot be completely cleaned.
9. Install a *new* gasket (4, **Figure 17**) over the inlet fitting and seat against the step on the fitting.
10. Install a *new* gasket (3, **Figure 17**) into the inlet fitting and center over the fuel passage.

11. Install the filter element (2, **Figure 17**) into the inlet fitting with the open end facing the smaller gasket. Install the spring (1, **Figure 17**) into the filter cavity in the carburetor.
12. Guide the filter element into the cavity. The closed end of the filter must contact the spring. Thread the inlet fitting into the cavity by hand. Securely tighten the inlet fitting.
13. Thread the fuel line fitting by hand into the inlet fitting. Engage a wrench onto the inlet fitting (A, **Figure 16**) to prevent rotation. Tighten the fuel line fitting (B, **Figure 16**) to 132-156 in.-lb. (15-18 N•m).
14. Install the flame arrestor and cover as described in this chapter.
15. Connect the battery cables.

WARNING
Stay clear of the propeller and propeller shaft while running an engine on a flush adapter.

CAUTION
If running the engine with the boat out of the water, use a flush device or other means to supply the engine with cooling water. The seawater pump is quickly damaged if operated without adequate cooling water.

16. Start the engine and immediately check for leaking fuel. Stop the engine and repair any leaks before operating the engine.

Two-Barrel Carburetor

The two-barrel Holley marine carburetor (**Figure 18**) can be identified by the single side-mounted float bowl. This section describes removal, repair and installation of the carburetor.

Removal

1. Remove the flame arrestor and cover as described in this chapter.
2. Disconnect the fuel inlet line and remove the fuel inlet filter as described in this section. Refer to *Fuel Inlet Filter Replacement*.
3. Carefully unplug the purple/white wire terminal (A, **Figure 19**) from the electric choke. It is not necessary to disconnect the black wire at this time.
4. On models using a mechanical fuel pump, unplug the sight tube from the carburetor fitting. Inspect the sight tube for brittle, deteriorated or cracked surfaces. Replace the sight tube with the manufacturer-recommended tube if these or other defects are evident. *Do not* replace the tube with any other type of hose or tubing.
5. If so equipped, unplug the PCV hose (B, **Figure 19**) from the fitting on the rear of the carburetor.
6. Remove the cotter pin (A, **Figure 20**) and then disconnect the throttle cable from the carburetor linkage.

18

7. Remove the carburetor mounting nuts (C, **Figure 19** and B, **Figure 20**) and then lift the carburetor off the intake manifold.
8. Remove the carburetor mounting gasket from the intake manifold or carburetor base. Discard the gasket. *Carefully* scrape residual gasket material from the surfaces. *Do not* damage the mating surfaces.
9. Use an aerosol carburetor cleaner to remove oily deposits from the mating surfaces.

Installation

1. Install a *new* carburetor mounting gasket over the studs and seat it against the intake manifold. Make sure the gasket openings match the intake manifold openings.
2. Align the openings with the studs and then install the carburetor onto the intake manifold. Fit the electric choke ground wire terminal (D, **Figure 19**) over the stud. Then thread the carburetor mounting nuts (C, **Figure 19** and B, **Figure 20**) onto the intake studs. Tighten the carburetor mounting nuts in a crossing pattern to 115-168 in.-lb. (13-19 N•m).
3. If removed, thread the flame arrestor stud into the carburetor opening and tighten to 65-80 in.-lb. (7-9 N•m).
4. On models with a mechanical fuel pump, push the sight tube fully over the carburetor fitting. Secure the tube to the fitting and the fuel line with plastic locking clamps. Do not tighten the clamp to the point that the tube begins to collapse.
5. If so equipped, plug the PCV hose (B, **Figure 19**) fully onto the fitting on the rear of the carburetor.
6. Plug the purple/white wire terminal (A, **Figure 19**) onto the electric choke terminal. The terminal must fit snugly on the choke terminal. If necessary, remove the terminal and squeeze *slightly* to achieve a snug fit.
7. Fit the opening in the end of the throttle cable over the throttle linkage post. Install the washer over the post and seat against the throttle cable end. Install a *new* cotter pin (A, **Figure 20**) into the opening in the post. Bend over both prongs to secure the cable.

8. Install the fuel inlet filter and connect the fuel line as described in this section. Refer to *Fuel Inlet Filter Replacement*.
9. Install the flame arrestor and cover as described in this chapter.

WARNING
Stay clear of the propeller and propeller shaft while running an engine on a flush adapter.

CAUTION
If running the engine with the boat out of the water, use a flush device or other means to supply the engine with cooling water. The seawater pump is quickly damaged if operated without adequate cooling water.

10. Connect the battery cable. Start the engine and immediately check for leaking fuel. Correct any leaks before operating the engine.
11. Adjust the throttle cable and idle speed as described in Chapter Five.

Disassembly

Refer to **Figure 21** for this procedure.
1. Remove the carburetor as described in this section.
2. Remove the four screws (**Figure 22**) securing the float bowl and metering block to the carburetor body. Remove the gaskets from the screws or float bowl. Discard the gaskets. Carefully pull the float bowl (**Figure 23**) and then the metering block (**Figure 24**) off the carburetor body. Pull the plastic baffle (69, **Figure 21**) off the metering block pins.
3. If so equipped, carefully pry the plugs out of the idle mixture screw openings on the sides of the metering block (**Figure 25**). Turn the idle mixture screws (**Figure 26**) clockwise and count the number of turns from the original position until the screws are lightly seated. Record the number of turns for each screw. Then remove the screws by turning them counterclockwise. Remove the seals from the screws. Discard the seals.
4. Use a wide-blade screwdriver to remove the main fuel jets (**Figure 27**) from the metering block. Mark each jet to indicate the mounting location in the carburetor. The opening on one side might be a different size.
5. Use a properly sized wrench to loosen and then remove the power valve (76, **Figure 21**) from the metering block. Remove the gasket (75, **Figure 21**) from the valve. Do not discard the gasket at this time. The rebuild kit might have several styles of gaskets. Use the style that matches the original.

6A. On earlier models with a center-hinged float (**Figure 28**), remove the float, inlet needle and seat as follows:
   a. Loosen the lockscrew (7, **Figure 28**) and then remove the inlet valve and seat assembly (3) by turning the adjusting nut (5) counterclockwise.
   b. Remove the gaskets from under the lockscrew and adjusting nut. Discard the gaskets.
   c. Remove the two screws (8, **Figure 28**) and then pull the retaining bracket (9) with float (10) out of the float bowl.
   d. Remove the pin (11, **Figure 28**) and then pull the float off the bracket.

6B. On later models with a side-hinged float, remove the float, inlet needle and seat as follows:
   a. Pull the baffle plate (61, **Figure 21**) out of the float bowl.
   b. Use needlenose pliers to pull the E-clip (63, **Figure 21**) off the float shaft.
   c. Pull the float (62, **Figure 21**) and spring assembly off the shaft. Pull the spring (64, **Figure 21**) off the float arm.
   d. Pull the inlet needle (48, **Figure 21**) and clip (49) out of the seat (47). Remove the clip from the needle.
   e. Use a suitable wrench to loosen and then remove the needle seat (47, **Figure 21**). Remove the gasket (46, **Figure 21**) from the seat. Discard the gasket.

7. Remove the four screws and lockwashers (**Figure 29**). Then lift the accelerator pump cover (**Figure 30**), diaphragm (**Figure 31**) and spring (**Figure 32**) from the float bowl.
8. Inspect the check ball in the float bowl (**Figure 33**) for free movement. For proper accelerator pump operation, the check ball must move freely. Replace the float bowl if the check ball does move freely after cleaning. The check ball is not available separately.

(21) **TWO-BARREL HOLLEY CARBURETOR**

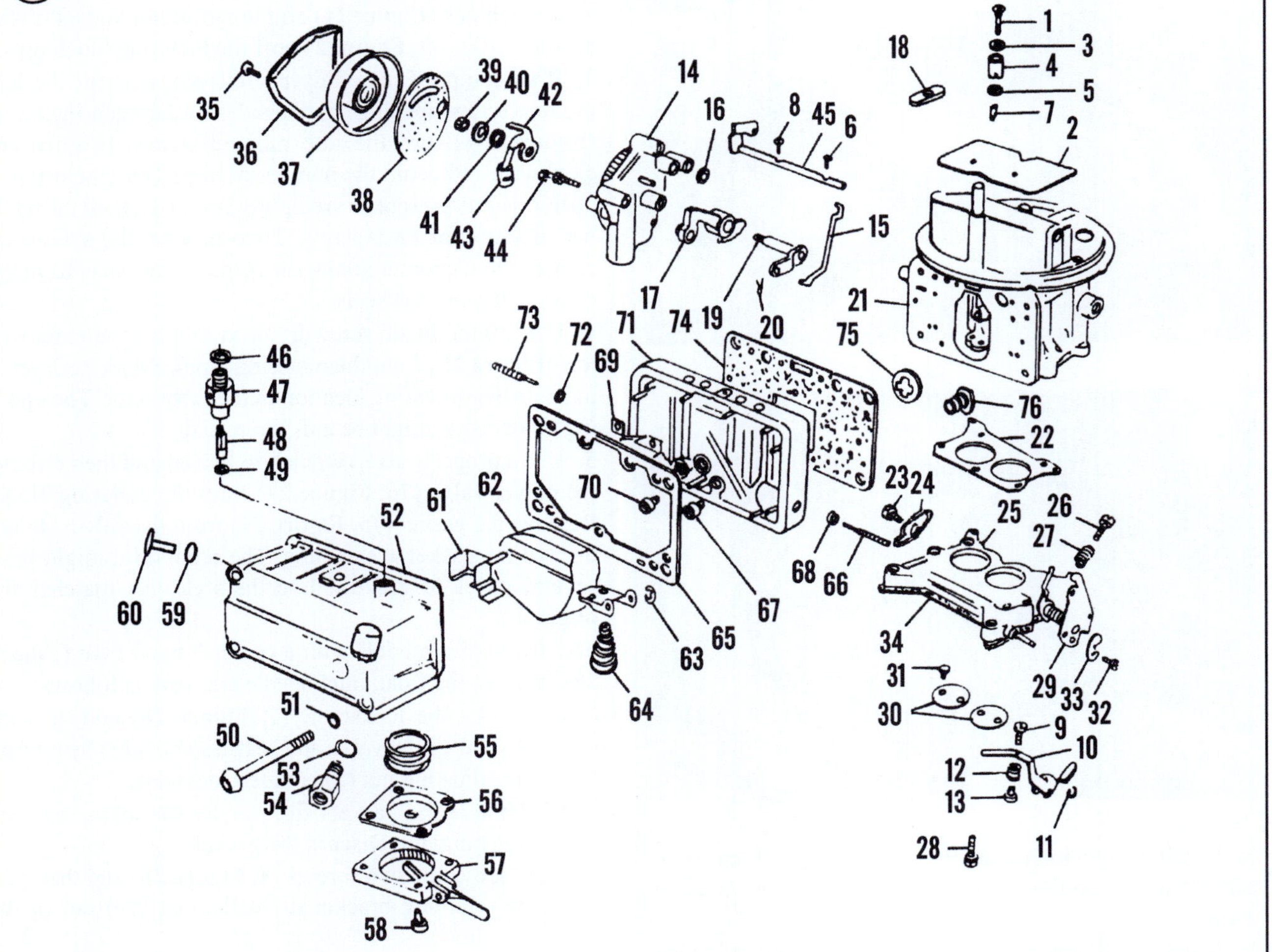

1. Screw
2. Choke plate
3. Gasket
4. Accelerator pump discharge nozzle
5. Gasket
6. Screw
7. Discharge needle
8. Screw
9. Screw
10. Accelerator pump lever
11. Retainer
12. Spring
13. Sleeve nut
14. Choke housing
15. Choke rod
16. Gasket
17. Fast idle cam
18. Choke rod seal
19. Choke shaft
20. Retainer
21. Carburetor body
22. Throttle body gasket
23. Screw
24. Diaphragm lever
25. Throttle body
26. Idle speed adjusting screw
27. Spring
28. Screw
29. Throttle shaft
30. Throttle plate
31. Screw
32. Screw
33. Accelerator pump cam
34. Gasket
35. Screw
36. Retainer
37. Choke cover
38. Gasket
39. Nut
40. Lockwasher
41. Spacer
42. Choke lever
43. Choke piston
44. Screw and washer
45. Choke valve shaft
46. Gasket*
47. Inlet needle seat*
48. Inlet needle*
49. Pull clip*
50. Screw
51. Gasket
52. Float bowl
53. Gasket
54. Fuel inlet filter nut
55. Spring
56. Diaphragm
57. Accelerator pump lever
58. Screw and lockwasher
59. Gasket
60. Fuel level sight plug
61. Baffle plate*
62. Float*
63. E-clip*
64. Float spring*
65. Float bowl gasket
66. Idle mixture screw
67. Main fuel jet
68. Seal
69. Baffle plate
70. Main fuel jet
71. Metering block
72. Seal
73. Idle mixture screw
74. Metering block gasket
75. Power valve gasket
76. Power valve

*Used on carburetor with side-hinged float assembly.

(28) CENTER-HINGED FLOAT (TWO-BARREL HOLLEY CARBURETOR)

1. Float bowl
2. O-ring
3. Inlet needle valve and seat assembly
4. Gasket
5. Fuel level adjusting nut
6. Gasket
7. Lockscrew
8. Screw
9. Float pin retaining bracket
10. Float
11. Float pin

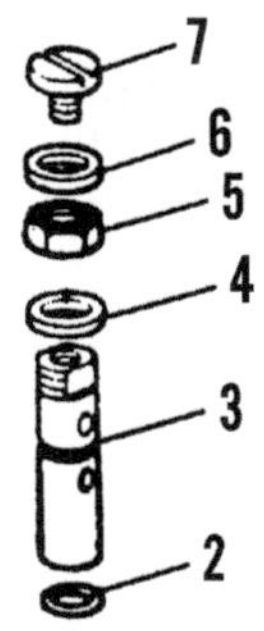

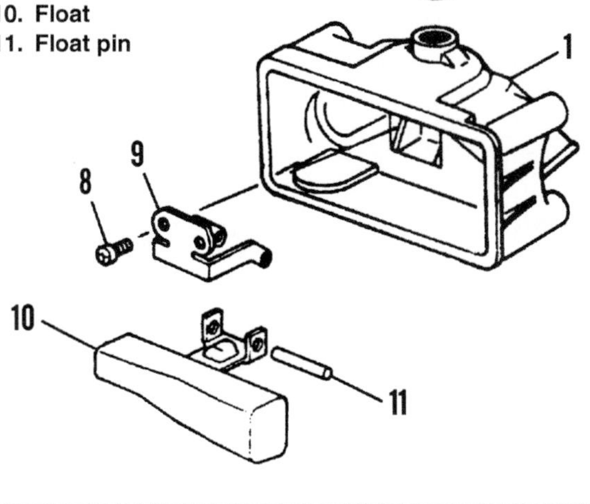

9. Remove the gaskets from the float bowl, metering block and carburetor body. Do not discard the gaskets at this time. The rebuild kit might contain several styles of gaskets. Use the style that matches the original. *Carefully* scrape residual gasket material off the mating surfaces. *Do not* damage the mating surfaces.
10. Pull the brass accelerator pump tube (**Figure 34**) out of the metering block or carburetor body. Remove the O-rings from the tube. Discard the O-rings.
11. Remove the retainer pin (20, **Figure 21**) from the choke rod (15) and separate the rod from the lever on the choke shaft (19).
12. Remove the screws (44, **Figure 21**) and then pull the electric choke off the carburetor. Remove the small gasket (16, **Figure 21**) from the port.
13. Remove the screw (1, **Figure 21**) and then remove the accelerator pump discharge nozzle and gasket. Remove the gaskets from the screw, nozzle or carburetor. Discard the gaskets.
14. Invert the carburetor and capture the discharge needle (7, **Figure 21**) from the nozzle bore.
15. Remove the retainer (11, **Figure 21**) and remove accelerator pump operating lever (10).
16. Disassemble the electric choke as follows:
   a. Make match marks on the choke cover and housing (**Figure 35**) to assist with proper assembly.
   b. Remove the screws (35, **Figure 21**) and retainer (36) and then pull the choke cover (37) off the choke housing (14).
   c. Remove the nut, lockwasher and spacer (39-41, **Figure 21**) and then pull the choke lever (42) and piston (43) out of the choke housing.
   d. Remove the choke rod (15, **Figure 21**) and seal (18). Remove the fast idle cam (17, **Figure 21**) and choke shaft (19).
17. Remove the screws (28, **Figure 21**) and then separate the throttle body (25) from the carburetor body (21). Remove the gasket (22, **Figure 21**). If necessary, carefully scrape gasket material from the mating surfaces. Do not damage the mating surfaces. Do not discard the gasket at this time. The repair kit might contain several styles of gaskets. Use the same gasket as the original.
18. Refer to *Servicing the Fuel System* in this chapter for procedures. Then clean and inspect the carburetor components.

Assembly

Refer to **Figure 21** for this procedure.

1. Install a new gasket (22, **Figure 21**) onto the throttle body (25). Then fit the throttle body onto the carburetor body. Align the openings and then thread the screws (28, **Figure 21**) into the openings. Tighten the throttle body-to-carburetor body screws in a crossing pattern to 50 in.-lb. (6 N•m).

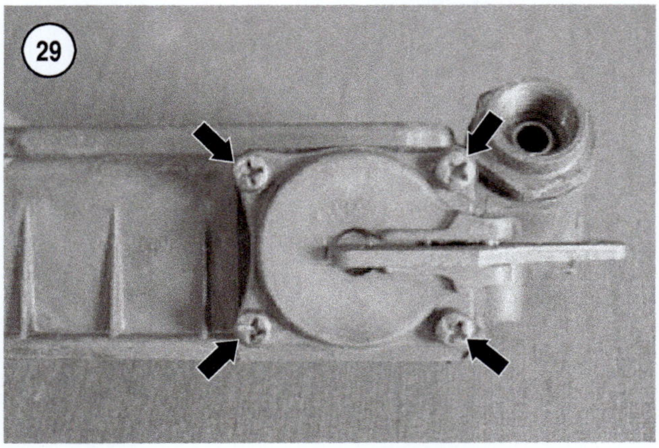

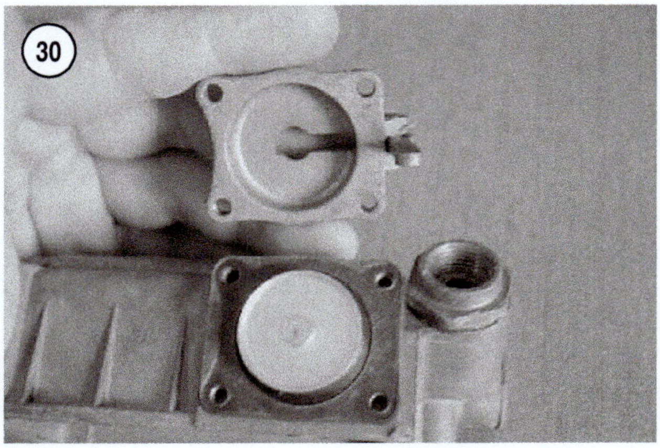

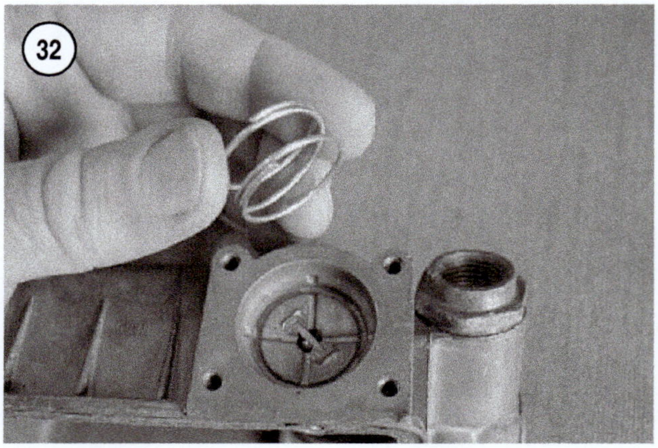

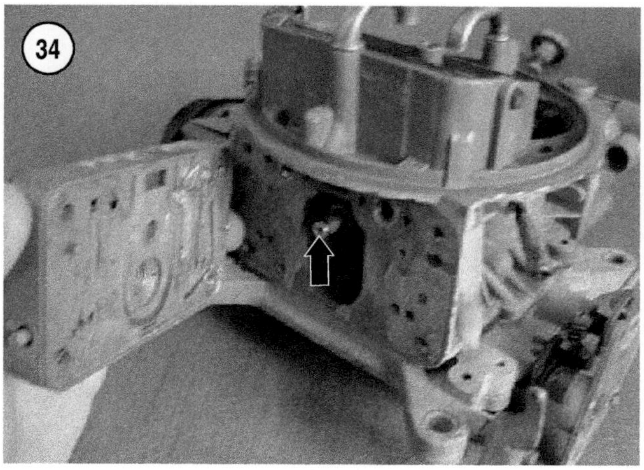

2. Attach the accelerator pump lever (10, **Figure 21**) to the carburetor body. Secure the lever with the retainer (11, **Figure 21**).

3. Install the accelerator pump discharge nozzle and related components as follows:

a. Drop the discharge needle (4, **Figure 36**) into the nozzle bore with the pointed tip facing downward.

b. Install a *new* gasket (2, **Figure 36**) onto the bottom of the pump nozzle (3). Then fit the nozzle onto the carburetor with the discharge openings pointing into the carburetor bore.

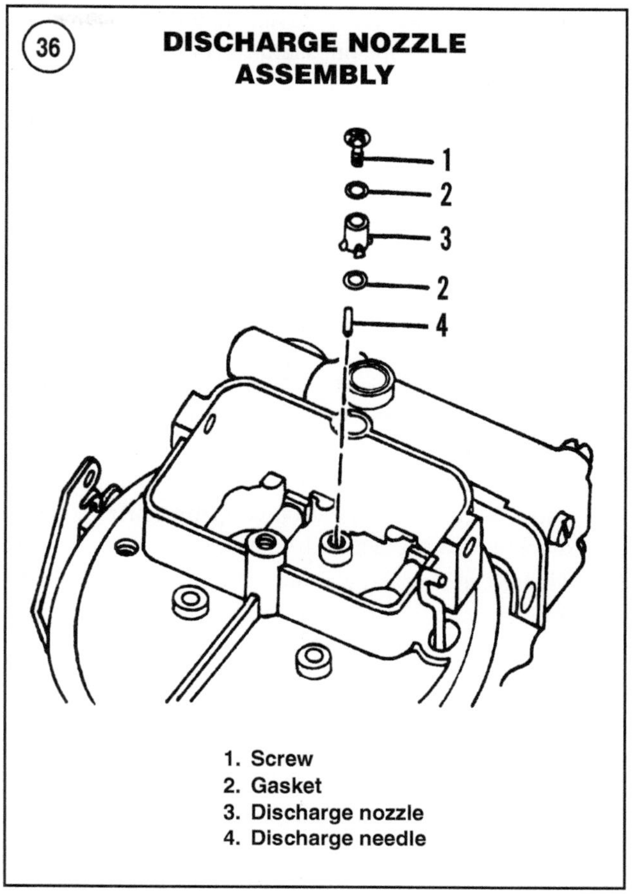

c. Install a *new* gasket (2, **Figure 36**) onto the top of the nozzle. Thread the screw (1, **Figure 36**) into the nozzle and carburetor opening. Securely tighten the screw.

d. Stake the screw lightly in place with a punch.

4. Assemble and install the electric choke as follows:

a. Insert the choke shaft (19, **Figure 21**) into the fast idle cam (17) opening. Guide the threaded end of the choke shaft into the choke housing.

b. Insert the piston (43, **Figure 21**) into the bore in the choke housing (14) and then fit the lever (42) over the choke shaft. Install the lockwasher and nut onto the shaft. Securely tighten the nut.

c. Fit the *new* gasket (16, **Figure 21**) into the carburetor body port that aligns with the choke housing port.

d. Insert the choke rod (15, **Figure 21**) into the choke shaft lever and then fit the choke housing onto the carburetor. The projection on the rod must be located under the fast idle cam to lift the cam when the choke is closed. Install a *new* retainer (20, **Figure 21**) onto the choke rod. Verify proper alignment and then install the screws and washers (44, **Figure 21**). Securely tighten the screws.

e. Install a *new* gasket (38, **Figure 21**) onto the choke housing. Make sure the tab on the lever fits into the slot in the gasket. Engage the spring loop over the tab on the lever (42, **Figure 21**) and then seat the choke cover (37) onto the housing. Fit the retainer (36, **Fig-**

7

ure 21) onto the choke cover. Rotate the retainer to align the screw openings and then install the screws (35, **Figure 21**). Rotate the choke cover to align the match marks made prior to disassembly (**Figure 35**). Then securely tighten the screws.

5. Install the accelerator pump components onto the float bowl as follows:

   a. Install the accelerator pump diaphragm onto the cover with the large lever disc facing the pump lever. Install the spring onto the diaphragm and center over the lever disc.
   b. Without dislodging the spring or diaphragm, fit the cover onto the float bowl. The pump lever must face toward the end of the float bowl.
   c. Make sure the screw openings in the diaphragm align with the cover and float bowl opening. Realign the diaphragm and cover as needed.
   d. Thread the four screws and lockwashers (**Figure 29**) into the cover and bowl. Do not tighten the screws at this time.
   e. Make sure the diaphragm is centered and then push the lever (**Figure 37**) to fully depress the diaphragm. Hold the lever in the compressed position and then securely tighten the cover screws.

6A. On earlier models with a center-hinged float (**Figure 28**), install the float, inlet needle and seat assembly and bracket into the float bowl as follows:

   a. Align the pin openings in the float arm and retaining bracket. Then insert the float pin (11, **Figure 28**) into the four openings.
   b. Guide the retaining bracket and float assembly into the float bowl. Make sure the float pin enters the recess in the float bowl.
   c. Thread the two screws (8, **Figure 28**) into the retainer and float bowl openings. Securely tighten the screws.
   d. Apply a light coat of engine oil to the O-ring (2, **Figure 28**) of the inlet needle and seat assembly (3). Then thread the assembly into the float bowl bore.
   e. Install a *new* gasket (4, **Figure 28**) onto the needle and seat assembly and then thread the adjusting nut (5) onto the needle and seat assembly.
   f. Install a *new* gasket (6, **Figure 28**) onto the needle and seat assembly and then thread the lockscrews (7) into the needle and seat assembly.

6B. On later models with a side-hinged float (62, **Figure 21**), install the float, inlet needle and seat into the float bowl as follows:

   a. Install a new gasket (46, **Figure 21**) and then thread the needle seat (47) into the float bowl opening. Tighten the seat to 10 in.-lb. (1 N•m).
   b. Install a new pull clip (49, **Figure 21**) onto the inlet needle (48). Insert the inlet needle and clip into the seat. Make sure the pull clip wire extends outward.

37

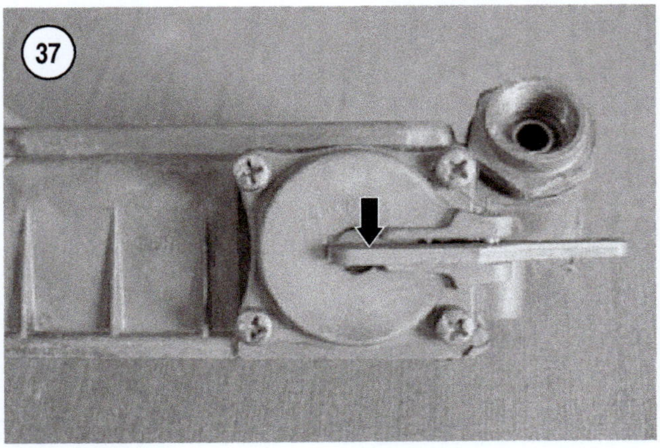

38

39

   c. Install the spring onto the float arm with the end with the small diameter fitting over the tabs as shown in **Figure 38**.
   d. Slide the float onto the float shaft. Make sure the spring is seated between the ridges on the boss in the float bowl floor as shown in **Figure 39**. Make sure the pull clip wire enters the slot in the float arm.
   e. Use needlenose pliers to install the E-clip (63, **Figure 21**) onto the float shaft.
   f. Insert the baffle plate (61, **Figure 21**) into the float bowl with the slot over the inlet needle seat.

7A. On models with a center-hinged float (**Figure 28**), adjust the float level as follows:

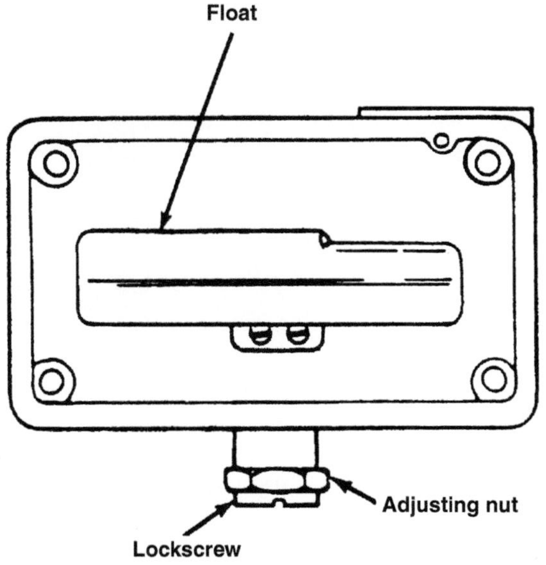

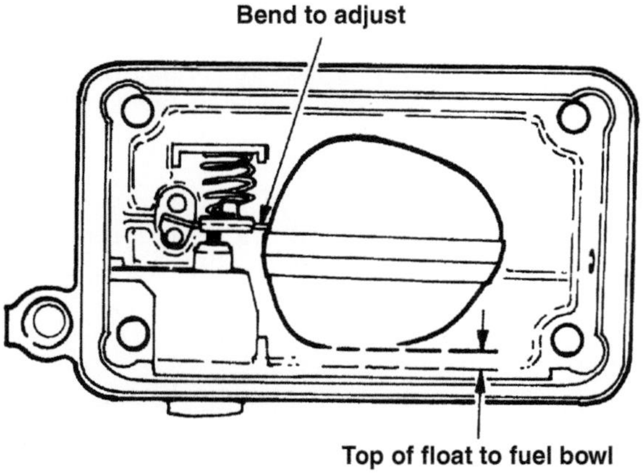

a. Invert the float bowl.

b. Loosen the lockscrew (**Figure 40**). Then turn the adjusting nut until the flat float surface is level with the flat float bowl (inner) surface.

c. Securely tighten the lockscrew.

7B. On models with a side-hinged float, place the float bowl on a level work surface with the inlet needle side facing downward. The top of the float should be parallel to the flat float bowl surface as shown in **Figure 41**. If adjustment is needed, carefully bend the curved part of the float arm that contacts the inlet needle to achieve parallel surfaces.

NOTE
Several different power valves and power valve gaskets might be included in the repair kit. The valves are identified by numbers stamped onto the faces of the valves. Install a valve with the same number as the original. Also, the valve must be installed with the correct gasket. If the power valve has round fuel openings, use the gasket with three internal projections. If the power valve has rectangular fuel openings, use the gasket without internal projections.

8. Install the *new* gasket (75, **Figure 21**) onto the power valve (76). Make sure the power valve and gasket are the same number and style as the originals. Thread the power valve by hand into the metering block opening. Tighten the power valve to 100 in.-lb. (11 N•m).

9. Thread the main fuel jets by hand into the metering block (**Figure 27**). Then tighten the jets to 10 in.-lb. (1 N•m).

10. Thread the idle mixture screws (**Figure 26**) into the metering block openings using *new* seals (68, **Figure 21**). Turn the needles in until lightly seated. Then back the screws out the number of turns recorded during disassembly. The setting is normally one-half to one turn out from lightly seated. If so equipped, install the plugs (**Figure 25**) into the screw openings to prevent tampering.

11. Install the metering block and float bowl as follows:

   a. Install the plastic baffle (69, **Figure 21**) onto the metering block and position over the dowels.

   b. Install a *new* inner gasket (74, **Figure 21**) onto the metering block (71). Use the dowels on the block to align and retain the gasket. Install a *new* O-ring onto the brass accelerator pump tube. Then insert the tube into the carburetor body opening (**Figure 34**).

   c. Fit the metering block and gasket onto the carburetor body (**Figure 24**). Install a *new* outer gasket (65, **Figure 21**) onto the metering block. Use the dowels on the block to align and retain the gasket.

   d. Install the float bowl onto the metering block (**Figure 23**). Use the gasket and dowel to align and retain the float bowl.

   e. Install a *new* gasket (51, **Figure 21**) onto each of the four screws (50). Thread the four screws (**Figure 22**) into the bowl, metering block and carburetor body openings. Verify proper fuel bowl, metering block and carburetor body alignment. Then tighten the screws securely in a crossing pattern.

12. Install the carburetor as described in this section.

13. Adjust the accelerator pump clearance as described in Chapter Five.

14. On models with a center-hinged float bowl, perform the final float level adjustment as described in Chapter Five.

Four-Barrel Carburetor

The four-barrel Holley marine carburetor (**Figure 42**) can be identified by the two side-mounted float bowls and

7

the two secondary openings. This section describes removal, repair and installation of the carburetor.

Removal

1. Remove the flame arrestor and cover as described in this chapter.
2. Disconnect the fuel inlet line and remove the fuel inlet filter as described in this section. Refer to *Fuel Inlet Filter Replacement.*
3. Carefully unplug the purple/white wire terminal (A, **Figure 19**) from the electric choke. It is not necessary to disconnect the black wire at this time.
4. On models using a mechanical fuel pump, unplug the sight tube from the carburetor fitting. Inspect the sight tube for brittle, deteriorated or cracked surfaces. Replace the sight tube with the manufacturer-recommended tube if these or other defects are evident. *Do not* replace the tube with any other type of hose or tubing.
5. If so equipped, unplug the PCV hose (B, **Figure 19**) from the fitting on the rear of the carburetor.
6. Remove the cotter pin (A, **Figure 20**) and then disconnect the throttle cable from the carburetor linkage.
7. Remove the carburetor mounting nuts (C, **Figure 19** and B, **Figure 20**) and then lift the carburetor off the intake manifold.
8. Remove the carburetor mounting gasket from the intake manifold or carburetor base. Discard the gasket. *Carefully* scrape residual gasket material from the surfaces. *Do not* damage the mating surfaces.
9. Use an aerosol carburetor cleaner to remove oily deposits from the mating surfaces.

Installation

1. Install a *new* carburetor mounting gasket over the studs and seat against the intake manifold. Make sure the gasket openings match the intake manifold openings.
2. Align the openings with the studs and then install the carburetor onto the intake manifold. Fit the electric choke ground wire terminal (D, **Figure 19**) over the stud. Then thread the carburetor mounting nuts (C, **Figure 19** and B, **Figure 20**) onto the intake studs. Tighten the nuts in a crossing pattern to 115-168 in.-lb. (13-19 N•m).
3. If removed, thread the flame arrestor stud into the carburetor opening and tighten to 65-80 in.-lb. (7.3-9.0 N•m).
4. On models with a mechanical fuel pump, push the sight tube fully over the carburetor fitting. Secure the tube to the fitting and the fuel line with plastic locking clamps. Do not tighten the clamps to the point that the tube begins to collapse.
5. If so equipped, plug the PCV hose (B, **Figure 19**) fully onto the fitting on the rear of the carburetor.
6. Plug the purple/white wire terminal (A, **Figure 19**) onto the electric choke terminal. The terminal must fit snugly on

42

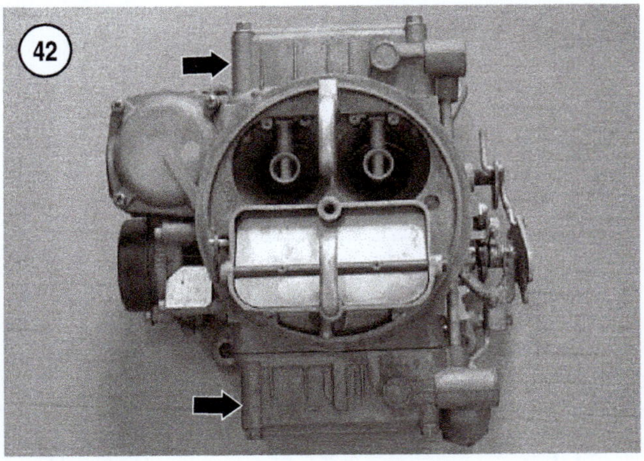

the choke terminal. If necessary, remove the terminal and squeeze *slightly* to achieve a snug fit.
7. Fit the opening in the end of the throttle cable over the throttle linkage post. Install the washer over the post and seat against the throttle cable end. Install a *new* cotter pin (A, **Figure 20**) into the opening in the post. Bend over both prongs to secure the cable.
8. Install the fuel inlet filter and connect the fuel line as described in section. Refer to *Fuel Inlet Filter Replacement.*
9. Install the flame arrestor and cover as described in this chapter.

WARNING
Stay clear of the propeller and propeller shaft while running an engine on a flush adapter.

CAUTION
If running the engine with the boat out of the water, use a flush device or other means to supply the engine with cooling water. The seawater pump is quickly damaged if operated without adequate cooling water.

10. Connect the battery cable. Start the engine and immediately check for leaking fuel. Correct any leaks before operating the engine.
11. Adjust the throttle cable and idle speed as described in Chapter Five.

Disassembly

Refer to **Figure 43** for this procedure.
1. Remove the carburetor as described in this section.
2. Remove the four screws (**Figure 22**) securing the primary float bowl (12, **Figure 43**) and metering block (18) to the carburetor body. The primary float bowl is on the front of the carburetor and on the same side as the choke valve. Remove the gaskets from the screws or float bowl. Discard the gaskets. Carefully pull the float bowl (**Figure 23**) and then metering block (**Figure 24**) off the carburetor body. Pull the plastic baffle (17, **Figure 43**) off the metering block pins.

(43) **FOUR-BARREL HOLLEY CARBURETOR**

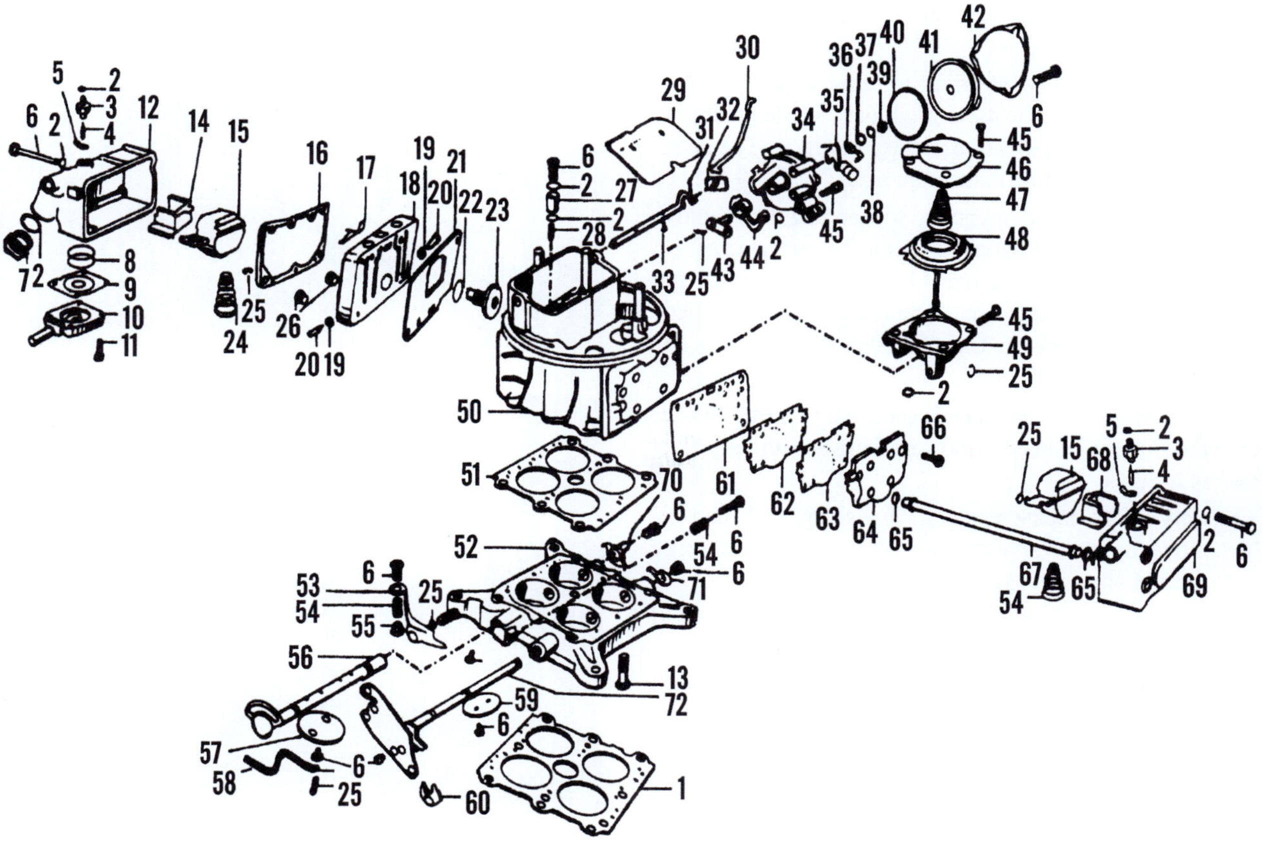

1. Mounting gasket
2. Gasket*
3. Inlet needle seat*
4. Inlet needle*
5. Pull off clip*
6. Screw
7. Fuel filter nut
8. Diaphragm spring
9. Diaphragm assembly
10. Accelerator pump cover
11. Screw and lockwasher
12. Primary float bowl
13. Screw
14. Baffle plate*
15. Float*
16. Float bowl gasket
17. Baffle plate
18. Primary metering block
19. Seal
20. Idle mixture screw
21. Metering block gasket
22. Power valve gasket
23. Power valve
24. Float spring*
25. E-clip*
26. Main fuel jet
27. Discharge nozzle
28. Discharge needle
29. Choke plate
30. Choke rod
31. Choke shaft
32. Choke rod seal
33. Screw
34. Choke housing
35. Choke link and piston
36. Choke lever
37. Spacer
38. Lockwasher
39. Nut
40. Gasket
41. Choke cover
42. Retainer
43. Choke shaft
44. Fast idle cam
45. Screw and washer
46. Cover
47. Diaphragm spring
48. Diaphragm assembly
49. Diaphragm housing
50. Carburetor body
51. Throttle body gasket
52. Throttle body
53. Accelerator pump operating lever
54. Spring
55. Sleeve nut
56. Secondary throttle shaft
57. Secondary throttle plate
58. Throttle connecting rod
59. Primary throttle plate
60. Accelerator pump cam
61. Gasket
62. Secondary plate
63. Metering body gasket
64. Secondary metering block
65. O-ring
66. Clutch screw
67. Fuel balance tube
68. Baffle plate
69. Secondary float bowl
70. Fast idle cam lever
71. Diaphragm lever assembly
72. Primary throttle shaft

*Used on carburetor with side-hinged float assembly.

3. Carefully pull the fuel balance tube from the primary or secondary float bowl (**Figure 44**). Remove the O-rings (65, **Figure 43**) from the balance tube or bowl openings. Discard the O-rings.

4. If so equipped, carefully pry the plugs out of the idle mixture screw openings on the sides of the metering block (**Figure 25**). Turn the idle mixture screws (**Figure 26**) *clockwise* and count the number of turns until the screws

are lightly seated. Record the number of turns for each screw. Then remove the screws by turning them counterclockwise. Remove the seals from the screws. Discard the seals.

5. Use a wide-blade screwdriver to remove the main fuel jets (**Figure 27**) from the metering block. Mark each jet to indicate the mounting location in the carburetor. The opening on one side might be a different size.

6. Use a properly sized wrench to loosen and then remove the power valve (23, **Figure 43**) from the metering block. Remove the gasket (22, **Figure 43**) from the valve. Do not discard the gasket at this time. The rebuild kit might have several styles of gaskets. Use the style that matches the original.

7A. On models with a center-hinged float (**Figure 28**), remove the float, inlet needle and seat from the primary float bowl as follows:

a. Loosen the lockscrew (7, **Figure 28**) and then remove the inlet valve and seat assembly (3) by turning the adjusting nut (5) counterclockwise.
b. Remove the gaskets from under the lockscrew and adjusting nut. Discard the gaskets.
c. Remove the two screws (8, **Figure 28**) and then pull the retaining bracket (9) with float (10) out of the float bowl.
d. Remove the pin (11, **Figure 28**) and then pull the float off the bracket.

7B. On models with a side-hinged float, remove the float, inlet needle and seat from the primary float bowl as follows:

a. Pull the baffle plate (14, **Figure 43**) out of the float bowl.
b. Use needlenose pliers to pull the E-clip (25, **Figure 43**) off the float shaft.
c. Pull the float (15, **Figure 43**) and spring assembly off the shaft. Pull the spring (24, **Figure 43**) off the float arm.
d. Pull the inlet needle (4, **Figure 43**) and clip (5) out of the seat (3). Remove the clip from the needle.
e. Use a suitable wrench to loosen and then remove the needle seat (3, **Figure 43**). Remove the gasket (2, **Figure 43**) from the seat. Discard the gasket.

8. Remove the four screws (**Figure 45**) securing the secondary float bowl and metering block to the carburetor body. Remove the gasket from the screws or float bowl. Discard the gaskets. Carefully pull the float bowl off the carburetor body (**Figure 46**).

9. Use a No. 2 clutch (double-D) screwdriver to remove the screws (**Figure 47**). Then separate the secondary metering block (**Figure 48**) and plate (**Figure 49**) from the carburetor body. Work carefully to avoid bending the plate.

10A. On models with a center-hinged float (**Figure 28**), remove the float, inlet needle and seat from the secondary float bowl as follows:

44

45

46

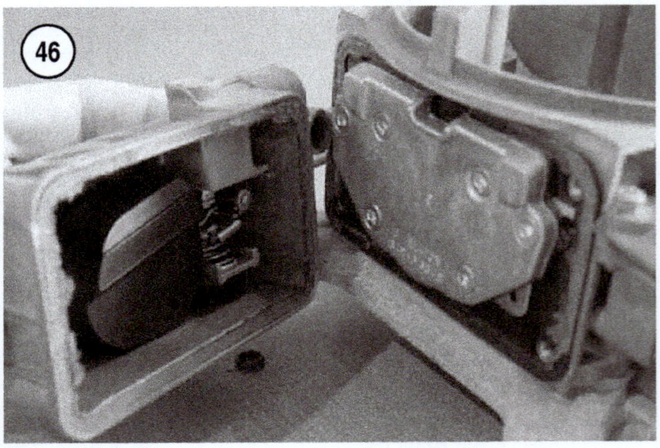

a. Loosen the lockscrew (7, **Figure 28**) and then remove the inlet valve and seat assembly (3) by turning the adjusting nut (5) counterclockwise.
b. Remove the gaskets from under the lockscrew and adjusting nut. Discard the gaskets.
c. Remove the two screws (8, **Figure 28**) and then pull the retaining bracket (9) with float (10) out of the float bowl.
d. Remove the pin (11, **Figure 28**) and then pull the float off the bracket.

10B. On models with a side-hinged float, remove the float, inlet needle and seat from the secondary float bowl as follows:

47

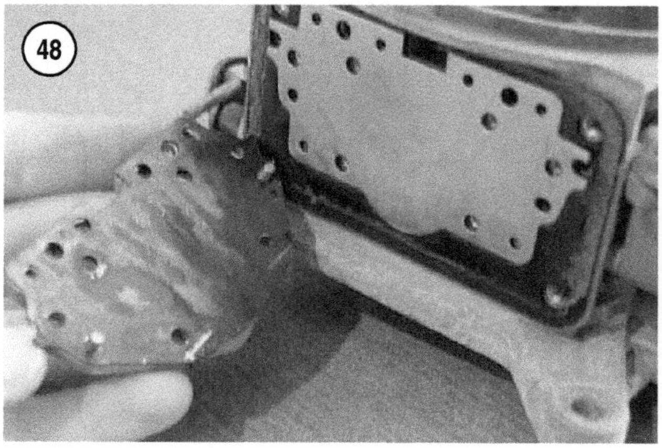

48

49

a. Pull the baffle plate (14, **Figure 43**) out of the float bowl.

b. Use needlenose pliers to pull the E-clip (25, **Figure 43**) off the float shaft.

c. Pull the float (15, **Figure 43**) and spring assembly off the shaft. Pull the spring (24, **Figure 43**) off the float arm.

d. Pull the inlet needle (4, **Figure 43**) and clip (5) out of the seat (3). Remove the clip from the needle.

e. Use a suitable wrench to loosen and then remove the needle seat (3, **Figure 43**). Remove the gasket (2, **Figure 34**) from the seat. Discard the gasket.

11. Remove the screw (6, **Figure 43**) and then lift the accelerator pump discharge nozzle (27) and gasket (2). Remove the gaskets from the screw, nozzle or carburetor. Discard the gaskets.

12. Invert the carburetor and capture the discharge needle (28, **Figure 43**) from the nozzle bore.

13. Remove the E-clip (25, **Figure 43**) that attaches the secondary diaphragm link to the secondary throttle shaft.

14. Remove the screws (13, **Figure 43**) that secure the throttle body to the carburetor body. Separate the throttle body (52, **Figure 43**) from the carburetor body (50). Remove the gasket (51, **Figure 43**) from the throttle body or carburetor body. If necessary, carefully scrape residual gasket material from the mating surfaces. Do not damage the mating surfaces. Do not discard the gasket at this time. The repair kit might contain several styles of gaskets. Use the same gasket as the original.

15. Remove the four screws and lockwashers (**Figure 29**). Then lift the accelerator pump cover (**Figure 30**), diaphragm (**Figure 31**) and spring (**Figure 32**) from the float bowl.

16. Inspect the check ball in the float bowl (**Figure 33**) for free movement. For proper accelerator pump operation, the check ball must move freely. Replace the float bowl if the check ball does move freely after cleaning. The check ball is not available separately.

17. Pull the brass accelerator pump tube (**Figure 34**) out of the metering block or carburetor body. Remove the O-rings from the tube. Discard the O-rings.

18. Disassemble the electric choke as follows:

a. Make match marks on the choke cover and housing (**Figure 35**) to assist with proper assembly.

b. Remove the three screws (6, **Figure 43**) and retainers (42) and then pull the choke cover (41) off the choke housing (34).

c. Remove the nut, lockwasher and spacer (37-39, **Figure 43**) and then pull the choke link and piston (35) out of the choke housing.

d. Remove the choke housing (34, **Figure 43**). Remove the small gasket (2, **Figure 43**) from the choke housing or port in the carburetor body.

e. Pull the choke shaft (43, **Figure 43**) and fast idle cam (44) out of the choke housing.

f. Remove the choke rod (30, **Figure 43**) and seal (32).

19. Remove the E-clip (25, **Figure 43**) and then remove accelerator pump operating lever (53).

20. Remove and disassemble the diaphragm housing as follows:

a. Remove the screws (45, **Figure 43**) and then separate the diaphragm housing assembly from the carburetor body (**Figure 50**).

b. Remove the small gasket (2, **Figure 43**) from the diaphragm assembly or carburetor body. Discard the gasket.

c. Remove the four screws (**Figure 51**) and then lift the cover off the diaphragm housing.

d. Remove the spring (47, **Figure 43**) and then pull the diaphragm and shaft out of the housing (**Figure 52**).
e. Inspect the diaphragm for torn, stretched or deteriorated surfaces. Replace the diaphragm and shaft assembly unless it is in excellent condition.

21. Refer to *Servicing the Fuel System* in this chapter and clean and inspect the carburetor components.

Assembly

Refer to **Figure 43** for this procedure.

1. Install a new gasket (51, **Figure 43**) onto the throttle body (52) and then fit the throttle body onto the carburetor body. Align the openings and then thread the screws (13, **Figure 43**) into the openings. Tighten the throttle body-to-carburetor body screws in a crossing pattern to 50 in.-lb. (6 N•m).
2. Attach the accelerator pump lever (53, **Figure 43**) onto the carburetor body. Secure the lever with the E-clip (25, **Figure 43**).
3. Install the accelerator pump discharge nozzle and related components as follows:
   a. Drop the discharge needle (4, **Figure 36**) into the nozzle bore with the pointed tip facing downward.
   b. Install a *new* gasket (2, **Figure 36**) onto the pump nozzle (3). Then fit the nozzle onto the carburetor with the discharge openings pointing into the carburetor bore.
   c. Install a *new* gasket (2, **Figure 43**) onto the top of the nozzle. Fit a new gasket (2, **Figure 43**) over the discharge passage in the carburetor air horn.
   d. Use needlenose pliers to install the nozzle and screw into the opening (**Figure 43**). Align the discharge openings to point into the throttle bores. Then securely tighten the screw.
   e. Stake the screw lightly in place with a punch.
4. Assemble and install the secondary diaphragm as follows:
   a. Install the secondary diaphragm and shaft into the housing (**Figure 52**). Rotate the diaphragm to align the screw openings. Make sure the corner with two openings has one opening positioned over the vacuum passage.
   b. Install the spring (47, **Figure 43**) onto the housing (49) with the side with the larger diameter against the diaphragm.
   c. Fit the cover onto the housing. Do not dislodge the spring or diaphragm.
   d. Make sure the openings align and then thread the four screws (**Figure 51**) into the cover and housing. *Do not* tighten the screws at this time.
   e. Support the housing and pull downward as far as possible on the diaphragm shaft. Hold the shaft in position and then securely tighten the screws. Release the diaphragm.

50

51

   f. Install a *new* gasket (2, **Figure 43**) over the diaphragm vacuum passage on the carburetor body.
   g. Install the secondary diaphragm assembly onto the carburetor body. Do not dislodge the gasket.
   h. Install the three screws (45, **Figure 43**) into the diaphragm assembly and carburetor body. Securely tighten the screws.
   i. Fit the opening in the diaphragm shaft over the secondary throttle lever. Install a *new* retainer onto the lever to secure the shaft.
5. Fit a new seal (32, **Figure 43**) onto the choke rod (30). Fit the seal with the rod into the grooves under the flame arrestor flange.
6. Assemble and install the electric choke as follows:
   a. Insert the choke shaft (43, **Figure 43**) into the fast idle cam (44) opening. Guide the threaded end of the choke shaft into the choke housing.
   b. Insert the piston (35, **Figure 43**) into the bore in the choke housing (34) and then fit the lever (36) over the choke shaft. Install the spacer lockwasher and nut onto the shaft. Securely tighten the nut.
   c. Fit the *new* gasket (3, **Figure 43**) into the carburetor body port that aligns with the choke housing port.
   d. Insert the choke rod (30, **Figure 43**) into the choke shaft lever, then fit the choke housing onto the carburetor. The projection on the rod must be located under the fast idle cam to lift the cam when the choke is

52

closed. Install a *new* retainer onto the choke rod. Verify proper alignment and then install the screws and washers (45, **Figure 43**). Securely tighten the screws.

e. Install a *new* gasket (40, **Figure 43**) onto the choke housing. Make sure the tab on the lever fits into the slot in the gasket. Engage the spring loop over the tab lever and then seat the choke cover (41, **Figure 43**) onto the housing. Fit the retainer (42, **Figure 43**) onto the choke cover. Rotate the retainer to align the screw openings and then install the screws (6, **Figure 43**). Rotate the choke cover to align the match marks made prior to disassembly (**Figure 35**). Then securely tighten the screws.

7. Install the accelerator pump components onto the primary float bowl as follows:

a. Install the accelerator pump diaphragm onto the cover with the large lever disc facing the pump lever. Install the spring onto the diaphragm and center over the lever disc.

b. Without dislodging the spring or diaphragm, fit the cover onto the float bowl. The pump lever must face toward the end of the float bowl.

c. Make sure the screw openings in the diaphragm align with the cover and float bowl opening. Realign the diaphragm and cover as needed.

d. Thread the four screws and lockwashers (**Figure 29**) into the cover and bowl. Do not tighten the screws at this time.

e. Make sure the diaphragm is centered and then push the lever (**Figure 37**) to fully depress the diaphragm. Hold the lever in the compressed position and then securely tighten the cover screws.

8. Identify the secondary float bowl as follows:

a. On models with a single fuel inlet fitting, the secondary float bowl does not have a threaded opening. The float bowl has an opening for the balance tube.

b. On models with twin fuel inlet fittings, the secondary float bowl has a threaded opening on the side opposite the throttle linkage.

9A. On models with a center-hinged float (**Figure 28**), install the float, inlet needle and seat assembly and bracket into the secondary float bowl as follows:

a. Align the pin openings in the float arm and retaining bracket and then insert the float pin (11, **Figure 28**) into the four openings.

b. Guide the retaining bracket and float assembly into the float bowl. Make sure the float pin enters the recess in the float bowl.

c. Thread the two screws (8, **Figure 28**) into the retainer and float bowl openings. Securely tighten the screws.

d. Apply a light coat of engine oil to the O-ring (2, **Figure 28**) of the inlet needle and seat assembly (3). Then thread the assembly into the float bowl bore.

e. Install a *new* gasket (4, **Figure 28**) onto the needle and seat assembly and then thread the adjusting nut (5) onto the needle and seat assembly.

f. Install a *new* gasket (6, **Figure 28**) onto the needle and seat assembly and then thread the lockscrews (7) into the needle and seat assembly.

9B. On models with a side-hinged float (15, **Figure 43**), install the inlet needle, seat and float into the secondary float bowl as follows:

a. Install a new gasket (2, **Figure 43**) and then thread the needle seat (3) into the float bowl opening. Tighten the seat to 10 in.-lb. (1 N•m).

b. Install a new pull clip (5, **Figure 43**) onto the inlet needle (4). Insert the inlet needle and clip into the seat. Make sure the pull clip wire extends outward.

c. Install the spring onto the float arm with the end with the small diameter fitting over the tabs as shown in **Figure 38**.

d. Slide the float onto the float shaft. Make sure the spring is seated between the ridges on the boss in the float bowl floor as shown in **Figure 39**. Make sure the pull clip wire enters the slot in the float arm.

e. Use needlenose pliers to install the E-clip (25, **Figure 43**) onto the float shaft.

f. Insert the baffle plate (68, **Figure 43**) into the float bowl with the slot over the inlet needle seat.

10A. On models with a center-hinged float (**Figure 28**), adjust the float level as follows:

a. Invert the float bowl.

b. Loosen the lockscrew (**Figure 40**) and then turn the adjusting nut until the flat float surface is level with the flat float bowl (inner) surface.

c. Securely tighten the lockscrew.

10B. On models with a side-hinged float, place the float bowl on a level work surface with the inlet needle side facing downward. The top of the float should be parallel to the flat float bowl surface as shown in **Figure 41**. If adjustment is needed, carefully bend the curved part of the float arm that contacts the inlet needle to achieve parallel surfaces.

11. Install the secondary metering block and float bowl as follows:

a. Install a *new* plate gasket (61, **Figure 43**) onto the secondary side of the carburetor body. The secondary side is opposite the choke valve. Install the plate onto

7

the gasket and carburetor body (**Figure 49**). Align the screw openings.

b. Install a *new* metering block gasket (63, **Figure 43**) onto the metering block. Install the metering block and gasket onto the plate (**Figure 48**).

c. Align all screw openings and then thread the screws (**Figure 47**) into the block, plate and carburetor body. Use a No. 2 clutch (double-D) screwdriver to securely tighten the screws in a crossing pattern.

d. Install the secondary float bowl onto the carburetor body (**Figure 46**). Install *new* gaskets (2, **Figure 43**) onto the screws and then thread them into the secondary float bowl and carburetor body (**Figure 45**). Tighten the screws securely in a crossing pattern.

12A. On models with a center-hinged float (**Figure 28**), install the float, inlet needle and seat assembly and bracket into the primary float bowl as follows:

a. Align the pin openings in the float arm and retaining bracket. Then insert the float pin (11, **Figure 28**) into the four openings.

b. Guide the retaining bracket and float assembly into the float bowl. Make sure the float pin enters the recess in the float bowl.

c. Thread the two screws (8, **Figure 28**) into the retainer and float bowl openings. Securely tighten the screws.

d. Apply a light coat of engine oil to the O-ring (2, **Figure 28**) on the inlet needle and seat assembly (3). Then thread the assembly into the float bowl bore.

e. Install a *new* gasket (4, **Figure 28**) onto the needle and seat assembly and then thread the adjusting nut (5) onto the needle and seat assembly.

f. Install a *new* gasket (6, **Figure 28**) onto the needle and seat assembly and then thread the lockscrews (7) into the needle and seat assembly.

12B. On models with a side-hinged float (15, **Figure 43**), install the inlet needle, seat and float into the primary float bowl as follows:

a. Install a new gasket (2, **Figure 43**) and then thread the needle seat (3) into the float bowl opening. Tighten the seat to 10 in.-lb. (1 N•m).

b. Install a new pull clip (5, **Figure 43**) onto the inlet needle (4). Insert the inlet needle and clip into the seat. Make sure the pull clip wire extends outward.

c. Install the spring onto the float arm with the end with the small diameter fitting over the tabs as shown in **Figure 38**.

d. Slide the float onto the float shaft. Make sure the spring is seated between the ridges on the boss in the float bowl floor as shown in **Figure 39**. Make sure the pull clip wire enters the slot in the float arm.

e. Use needlenose pliers to install the E-clip (25, **Figure 43**) onto the float shaft.

f. Insert the baffle plate (68, **Figure 43**) into the float bowl with the slot over the inlet needle seat.

13A. On models with a center-hinged float (**Figure 28**), adjust the float level as follows:

53

a. Invert the float bowl.

b. Loosen the lockscrew (**Figure 40**) and then turn the adjusting nut until the flat float surface is level with the flat float bowl (inner) surface.

c. Securely tighten the lockscrew.

13B. On models with a side-hinged float, place the float bowl on a level work surface with the inlet needle side facing downward. The top of the float should be parallel to the flat float bowl surface as shown in **Figure 41**. If adjustment is needed, carefully bend the curved part of the float arm that contacts the inlet needle to achieve parallel surfaces.

14. Make sure the *used* balance tube O-rings (65, **Figure 43**) are removed from the openings in the float bowls. Install *new* O-rings onto each end of the tube. Insert the tube into the opening in the secondary float bowl.

NOTE

Several different power valves and power valve gaskets might be included in the repair kit. The valves are identified by numbers stamped onto the faces of the valves. Install a valve with the same number as the original. Also, the valve must be installed with the correct gasket. If the power valve has round fuel openings, use the gasket with three internal projections. If the power valve has rectangular fuel openings, use the gasket without internal projections.

15. Install the *new* gasket (22, **Figure 43**) onto the power valve (23). Make sure the power valve and gasket are the same number and style as the originals. Thread the power valve by hand into the metering block opening. Tighten the power valve to 100 in.-lb. (11 N•m) with a suitable wrench.

16. Thread the main fuel jets by hand into the metering block (**Figure 27**). Then use a wide-blade screwdriver to tighten the jets to 10 in.-lb. (1 N•m).

17. Thread the idle mixture screws (**Figure 26**) into the metering block openings using *new* seals (66, **Figure 21**). Turn in the needles until lightly seated. Then back the screws out the number of turns recorded during disassembly. The setting is normally one-half to one turn out from

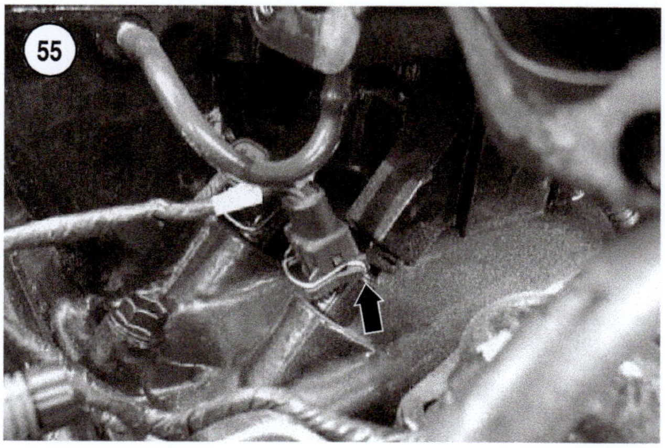

lightly seated. If so equipped, install the plugs (**Figure 25**) into the screw openings to prevent tampering.

18. Install the primary metering block and float bowl as follows:

a. Install the baffle (17, **Figure 43**) onto the metering block and position over the dowels.
b. Install a *new* inner gasket (21, **Figure 43**) onto the metering block (18). Use the dowels on the block to align and retain the gasket. Install a *new* O-ring onto the brass accelerator pump tube and then insert the tube into the carburetor body opening (**Figure 34**).
c. Fit the metering block and gasket onto the carburetor body (**Figure 24**). Install a *new* outer gasket (16, **Figure 43**) onto the metering block. Use the dowels on the block to align and retain the gasket.
d. Guide the balance tube, if so equipped, into the opening in the primary float bowl. Then install the bowl onto the metering block (**Figure 23**). Use the dowels to align and retain the float bowl.
e. Install a *new* gasket (2, **Figure 43**) onto each of the four screws (6). Thread the four screws (**Figure 22**) into the bowl, metering block and carburetor body openings. Verify proper fuel bowl, metering block and carburetor body alignment. Then tighten the screws securely in a crossing pattern.

19. Install the carburetor as described in this section.

20. Adjust the accelerator pump clearance as described in Chapter Five.

21. On models with a center-hinged float bowl, perform the final float level adjustment as described in Chapter Five.

ELECTRONIC FUEL INJECTION (EFI)

This section describes service procedures for the electronic fuel injection (EFI) components. Two different types of fuel injection systems are used on the engines covered in this manual. The throttle body injection (TBI) system is used on GM 4.3L, 5.0L and 5.7L models. The multiport injection (MPI) system is used on Ford 5.0L and 5.8L and GM 7.4L and 8.2L models.

On the TBI system, the throttle body is located on the intake manifold in the same location as the carburetor on models so equipped. The TBI throttle body houses the fuel injectors (**Figure 53**). On MPI systems, the throttle body (**Figure 54**, typical) is located on the front or side of the intake plenum. The plenum is located directly over the intake manifold. On MPI systems, the fuel injectors (**Figure 55**) fit into openings in the intake manifold that align with the intake passages for the individual cylinders. 7

Replacement of the injector filters used on TBI and MPI models is described in the injector replacement procedure.

Always relieve the fuel system pressure as described in this section *before* disconnecting any fuel line or removing any EFI component. The throttle body used on TBI systems is equipped with an internal bleed system that automatically relieves the system pressure within a few seconds of stopping the engine. However, as a precaution, always relieve the fuel system pressure as described in this section.

Take all precautions to prevent contamination and work only in a clean environment when servicing the EFI system. Use only clean solvent and lint-free shop towels when cleaning components. Dry components with compressed air and promptly cover them. Replace all gaskets, seals and O-rings if disturbed. Unless specified otherwise, apply liquid dishwashing soap to all O-rings before installing them.

WARNING

Use caution when working with the fuel system. Never smoke around fuel or fuel vapor. Make sure no flame or source of ignition is present in the work area. Flame or sparks can ignite fuel or fuel vapor and cause a fire and explosion.

CAUTION

Do not use solvents containing methyl ethyl ketone to clean the fuel rail, throttle body and related components. Harsh solvents can dissolve protective coatings and damage some fuel system components. Use only mild solvents designed to clean fuel system components.

Relieving Fuel System Pressure

Throttle body injection (TBI) models

1. Disconnect the battery cables.
2. Remove the flame arrestor and cover as described in this chapter.
3. Locate the fuel delivery (pressure) line (A, **Figure 56**) on the fitting (B) at the rear port side of the throttle body assembly. The fuel return line threads into the fitting on the rear starboard side. Use compressed air to remove debris from these fittings and the throttle body surfaces.
4. Wrap a suitable shop towel around the fitting to capture spilled fuel.
5. Hold the throttle body fitting with a suitable wrench to prevent rotation. Then use a flare nut wrench to loosen the fuel line fitting.
6. *Slowly* loosen the fuel line fitting by hand until fuel just starts dripping from the connection. Wait until fuel stops dripping and then unthread the fitting.
7. Drain residual fuel in the line into a suitable container. Clean up any spilled fuel at once.
8. Remove the O-ring from the groove at the end of the fuel line fitting. Discard the O-ring.
9. Install a *new* O-ring into the fitting groove. Then thread the line fitting by hand into the throttle body fitting.
10. Hold the throttle body fitting (B, **Figure 56**) with a suitable wrench. Then tighten the fuel line (A, **Figure 56**) to 204 in.-lb. (23 N•m). Do not allow the fuel line to twist during the tightening procedure. Twisting damages the line.

WARNING
Stay clear of the propeller and propeller shaft while running an engine on a flush adapter.

CAUTION
If running the engine with the boat out of the water, use a flush device or other means to supply the engine with cooling water. The seawater pump is quickly damaged if operated without adequate cooling water.

11. After performing the fuel system service procedure, start the engine and immediately check for fuel leakage at the fuel line-to-throttle body connection. Correct any fuel leakage before operating the engine.
12. Install the flame arrestor and cover as described in this chapter.

Multiport injection (MPI) models

A suitable pressure gauge with a bleed-off hose (**Figure 57**) available from Kent-Moore (part No. J-34730-1A) is required for this procedure.

WARNING
Stay clear of the propeller and propeller shaft while running an engine on a flush adapter.

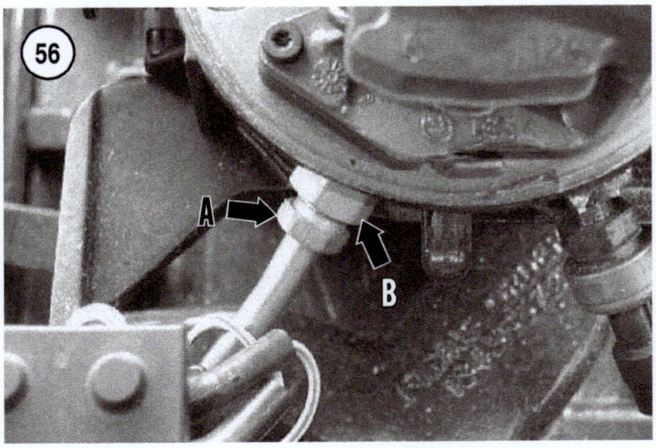

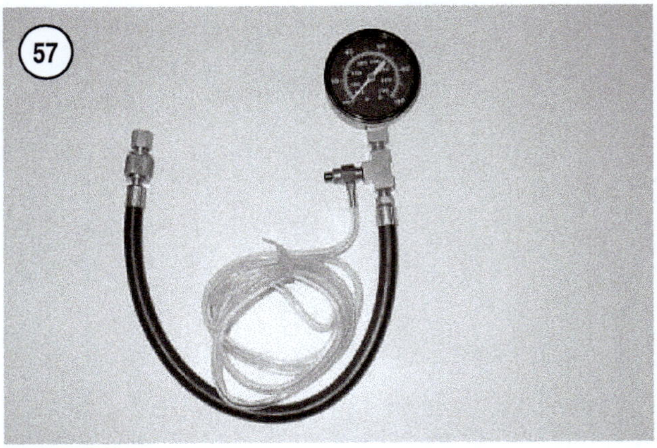

CAUTION
If running the engine with the boat out of the water, use a flush device or other means to supply the engine with cooling water. The seawater pump is quickly damaged if operated without adequate cooling water.

1. Remove the fuel pump relay as described in Chapter Nine.
2. Supply the engine with cooling water as described in Chapter Three. Refer to *Flushing the Cooling System*.
3. Attempt to start the engine.
   a. If the engine starts, allow it to run until it stops. Then crank the engine for an additional 20 seconds to relieve any residual pressure.
   b. If the engine does not start, crank the engine for two 20-second intervals to relieve any residual pressure.
4. Disconnect the negative battery cable.
5. Refer to *Fuel Pressure Test* in Chapter Two to locate the fuel pressure test port for the high-pressure electric fuel pump.
6. Remove the cap from the test port fitting. Wrap a suitable shop towel around the test port to capture spilled fuel.
   a. If the test gauge uses a lever to lock the hose fitting onto the test port, pull up on the locking lever and then *quickly* push the connector onto the test port and release the locking lever.

58
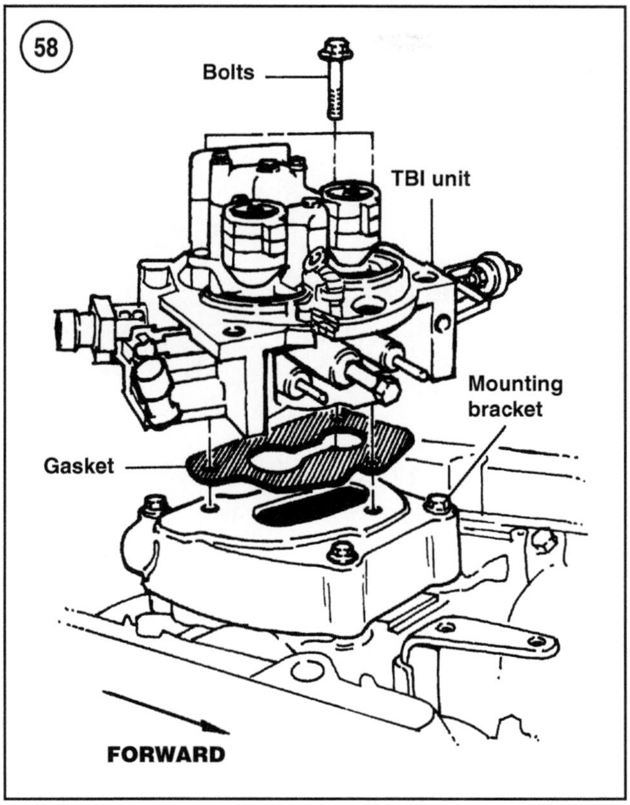

59

b. If the test gauge threads onto the test port, hold the hose fitting in firm contact with the test port fitting and then *quickly* thread the hose fitting onto the test port.

7. Direct the bleed-off hose into a container suitable for holding fuel. Open the valve on the gauge until the pressure is completely relieved. Allow all fuel to drain from the hose.

a. If the test gauge uses a lever to lock the hose fitting onto the test port, lift the locking lever and pull the connector off the test port.

b. If the test gauge threads onto the test port, wrap a shop towel around the test port and then *slowly* unthread the fitting.

8. Clean up any spilled fuel and replace the cap. Open the valve and allow all fuel to drain from the bleed-off hose.

60

9. Inspect the fuel for contamination. If it is not contaminated, empty the fuel into the boat fuel tank. Dispose of contaminated fuel in an environmentally responsible manner. Contact a marine dealership or local government official for information on the proper disposal of fuel.

10. Install the fuel pump relay as described in Chapter Nine.

WARNING
Stay clear of the propeller and propeller shaft while running an engine on a flush adapter.

CAUTION
If running the engine with the boat out of the water, use a flush device or other means to supply the engine with cooling water. The seawater pump is quickly damaged if operated without adequate cooling water.

11. Connect the battery cable after performing the fuel system repair or other service procedure. Start the engine and immediately check for fuel leakage at all fuel system components. Correct any fuel leakage before operating the engine.

12. Install the flame arrestor as described in this chapter.

THROTTLE BODY

This section describes throttle body removal, repair and installation. Replace the mounting gasket anytime the throttle body mounting fasteners are removed.

Removal (GM 4.3L, 5.0L and 5.7L TBI Models)

Refer to **Figure 58** for this procedure.

1. Disconnect the negative battery cable.
2. Relieve the fuel system pressure as described in this section.
3. Remove the flame arrestor (A, **Figure 59**) and cover as described in this chapter.
4. Disconnect the wire harness connectors from the throttle position sensor (A, **Figure 60**), idle air control motor

(B, **Figure 60**) and both fuel injectors (A, **Figure 61**). Pinch the two white plastic arms to unplug the fuel injector connectors. Pull the injector harness grommet (B, **Figure 61**) out of the notch in the throttle body.

5. Disconnect the pressure line from the throttle body as follows:
   a. To prevent rotation, engage a wrench onto the throttle body pressure line fitting (B, **Figure 59**) at the rear of the throttle body. The pressure line fitting is on the port side.
   b. Wrap a shop towel around the fitting to capture spilled fuel.
   c. Use another wrench to loosen the pressure line fitting. Allow the line to drain completely and then unthread the pressure line. Remove the O-ring from the line or throttle body fitting. Discard the O-ring.
6. Disconnect the return line from the throttle body as follows:
   a. To prevent rotation, engage a wrench onto the throttle body return line fitting (C, **Figure 59**) at the rear of the throttle body. The return line fitting is on the starboard side.
   b. Wrap a shop towel around the fitting to capture spilled fuel.
   c. Use another wrench to loosen the return line fitting. Allow the line to drain completely and then unthread the return line. Remove the O-ring from the line or throttle body fitting. Discard the O-ring.
7. Drain any residual fuel in the pressure and return lines into a suitable container. Plug the lines and throttle body fittings to prevent contamination.
8. On models using a mechanical fuel pump, disconnect the sight tube from the throttle body.
9. Disconnect the throttle cable from the throttle arm on the port side of the throttle body.
10. Remove the three bolts (**Figure 58**) and then lift the TBI unit off the mounting bracket.
11. Remove the gasket (**Figure 58**) from the mounting bracket or throttle body. Carefully scrape residual gasket material from the mating surfaces. Do not damage the surfaces. Residual gasket material or damaged surfaces will allow air leakage and cause poor idle characteristics and a lean operating condition. Use aerosol carburetor cleaner to remove oily deposits or other material from the mating surfaces.
12. If necessary, repair the throttle body as described in this section.

Disassembly/Assembly (GM 4.3L, 5.0L and 5.7L TBI Models)

Failure of the fuel pressure regulator is almost always the result of debris under the diaphragm and housing seats. Cleaning the seats usually corrects the fault. Replace the throttle body assembly if the regulator diaphragm or seating surfaces are damaged. The housing, cover, spring and diaphragm are not available separately.

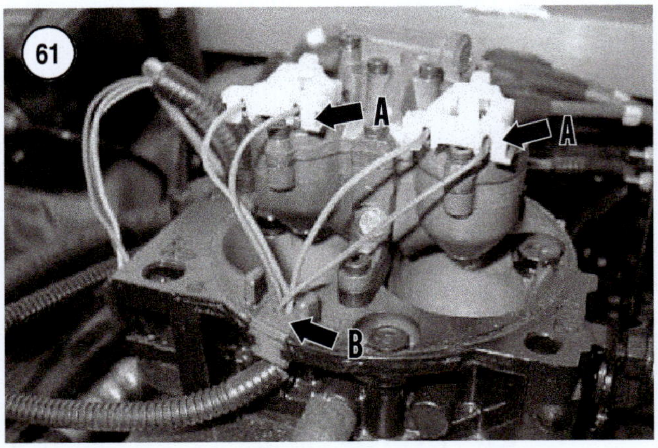

Refer to **Figure 62** for this procedure.

1. Disconnect the negative battery cable.
2. Relieve the fuel system pressure as described in this section.
3. Remove the throttle body as described in this section.
4. Remove the screws and then *carefully* lift the injector cover and regulator housing (1, **Figure 62**) off the injector housing (9). Do not remove the regulator unless it must be replaced. Refer to *Fuel Pressure Regulator Replacement* in this section.
5. To protect the injector housing, leave the gasket (3, **Figure 62**) in place during injector removal.
6. Use a screwdriver as a fulcrum as shown in **Figure 63** and pry the injectors from the housing.
7. Remove the gaskets (3 and 5, **Figure 62**) from the injector housing or injector cover and regulator housing. Discard the gaskets.
8. Remove the O-rings (4 and 8, **Figure 62**) from the fuel injectors (6) or injector bores in the housing (9). Discard the O-rings.
9. Carefully slide the filters (7, **Figure 62**) off the injectors.
10. Remove the pressure and return fittings (12, **Figure 62**) and gaskets (11) from the housing. Discard the gaskets.
11. Remove the flame arrestor stud from the throttle body.
12. Remove the screws and then lift the injector housing out of the throttle body (13, **Figure 62**).
13. Remove the throttle position sensor (17, **Figure 62**) and idle air control motor (15) from the throttle body as described in Chapter Nine.
14. Remove the gasket (10, **Figure 62**) from the throttle body or injector housing. Carefully scrape residual gasket material from the surfaces. Do not damage the surfaces. Surface damage will allow air or fuel leakage.
15. Soak the injector housing and throttle body in mild solvent and dry in compressed air. *Do not* soak the injector cover and regulator housing or injectors in solvent. Solvent can damage these components.
16. Inspect the passages in the injector cover and regulator housing for gummy deposits or other contaminants. If contaminated, remove the fuel pressure regulator as described in this section.

62

THROTTLE BODY ASSEMBLY (GM 4.3L, 5.0L AND 5.7L ELECTRONIC FUEL INJECTION MODELS)

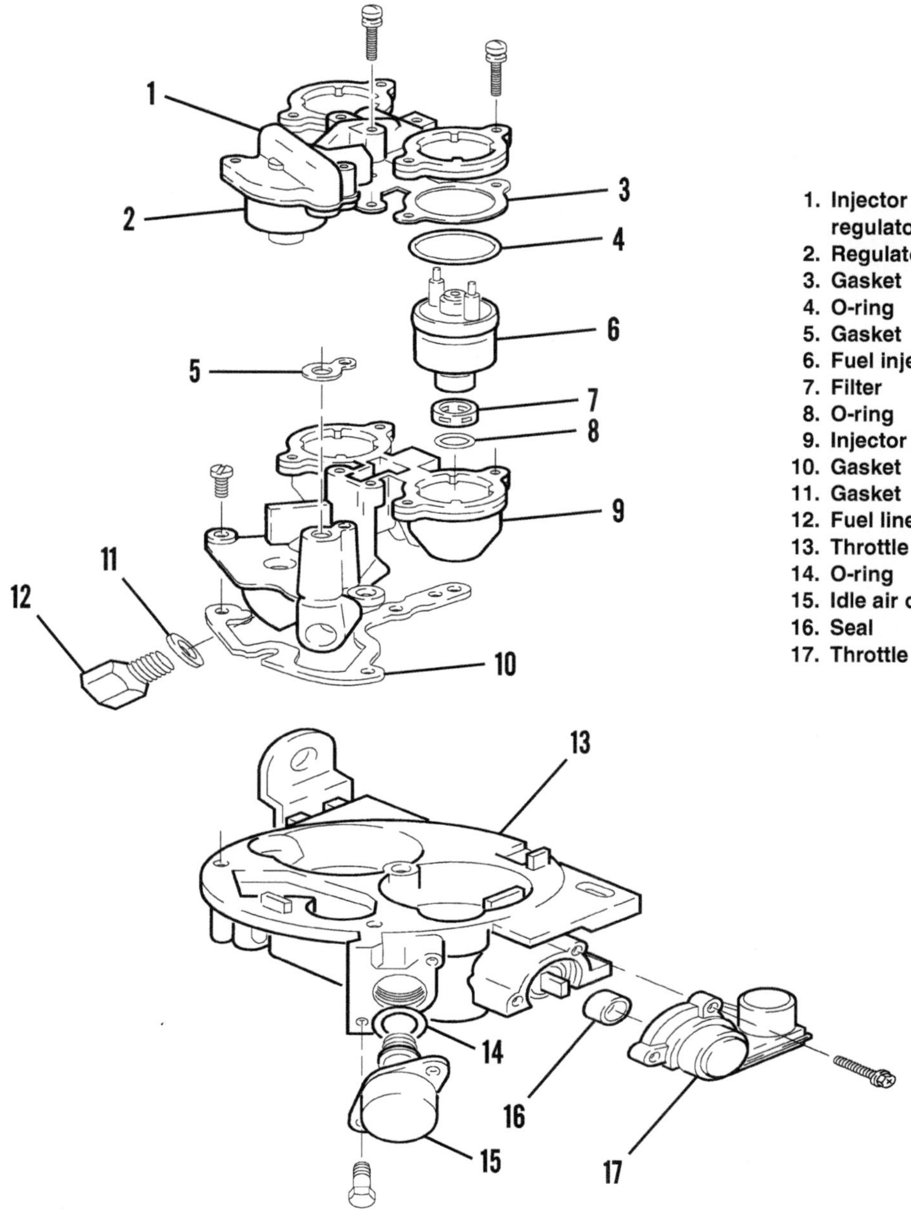

1. Injector cover and regulator housing
2. Regulator cover
3. Gasket
4. O-ring
5. Gasket
6. Fuel injector
7. Filter
8. O-ring
9. Injector housing
10. Gasket
11. Gasket
12. Fuel line fitting
13. Throttle body
14. O-ring
15. Idle air control motor (IAC)
16. Seal
17. Throttle position sensor (TPS)

63

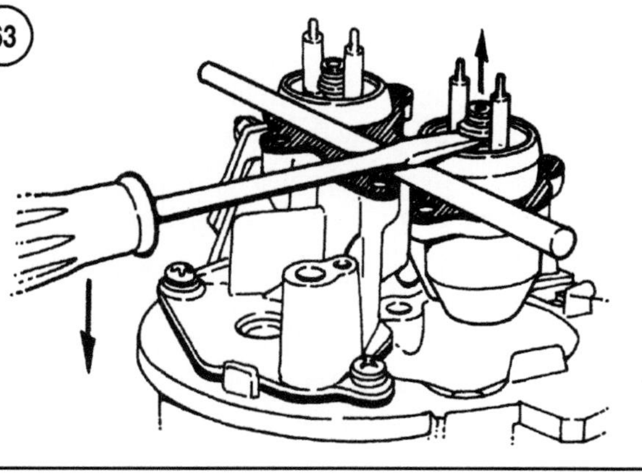

17. Inspect the filters for blockage or damage. Replace filters that are damaged or cannot be completely cleaned with solvent.

18. Inspect the throttle body for cracks, looseness of the throttle shaft or plate, or other defects. Replace the throttle body assembly if these or other defects are evident.

19. Inspect the injector housing and injector cover and regulator housing for cracks or damaged mating surfaces. Replace the throttle body assembly if these or other defects are evident.

20. If removed, install the fuel pressure regulator as described in this section.

21. Install a *new* gasket (10, **Figure 62**) onto the throttle body (13). Install the injector housing (9, **Figure 62**) into the throttle body and seat against the gasket.
22. Apply threadlocking compound (OMC part No. 1161053) onto the screw threads. Align the openings and then thread the screws into the housing, gasket and throttle body. Tighten the injector housing screws in a crossing pattern to 35 in.-lb. (4 N•m).
23. Carefully slide the filters over the injector body and position over the openings. Lubricate with dishwashing soap and fit *new* O-rings (4 and 8, **Figure 62**) into the injector housing. Rest the smaller O-rings in the recess at the bottom of the injector bore. Rest the larger O-ring in the step at the top of the bore.
24. Without dislodging the O-rings, insert the injectors into the housing. Push down on the injectors until fully seated.
25. Install *new* gaskets (3 and 5, **Figure 62**) onto the injector housing (9) and then install the injector cover and regulator housing.
26. Align the openings and then install the screws. Tighten the injector cover and regulator housing screws in a crossing pattern to 27 in.-lb. (3 N•m).
27. Install the throttle position sensor (17, **Figure 62**) and idle air control motor (15) onto the throttle body as described in Chapter Nine.
28. Fit a *new* gasket (11, **Figure 62**) onto the pressure and return fittings (12). Thread the fittings into the injector housing. Tighten the pressure line fitting (B, **Figure 56**) to 30 ft.-lb. (41 N•m). Tighten the return line fitting to 21 ft.-lb. (29 N•m).
29. Thread the flame arrestor stud into the throttle body and tighten until the offset end aligns with the center of the assembly.
30. Install the throttle body as described in this section.

Installation (4.3L, 5.0L and 5.7L TBI Models)

Refer to **Figure 58** for this procedure.

1. Fit a new gasket (**Figure 58**) onto the mounting plate. Align the mounting bolt and throttle bore openings.
2. Align the mounting bolt and throttle bore openings and then install the throttle body (**Figure 58**) onto the mounting bracket.
3. Thread the three mounting bolts (**Figure 58**) by hand into the throttle body and mounting bracket openings. Tighten the throttle body mounting bolts evenly to 142 in.-lb. (16 N•m).
4. Connect the pressure line (B, **Figure 59**) to the throttle body as follows:
   a. Lubricate with dishwashing soap and then install a *new* O-ring onto the pressure line fitting. Do not cut the O-ring on the fitting threads during installation.
   b. Remove the caps and plugs from the fittings and then thread the pressure line fitting by hand into the port side throttle body fitting.

64

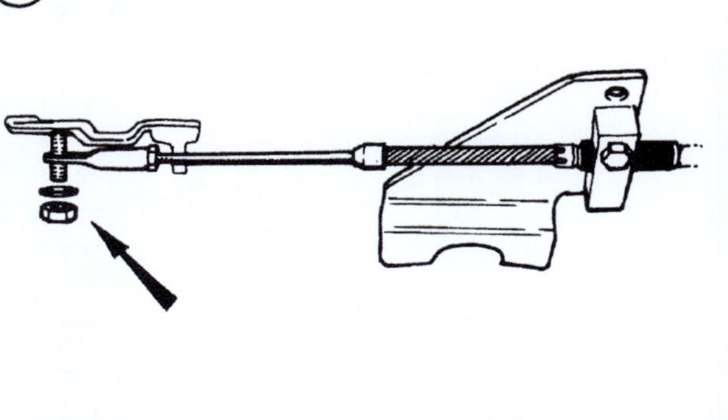

65

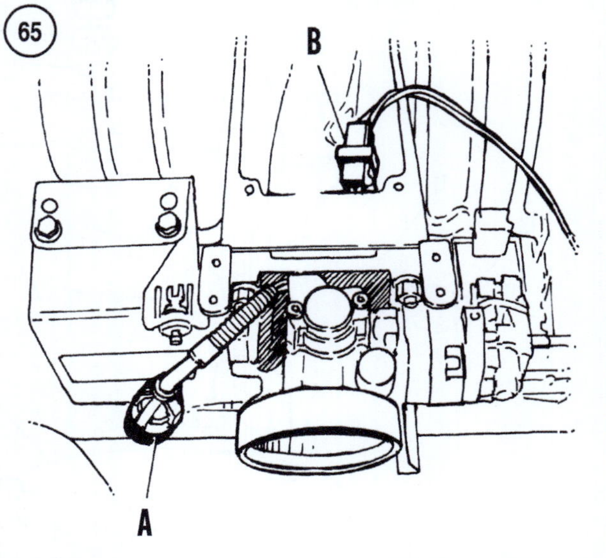

   c. Rotate the fitting to prevent line stress or kinking. Engage a wrench onto the throttle body fitting to prevent rotation.
   d. Hold the elbow section of the line in position and then tighten the pressure line (A, **Figure 56**) to 204 in.-lb. (23 N•m). Do not allow the throttle body fitting to rotate while tightening the pressure line fitting.
5. Connect the return line (C, **Figure 59**) to the throttle body as follows:
   a. Lubricate with dishwashing soap and then install a *new* O-ring onto the return line fitting. Do not cut the O-ring on the fitting threads during installation.
   b. Remove the caps and plugs from the fittings and then thread the return line fitting by hand into the starboard side throttle body fitting.
   c. Rotate the fitting to prevent line stress or kinking. Engage a wrench onto the throttle body fitting to prevent rotation.
   d. Hold the elbow section of the line in position and then tighten the return line to 204 in.-lb. (23 N•m).

Do not allow the throttle body fitting to rotate while tightening the return line fitting.

6. On models with a mechanical fuel pump, connect the sight tube to the throttle body fitting and secure with a clamp.

7. Fit the injector wire harness grommet (B, **Figure 61**) into the notch in the throttle body. Pinch the white plastic arms and then plug the injector harness connectors onto the fuel injectors (A, **Figure 61**). The connectors can attach to either injector.

8. Connect the engine wire harness connectors to throttle position sensor (A, **Figure 60**) and idle air control motor (B).

9. Attach the throttle cable to the throttle arm on the port side of the throttle body.

10. Route the wiring to prevent interference with the throttle linkage and other moving components. Secure the wiring with plastic locking clamps as needed.

11. Adjust the throttle cable as described in Chapter Five.

WARNING
Stay clear of the propeller and propeller shaft while running an engine on a flush adapter.

CAUTION
If running the engine with the boat out of the water, use a flush device or other means to supply the engine with cooling water. The seawater pump is quickly damaged if operated without adequate cooling water.

12. Connect the battery.

13. Install the flame arrestor and cover as described in this chapter.

14. Start the engine and immediately check for fuel leakage at all fuel line and hose connections, throttle body injector housing mating surfaces and fuel pressure regulator cover. Stop the engine and correct any leaks before operating the engine.

Removal, Inspection and Installation (Ford 5.0L and 5.8L Models)

The throttle body is located on the port side of the intake plenum/manifold.

1. Disconnect the battery cable.
2. Remove the flame arrestor as described in this chapter.
3. Remove the locknut and flat washer (**Figure 64**) and then disconnect the throttle cable end from the throttle lever.
4. Pinch the locking tabs to release the connectors and then disconnect the engine wire harness connectors from the throttle position sensor (TPS) harness connector (A, **Figure 65**) and idle speed control solenoid (ISC) harness connector (B).
5. Remove the four nuts (**Figure 66**) and then carefully pull the throttle body off the intake plenum.
6. Remove the gasket from the throttle body or intake plenum surface. Discard the gasket.
7. Carefully scrape residual gasket material from the mating surfaces. Do not damage the surfaces. Residual gasket material or damaged surfaces will allow air leakage and cause poor idle characteristics and a lean operating condition. Use aerosol carburetor cleaner to remove oily deposits or other material from the mating surfaces.
8. Remove the TPS and ISC as described in Chapter Nine.
9. Soak the throttle body in a mild cool solvent and blow dry with compressed air. Inspect the throttle body for cracks, looseness of the throttle shaft or plate, or other defects. Replace the throttle body if these or other defects are evident.
10. Install the TPS and ISC as described in Chapter Nine.
11. Fit a new gasket over the mounting studs and seat it onto the intake plenum. The tab on the gasket must face down and toward the rear.
12. Align the stud and throttle bore openings and then install the throttle body onto the intake plenum.
13. Thread the four nuts (**Figure 66**) by hand onto the mounting studs. Tighten the throttle body mounting nuts evenly to 144-216 in.-lb. (16-24 N•m).
14. Pinch the locking tabs to release the locking connector. Then connect the engine wire harness connectors to the TPS harness connector (A, **Figure 65**) and ISC harness connector (B).
15. Connect the throttle cable to the throttle lever. Secure the cable end with the flat washer and locknut (**Figure 64**).
16. Install the flame arrestor as described in this chapter.
17. Connect the battery cable.

Removal, Inspection and Installation (1998 74GI and 82GI Models)

Refer to **Figure 67** for this procedure.

1. Disconnect the battery cables.
2. Remove the flame arrestor as described in this chapter.

(67) INTAKE MANIFOLD, FUEL RAIL, PLENUM AND THROTTLE BODY (1998 GM MPI 74GI AND 82GI MODELS)

1. Torx screw
2. Plastic cover
3. Plenum
4. Bolt
5. Plastic cover
6. Plenum gasket
7. Injector harness
8. Stud bolt
9. Manifold air pressure (MAP) sensor
10. Seal
11. Bolt
12. Intake manifold
13. Fitting
14. Clamp
15. Bypass hose
16. Clamp
17. Fitting
18. Intake gasket
19. Flame arrestor
20. Screw
21. Throttle lever hardware
22. Nut
23. Bracket
24. Screw
25. PCV hose
26. Fitting
27. Adapter
28. Gasket
29. Nut
30. Throttle body
31. Clamp
32. Vacuum hose fitting (pressure regulator hose)
33. Fuel rail assembly
34. Fuel injector
35. O-ring
36. Plug
37. Screw
38. Gasket
39. Stud

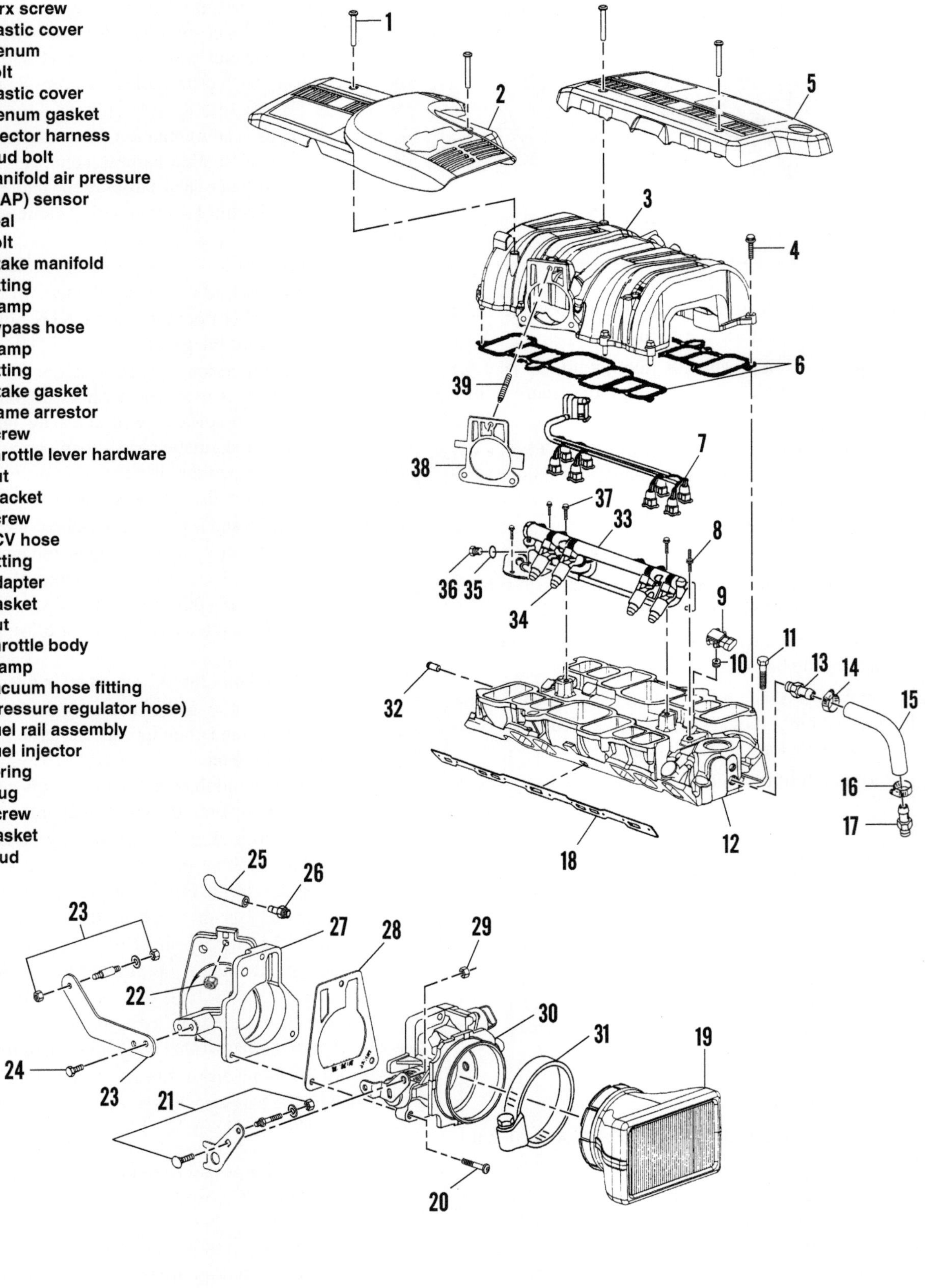

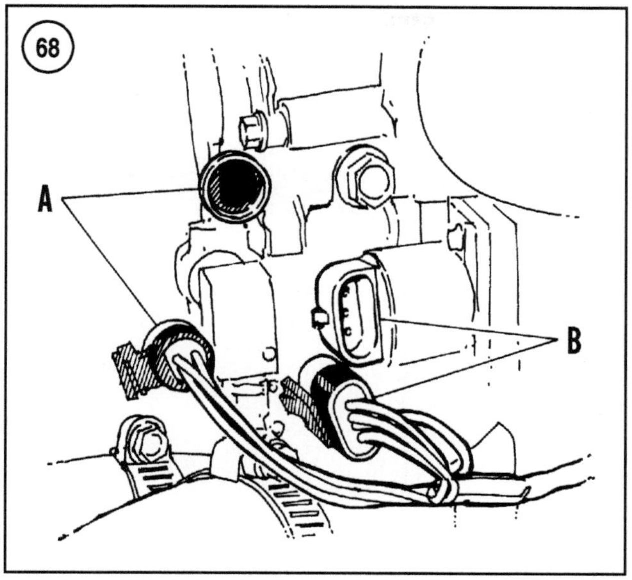

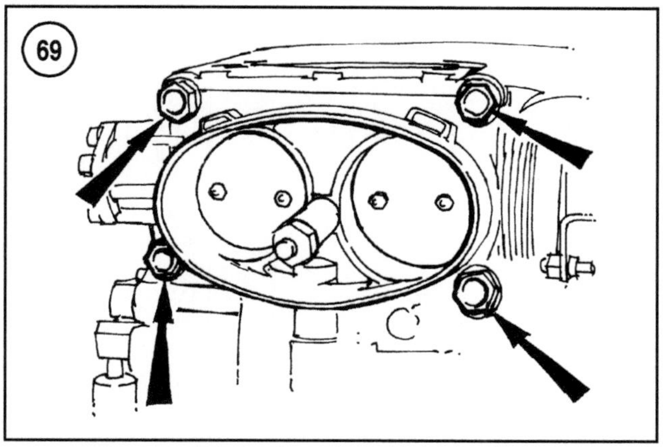

3. Disconnect the wire harness connectors from the throttle position sensor (TPS) (A, **Figure 60**, typical) and idle air control motor (IAC) (B, **Figure 60**). These components are on the side of the throttle body opposite the throttle lever.

4. Remove the locknut and flat washer (**Figure 64**) and then disconnect the throttle cable end from the throttle lever.

5. Support the throttle body and then remove the three screws (20, **Figure 67**). Pull the throttle body (30, **Figure 67**) off the plenum (3).

6. Remove the gasket (28, **Figure 67**) from the plenum or throttle body. Carefully scrape residual gasket material from the mating surfaces. Do not damage the surfaces. Residual gasket material or damaged surfaces will allow air leakage and cause poor idle characteristics and a lean operating condition. Use aerosol carburetor cleaner to remove oily deposits or other material from the mating surfaces.

7. Remove the TPS and IAC as described in Chapter Nine.

8. Soak the throttle body in a mild solvent and dry with compressed air. Inspect the throttle body for cracks, looseness of the throttle shaft or plate, or other defects. Replace the throttle body if these or other defects are evident.

9. After cleaning and inspection, install the TPS and IAC as described in Chapter Nine.

10. Insert the three screws (20, **Figure 67**) into the throttle body openings and then fit the *new* gasket (28) onto the throttle body. Make sure the openings in the gasket match the openings in the throttle body and plenum.

11. Align the mounting screw and throttle bore openings and then guide the throttle body (30, **Figure 67**) onto the plenum.

12. Thread the three mounting screws by hand into the threaded openings. Tighten the three throttle body mounting screws evenly to 132 in.-lb. (15 N•m).

13. Lift the locking tabs and then connect the engine wire harness connectors to the TPS and IAC connectors. Both components are located on the side of the throttle body opposite the throttle cable.

14. Connect the throttle cable to the throttle lever. Secure the cable end with the flat washer and locknut (**Figure 64**).

15. Adjust the throttle cable as described in Chapter Five.

16. Install the flame arrestor as described in this chapter.

17. Route the wiring to prevent interference with the throttle linkage and other moving components. Secure the wiring with plastic locking clamps as needed.

18. Connect the battery cables.

Removal, inspection and installation (74GSI and 82GSI models)

1. Disconnect the battery cables.
2. Remove the flame arrestor as described in this chapter.
3. Disconnect the engine wire harness connectors from the throttle position sensor (TPS) (A, **Figure 68**) and idle air control motor (IAC) (B). These components are on the side of the throttle body opposite the throttle lever.
4. Remove the locknut and flat washer (**Figure 64**) and then disconnect the throttle link arm from the throttle lever.
5. Support the throttle body and then remove the four screws (**Figure 69**). Pull the throttle body off the plenum.

CAUTION

Do not soak the throttle body in solvent or allow solvent onto the throttle bore or throttle plate surfaces. Solvent can damage or remove the special sealing compound applied at the time of manufacture.

6. Remove the gasket from the plenum or throttle body. Carefully scrape residual gasket material from the mating surfaces. Do not damage the surfaces. Residual gasket material or damaged surfaces will allow air leakage and cause poor idle characteristics and a lean operating condition. Use a clean shop towel to remove oily deposits or other material from the mating surfaces.

7. Remove the TPS and IAC as described in Chapter Nine.

8. Remove the small screws and then carefully remove the top cover from the throttle body. Remove the gasket from the cover. Discard the gasket. Carefully scrape residual gas-

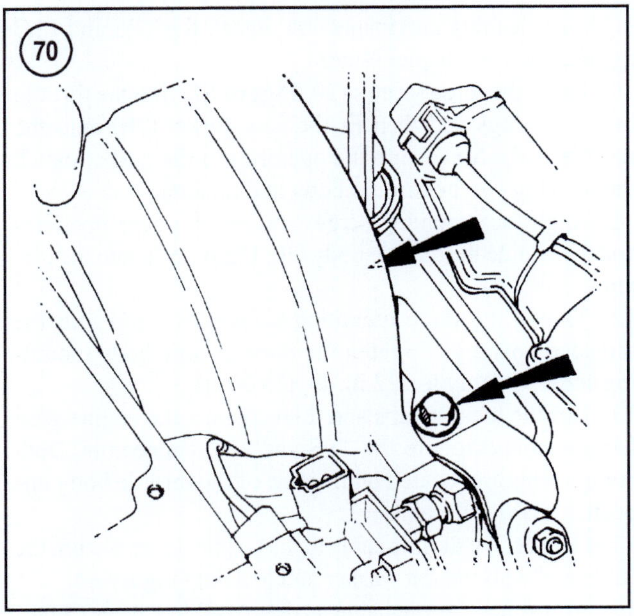

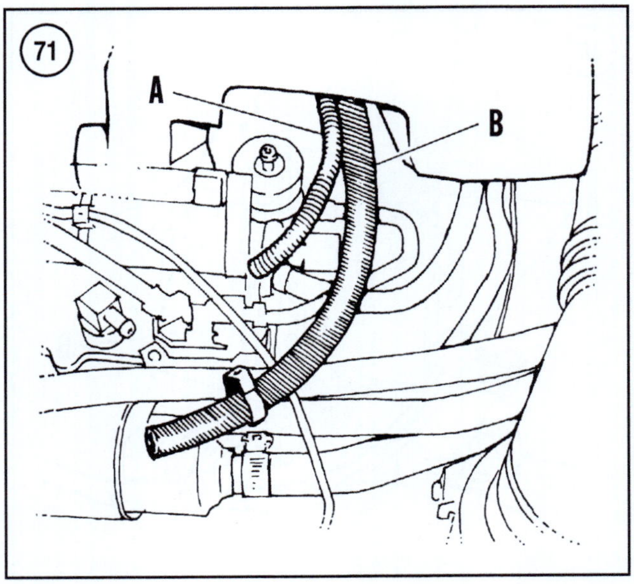

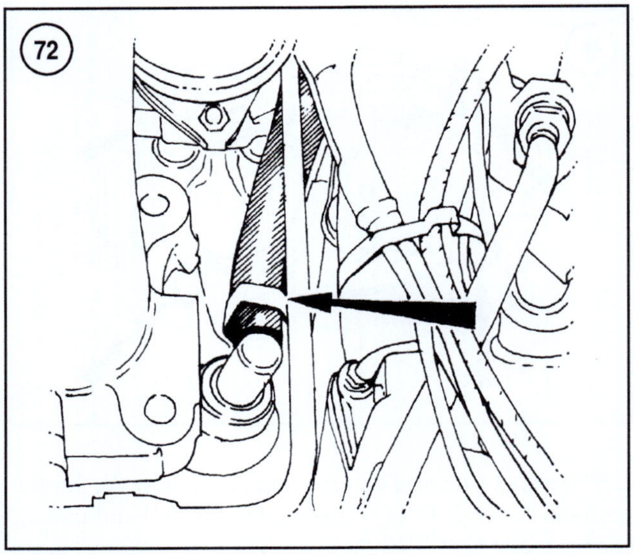

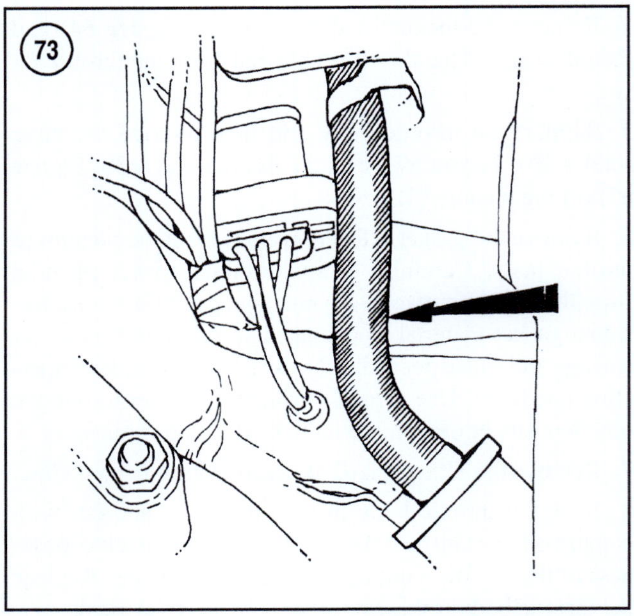

ket material from the mating surfaces. Do not damage the surfaces. Residual gasket material or damaged surfaces will allow water leakage.

9. Use compressed air to remove loose material from the throttle body. Inspect the throttle body for cracks, loose throttle shaft or plate, or other defects. Replace the throttle body if these or other defects are evident.

10. After cleaning and inspection, install the TPS and IAC as described in Chapter Nine.

11. Install a *new* gasket onto the throttle body top cover and then fit the cover and gasket onto the top of the throttle body. Thread the screws into the cover, gasket and throttle body openings. Tighten the top cover screws evenly to 28 in.-lb. (3 N•m).

12. Insert the four screws (**Figure 69**) into the throttle body openings and then fit the *new* gasket onto the throttle body. Make sure the openings in the gasket match the openings in the throttle body and plenum.

13. Align the mounting screws and throttle bore openings and then guide the throttle body onto the plenum.

14. Thread the three mounting screws by hand into the threaded openings. Tighten the three throttle body mounting screws evenly to 132 in.-lb. (15 N•m).

15. Lift the locking tabs and then connect the engine wire harness connectors to the TPS (A, **Figure 68**) and IAC (B) connectors. Both components are located on the side of the throttle body opposite the throttle cable.

16. Connect the throttle cable to the throttle linkage. Secure the cable end with the flat washer and locknut (**Figure 64**).

17. Adjust the throttle cable as described in Chapter Five.

18. Install the flame arrestor as described in this chapter.

19. Route the wiring to prevent interference with the throttle linkage and other moving components. Secure the wiring with plastic locking clamps as needed.

20. Connect the battery cables.

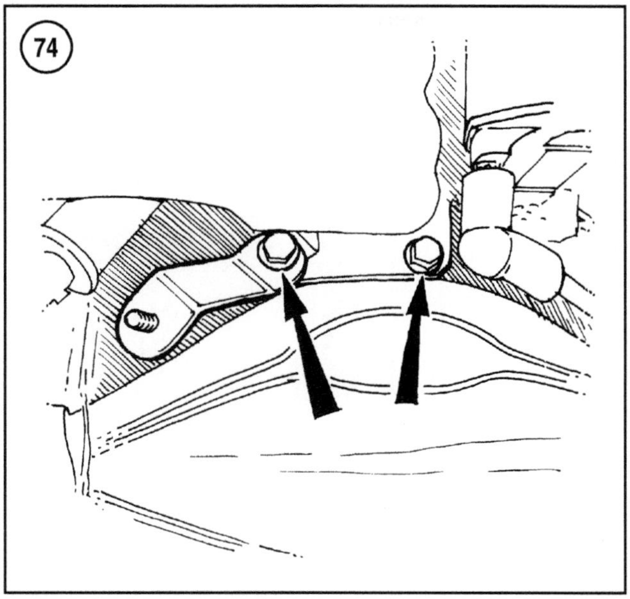

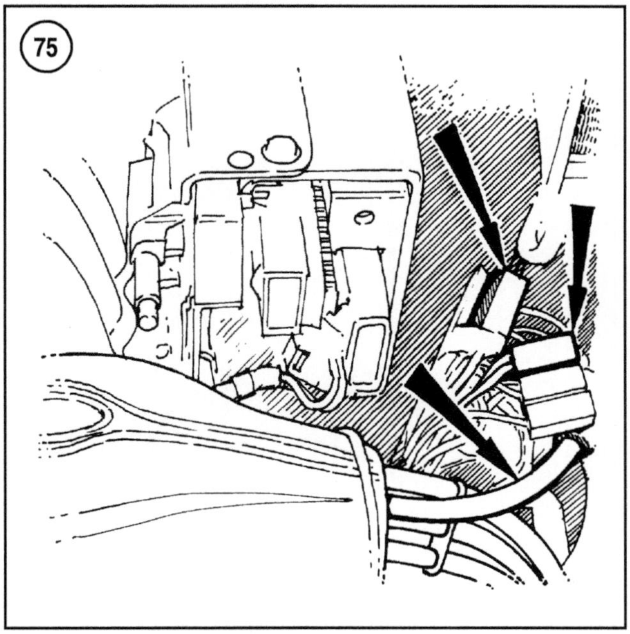

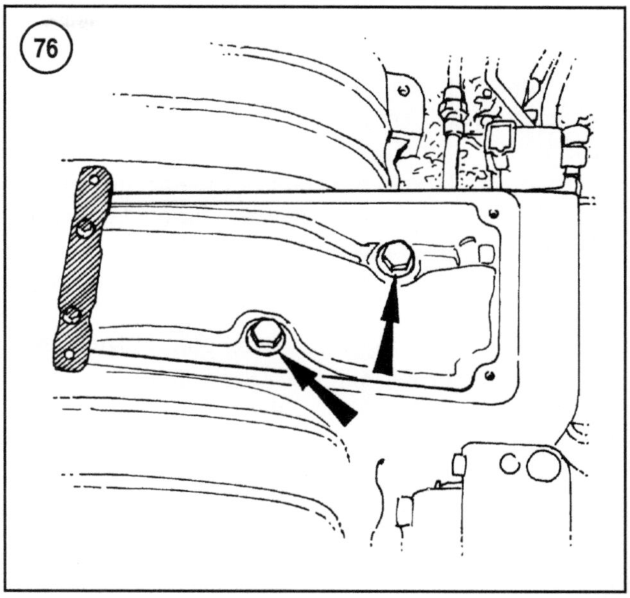

INTAKE PLENUM, FUEL RAIL AND INJECTORS

This section describes removal and installation of the intake plenum, fuel rail and fuel injectors. Replace all O-rings, seals and gaskets anytime they are removed. Consider reusing a gasket only if it is in excellent condition and a *new* replacement is not available. Never reuse O-rings.

Ford 5.0L and 5.8L Models

Removal, inspection and installation

1. Relieve the fuel system pressure as described in this chapter.
2. Disconnect the battery cable.
3. Remove the flame arrestor as described in this chapter.
4. Remove the throttle body as described in this chapter.
5. Remove the two screws (**Figure 70**) attaching the cable bracket and then position the cable away from the intake plenum.
6. Remove the clamps. Then disconnect the fuel pressure regulator vacuum hose (A, **Figure 71**) and fuel reservoir vent hose (B) from the plenum fittings.
7. Remove the hose clamps, the disconnect the vacuum hose from the PVC valve (**Figure 72**).
8. Remove the hose clamps. Then disconnect the manifold air pressure (MAP) sensor hose (**Figure 73**) from the plenum fitting.
9. Remove the two screws (**Figure 74**) and then position the shift bracket support away from the plenum.
10. Disconnect the high-tension lead and engine wire harness connector (**Figure 75**) from the ignition coil.
11. Remove the two center screws (**Figure 76**) from the top of the plenum. Carefully lift the plenum from the intake manifold.
12. Cover all of the openings in the intake manifold with masking tape to prevent screws or other objects from falling into the intake passages.
13. Clean and inspect the mating surfaces of the intake manifold and intake plenum. Deep scratches or other surface imperfections will allow air leakage and cause a poor idle quality.
14. Use compressed air to remove any loose material from the area surrounding the fuel injectors.
15. Disconnect the fuel delivery (pressure) line fitting (A, **Figure 77**) and return line fitting (B) from the fuel rail and fuel pressure regulator fittings.
16. Note the wire routing, and clamp positioning. Then cut the plastic locking clamps that secure the injector wire harness to the fuel rail.
17. Push the locking clip and then disconnect the wire harness connector from each of the fuel injectors.

18. Remove the four screws (A, **Figure 78**). Then while pulling up slightly on the fuel rail, disengage each fuel injector from the fuel rail.
19. Carefully pull each of the injectors out of the intake manifold.
20. Use aerosol spray carburetor cleaner to remove all oily deposits from the injector openings in the intake manifold. Dry the openings with compressed air. Then cover the injector openings in the intake manifold to prevent contamination.
21. Remove the O-rings from the injectors (**Figure 79**).
22. Use mild solvent followed by compressed air to remove all foreign material from the fuel rail passages.
23. Clean the fuel injector terminals and harness connector terminals with electrical contact cleaner.
24. Apply liquid dishwashing soap to the *new* O-rings. Then install the O-rings onto the injectors (**Figure 79**).
25. Remove the covers and then fit the injectors into the corresponding openings in the intake manifold. Rotate the injectors slightly during installation to prevent damage to the O-rings.
26. Align the injector openings in the rail with the injectors and then lower the fuel rail onto the intake manifold. Push down lightly while slightly rotating the injectors. When all injectors are properly aligned, the O-rings will snap into position in the opening. When all injectors are fully seated and the rail contacts the intake manifold, thread the four screws (A, **Figure 78**) by hand into the rail and intake openings. Tighten the fuel rail-to-intake manifold screws in a crossing pattern to 70-105 in.-lb. (8-12 N•m).
27. Use plastic locking clamps to secure the injector harness wiring to the fuel rail in the original locations. The six clamping points (B, **Figure 78**) are the usual clamping positions.
28. Thread the fuel delivery (pressure) line (A, **Figure 77**) and return line (B) by hand onto the fuel rail and fuel pressure regulator fittings. Tighten the fuel delivery line and return line fittings to 120-144 in.-lb. (14-16 N•m).
29. Apply electric terminal grease (Ford part No. F2AZ-19584-A) to the terminals and then plug each of the injector harness connectors onto the corresponding fuel injector. Tug lightly on each harness to verify a secure connection.
30. Verify the vacuum hose routing on the bottom of the plenum by comparing to the diagram in **Figure 80**. Use plastic locking clamps to retain the hose and prevent pinching during plenum installation.
31. Install the *new* gasket onto the intake manifold. Align the intake passage and screw openings.
32. Carefully guide the intake plenum onto the intake manifold. Work carefully to avoid dislodging the gaskets or pinching the vacuum hoses.
33. Verify proper gasket alignment and hose routing and then thread the two center screws (**Figure 76**) into the plenum and intake manifold. *Do not* tighten the screws at this time.

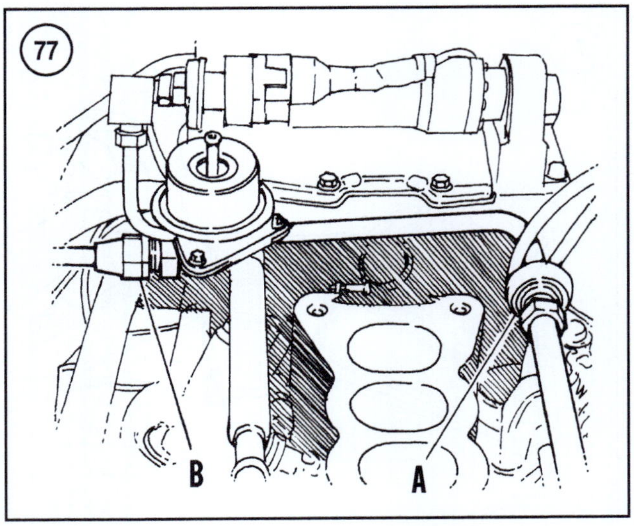

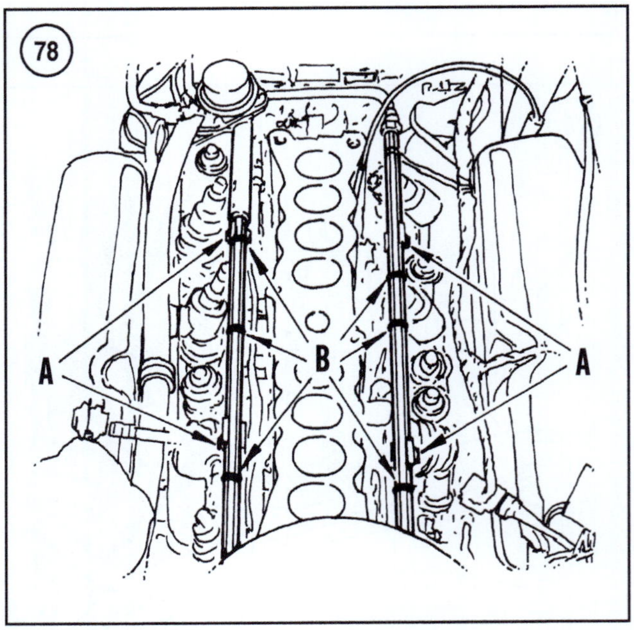

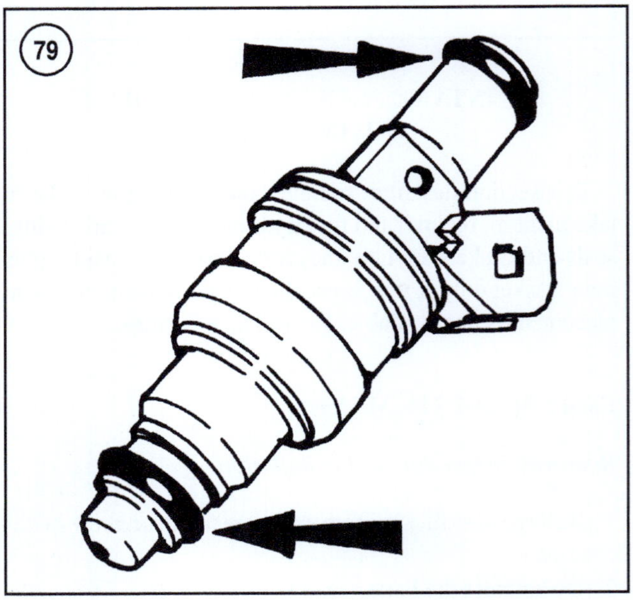

(80) **VACUUM HOSE ROUTING AND CONNECTION POINTS (FORD 5.0L AND 5.8L ELECTRONIC FUEL INJECTION [EFI] MODELS)**

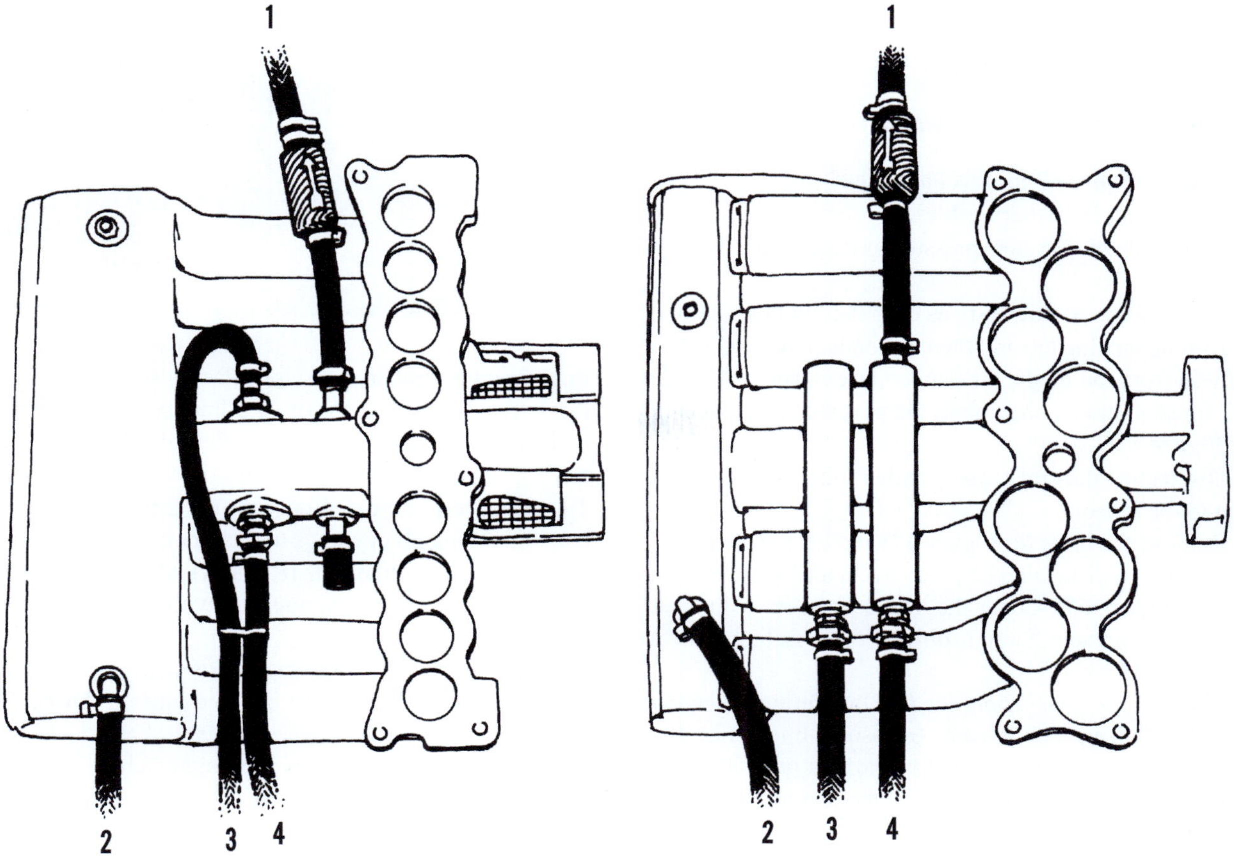

1. To manifold air pressure sensor (MAP)
2. To PCV valve
3. To fuel pressure regulator
4. To fuel reservoir vent fitting

34. Apply electric terminal grease (Ford part No. F2AZ-19584-A) to the terminals and then connect the high-tension lead and engine wire harness connector (**Figure 75**) to the ignition coil.

35. Position the shift support bracket onto the intake plenum. Align the openings and then thread the two screws (**Figure 74**) by hand into the bracket, plenum and intake manifold openings. *Do not* tighten the screws at this time.

36. Position the throttle cable attaching bracket onto the intake plenum. Align the opening and then thread the two screws (**Figure 70**) by hand into the bracket, plenum and intake manifold openings. *Do not* tighten the screws at this time.

37. Tighten the six intake plenum-to-manifold screws installed in Steps 33-36 in a crossing pattern to 144-216 in.-lb. (16-24 N•m).

38. Connect the MAP sensor hose (**Figure 73**) to the plenum fitting. Secure the hose with the reusable clamp.

39. Connect the vacuum hose to the PCV valve (**Figure 72**). Secure the hose with the reusable clamp.

40. Connect the fuel pressure regulator vacuum hose (A, **Figure 71**) and fuel reservoir vent hose (B) to the plenum fittings. Secure the hoses with the reusable clamps.

41. Install the throttle body as described in this chapter.

42. Install the flame arrestor as described in this chapter.

43. Connect the battery cable.

WARNING

Stay clear of the propeller and propeller shaft while running an engine on a flush adapter.

CAUTION

If running the engine with the boat out of the water, use a flush device or other means to supply the engine with cooling water. The seawater pump is quickly damaged if operated without adequate cooling water.

44. Start the engine and immediately check for fuel leakage at all fuel rail and hose connections, fuel injector mounting surfaces and at the fuel pressure regulator. Stop

the engine and correct any leaks before operating the engine.

GM 7.4L and 8.2L (1998 74GI and 82GI) Models

Removal, inspection and installation

Refer to **Figure 67** for this procedure.

1. Disconnect the battery cables.
2. Relieve the fuel system pressure as described in this section.
3. Remove the throttle body as described in this section.
4. Lift the locking tab and then disconnect the engine wire harness from the intake air temperature sensor (IAT). The IAT threads into an opening in the port side of the plenum (3, **Figure 67**).
5. Disconnect the crankcase breather tube from the fitting next to the sensor.
6. Remove the bolts (4, **Figure 67**) and lift the plenum (3) off the intake manifold (12). Remove and discard the gaskets (6, **Figure 67**). Cover the intake passages with masking tape to prevent foreign objects from falling into the openings.
7. Place a shop towel under the fuel line fittings to capture spilled fuel. The fuel delivery (pressure) line connects to the lower fitting on the aft end of the fuel rail. The return line connects to the upper fitting at the end of the fuel rail. Do not loosen the fittings at this time.
8. Unplug the vacuum hose from the regulator. Replace the regulator as described in this section if fuel is found in the vacuum hose. This indicates the diaphragm in the regulator is leaking.
9. Mark the cylinder number on each fuel injector harness connector (**Figure 81**, typical). The fuel injectors are located above the respective cylinders. Odd-numbered cylinders are located on the port bank with the No. 1 cylinder at the front followed by No. 3, 5 and 7. Even-numbered cylinders are located on the starboard bank with the No. 2 cylinder at the front followed by No. 4, 6 and 8. Depress the locking clip and then unplug the connector from each injector. Route the injector harness (7, **Figure 67**) to a location that prevents interference during fuel rail removal and installation.
10. Use compressed air to blow all debris from the intake manifold, fuel rail and hose fittings. Mark the fuel line fittings to identify the connection points prior to removal.
11. To prevent rotation, engage a wrench onto the fuel rail fitting for the delivery (pressure) line. Wrap a shop towel around the fitting to capture spilled fuel. Use a flare nut wrench to loosen the fuel delivery line fitting. Allow the line to drain completely and then unthread the pressure line. Remove the O-ring from the line or fuel rail fitting. Discard the O-ring.
12. To prevent rotation, engage a wrench onto the fuel rail fitting for the return line. Wrap a shop towel around the fitting to capture spilled fuel. Use a flare nut wrench to loosen the return line fitting. Allow the line to drain completely and then unthread the return line. Remove the O-ring from the line or fuel rail fitting. Discard the O-ring.

13. Drain residual fuel from the fuel lines into a suitable container. Plug all openings to prevent contamination.
14. Remove the stud bolts (8, **Figure 67**) and screws (37) and then carefully lift the fuel rail with injectors from the intake manifold.
15. Remove the seals (**Figure 82**) from the injectors or corresponding openings in the intake manifold. Discard the seals.
16. Use aerosol solvent to thoroughly clean the intake manifold openings and dry with compressed air. Cover the openings to prevent contamination.

CAUTION

*Remove the fuel injector filter (**Figure 82**) only if it has a flange of sufficient diameter to rest against the top of the injector. Do not remove the filter if it does not have the described flange or is located completely within the opening. The filter without the flange is not designed for removal.*

17. Remove the injectors from the fuel rail as follows:
    a. Expand the arms of the metal retaining clip to disengage it from the tabs on the fuel rail and then pull the injector out of the fuel rail opening.
    b. Remove the grommet (**Figure 82**) from the injector. Some models do not use the grommet.
    c. Remove the O-ring (**Figure 82**) from the groove at the top of the injector. Discard the O-ring.
    d. If the flange at the top of the filter (**Figure 82**) extends from the opening, carefully pull the filter out of the injector. Discard the filter.
    e. Repeat this step for the remaining injectors. Place the injectors in a sealed container to prevent contamination.
18. Remove the fuel pressure regulator as described in this section.
19. Use mild solvent to flush the fuel rail and fuel line passages. Do not apply solvent to the injectors or fuel pressure

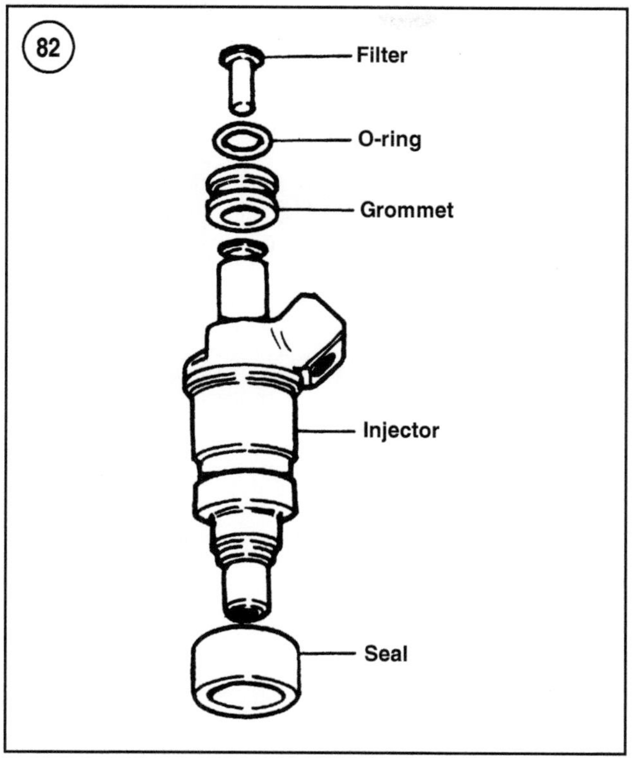

regulator. Solvent can damage these components. Direct pressurized air into the passages to remove solvent and contaminants.

20. Remove the IAT from the plenum as described in Chapter Nine.

21. Use soap and water to remove oily deposits or other contamination from the plenum (3, **Figure 67**). Dry the plenum with compressed air. Direct air into all passages to remove residual moisture.

22. Remove the tape from the intake openings. Clean any residual gasket material from the intake-to-plenum and throttle body-to-plenum mating surfaces.

23. Install the IAT into the plenum as described in Chapter Nine.

24. Install the fuel pressure regulator as described in this section.

25. Remove any covers or caps from fuel rail openings.

26. Install the fuel injectors into the fuel rail as follows:

   a. If removed, insert the new filter (**Figure 82**) into the injector opening. Seat the filter flange against the top of the injector.
   b. Lubricate with engine oil and then install the *new* O-ring (**Figure 82**) into the groove near the top of the fuel injector.
   c. If used, install the grommet (**Figure 82**) over the inlet end of the fuel injector and seat against the step in the injector body.
   d. Without dislodging or pinching the O-ring, insert the injector into the fuel rail opening.
   e. Hold the injector in position and then push the injector toward the rail until the metal arms on the retaining clip snap over the tabs on the fuel rail. Tug on the injector to verify a secure connection.
   f. Lubricate with engine oil and then fit the *new* seal (**Figure 82**) over the lower end of the injector. Seat the seal against the step on the injector body.
   g. Repeat this step for the remaining injectors.

27. *Slowly* rotate the injector to position the harness connector side of the injector facing upward.

28. Remove the covering and then apply a light coat of oil into the injector openings in the intake manifold.

29. Carefully guide the fuel rail onto the throttle body and then align all of the injectors with the respective intake manifold openings. Verify that all injectors have entered the openings and then seat the fuel rail.

30. Thread the stud bolts (8, **Figure 67**) and screws (37) by hand into the respective fuel rail and intake manifold openings. Tighten the fuel rail-to-intake manifold stud bolts and screws to 28 in.-lb. (3 N•m).

31. Connect the fuel delivery (pressure) line to the fuel rail as follows:

   a. Lubricate the surfaces with liquid dish soap and then install a *new* O-ring onto the pressure line fitting. Do not cut the O-ring on the fitting threads during installation.
   b. Remove the caps and plugs from the fittings and then thread the pressure line fitting into the fuel rail fitting by hand.
   c. Rotate the fitting to prevent line stress or kinking. Engage a wrench onto the fuel rail fitting to prevent rotation.
   d. Have an assistant hold the fuel line in position to prevent twisting and then tighten the pressure line fitting to 216 in.-lb. (24 N•m). Do not allow the fuel rail fitting to rotate while tightening the pressure line fitting.

32. Connect the return line to the fuel rail as follows:

   a. Lubricate with liquid dish soap and then install a *new* O-ring onto the return line fitting. Do not cut the O-ring on the fitting threads during installation.
   b. Remove the caps and plugs from the fittings and then thread the return line fitting into the fuel rail fitting by hand.
   c. Rotate the fitting to prevent line stress or kinking. Engage a wrench onto the fuel rail fitting to prevent rotation.
   d. Have an assistant hold the fuel line in position to prevent twisting and then tighten the return line fitting to 216 in.-lb. (24 N•m). Do not allow the fuel rail fitting to rotate while tightening the return line fitting.

33. Depress the locking clip and then plug the injector harness connectors onto the respective injectors. Route the injector harness wiring to prevent pinching or interference with other components.

34. Plug the vacuum hose onto the fuel pressure regulator fitting.

35. Install new gaskets (6, **Figure 67**) onto the intake manifold (12). Align the intake port and bolt openings and then install the plenum (3, **Figure 67**) onto the intake manifold. Make sure the gaskets are aligned and then thread the bolts (4, **Figure 67**) into the plenum, gaskets and intake manifold openings. Tighten the intake plenum-to-manifold bolts in a crossing pattern to 30 ft.-lb. (41 N•m).

36. Lift the locking tab and then plug the engine wire harness onto the IAT.

37. Connect the crankcase breather tube to the fitting next to the IAT. Secure the tube with a plastic locking clamp.

38. Install the throttle body as described in this section.

39. Install the flame arrestor as described in this chapter.

40. Connect the battery cables.

WARNING

Stay clear of the propeller and propeller shaft while running an engine on a flush adapter.

CAUTION

If running the engine with the boat out of the water, use a flush device or other means to supply the engine with cooling water. The seawater pump is quickly damaged if operated without adequate cooling water.

41. Start the engine and immediately check for fuel leakage at all fuel line and hose connections, throttle body injector housing mating surfaces and fuel pressure regulator cover. Stop the engine and correct any leaks before operating the engine.

GM 7.4L and 8.2L (74GSI and 82GSI) Models

Removal, inspection and installation

1. Disconnect the battery cables.
2. Relieve the fuel system pressure as described in this section.
3. Remove the throttle body as described in this section.
4. Disconnect the J-1 and J-2 connectors from the engine control unit (ECU). Refer to Chapter Nine.
5. Disconnect the engine wire harness connectors from the knock sensor, manifold air pressure sensor (MAP) and intake air temperature sensor (IAT) as described in the removal instructions in Chapter Nine.
6. Disconnect the fuel pressure regulator vacuum hose from the intake plenum fitting (**Figure 83**).
7. Disconnect the fuel reservoir vent hose from the pulse limiter fitting (**Figure 84**) that threads into the plenum.
8. Remove the 12 screws (A, **Figure 85**) that secure the intake plenum to the intake manifold. Note the position of the throttle cable and fuel line brackets (B and C, **Figure 85**). Then carefully lift the plenum off the intake manifold.
9. Remove the eight O-ring seals from between the intake plenum and intake manifold.

83

84

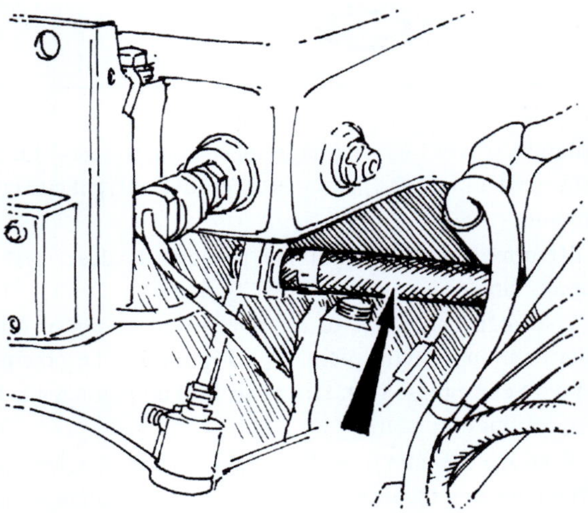

10. Cover the intake passages in the intake manifold with masking tape to prevent any foreign objects from falling into the passages.
11. Use compressed air to blow all debris from the intake manifold, fuel rail and hose fittings. Mark the fuel line fittings to identify the connection points prior to removal.
12. To prevent rotation, engage a wrench onto the fuel rail fitting for the return line. Wrap a shop towel around the fitting to capture spilled fuel. Use a flare nut wrench to loosen the line fitting at the fuel reservoir connection. Allow the line to drain completely and then unthread the return line (**Figure 86**).
13. To prevent rotation, engage a wrench onto the fuel rail fitting for the delivery (pressure) line. Wrap a shop towel around the fitting to capture spilled fuel. Use a flare nut wrench to loosen the delivery line fitting at the fuel rail connection (A, **Figure 87**). Allow the line to drain completely and then unthread the delivery line.

85

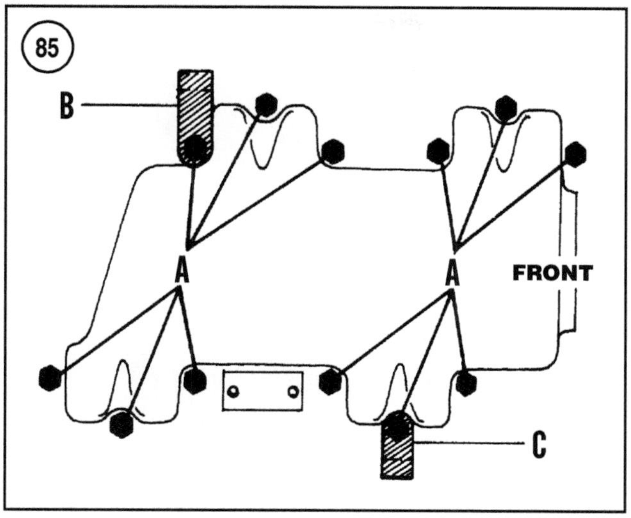

86

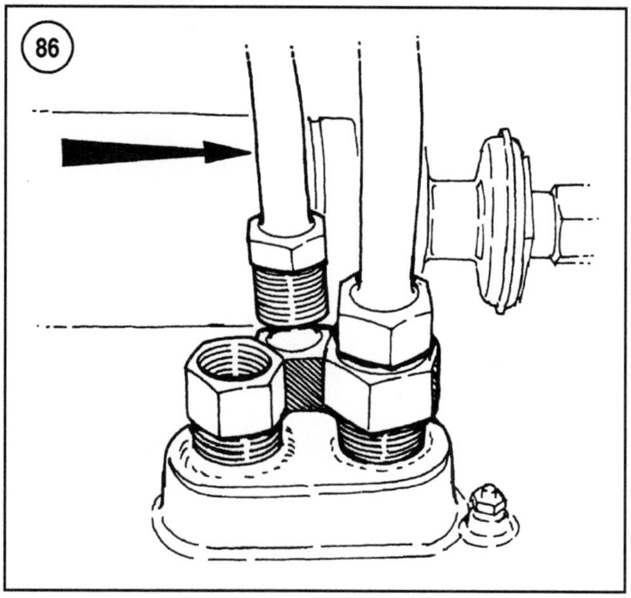

87

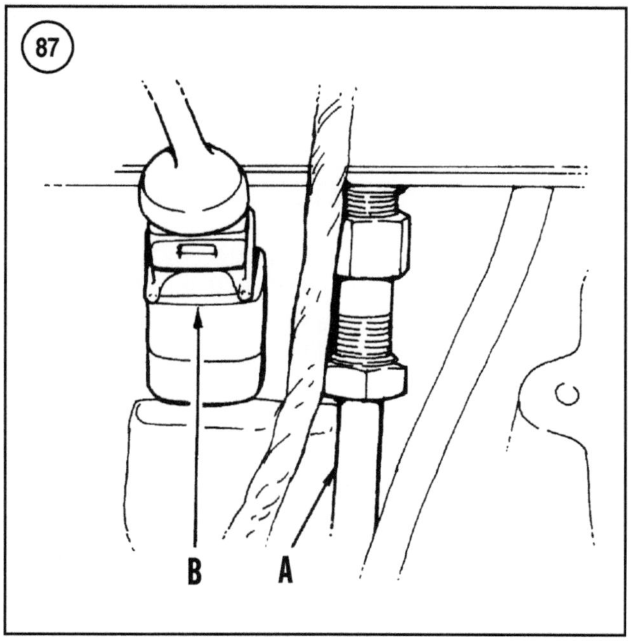

14. Drain residual fuel from the fuel lines into a suitable container. Plug all openings to prevent contamination.
15. Note the locations of the plastic locking clamps that secure the engine wire harness to the fuel rail. Then cut the clamps.
16. Push in on the locking clip and then disconnect the harness connector (B, **Figure 87**) from each of the fuel injectors.
17. Remove the four screws that secure the fuel rail to the intake manifold. Then carefully lift the fuel rail with the injectors from the intake manifold.
18. Remove the seals (**Figure 82**) from the injectors or corresponding openings in the intake manifold. Discard the seals.
19. Use aerosol solvent to thoroughly clean the intake manifold openings and dry with compressed air. Cover the openings to prevent contamination.
20. Slide the retaining clips to the side and then pull each of the fuel injectors out of the respective fuel rail opening.
21. Remove the fuel pressure regulator as described in this section.
22. Use mild solvent to flush the fuel rail and fuel line passages. Do not apply solvent to the injectors or fuel pressure regulator. Solvent can damage these components. Direct pressurized air into the passages to remove solvent and contaminants.
23. Remove the IAT and MAP from the plenum as described in Chapter Nine.
24. Use soap and water to remove oily deposits or other contamination from the plenum (3, **Figure 67**). Dry the plenum with compressed air. Direct air into all passages to remove residual moisture.
25. Clean the fuel injector terminals and harness connector terminals with electrical contact cleaner.
26. Remove the grommet (**Figure 82**) from the injectors. Some models do not use the grommet. Remove the O-rings (**Figure 82**) from the groove at the top of the injectors. Discard the O-rings. If the flange at the top of the filter (**Figure 82**) extends from the opening, carefully pull the filters out of the injectors. Discard the filters.
27. Remove any covers or caps from the fuel rail openings. Install the fuel injectors into the fuel rail as follows:
   a. If removed, insert the new filter (**Figure 82**) into the injector opening. Seat the filter flange against the top of the injector.
   b. Lubricate with engine oil and then install the *new* O-ring (**Figure 82**) into the groove near the top of the fuel injector.
   c. If used, install the grommet (**Figure 82**) over the inlet end of the fuel injector and seat against the step in the injector body.
   d. Without dislodging or pinching the O-ring, insert the injector into the fuel rail opening.
   e. Hold the injector in position and then push the locking clip over the injector body. Tug on the injector to verify a secure connection.

7

f. Lubricate with engine oil and then fit the *new* seal (**Figure 82**) over the lower end of the injector. Seat the seal against the step on the injector body.

g. Repeat this step for the remaining injectors.

28. *Slowly* rotate the injector to position the harness connector side of the injector facing upward.

29. Remove the covering and then apply a light coat of oil into the injector openings in the intake manifold.

30. Carefully guide the fuel rail onto the throttle body and then align all of the injectors into their respective intake manifold openings. Verify that all injectors have entered the openings and then seat the fuel rail.

31. Thread the screws by hand into the respective fuel rail and intake manifold openings. Tighten the fuel rail-to-intake manifold screws to 88 in.-lb. (10 N•m).

32. Secure the injector wire harness to the fuel rail with plastic locking clamps. Position the clamps in the locations noted prior to removal.

33. Plug the injector wire harness connectors onto each of the fuel injectors. Make sure the locking clip fully engages the body of the injector. Tug lightly on the harness to verify a secure connection.

34. Install the fuel pressure regulator as described in this section.

35. Thread the fuel return line fitting by hand into the fuel reservoir fitting (**Figure 86**). To prevent rotation, engage a wrench onto the reservoir fitting for the return line. Tighten the return line fitting to 156 in.-lb. (18 N•m).

36. Thread the fuel delivery (pressure) line fitting into the fuel rail fitting (A, **Figure 87**). To prevent rotation, engage a wrench onto the rail fitting for the delivery line. Tighten the delivery line fitting to 156 in.-lb. (18 N•m).

37. Remove the tape from the intake openings. Clean any residual gasket material from the intake-to-plenum and throttle body-to-plenum mating surfaces.

38. Install the IAT and MAP into the plenum as described in Chapter Nine.

39. Apply a light coat of water-resistant grease to the surfaces and then fit the eight *new* O-rings into the respective grooves in the (lower) intake manifold.

40. Without dislodging the O-rings, carefully lower the plenum onto the intake manifold. Verify that all eight O-rings are in position and then seat the plenum on the intake manifold.

41. Apply a coat of Lubriplate 777 or equivalent to the threads and shafts of the 12 intake plenum screws. Thread the screws by hand into the plenum and intake manifold. Tighten the intake plenum-to-manifold screws in the sequence shown in **Figure 88** to 124 in.-lb. (14 N•m).

42. Connect the fuel reservoir vent hose to the pulse limiter fitting (**Figure 84**) that threads into the plenum. Secure the hose to the fitting with a plastic locking clamp.

43. Connect the fuel pressure regulator vacuum to the intake plenum fitting (**Figure 83**). Secure the hose to the fitting with a plastic locking clamp.

88

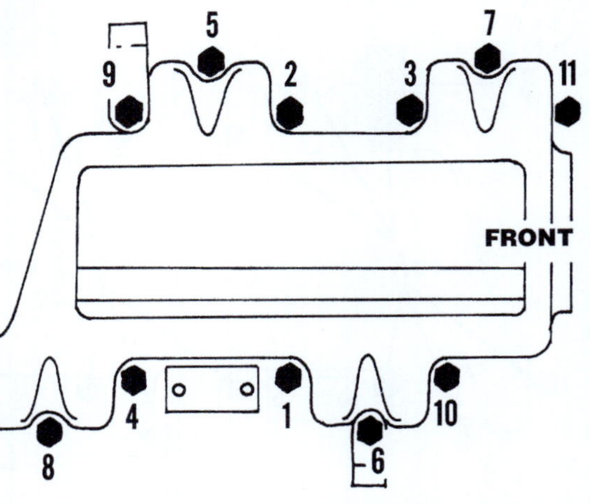

44. Connect the engine wire harness connectors to the knock sensor, MAP and IAT as described in the installation instructions in Chapter Nine.

45. Connect the J-1 and J-2 connectors to the ECU. Refer to Chapter Nine.

46. Install the throttle body as described in this section.

47. Install the flame arrestor as described in this chapter.

48. Connect the battery cables.

WARNING
Stay clear of the propeller and propeller shaft while running an engine on a flush adapter.

CAUTION
If running the engine with the boat out of the water, use a flush device or other means to supply the engine with cooling water. The seawater pump is quickly damaged if operated without adequate cooling water.

49. Start the engine and immediately check for fuel leakage at all fuel rail and hose connections, fuel injector mounting surfaces and at the fuel pressure regulator. Stop the engine and correct any leaks before operating the engine.

FUEL PRESSURE REGULATOR

On GM 4.3L, 5.0L and 5.7L models with throttle body injection (TBI), the fuel pressure regulator is incorporated into the throttle body. On multiport injection (MPI) models, the fuel pressure regulator is located on the fuel rail assembly.

GM 4.3L, 5.0L and 5.7L Throttle Body Injection (TBI) Models

The fuel pressure regulator is integrated into the throttle body. The regulator can be replaced without removing the throttle body from the intake manifold.

Failure of the fuel pressure regulator is almost always the result of debris under the diaphragm and housing seats. Cleaning the seats usually corrects the fault. Replace the throttle body assembly if the regulator diaphragm or seating surfaces are damaged. The housing, cover, spring and diaphragm are not available separately.

Removal and installation

1. Disconnect the battery cables.
2. Remove the flame arrestor and cover as described in this chapter.
3. Relieve the fuel system pressure as described in this section.
4. Pinch the two white plastic arms (A, **Figure 61**) and then unplug the engine wire harness from the fuel injector connectors. Pull the injector harness grommet (B, **Figure 61**) out of the notch in the throttle body.
5. Place shop towels into the throttle bores to capture spilled fuel.
6. Remove the screws and then *carefully* lift the injector cover and regulator housing (1, **Figure 62**) off the injector housing (9).
7. Remove the gaskets (3 and 5, **Figure 62**) from the injector housing or injector cover and regulator housing. Discard the gaskets.
8. Fit a C-clamp against the regulator cover (2, **Figure 62**) and the top side of the injector cover and regulator housing. Tighten the C-clamp just enough to lightly clamp the components.
9. Place the injector cover and regulator housing on a clean work surface with the regulator cover facing upward. Remove the screws that secure the regulator cover to the injector cover. Then *slowly* loosen the clamp to relieve the internal spring tension.
10. Remove the clamp and then lift the cover, spring and diaphragm from the injector housing.
11. Soak the cover, housing and spring in a mild solvent. Use mild dishwashing soap to clean the diaphragm. Dry the components with compressed air.
12. Inspect the diaphragm for tears, missing sections or other defects. Inspect the seating surfaces on the diaphragm and housing for pitting or other defects. Make sure the pivoting seat on the diaphragm moves freely. Inspect the spring for corrosion damage or broken loops. Replace the throttle body assembly if the diaphragm, spring or the seating surfaces are defective. The parts are not available separately.
13. Fit the diaphragm onto the injector cover and regulator housing. The pivoting seat on the diaphragm must fit against the seat in the cover.
14. Install the spring onto the diaphragm and center over the metal support. Fit the cover (2, **Figure 62**) over the spring. Rotate the spring and diaphragm to align the screw openings.
15. Fit a C-clamp against the regulator cover (2, **Figure 62**) and the top side of the injector cover and regulator housing. *Slowly* tighten the C-clamp to compress the spring and bring the regulator cover in contact with the housing. Make sure the screw opening in the regulator cover, diaphragm and housing remain aligned while tightening the clamp.
16. Verify proper alignment of the openings. Apply a light coat of Loctite 271 to the threads and then thread the screws into the regulator cover and housing. Securely tighten the screws in a crossing pattern. Lightly stake the screws to prevent loosening.
17. Install *new* gaskets (3 and 5, **Figure 62**) onto the injector housing (9) and then install the injector cover and regulator housing.
18. Align the screw openings in the cover and throttle body. Apply a light coat of Loctite 262 to the threads and then install the screws. Tighten the injector cover and regulator housing screws in a crossing pattern to 27 in.-lb. (3 N•m).
19. Fit the injector wire harness grommet into the notch in the throttle body (B, **Figure 61**). Pinch the white plastic arms and then plug the injector harness connectors (A, **Figure 61**) onto the fuel injectors. The connectors can attach to either injector.
20. Remove the shop towels from the throttle bores.
21. Connect the battery cables.

WARNING
Stay clear of the propeller and propeller shaft while running an engine on a flush adapter.

CAUTION
If running the engine with the boat out of the water, use a flush device or other means to supply the engine with cooling water. The seawater pump is quickly damaged if operated without adequate cooling water.

22. Start the engine and immediately check for fuel leakage at all fuel line and hose connections, throttle body injector housing mating surfaces and fuel pressure regulator cover. Stop the engine and repair any leaks before operating the engine.
23. Install the flame arrestor and cover as described in this chapter.

Ford 5.0L and 5.8L Electronic Fuel Injection (EFI) Models

Removal and installation

The fuel pressure regulator is attached to the rear starboard side of the fuel rail and can be replaced without removing the fuel rail or injectors from the intake manifold.

1. Relieve the fuel system pressure as described in this chapter.
2. Disconnect the negative battery cable.
3. Remove the throttle body and intake plenum as described in this chapter. Do not remove the fuel rail or fuel injectors from the intake manifold for this procedure.
4. Disconnect the fuel return line fitting (B, **Figure 77**) from the fuel pressure regulator fitting.

5. Remove the three screws (**Figure 89**) and then carefully pull the regulator off the fuel rail.

6. Remove the O-ring and gasket from the regulator or fuel rail mating surface.

7. Carefully scrape any gasket residue from the regulator and fuel rail mating surfaces. Work carefully to avoid damaging the delicate sealing surfaces.

8. Clear the regulator passages with compressed air. Do not apply air pressure into the openings. Excess pressure can damage the diaphragm within the regulator.

9. Lubricate the *new* O-ring and the contact surfaces with light assembly oil (Ford part No. ESE-M2C39-F) or equivalent. Do not use silicone grease, as it will clog the fuel injectors. Install the O-ring onto the regulator.

10. Install a *new* gasket onto the fuel rail at the regulator mating surface and then install the regulator onto the rail. Rotate the regulator and gasket to align the openings with the fuel rail screw openings.

11. Thread the three screws (**Figure 89**) into the regulator, gasket and fuel rail. Tighten the pressure regulator-to-fuel rail screws evenly to 27-40 in.-lb. (3-5 N•m).

12. Thread the fuel return line fitting (B, **Figure 77**) by hand onto the fuel pressure regulator fittings. Tighten the fuel return line fitting to 120-144 in.-lb. (14-16 N•m).

13. Install the intake plenum and throttle body as described in this chapter.

14. Connect the battery cables.

WARNING
Stay clear of the propeller and propeller shaft while running an engine on a flush adapter.

CAUTION
If running the engine with the boat out of the water, use a flush device or other means to supply the engine with cooling water. The seawater pump is quickly damaged if operated without adequate cooling water.

15. Start the engine and immediately check for fuel leakage at all fuel line connections and the fuel pressure regulator-to-fuel rail mating surfaces. Stop the engine and repair any leaks before operating the engine.

16. Install the flame arrestor and cover as described in this chapter.

GM 7.4L and 8.2L (1998 74GI and 82GI) Models

The fuel pressure regulator is integrated into the fuel rail and can be replaced without removing the fuel rail from the intake manifold. Replace all O-rings and gaskets anytime they are removed. Consider reusing a gasket only if it is in excellent condition and a *new* replacement is not available. Never reuse O-rings.

89

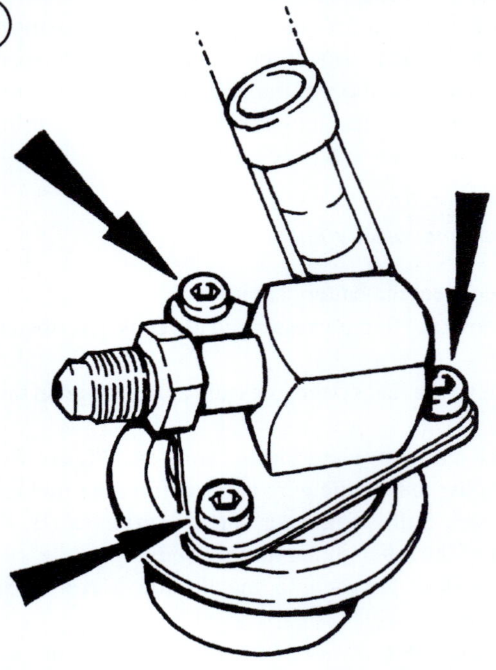

Removal and installation

1. Relieve the fuel system pressure as described in this chapter.

2. Disconnect the negative battery cable.

3. Remove the throttle body and intake plenum as described in this chapter. Do not remove the fuel rail or fuel injectors from the intake manifold for this procedure.

4. Use compressed air to blow debris or loose material from the intake manifold and fuel rail. Place a shop towel under the regulator housing to capture spilled fuel. The regulator housing is located on the port side of the fuel rail assembly. Identify the regulator housing by the vacuum hose fitting.

5. Unplug the vacuum hose from the pressure regulator fitting. Replace the regulator if fuel is found in the vacuum hose. This indicates the diaphragm in the regulator has failed.

6. Push the regulator body into the regulator housing to compress the spring and then remove the snap ring from the groove in the housing.

7. Pull the regulator, spacer, large O-ring, filter and small O-ring out of the regulator housing. Separate the components. Discard the O-rings.

8. Soak the spacer and filter in mild solvent and dry in compressed air. *Do not* soak the regulator. Solvent can damage the regulator.

9. Inspect the filter for blockage or damage. Replace the filter if damaged or if it cannot be completely cleaned with solvent.

10. Inspect the regulator housing for gummy deposits or other contaminants. If contaminated, remove the fuel rail

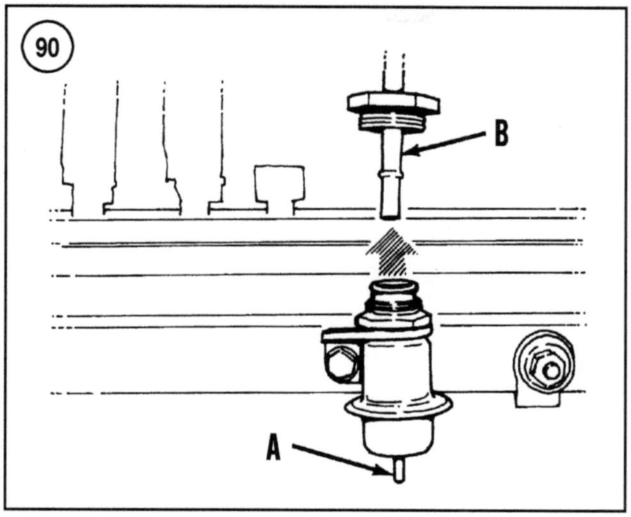

and injectors and then clean the passages as described in this chapter.

11. Fit the spacer over the regulator body and seat against the flange. Install a *new* large O-ring over the regulator body and rest it on the spacer.

12. Fit the filter over the tube on the regulator body and rest it against the O-ring and spacer. Fit the *new* small O-ring over the tube on the regulator and seat it against the filter.

13. Apply a light coat of engine oil to the O-rings, regulator, filter and spacer.

14. Insert the regulator and filter assembly into the housing. Guide the regulator tube into the opening at the bottom of the housing and then seat the assembly into the housing.

15. Push the regulator into the regulator housing to compress the spring and then install the snap ring into the groove in the regulator housing. Make sure the snap ring is seated fully in the groove. Release the regulator. Tug on the regulator to verify a secure connection.

16. Plug the vacuum hose onto the regulator fitting.

17. Install the intake plenum and throttle body as described in this chapter.

18. Connect the battery cables.

WARNING

Stay clear of the propeller and propeller shaft while running an engine on a flush adapter.

CAUTION

If running the engine with the boat out of the water, use a flush device or other means to supply the engine with cooling water. The seawater pump is quickly damaged if operated without adequate cooling water.

19. Start the engine and immediately check for fuel leakage at all fuel line connections and the fuel pressure regulator on the fuel rail. Stop the engine and repair any leaks before operating the engine.

20. Install the flame arrestor and cover as described in this chapter.

GM 7.4L and 8.2L (74GSI and 82GSI) Models

The fuel pressure regulator is located on the port side fuel rail and can be replaced without removing the fuel rail or injectors from the intake manifold.

Removal and installation

1. Relieve the fuel system pressure as described in this chapter.

2. Disconnect the negative battery cable.

3. Remove the throttle body and intake plenum as described in this chapter. Do not remove the fuel rail or fuel injectors from the intake manifold for this procedure.

4. Use compressed air to blow debris or loose material from the intake manifold and fuel rail. Place a shop towel under the regulator housing to capture spilled fuel. The regulator housing is located on the port side fuel rail. Identify the regulator housing by the vacuum hose fitting.

5. Disconnect the vacuum hose from the regulator fitting (A, **Figure 90**). Replace the regulator if fuel is found in the vacuum hose. This indicates the diaphragm in the regulator has failed.

6. To prevent rotation, engage a wrench onto the hex-shaped section on the upper end of the fuel pressure regulator. Wrap a shop towel around the fitting to capture spilled fuel. Use a flare nut wrench to loosen the return line fitting (B, **Figure 90**) from the fuel pressure regulator. Allow the line to drain completely and then unthread the return line.

7. Drain residual fuel from the fuel line into a suitable container. Plug all openings to prevent contamination.

8. Remove the single retaining screw and then carefully pull the regulator off the fuel rail.

9. Remove the O-ring from the regulator or fuel rail. Discard the O-ring.

10. Use a small pick to remove the screen from the rail opening. Discard the screen.

11. Carefully push the *new* screen into the opening. Make sure the screen is fully seated in the opening.

12. Lubricate with engine oil and then fit the *new* O-ring onto the fuel pressure regulator.

13. Rotate the regulator slightly while carefully guiding the regulator into the opening in the fuel rail. Work carefully to avoid damaging the O-ring.

14. Seat the regulator on the fuel rail and then rotate the regulator to align the screw openings. Thread the screw into the regulator and fuel rail. Verify that the O-ring has entered the opening and then tighten the pressure regulator-to-fuel rail screws to 84 in.-lb. (10 N•m).

15. Thread the fuel return line fitting (B, **Figure 90**) by hand into the regulator fitting. To prevent rotation, engage a wrench onto the hex-shaped section on the upper end of the fuel pressure regulator. Use a flare nut wrench to tighten the return line fitting to 156 in.-lb. (18 N•m).

7

16. Connect the vacuum hose to the fuel pressure regulator fitting (A, **Figure 90**). Secure the hose with a plastic locking clamp.
17. Install the intake plenum and throttle body as described in this chapter.
18. Connect the battery cables.

WARNING
Stay clear of the propeller and propeller shaft while running an engine on a flush adapter.

CAUTION
If running the engine with the boat out of the water, use a flush device or other means to supply the engine with cooling water. The seawater pump is quickly damaged if operated without adequate cooling water.

19. Start the engine and immediately check for fuel leakage at all fuel line connections and the fuel pressure regulator-to-fuel rail mating surfaces. Stop the engine and repair any leaks before operating the engine.
20. Install the flame arrestor and cover as described in this chapter.

HIGH-PRESSURE ELECTRIC FUEL PUMP

The high-pressure electric fuel pump is used on all electronic fuel injection (EFI) models. The pump is mounted on the top of the fuel reservoir at the rear of the engine.

Removal and Installation

1. Relieve the fuel system pressure as described in this chapter.
2. Disconnect the negative battery cable.
3. Cut the plastic locking clamp that secures the engine wire harness (for the fuel pump) to the fuel line.
4. Push in on the locking clip and then unplug the engine wire harness connector from the high-pressure electric fuel pump (**Figure 91**).
5. Note the position of the fuel pump elbow (**Figure 92**) so the elbow can be installed onto the pump in same position during assembly.
6. Engage a wrench onto the fuel pump elbow to prevent rotation. Wrap a shop towel around the fitting to capture spilled fuel. Use a flare nut wrench to loosen the fuel line fitting (C, **Figure 93**). Allow the residual fuel to fully drain and then unthread the line from the elbow.
7. Engage a wrench onto the hex end of the fuel pump to prevent rotation. Then loosen the locknut on the back of the elbow (**Figure 94**). Unthread the elbow from the pump. Remove the washer and O-ring from the elbow. Discard the O-ring.
8. Remove the two screws that secure the pump end or filter support bracket to the fuel reservoir. Remove the bracket from the pump.

91

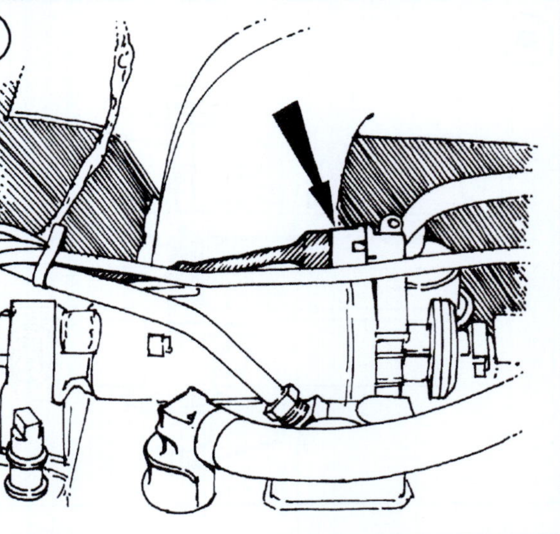

92

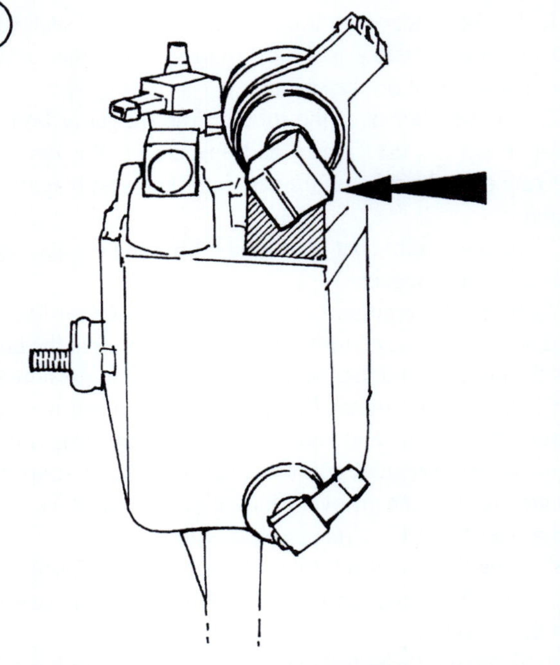

9. Place a shop towel around the fitting to capture spilled fuel. Then unthread the banjo fitting (**Figure 95**) that attaches the end of the pump to the fuel reservoir. Remove the pump from the reservoir opening.
10. Remove the O-ring from the threaded end of the banjo bolt (**Figure 95**). Then remove the banjo bolt from the reservoir opening.
11. Remove the O-ring from the banjo bolt. Discard both O-rings.
12. Lubricate the *new* O-ring with engine oil and then guide it over the banjo bolt. Fit the O-ring into the groove near the head of the banjo bolt.
13. Install the banjo bolt into the reservoir opening as shown in **Figure 94**. Lubricate with engine oil and then in-

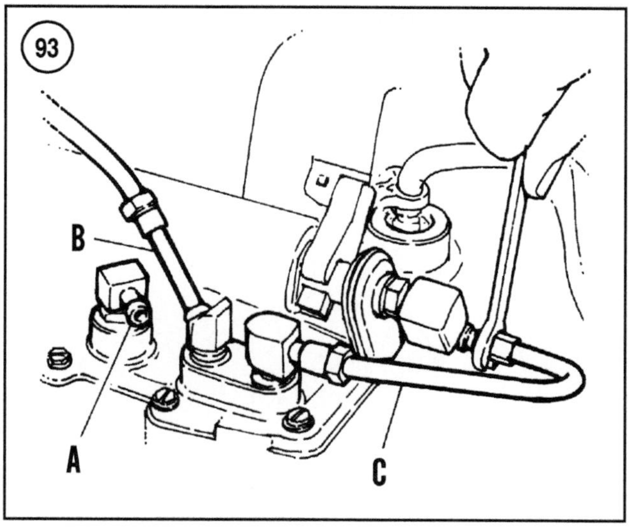

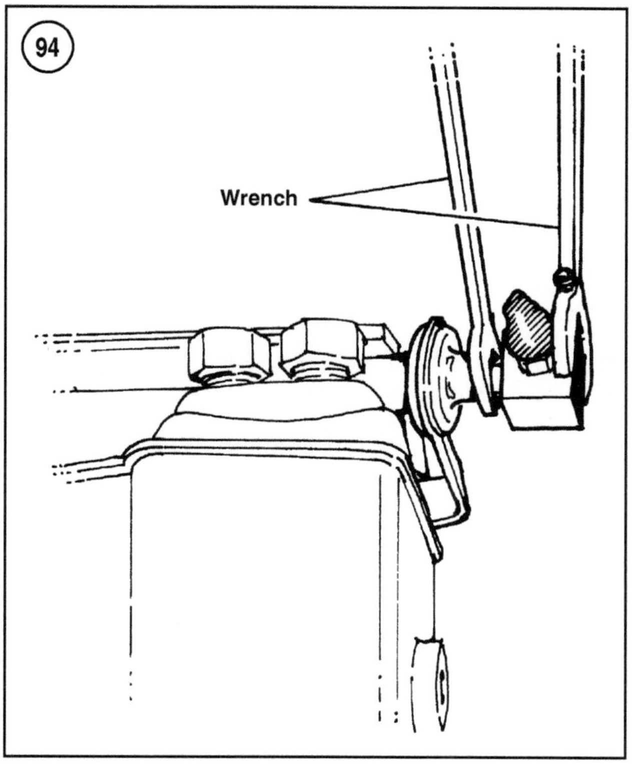

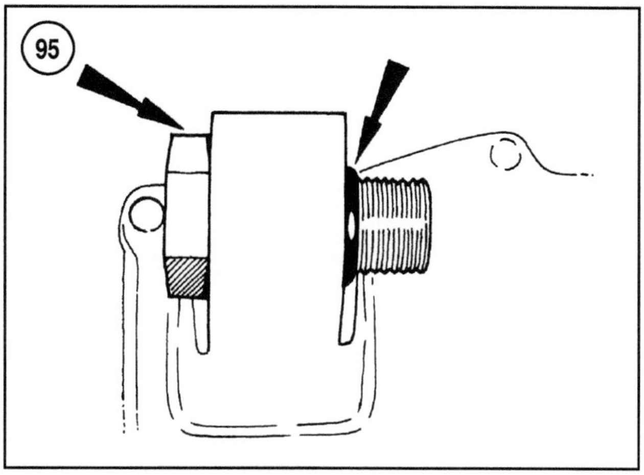

stall the *new* O-ring over the threaded end of the banjo bolt and position next to the reservoir opening.

14. Fit the high-pressure electric fuel pump onto the top of the fuel reservoir with the opening aligned with the banjo bolt. Rotate the pump to position the harness connector at 45° as shown in **Figure 92**. Hold the pump in this position and then thread the banjo bolt by hand into the pump opening.

15. Engage a wrench onto the hex-shaped end of the pump to prevent rotation. Then tighten the banjo bolt to 18-22 ft.-lb. (24-30 N•m).

16. Fit the pump end support bracket over the pump and against the fuel reservoir. Thread the two screws into the bracket and reservoir. Tighten the pump support bracket screws to 27-35 in.-lb. (3-4 N•m).

17. Lubricate with engine oil and then install a *new* O-ring onto the elbow. Install the washer onto the elbow and then thread the elbow into the pump opening. Tighten the elbow fully into the pump opening and then loosen the elbow just enough to position the elbow as noted prior to removal (**Figure 92**).

18. Thread the fuel line (C, **Figure 93**) by hand into the elbow. Engage a wrench onto the elbow to prevent rotation. Then tighten the fuel line fitting to 156 in.-lb. (18 N•m).

19. Engage a wrench onto the elbow to prevent rotation. Then use another wrench (**Figure 94**) to tighten the locknut to 60-84 in.-lb. (7-10 N•m).

20. Push in on the locking clip and then plug the engine wire harness connector onto the high-pressure electric fuel pump (**Figure 91**). Release the locking clip and then tug on the harness to verify a secure connection.

21. Secure the engine wire harness for the high-pressure electric fuel pump to the fuel line with a plastic locking clamp.

22. Connect the battery cables.

WARNING
Stay clear of the propeller and propeller shaft while running an engine on a flush adapter.

CAUTION
If running the engine with the boat out of the water, use a flush device or other means to supply the engine with cooling water. The seawater pump is quickly damaged if operated without adequate cooling water.

23. Start the engine and immediately check for fuel leakage at all fuel line connections and the fuel pump-to-reservoir banjo bolt connection. Stop the engine and repair any leaks before operating the engine.

FUEL RESERVOIR

The fuel reservoir is located on the rear of the engine just above the flywheel housing. The reservoir houses the float, inlet needle and needle seat.

Removal and Installation

1. Relieve the fuel system pressure as described in this chapter.
2. Disconnect the negative battery cable.
3. Remove the high-pressure electric fuel pump as described in this chapter.
4. Loosen the clamps and then disconnect the cooling water hoses from the fuel reservoir.
5. Remove the nut from the clamp and then separate the power steering fluid cooler from the fuel reservoir.
6. Cut the plastic locking clamp that secures the engine wire harness (for the fuel pump) to the fuel line. Push in on the locking clip and then unplug the engine wire harness connector from the high-pressure electric fuel pump (**Figure 91**).
7. Loosen the clamp and then disconnect the vent hose from the fuel reservoir fitting.
8. Engage a wrench onto the fuel inlet fitting to prevent rotation. Wrap a shop towel around the fitting to capture spilled fuel. Use a flare nut wrench to loosen the fuel inlet hose fitting. Allow residual fuel to drain from the hose.
9. Using the procedure described in Step 8, disconnect the remaining fuel pressure regulator return hose from the reservoir.
10. Remove the two screws that attach the reservoir to the bracket and then remove the reservoir.
11. To install, fit the reservoir onto the bracket. Align the openings and then thread the two screws into the reservoir and bracket. Tighten the reservoir-to-bracket mounting screws to 24-36 ft.-lb. (33-49 N•m).
12. Connect the cooling water hoses to the corresponding reservoir fittings. Secure the hoses to the fittings with the clamps.
13. Thread the pressure regulator return hose by hand onto the reservoir fitting. Engage a wrench onto the reservoir fitting to prevent rotation. Then use a flare nut wrench to tighten the fuel return hose fitting to 156 in.-lb. (18 N•m).
14. Thread the fuel inlet hose by hand onto the reservoir fitting. Engage a wrench onto the reservoir fitting to prevent rotation. Then use a flare nut wrench to tighten the inlet hose fitting to 156 in.-lb. (18 N•m).
15. Connect the vent hose to the reservoir fitting and secure with the clamp.
16. Install the high-pressure electric fuel pump as described in this section.
17. Fit the power steering fluid cooler onto the reservoir. Secure the cooler clamp to the reservoir with the nut and washer.
18. Push in on the locking clip and then plug the engine wire harness connector onto the high-pressure electric fuel pump (**Figure 91**). Release the locking clip and then tug on the harness to verify a secure connection.
19. Secure the engine wire harness for the high-pressure electric fuel pump to the fuel line with a plastic locking clamp.
20. Connect the battery cables.

96

INTAKE MANIFOLD (TYPICAL V-6 AND V-8 MODELS)

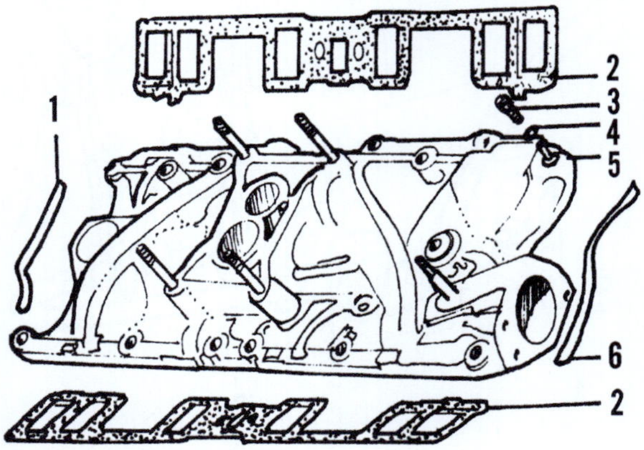

1. Rear manifold seal
2. Intake to cylinder head gasket
3. Bolt
4. Sealing washer
5. Intake manifold
6. Front manifold seal

WARNING

Stay clear of the propeller and propeller shaft while running an engine on a flush adapter.

CAUTION

If running the engine with the boat out of the water, use a flush device or other means to supply the engine with cooling water. The seawater pump is quickly damaged if operated without adequate cooling water.

21. Start the engine and immediately check for fuel leakage at all fuel line connections. Stop the engine and repair any leaks before operating the engine.
22. Check for a flooding reservoir as described in Chapter Two.

Disassembly/Assembly

1. Remove the fuel reservoir as described in this section.
2. Remove the screws that secure the cover to the reservoir.
3. Lift the cover off the reservoir. Remove the gasket from the cover or reservoir. Discard the gasket.
4. Remove the pin and then lift the float from the cover. Unthread the inlet needle seat from the cover.
5. Clean all components with clean solvent and dry with compressed air. Thoroughly clean all passages and fittings. Direct solvent and compressed air through all passages.
6. Inspect the inlet needle for wear or deterioration of the tip. Flooding is almost always due to deteriorated surfaces or dirt or debris on the inlet needle and seat. Replace the needle and seat unless in excellent condition.

(97) **INTAKE MANIFOLD TORQUE SEQUENCE (GM 4.3L, 5.0L AND 5.7L MODELS WITH A 12-BOLT INTAKE MANIFOLD)**

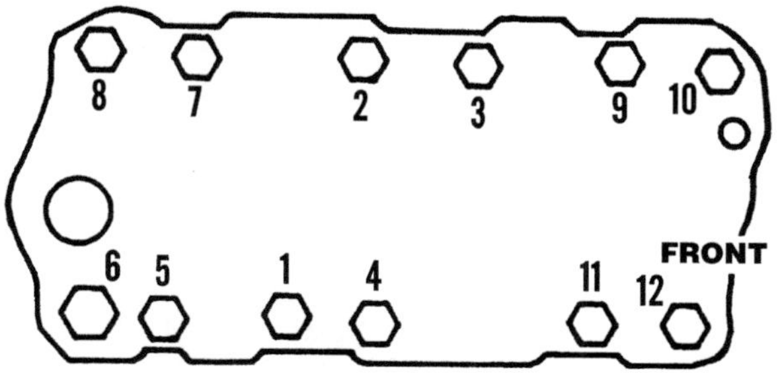

7. Inspect the float for damage or fuel saturation and replace as needed.

8. Thread the needle seat onto the cover and securely tighten. Insert the needle into the seat. Fit the float onto the cover. Verify proper alignment of the tab on the float with the needle. Then insert the float pin through the float arm and cover bosses.

9. Inspect the float for proper adjustment. The flat upper surface of the float must be approximately 1/4 in. (6.35 mm) from and parallel with the flat surface of the inner cover. Bend the metal tab on the float arm as needed to adjust.

10. Install a new cover gasket onto the reservoir. Guide the float into the reservoir and then seat the cover onto the reservoir. Align the openings and then thread the screws into the cover and reservoir. Securely tighten the screws.

11. Install the reservoir as described in this section.

INTAKE MANIFOLD

This section describes removal and installation of the intake manifold. Remove the intake manifold if leaking or as necessary to remove the cylinder head(s), camshaft, lifters or pushrods. Always replace the manifold-to-cylinder head gaskets (2, **Figure 96**) and seals (1 and 6) anytime the manifold is removed. Some models do not use front and rear seals on the intake manifold. Although the used gaskets and seals might appear to be in excellent condition, small tears will allow exhaust, oil, air, water or coolant leakage. The resulting damage will exceed the cost of the gaskets. *Never* use automotive gaskets on the intake manifold. Purchase the new gasket set from an OMC dealership. RTV sealant is required to seal the front and rear intake manifold mating surfaces in this procedure. Purchase the RTV sealant from an automotive parts store or GM automobile dealership (GM part No. 12346141). Make sure the sealant is fresh. Old sealant might not cure properly and, therfore, might allow oil leakage.

NOTE

*On GM 3.0L models, refer to **Exhaust Manifold** in Chapter Eight for removal/installation procedures and specifications.*

Intake Manifold Removal

1. Disconnect the battery cables.
2. Use compressed air to blow debris and loose material from the intake manifold and attached components.

3A. On GM and Ford carburetor-equipped models, remove the carburetor as described in this chapter.

3B. On GM 4.3L, 5.0L and 5.7L electronic fuel injection (EFI) models, remove the throttle body as described in this chapter.

3C. On Ford 5.0L and 5.8L EFI models, remove the intake plenum, fuel rail and injectors as described in this chapter.

3D. On GM 7.4L and 8.2L EFI models, remove the intake plenum, fuel rail and injectors as described in this chapter. 7

4. On models with a standard (open) cooling system, drain water from the block as described in *Winterizing* in Chapter Four.
5. On models with a closed cooling system, drain coolant from the block as described in *Engine Coolant* in Chapter Three.
6. Remove the distributor as described in Chapter Nine.
7. Remove the thermostat housing and thermostat as described in Chapter Eight.
8. Where applicable, remove the coolant temperature sender and engine temperature sensor from the intake manifold as described in Chapter Nine.
9. If the ignition coil is mounted on the intake manifold, remove it as described in Chapter Nine.
10. Remove any remaining brackets or components that might interfere with manifold removal. Route wiring away from the manifold.
11. Loosen the intake manifold bolts in the opposite of the tightening sequence.
    a. On earlier (1994-1996) GM 4.3L, 5.0L and 5.7L models with a 12-bolt intake, refer to **Figure 97**.
    b. On later (1997-2000) GM 4.3L, 5.0L and 5.7L models with an eight-bolt intake refer to **Figure 98**.
    c. On Ford 5.0L and 5.8L carburetor-equipped models, refer to **Figure 99**.
    d. On Ford 5.0L and 5.8L EFI models, refer to **Figure 100**.
    e. On 7.4L and 8.2L models with a 12-bolt intake, refer to **Figure 101**.
    f. On 7.4L and 8.2L models with a 15-bolt intake, refer to **Figure 102**.
12. Fit a pry bar into the gap at the front of the manifold and carefully pry the manifold from the cylinder heads and block. Remove and discard the gaskets.
13. Remove the intake port and water crossover gaskets (2, **Figure 96**, typical) from the intake manifold or cylinder head surfaces.

98 **INTAKE MANIFOLD TORQUE SEQUENCE (GM 4.3L, 5.0L AND 5.7L MODELS WITH AN 8-BOLT INTAKE MANIFOLD)**

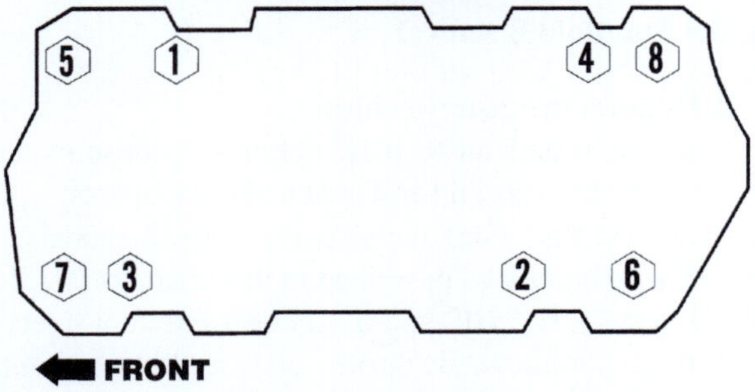

99 **INTAKE MANIFOLD TORQUE SEQUENCE (FORD 5.0L AND 5.8L CARBURETOR-EQUIPPED MODELS)**

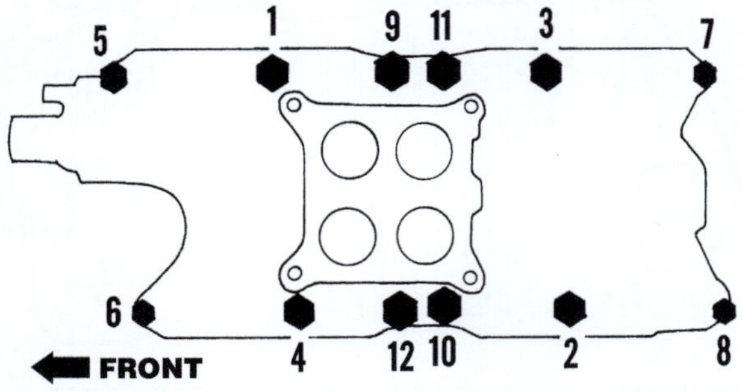

14. On GM 7.4L and 8.2L models and Ford 5.0L and 5.8L models, remove the neoprene manifold seals (1 and 6, **Figure 96**) from the front and rear surfaces of the cylinder block.
15. Cover the cylinder head openings and place clean shop towels into the valley between the cylinder heads. Then carefully scrape gasket material and sealant from the cylinder head and cylinder head surfaces.
16. Use clean solvent and a brush with plastic bristles to remove carbon and oil residue from the intake manifold. Dry the manifold and bracket with compressed air. Direct the air into all passages to remove debris and contaminants.
17. Inspect the mating surfaces for deep corrosion pitting, cracks or gouges. Defective mating surfaces will allow water, coolant, air, oil or exhaust leakage.

CAUTION
Never apply muriatic acid (pool acid) to aluminum intake manifolds or other aluminum components. The acid will rapidly destroy exposed surfaces.

18. Inspect the water or coolant passages in the manifold for corrosion or mineral deposit buildup. Have a machine shop soak the manifold in a special cleaning vat to remove the deposits. The manifolds used on some models are constructed of aluminum, and some acids will damage the material.

Intake Manifold Installation

1. Remove any shop towels or other covers from the cylinder head ports or valley between the cylinder heads. Remove any gasket material from these areas.
2A. On earlier GM 4.3L, 5.0L and 5.7L models with a 12-bolt intake manifold, install the gaskets and apply sealant as follows:
   a. Apply a 1/8 in. (3.2 mm) bead of RTV into the front and rear corners where the cylinder head and cylinder block mate (**Figure 103**). Apply the bead the full width of the seam.
   b. Coat the gasket mating surfaces on the cylinder heads with Permatex Gasket Sealing Compound.

100 **INTAKE MANIFOLD TORQUE SEQUENCE (FORD 5.0L AND 5.8L ELECTRONIC FUEL INJECTION MODELS)**

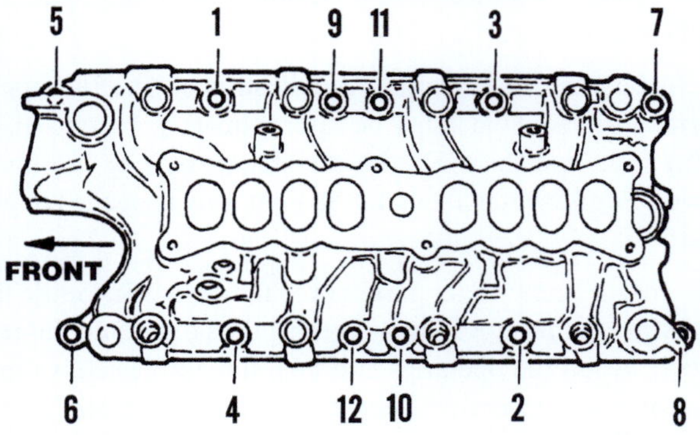

101 **INTAKE MANIFOLD TORQUE SEQUENCE (GM 7.4L AND 8.2L MODELS WITH A 12-BOLT INTAKE MANIFOLD)**

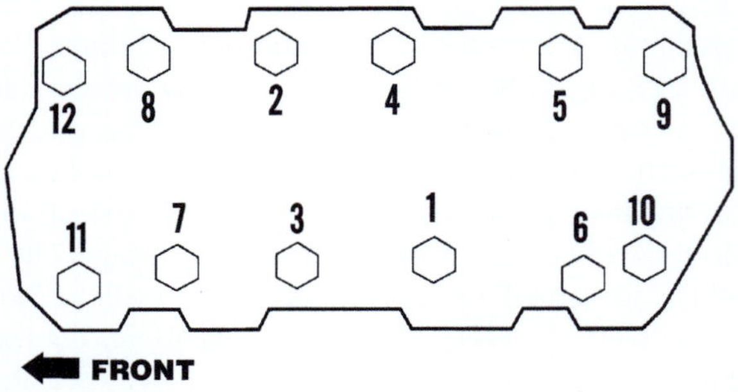

   c. Install the *new* intake gaskets (2, **Figure 96**, typical) onto the cylinder heads. The *TOP* marks on the gaskets must face up. Position the gaskets to perfectly align the gasket and cylinder head ports.

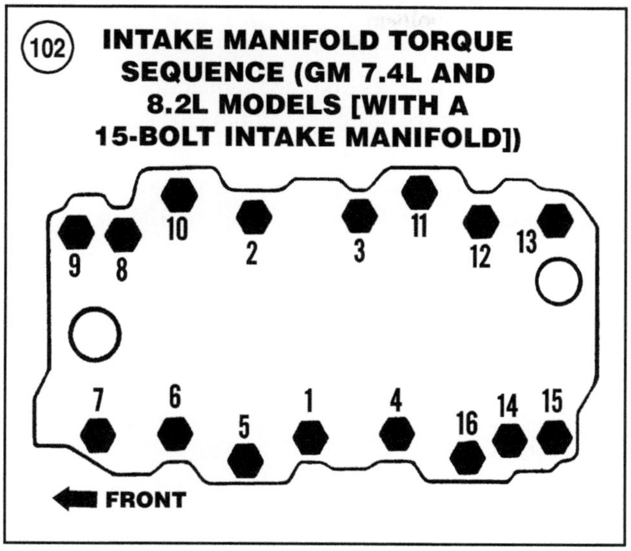

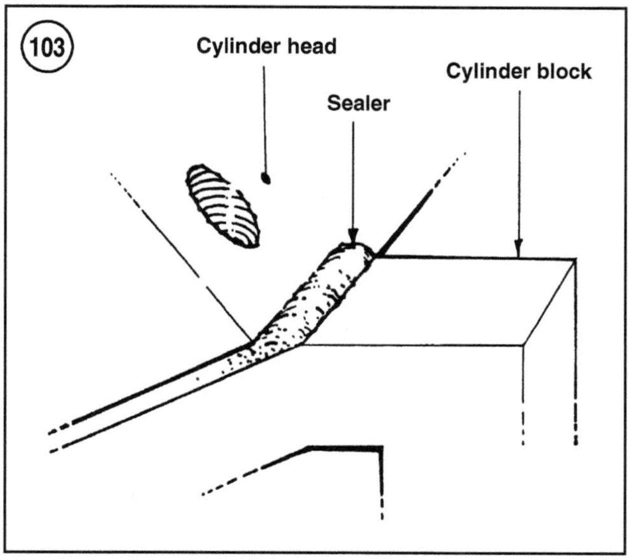

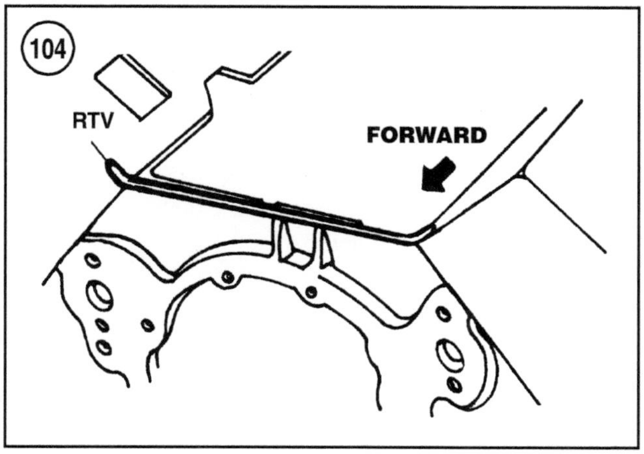

d. Apply a 3/16 in. (4.8 mm) diameter bead of RTV sealant to the front and rear manifold-to-cylinder block mating surfaces. (**Figure 104**). Extend the bead up approximately 1/2 in. (13 mm) onto the intake gaskets.

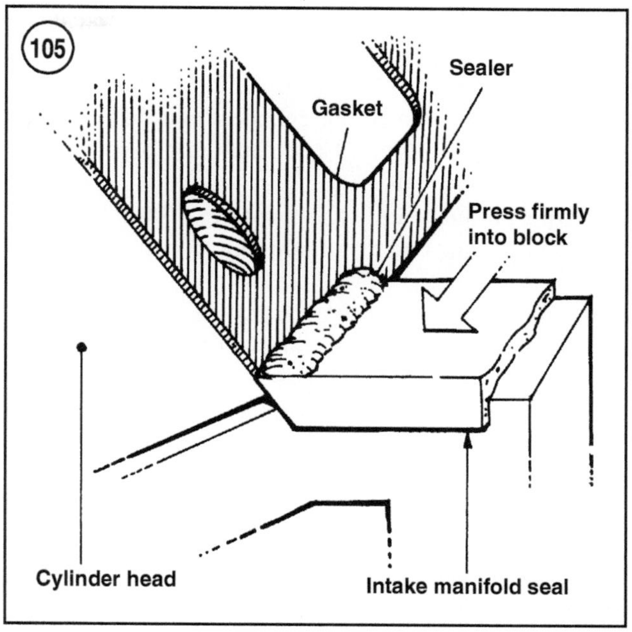

e. Apply a coat of Permatex Gasket Sealing Compound to the gasket mating surfaces of the intake manifold. *Do not* apply the compound to the manifold surfaces that will contact the bead of RTV sealant.

2B. On later GM 4.3L 5.0L and 5.7L models with an eight-bolt intake manifold, install the gaskets and apply the sealant as follows:

a. Apply a 1/8 in. (3.2 mm) bead of RTV to the front and rear corners where the cylinder head and cylinder block mate (**Figure 103**). Apply the bead the full width of the seam.

b. Fit the locating pins (molded into the gaskets) into the cylinder head openings while installing the *new* intake manifold gaskets onto the cylinder heads. The *TOP* marks on the gaskets must face up.

c. Apply a 3/16 in. (4.8 mm) diameter bead of RTV sealant to the front and rear manifold-to-cylinder block mating surfaces. (**Figure 104**). Extend the bead up approximately 1/2 in. (13 mm) onto the intake gaskets.

2C. On Ford 5.0L and 5.8L models, install the gaskets and apply sealant as follows:

a. Apply a 1/8 in. (3.2 mm) bead of RTV into the front and rear corners where the cylinder head and cylinder block mate (**Figure 103**). Apply the bead the full width of the seam.

b. Install the *new* intake manifold gaskets onto the cylinder heads. Position the gaskets to perfectly align the gasket and cylinder head ports.

c. Install the front and rear seals (1 and 6, **Figure 96**, typical) onto the cylinder block surfaces. The slot in the bottom of the seals must fit over the cylinder block ridge as shown in **Figure 105**.

d. Apply a 1/16 in. (1.6 mm) bead of RTV at the joint of the gaskets and end seals as shown in **Figure 105**. Apply the bead the full width of the seam.

2D. On 7.4L, 8.1L and 8.2L models, install the intake manifold gaskets and apply sealant as follows:

a. Apply a 1/8 in. (3.2 mm) bead of RTV to the front and rear corners where the cylinder head and cylinder block mate (**Figure 103**). Apply the bead the full width of the seam.
b. Apply a small bead of Scotch-Grip 1300 adhesive to the front and rear manifold-to-cylinder block mating surfaces.
c. *Quickly* install the neoprene gaskets (1 and 6, **Figure 96**, typical) onto the front and rear cylinder block surfaces. Make sure slot in the bottom fits over the cylinder block as shown in **Figure 105** and the square recesses for the intake manifold gaskets face upward. Hold the gaskets in position until the adhesive dries. Apply a small bead of RTV sealant into the square recesses.
d. Install the *new* intake port and water crossover gaskets (2, **Figure 96**, typical) onto the cylinder heads with *TOP* marks facing upward.
e. Fit the tabs at the lower front and rear of the gaskets into the square recesses in the neoprene gaskets.
f. Apply a coat of Permatex Gasket Sealing Compound to the gasket mating surfaces of the intake manifold. *Do not* apply the compound to the manifold surfaces that will contact the bead or RTV sealant.
g. Apply a 3/16 in. (4.8 mm) bead of RTV sealant that extends approximately 1/2 in. (13 mm) over the ends of the cylinder head-mounted gaskets and down onto the neoprene cylinder block-mounted gaskets.

3. Use aerosol carburetor cleaner to remove all oily deposits from the manifold mating surfaces. Deposits can prevent the RTV sealant from bonding. Wipe the surfaces dry with a clean shop towel.

4. Carefully lower the manifold into position on the cylinder heads. Do not remove the manifold after contacting the sealant because the manifold must be removed, and the RTV sealant must be reapplied.

5. On GM V-6 and V-8 models, apply a coat of GM Threadlock (GM part No. 12345382) to the manifold bolts. Purchase the Threadlock from a GM automobile dealership.

6. If necessary to align the openings, move the manifold or gaskets *slightly*. Then thread the bolts by hand into the manifold and cylinder head. *Do not* dislodge the end seals used on some models.

7A. On earlier GM 4.3L, 5.0L and 5.7L models with a 12-bolt intake manifold, tighten the intake manifold bolts as follows:

a. Tighten the intake manifold bolts in the sequence shown in **Figure 97** to 180 in.-lb. (20 N•m).
b. Tighten the intake manifold bolts in sequence (**Figure 97**) a final time to 30 ft.-lb. (41 N•m).

7B. On later GM 4.3L, 5.0L and 5.7L models with an eight-bolt intake manifold, tighten the intake manifold bolts in three steps as follows:

a. Tighten the bolts in the sequence shown in **Figure 98** to 27 in.-lb. (3 N•m).
b. Wait a few minutes and then tighten the bolts in sequence (**Figure 98**) a second time to 106 in.-lb. (12 N•m).
c. Wait a few minutes and then tighten the bolts in sequence (**Figure 98**) a final time to 133 in.-lb. (15 N•m).

7C. On Ford 5.0L and 5.8L carburetor-equipped models, tighten the intake manifold bolts in sequence (**Figure 99**) to 23-25 ft.-lb. (31-34 N•m).

7D. On Ford 5.0L and 5.8L EFI models, tighten the intake manifold bolts in sequence (**Figure 100**) to 23-25 ft.-lb. (31-34 N•m).

7E. On GM 7.4L and 8.2L models with a 12-bolt intake, tighten the manifold bolts as follows:

a. Tighten the intake manifold bolts in the sequence shown in **Figure 101** to 180 in.-lb. (20 N•m).
b. Tighten the intake manifold bolts in sequence (**Figure 101**) a final time to 30 ft.-lb. (41 N•m).

7F. On GM 7.4L and 8.2L models with a 15-bolt intake, tighten the manifold bolts as follows:

a. Tighten the intake manifold bolts in the sequence shown in **Figure 102** to 210 in.-lb. (24 N•m).
b. Tighten the intake manifold bolts in sequence (**Figure 102**) a final time to 35 ft.-lb. (47 N•m).

8. Install any brackets and clamps that were removed for manifold removal.

9. If the ignition coil is mounted on the intake manifold, install it as described in Chapter Nine.

10. Install the coolant temperature sender as described in Chapter Nine.

11. Install the engine temperature sensor as described in Chapter Nine.

12. Install the thermostat housing and thermostat as described in Chapter Eight.

13. Install the distributor as described in Chapter Nine.

14. On models with a closed cooling system, fill the system with the recommended coolant as described in *Changing the Coolant* in Chapter Three.

15A. On GM and Ford carburetor-equipped models, install the carburetor as described in this chapter.

15B. On GM 4.3L, 5.0L and 5.7L EFI models, install the throttle body as described in this chapter.

15C. On Ford 5.0L and 5.8L EFI models, install the intake plenum, fuel rail and injectors as described in this chapter.

15D. On GM 7.4L and 8.2L EFI models, install the intake plenum, fuel rail and injectors as described in this chapter.

16. Route all wiring to prevent interference with moving components. Clamp the wiring as needed. Do not install the plastic engine cover at this time.

WARNING

Stay clear of the propeller and propeller shaft while running an engine on a flush adapter.

CAUTION
If running the engine with the boat out of the water, use a flush device or other means to supply the engine with cooling water. The seawater pump is quickly damaged if operated without adequate cooling water.

17. Connect the battery cables.
18. Start the engine and immediately check for water, coolant, fuel, air or oil leakage. Stop the engine and correct any leaks before proceeding.
19. Adjust the ignition timing as described in Chapter Five.

Table 1 FUEL SYSTEM TORQUE SPECIFICATIONS

Fastener	in.-lb.	ft.-lb.	N•m
Carburetor			
Mounting nuts	115-168	–	13-19
Flame arrestor stud	65-80	–	7-9
Fuel filter inlet line	132-156	–	15-18
Fuel jets	10	–	1
Inlet needle seat	10	–	1
Power valve	100	–	11
Throttle body to carburetor body screws	50	–	6
Electric fuel pump			
Low-pressure pump (carburetor models)			
Fuel pump to filter bracket fitting	97-124	–	11-14
Fuel line fittings	60-84	–	7-10
Pump mounting bracket bolts	–	20-25	27-34
High-pressure pump (EFI) models			
Banjo bolt	–	18-22	24-30
Elbow locknut	60-84	–	7-10
Pump end support bracket screws	27-35	–	3-4
Fuel line to elbow fitting	156	–	18
Flame arrestor			
Carburetor and TBI-equipped models (nuts)	24-35	–	3-4
Ford 5.0L and 5.8L (EFI) models (screws)	24-48	–	3-5
Plastic top cover screws	24-36	–	3-4
Fuel pressure regulator to fuel rail screws			
Ford 5.0L and 5.8L (EFI) models	27-40	–	3-5
GM 7.4L and 8.2L (74GSI and 82GSI) models	84	–	10
Fuel pressure regulator return line			
GM 7.4L and 8.2L (74GSI and 82GSI) models	156	–	17.6
Fuel rail to intake manifold fasteners			
Ford 5.0L and 5.8L (EFI) models	70-105	–	8-12
GM 7.4L and 8.2L			
74GI and 82GI models	28	–	3
74GSI and 82GSI models	88	–	10
Fuel rail delivery (pressure) line fitting			
Ford 5.0L and 5.8L (EFI) models	120-144	–	14-16
GM 7.4L and 8.2L			
74GI and 82GI models	216	–	24
74GSI and 82GSI models	156	–	18
Fuel rail return line fitting			
Ford 5.0L and 5.8L (EFI) models	120-144	–	14-16
GM 7.4L and 8.2L			
74GI and 82GI models	216	–	24
74GSI and 82GSI models	156	–	18
Fuel reservoir			
Bracket mounting screws	–	24-36	33-49
Fuel inlet hose fitting	156	–	18
Regulator return hose fitting	156	–	18
Intake manifold			
GM 4.3L, 5.0L and 5.7L models			
With a 12-bolt intake manifold			
First step	180	–	20
Final step	–	30	41
With an 8-bolt intake manifold			
First step	27	–	3

(continued)

7

Table 1 FUEL SYSTEM TORQUE SPECIFICATIONS (continued)

Fastener	in.-lb.	ft.-lb.	N•m
Intake manifold			
GM 4.3L, 5.0L and 5.7L models			
With an 8-bolt intake manifold (continued)			
Second step	106	–	12
Final step	133	–	15
Ford 5.0L and 5.8L models	–	23-25	31-34
GM 7.4L and 8.2L models			
With a 12-bolt intake manifold			
First step	180	–	20
Final step	–	30	41
With a 15-bolt intake manifold			
First step	210	–	24
Final step	–	35	47
Intake plenum to intake manifold fasteners			
Ford 5.0L and 5.8L (EFI) models	144-216	–	16-24
GM 7.4L and 8.2L			
74GI and 82GI models	–	30	41
74GSI and 82GSI models	124	–	14
Mechanical fuel pump			
GM 3.0L four-cylinder models			
Mounting bolts	–	20-25	27-34
Inlet barb fitting	48-59	–	5-7
Outlet elbow fitting	48-59	–	5-7
Outlet flare nut fitting	177-212	–	20-24
GM 5.0L and 5.7L models			
Mounting bolts	–	20	27
Mounting plate bolts	–	15	20
Inlet barb fitting	48-59	–	5-7
Outlet elbow fitting	48-59	–	5-7
Inlet flare nut fitting	177-212	–	20-24
Outlet flare nut fitting	177-212	–	20-24
Ford 5.0L and 5.8L models			
Mounting bolts	–	18-20	24-27
Fuel line fittings	120-144	–	13-16
Throttle body			
GM 4.3L, 5.0L and 5.7L (TBI) models			
Fuel delivery (pressure) line	204	–	23
Fuel delivery (pressure) line fitting	–	30	41
Fuel return line	204	–	23
Fuel return line fitting	–	21	29
Injector cover/regulator housing	27	–	3
Injector housing screws	35	–	4
Mounting bolts	142	–	16
Ford 5.0L and 5.8L models			
Mounting nuts	144-216	–	16-24
GM 7.4L and 8.2L models			
Top cover screws	28	–	3
Mounting screws	132	–	15

Chapter Eight

Cooling and Exhaust Systems

This chapter covers service procedures for the thermostat, engine recirculation pump, hoses, seawater pump, oil and power steering fluid coolers, exhaust elbow, flapper valve and exhaust manifold. Procedures for flushing the cooling system and maintaining closed cooling systems are described in Chapter Three.

On models equipped with a standard (open) cooling system, water in which the boat is operating is picked up at the drive unit and circulated throughout the engine to absorb heat. The heated water is transferred to the exhaust manifold(s) where it is expelled overboard with the exhaust gasses.

Some engines are equipped with a closed cooling system that is divided into two separate subsystems. One subsystem is a closed system of coolant (antifreeze-and-water mixture) circulating in the engine cooling passages and absorbing the engine heat. Heat absorbed by the coolant is then transferred (exchanged) to the seawater in the heat exchanger. In the other subsystem, water in which the boat is operating is picked up at the drive unit and transferred through the passages in the heat exchanger where it absorbs heat from the coolant. The heated seawater from the exchanger is transferred to the exhaust manifold(s) where it is expelled overboard with the exhaust gasses. **Figure 1** shows a typical standard (open) cooling system. **Figure 2** shows a typical closed cooling system. The location of the pumps and path of flow might differ slightly from those shown in the illustrations.

Table 1 lists torque specifications for most cooling and exhaust system fasteners. Use the general torque recommendations provided in Chapter One for fasteners not listed in **Table 1**. **Table 1** is located at the end of this chapter.

WARNING
Stay clear of the propeller shaft while running the engine on a flush/test adapter. Remove the propeller(s) before running the engine to prevent injury or death. Always disconnect the battery cables before removing or installing the propeller.

CAUTION
Never run the engine without first providing cooling water. Use either a test tank or a flush/test adapter.

THERMOSTAT

The thermostat controls water or coolant circulation to provide quick engine warm-up and maintain a consistent operating temperature.

① **STANDARD (OPEN) COOLING SYSTEM (TYPICAL FOUR-CYLINDER MODELS)**

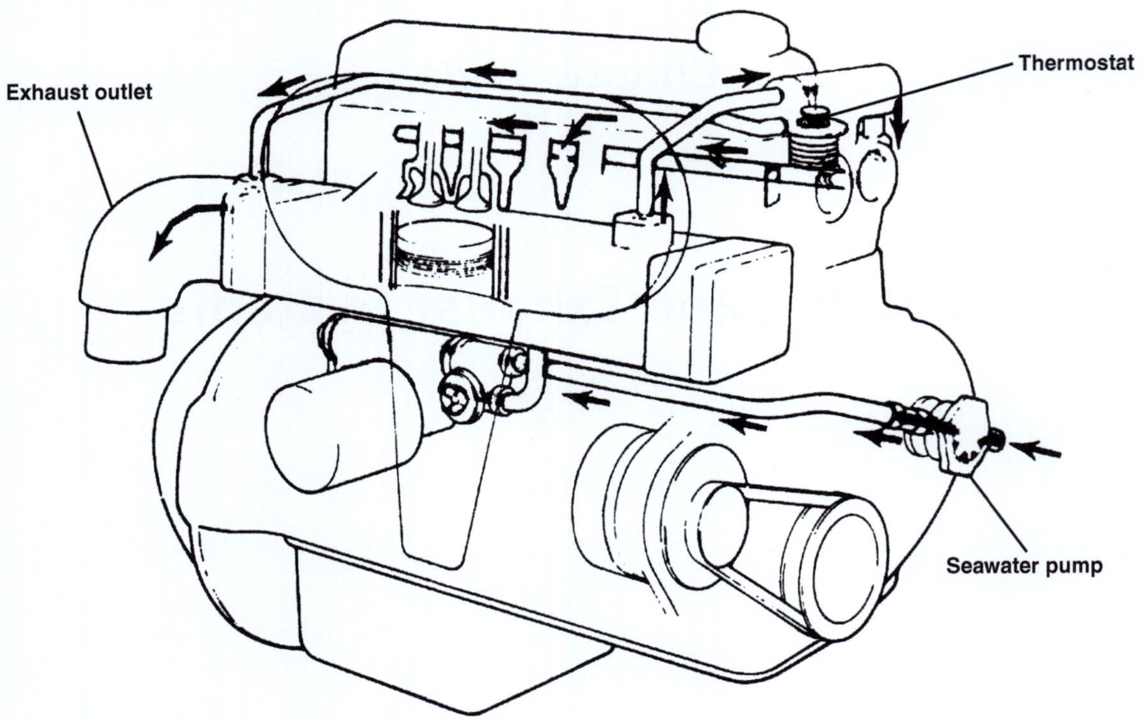

On a standard (open) cooling system, the thermostat controls seawater flow through the cylinder block to control engine temperature. When the engine is cold, the thermostat is closed, and water bypasses the cylinder block and is sent directly to the exhaust manifold where it is expelled. When the engine warms up, the thermostat opens, and water flows through the cylinder block to absorb excess heat before flowing to the exhaust manifold.

On a closed cooling system, the thermostat controls coolant flow through the cylinder block and heat exchanger to control engine temperature. When the engine is cold, the thermostat is closed, and the coolant bypasses the heat exchanger. This allows the cylinder block to warm to normal operating temperature. When the engine warms up, the thermostat opens, and coolant flows through the heat exchanger to remove excess heat.

CAUTION

Never operate the engine with a faulty or missing thermostat. Either condition can cause the engine to overheat or fail to reach normal operating temperature. Operating at too low of a temperature for an extended period of time can result in oil dilution and possible engine damage from inadequate lubrication. Operating the engine at too high of a temperature can result in serious damage to all engine components.

Thermostats are rated according to the opening temperature. The opening temperature value is stamped in the thermostat flange (**Figure 3**) or another area. The thermostat should start to open at the temperature stamped on the thermostat and be fully opened at 45° F (25° C) above that temperature. The manufacturer does not recommend installing a thermostat with a higher or lower temperature than the original equipment. Doing so can adversely affect engine operation, durability and exhaust emissions.

CAUTION

Do not substitute an automotive thermostat for the original equipment. The higher rating will cause the engine to overheat and can cause serious engine damage. Make sure any thermostat installed in the engine is approved for use in the indicated OMC engine model.

Thermostat Removal

1A. On models with a standard cooling system, drain all water from the cylinder block as described in *Winterizing* in Chapter Four.

1B. On models with a closed cooling system, drain coolant from the cylinder block as described in Chapter Three. Refer to *Changing the Engine Coolant*.

2. Loosen the hose clamps and disconnect all hoses from the thermostat housing (**Figure 4**, typical).

3. Note the wire routing and connection points, then disconnect the engine wire harness terminal from the engine

2

OPTIONAL CLOSED COOLING SYSTEM (TYPICAL V-8 OR V-8 ENGINE)

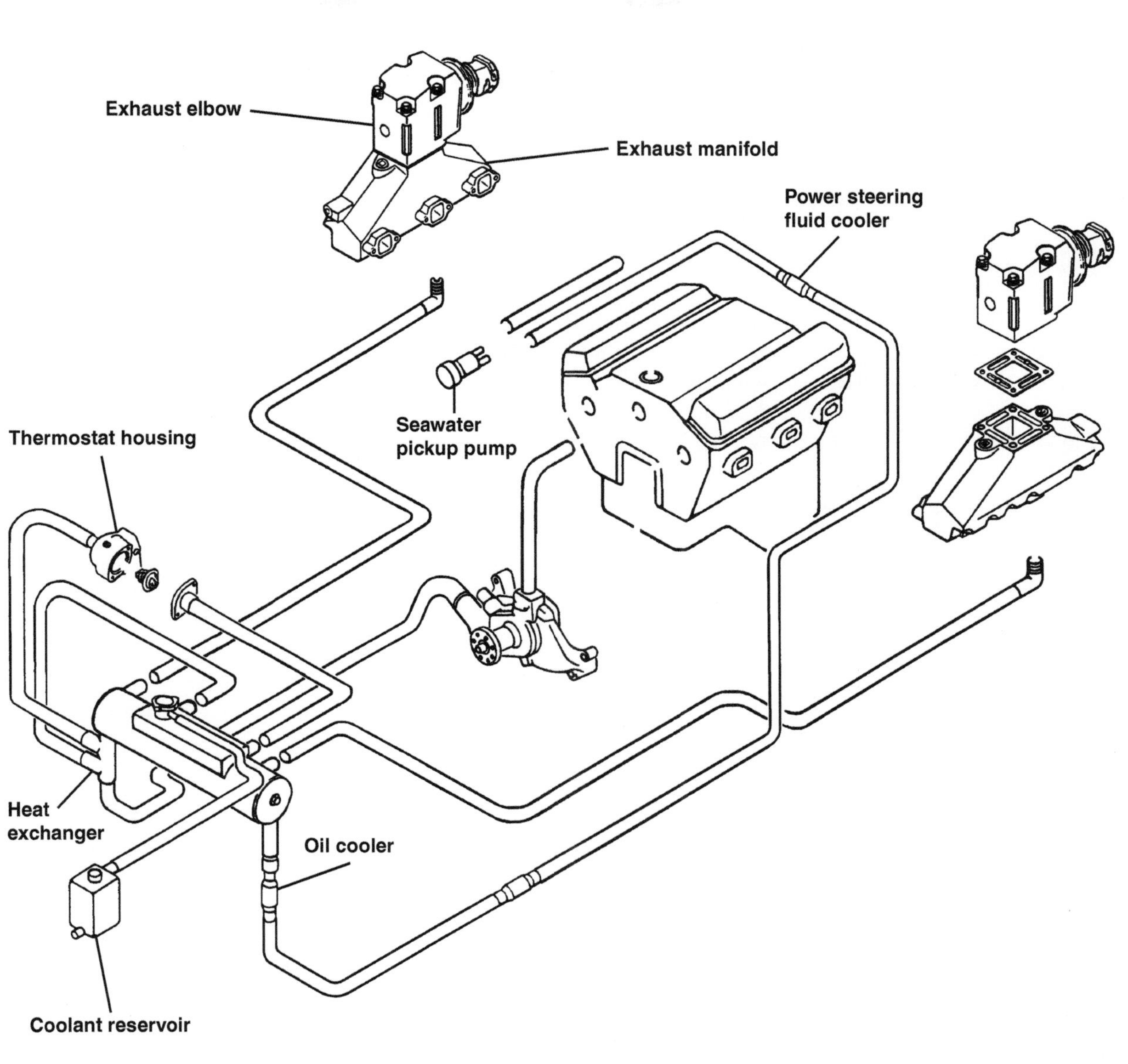

3

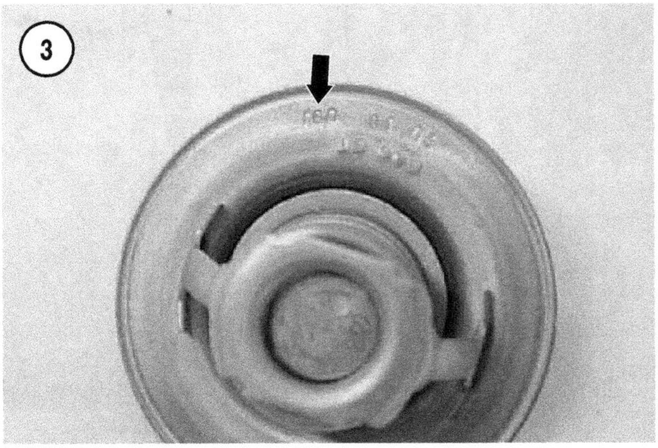

4

temperature sender, overheat switch or engine temperature sensor mounted in the thermostat/distribution housing.

4. Remove the bolts (**Figure 4**, typical) attaching the thermostat housing or thermostat housing cover to the cylinder head or intake manifold.

5. Carefully remove the housing or cover and the gasket. If necessary, carefully pry the housing or cover loose. Remove and discard the gasket.

NOTE
Some models might have an O-ring that retains the thermostat into the housing. Carefully pull the O-ring out of the groove and then remove the thermostat from the housing.

6. If an O-ring rests in the groove next to the thermostat, use a pick and carefully remove the thermostat from the groove. Replace the O-ring unless it is in excellent condition. Remove the thermostat from the opening. If necessary, grip the thermostat with locking pliers and twist lightly to free the thermostat.

7. Scrape all remaining gasket material from the cover or cover mating surface. Any residual material on the mating surfaces will allow water or coolant leakage.

8. Remove all corrosion, salt or mineral deposits from the thermostat flange recess in the thermostat/distribution housing. Any material on the flange surface will allow water or coolant to bypass the thermostat and greatly increase the warm-up time.

9. Use a wire brush to remove all corrosion or mineral deposits from the thermostat housing bolts. Use a tap or thread chaser to remove corrosion or deposits from the bolt openings in the housing. Use compressed air to clear debris from the openings.

10. Test the thermostat as described in Chapter Two. Refer to *Cooling System*.

Thermostat Installation

1. Install the thermostat into the housing with the element end (**Figure 5**) facing toward the cylinder head (four-cylinder models) or intake manifold (V-6 and V-8 models). Make sure the flange fully seats into the recess in the housing. If the housing uses an O-ring to retain the thermostat, carefully fit the O-ring into the groove next to the thermostat.

2. Apply a light coat of gasket sealing compound to the mating surfaces and then fit a new gasket onto the cover.

3. Install the cover and gasket to the housing. Apply a light coat of gasket sealing compound to the threads and then thread the bolts into the cover and housing. Verify proper alignment of the cover to housing and then tighten the thermostat housing bolts evenly to 144-192 in.-lb. (16-22 N•m).

4. Connect all hoses to the thermostat housing. Securely tighten the hose clamps.

5

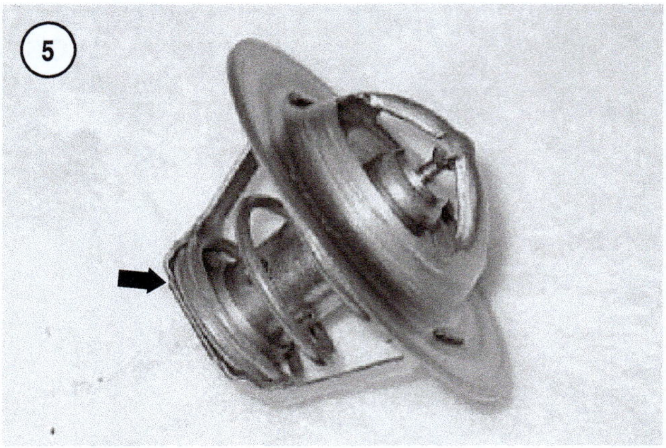

5. Connect the corresponding wire harness to the coolant temperature sender, overheat switch or coolant temperature sensor. Route the wiring to prevent entanglement in the drive belt(s) or other moving components.

6. On models with a closed cooling system, refill the coolant as described in Chapter Three. Refer to *Changing the Engine Coolant*.

7. Start the engine and immediately check for water or coolant leakage from the thermostat housing. Correct the leakage before continued operation.

8. Monitor the engine and temperature gauge while the engine warms to operating temperature. Stop the engine if overheating or continued low temperature is evident. Correct the cause of overheating or overcooling before putting the engine into service.

HOSE REPLACEMENT

Replace any hoses that are cracked, brittle, mildewed or very soft and spongy. Always replace hoses in questionable condition. Attention to hose condition can prevent failure while the boat is offshore. Some of the hoses are difficult to access under normal working conditions and with all of the necessary tools available. It might be impossible to change the hose while offshore with a limited supply of tools.

Hose manufacturers generally recommend replacing the hoses every 2 years. How long the hoses last depends a great deal on how much the boat is used and how well it is maintained. Regardless, it is good to replace hoses at recommended 2-year intervals. Always replace cooling system hoses with the same types as the originals. Pleated rubber hoses do not allow the same amount of water or coolant flow and do not have the same strength as reinforced molded hoses. Straight hose is not an acceptable substitute for curved or molded hose. In such applications, the straight hose might kink and prevent adequate water or coolant flow.

Always check the condition of the hose clamps and install new worm screw clamps if necessary. A damaged clamp might loosen while underway and result in coolant loss, overheating and serious engine damage.

6

7

If replacing hoses that connect only to the upper part of the engine, it is acceptable to partially drain the seawater or coolant. Always fully drain the system if replacing hoses connecting to the recirculation pump, heat exchanger or exhaust manifold(s).

1. Loosen the clamp at each end of the hose. Grasp the hose and twist it off the fitting with a pulling motion. Avoid using a sharp instrument to remove the hose. Doing so can damage the reinforcement material in the hose in a manner that might not be visually apparent. The hose might fail while offshore.

2. If the hose is corroded on the fitting and will not twist free, cut it off with a sharp knife about 1.0 in. (25.4 mm) beyond the fitting. Remove the clamp and slit the remaining piece of hose lengthwise. Then peel it off the fitting.

3. Clean any rust or corrosion from the fitting by wrapping a piece of medium grit sandpaper around it and rotating until the fitting is clean and smooth. Be careful to avoid removing too much material when cleaning composite or plastic fittings.

4. Wipe the inside diameter of the replacement hose with a soap-and-water solution. Fit the hose over the fittings with a twisting motion. Align the hose to prevent kinking or twisting and then push the hose fully onto the fittings.

5. Position the clamps approximately 1/4 in. (6.3 mm) from the end of the hose. Make sure the clamp screw is oriented for easy access and does not contact other engine components. Tighten the clamp securely.

6. If the engine was drained to replace the hose, refill the system with coolant as described in Chapter Four.

7. Start the engine and immediately check for leakage at the fittings. Tighten the clamps if leaking. If leakage cannot be corrected by tightening the clamps, check for residual debris on the fittings.

RECIRCULATION PUMP REPLACEMENT

If the recirculation pump is making noise, it might be an indication of impending failure. If the seal is defective, seawater or coolant can leak from behind the pump pulley. The pump is serviced as an assembly and can be replaced with the engine in the boat. Marine recirculation pumps contain stainless steel components and use a special shaft seal. *Do not* replace it with an automotive pump.

1. Disconnect the negative battery cable.

2. Loosen but do not remove the four bolts (**Figure 6**, typical) that secure the pulley to the pump shaft. The tension of the drive belt(s) will help limit pulley rotation while loosening the bolts.

3A. On models with a standard (open) cooling system, drain seawater from the cylinder block as described in *Winterizing* in Chapter Four. It is not necessary to drain the exhaust manifold(s) for this procedure.

3B. On models with a closed cooling system, drain the coolant as described in Chapter Three. Refer to *Changing the Coolant*. It is not necessary to drain the exhaust manifold(s) for this procedure.

4. Remove the alternator drive belt as described in Chapter Nine. Refer to *Alternator Removal*.

5. Remove any accessory brackets or clamps attached to the recirculation pump fasteners or those which can interfere with pump removal.

6. Remove the pump pulley bolts (**Figure 6**, typical) and carefully pull the pulley from the pump shaft. Clean rust or corrosion from the pulley surfaces. Rust or corrosion pitting on the belt contact surfaces will allow slippage and quickly wear out a new belt. Clean and repaint the pulley as necessary.

7. Loosen the clamps and disconnect the large hose (**Figure 7**) from the pump.

8. On Ford 5.0L and 5.8L EFI models, remove the engine temperature sender as described in Chapter Nine.

9A. On GM 7.4L and 8.2L carburetor-equipped models, remove the coolant temperature sender from the recirculation pump as described in Chapter Nine.

9B. On GM 7.4L and 8.2L EFI models, loosen the clamp and then disconnect the bypass hose (**Figure 8**) from the fitting on the top of the recirculation pump.

10. If the boat is equipped with a hot water heater, loosen the clamp and then disconnect the water heater hose (**Figure 9**, typical) from the recirculation pump fitting.

11. Note the location of the stud bolts used on GM V-6 and V-8 models. The stud bolts must be installed in the original locations for proper accessory bracket and clamp alignment.

12. Support the pump while removing the pump fasteners (**Figure 10**, typical). Tap on the body with a rubber mallet to free the pump from the cylinder block.

13. Carefully scrape all gasket material and old sealant from the cylinder block and recirculation pump mating surfaces. Be thorough. Residual gasket material is easy to overlook and will cause the pump to leak after installation. Use a thread chaser to remove corrosion and old sealant from the pump fastener openings. Use a wire brush to clean the fastener threads.

14. Apply gasket sealing compound to the mating surfaces and then install a new gasket onto the pump. The sealant should hold the gaskets in position. On GM 4.3L, 5.0L and 5.7L models, the replacement gasket does not perfectly match the openings on the port side. Although the gasket will protrude on the bottom, the gasket will seal satisfactorily.

15. Insert the pump bolts through the pump body and gasket(s) to help retain the gasket(s). On GM V-6 and V-8 models, make sure the stud bolts are in the proper openings as noted in Step 10. Apply a light coat of gasket sealing compound to the bolt threads.

16. Fit the pump onto the cylinder block. Thread the bolts into the cylinder block. Make sure the gaskets are properly positioned before tightening the bolts. Tighten the recirculation pump bolts in a crossing pattern to the following specification:

   a. On GM 3.0L models, tighten the bolts to 160-203 in.-lb. (18-23 N•m).
   b. On GM 4.3L, 5.0L, 5.7L, 7.4L and 8.2L models, tighten the bolts to 30 ft.-lb. (41 N•m).
   c. On Ford 5.0L and 5.8L models, tighten the bolts to 120-180 in.-lb. (14-20 N•m).

17. Align the center opening and then fit the pulley onto the water pump shaft. Rotate the pulley to align the four bolt openings. The replacement pump may have eight bolt openings, but only four will align simultaneously. Rotate the pulley until all four openings align. Thread the four bolts into the openings and tighten by hand. The bolts are easier to tighten with the drive belt(s) installed. Make sure the pulley is fully seated on the shaft flange. Tighten the bolts an additional amount if necessary.

18. Connect the large hose to the water pump (**Figure 7**) and secure with the worm screw clamp.

19. On Ford 5.0L and 5.8L EFI models, install the engine temperature sender as described in Chapter Nine.

20A. On GM 7.4L and 8.2L carburetor-equipped models, install the coolant temperature sender from the recirculation pump as described in Chapter Nine.

20B. On GM 7.4L and 8.2L EFI models, connect the bypass hose (**Figure 8**) to the fitting on the top of the recirculation pump. Secure the hose to the fitting with a worm screw clamp.

8

9

10

21. If the boat is equipped with a hot water heater, connect the water heater hose (**Figure 9**, typical) to the recirculation pump fitting. Secure the hose to the fitting with a worm screw clamp.

22. Install any accessory bracket or clamp removed to access the pump.

23. Install the alternator drive belt as described in Chapter Nine. Refer to *Alternator Installation*.

24. Tighten the four recirculation pump pulley bolts in a crossing pattern. On GM models, tighten the bolts to the torque recommendation provided in the table at the end of

Chapter One, and for Ford 5.0L and 5.8L models, tighten to 168-240 in.-lbs. (19-27 N•m). Check and readjust the drive belt tension. Refer to *Alternator Installation* in Chapter Nine.

25. On models with a closed cooling system, refill the coolant as described in Chapter Three.

26. Connect the battery cable.

WARNING

Stay clear of the propeller shaft while running the engine on a flush/test adapter. Remove the propeller(s) before running the engine to prevent injury or death. Always disconnect the battery cables before removing or installing the propeller.

CAUTION

Never run the engine without first providing cooling water. Use either a test tank or a flush/test adapter.

27. Start the engine and immediately check for leakage at the hose connection and water pump-to-cylinder block mating surfaces. If leaking from the hose connection, remove the hose, clean the hose and fitting, and install the hose. If leaking from the mating surfaces, remove the pump and remove residual gasket material from the pump and cylinder block mating surfaces.

SEAWATER PUMP

A seawater pump with a replaceable rubber impeller is used on all models. Two different types of pumps are used on the models covered in this manual. Earlier models use a belt-driven seawater pump (**Figure 11**) that mounts on the lower front starboard side of the engine. Later models use a crankshaft-driven seawater pump (**Figure 12**) located on the lower front of the engine. The pump attaches to and is driven by the crankshaft pulley. A bracket attaches to the cylinder block and the pump body to provide support and prevent the body from rotating with the crankshaft. This section describes pump removal, repair and installation for both types of pumps.

Belt-Driven Seawater Pump

Removal

1. Loosen the clamps and then carefully pull the inlet (A, **Figure 13**) and outlet (B) hoses from the pump.

2. Rotate the pulley if necessary to align the bolt access holes in the pulley with the mounting bolts (**Figure 14**). Loosen but do not remove the mounting bolts.

3. Slide the pump toward the port side to relieve belt tension and then slip the belt off the pump pulley. Inspect the belt for worn, damaged or deteriorated surfaces and replace as needed.

4. Remove the mounting bolts and carefully slide the pump assembly out of the mounting bracket.

Pump repair

The water pump repair kit includes a new impeller, gaskets, glycerin water pump lubricant and cam element. Several styles of gaskets are included in the kit. Use the style that perfectly matches the original gasket.

1. Remove the belt-driven water pump as described in this chapter.
2. Place the water pump on a suitable work surface with the pulley side facing down.
3. Remove the three brass screws (**Figure 15**) and then carefully pull the cover off the pump body (**Figure 16**). Do not pry off the cover. The gasket surfaces are easily damaged. If necessary, carefully twist the cover loose.
4. Use two blunt screwdrivers to pry the impeller out of the pump body (**Figure 17**). Work carefully to avoid damaging the pump body or gasket mating surfaces.
5. Remove the single screw on the side of the pump body (**Figure 18**). Carefully remove the cam element from the pump body (**Figure 19**). Always replace the cam element with the new element included in the repair kit. The design of the element is sometimes changed to match the impeller in the kit.
6. Use compressed air to clear residual water and debris from the pump body.
7. Inspect the impeller contact surfaces in the pump body and cover (**Figure 20**) for deep corrosion pitting, deep scratches, cracking or roughness. Replace the water pump assembly if these defects are evident.
8. Grasp the water pump body and rotate the pump pulley to check for roughness or excessive play in the pump bearings. Replace the pump body if these defects are evident.
9. Inspect the shaft seal (**Figure 21**) for cracking at the seal lip, missing sections or other defects. Replace the seal if these or other defects are evident. Make sure the seal is available before removing it from the housing. The seal is not available separately on some models.
10. Inspect the impeller for worn, charred or missing sections. Replace the impeller unless a new replacement is not available and the impeller is in like-new condition. Charred surfaces indicate the water pump was operated without sufficient water. Check for blocked passages, loose hose clamps, damaged or kinked hoses or other defects that restrict water flow or allow air to flow into the pump. If sections are missing, remove all hoses and the thermostat and then check all cooling system passages for remnants. Missing sections are usually found in the water pump-to-thermostat hose or in the thermostat/distribution housing.
11. Fit the cam element into the pump body. Work carefully to avoid bending the element fingers. Reposition the element if necessary to align the screw opening with the corresponding opening in the pump body.
12. Apply a light coat of gasket sealing compound to the threads and then install and carefully tighten the screw (**Figure 18**). Do not overtighten the screw. Overtightening will break the screws.

15

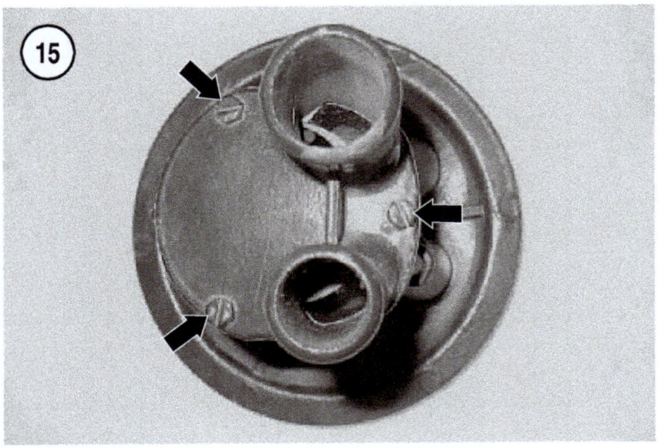

16

17

18

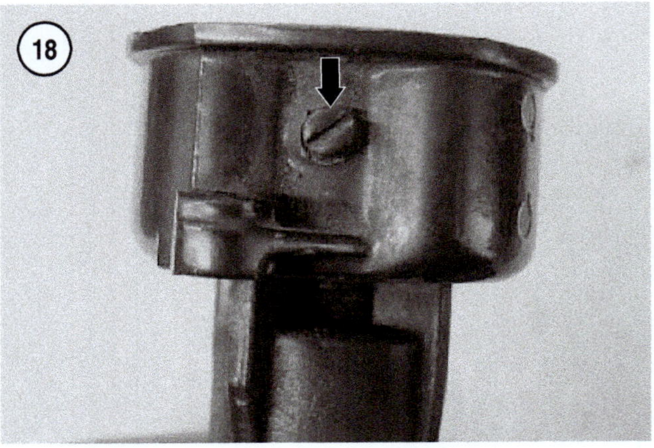

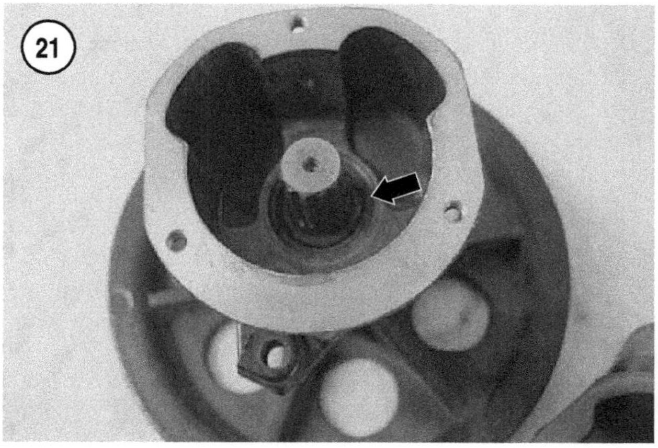

13. Coat the inner surfaces of the pump body and the replacement impeller with the glycerin water pump lubricant included in the repair kit.
14. Align the shaft and impeller splines and then carefully fit the impeller into the pump body. Rotate the pump pulley clockwise, as viewed from the pulley side, while pushing the impeller into the body to make sure all of the impeller vanes curl in the clockwise direction as shown in **Figure 22**. Remove and reinstall the impeller if necessary.
15. Carefully match the original gasket with the correct replacement gasket from the repair kit. Some of the gaskets are very similar. Fit the new gasket onto the pump body (**Figure 23**). Do not apply sealant to the gasket. Sealant will cause the gasket to slide out of position while tightening the cover screws.
16. Align the screw openings while carefully fitting the cover onto the pump body. Do not dislodge the gasket.
17. Apply a light coat of gasket sealing compound to the threads and then carefully thread the three screws into the cover, gasket and pump body openings. Verify that the gasket has not slipped out of position and then tighten the cover screws. Do not overtighten the screws. Overtightening will break the screws.
18. Rotate the water pump pulley clockwise, as viewed from the pulley side, to check for smooth operation. Do not rotate the pulley counterclockwise. Doing so might cause one or more of the impeller vanes to curl in the wrong direction, which can cause inefficient water pump operation, a damaged impeller and/or overheating. Disassemble the pump and check for improper assembly or damaged components if binding or roughness is evident.
19. Install the belt-driven water pump as described in this chapter.

Installation

1. Carefully slide the pump into the mounting bracket with the pulley facing forward.
2. Thread the mounting bolts (**Figure 14**) into the bracket and pump body openings. Do not tighten the bolts at this time.
3. Move the pump toward the crankshaft pulley. Then slip the drive belt onto the pump and crankshaft pulleys. Make

sure the belt is not twisted and fits into the groove in each pulley.

4. Rotate the pump pulley to align the openings in the pulley with the mounting bolts.

5. Move the pump away from the crankshaft pulley to apply tension to the drive belt. When properly tensioned, the belt will deflect approximately 1/4-1/2 in. (6-13 mm) with thumb pressure applied at a point midway between the pulleys (**Figure 24**).

6. Hold the pump in position to maintain belt tension while tightening the mounting bolts to 30 ft.-lb. (41 N•m).

7. Fit the clamps over the ends and then push the inlet (A, **Figure 13**) and outlet (B) hoses onto the pump fittings. The inlet hose leads to the transom bracket fitting. The outlet hose leads to the thermostat housing. Position the hose clamps approximately 1/4 in. (6 mm) from the end of the hoses and then securely tighten the clamps.

WARNING
Stay clear of the propeller shaft while running the engine on a flush/test adapter. Remove the propeller(s) before running the engine to prevent injury or death. Always disconnect the battery cables before removing or installing the propeller.

CAUTION
Never run the engine without first providing cooling water. Use either a test tank or a flush/test adapter.

8. Start the engine and immediately check for leakage from the hose fittings and body-to-end plate mating surfaces. If there is leakage from the hoses, remove the hoses, clean the fittings and reinstall the hoses. If there is leakage from the mating surfaces, the gasket probably slipped out of position during assembly. Replace the gasket to correct the leak.

9. Monitor the engine temperature while allowing the engine to reach normal operating temperature. If the engine begins to overheat, refer to Chapter Two for troubleshooting procedures.

10. Stop the engine and disconnect the negative battery cable. Check the belt tension as described in Step 5 and readjust as necessary.

11. Reconnect the battery cable.

Crankshaft-Driven Seawater Pump

The pump mounts on the front of the crankshaft pulley. The pump must be partially disassembled for removal. The pump repair kit includes a new impeller, O-ring, sealing ring and a tube of glycerin water pump lubricant.

Removal, repair and installation

1. Disconnect the negative battery cable.

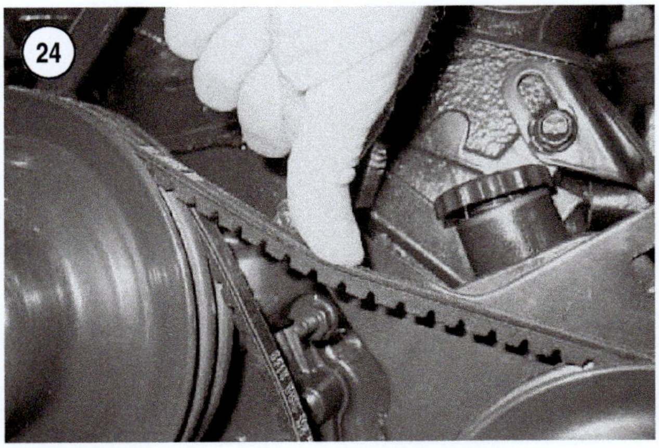

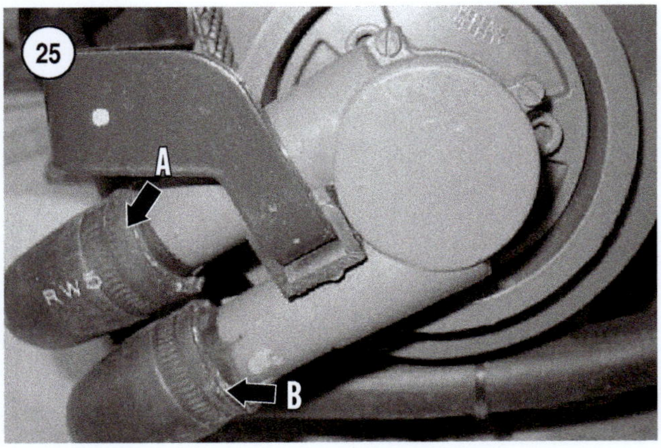

2. Loosen the hose clamps and then carefully pull the inlet (A, **Figure 25**) and outlet (B) hoses from the pump fittings. Do not pry the hoses off the fittings. Doing so will damage the fittings. If necessary, carefully twist the hoses to break the bond and then push the hoses off with a blunt instrument.

3. Remove the two bolts and washers (A, **Figure 26**) and then pull the support bracket (B) from the cylinder block and pump body. Reposition the rubber bumper (C, **Figure 26**) onto the tip of the support arm if dislodged or stuck in the pump body recess.

4. Remove the four screws (1, **Figure 27**) that secure the water pump body (2) to the pulley (6). Then carefully pull the body off the pulley. If the impeller (3, **Figure 27**) remains on the pulley shaft, carefully pry it loose with two blunt instruments. If the impeller remains in the pump body, carefully pry it out with two blunt screwdrivers. Work carefully to avoid damaging the pump body.

5. Remove the O-ring (4, **Figure 27**) from the groove in the pump body or step on the pulley. Discard the O-ring.

6. Remove the sealing ring (5, **Figure 27**) from the pump pulley. Discard the sealing ring.

7. Use compressed air to clear water and debris from the pump body and pulley.

8. Inspect the impeller contact surfaces in the pump body and pulley for deep corrosion pitting, deep scratches, crack-

26

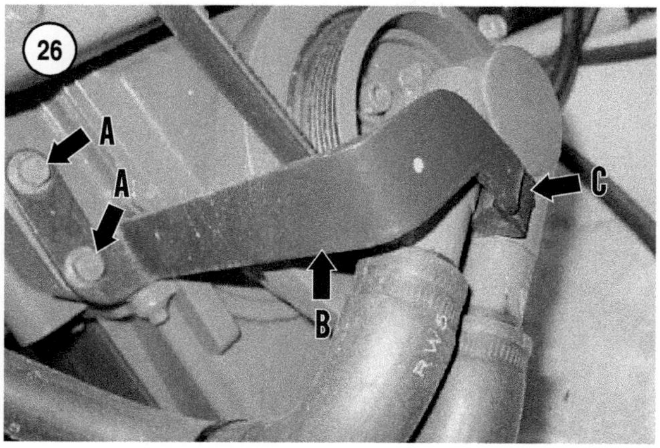

27

CRANKSHAFT-DRIVEN SEAWATER PUMP

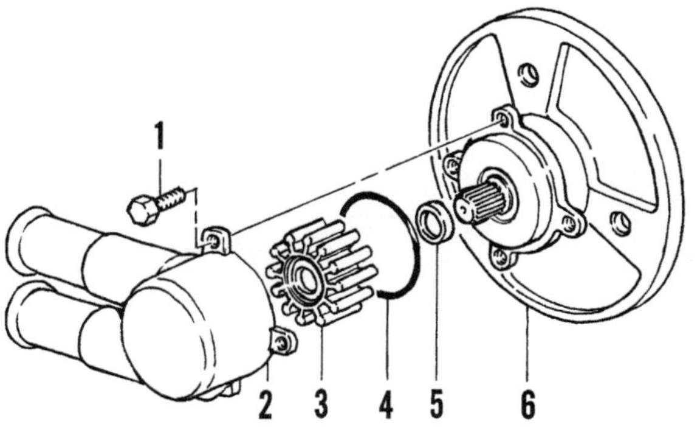

1. Screw
2. Pump body
3. Impeller
4. O-ring
5. Sealing ring
6. Pulley

ing or roughness. Replace the pump body or pulley if these or other defects are evident.

9. Inspect the impeller for worn, charred or missing sections. Replace the impeller unless a new replacement is not available and the impeller is in like-new condition. Charred surfaces indicate the water pump was operated without sufficient water. Check for blocked passages, loose hose clamps, damaged or kinked hoses or other defects that allow air to flow into the pump or restrict water flow. If sections of the impeller are missing, remove all hoses and the thermostat and then check all cooling system passages for remnants. Missing sections are usually found in the water pump-to-thermostat hose or in the thermostat/distribution housing.

10. Coat the inner surfaces of the pump body and the replacement impeller with the glycerin water pump lubricant included in the repair kit.

11. Coat the surfaces with grease and then slide the new sealing ring (5, **Figure 27**) over the pulley shaft. Seat the ring into the recess in the pulley.

12. Align the pulley shaft and impeller splines, then fit the replacement impeller onto the shaft. Seat the impeller against the pulley.

13. Lubricate with grease and then fit the new O-ring (4, **Figure 27**) onto the step on the pulley. Do not fit the O-ring into the pump body. It might become dislodged during pump assembly.

14. Align the opening in the pump body with the impeller. Then rotate the pump body counterclockwise, as viewed from the front, while carefully pushing the body over the impeller. This will cause the impeller vanes to curl in the proper direction as they enter the body. Continue until the body mates against the pulley and the hose fittings are oriented in the direction shown in **Figure 25**.

15. Temporarily position the support bracket (B, **Figure 26**) on the cylinder block. Thread the two bolts two to three turns into the bracket and cylinder block. The bracket must be free to move.

16. Rotate the pump body counterclockwise until the four screw openings align and the rubber bumper (C, **Figure 26**) on the support can be inserted between the pump fittings. Do not rotate the pump clockwise to align the screw openings and fittings. Doing so might cause one or more of the impeller vanes to curl in the wrong direction, which can cause in inefficient water pump operation, a damaged impeller and/or overheating.

17. Hold the pump body in position and then thread the four screws (1, **Figure 27**) into the body and pulley openings. Tighten the pump body screws in a crossing pattern to 19-25 in.-lb. (2-3 N•m).

18. Fit the rubber bumper on the tip of the support bracket into the recess between the pump fittings, then tighten the two support bracket bolts (A, **Figure 26**) to 20-25 ft.-lb. (27-34 N•m).

19. Fit the clamps over the ends, then push the inlet (A, **Figure 25**) and outlet (B) hoses onto the pump fittings. The inlet hose leads to the transom bracket fitting. The outlet hose leads to the thermostat housing. Position the hose clamps approximately 1/4 in. (6 mm) from the ends of the hoses and then securely tighten the clamps.

WARNING

Stay clear of the propeller shaft while running the engine on a flush/test adapter. Remove the propeller(s) before running the engine to prevent injury or death. Always disconnect the battery cables before removing or installing the propeller.

CAUTION

Never run the engine without first providing cooling water. Use either a test tank or a flush/test adapter.

20. Connect the battery cable. Start the engine and immediately check for leakage from the hose fittings and body-to-pulley mating surfaces. If there is leakage from the hoses, remove the hoses, clean the fittings and reinstall the

hoses. If there is leakage from the mating surfaces, remove the pump and check for improper seating of the pump body to the pulley or damage to the O-ring.

21. Monitor the engine temperature while allowing the engine to reach normal operating temperature. If the engine begins to overheat, refer to Chapter Two for troubleshooting procedures.

OIL AND POWER STEERING FLUID COOLERS (REPLACEMENT)

The power steering fluid cooler (**Figure 28**, typical) is located along the hose connecting the seawater pump to the transom bracket fitting and is used on all models equipped with power steering. The power steering fluid hoses connect to nipple fittings on the side of the cooler and are secured with worm gear hose clamps or crimp clamps.

The oil cooler (**Figure 29**, typical) is located on the lower port side of the engine. The oil cooler is used only on GM 7.4L and 8.2L models. The oil hoses connect to the cooler with flare nut fittings.

Both the steering fluid and oil cooler are a one-piece welded design. Cooler replacement is usually required due to freeze or corrosion damage. Internal pressure from freezing water usually pushes the end piece off the cooler. In most instances, the end can be reinstalled and soldered in place at a reputable radiator repair shop. Always have the repair shop pressure test the cooler after the repair to make sure internal parts of the cooler are not damaged. Replace the cooler if leaking due to corrosion damage.

1. Disconnect the negative battery cable.
2. Loosen the clamps and then carefully pull the larger seawater hoses from the cooler fittings. Do not attempt to pry or cut the hoses off the fittings. Doing so will damage the thin fitting material. If necessary, use a blunt instrument to carefully push the hoses off the fittings. If the boat is stored in the water, plug the disconnected hoses to prevent water from siphoning into the bilge.
3. Place a suitable container under the cooler to capture spilled fluid or oil. Note the hose routing and connection points and then disconnect the fluid or oil hoses from the cooler fittings. Drain the hoses and then plug the ends to prevent contamination.
4. Support the cooler while removing the mounting bracket bolt. Remove the cooler. Drain residual water, fluid or oil from the cooler.
5. Fit the replacement cooler onto the engine. Make sure the cooler is oriented properly and is not in contact with any moving components, electrical terminals or sharp or pointed surfaces. The power steering fluid cooler usually tilts down on one end to allow water to drain. Install the mounting bolt and tighten to the general torque specification provided in Chapter One.
6. Connect the fluid or oil hoses to the pump fittings. Secure the fluid or oil hoses with clamps or by securely tightening the fittings. Make sure the hoses are not kinked or in contact with sharp or pointed objects.

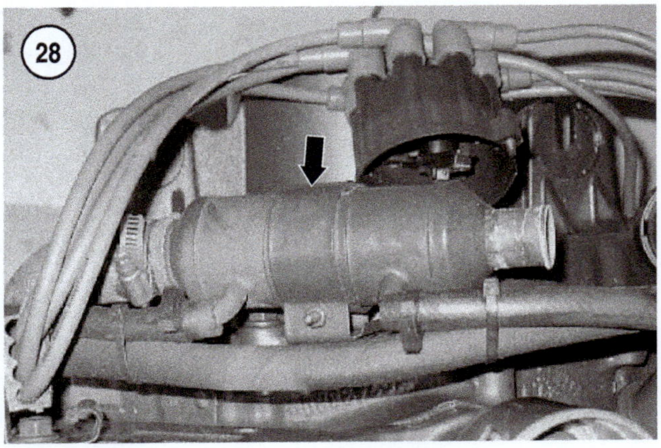

7. Fit the clamps over the ends and then push the seawater hoses onto the cooler fittings. Position the clamps approximately 1/4 in. (6 mm) from the end of the hose and then securely tighten the clamps. Make sure the clamps are not in contact with moving components, wiring or wire terminals.
8. Correct the power steering or engine oil level as described in Chapter Three.

WARNING

Stay clear of the propeller shaft while running the engine on a flush/test adapter. Remove the propeller(s) before running the engine to prevent injury or death. Always disconnect the battery cables before removing or installing the propeller.

CAUTION

Never run the engine without first providing cooling water. Use either a test tank or a flush/test adapter.

9. Connect the battery cable. Start the engine and immediately check for water, power steering fluid or oil leakage from the cooler or connecting hoses. Correct any leakage before operating the engine.
10. Check and correct the power steering fluid or engine oil level.

30

EXHAUST SYSTEM (GM 3.0L MODELS)

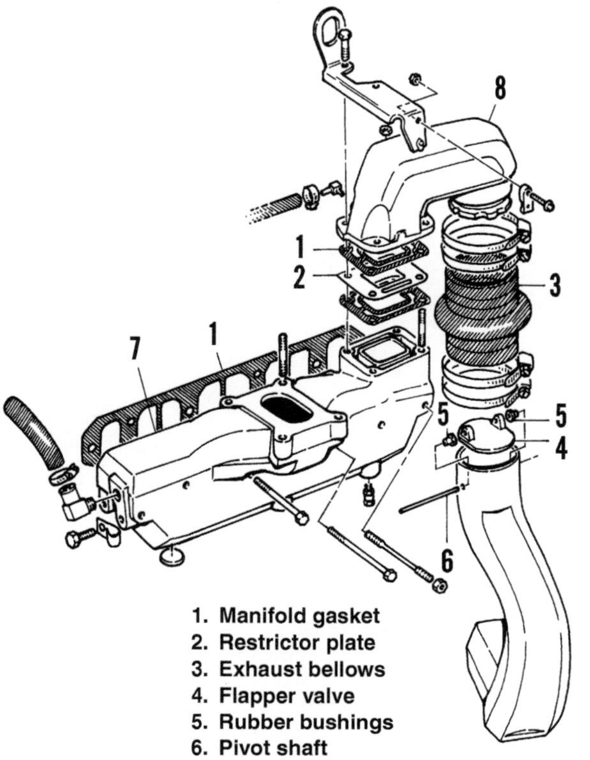

1. Manifold gasket
2. Restrictor plate
3. Exhaust bellows
4. Flapper valve
5. Rubber bushings
6. Pivot shaft
7. Exhaust manifold
8. Exhaust elbow

31

EXHAUST ELBOW

GM 3.0L Models

The intake and exhaust manifolds are combined into a single unit. **Figure 30** shows the exhaust system components. The exhaust elbow, exhaust bellows and flapper valve can be replaced without removing the engine from the boat.

Removal/installation

The exhaust elbow can be removed without removing the manifold from the engine provided that the boat allows adequate clearance. If clearance above the manifold is limited, remove the manifold or engine before removing the elbow.

Refer to **Figure 30** for this procedure.

1. Disconnect the negative battery cable.
2. Drain water from the exhaust manifold as described in *Winterizing* in Chapter Four.
3. Loosen the two clamps that secure the exhaust bellows (3, **Figure 30**) to the exhaust elbow outlet.
4. On later models with the bypass system, loosen the hose clamp and disconnect the bypass hose from the small fitting on the starboard side of the exhaust elbow.
5. Insert a blunt awl or other suitable instrument between the bellows and elbow outlet. Carefully work the instrument around the bellows to break the bond between the metal and rubber. Work carefully to avoid damaging the bellows.
6. Note the orientation of any attached brackets or clamps and remove the four bolts or nuts and washers that secure the elbow to the exhaust manifold. Tap lightly on the elbow with a rubber mallet to break the bond and then lift the elbow off the manifold. If necessary, carefully pry the elbow loose. Work carefully to avoid contact with the elbow and manifold mating surfaces. Make sure the pry bar contacts only the gasket or restrictor plate surfaces.
7. Remove the gaskets (1, **Figure 30**) and restrictor plate (2) from the exhaust manifold or exhaust elbow mating surfaces. Discard the gaskets. Thoroughly clean all residual gasket material from the mating surfaces with a Scotch-Brite pad.
8. Peer into the exhaust bellows to check the condition of the exhaust flapper (**Figure 31**). Replace the flapper valve as described in this chapter if missing or if the surfaces appeared burned.
9. Inspect the restrictor plate for scratched, pitted or damaged surfaces. Replace the plate if defects are evident on any of the gasket mating surfaces. Damaged surfaces will allow water leakage into the manifold, cylinder head and cylinder bore(s).
10. Inspect the mating surfaces of the manifold and exhaust elbow for deep pitting, scratches, cracks or other defects. Have the mating surfaces machined to remedy surface defects. Do not remove more than 0.020 in. (0.5 mm) from the mating surfaces. Replace the manifold and/or elbow if surface defects cannot be corrected by machining or if cracks are evident on either component.
11. Use an appropriate thread chaser to remove corrosion, old sealant or other foreign material from the bolt openings.
12. Inspect the exposed water passages in the elbow and manifold for excessive salt, mineral or rust deposits. If heavy deposits are present, remove the manifold as described in this chapter and have it and the elbow cleaned in an acid vat at a reputable machine shop. Deposits in the passages will cause the engine to overheat by restricting water

flow and insulating the water from the heat-producing components.

13. Inspect the exhaust bellows (3, **Figure 30**) for cut, charred, cracked or hardened surfaces. Replace the bellows if these or other defects are evident.

14. Apply a light coat of gasket sealing compound to the gasket mating surfaces. Fit the first gasket (1, **Figure 30**), restrictor plate (2) and then second gasket over the manifold studs. Then seat them against the manifold. Make sure the bolt, exhaust and water passages align with the manifold openings.

15. Apply a soap-and-water solution to the inner exhaust bellows surfaces. Guide the outlet of the exhaust elbow into the exhaust bellows and then carefully lower the elbow onto the manifold. Work carefully to avoid dislodging or damaging the gaskets.

16. Install any attached brackets or clamps and then thread the fasteners onto the studs or into the elbow and manifold openings. Rotate the elbow to achieve proper alignment of the gaskets, restrictor plate, manifold and elbow. Then tighten the nuts and bolts in a crossing pattern to 142-168 in.-lb. (16-19 N•m).

17. Position the two hose clamps into the recesses in the exhaust bellows. Make sure the clamp surface is positioned directly over the clamping surface on the elbow. Reposition the bellows if necessary. Then securely tighten all four hose clamps.

18. On later models using a water bypass system, connect the water bypass hose to the fitting on the starboard side of the elbow. Connect the water hose to the fitting on the front of the manifold. Secure both hoses with screw clamps.

WARNING

Stay clear of the propeller shaft while running the engine on a flush/test adapter. Remove the propeller(s) before running the engine to prevent injury or death. Always disconnect the battery cables before removing or installing the propeller.

CAUTION

Never run the engine without first providing cooling water. Use either a test tank or a flush/test adapter.

19. Connect the battery cable. Start the engine and immediately check for water leakage at the hose fitting, exhaust bellows and elbow-to-manifold mating surfaces. Tighten the hose clamps and check for damaged hoses or bellows if there is leakage from the fittings. If there is leakage from the mating surfaces, remove the elbow and check for damaged gaskets or overlooked mating surface defects.

20. Allow the engine to completely cool and then retighten the elbow mounting fasteners as described in Step 16.

32 **EXHAUST SYSTEM (GM 4.3L MODELS)**

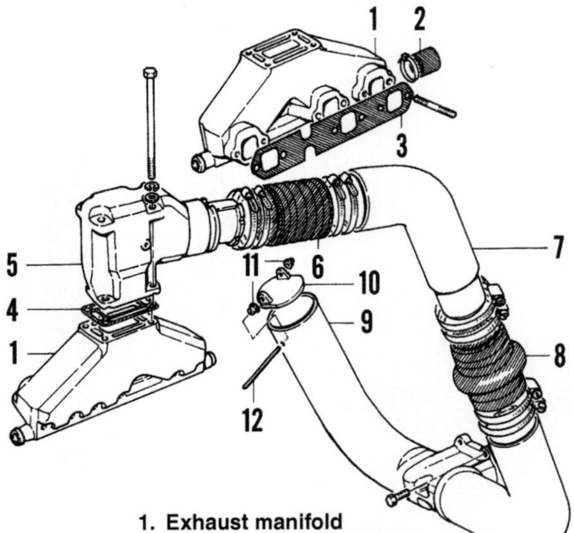

1. Exhaust manifold
2. Drain cap (earlier models) or plug (later models)
3. Exhaust manifold-to-cylinder head gasket
4. Exhaust elbow-to-exhaust manifold gasket
5. Exhaust elbow
6. Rubber exhaust tube
7. Curved exhaust tube (metal)
8. Rubber bellows
9. Exhaust collector
10. Flapper valve
11. Bushings
12. Pivot shaft

GM 4.3L Models

These models use an exhaust manifold and elbow for each cylinder bank. Rubber exhaust tubes connect each exhaust elbow and tube to the single exhaust collector. **Figure 32** shows the exhaust system components. The exhaust elbow(s) can be removed without removing the manifold from the engine provided that the boat allows adequate clearance. If clearance above the manifold is limited, remove the manifold before removing the elbow.

Removal/installation

Refer to **Figure 32** for this procedure.

1. Disconnect the negative battery cable.
2. Drain water from the exhaust manifold as described in *Winterizing* in Chapter Four.
3. Loosen the two clamps that secure the exhaust elbow outlet to the rubber exhaust tube (6, **Figure 32**).
4. On later carburetor-equipped models with hydraulic lock prevention hoses, remove the plastic flame arrestor cover (**Figure 33**). Remove the thin nut and lift the hose retainers (**Figure 34**) from the flame arrestor stud.

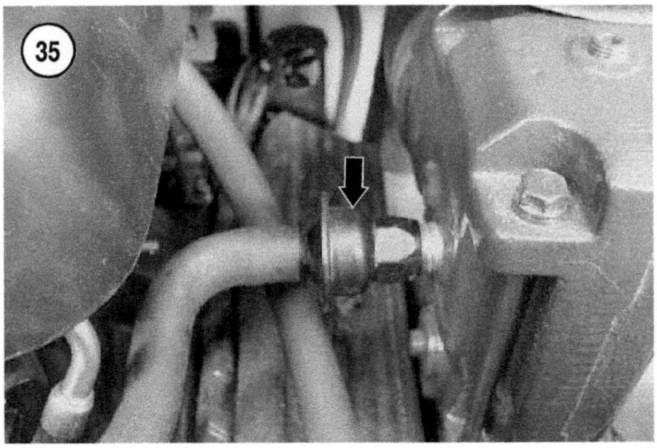

5. Note the orientation and then remove any clamps or brackets that attach to the exhaust elbow.

6. Insert a blunt awl or other suitable instrument between the elbow outlet and the rubber exhaust tube. Carefully work the instrument around the tube to break the bond between the rubber and metal. Work carefully to avoid damaging the bellows.

7. Remove the four long bolts and washers that secure the elbow to the manifold. Tap lightly on the elbow with a rubber mallet to break the bond and then lift the elbow off the manifold. Carefully pull the outlet end of the elbow out of the rubber exhaust tube. Work carefully to avoid dragging the elbow across the manifold mating surface.

8. Remove the gasket (4, **Figure 32**) from the manifold or elbow. Discard the gasket. Thoroughly clean all residual gasket material from the mating surfaces with a Scotch-Brite pad.

9. If the curved exhaust tube (7, **Figure 32**) is removed, peer into the exhaust bellows to check the condition of the exhaust flapper (**Figure 31**). Replace the flapper valve as described in this chapter if it is missing or if the surfaces appeared burned.

10. Inspect the mating surfaces of the manifold and exhaust elbow for deep pitting, scratches, cracks or other defects. Have the mating surfaces machined to remedy surface defects. Do not remove more than 0.020 in. (0.5 mm) from the mating surfaces. Replace the manifold and/or elbow if surface defects cannot be corrected by machining or if cracks are evident on either component.

11. Use an appropriate thread chaser to remove corrosion, old sealant or other foreign material from the bolt openings.

12. Inspect the rubber exhaust tube (6, **Figure 32**) for cut, charred, cracked or hardened surfaces. Replace the tube if these or other defects are evident.

8

13. Inspect the exposed water passages in the elbow and manifold for excessive salt, mineral or rust deposits. If heavy deposits are present, remove the manifold as described in this chapter and have it and the elbow cleaned in an acid vat at a reputable machine shop. Deposits in the passages will cause the engine to overheat by restricting water flow and insulating the water from the heat-producing components. Always remove any clamps, fittings or plugs from the elbow or manifold before acid-vat cleaning.

14. If the engine is equipped with hydraulic lock prevention hoses and the elbow must be replaced, clamp the elbow in a vise and unthread the check valve and hose (**Figure 35**). Apply sealant to the threads and then thread the hose and valve into the replacement elbow. Securely tighten the fitting.

15. If necessary, remove the elbow from the other cylinder bank as described in Steps 3-14. Any defect affecting one elbow is probably affecting the other.

16. Apply a light coat of gasket sealing compound to the gasket mating surfaces. Install the new gasket onto the manifold surface.

17. Apply a soap-and-water solution to the inner surfaces of the rubber exhaust tube. Guide the outlet of the exhaust elbow into the exhaust bellows and then carefully lower the elbow onto the manifold. Work carefully to avoid dislodging or damaging the gaskets.

NOTE

Some replacement gaskets are made of a material different from the original gaskets. Often, the torque specification and tightening sequence is changed for the different material. Refer to the instructions supplied with the replacement gaskets for special instructions.

18. Align the bolt openings in the elbow, gasket and manifold. Install any bracket or clamp that attaches with the elbow mounting bolts. Then thread the four bolts into the elbow and manifold. Tighten the bolts in a crossing pattern to 144-216 in.-lb. (16-24 N•m).

19. Position the two hose clamps into the recesses in the rubber exhaust tube. Make sure the clamp surface is positioned directly over the clamping surface on the elbow. Reposition the tube if necessary. Then securely tighten the four hose clamps.

20. If removed, install the remaining exhaust elbow as described in Steps 16-19.

21. On carburetor-equipped models with hydraulic lock prevention hoses, fit the hose retainers (**Figure 34**) onto the flame arrestor and secure with the thin nut. Install the plastic flame arrestor cover (**Figure 33**).

WARNING
Stay clear of the propeller shaft while running the engine on a flush/test adapter. Remove the propeller(s) before running the engine to prevent injury or death. Always disconnect the battery cables before removing or installing the propeller.

CAUTION
Never run the engine without first providing cooling water. Use either a test tank or a flush/test adapter.

22. Connect the battery cable. Start the engine and immediately check for water leakage at the hose fittings, rubber exhaust tube and elbow-to-manifold mating surfaces. Tighten the hose clamps and check for damaged hoses if there is leakage from the fittings. If there is leakage from the mating surfaces, remove the elbow and check for a damaged gasket mating surface defects.

23. Allow the engine to completely cool and retighten the elbow mounting bolts as described in Step 18.

GM 5.0L and 5.7L Models

These models use an exhaust manifold and elbow for each cylinder bank. Rubber exhaust tubes connect each exhaust elbow and tube to the single exhaust collector. **Figure 36** shows the exhaust system components. The exhaust elbow(s) can be removed without removing the manifold from the engine provided that the boat allows adequate clearance. If clearance above the manifold is limited, remove the manifold before removing the elbow.

Removal/installation

Refer to **Figure 36** for this procedure.

1. Disconnect the negative battery cable.
2. Drain water from the exhaust manifold as described in *Winterizing* in Chapter Four.

(36) **EXHAUST SYSTEM (GM 5.0L AND 5.7L MODELS)**

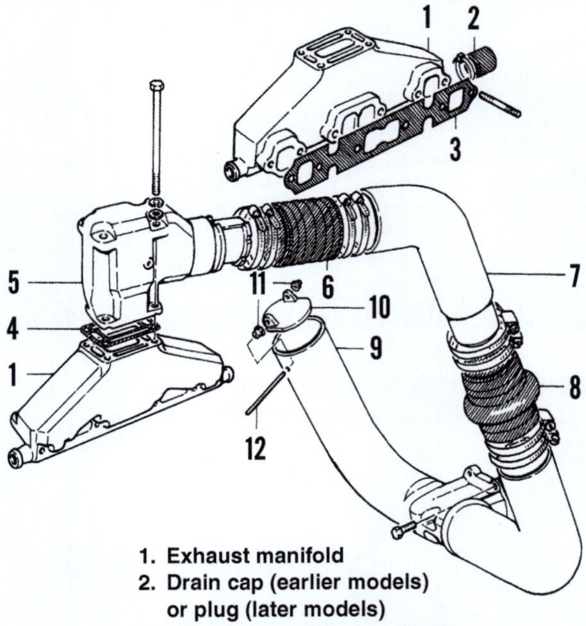

1. Exhaust manifold
2. Drain cap (earlier models) or plug (later models)
3. Exhaust manifold-to-cylinder head gasket
4. Exhaust elbow-to-exhaust manifold gasket
5. Exhaust elbow
6. Rubber exhaust tube
7. Curved exhaust tube (metal)
8. Rubber bellows
9. Exhaust collector
10. Flapper valve
11. Bushings
12. Pivot shaft

3. Loosen the two clamps that secure the exhaust elbow outlet to the rubber exhaust tube (6, **Figure 36**).
4. On later carburetor-equipped models with hydraulic lock prevention hoses, remove the plastic flame arrestor cover (**Figure 33**). Remove the thin nut and lift the hose retainers (**Figure 34**) from the flame arrestor stud.
5. Note the orientation and then remove any clamps or brackets that attach to the exhaust elbow.
6. Insert a blunt awl or other suitable instrument between the elbow outlet and the rubber exhaust tube. Carefully work the instrument around the tube to break the bond between the rubber and metal. Work carefully to avoid damaging the bellows.
7. Remove the four long bolts and washers that secure the elbow to the manifold. Tap lightly on the elbow with a rubber mallet to break the bond and then lift the elbow off the manifold. Carefully pull the outlet end of the elbow out of the rubber exhaust tube. Work carefully to avoid dragging the elbow across the manifold mating surface.
8. Remove the gasket (4, **Figure 36**) from the manifold or elbow. Discard the gasket. Thoroughly clean all residual

gasket material from the mating surfaces with a Scotch-Brite pad.

9. If the curved exhaust tube (7, **Figure 36**) is removed, peer into the exhaust bellows to check the condition of the exhaust flapper (**Figure 31**). Replace the flapper valve as described in this chapter if it is missing or if the surfaces appeared burned.

10. Inspect the mating surfaces of the manifold and exhaust elbow for deep pitting, scratches, cracks or other defects. Have the mating surfaces machined to remedy surface defects. Do not remove more than 0.020 in. (0.5 mm) from the mating surfaces. Replace the manifold and/or elbow if surface defects cannot be corrected by machining or if cracks are evident on either component.

11. Use an appropriate thread chaser to remove corrosion, old sealant or other foreign material from the bolt openings.

12. Inspect the rubber exhaust tube (6, **Figure 36**) for cut, charred, cracked or hardened surfaces. Replace the tube if these or other defects are evident.

13. Inspect the exposed water passages in the elbow and manifold for excessive salt, mineral or rust deposits. If heavy deposits are present, remove the manifold as described in this chapter and have it and the elbow cleaned in an acid vat at a reputable machine shop. Deposits in the passages will cause the engine to overheat by restricting water flow and insulating the water from the heat-producing components. Always remove any clamps, fittings or plugs from the elbow or manifold before acid-vat cleaning.

14. If the engine is equipped with hydraulic lock prevention hoses and the elbow must be replaced, clamp the elbow in a vise and unthread the check valve and hose (**Figure 35**). Apply sealant to the threads and then thread the hose and valve into the replacement elbow. Securely tighten the fitting.

15. If necessary, remove the elbow from the other cylinder bank as described in Steps 3-14. Any defect affecting one elbow is likely affecting the other.

16. Apply a light coat of gasket sealing compound to the gasket mating surfaces. Install the new gasket onto the manifold surface.

17. Apply a soap-and-water solution to the inner surfaces of the rubber exhaust tube. Guide the outlet of the exhaust elbow into the exhaust bellows and then carefully lower the elbow onto the manifold. Work carefully to avoid dislodging or damaging the gaskets.

NOTE

Some replacement gaskets are made of a material different from the original gaskets. Often, the torque specification and tightening sequence is changed for the different material. Refer to the instructions supplied with the replacement gaskets for special instructions.

18. Align the bolt openings in the elbow, gasket and manifold. Install any bracket or clamp that attaches with the elbow mounting bolts. Then thread the four bolts into the elbow and manifold. Tighten the bolts in a crossing pattern to 144-216 in.-lb. (16-24 N•m).

19. Position the two hose clamps into the recesses in the rubber exhaust tube. Make sure the clamp surface is positioned directly over the clamping surface on the elbow. Reposition the tube if necessary. Then securely tighten the four hose clamps.

20. If removed, install the remaining exhaust elbow as described in Steps 16-19.

21. On later carburetor-equipped models with hydraulic lock prevention hoses, fit the hose retainers (**Figure 34**) onto the flame arrestor and secure with the thin nut. Install the plastic flame arrestor cover (**Figure 33**).

WARNING

Stay clear of the propeller shaft while running the engine on a flush/test adapter. Remove the propeller(s) before running the engine to prevent injury or death. Always disconnect the battery cables before removing or installing the propeller.

CAUTION

Never run the engine without first providing cooling water. Use either a test tank or a flush/test adapter.

22. Connect the battery cable. Start the engine and immediately check for water leakage at the hose fittings, rubber exhaust tube and elbow to manifold mating surfaces. Tighten the hose clamps and check for damaged hoses if there is leakage from the fittings. If there is leakage from the mating surfaces, remove the elbow and check for a damaged gasket or overlooked mating surface defects.

23. Allow the engine to completely cool and retighten the elbow mounting bolts as described in Step 18.

Ford 5.0L and 5.8L Models

These models use an exhaust manifold and elbow for each cylinder bank. Rubber exhaust tubes connect each exhaust elbow and tube to the single exhaust collector. **Figure 37** shows the exhaust system components. The exhaust elbow(s) can be removed without removing the manifold from the engine provided that the boat allows adequate clearance. If clearance above the manifold is limited, remove the manifold before removing the elbow.

Removal/installation

Refer to **Figure 37** for this procedure.

1. Disconnect the negative battery cable.

2. Drain water from the exhaust manifold as described in *Winterizing* in Chapter Four.

3. Loosen the two clamps that secure the exhaust elbow outlet to the rubber exhaust tube (6, **Figure 37**).

37 **EXHAUST SYSTEM (FORD 5.0L AND 5.8L MODELS)**

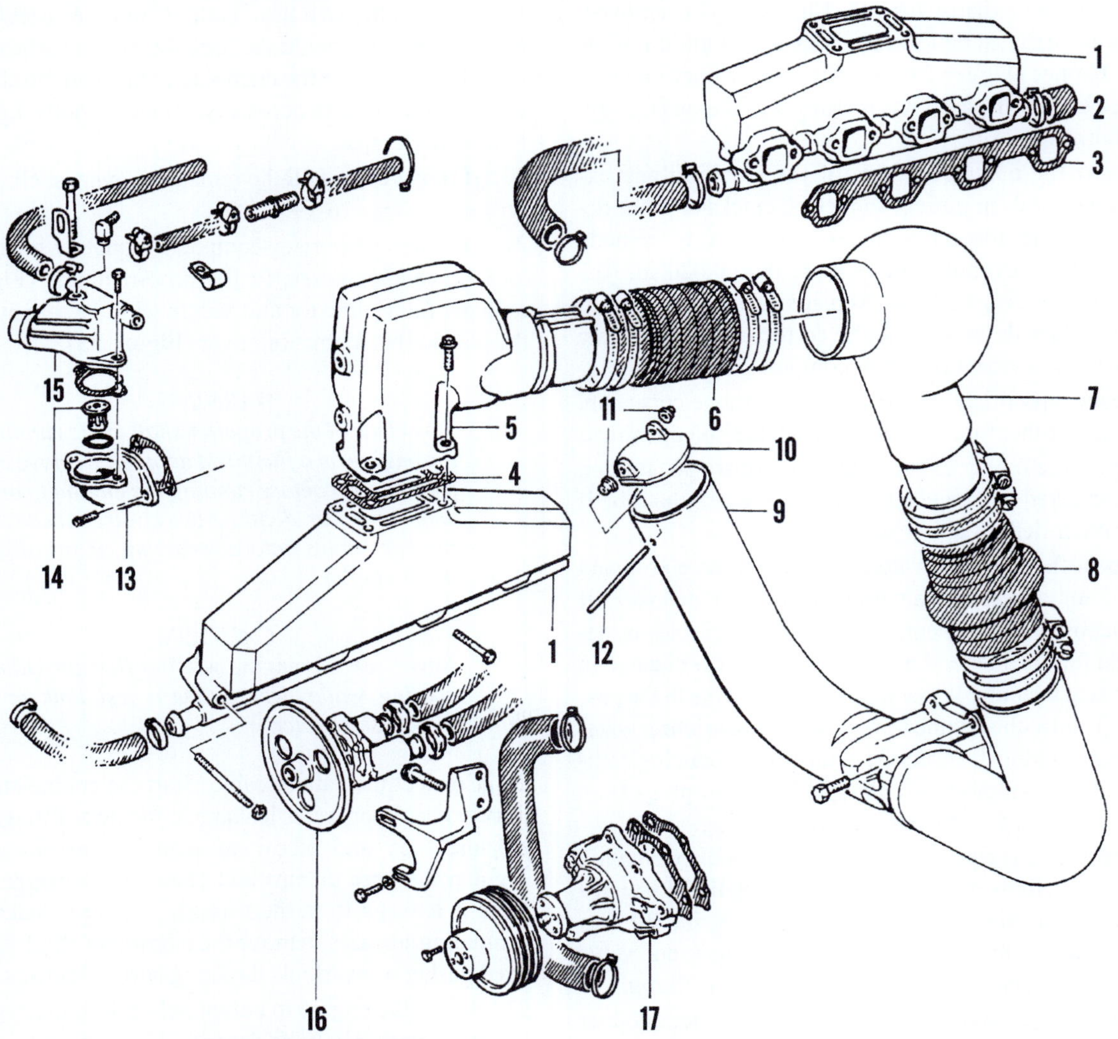

1. Exhaust manifolds
2. Drain cap
3. Exhaust manifold-to-cylinder head gasket
4. Exhaust elbow-to-exhaust manifold gasket
5. Exhaust elbow
6. Rubber exhaust tube
7. Curved exhaust tube (metal)
8. Exhaust bellows
9. Exhaust collector
10. Flapper valve
11. Bushings
12. Pivot shaft
13. Thermostat housing elbow
14. Thermostat
15. Thermostat housing
16. Seawater pump
17. Recirculation pump

4. Note the orientation and then remove any clamps or brackets that attach to the exhaust elbow.

5. Insert a blunt awl or other suitable instrument between the elbow outlet and the rubber exhaust tube. Carefully work the instrument around the tube to break the bond between the rubber and metal. Work carefully to avoid damaging the bellows.

6. Remove the four long bolts and washers that secure the elbow to the manifold. Tap lightly on the elbow with a rubber mallet to break the bond and then lift the elbow off the manifold. Carefully pull the outlet end of the elbow out of the rubber exhaust tube. Work carefully to avoid dragging the elbow across the manifold mating surface.

7. Remove the gasket (4, **Figure 37**) from the manifold or elbow. Discard the gasket. Thoroughly clean all residual gasket material from the mating surfaces with a Scotch-Brite pad.

8. If the curved exhaust tube (7, **Figure 37**) is removed, peer into the exhaust bellows to check the condition of the exhaust flapper (**Figure 31**). Replace the flapper valve as described in this chapter if it is missing or if the surfaces appeared burned.

9. Inspect the mating surfaces of the manifold and exhaust elbow for deep pitting, scratches, cracks or other defects. Have the mating surfaces machined to remedy surface defects. Do not remove more than 0.020 in. (0.5 mm) from the mating surfaces. Replace the manifold and/or elbow if surface defects cannot be corrected by machining or if cracks are evident on either component.
10. Use an appropriate thread chaser to remove corrosion, old sealant or other foreign material from the bolt openings.
11. Inspect the rubber exhaust tube (6, **Figure 37**) for cut, charred, cracked or hardened surfaces. Replace the tube if these or other defects are evident.
12. Inspect the exposed water passages in the elbow and manifold for excessive salt, mineral or rust deposits. If heavy deposits are present, remove the manifold as described in this chapter and have it and the elbow cleaned in an acid vat at a reputable machine shop. Deposits in the passages will cause the engine to overheat by restricting water flow and insulating the water from the heat-producing components. Always remove any clamps, fittings or plugs from the elbow or manifold before acid-vat cleaning.
13. If necessary, remove the elbow from the other cylinder bank as described in Steps 3-12. Any defect affecting one elbow is probably affecting the other.
14. Apply a light coat of gasket sealing compound to the gasket mating surfaces. Install the new gasket onto the manifold surface.
15. Apply a soap-and-water solution to the inner surfaces of the rubber exhaust tube. Guide the outlet of the exhaust elbow into the exhaust bellows and then carefully lower the elbow onto the manifold. Work carefully to avoid dislodging or damaging the gaskets.

NOTE
Some replacement gaskets are made of a material different from the original gaskets. Often, the torque specification and tightening sequence is changed for the different material. Refer to the instructions supplied with the replacement gaskets for special instructions.

16. Align the bolt openings in the elbow, gasket and manifold. Install any bracket or clamp that attaches with the elbow mounting bolts. Then thread the four bolts into the elbow and manifold. Tighten the bolts in a crossing pattern to 144-216 in.-lb. (16-24 N•m).
17. Position the two hose clamps into the recesses in the rubber exhaust tube. Make sure the clamp surface is positioned directly over the clamping surface on the elbow. Reposition the tube if necessary. Then securely tighten the four hose clamps.
18. If removed, install the remaining exhaust elbow as described in Steps 14-17.

WARNING
Stay clear of the propeller shaft while running the engine on a flush/test adapter. Remove the propeller(s) before running the engine to prevent injury or death. Always disconnect the battery cables before removing or installing the propeller.

CAUTION
Never run the engine without first providing cooling water. Use either a test tank or a flush/test adapter.

19. Connect the battery cable. Start the engine and immediately check for water leakage at the hose fittings, rubber exhaust tube and elbow-to-manifold mating surfaces. Tighten the hose clamps and check for damaged hoses if there is leakage from the fittings. If there is leakage from the mating surfaces, remove the elbow and check for a damaged gasket or mating surface defects.
20. Allow the engine to completely cool and retighten the elbow mounting bolts as described in Step 18.

GM 7.4L and 8.2L Models

These models use an exhaust manifold and elbow for each cylinder bank. Rubber exhaust tubes connect each exhaust elbow and tube to the single exhaust collector. **Figure 38** shows the exhaust system components. The exhaust elbow(s) can be removed without removing the manifold from the engine provided that the boat allows adequate clearance. If clearance above the manifold is limited, remove the manifold before removing the elbow.

Removal/installation

Refer to **Figure 38** for this procedure.
1. Disconnect the negative battery cable.
2. Drain water from the exhaust manifold as described in *Winterizing* in Chapter Four.
3. Loosen the two clamps that secure the exhaust elbow outlet to the rubber exhaust tube (6, **Figure 38**).
4. Note the orientation and then remove any clamps or brackets that attach to the exhaust elbow.
5. Insert a blunt awl or other suitable instrument between the elbow outlet and the rubber exhaust tube. Carefully work the instrument around the tube to break the bond between the rubber and metal. Work carefully to avoid damaging the bellows.
6. Remove the four long bolts and washers that secure the elbow to the manifold. Tap lightly on the elbow with a rubber mallet to break the bond and then lift the elbow off the manifold. Carefully pull the outlet end of the elbow out of the rubber exhaust tube. Work carefully to avoid dragging the elbow across the manifold mating surface.
7. Remove the gasket (4, **Figure 38**) from the manifold or elbow. Discard the gasket. Thoroughly clean all residual gasket material from the mating surfaces with a Scotch-Brite pad.

8. If the curved exhaust tube (7, **Figure 38**) is removed, peer into the exhaust bellows to check the condition of the exhaust flapper (**Figure 31**). Replace the flapper valve as described in this chapter if it is missing or if the surfaces appeared burned.

9. Inspect the mating surfaces of the manifold and exhaust elbow for deep pitting, scratches, cracks or other defects. Have the mating surfaces machined to remedy surface defects. Do not remove more than 0.020 in. (0.5 mm) from the mating surfaces. Replace the manifold and/or elbow if surface defects cannot be corrected by machining or if cracks are evident on either component.

10. Use an appropriate thread chaser or die to remove corrosion, old sealant or other foreign material from the bolt openings or stud threads.

11. Inspect the rubber exhaust tube (6, **Figure 38**) for cut, charred, cracked or hardened surfaces. Replace the tube if these or other defects are evident.

12. Inspect the exposed water passages in the elbow and manifold for excessive salt, mineral or rust deposits. If heavy deposits are present, remove the manifold as described in this chapter and have it and the elbow cleaned in an acid vat at a reputable machine shop. Deposits in the passages will cause the engine to overheat by restricting water flow and insulating the water from the heat-producing components. Always remove any clamps, fittings or plugs from the elbow or manifold before acid-vat cleaning.

13. If necessary, remove the elbow from the other cylinder bank as described in Steps 3-12. Any defect affecting one elbow is probably affecting the other.

14. Apply a light coat of gasket sealing compound onto the gasket mating surfaces. Install the new gasket onto the manifold surface.

15. Apply a soap-and-water solution to the inner surfaces of the rubber exhaust tube. Guide the outlet of the exhaust elbow into the exhaust bellows and then carefully lower the elbow onto the manifold. Work carefully to avoid dislodging or damaging the gaskets.

NOTE

Some replacement gaskets are made of a material different from the original gaskets. Often, the torque specification and tightening sequence is changed for the different material. Refer to the instructions supplied with the replacement gaskets for special instructions.

16. Align the bolt openings in the elbow, gasket and manifold. Install any bracket or clamp that attaches with the elbow mounting bolts. Then thread the four bolts into the elbow and manifold. Tighten the bolts in a crossing pattern to 144-216 in.-lb. (16-24 N•m).

17. Position the two hose clamps into the recesses in the rubber exhaust tube. Make sure the clamp surface is positioned directly over the clamping surface on the elbow. Reposition the tube if necessary. Then securely tighten the four hose clamps.

38

EXHAUST SYSTEM (GM 7.4L AND 8.2L MODELS)

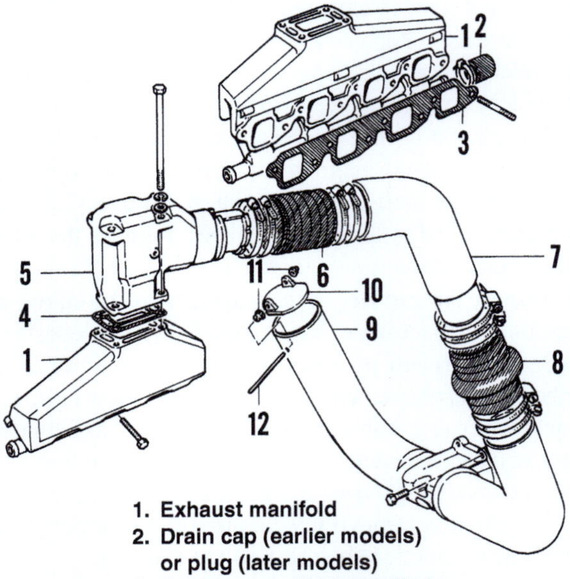

1. Exhaust manifold
2. Drain cap (earlier models) or plug (later models)
3. Exhaust manifold-to-cylinder head gasket
4. Exhaust elbow-to-exhaust manifold gasket
5. Exhaust elbow
6. Rubber exhaust tube
7. Curved exhaust tube (metal)
8. Rubber bellows
9. Exhaust collector
10. Flapper valve
11. Bushings
12. Pivot shaft

18. If removed, install the remaining exhaust elbow as described in Steps 14-17.

WARNING

Stay clear of the propeller shaft while running the engine on a flush/test adapter. Remove the propeller(s) before running the engine to prevent injury or death. Always disconnect the battery cables before removing or installing the propeller.

CAUTION

Never run the engine without first providing cooling water. Use either a test tank or a flush/test adapter.

19. Connect the battery cable. Start the engine and immediately check for water leakage at the hose fittings, rubber exhaust tube and elbow-to-manifold mating surfaces. Tighten the hose clamps and check for damaged hoses if there is leakage from the fittings. If there is leakage from the mating surfaces, remove the elbow and check for a damaged gasket or mating surface defects.

20. Allow the engine to completely cool and retighten the elbow mounting bolts as described in Step 16.

EXHAUST BELLOWS, EXHAUST TUBE AND FLAPPER VALVE

GM 3.0L Models

Figure 30 shows the exhaust system components. The exhaust elbow, exhaust bellows and flapper valve can be replaced without removing the engine from the boat.

Removal/installation

The exhaust bellows and flapper valve can be removed without removing the manifold from the engine. Refer to **Figure 30** for this procedure.

1. Disconnect the negative battery cable.
2. Remove the exhaust elbow as described in this chapter.
3. Loosen the two hose clamps and then pull the exhaust bellows from the lower exhaust tube. Grasp the exhaust flapper valve (4, **Figure 30**) with locking pliers to prevent it from dropping into the exhaust tube during removal.

NOTE
Work carefully to keep the bushings from falling into the exhaust tube during flapper valve removal. The bushing flanges face toward the inside of the tube.

4. Use a punch to push the pivot shaft (6, **Figure 30**) out of the bushings (5) and flapper valve openings.
5. Lift the flapper valve out of the opening. Carefully remove the bushings from the lower exhaust tube.
6. Inspect the flapper valve for worn, burned or damaged surfaces and wear in the pivot shaft openings. Replace the valve if these or other defects are evident. If the flapper valve is missing, remove the drive unit as described in Chapter Ten and inspect the exhaust passages in the drive and transom bracket. The remnant might block the exhaust passage and cause poor performance, overheating and possible engine damage due to increased exhaust back pressure.

CAUTION
Never reuse a worn flapper valve pivot shaft. The shaft will likely break at the worn area and drop the valve and shaft into the lower exhaust tube. Exhaust pressure pulses will move the and cause the sharp end of the broken shaft to wear a hole in the lower exhaust tube. The hole will allow exhaust and water to leak into the engine compartment.

7. Inspect the pivot shaft for wear at the flapper valve pivoting surfaces. Replace the shaft unless it is in excellent condition.
8. Inspect the rubber shaft bushings (5, **Figure 30**) for missing sections and burned or worn surfaces and replace as needed.
9. Apply 3M Scotch-Grip rubber adhesive to the surfaces and then carefully push the bushings into the exhaust tube openings. The flange side of the bushings must face the inside of the tube.
10. Grasp the replacement flapper valve with locking pliers and carefully lower it into the exhaust tube. The long end of the flapper valve must face downward. Align the pivot shaft openings in the flapper valve with the corresponding openings in the bushings. Then carefully slide the pivot shaft through the openings. Verify that the shaft is passing through both flapper valve openings and then remove the locking pliers. Pivot the flapper valve on the shaft to verify free movement. Check for improper installation if binding is evident.

NOTE
The short end of the exhaust bellows has only two recesses for the hose clamps and fits over the lower exhaust tube. The longer end has three recesses and fits over the discharge end of the exhaust elbow.

11. Lubricate the inner surface of the exhaust bellows (3, **Figure 30**) with a soap-and-water solution. Then, without dislodging the bushings, carefully slide the bellows over the lower exhaust tube. The short end of the bellows, with only two clamp recesses, must fit over the exhaust tube. Make sure the bellows flange fits fully over both bushings.
12. Fit the two lower clamps into the recesses in the exhaust bellows. Securely tighten the clamps.
13. Fit the two upper hose clamps into the two upper recesses in the exhaust bellows. The clamps must fit loosely at this time to allow installation of the exhaust elbow.
14. Install the exhaust elbow as described in this chapter.
15. Connect the negative battery cable.

GM 4.3L Models

The exhaust bellows, exhaust tube and flapper valve can be removed without removing the manifold from the engine; however, removal is usually easier and quicker with the elbow removed. It can be difficult to remove and reinstall the curved exhaust tube with the elbow installed.

Removal/installation

Refer to **Figure 32** for this procedure.

1. Disconnect the negative battery cable.
2. Remove the exhaust elbow as described in this chapter.
3. Loosen the two lower clamps that secure the rubber exhaust bellows (8, **Figure 32**) to the exhaust collector (9). Insert a blunt awl or other suitable object between the rubber exhaust bellows and the collector. Then work the tool around the connector to break the rubber-to-metal bond.
4. Carefully pull the curved exhaust tube (7, **Figure 32**) and bellows (8) as an assembly off the collector. Work care-

fully to avoid dropping the screw clamps or dislodging the bushings (11, **Figure 32**).

5. Grasp the exhaust flapper valve (10, **Figure 32**) with locking pliers to prevent it from dropping into the exhaust collector during removal.

NOTE
Work carefully to keep the bushings from falling into the exhaust collector during flapper valve removal. The bushing flanges face toward the inside of the collector.

6. Use a punch to push the pivot shaft (12, **Figure 32**) out of the bushings (11) and flapper valve openings.
7. Lift the flapper valve out of the opening. Carefully remove the bushings from the lower exhaust collector.
8. Loosen the clamps and then remove the rubber exhaust tube (6, **Figure 32**) from the curved exhaust tube (7).
9. Inspect the flapper valve for worn, burned or damaged surfaces and wear in the pivot shaft openings. Replace the valve if these or other defects are evident. If the flapper valve is missing, remove the drive unit as described in Chapter Ten and inspect the exhaust passages in the drive and transom bracket. The remnant might block the exhaust passage and cause poor performance, overheating and possible engine damage due to increased exhaust back pressure.

CAUTION
Never reuse a worn flapper valve pivot shaft. The shaft will probably break at the worn surface and drop the valve and remaining shaft into the lower exhaust collector. Exhaust pressure pulses will move the flapper and cause the sharp end of the broken shaft to wear a hole in the lower exhaust collector. The hole will allow exhaust and water to leak into the engine compartment.

10. Inspect the pivot shaft for wear at the flapper valve pivoting surfaces. Replace the shaft unless it is in excellent condition.
11. Inspect the rubber shaft bushings (11, **Figure 32**) for missing sections and burned or worn surfaces and replace as needed.
12. Inspect the rubber exhaust tube (6, **Figure 32**) and bellows (8) for cut, charred, cracked or hardened surfaces. Replace the tube and bellows if these or other defects are evident. The condition affecting one of these components is probably affecting the other.
13. Apply 3M Scotch-Grip rubber adhesive to the outer diameter of the bushings and then carefully push the bushings into the exhaust collector openings. The flange side of the bushings must face the inside of the collector.
14. Grasp the replacement flapper valve with locking pliers and carefully lower it into the exhaust collector. The long end of the flapper valve must face downward. Align the pivot shaft openings in the flapper valve with the corresponding openings in the bushings. Then carefully slide the pivot shaft through the openings. Verify that the shaft is passing through both flapper valve openings and then remove the locking pliers. Pivot the flapper valve on the shaft to verify free movement. Check for improper installation if binding is evident.

NOTE
The short end of the exhaust bellows has only two recesses for the hose clamps and fits over the lower exhaust collector. The longer end has three recesses and fits over the discharge end of the curved exhaust tube.

15. Lubricate the inner surfaces of the rubber exhaust bellows (8, **Figure 32**) with a soap-and-water solution. Then, without dislodging the bushings, carefully slide the bellows over the lower exhaust collector. The short end of the bellows, with only two clamp recesses, must fit over the exhaust collector. Make sure the bellows flange fits fully over both bushings.
16. Fit the two lower clamps into the recesses in the exhaust bellows. Securely tighten the clamps.
17. Fit the two upper hose clamps into the two upper recesses in the exhaust bellows. The clamps must fit loosely at this time to allow installation of the curved exhaust tube.
18. Lubricate the inner surface with a soap-and-water solution. Then fit the rubber exhaust tube (6, **Figure 32**) over the larger-diameter end of the curved exhaust tube until the clamp grooves are positioned directly over the clamping surfaces of the tube. Fit the two clamps into the recesses. Do not tighten the clamps at this time.
19. Guide the smaller-diameter end of the tube into the rubber exhaust bellows (8, **Figure 32**) until the clamp recesses in the bellows are directly over the clamping surface on the curved tube. Then securely tighten the two clamps.
20. Install the exhaust elbow as described in this chapter.
21. If necessary, replace the exhaust bellows, tube and flapper on the other cylinder bank as described in Steps 2-20. Any defect affecting the exhaust components on one cylinder bank is probably affecting the other.
22. Connect the battery cable.

GM 5.0L and 5.7L Models

The exhaust bellows, exhaust tube and flapper valve can be removed without removing the manifold from the engine; however, removal is usually easier and quicker with the elbow removed. It can be difficult to remove and reinstall the curved exhaust tube with the elbow installed.

Removal/installation

Refer to **Figure 36** for this procedure.

1. Disconnect the negative battery cable.
2. Remove the exhaust elbow as described in this chapter.
3. Loosen the two lower clamps that secure the rubber exhaust bellows (8, **Figure 36**) to the exhaust collector (9). In-

sert a blunt awl or other suitable object between the rubber exhaust bellows and the collector. Then work the tool around the connector to break the rubber-to-metal bond.

4. Carefully pull the curved exhaust tube (7, **Figure 6**) and bellows (8) as an assembly off the collector. Work carefully to avoid dropping the screw clamps or dislodging the bushings (11, **Figure 36**).

5. Grasp the exhaust flapper valve (10, **Figure 36**) with locking pliers to prevent it from dropping into the exhaust collector during removal.

NOTE
Work carefully to keep the bushings from falling into the exhaust collector during flapper valve removal. The bushing flanges face toward the inside of the collector.

6. Use a punch to push the pivot shaft (12, **Figure 36**) out of the bushings (11) and flapper valve openings.

7. Lift the flapper valve out of the opening. Carefully remove the bushings from the lower exhaust collector.

8. Loosen the clamps and then remove the rubber exhaust tube (6, **Figure 36**) from the curved exhaust tube (7).

9. Inspect the flapper valve for worn, burned or damaged surfaces and wear in the pivot shaft openings. Replace the valve if these or other defects are evident. If the flapper valve is missing, remove the drive unit as described in Chapter Ten and inspect the exhaust passages in the drive and transom bracket. The remnant might block the exhaust passage and cause poor performance, overheating and possible engine damage due to increased exhaust back pressure.

CAUTION
Never reuse a worn flapper valve pivot shaft. The shaft will probably break at the worn area and drop the valve and remaining shaft into the lower exhaust collector. Exhaust pressure pulses will move the flapper and cause the sharp end of the broken shaft to wear a hole in the lower exhaust collector. The hole will allow exhaust and water to leak into the engine compartment.

10. Inspect the pivot shaft for wear at the flapper valve pivoting surfaces. Replace the shaft unless it is in excellent condition.

11. Inspect the rubber shaft bushings (11, **Figure 36**) for missing sections and burned or worn surfaces and replace as needed.

12. Inspect the rubber exhaust tube (6, **Figure 36**) and bellows (8) for cut, charred, cracked or hardened surfaces. Replace the tube and bellows if these or other defects are evident. The condition affecting one of these components is probably affecting the other.

13. Apply 3M Scotch-Grip rubber adhesive to the outer surface of the bushings and then carefully push the bushings into the exhaust collector openings. The flange side of the bushings must face the inside of the collector.

14. Grasp the replacement flapper valve with locking pliers and carefully lower it into the exhaust collector. The long end of the flapper valve must face downward. Align the pivot shaft openings in the flapper valve with the corresponding openings in the bushings. Then carefully slide the pivot shaft through the openings. Verify that the shaft is passing through both flapper valve openings and then remove the locking pliers. Pivot the flapper valve on the shaft to verify free movement. Check for improper installation if binding is evident.

NOTE
The short end of the exhaust bellows has only two recesses for the hose clamps and fits over the lower exhaust collector. The longer end has three recesses and fits over the discharge end of the curved exhaust tube.

15. Lubricate the inner surface of the rubber exhaust bellows (8, **Figure 36**) with a soap-and-water solution. Then, without dislodging the bushings, carefully slide the bellows over the lower exhaust collector. The short end of the bellows, with only two clamp recesses, must fit over the exhaust collector. Make sure the bellows fits fully over both bushings.

16. Fit the two lower clamps into the recesses in the exhaust bellows. Securely tighten the clamps.

17. Fit the two upper hose clamps into the two upper recesses in the exhaust bellows. The clamps must fit loosely at this time to allow installation of the curved exhaust tube.

18. Lubricate the inner surface with a soap-and-water solution. Then fit the rubber exhaust tube (6, **Figure 36**) over the larger-diameter end of the curved exhaust tube until the clamp groove are positioned directly over the clamping surfaces of the tube. Fit the two clamps into the recess. Do not tighten the clamps at this time.

19. Guide the smaller-diameter end of the tube into the rubber exhaust bellows (8, **Figure 36**) until the clamp recesses in the bellows are directly over the clamping surface on the curved tube. Then securely tighten the two clamps.

20. Install the exhaust elbow as described in this chapter.

21. If necessary, replace the exhaust bellows, tube and flapper on the other cylinder bank as described in Steps 2-20. Any defect affecting the exhaust components on one cylinder bank is probably affecting the other.

22. Connect the battery cable.

Ford 5.0L and 5.8L Models

The exhaust bellows, exhaust tube and flapper valve can be removed without removing the manifold from the engine; however, removal is usually easier and quicker with the elbow removed. It can be difficult to remove and reinstall the curved exhaust tube with the elbow installed.

Removal/installation

Refer to **Figure 37** for this procedure.

1. Disconnect the negative battery cable.
2. Remove the exhaust elbow as described in this chapter.
3. Loosen the two lower clamps that secure the rubber exhaust bellows (8, **Figure 37**) to the exhaust collector (9). Insert a blunt awl or other suitable object between the rubber exhaust bellows and the collector. Then work the tool around the connector to break the rubber-to-metal bond.
4. Carefully pull the curved exhaust tube (7, **Figure 37**) and bellows (8) off the collector as an assembly. Work carefully to avoid dropping the clamps or dislodging the bushings (11, **Figure 37**).
5. Grasp the exhaust flapper valve (10, **Figure 37**) with locking pliers to prevent it from dropping into the exhaust collector during removal.

NOTE
Work carefully to keep the bushings from falling into the exhaust collector during flapper valve removal. The bushing flanges face toward the inside of the collector.

6. Use a punch to push the pivot shaft (12, **Figure 37**) out of the bushings (11) and flapper valve openings.
7. Lift the flapper valve out of the opening. Carefully remove the bushings from the lower exhaust collector.
8. Loosen the clamps and then remove the rubber exhaust tube (6, **Figure 37**) from the curved exhaust tube (7).
9. Inspect the flapper valve for worn, burned or damaged surfaces and wear in the pivot shaft openings. Replace the valve if these or other defects are evident. If the flapper valve is missing, remove the drive unit as described in Chapter Ten and inspect the exhaust passages in the drive and transom bracket. The remnant might block the exhaust passage and cause poor performance, overheating and possible engine damage due to increased exhaust back pressure.

CAUTION
Never reuse a worn flapper valve pivot shaft. The shaft will probably break at the worn area and drop the valve and remaining shaft into the lower exhaust collector. Exhaust pressure pulses will move the flapper and cause the sharp end of the broken shaft to wear a hole in the lower exhaust collector. The hole will allow exhaust and water to leak into the engine compartment.

10. Inspect the pivot shaft for wear at the flapper valve pivoting surfaces. Replace the shaft unless it is in excellent condition.
11. Inspect the rubber shaft bushings (11, **Figure 37**) for missing sections and burned or worn surfaces and replace as needed.
12. Inspect the rubber exhaust tube (6, **Figure 37**) and bellows (8) for cut, charred, cracked or hardened surfaces. Replace the tube and bellows if these or other defects are evident. The condition affecting one of these components is probably affecting the other.
13. Apply Scotch-Grip rubber adhesive to the outer surface of the bushings and then carefully push the bushings into the exhaust collector openings. The flange side of the bushings must face the inside of the collector.
14. Grasp the replacement flapper valve with locking pliers and carefully lower it into the exhaust collector. The long end of the flapper valve must face downward. Align the pivot shaft openings in the flapper valve with the corresponding openings in the bushings. Then carefully slide the pivot shaft through the openings. Verify that the shaft is passing through both flapper valve openings and then remove the locking pliers. Pivot the flapper valve on the shaft to verify free movement. Check for improper installation if binding is evident.

NOTE
The short end of the exhaust bellows has only two recesses for the hose clamps and fits over the lower exhaust collector. The longer end has three recesses and fits over the discharge end of the curved exhaust tube.

15. Lubricate the inner surface of the rubber exhaust bellows (8, **Figure 37**) with a soap-and-water solution. Then, without dislodging the bushings, carefully slide the bellows over the lower exhaust collector. The short end of the bellows, with only two clamp recesses, must fit over the exhaust collector. Make sure the bellows fits fully over both bushings.
16. Fit the two lower clamps into the recesses in the exhaust bellows. Securely tighten the clamps.
17. Fit the two upper hose clamps into the two upper recesses in the exhaust bellows. The clamps must fit loosely at this time to allow installation of the curved exhaust tube.
18. Lubricate the inner surface with a soap-and-water solution. Then fit the rubber exhaust tube (6, **Figure 37**) over the larger diameter end of the curved exhaust tube until the clamp grooves are positioned directly over the clamping surfaces of the tube. Fit the two clamps into the recess. Do not tighten the clamps at this time.
19. Guide the smaller-diameter end of the tube into the rubber exhaust bellows (8, **Figure 37**) until the clamp recesses in the bellows are directly over the clamping surface on the curved tube. Then securely tighten the two clamps.
20. Install the exhaust elbow as described in this chapter.
21. If necessary, replace the exhaust bellows, tube and flapper on the other cylinder bank as described in Steps 2-20. Any defect affecting the exhaust components on one cylinder bank is probably affecting the other.
22. Connect the battery cable.

GM 7.4L and 8.2L Models

The exhaust bellows, exhaust tube and flapper valve can be removed without removing the manifold from the engine; however, removal is usually easier and quicker with the elbow removed. It can be difficult to remove and reinstall the curved exhaust tube with the elbow installed.

Removal/installation

Refer to **Figure 38** for this procedure.

1. Disconnect the negative battery cable.
2. Remove the exhaust elbow as described in this chapter.
3. Loosen the two lower clamps that secure the rubber exhaust bellows (8, **Figure 38**) to the exhaust collector (9). Insert a blunt awl or other suitable object between the rubber exhaust bellows and the collector. Then work the tool around the connector to break the rubber-to-metal bond.
4. Carefully pull the curved exhaust tube (7, **Figure 38**) and bellows (8) as an assembly off the collector. Work carefully to avoid dropping the clamps or dislodging the bushings (11, **Figure 38**).
5. Grasp the exhaust flapper valve (10, **Figure 38**) with locking pliers to prevent it from dropping into the exhaust collector during removal.

NOTE
Work carefully to keep the bushings from falling into the exhaust collector during flapper valve removal. The bushing flanges face toward the inside of the collector.

6. Use a punch to push the pivot shaft (12, **Figure 38**) out of the bushings (11) and flapper valve openings.
7. Lift the flapper valve out of the opening. Carefully remove the bushings from the lower exhaust collector.
8. Loosen the clamps and then remove the rubber exhaust tube (6, **Figure 38**) from the curved exhaust tube (7).
9. Inspect the flapper valve for worn, burned or damaged surfaces and wear in the pivot shaft openings. Replace the valve if these or other defects are evident. If the flapper valve is missing, remove the drive unit as described in Chapter Ten and inspect the exhaust passages in the drive and transom bracket. The remnant might block the exhaust passage and cause poor performance, overheating and possible engine damage due to increased exhaust back pressure.

CAUTION
Never reuse a worn flapper valve pivot shaft. The shaft will probably break at the worn area and drop the valve and remaining shaft into the lower exhaust collector. Exhaust pressure pulses will move the flapper and cause the sharp end of the broken shaft to wear a hole in the lower exhaust collector. The hole will allow exhaust and water to leak into the engine compartment.

10. Inspect the pivot shaft for wear at the flapper valve pivoting surfaces. Replace the shaft unless it is in excellent condition.
11. Inspect the rubber shaft bushings (11, **Figure 38**) for missing sections and burned or worn surfaces and replace as needed.
12. Inspect the rubber exhaust tube (6, **Figure 38**) and bellows (8) for cut, charred, cracked or hardened surfaces. Replace the tube and bellows if these or other defects are evident. The condition affecting one of these components is probably affecting the other.
13. Apply Scotch-Grip rubber adhesive to the outer surface of the bushings and then carefully push the bushings into the exhaust collector openings. The flange side of the bushings must face the inside of the collector.
14. Grasp the replacement flapper valve with locking pliers and carefully lower it into the exhaust collector. The long end of the flapper valve must face downward. Align the pivot shaft openings in the flapper valve with the corresponding openings in the bushings. Then carefully slide the pivot shaft through the openings. Verify that the shaft is passing through both flapper valve openings, then remove the locking pliers. Pivot the flapper valve on the shaft to verify free movement. Check for improper installation if binding is evident.

NOTE
The short end of the exhaust bellows has only two recesses for the hose clamps and fits over the lower exhaust collector. The longer end has three recesses and fits over the discharge end of the curved exhaust tube.

15. Lubricate the inner surface of the rubber exhaust bellows (8, **Figure 38**) with a soap-and-water solution. Then, without dislodging the bushings, carefully slide the bellows over the lower exhaust collector. The short end of the bellows, with only two clamp recesses, must fit over the exhaust collector. Make sure the bellows fits fully over both bushings.
16. Fit the two lower clamps into the recesses in the exhaust bellows. Securely tighten the clamps.
17. Fit the two upper hose clamps into the two upper recesses in the exhaust bellows. The clamps must fit loosely at this time to allow installation of the curved exhaust tube.
18. Lubricate the inner surface with a soap-and-water solution. Then fit the rubber exhaust tube (6, **Figure 38**) over the larger-diameter end of the curved exhaust tube until the clamp grooves are positioned directly over the clamping surfaces of the tube. Fit the two clamps into the recess. Do not tighten the clamps at this time.
19. Guide the smaller-diameter end of the tube into the rubber exhaust bellows (8, **Figure 38**) until the clamp recesses in the bellows are directly over the clamping surface on the curved tube. Then securely tighten the two clamps.
20. Install the exhaust elbow as described in this chapter.

21. If necessary, replace the exhaust bellows, tube and flapper on the other cylinder bank as described in Steps 2-20. Any defect affecting the exhaust components on one cylinder bank is probably affecting the other.
22. Connect the battery cable.

EXHAUST MANIFOLD

GM 3.0L Models

On this model, the intake and exhaust manifolds are combined into a single unit. **Figure 30** shows the exhaust system components. The exhaust elbow, exhaust bellows and flapper valve can be replaced without removing the engine from the boat; however, the manifold requires clear access from the port side for removal. The 3.0L models are typically used on smaller boats and many of them have limited access due to integrated seating or other boat structure. In such instances, the engine must be removed before removing the manifold.

Removal/installation

Refer to **Figure 30** for this procedure.
1. Disconnect the negative battery cable.
2. Drain water from the exhaust manifold as described in *Winterizing* in Chapter Four.
3. Loosen the four hose clamps that secure the exhaust bellows (3, **Figure 30**) to the exhaust elbow and lower exhaust tube. Carefully work the bellows down the exhaust tube until the top of the bellows is below the top of the tube. Work carefully to avoid dislodging the bushings. (5, **Figure 30**). This is necessary to allow the manifold to slide off the mounting studs.
4. On later models using a water bypass system, loosen the clamp and then disconnect the water bypass hose from the fitting on the starboard side of the exhaust elbow.
5. Loosen the clamp and disconnect the water hose from the fitting on the front of the manifold.
6. Disconnect any wire clamps, brackets or other components connected onto the manifold.
7. Provide the manifold with overhead support and then remove the bolts and nuts that secure the manifold to the cylinder head. Slide the manifold toward the port side until it is free from the studs. Then lift the manifold out of the boat.
8. Remove the gasket from the manifold or cylinder block surfaces. Carefully scrape residual gasket material from the surfaces.

CAUTION
Always thoroughly clean the manifold-to-cylinder head mating surfaces after removing the manifold. Debris or residual gasket material can allow an intake air leak and/or exhaust leakage that leads to serious engine damage.

9. Inspect the manifold for cracks, corrosion and other defects. Replace the manifold if these or other defects are evident.
10. Use a straightedge to check for warp at the cylinder head mating surface. If a gap is visually apparent, measure it with a feeler gauge. Replace the manifold if the surface warp exceeds 0.006 in. (0.15 mm) within any 12 in. (30.5 cm) span.
11. If replacing the manifold, remove the exhaust elbow and install it onto the replacement manifold as described in this chapter.
12. Align the bolt and port openings and then fit the new gasket (1, **Figure 30**) onto the manifold. Insert the mounting bolts into the manifold openings to help hold the gasket into position.
13. Using overhead support, carefully guide the manifold over the corresponding stud openings while sliding the manifold into position on the cylinder head. Thread the six fasteners onto the studs or bolt openings. Starting from the middle and working toward the ends, tighten the manifold fasteners to 20-25 ft.-lb. (27-34 N•m).
14. Reinstall any bracket, wire or hose clamp removed for manifold removal.
15. Without dislodging the bushings (5, **Figure 31**), carefully move the bellows (3) up the exhaust tube and onto the discharge end of the elbow. Make sure the upper end of the bellows fits fully over the elbow outlet.
16. Position the hose clamps into the upper two recesses in the bellows and then securely tighten the clamps.
17. Connect the water hose onto the fitting on the front of the manifold and secure with the screw clamp.
18. On later models with a water bypass system, connect the water bypass hose to the fitting on the starboard side of the elbow. Connect the water hose to the fitting on the front of the manifold. Secure both hoses with screw clamps.

WARNING
Stay clear of the propeller shaft while running the engine on a flush/test adapter. Remove the propeller(s) before running the engine to prevent injury or death. Always disconnect the battery cables before removing or installing the propeller.

CAUTION
Never run the engine without first providing cooling water. Use either a test tank or a flush/test adapter.

19. Connect the battery cable. Start the engine and immediately check for water leakage at the hose fitting, exhaust bellows and elbow-to-manifold mating surfaces. If there is leakage, tighten the hose clamps and check for damaged or improperly installed hoses or bellows. If there is leakage from the mating surfaces, remove the elbow and check for damaged gaskets or overlooked mating surface defects.

GM 4.3L Models

This model uses an exhaust manifold and elbow for each cylinder bank. Rubber exhaust tubes connect each exhaust elbow and tube to the single exhaust collector. **Figure 32** shows the exhaust system components. The exhaust manifold(s), exhaust elbow, exhaust tube and flapper valve can be replaced without removing the engine from the boat; however, the manifold requires clear access from the top and sides for removal. If the boat provides limited access due to seating or other boat structure, remove the engine before removing the manifold(s).

Removal/installation

Refer to **Figure 32** for this procedure.

1. Disconnect the negative battery cable.
2. Drain water from the exhaust manifold as described in *Winterizing* in Chapter Four.
3. Loosen the two clamps that secure the exhaust elbow outlet onto the rubber exhaust tube (6, **Figure 32**).
4. Note the orientation and then remove any clamps or brackets that attach to the exhaust elbow.
5. On later carburetor-equipped models with hydraulic lock prevention hoses, remove the plastic flame arrestor cover (**Figure 33**). Remove the thin nut and lift the hose retainers (**Figure 34**) from the flame arrestor stud.
6. Insert a blunt awl or other suitable instrument between the elbow outlet and the rubber exhaust tube (6, **Figure 32**). Carefully work the instrument around the tube to break the bond between the rubber and metal. Work carefully to avoid damaging the tube.
7. Remove the spark plugs to prevent damage if the manifold should drop during the removal or installation process.
8. Loosen the clamp and disconnect the water hose from the fitting on the front end of the manifold.
9. Provide overhead support or have an assistant support the manifold before removing the mounting fasteners.
10. Remove the mounting bolts and then carefully pull the manifold forward until the elbow outlet is clear of the rubber exhaust tube. Then lift the manifold and elbow assembly out of the boat.
11. Remove the gasket (3, **Figure 32**) from the manifold or cylinder block. Carefully scrape residual gasket material from the mating surfaces.

CAUTION
Always thoroughly clean the manifold-to-cylinder head mating surfaces after removing the manifold. Debris or residual gasket material can allow exhaust leakage.

12. If necessary, remove the exhaust manifold from the other cylinder bank as described in Steps 2-11.
13. Inspect the manifold for cracks, corrosion and other defects. Replace the manifold if these or other defects are evident.
14. Use a straightedge to check for warpage at the cylinder head mating surface. If a gap is visually apparent, measure it with a feeler gauge. Replace the manifold if the surface warp exceeds 0.006 in. (0.15 mm) within any 12 in. (30.5 cm) span.
15. If replacing the manifold, remove the exhaust elbow and install it onto the replacement manifold as described in this chapter.
16. Fit the manifold onto the cylinder head as follows:
    a. Align the bolt and port openings and then fit the new gasket (3, **Figure 32**) onto the manifold. Insert the mounting bolts into the manifold openings to help hold the gasket into position.
    b. Using overhead support or an assistant, carefully guide the exhaust elbow outlet into the rubber exhaust tube (6, **Figure 32**). Work carefully to avoid dislodging the gasket (3, **Figure 32**).
    c. Align the port and bolt openings and then seat the manifold and gasket against the cylinder head. Thread the mounting bolts into the manifold and cylinder head.
17. Starting from the center and working toward the ends, tighten the manifold bolts as follows:
    a. On 1994-1996 models, tighten the bolts to 20-26 ft.-lb. (27-35 N•m).
    b. On 1997-2000 models, tighten the bolt to 132 in.-lb. (15 N•m). Tighten the bolts a final time to 22 ft.-lb. (30 N•m).
18. Reinstall any bracket, wire or hose clamp removed for manifold removal.
19. Connect the water hose to the fitting at the end of the manifold and secure with the screw clamp.
20. Position the two hose clamps into the recesses in the rubber exhaust tube. Make sure the clamps are positioned directly over the clamping surface on the elbow. Reposition the tube if necessary. Then securely tighten the hose clamps.
21. If removed, install the manifold onto the other cylinder bank as described in Steps 16-20.
22. On carburetor-equipped models with hydraulic lock prevention hoses, fit the hose retainers (**Figure 34**) onto the flame arrestor and secure with the thin nut. Install the plastic flame arrestor cover (**Figure 33**).
23. Install the spark plugs and tighten to 22 ft.-lb. (30 N•m) if the cylinder head is new and 15 ft.-lb. (20 N•m) if the cylinder head is used.
24. Connect the spark plug leads.

WARNING
Stay clear of the propeller shaft while running the engine on a flush/test adapter. Remove the propeller(s) before running the engine to prevent injury or death. Always disconnect the battery cables before removing or installing the propeller.

CAUTION
Never run the engine without first providing cooling water. Use either a test tank or a flush/test adapter.

25. Connect the battery cable. Start the engine and immediately check for water leakage at the hose fittings, rubber exhaust tube and elbow-to-manifold mating surfaces. Tighten the hose clamps and check for damaged hoses if there is leakage from the fittings. If there is leakage from the mating surfaces, remove the elbow and check for a damaged gasket or mating surface defects.

GM 5.0L and 5.7L Models

These models use an exhaust manifold and elbow for each cylinder bank. Rubber exhaust tubes connect each exhaust elbow and tube to the single exhaust collector. **Figure 36** shows the exhaust system components. The exhaust manifold(s), exhaust elbow, exhaust tube and flapper valve can be replaced without removing the engine from the boat; however, the manifold requires clear access from the top and sides for removal. If the boat provides limited access due to seating or other boat structure, remove the engine before removing the manifold(s).

Removal/installation

Refer to **Figure 36** for this procedure.

1. Disconnect the negative battery cable.
2. Drain water from the exhaust manifold as described in *Winterizing* in Chapter Four.
3. Loosen the two clamps that secure the exhaust elbow outlet to the rubber exhaust tube (6, **Figure 36**).
4. Note the orientation and then remove any clamps or brackets that attach to the exhaust elbow.
5. On later carburetor-equipped models with hydraulic lock prevention hoses, remove the plastic flame arrestor cover (**Figure 33**). Remove the thin nut and lift the hose retainers (**Figure 34**) from the flame arrestor stud.
6. Insert a blunt awl or other suitable instrument between the elbow outlet and the rubber exhaust tube (6, **Figure 36**). Carefully work the instrument around the tube to break the bond between the rubber and metal. Work carefully to avoid damaging the tube.
7. Remove the spark plugs to prevent damage if the manifold should drop during the removal or installation process.
8. Loosen the clamp and disconnect the water hose from the fitting on the front end of the manifold.
9. Provide overhead support or have an assistant support the manifold before removing the mounting fasteners.
10. Remove the mounting bolts and then carefully pull the manifold forward until the elbow outlet is clear of the rubber exhaust tube. Then lift the manifold and elbow assembly out of the boat.
11. Remove the gasket (3, **Figure 36**) from the manifold or cylinder block surfaces. Carefully scrape residual gasket material from the surfaces.

CAUTION
Always thoroughly clean the manifold-to-cylinder head mating surfaces after removing the manifold. Debris or residual gasket material can allow exhaust leakage.

12. If necessary, remove the exhaust manifold from the other cylinder bank as described in Steps 2-11.
13. Inspect the manifold for cracks, corrosion and other defects. Replace the manifold if these or other defects are evident.
14. Use a straightedge to check for warpage at the cylinder head mating surface. If a gap is visually apparent, measure it with a feeler gauge. Replace the manifold if the surface warpage exceeds 0.006 in. (0.15 mm) within any 12 in. (30.5 cm) span.
15. If replacing the manifold, remove the exhaust elbow and install it onto the replacement manifold as described in this chapter.
16. Fit the manifold onto the cylinder head as follows:
    a. Align the bolt and port openings and then fit the new gasket (3, **Figure 36**) onto the manifold. Insert the mounting bolts into the manifold openings to help hold the gasket into position.
    b. Using overhead support or an assistant, carefully guide the exhaust elbow outlet into the rubber exhaust tube (6, **Figure 36**). Work carefully to avoid dislodging the gasket (3, **Figure 36**).
    c. Align the port and bolt openings and then seat the manifold and gasket against the cylinder head. Thread the mounting bolts into the manifold and cylinder head.
17. Starting from the center and working toward the ends, tighten the manifold bolts as follows:
    a. On 1994-1997 models, tighten the bolts to 20-26 ft.-lb. (27-35 N•m).
    b. On 1998-2000 models, tighten the bolts to 132 in.-lb. (15 N•m). Tighten the bolts a final time to 22 ft.-lb. (30 N•m).
18. Reinstall any bracket, wire or hose clamp removed for manifold removal.
19. Connect the water hose to the fitting at the end of the manifold and secure with the screw clamp.
20. Position the two hose clamps into the recesses in the rubber exhaust tube. Make sure the clamps are positioned directly over the clamping surface on the elbow. Reposition the tube if necessary. Then securely tighten the hose clamps.
21. If removed, install the manifold onto the other cylinder bank as described in Steps 16-20.
22. On carburetor-equipped models with hydraulic lock prevention hoses, fit the hose retainers (**Figure 34**) onto the flame arrestor and secure with the thin nut. Install the plastic flame arrestor cover (**Figure 33**).
23. Install the spark plugs and tighten to 22 ft.-lb. (30 N•m) if the cylinder head is new and 15 ft.-lb. (20 N•m) if the cylinder head is used.
24. Connect the spark plug leads.

WARNING
Stay clear of the propeller shaft while running the engine on a flush/test adapter. Remove the propeller(s) before running the engine to prevent injury or death. Always disconnect the battery cables before removing or installing the propeller.

CAUTION
Never run the engine without first providing cooling water. Use either a test tank or a flush/test adapter.

25. Connect the battery cable. Start the engine and immediately check for water leakage at the hose fittings, rubber exhaust tube and elbow-to-manifold mating surfaces. Tighten the hose clamps and check for damaged hoses if there is leakage from the fittings. If there is leakage from the mating surfaces, remove the elbow and check for a damaged gasket or mating surface defects.

Ford 5.0L and 5.8L Models

These models use an exhaust manifold and elbow for each cylinder bank. Rubber exhaust tubes connect each exhaust elbow and tube to the single exhaust collector. **Figure 37** shows the exhaust system components. The exhaust manifold(s), exhaust elbow, exhaust tube and flapper valve can be replaced without removing the engine from the boat; however, the manifold requires clear access from the top and sides for removal. If the boat provides limited access due to seating or other boat structure, remove the engine before removing the manifold(s).

Removal/installation

Refer to **Figure 37** for this procedure.
1. Disconnect the negative battery cable.
2. Drain water from the exhaust manifold as described in *Winterizing* in Chapter Four.
3. Loosen the two clamps that secure the exhaust elbow outlet to the rubber exhaust tube (6, **Figure 37**).
4. Note the orientation and then remove any clamps or brackets that attach to the exhaust elbow.
5. Insert a blunt awl or other suitable instrument between the elbow outlet and the rubber exhaust tube (6, **Figure 37**). Carefully work the instrument around the tube to break the bond between the rubber and metal. Work carefully to avoid damaging the tube.
6. Remove the spark plugs to prevent damage if the manifold should drop during the removal or installation process.
7. Loosen the clamp and disconnect the water hose from the fitting on the front end of the manifold.
8. Provide overhead support or have an assistant support the manifold before removing the mounting fasteners.
9. Remove the mounting bolts and then carefully pull the manifold forward until the elbow outlet is clear of the rubber exhaust tube. Then lift the manifold and elbow assembly out of the boat.
10. Remove the gasket (3, **Figure 37**) from the manifold or cylinder block. Carefully scrape residual gasket material from the mating surfaces.

CAUTION
Always thoroughly clean the manifold-to-cylinder head mating surfaces after removing the manifold. Debris or residual gasket material may allow exhaust leakage.

11. If necessary, remove the exhaust manifold from the other cylinder bank as described in Steps 2-10.
12. Inspect the manifold for cracks, corrosion and other defects. Replace the manifold if these or other defects are evident.
13. Use a straightedge to check for warpage at the cylinder head mating surface. If a gap is visually apparent, measure it with a feeler gauge. Replace the manifold if the surface warpage exceeds 0.006 in. (0.15 mm) within any 12 in. (30.5 cm) span.
14. If replacing the manifold, remove the exhaust elbow and install it onto the replacement manifold as described in this chapter.
15. Fit the manifold onto the cylinder head as follows:
   a. Align the bolt and port openings and then fit the new gasket (3, **Figure 37**) onto the manifold. Insert the mounting bolts into the manifold openings to help hold the gasket into position.
   b. Using overhead support or an assistant, carefully guide the exhaust elbow outlet into the rubber exhaust tube (6, **Figure 37**). Work carefully to avoid dislodging the gasket (3, **Figure 37**).
   c. Align the port and bolt openings and then seat the manifold and gasket against the cylinder head. Thread the mounting bolts into the manifold and cylinder head.
16. Starting from the center and working toward the ends, tighten the manifold bolts to 20-26 ft.-lb. (27-35 N•m).
17. Reinstall any bracket, wire or hose clamp removed for manifold removal.
18. Connect the water hose to the fitting at the end of the manifold and secure with the screw clamp.
19. Position the two hose clamps into the recesses in the rubber exhaust tube. Make sure the clamps are positioned directly over the clamping surface on the elbow. Reposition the tube if necessary. Then securely tighten the hose clamps.
20. If removed, install the manifold onto the other cylinder bank as described in Steps 15-19.
21. Install the spark plugs and tighten to 120-180 in.-lb. (14-20 N•m).
22. Connect the spark plug leads.

WARNING
Stay clear of the propeller shaft while running the engine on a flush/test adapter. Remove the propeller(s) before running the engine to pre-

vent injury or death. Always disconnect the battery cables before removing or installing the propeller.

CAUTION
Never run the engine without first providing cooling water. Use either a test tank or a flush/test adapter.

23. Connect the battery cable. Start the engine and immediately check for water leakage at the hose fittings, rubber exhaust tube and elbow-to-manifold mating surfaces. Tighten the hose clamps and check for damaged hoses if there is leakage from the fittings. If there is leakage from the mating surfaces, remove the elbow and check for a damaged gasket or mating surface defects.

GM 7.4L and 8.2L Models

These models use an exhaust manifold and elbow for each cylinder bank. Rubber exhaust tubes connect each exhaust elbow and tube to the single exhaust collector. **Figure 38** shows the exhaust system components. The exhaust manifold(s), exhaust elbow, exhaust tube and flapper valve can be replaced without removing the engine from the boat; however, the manifold requires clear access from the top and sides for removal. If the boat provides limited access due to seating or other boat structure, remove the engine before removing the manifold(s).

Removal/installation

Refer to **Figure 38** for this procedure.

1. Disconnect the negative battery cable.
2. Drain water from the exhaust manifold as described in *Winterizing* in Chapter Four.
3. Loosen the two clamps that secure the exhaust elbow outlet to the rubber exhaust tube (6, **Figure 38**).
4. Note the orientation and then remove any clamps or brackets that attach to the exhaust elbow.
5. Insert a blunt awl or other suitable instrument between the elbow outlet and the rubber exhaust tube (6, **Figure 38**). Carefully work the instrument around the tube to break the bond between the rubber and metal. Work carefully to avoid damaging the tube.
6. Remove the spark plugs to prevent damage if the manifold should drop during the removal or installation process.
7. Loosen the clamp and disconnect the water hose from the fitting on the front end of the manifold.
8. Provide overhead support or have an assistant support the manifold before removing the mounting fasteners.
9. Remove the mounting bolts and nuts. Pivot the bottom of the manifold away from the block until free from the studs. Then pull the manifold forward until the elbow outlet is clear of the rubber exhaust tube. Lift the manifold and elbow assembly out of the boat.
10. Remove the gasket (3, **Figure 38**) from the manifold or cylinder block. Carefully scrape residual gasket material from the mating surfaces.

CAUTION
Always thoroughly clean the manifold-to-cylinder head mating surfaces after removing the manifold. Debris or residual gasket material can allow exhaust leakage.

11. If necessary, remove the exhaust manifold from the other cylinder bank as described in Steps 3-10.
12. Inspect the manifold for cracks, corrosion and other defects. Replace the manifold if these or other defects are evident.
13. Use a straightedge to check for warpage at the cylinder head mating surface. If a gap is visually apparent, measure it with a feeler gauge. Replace the manifold if the surface warpage exceeds 0.006 in. (0.15 mm) within any 12 in. (30.5 cm) span.
14. If replacing the manifold, remove the exhaust elbow and install it onto the replacement manifold as described in this chapter.
15. Fit the manifold onto the cylinder head as follows:
    a. Align the stud and port openings and then fit the new gasket (3, **Figure 38**) onto the cylinder head studs.
    b. Pivot the lower end of the manifold away from the cylinder head while guiding the outlet end of the exhaust elbow into the rubber exhaust tube.
    c. Pivot the manifold down enough to align the studs with the openings in the manifold.
    d. Fit the manifold over the studs and then seat the manifold against the gasket and head.
    e. Thread the washers and nuts onto the studs. Then thread the bolts into the manifold and cylinder head openings. Starting from the center and working toward the ends, tighten the fasteners to 20-26 ft.-lb. (27-35 N•m) for the nuts and 24-28 ft.-lb. (33-38 N•m) for the bolts.
16. Reinstall any bracket, wire or hose clamp removed for manifold removal.
17. Connect the water hose to the fitting at the end of the manifold and secure with the screw clamp.
18. Position the two hose clamps into the recesses in the rubber exhaust tube. Make sure the clamps are positioned directly over the clamping surface on the elbow. Reposition the tube if necessary. Then securely tighten the hose clamps.
19. If removed, install the manifold onto the other cylinder bank as described in Steps 15-18.
20. Install the spark plugs and tighten to 22 ft.-lb. (30 N•m).
21. Connect the spark plug leads.

WARNING
Stay clear of the propeller shaft while running the engine on a flush/test adapter. Remove the propeller(s) before running the engine to prevent injury or death. Always disconnect the

battery cables before removing or installing the propeller.

CAUTION

Never run the engine without first providing cooling water. Use either a test tank or a flush/test adapter.

22. Connect the battery cable. Start the engine and immediately check for water leakage at the hose fittings, rubber exhaust tube and elbow-to-manifold mating surfaces. Tighten the hose clamps and check for damaged hoses if there is leakage from the fittings. If there is leakage from the mating surfaces, remove the elbow and check for a damaged gasket or mating surface defects.

Table 1 COOLING AND EXHAUST SYSTEMS TORQUE SPECIFICATIONS

Fastener	in.-lb.	ft.-lb.	N•m
Belt-driven seawater pump			
Mounting bolts	–	30	41
Crankshaft-driven seawater pump			
Pump body screws	19-25	–	2-3
Support bracket bolts	–	20-25	27-34
Exhaust elbow to manifold fasteners			
GM 3.0L models	142-168	–	16-19
GM 4.3L models	144-216	–	16-24
GM 5.0L, 5.7L, 7.4L and 8.2L models	144-216	–	16-24
Ford 5.0L and 5.8L models	144-216	–	16-24
Exhaust manifold to cylinder head fasteners			
GM 3.0L models	–	20-25	27-34
GM 4.3L models			
1994-1996	–	20-26	27-35
1997-2000			
First step	132	–	15
Final step	–	22	30
GM 5.0L and 5.7L models			
1994-1997	–	20-26	27-35
1998-2000			
First step	132	–	15
Final step	–	22	30
Ford 5.0L and 5.8L models	–	20-26	27-35
GM 7.4L and 8.2L models			
Nuts	–	20-26	27-35
Bolts	–	24-28	33-38
Recirculation pump mounting bolts			
GM 3.0L models	160-203	–	18-23
GM 4.3L, 5.0L, 5.7L, 7.4L and 8.2L models	–	30	41
Ford 5.0L and 5.8L models	120-180	–	14-20
Recirculation pump pulley			
Ford 5.0L and 5.8L models	168-240	–	19-27
Spark plug			
GM 3.0L models	–	20	27
GM 4.3L, 5.0L and 5.7L models			
New cylinder head	–	22	30
Used cylinder head	–	15	20
Ford 5.0L and 5.8L models	120-180	–	14-20
GM 7.4L and 8.2L models	–	22	30
Thermostat housing bolts	144-192	–	16-22

Chapter Nine

Electrical System

This chapter provides service procedures for the battery, charging system, starting, ignition and warning system components. Wiring diagrams are located at the end of the manual. Torque specifications are listed in **Table 1**. Use the general torque recommendations listed in Chapter One for fasteners not listed in **Table 1**. Battery capacity, battery cable size recommendations, battery charge percentage and engine firing order specifications are listed in **Tables 2-5**. **Tables 1-5** are located at the end of this chapter.

BATTERY

Batteries used in marine applications endure far more rigorous treatment than those used in automotive electrical systems. Marine batteries (**Figure 1**) have thicker exterior cases to cushion the plates during tight turns and rough water operation. Thicker plates are also used, and each one is individually fastened within the case to prevent premature failure. Spill-proof caps on the battery cells prevent electrolyte from spilling into the bilge.

Automotive batteries should be used in a boat *only* during an emergency situation when a suitable marine battery is not available.

Battery Rating Methods

The battery industry has developed specifications and performance standards to evaluate batteries and the energy potential. Several rating methods are available to provide information on battery selection.

Cold cranking amps (CCA)

Cold cranking amps is the amperage the battery can deliver for 30 seconds at 0° F (-18° C) without dropping below 1.2 volts per cell (7.2 volts on a standard 12-volt battery). The higher the number, the more amps the battery can deliver to crank the engine. CCA × 1.3 = MCA.

Marine cranking amps (MCA)

Marine cranking amps is similar to the CCA rating, except the test occurs at 32° F (0° C) instead of 0° F (-18° C). This is closer to actual operating environments. MCA × 0.77 = CCA.

Reserve capacity

Reserve capacity represents the time in minutes that a fully charged battery at 80° F (27° C) can deliver 25 amps,

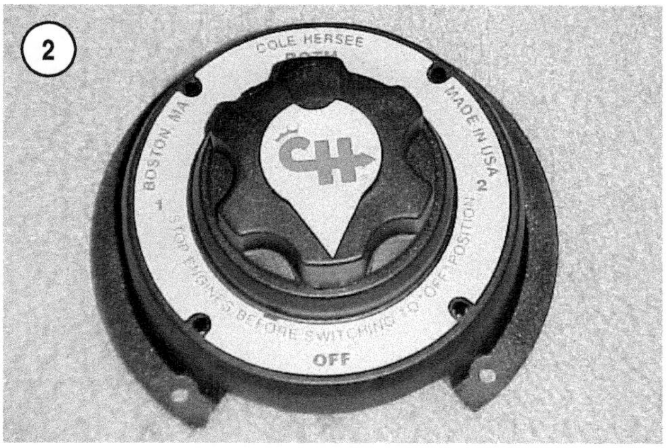

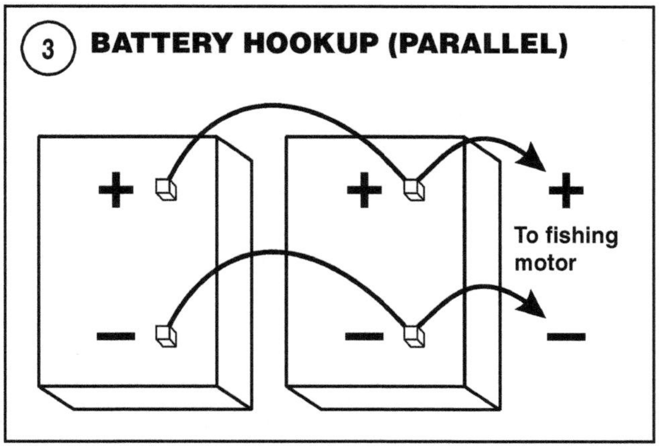

without dropping below 1.75 volts per cell (10.5 volts on a standard 12-volt battery). The reserve capacity rating defines the length of time that a typical vehicle can be operated after the charging system fails. The 25-amp rating takes into account the power required by the ignition system, lighting and other accessories. The higher the reserve capacity, the longer the vehicle could be operated after a charging system failure.

Ampere hour rating

The ampere hour rating is also called the 20-hour rating. This rating represents the steady current flow that the bat-

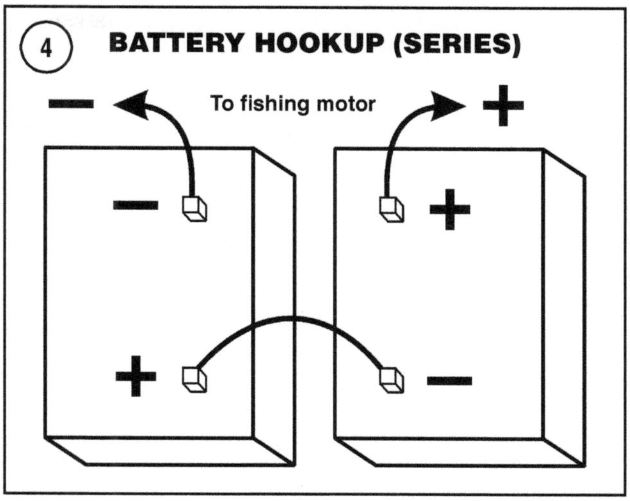

tery will deliver for 20 hours while at 80° F (27° C) without dropping below 1.75 volts per cell (10.5 volts on a standard 12-volt battery). Example: A 60 A.h. battery will deliver 3 amps continuously for 20 hours. This rating method has been largely discontinued by the battery industry. Cold cranking amps or marine cranking amps and reserve capacity are now the most common battery rating methods.

Battery Recommendations

A battery with inadequate capacity can cause hard starting or an inability to start the engine. For GM 3.0L and GM 4.3L carburetor-equipped models, use a battery with a minimum rating of 360 cold cranking amps (CCA) and 115 minutes or more of reserve capacity. For all other models covered in this manual, use a battery with a minimum rating of 650 CCA and 165 minutes or more of reserve capacity.

A battery with a capacity exceeding the minimum requirement is acceptable and is highly recommended if the boat is equipped with numerous electrical accessories. Consider adding an additional battery and installing a battery switch (**Figure 2**) on such applications. The switch allows starting and charging operations to use one or both batteries. To prevent discharge caused by some on-board accessories, the switch can be turned off if the boat is at rest or in storage.

Battery Usage

Separate batteries may be used to provide power for accessories such as lighting, fish finders, depth finders and radios. To determine the required capacity of such batteries, calculate the accessory current amperage draw and refer to **Table 2**.

Two batteries may be connected in parallel to double the ampere hour capacity while maintaining the required 12 volts. See **Figure 3**. For accessories that require 24 volts, batteries may be connected in series (**Figure 4**), but only accessories specifically requiring 24 volts should be connected to the system. If charging becomes necessary, individually disconnect and charge batteries connected in a parallel or series circuit.

Safe Battery Mounting

The battery must be securely fastened in the boat to prevent the battery from shifting or moving in the bilge area. The positive battery terminal or the entire top of the battery must be covered with a nonconductive shield or boot.

An improperly secured battery can contact the hull or metal fuel tank in rough water or while trailering the boat. If the battery terminals short against the metal hull or fuel tank, the short circuit can create sparks and cause an electrical fire. An explosion could occur if the fuel tank or battery case is compromised.

If the battery is not properly grounded and the battery contacts the metal hull, the battery will seek a ground through the control cables or the boat wiring harness. Again, the short circuit can create sparks and an electrical fire. The control cables and boat wiring can be irreparably damaged.

Observe the following preventive steps when installing a battery in a boat, especially a metal boat or a boat with a metal fuel tank.

1. Choose a mounting location that provides good access to the battery for maintenance and is as far as practical from the fuel tank. If possible, choose a location that is close enough to the engine to use the original equipment battery cable. If longer cables are required, refer to the battery cable recommendations in **Table 3**. It is always preferable to use the shortest cable possible.
2. Secure the battery to the hull with a plastic battery box and tie-down strap (**Figure 5**) or a battery tray (**Figure 6**) with a nonconductive shield or boot covering the positive battery terminal. Install an additional shield or boot onto the negative battery terminal for added protection against arcing between the terminals.
3. Make sure all battery cable connections (two at the battery and two at the engine) are clean and tight. *Do not* use wingnuts to secure battery cables. If wingnuts are present, discard them and replace them with corrosion-resistant hex nuts and lockwashers to ensure good electrical connections. Loose or dirty battery connections can cause engine malfunction and failure of expensive components.
4. Periodically inspect the installation to make sure the battery is physically secured to the hull and the battery cable connections are clean and tight.

5

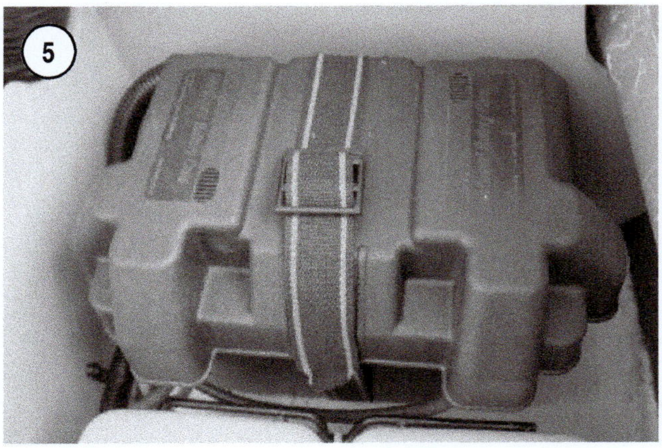

6

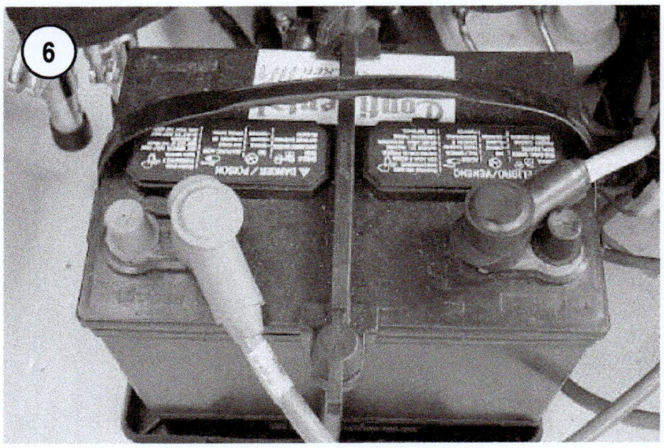

7

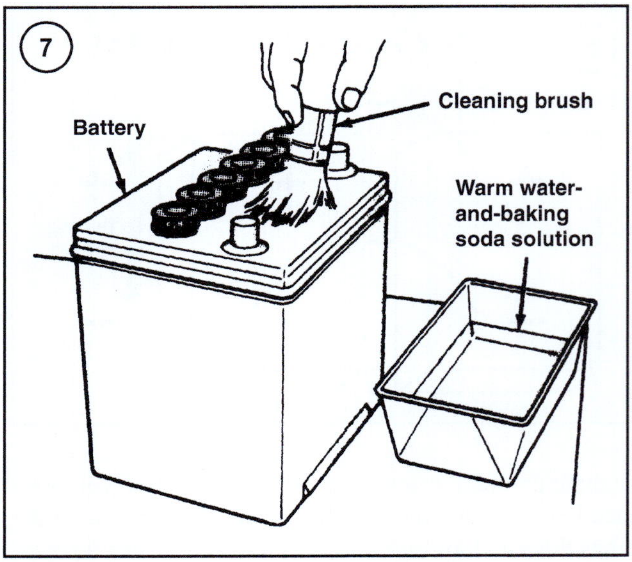

Battery Care and Inspection

For reliable starting and maximum battery life, inspect the battery case and terminals on a frequent basis.

1. Remove the battery box cover or battery tray cover. See **Figure 5** or **Figure 6**.
2. Disconnect the negative battery cable and then the positive battery cable.

NOTE
*Some batteries have a built-in lifting strap (**Figure 1**).*

3. Attach a battery carry strap to the terminal post. Remove the battery from the boat.
4. Inspect the entire battery case for cracks, holes, deep abrasion or other damage. Replace the battery if these or other defects are evident.

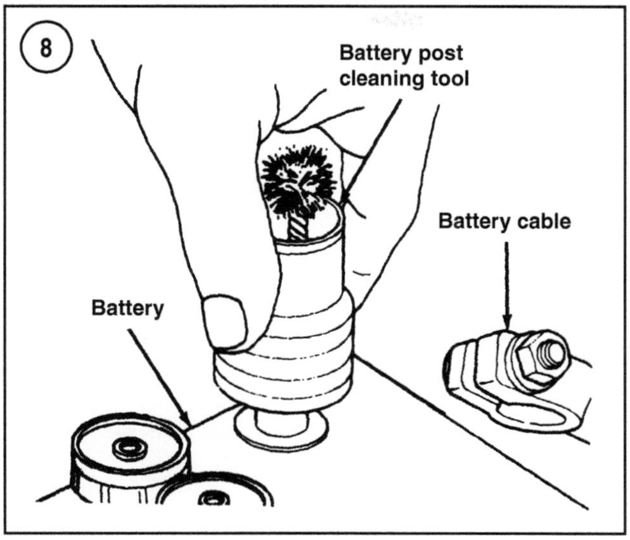

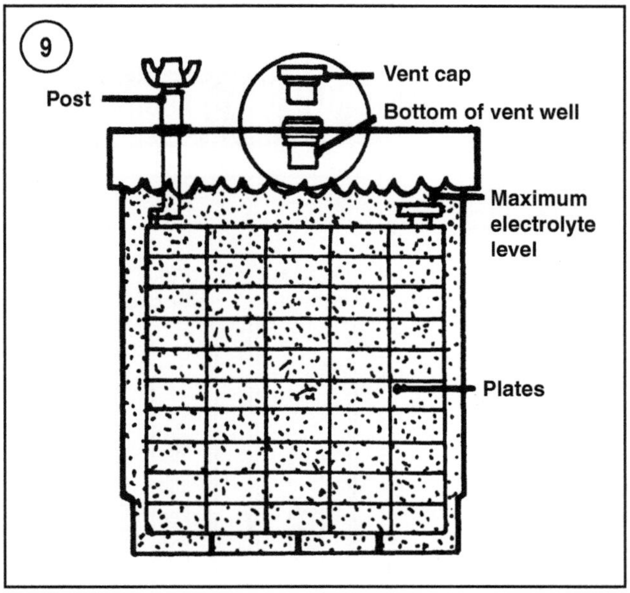

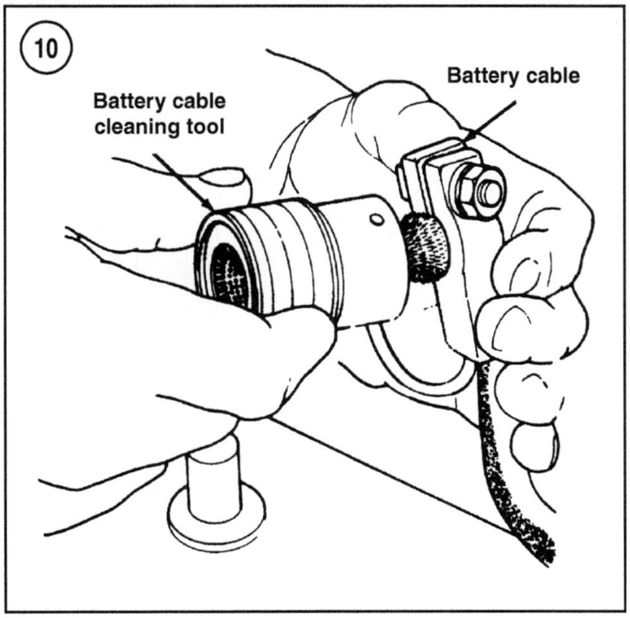

5. Inspect the battery box or battery tray for cracks, corrosion, deterioration or other damage. Replace the box or tray if these other defects are evident.

NOTE
Do not allow the baking soda solution to enter the battery cells because the electrolyte will be severely weakened.

6. Using a baking soda-and-water solution, clean the top of the battery with a stiff bristle brush (**Figure 7**). Rinse the battery case with clear water and wipe it dry with a clean cloth or paper towel.
7. Clean the battery terminal post with a stiff wire brush or battery terminal cleaning tool (**Figure 8**).

NOTE
Do not overfill the battery cells. The electrolyte expands due to heat from the charging system and will overflow if the level is more than 3/16 in. (4.8 mm) above the battery plates.

8. Remove the filler caps and check the electrolyte level. Add distilled water, if needed, to bring the level up to approximately 3/16 in. (4.8 mm) above the plates in the battery case. See **Figure 9**.
9. Clean the battery cable clamps with a stiff wire brush (**Figure 10**).
10. Place the battery back into the boat and into the battery box or tray. When using a battery tray, install and secure the retaining bracket.
11. Connect the positive battery cable first and then connect the negative cable.
12. Securely tighten the battery connections. Coat the connections with petroleum jelly or a light grease to minimize corrosion. When using a battery box, install the cover and secure the assembly with a tie down strap.

9

Battery Testing

Hydrometer test

On batteries with removable vent caps, the best way to check the state of charge is to check the specific gravity of the electrolyte with a hydrometer. Use a hydrometer with numbered graduations from 1.100-1.300 points rather than one with color-coded bands. To use the hydrometer, squeeze the rubber bulb, insert the tip fully into the cell and then release the bulb to fill the hydrometer. See **Figure 11**.

NOTE
Do not test specific gravity immediately after adding water into the battery cells because the water will dilute the electrolyte and lower the specific gravity. To obtain an accurate hydrometer reading, charge the battery after adding water and before testing with a hydrometer.

Draw enough electrolyte to raise the float inside the hydrometer. When using a temperature-compensated hydrometer, discharge the electrolyte back into the battery cell and repeat the process several times to adjust the temperature of the hydrometer to the electrolyte.

Hold the hydrometer upright and note the number on the float that is even with the surface of the electrolyte (**Figure 12**). This number is the specific gravity for the cell. Discharge the electrolyte into the cell from which it came.

The specific gravity of a cell is the indicator of the state of charge of the cell. A fully charged cell will read 1.260 or higher at 80° F (27° C). A cell that is 75 percent charged will read from 1.220-1.230 while a cell with a 50 percent charge will read from 1.170-1.180. A cell reading 1.120 or lower is discharged. All cells should be within 30 points specific gravity of each other. If there are more than 30 points of variation, the battery condition is questionable. Charge the battery and recheck the specific gravity. If 30 points or more of variation remain between the cells after charging, the battery has failed and must be replaced. Refer to **Table 4** for battery charge level based on specific gravity readings.

NOTE

*If a temperature-compensated hydrometer is **not** used, add 4 points of specific gravity to the actual reading for every 10° F (6° C) above 80° F (27° C). Subtract 4 points of specific gravity for every 10° F (6° C) below 80° F (27° C).*

Open-circuit voltage test

On sealed or maintenance-free batteries, check the state of charge by measuring the open-circuit (no load) voltage of the battery. For best results, use a digital voltmeter. For the most accurate results, allow the battery to rest for at least 30 minutes to allow the battery to stabilize. Then, observing the correct polarity, connect the meter to the battery terminals and note the meter reading. If the open-circuit voltage is 12.7 volts or higher, the battery is fully charged. A reading of 12.4 volts means the battery is approximately 75 percent charged, a reading of 12.2 means the battery is approximately 50 percent charged, and a reading of 12.1 means the battery is approximately 25 percent charged.

Load testing

Two common methods are used to load-test batteries. A load tester (**Figure 13**) measures the battery voltage as it applies a load across the terminals. If a load tester is not available, measure the cranking voltage by following the instructions in this section.

1. Attach a voltmeter to the battery terminals as shown in **Figure 14**.
2. To prevent accidental starting, remove and ground the spark plug lead onto the power head.

11

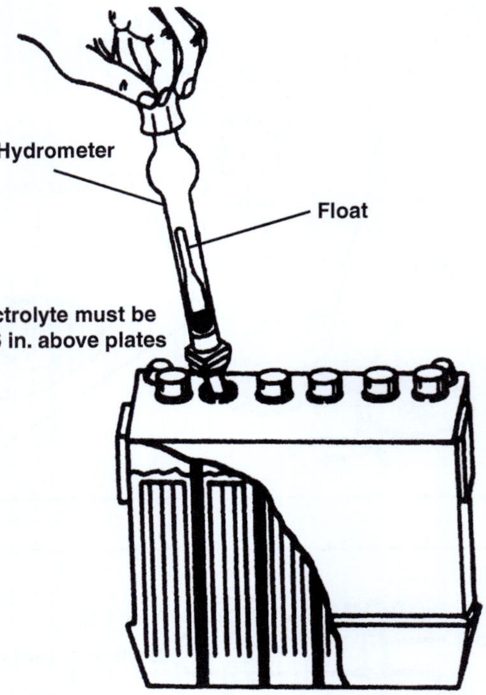

12

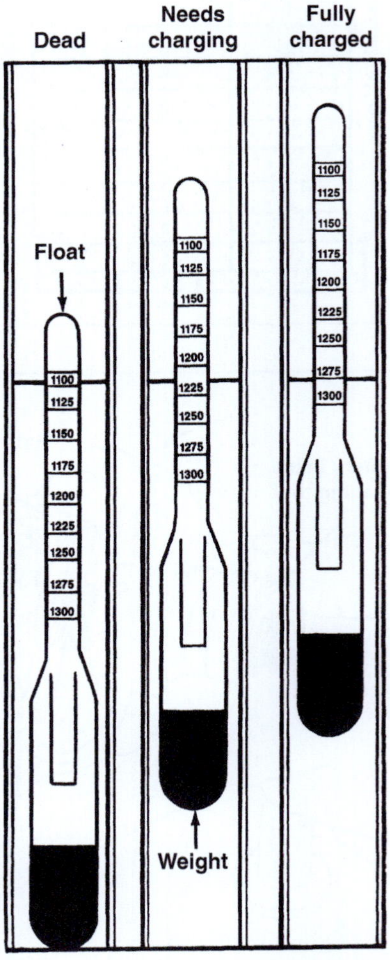

13

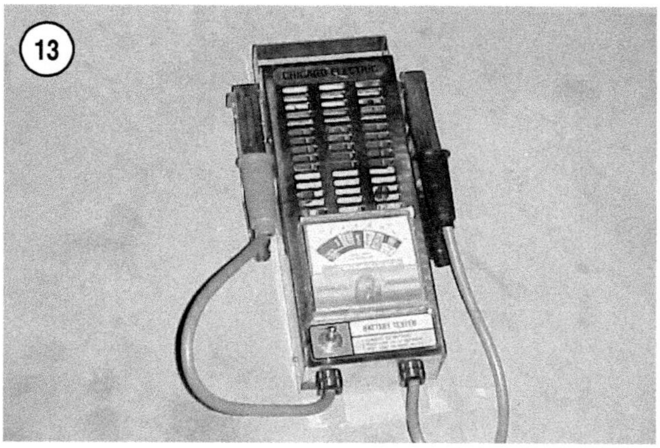

14

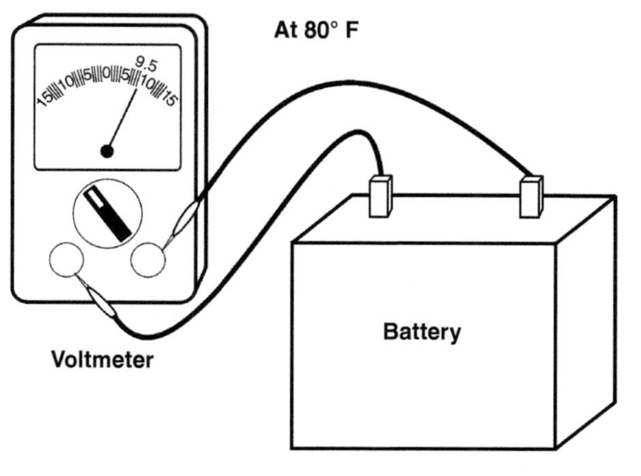

3. Crank the engine for approximately 15 seconds while noting the meter reading. Note the voltage reading at the end of the 15-second period.

4A. If the voltage is 9.5 volts or higher, the battery is sufficiently charged and of sufficient capacity for the outboard.

4B. If the voltage is below 9.5 volts, one of the following conditions is present.

   a. The battery is discharged or defective. Charge the battery and retest.
   b. The battery capacity is too small for the engine. Refer to *Battery Recommendations* in this chapter.
   c. The starting system is drawing excessive current and causing the battery voltage to drop. Refer to Chapter Two for starting system troubleshooting procedures.
   d. A mechanical defect is creating excessive load and current draw on the starting system.

Battery Storage

Wet cell batteries slowly discharge when stored. Before storing a battery, clean the case with a solution of baking soda and water. Rinse it with clear water and wipe dry. Fully charge the battery and then store it in a cool, dry location. Check the electrolyte and state of charge frequently during storage. If the specific gravity falls to 40 points or more below full charge (1.260), or if the open circuit voltage falls below 12.4 volts, recharge the battery.

Battery Charging

Check the state of charge with a hydrometer or digital voltmeter as described in the previous section.

Remove the battery from the boat for charging. A charging battery releases highly explosive hydrogen gas. In many boats, the area around the battery is not ventilated well unless the vessel is moving, and the gas can remain in the area for hours after the charging process has completed. Sparks or flames occurring near the battery can cause it to explode and spray battery acid over a wide area.

If the battery cannot be removed for charging, make sure the bilge access hatches, doors or vents are fully open to allow adequate ventilation. Observe the following precautions when charging batteries:

1. Never smoke in close proximity to a battery.
2. Make sure all accessories are turned off before disconnecting the battery cables. Disconnecting a circuit that is electrically active will create a spark that can ignite explosive gas that might be present.
3. Always disconnect the negative battery cable first and then the positive cable.
4. On batteries with removable vent caps, always check the electrolyte level before charging the battery. Maintain the correct level throughout the charging process.
5. Never attempt to charge a battery that is frozen.

WARNING
Be extremely careful not to create any sparks around the battery when connecting the battery charger.

6. Connect the negative charger lead to the negative battery terminal and the positive charger lead to the positive terminal. If the charger output is variable, select a setting of approximately 4 amps. Charge the battery slowly at low amp settings, rather than quickly at high amp settings.
7. If the charger has a dual voltage setting, set the voltage to 12 volts and then switch the charger *ON*.
8. If the battery is severely discharged, allow it to charge for at least 8 hours. Check the charging process with a hydrometer.

Jump-Starting

If the battery becomes severely discharged, the engine can be jump-started from another battery in or out of the boat. Jump-starting can be dangerous if the proper procedure is not followed. Always use caution when jump-starting.

Check the electrolyte level of the discharged battery before attempting the jump start. If the electrolyte is not visible or it appears to be frozen, do not jump-start the discharged battery.

WARNING
To avoid personal injury or damage to the electrical system, use extreme caution when connecting the booster battery to the discharged battery. ***Make sure*** *the jumper cables are connected to the correct polarity.*

1. Connect the jumper cables in the order shown in **Figure 15**.

WARNING
An electrical arc can occur when the final connection is made. This could cause an explosion if it occurs near the battery. For this reason, make the final connection to a good engine ground away from the battery and not to the battery itself.

2. Make sure all jumper cables are out of the way of moving engine parts.

WARNING
Stay clear of the propeller and propeller shaft while running an engine on a flush adapter.

CAUTION
If running the engine with the boat out of the water, use a flush device or other means to supply the engine with cooling water. The seawater pump is quickly damaged if operated without adequate cooling water.

3. Start the engine. Once it starts, run it at a moderate speed (fast idle).

CAUTION
Running the engine at high speed with a discharged battery can damage the charging system.

4. Remove the jumper cables in the exact reverse of the order shown in **Figure 15**.

15

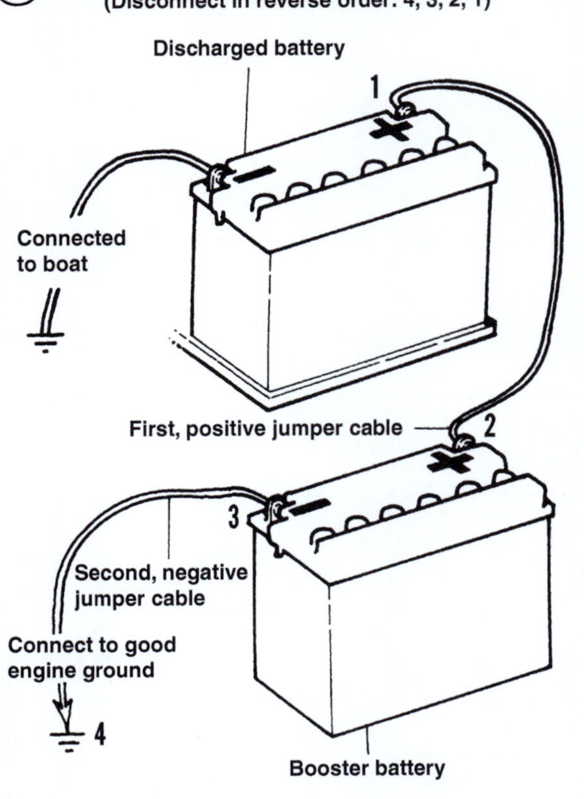

16

ALTERNATOR

NOTE
Alternator repair for the home mechanic is neither practical or economical. In some cases, the necessary bushings, bearings or other components are not available separately from OMC. If testing indicates a fault with the alternator, install a new or rebuilt marine alternator. Make sure, however, that the new or rebuilt alternator is an exact replacement for the one removed.

Isolate the cause of failure before installing the replacement alternator. Alternator failure is usually caused by exposure to water or faulty wiring and connections. Dirty or corroded battery terminals are a common cause of alternator failure. Another common cause is disconnecting the battery cables/wiring or turning the battery switch while the engine is operating. The alternator output and ground terminals must have a good circuit to the battery terminals. Wiring, fuses, switches and block ground connections are incorporated into the circuit. To prevent repeat failures, refer to the wiring diagrams at the end of the manual and check the entire circuit for faulty wiring, terminals, fuses or circuit breakers.

Drive belt removal, installation and tension adjustment is described in the alternator removal and installation procedures.

17

18

19

Removal

1. Disconnect the negative battery cable.

2. Note the belt routing around all of the pulleys prior to removing the belt. Depending on the model, one or more of the other belts must also be removed. Refer to Chapter Eight for the seawater pump belt removal procedure. Refer to Chapter Thirteen for the power steering belt removal procedure. If necessary, make a sketch indicating the belt routing around *all* of the pulleys.

3. Loosen but *do not* remove the pivot (C, **Figure 16**, typical) and adjusting bracket bolt (B). Carefully pivot the alternator toward the crankshaft and recirculation pump pulleys.

4. Slip the belt (A, **Figure 16**, typical) off the power steering pulley. Inspect the belt for wear, damage or deterioration and replace as needed.

5. Support the alternator and then remove the fasteners attaching the alternator to the mounting bracket. Carefully pull the alternator away from the bracket. Support the alternator until the wiring is disconnected in the next step.

6. Make a sketch of the wire terminal orientation and connections on the alternator (**Figure 17**, typical). Disconnect the engine wire harness from the alternator. Remove the alternator from the engine. If the alternator will be rebuilt, remove the jumper strap or wire (**Figure 18**) from the alternator. The strap or wire is often lost during the rebuild.

7. Clean rust or corrosion from the pulley surfaces. Rust or corrosion pitting on the belt contact surfaces will allow slippage and quickly wear out a new belt. Clean, sand and repaint the pulley as necessary. Replace the pulley if a smooth belt surface cannot be achieved by sanding and painting.

Installation

1. Connect each of the wire harness terminals to the corresponding terminals on the back of the alternator (**Figure 17**, typical). Rotate the terminals to prevent contact with other terminals or inadvertent contact with the alternator housing. Install and securely tighten the terminal nuts. If removed, install the jumper strap or wire (**Figure 18**) onto the alternator. To prevent terminal corrosion and possible repeat alternator failure, thoroughly coat the terminals with OMC Black Neoprene Dip. Purchase the neoprene dip (OMC part No. 909570) from an OMC dealership.

2. Carefully mount the alternator into the mounting bracket. Install and tighten the mounting fasteners by hand.

3. Pivot the alternator toward the crankshaft and recirculation pump pulleys amd then slip the drive belt (A, **Figure 16**) onto pulleys. Make sure the belt is not twisted and fits into the groove in each pulley.

4. Pivot the alternator away from the crankshaft and recirculation pump pulleys to apply tension on the drive belt. When properly tensioned, the belt will deflect approximately 1/4-1/2 in. (6-13 mm) with thumb pressure applied at a point midway between the pulleys (**Figure 19**).

5. Hold the alternator in position to maintain belt tension and tighten the mounting bolts to 26-30 ft.-lb. (35-41 N•m).

WARNING

Stay clear of the propeller and propeller shaft while running an engine on a flush adapter.

CAUTION

If running the engine with the boat out of the water, use a flush device or other means to supply the engine with cooling water. The seawater pump is quickly damaged if operated without adequate cooling water.

NOTE
Carburetor-equipped models are equipped with an electric choke. The design of the alternator can prevent charging output until the idle speed is raised to approximately 1500 rpm for a few seconds after starting. This allows the internal circuits to build enough field strength to switch on the alternator and begin charging. The alternator will subsequently provide charging output at normal idle speed.

6. Connect the battery. Start the engine and check the dash-mounted voltmeter for alternator output. The gauge should indicate 12.6-13.8 volts at idle speed. If the engine is equipped with a carburetor and electric choke heater, raise the NEUTRAL idle speed to approximately 1500 rpm for a few seconds to switch on the alternator. Check for a slipping belt and faulty wiring or terminals if the voltmeter does not indicate charging. If the belt is slipping, check for a faulty belt, incorrect belt routing, dirty or damaged pulleys, incorrect belt tension adjustment, or seizure of the water pump or power steering pump.
7. Stop the engine.
8. If the belt is new, seat the belt and readjust the tension as follows:
   a. Start the engine and allow it to run for approximately 10 minutes at idle speed.
   b. Stop the engine and disconnect the negative battery cable.
   c. Apply thumb pressure at a point midway between the pulleys (**Figure 19**). The belt must deflect approximately 1/4-1/2 in. (6-13 mm). If necessary, loosen the mounting bolts (B and C, **Figure 16**, typical) and pivot the alternator to correct the tension. Hold the alternator in position and tighten the mounting bolts to 26-30 ft.-lb. (35-41 N•m).
   d. Connect the battery cable.

STARTER AND STARTER SOLENOID

This section describes removal, repair and installation of the starter and replacement of the starter solenoid. Although the starter solenoid can be replaced without removing the starter from the engine, it is usually easier to first remove the starter. Consider replacing the starter solenoid with the starter installed only if you have clear access to the solenoid wire terminals and mounting screws.

Earlier GM models use the Delco electric coil frame starter (**Figure 20**) that is equipped with an electric coil frame and an armature shaft drive system.

Later GM models use a Delco PMGR starter (**Figure 21**) that is equipped with a permanent magnet frame and a planetary gear reduction drive system. The Delco PMGR starter is much smaller and lighter than the previous Delco starter.

20

21

22

The starter solenoid is integrated into the starter housing, and the housing assembly must be replaced if the solenoid fails.

The latest production GM models use a Delco PG260 starter (**Figure 22**) that is also equipped with a permanent magnet frame and a planetary gear reduction drive system. The PG260 starter is slightly smaller than the Delco PMGR. Unlike the Delco PMGR starter, the starter solenoid can be replaced separately on the PG260 starter.

All of the Ford 5.0L and 5.8L models covered in this manual, use a Motorcraft marine starter (**Figure 23**) that is equipped with a permanent magnet frame and a planetary gear reduction drive system that are very similar to those on

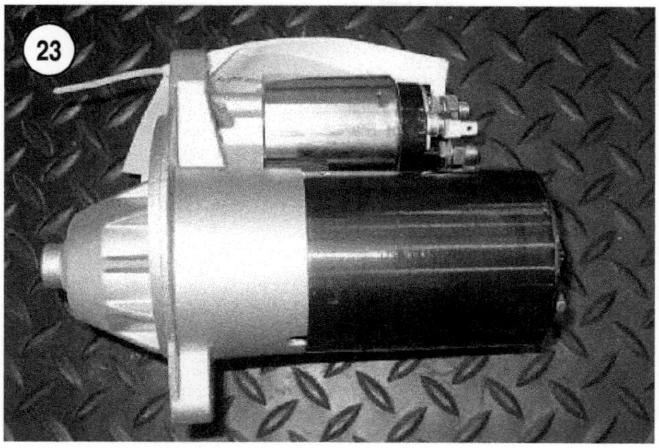

the Delco PG260 starter. The starter solenoid can be replaced separately on this starter.

On GM models, the later-style starters may be used to replace earlier-style starters used on some models. Always refer to **Figures 20-22** and identify the starter before performing any service procedure.

Starter removal and installation is a fairly simple operation provided you have good access to the mounting bolts. In some instances, engine removal is required to gain access to the bolts. Engine removal and installation are described in Chapter Six. The manufacturer does not offer internal replacement parts for the Delco PMGR and PG260 starters. If internal components require replacement, purchase a new or rebuilt marine starter or check with a reputable marine starter and alternator repair shop for a source of suitable aftermarket replacement parts. Internal parts are available for the Delco electric coil frame starter and the Motorcraft marine starter.

CAUTION
Some GM engines use a shim between the starter housing and the cylinder block. If so equipped, the shim must be reinstalled to provide adequate pinion gear-to-flywheel clearance. Otherwise, the starter will be noisy, wear rapidly or fail. Inadequate clearance can cause rapid wear and eventual failure of the flywheel teeth as well as starter failure.

Removal

1. Disconnect the negative battery cable.
2. Note the wire routing and connection points. Then disconnect all wire terminals (A, B and C, **Figure 24**) and the terminal cap (if so equipped) from the starter solenoid.
3. On models using a support brace on the starter (**Figure 25**), loosen the nut (A) that secures the starter brace to the starter throughbolt. Loosen the bolt that secures the upper end of the brace to the cylinder block and then swing the starter brace (B, **Figure 25**) away from the starter.
4. Support the starter while removing the two mounting bolts located near the flywheel. Back the nose of the starter out of the flywheel housing and then remove the starter. Retrieve the shim (if so equipped) from the starter or cylinder block mating surface.
5. Clean corrosion or debris from the starter mating surfaces. Use the proper thread chaser to clean and dress the mounting bolt holes.

Installation

1. Guide the nose of the starter into the flywheel housing. Align the starter with the cylinder block mounting holes. Insert the shim (if so equipped) between the starter and the cylinder block. Align the starter and cylinder block mounting bolt holes.
2. Thread the mounting bolts into the starter and cylinder block by hand. Do not tighten the bolts at this time. On GM models, the mounting bolts thread in from the bottom. On Ford models, the mounting bolts thread in from the side.
3. On models using a support brace on the starter, swing the brace (B, **Figure 25**) toward the starter and position the opening in the brace over the starter through-bolt stud. Install the spacer, washer and nut. Do not tighten the nut at this time.
4. Connect the yellow/red wire terminal to the solenoid terminal (B, **Figure 24**) next to the *S* molded into the solenoid housing. Install and securely tighten the terminal nut. Some Motorcraft starters use push-on terminals in place of the ring terminal and nut.

9

26

DELCO STARTER COMPONENTS (WITH ELECTRIC COIL FRAME AND ARMATURE DRIVE SYSTEM)

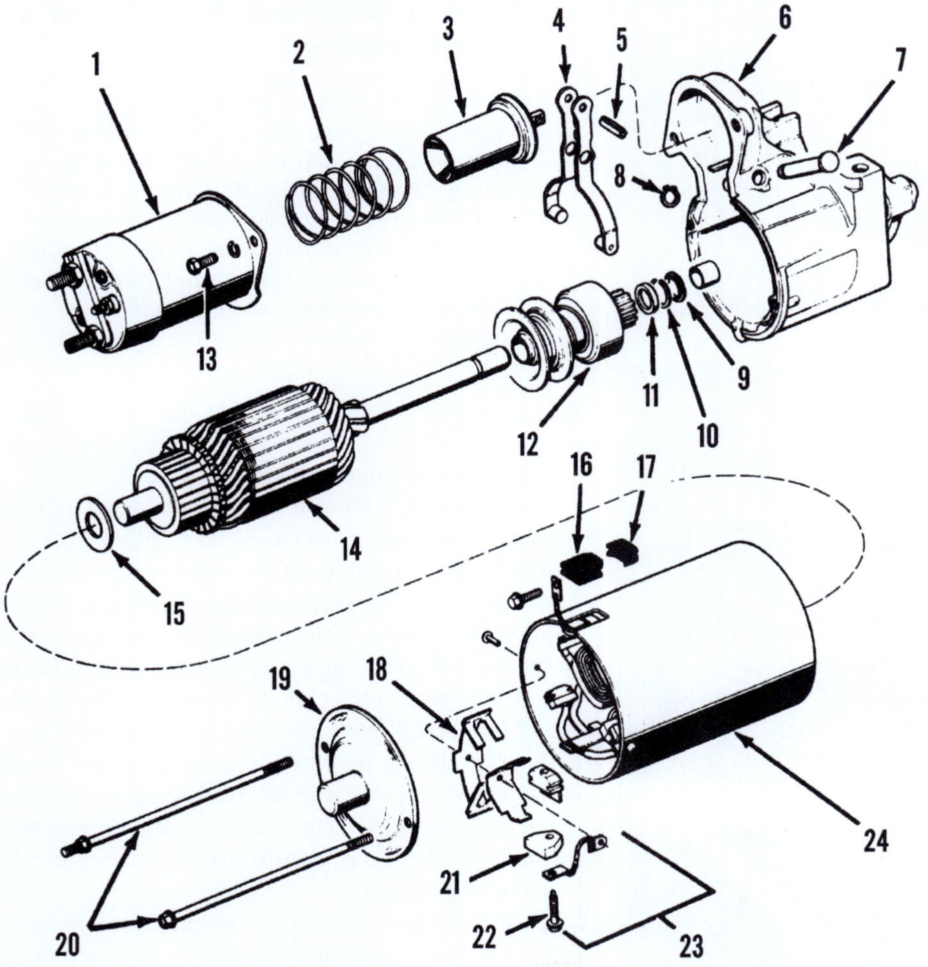

1. Starter solenoid
2. Plunger return spring
3. Plunger
4. Shift lever
5. Plunger pin
6. Starter housing
7. Shift lever pin
8. Retainer
9. Thrust collar
10. Lockring
11. Collar
12. Pinion drive
13. Screw and washer
14. Armature
15. Washer
16. Grommet
17. Grommet
18. Brush holder
19. End cover
20. Throughbolt
21. Brush
22. Screw
23. Brush holder assembly
24. Electric coil frame assembly

5A. On GM 3.0L models, install the terminal cap over the solenoid terminal (C, **Figure 24**) next to the *I* molded into the solenoid. If the cap is not available, use another means to completely insulate the terminal. Use a heavy wrap of electrical tape if necessary.

5B. On GM V-6 and V-8 models, connect the purple wire terminal to the solenoid terminal (C, **Figure 24**) next to the *I* molded into the solenoid. Install and securely tighten the terminal nut.

6. Connect the battery cable and, where applicable, the engine harness battery feed wiring to the large solenoid terminal (A, **Figure 24**). Thread the nut onto the terminal. Inspect the cable terminal to make sure it does not contact other terminals or the starter case. Reposition the cable as needed and then securely tighten the nut. Do not overtighten the nut. Overtightening can crack the solenoid housing.

7. Position the battery cable insulator over the cable end and terminal nut. To prevent terminal corrosion, thoroughly coat the terminals with OMC Black Neoprene Dip. Purchase the neoprene dip (OMC part No. 909570) from an OMC dealership.

8. On GM models, evenly tighten the two starter mounting bolts to 30-36 ft.-lb. (41-49 N•m). On Ford 5.0L and 5.8L models, tighten the bolts to 24-30 ft.-lb. (33-41 N•m).

9. On models using a support brace on the starter, securely tighten the starter brace nut (A, **Figure 25**) and the bolt that secures the brace to the cylinder block.

WARNING

Stay clear of the propeller and propeller shaft while running an engine on a flush adapter.

CAUTION

If running the engine with the boat out of the water, use a flush device or other means to supply the engine with cooling water. The seawater pump is quickly damaged if operated without adequate cooling water.

10. Connect the battery cable. Check for proper starter operation. Refer to *Starting System* in Chapter Two if the starter fails to operate or operates improperly.

Starter Solenoid Removal and Installation

NOTE
The starter solenoid cannot be replaced separately on the Delco PMGR starter. If the solenoid fails, disassemble the starter as described in this section and replace the starter housing and solenoid assembly.

Delco starter with electric coil frame

Refer to **Figure 26** for this procedure.

1. Remove the starter as described in this section.
2. Remove the screw from the solenoid terminal (A, **Figure 27**) and then pull the spacer (B) from between the solenoid terminal and brush holder terminal.
3. Remove the two screws and washers (13, **Figure 26**). Then while pushing the solenoid toward the starter housing to collapse the spring, rotate the solenoid to disengage the solenoid flange from the slot in the starter housing.
4. Carefully relieve the pressure and remove the solenoid and spring from the starter (**Figure 28**). Inspect the spring for corrosion damage, bent loops or other damage and replace as needed.
5. Inspect the plunger for corrosion pitting or other imperfections. Remove the pin and replace the plunger if imperfections cannot be removed by polishing with emery paper. Secure the replacement plunger to the shift lever with the pin. Replace the pin if it does not fit securely into the plunger and lever openings.
6. Guide the return spring into the solenoid opening.
7. Align the solenoid with the plunger and then rotate the solenoid to position the solenoid flange away from the slot in the starter housing. Guide the plunger into the spring and solenoid opening and then seat the solenoid against the starter housing.
8. Maintain pressure to collapse the spring and then rotate the solenoid until the flange enters the slot in the starter housing. Release the pressure. The flange engagement in the slot will hold the solenoid in place.
9. Rotate the solenoid slightly to align the openings and then thread the two screws and washers (13, **Figure 26**) into the solenoid and starter housing. Securely tighten the screws.
10. Fit the spacer (B, **Figure 27**) between the solenoid terminal and brush holder terminal. Thread the screw into the brush holder terminal, spacer and starter solenoid terminal and securely tighten.
11. Install the starter as described in this section.

Delco PG260 starter

Refer to **Figure 29** for this procedure.

1. Remove the starter as described in this chapter.
2. Remove the nut (**Figure 30**) and then disconnect the brush plate lead (**Figure 31**) from the solenoid terminal.
3. Remove the three inverted Torx screws (**Figure 32**) from the starter housing and solenoid.
4. Lift the plunger out of the slot in the pinion yoke (19, **Figure 29**) and then remove the starter solenoid (**Figure 33**).
5. Clean debris or corrosion from the solenoid mating surfaces.
6. Engage the plunger of the solenoid into the slot in the pinion yoke. Seat the body of the solenoid against the starter housing (17, **Figure 29**).
7. Rotate the solenoid to position the *R* terminal (B, **Figure 24**) toward the side of the starter that faces the cylinder block when installed.
8. Rotate the solenoid on the mount to align the openings and then install the three screws (**Figure 33**). Securely tighten the screws.
9. Attach the brush plate lead to the bottom terminal (**Figure 31**) of the solenoid. Install the nut (**Figure 30**) and tighten securely. Do not overtighten the nut. Overtightening can crack the solenoid housing.
10. Install the starter as described in this chapter.

(29)

DELCO PG260 STARTER COMPONENTS (WITH PERMANENT MAGNET FRAME AND REPLACEABLE SOLENOID)

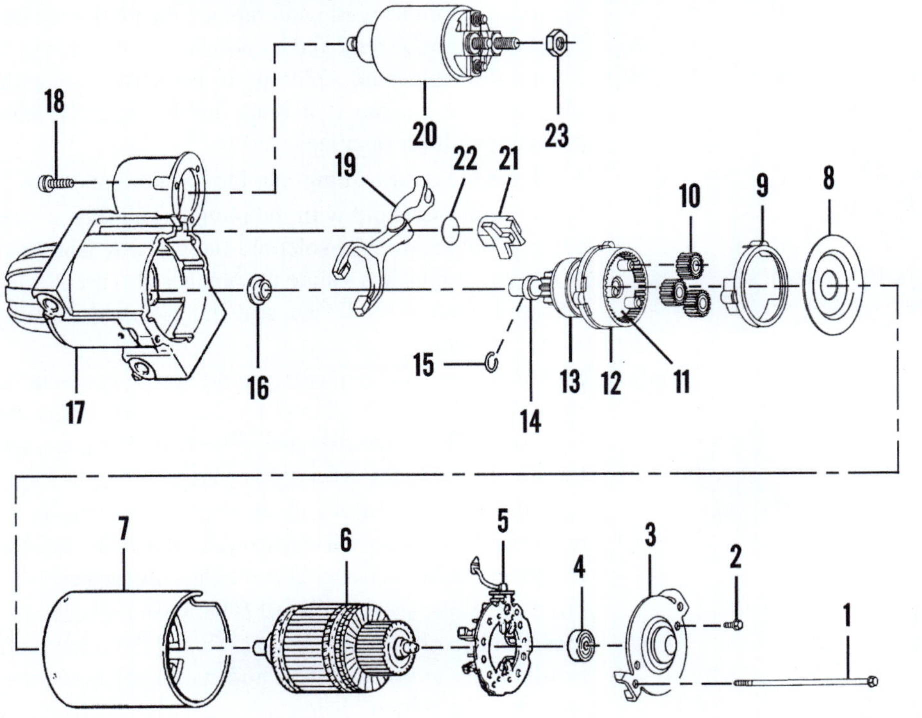

Motorcraft starter

Refer to **Figure 34** for this procedure

1. Remove the starter as described in this chapter.

2. Remove the nut (**Figure 30**) and then disconnect the brush plate lead (**Figure 31**) from the solenoid terminal.

3. Remove the two screws (6, **Figure 34**) from the starter housing and solenoid.

4. Lift the plunger out of the slot in the pinion yoke (3, **Figure 34**) and then remove the starter solenoid (**Figure 33**).

5. Clean debris or corrosion from the solenoid mating surfaces.

6. Engage the plunger of the solenoid into the slot in the pinion yoke. Seat the body of the solenoid against the starter housing (5, **Figure 34**).

7. Rotate the solenoid to position the *R* terminal (B, **Figure 24**) toward the side of the starter that faces the cylinder block when installed.

8. Rotate the solenoid on the mount to align the openings and then install the two screws (6. **Figure 34**). Tighten the solenoid mounting screws to 45-85 in.-lb. (5-10 N•m).

9. Attach the brush plate lead to the bottom terminal (**Figure 31**) of the solenoid. Install the terminal nut (**Figure 30**) and tighten to 80-120 in.-lb. (9-14 N•m).

10. Install the starter as described in this chapter.

(30)

(31)

32

33

34

FORD MARINE STARTER
(PERMANENT MAGNET FRAME AND PLANETARY GEAR REDUCTION)

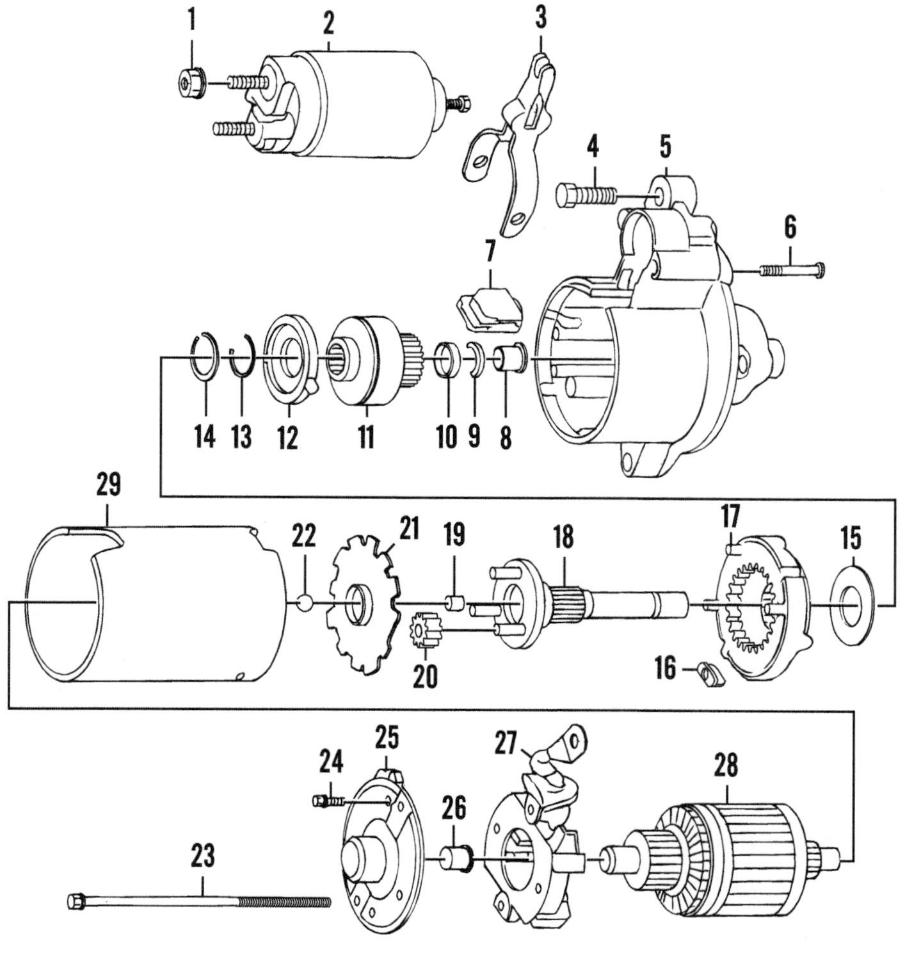

1. Nut
2. Starter solenoid
3. Pinion yoke
4. Starter mounting bolt
5. Starter housing
6. Screw
7. Seal
8. Bearing
9. Stop ring
10. Retaining ring
11. Pinion drive
12. Yoke collar
13. Lockring
14. E-ring
15. Thrust washer
16. Wedge block
17. Planetary gear housing
18. Pinion/planetary shaft
19. Ball bearing
20. Planetary gear
21. Plate
22. Armature thrust bearing
23. Throughbolt
24. Screw
25. End plate
26. Bearing
27. Brush plate
28. Armature
29. Frame assembly

Starter Disassembly and Assembly

Delco starter with electric coil frame

Apply a light coat of OMC EP/Wheel Bearing Grease or other suitable water-resistant grease to the bushings, helical splines and armature bearing surfaces during assembly. Purchase the grease from an OMC dealership. Do not apply grease or other lubricants to the brushes or commutator portion of the armature.

Refer to **Figure 26** for this procedure.

1. Remove the starter and then the starter solenoid as described in this section.

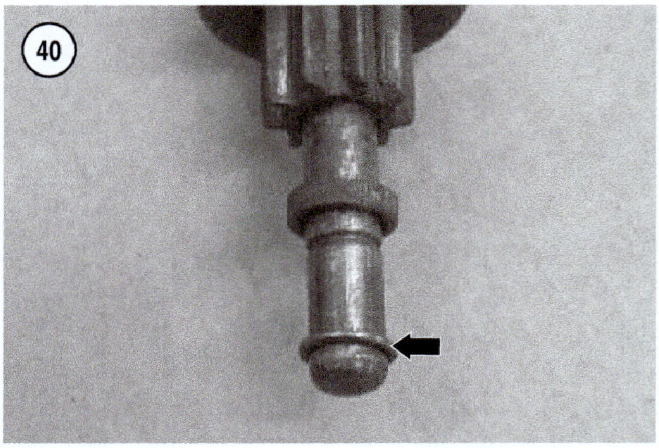

2. Make match marks on the end cover, frame and starter housing (**Figure 35**).

3. Note the location of the throughbolt and stud bolt (**Figure 36**) prior to removal. The stud bolt must be installed into the correct opening during assembly to align with the starter brace.

4. Remove the stud bolt and throughbolt. Then carefully pry the end cover off the starter frame (**Figure 37**).

5. Carefully pull the frame off the armature (**Figure 38**). Remove the washer (15, **Figure 26**) from the commutator end of the armature.

6. Tilt the armature in the housing to disengage it from the shift lever (4, **Figure 26**). Then pull the armature out of the starter housing (**Figure 39**).

7. Remove the thrust collar (9, **Figure 26**) from the armature drive shaft.

8. Using a punch, drive the collar toward the pinion gear until the lockring is exposed. Carefully pry the lockring out of the groove in the armature shaft (**Figure 40**). Then pull the lockring and collar off the armature shaft. Rotate the pinion drive counterclockwise to disengage it from the helical splines and then pull the pinion drive off the armature (**Figure 41**).

9. Remove the grommets (16 and 17, **Figure 26**) from the brush holder terminal and armature slot. Replace the grommets if torn, deteriorated or in questionable condition.

10. Remove the pins (**Figure 42**), then carefully pull the brush holders (18, **Figure 26**) out of the frame. Note the

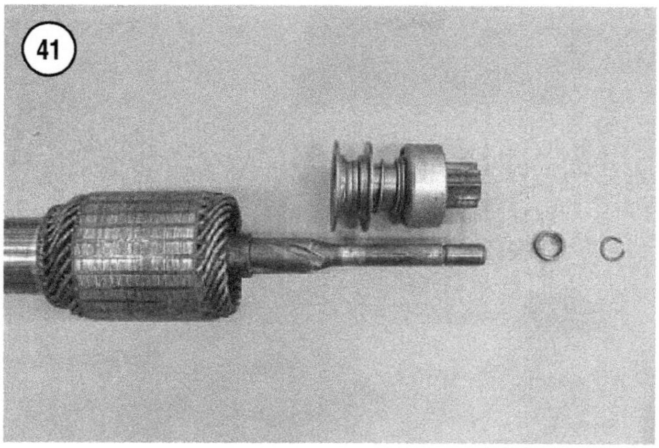

brush lead routing and connection points, then remove the screws (22, **Figure 26**) that retain the brush leads onto the brushes (21).

11. Remove the retainer (8, **Figure 26**) and then pull the lever pin (7) and shift lever (4) out of the starter housing.

12. Remove the pin (5, **Figure 26**) and then pull the plunger (3) out of the shift lever.

13. Clean and inspect the starter components as described in this section.

14. Fit the plunger shaft into the slot in the shift lever (4, **Figure 26**). Align the openings and then insert the pin (5, **Figure 26**) through the corresponding openings in the lever and plunger. Drive the pin into the openings until it reaches the same depth on each side of the lever.

15. Fit the shift lever (4, **Figure 26**) into the starter housing (7) as shown in **Figure 43**. Guide the shift lever pin (7, **Figure 26**) through the openings in the starter housing and shift lever yoke. Install the retainer (8, **Figure 26**) into the groove to retain the pin.

16. Apply grease to the helical splines of the armature shaft and then thread the pinion drive clockwise onto the splines.

17. Slide the collar over the shaft with the open end facing away from the pinion gear (**Figure 44**). Carefully fit the lockring onto the end of the shaft (**Figure 40**). Work the ring down the shaft and into the groove (**Figure 45**). Use two pliers to pull the collar fully over the lockring (**Figure 46**) . The ring must fully seat into the collar (**Figure 47**).

47

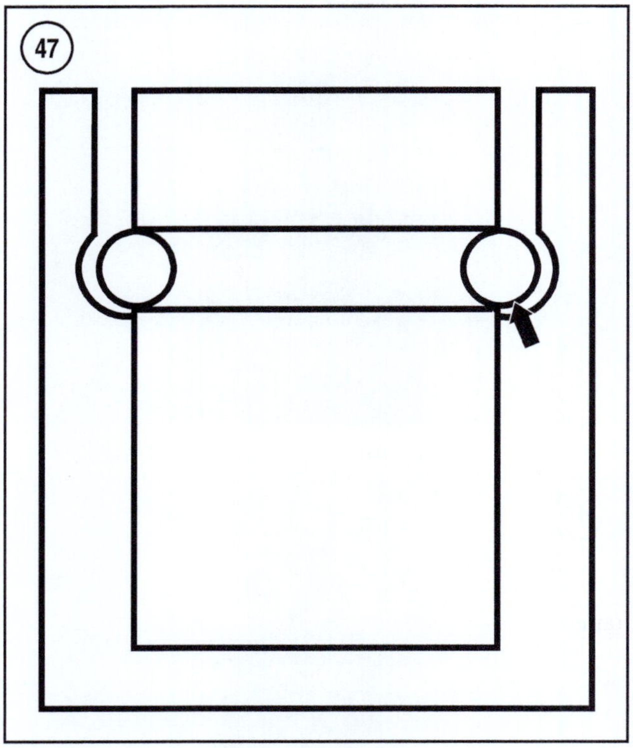

18. Fit the thrust collar (9, **Figure 26**) over the end of the shaft with the larger side facing away from the pinion gear. Seat the collar against the lockring and shaft collar.

19. Connect the brush leads to the brushes (21, **Figure 26**) and holder (18) with the screws (22). Securely tighten the screws.

20. Fit the brush holders and brushes into the frame assembly as shown in **Figure 42**. Insert the pins into the brush holders and openings in the frame brackets to secure the brush holders. Make sure the springs are installed behind the holders to ensure they press the brushes against the commutator when installed.

21. Fit the brush holder terminal into the opening in the frame. Fit the grommets (16 and 17, **Figure 26**) over the terminal and slide them into the opening. Make sure the grooves in the grommets engage the opening. Otherwise, the grommets will not seal properly, or they will work out of the opening and short the lead against the frame.

22. Guide the armature and pinion drive assembly into the starter housing. Tilt the armature as needed to engage the shift lever yoke into the groove in the pinion drive (**Figure 48**). Then fit the end of the armature into the bushing in the nose of the starter housing. Seat the armature shaft and pinion drive assembly into the housing.

23. Install the washer (15, **Figure 26**) over the end of the armature and seat against the commutator step.

24. Without disturbing the armature or starter housing, guide the frame assembly over the armature. Move the brush holder to compress the springs enough to fit the brushes over the commutator as shown in **Figure 49**. This will allow the frame to seat against the starter housing.

48

49

50

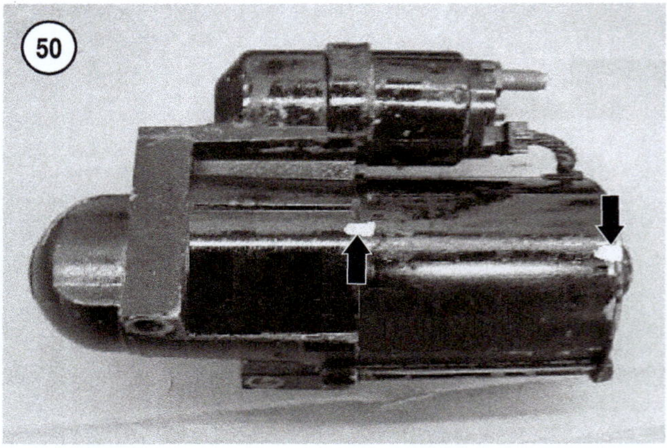

51

52

53

54

55

25. Apply a light coat of water-resistant marine grease to the bushing and then install the end cover (19, **Figure 26**) onto the frame.

26. Rotate the armature and frame assembly and cover to align the match marks (**Figure 35**).

27. Insert the throughbolts (**Figure 36**) into the cover, frame and starter housing openings. Tighten the throughbolts by hand.

28. Verify proper alignment and seating of the housing, frame and cover. Disassemble the starter and correctly assemble if necessary.

29. Securely tighten the throughbolts. The starter housing and cover must fully seat onto the frame.

30. Install the starter solenoid and then the starter as described in this section.

Delco PMGR starter

NOTE

Except for the throughbolts and starter housing, the manufacturer does not offer individual starter components. Many of the components are available from starter and alternator repair shops that are experienced with marine starter repair. It is usually more economical to replace the starter if several components require replacement.

Apply a light coat of OMC EP/Wheel Bearing Grease or other suitable water-resistant grease to the needle bearings, bushings, planetary shaft and gear teeth during assembly. Purchase the grease from an OMC dealership. Do not apply grease or other lubricants to the brushes or commutator portion of the armature.

1. Remove the starter as described in this section.

2. Make match marks on the end cover, frame and starter housing (**Figure 50**).

3. Remove the nut (**Figure 30**) and then disconnect the braided brush terminal wire from the solenoid terminal (**Figure 31**).

4. Remove the throughbolts (**Figure 51**) and pull the frame and cover assembly from the starter housing (**Figure 52**).

5. Remove the metal shield from the frame assembly or starter housing. Push in on the shaft to hold the armature into the rear cover and brush holder. Then pull the frame off the armature (**Figure 53**).

6. Carefully pull the armature out of the rear cover and brush holder.

7. Pull the planetary drive and pinion shaft assembly out of the starter housing bore enough to access the plastic pinion yoke shaft. Use a thin screwdriver to pry the pinion yoke arms off the pivots on the pinion shaft. Then pull the planetary drive and pinion shaft assembly out of the starter housing (**Figure 54**).

8. Remove the thrust collar from the pinion shaft (**Figure 55**).

9

9. Using a punch, drive the collar toward the pinion gear until the lockring is exposed. Carefully pry the lockring out of the groove in the pinion shaft (**Figure 40**). Then pull the lockring and collar off the armature shaft (**Figure 56**). Rotate the pinion drive counterclockwise to disengage it from the helical splines and then pull the pinion drive and planetary ring gear off the pinion shaft (**Figure 57**).

10. Clean and inspect the starter components as described in this section.

11. Apply a light coat of water-resistant marine grease to the roller bearing in the planetary ring gear and then guide the ring gear (gear end first) over the pinion shaft. Align the planetary gear teeth with the ring gear teeth as shown in **Figure 58** and then seat the ring gear onto the shaft.

12. Apply a light coat of marine grease to the helical splines of the pinion shaft. Then thread the pinion drive clockwise onto the splines.

13. Slide the lockring collar over the shaft with the open end facing away from the pinion gear (**Figure 44**). Carefully fit the lockring onto end of the shaft (**Figure 40**). Work the ring down the shaft and into the groove (**Figure 45**). Use two pliers to pull the collar over the lockring (**Figure 46**). The ring must fully seat into the collar (**Figure 47**).

14. Fit the thrust collar over the end of the shaft with the larger side facing away from the pinion gear (**Figure 55**). Seat the collar against the lockring and shaft collar.

15. Align the planetary drive and pinion shaft with the opening in the starter housing (**Figure 54**). Hold the shaft toward the side opposite the starter solenoid. Then guide the shaft into the housing until the yoke arms align with the pivots on the shaft. Then hold the shaft parallel to the solenoid while moving it toward the solenoid. This will cause the pivots to spread the yoke arm shaft enough to snap the pivots into the yoke. The shaft will make two distinct snapping noises as the pivots enter the openings. Pull on the shaft to verify proper engagement.

16. Make sure the thrust collar was not dislodged from the shaft. Reposition the collar as needed.

17. Apply a light coat of water-resistant marine grease to the needle bearing in the nose of the starter housing. Then guide the pinion shaft into the bearing. Rotate the planetary ring to align the alignment bosses and then seat the planetary ring and pinion shaft into the starter housing.

18. Apply a light coat of water-resistant marine grease to the armature shaft needle bearing and gears in the planetary assembly. Then install the metal shield onto the planetary assembly with the concave side toward the planetary drive assembly.

19. Collapse the brush springs and guide the commutator end of the armature into the rear cover and brush assembly. Release the brush springs. The brush surfaces must contact the commutator (**Figure 59**, typical).

20. Carefully guide the frame assembly over the armature, rear cover and brush assembly. Do not allow the magnets to inadvertently pull the armature out of the brushes. If necessary, insert a wooden dowel through the frame and use it to push the armature firmly into the cover while guiding the frame into position.

56

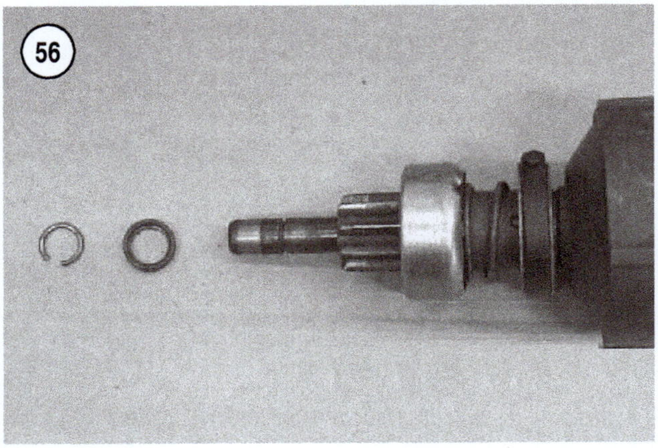

57

58

21. Apply a light coat of water-resistant marine grease to the gear teeth on the end of the armature shaft. Then guide the frame, armature and rear cover assembly into position on the starter housing. Rotate the frame to align the gear teeth and then seat the assembly against the starter housing.

22. Rotate the armature and frame assembly and cover to align the match marks (**Figure 50**).

23. Insert the throughbolts (**Figure 51**) into the cover, frame and starter housing openings. Tighten the throughbolts by hand.

59

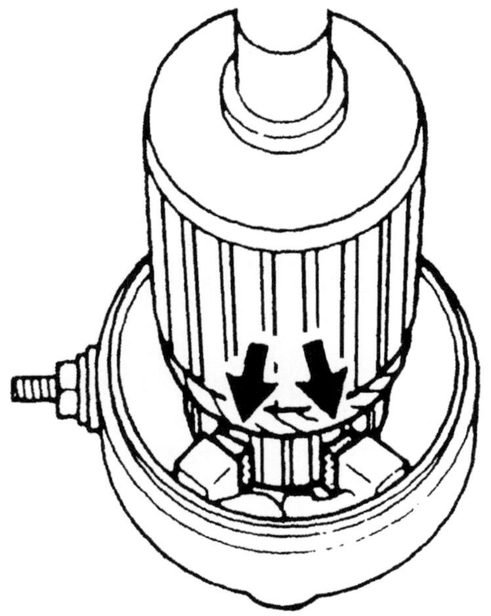

60

61

24. Verify proper alignment and seating of the housing, frame and cover. Disassemble the starter and correctly assemble if necessary.
25. Securely tighten the throughbolts. The starter housing and cover must fully seat onto the frame.

62

26. Connect the braided brush terminal wire to the starter solenoid terminal (**Figure 31**). Thread the nut onto the terminal (**Figure 30**). Verify that the brush terminal and wire are not contacting other terminals or the starter frame and then securely tighten the nut.
27. Install the starter as described in this section.

Delco PG1 starter

NOTE

Except for the starter solenoid, the manufacturer does not offer individual starter components. Many of the components are available from starter and alternator repair shops that are experienced with marine starter repair. It is usually more economical to replace the starter if several components require replacement.

Apply a light coat of OMC EP/Wheel Bearing Grease or other suitable water-resistant grease to the needle bearings, bushings, planetary shaft and gear teeth during assembly. Purchase the grease from an OMC dealership. Do not apply grease or other lubricants to the brushes or commutator portion of the armature.

Refer to **Figure 29** for this procedure.

1. Remove the starter and starter solenoid (**Figure 60**) as described in this section.
2. Make match marks on the end cover, frame and starter housing (**Figure 61**).
3. Remove the throughbolts (**Figure 62**) and pull the frame and cover assembly from the starter housing (**Figure 63**).
4. Push on the shaft to hold the armature into the cover and then pull the frame off the armature (**Figure 64**).
5. Carefully pull the armature out of the cover (**Figure 65**). Remove the two inverted Torx screws (**Figure 66**) and then pull the brush plate out of the cover (**Figure 67**).
6. Remove the plate from the starter housing (**Figure 68**). Remove the grommet (**Figure 69**) and then pull the planetary shaft and related components from the starter housing (**Figure 70**).
7. Remove the thrust collar from the pinion shaft (**Figure 71**) and then lift the yoke and disc off the planetary assembly (**Figure 72**).

63

67

64

68

65

69

66

70

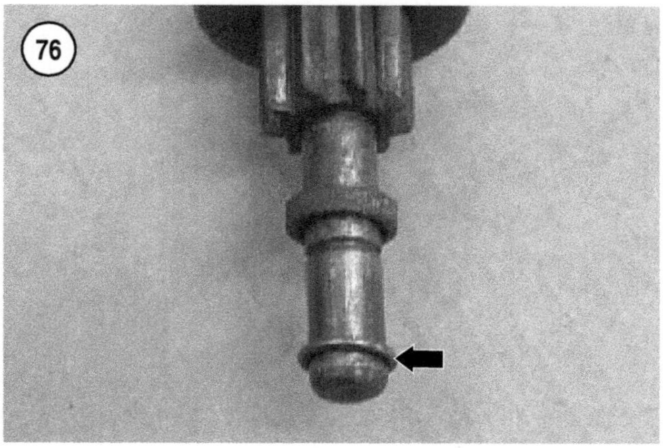

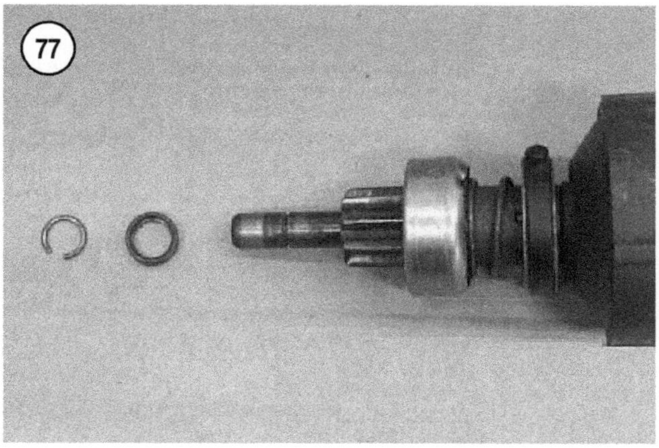

8. Push the ring gear toward the pinion gear to expose the planetary gears (**Figure 73**). Remove the three planetary gears from the pinion shaft (**Figure 74**).

9. Using a punch, drive the collar toward the pinion gear until the lockring is exposed (**Figure 75**). Carefully pry the lockring out of the groove in the pinion shaft (**Figure 76**). Then pull the lockring and collar off the pinion shaft (**Figure 77**).

10. Rotate the pinion drive counterclockwise to disengage it from the helical splines and then pull the pinion gear and planetary ring gear off the pinion shaft (**Figure 78**).

11. Clean and inspect the starter components as described in this section.

12. Carefully guide the frame assembly over the armature.
13. Collapse the brush springs and fit the brush plate over the commutator (**Figure 79**). Release the brush springs. The brush surfaces must contact the commutator (**Figure 59**, typical).
14. Align the brush lead grommet with the notch and fit the cover onto the brush plate, armature and frame. Rotate the cover as needed to align the opening and then install and securely tighten the screws (**Figure 66**).
15. Guide the planetary ring gear over the helical-splined end of the pinion shaft. The open side of the ring gear must face away from the helical splines when installed.
16. Apply a light coat of marine grease to the helical splines of the pinion shaft and then thread the pinion drive clockwise onto the splines.
17. Slide the lockring collar over the shaft with the open end facing away from the pinion gear (**Figure 80**). Carefully fit the lockring onto the end of the shaft (**Figure 76**). Work the ring down the shaft and into the groove (**Figure 75**). Use two pliers to pull the collar over the lockring (**Figure 81**). The ring must fully seat into the collar (**Figure 82**).
18. Fit the thrust collar over the end of the shaft with the larger side facing away from the pinion gear (**Figure 71**). Seat the collar against the lockring and shaft collar.
19. Apply a light coat of water-resistant marine grease to the surfaces and then install the gears onto the pinion shaft (**Figure 74**).
20. Align the planetary gear teeth with the ring gear teeth as shown in **Figure 58** and then seat the ring gear onto the shaft.
21. Apply a light coat of water-resistant marine grease onto the needle bearing in the nose of the starter housing.
22. Fit the yoke arm into the pinion gear spool (**Figure 83**). Align the yoke with the solenoid bore and then carefully guide the planetary assembly into the starter housing (**Figure 70**). Align the end of the shaft with the needle bearing in the nose of the housing. Seat the planetary assembly into the housing.
23. Insert the disc (22, **Figure 60**) into the opening and then seat it against the pinion yoke (19). Insert the grommet into the starter housing (**Figure 69**). Make sure the lips on the side of the grommet fit over the slot in the starter housing.
24. Apply a coat of water-resistant marine grease to the planetary gear teeth, the needle bearing in the ring gear and the gear on the end of the armature shaft.
25. Install the plate into the starter housing with the open side facing outward (**Figure 68**). Seat the plate against the planetary ring gear.
26. Align the gear teeth on the armature shaft with the planetary gear teeth (**Figure 63**) and then guide the armature, frame and cover assembly into the planetary assembly. Rotate the armature to mesh the gear teeth and then seat the frame assembly onto the starter housing. Rotate the armature and frame assembly to align the match marks (**Figure 61**).

78

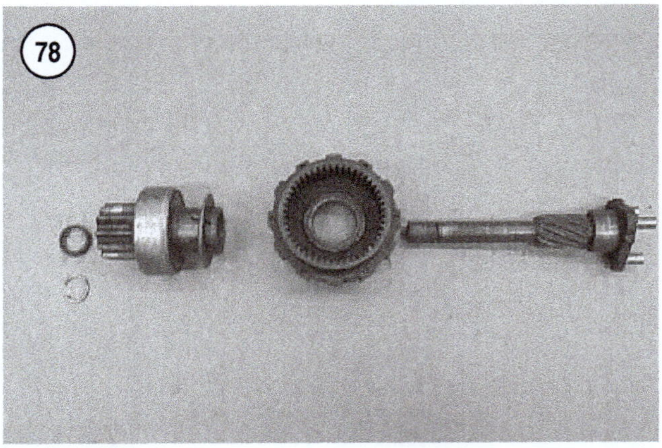

79

80

81

82

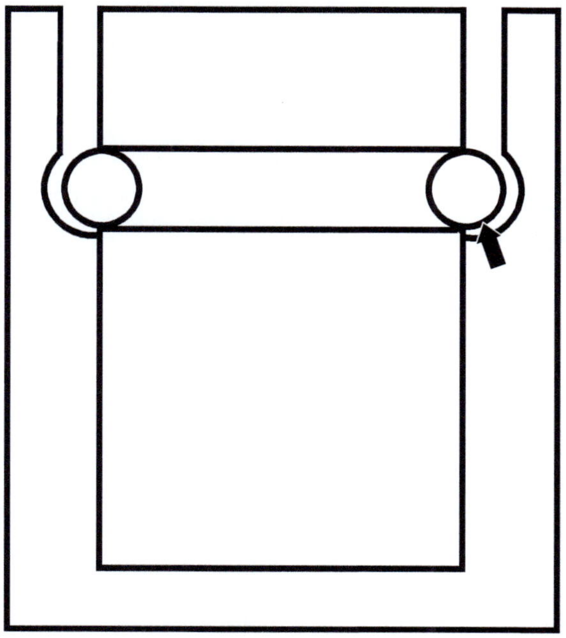

83

27. Insert the through bolts (**Figure 62**) into the cover, frame and starter housing openings. Tighten the throughbolts by hand.
28. Verify proper alignment and seating of the housing, frame and cover. Disassemble the starter and correctly assemble if necessary.
29. Securely tighten the throughbolts. The starter housing and cover must fully seat onto the frame.
30. Install the starter solenoid and then the starter as described in this section.

Motorcraft starter

Apply a light coat of OMC EP/Wheel Bearing Grease or other suitable water-resistant grease to the needle bearings, bushings, planetary shaft and gear teeth during assembly. Purchase the grease from an OMC dealership. Do not apply grease or other lubricants to the brushes or commutator portion of the armature.

Refer to **Figure 84** for this procedure.

1. Remove the starter and starter solenoid (**Figure 60**, typical) as described in this section.
2. Make match marks on the end cover, frame and starter housing (**Figure 61**, typical).
3. Remove the through bolts (**Figure 62**, typical) and pull the frame and planetary assembly off the starter housing (5, **Figure 84**).
4. Pull the pinion yoke off the pinion drive assembly. Then pull the seal (7, **Figure 84**) out of the starter housing.
5. Remove the stop (9, **Figure 84**) and retaining ring (10) from the planetary shaft (18).
6. Rotate the pinion drive assembly counterclockwise to disengage it from the helical splines and then pull the pinion gear assembly off the pinion shaft.
7. Remove the E-clip (14, **Figure 84**) and washer (15).
8. Carefully pull the plate (21, **Figure 84**) from the planetary gear housing. Remove the three planetary gears (20, **Figure 84**) from the gear housing. Tap the end of the planetary shaft against the work surface to remove the ball (19, **Figure 84**). Remove the planetary gear housing (17, **Figure 84**) from the pinion shaft (18).
9. Remove the two screws (24, **Figure 84**) that retain the brush plate to the end plate. Carefully pull the end plate from the frame and armature assembly.
10. Pull the brush plate out of the frame assembly. Then pull the armature out of the frame.
11. Clean and inspect the starter components as described in this section.
12. Carefully guide the frame assembly over the armature.
13. Collapse the brush springs and fit the brush plate over the commutator (**Figure 79**). Release the brush springs. The brush surfaces must contact the commutator (**Figure 59**, typical).
14. Apply water-resistant grease to the bearing within the end plate and the corresponding bearing surface on the armature. Do not allow the grease to contaminate the brushes or commutator. Guide the armature shaft into the end plate bearing. Rotate the end plate to align the brush lead grommet with the notch and fit the end plate (25, **Figure 84**) onto the brush plate, armature and frame. Rotate the cover as needed to align the opening and then install the screws (**Figure 66**, typical). Tighten the brush plate screws to 20-30 in.-lb. (2-3 N•m).
15. Apply water-resistant grease to the helical splines and all bearing surfaces. Then guide the planetary shaft (18, **Figure 84**) into the planetary gear housing (17).
16. Apply a light coat of water-resistant grease to the surfaces and then install the gears onto the pinion shaft and planetary gear housing (**Figure 74**). Align the planetary gear teeth with the housing teeth as shown in **Figure 58**.
17. Apply a light coat of water-resistant grease to the ball bearing (19, **Figure 84**) and then fit the bearing into the opening in the planetary shaft (18).

9

84

FORD MARINE STARTER
(PERMANENT MAGNET FRAME AND PLANETARY GEAR REDUCTION)

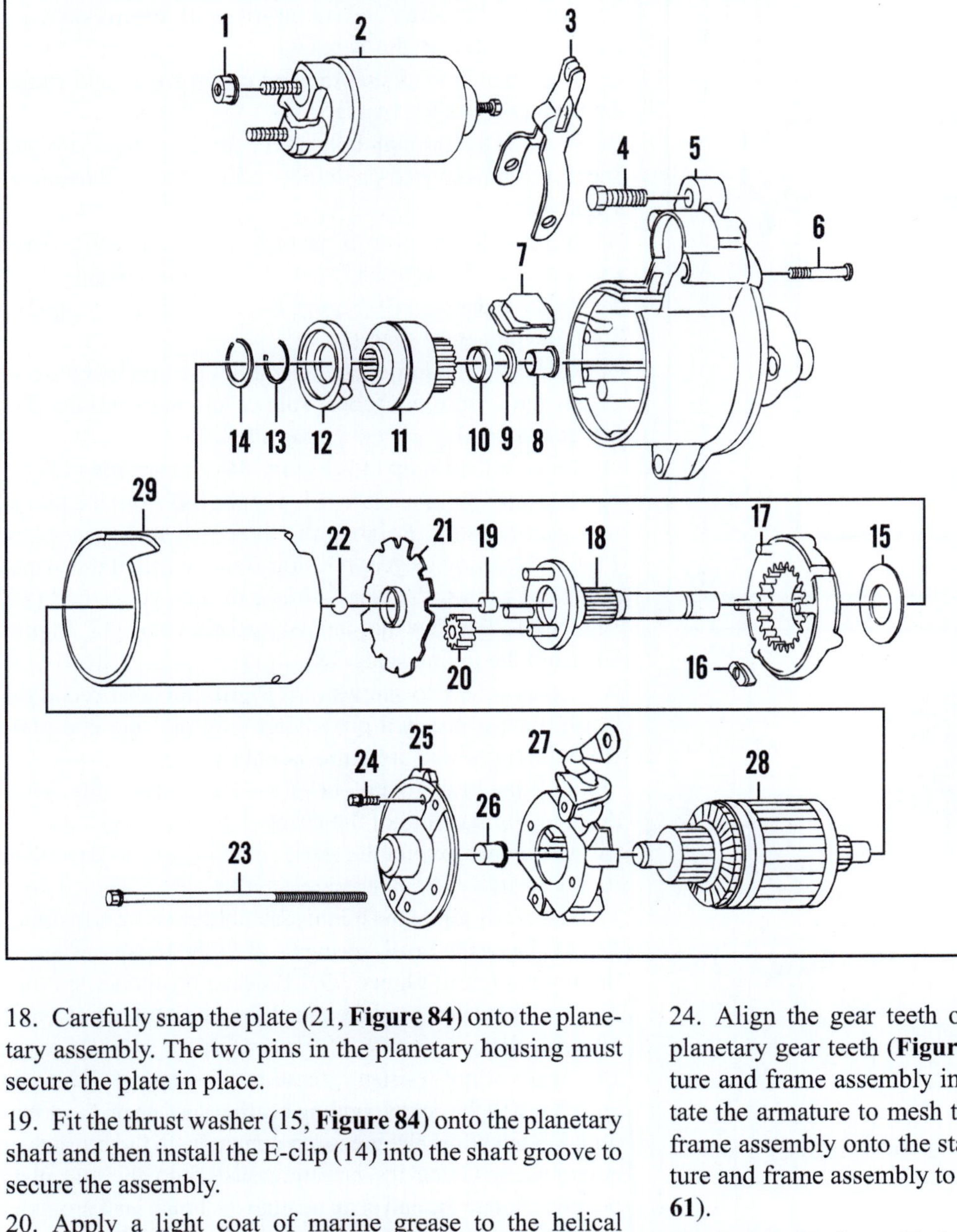

1. Nut
2. Starter solenoid
3. Pinion yoke
4. Starter mounting bolt
5. Starter housing
6. Screw
7. Seal
8. Bearing
9. Stop ring
10. Retaining ring
11. Pinion drive
12. Yoke collar
13. Lockring
14. E-ring
15. Thrust washer
16. Wedge block
17. Planetary gear housing
18. Pinion/planetary shaft
19. Ball bearing
20. Planetary gear
21. Plate
22. Armature thrust bearing
23. Throughbolt
24. Screw
25. End plate
26. Bearing
27. Brush plate
28. Armature
29. Frame assembly

18. Carefully snap the plate (21, **Figure 84**) onto the planetary assembly. The two pins in the planetary housing must secure the plate in place.

19. Fit the thrust washer (15, **Figure 84**) onto the planetary shaft and then install the E-clip (14) into the shaft groove to secure the assembly.

20. Apply a light coat of marine grease to the helical splines of the pinion shaft and then thread the pinion drive clockwise onto the splines.

21. Install the stop ring (9, **Figure 84**) and retaining ring (10) onto the planetary shaft (18).

22. Carefully snap the openings in the pinion yoke (3, **Figure 84**) over the pins of the planetary yoke collar (12).

23. Apply water-resistant grease into the bearing (8, **Figure 84**) within the starter housing. Guide the planetary assembly into the starter housing. Make sure the pinion yoke fits into the recess in the housing. Then fit the seal (7, **Figure 84**) into the motor housing. Fit the wedge block (16, **Figure 84**) between the planetary assembly and the starter housing. The long end must fit into the starter housing, and the opening must fit over the post on the planetary housing.

24. Align the gear teeth on the armature shaft with the planetary gear teeth (**Figure 63**) and then guide the armature and frame assembly into the planetary assembly. Rotate the armature to mesh the gear teeth and then seat the frame assembly onto the starter housing. Rotate the armature and frame assembly to align the match marks (**Figure 61**).

25. Insert the throughbolts (**Figure 62**) into the cover, frame and starter housing openings. Tighten the throughbolts by hand. Verify proper alignment and seating of the housing, frame and end plate. Then tighten the throughbolts to 45-84 in.-lb. (5-10 N•m).

26. Install the starter solenoid and then the starter as described in this section.

Starter Component Cleaning and Inspection

1. Clean all components in a mild solvent. Do not use solvents that are damaging to plastic or rubber or apply any solvent to the starter solenoid.

85

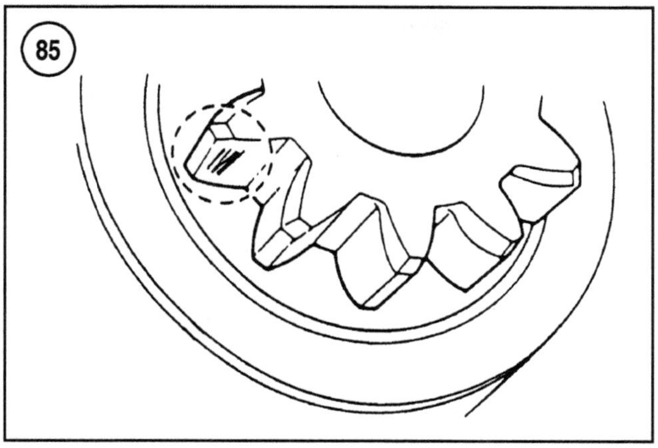

86

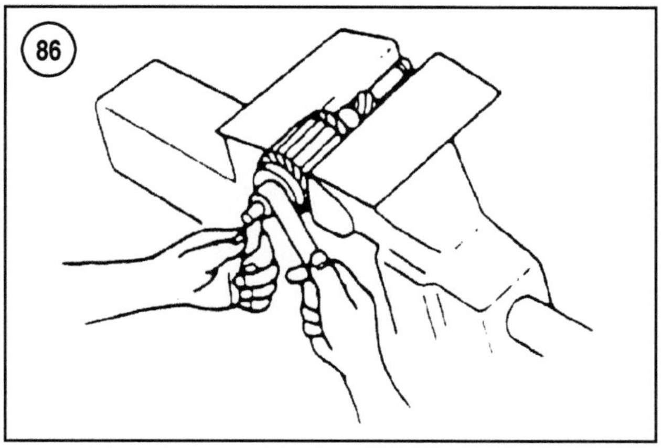

87

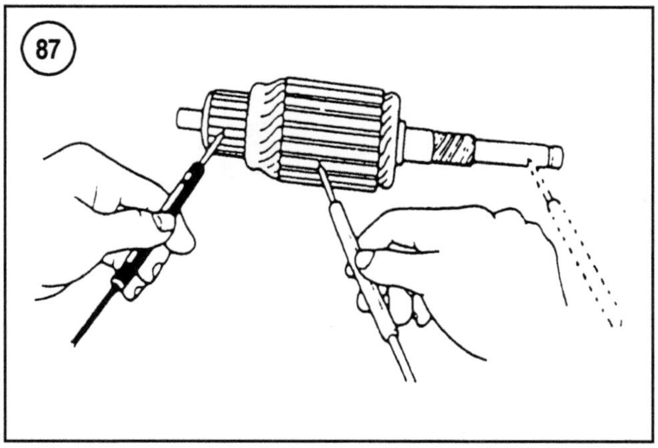

88

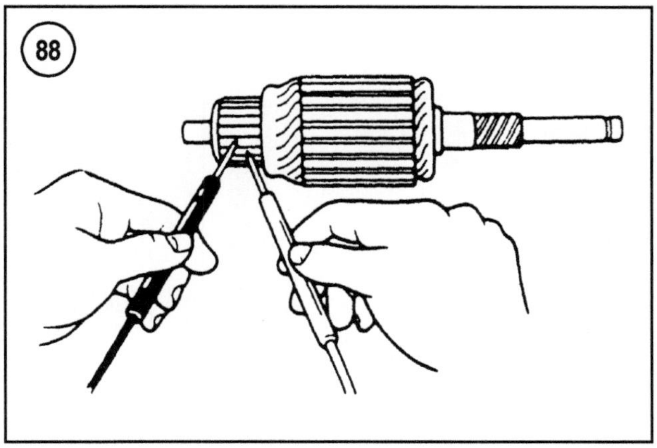

89

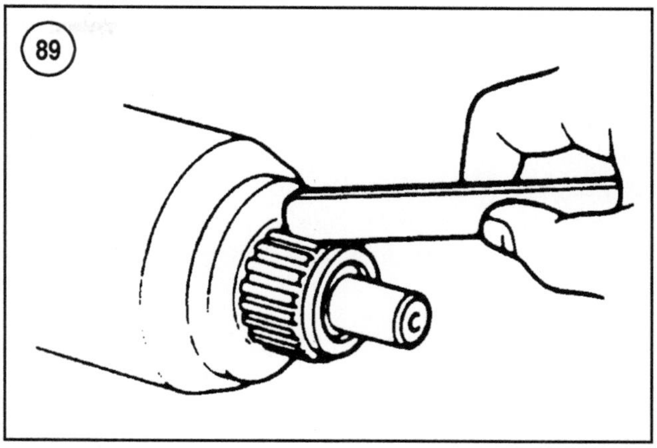

2. Inspect the pinion gear for chipped, cracked or worn teeth (**Figure 85**). Replace the gear if these or other defects are evident.
3. Inspect the helical splines at the pinion end of the planetary shaft for excessive wear or damaged splines and replace as needed.
4. Inspect the armature for deep corrosion pitting on the commutator shaft or excessive wear on the bearing surfaces. Replace the armature if these or other defects are evident.
5. Carefully secure the armature in a vise with protective jaws (**Figure 86**). Tighten the vise just enough to secure the armature. Do not overtighten the vise. Carefully polish away any corrosion deposits and glazed surfaces from the commutator using 600-grit Carborundum cloth. Avoid removing too much material. Rotate the armature often to polish the commutator evenly.
6. Connect an ohmmeter between any commutator segment and the armature laminations (**Figure 87**). Connect the ohmmeter between any commutator segment and the armature shaft. The meter must indicate *no continuity* at both connection points. If otherwise, the armature is shorted and must be replaced.
7. Connect the ohmmeter between commutator segments (**Figure 88**) and note the reading. Repeat this test with the meter connected to each commutator segment. The meter must indicate *continuity* between any pair of segments. If otherwise, the commutator has failed open and must be replaced.
8. On models using a Delco starter with an electric coil frame, have the coil frame tested by a reputable starter repair shop.
9. Using a small file, remove the mica particles from the undercut between the commutator segments (**Figure 89**).
10. Inspect the brushes for grooved, discolored or corroded surfaces. Inspect the brush springs for corrosion or lost spring tension. Replace the brush and lead or plate assembly if these or other defects are evident.
11. Measure the length of the four brushes (**Figure 90**).
    a. On the Delco starter with the electric coil frame, replace the brush plate assembly if any of the brushes are worn to less than 1/4 in. (6 mm).
    b. On the Delco PMGR or Delco PG260 permanent magnet starter, replace the brush plate assembly if any of the brushes are worn to less than 0.30 in. (8 mm).

9

c. On the Motorcraft permanent magnet gear reduction starter, replace the brush plate if any of the brushes are worn to less than 1/4 in. (6 mm).

12. Inspect the bearing surfaces on the armature and planetary shaft for excessive or uneven wear. Inspect the planetary gears for worn, cracked or missing teeth. Inspect the needle bearing in the starter housing for wear, corrosion or discoloration. Replace all worn or damaged components.

90

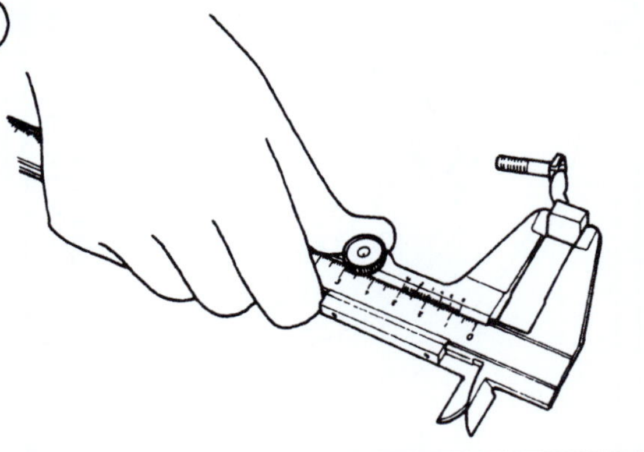

IGNITION SYSTEM

Distributors

Types and locations

The models covered in this manual use several different types of ignition systems and distributors.

All GM 3.0L models use a Delco EST (electronic spark timing) ignition system and use the type of distributor shown in **Figure 91**. The distributor mounts on the starboard side of the cylinder block (**Figure 92**) and houses the ignition module and pickup coil. The ignition coil mounts on the starboard side of the cylinder block (near the distributor).

All GM V-6 and V-8 electronic fuel injection (EFI) models use a Delco EST ignition system. All carburetor-equipped GM 4.3L and later carburetor-equipped GM V-8 models use a Delco EST ignition system. The Delco EST ignition system uses the type of distributor shown in **Figure 93**. The distributor is located on the rear of the intake manifold and houses the ignition module and pickup coil. The ignition coil mounts on the rear of the cylinder head or the intake manifold.

All carburetor-equipped Ford 5.0L and 5.8L models, and earlier GM V-8 models, use a Prestolite BID (breakerless integral distributor) ignition system. The BID ignition system uses the type of distributor shown in **Figure 94**. On Ford 5.0L and 5.8L models, the distributor is located on the front of the intake manifold. On GM V-8 models, the distributor is located on the rear of the intake manifold. The distributor houses the ignition module and centrifugal advance mechanism. The ignition coil mounts onto the intake manifold or rear of the cylinder head.

91

DELCO EST DISTRIBUTOR ASSEMBLY (3.0L MODELS)

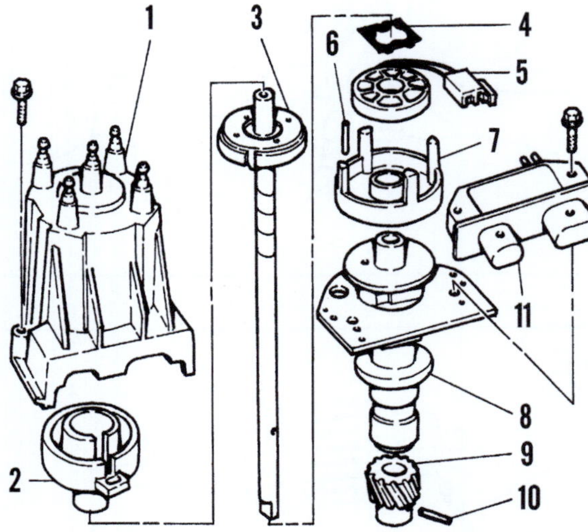

1. Distributor cap
2. Rotor
3. Shaft
4. Retainer
5. Pickup coil
6. Locating pin
7. Pole piece
8. Housing
9. Driven gear
10. Roll pin
11. Ignition module

Cap and rotor replacement

Refer to **Figures 91-95** for this procedure.

1. Disconnect the negative battery cable.

2. On models using a thick film integrated (TFI) ignition system, note which spark plug wires exited the side openings and then carefully remove the rubber boot (1, **Figure 95**) from the distributor.

3. Mark the cylinder numbers on the plug wire caps and next to the corresponding terminals on the distributor cap. Carefully disconnect the spark plug leads and ignition coil secondary lead from the cap.

92

93

DELCO EST IGNITION SYSTEM (GM ELECTRONIC FUEL INJECTION MODELS AND LATER GM V-6/V-8 CARBURETOR-EQUIPPED MODELS)

1. Ignition coil
2. Distributor shaft
3. Rotor
4. Distributor cap
5. Ignition module
6. Retainer
7. Washer
8. Pickup coil
9. Pole piece
10. Locating pin
11. Housing
12. Bolt
13. Hold-down clamp
14. Washer
15. Washer
16. Driven gear
17. Gasket
18. Intake manifold

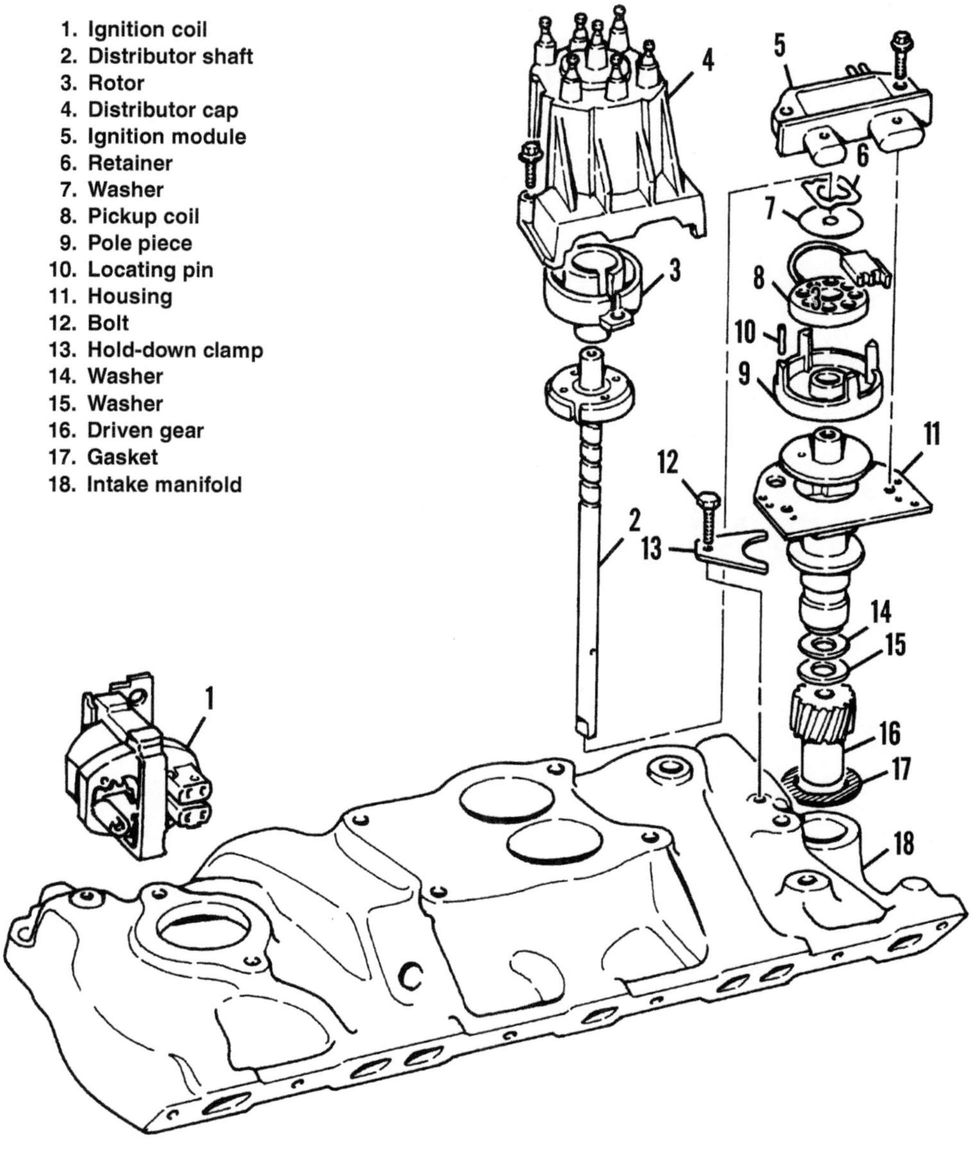

9

4A. On models using a Delco EST or Prestolite BID ignition system, make match marks on the side of the cap and distributor housing to aid in installation. Use two side-by-side marks on each component to avoid confusion with other alignment marks.

4B. On models using a TFI ignition system, make match marks on the distributor housing (8, **Figure 95**), distributor base (4) and distributor cap (2). Use two side-by-side marks on each component to avoid confusion with other alignment markings.

5. Fully loosen the screws and then lift the cap off the distributor. Use compressed air to blow loose material from the distributor and cap.

6. Scribe a *single* mark on the distributor housing that aligns with the rotor tip (**Figure 96**, typical).

7. Grip the rotor and carefully pull the rotor off the distributor shaft (**Figure 97**, typical). Do not twist the rotor during removal.

8. Inspect the rotor for worn or burned contacts. Inspect the cap (**Figure 98**, typical) for hairline cracks, carbon tracking, and worn, burned or corroded contacts. Replace the cap and rotor if these or other defects are evident. Transfer the mark made in Step 6 and the cylinder number markings (Step 3) to the replacement cap.

9. Fit the rotor onto the distributor shaft with the tip aligned with the *single* scribe mark (**Figure 96**). Do not

(94)

PRESTOLITE IGNITION SYSTEM (CARBURETOR-EQUIPPED FORD 5.0L/5.8L MODELS AND EARLIER CARBURETOR-EQUIPPED GM V-8 MODELS)

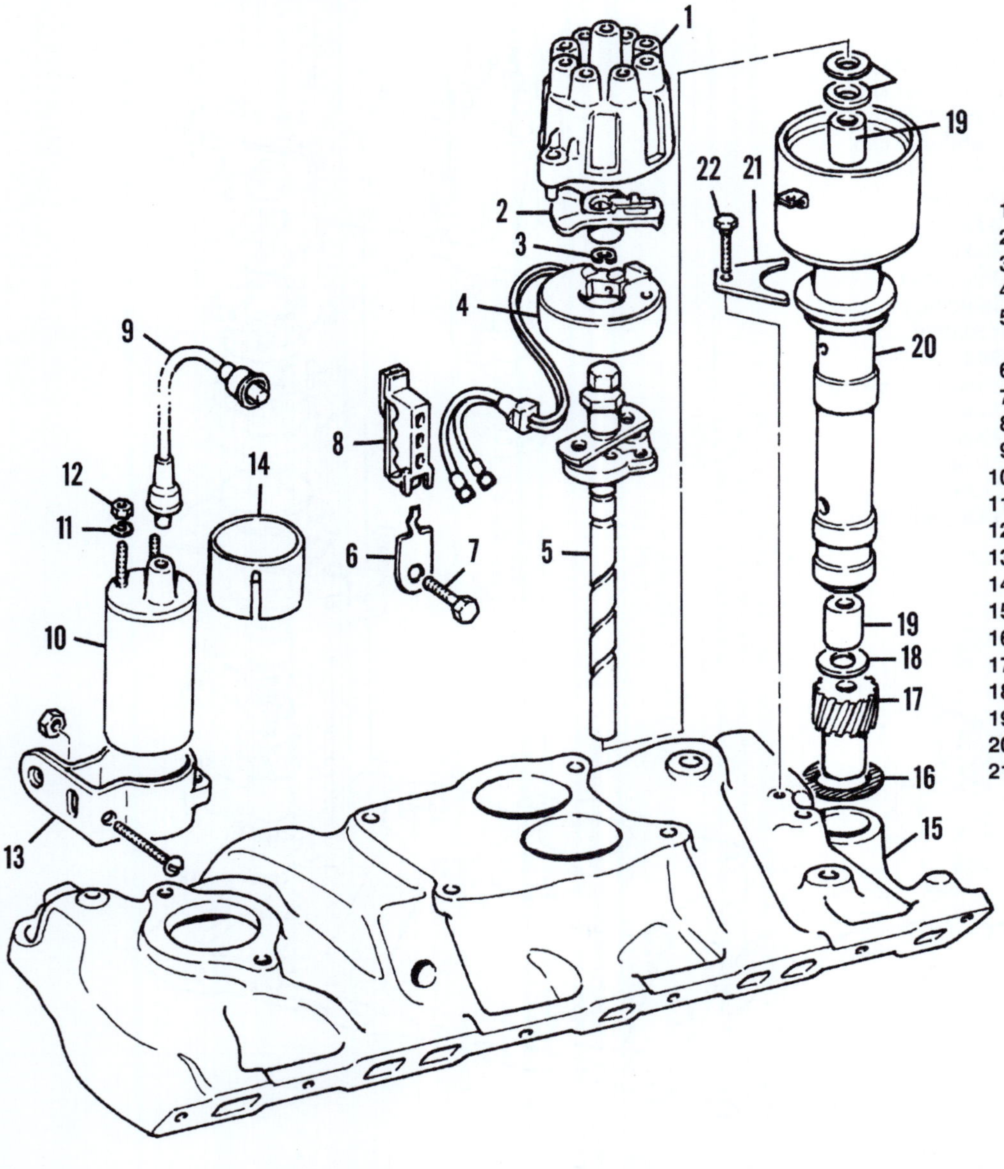

1. Distributor cap
2. Rotor
3. E-clip
4. Ignition module
5. Distributor shaft assembly
6. Bracket
7. Bolt
8. Retainer
9. High-tension lead
10. Ignition coil
11. Lockwasher
12. Nut
13. Bracket
14. Sleeve
15. Intake manifold
16. Gasket
17. Driven gear
18. Washer
19. Bushing
20. Distributor housing
21. Hold-down clamp

confuse the scribe mark with the distributor cap and housing match marks made in Step 3. Turn the rotor slightly to align the raised boss (inside the rotor opening) with the slot (at the top of the distributor shaft). Then push down until the rotor fully seats on the shaft. On models using a TFI ignition system, the pin on the bottom of the rotor must align with the opening in the switch wheel.

10A. On models using a Delco EST ignition system, align the match marks (Step 4) and fit the distributor cap onto the distributor housing. Fit the slot in the cap (1, **Figure 91** or 4, **Figure 93**) over the ridge on the ignition module (11, **Figure 91** or 5, **Figure 93**). Rotate the cap slightly to align the screws with the openings in the housing and then tighten the screws securely.

10B. On models using a Prestolite BID ignition system, align the match marks (Step 4) and fit the distributor cap (1, **Figure 94**) over the distributor housing (20). Rotate the cap slightly to align the screw boss on the cap over the threaded boss on the housing. Then seat the cap onto the distributor. Thread the screws into the boss and tighten securely.

10C. On models using a TFI ignition system, fit the distributor cap (2, **Figure 95**) over the distributor base (4). Rotate the cap slightly to align the screw boss on the cap over the threaded boss and then seat the cap onto the distributor base. Thread the two screws into the cap and base. Tighten the distributor cap screws to 18-23 in.-lb. (2-3 N•m).

11. Starting with the No. 1 cylinder, plug the spark plug wire cap onto the corresponding cap terminals. The cylin-

95

TFI IGNITION SYSTEM
(FORD 5.0L AND 5.8L ELECTRONIC FUEL INJECTION MODELS)

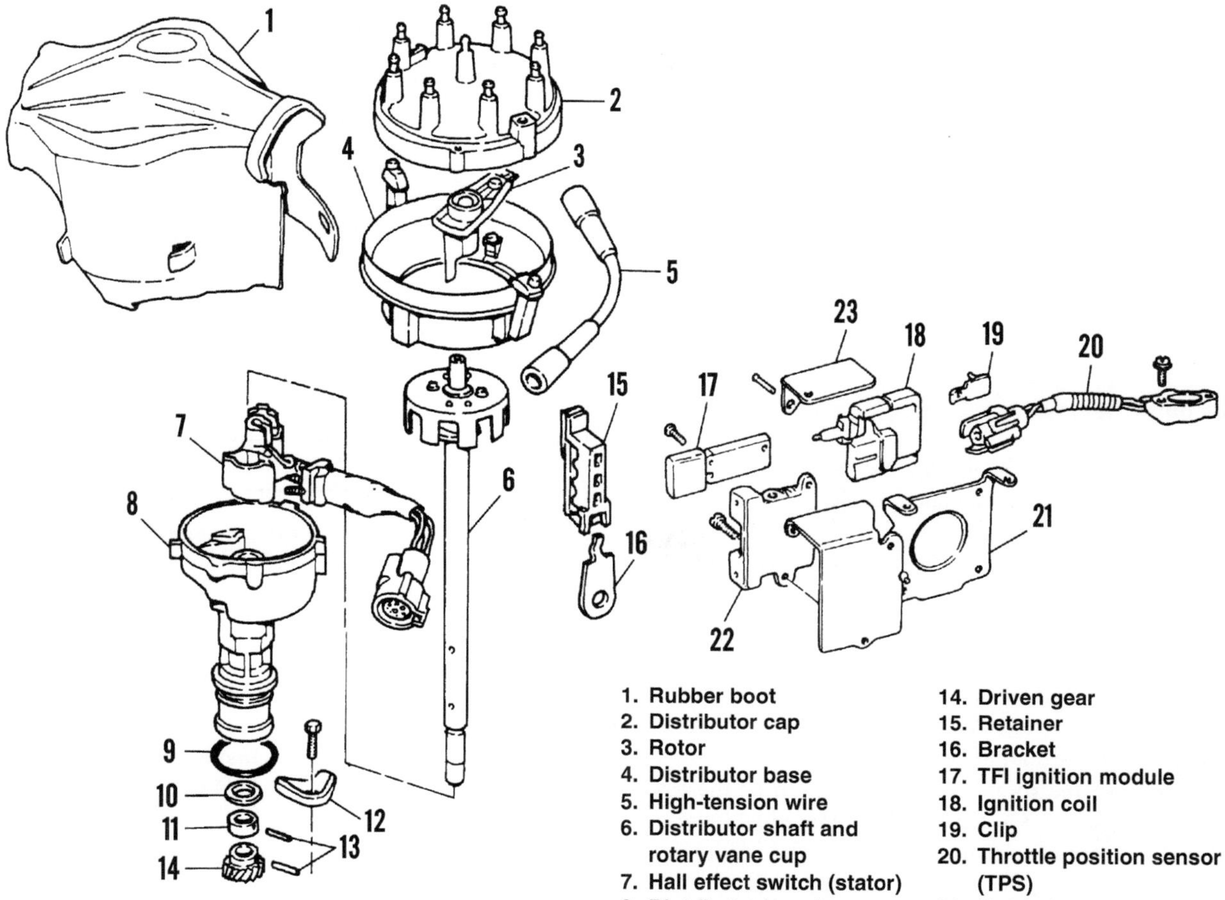

1. Rubber boot
2. Distributor cap
3. Rotor
4. Distributor base
5. High-tension wire
6. Distributor shaft and rotary vane cup
7. Hall effect switch (stator)
8. Distributor housing
9. O-ring
10. Spacer
11. Spacer
12. Hold-down clamp
13. Roll pin
14. Driven gear
15. Retainer
16. Bracket
17. TFI ignition module
18. Ignition coil
19. Clip
20. Throttle position sensor (TPS)
21. Bracket
22. Heat sink
23. Bracket

96

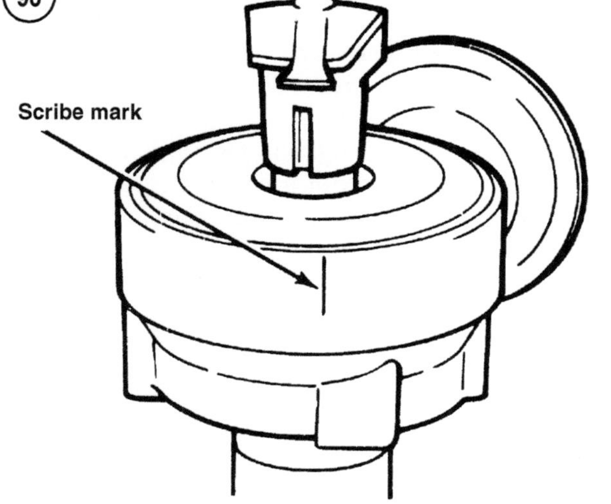

97

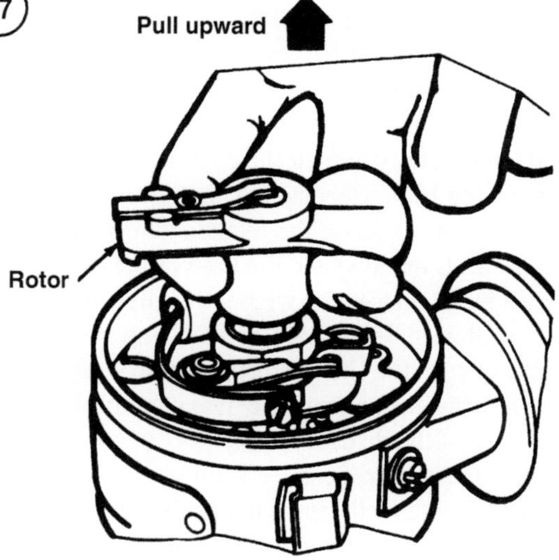

der number sequence must match the firing order sequence listed in **Table 5**.

a. On GM models, the firing order progresses in the clockwise direction around the cap.

b. On Ford models, the firing order progresses in the counterclockwise direction around the cap.

12. On models using a TFI ignition system, guide the spark plug wires into the openings and then fit the rubber boot (1, **Figure 95**) onto the distributor. Fit the tab on the side of the boot into the corresponding openings on the opposite side of the boot.

13. Connect the battery cable.

14. Adjust the ignition timing as described in Chapter Five.

Distributor Removal

Refer to **Figures 91-95** for this procedure.

1. Disconnect the negative battery cable.
2. Remove the spark plugs.
3. Place No. 1 piston at TDC firing position as described in Chapter Six.
4. Mark the cylinder numbers on the plug wire caps and next to the corresponding terminals on the distributor cap. Carefully disconnect the spark plug leads and ignition coil secondary lead from the cap.
5. Remove the distributor cap as described in this chapter. Do not remove the rotor. Verify that the tip of the rotor would align with the No. 1 cylinder wire terminal if the cap were installed onto the distributor. If it would not, refer to Step 3.
6. Scribe a *single* mark on the distributor housing that aligns with the rotor tip (**Figure 96**).

7A. On models using a Delco EST ignition system, lift up on the locking tab and then unplug the engine wire harness connector(s) from the ignition module (11, **Figure 91** or 5, **Figure 93**). A single harness and connector are used on carburetor-equipped models. Two harnesses and connectors are used on electronic fuel injection (EFI) models.

7B. On models using a Prestolite BID ignition system, disconnect the engine wire harness connectors from the ignition module (4, **Figure 94**) wire connectors.

7C. On models using a Thick Film Integrated (TFI) ignition system, carefully unplug the engine wire harness connector from the Hall effect switch (7, **Figure 95**).

8. Scribe match marks on the intake manifold or cylinder block (GM 3.0L models) and the base of the distributor.
9. Use a special distributor wrench (**Figure 99**) to loosen the bolt and then slide the hold-down clamp away from the base of the distributor. Lift the distributor from the engine.
10. Remove the gasket from the intake manifold, cylinder block or base of the distributor.
11. Inspect the driven gear teeth for excessive wear and cracked or missing teeth. Have a machine shop replace the driven gear if excessively worn or damaged. The replacement gear must be drilled to match the distributor shaft and requires special equipment and expertise. If gear teeth are missing, remove the oil pan and retrieve the remnants. Oil pan removal and installation is described in Chapter Six. Look into the distributor opening and check for damage to the camshaft teeth. If the camshaft gear teeth are damaged, replace the camshaft and lifters as described in Chapter Six.

98

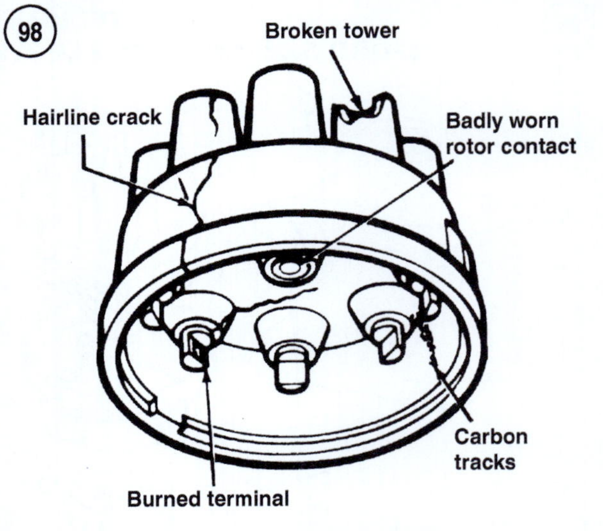

99

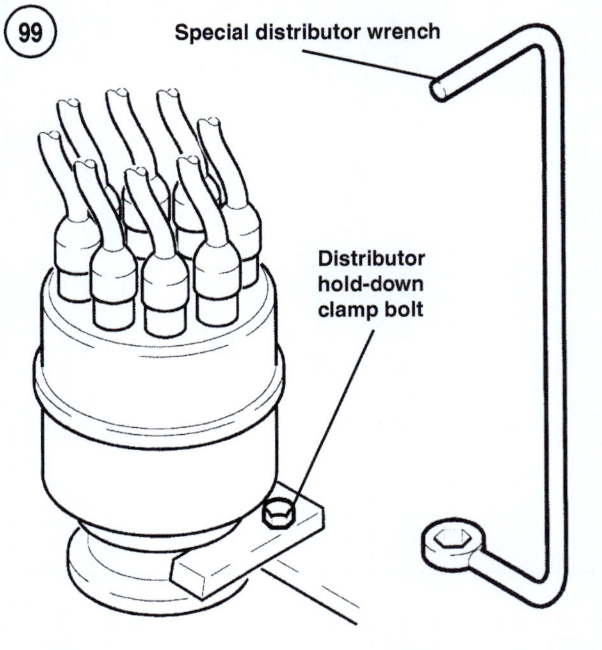

Distributor Installation

1. If the crankshaft was rotated with the distributor removed, place the No. 1 piston at TDC firing position as described in Chapter Six.
2. If the distributor, intake manifold or cylinder block (GM 3.0L models) was replaced, transfer any marks made during

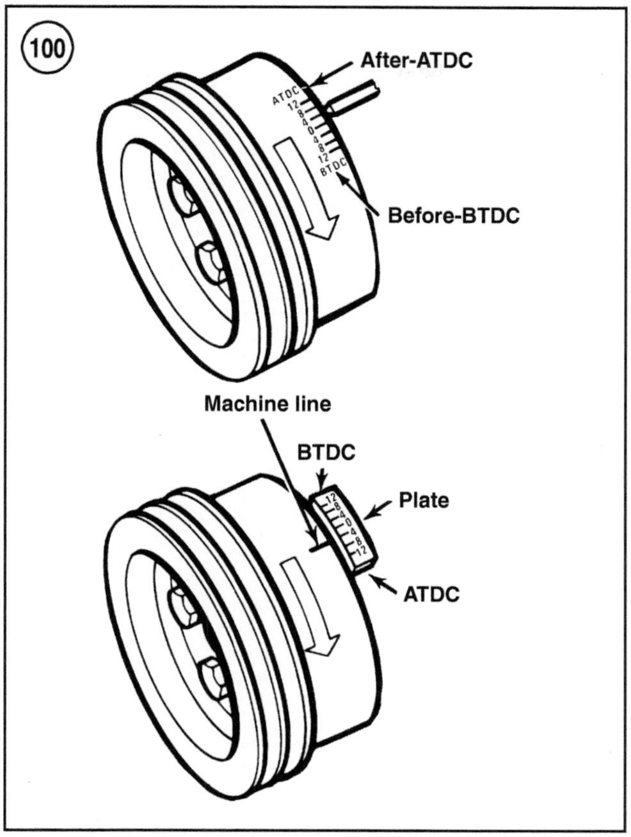

the distributor removal procedures to the replacement components.

3. Fit a *new* gasket over the driven gear and fit onto the distributor base.

4. Turn the driven gear on the distributor until the rotor tip aligns with the *single* scribe mark (**Figure 96**). This mark was made in Step 6 of the distributor removal procedure.

5. Carefully guide the distributor into the intake manifold or cylinder block opening. At this time, do not insert the distributor into the opening enough to engage the driven gear teeth onto the camshaft teeth.

6. Rotate the distributor housing to align the distributor base to the intake manifold or cylinder block (3.0L models) match marks made during the distributor removal procedures (Step 8).

7. Verify that the rotor is aligned with the *single* scribe mark as described in Step 4.

  a. On GM models, turn the rotor counterclockwise (as viewed from the top) approximately 45° (one-eighth of a turn).

  b. On Ford models, turn the rotor clockwise (as viewed from the top) approximately 45° (one-eighth of a turn).

8. Carefully guide the driven gear into the camshaft gear while installing the distributor. The rotor will move toward alignment with the *single* scribe mark as the gear teeth mesh. The distributor base might not fully seat at this time because the oil pump drive shaft might not be aligned with the slot in the driven gear. Refer to the following:

  a. If the distributor is seated, proceed to Step 10.

  b. If the distributor is not seated, proceed to Step 9.

9. Hold the distributor in position with the match marks aligned. Apply light downward pressure while an assistant *slowly* rotates the harmonic balancer clockwise while counting the turns. The distributor base will drop onto the intake manifold or cylinder block when the oil pump drive shaft aligns with the slot in the gear. Have the assistant stop when the harmonic balancer has rotated *exactly* two revolutions and the timing mark aligns with the 0° TDC marking on the timing scale (**Figure 100** typical). This should again place the No. 1 piston at TDC firing position as described in Chapter Six.

10. Turn the distributor to align the match marks as described in Step 6. The rotor tip should align with or be within a few degrees of the *single* scribe mark (**Figure 96**). Verify that the tip of the rotor would align with the No. 1 cylinder wire terminal if the cap were installed onto the distributor. If not, remove the distributor and repeat the installation procedure.

11. Hold the distributor in position while placing the hold-down clamp over the distributor base. Tighten the hold-down clamp bolt to 18 ft.-lb. (24 N•m).

12. Install the distributor cap as described in this section.

13. Starting with the No. 1 cylinder, plug the spark plug wire caps onto the corresponding cap terminals. The cylinder number sequence, must match the firing order sequence listed in **Table 5**.

  a. On GM models, the firing order progresses in the clockwise direction around the cap.

  b. On Ford models, the firing order progresses in the counterclockwise direction around the cap.

14A. On GM 3.0L models, install the spark plug and tighten to 20 ft.-lb. (27 N•m).

14B. On GM 4.3L, 5.0L and 5.7L models, install the spark plugs and tighten to 22 ft.-lb. (30 N•m) if the cylinder head is new and 15 ft.-lb. (20 N•m) if the cylinder head is used.

14C. On Ford 5.0L and 5.8L models, install the spark plugs and tighten to 120-180 in.-lb. (14-20 N•m).

14D. On GM 7.4L and 8.2L models, install the spark plugs and tighten to 22 ft.-lb. (30 N•m).

15A. On models using a Delco EST ignition system, lift up on the locking tab and then carefully plug the engine wire harness connector(s) onto the ignition module (11, **Figure 91** or 5, **Figure 93**). A single harness and connector is used on carburetor-equipped models. Two harnesses and connectors are used on electronic fuel injection (EFI) models.

15B. On models using a Prestolite BID ignition system, connect the engine wire harness connectors to the ignition module (4, **Figure 94**).

15C. On models using a thick film integrated (TFI) ignition system, plug the engine wire harness connector onto the Hall effect switch (7, **Figure 95**).

16. To avoid electrical interference, route the wiring away from the ignition coil or spark plug secondary leads.

17. Connect the battery cable.

18. Adjust the ignition timing as described in Chapter Five.

Ignition Module Replacement

The ignition module can be replaced without removing the distributor from the engine. On Ford 5.0L and 5.8L electronic fuel injection (EFI) models using thick film integrated (TFI) ignition, the ignition module is located on a bracket near the distributor. On all other models using electronic spark timing (EST) or Prestolite BID ignition, the ignition module is located inside the distributor housing.

Models with a Delco EST ignition system

The ignition module is located in the distributor housing. To prevent overheating and module failure, apply heat transfer compound to the bottom of the module before installation. A small tube of compound is supplied with the replacement module. If the compound is removed and the module must be reused, purchase the compound from a GM automotive dealership.

Refer to **Figure 91** or **Figure 93** for this procedure.

1. Disconnect the negative battery cable.
2. Remove the distributor cap as described in this section.
3. Pull up on the locking tab and then disconnect the pickup coil connector (**Figure 101**) from the ignition module.
4. Pull up the locking tab and then unplug the engine wire harness connector(s) from the ignition module (11, **Figure 91** or 5, **Figure 93**). A single harness and connector is used on carburetor-equipped models. Two harnesses and connectors are used on electronic fuel injection (EFI) models.
5. Remove the two screws and then lift the module off the distributor.
6. Use compressed air to blow loose material off the distributor.
7. Wipe the heat transfer compound from the distributor and the bottom of the module.
8. Apply an even film of the compound to the bottom of the module. The bottom of the module is metal.
9. Align the tabs on the bottom of the module with the corresponding openings in the distributor. Firmly seat the module on the distributor to spread the compound.
10. Install the module mounting screws and securely tighten.
11. Plug the pickup coil connector (**Figure 101**) onto the module. The locking tab must engage the tab on the module.
12. Pull up on the locking tab and then carefully plug the engine wire harness connector(s) onto the ignition module (11, **Figure 91** or 5, **Figure 93**).
13. To avoid electrical interference, route the wiring away from the ignition coil or spark plug secondary leads.
14. Install the distributor cap as described in this chapter.
15. Connect the battery cable.

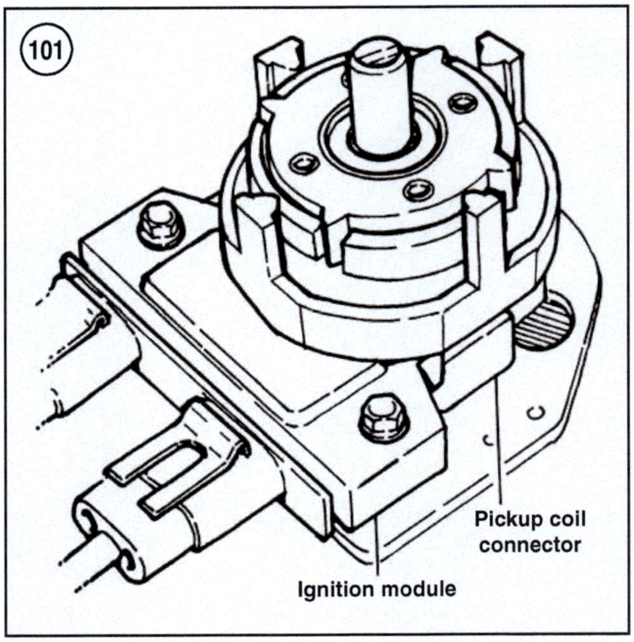

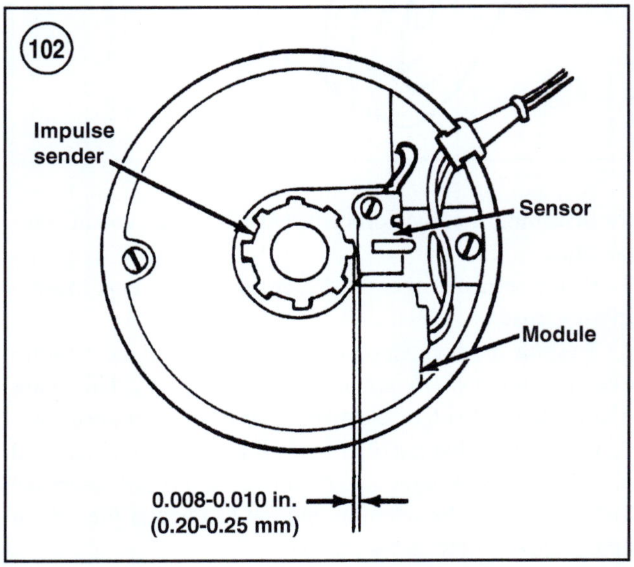

16. Adjust the ignition timing as described in Chapter Five.

Models with a Prestolite BID ignition system

The ignition module is located in the distributor housing. To prevent an ignition system malfunction, adjust the air gap between the sensor and the impulse sender (incorporated into the module and distributor shaft) during module installation. Refer to **Figure 94** for this procedure.

1. Disconnect the negative battery cable.
2. Remove the distributor cap and rotor as described in this section.
3. Disconnect the ignition module leads coming out of the distributor housing from the engine wire harness.

103

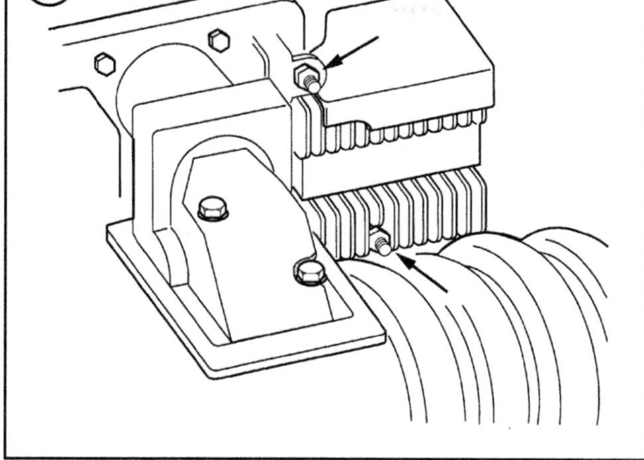

104

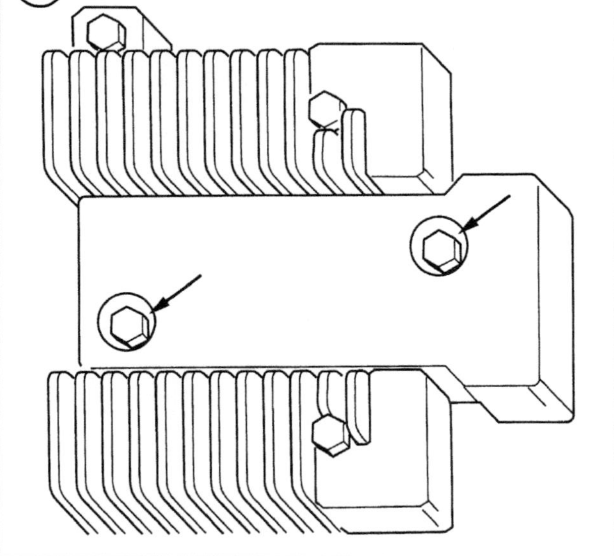

4. Remove the two screws and lift the ignition module (4, **Figure 94**) out of the distributor.

5. Use compressed air to blow loose material off the module and distributor.

6. Wipe any oil film or debris from the sensor or mount in the distributor.

7. Guide the wiring through the opening and then install the replacement module into the distributor. Seat the grommet on the wiring into the distributor opening.

8. Align the openings and then thread the two screws into the module and distributor. Do not tighten the screws at this time.

9. Remove the spark plugs. Rotate the harmonic balancer until the sensor (incorporated into the module) aligns with a finger on the impulse sender (incorporated into the distributor shaft) as shown in **Figure 102**.

10. Insert a 0.009 in. (0.22 mm) feeler gauge into the gap between the sensor and sender. Hold the module in light contact against the feeler gauge and then securely tighten the module mounting screws. After tightening the screws, the gap must be 0.008-0.010 in. (0.20-0.25 mm). Reposition the module as needed.

11. Connect the ignition module leads to the engine wire harness. To avoid electrical interference, route the wiring away from the ignition coil or spark plug secondary leads.

12. Install the rotor and distributor cap as described in this section.

13A. On GM 5.0L and 5.7L models, install the spark plugs and tighten to 22 ft.-lb. (30 N•m) if the cylinder head is new and 15 ft.-lb. (20 N•m) if the cylinder head is used.

13B. On Ford 5.0L and 5.8L models, install the spark plugs and tighten to 120-180 in.-lb. (14-20 N•m).

13C. On GM 7.4L and 8.2L models, install the spark plugs and tighten to 22 ft.-lb. (30 N•m).

14. Connect the battery cables and spark plug leads.

15. Adjust the ignition timing as described in Chapter Five.

Models with a Thick Film Integrated (TFI) ignition system

The ignition module is located near the ignition coil on the ignition system mounting bracket. To prevent overheating and module failure, apply heat transfer compound to the mating surface of the module before installation. A small tube of compound is supplied with the replacement module. If the compound is removed and the module must be reused, purchase the compound from a Ford automotive dealership.

Refer to **Figure 95** for this procedure.

1. Disconnect the negative battery cable.

2. Remove the ignition coil (18, **Figure 95**) as described in this chapter.

3. Push in on the side clips and then disconnect the wire harness connector from the ignition module (17, **Figure 95**).

4. Remove the two nuts (**Figure 103**) that secure the heat sink (22, **Figure 95**) to the bracket. Remove the heat sink and module from the bracket.

5. Remove the two screws (**Figure 104**) that secure the module to the heat sink. Carefully pull the module off the heat sink.

6. Use compressed air to blow loose material off the heat sink and module.

7. Wipe the heat transfer compound from the heat sink and module.

8. Apply an even film of the compound to the metal side of the module. The metal side must contact the heat sink.

9. Align the screw openings in the module and heat sink and then firmly seat the module on the heat sink to spread the compound.

10. Thread the two screws (**Figure 104**) into the module and heat sink. Tighten the screws to 24-36 in.-lb. (3-4 N•m).

9

11. Install the heat sink and module assembly onto the bracket. Secure the assembly to the bracket with the two nuts (**Figure 103**). Tighten the nuts to 60-84 in.-lb. (7-10 N•m).
12. Push in on the side clips and then plug the wire harness connector onto the ignition module. Release the side clips and then tug on the connector to verify a secure connection.
13. Install the ignition coil as described in this chapter.
14. Connect the battery cable.
15. Adjust the ignition timing as described in Chapter Five.

Hall Effect Switch

Removal and installation

The Hall effect switch (stator) is used on Ford 5.0L and 5.8L models with electronic fuel injection (EFI). The switch is located within the distributor housing.

1. Disconnect the negative battery cable.
2. Remove the distributor cap and rotor as described in this chapter.
3. Remove the two screws and then lift the distributor base (4, **Figure 95**) off the distributor housing (8).
4. Remove the two screws (1, **Figure 105**) and then lift the switch wheel (2) off the distributor shaft.
5. Remove the two screws (3, **Figure 105**) and then carefully pull the switch and harness (5) out of the distributor housing.
6. Use compressed air to remove any debris from the distributor housing, distributor base and switch. Use aerosol carburetor cleaner to remove any oil film from the distributor housing and switch wheel. Do not apply aerosol cleaner to the switch.
7. Install the Hall effect switch into the distributor housing as follows:
   a. Carefully guide the switch (5, **Figure 105**) into the distributor housing.
   b. Move the switch to align the harness connector tab on the switch with the notch in the housing. When aligned, the mounting tabs will align with the screw openings in the housing.
   c. Press the switch fully into the housing.
   d. Install the two mounting screws (4, **Figure 105**) and tighten to 15-35 in.-lb. (2-4 N•m).
8. Install the switch wheel (2, **Figure 105**) over the distributor shaft and seat against the step. Rotate the switch wheel to align the screw openings. Thread the two screws (1, **Figure 105**) into the wheel and distributor shaft openings. Verify proper seating of the wheel onto the shaft step and then tighten the two screws to 25-35 in.-lb. (3-4 N•m).
9. Install the distributor base (4, **Figure 95**) onto the distributor housing (8). Pivot the base to align the marks made prior to cap and rotor removal. The locating tab on the base will align with the notch in the distributor housing when aligned. Make sure the base is fully seated on the housing and the screw openings align. Thread the screws into the distributor base and housing. Securely tighten the screws.

(105) **THICK FILM INTEGRATED DISTRIBUTOR (FORD 5.0L AND 5.8L ELECTRONIC FUEL INJECTION MODELS)**

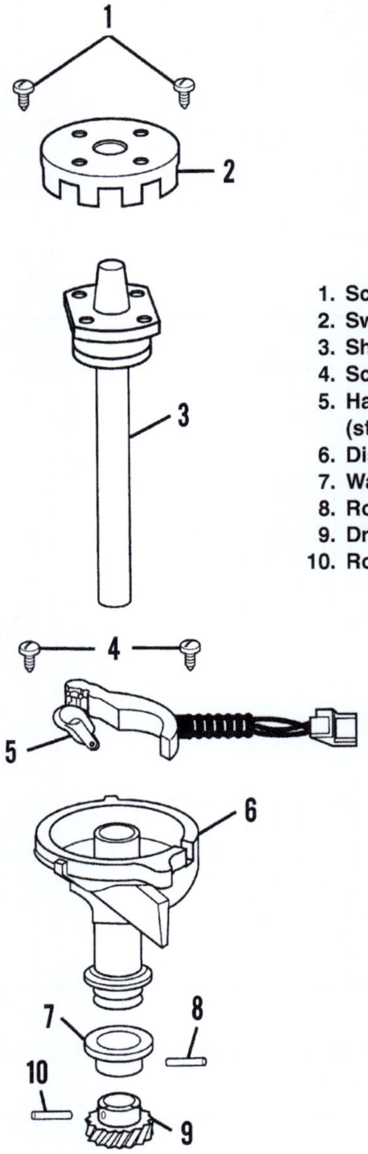

1. Screws
2. Switch wheel
3. Shaft
4. Screw
5. Hall effect switch (stator)
6. Distributor housing
7. Washer/spacer
8. Roll pin
9. Driven gear
10. Roll pin

10. Install the rotor and distributor cap as described in this chapter.
11. Connect the battery cable.
12. Adjust the ignition timing as described in Chapter Five.

Pickup Coil Replacement

The pickup coil is used on all models equipped with a Delco EST (electronic spark timing) ignition system. The

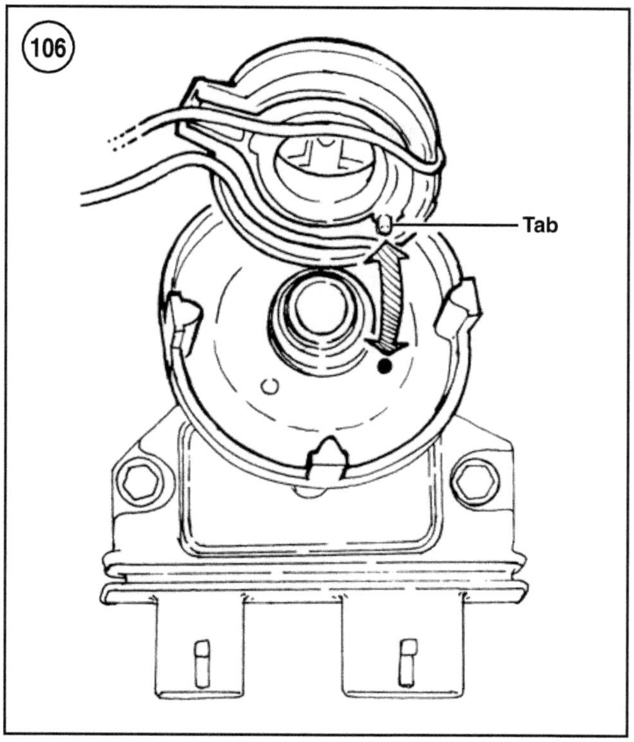

pickup coil is located in the distributor body. The coil can be replaced without removing the distributor.

Refer to **Figure 91** or **Figure 93** for this procedure.

1. Disconnect the negative battery cable.
2. Remove the distributor cap and rotor as described in this section.
3. Lift the locking tab and then disconnect the pickup coil connector (**Figure 101**) from the module.
4. Grip two opposing corners of the retainer (4, **Figure 91** or 6, **Figure 93**) with needlenose pliers. Pull upward on the corners until the retainer slips off the distributor bushing. Discard the retainer.
5. On GM V-6 and V-8 models, lift the copper shield off the pole piece (8, **Figure 93**). Remove the washer (7, **Figure 93**) from the pickup coil (8).
6. Lift the pickup coil out of the distributor. Use compressed air to blow debris and loose material from the distributor.

7. Guide the tab on the pickup coil into the opening (**Figure 106**) and seat the pickup coil into the pole piece.
8. On GM V-6 and V-8 models, align the notch over the pickup coil wiring and insert the copper shield onto the pole piece (8, **Figure 93**). Install the washer (7, **Figure 93**) over the distributor shaft and seat it on the pickup coil (8).
9. Fit a *new* retainer over the distributor bushing with the teeth side facing outward. Position a deep 5/8 in. socket over the bushing and seat against the retainer. Tap on the socket until the teeth on the retainer fit into the groove in the bushing. The retainer must hold the shield and pickup coil firmly.
10. Plug the pickup coil connector (**Figure 101**) onto the module. The locking tab must engage the tab on the module.
11. Install the rotor and distributor cap as described in this chapter.
12. Connect the battery cable. Adjust the ignition timing as described in Chapter Five.

9

Ignition Coil Replacement

On 3.0L models, the electronic spark timing (EST) ignition coil (**Figure 107**) is located on the starboard side of the cylinder block.

On GM V-6 and V-8 models with an Delco EST ignition system, the ignition coil (1, **Figure 93**) is located on the intake manifold or back of the port cylinder head.

On carburetor-equipped V-8 models using a Prestolite BID ignition system, the ignition coil (**Figure 108**) is located on the rear of the port side cylinder head.

On Ford 5.0L and 5.8L electronic fuel injection (EFI) models, using a thick film integrated (TFI) ignition system, the ignition coil (18, **Figure 95**) mounts onto the bracket near the front starboard side of the engine.

Models using a Delco EST ignition system

1. Disconnect the negative battery cable.

2. Pull up on the locking tabs and remove the black and gray wire harness connectors as an assembly. Separate the connectors after removal.
3. Grip the rubber boot and then carefully pull the secondary wire off the coil tower. Do not tug on the wire.
4. Remove the two mounting bolts and lift the coil off the engine.
5. Use compressed air to blow debris and loose material off the coil mounting surface. Remove rust or other contaminants from the mating surface. To operate at full efficiency, the coil must have good contact with an engine ground.
6. Wire all material off the coil tower. Inspect the coil tower for hairline cracks or carbon tracking. Replace the coil if either of these defects are evident.
7. Install the coil onto the engine and secure with the two bolts. Tighten the bolts evenly to 20-25 ft.-lb. (27-34 N•m).
8. Plug the coil secondary lead onto the coil tower.
9. Fit the gray wire harness connector into the black harness connector. Match the shape of the openings with the shape of the coil terminal. Carefully plug both connectors as an assembly onto the coil. Make sure the locking tab engages the tab on the coil.
10. Connect the battery cable.

Models using a Prestolite BID ignition system

1. Disconnect the negative battery cable.
2. Grip the rubber boot and then carefully pull the secondary wire off the coil tower. Do not tug on the wire.
3. Remove the terminal nuts and then lift the gray and purple wires off the coil terminals.
4. Loosen the screw and nut that secures the coil into the mounting bracket. If the screw is not accessible, remove the bolt that secures the coil and bracket to the cylinder head.
5. Pull the coil out of the mounting bracket. Remove the sleeve from the coil or mount bracket. Replace the sleeve if it is torn or deteriorated.
6. Wipe all material off the coil tower. Inspect the coil tower for hairline cracks or carbon tracking. Replace the coil if either of these defects are evident.
7. Apply a soap-and-water solution to the sleeve and then slide it over the ignition coil.
8. Insert the coil into the mounting bracket and carefully slide it into position. Securely tighten the screw and nut that secures the coil into the bracket. If removed for access, install the coil and bracket assembly onto the cylinder head. Install the bolt into the bracket and cylinder head openings and tighten securely.
9. Plug the coil secondary lead onto the coil tower.
10. Attach the gray harness wire to the negative coil terminal, and the purple wires connect to the positive coil terminal. Install and securely tighten the terminal nuts.
11. To prevent terminal corrosion and reduce arcing to the terminals, thoroughly coat the terminals with OMC Black Neoprene Dip (OMC part No. 909570). Purchase the neoprene dip from an OMC dealership.

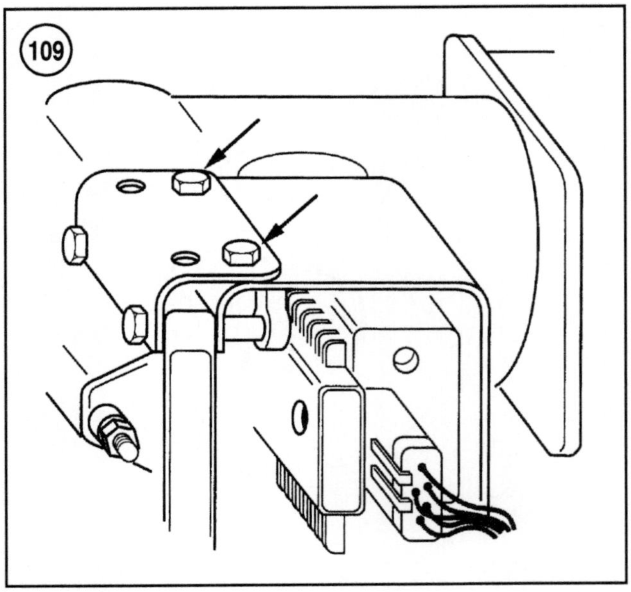

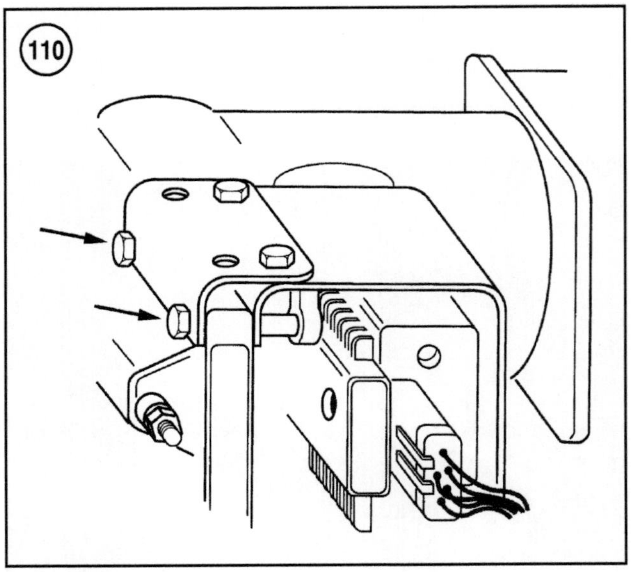

12. Connect the battery cable.

Models using a Thick Film Integrated (TFI) ignition system

Refer to **Figure 95** for this procedure.

1. Disconnect the negative battery cable.
2. Remove the plastic cover from the front of the engine.
3. Grip the rubber boot and then carefully pull the secondary wire off the coil tower. Do not tug on the wire.
4. Use a small screwdriver to release the tab that locks the engine harness connector onto the ignition coil terminal. Then pull the connector off the coil.
5. Remove the two screws (**Figure 109**) that secure the coil mounting bracket (23, **Figure 95**) to the ignition module mounting bracket (21).
6. Remove the two screws (**Figure 110**) that secure the ignition coil mounting bracket and coil to the ignition module

(111)

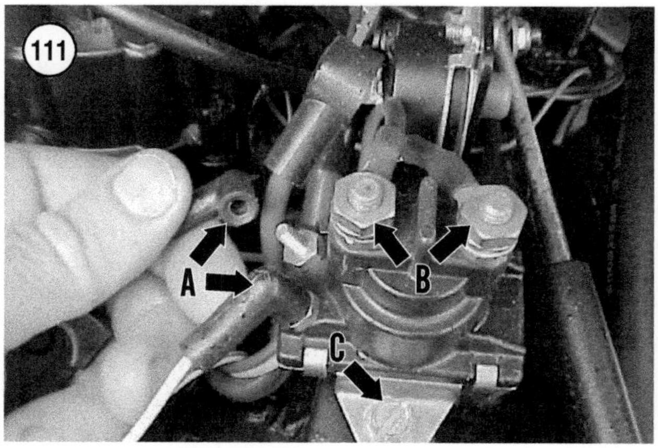

(112)

mounting bracket. Then remove the coil (18, **Figure 95**) from the bracket.

7. Use compressed air to blow debris and loose material off the coil mounting surface. Remove rust or other contaminants from the mating surface.

8. Wipe all material off the coil tower. Inspect the coil tower for hairline cracks or carbon tracking. Replace the coil if either of these defects are evident.

9. Fit the coil (18, **Figure 95**) onto the mounting bracket. Make sure the four openings are on the top side of the bracket.

10. Thread the two longer screws (**Figure 110**) into the coil mounting bracket, ignition coil and module mounting bracket. Do not tighten the screws at this time.

11. Thread the two shorter screws (**Figure 109**) into the coil mounting bracket and module mounting bracket. Tighten the longer screws to 24-36 in.-lb. (3-4 N•m). Tighten the shorter screws to 60-84 in.-lb. (7-10 N•m).

12. Apply a light coat of electrical terminal grease (Ford part No. F2AZ-19584-A) to the connector plug and then plug the engine harness connector onto the ignition coil terminal. Tug on the connector to verify a secure connection.

13. Apply a coat of OMC EP Wheel Bearing Grease onto the ignition coil tower and then push the secondary lead fully over the coil tower.

14. Install the plastic cover onto the top of the engine. Thread the screws into the cover. Then tighten them to 24-36 in.-lb. (3-4 N•m).

15. Connect the battery cable.

STARTER SLAVE SOLENOID

Removal and Installation

The starter slave solenoid (**Figure 111**, typical) is used on Ford 5.0L and 5.8L models. On most models, the solenoid is located within the circuit breaker housing (**Figure 112**). On carburetor-equipped models, the solenoid is mounted on the ignition system bracket on the top and port side of the engine. On electronic fuel injection (EFI) models, the solenoid is located on a bracket on the front starboard side of the engine.

Identify the slave solenoid by the yellow/red and red wires connecting to the terminals. The manufacturer has used different brands and types of solenoids, and the type used on later models can be installed on some of the earlier models.

1. Disconnect the battery cables.

2. If so equipped, remove the terminal protector from the top of the solenoid. 9

3. If the solenoid is not mounted on the ignition system bracket, remove the screws or pushpins. Then pull the plastic cover (**Figure 112**) off the circuit breaker housing. The solenoid mounts on the back side of the housing bracket.

4. Disconnect the yellow/red and black wires from the small terminals (A, **Figure 111**, typical).

5. Remove the terminal nuts and then disconnect the yellow/red and red/purple wires from the larger terminals (B, **Figure 111**, typical).

6. Remove the mounting screws or nuts (C, **Figure 111**, typical) and then lift the solenoid from the mount.

7. Clean all corrosion, belt material or oily deposits from the wire terminals and the solenoid mount.

8. Install the replacement solenoid onto the mount. Install the mounting screws or nuts and tighten securely.

9. Connect the yellow/red and red/purple wires to the larger terminals (B, **Figure 111**, typical). The larger wires can be installed onto either large terminal.

10. Connect the yellow/red and black wires onto the small terminals (A, **Figure 111**, typical). The smaller wires can be installed onto either terminal.

11. Make sure the wire terminals do not contact other terminals or other components on the engine. Then securely tighten the terminals. To prevent terminal corrosion and possible repeat solenoid failure, thoroughly coat the terminals with OMC Black Neoprene Dip (OMC part No. 909570). Purchase the neoprene dip from an OMC dealership.

12. If so equipped, install the terminal protector onto the solenoid.

13. If the solenoid is mounted within the circuit breaker housing, push the plastic cover fully over the housing bracket. Push the plastic pins into the openings or install the screws to secure the cover.
14. Connect the battery cables.

STARTER RELAY

Removal and Installation

The starter relay (**Figure 113**, typical) is used on all GM models. The starter relay is physically identical to the electronic fuel injection (EFI) system and electric fuel pump relays and is mounted in the same general location on some models.

On GM 3.0L models, the starter relay (**Figure 114**) is located on the rear starboard side of the engine.

On most V-6 and V-8 models, the starter relay is mounted on an engine-mounted bracket (**Figure 113**). On some models, the relay is located near or within the circuit breaker housing (**Figure 112**).

NOTE
Identify the starter relay by the two yellow/red wires leading to the harness connector. The system relay and fuel pump relay do not use two yellow/red wires; however, on some models, the fuel pump relay will use a single yellow/red wire.

113

114

1. Disconnect the battery cables.
2. If the relay is not mounted on the cylinder block (3.0L models) or the ignition system bracket, remove the screws or pushpins. Then pull the plastic cover (**Figure 112**) off the circuit breaker housing. The relay mounts on the back side of the housing bracket.
3A. If the relay is mounted in a bracket, work the relay back and forth while carefully pulling it out of the mounting bracket.
3B. If the relay plugs into the engine harness, depress the locking tab and then pull the engine harness connector out of the bottom of the starter relay.
4. Use compressed air to remove any debris from the terminals within the mounting bracket or engine harness connector. Inspect the terminals for corroded or bent terminals. Replace or repair the terminals if needed.
5A. If the relay is mounted in a bracket, align the spade terminals in the bottom of the relay with the openings in the bracket. The relay will only fit one way. Carefully push the starter relay fully into the terminals.
5B. If the relay plugs into the engine harness connector, align the spade terminals in the bottom of the relay with the openings in the harness plug. Depress the locking tab and then push the engine harness connector onto the starter relay.
6. If the relay is mounted within the circuit breaker housing, push the plastic cover fully over the housing bracket. Push the plastic pins into the openings or install the screws to secure the cover.
7. Connect the battery cables.

FUEL PUMP RELAY

Removal and Installation

The fuel pump relay is used on all electronic fuel injection (EFI) models and later carburetor-equipped V-6 and V-8 models. The fuel pump relay is physically identical to the starter relay and EFI system relay used on some models.

On carburetor-equipped V-6 and V-8 models, the fuel pump relay (**Figure 115**) mounts on a bracket at the front of the engine near the electric fuel pump. On GM 4.3L, 5.0L and 5.7L electronic fuel injection (EFI) models, the fuel pump relay mounts alongside the EFI system relay on a bracket between the thermostat housing and the throttle body assembly. On Ford 5.0L and 5.8L EFI models, the relay is located on the bracket at the front starboard side of the engine. On GM 7.4L and 8.2L EFI models, the relay is located on a bracket at the front starboard side of the engine.

NOTE
Identify the fuel pump relay used on carburetor-equipped models by the single yellow/red and single orange wires leading into the harness connector. On GM V-6 and V-8 EFI mod-

els, identify the fuel pump relay by the single orange/white and two yellow/blue wires leading into the harness connector. On Ford 5.0L and 5.8L EFI models, identify the fuel pump relay by the single blue/orange and the two green/yellow wires leading into the harness connector. The starter relay and EFI system relays do not use these color combinations.

1. Disconnect the negative battery cable.

2A. On V-6 and V-8 carburetor-equipped models, remove the single screw that secures the guard to the fuel pump bracket. Pull the tab on the guard out of the bracket opening and then remove the guard. Depress the locking tab and carefully lift the relay (**Figure 115**) out of the mounting bracket.

2B. On GM 4.3L, 5.0L and 5.7L EFI models, remove the flame arrestor cover as described in Chapter Seven. The fuel pump relay is located on the port side of the mounting bracket. The EFI system relay is located on the starboard side. Depress the locking tab and then pull the relay out of the mounting bracket.

2C. On Ford 5.0L and 5.8L EFI models, lift the locking tab and then unplug the engine wire harness from the relay. Remove the screw and then remove the relay from the mounting bracket.

2D. On GM 7.4L and 8.2L electronic fuel injection (EFI) models, lift up on the locking tab, then unplug the engine wire harness from the relay. Remove the screw then remove the relay from the mounting bracket.

3. Use compressed air to remove any debris from the terminals within the mounting bracket or engine harness connector. Check for corroded or bent terminals. Replace or repair the terminals if needed.

4A. If the relay is mounted in a bracket, align the spade terminals in the bottom of the relay with the openings in the bracket. The relay will only fit one way. Carefully push the starter relay fully into the terminals.

4B. If the relay plugs into the engine harness connector, align the spade terminals in the bottom of the relay with the openings in the harness plug. Depress the locking tab and then push the engine harness connector onto the starter relay. Thread the screw into the relay tab and bracket opening. Securely tighten the screw.

5. On GM 4.3L, 5.0L and 5.7L EFI models, install the flame arrestor cover as described in Chapter Seven.

6. Connect the battery cable.

ELECTRONIC FUEL INJECTION (EFI) SYSTEM RELAY

Removal and Installation

The electronic fuel injection (EFI) system relay is used on all EFI models. The system relay is physically identical to the fuel pump relay and starter relay used on some models.

On GM 4.3L, 5.0L and 5.7L EFI models, the system relay mounts alongside the fuel pump relay on a bracket between the thermostat housing and the throttle body assembly. On Ford 5.0L and 5.8L EFI models, the relay is located on the engine control assembly (ECA) bracket at the front and starboard side of the engine. On GM 7.4L and 8.2L EFI models, the relay is located on a bracket at the front starboard side of the engine.

NOTE

Identify the EFI system relay used on GM V-6 and V-8 EFI models by the single red/purple and single pink/white wires leading into the harness connector. On Ford 5.0L and 5.8L EFI models, identify the system relay by the single red/purple, single purple and single purple/black wires leading into the harness connector. The starter relay and fuel pump relay do not use these color combinations.

1. Disconnect the negative battery cable.

2A. On GM 4.3L, 5.0L and 5.7L models, remove the flame arrestor cover as described in Chapter Seven. The EFI system relay is located on the starboard side. The fuel pump relay is located on the port side of the mounting bracket. Depress the locking tab and then pull the EFI system relay out of the mounting bracket.

2B. On Ford 5.0L and 5.8L EFI models, cut the plastic locking clamp that secures the lower part of the relay to the bracket. Lift the locking tab and then unplug the engine wire harness from the relay. Remove the screw and then remove the relay from the mounting bracket.

2C. On GM 7.4L and 8.2L EFI models, lift up on the locking tab and then unplug the engine wire harness from the relay. Remove the screw then remove the relay from the mounting bracket.

3. Use compressed air to remove any debris from the terminals within the mounting bracket or engine harness connector. Check for corroded or bent terminals. Replace or repair the terminals if needed.

4A. If the relay is mounted in a bracket, align the spade terminals in the bottom of the relay with the openings in the bracket. The relay will only fit one way. Carefully push the starter relay fully into the terminals.

4B. If the relay plugs into the engine harness connector, align the spade terminals in the bottom of the relay with the openings in the harness plug. Depress the locking tab and then push the engine harness connector onto starter relay. Thread the screw into the relay tab and bracket opening. Securely tighten the screw.

5. On Ford 5.0L and 5.8L models, secure the lower part to the bracket using a suitable plastic locking clamp passing through the openings in the bracket and around the relay body.

6. On GM 4.3L, 5.0L and 5.7L EFI models, install the flame arrestor cover as described in Chapter Seven

7. Connect the battery cable.

OIL PRESSURE SENDER

Removal and Installation

NOTE
The oil pressure sender and oil pressure switch are similar in appearance and mount in the same area on the engine. Refer to the wiring diagrams at the end of the manual to identify the wires for the sender. A blue wire connects to the oil pressure sender. A tan/blue or tan/black wire connects to the oil pressure switch.

An oil pressure sender is used on all models. The sender operates the oil pressure gauge and is not part of the ignition or fuel injection system.

On GM 3.0L models, the sender is located on the starboard side of the cylinder block (**Figure 116**). On GM 4.3L, 5.0L and 5.7L models, the sender is located on the top rear of the cylinder block next to the distributor (**Figure 117**). On Ford 5.0L and 5.8L models, the sender is located on the port side of the engine near the oil filter. On GM 7.4L and 8.2L models, the sender is located on the rear port side of the cylinder block (A, **Figure 118**).

1. Disconnect the negative battery cable.
2. Remove the nut or unplug the push-on terminal to disconnect the blue wire from the sender terminal.
3. Place a shop towel under the sender to capture spilled oil.
4. Attach a wrench to the brass fitting to prevent rotation and then use another wrench to remove the sender from the fitting.
5. Clean all old sealant from the brass fitting and sender threads.
6. Apply a light coat of Loctite Pipe Sealant with Teflon to the sender threads. Purchase the sealant from an OMC dealership (OMC part No. 1141570), automotive parts store or GM automobile dealership (GM part No. 12346004).

116

117

7. Thread the sender into the fitting by hand. Attach a wrench to the brass fitting to prevent rotation and then tighten the sender to 124-168 in.-lb. (14-19 N•m).
8. Connect the blue wire to the sender terminal.
   a. If the blue wire uses a ring terminal, thread the nut onto the terminal and securely tighten. To prevent corrosion and loosening, thoroughly coat the terminal with OMC Black Neoprene Dip. Purchase the neoprene dip (OMC part No. 909570) from an OMC dealership.
   b. If the blue wire uses a push-on terminal, make sure the terminal fits snugly on the sender terminal. If the fit is loose, unplug the terminal and carefully squeeze the terminal to achieve a snug fit.
9. Connect the battery cable.

WARNING
Stay clear of the propeller and propeller shaft while running an engine on a flush adapter.

CAUTION
If running the engine with the boat out of the water, use a flush device or other means to supply the engine with cooling water. The seawater pump is quickly damaged if operated without adequate cooling water.

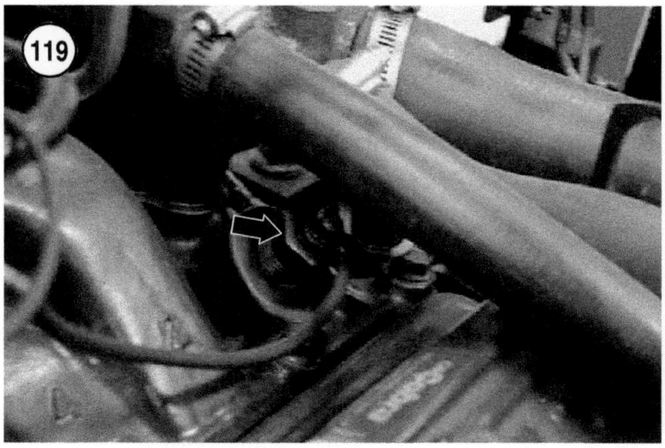

10. Start the engine and immediately check for oil leakage. Make sure the oil pressure gauge is operating properly. If the gauge is not operating, the sender might not have a good connection to engine ground. Temporarily attach a jumper wire to the sender body and a good ground. If the gauge now operates correctly, remove the sender, clean the threads and reinstall as described in this procedure.

OIL PRESSURE SWITCH

Removal and Installation

NOTE
The oil pressure switch and oil pressure sender are similar in appearance and mount in the same area on the engine. Refer to the wiring diagrams at the end of the manual to identify the wires for the sender. A tan/blue or tan/black wire connects to the oil pressure switch. A blue wire connects to the oil pressure sender.

An oil pressure switch (B, **Figure 118**, typical) is used on all electronic fuel injection (EFI) models and carburetor-equipped models with the optional audio warning system. The switch controls operation of the warning horn by completing the ground circuit if the oil pressure drops below a preset limit. The switch threads into a cylinder block T-fitting along with the oil pressure sender.

1. Disconnect the negative battery cable.
2. Pull the tan/blue wire off the switch terminal.
3. Place a shop towel under the switch to capture spilled oil.
4. Attach a wrench to the brass fitting to prevent rotation and then use another wrench to remove the switch from the fitting.
5. Clean all old sealant from the brass fitting and switch threads.
6. Apply a light coat of Loctite Pipe Sealant with Teflon to the switch threads. Purchase the sealant from an OMC dealership (OMC part No. 1141570), automotive parts store or GM automobile dealership (GM part No. 12346004).
7. Thread the replacement switch into the fitting by hand. Attach a wrench onto the brass fitting to prevent rotation and then securely tighten the switch.
8. Connect the tan/blue wire to the switch terminal. The terminal must fit snugly. If necessary, remove the connector and pinch the connector to achieve a snug fit.
9. Touch one ohmmeter lead to the metal body of the switch. Touch the other lead to an engine ground. The meter must indicate continuity. If not, remove the switch, clean the threads and reinstall the switch as described in this procedure. The switch cannot operate properly with a faulty ground connection.
10. Connect the battery cable.
11. Start the engine and immediately check for oil leakage. If there is leakage, remove the switch, clean the threads and reinstall the switch as described in this procedure.

COOLANT TEMPERATURE SENDER

Removal and Installation

A coolant temperature sender (**Figure 119**, typical) is used on all models. The sender operates the temperature gauge and is not part of the ignition or fuel injection system.

On GM 3.0L models, the sender threads into the port side of the thermostat housing at the front of the cylinder head.

On 4.3L, 5.0L and 5.7L models, the sender threads into the front starboard side of the intake manifold.

On Ford 5.0L and 5.8L models, the sender threads into the front starboard side of the intake manifold on carburetor-equipped models and the starboard side of the recirculation pump on electronic fuel injection (EFI) models.

On GM 7.4L and 8.2L models, the sender threads into the top of the recirculation pump on carburetor-equipped models and the front of the intake manifold on EFI models.

NOTE
The coolant temperature sender, coolant temperature sensor and engine overheat switch are very similar in appearance and might

mount in the same area on the engine. A tan wire connects to the engine temperature sender. A tan/blue or tan/black wire connects to the overheat switch.

1. Disconnect the negative battery cable.

2A. On models using an standard open cooling system, drain all water from the cylinder block as described in *Winterizing* in Chapter Four.

2B. On models with a closed cooling system, drain coolant from the cylinder block as described in Chapter Three. Refer to *Changing the Engine Coolant*.

3. Reinstall the drain plugs after draining the water or coolant.

4. Remove the terminal nut and then disconnect the tan wire from the sender terminal.

5. Use a deep socket to remove the sender. Thoroughly clean the threads on the sender and in the thermostat housing or intake manifold.

6. Apply a light coat of Loctite Pipe Sealant with Teflon to the sender threads. Purchase the sealant from an OMC dealership (OMC part No. 1141570), automotive parts store or GM automobile dealership (GM part No. 12346004).

7. Thread the replacement sender into the thermostat housing or intake manifold by hand and then use a deep socket to securely tighten the sender.

8. Connect the tan wire to the sender terminal. Install the brass terminal nut and tighten securely. To prevent corrosion and loosening, thoroughly coat the terminal nut with OMC Black Neoprene Dip. Purchase the neoprene dip (OMC part No. 909570) from an OMC dealership.

9. On models with a closed cooling system, refill the coolant as described in Chapter Three. Refer to *Changing the Engine Coolant*.

10. Connect the battery cable.

WARNING

Stay clear of the propeller and propeller shaft while running an engine on a flush adapter.

CAUTION

If running the engine with the boat out of the water, use a flush device or other means to supply the engine with cooling water. The seawater pump is quickly damaged if operated without adequate cooling water.

11. Start the engine and immediately check for water or coolant leakage at the sender threads. Allow the engine to operate long enough to make sure the temperature gauge is operating properly. If the gauge is not operating, the sender might not have a good connection to an engine ground. Temporarily attach a jumper wire to the sender body and a good ground. If the gauge now reads, remove the sender, clean the threads and reinstall as described in this procedure.

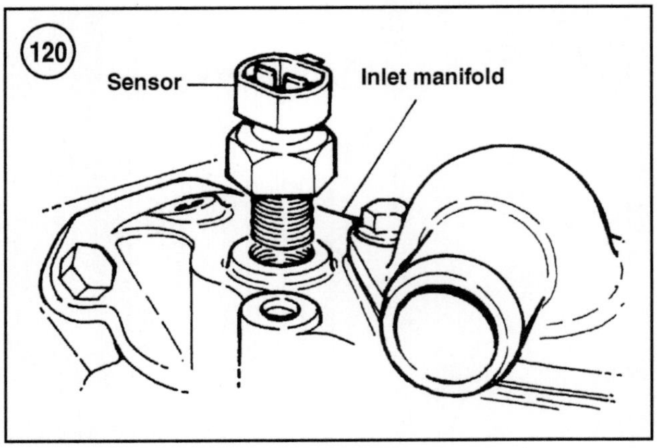

COOLANT TEMPERATURE SENSOR

Removal and Installation

A coolant temperature sensor (**Figure 120**, typical) is used on all electronic fuel injection (EFI) models. The sensor provides a varying resistance value to the engine control unit (ECU) or engine control assembly (ECA) to indicate the water or coolant temperature in the cooling system. The ECU or ECA adjusts the fuel delivery and ignition timing to the optimum setting for the given temperature. On GM models, the ECU also uses the input to operate the warning horn if the engine overheats. On Ford models, an overheat switch is used to operate the warning horn if the engine overheats. The coolant temperature sensor is not part of the circuit that operates the engine temperature gauge.

On GM 4.3L, 5.0L and 5.7L models, the sensor threads into the front side of the thermostat housing.

On Ford 5.0L and 5.8L models, the sensor threads into the front port side of the intake manifold.

On GM 7.4L and 8.2L models, the sensor threads into the front of the intake manifold and near the engine temperature sender. The sensor is close to the starboard side, and the sender is close to the port side.

NOTE

The coolant temperature sender, coolant temperature sensor and engine overheat switch are very similar in appearance and might mount in the same area on the engine. Identify the coolant temperature sensor by the two red/black and yellow/blue or red/black and green/pink wires leading into the harness connector for the sensor. The coolant temperature sender and overheat switch use a single wire connector.

1. Disconnect the negative battery cable.

2A. On models with a standard open cooling system, drain all water from the cylinder block as described in *Winterizing* in Chapter Four.

2B. On models with a closed cooling system, drain coolant from the cylinder block as described in Chapter Three. Refer to *Changing the Engine Coolant*.

3. Reinstall the drain plugs after draining the water or coolant.

4. Lift the locking tab, then unplug the engine wire harness connector from the sensor. The engine temperature *sensor* uses two wires use a locking type plug on connector. Do not confuse the engine temperature sensor with the engine temperature sender or engine overheat switch which use single wire with a push-on or ring type terminals connector.

5. Use a deep socket to remove the sensor. Thoroughly clean the threads on the sensor and in the thermostat housing or intake manifold.

6. Apply a light coat of Loctite Pipe Sealant with Teflon to the sender threads. Purchase the sealant from an OMC dealership (OMC part No. 1141570), automotive parts store or GM automobile dealership (GM part No. 12346004).

7. Thread the replacement sensor into the opening by hand. Using a deep socket, tighten the sensor to 106 in.-lb. (12 N•m) for GM V-6 and V-8 models. For Ford V-8 models, tighten the sensor to 120-180 in.-lb. (14-20 N•m).

8. Lift the tab and then plug the engine harness connector fully onto the sensor. The tab must engage the tab on the sensor.

9. On models with a closed cooling system, refill the coolant as described in Chapter Three. Refer to *Changing the Engine Coolant*.

10. Connect the battery cable.

WARNING

Stay clear of the propeller and propeller shaft while running an engine on a flush adapter.

CAUTION

If running the engine with the boat out of the water, use a flush device or other means to supply the engine with cooling water. The seawater pump is quickly damaged if operated without adequate cooling water.

11. Start the engine and immediately check for water or coolant leakage at the sensor threads. If leaking, remove the sensor, clean the threads and reinstall the sensor as described in this procedure.

ENGINE OVERHEAT SWITCH

Removal and Installation

An engine overheat switch (**Figure 121**, typical) is used on carburetor-equipped models with the optional audible warning system and Ford 5.0L and 5.8L electronic fuel injection (EFI) models. The switch controls operation of the warning horn by completing the ground circuit if the engine temperature exceeds a preset limit.

On carburetor-equipped models, the switch threads into the thermostat housing. On Ford 5.0L and 5.8L EFI models, the switch threads into the front starboard side of the intake manifold and near the thermostat housing.

NOTE

The engine overheat switch, coolant temperature sender and coolant temperature sensor are very similar in appearance and might mount in the same area on the engine. Identify the overheat switch by the single tan/blue or tan/black wire leading into the connector. The coolant temperature sender has a single tan wire leading into the connector. The coolant temperature sensor has the two red/black and yellow/blue or red/black and green/pink wires leading into the connector.

1. Disconnect the negative battery cable.

2A. On models with a standard open cooling system, drain all water from the cylinder block as described in *Winterizing* in Chapter Four.

2B. On models with a closed cooling system, drain coolant from the cylinder block as described in Chapter Three. Refer to *Changing the Engine Coolant*.

3. Reinstall the drain plugs after draining the water or coolant.

4. Disconnect the tan/blue or tan/black wire terminal from the switch.

5. Use a deep socket to remove the switch. Thoroughly clean the threads on the switch and in the thermostat housing.

6. Apply a light coat of Loctite Pipe Sealant with Teflon to the switch threads. Purchase the sealant from an OMC dealership (OMC part No. 1141570), automotive parts store or GM automobile dealership (GM part No. 12346004).

7. Thread the replacement switch into the opening by hand and then use a deep socket to securely tighten the switch.

8. Plug the tan/blue wire terminal onto the switch terminal. The terminal must fit securely. If not, disconnect the terminal and squeeze the terminal to provide a secure fit.

9. Touch one ohmmeter lead to the metal body of the switch. Touch the other lead to an engine ground. The meter must indicate continuity. If not, remove the switch, clean the threads and reinstall the switch as described in this procedure. The switch cannot operate properly with a faulty ground connection.

10. On models with a closed cooling system, refill the coolant as described in Chapter Three. Refer to *Changing the Engine Coolant*.
11. Connect the battery cable.

WARNING
Stay clear of the propeller and propeller shaft while running an engine on a flush adapter.

CAUTION
If running the engine with the boat out of the water, use a flush device or other means to supply the engine with cooling water. The seawater pump is quickly damaged if operated without adequate cooling water.

12. Start the engine and immediately check for water or coolant leakage at the switch threads. If leaking, remove the switch, clean the threads and reinstall the switch as described in this procedure.

MANIFOLD AIR PRESSURE SENSOR

The manifold air pressure sensor (MAP) is used on all electronic fuel injection (EFI) models. The sensor provides data to the engine control unit (ECU) or engine control assembly (ECA) that is used to compute the correct air-to-fuel ratio during all running conditions.

GM 4.3L, 5.0L and 5.7L V-8 Models

The manifold air pressure sensor (MAP) mounts within the circuit breaker housing on the starboard side exhaust manifold. A hose (**Figure 122**) connects to the sensor and throttle body fittings.

Removal and installation

1. Disconnect the negative battery cable. Remove the two screws or pull out on the plastic pushpins and then pull the plastic cover off the circuit breaker housing.
2. Disconnect the hose from the sensor fitting.
3. Lift the locking tab and then unplug the engine harness connector from the sensor.
4. Remove the two screws and lift the sensor off the bracket.
5. Inspect the hose for torn, pinched or deteriorated surfaces. Replace the hose if these or other defects are evident.
6. Install the replacement sensor onto the mounting bracket and align the mounting screw openings.
7. Install the mounting screws and tighten evenly to 44-62 in.-lb. (5-7 N•m).
8. Lift the locking tab and plug the engine harness connector fully onto the sensor. The locking tab must engage the tab on the sensor.

122

123

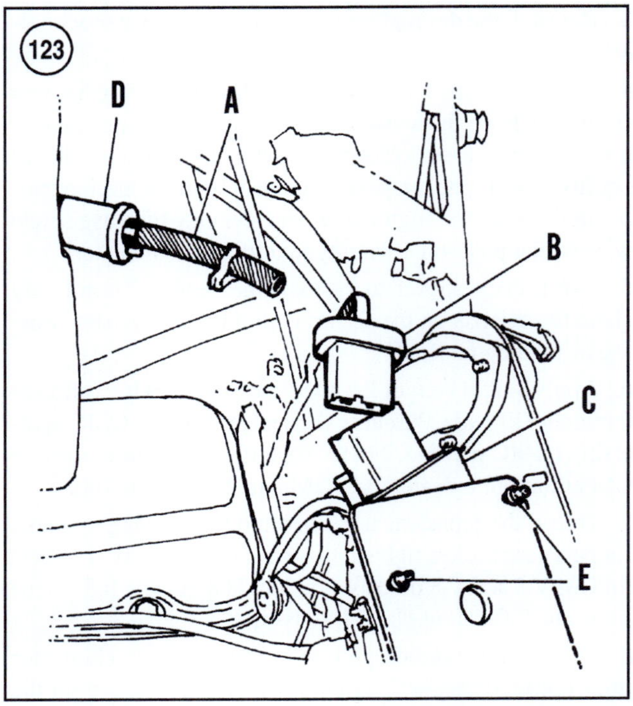

9. Plug the hose onto the sensor fitting. Install the plastic cover onto the housing. Secure the cover by pushing in the two plastic pushpins or with the two screws.
10. Connect the battery cable.

Ford 5.0L and 5.8L Models

The manifold air pressure sensor (MAP) mounts on the engine control assembly (ECA) bracket on the front starboard side of the engine. A hose with an inline filter connects the sensor and intake manifold fittings.

Removal and installation

1. Disconnect the negative battery cable.
2. Remove the self-test connector and plastic cover from the top of the engine control assembly (ECA) mounting bracket.

124

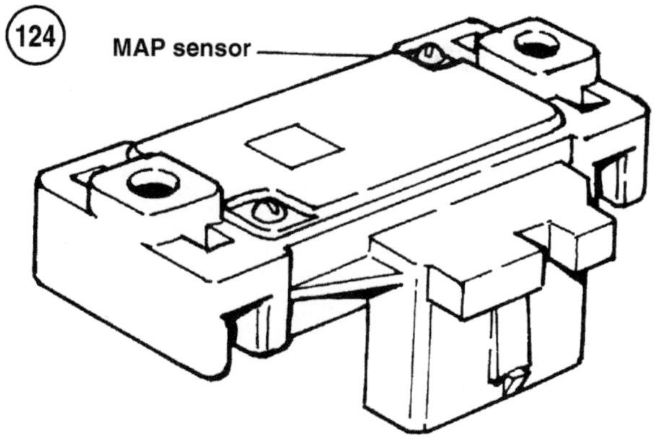

3. Remove the three screws that secure the ECA bracket to the engine.
4. Loosen the clamp and then disconnect the vacuum hose (A, **Figure 123**) from the MAP sensor (C).
5. Push in on the locking tab and then disconnect the engine wire harness connector (B, **Figure 123**) from the sensor.
6. Remove the two screws (E, **Figure 123**) and then pull the sensor off the bracket.
7. Cut the plastic locking clamps and then pull the inline filter (D, **Figure 123**) out of the vacuum hose. Clean the filter with mild solvent and dry with compressed air. Replace the filter if debris is noted within the translucent filter.
8. Use compressed air to remove debris from the sensor mounting bracket.
9. Insert the filter into the vacuum hose. The arrow cast into the filter body must face the hose connected to the sensor. Secure the filter to each hose with plastic locking clamps.
10. Fit the MAP sensor onto the bracket with the flat side facing upward. Thread the two screws into the sensor and bracket openings. Tighten the screws to 24-36 in.-lb. (3-4 N•m).
11. Apply a light coat of electric terminal grease (Ford part No. F2AZ-19584-A) to the terminals and then connect the engine wire harness connector (B, **Figure 123**) to the sensor connector. Tug on the connector to verify a secure connection.
12. Connect the vacuum hose to the sensor fitting. Secure the hose with the clamp.
13. Fit the ECA bracket onto the engine. Thread the three screws into the bracket and engine. Tighten the screws to 60-84 in.-lb. (7-10 N•m).
14. Install the plastic cover onto the ECA bracket. Install the self-test connector onto the ECA bracket. Tighten the cover screws to 24-36 in.-lb. (3-4 N•m).
15. Connect the battery cable.

GM 7.4L and 8.2L Models

The manifold air pressure (MAP) sensor (**Figure 124**) mounts on the top starboard side of the intake plenum. The sensor tip fits in an opening in the intake plenum. Two screws secure the sensor to the plenum.

Removal and installation

1. Disconnect the negative battery cable.
2. Use compressed air to blow debris or loose material from the intake plenum and sensor.
3. Lift the locking tab and then unplug the engine harness connector from the sensor.
4. Remove the two screws and then carefully pull the sensor out of the intake manifold.
5. Inspect the seal on the sensor tip for distorted or damaged surfaces and replace as needed.
6. Wipe oily deposits or other material off the manifold opening and the sensor mating surface.
7. Carefully guide the sensor tip into the manifold opening. Rotate the sensor *slightly* to prevent seal damage while guiding the sensor into the manifold.
8. When seated, rotate the sensor to align the screw openings. Thread the screws into the sensor and intake plenum openings. Tighten the screws to 44-62 in.-lb. (5-7 N•m).
9. Lift the locking tab and plug the engine harness connector fully onto the sensor. The locking tab must engage the tab on the sensor. Tug on the harness to verify a secure connection.
10. Connect the battery cable.

INTAKE AIR TEMPERATURE (IAT) SENSOR

The intake air temperature sensor is used only on electronic fuel injection (EFI) models that are equipped with multiport injection (MPI) systems. This includes Ford 5.0L, Ford 5.8L, GM 7.4L and GM 8.2L models.

Ford 5.0L and 5.8L Models

The IAT sensor threads into an opening on the port side of the intake manifold.

Removal and installation

1. Disconnect the negative battery cable.
2. Remove the four screws and then lift the plastic cover off the engine.
3. Remove the ignition coil as described in this chapter.
4. Push in the locking tab and then unplug the engine wire harness connector from the sensor (**Figure 125**).
5. Use compressed air to remove loose material from the sensor and surrounding areas on the intake manifold.
6. Use a deep socket to unthread the sensor from the intake manifold.
7. Clean any contaminants from the sensor threads and intake manifold threads. Do not allow solvent onto the tip of the sensor.
8. Thread the sensor by hand into the intake manifold. Tighten the sensor to 144-216 in.-lb. (16-24 N•m) with a deep socket.

9. Apply a light coat of electric terminal grease (Ford part No. F2AZ-19584-A) to the terminals and then connect the engine wire harness connector (B, **Figure 123**) to the sensor connector. Tug on the connector to verify a secure connection.
10. Install the ignition coil as described in this chapter.
11. Install the plastic cover onto the top of the engine. Install the four cover screws. Tighten to 24-36 in.-lb. (3-4 N•m).
12. Connect the battery cable.

GM 7.4L and 8.2L Models

The IAT sensor threads into an opening on the rear port side of the intake plenum.

Removal and installation

1. Disconnect the negative battery cable.
2. Push in the locking tab and then unplug the engine wire harness connector from the sensor (**Figure 125**).
3. Use compressed air to remove loose material from the sensor and surrounding areas on the intake manifold.
4. Use a deep socket to unthread the sensor from the intake manifold.
5. Clean the sensor threads and intake manifold threads. Do not allow solvent onto the tip of the sensor.
6. Thread the sensor by hand into the intake manifold opening. Use a deep socket to securely tighten the sensor.
7. Push in the locking tab and then connect the engine wire harness connector to the sensor connector. Release the locking tab and then tug on the connector to verify a secure connection.
8. Connect the battery cable.

THROTTLE POSITION SENSOR (TPS)

The throttle position sensor (TPS) is used on all electronic fuel injection (EFI) models.

GM V-6 and V-8 Models

Two screws secure the TPS (**Figure 126**) onto the starboard side of the throttle body.
1. Disconnect the negative battery cable. Verify that the throttle valve is completely closed.
2. Pull up the locking tab and then unplug the engine wire harness from the sensor.
3. Use compressed air to blow debris or loose material off the sensor and throttle body.
4. Remove the two screws (**Figure 127**) and then pull the sensor off the throttle body.
5. Retrieve the seal (**Figure 127**) from the sensor shaft or throttle body. Replace the seal if deteriorated or damaged.
6. Fit the seal onto the replacement TPS. Do not apply any sealant or other material to the seal.

125

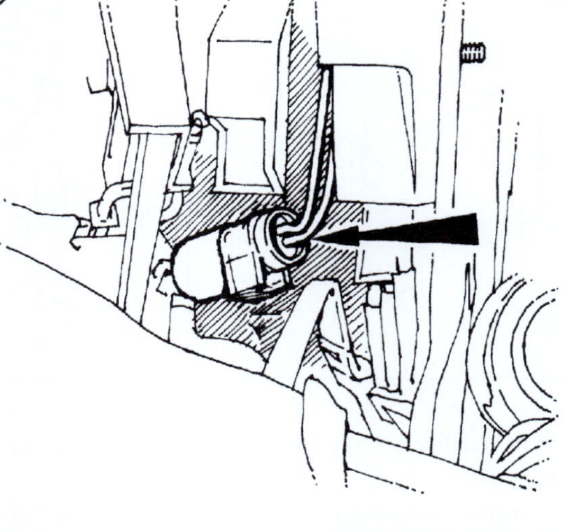

126

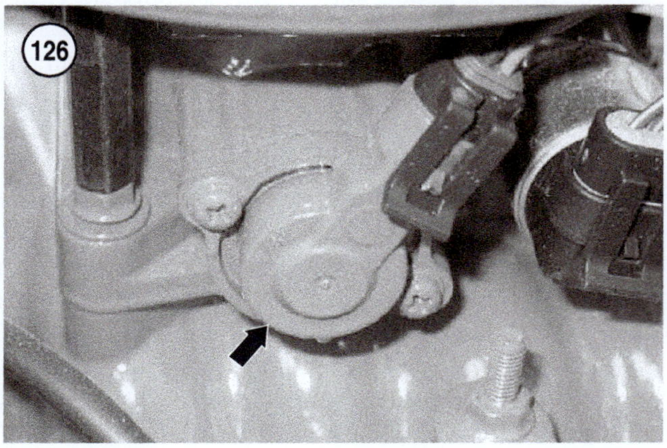

127

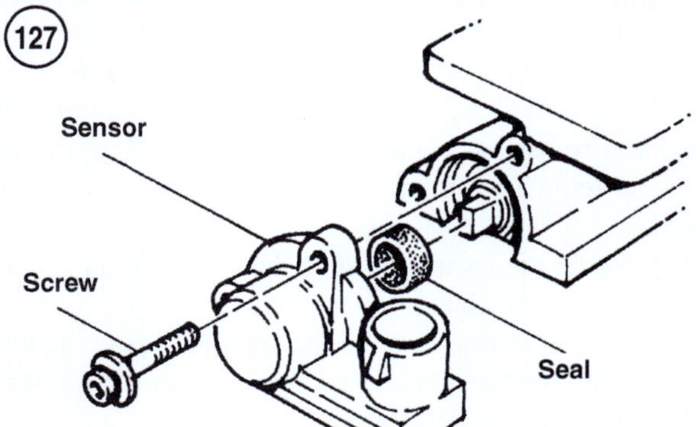

7. Rotate the sensor to align the flat surfaces on the sensor and throttle shaft and then fit the replacement sensor onto the throttle body.

8. Verify that the throttle valve is completely closed.

9. Rotate the sensor *slightly* counterclockwise on the throttle shaft to align the mounting screw openings and then seat the sensor on the throttle body.

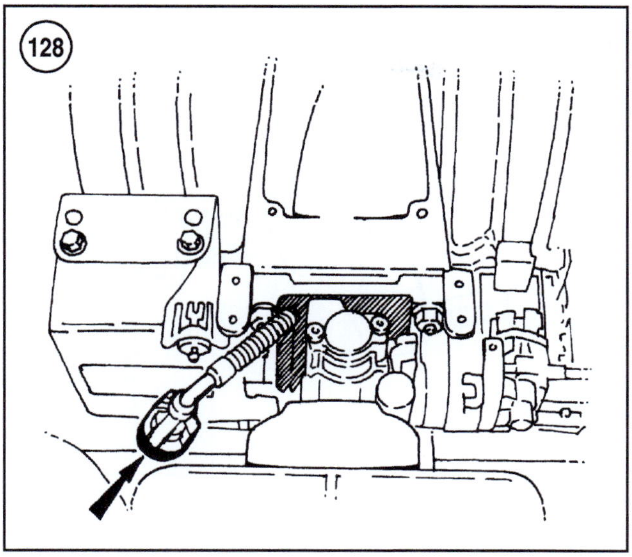

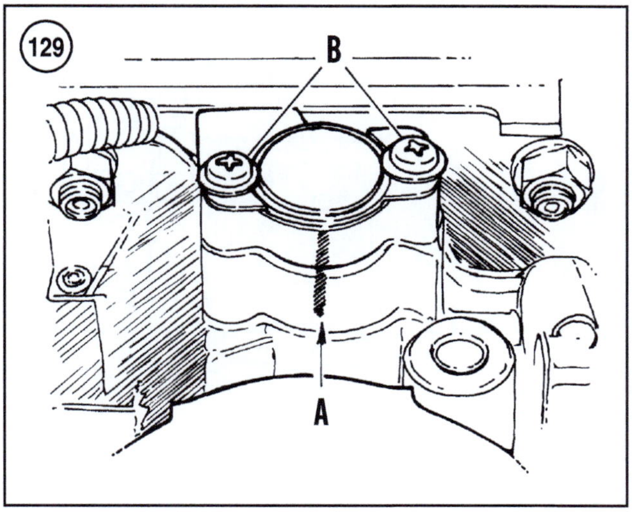

10. Thread the two screws into the sensor and throttle body. If new screws are supplied with the replacement sensor, use them during the installation. Tighten the two screws evenly to 18 in.-lb. (2.0 N•m).

11. Pull up the locking tab and then plug the engine wire harness connector onto the TPS. The tab on the harness connector must engage the tab on the sensor. Tug on the harness to verify a secure connection.

12. Connect the battery cable.

Ford 5.0L and 5.8L Models

Two screws secure the TPS onto the top of the throttle body.

1. Disconnect the negative battery cable.
2. Remove the four screws and then lift the plastic cover off the engine.
3. Use compressed air to remove loose material from the TPS and throttle body.
4. Disconnect the TPS three-wire harness connector (**Figure 128**) from the engine wire harness connector.
5. Make match marks (A, **Figure 129**) across the TPS and throttle body.
6. Remove the two screws (B, **Figure 129**) and then lift the sensor out of the throttle bore.
7. Fit the replacement TPS onto the throttle body. The sensor wire harness must be parallel to the throttle bore. Rotate the sensor to align the match marks (A, **Figure 129**). Rotate the sensor as needed to align the flat surfaces on the throttle shaft and sensor shaft. When aligned, the sensor will seat against the throttle body.
8. Thread the two screws (B, **Figure 129**) into the sensor and throttle body openings. Then tighten the screws to 18-26 in.-lb. (2-3 N•m).
9. Apply a light coat of electric terminal grease (Ford part No. F2AZ-19584-A) to the terminals. Then connect the TPS wire harness connector (**Figure 128**) to the engine wire harness connector. Tug on the harness to verify a secure connection.
10. Install the plastic cover onto the top of the engine. Install the four cover screws. Tighten to 24-36 in.-lb. (3-4 N•m).
11. Connect the battery cable.

9

IDLE AIR CONTROL MOTOR (IAC)

The idle air control motor is used on GM V-6 and V-8 electronic fuel injection (EFI) models. Two screws secure the IAC (**Figure 130**) to the throttle body. Refer to Chapter Seven to locate the throttle body on the engine.

Removal and installation

1. Disconnect the negative battery cable.
2. Pull up the locking tab and then unplug the engine wire harness from the IAC.
3. Use compressed air to blow debris or loose material off the motor and throttle body.
4. Remove the two screws and then pull the motor off the throttle body. If the IAC will be reused, *do not* remove the threadlocking compound from the screws.

5. Retrieve the O-ring (**Figure 131**) from the motor housing or throttle body opening. Discard the O-ring.

CAUTION
Do not use carburetor cleaner containing methyl ethyl ketone to clean the IAC. This type of cleaner removes the protective coat and can damage the component.

6A. If the IAC will be reused, wipe debris or contaminants from the motor body, spring and pintle shaft. Use aerosol type carburetor cleaner to remove stubborn deposits from the pintle shaft and spring. Work carefully to avoid moving the pintle shaft in or out. Moving the pintle will disturb the computer controlled setting and moving the pintle shaft with excessive force will damage the internal components.

6B. If a new IAC will be installed, *carefully* remove the wire retainer, then measure the distance from the tip of the pintle to the mating surface (**Figure 131**). The distance must be 1.102 in. (28 mm). If necessary, use light finger pressure and push on the pintle tip to achieve the desired distance. Light pressure will not damage the internal components. Measure the pintle diameter and compare with the diameter of the used IAC. The diameter must be the same. If not, the part is not correct for the engine model. Install the correct part.

7. Lubricate the surfaces with engine oil then fit a *new* O-ring onto the idle air control motor body (**Figure 131**).

8. *Carefully* guide the pintle shaft into the throttle body opening and seat the motor onto the throttle body. Work carefully to avoid contact and inadvertently moving the pintle shaft.

9. Rotate the motor on the throttle body to align the screw openings. Install the two screws and tighten evenly to 28 in.-lb. (3 N•m).

10. Lift up on the locking tab and then plug the engine wire harness connector onto the motor. The tab on the harness connector must engage the tab on the motor. Tug on the harness to verify a secure connection.

11. Connect the battery cable.

12. Start the engine and then put the engine into service mode as described under the ignition timing adjustment procedure in Chapter Five. Take the engine out of service mode after operating at idle speed for a few minutes. Operating the engine in service mode calibrates the IAC position to the engine control unit (ECU) setting. Improper calibration can cause hard starting (particularly when cold), excessively fast or excessively slow idle speed or stalling when returned to the idle position.

IDLE SPEED CONTROL SOLENOID (ISC)

The idle speed control solenoid is used on Ford 5.0L and 5.8L electronic fuel injection models. The ISC is located on the rear of the throttle body. The throttle body is located on the port side of the intake manifold and air plenum.

131

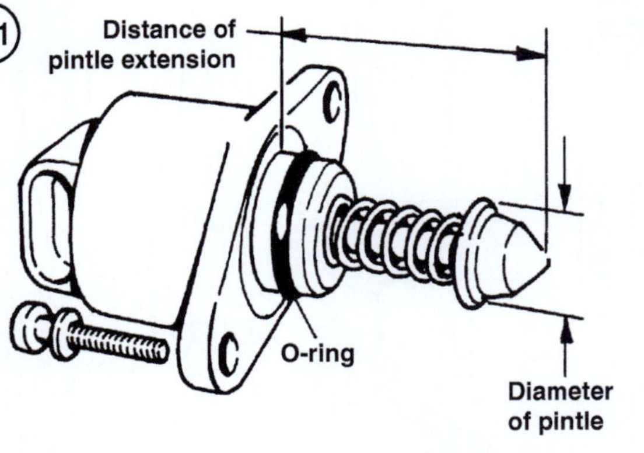

132

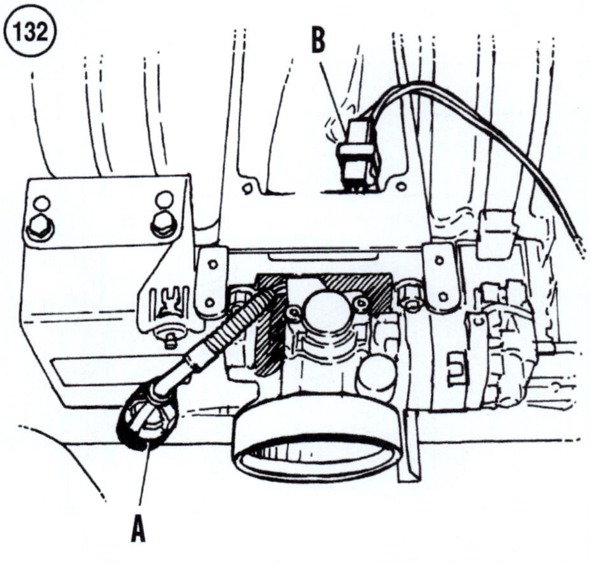

Removal and Installation

1. Disconnect the negative battery cable.

2. Remove the four screws and then lift the plastic cover off the engine.

3. Use compressed air to remove loose material from the idle speed control solenoid (ISC) and throttle body surfaces.

4. Disconnect the ISC three-wire harness connector (B, **Figure 132**) from the engine wire harness connector. Then disconnect the throttle position sensor (TPS) two-wire harness connector (A, **Figure 132**) from the engine harness connectors.

5. Remove the two screws (**Figure 133**) and then carefully pull the ISC out of the throttle body.

6. Remove the two screws and then pull the solenoid adapter off the throttle body.

7. Remove the foam muffler from the adapter. Remove the gaskets from the adapter and throttle body mating surfaces. Discard the gaskets.

8. Use mild solvent to clean oil and carbon deposits from the adapter and corresponding throttle body mating sur-

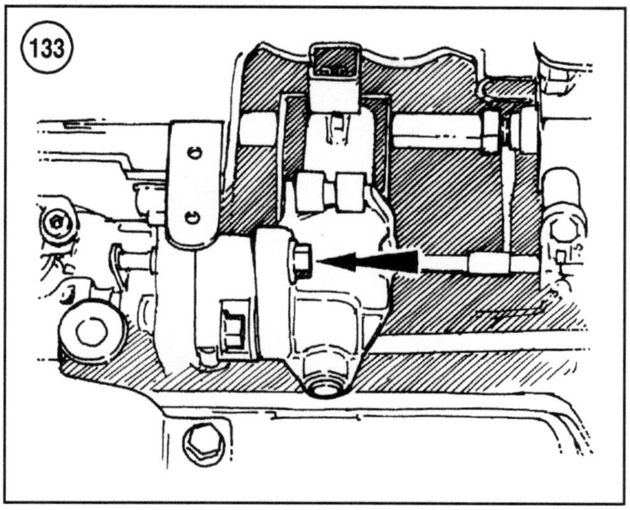

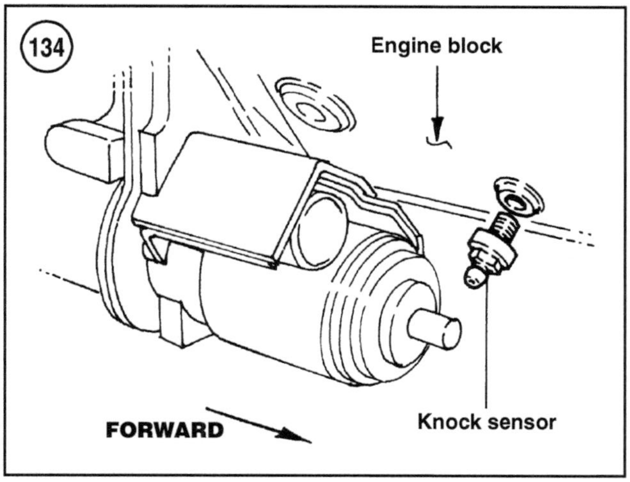

faces. Do not apply solvent to the ISC or foam muffler. Use compressed air to clean the foam muffler.

9. Insert the foam muffler into the adapter bore until flush or slightly below the adapter mating surface.

10. Place a *new* gasket onto the adapter and then fit the adapter onto the throttle body. Align the gasket, adapter and throttle body openings. Then thread the two screws into the openings. Tighten the adapter-to-throttle body screws evenly to 71-102 in.-lb. (8-12 N•m).

11. Place a *new* gasket onto the ISC and then fit the ISC onto the adapter. Align the gasket, ISC and adapter openings and then thread the two screws (**Figure 133**) into the openings. Tighten the ISC-to-adapter screws evenly to 71-102 in.-lb. (8-12 N•m).

12. Apply a light coat of electric terminal grease (Ford part No. F2AZ-19584-A) to the terminals and then connect the ISC three-wire harness connector (B, **Figure 132**) and the TPS two-wire harness connector (A) to the engine harness connectors. Tug on the harnesses to verify a secure connection.

13. Install the plastic cover onto the top of the engine. Install the four cover screws. Tighten to 24-36 in.-lb. (3-4 N•m).

14. Connect the battery cable.

KNOCK SENSOR (KS)

A knock sensor (KS) is used on all electronic fuel injection models.

NOTE
The KS can be easily damaged if dropped or struck with a hard object. Replace the sensor if dropped or struck. A damaged sensor might deliver a false detonation signal or fail to detect engine damaging detonation.

9

GM V-6 and V-8 Models

On all GM V-6 and V-8 EFI models, the knock sensor (KS) threads into the starboard side of the cylinder block just forward of the starter (**Figure 134**). Some later 7.4L and 8.2L models have a second KS that threads in the port side of the cylinder block (**Figure 135**).

Removal and installation

1. Disconnect the negative battery cable.

2A. On models with a standard cooling system, drain all water from the cylinder block as described in *Winterizing* in Chapter Four.

2B. On models with a closed cooling system, drain coolant from the cylinder block as described in Chapter Three. Refer to *Changing the Engine Coolant*.

3. Pinch the longer sides of the harness connector and then pull the engine harness connector off the knock sensor.

4. Use a deep six-point socket to remove the sensor from the cylinder block. Thoroughly clean the threads on the sensor and in the cylinder block.

5. Apply a light coat of Loctite Pipe Sealant with Teflon onto the sensor threads. Purchase the sealant from an OMC dealership (OMC part No. 1141570), automotive parts store or GM automobile dealership (GM part No. 12346004).

6. Thread the replacement sensor into the opening by hand and then use a deep socket to tighten the sensor to 168 in.-lb. (19 N•m).
7. Touch one ohmmeter lead to the metal body of the sensor. Touch the other lead to an engine ground. The meter must indicate continuity. If not, remove the sensor, clean the threads and reinstall the sensor as described in this procedure. The sensor cannot operate properly with a faulty ground connection.
8. On models with a closed cooling system, refill the coolant as described in Chapter Three. Refer to *Changing the Engine Coolant*.
9. Pinch the longer side of the connector and then plug the engine harness connector onto the sensor. Release the grip and then tug *lightly* on the wire to verify a secure connection.
10. Connect the battery cable.
11. Start the engine and immediately check for water or coolant leakage at the sensor threads. If thre is leakage, remove the sensor, clean the threads and reinstall the sensor as described in this procedure.

Ford 5.0L and 5.8L Models

The knock sensor threads into the rear of the port side cylinder head.
1. Disconnect the negative battery cable.
2A. On models with a standard cooling system, drain all water from the cylinder block as described in *Winterizing* in Chapter Four.
2B. On models with a closed cooling system, drain coolant from the cylinder block as described in Chapter Three. Refer to *Changing the Engine Coolant*.
3. Pull the engine harness connector (**Figure 136**) off the KS.
4. Use a deep six-point socket to remove the sensor from the cylinder head.
5. Use an appropriate tap to remove corrosion or sealant from the opening in the cylinder head. Use a wire brush to thoroughly clean the threads on the KS.
6. Apply a light coat of Loctite Pipe Sealant with Teflon to the sensor threads. Purchase the sealant from an OMC dealership (OMC part No. 1141570), automotive parts store or GM automobile dealership (GM part No. 12346004).
7. Thread the replacement sensor into the opening by hand and then use a deep socket to tighten the sensor to 84-120 in.-lb. (10-14 N•m).
8. Touch one ohmmeter lead to the metal body of the sensor. Touch the other lead to an engine ground. The meter must indicate continuity. If not, remove the sensor, clean the threads and reinstall the sensor as described in this procedure. The sensor cannot operate properly with a faulty ground connection.
9. On models with a closed cooling system, refill the coolant as described in Chapter Three. Refer to *Changing the Engine Coolant*.

136

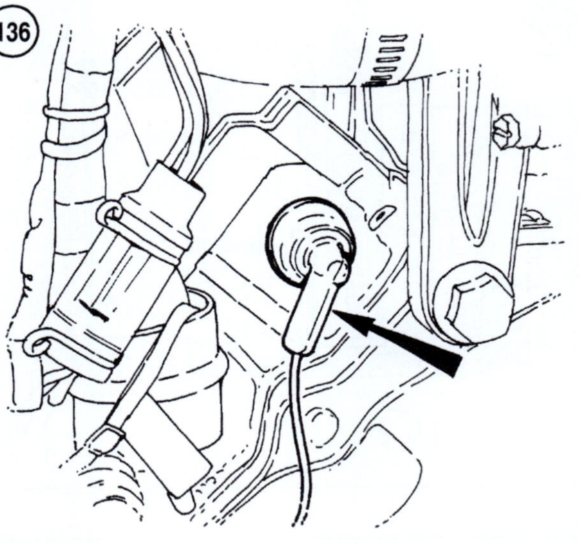

137

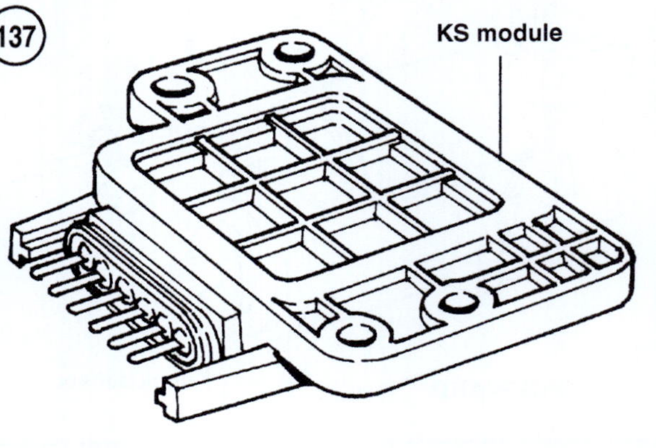

10. Push the engine harness connector onto the KS terminal (**Figure 136**). Then tug *lightly* on the wire to verify a secure connection.
11. Connect the battery cable.
12. Start the engine and immediately check for water or coolant leakage at the sensor threads. If there is leakage, remove the sensor, clean the threads and reinstall the sensor as described in this procedure.

KNOCK SENSOR MODULE

A knock sensor module (**Figure 137**) is used on all GM electronic fuel injection (EFI) models.

The knock sensor module amplifies the knock sensor signal and signals the engine control unit (ECU) if spark knock is detected.

Removal and Installation

The knock sensor module mounts on the circuit breaker housing on the starboard side exhaust elbow.
1. Disconnect the negative battery cable.

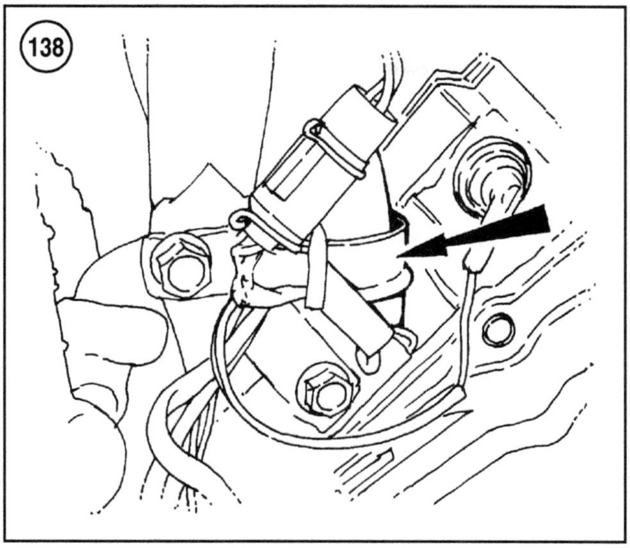

2. Lift the locking tab and then unplug the engine harness connector from the module.

3. Remove the two mounting screws and then lift the module off the bracket.

4. Use compressed air to blow debris and loose material off the module and mounting bracket. Wipe oily deposits or other contaminants from the mating surfaces.

5. Fit the replacement module onto the bracket. Then install the mounting screws. Make sure to fit the ground wire ring terminal onto the mounting screw. Tighten the screws to 88-124 in.-lb. (10-14 N•m).

6. Lift the locking tab and then plug the engine harness connector onto the module. The harness locking tab must engage the tab on the module. Tug on the harness to verify a secure connection.

7. Connect the battery cable.

KNOCK SENSOR AMPLIFIER

The knock sensor amplifier is used on Ford 5.0L and 5.8L electronic fuel injection (EFI) models.

Removal and Installation

1. Disconnect the negative battery cable.

2. Cut the plastic locking clamp to free the wire harness and amplifier from the mounting bracket (**Figure 138**). Lift the amplifier out of the bracket.

3. Pull the retaining tab out of the amplifier wire harness connector and then unplug the amplifier harness from the engine wire harness.

4. Apply a light coat of electric terminal grease (Ford part No. F2AZ-19584-A) to the terminals and then plug the amplifier harness connector into the engine wire harness connector. Push the retaining tab into the slot to secure the connectors. Tug on the harness to verify a secure connection.

5. Slide the amplifier into the mounting bracket. Secure the amplifier and harness into the bracket with a plastic locking clamp.

6. Connect the battery cable.

ENGINE CONTROL UNIT (ECU)

The engine control unit (**Figure 139**, typical) is used on all GM V-6 and V-8 electronic fuel injection models.

Removal and Installation

Three screws secure the ECU (**Figure 139**, typical) to the port side exhaust elbow bracket.

1. Disconnect the negative battery cable.

2. Mark the two harness connectors and ECU to identify the J1 and J2 harness connectors. The connectors are similar in appearance.

3. Depress the locking tab (**Figure 139**, typical) and then pull the J1 and J2 harness connectors off the ECU.

4. Remove the three screws and then lift the ECU off the mounting bracket.

5. Use compressed air to blow debris and loose material off the ECU and mounting bracket. Wipe oily deposits or other contaminants from the mating surfaces.

6. Fit the replacement ECU onto the bracket and then install the mounting screws. Make sure to fit the ground wire ring terminal onto the mounting screw. Tighten the mounting screws to 88-124 in.-lb. (10-14 N•m).

7. Depress the locking tab (**Figure 139**, typical) and then carefully plug the J1 and J2 harness connectors onto the ECU. Do not force the connectors into the openings. The connectors are similar in appearance, but the shapes of the connectors are different. Forcing the connector into the wrong opening can damage the connector, ECU or both components.

8. Connect the battery cable.

ENGINE CONTROL ASSEMBLY (ECA)

The engine control assembly (ECA) is used on Ford 5.0L and 5.8L electronic fuel injection (EFI) models.

Removal and Installation

The ECA is located on the front and starboard side of the engine.

1. Make sure the ignition switch is in OFF. Disconnect the negative battery cable.
2. Cut the plastic locking clamp and then remove the rubber cover from the plastic ECA cover.
3. Move the rubber cover away from the ECA to access the engine harness connector to the ECA.
4. Move the wire to access the screw that secures the harness connector to the ECA. Loosen the screw while pulling lightly on the harness until the connector is free from the ECA.
5. Unthread the four screws that secure the ECA to the plastic cover and then pull the ECA out of the cover.
6. Use compressed air to remove loose material from the ECA, cover and mounting bracket.
7. Fit the replacement ECA into the plastic cover. Fit the cover and ECA onto the mounting bracket and then thread the four screws into the cover and bracket. Tighten the screws to 24-36 in.-lb. (3-4 N•m).
8. Apply a light coat of electric terminal grease (Ford part No. F2AZ-19584-A) to the terminals. Connect the engine wire harness connector to the ECA. Verify that the connector is fully seated. Thread the connector retaining screw into the ECA. Tighten the screw to 46-64 in.-lb. (5-7 N•m).
9. Fit the rubber cover over the ECA cover and harness connection. Secure the rubber cover to the ECA cover with plastic locking clamps.
10. Connect the battery cable.

Table 1 ELECTRICAL SYSTEM TORQUE SPECIFICATIONS

Fastener	in.-lb.	ft.-lb.	N•m
Alternator mounting bolts	–	26-30	35-41
Distributor cap screws			
Ford 5.0L and 5.8L EFI models	18-23	–	2-3
Distributor hold-down clamp bolt	–	18	24
Engine control assembly (ECA) bracket			
Ford 5.0L and 5.8L EFI models	60-84	–	7-10
Engine control assembly (ECA) connector screw			
Ford 5.0L and 5.8L EFI models	46-64	–	5-7
Engine control assembly (ECA) cover screws			
Ford 5.0L and 5.8L EFI models	24-36	–	3-4
Engine control unit (ECU) mounting screws			
GM V-6 and V-8 models	88-124	–	10-14
Engine coolant temperature sensor (EFI models)			
GM V-6 and V-8 models	106	–	12
Ford 5.0L and 5.8L EFI models	120-180	–	14-20
Hall effect sensor (stator) mounting screws			
Ford 5.0L and 5.8L EFI models	15-35	–	2-4
Heat sink to ignition module bracket mounting nuts			
Ford 5.0L and 5.8L EFI models	60-84	–	7-10
Idle air control motor (IAC)			
GM V-6 and V-8 models	28	–	3
Idle air control solenoid adapter-to-throttle body screws			
Ford 5.0L and 5.8L EFI models	71-102	–	8-12
Idle air control solenoid to adapter screws			
Ford 5.0L and 5.8L EFI models	71-102	–	8-12
Ignition coil mounting			
Delco EST ignition coil	–	20-25	27-34
Ford 5.0L and 5.8L EFI models			
Shorter screws	60-84	–	7-10
Longer screws	24-36	–	3-4
Ignition module-to-heat sink screws			
Ford 5.0L and 5.8L EFI models	24-36	–	3-4
Intake air temperature (IAT) sensor			
Ford 5.0L and 5.8L EFI models	144-216	–	16-24
Knock sensor (KS)			
GM V-6 and V-8 models	168	–	19
Ford 5.0L and 5.8L EFI models	84-120	–	10-14

(continued)

Table 1 ELECTRICAL SYSTEM TORQUE SPECIFICATIONS (continued)

Fastener	in.-lb.	ft.-lb.	N•m
Knock sensor module			
GM V-6 and V-8 models	88-124	–	10-14
Manifold air pressure (MAP) sensor			
GM V-6 and V-8 models	44-62	–	5-7
Ford 5.0L and 5.8L EFI models	24-36	–	3-4
Oil pressure sender	124-168	–	14-19
Plastic engine cover			
Ford 5.0L and 5.8L EFI models	24-36	–	3-4
Sensor wheel mounting screws			
Ford 5.0L and 5.8L EFI models	25-35	–	3-4
Spark plug			
GM 3.0L models	–	20	27
GM 4.3L, 5.0L and 5.7L models			
New cylinder head	–	22	30
Used cylinder head	–	15	20
Ford 5.0L and 5.8L V-8 models	120-180		14-20
GM 7.4L and 8.2L V-8 models	–	22	30
Starter brush plate screws			
Ford 5.0L and 5.8L models	20-30	–	2-3
Starter mounting bolts			
GM four-cylinder, V-6 and V-8 models	–	30-36	41-49
Ford 5.0L and 5.8L V-8 models	–	24-30	33-41
Starter solenoid mounting screws			
Ford starter	45-85	–	5-10
Starter terminal nut	80-120	–	9-14
Starter throughbolts			
Ford 5.0L and 5.8L models	45-84	–	5-10
Throttle position sensor (TPS)			
GM V-6 and V-8 EFI models	18	–	2
Ford 5.0L and 5.8L EFI models	18-26	–	2-3

9

Table 2 BATTERY CAPACITY (HOURS)

Accessory draw	Provides continuous power for:	Approximate recharge time
80 ampere hour battery		
5 amps	13.5 hours	16 hours
15 amps	3.5 hours	13 hours
25 amps	1.6 hours	12 hours
105 ampere hour battery		
5 amps	15.8 hours	16 hours
15 amps	4.2 hours	13 hours
25 amps	2.4 hours	12 hours

Table 3 BATTERY CABLE RECOMMENDATIONS

Cable length	Minimum cable gauge size (AWG)
0 to 10 ft.	0
10 to 15 ft.	00
15 to 20 ft.	0000

Table 4 BATTERY STATE OF CHARGE

Specific Gravity Reading	Percentage of Charge Remaining
1.120-1.140	0
1.135-1.155	10

(continued)

Table 4 BATTERY STATE OF CHARGE (continued)

Specific Gravity Reading	Percentage of Charge Remaining
1.150-1.170	20
1.160-1.180	30
1.175-1.195	40
1.190-1.210	50
1.205-1.225	60
1.215-1.235	70
1.230-1.250	80
1.245-1.265	90
1.260-1.280	100

Table 5 ENGINE FIRING ORDER

Model	Firing order
GM 3.0L models	1-3-4-2
GM 4.3L models	1-6-5-4-3-2
GM 5.0L and 5.7L models	1-8-4-3-6-5-7-2
Ford 5.0L and 5.8L models	1-3-7-2-6-5-4-8
GM 7.4L and 8.2L models	1-8-4-3-6-5-7-2

Chapter Ten

Drive Unit

This chapter provides drive unit identification information and removal/installation and repair procedures. **Table 1** lists torque specifications for the drive unit. Use the general torque recommendations provided in Chapter One for fasteners not listed in **Table 1**. **Table 2** lists drive unit bearing preload (rolling torque) specifications. **Table 3** lists gear lash specifications. **Table 4** lists gear lash and gear pattern correction guides. **Table 5** lists drive unit identification and available ratios. **Tables 1-5** are located at the end of this chapter.

Exploded-view illustrations of each drive unit are located in the appropriate disassembly and assembly section and are helpful references for many service procedures.

DRIVE UNIT IDENTIFICATION

Volvo Penta and Outboard Marine Corporation formed a joint venture in 1993 to develop a new series of drive units. The joint venture is called Volvo Penta Marine Products, Limited Partnership. The outcome of the joint venture is product incorporating the best features of both the OMC Cobra and Volvo Penta drive units. The engines covered in this manual use either the Cobra SX or the DP-S DuoProp drive units.

OMC Cobra SX

The OMC Cobra SX drive unit (**Figure 1**) is a product of the joint venture and incorporates the best features of the Volvo Penta and previous OMC Cobra drive units. Except for the engine-mounted seawater pump (replacing the pump incorporated into the drive unit), the engine and transom bracket are generally of original OMC Cobra design. The transom cutout is the industry standard. The upper drive shaft housing assembly is an OMC Cobra-style casting modified to use Volvo Penta gears, shafts and cone clutch shift mechanism. The lower gearcase housing, propeller shaft bearing carrier and all bearings are from the original OMC King Cobra drive unit. The shafts and gears are from the design used on Volvo Penta DP-E drive units. The exhaust exits through the propeller hub with additional exhaust relief ports in the transom bracket (**Figure 2**) and the lower gearcase (**Figure 3**) (above the antiventilation plate). This drive uses a single propeller with a spline pattern unique to this drive unit. Normal rotation of the drive unit is clockwise (as viewed from the rear) when in forward gear. For twin engine applications, a simple shift linkage adjustment at the upper gearcase can change rotation to counterclockwise in forward.

OMC DP-S DuoProp

The OMC DP-S DuoProp drive unit (**Figure 4**) is a product of the joint venture. This drive unit uses the same upper drive shaft housing assembly and transom bracket as the Volvo Penta SX drive unit. The lower gearcase features internal components similar to the Volvo Penta DP-E drive units with an improved hydrodynamic shape. Like the Cobra SX drive unit, the exhaust exits through the propeller hubs and exhaust relief ports in the transom bracket (**Figure 2**). An additional exhaust passage is incorporated into the antiventilation plate (**Figure 5**). This drive uses two counter-rotating OMC Cobra or Volvo Penta D series aluminum or F series stainless steel propellers in matched pitch sets only.

DRIVE UNIT OPERATION

Both the Cobra SX and Cobra DP-S DuoProp drive units use the Volvo Penta cone clutch shift mechanism in the upper gearcase. The upper drive shaft housing contains three gears: the drive gear and the upper and lower output gears. All gears are continuously meshed and turn anytime the engine is running. A sliding cone clutch is positioned between the upper and lower output gears and is connected to the output shaft with helical splines. A spring-loaded shift shoe moves the sliding cone up or down to engage the appropriate gear.

Whenever the cone clutch contacts the spinning sleeve of an output gear, friction begins to turn the cone clutch in the same direction as the output gear. Because of the helical splines on the output shaft, the cone clutch is forced tighter into the output gear sleeve, and power is transferred to the lower gearcase through the output shaft. The more power applied, the tighter the wedging action and the more positive the engagement.

To shift out of gear, it is necessary to break the bond between the cone clutch and the gear sleeve. An eccentric is machined into the shift shoe groove on the cone clutch. This causes the shoe to move in and out slightly as the cone clutch rotates, much like a lifter on a camshaft. As the shift shoe attempts to lift the cone clutch out of the gear, the shoe will begin to compress the internal spring and ride up and away from the engaged side of the cone clutch until it bottoms against the shimmed stop bolt in the shift housing arm. When the shoe bottoms and is pinched between the stop bolt and the eccentric groove on the cone clutch, an impact is transferred to the cone clutch eccentric groove and breaks the clutch free from the gear sleeve. Once the cone clutch is free, the shift shoe returns to the bottom of the eccentric groove and rides the groove normally.

For proper operation, the shift shoe must be shimmed properly. Too many shims under the stop bolt can make it difficult to shift out of gear. Too few shims will cause mechanical binding resulting in shift shoe or cone clutch failure.

1

COBRA SX DRIVE UNIT (SINGLE PROPELLER)

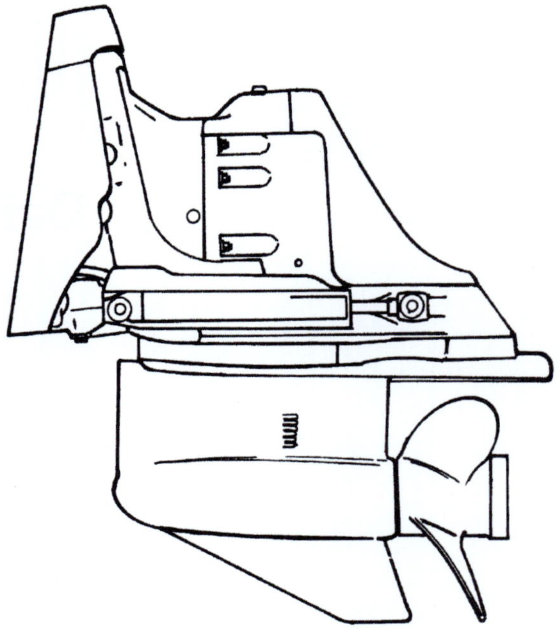

2

3

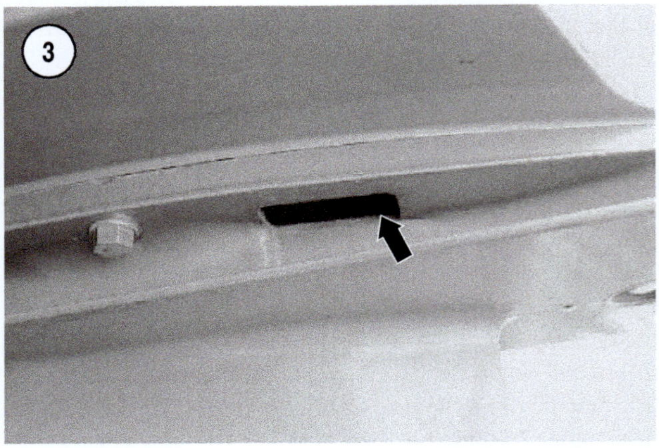

4 **COBRA DP-S DUOPROP DRIVE UNIT (TWIN PROPELLERS)**

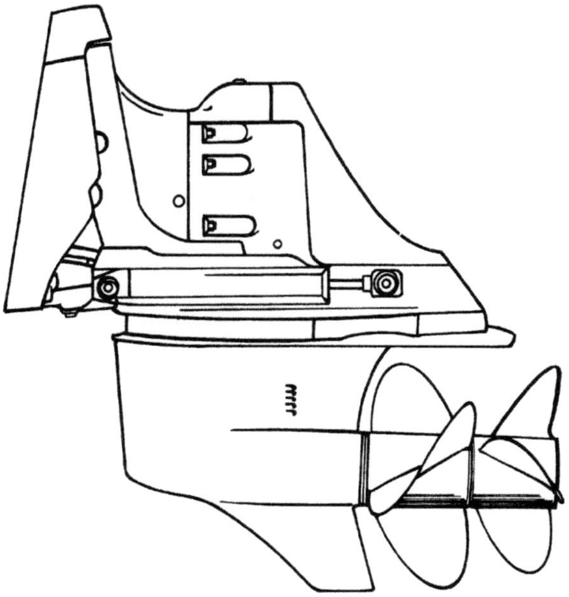

5

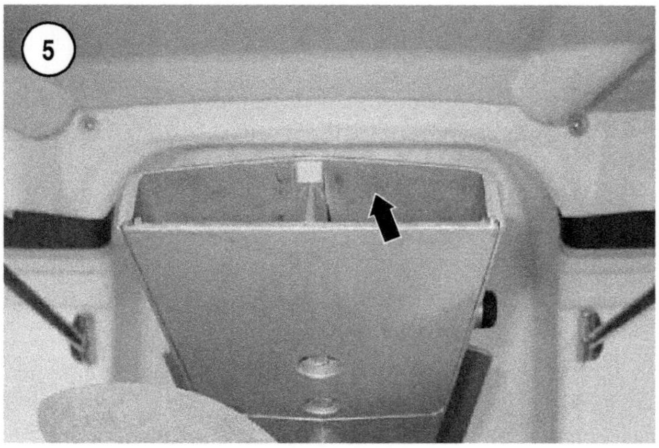

6

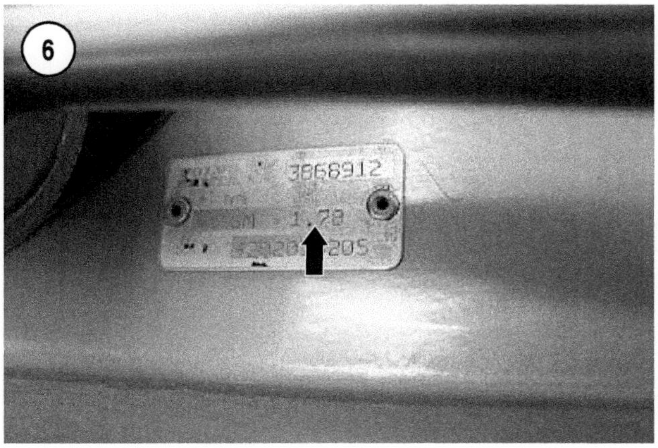

Engine torque transfers from the upper drive shaft housing to the lower gearcase via a female splined drive shaft coupler. The coupler has a deep groove machined into the middle of the shaft that is designed to provide a breaking point should the gearcase encounter an extreme impact load. This sacrificial feature can prevent damage to other drive unit components. The shallow groove in one end of the shaft is for orientation and should always face upward when installed.

A pinion gear mounted on the bottom of the lower drive shaft is in constant mesh with the driven gear(s). All OMC cobra gearcases are considered direct drive because all shifting is accomplished in the upper drive shaft housing.

Cobra SX drives use a single propeller, and the drive unit is considered birotational; it can rotate the propeller shaft either clockwise or counterclockwise for forward gear. The propeller is powered by the single driven gear mounted in the front of the gearcase.

Cobra DP-S DuoProp drives incorporate twin counter-rotating propellers. The front propeller is powered by the aft driven gear and turns counterclockwise (in forward) when viewed from the rear. The rear propeller is powered by the front gear and turns clockwise (in forward) when viewed from the rear.

The gearcase incorporates a feature that circulates the lubricant whenever the drive unit is shifted into gear. The lubricant cools as it flows past parts of the housing that are cooled by water flowing past cooling water passages. Both the SX and DP-S drives incorporate a lubricant screen in the upper drive shaft housing- or intermediate housing-to-gearcase mating surface to capture impurities before they reach the upper drive shaft housing. Replace or at least clean the screen each time the lower gearcase is removed.

DRIVE UNIT GEAR RATIO

The overall ratio refers to the amount of gear reduction between the crankshaft and the propeller shaft and includes the ratios of both the upper and lower gearcases. Overall gear ratios range from as low as 2.30:1 on DP-S drives to as high as 1.43:1 on SX drives. The factory-installed overall gear ratio is stamped on the model identification plate (**Figure 6**). An overall gear ratio of 2.30:1 means that the crankshaft turns 2.3 times for every one turn of the propeller shaft. Higher gear ratios are easier for the engine to turn.

If the model identification is missing or illegible, or if the gear ratio is suspected of being incorrect, the gear ratio can be determined by two different methods. The first method requires removing the drive unit. Mark the input shaft and the propeller blades for counting purposes. Manually shift the drive unit into either gear. While counting, turn the input shaft in the normal direction of rotation (counterclockwise as viewed from the rear of the drive unit) until the propeller has made exactly 10 turns. Divide the number of drive shaft rotations by 10 to determine the overall ratio.

Compare the results with the list of gear ratios in **Table 5**. Round the result to the nearest ratio listed.

The second method of determining the gear ratio involves counting the actual number of teeth on the drive unit gears. This method requires at least partial disassembly of the drive unit. Because a drive unit has two sets of gears (upper and lower gear sets) both gear sets must be counted. To determine the overall gear ratio, first divide the upper driven gear tooth count by the upper drive gear tooth count. Divide the lower driven gear tooth count by the lower drive gear tooth count. Multiply the upper ratio by the lower ratio.

For example, an upper gearcase has a 22:23 (drive to driven) tooth count, and a lower gearcase has a 17:30 (drive to driven) tooth count. Divide 23 (driven) by 22 (drive). The upper ratio is 1.05. Divide 30 (driven) by 17 (drive). The lower ratio is 1.76. Multiply 1.05 (upper ratio) by 1.76 (lower ratio). The overall gear ratio of the drive unit is 1.85.

It is very important that the engine is operated with the factory-recommended gear ratio. Running the engine with an incorrect gear ratio can cause poor performance and poor fuel economy and make it difficult or impossible to obtain the correct wide open throttle engine speed, which can lead to engine failure.

Some engines have a choice of several factory ratios depending on boat size and weight. Light, fast boats should use the lower ratio, and heavy, slower boats should use the higher ratio. If there is any doubt about which ratio should be used, contact the boat manufacturer or reputable marine dealership.

No matter which ratio is used, it is imperative that the engine operates within the recommended speed at wide open throttle. Change propeller(s) pitch and diameter as necessary to adjust engine speed. Increasing pitch or diameter increases load on the engine and reduces wide open throttle speed. Decreasing pitch or diameter reduces the load on the engine and increases wide open throttle speed. Always use an accurate shop tachometer to verify wide open throttle speed.

HIGH ALTITUDE OPERATION

The factory-recommended gear ratio is adequate for operation at altitudes up to 3500 ft. (1067 m). At higher altitudes, it is recommended to change the gear ratio to a higher number ratio (more gear reduction) to compensate for the loss of horsepower caused by thinner air.

NOTE

If the boat is operated in a high-altitude environment temporarily (such as a vacation trip), it is recommended to change only the propeller(s) to achieve the correct wide open throttle speed during the stay at higher elevation. Although this change will not usually achieve the same efficiency as a gear change, it will provide satisfactory performance during the temporary visit. Change back to the original propeller(s) when the boat returns to normal altitude.

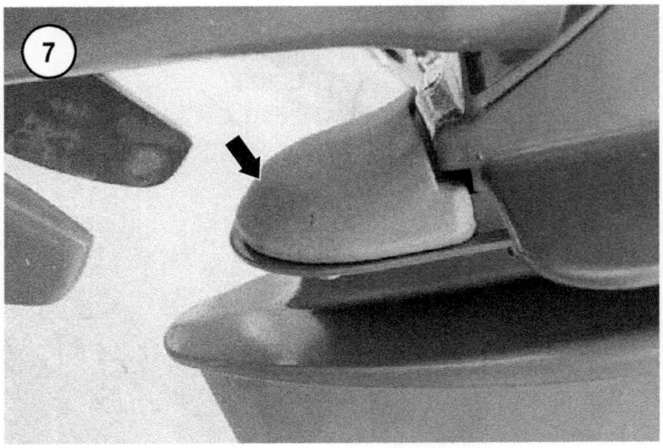

OUTDRIVE SERVICE PRECAUTIONS

When working on a drive unit, there are several precautions to keep in mind that will make your work easier, faster and more accurate.

1. Whenever possible, use an appropriate fixture to hold the gearcase. Use a vise with protective jaws to hold smaller housings or individual components. If protective jaws are not available, insert blocks of wood or similar padding on each side of the housing or components before clamping.
2. To help reduce corrosion, especially in saltwater areas, apply OMC or Volvo Penta White Sealing Compound (part No. 1141570-0) or equivalent to all external surfaces of bearings carriers, housing mating surfaces and fasteners unless otherwise specified. Do not apply sealing compound where it can get into gears and bearings.
3. Discard all O-rings, seals and gaskets during disassembly. To provide initial lubrication, apply OMC Triple Guard Grease or equivalent to the new O-rings and seal lips during assembly.
4. All drive units use precision shimmed gears. Tag all shims with the location and measured thickness of each shim as it is removed from the gearcase. Shims are reusable as long as they are not physically damaged or corroded. Follow shimming procedures carefully. Shims control gear location and bearing preload. Incorrect shimming can lead to gear and/or bearing failure.
5. Refer to *Shop Supplies, Fasteners, Special Tools* and *Basic Mechanical Skills* in Chapter One for appropriate service procedures that apply to drive unit repair.

SACRIFICIAL ANODES

Sacrificial zinc or aluminum anodes are standard equipment on all models. The anodes must have good electrical continuity to ground or they will not function. Never paint or coat the anodes with any material. Inspect the anodes visually and test them electrically anytime the drive is ser-

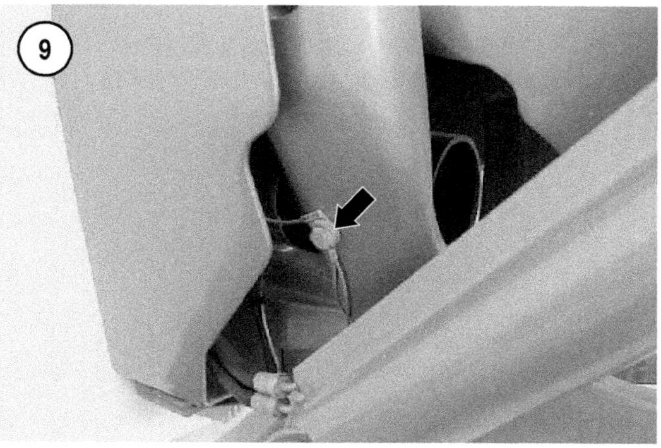

viced. Anodes are mounted on top of the leading edge of the lower gearcase (**Figure 7**) and on the bottom of the transom bracket (**Figure 8**). The Cobra SX drive has an additional anode mounted in the propeller shaft bearing carrier area of the lower gearcase.

Sacrificial Anode Visual Inspection

Check for loose mounting hardware, verify that the anodes are not painted, and check the amount of deterioration. Replace anodes that have lost more than one-third of the material. After installation, test the electrical continuity of each anode as described in this section.

Sacrificial Anode Electrical Testing

This test requires a common ohmmeter.

1. Calibrate the meter to the lowest scale available.
2. On Cobra SX drive units, remove the propeller as described in this chapter to access the bearing carrier-mounted anode.
3. Connect one ohmmeter lead to the anode being tested.
4. Connect the other test lead to a good ground point on the housing in which the anode is mounted.
5. The meter must indicate continuity. If otherwise, remove the anode and clean the mounting surface of the anode, gearcase and mounting hardware. Reinstall the anode and repeat the test.
6. Test the continuity of the gearcase to the engine and negative battery terminal by connecting one of the test leads to the negative battery terminal and the other terminal to a ground point on the gearcase. The meter must indicate continuity. If otherwise, check electrical continuity of the lower gearcase to the upper drive shaft housing, the upper drive shaft housing to each metal component on the transom bracket and the transom bracket to the engine block. Check for loose mounting hardware, broken or missing ground straps or excessive corrosion. Repair or replace wiring or fasteners (**Figure 9**, typical) to establish a good electrical path.
7. On Cobra SX drive units, install the propeller as described in this chapter.

Sacrificial Anode Replacement

1. Remove the fasteners and then carefully tap the anode from the mounting surface with a plastic mallet.

CAUTION
Do not use a wire brush to clean the surfaces because too much material might be removed, and steel particles can become embedded into the aluminum material and promote increased galvanic corrosion.

2. Use sandpaper to remove all corrosion and other material from the mounting surface. Use a plastic parts-cleaning brush to remove loose material.
3. Clean the mounting fastener openings with a suitable thread chaser.
4. Use aerosol carburetor cleaner to remove all residue from the mounting surface and threaded openings. Dry the threads and mounting surfaces with compressed air.
5. Fit the anode onto the mounting surface and secure with the fasteners. Do not apply sealant to the fastener threads. Securely tighten the fasteners.

SHIFTING AND PROPELLER ROTATION

NOTE
The drive unit requires a remote control setup that retracts the shift cable at the drive unit when the remote control is shifted into forward and extends the cable when the remote control is shifted into reverse.

On Cobra DP-S DuoProp drive units, the front propeller must rotate counterclockwise, and the rear propeller rotates clockwise when in forward. To accomplish this, the shift arm (**Figure 10**, typical) must rotate clockwise when the remote control is shifted into forward. This engages the cone clutch with the lower output gear. If the input from the remote control is correct, the shift linkage will connect to the opening (A, **Figure 10**) on the port side of the shift arm.

If the remote control is set up correctly, the Cobra SX drive unit is factory-set to provide right-hand (clockwise) rotation in forward. The shift linkage is factory-connected to the opening on the port side (A, **Figure 10**) of the shift arm. This engages the cone clutch to the lower output gear to provide clockwise propeller rotation.

If the boat uses two SX drive units, the starboard drive unit should remain standard (clockwise) rotation, and the port drive unit should be changed to left-hand (counter-clockwise) rotation with a matching left-hand (counter-rotation) propeller installed. Moving the shift linkage to connect to the starboard side opening (B, **Figure 10**) on the shift arm is the only adjustment required to change drive unit rotation. The port (counter-rotating) drive unit will then use the upper output gear for forward. Running two drive units with counter-rotating propellers reduces steering torque, is generally more efficient and reduces torque roll (the tendency of the boat to lean to one side while underway).

10

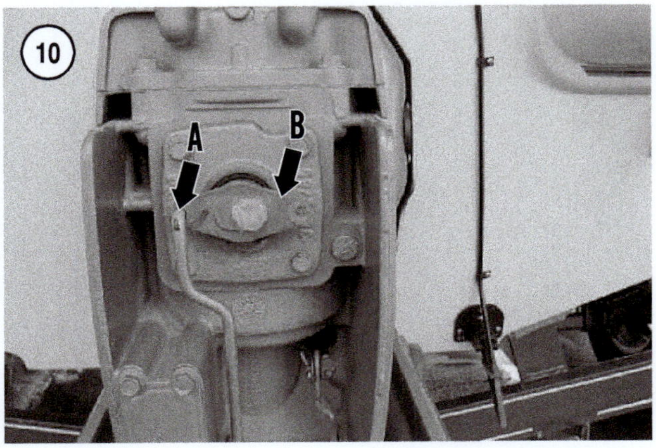

11

PROPELLER INSTALLATION (OMC COBRA SX DRIVE UNIT)

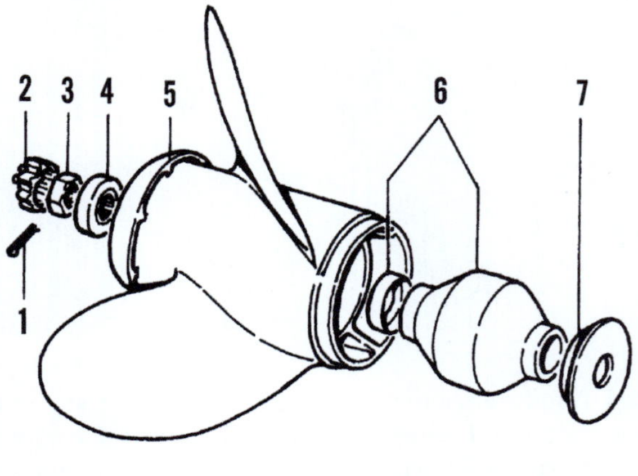

1. Cotter pin
2. Castellated retainer
3. Propeller nut
4. Splined washer
5. Propeller
6. Rubber drive hub and sleeve
7. Thrust washer

PROPELLER

Removal and Installation

Cobra SX (single-propeller drive)

This drive unit uses a single propeller with a unique spline pattern. Propellers are available in both right- and left-hand rotations and in a variety of aluminum and stainless steel designs. The rubber hub and sleeve (6, **Figure 11**) incorporated into the propeller are not consumer-serviceable. Have the hub and sleeve replaced at a reputable propeller repair shop if they should fail.

Refer to **Figure 11** for this procedure.

1. Disconnect the negative battery cable. Shift the drive into NEUTRAL.
2. Wedge a block of wood between the propeller and the antiventilation plate (**Figure 12**, typical) to prevent propeller rotation. Never attempt to hold the propeller by hand.
3. Remove the cotter pin and retainer (1 and 2, **Figure 11**). Discard the cotter pin.
4. Remove the splined washer (3, **Figure 11**). Then pull the propeller off the propeller shaft. If seized, use a block of wood as a cushion and carefully drive rearward on the propeller blade. Do not use excessive force. The propeller blade or propeller shaft might bend from excessive force.
5. Remove the thrust washer (7, **Figure 11**) from the propeller shaft. If seized, carefully tap on the sides of the washer until free.
6. Clean all old grease, salt crystals or other contaminants from the propeller shafts.
7. Apply a coat of OMC Triple Guard or Volvo Penta Propeller Shaft Grease (part No. 1141644-3) to the splined and tapered surfaces of each propeller shaft.
8. Fit the thrust washer onto the propeller shaft. The tapered bore in the washer must mate with the tapered surface on the propeller shaft.
9. Align the propeller splines with the propeller shaft splines and then fit the propeller onto the shaft. Seat the propeller against the thrust washer.
10. Fit the splined washer onto the propeller shaft. Align the splines and then seat the washer against the propeller.
11. Wedge a block of wood between the propeller and the antiventilation plate (**Figure 12**) to prevent propeller rotation.
12. Thread the propeller nut (3, **Figure 11**) onto the propeller shaft. Tighten the nut to 70-80 ft.-lb. (95-108 N•m).
13. Fit the castellated retainer (2, **Figure 11**) onto the propeller nut. Index the retainer as necessary to align the slots with the cotter pin opening in the propeller shaft. If necessary, remove the retainer and tighten the nut an additional amount. Never loosen the nut to align the openings.

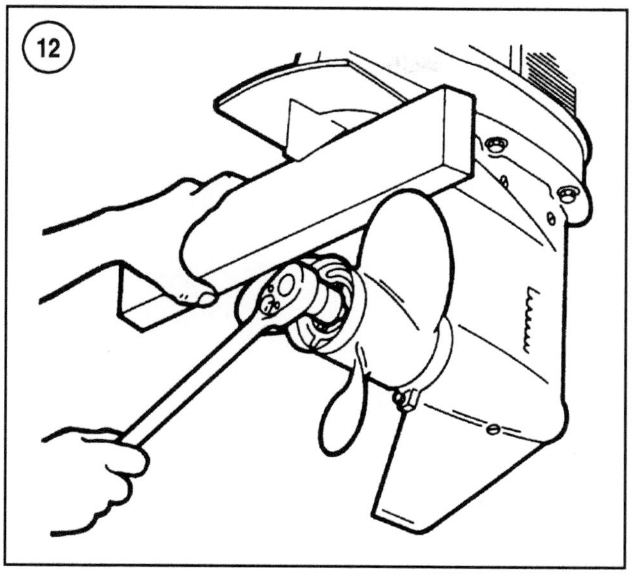

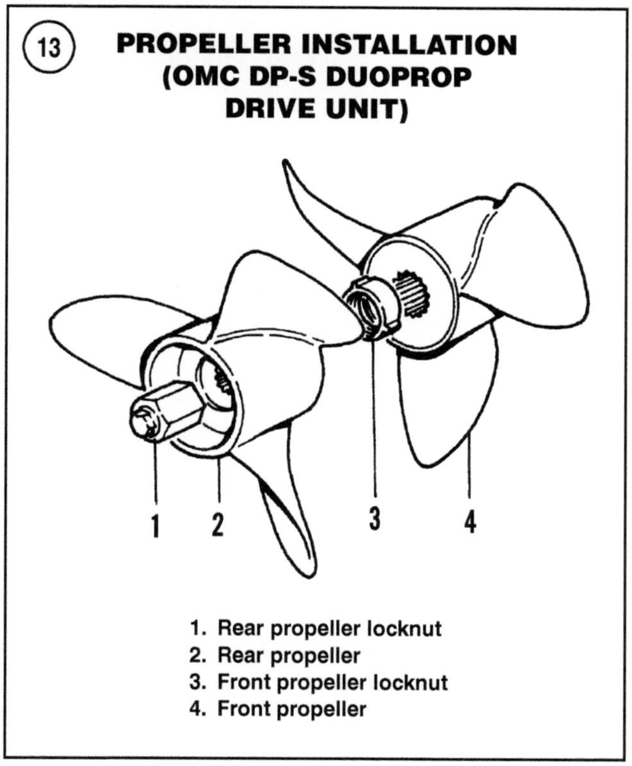

14. Insert the new cotter pin and bend the ends over to secure the retainer and nut.

15. Remove the block of wood. Carefully rotate the rear propeller to check for binding. Remove the propeller and check for improper installation if binding is evident.

16. Connect the battery cable.

Cobra DP-S DuoProp (twin-propeller drive)

The Cobra DP-S DuoProp drive unit uses two counter-rotating Volvo Penta D series aluminum or F series stainless steel propellers in matched pitch sets only. Use the propeller tool kit (OMC or Volvo Penta part No. 3855516-5) to remove and install the propeller nuts. The front and rear propeller nuts are elastic locknuts. Replace the nut if it can be turned by hand (without the aid of a wrench). These propellers do not use fish line cutters. The thrust washers are built into the propeller or propeller hub.

Refer to **Figure 13** for this procedure.

1. Disconnect the negative battery cable. Shift the drive into NEUTRAL.

2. Wedge a block of wood between the rear propeller and the antiventilation plate (**Figure 12**, typical) to prevent propeller rotation. Never attempt to hold the propeller by hand.

3. Rotate the rear propeller nut (1, **Figure 13**) counterclockwise until free from the propeller shaft. Then pull the rear propeller (2, **Figure 13**) off the propeller shaft.

4. Wedge the block of wood between the front propeller and the antiventilation plate (**Figure 12**, typical).

5. Engage the slots in the propeller nut removal tool (part No. 3855516-5) over the lugs on the front propeller nut (3, **Figure 13**). Rotate the nut counterclockwise until free from the shaft.

6. Pull the front propeller (4, **Figure 13**) from the shaft.

7. Clean all old grease, salt crystals or other contaminants from the propeller shafts.

8. Apply a coat of OMC Triple Guard or Volvo Penta Propeller Shaft Grease (part No. 1141644-3) to the splined and tapered surfaces of each propeller shaft.

9. Fit the front propeller (4, **Figure 13**) over the smaller-diameter propeller shaft. Align the propeller shaft splines with the splines on the larger-diameter propeller shaft and then seat the propeller against the tapered surface on the propeller shaft.

10. Wedge a suitable block of wood between the front propeller blade and the antiventilation plate to prevent propeller rotation. Never attempt to hold the propeller by hand.

11. Thread the front propeller nut (3, **Figure 13**) onto the larger-diameter propeller shaft with the tapered end facing inward.

12. Engage the slots in the propeller nut removal tool (part No. 38555-16) over the lugs on the front propeller nut. Use the tool to tighten the front propeller nut to 44 ft.-lb. (60 N•m).

13. Align the rear propeller splines with the propeller shaft splines and then fit the rear propeller (2, **Figure 13**) onto the propeller shaft. Seat the propeller against the tapered surface on the propeller shaft.

14. Wedge a block of wood between the rear propeller and the antiventilation plate (**Figure 12**, typical) to prevent propeller rotation.

15. Make sure the inner bolt in the rear propeller nut (1, **Figure 13**) is not protruding at the front of the nut. Then thread the rear nut onto the propeller shaft. Tighten the nut to 52 ft.-lb. (71 N•m).

16. Remove the block of wood. Carefully rotate the rear propeller to check for binding. Remove the propellers and check for improper installation if binding is evident.

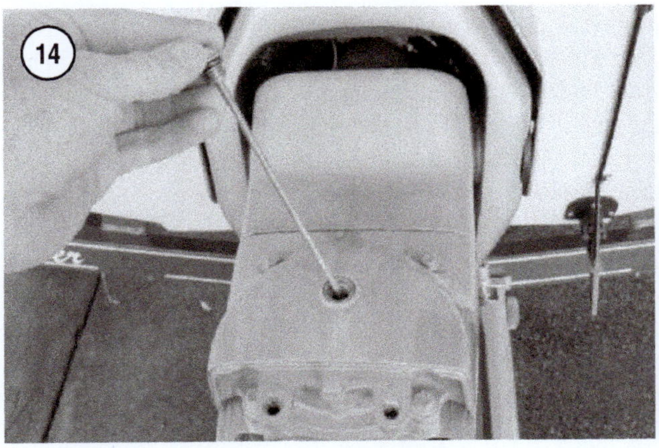

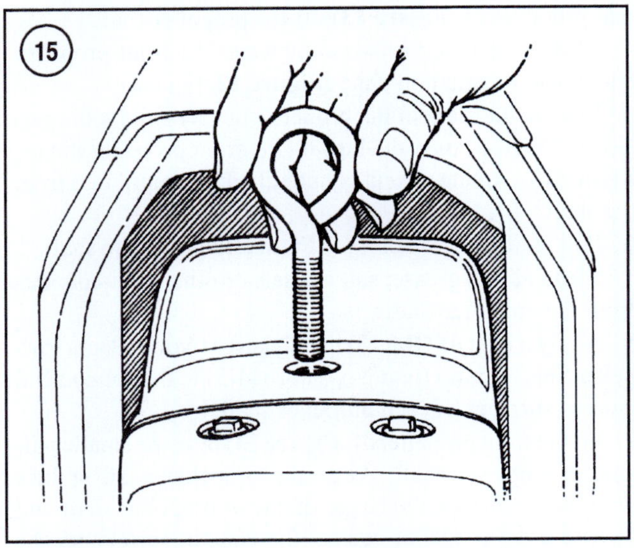

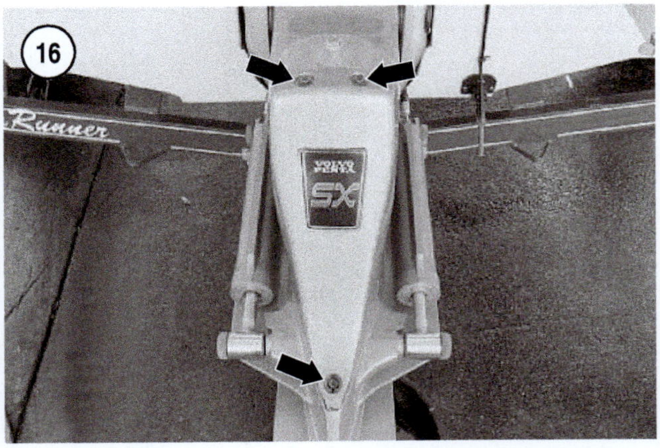

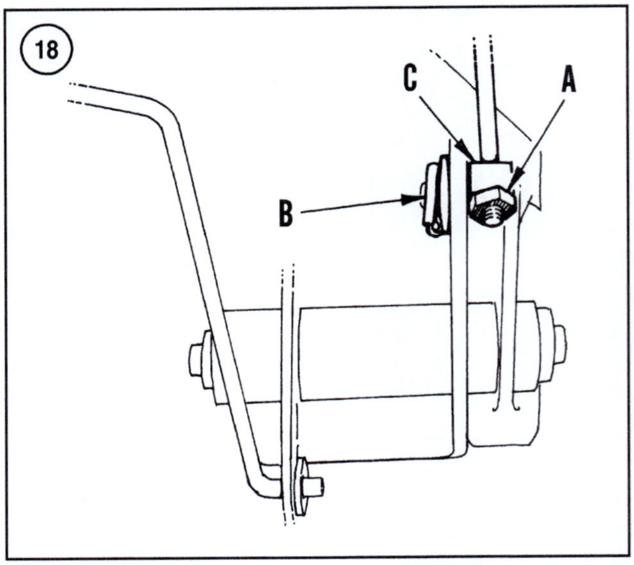

17. Connect the battery cable.

DRIVE UNIT

WARNING
Do not attempt to remove the drive unit from the boat without the aid of a suitable hoist. The drive unit is heavy and can slip and cause damage to the drive and possible injury.

Removal

1. Disconnect the negative battery cable.
2. Remove the propeller(s) as described in this chapter.

CAUTION
Make sure the lifting eye installed in Step 3 fits the threads properly and is threaded completely into the opening.

3. Remove the dipstick from the drive unit (**Figure 14**). Thread a 1/2 × 13 lifting eye (available at most hardware stores) into the dipstick opening (**Figure 15**). Make sure the lifting eye threads match the dipstick threads.
4. Remove the three screws (**Figure 16**) that secure the plastic cover to the upper housing. Then pull the cover off the housing. (**Figure 17**).
5. Disconnect the shift cable from the drive unit as follows:
   a. Locate the shift cable and linkage at the rear of the upper housing. Remove the jam nut (A, **Figure 18**) from the shift cable.
   b. Remove the cotter pin (B, **Figure 18**) and washer from the brass cube nut (C). Pull the cube nut out of

19
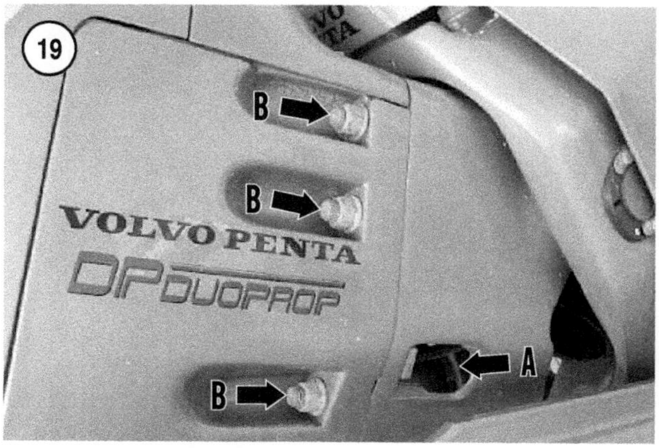

22

20
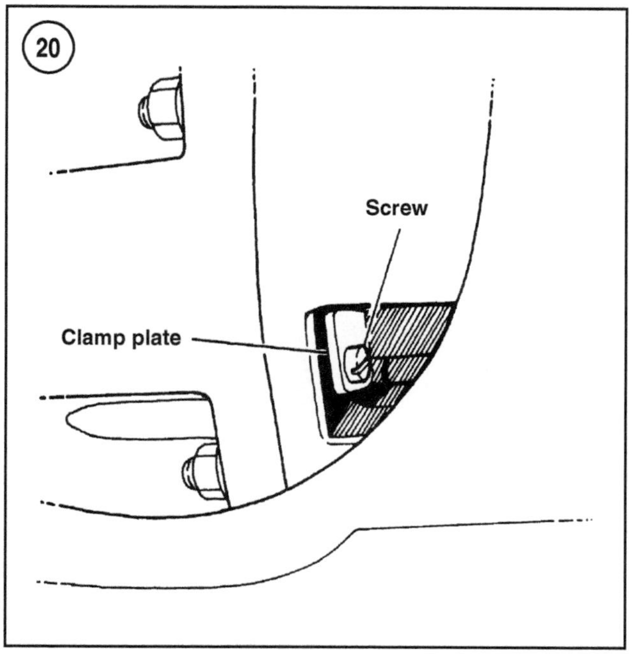

21
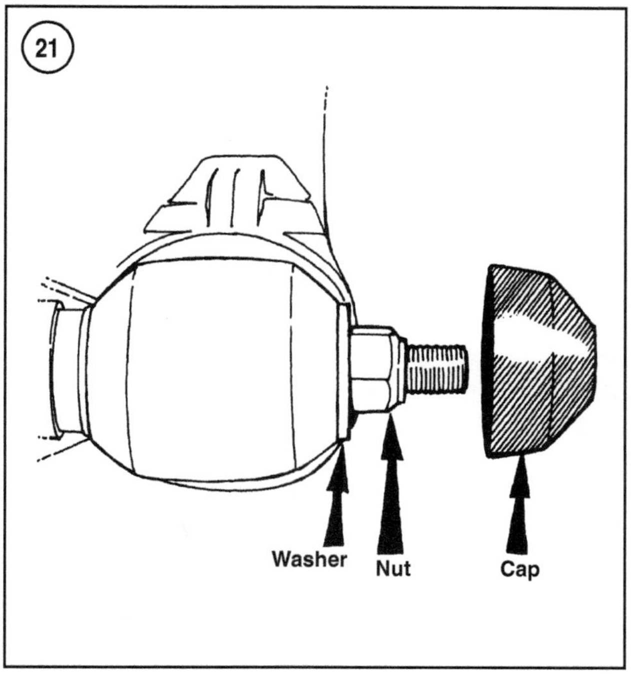

the bellcrank opening. Unthread the brass cube nut from the shift cable.

c. Locate the shift cable clamp plate (A, **Figure 19**) in the opening on the starboard side of the drive unit. Loosen the screw (**Figure 20**) securing the clamp plate to the pivot housing. Slide the clamp plate to the starboard side to free the cable.

d. Turn the drive unit to port and pull the shift cable out of the pivot housing opening. Make sure both shift cable rubber grommets are present on the shift cable. Locate the grommets if missing. Allow the cable to hang beneath the transom bracket during drive removal and installation.

6. Disconnect the hydraulic trim/tilt cylinder from the drive unit as follows:

a. Remove the protective cap, locknut and washer (**Figure 21**) from one end of the cylinder rod.

b. While supporting the drive unit, tap on the threaded rod end with a soft-faced hammer and brass or aluminum punch. Do not damage the threads on the cylinder rod.

c. Pull the cylinder rod out from the opposite end and then remove the bushings, washers and rubber grommets from the trim cylinders.

7. Attach a suitable hoist to the lifting eye (**Figure 15**) installed in Step 3. Apply enough lifting force with the hoist just to support the weight of the drive unit.

8. Remove and discard the six washer-faced elastic locknuts (three on each side [B, **Figure 19**]) and carefully pull the drive unit away from the transom. To prevent it from falling and damaging the housing or drive unit, support the universal joint shaft as it is removed from the pivot housing. Support and lower the hydraulic trim/tilt cylinders as the drive is removed to prevent hydraulic hose damage.

9. Once the drive is removed, tie the trim/tilt cylinders in the normal running position with rope.

10. Secure the drive unit in a suitable holding fixture. Remove and discard the two O-rings (**Figure 22**) from the universal joint drive shaft.

11. Use solvent to clean all old grease, corrosion or other debris from the universal shaft splines.

12. Reach into the universal shaft bellows and clean all old grease, corrosion or other contaminants from the bellows. If water is found in the bellows, check for torn, punctured or worn-through surfaces. If any defects are present, replace the bellows as described in Chapter Eleven. If water is present but no defects are found in the bellows, check for evidence of a high water level inside the engine compartment, loose drive unit attaching nuts and dirty or damaged mating surfaces.

13. Use a suitable solvent to thoroughly clean both the drive unit and pivot housing mating surfaces.

14. Remove the water tube seal (**Figure 23**) from the pivot housing. Discard the seal.

15. Check the engine alignment as described in Chapter Six. Correct any misalignment before installing the drive unit.

CAUTION

Failure to check and properly adjust the engine alignment will lead to premature engine coupler failure, excessive spline wear and difficult drive unit installation and removal. Misalignment can also cause universal joint knocking noises and vibration when the drive unit is turned under load.

Installation

1. Install two new O-rings into the grooves on the universal drive shaft. Work carefully to avoid tearing the O-rings as they slide over the splined section.

2. Apply a coat of OMC Moly Lube (part No. 175356) or OMC Triple Guard Grease to the universal joint shaft splines. Do not apply the grease to the two O-rings.

3. Grease the gimbal bearing as described in Chapter Four.

4. Using a suitable grease gun, pump OMC EP Marine Wheel Bearing Grease into the fittings until grease just appears at the U-joint seals (**Figure 24**).

5. Apply a light coat of OMC or Volvo Penta White Sealing Compound (part No. 1141570-0) to the threads of the six studs and drive mating surfaces on the pivot housing.

6. Use OMC Type M adhesive (part No. 318535) to glue the new water tube seal (**Figure 23**) to the groove in the pivot housing.

7. Apply a light coat of OMC Triple Guard Grease to the water tube seal and the V-shaped lip of the universal joint bellows (**Figure 25**).

8. Shift the drive unit into either gear by moving the shift arm (**Figure 26**) either up or down.

9. With the aid of an assistant, suitable hoist and lifting eye or other suitable means of support, position the drive unit at the correct installation height with the universal drive shaft even with the bellows opening.

10. Carefully guide the universal drive shaft into the bellows while moving the drive unit toward the pivot housing. Rotate the propeller shaft as necessary to align the universal drive shaft splines with the splines in the engine coupler. When aligned, seat the drive unit against the pivot housing.

23

24

25

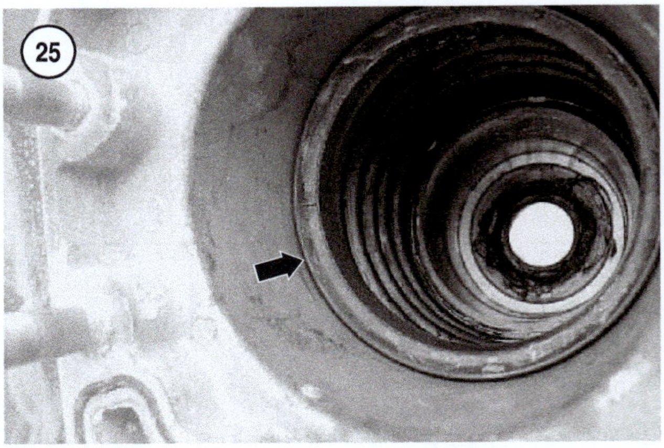

11. Thread the six new washer-faced elastic locknuts (B, **Figure 19**) onto the pivot housing studs. Working from the center to the corners, tighten the nuts in a crossing pattern to 25 ft.-lb. (34 N•m). Then tighten the nuts a second time in the same pattern to 50 ft.-lb. (68 N•m).

12. Remove the lifting eye (**Figure 15**) and install the dipstick and tighten to 48-72 in.-lb. (5-8 N•m).

13. Apply a coat of OMC Triple Guard Grease to the bushings and grommets. Then position the rubber grommet (1, **Figure 27**) on each bushing with the beveled edge facing away from the bushing flange.

26

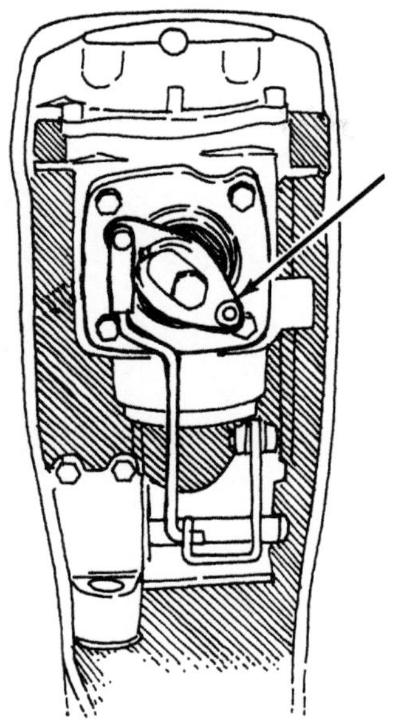

27

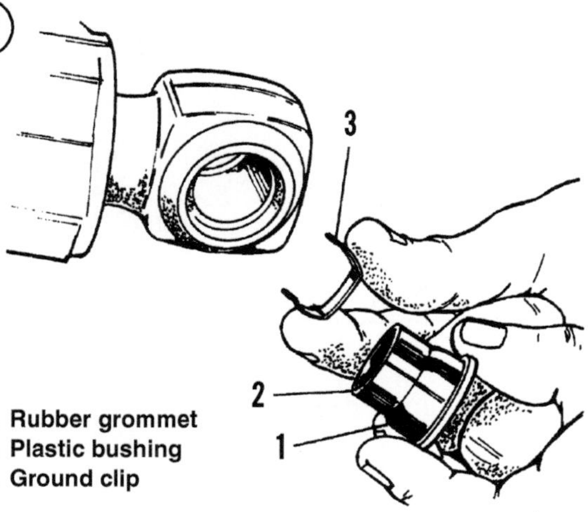

1. Rubber grommet
2. Plastic bushing
3. Ground clip

14. Install a bushing assembly into the opening on each side of the upper drive shaft housing assembly. Install two bushing assemblies into each opening in each trim/tilt cylinder with a ground clip (3, **Figure 27**) between the two bushing assemblies.

15. Align the openings in the trim/tilt cylinder cylinders with the openings in the upper drive shaft housing assembly. Apply a coat of OMC Triple Guard Grease to the entire length of the cylinder pivot rod. Insert the rod through one of the trim/tilt cylinder openings, through the opening in the upper drive shaft housing and out the other trim cylinder opening. Be careful not to displace any ground clips or damage any bushing assemblies. Use a rubber or plastic mallet to carefully tap the rod into position.

16. Install the washer and thread the elastic locknut onto the cylinder rod. Tighten both locknuts until seated and then tighten the nuts to 124-142 in.-lb. (14-16 N•m).

17. Turn the drive unit to full port. Then guide the shift cable into the opening in the pivot housing and upper drive shaft housing. Verify the rubber grommets are still in position and then position the cable to align the groove in the cable with the clamp plate (**Figure 20**). Slide the clamp toward the port side until fully engaged into the cable groove. Hold the clamp in position and securely tighten the clamp screw (**Figure 20**).
18. Connect the remote control cable as follows:
   a. Place the remote control into NEUTRAL.
   b. Place the shift arm (**Figure 26**) in the NEUTRAL position detent. The shift arm should be horizontal.
   c. Thread the brass cube nut (C, **Figure 18**) onto the shift cable and position it so that when all play in both the remote control cable and the shift linkage is centered, the brass cube nut will slide freely into the opening in the shift linkage.
   d. Secure the brass cube nut into the bellcrank arm with a washer and new stainless steel cotter pin (B, **Figure 18**). Bend over both prongs of the cotter pin to secure the nut.
   e. Thread the jam nut (A, **Figure 18**) onto the shift cable until it contacts the brass cube nut. Hold the cube nut with pliers and tighten the jam nut against the cube nut.
19. Fit the plastic cover onto the upper drive shaft housing and secure with the three screws. Securely tighten the screws.
20. Check the gearcase lubricant level as described in Chapter Three.
21. Adjust the shift cable as described in Chapter Five.
22. Install the propeller(s) as described in this chapter.
23. Connect the negative battery cable.

Upper Drive Shaft Housing

Removal

1. Disconnect the negative battery cable.
2. Remove the propeller(s) as described in this chapter.
3. Drain the gearcase lubricant as described in Chapter Three.
4. Remove the drive unit as described in this chapter. Secure the drive unit in a suitable holding fixture.
5. Remove the four bolts or nuts (**Figure 28**), two on each side, and the two bolts under the antiventilation plate.
6. With the aid of an assistant, lift the upper drive shaft housing assembly off the lower gearcase. Be ready to catch the intermediate drive shaft coupler and water tube should they become dislodged. If the coupler is dislodged, reinstall it onto the lower gearcase drive shaft with the grooved end facing upward.

7A. On earlier Cobra SX drives, remove the copper water tube and grommet (1 and **Figure 29**) from the upper drive shaft housing. Do not remove the plastic water tube adapter (1, **Figure 30**) from the gearcase unless it must be replaced.

7B. On later Cobra SX and all DP-S DuoProp drives, remove the copper water tube (3, **Figure 31**), upper grommet (4), lower grommet (2) and plastic water guide (1) from the gearcase.

8. Hook the screen (3, **Figure 29**) and remove it from the upper drive shaft housing. If a replacement screen is not available, clean the used screen with solvent and dry with compressed air. Replace the screen if damaged or if it cannot be satisfactorily cleaned.

9. Remove the drive shaft O-ring and lubricant passage O-rings. Discard the O-rings.

10. Thoroughly clean the mating surfaces on the upper drive shaft housing and lower gearcase.

11. Inspect the water tube grommets (2 and 4, **Figure 31**) for burned, melted or cracked surfaces. Replace the grommet(s) unless it is in excellent condition.

28

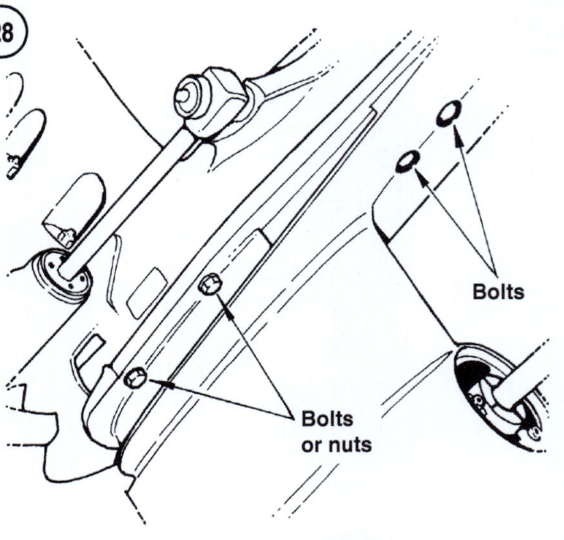

29

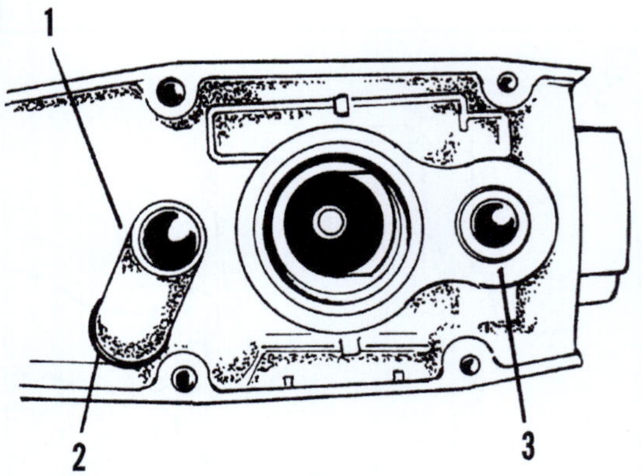

1. Copper water tube
2. O-ring
3. Lubricant screen

Disassembly (SX and DP-S models)

Record the measurement of all shims as they are removed. Shims can be reused if not damaged. Tag the shims by location for reassembly purposes.

All components that are reused must be installed in the original locations and orientations. Mark components as necessary to ensure correct assembly.

The gearcase holding fixture (OMC or Volvo Penta part No. 3850605-1) is designed to hold the upper drive shaft housing securely during all service procedures. This tool fits the standard Bob Kerr Tool Company workbench base or floor stand. It is highly recommended to use only the holding fixture during the disassembly and assembly procedures. Many of the procedures are difficult to perform without the fixture, even with the aid of an assistant. Purchase the holding fixture from an OMC or Volvo Penta dealership. Purchase the workbench base or floor stand from Bob Kerr Tool Company, P.O. Box 771135, Winter Garden, Florida 34777.

Refer to **Figure 32** for this procedure.

1. Remove and discard the cotter pin securing the clevis pin (1, **Figure 33**) to the eccentric piston arm. Remove the clevis pin and clevis end (3, **Figure 33**) from the arm.

NOTE

The shift shoe is spring-loaded and can easily fall free from the shift housing.

2. Disconnect the linkage from the shift arm on the shift housing. Remove the four screws (2, **Figure 33**) securing the shift housing to the housing. Remove the shift housing and shift shoe assembly. Remove and discard the O-ring (49, **Figure 32**).

3. Refer to *Shift Housing Service* in this chapter if the shift housing seal or any shift housing components require service.

4. Remove the drive gear bearing carrier as follows:

a. Remove the four bolts (29, **Figure 32**) securing the carrier and universal drive shaft assembly into the housing.

b. Carefully pull the carrier assembly out of the housing. If the carrier binds, tap on the side of the carrier with a plastic or rubber mallet while an assistant pulls on the universal drive shaft. If necessary, carefully pry the carrier out of the housing with two blunt pry bars inserted into the relief slots on each side of the carrier.

c. Remove and discard the O-ring (26, **Figure 32**). Remove the plastic shims (27, **Figure 32**) from the carrier or housing mating surfaces.

5. Remove the four bolts (2, **Figure 32**) securing the top cover (3) to the housing. Carefully lift the cover from the housing. If necessary, tap on the sides of the cover with a plastic or rubber mallet to free the cover.

(30) **GEARCASE WATER TUBE ADAPTER (EARLIER COBRA SX-S DRIVE UNITS)**

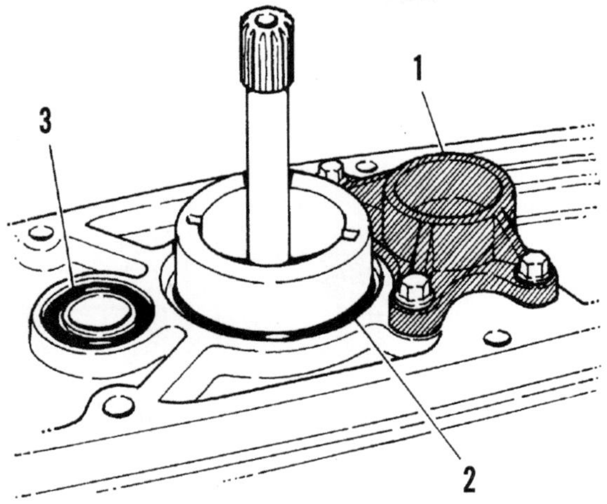

1. Plastic water tube adapter
2. O-ring
3. O-ring

(31) **GEARCASE WATER TUBE AND GROMMETS (LATER COBRA SX AND DP-S DRIVE UNITS)**

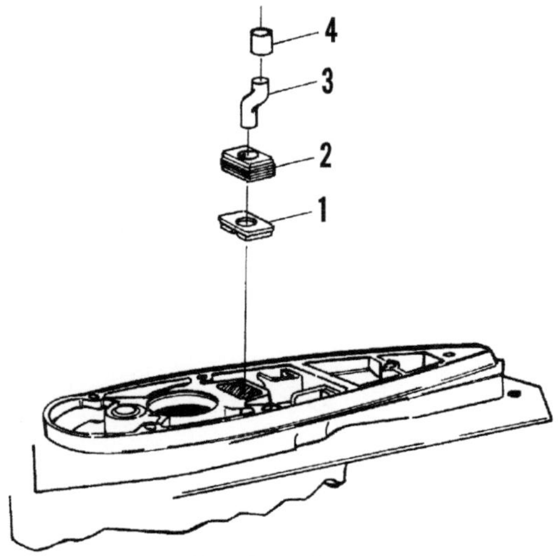

1. Plastic guide
2. Lower grommet
3. Copper water tube
4. Upper drive shaft housing grommet

6. Remove and discard the cover seal (5, **Figure 32**). Remove the cover shims (4, **Figure 32**) from the cover or upper output gear bearings.

NOTE
*The output shaft nut (**Figure 34**) has left-hand threads and must be turned clockwise for removal. The nut also is available in four different sizes (thicknesses) to control output shaft end play.*

7A. If using the recommended holding fixture, loosen the nut as follows:

a. Attach the splined adapter (OMC or Volvo Penta part No. 3895098-1) to the shaft of the holding fixture. Secure the adapter with the setscrew. Push up on the shaft while rotating to engage the adapter splines onto the output shaft splines. Rotate the shaft to align the flat shaft surface with the lock bolt. Hold the shaft while securely tightening the lock bolt.
b. Remove the output shaft nut by turning it clockwise with a 30 mm socket and breaker bar. Record the number stamped onto the nut (**Figure 35**) for later reference.

7B. If not using the recommended holding fixture, proceed as follows:

a. Lay the housing on the side. Install the splined socket (OMC or Volvo Penta part No. 3855098) onto the lower splines of the output shaft.
b. Have an assistant engage a 1/2 in. breaker bar and socket onto the splined socket.
c. Have the assistant secure the breaker bar to prevent shaft rotation. Remove the output shaft nut by turning it clockwise with a 30 mm socket and breaker bar. Record the number stamped onto the nut (**Figure 35**) for later reference.
d. Stuff several shop towels into the drive shaft bore in the housing to prevent the output shaft from dropping down and dislodging the split rings and thrust washer.

8. Carefully lift the upper output gear and bearing assembly (1, **Figure 36**) from the housing. Remove the shims (3, **Figure 36**) from the bearing or housing.

NOTE
If the double-stacked bearings removed in Step 9 are reused, mark the bearing locations and orientations for reference during assembly.

9. Remove the double-stacked needle bearings (2, **Figure 36**) from the output shaft or gear bore.

NOTE
The cone clutch is stamped with the word TOP on the upper surface. If the mark is not visible and the clutch is reused, make sure to mark the top of the clutch before removal.

10. Rotate counterclockwise while lifting the cone clutch and spring from the output shaft and housing (**Figure 37**). Mark the top side of the cone clutch and spring for reference during assembly.

CAUTION
*The lower output gear retaining ring (17, **Figure 32**) is tightened to 145-165 ft.-lb.*

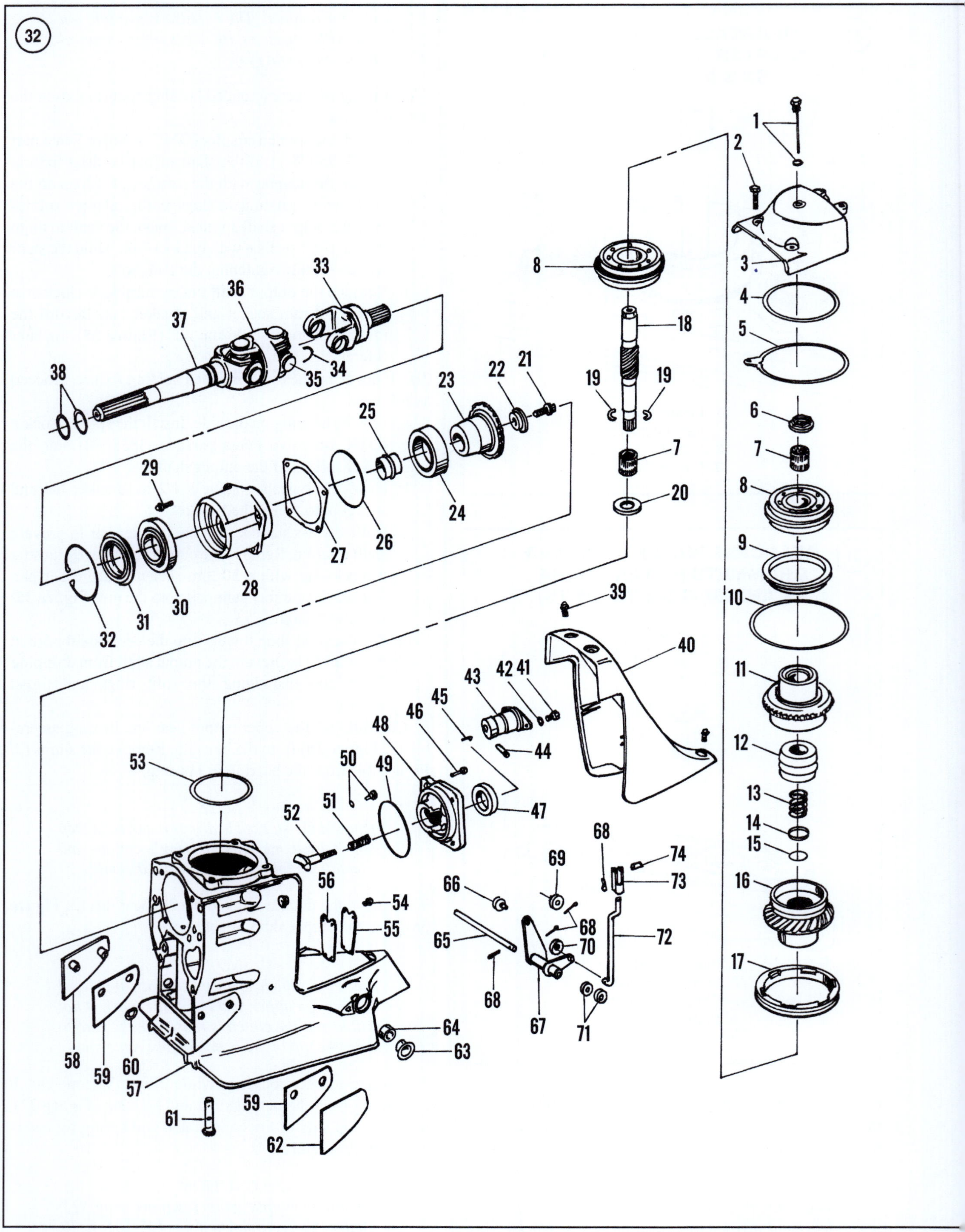
32
1
2
3
4
5
6
7
8
9
10
11
12
13
14
15
16
17
18
19
20
21
22
23
24
25
26
27
28
29
30
31
32
33
34
35
36
37
38
39
40
41
42
43
44
45
46
47
48
49
50
51
52
53
54
55
56
57
58
59
60
61
62
63
64
65
66
67
68
69
70
71
72
73
74

UPPER DRIVE SHAFT HOUSING

1. Lubricant level dipstick and O-ring
2. Bolt
3. Top cover
4. Top cover shims
5. Top cover seal
6. Output shaft nut
7. Double-stacked needle bearings
8. Output gear main bearing
9. Adapter ring (upper gear only)
10. Upper output gear shims
11. Upper output gear
12. Cone clutch
13. Cone clutch support spring
14. Spring support washer
15. C-clip
16. Lower output gear
17. Lower gear retaining ring
18. Output shaft
19. Split lockrings
20. Thrust washer
21. Torx head bolt
22. Tapered washer
23. Drive gear
24. Large tapered roller bearing
25. Pretension (crush) sleeve
26. O-ring
27. Shims (plastic)
28. Bearing carrier
29. Bolt
30. Small tapered roller bearing
31. Seal
32. Snap ring
33. Yoke shaft
34. Lockring
35. Bearing cap
36. Center coupler
37. Splined input yoke
38. O-rings
39. Screw
40. Rear cover (plastic)
41. Shift shoe stop bolt
42. Shim(s)
43. Eccentric piston
44. Ramp follower pin
45. Roll pin
46. Screw
47. Seal
48. Shift housing
49. O-ring
50. Oil level/vent plug and seal
51. Shift shoe spring
52. Shift shoe
53. Lower output gear shim
54. Screw
55. Water cover
56. Gasket
57. Gear housing
58. Starboard thrust pad
59. Thrust plate shim(s)
60. Retaining clip
61. Lubricant screen
62. Port thrust pad
63. Pivot pin bushing
64. Rubber grommet
65. Bellcrank shaft
66. Brass cube nut
67. Bellcrank
68. Cotter pin
69. Washer
70. Washer
71. Washer
72. Shift link
73. Clevis end
74. Clevis pin

33

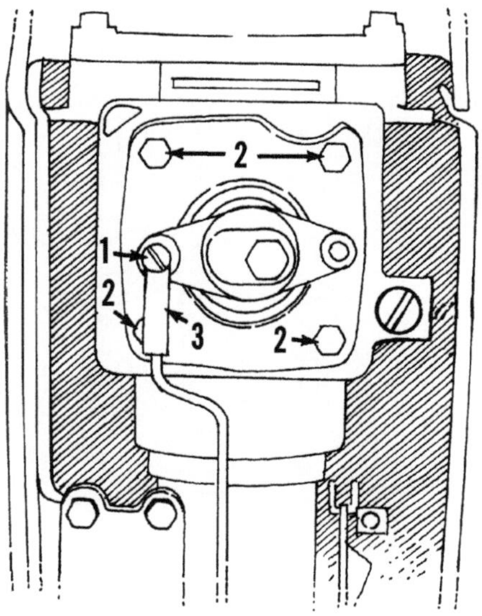

1. Clevis pin
2. Screws (4)
3. Clevis end/linkage

34

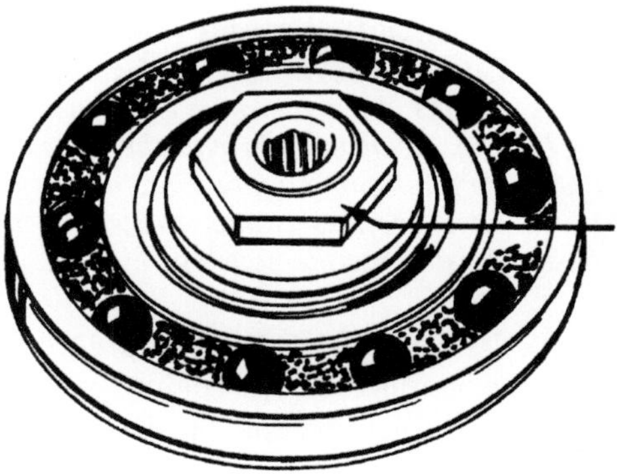

35

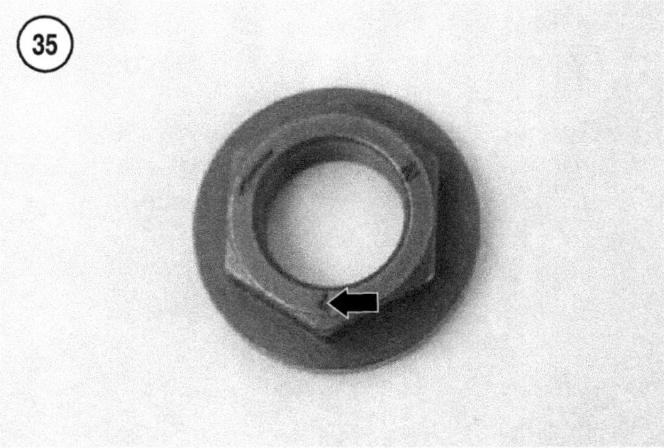

(197-224 N•m). The housing must be secured to a holding fixture (OMC or Volvo Penta part No. 3850605-1), a strong bench or floor mount, and the spanner wrench (OMC or Volvo Penta part No. 3850604-1) must be

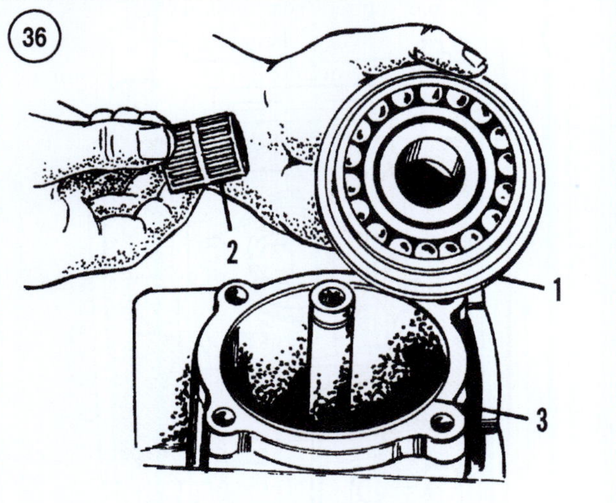

1. Upper output gear and bearing
2. Double-stacked needle bearings
3. Shim(s)

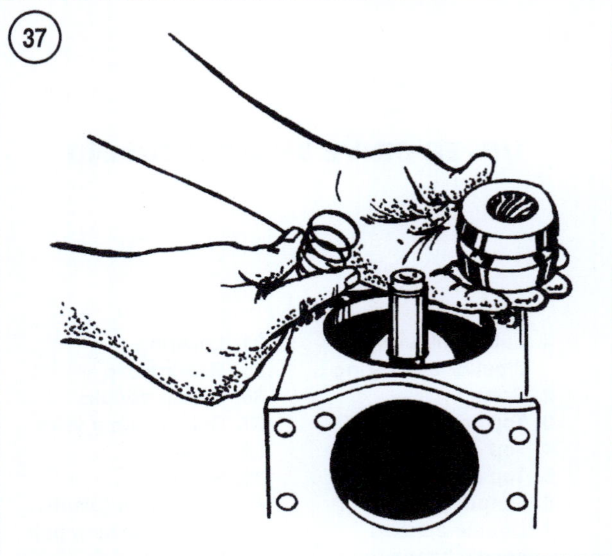

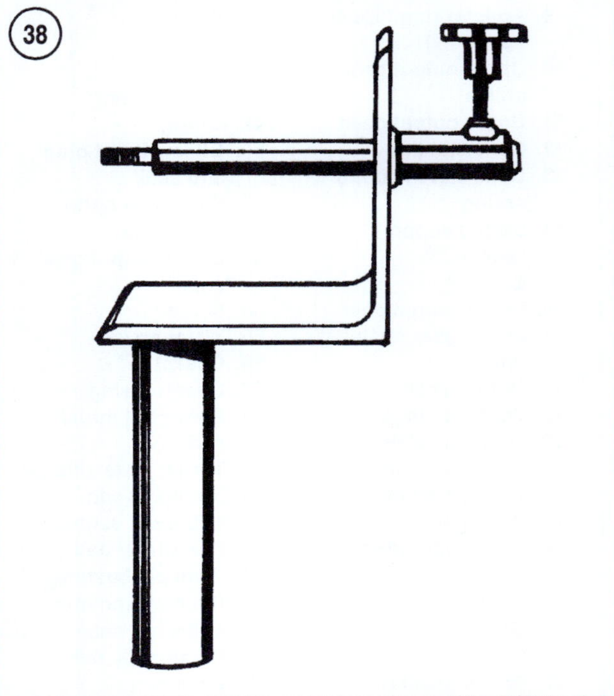

used when loosening or tightening the retainer. Make sure that any debris from gear or bearing failure is cleaned from the retainer and housing threads. If necessary, scrape the threads clean with a suitable pick tool and liberally wash the area with lubricant. If the retainer binds during removal, thread the ring back in, clean and lubricate the threaded area and try loosening again. Forcing the retainer will cause the retainer to lock in the housing, destroying the housing.

11. Remove the lower output gear retaining ring as follows:
   a. Mount the housing onto the holding fixture (**Figure 38**). Secure the housing onto the cover with the four bolts.
   b. Fit the spanner wrench (**Figure 39**) over the output shaft and engage the wrench onto the lugs of the retainer ring. Thread two of the cover bolts into the spanner wrench and housing openings. Securely tighten the bolts.
   c. Using a 3/4 in. breaker bar, break the retainer loose with a sudden, quick movement. Do not turn the ring more than a quarter turn with the breaker bar.
   d. Remove the breaker bar and two bolts securing the spanner wrench to the housing.
   e. Unthread the retainer by hand until free.

12. Grasp the output shaft and lift the shaft and lower output gear assembly out of the housing. Place the assembly on a clean work surface.

13. Remove the lower output gear shims from the housing or output gear bearing.

14. Remove the split lockrings (**Figure 40**) and thrust washer (20, **Figure 32**) from the output shaft.

15. Pull the output shaft (18, **Figure 32**) out of the lower output gear and bearing assembly.

NOTE
If the double-stacked bearings removed in Step 16 is reused, mark the bearing location and orientation for reference during assembly.

16. Remove the double-stacked needle bearings (7, **Figure 32**) from the output shaft or gear bore.

NOTE
Do not remove the output gear main bearing(s) unless they must be replaced. The removal process damages the bearings.

17. If the upper or lower output gear main bearing must be replaced, proceed as follows:

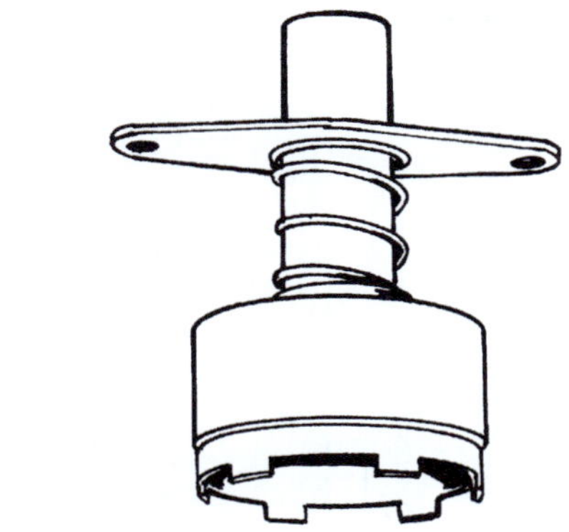
39

40

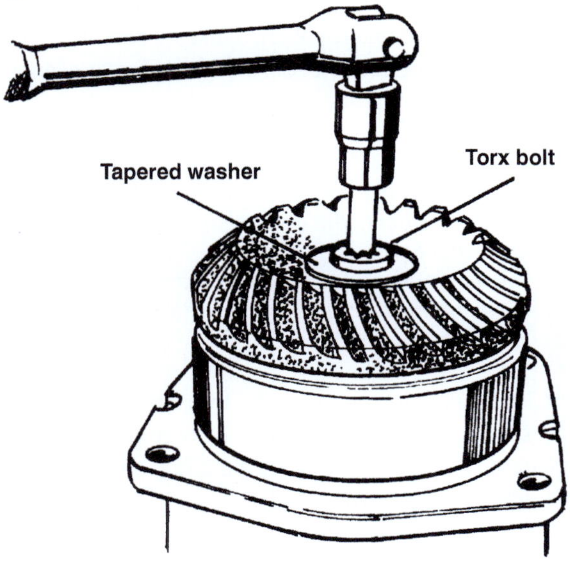

41

a. For the upper output gear only, place the support sleeve (OMC or Volvo Penta part No. 3850606-9) into an arbor or hydraulic press with the large end facing upward.
b. For the upper output gear only, place the upper output gear assembly into the support with the gear teeth facing upward.
c. For the upper output gear only, place the sleeve (OMC or Volvo Penta part No. 884263-5) against the gear.
d. For the upper output gear only, press the adapter ring (9, **Figure 32**) off the output gear.
e. Place the support sleeve (OMC or Volvo Penta part No. 3850606-9) into an arbor or hydraulic press with the large end facing upward. Place a shop towel into the support sleeve to protect the output gear.
f. Place the output gear and bearing assembly into the sleeve with the gear teeth facing downward.
g. Place the sleeve (OMC or Volvo Penta part No. 884263-5) against the gear hub. Press the gear out of the bearing.
h. Repeat Steps 17e-17g for the lower output gear.
i. Discard the output gear bearings.

18. Remove the C-clip (15, **Figure 32**) from the output shaft. Replace the C-clip if excessive wear, rough surfaces, distortion, discoloration or lost spring tension is evident.
19. Remove the support washer (14, **Figure 32**) from the output shaft. Replace the washer if it is excessively worn, distorted or discolored.
20. To service the drive gear bearing carrier, remove the two caps (35, **Figure 32**) and separate the yoke (33) from the universal drive shaft as described in this chapter. Refer to *Universal Joint Service*.
21. Fabricate a universal joint yoke support by cutting a steel or cast-iron pipe with a 1.0 in. (25.4 mm) diameter to a length of 2 7/8 in. (73 mm). Insert the pipe through the cap bores in the yoke. Then clamp the pipe into a vise. The pipe will absorb the compression loads and prevent damage to the yoke during service.

NOTE
Do not remove the drive gear unless it or the bearings must be replaced. The pretension sleeve and bearings must be replaced each time the gears or bearings are removed.

22. With the yoke secured, remove the Torx bolt (**Figure 41**) from the center of the drive gear. The bolt is sealed with threadlocking compound and might be difficult to remove. Discard the bolt upon removal. Remove the tapered washer (**Figure 41**) from the center of the gear.
23. Pull the drive gear bearing carrier off the yoke. Place the carrier on a suitable work surface with the gear teeth facing down. Use robust snap ring pliers to remove the snap ring (32, **Figure 32**) from the carrier. Do not remove the seal (31, **Figure 32**) at this time.
24. To remove the drive gear, proceed as follows:

a. Place the support sleeve (OMC or Volvo Penta part No. 884938-2) onto an arbor or hydraulic press with the large end facing upward. Place a shop towel in the bottom of the sleeve to protect the gear.
b. Set the drive gear bearing carrier into the sleeve with the drive gear facing downward.
c. Place the sleeve (OMC or Volvo Penta part No. 884266-8) against the hub of the gear. Press the gear out of the bearings.
d. Remove the pretension sleeve (25, **Figure 32**) from the gear or bearing bores. Discard the sleeve.

25. Turn the bearing carrier over and press the roller bearing (30, **Figure 32**) and seal (31) out using the recommended sleeve (OMC or Volvo Penta part No. 884263-5). The bearing and seal can usually be pressed out with light pressure. Discard the seal upon removal.

26. To remove the smaller bearing race from the drive gear bearing carrier, proceed as follows:
a. Thread the driver handle (OMC or Volvo Penta part No. 9991801-3) fully into the mandrel (OMC or Volvo Penta part No. 884933-3).
b. Place the bearing carrier onto a press with the large diameter facing downward.
c. Seat the mandrel and handle against the smaller bearing race. Press the race out of the carrier. Discard the bearing and race.

27. To remove the larger-diameter bearing race from the bearing carrier, proceed as follows:
a. Place the support sleeve (OMC or Volvo Penta part No. 884938-2) on a press with the large end facing upward. Set the bearing carrier into the support sleeve with the large end facing upward.
b. Thread the driver handle (OMC or Volvo Penta part No. 9991801-3) fully into the mandrel (OMC or Volvo Penta part No. 884933-3).
c. Seat the mandrel and handle against the larger bearing race. Press the race out of the carrier. Discard the bearing and race upon removal.

28. To remove the tapered roller bearing from the drive gear, proceed as follows:
a. Engage a bearing separator between the gear and bearing.
b. Place the separator and gear on a press. Place supports under the separator to ensure adequate space for the gear after removal.
c. Seat the sleeve (OMC or Volvo Penta part No. 884263-5) against the hub of the gear as shown in **Figure 42**.
d. Press the gear out of the bearing. Discard the bearing.

29. Remove the housing from the holding fixture.

30. Wash all components with clean solvent and dry with compressed air. Apply a light coat of oil to all cleaned components to prevent rusting.

31. Inspect the gear teeth (A, **Figure 43**) and the internal needle bearing surfaces (B) for excessive wear, chips, cracks, nicks or other damage. Inspect the cone clutch sleeve of the gears (C, **Figure 43**) for scoring, glazing or metal transfer. Use emery cloth to lightly dress superficial imperfections on the sleeve. Replace all three gears if any of the gears have damaged teeth, damaged bearing surfaces or if surface imperfections on the cone sleeve cannot be corrected by dressing with emery cloth.

42

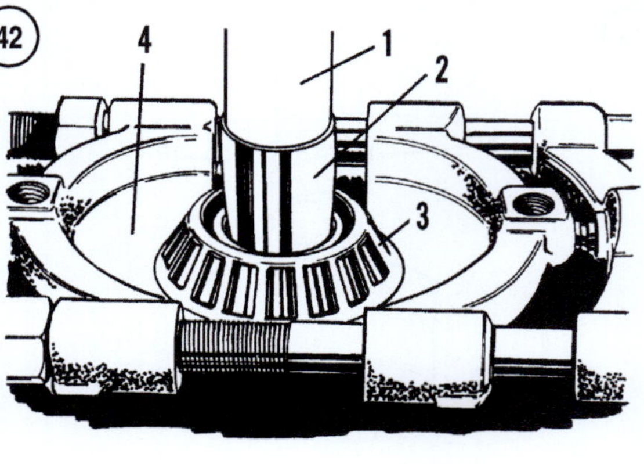

1. Press ram
2. Removal sleeve
3. Tapered roller bearing
4. Bearing separator

43

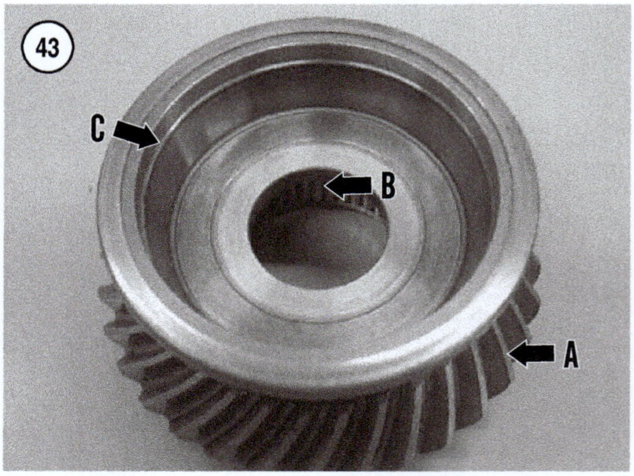

CAUTION

A tapered roller bearing consists of the caged tapered rollers, the cone-shaped inner race and the tapered outer race. The bearings must be replaced as a set, and parts from one bearing must never be used with parts from another.

32. Check all bearings and races for excessive or uneven wear, scoring, roughness, heat discoloration and binding. Replace any bearing and corresponding race if these or other defects are evident.

33. Check the splines in the input drive shaft coupler, universal joint shaft yoke and output drive shaft for twisting, burrs, discoloration, cracks and excessive wear. Replace

the shaft coupler, yoke or drive shaft if these or other defects are evident.

34. Check the helical splines and bearing contact surfaces on the output drive shaft (18, **Figure 32**) for excessive wear, burring, discoloration, uneven surfaces, scoring and other surface imperfections. Replace the output drive shaft if these or other defects are evident.

35. Inspect the cone clutch splines (B, **Figure 44**) for excessive wear, dislocation, cracks or missing sections. Check the contact surfaces (A, **Figure 44**) for excessive wear, discoloration, embedded debris or damage to the clutch-to-output gear sleeve surfaces. Replace the cone clutch if these or other defects are evident.

36. Check the cone clutch support spring (13, **Figure 32**) for discoloration, distortion or lost spring tension. Replace the spring if it is in questionable condition.

37. Check the double-stacked needle bearings (7, **Figure 32**) for roughness, discoloration, binding, missing or damaged needles and other defects. Replace both bearings in the stack if these or other defects are found with either section.

38. Inspect all shims and replace any that are damaged. If shimming adjustments are not required for the repair, damaged shims can be replaced with new shims of the same thickness.

39. Inspect the water passage cover (55, **Figure 32**) for gasket deterioration, loose fasteners and evidence of leakage. If defects are evident, proceed as follows:

a. Remove the four screws (54, **Figure 32**) securing the cover to the housing.

b. Remove the cover and all gasket material from the cover and housing surfaces. Replace the cover if it is warped, corroded, cracked or otherwise defective.

c. Check the water passage in the gearcase for corrosion, debris or deposit buildup. Clean the passage as required.

40. Inspect the thrust plates (58 and 62, **Figure 32**) for excessive wear, cracks or missing sections. Remove the retaining clips (60, **Figure 32**) and replace worn or damaged thrust plates. Always install new retaining clips during assembly.

Assembly (SX and DP-S models)

For maximum durability and quiet operation, all of the gears must be properly located in the housing through shimming procedures. When properly located, the correct gear lash and gear contact pattern will be obtained.

Additionally, the tapered bearings in the bearing carrier must be properly tensioned or *preloaded*. Preload refers to the tension applied to hold the tapered roller bearing against the bearing race. Insufficient preload will allow the bearing to move in relation to the race. Any movement will result in premature failure of the bearing and supported gears or shaft. Excessive preload will cause the bearings to overheat and fail.

10

During assembly, lubricate all gears, bearings, seals, shafts and O-rings with OMC, Volvo Penta or Mobil One GL5 Synthetic Gear Lube.

Refer to **Figure 32** for this procedure.

1. If the small cover (55, **Figure 32**) was removed from the rear of the housing, install the new gasket (56) and cover onto the upper drive shaft housing. Do not apply sealant to the gasket. Thread the four screws (54, **Figure 32**) into the water cover and housing openings. Tighten the screws in a crossing pattern to 60-84 in.-lb. (7-10 N•m).

NOTE

*Thrust plate shims (59, **Figure 32**) come in one size only, 0.015 in. (0.381 mm). Shims must always be installed or removed equally on each side with a minimum of one shim and a maximum of four shims per side. Each side must have the same number of shims to keep the drive centered. If more than four shims are required on each side to obtain the required dimension, replace both thrust plates and recheck the dimension.*

2. Using a caliper, measure the distance from the outside edge of one thrust plate to the outside edge of the other thrust plate (**Figure 45**). Make sure to press the thrust plates against the shims and housing to eliminate any air gap during the measurement. Take several measurements along the length of the plate to locate the widest dimension. Record

the measurement. The thrust plate measurement must be 5.007-5.036 in. (12.72-12.79 cm) at the widest point. Refer to the following:

a. If the thrust plate measurement is within the specification, no shim change is required.
b. If the thrust plate measurement exceeds 5.036 in. (12.79 cm), remove the retaining clips (60, **Figure 32**) and carefully pry the thrust plates (58 and 62) and shims (59) from the housing. Install one fewer shim onto each side of the housing. Reinstall the thrust plates and secure with new retaining clips. Recheck the dimension. Remove additional shims if necessary to obtain the correct dimension.
c. If the thrust plate measurement is less than 5.007 in. (12.72 cm), remove the retaining clips (60, **Figure 32**) and carefully pry the thrust plates (58 and 62) and shims (59) from the housing. Install one additional shim onto each side of the housing. Reinstall the thrust plates and secure with new retaining clips. Recheck the dimension. Add additional shims if necessary to obtain the correct dimension. If more than four shims are required on each side to obtain the required dimension, replace both thrust plates and recheck the dimension. Always replace the retaining clips with new ones upon removal.

3. If the bearing races were removed from the drive gear bearing carrier, install new bearing races as follows:

a. Place the bearing carrier onto a press with the large end facing downward.
b. Lubricate the larger-diameter bearing race and place it into the bearing carrier with the tapered side facing upward.
c. Seat the larger-diameter side of the installation mandrel (OMC or Volvo Penta part No. 884932-5) against the race. Press the race into the housing until fully seated.
d. Place the bearing carrier onto a press with the small end facing downward.
e. Lubricate the smaller-diameter bearing race and place it into the bearing carrier with tapered side facing upward.
f. Seat the smaller-diameter side of the installation mandrel (OMC or Volvo Penta part No. 884932-5) against the race. Press the race into the carrier until fully seated.

4. If the bearing was removed from the drive gear, install the new bearing as follows:

a. Place a block of wood or other suitable material on the press to protect the gear teeth. Place the gear on the block with the teeth facing downward.
b. Lubricate the gear hub and inner bore of the new bearing.
c. Place the bearing onto the gear hub with the rollers facing upward.
d. Seat the installation sleeve (OMC or Volvo Penta part No. 884263-5) against the inner bearing race.

46

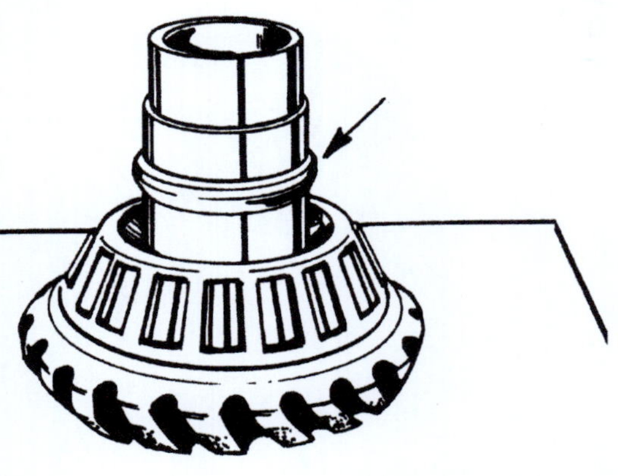

47

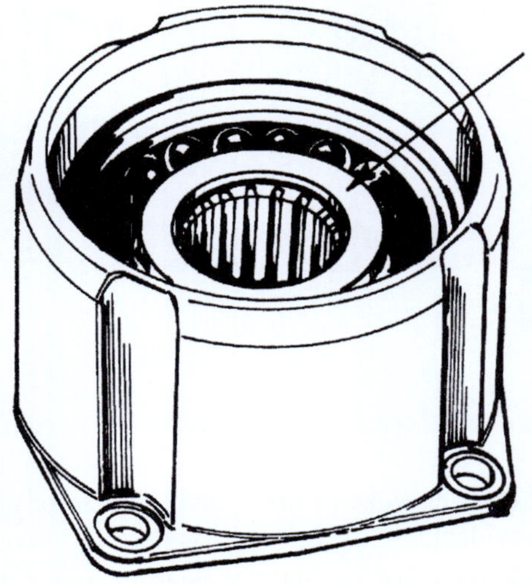

e. Press the bearing onto the gear until fully seated.

CAUTION
A new pretension (crush) sleeve must be installed each time the bearing carrier is disassembled. Failure to install a new sleeve will result in premature bearing and/or gear failure.

NOTE
Protect the gear teeth from damage when pressing the bearing onto the gear hub by placing a block of wood or other suitable material against the gear.

5. Position a *new* pretension sleeve (**Figure 46**) over the hub of the gear and rest it against the inner bearing race.

6. With the large end facing upward, position the bearing carrier over the gear. Rest the bearing race in the carrier against the roller bearing on the gear.

7. Lubricate the smaller-diameter roller bearing and position it onto the gear hub with the rollers facing downward. See **Figure 47**.

48

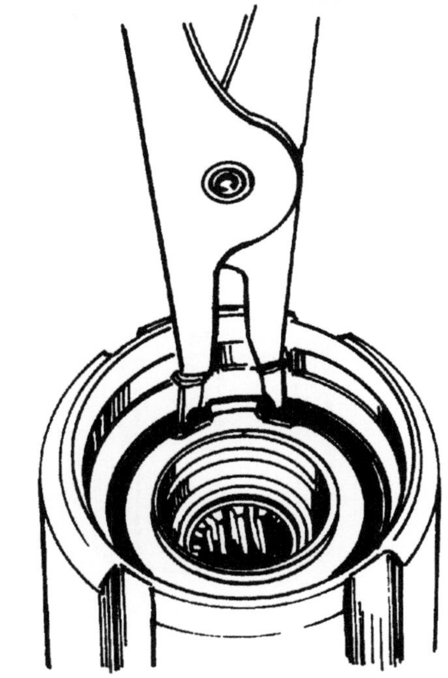

49

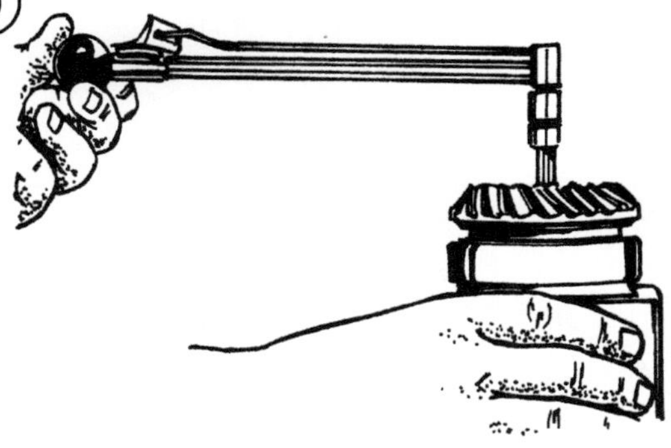

CAUTION
Do not press the bearing too far in Step 8. If the new pretension sleeve is overcrushed, the bearing carrier must be disassembled, and another new pretension sleeve must be installed.

8. Press the small bearing into the carrier as follows:
   a. Place a block of wood or other suitable material on the press to protect the gear teeth. Place the bearing carrier and gear on the block with the teeth facing downward.
   b. Seat the installation sleeve (OMC or Volvo Penta part No. 884263-5) against the inner bearing race.
   c. *Slowly* press the small bearing onto the gear hub until there is almost no play (approximately 0.40 in. [1 cm]) between the bearing carrier, gears and bearings. Wiggle the bearing carrier up and down while pressing to monitor the bearing progress.
9. If removed, install the *new* universal joint shaft seal (31, **Figure 32**) as follows:
   a. Place the support sleeve (OMC or Volvo Penta part No. 884938-2) onto the press with the large end facing upward.
   b. Set the bearing carrier into the sleeve with the large end facing upward.
   c. Coat the outer diameter of the new seal with OMC or Volvo Penta White Sealing Compound (part No. 1141570-0). Fit the seal into the carrier with the lip side facing upward.
   d. Seat the installation sleeve (OMC or Volvo Penta part No. 3850607-7) against the seal casing. Press the seal into the bore until seated against the bearing carrier shoulder.
   e. Secure the seal by installing the snap ring (32, **Figure 32**) with robust snap ring pliers (**Figure 48**). Make sure the snap ring is fully expanded into the groove in the carrier.
   f. Coat the seal lip with OMC Triple Guard Grease.
10. Insert the pipe used during disassembly through the cap bores in the yoke. Then clamp the pipe into a vise. The pipe will absorb the compressionloads and prevent damage to the yoke during service.
11. Slide the drive gear bearing carrier over the yoke shaft. Rotate the yoke shaft as necessary to align the splines and then seat the bearing carrier on the yoke shaft.
12. Insert the tapered washer (22, **Figure 32**) into the gear face with the taper facing the gear. Apply a light coat of Loctite 271 threadlocking adhesive to the threads and then thread the Torx head bolt (21, **Figure 32**) into the gear and yoke. Tighten the bolt by hand.

NOTE
The next step sets the tension (preload) on the tapered roller bearings. Do not overtighten the Torx head bolt because the pretension sleeve will be damaged and require replacement. The preload is checked by measuring the torque required to turn the yoke shaft as the bearing carrier is held stationary. For accurate measurement, the yoke must be turned slowly and smoothly and the bearings must be well-lubricated. Use a 0-30 in.-lb. (0-3.4 N•m) dial torque wrench for this procedure.

13. Slowly tighten the Torx head bolt while rotating the bearing carrier (**Figure 49**). Tightening the bolt forces the bearings close together as they compress the pretension sleeve. Stop tightening the bolt when the effort required to rotate the carrier increases. This indicates the bearings have started to preload.

CAUTION
It is common to tighten the bolt and see no immediate increase in preload. Do not tighten the bolt in large increments. Be patient, tighten the bolt in small amounts, and check the preload each time.

10

14. Check the bearing preload by gently clamping the bearing carrier into a vise with protective jaws. Use a T50 socket and dial torque wrench to rotate the drive gear several revolutions to seat the bearings. Then rotate the drive gear *slowly* and *smoothly* while noting the torque reading. The torque wrench must indicate 9-14 in-lb. (1.0-1.6 N•m) of rolling torque (preload). Refer to the following instructions:

a. If the preload is within the specification, remove the torque wrench. Proceed to Step 15.

b. If the preload is less than the specification, clamp the bearing carrier into a vise. Tighten the Torx bolt *slightly* as described in Step 13. Do not use the torque wrench to tighten the bolt. Recheck the preload. Continue tightening and measuring until the preload is correct.

c. If the preload exceeds the specification, remove the drive gear and replace the pretension sleeve as described in the disassembly and assembly procedures.

CAUTION

If the bearing preload exceeds the specification, the bearing carrier must be disassembled to replace the pretension sleeve. Do not continue with the assembly until the preload is correct.

15. Install the universal drive shaft onto the yoke as described in this chapter. Refer to *Universal Joint Service*.

16. If the bearings (8, **Figure 32**) were removed from the output gears (11 and 16), install the new bearings as follows:

a. Place a block of wood or other suitable material on the press table to protect the gear.

b. Place the output gear on a press with the gear sleeve and teeth facing downward.

c. Lubricate the new bearing and output gear hub. Place the bearing onto the gear hub with the numbered side facing upward.

d. Seat the sleeve (OMC or Volvo Penta part No. 884168-6) against the inner bearing race. Press the bearing onto the gear hub until fully seated.

e. Repeat this step for the remaining output gear.

17. If removed, install the adapter ring (9, **Figure 32**) onto the upper output gear (11) as follows:

a. Place the support sleeve (OMC or Volvo Penta part No. 884938-2) onto a press with the larger opening facing upward as shown in **Figure 50**.

b. Lubricate the adapter ring (2, **Figure 50**) on the bearing with the larger diameter (flanged side) facing upward.

c. Set the upper output gear and bearing (3, **Figure 50**) onto the adapter ring with the gear side facing downward.

50

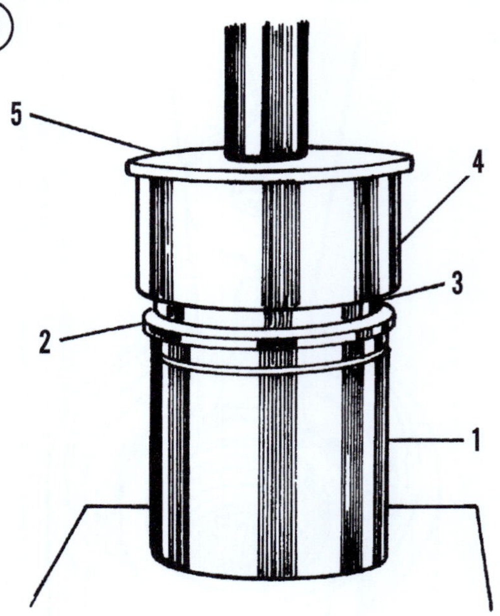

1. Support sleeve
2. Adapter ring (flange side up)
3. Upper output gear and bearing (gear side down)
4. Mandrel
5. Suitable mandrel or plate

51

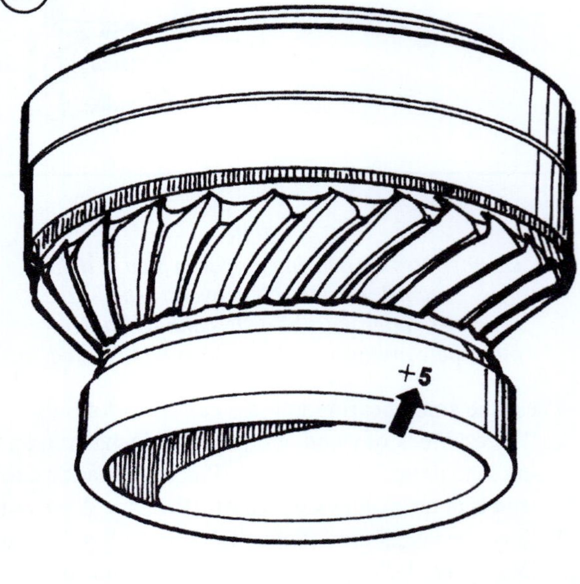

d. Seat the mandrel (4, **Figure 50**) onto the bearing. Place a suitable mandrel or plate (5, **Figure 50**) on the mandrel.

e. Press the bearing and gear into the adapter ring until the bearing lip seats against the adapter ring.

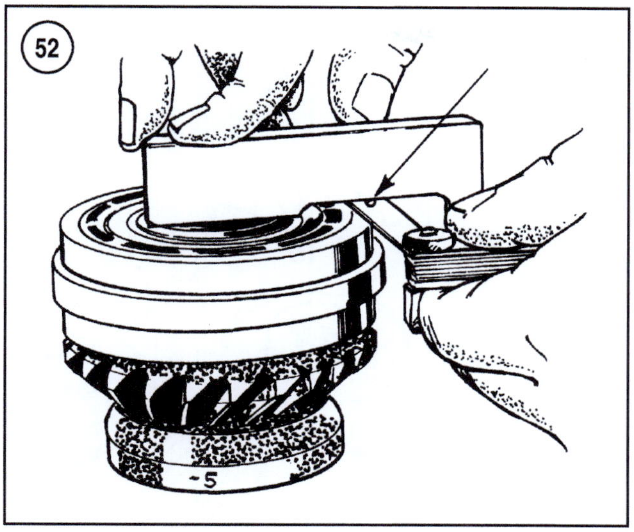

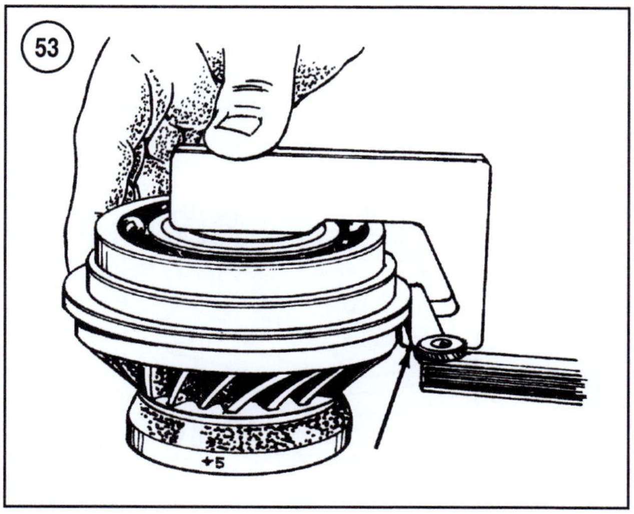

CAUTION
*Due to the mounting location of the lower output gear and shims, a negative (–) correction number must be **added** to the feeler gauge reading, and a positive (+) correction number must be **subtracted** from the feeler gauge reading.*

NOTE
When the correct feeler gauge is selected for the measured clearance, a slight drag is felt as the feeler gauge passes between the measuring points. To verify, insert the next larger gauge and check for a noticeable increase in drag. Then insert the next smaller gauge and check for the absence of drag.

18. If a new lower output gear (16, **Figure 32**) or bearing (8) was installed, select the proper shims as follows:

a. Find the correction number etched or stamped onto the lower output gear sleeve (**Figure 51**). Record this number in thousandths of an inch. For example, if +5 is etched into the gear, write it as +0.005 in. If -5 is etched into the gear, write is as -0.005 in.

b. Set the lower output gear on a clean work surface with the gear teeth facing downward.

c. Position the machined flat on the shim gauge (OMC or Volvo Penta part No. 3850701-8) onto the gear hub with the measuring tip positioned directly over the outer bearing race.

d. While holding the shim gauge firmly in place, take a feeler gauge reading between the shim gauge upper tip and the outer bearing race (**Figure 52**). Record the feeler gauge reading.

e. Add or subtract the correction number (Step 18a) to or from the feeler gauge reading (Step 18d). For example, if the feeler gauge reading is 0.011 in. and the correction number is +5, *subtract* 0.005 in. from 0.011 in. The result is 0.006 in. If the feeler gauge reading is 0.011 in. and the correction number is -5, *add* 0.005 in. to 0.011 in. The result is 0.016 in. In the last example, a 0.016 in. shim would be used.

f. Select the correct number of shims for the lower output gear based on the results in this step. Install this number of shims into the upper drive shaft housing and rest on the step in the lower output bearing bore.

CAUTION
*Due to the mounting location of the upper output gear and shims, a negative (–) correction number must be **subtracted** from the feeler gauge reading, and a positive (+) correction number must be **added** to the feeler gauge reading.*

19. If a new upper output gear (11, **Figure 32**) or bearing (8) was installed, select the proper shims as follows:

a. Find the correction number etched or stamped onto the lower output gear sleeve (**Figure 51**). Record this number in thousandths of an inch. For example, if +5 is etched into the gear, write it as +0.005 in. If -5 is etched into the gear, write it as -0.005 in.

b. Set the upper output gear on a clean work surface with the gear teeth facing downward.

c. Position the machined flat on the shim gauge (OMC or Volvo Penta part No. 3850701-8) onto the gear hub with the lower measuring tip positioned directly under the adapter ring lip (**Figure 53**).

d. While holding the shim gauge firmly in place, take a feeler gauge reading between the shim gauge lower tip and the adapter ring on the bearing (**Figure 53**). Record the feeler gauge reading.

e. Add or subtract the correction number (Step 19a) to or from the feeler gauge reading (Step 19d). For example, if the feeler gauge reading is 0.011 in. and the correction number is +5, *add* 0.005 in. to 0.011 in. The result is 0.016 in. If the feeler gauge reading is 0.011 in. and the correction number is -5, *subtract*

10

0.005 in. from 0.011 in. The result is 0.006 in. In the last example, a 0.006 in. shim would be used.

f. Select the correct number of shims for the upper output gear based on the results in this step. Place the shims with the upper output gear assembly.

NOTE

Shim selection for the driver gear can be performed with or without the shim plate (OMC or Volvo Penta part No. 3850600-2). The shim plate is designed to sit on the gear teeth and provides a stable platform for measurement. If the shim plate is not available, a wide-based depth micrometer may be used. Measurement using the shim plate is easier and generally more accurate.

NOTE

*All three gears are stamped with a correction number and match number. The match number always starts with a T (A, **Figure 54**). All three gears in a set have the same T number. Though possible, it is uncommon for all three gears to have the same correction number (B, **Figure 54**). Do not confuse the horizontal line on the T mark with a negative (–) mark.*

20A. If using a shim plate, and if the drive gear, drive gear bearings or drive gear bearing carrier was replaced, select drive gear shims (27, **Figure 32**) as follows:

a. Find the correction number etched or stamped onto the drive gear face (B, **Figure 54**). Record this figure in thousandths of an inch. For example, if +5 is etched into the gear, write it down as +0.005 in. If -5 is etched into the gear, write it down as -0.005 in.

b. Lightly clamp the drive gear bearing carrier in a vise with protective jaws. The gear teeth must face upward.

c. Place the shim plate (OMC or Volvo Penta part No. 3850600-2) onto the gear with the recessed side facing upward. See **Figure 55**. Center the plate on the gear teeth.

d. Use a depth micrometer to measure the distance from the top of the shim plate (A, **Figure 55**) to the flange on the bearing carrier (B). Do not measure near bolt openings. Take several measurements to ensure accuracy. Record the average dimension.

e. Subtract the thickness of the shim plate (0.500 in. [12.70 mm]) from the recorded measurement. Record the results. This is the actual distance from the gear teeth to the carrier flange.

f. Add or subtract the etched correction number determined in Step 20a to or from the actual distance recorded in Step 20e. For example, if the actual distance is 1.999 in. and the correction number is +4, *add* 0.004 in. to 1.999 in. The result is 2.003. If the actual distance is 1.999 in. and the correction number is -4, *subtract* 0.004 in. from 1.999 in. The result is 1.995 in. Record the results.

54

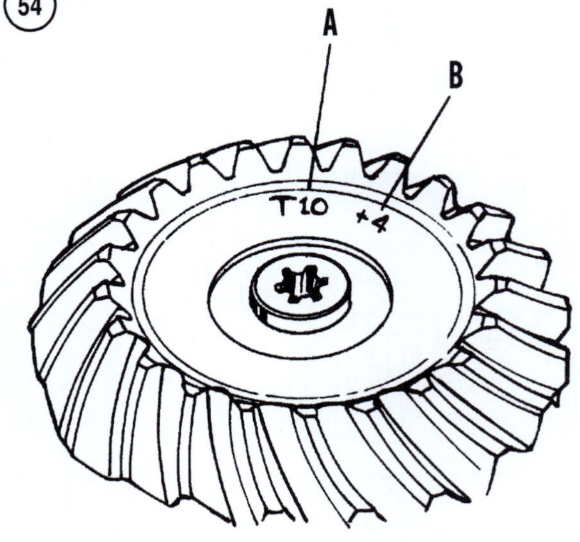

55

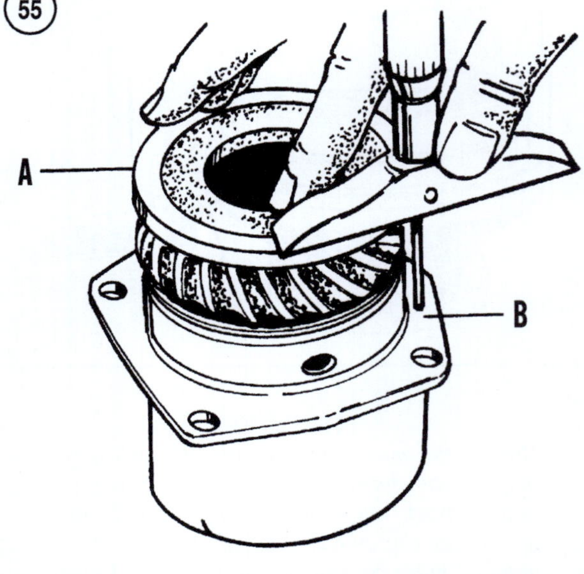

g. If the drive gear has 21 teeth, *subtract* 1.898 in. from the actual distance and correction number calculation recorded in Step 20f. If the drive gear has any other tooth count, *subtract* 1.971 in. from the actual distance and correction number calculation recorded in Step 20f. For example, if the calculation recorded in Step 20f is 1.995 and the gear has 22 teeth, *subtract* 1.971 in. from 1.995 in. The result is 0.024 in. In this example, a 0.024 in. shim is used for the drive gear, bearing and housing combination. Select the correct thickness of shims and place them on the drive gear bearing carrier.

20B. If not using a shim plate, and if the drive gear, drive gear bearings or drive gear bearing carrier was replaced, select drive gear shims as follows:

56

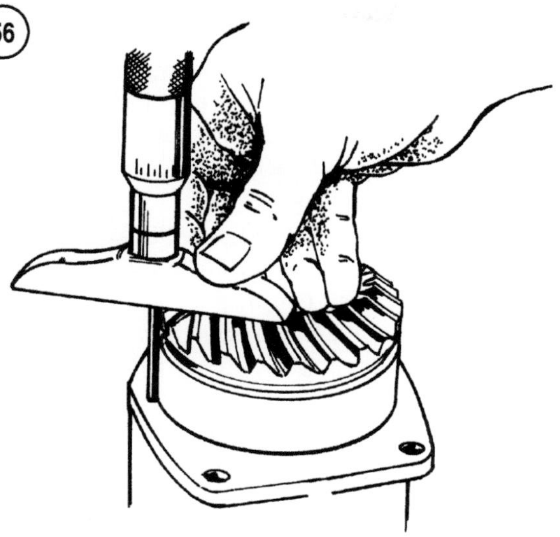

a. Find the correction number etched or stamped onto the drive gear face (B, **Figure 54**). Record this figure in thousandths of an inch. For example, if +5 is etched into the gear, write it down as +0.005 in. If -5 is etched into the gear, write it down as -0.005 in.
b. Lightly clamp the drive gear bearing carrier in a vise with protective jaws. The gear teeth must face upward.
c. Use a depth micrometer to measure the distance from the top of the gear teeth to the flange on the bearing carrier (**Figure 56**). Do not measure near bolt openings. Take several measurements to ensure accuracy. Record the average dimension as the actual distance from the gear teeth to the carrier flange.
d. Add or subtract the etched correction number determined in Step 20a to or from the actual distance recorded in Step 20c. For example, if the actual distance is 1.999 in. and the correction number is +4, *add* 0.004 in. to 1.999 in. The result is 2.003. If the actual distance is 1.999 in. and the correction number is -4, *subtract* 0.004 in. from 1.999 in. The result is 1.995 in. Record the results.
e. If the drive gear has 21 teeth, *subtract* 1.898 in. from the actual distance and correction number calculation recorded in Step 20d. If the drive gear has any other tooth count, *subtract* 1.971 in. from the actual distance and correction number calculation recorded in Step 20d. For example, if the calculation recorded in Step 20d is 1.995 and the gear has 22 teeth, *subtract* 1.971 in. from 1.995 in. The result is 0.024 in. In this example, a 0.024 in. shim is used for the drive gear, bearing and housing combination. Select the correct thickness of shims and place them on the drive gear bearing carrier.

21. Install the C-clip (15, **Figure 32**) into the output shaft groove. Slide the support washer (14, **Figure 32**) over the top of the output shaft and rest on the C-clip.

22. Lubricate and then insert the double-stacked bearings (7, **Figure 32**) into the bore in the lower output gear and bearing assembly. If reused, the bearings must be installed in the original locations and orientations.

23. Guide the lower end of the output shaft gear teeth side first through the gear and needle bearing bore. Slide the thrust washer (20, **Figure 32**) over the lower end of the output shaft with the groove recess facing away from the gear. Coat the two split lockrings with OMC Needle Bearing Assembly Grease (part No. 378642). Move the output shaft to expose the groove and then fit the split lockrings into the shaft groove (**Figure 40**). Push the lower end of the shaft toward the gear to seat the split lockrings into the thrust washer recess.

CAUTION

When the output shaft and lower output gear and bearing assembly is installed in the housing, the output shaft must be supported to prevent it from dropping and dislodging the thrust washer and split lockrings.

CAUTION

*The lower output gear retaining ring (17, **Figure 32**) is tightened to 145-165 ft.-lb. (197-224 N•m). The housing must be secured to a holding fixture (OMC or Volvo Penta part No. 3850605-1), a strong bench or floor mount, and the spanner wrench (OMC or Volvo Penta part No. 3850604-1) must be used when loosening or tightening the retainer. Make sure that any debris from gear or bearing failure is cleaned from the retainer and housing threads prior to installation. If necessary, scrape the threads clean with a suitable pick tool and liberally wash the area with lubricant. If the retainer binds on installation, thread the ring back out, clean and lubricate the threaded area, and try tightening again. Forcing the retainer will cause the retainer to lock in the housing and destroy the housing.*

NOTE

The gearcase holding fixture (OMC or Volvo Penta part No. 3850605-1) is designed to hold the upper drive shaft housing securely during all service procedures. This tool fits the standard Bob Kerr Tool Company workbench base or floor stand. It is highly recommended to use only the holding fixture during the disassembly and assembly procedures. Many of the procedures are difficult to perform without the fixture, even with the aid of an assistant. Purchase the holding fixture from a Volvo Penta dealership. Purchase the workbench base or floor stand from Bob Kerr Tool Company, P.O. Box 771135, Winter Garden, Florida 34777.

10

24A. If using the recommended holding fixture, install the output shaft and lower gear retaining ring as follows:

a. Attach the splined adapter (OMC or Volvo Penta part No. 3895098-1) to the shaft of the holding fixture and secure the adapter with the setscrew.
b. Verify that the predetermined shim(s) (10, **Figure 32**) for the lower output gear are installed into the bottom of the housing bore.
c. Grasp the output shaft at the upper end and carefully lower the shaft and lower output gear assembly into the housing. While maintaining upward lifting on the shaft, seat the lower output gear bearing into the bore until it contacts the shim(s). The threads for the retaining ring will be clearly visible when the bearing seats.
d. While maintaining upward lifting on the shaft, loosen the lock bolt and move the fixture shaft upward until the splined adapter contacts the splines of the output shaft. Rotate the shaft until the splines align and the output shaft rotates with the fixture shaft. Rotate the fixture shaft to align the flat shaft surface with the lock bolt. Hold the shaft while securely tightening the lock bolt. Release the upper drive shaft.
e. Lubricate the threads of the retaining ring (17, **Figure 32**) and the corresponding threads in the housing. Install and thread the ring in as far as possible by hand. The threads of the ring should not be visible when viewed through the drive gear opening.
f. Fit the spanner wrench (**Figure 39**) over the output shaft and engage the wrench onto the lugs of the retainer ring. Thread two of the cover bolts into the spanner wrench and housing openings. Securely tighten the bolts.
g. Using a 1/2 in.-to-3/4 in. adapter and a suitable torque wrench, tighten the retaining nut to 145-165 ft.-lb. (197-224 N•m).
h. *Slowly* loosen the two cover bolts used to secure the spanner wrench. Lift the spanner wrench out of the housing.

24B. If not using the recommended holding fixture, install the output shaft and lower gear retaining ring as follows:

a. Verify that the predetermined shim(s) (10, **Figure 32**) for the lower output gear are installed into the bottom of the housing bore.
b. Stuff several shop towels in the drive shaft bore from the bottom of the housing to support the output shaft during installation.
c. Grasp the output shaft at the upper end and carefully lower the shaft and lower output gear assembly into the housing. While maintaining upward lifting of the shaft, seat the lower output gear bearing into the bore until it contacts the shim(s). The threads for the retaining ring will be clearly visible when the bearing seats.
d. While still lifting the shaft, make sure the shop towels are pressed up against the lower end of the output shaft to prevent it from dropping and dislodging the thrust washer and split lockrings.
e. Lubricate the threads of the retaining ring (17, **Figure 32**) and the threads in the housing. Install and thread the ring in as far as possible by hand. The threads of the ring should not be visible when viewed through the drive gear opening.
f. Fit the spanner wrench (**Figure 39**) over the output shaft and engage the wrench onto the lugs of the retainer ring. Thread two of the cover bolts into the spanner wrench and housing openings. Securely tighten the bolts.
g. Attach a 1/2 in.-to-3/4 in. adapter and a suitable torque wrench onto the spanner wrench. With the aid of an assistant, tighten the retaining nut to 145-165 ft.-lb. (197-224 N•m).
h. *Slowly* loosen the two cover bolts used to secure the spanner wrench. Lift the spanner wrench out of the housing.

25. Install the cone clutch support spring (13, **Figure 32**) over the output shaft and rest it on the support washer (14).

26. Install the cone clutch onto the output shaft. Rotate the clutch clockwise to engage the helical splines and then rest the clutch against the spring. Make sure the *TOP* mark on the clutch faces upward.

27. Install the predetermined shims (10, **Figure 32**) onto the upper output gear adapter ring. Secure the shims in place with OMC Needle Bearing Assembly Grease. When assembled, the shims will fit between the adapter ring and the housing.

28. Carefully install the upper output gear assembly and shims into the housing. Make sure all of the shims are in position on the output gear adapter ring.

NOTE

If using the holding fixture, it might be necessary to loosen the mounting fasteners and move the housing slightly to align the output shaft with the bore center to install the double-stacked needle bearings. Retighten the mounting fasteners after installing the bearings.

29. Lubricate the surfaces and then insert the double-stacked bearings (7, **Figure 32**) into the bore in the upper output gear and bearing assembly. If reused, the bearings must be installed in the original locations and orientations.

NOTE

Output shaft nuts are available in four different thicknesses in 0.008 in. (0.2 mm) increments. The nuts are numbered 0, 2, 4 and 6, and 0 is the thinnest. The nuts use left-hand threads.

30A. If using the recommended holding fixture, install the output shaft nut and measure the shaft end play as follows:

57

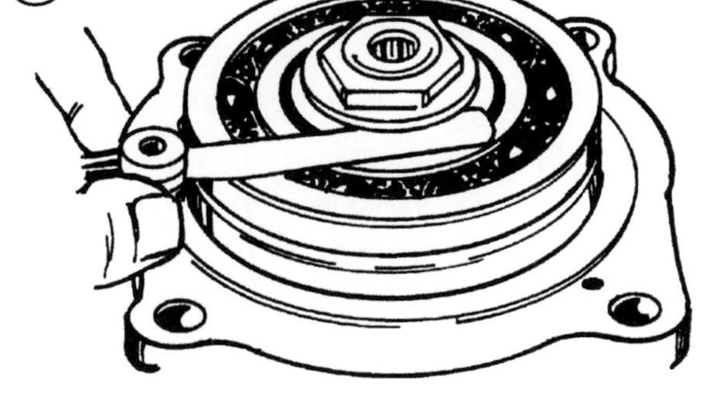

a. Rotate the upper drive shaft to verify full engagement of the splined adapter on the fixture shaft to the output shaft. If necessary, loosen the lock bolt and engage the splines as described in Step 24A.
b. Lubricate the output shaft and nut threads and then thread the output shaft nut (**Figure 34**) onto the shaft. The nut uses left-hand threads.
c. Using a 30 mm socket and suitable torque wrench, tighten the nut to 96-110 ft.-lb. (130-149 N•m).
d. Loosen the lock bolt on the holding fixture and then push up firmly on the holding fixture shaft. Hold up on the shaft while securely tightening the lock bolt.
e. Apply downward pressure on the upper driven gear bearing. Then use feeler gauges to measure the clearance between the upper output gear hub and the flat surface of the nut (**Figure 57**) to determine the output shaft end play. The end play must be 0.002-0.010 in. (0.05-0.25 mm). If the end play is not within the specification, install a thicker or thinner output shaft nut and recheck the clearance. If the required nut thickness is not available, check for improper seating of the upper and lower output gears. If the gears are seated properly and the clearance cannot be corrected with a nut change, disassemble and reassemble the housing and check all clearances, shim calculations and gear lash measurements.
f. After correcting the output shaft end play, loosen the lock bolt and lower the fixture shaft to disengage the splined adapter.

30B. If not using the recommended holding fixture, install the output shaft nut and measure the shaft end play as follows:

a. Lubricate the output shaft and nut threads and then thread the output shaft nut (**Figure 34**) onto the shaft. Tighten the nut by hand at this time. The nut uses left-hand threads.
b. Remove the shop towels from the lower drive shaft bore. Using a flashlight, verify that the two split lockrings remain seated within the thrust washer recess.
c. Insert the splined adapter (OMC or Volvo Penta part No. 3850598-8) into the drive shaft bore and engage the adapter fully onto the output shaft splines. Engage a socket, extension and 1/2 in. breaker bar onto the adapter.
d. Have the assistant hold the breaker bar and housing to prevent shaft rotation. Then use a 30 mm socket and suitable torque wrench to tighten the output shaft nut to 96-110 ft.-lb. (130-149 N•m).
e. Have an assistant push up on the output shaft with an extension, nut and the adapter. At the same time, apply downward pressure on the upper driven gear bearing. While maintaining pressure, use feeler gauges to measure the clearance between the upper output gear hub and the flat surface of the nut (**Figure 57**) to determine the output shaft end play. The end play must be 0.002-0.010 in. (0.05-0.25 mm). If the end play is not within the specification, install a thicker or thinner output shaft nut and recheck the clearance. If the required nut thickness is not available, check for improper seating of the upper and lower output gears. If the gears are seated properly and the clearance cannot be corrected with a nut change, disassemble and reassemble the housing and check all clearances, shim calculations and gear lash measurements.
f. After correcting the end play, remove the socket, extension and adapter.

31. Set the top cover (3, **Figure 32**) onto the housing *without* the seal (5) and with the original shims (4) or new shims of the same thickness. Apply a coat of OMC Needle Bearing Assembly Grease to hold the shims in position during installation. Press down with firm hand pressure and use feeler gauges to measure the clearance between the top cover and the housing. If there is no clearance, remove the cover and add enough shims to achieve a measurable clearance. Take several measurements and compute the average clearance. The average clearance must be 0.002-0.004 in. (0.05-0.10 mm). Add or subtract shims to achieve the required clearance.

32. Thread the top cover bolts (2, **Figure 32**) into the cover and housing. Do not install the top cover seal at this time. Tighten the bolts in a crossing pattern to 195-212 in.-lb. (22-24 N•m).

33. Lubricate the mating surfaces and then insert the drive gear bearing carrier and universal drive shaft assembly into the shift housing with the predetermined shims (27, **Figure 32**). Do not install the O-ring (26, **Figure 32**) at this time. Rotate the universal shaft as necessary to mesh the gear teeth. Make sure the shims align with the bolt openings and then thread the four bolts (29, **Figure 32**) into the bearing carrier and shift housing. Tighten the bolts in a crossing pattern to 142-168 in.-lb. (16-19 N•m).

NOTE

Gear lash is a measurement of the clearance between the drive gear and driven gear teeth. The drive gear must not move during the measurement and must be clean. The gears should make a distinct clicking noise as the output gear is rotated back and forth against the drive gear.

34. Mount a dial indicator capable of measuring lateral movement into a stable mount. Place the ball tip of the dial indicator on the horizontal and vertical midpoint of one of the lower output gear teeth. Make sure the dial indicator arm is pointing directly toward the imaginary center of the output shaft. See **Figure 58**, typical. Zero the dial and *lightly* rock the output gear back and forth to read the gear lash. Record the reading. Lift the dial indicator from the gear tooth, turn the gear one-half revolution, reposition the indicator, and recheck the gear lash.

35. Reposition the dial indicator and repeat Step 34 for the upper output gear. Record the gear lash measurement. Remove the dial indicator when finished.

NOTE

If the gear lash requires correction, change approximately 0.001 in. (0.03 mm.) of shim thickness for every 0.001 in. (0.03 mm) of gear lash correction required. Move the appropriate gear toward the other gears to decrease gear lash and move the appropriate gear away from the other gears to increase gear lash.

36. The gear lash for either gear must be 0.006-0.010 in. (0.152-0.254 mm). Refer to the following:

a. If both gears are within the specification, proceed to Step 37.

b. If the gear lash for both gears is greater than 0.010 in. (0.254 mm), remove the drive gear bearing carrier and install thinner shims (27, **Figure 32**) between the carrier and housing. Reinstall the carrier and recheck the gear lash as described in Steps 33-35.

c. If the gear lash for both gears is less than 0.006 in. (0.152 mm), remove the drive gear bearing carrier and install thicker shims (27, **Figure 32**) between the carrier and housing. Reinstall the carrier and recheck the gear lash as described in Steps 33-35.

d. If the gear lash for one gear is within the specification (0.006-0.010 in. [0.152-0.254 mm]) and the gear lash for the other gear is not within the specification, check the calculations used to determine shim thickness as described in Step 15. If the calculation is correct, remove the gear and bearings assembly, add or subtract shims as necessary to bring the suspect output gear into the specification, and install the gear and bearing. Recheck the gear lash as described in Steps 33-35. If the upper output gear shims are changed, check the top cover clearance and output shaft end play during the assembly and correct as needed. If the lower output gear shims are changed, check the output shaft end play during the assembly and correct as needed.

58

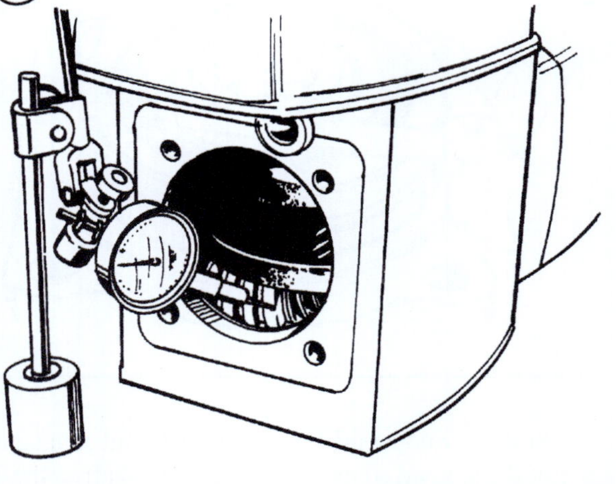

59

CHECKING DRIVEN GEAR PATTERN (UPPER DRIVE SHAFT HOUSING)

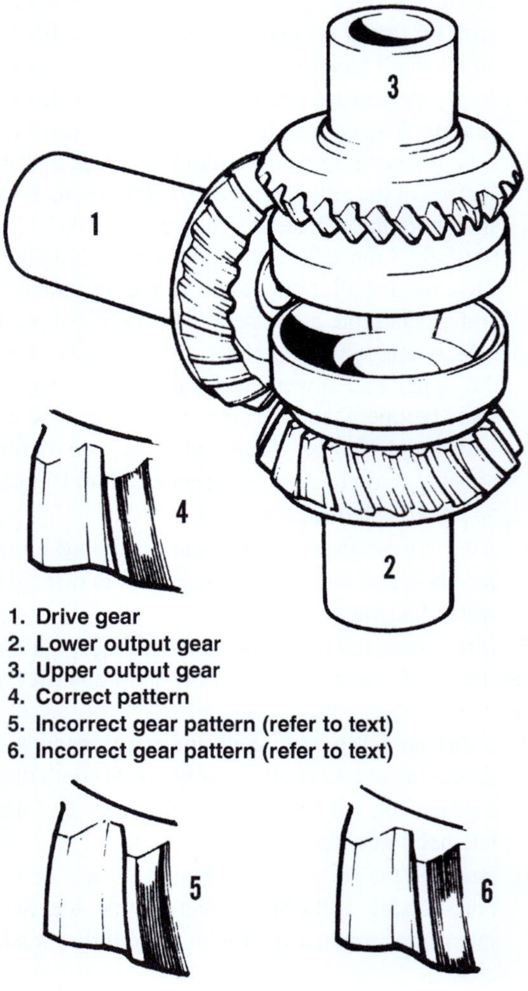

1. Drive gear
2. Lower output gear
3. Upper output gear
4. Correct pattern
5. Incorrect gear pattern (refer to text)
6. Incorrect gear pattern (refer to text)

60

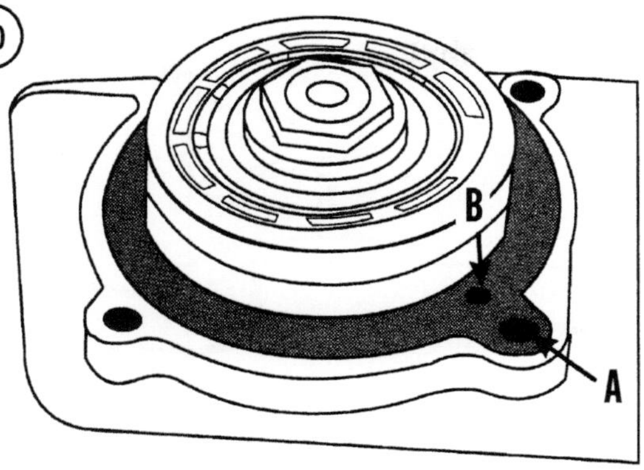

61

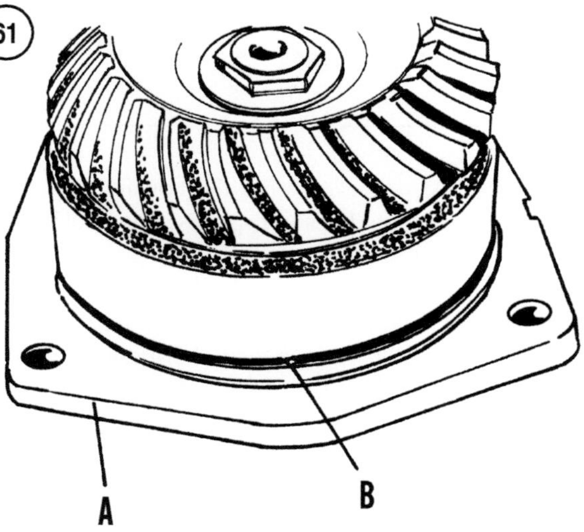

37. Coat three teeth on both output gears with a thin coat of GM Gear Marking Compound available from any General Motors (GM) automotive dealership.

38. Wedge a wooden hammer handle against the gear sleeves and rotate the gear by turning the universal drive shaft counterclockwise (as viewed from the rear of the drive) several revolutions to produce a tooth pattern on the output gears. The gears must be loaded to produce a pattern.

39. Withdraw the hammer handle and observe the pattern on the output gear teeth through the shift housing bore. Compare the pattern with the patterns shown in **Figure 59**. Refer to the following:

a. If the pattern is like the pattern shown in 4, **Figure 59**, the pattern is correct. Proceed to Step 40.

b. If the pattern is more like the pattern shown in 5, **Figure 59**, remove the drive gear bearing carrier as described in the disassembly procedures. Do not disassemble the carrier. Add thicker shims (27, **Figure 32**) between the bearing carrier (28) and housing. Remove the same thickness from the shims (10 and 53, **Figure 32**) between the output gear bearings and housing. The new shims should move the contact pattern toward the center of the tooth. Reinstall the carrier and gears and recheck the pattern.

c. If the pattern is more like the pattern shown in 6, **Figure 59**, remove the drive gear bearing carrier as described in the disassembly procedures. Do not disassemble the carrier. Add thinner shims (27, **Figure 32**) between the bearing carrier (28) and housing. Add the same thickness to the shims (10 and 53, **Figure 32**) between the output gear bearings and housing. The new shims should move the contact pattern toward the center of the tooth. Reinstall the carrier and gears and recheck the pattern.

CAUTION

If the gear pattern must be corrected, gear lash must again by verified after the pattern correction and before continuing with final assembly.

40. Remove the top cover as described in the disassembly procedures. Install a new seal (5, **Figure 32**) onto the top cover. Make sure the protrusion on the seal fits over the starboard front bolt opening (A, **Figure 60**). Also make sure the opening in the seal (B, **Figure 60**) aligns with the lubricant passage in the top cover.

41. Install the top cover onto the housing with the predetermined shims installed onto the cover. Apply a coat of OMC Needle Bearing Assembly Grease to hold the shims in position during the installation. Make sure the shims and seal are not dislodged during the cover installation. Reposition the shims and seals if necessary.

42. Apply a coat of OMC or Volvo Penta White Sealing Compound (part No. 1141570-0) to the threads. Then thread the top cover bolts (2, **Figure 32**) into the cover and housing openings. Tighten the bolts in a crossing pattern to 195-212 in.-lb. (22-24 N•m).

43. Remove the drive gear bearing carrier and universal drive shaft assembly as described in the disassembly procedures. Do not disassemble the carrier.

44. Make sure the predetermined shims are installed onto the flange of the carrier (A, **Figure 61**). Install a new O-ring into the groove in the carrier (B, **Figure 61**).

45. Apply a light coat of OMC or Volvo Penta White Sealing Compound (part No. 1141570-0) to the mating surfaces and then insert the drive gear bearing carrier and universal drive shaft assembly into the housing. Rotate the universal shaft as necessary to mesh the gear teeth. Make sure the shims align with the bolt openings.

46. Apply a light coat of Volvo Penta White Sealing Compound (part No. 1141570-0) to the threads and then thread the four bolts (29, **Figure 32**) into the bearing carrier and shift housing. Tighten the bolts in a crossing pattern to 142-168 in.-lb. (16-19 N•m).

47. Install a new O-ring (49, **Figure 32**) into the shift housing groove. Coat the O-ring, shift housing and upper drive shaft housing mating surfaces with OMC or Volvo Penta White Sealing Compound (part No. 1141570-0). Carefully

guide the shift shoe (52, **Figure 32**) into the clutch groove while installing the assembly into the housing. The big end of the shoe must face starboard as shown in **Figure 62**. Reposition the housing to align the screw openings. Apply OMC or Volvo Penta White Sealing Compound (part No. 1141570-0) to the threads and then thread the four screws (46, **Figure 32**) into the housing. Tighten the shift housing screws in a crossing pattern to 60-84 in.-lb. (7-10 N•m).
48. Refer to *Shift Shoe Shimming* in this chapter and determine the shims required to properly locate the shift shoe stop bolt. If desired, this procedure can be performed after the upper gearcase housing is installed onto the drive unit.
49. Reconnect the shift linkage to the shift arm on the shift housing.
50. Install the upper drive shaft housing as described in this chapter.
51. Pressure and vacuum test the drive unit as described in this chapter.

Installation (SX and DP-S models)

1. Lubricate the surfaces and then install a new O-ring into the groove surrounding the lubricant screen passage (3, **Figure 29**).
2. Lubricate and then install a new O-ring (2, **Figure 29**) onto the drive shaft retainer. Seat the O-ring against the housing.
3. Insert a new screen into the lubricant passage (3, **Figure 29**) with the cone end facing upward. Seat the screen flush with the mating surface.
4. Install the water tube grommet into the upper drive shaft housing with the larger-diameter opening facing outward. To ease installation of the copper water tube, apply a light coat of grease to the inner surfaces.
5. Install the water tube into the upper water tube grommet. Rotate the lower end of the water tube to align it with the lubricant screen bore and the lower drive shaft.
6A. On earlier Cobra SX drives, verify that the plastic water tube guide (1, **Figure 30**) is secured to the gearcase housing. Apply a light coat of OMC Type M adhesive to the seal in the guide bore.
6B. On later Cobra SX and all DP-S DuoProp drives, insert the plastic guide (1, **Figure 31**) into the gearcase with the protruding edges facing downward. Apply a coat of OMC Gasket Sealing Compound (part No. 508235) to the outer surfaces of the lower grommet (2, **Figure 31**). Insert the lower grommet into the opening with the tapered end facing downward. Seat the grommet against the guide.
7. Install the intermediate drive shaft coupler onto the lower drive shaft with the grooved end facing upward.
8. To reduce corrosion damage, apply a coat of OMC or Volvo Penta White Sealing Compound to the upper drive shaft housing and lower gearcase mating surfaces.
9. With the aid of an assistant, set the upper drive shaft housing onto the lower gearcase housing. Be sure to align the copper water tube with the lower grommet or plastic water tube guide while lowering the housing. Rotate the propeller shaft to align the shaft splines. The housings will mate when the splines align.

62

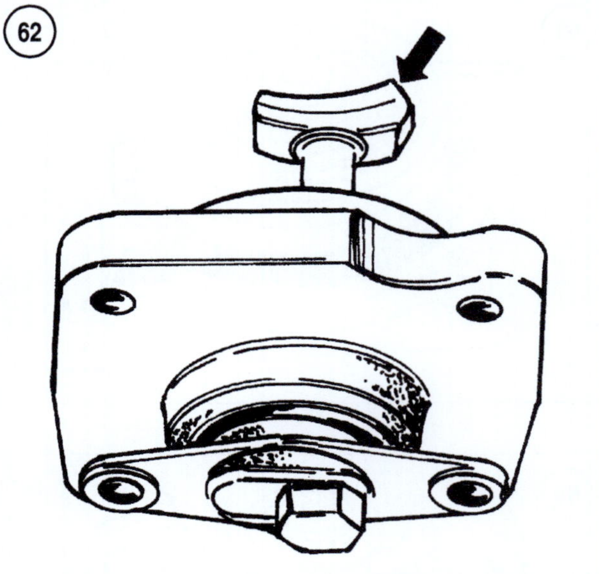

10. Apply a coat of OMC Gasket Sealing Compound to the threads of the six fasteners. Install the two bolts (**Figure 28**) and four nuts or bolts. Tighten the bolts and nuts evenly to 22-34 ft.-lb. (30-46 N•m) for 3/8-16 fasteners and 32-40 ft.-lb. (43-54 N•m) for 7/16-14 fasteners.
11. Pressure and vacuum test the drive unit as described in this chapter.
12. Install the drive unit as described in this chapter.
13. Install the propeller(s) as described in this chapter.
14. Adjust the shift cable as described in Chapter Five.
15. Fill the drive unit with lubricant as described in Chapter Three.
16. Connect the negative battery cable

UNIVERSAL JOINT SERVICE

This section describes removal and replacement of the universal joints. If the universal joint is removed to service the universal joint bearing carrier, only two caps require removal. These are the two cap on the yoke shaft of the universal joint bearing carrier. The universal joints can be serviced without removing the yoke shaft from the drive gear bearing carrier and without removing the drive gear bearing carrier from the upper drive shaft housing.

Inspecting the Universal Joints

Inspect the universal joints anytime the drive or upper drive shaft housing is removed.
1. Remove the drive unit or upper drive shaft housing as described in this chapter.
2. Wipe all grease, corrosion or other residue from the universal joint shaft. Carefully scrape all grease and other ma-

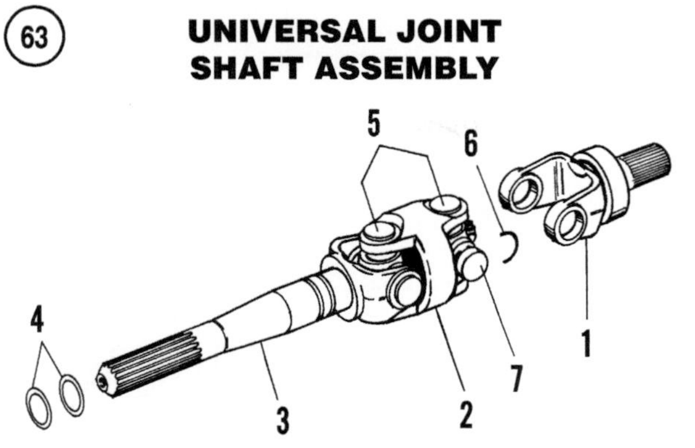

1. Yoke shaft (upper drive shaft housing assembly)
2. Center coupler
3. Input drive shaft
4. O-rings
5. Universal joint cross
6. Lockring
7. Universal joint cap

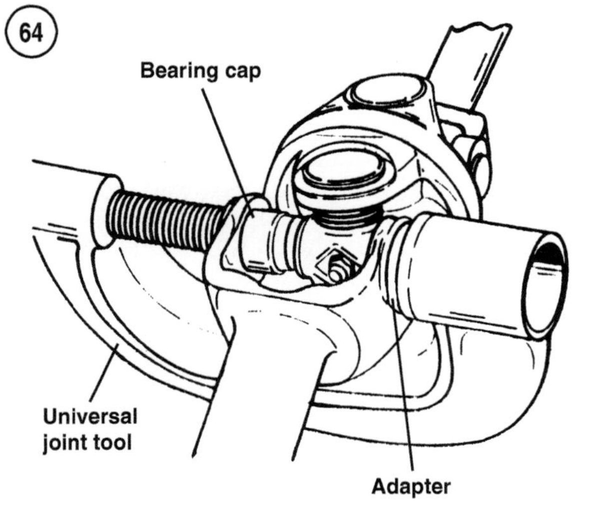

terial from the splined end of the shaft. Use an aerosol carburetor cleaner to remove residual material from the splines.

3. Remove and discard the O-rings from the input drive shaft.

4. Manipulate the universal joint shaft in all directions while checking for binding or looseness of the universal joints. If binding or looseness is evident, disassemble the shaft and replace the universal joint as described in this section.

5. Inspect the shaft splines for excessive wear or sharp edges. If excessive wear or sharp edges are evident, replace the defective shaft.

6. Install *new* O-rings into the groove on the input drive shaft.

Universal Joint Disassembly

NOTE

Service the universal joints with Owatonna Tool Company part No. T74-4665535-C (or OEM-4635) used in conjunction with the OMC or Volvo Penta adapter kit (part No. 3850628-3). Under no circumstances should the joints be beaten apart with a hammer. A pair of suitably sized sockets and a vise or arbor press may be used, if necessary.

Refer to **Figure 63** for this procedure.

1. Remove the drive unit as described in this chapter.
2. Remove the lockrings (6, **Figure 63**) from all universal joints that must be replaced. Drive the lockrings from the caps with an appropriate punch and small hammer. Do not strike directly on the bearing cap. Discard the lockrings upon removal.

NOTE

Once a lockring has been removed, it must be discarded. Replacement lockrings are available separately.

3. Clamp the universal joint tool into an appropriate vise. Install the *open* adapter from the adapter kit (OMC or Volvo Penta part No. 3850628-3) into the open end of the universal joint tool.
4. Install the universal joint shaft assembly into the tool. Insert the cap to be removed into the adapter as shown in **Figure 64**.
5. Tighten the pressing screw to push the universal joint cap out of the yoke (1, **Figure 63**) or center coupler (2). Tighten the pressing screw only until the end of cap is sufficiently exposed. Do not press to the point that the cross contacts the yoke or center coupler. Remove the shaft from the tool and pull the cap from the yoke or coupler with slip-joint pliers.

NOTE

If the universal joint is reused, keep each bearing cap and loose needle bearing separate from other caps and bearings. Mark the components upon removal. If reused, the caps and bearings must be installed into the original locations on the cross.

6. Rotate the universal joint 180° and reinstall the shaft assembly into the universal joint tool as shown in **Figure 64**.
7. Tighten the pressing screw against the exposed end of the cross to press the remaining cap free from the yoke or center coupler.
8. Carefully work the cross out of the yoke or coupler. If removal is difficult, remove the seals from the cross to provide additional clearance. Reposition the seals onto the caps.
9. Repeat Steps 3-8 as necessary to remove any remaining parts requiring service or replacement.

10

10. Clean all removed parts in solvent and dry with compressed air. Replace any parts that are worn, grooved, corroded or otherwise damaged. Inspect the rubber seals on the caps for damage. Replace the universal joint if the caps or other components are defective.

Universal Joint Assembly

The universal joint grease fittings must be installed with the fitting facing away from the drive unit at a 45° angle to any joint cap.

NOTE

Make sure that none of the needle rollers fall into the bottom of the bearing cap when pressing the cap onto the cross. If a roller is in the bottom of the cap, it will be difficult if not impossible to install the cap, and the roller, cross and cap will be damaged.

Refer to **Figure 63** for this procedure.

1. Pack all crosses, loose needle rollers and the caps with OMC EP/Wheel Bearing Grease. Make sure the seals are installed on each cap. The seals help hold the rollers in position.
2. Insert the cross into the appropriate yoke or coupler (**Figure 65**). Make sure the grease fitting faces away from the drive unit at a 45° angle to any joint cap.
3. Install the two bearing caps (1, **Figure 65**) onto opposite ends of the cross and press in as far as possible with hand pressure. Slide the cross assembly back and forth between the caps to make sure the caps are aligned correctly.
4. Position the universal joint assembly with one cap piloted into the open adapter and the other cap against the pressing screw (**Figure 64**). Tighten the pressing screw until the lockring groove is just visible at the yoke or coupler inner surface.
5. Fit the ends of a new lockring (6, **Figure 63**) into the groove. Use a hammer to tap the lockring completely into the groove (**Figure 66**).
6. Rotate the universal joint assembly 180° in the tool. Make sure the cap with the lockring installed fits into the open adapter (**Figure 64**) and the pressing screw contacts the opposite cap.
7. Tighten the pressing screw until the lockring groove is just visible at the yoke or coupler inner surface. If the cap cannot be pressed in far enough to expose the groove without excessive pressure, remove the caps and check for dislodged needle bearings.
8. Fit the ends of a new lockring (6, **Figure 63**) into the groove. Use a hammer to tap the lockring completely into the groove (**Figure 66**).
9. Once both lockrings are installed, make sure the cross assembly pivots freely in the bearing caps. If the cross assembly feels tight, tap the cross assembly sideways by tapping on each side of the yoke or coupler to seat the caps against the yoke.

65

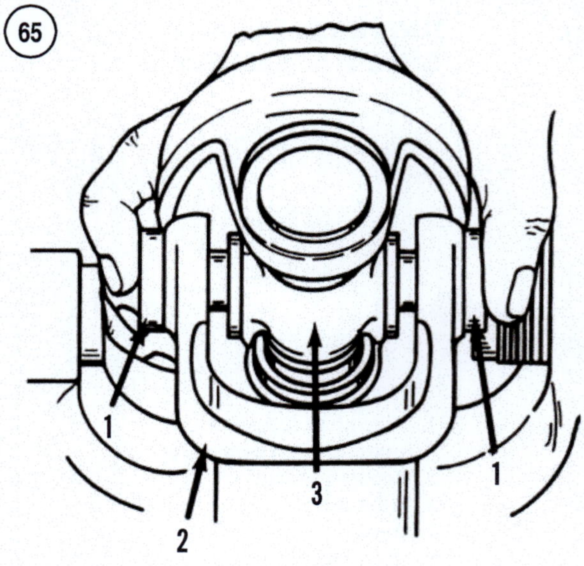

1. Bearing caps
2. Yoke or coupler
3. Universal joint cross

66

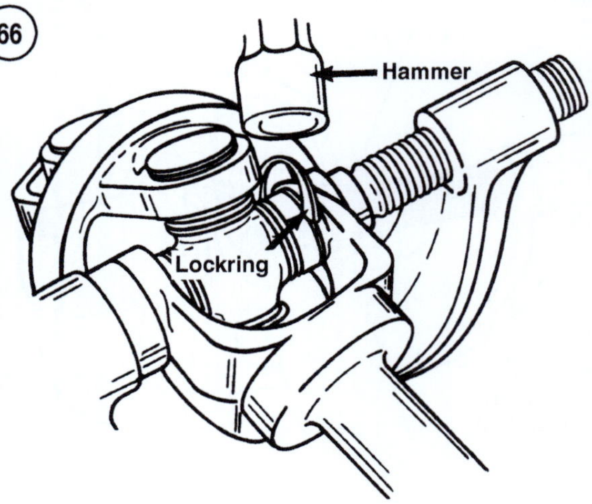

10. Repeat Steps 2-9 for all remaining caps and cross assemblies.
11. *Slowly* pump OMC EP/Wheel Bearing Grease or equivalent into the fitting until fresh grease appears at all four bearing cap seals.

SHIFT HOUSING SERVICE

The shift housing can be serviced without removing the upper drive shaft housing from the drive unit or removing the drive unit from the transom shield or bracket. However, if the shift shoe is damaged or broken, the drive unit (or at least the upper drive shaft housing) must be disassembled and inspected for damage. The shift housing is mounted on the rear of the upper drive shaft housing, and the shift hous-

67 **SHIFT HOUSING ASSEMBLY**

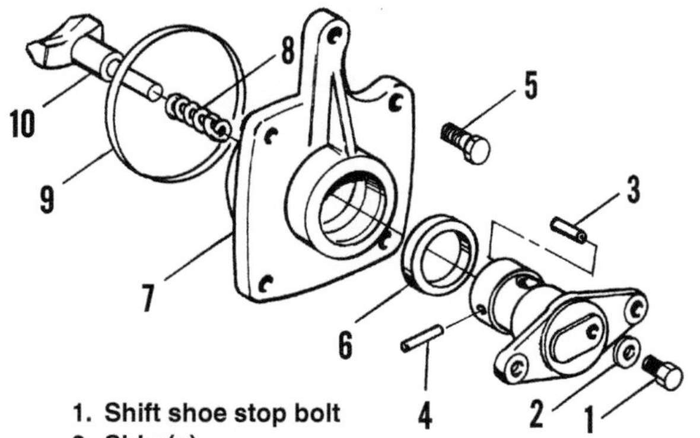

1. Shift shoe stop bolt
2. Shim(s)
3. Roll pin
4. Ramp follower pin
5. Screw
6. Seal
7. Shift housing
8. Spring
9. O-ring
10. Shift shoe

68

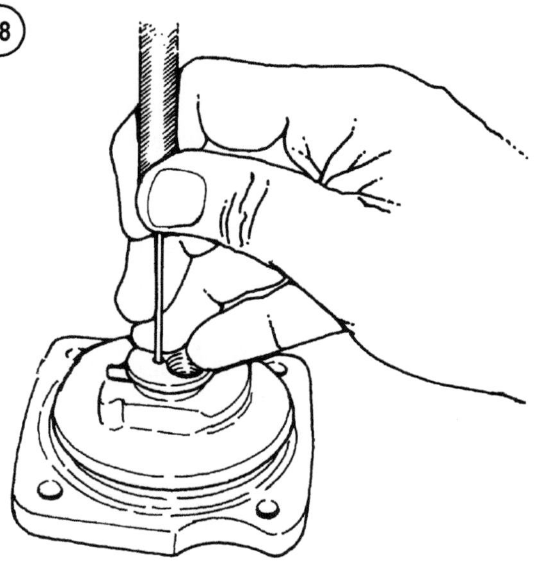

69

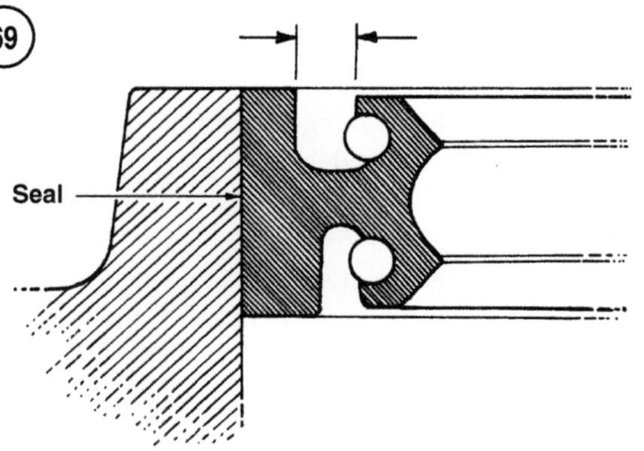

ing has two external shift arms. The shift shoe stop bolt requires shimming to ensure smooth, positive gear disengagement. If shifting out of gear is difficult, shim the shift shoe as described in this section.

Removal and Disassembly

Refer to **Figure 67** for this procedure.

1. Remove the rear cover as described in the drive unit removal procedures described in this chapter.
2. Disconnect the shift linkage from the shift arm.
3. Drain the drive unit lubricant as described in Chapter Three.

NOTE
The shift shoe is spring-loaded and falls free when the housing is removed.

4. Remove the four screws (5, **Figure 67**). Carefully pull the shift housing assembly from the upper drive shaft housing. Remove and discard the O-ring (9, **Figure 67**).
5. Clamp the shift housing into a vise with protective jaws.
6. Remove the shift shoe (10, **Figure 67**) and spring (8).
7. Use a 5/64 in. (1.98 mm) pin punch to drive the roll pin into the eccentric cavity piston (**Figure 68**).
8. Remove the ramp follower pin (4, **Figure 67**) from the eccentric piston with pliers. Shake the eccentric piston until the roll pin (3, **Figure 67**) falls out. Discard the roll pin.

CAUTION
The roll pin must be removed from the eccentric piston as described in Step 8. Failure to remove the pin at this time will prevent future disassembly.

9. Pull the eccentric piston out of the shift housing.
10. Carefully pry the seal out of the cover bore. Use a blunt pry bar and work carefully to prevent damage to the seal bore. Discard the seal.
11. Clean all shift housing components with solvent and dry with compressed air. Check all parts for excessive wear, cracks or other defects and replace as needed.
12. Clean the inside of the eccentric piston to remove all corrosion and deposits that could cause binding.

Assembly and Installation

Refer to **Figure 67** for this procedure.

1. Coat the outside diameter of the new seal (6, **Figure 67**) with Loctite 271. The inner side of the seal is fitted with a steel ring and can be identified by the narrower gap (**Figure 69**). Use the larger end of the installation tool (OMC or Volvo Penta part No. 884359-3) to press the seal into the bore with the narrower gap side facing down. Press until the tool contacts the housing or the seal surface is flush as shown in **Figure 69**.

10

2. Apply a coat of OMC Triple Guard to both seal lips. Lubricate the eccentric piston with gearcase lubricant. Then insert the piston into the seal opening with a rotating motion. Seat the piston against the housing.
3. Insert the ramp follower pin (4, **Figure 67**) into the eccentric piston opening. Align the roll pin opening in the follower pin with the corresponding opening in the eccentric piston. Drive the *new* roll pin (3, **Figure 67**) into the openings until flush with the eccentric piston surface. Rotate the eccentric piston in the housing to verify free movement.
4. Lubricate the spring (8, **Figure 67**) and shift shoe (8) with gearcase lubricant. Install the spring and then the shift shoe into the eccentric piston opening.
5. Install a *new* O-ring (9, **Figure 67**) into the groove on the shift housing. Apply a coat of OMC or Volvo Penta White Sealing Compound (part No. 1141570-0) to the O-ring and mating surfaces of the upper drive shaft housing.
6. Carefully guide the shift shoe (10, **Figure 67**) into the clutch groove while installing the assembly into the housing. The big end of the shift shoe (**Figure 62**) must face toward the starboard side of the drive unit.
7. Reposition the housing to align the screw openings. Coat the threads of the four screws (5, **Figure 67**) with Volvo Penta White Sealing Compound (part No. 1141570-0). Thread the screws into the shift housing and tighten in a crossing pattern to 60-84 in.-lb. (7-10 N•m).
8. Connect the shift linkage to the shift arm.
9. Install the rear cover as described in the appropriate drive unit installation procedures in this chapter.
10. Fill the drive unit with lubricant as described in Chapter Three.

Shift Shoe Shimming

The shift shoe must be shimmed anytime the cone clutch, shift shoe or any of the shift housing components are replaced. The shift shoe should also be shimmed anytime the drive unit is hard to shift out of gear. This operation can be performed with the drive unit installed or removed.
1. Disconnect the shift linkage from the shift arm on the shift housing. Remove the rear cover as described in the appropriate drive unit removal procedure described in this chapter.
2. Support the shift arm with an adjustable wrench and remove the shift shoe stop bolt (1, **Figure 67**).
3. Carefully remove the shim(s) (2, **Figure 67**) from the stop bolt or face of the eccentric piston.
4. Clean all sealant from the stop bolt, shims, eccentric piston face and threads.
5. Install the stop bolt *without* shims and *ital)without* any sealant.
6. Rotate the shift arm to position the internal ramp follower pin (A, **Figure 70**) on a *flat* surface (B) on the shift housing ramp. The flat is located just past either side of the neutral detent (C, **Figure 70**). Locate the flat by *feel* while slightly moving the arm. The shift arm should be just above or just the below the horizontal position. With the pin on the flat, the eccentric piston and stop bolt should be as close as possible to the cone clutch (if installed onto the drive).

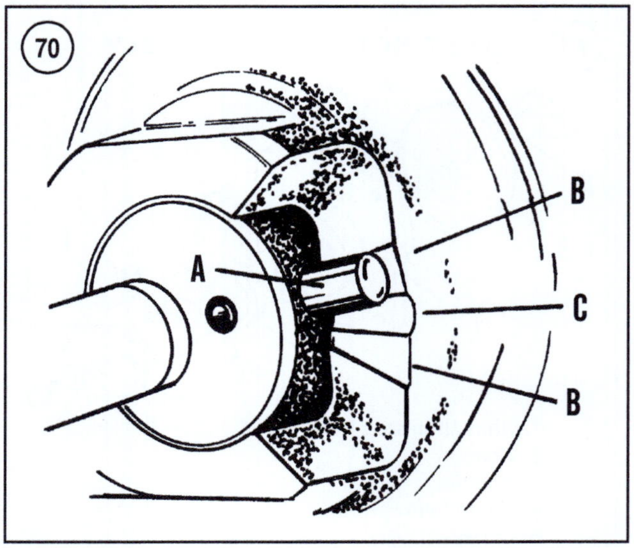

WARNING
Shim the shift shoe by rotating the propeller shaft(s) or the upper drive shaft housing output shaft, depending on whether the drive shaft housing is installed. If the drive unit is installed onto the transom bracket, disconnect the negative battery cable to prevent accidental starting.

NOTE
The shift shoe rides on an eccentric machined into the cone clutch groove. During operation, the shift shoe moves in and out, much like a lifter on a camshaft. There is only one eccentric high point per revolution of the cone clutch.

7. If the drive unit is installed onto the transom bracket, disconnect the negative battery cable to prevent accidental starting.
8. While rotating the propeller shaft, slowly turn the stop bolt inward in small increments until the propeller or output shaft binds from the shift shoe being pinched between the shift shoe stop bolt and the cone clutch eccentric groove. Wait at least one revolution of the cone clutch for each incremental turn of the stop bolt.
9. When you just feel binding of the shift shoe, verify that the shift arm has not moved from the correct position as described in Step 6. If the arm has moved, reposition the arm and readjust the stop bolt as described in Steps 6-8.
10. With the shift arm in position, slowly turn the stop bolt outward while rotating the propeller until the eccentric on the cone clutch just *cannot* be felt. Do not confuse the resulting friction from the shift shoe spring compressing and relaxing with the mechanical binding of the shoe to the clutch and stop bolt.

71

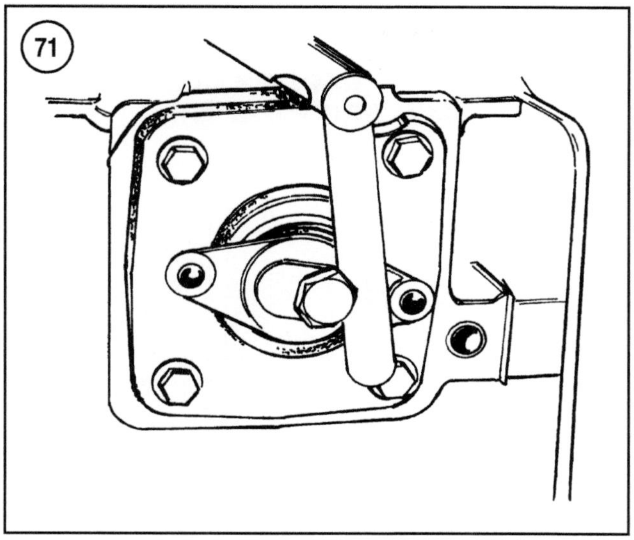

72

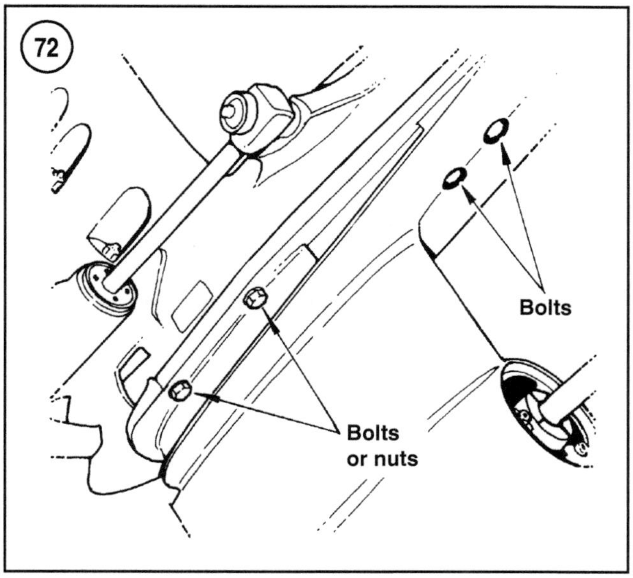

11. Use feeler gauges to measure the gap between the eccentric piston face and the underside of the stop bolt head (**Figure 71**). If the bolt is slightly cocked, measure the smallest and largest gaps and average the measurements. Shift shoe shims come in 0.008 in. (0.203 mm) size only.

NOTE
Always install shift shoe shims with a total thickness at least equal to the thickness of the feeler gauge. Never install shims with a thickness less than the feeler gauge.

12. Compare the feeler gauge reading to the total thickness of the original shims (2, **Figure 67**). Select a quantity of shims with a total thickness that equals or slightly exceeds the feeler gauge thickness. For example, if the feeler gauge thickness is 0.019 in. (0.483 mm), install three shims for a total thickness of 0.024 in. (0.610 mm). Select the appropriate number of shims based upon the feeler gauge reading and install them onto the stop bolt (1, **Figure 67**).
13. Support the shift arm with an adjustable wrench while tightening the stop bolt to 124-142 in.-lb. (14-16 N•m).
14. Reposition the shift arm as described in Step 6. Rotate the propeller shaft or upper output shaft and feel for any binding of the shift shoe against the cone clutch groove and stop bolt. The shift shoe is shimmed correctly if *no* binding is evident. If binding is evident, remove the stop bolt and install one additional shim. Rotate the propeller shaft or output shaft and recheck for binding.

CAUTION
Do not continue assembly if binding is evident. Operating the drive with binding will damage the shift shoe and cone clutch. Generally, no fewer than one shim and no more than five shims are required to correctly shim the shoe.

CAUTION
The shift shoe stop bolt must be sealed correctly during the final assembly. Otherwise, the drive unit will not hold lubricant or keep out water.

15. With the shift shoe correctly shimmed, support the arm and remove the stop bolt and shims. Coat the threads of the stop bolt with Loctite PST pipe sealant (OMC part No. 9100048). Support the shift arm with an adjustable wrench while tightening the stop bolt to 124-142 in.-lb. (14-16 N•m).
16. Connect the shift linkage to the shift arm (on the shift housing). Always use a new cotter pin and bend the ends completely over the connector.
17. Install the rear cover as described in the appropriate drive unit installation procedures described in this chapter.

LOWER GEARCASE

Removal

The lower gearcase can be removed without removing the drive unit from the transom bracket.
1. Remove the propeller(s) as described in this chapter.
2. Drain the gearcase lubricant as described in Chapter Three.
3. Raise the drive unit to the full UP position.
4. Disconnect the negative battery cable to prevent accidental starting.

NOTE
Support the lower gearcase while removing the fasteners in Step 5. The gearcase will fall free when the last fastener is removed.

5. Remove the four bolts or nuts (**Figure 72**), two on each side, and the two bolts under the antiventilation plate.

NOTE
If the lower gearcase does not drop free from the upper drive shaft housing, use a razor

73

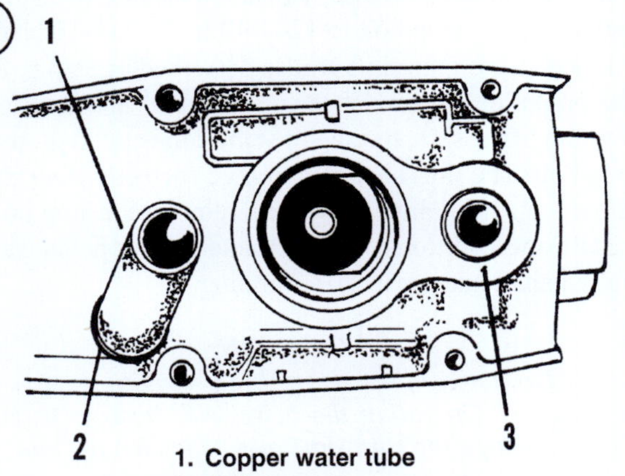

1. Copper water tube
2. O-ring
3. Lubricant screen

blade or other sharp instrument to cut the paint film that might be spanning the parting line for the housing mating surfaces

6. With the aid of an assistant, carefully lower the gearcase off the drive shaft housing. Be prepared to catch the drive shaft coupler and copper water tube if they fall free. Place the gearcase in a suitable holding fixture or clean work surface. If the coupler was dislodged, reinstall it onto the lower gearcase drive shaft with the grooved end facing upward.

7A. On earlier SX drives, remove the copper water tube and grommet (1 and **Figure 73**) from the upper drive shaft housing. Do not remove the plastic water tube adapter (1, **Figure 74**) from the gearcase unless it must be replaced.

7B. On later Cobra SX and all DP-S DuoProp drives, remove the copper water tube (3, **Figure 75**), upper grommet (4), lower grommet (2) and plastic water guide (1) from the gearcase.

8. Hook the screen (3, **Figure 73**) and remove it from the upper drive shaft housing. If a replacement screen is not available, clean the used screen with solvent and dry with compressed air. Replace the screen if damaged or if it cannot be satisfactorily cleaned.

9. Remove the drive shaft O-ring and lubricant passage O-rings. Discard the O-rings.

10. Thoroughly clean the mating surfaces on the upper drive shaft housing and lower gearcase.

11. Inspect the water tube grommets (2 and 4, **Figure 75**) for burned, melted or cracked surfaces. Replace the grommet(s) unless in excellent condition.

74

GEARCASE WATER TUBE ADAPTER (EARLIER COBRA SX-S DRIVE UNITS)

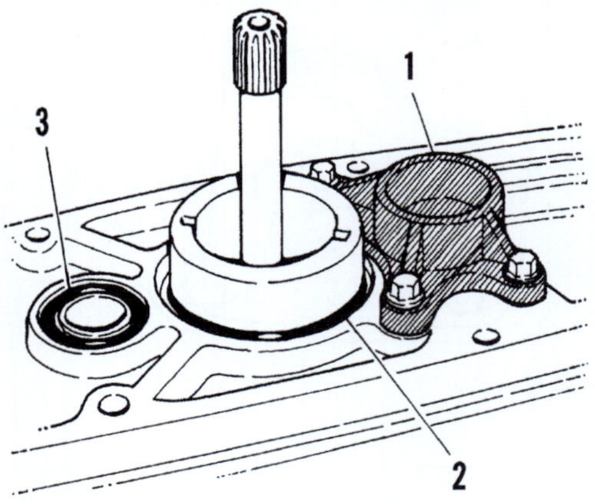

1. Plastic water tube guide
2. O-ring
3. O-ring

75

GEARCASE WATER TUBE AND GROMMETS (LATER COBRA SX AND DP-S DRIVE UNITS)

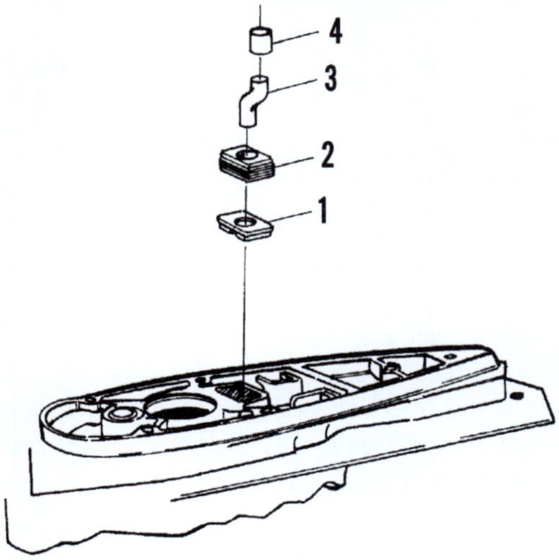

1. Plastic guide
2. Lower grommet
3. Copper water tube
4. Upper drive shaft housing grommet

SX (Single-Propeller) Drives

Disassembly

Upon removal, secure the gearcase into a suitable holding fixture or place the gearcase on a clean work surface.

Record the measurement of all shims as they are removed. Tag the shims by location for reassembly purposes. Do not lose or discard any shim.

All components that may be reused must be installed into the original locations and orientations. Refer to **Figure 76** for this procedure.

(76) **LOWER GEARCASE COMPONENTS (COBRA SX DRIVES)**

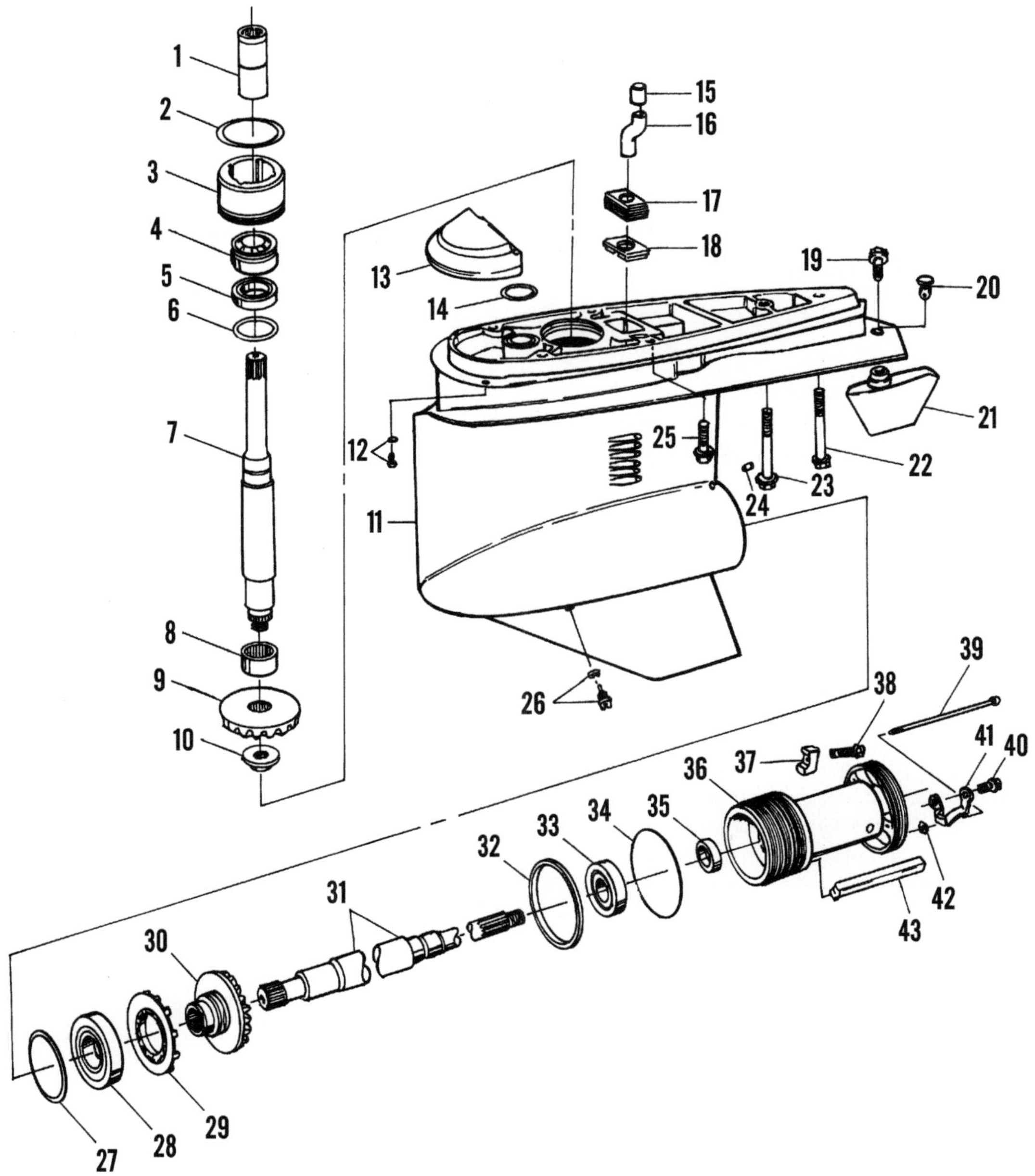

1. Drive shaft coupler
2. Drive shaft O-ring
3. Drive shaft bearing retainer
4. Larger tapered roller bearing
5. Smaller tapered roller bearing
6. Shim (pinion gear location)
7. Lower drive shaft
8. Lower drive shaft (pinion) bearing
9. Pinion gear
10. Pinion nut
11. Lower gearcase housing
12. Screw and washer
13. Anode
14. O-ring
15. Water tube grommet (upper drive shaft housing)
16. Copper water tube
17. Water tube grommet (lower gearcase)*
18. Plastic water tube guide*
19. Trim tab bolt (manual steering models)
20. Plug (power steering models)
21. Trim tab (manual steering models)
22. Bolt
23. Bolt
24. Setscrew
25. Bolt
26. Drain/fill plug and sealing washer
27. Shim (propeller shaft location)
28. Tapered roller bearing
29. Oil slinger
30. Driven gear
31. Propeller shaft
32. Shim ring (bearing preload)
33. Tapered roller bearing
34. O-ring
35. Seal (double-lip, single casing)
36. Propeller shaft bearing carrier
37. Retainer block
38. Retainer screw
39. Screw
40. Screw and washer
41. Anode bracket
42. Wave washer
43. Anode

*Used only on later Cobra SX drive units

1. Remove and clean the screen (if not already removed). Replace the strainer if damaged or if it cannot be satisfactorily cleaned.

2. On manual steering models, use a felt-tipped marker to mark the position of the trim tab (**Figure 77**). Remove the plug (20, **Figure 76**) and bolt (19) and then pull the trim tab (21) off the gearcase.

3. Use 1/4 in. drive tools to remove the screw (38, **Figure 76**) securing the retainer block (37) to the propeller shaft bearing carrier. The screw is located inside the exhaust passage. Invert the gearcase and catch the block and screw as they fall out the exhaust passage.

4. Use a T-27 Torx bit to remove the bearing carrier setscrew (24, **Figure 76**).

5. Use a T-40 Torx bit to remove the two anode bracket screws and washers (40, **Figure 76**). Use a T-30 Torx bit to remove the anode screw (39, **Figure 76**). Pull the anode (43, **Figure 76**), wave washer (42) and anode bracket (41) as an assembly from the carrier.

CAUTION

Work carefully to prevent damage to the bearing carrier and lower gearcase threads when removing the bearing carrier. Forcing the carrier out will destroy the carrier and lower gearcase housing. Before removal, remove any corrosion from the carrier flange and gearcase bore. Liberally lubricate the front and rear flanges of the carrier. If the carrier begins to bind during removal, stop and apply additional lubricant. Let the carrier set for a few minutes to allow the lubricant to penetrate the threads and mating surfaces. Then turn the carrier back into the gearcase and apply additional lubricant.

6. Fit the carrier removal tool (OMC or Volvo Penta part No. 3850707-5) over the propeller shaft. Insert the protrusions on the tool into the slots in the carrier. With the aid of an assistant, rotate the tool counterclockwise (**Figure 78**) until the carrier is free from the gearcase. Remove the O-ring (34, **Figure 76**) from the carrier (36). Discard the O-ring.

7. Remove the thick shim ring (32, **Figure 76**) from the carrier shoulder. Measure the shim ring and record the thickness for reference during assembly. Tag the shim ring for location purposes.

8. Remove the propeller shaft seal (35, **Figure 76**) by driving it out with a small punch and hammer (**Figure 79**). Work carefully to avoid damaging the seal bore.

9. Pull the propeller shaft (31, **Figure 76**) out of the gearcase. If necessary, use a slide hammer threaded onto the propeller nut threads to pull the shaft out of the driven gear (**Figure 80**).

10. Remove the drive shaft coupler (1, **Figure 76**) from the drive shaft.

11. Remove the two screws and washers (12, **Figure 76**) and then lift the anode (13) off the gearcase housing.

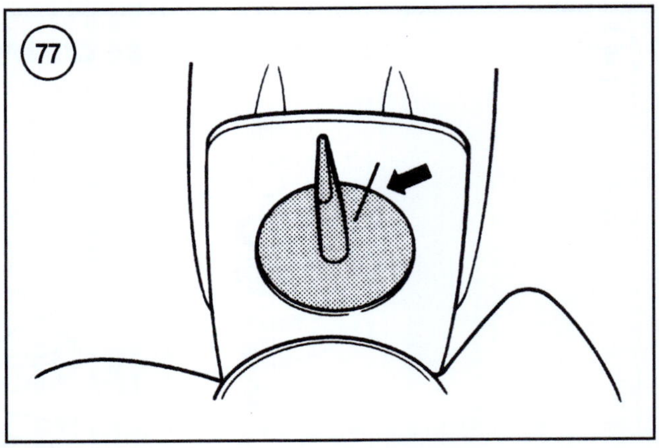

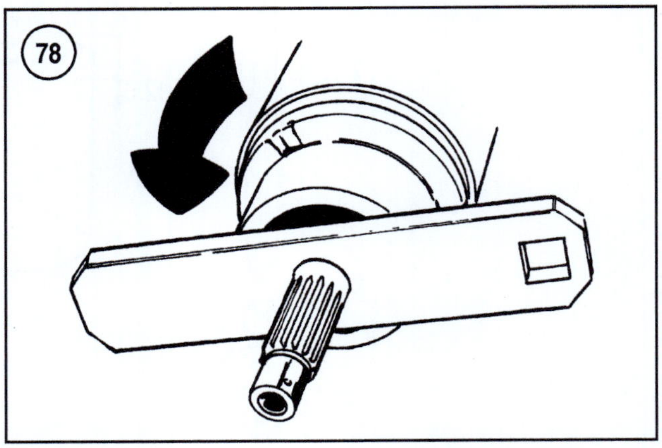

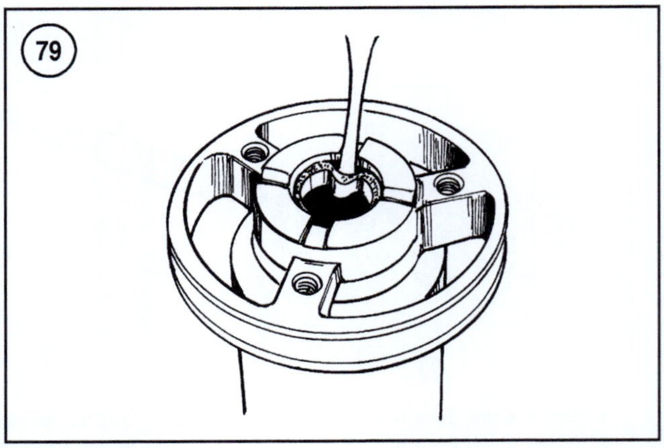

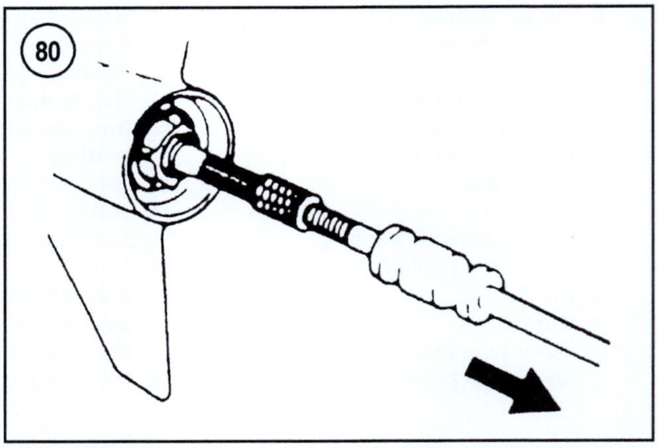

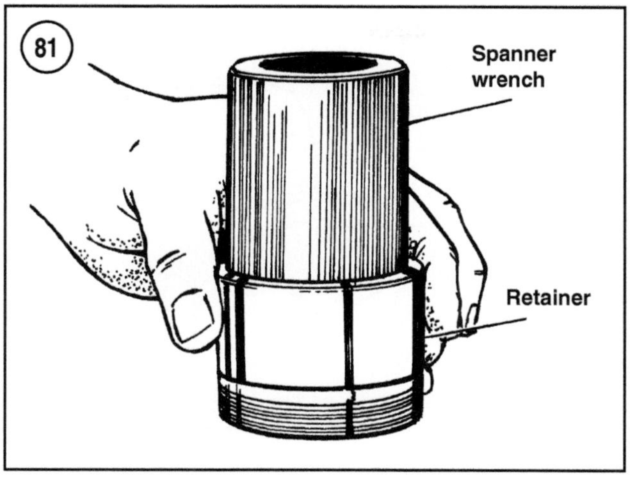

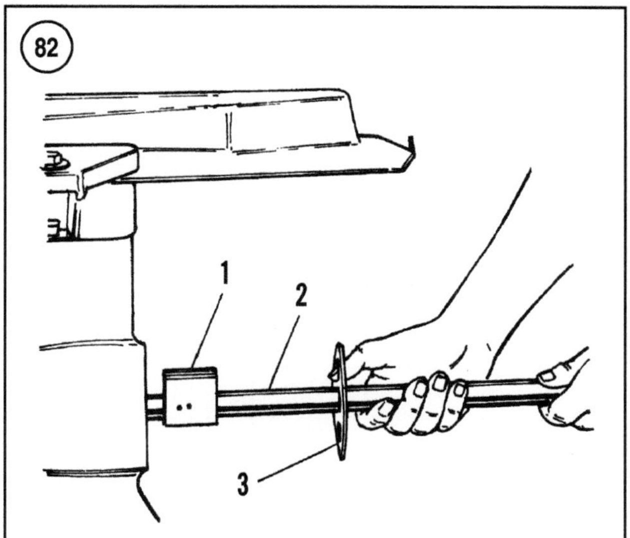

1. Nut holder (OMC or Volvo Penta part No. 3854864-0)
2. Drive handle (OMC or Volvo Penta part No. 3850609-3)
3. Alignment plate (OMC or Volvo Penta part No. 3850613-5)

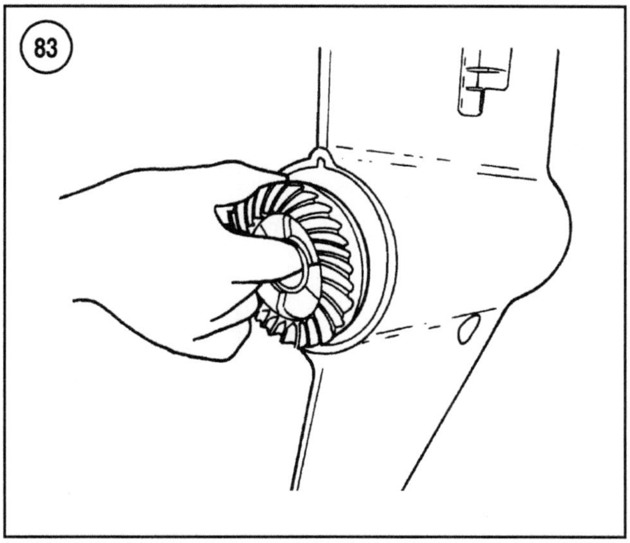

12. Use the spanner wrench (OMC or Volvo Penta part No. 3850601-0) to unthread the drive shaft bearing retainer from the gearcase (**Figure 81**). The carrier can be removed with hand pressure.

13. Thread the nut holder (OMC or Volvo Penta part No. 3854864-0) onto the driver handle (OMC or Volvo Penta part No. 3850609-3) as shown in **Figure 82**. Slide the alignment plate (3, **Figure 82**) over the driver handle. Insert the tool into the carrier bore and engage the nut holder onto the pinion nut (10, **Figure 76**).

14. Install the splined adapter (OMC or Volvo Penta part No. 3850598-8) onto the lower drive shaft. Using a breaker bar and socket, rotate the adapter counterclockwise until loose. Then remove the pinion nut holding tools from the gearcase.

NOTE
The 19 rollers will fall out of the lower drive shaft (pinion) bearing case when the lower drive shaft is removed.

15. Remove the pinion nut by hand and carefully lift the lower drive shaft out of the gearcase. Pull the pinion gear (9, **Figure 76**) out of the gearcase.

16. Collect and remove the 19 rollers from the lower drive shaft (pinion) bearing (8, **Figure 76**). Pull the driven gear out of the gearcase (**Figure 83**).

NOTE
The drive shaft bearing race in the top of the drive shaft bore and the driven gear bearing in the bottom of the propeller shaft bore are shimmed bearings. They must be removed to perform the shimming procedures. All other bearings are not shimmed and should not be removed unless the race or roller bearing must be replaced.

17. Remove the bearing race and shims from the drive shaft bore as follows:

a. Thread the bearing remover (OMC or Volvo Penta part No. 3855859) onto the threaded rod (OMC or Volvo Penta part No. 3855860-7) as shown in **Figure 84**. Fit the guide plate (OMC or Volvo Penta part No. 3850218-3) over the top end of the threaded rod and then thread the nut and washer (5, **Figure 84**) onto the threaded rod.

b. Fit the bearing remover behind the bearing race (4, **Figure 84**). Hold up the threaded shaft while seating the guide plate into the housing.

c. Tighten the nut to remove the bearing race (5, **Figure 76**) from the housing.

d. Remove the shims (6, **Figure 76**) from the bore.

18. Using a slide hammer and a suitable two-jaw puller (**Figure 85**), remove the driven gear bearing race (28, **Figure 76**). Remove the shims (**Figure 85**) from the bearing race bore.

84

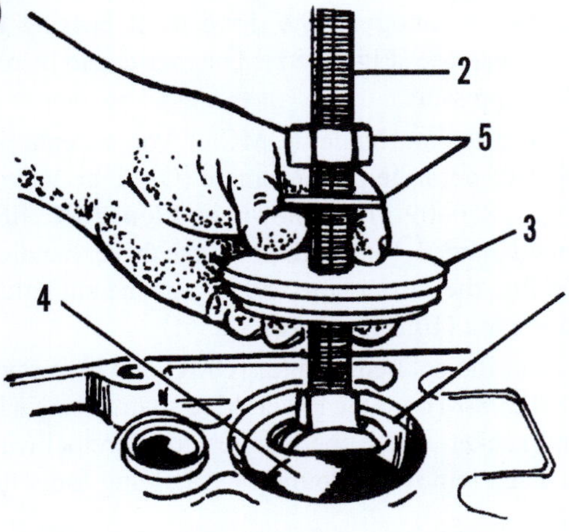

1. Bearing remover (OMC or Volvo Penta part No. 3855859-9)
2. Threaded rod (OMC or Volvo Penta part No. 3855860-7)
3. Guide plate (OMC or Volvo Penta part No. 3850218-3)
4. Bearing race (in drive shaft bore)
5. Nut and washer

85

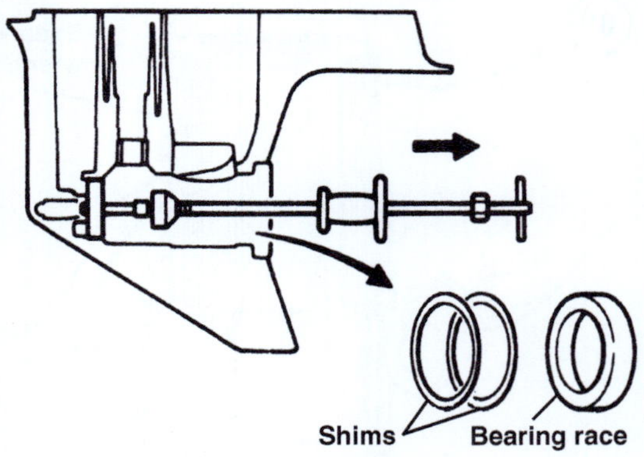

86

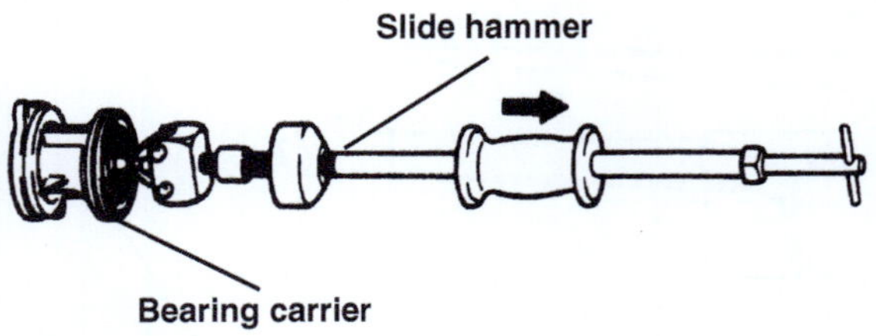

87

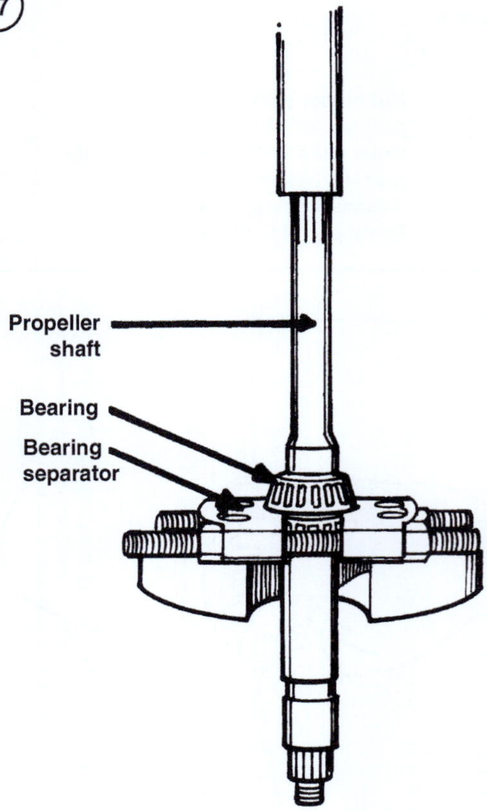

19. If the bearing carrier (36, **Figure 76**), roller bearing (33) or corresponding race in the bearing carrier must be replaced, proceed as follows:
   a. Clamp the bearing carrier into a vise with protective jaws. Engage the jaws of a slide hammer under the bearing race. Use short hammer strokes to remove the race (**Figure 86**). Discard the race upon removal.
   b. Using a bearing separator, support the propeller shaft on a press. The propeller nut threads must face upward, and the bearing separator must engage the inner bearing race. See **Figure 87**, typical.
   c. Thread a propeller nut onto the propeller shaft to protect the threads.
   d. Press against the nut until the bearing is free from the propeller shaft. Discard the bearing.

20. If the driven gear roller bearing (28, **Figure 76**), gear (30) or oil slinger (29) must be replaced, proceed as follows:
   a. Support the gear and bearing in a press with a bearing separator engaging the bearing (**Figure 88**). Block the separator to provide clearance for the gear and slinger to exit.
   b. Thread the removal tool (OMC or Volvo Penta part No. 3850625-9) onto the driver handle (OMC or Volvo Penta part No. 3850610-1).
   c. Install the tool assembly onto the hub of the gear. Press the gear and slinger out of the bearing. Discard the bearings.
   d. Support the gear and slinger in a press with a bearing separator engaging the slinger. Using the removal tool, press the gear out of the slinger. Discard the slinger.

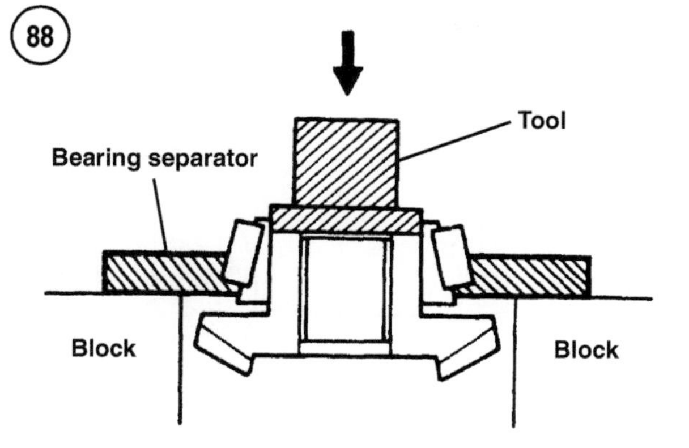

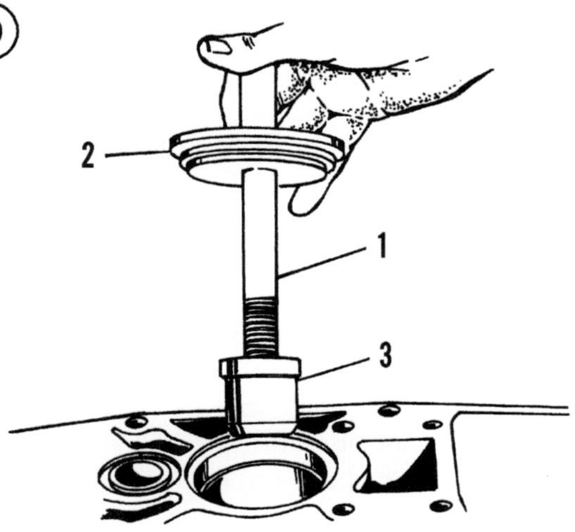

1. Driver rod (OMC or Volvo Penta part No. 3850623-4)
2. Guide plate (OMC or Volvo Penta part No. 3850218)
3. Bearing remover (OMC or Volvo Penta part No. 3850622)

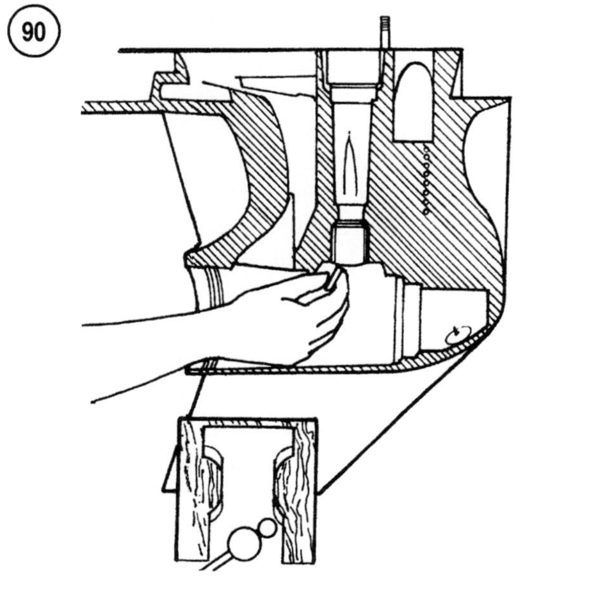

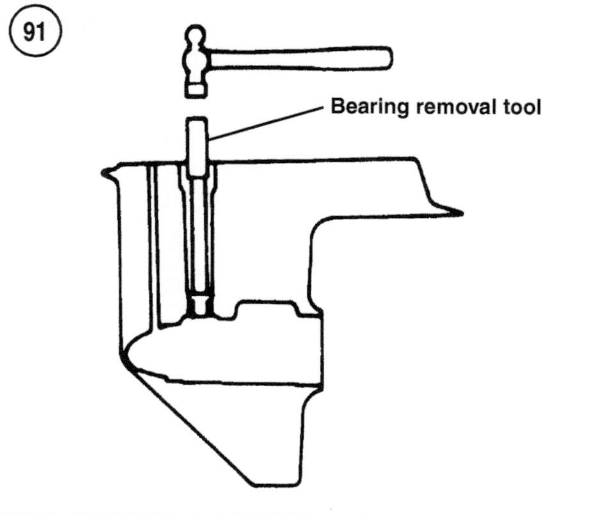

21. If the lower drive shaft (7, **Figure 76**) or either bearing (4 or 5) must be replaced, remove the bearings as follows:

a. Support the lower drive shaft in a press with a bearing separator. The separator must engage the inner race for the small tapered roller bearing (5, **Figure 76**). The drive shaft splines must face upward.
b. Press on the splined end until both bearings are free from the shaft. Discard the bearings.

22. If the lower drive shaft (pinion) bearing (8, **Figure 76**) or lower drive shaft must be replaced, remove the lower drive shaft bearings as follows:

a. Slide the guide plate (OMC or Volvo Penta part No. 3850218) over the driver rod (OMC or Volvo Penta part No. 3850623-4) with the small diameter side facing the threaded end. See **Figure 89**. Thread the bearing remover (OMC or Volvo Penta part No. 3850622) onto the driver rod with the smaller-diameter side facing away from the guide. See **Figure 89**.
b. Lubricate the 19 rollers with OMC Needle Bearing Assembly Grease (OMC part No. 378642) and then insert them into the bearing case (**Figure 90**).
c. Insert the bearing removal tool into the bearing and seat it against the rollers (**Figure 91**, typical). Seat the guide plate into the drive shaft bore.
d. Place shop towels into the gearcase and position them directly under the bearing.
e. Carefully drive the bearing into the gearcase. Avoid using excessive force. Excessive force can break the bearing case and make removal very difficult. If the bearing proves difficult to remove, apply mild heat to the outside of the gearcase in the area around the bearing with a heat gun or heat lamp. Do not use a torch or open flame. Discard the bearing upon removal.

23. On earlier Cobra SX models, remove the four screws retaining the water tube guide (1, **Figure 74**) onto the gearcase housing. Discard the water tube guide.

24. Clean all parts in clean solvent and dry with compressed air. Make sure all sealant and gasket residue is removed and blow out all threaded openings, lubricant passages and water intake passages.
25. Apply a light coat of oil to all cleaned components to prevent rusting.
26. Inspect the gear teeth for chipped, excessively worn, broken, missing, rough or discolored surfaces. If one gear is defective and requires replacement, both gears must be replaced.

CAUTION
A tapered roller bearing set consists of the caged rollers, cone-shaped inner bearing race and a tapered outer race. The bearing components must be replaced as a set. Never use parts from one bearing set with parts from a different bearing set.

27. Check all bearings for scoring, pitted rollers, roughness, heat discoloration and binding. Replace the bearings and corresponding race if these or other defects are evident. Do not remove pressed-on bearings unless replacement is required or the removal procedure specifically directs you to do so. Many of the bearings cannot be reused after removal.
28. Check the splines in the drive shaft coupler (1, **Figure 76**) for twisted, excessively worn or damaged splines. Replace the drive coupler if these or other defects are evident.
29. Inspect all shims and replace any that are rusted, bent or not in like-new condition with new shims of the same thickness.
30. Check all threads, sealing surfaces and splines on the propeller and drive shafts. Replace any shaft with deep grooving at the seal contact surfaces, roughness at the bearing contact surfaces, twisted splines or mechanical damage. Replace the shaft(s) if surface imperfections cannot be removed with emery paper.
31. Check the gearcase housing and propeller shaft bearing carrier for cracks around bolt or screw openings and bearing bores. Also check for cracks, nicks or burrs on the housing and O-ring mating surfaces. Replace the housing or bearing carrier if cracked or if surface imperfections cannot be removed with emery paper.

Assembly

For maximum durability and quiet operation, all of the gears must be properly located in the housing through shimming procedures. When properly located, the correct gear lash and gear contact pattern will be obtained.

Additionally, the tapered bearings for the propeller and lower drive shaft must be properly tensioned or *preloaded*. Preload refers to the tension applied to hold the tapered roller bearing against the bearing race. Insufficient preload will allow the bearing to move in relation to the race. Any movement will result in premature failure of the bearing

92

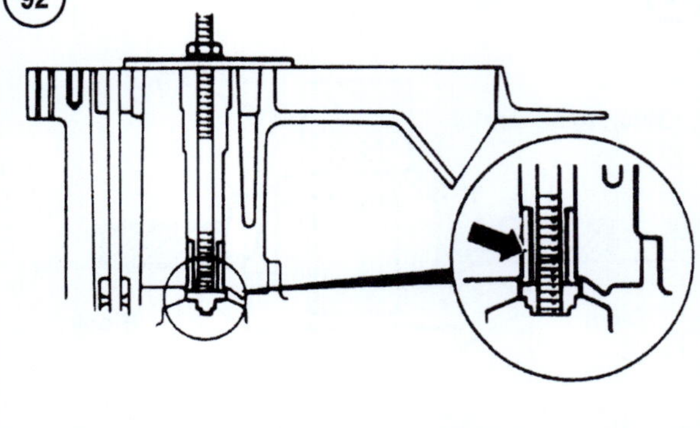

and supported gears or shaft. Excessive preload will cause the bearings to overheat and fail.

Do not install new seals and O-rings until the gear lash, gear pattern and bearing preload are corrected. Seal and O-rings installation is performed during the final assembly.

During assembly, lubricate all gears, bearings, seals, shafts and O-rings with GL5 Synthetic Gear Lube.

Refer to **Figure 76** for this procedure.

1. If the roller bearing (33, **Figure 76**) was removed from the propeller shaft (31), install the *new* bearing as follows:
   a. Lubricate the new bearing and slide it over the propeller end of the shaft with the rollers facing the threaded end of the shaft.
   b. Slide the bearing installer (OMC or Volvo Penta part No. 3850617-6) over the propeller end of the propeller shaft. The raised edge of the tool must face toward the bearing. Seat the tool against the inner race of the bearing.
   c. Support the tool and shaft in a press with the propeller end of the shaft facing downward.
   d. Place a piece of soft metal on the end of the shaft to protect the splines from the press ram.
   e. Press on the propeller shaft until the bearing fully seats against the propeller shaft shoulder.
2. If the lower drive shaft (pinion) bearing (8, **Figure 76**) was removed, install the *new* bearing as follows:
   a. Lubricate the new bearing and place it onto the installation tool (OMC or Volvo Penta part No. 3850620-0) with the lettered side of the bearing casing facing the hex end of the tool. Make sure all 19 rollers are installed into the casing.
   b. Place the tool and bearing into the drive shaft bore from the bearing carrier opening (**Figure 92**, typical). The lettered side of the bearing must face downward when installed into the drive shaft bore.
   c. Guide the threaded rod (3855860-7) into the drive shaft bore. Thread the bearing installation tool fully onto the threaded rod.
   d. Place the guide plate (OMC or Volvo Penta part No. 3850619-2) over the threaded rod with the small-diameter side facing the gearcase. Seat the guide plate

93

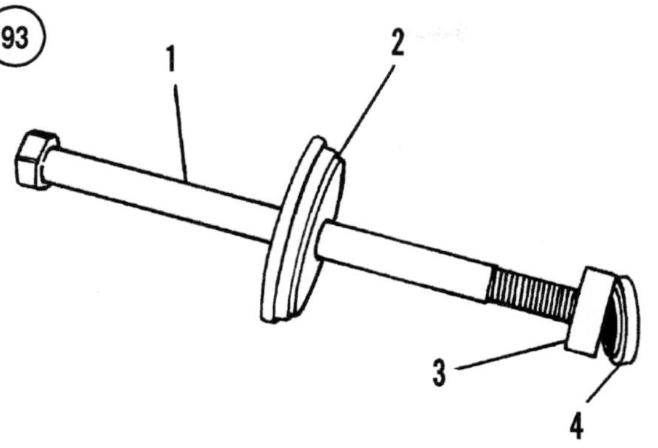

1. Large puller bolt
2. Guide plate
3. Bearing installer or seal installer
4. Propeller shaft seal

into the bore. Install the washer and then the nut onto the threaded rod.

e. Hold the threaded rod and tighten the nut to pull the bearing into the bore. Stop tightening when a marked increase in effort is required to turn the rod. This indicates the bearing is seated in the bore. Do not overtighten the rod. Overtightening damages the bearing.

f. Remove the installation tools. Look into the drive shaft bore and inspect the bearing for proper seating or damage. Remove and replace the bearing if damaged or improperly seated.

NOTE

When installed correctly, the drive shaft tapered bearing rollers face away from each other. Install the smaller bearing first and the larger bearing second.

3. If the tapered roller bearings (4 and 5, **Figure 76**) were removed from the lower drive shaft, install the *new* bearings as follows:

a. Lubricate the new smaller-diameter bearing (5, **Figure 76**) and place it over the splined end of the drive shaft with the rollers facing the pinion gear end of the shaft.

b. Slide the bearing installer (OMC or Volvo Penta part No. 3850617-6) over the splined end of the drive shaft. The raised edge of the tool must face toward the bearing. Seat the tool against the inner race of the bearing.

c. Thread a used pinion nut onto the drive shaft to protect the threads. Support the tool and shaft in a press with the pinion gear end of the shaft facing upward. Place wooden blocks under the bearing installer for support.

d. Press against the pinion nut until the bearing fully seats against the drive shaft shoulder.

e. Lubricate the new larger-diameter bearing (4, **Figure 76**) and place it over the splined end of the drive shaft with the rollers facing the smaller bearing.

f. Slide the bearing installer (OMC or Volvo Penta part No. 3850617-6) over the splined end of the drive shaft. The raised edge of the tool must face toward the bearing. Seat the tool against the inner race of the bearing.

g. Support the tool and shaft in a press with the pinion gear end of the shaft facing upward. Place blocks under the bearing installer for support.

h. Press against the pinion nut until the bearing fully seats against the inner bearing race of the smaller bearing. Remove the used pinion nut.

4. If removed, install the race for the *new* bearing carrier tapered roller bearing (33, **Figure 76**) as follows:

a. Lubricate the race and place it into the bearing carrier with the tapered side facing outward (away from the propeller end of the carrier).

b. Insert the large puller bolt (OMC or Volvo Penta part No. 3850623-4) into the guide plate (OMC or Volvo Penta part No. 3850619-2) with the larger-diameter side of the plate facing the bolt head. Thread the bearing installer (OMC or Volvo Penta part No. 3830621-8) onto the puller bolt with the beveled edge facing away from the bolt head. See **Figure 93**, typical.

c. Place the bearing carrier into a press with the bearing race opening facing upward. Place a piece of wood, plastic or soft metal under the carrier to protect the surfaces.

d. Guide the bearing installer into the bearing race while installing the tool assembly onto the carrier. Fit the guide plate into the carrier opening.

e. Press on the puller bolt until the bearing race fully seats in the carrier bore. Remove the tool.

5. Install the new propeller shaft seal (35, **Figure 76**) as follows:

a. Insert the large puller bolt (OMC or Volvo Penta part No. 3850623-4) into the guide plate (OMC or Volvo Penta part No. 3850619-2) with the larger-diameter side of the plate facing the bolt head. Thread the seal installer (OMC or Volvo Penta part No. 3830627-5) onto the puller bolt with the recessed side facing away from the bolt head. See **Figure 93**.

b. Apply a coat of OMC Gasket Sealing Compound (part No. 508235) onto the outer diameter new seal casing. Fit the seal (4, **Figure 93**) onto the installer (3) with the protruding lip fitting into the recess.

c. Guide the seal into the bore while fitting the tool into the bearing carrier. Invert the carrier and place it in a press with the puller bolt facing upward. Place a piece of wood, plastic or soft metal under the carrier to protect the surfaces.

d. Press on the puller bolt until the seal fully seats in the bore. Do not press with excessive force. Excessive force damages the seal.

e. Coat the seal lips with OMC Triple Guard Grease.

6. If removed, install the new oil slinger (29, **Figure 76**) and tapered roller bearing (28) onto the driven gear (30).

a. Place the gear and bearing onto a press with the gear teeth facing downward. Place a block of wood, plastic or soft metal under the gear to protect the teeth.

b. Lubricate the new slinger and fit the slinger onto the gear with the aligning pins on the slinger aligned with the corresponding openings in the gear.

c. Place the support tool (OMC or Volvo Penta part No. 3850614-3) over the oil slinger. Place the installer (OMC or Volvo Penta part No. 3850617-6) over the support tool. Press against the mandrel until the slinger fully seats against the gear.

d. Lubricate the new roller bearing and place it over the gear hub with the roller facing away from the slinger.

e. Place the installer (OMC or Volvo Penta part No. 3850624-2) onto the bearing and center over the inner race. Press on the mandrel until the bearing fully seats against the gear (**Figure 94**).

NOTE

When the correct feeler gauge is selected for the measured clearance, a slight drag is felt as the feeler gauge passes between the measuring points. To verify, insert the next larger gauge and check for a noticeable increase in drag. Then insert the next smaller gauge and check for the absence of drag.

7. If the pinion gear, lower drive shaft or drive shaft bearings were replaced, select the correct pinion gear shim thickness as follows:

a. Clamp the splined adapter (OMC or Volvo Penta part No. 3850598-8) into a vise. Make sure the opening is facing vertical. Align the splines and insert the lower drive shaft into the vise.

b. Guide the small tapered roller bearing over the drive shaft with the tapered side facing downward and seat it against the small roller bearing. Guide the shim fixture (OMC or Volvo Penta part No. 3855098-4) over the drive shaft, with the slotted end facing upward, and seat it against the bearing race.

c. Align the splines and install the pinion gear onto the drive shaft. Thread the *used* pinion nut (10, **Figure 76**) onto the drive shaft. Tighten the pinion nut to 150-160 ft.-lb. (203-217 N•m). Rotate the shim tool several times to seat the bearing rollers.

d. Use a feeler gauge to measure the distance between the shim tool and the pinion gear at each of the three slots (**Figure 95**). Record the average of the three clearances as the *pinion gear feeler gauge measurement*. Remove the pinion gear, shim tool and small bearing race.

94

95

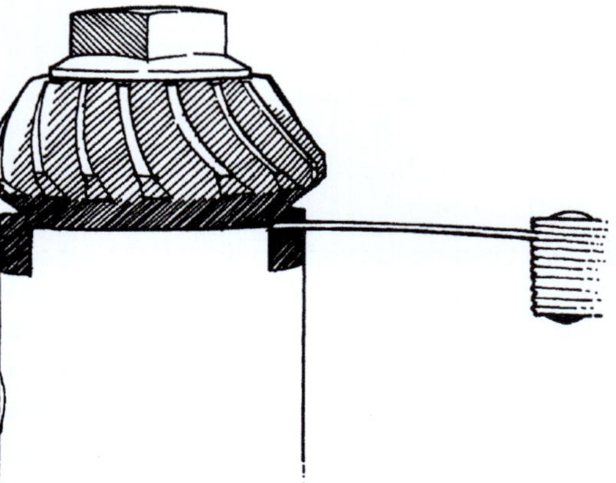

96

e. Locate the correction number stamped or etched into the pinion gear (**Figure 96**). Record the correction number in thousandths of an inch. For example, if +10 is etched or stamped into the gear, write it down as +0.010 in. If -10 is etched or stamped into the gear, write it down as -0.010 in.

f. Add or subtract (as indicated by the plus or minus signs) the correction number value recorded in Step 7F to or from *pinion gear feeler gauge measurement* recorded in Step 7d. Perform the calculation for the actual pinion gear and record the results as *pinion gear shims*.

97

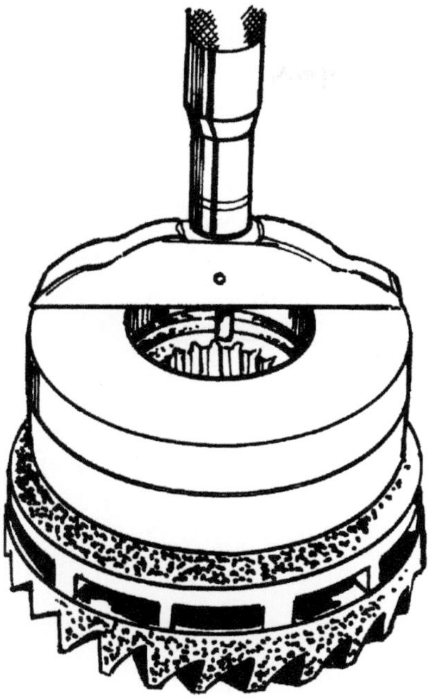

98

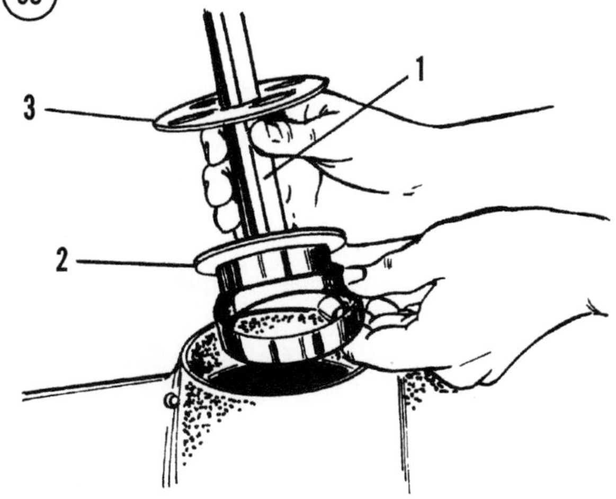

1. Driver handle part
2. Bearing installer
3. Alignment plate

8. Install the pinion gear shims and bearing race into the gearcase as follows:

   a. Select pinion gear shims (6, **Figure 76**) with a total thickness equal to the predetermined value recorded in Step 7f. Insert the shims into the drive shaft bore and rest them on the shoulder in the bore.

   b. Insert the large puller bolt (OMC or Volvo Penta part No.3850623-4) into the guide plate (OMC or Volvo Penta part No. 3850619-2) with the larger-diameter side of the plate facing the bolt head. Thread the bearing installer (OMC or Volvo Penta part No. 3830621-8) onto the puller bolt with the beveled edge facing away from the bolt head. See **Figure 93**, typical.

   c. Install the bearing race into the drive shaft bore with the tapered side facing upward.

   d. Seat the bearing installer into the bearing race. Fit the guide plate into the drive shaft bore.

   e. Drive against the puller bolt with a plastic mallet until the race seats against the step in the bore.

9. If the driven gear (30, **Figure 76**), oil slinger (29) or tapered roller bearing (28) was replaced, select the correct driven gear shims as follows:

   a. Place the driven gear on a clean work surface with the gear teeth facing downward. Install the race onto the tapered roller bearing. Rotate the bearing race several revolutions to seat the rollers.

   b. Place the shim fixture (OMC or Volvo Penta part No. 3850600-2) onto the roller bearing race with the recessed side facing the bearing.

   c. Use a depth micrometer to measure the distance from the top surface of the shim fixture to the end of the driven gear stub shaft (hub). See **Figure 97**. Record the measurement. Subtract the shim fixture thickness (0.500 in.) from the measurement. For example, if the measurement is 0.599 in., subtract 0.500 in. from 0.599 in., which equals 0.099 in. Perform the calculation using the actual measurement and record the results as *actual driven gear and bearing dimension*.

   d. Locate the correction number stamped or etched into the outer diameter of the driven gear. Record the correction number in thousandths of an inch. For example, if +5 is etched or stamped into the gear, write it down as +0.005 in. If -5 is etched or stamped into the gear, write it down as -0.005 in.

   e. Add or subtract (as indicated by the plus or minus signs) the correction number value recorded in Step 9d to or from the *driven gear and bearing dimension* recorded in Step 9c. Perform the calculation, using the driven gear correction number, and record the results as *corrected driven gear dimension*.

   f. Subtract the *corrected driven gear dimension*, recorded in Step 9e, from the fixed nominal dimension of 0.106 in. (2.692 mm). Perform the calculation and record the results as *driven gear shims*.

10. Install the driven gear shims (27, **Figure 76**) and bearing race as follows:

   a. Select driven gear shims with a total thickness equal to the predetermined value recorded in Step 9f. Install the shims into the gearcase opening and rest them on the step in the bore.

   b. Thread the driver handle (OMC or Volvo Penta part No. 3850609-3) into the bearing installer (OMC or Volvo Penta part No. 3850616-8) with the larger diameter of the installer facing the driver handle. Install the alignment plate (OMC or Volvo Penta part No. 3850613-5) over the handle as shown in **Figure 98**.

10

c. Lubricate the bearing race and place it into the recess with the tapered side facing outward.
d. Guide the bearing installer into the bearing race while installing the tool assembly into the housing. Fit the alignment plate into the bearing carrier bore of the housing and seat it against the step in the bore.
e. If the housing is not mounted into a suitable holding fixture, rest the leading edge of the gearcase against a block of wood to prevent damage to the gearcase.
f. Drive against the tool until the race seats against the shim(s) and the step in the bore. Remove the tool and inspect the bearing for proper installation. Remove and replace the bearing if found to be damaged or tilted in the bore.

11. Verify that all 19 bearing rollers are still located in the lower drive shaft bearing case. If any are dislodged, lubricate them with OMC Needle Bearing Assembly Grease (part No. 378642) and then insert them into the bearing case (**Figure 90**).
12. Coat at least three teeth of the driven gear with GM Gear Marking Compound, available from any General Motor automobile dealership.
13. Position the gearcase with the bearing carrier opening facing upward. Lubricate the driven gear and bearing assembly and then install it into the gearcase housing (**Figure 83**). Seat the tapered roller bearing into the race at the front of the housing.
14. Mesh the gear teeth and hold the pinion gear in position directly aligned with the drive shaft bore. Carefully guide the lower drive shaft assembly into the bore. Work carefully to avoid dislodging the rollers in the lower drive shaft bearing. Rotate the drive shaft to align the shaft and pinion gear splines. Then seat the smaller roller bearing against the race in the bore.

NOTE
Use the original pinion nut during assembly at this time to check the gear lash and gear pattern. Install a new pinion nut during the final assembly.

15. Thread the *used* pinion nut onto the drive shaft and tighten by hand only at this time.
16. Install the larger drive shaft roller bearing race (4, **Figure 76**) over the drive shaft with the tapered side facing the housing. Seat the bearing race onto the tapered roller bearing on the drive shaft.
17. Lubricate the drive shaft bearing retainer (3, **Figure 76**) and install it over the drive shaft with the threaded end facing the housing. Thread the retainer into the housing by hand only at this time.
18. Thread the nut holder (OMC or Volvo Penta part No. 3854864-0) onto the driver handle (OMC or Volvo Penta part No. 3850609-3) as shown in **Figure 82**. Slide the alignment plate (3, **Figure 82**) over the driver handle. Insert the tool into the carrier bore and engage the nut holder onto the pinion nut (10, **Figure 76**).

99

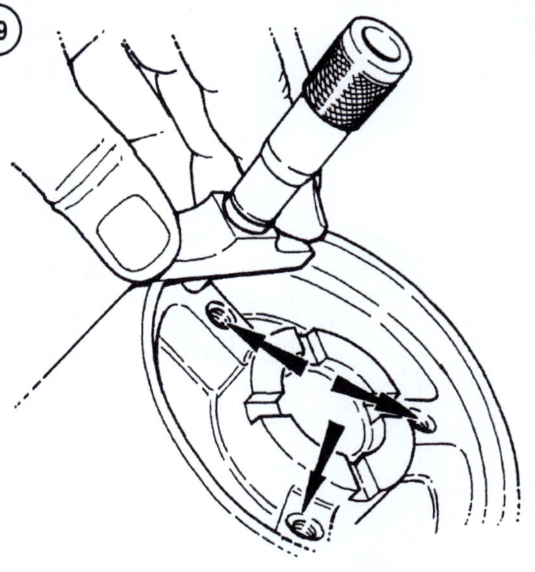

19. Install the splined adapter (OMC or Volvo Penta part No. 3850598-8) onto the lower drive shaft. Attach a suitable torque wrench to the splined adapter and tighten the pinion nut to 150-160 ft.-lb. (203-217 N•m).
20. Rotate the gearcase assembly to position the bearing carrier bore facing straight up. Adjust the drive shaft rolling torque as follows:

a. Attach a dial torque wrench capable of reading 2-4 in.-lb. (0.23-0.45 N•m) onto the splined adapter. Turn the drive shaft clockwise several turns to seat the bearing rollers.
b. Note the torque wrench reading while rotating the drive shaft at approximately 3 rpm.
c. If the reading is less than 2 in.-lb. (0.23 N•m), tighten the drive shaft retainer (3, **Figure 76**) *slightly* with the spanner wrench (**Figure 81**) and recheck the rolling torque.
d. If the reading exceeds 4 in.-lb. (0.45 N•m), loosen the drive shaft retainer (3, **Figure 76**) *slightly* with the spanner wrench (**Figure 81**) and recheck the rolling torque.
e. Do not proceed with assembly until the drive shaft rolling torque is adjusted to 2-4 in.-lb. (0.23-0.45 N•m). Remove the tools. Do not disturb the retainer position.

21. Measure the shim ring tool (OMC or Volvo Penta part No. 3850626-7) to verify that it is 0.160 in. (4.064 mm) thick. Record the measurement. Apply a coat of OMC Needle Bearing Assembly Grease to the shim ring tool and install it onto the bearing carrier flange.
22. Lubricate the threads and flange on the bearing carrier with gearcase lubricant. Install the bearing carrier into the housing. Thread the carrier into the housing as far as possible by hand. Work carefully to avoid dislodging the shim ring tool.

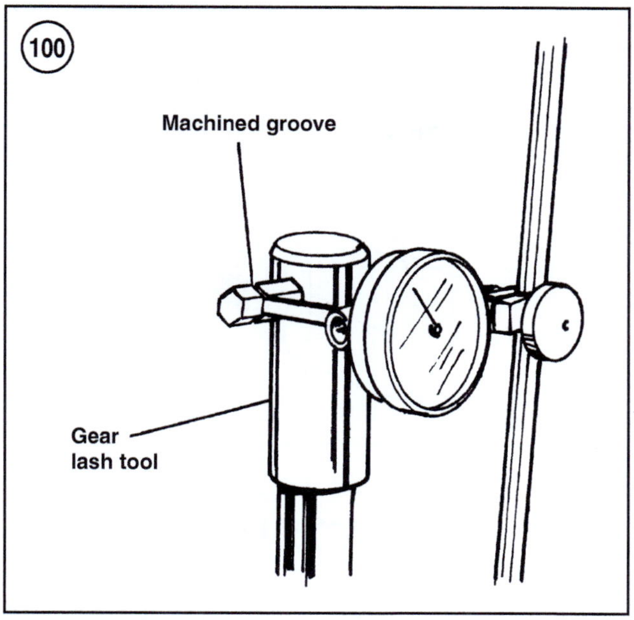

23. Fit the carrier removal tool (OMC or Volvo Penta part No. 3850707-5) over the propeller shaft (**Figure 78**). Insert the protrusions on the tool into the slots in the carrier. Attach a torque wrench to the removal tool at a 90° angle to the bar of the tool. Use the tool to temporarily tighten the bearing carrier to 80 ft.-lb. (109 N•m).

24. Use a depth micrometer to measure the distance from the trailing edge at the rear of gearcase to the flange on the bearing carrier. Take three measurements at each of the carrier support ribs. See **Figure 99**. Record the average measurement as *initial carrier measurement.*

25. Add the *initial carrier measurement* recorded in Step 24 to the shim ring tool measurement recorded in Step 21. Record the results as *total initial carrier measurement.*

26. Fit the carrier removal tool (OMC or Volvo Penta part No. 3850707-5) over the propeller shaft (**Figure 78**). Insert the protrusions on the tool into the slots in the carrier. Attach a breaker bar to the tool. Rotate the tool counterclockwise until free. Remove the tool and then remove the carrier by hand.

27. Remove the shim ring tool (OMC or Volvo Penta part No. 3850626-7) from the carrier or housing bore. Put the tool away so it will not be confused with any other shim ring.

28. Lubricate the propeller shaft tapered roller bearing and driven gear splines. Align the propeller shaft and driven gear splines and then insert the shaft into the gear.

29. Lubricate the threads and flange on the bearing carrier with gearcase lubricant. Install the bearing carrier *without* the shim ring or O-ring over the propeller shaft and into the housing. Thread the carrier into the housing as far as possible by hand.

NOTE

The overall rolling torque varies depending on the drive unit gear ratio. Inspect the identification plate on the side of the upper drive shaft housing to determine the drive unit ratio. If the plate is missing or illegible, refer to ***Drive Unit Ratio*** *in this chapter to determine the ratio.*

30. Set the overall rolling torque for the gear lash and gear pattern check procedures as follows:

a. Install the splined adapter (OMC or Volvo Penta part No. 3850598-8) onto the lower drive shaft. Attach a suitable torque wrench to the splined adapter.
b. Rotate the gearcase assembly to position the drive shaft facing straight up. Attach a dial torque wrench capable of reading 10-30 in.-lb. (1.13-3.39 N•m) onto the splined adapter. Turn the drive shaft clockwise several turns to seat the bearing rollers.
c. Note the torque wrench reading while rotating the drive shaft at approximately 3 rpm.
d. If the reading is less than 14 in.-lb. (1.58 N•m) for 1.43:1, 1.51:1 and 1.60:1 gear ratios, or 11 in.-lb. (1.24 N•m) for all other gear ratios, tighten the bearing carrier *slightly* with the carrier removal tool (OMC or Volvo Penta part No. 3850707-5) and recheck the rolling torque.
e. If the reading exceeds 23 in.-lb. (2.60 N•m) for 1.43:1, 1.51:1 and 1.60:1 gear ratio units, or 18 in.-lb. (2.03 N•m) for all other gear ratio units, loosen the bearing carrier *slightly* with the carrier removal tool (OMC or Volvo Penta part No. 3850707-5) and recheck the rolling torque.
f. Do not proceed with assembly until the overall rolling torque is adjusted to 14-23 in.-lb. (1.58-2.60 N•m) for 1.43:1, 1.51:1 and 1.60:1 gear ratios, or 11-18 in.-lb. (1.24-2.03 N•m) for all other gear ratios. Remove the tools. Do not disturb the bearing carrier position.

10

31. Attach the gear lash tool (OMC or Volvo Penta part No. 3850602-8) onto the drive shaft splines (**Figure 100**). Install the tool with hand pressure only. Do not drive the tool onto the shaft. Securely attach a dial indicator to the gearcase housing. Position the dial indicator plunger in line with the machined groove (**Figure 100**) on the gear lash tool and at a 90° angle to the tool arm. The plunger must extend perfectly horizontally to the tool arm as shown in **Figure 100**. Zero the dial indicator.

NOTE

Gear lash is a measurement of the clearance between the pinion gear and the teeth of the lower gearcase driven gear. The driven gear must not move during the measurement and must be clean. The gears should make a distinct clicking noise as the pinion gear is rotated back and forth against the drive gear.

32. Check the gear lash by *lightly* rotating the drive shaft back and forth while holding the propeller shaft stationary. Record the reading as *gear lash.* Lift the dial indicator and

rotate the drive shaft one-half turn. Reposition the dial indicator and repeat Steps 31-32 to confirm the first reading.

NOTE

If the gear lash requires correction, change approximately 0.001 in. (0.025 mm) in shim thickness for every 0.001 in. (0.025 mm) of correction required. Move both gears (pinion and driven) toward each other to decrease the gear lash. Move both gears away from each other to increase gear lash. Make one-half of the shim change at the pinion gear shim and the other half of the shim change at the driven gear shim. Always change equal amounts of shim thickness at both locations.

33. The gear lash measured in Step 32 must be 0.005-0.012 in. (0.13-0.30 mm). Refer to the following:
   a. If the measured gear lash is 0.005-0.012 in. (0.13-0.30 mm), proceed to Step 34.
   b. If the measured gear lash is less than 0.005 in. (0.13 mm), disassemble the gearcase as described in this chapter. Subtract shims (27, **Figure 76**) from behind the driven gear race and add shims (6, **Figure 76**) below the small bearing race. Reassemble the gearcase. Recheck the rolling torque and recheck the gear lash. Do not proceed with assembly until the gear lash is correct.
   c. If the measured gear lash exceeds 0.012 in. (0.30 mm), disassemble the gearcase as described in this chapter. Add shims (27, **Figure 76**) behind the driven gear race and subtract shims (6, **Figure 76**) from below the small bearing race. Reassemble the gearcase. Recheck the rolling torque and recheck the gear lash. Do not proceed with assembly until the gear lash is correct.

34. Install the splined adapter (OMC or Volvo Penta part No. 3850598-8) onto the drive shaft . Attach a breaker bar and suitable socket to the adapter (**Figure 101**).

NOTE

The gears must be loaded to produce a contact pattern. A simple clamping brake can be fabricated from two 2 × 4 pieces of wood approximately 18 in. (457 mm) long. Hinge the boards together so they can be folded upon each other and pinch the propeller shaft between them.

35. Place an appropriate clamping brake over the propeller shaft (**Figure 101**). Apply enough clamping load to resist rotation and load the aft driven gear. Rotate the drive shaft 6-8 revolutions in each direction.

36. Remove the propeller shaft bearing carrier and propeller shaft as described in this chapter. Do not remove the drive shaft or pinion gear.

37. Compare the pattern on the driven gear teeth with the following patterns.

101

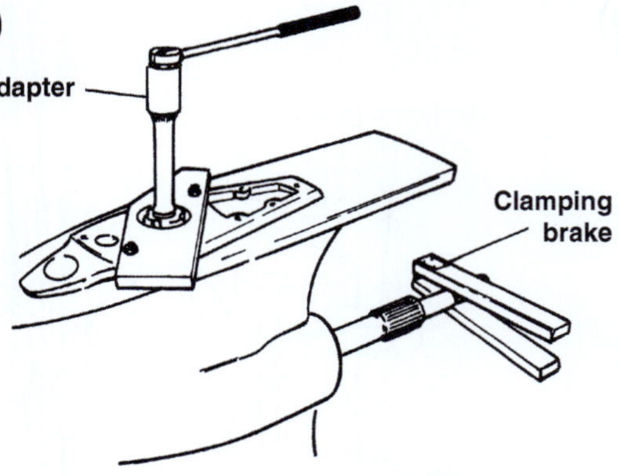

102

CORRECT PATTERN

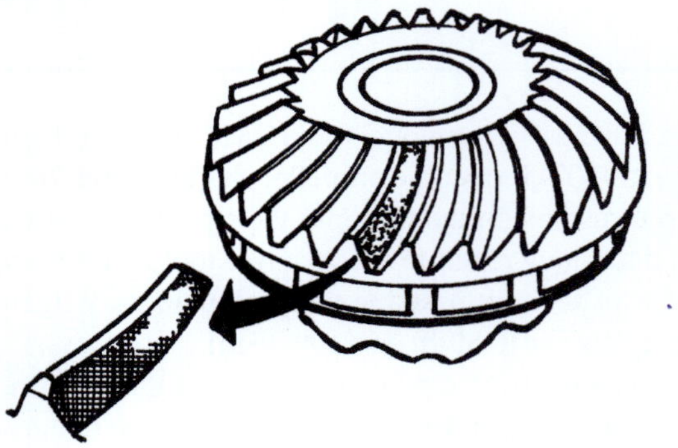

   a. **Figure 102** shows a satisfactory contact pattern that is completely on the tooth with a *slight* bias toward the inner diameter of the gear. Correction is not required. Proceed to Step 38.
   b. **Figure 103** shows an unsatisfactory contact pattern too close to the inner diameter or running off the inner diameter of the gear. Disassemble the gearcase as described in this chapter. Add equal amounts of shim thickness (approximately 0.004 in. [0.102 mm]) under the driven gear race (27, **Figure 76**) and under the small bearing race (6) in the drive shaft bore. Reassemble the gearcase. Recheck the rolling torque, gear lash and gear pattern. Do not proceed with the assembly until the pattern is correct.
   c. **Figure 104** shows an unsatisfactory contact pattern too close to the outer diameter or running off the outer diameter of the gear. Disassemble the gearcase as described in this chapter. Subtract equal amounts of shim thickness (approximately 0.004 in. [0.102 mm]) from under the driven gear race (27, **Figure 76**) and under the small bearing race (6) in the drive shaft bore. Reassemble the gearcase. Recheck the rolling torque, gear lash and gear pattern. Do not proceed with the assembly until the pattern is correct.

38. Once the gear pattern is satisfactory, disassemble the gearcase as necessary to remove all of the gear marking

(103) **INCORRECT PATTERN (TOO CLOSE TO INNER DIAMETER)**

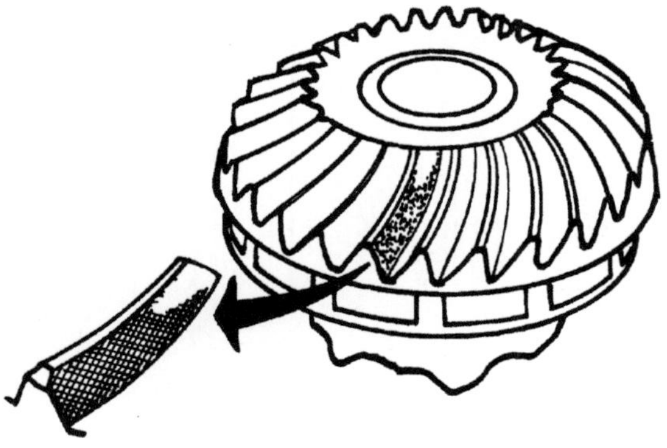

(104) **INCORRECT PATTERN (TOO CLOSE TO OUTER DIAMETER)**

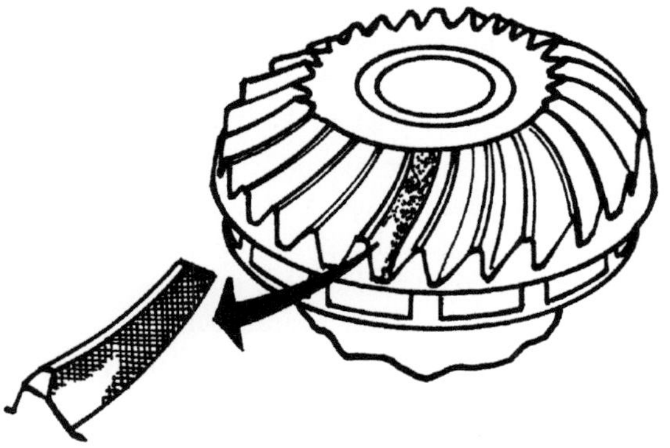

compound from the gear teeth. Discard the *used* pinion nut. Do not remove any pressed-in bearings or races.

39. Verify that all 19 bearing rollers are still located in the lower drive shaft bearing case. If any are dislodged, lubricate them with OMC Needle Bearing Assembly Grease (part No. 378642) and then insert them into the bearing case (**Figure 90**). Purchase the grease from an OMC dealership.

40. Position the gearcase with the bearing carrier opening facing upward. Lubricate the driven gear and bearing assembly and then install it into the gearcase housing (**Figure 83**). Seat the tapered roller bearing into the race at the front of the housing.

41. Mesh the gear teeth and hold the pinion gear in position directly in line with the drive shaft bore. Carefully guide the lower drive shaft assembly into the bore. Work carefully to avoid dislodging the rollers in the lower drive shaft bearing. Rotate the drive shaft to align the shaft and pinion gear splines. Then seat the smaller roller bearing against the race in the bore.

42. Apply a coat of Loctite 271 to the threads of the *new* pinion nut. Thread the pinion nut onto the drive shaft and tighten by hand only at this time.

43. Install the larger drive shaft roller bearing race (4, **Figure 76**) over the drive shaft with the tapered side facing the housing. Seat the bearing race onto the tapered roller bearing on the drive shaft.

44. Lubricate the drive shaft bearing retainer (3, **Figure 76**) and install it over the drive shaft with the threaded end facing the housing. Thread the retainer into the housing by hand only at this time.

45. Thread the nut holder (OMC or Volvo Penta part No. 3854864-0) onto the driver handle (OMC or Volvo Penta part No. 3850609-3) as shown in **Figure 82**. Slide the alignment plate (3, **Figure 82**) over the driver handle. Insert the tool into the carrier bore and engage the nut holder onto the pinion nut (10, **Figure 76**).

46. Install the splined adapter (OMC or Volvo Penta part No. 3850598-8) to the lower drive shaft. Attach a suitable torque wrench onto the splined adapter and tighten the pinion nut to 150-160 ft.-lb. (203-217 N•m).

47. Rotate the gearcase assembly to position the bearing carrier bore facing straight up. Adjust the drive shaft rolling torque as follows:

a. Attach a dial torque wrench capable of reading 2-4 in.-lb. (0.23-0.45 N•m) to the splined adapter. Turn the drive shaft clockwise several turns to seat the bearing rollers.
b. Note the torque wrench reading while rotating the drive shaft at approximately 3 rpm.
c. If the reading is less than 2 in.-lb. (0.23 N•m), tighten the drive shaft retainer (3, **Figure 76**) *slightly* with the spanner wrench (**Figure 81**) and recheck the rolling torque.
d. If the reading exceeds 4 in.-lb. (0.45 N•m), loosen the drive shaft retainer (3, **Figure 76**) *slightly* with the spanner wrench (**Figure 81**) and recheck the rolling torque.
e. Do not proceed with assembly until the drive shaft rolling torque is adjusted to 2-4 in.-lb. (0.23-0.45 N•m). Remove the tools. Do not disturb the retainer position.

48. Lubricate the propeller shaft tapered roller bearing and driven gear splines. Align the propeller shaft and driven gear splines and then insert the shaft into the gear.

49. Lubricate the threads and flange on the bearing carrier with gearcase lubricant. Do not install the shim ring (32, **Figure 76**) or O-ring (34) at this time. Install the bearing carrier into the housing. Thread the carrier into the housing as far as possible by hand.

50. Adjust the overall preload for the final shim ring adjustment as follows:

a. Fit the carrier removal tool (OMC or Volvo Penta part No. 3850707-5) over the propeller shaft (**Figure 78**). Insert the protrusions on the tool into the slots in the carrier.
b. Install the splined adapter (OMC or Volvo Penta part No. 3850598-8) onto the lower drive shaft. Attach a suitable torque wrench to the splined adapter.

10

c. Rotate the gearcase assembly to position the drive shaft facing straight up. Attach a dial torque wrench capable of reading 10-30 in.-lb. (1.13-3.39 N•m) to the splined adapter. Turn the drive shaft clockwise several turns to seat the bearing rollers.
d. Note the torque wrench reading while rotating the drive shaft at approximately 3 rpm.
e. If the reading is less than 14 in.-lb. (1.58 N•m) for 1.43:1, 1.51:1 and 1.60:1 gear ratios, or 11 in.-lb. (1.24 N•m) for all other gear ratios, tighten the bearing carrier *slightly* with the carrier removal tool (OMC or Volvo Penta part No. 3850707-5) and recheck the rolling torque.
f. If the reading exceeds 23 in.-lb. (2.60 N•m) for 1.43:1, 1.51:1 and 1.60:1 gear ratios, or 18 in.-lb. (2.03 N•m) for all other gear ratios, loosen the bearing carrier *slightly* with the carrier removal tool (OMC or Volvo Penta part No. 3850707-5) and recheck the rolling torque.
g. Do not proceed with assembly until the overall rolling torque is adjusted to 14-23 in.-lb. (1.58-2.60 N•m) for 1.43:1, 1.51:1 and 1.60:1 gear ratios, or 11-18 in.-lb. (1.24-2.03 N•m) for all other gear ratios. Remove the tools. Do not disturb the bearing carrier position.

51. Use a depth micrometer to measure the distance from the trailing edge at the rear of the gearcase to the flange on the bearing carrier. Take three measurements at each of the carrier support ribs. See **Figure 105**. Record the average measurement as *final carrier measurement.*

52. Subtract the *final carrier measurement* recorded in Step 51 from the *total initial carrier measurement* recorded in Step 25. Record the results as *bearing carrier calculated shim ring*.

53. Fit the carrier removal tool (OMC or Volvo Penta part No. 3850707-5) over the propeller shaft (**Figure 78**). Insert the protrusions on the tool into the slots in the carrier. Attach a breaker bar to the tool. Rotate the tool counterclockwise until free. Remove the tool and then remove the carrier by hand.

54. Select a shim ring (32, **Figure 76**) with a thickness that comes closest to the measurement recorded in Step 52. Apply a coat of OMC Needle Bearing Assembly Grease to the shim ring and install it onto the bearing carrier flange. Lubricate and then install a new O-ring (34, **Figure 76**) onto the bearing carrier.

55. Lubricate the threads and flange on the bearing carrier with gearcase lubricant. Install the bearing carrier over the propeller shaft and into the housing. Thread the carrier into the housing as far as possible by hand. Work carefully to avoid dislodging the shim ring or O-ring. The O-ring will provide substantial resistance. Fit the carrier removal tool (OMC or Volvo Penta part No. 3850707-5) over the propeller shaft (**Figure 78**). Insert the protrusions on the tool into the slots in the carrier. Attach a torque wrench to the removal tool at a 90° angle to the bar of the tool (**Figure 106**).

105

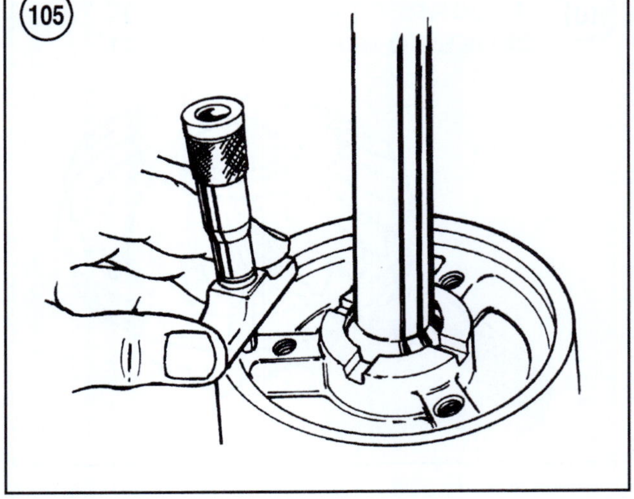

With the aid of the assistant, tighten the propeller shaft bearing carrier to 200-225 ft.-lb. (271-305 N•m).

56. Set the final overall rolling torque as follows:
a. Install the splined adapter (OMC or Volvo Penta part No. 3850598-8) onto the lower drive shaft. Attach a suitable torque wrench to the splined adapter.
b. Rotate the gearcase assembly to position the drive shaft facing straight up. Attach a dial torque wrench capable of reading 10-30 in.-lb. (1.13-3.39 N•m) to the splined adapter. Turn the drive shaft clockwise several turns to seat the bearing rollers.
c. Note the torque wrench reading while rotating the drive shaft at approximately 3 rpm.
d. If the reading is less than 14 in.-lb. (1.58 N•m) for 1.43:1, 1.51:1 and 1.60:1 gear ratios, or 11 in.-lb. (1.24 N•m) for all other gear ratios, remove the bearing carrier and install the next *thinner* shim ring. Reinstall the bearing carrier and recheck the rolling torque.
e. If the reading exceeds 23 in.-lb. (2.60 N•m) for 1.43:1, 1.51:1 and 1.60:1 gear ratios, or 18 in.-lb. (2.03 N•m) for all other gear ratios, remove the bearing carrier and install the next *thicker* shim ring. Reinstall the bearing carrier and recheck the rolling torque.
f. Do not proceed with assembly until the overall rolling torque is adjusted to 14-23 in.-lb. (1.58-2.60 N•m) for 1.43:1, 1.51:1 and 1.60:1 gear ratios, or 11-18 in.-lb. (1.24-2.03 N•m) for all other gear ratios.

CAUTION
Do not overtighten the bearing carrier setscrews. Overtightening will damage the bearing carrier or gearcase housing.

57. Apply a coat of Loctite 271 to the threads and then install the bearing carrier setscrew (24, **Figure 76**) into the opening at the rear port side of the gearcase. Use a T-27

106

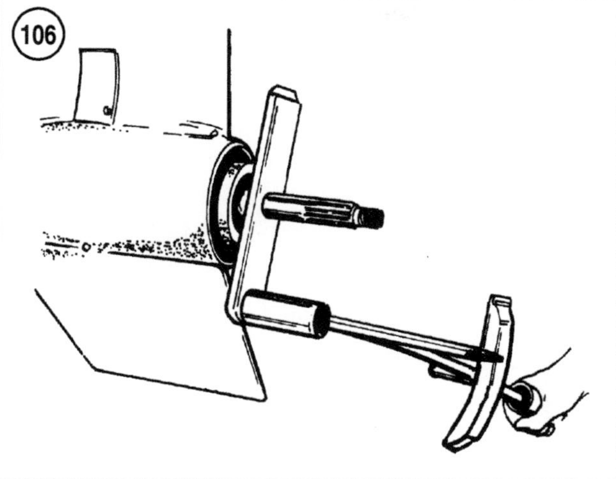

107

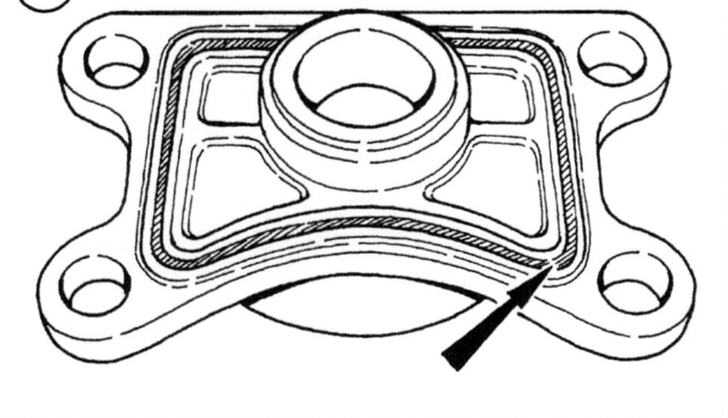

Torx bit to tighten the screw to 42-60 in.-lb. (5-7 N•m). Do not overtighten the setscrew.

58. Coat the threads with Volvo Penta White Sealing Compound (part No. 1141570-0). Then use mechanical fingers to install the retainer block (37, **Figure 76**) and screw (38) into the bearing carrier. The retainer block screw angles up slightly toward the gearcase mating surface. Make sure the retainer is positioned to allow the screw to angle upward. Tighten the screw to 20-25 ft.-lb. (27-34 N•m).

59. Assemble the anode screw (39, **Figure 76**), bracket (41), wave washer (42) and anode (43). Coat the threads of the screw with Volvo Penta White Sealing Compound (part No. 1141570-0). Match the curve of the anode to the curve in the bearing carrier and then install the anode into the lowest possible carrier opening. Thread the anode screw into the carrier and tighten by hand at this time.

60. Coat the threads of the two anode bracket screws (40, **Figure 76**) with Volvo Penta White Sealing Compound (part No. 1141570-0) and install the screws and washers into the carrier. Use a T-40 Torx bit to tighten the screws to 144-168 in.-lb. (16-19 N•m). Tighten the anode screw to 50-84 in.-lb. (6-10 N•m).

61. Install the drive shaft coupler (1, **Figure 76**) onto the drive shaft with the groove end facing upward (away from the lower gearcase).

62. On manual steering models, apply a coat of Volvo Penta White Sealing Compound (part No. 1141570-0) to the threads and then insert the trim tab bolt (19, **Figure 76**) into the opening at the upper rear of the gearcase housing. Install the trim tab (21, **Figure 76**) onto the gearcase. Align the trim tab with the mark (**Figure 77**) made prior to removal. Hold the tab in position and tighten the trim tab bolt to 168-195 in.-lb. (19-22 N•m).

63. On earlier Cobra SX models, install the water tube guide as follows:

a. Apply a coat of OMC Type M adhesive to the sealing ribs of the *new* water tube guide (**Figure 107**).

b. Fit the water tube guide onto the gearcase. Make sure the curved side aligns with the drive shaft. Rotate the guide to align the screw openings.

c. Apply OMC or Volvo Penta White Sealing Compound (part No.1141570-0) to the threads and then thread the four screws by hand into the guide and gearcase.

d. Tighten the screws in a crossing pattern to 120-144 in.-lb. (14-16 N•m).

64. Install the lower gearcase as described in this chapter.

65. Pressure and vacuum test the drive unit as described in this chapter.

66. If removed, install the drive unit as described in this chapter. 10

67. Install the propeller as described in this chapter.

68. Fill the drive unit with lubricant as described in Chapter Three.

DP-S DuoProp (Twin-Propeller) Drives

Disassembly

Upon removal, secure the gearcase into a suitable holding fixture or place the gearcase on a clean work surface.

Record the measurement of all shims as they are removed. Tag the shims by location for reassembly purposes. Do not lose or discard any shim.

All components that will be reused must be installed into the original locations and orientations. Refer to **Figure 108** for this procedure.

1. Remove and clean the screen from the upper drive shaft housing (if not already removed). Replace the strainer if damaged or if it cannot be satisfactorily cleaned.

2. Remove the drain/fill plug (41, **Figure 108**) from the bearing carrier. Remove and discard the O-rings (40, **Figure 108**).

3. Fit the removal tool (OMC or Volvo Penta part No. 3885877-1) over the outer propeller shaft (**Figure 109**). Engage the lugs on the tool into the retaining ring lugs. Rotate the tool counterclockwise to remove the retaining ring (46, **Figure 108**). Remove the O-rings (44 and 47, **Figure 108**) from the retaining ring. Discard the O-rings.

108

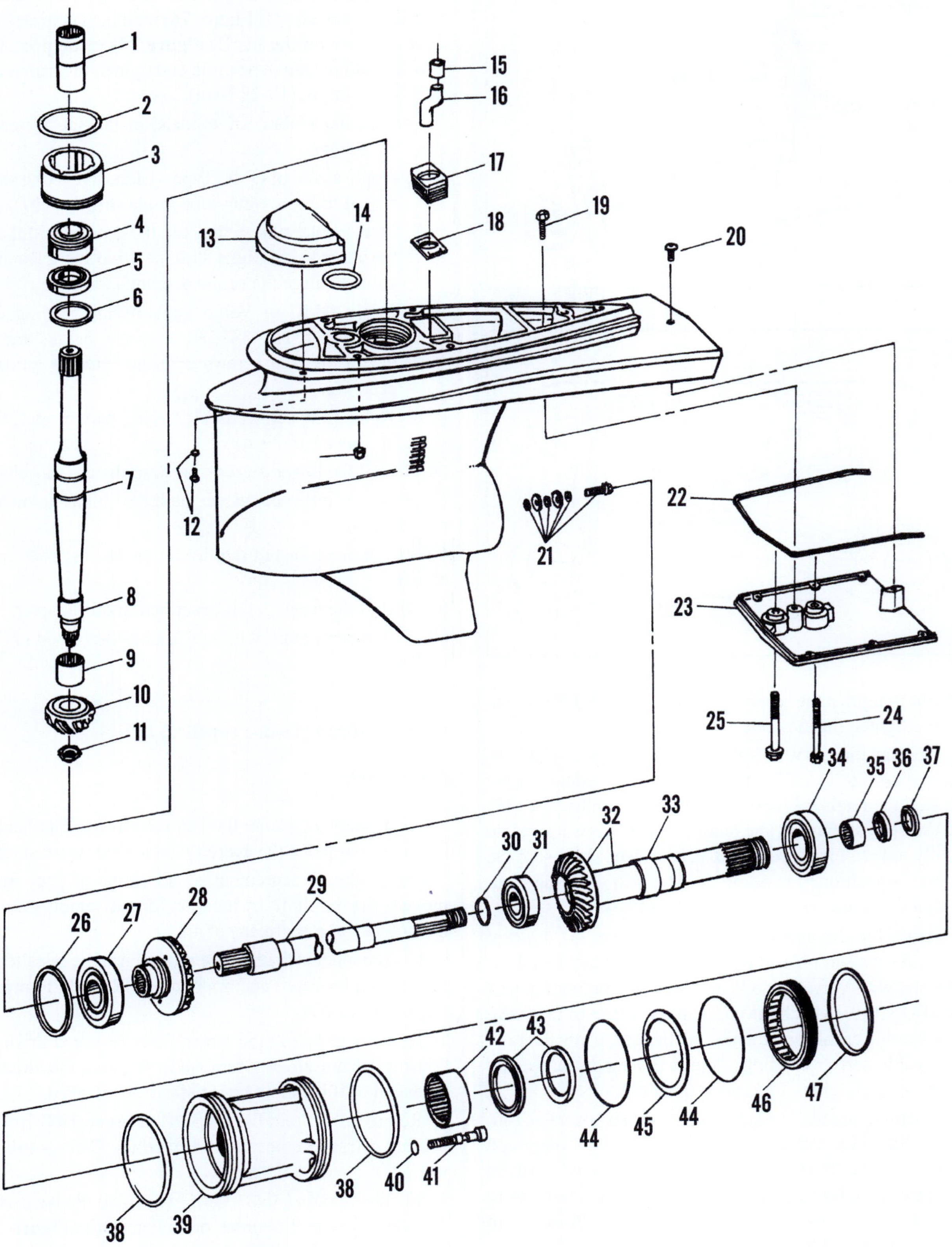
1
2
3
4
5
6
7
8
9
10
11
12
13
14
15
16
17
18
19
20
21
22
23
24
25
26
27
28
29
30
31
32
33
34
35
36
37
38
39
38
40
41
42
43
44
45
44
46
47

LOWER GEARCASE COMPONENTS (DP-S DRIVES)

1. Drive shaft coupler
2. Drive shaft O-ring
3. Drive shaft bearing retainer
4. Larger tapered roller bearing
5. Smaller tapered roller bearing
6. Shim (pinion gear location)
7. Lower drive shaft
8. Bearing sleeve
9. Lower drive shaft (pinion) bearing
10. Pinion gear
11. Pinion nut
12. Anode screw and washer
13. Anode
14. O-ring
15. Water tube grommet (upper drive shaft housing)
16. Copper water tube
17. Water tube grommet (lower gearcase housing)
18. Plastic water tube guide
19. Exhaust plate screw (hex head)
20. Exhaust plate screw (Torx head)
21. Magnet assembly
22. Exhaust plate seal
23. Exhaust plate
24. Bolt (3/8 in. shank)
25. Bolt (7/16 in. shank)
26. Shim (aft driven gear location)
27. Tapered roller bearing
28. Front driven gear
29. Inner propeller shaft
30. Shim (aft driven gear location)
31. Tapered roller bearing
32. Aft driven gear and outer propeller shaft
33. Bearing sleeve
34. Tapered roller bearing
35. Needle bearing
36. Seal (double-lip, single case)
37. Seal (single lip)
38. O-ring
39. Propeller shaft bearing carrier
40. O-ring (drain/fill plug)
41. Drain/fill plug
42. Needle bearing
43. Seals
44. O-ring
45. Tab washer
46. Retaining ring
47. O-rings (2)

109

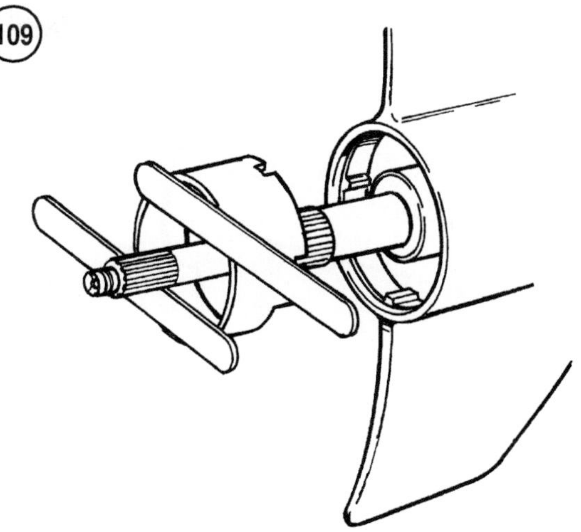

110

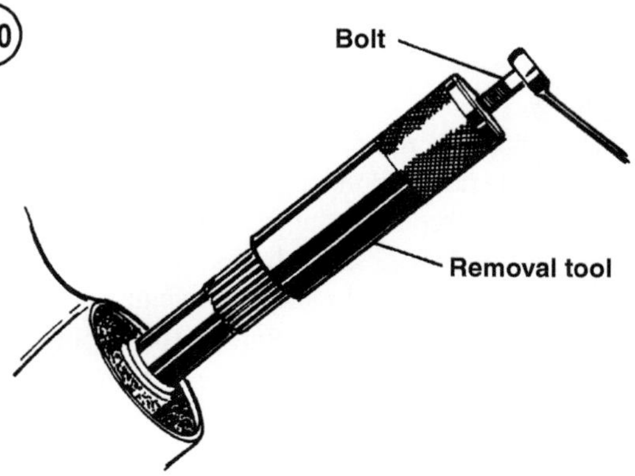

4. Remove the tabbed retaining washer (45, **Figure 108**). Remove the O-ring (44, **Figure 108**) from beneath the washer. Discard the O-rings.

5. Thread the removal tool (OMC or Volvo Penta part No. 884789-9) fully onto the outer propeller shaft (**Figure 110**). Tighten the bolt (**Figure 110**) to pull the outer propeller shaft and bearing carrier out of the gearcase. Remove the tool from the propeller shaft.

6. Remove the inner propeller shaft and loosen the pinion nut as follows:

a. Thread the adapter (OMC or Volvo Penta part No. 3855919-1) onto the inner propeller shaft threads (A, **Figure 111**).

b. Attach the pusher tip (OMC or Volvo Penta part No. 3855921-7) to the tube (OMC or Volvo Penta part No. 3855922-5). Securely tighten the pusher tip set-screw.

c. Guide the pusher tip and tube assembly over the inner propeller shaft. Rotate the drive shaft to align the slot in the pusher tip with the flats on the pinion nut. Seat the tool (B, **Figure 111**) into the gearcase.

d. Install the splined adapter (OMC or Volvo Penta part No. 3850598-8) over the drive shaft splines. Attach a

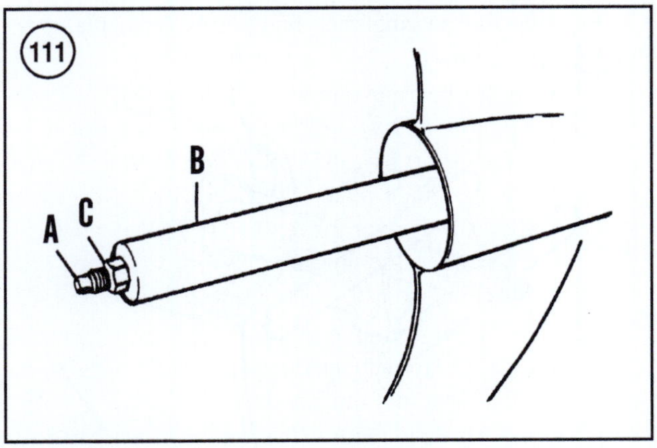

breaker bar and socket to the adapter. Turn the adapter counterclockwise and loosen the pinion nut three turns.

e. Lubricate the special nut (OMC or Volvo Penta part No. 3855920-9) and then thread the nut (C, **Figure 111**) onto the adapter and seat it against the tube assembly.

f. While holding the tube with an adjustable wrench, tighten the nut until the inner propeller shaft (29, **Figure 108**) is free from the front driven gear (28). Lift the shaft and tool assembly out of the gearcase. Remove the tool from the propeller shaft.

7. Unthread the pinion nut (11, **Figure 108**) from the drive shaft. Retain the used pinion nut for use during the assembly procedures.

8. Remove the two screws and washers (12, **Figure 108**) and then lift the anode (13) off the gearcase housing.

9. Use the spanner wrench (OMC or Volvo Penta part No. 3850601-0) to unthread the drive shaft bearing retainer from the gearcase (**Figure 112**). The carrier can be removed by hand pressure.

10. Remove the lower drive shaft as follows:

a. Install the drive shaft puller plate (OMC or Volvo Penta part No. 3855923-3) onto the drive shaft as shown in **Figure 113**. Tighten the clamp screws (A, **Figure 113**) to hold the tool securely to the drive shaft.

b. Rotate the drive shaft to position the bolts (B, **Figure 113**) over a flat surface on the gearcase. Tighten the bolts evenly until the drive shaft is free from the pinion gear.

c. Remove the tool from the drive shaft. Carefully lift the lower drive shaft assembly out of the housing.

d. Reach into the housing and remove the pinion gear and then the front driven gear (**Figure 114**).

11. Remove the screw, magnets and spacers (21, **Figure 108**) from the very bottom of the propeller shaft bore.

NOTE
The drive shaft bearing race in the top of the drive shaft bore and the front driven gear bearing in the bottom of the propeller shaft bore are shimmed bearings. They must be removed to perform the shimming procedures. All other bearings are not shimmed and should not be removed unless the race or roller bearing must be replaced.

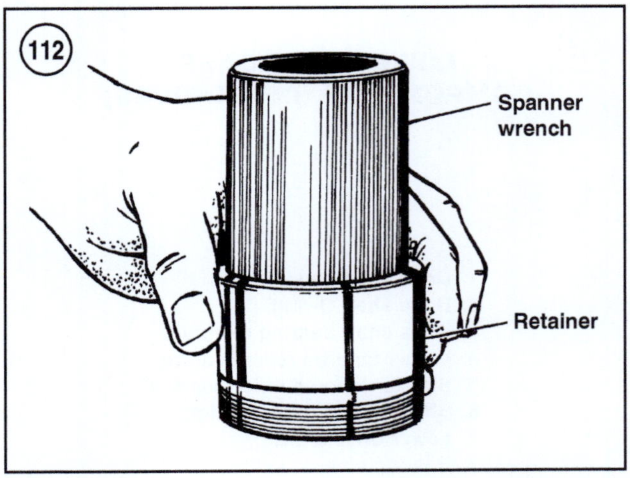

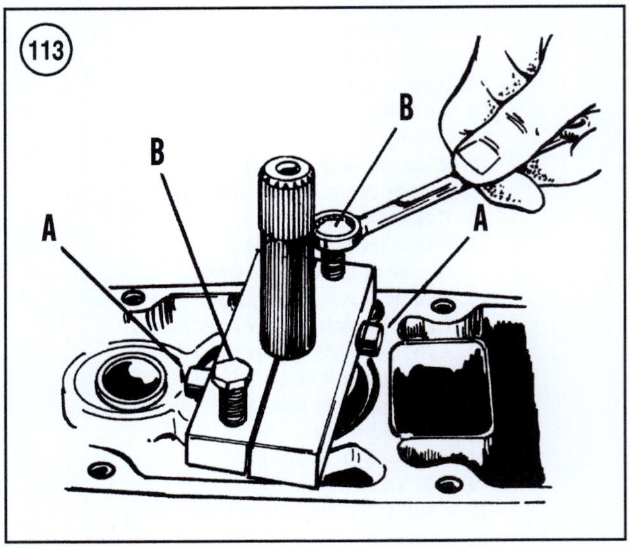

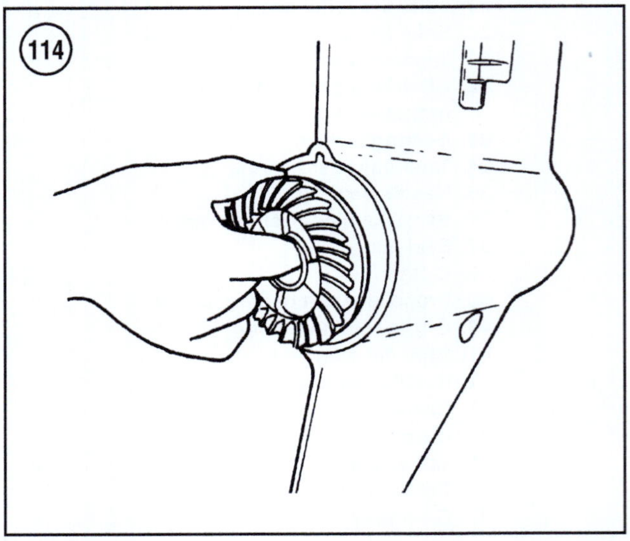

(115)

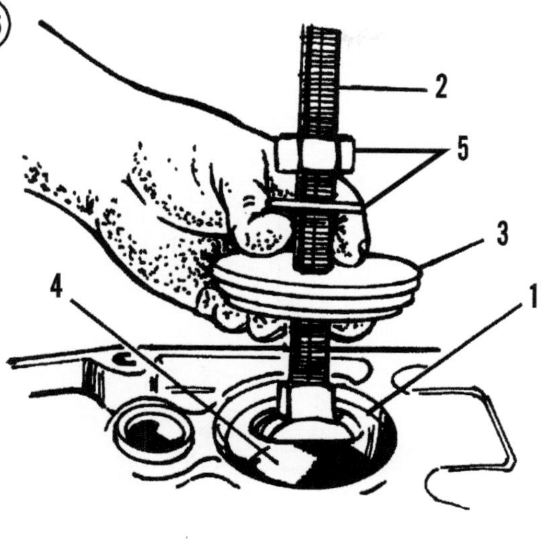

1. Bearing remover
2. Threaded rod
3. Guide plate
4. Bearing race (in drive shaft bore)
5. Nut and washer

(116)

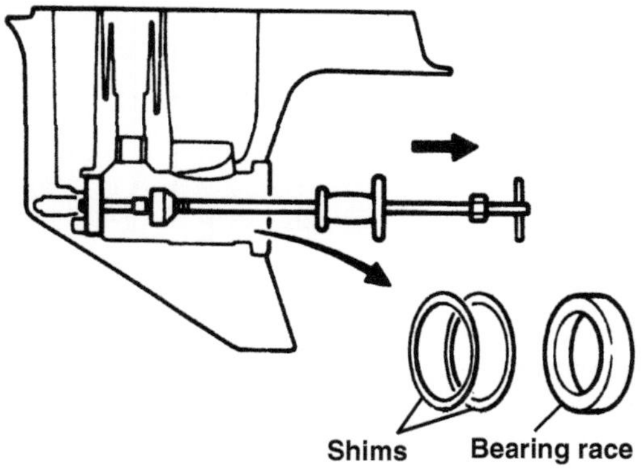

(117)

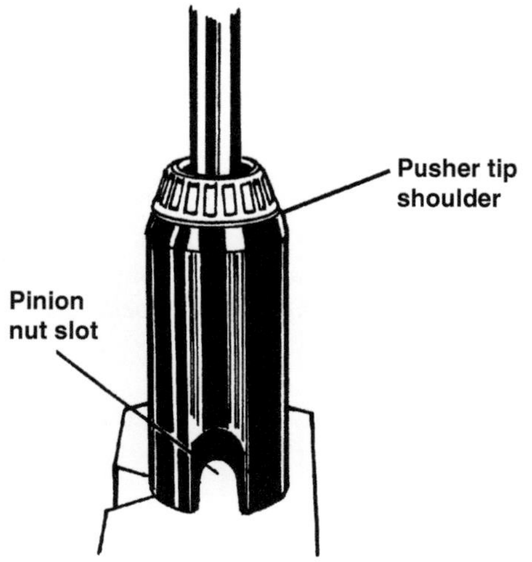

12. Remove the bearing race and shims from the drive shaft bore as follows:

   a. Thread the bearing remover (OMC or Volvo Penta part No. 3855859-9) onto the threaded rod (OMC or Volvo Penta part No. 3855860-7) as shown in **Figure 115**. Fit the guide plate (OMC or Volvo Penta part No. 3850619-2) over the top end of the threaded rod and then thread the nut and washer (5, **Figure 115**) onto the threaded rod.
   b. Fit the bearing remover behind the bearing race (4, **Figure 115**). Hold up the threaded shaft while seating the guide plate into the housing.
   c. Tighten the nut to remove the bearing race (5, **Figure 108**) from the housing.
   d. Remove the shims (6, **Figure 108**) from the bore.

13. Using a slide hammer and a suitable two-jaw puller (**Figure 116**), remove the front driven gear bearing race (27, **Figure 108**). Remove the shims (**Figure 116**) from the bearing race bore.

NOTE
Remove the tapered roller bearing and shims from the inner propeller shaft only if the shims or bearing must be replaced.

14. If the tapered roller bearing (31, **Figure 108**) must be removed from the inner propeller shaft (29), proceed as follows:

   a. Support the removal tool (OMC or Volvo Penta part No. 3855921-7) in a press with the setscrew end facing upward and the pinion nut slot (**Figure 117**) facing downward.
   b. Thread a used propeller nut onto the inner propeller shaft to protect the threads.
   c. Guide the gear end of the inner propeller shaft into the removal tool and seat the roller bearing against the pusher tip shoulder (**Figure 117**).
   d. Press against the propeller nut until the bearing is free from the propeller shaft.
   e. Remove the shim(s) (30, **Figure 108**) from the propeller shaft or bearing. Measure and record the shim(s) thickness for reference during assembly. Tag the shim(s) for location purposes. Remove the propeller nut from the shaft.

15. Carefully pull the bearing carrier (39, **Figure 108**) off the outer propeller shaft and aft driven gear (32).

NOTE
If bearing replacement is required, the needle bearing and inner propeller shaft seals can be removed from the outer propeller shaft at the same time. If bearing replacement is not required, pull the seals from the outer propeller shaft with a suitable two-jaw puller and slide hammer. If replacing the seals only, work carefully to avoid damaging the seal bore or needle bearing in the propeller shaft.

16A. For seal removal only, remove the seals (36 and 37, **Figure 108**) from the outer propeller shaft as follows:

a. Clamp the outer propeller shaft into a vise with protective jaws.
b. Engage the puller jaws into the inner seal case. Do not allow the jaws to contact the seal bore in the propeller shaft.
c. Use short hammer strokes to remove the seals. Discard the seals.

16B. For seal and bearing removal, remove the seals (36 and 37, **Figure 108**) and needle bearing (35) from the outer propeller shaft as follows:

a. Thread the bearing removal tool (OMC or Volvo Penta part No. 884803-8) onto the driver handle (OMC or Volvo Penta part No. 884143-9).
b. Thread a used propeller nut onto the outer propeller shaft to protect the threads.
c. Guide the removal tool and handle into the gear end of the propeller shaft and seat the tool against the needle bearing.
d. Place the tool and outer propeller shaft assembly onto a press with the propeller nut supported by the table of the press. Make sure the table provides an opening for the seals and bearing to exit.
e. Press on the removal tool until the seals and bearing are free from the propeller shaft.

17. Remove the two O-rings (38, **Figure 108**) from the bearing carrier (39). Discard the O-rings.

CAUTION
Remove the lower drive shaft bearing (pinion bearing) only if it, the drive shaft or bearing sleeve on the drive shaft must be replaced. The removal process damages the bearing.

18. If the lower drive shaft (pinion) bearing (9, **Figure 108**) or lower drive shaft (7) must be replaced, remove the lower drive shaft bearings as follows:

a. Slide the guide plate (OMC or Volvo Penta part No. 3850619-2) over the driver rod (OMC or Volvo Penta part No. 3855860-7) with the small-diameter side facing the threaded end. See **Figure 118**. Thread the bearing remover (OMC or Volvo Penta part No. 3855898-0) onto the driver rod with the smaller-diameter side facing away from the guide. See **Figure 118**.
b. Insert the bearing removal tool into the bearing and seat it against the rollers. Seat the guide plate into the drive shaft bore.
c. Place shop towels into the gearcase and position them directly under the bearing.
d. Carefully drive the bearing into the gearcase. Avoid using excessive force. Excessive force can break the bearing case and make removal very difficult. If the bearing proves difficult to remove, apply mild heat to the outside of the gearcase in the area around the bearing with a heat gun or heat lamp. Do not use a torch or open flame. Discard the bearing upon removal.

118

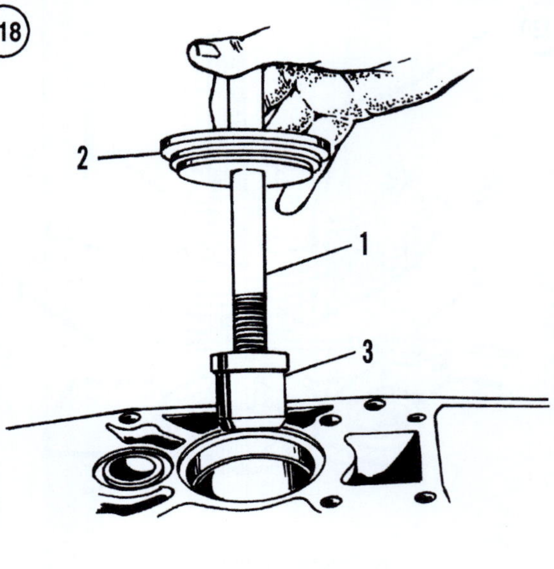

1. Driver rod
2. Guide plate
3. Bearing remover

119

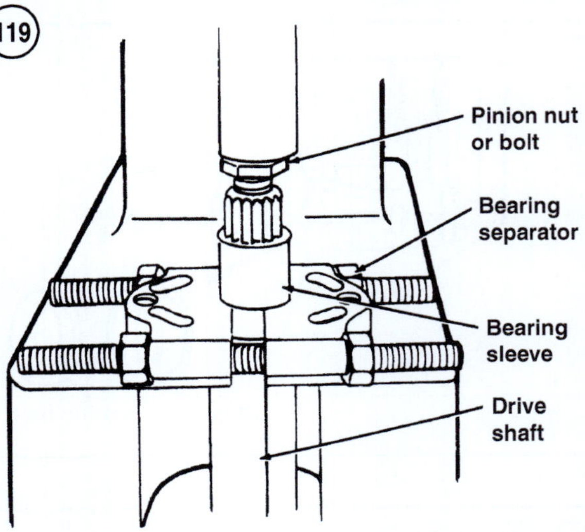

CAUTION
Do not remove the tapered roller bearing from the lower drive shaft unless the bearing or drive shaft must be replaced. The removal process damages the bearings.

19. If the lower drive shaft (7, **Figure 108**) or either bearing (4 or 5) must be replaced, remove the bearings as follows:

a. Support the lower drive shaft in a press with a bearing separator. The separator must engage the inner race

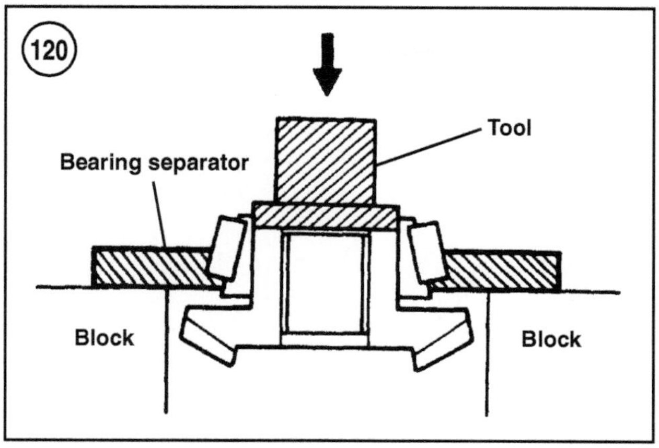

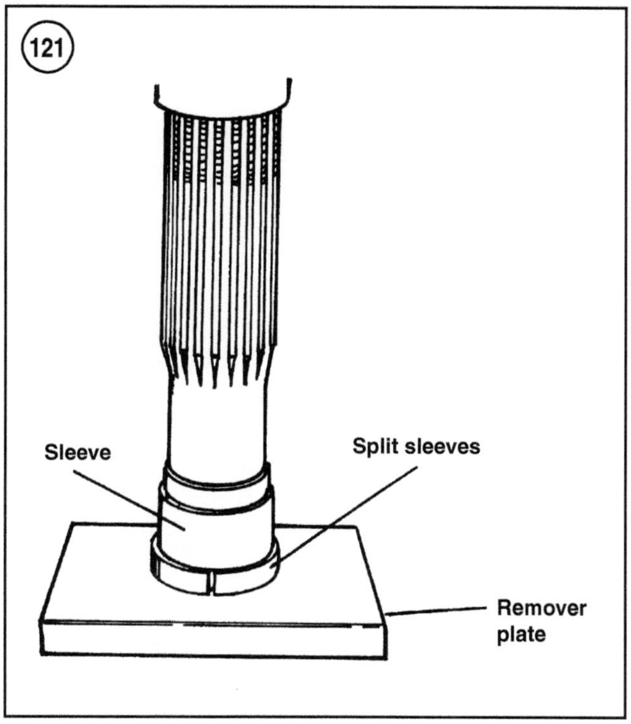

for the small tapered roller bearing (5, **Figure 108**). The drive shaft splines must face upward.

b. Press on the splined end until both bearings are free from the shaft. Discard the bearings.

20. If the lower drive shaft bearing sleeve (8, **Figure 108**) requires replacement, proceed as follows:

a. Thread the *used* pinion nut (11, **Figure 108**) onto the lower end of the drive shaft.

b. Support the drive shaft sleeve in a press with a bearing separator (**Figure 119**). The pinion nut must face upward.

c. Press against the pinion nut until the sleeve is free from the drive shaft. Discard the sleeve. Remove the pinion nut from the shaft.

CAUTION

Remove the tapered roller bearing from the front driven gear only if it must be replaced. The removal process damages the bearing.

21. If the front driven gear bearing (27, **Figure 108**) requires replacement, proceed as follows:

a. Support the bearing and gear assembly in a press with a bearing separator. The gear teeth must face downward. The separator must fit between the bearing and gear (**Figure 120**). Place blocks under the separator to provide adequate travel for the gear.

b. Use a socket, section of tubing or other suitable tool with the proper diameter to press only against the hub of the gear.

c. Align the tool with the hub of the gear and then press the gear out of the bearing. Discard the bearing.

CAUTION

Do not remove the tapered roller bearing from the aft gear unless it must be replaced. The removal process damages the bearings. If the tapered roller bearing is replaced, the bearing sleeve (inner bearing race) for the propeller shaft bearing carrier must first be removed.

NOTE

If the sleeve proves difficult to remove, apply mild heat to the sleeve. Use a heat gun, heat lamp or propane torch.

22. If the tapered roller bearing (34, **Figure 108**) on the aft gear and outer propeller shaft assembly (32) must be replaced, proceed as follows:

a. Insert the outer propeller shaft into the sleeve removal tool (OMC or Volvo Penta part No. 884831-9). Position the split sleeves onto the gear side of the bearing sleeve and rest them against the remover plate as shown in **Figure 121**.

b. Thread a used propeller nut onto the propeller shaft to protect the threads. Support the tool and propeller shaft on a press with the propeller nut facing upward.

c. Press on the propeller nut until the sleeve is free from the propeller shaft. Discard the sleeve.

d. Support the propeller shaft and gear assembly in a press with a bearing separator. The splined end of the propeller shaft must face upward, and the bearing separator must fit between the gear and the tapered roller bearing. The race for the bearing (in the bearing carrier) must also be removed.

e. Thread a used propeller nut onto the propeller shaft to protect the threads. Press on the propeller nut until the tapered roller bearing is free from the propeller shaft and gear.

23. If the small tapered roller bearing race in the aft gear face must be replaced, proceed as follows:

a. Insert the split remover halves (OMC or Volvo Penta part No. 884832-7) through the propeller end of the shaft and position the tool lips behind the bearing race as shown. See **Figure 122**.

10

b. Insert the driver handle (OMC or Volvo Penta part No. 884143-9) through the propeller end of the shaft and guide it between the split halves.
c. Support the gear and shaft in a press with the gear resting on the table of the press. Make sure an opening is provided in the table for the race to exit. Protect the gear teeth with a block of wood, plastic or soft metal.
d. Press on the driver handle until the race is free from the gear. Discard the bearing race.

NOTE
If the needle bearing must be replaced, the propeller shaft bearing carrier inner needle bearing and outer propeller shaft seals can be removed from the bearing carrier at the same time. If bearing replacement is not required, pull the seals from the carrier with a suitable two-jaw puller and slide hammer. If replacing the seals only, work carefully to avoid damaging the seal bore or needle bearing in the carrier.

24A. For seal removal only, remove the seals (43, **Figure 108**) from the propeller shaft bearing carrier as follows:
a. Clamp the outer propeller shaft into a vise with protective jaws.
b. Engage the puller jaws onto the inner seal case. Do not allow the jaws to contact the seal bore in the propeller shaft.
c. Use short hammer strokes to remove the seals. Discard the seals.

24B. For seal and bearing removal, remove the seals (43, **Figure 108**) and needle bearing (42) from the propeller shaft bearing carrier (39) only if the bearing must be replaced. Remove the seals and bearing as follows:
a. Thread the bearing removal tool (OMC or Volvo Penta part No. 3855924-1) onto the driver handle (OMC or Volvo Penta part No. 9991801-3).
b. Insert the removal tool into the bearing carrier from the gear side and seat it against the needle bearing.
c. Place the bearing carrier and tool on a press with the tool side facing upward. Support the carrier with blocks to provide enough room for the seals and bearing to exit. See **Figure 123**.
d. Press against the removal tool until the seals and needle bearings are free from the carrier (**Figure 123**).

NOTE
Remove the bearing race from the propeller shaft bearing carrier only if the bearing carrier or corresponding roller bearing must be replaced.

25. If the tapered roller bearing race or shims in the propeller shaft bearing carrier must be replaced, proceed as follows:
a. Clamp the bearing carrier into a vise with protective jaws.

122
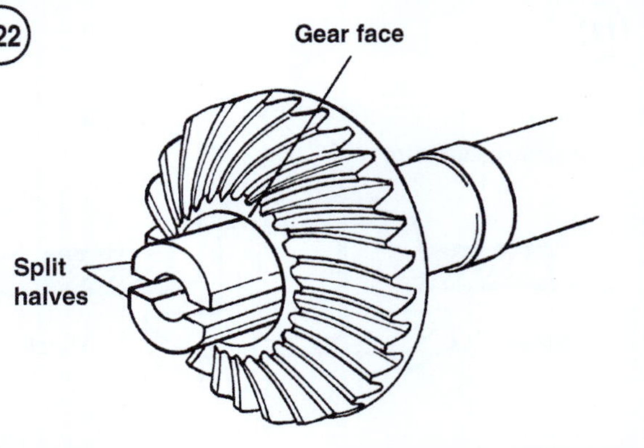

123
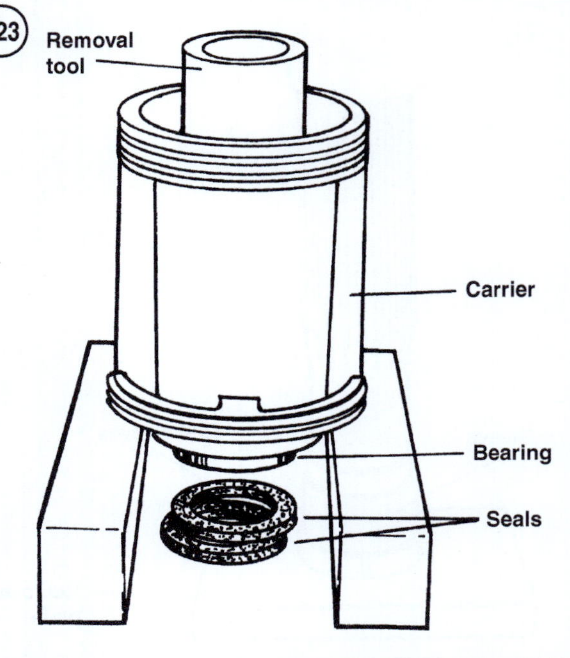

b. Engage the jaws of a slide hammer against the back lip of the race.
c. Use short hammer strokes to pull the race out of the carrier (**Figure 124**, typical).
d. Discard the bearing race.

26. Clean all parts in fresh solvent and dry with compressed air. Make sure all sealant and gasket residue is removed and blow out all threaded openings, lubricant passages and water intake passages.

27. Apply a light coat of oil to all cleaned components to prevent rusting.

28. Inspect the gear teeth for chipped, excessively worn, broken, missing, rough or discolored surfaces. If one gear requires replacement, both gears and the outer propeller shaft and gear assembly must be replaced.

CAUTION
A tapered roller bearing set consists of the caged rollers, cone-shaped inner bearing race and a tapered outer race. The bearing

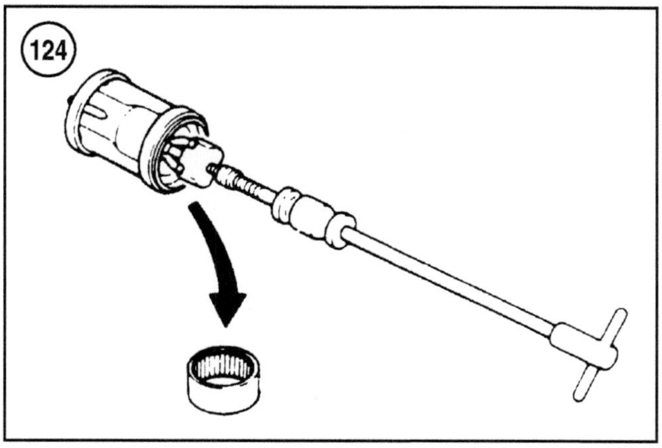

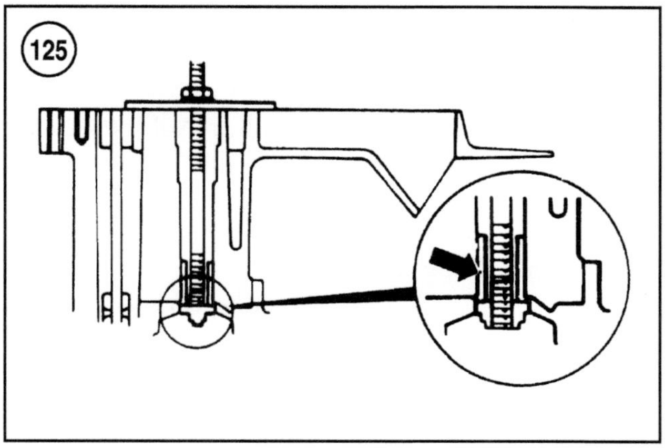

components must be replaced as a set. Never use parts from one bearing set with parts from a different bearing set.

29. Check all bearings for scoring, pitted rollers, roughness, heat discoloration and binding. Replace the bearing and corresponding race if these or other defects are evident. Do not remove pressed-on bearings unless replacement is required or the removal procedures specifically direct you to do so. Many of the bearings cannot be reused after removal.

30. Check the splines in the drive shaft coupler (1, **Figure 108**) for twisted, excessively worn or damaged splines. Replace the drive coupler if these or other defects are evident.

31. Inspect all shims and replace any that are rusted, bent or not in like-new condition with shims of the same thickness.

32. Check all threads, sealing surfaces and splines on the propeller and drive shafts. Replace any shaft with deep grooving at the seal contact surface, roughness at the bearing contact surfaces, twisted splines or mechanical damage. Replace the shaft(s) if surface imperfections cannot be removed with emery paper.

33. Check the gearcase housing and propeller shaft bearing carrier for cracks around bolt or screw openings and bearing bores. Also check for cracks, nicks or burrs on the housing and O-ring mating surfaces. Replace the housing or bearing carrier if cracked or if surface imperfections cannot be removed with emery paper.

Assembly

For maximum durability and quiet operation, all of the gears must be properly located in the housing through shimming procedures. When properly located, the correct gear lash and gear contact pattern will be obtained.

Additionally, the tapered bearings for the propeller and lower drive shaft must be properly tensioned or *preloaded*. Preload refers to the tension applied to hold the tapered roller bearing against the bearing race. Insufficient preload will allow the bearing to move in relation to the race. Any movement will result in premature failure of the bearing and supported gears or shaft. Excessive preload will cause the bearings to overheat and fail.

Do not install new seals and O-rings until the gear lash, gear pattern and bearing preload are corrected. Seal and O-ring installation is performed during the final assembly.

During assembly, lubricate all gears, bearings, seals, shafts and O-rings with OMC, Volvo Penta or Mobil One GL5 Synthetic Gear Lube.

Refer to **Figure 108** for this procedure.

1. If the lower drive shaft (pinion) bearing (9, **Figure 108**) was removed, install the *new* bearing as follows:
   a. Lubricate the new bearing and place it onto the installation tool (OMC or Volvo Penta part No. 3855868-0) with the lettered side of the bearing casing facing the flange of the tool.
   b. Place the tool and bearing into the drive shaft bore from the bearing carrier opening (**Figure 125**, typical). The lettered side of the bearing must face downward when installed into the drive shaft bore.
   c. Guide the threaded rod (3855860-7) into the drive shaft bore. Thread the bearing installation tool fully onto the threaded rod.
   d. Slide the spacer (OMC or Volvo Penta part No. 3855867-2) over the threaded rod and into the drive shaft bore.
   e. Place the guide plate (OMC or Volvo Penta part No. 3850619-2) over the threaded rod with the small-diameter side facing the gearcase. Seat the guide plate into the bore. Install the washer and then the nut onto the threaded rod.
   f. Hold the threaded rod and tighten the nut to pull the bearing into the bore. Stop tightening when a marked increase in effort is required to turn the rod. This indicates the bearing is seated in the bore. Do not overtighten the rod. Overtightening damages the bearing.
   g. Remove the installation tools. Look into the drive shaft bore and inspect the bearing for proper seating or damage. Remove and replace the bearing if damaged or improperly seated.
2. If the bearing sleeve (8, **Figure 108**) was removed from the lower drive shaft, install the *new* sleeve as follows:

10

a. Lightly coat the sleeve contact area on the lower drive shaft with Loctite 271 threadlocking compound. Fit the new sleeve onto the pinion gear end of the lower drive shaft. Place the sleeve installer (OMC or Volvo Penta part No. 3855869-8) over the shaft and seat it against the sleeve.

b. Place the drive shaft onto a press with the sleeve side facing upward. Place a block of wood, plastic or soft metal under the drive shaft to protect the splines.

c. Press against the installer until the sleeve fully seats on the shaft.

NOTE
When installed correctly, the drive shaft tapered bearing rollers face away from each other. Install the smaller bearing first and the larger bearing second.

3. If the tapered roller bearings (4 and 5, **Figure 108**) were removed from the lower drive shaft, install the *new* bearings as follows:

a. Lubricate the new smaller-diameter bearing (5, **Figure 108**) and place it over the splined end of the drive shaft with the rollers facing the pinion gear end of the shaft.

b. Slide the bearing installer (OMC or Volvo Penta part No. 3850617-6) over the splined end of the drive shaft. The raised edge of the tool must face toward the bearing. Seat the tool against the inner race of the bearing.

c. Thread a used pinion nut onto the drive shaft to protect the threads. Support the tool and shaft in a press with the pinion gear end of the shaft facing upward. Place blocks under the bearing installer for support.

d. Press against the pinion nut until the bearing fully seats against the drive shaft shoulder.

e. Lubricate the new larger-diameter bearing (4, **Figure 108**) and place it over the splined end of the drive shaft with the rollers facing the smaller bearing.

f. Slide the bearing installer (OMC or Volvo Penta part No. 3850617-6) over the splined end of the drive shaft. The raised edge of the tool must face toward the bearing. Seat the tool against the inner race of the bearing.

g. Support the tool and shaft in a press with the pinion gear end of the shaft facing upward. Place blocks under the sleeve installer for support.

h. Press against the pinion nut until the bearing fully seats against the inner bearing race of the smaller bearing. Remove the used pinion nut.

4. If the needle bearing (35, **Figure 108**) was removed from the outer propeller shaft and driven gear (32), install the *new* bearing as follows:

a. Insert the new needle bearing into the propeller end of the shaft with the numbered side facing outward.

126

b. Place the propeller shaft onto a press with the gear teeth facing downward. Place a block of wood, plastic or soft metal under the gear to protect the teeth.

c. Place the bearing installer (OMC or Volvo Penta part No. 3855928-2) onto the shaft with the smaller end against the bearing.

d. Press against the installer until the tool seats against the shaft. Remove the installer.

5. If the large tapered roller bearing (34, **Figure 108**) was removed from the aft driven gear and outer propeller shaft, install the *new* roller bearing as follows:

a. Lubricate the new roller bearing and place it over the propeller shaft with the rollers facing away from the gear.

b. Place the shaft and bearing onto a press with the bearing facing upward. Place a block of wood, plastic or soft metal under the gear to protect the teeth.

c. Install the installation tool (OMC or Volvo Penta part No. 3855866-4) onto the bearing and center over the inner bearing race.

d. Press against the tool until the bearing fully seats against the gear.

6. If the bearing sleeve (33, **Figure 108**) was removed from the outer propeller shaft, install the *new* sleeve as follows:

a. Lightly coat the inner diameter of the new sleeve with Loctite 271 threadlocking compound. Guide the sleeve over the propeller shaft splines. Guide the installation tool (OMC or Volvo Penta part No. 3855866-4) over the propeller shaft splines and seat it against the sleeve.

b. Place the outer propeller shaft assembly onto a press with the gear teeth facing downward. Place a block of wood, plastic or soft metal under the gear to protect the teeth.

c. Press against the installation tool until it seats against the inner race of the tapered roller bearing.

7. If the small tapered roller bearing race was removed from the face of the aft driven gear, install the *new* bearing race as follows:

127

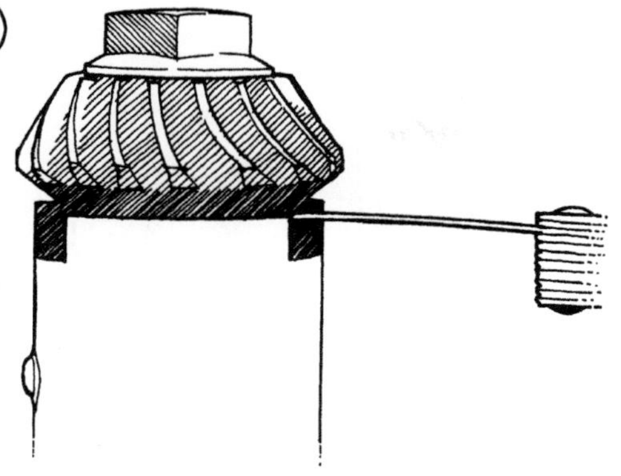

a. Guide the installation tool (OMC or Volvo Penta part No. 3855866-4) over the propeller shaft and seat against the inner race of the tapered roller bearing.

b. Place the outer propeller shaft assembly onto a press with the gear teeth facing upward.

c. Lubricate the new bearing race and place it into the recess in the gear face with the tapered side facing upward.

d. Place the installation tool (OMC or Volvo Penta part No. 3855865-6) onto the race. Center the tool over the race.

e. Press against the tool until the race fully seats in the recess.

8. If the tapered roller bearing (27, **Figure 108**) was removed from the front driven gear (28), install the *new* bearing as follows:

a. Lubricate the new roller bearing and place it over the gear hub with the roller facing away from the gear.

b. Place the gear onto a press with the gear teeth facing downward. Place a block of wood, plastic or soft metal under the gear to protect the teeth.

c. Place the mandrel (OMC or Volvo Penta part No. 3855861-5) onto the bearing and center over the inner race.

d. Press on the mandrel until the bearing fully seats against the gear (**Figure 126**).

9. If the needle bearing (42, **Figure 108**) was removed from the propeller shaft bearing carrier (39), install the *new* needle bearing as follows:

a. Lubricate the new needle bearing and insert it into the propeller end of the carrier with the numbered side facing outward.

b. Place the bearing carrier onto a press with the propeller end facing upward. Place a block of wood or plastic under the carrier to protect the surfaces.

c. Insert the bearing installer (OMC or Volvo Penta part No. 3855925-8) into the carrier with the larger end against the bearing.

d. Press against the bearing installer until it seats against the carrier.

10. If the tapered roller bearing race (34, **Figure 108**) in the bearing carrier (39) was removed, install the *new* race as follows:

a. Lubricate the tapered roller bearing race and place it into the bore with the tapered side facing outward.

b. Place the bearing carrier onto a press with the propeller end facing downward. Place a block of wood or plastic under the carrier to protect the surfaces.

c. Place the installation tool (OMC or Volvo Penta part No. 3855864-9) onto the bearing race. Center the tool over the race.

d. Press against the tool until the race fully seats in the bore.

NOTE

When the correct feeler gauge is selected for the measured clearance, a slight drag is felt as the feeler gauge passes between the measuring points. To verify, insert the next larger gauge and check for a noticeable increase in drag. Then insert the next smaller gauge and check for the absence of drag.

11. If the pinion gear, lower drive shaft or drive shaft bearings were replaced, select the correct pinion gear shim thickness as follows:

a. Clamp the splined adapter (OMC or Volvo Penta part No. 3850598-8) into a vise. Make sure the opening is vertical. Align the splines and insert the lower drive shaft into the vise.

b. With the tapered side facing downward, guide the small tapered roller bearing over the drive shaft and seat it against the small roller bearing. With the slotted end facing upward, guide the shim fixture (OMC or Volvo Penta part No. 3855870-6) over the drive shaft and seat it against the bearing race.

c. Align the splines and install the pinion gear onto the drive shaft. Thread the *used* pinion nut (11, **Figure 108**) onto the drive shaft. Tighten the pinion nut to 72-87 ft.-lb. (98-118 N•m). Record the actual torque applied to the nut. Rotate the shim tool several times to seat the bearing rollers.

d. Use a feeler gauge to measure the distance between the shim tool and the pinion gear at each of the three slots (**Figure 127**). Record the average of the three clearances as the *pinion gear feeler gauge measurement*. Remove the pinion gear, shim tool and small bearing race.

e. Locate the correction number stamped or etched into the pinion gear (**Figure 128**). Record the correction number in thousandths of an inch. For example, if +10 is etched or stamped into the gear, write it down as +0.010 in. If -10 is etched or stamped into the gear, write it down as -0.010 in.

f. Add or subtract (as indicated by the plus or minus sign) the correction number value recorded in Step 11e to or from the *pinion gear feeler gauge measure-*

ment recorded in Step 11d. Perform the calculation for the pinion gear and record the results as *pinion gear shims*.

12. Install the pinion gear shims and bearing race into the gearcase as follows:

a. Select pinion gear shims (6, **Figure 108**) with a total thickness equal to the predetermined value recorded in Step 11f. Insert the shims into the drive shaft bore and rest them on the shoulder in the bore.
b. Insert the large puller bolt (OMC or Volvo Penta part No.3850623-4) into the guide plate (OMC or Volvo Penta part No. 3850619-2) with the larger-diameter side of the plate facing the bolt head. Thread the bearing installer (OMC or Volvo Penta part No. 3830621-8) onto the puller bolt with the beveled edge facing away from the bolt head. See **Figure 129**, typical.
c. Install the bearing race into the drive shaft bore with the tapered side facing upward.
d. Seat the bearing installer into the bearing race. Fit the guide plate into the drive shaft bore.
e. Drive against the puller bolt with a plastic mallet until the race seats against the step in the bore.

13. If the front driven gear (28, **Figure 108**) or tapered roller bearing (27) was replaced, select the correct front driven gear shims as follows:

a. Place the front driven gear on a clean work surface with the gear teeth facing downward. Install the race (**Figure 130**) onto the tapered roller bearing. Rotate the bearing race several revolutions to seat the rollers.
b. Place the shim fixture (OMC or Volvo Penta part No. 3850600-2) onto the roller bearing race with the recessed side facing the bearing. See **Figure 130**.
c. Use a depth micrometer to measure the distance from the top surface of the shim fixture to the end of the driven gear stub shaft (hub). See **Figure 130**. Record the measurement. Subtract the shim fixture thickness (0.500 in.) from the measurement. For example, if the measurement is 0.544 in., subtract 0.500 in. from 0.544 in., which equals 0.044 in. Perform the calculation using the actual measurement and record the results as *front driven gear and bearing dimension*.
d. Locate the correction number stamped or etched into the outer diameter of the front driven gear (**Figure 131**, typical). Record the correction number in thousandths of an inch. For example, if +5 is etched or stamped into the gear, write it down as +0.005 in. If -5 is etched or stamped into the gear, write it down as -0.005 in.
e. Add or subtract (as indicated by the plus or minus sign) the correction number value recorded in Step 13c to or from the *front driven gear and bearing dimension* recorded in Step 13c. Using the driven gear correction number, perform the calculation and record the results as *corrected front driven gear dimension*.
f. Subtract the *corrected front driven gear dimension*, recorded in Step 13e, from the fixed nominal dimen-

128

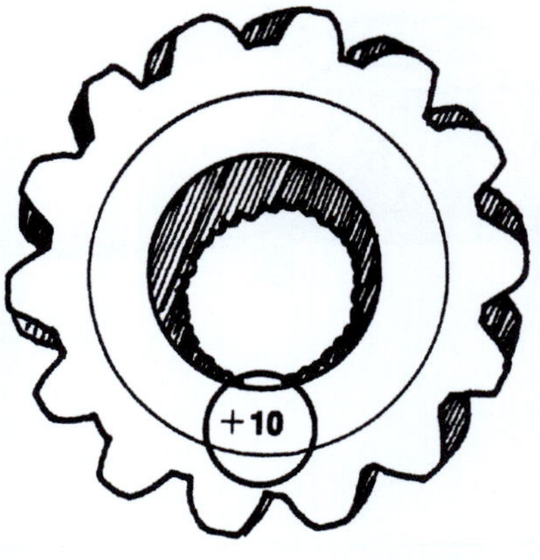

129

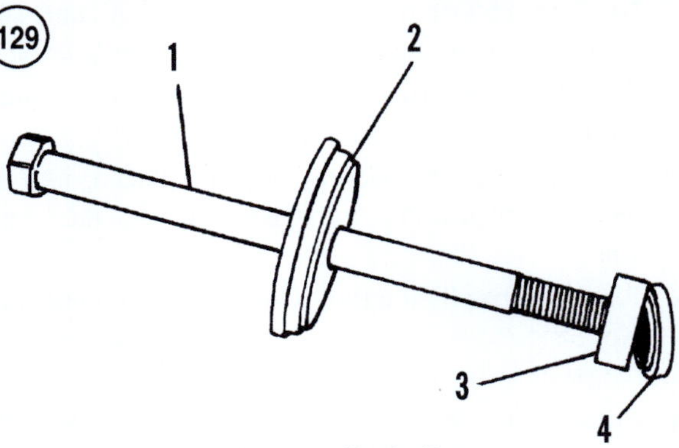

1. Large puller bolt
2. Guide plate
3. Bearing installer
4. Propeller shaft seal

130

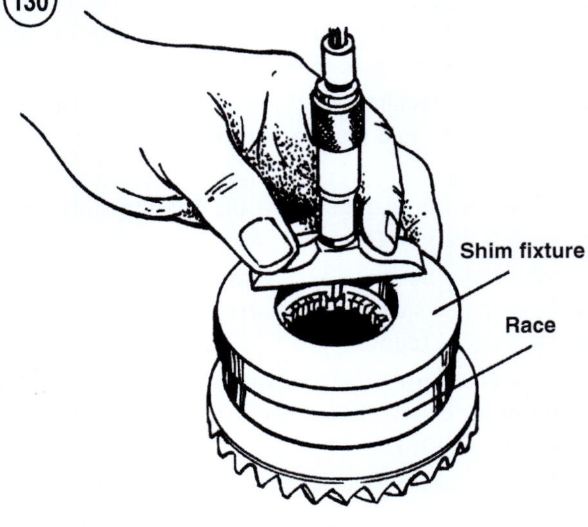

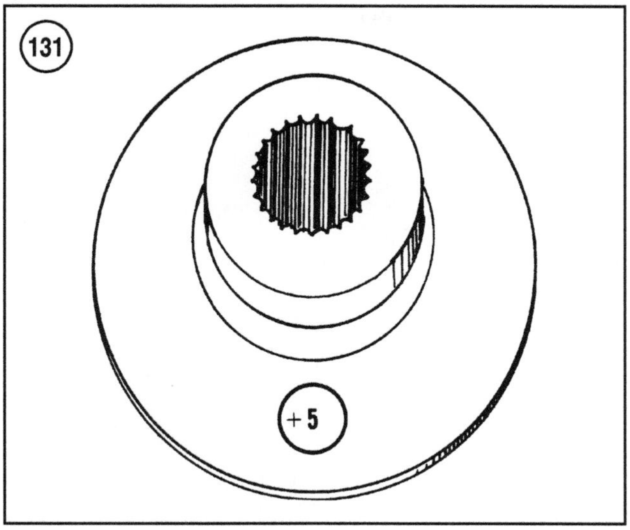

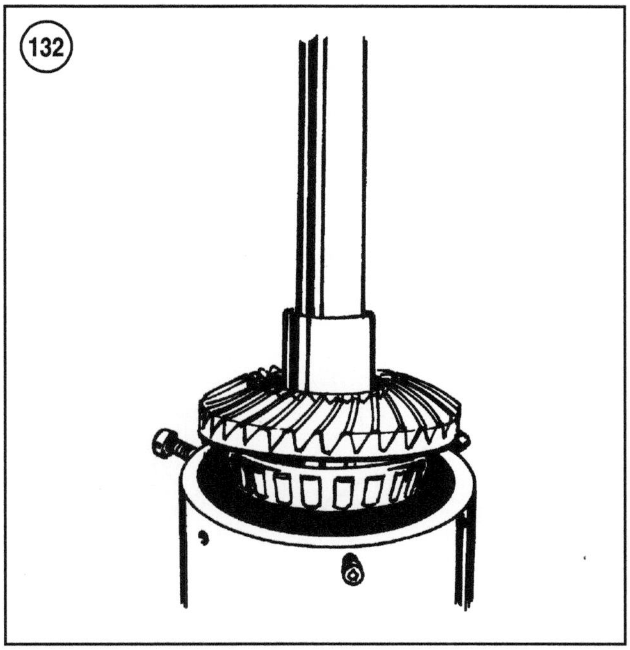

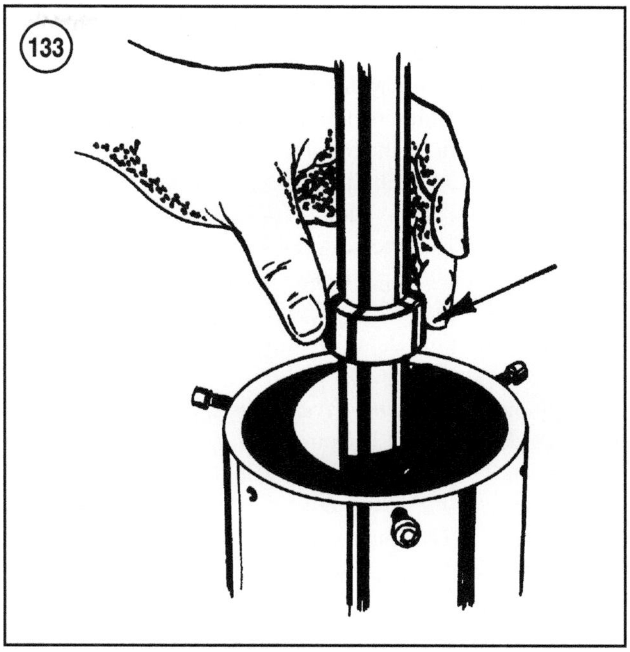

sion of 0.055 in. (1.397 mm). Perform the calculation and record the results as *front driven gear shims*.

14. Assemble the two magnets and three washers over the mounting screw (21, **Figure 108**). Coat the threads of the mounting screw with Loctite 242 threadlocking compound. Thread the assembly into the mounting boss at the very bottom of the gearcase bore. Securely tighten the mounting screw.

15. Install the inner propeller shaft (29, **Figure 108**) into the front driven gear (28) as follows:

a. Place the bearing installer (OMC or Volvo Penta part No. 3855861-5) onto the table of a press.

b. Set the front driven gear onto the installer with the gear teeth facing upward. Make sure the inner bearing race seats against the installer and the rollers or cage do not contact the installer.

c. Thread a used propeller nut onto the inner propeller shaft to protect the threads. Lubricate the splines on the propeller shaft and in the front driven gear.

d. Rotate the propeller shaft to align the splines while inserting the shaft into the gear.

e. Hold the propeller shaft perpendicular while lowering the press ram against the propeller nut. Verify proper alignment and then press against the nut until the shaft fully seats in the gear.

f. Remove the propeller nut and tools.

16. If the outer propeller shaft and aft driven gear (32, **Figure 108**) or bearing (34) were replaced, select the correct aft driven gear locating shim(s) (30) as follows:

a. Place the shim fixture (OMC or Volvo Penta part No. 3855872-2) on a clean workbench with the tapered side facing upward. Set the front driven gear bearing race into the fixture with the tapered surface facing upward. Rest the race on the three bosses in the bottom of the fixture bore.

b. Guide the front driven gear and inner propeller shaft assembly into the holding fixture (**Figure 132**) and then seat the driven gear roller bearing against the bearing race in the fixture.

c. Slide the shim spacer (OMC or Volvo Penta part No. 3855871-4) over the propeller shaft and seat it against the propeller shaft shoulder. See **Figure 133**. Slide the inner propeller shaft small roller bearing (31, **Figure 108**) with the bearing rollers facing upward. Seat the bearing against the shim fixture.

d. Guide the outer propeller shaft and aft driven gear assembly over the inner propeller shaft. Seat the race in the aft driven gear face onto the roller bearing on the inner propeller shaft. Rotate the outer propeller shaft several turns to seat the bearing rollers. Press down on the outer propeller shaft while tightening the three fixture setscrews.

e. Use a depth micrometer to measure the distance from the top surface of the large shim fixture to the machined ring on the rear face of the aft driven gear. See **Figure 134**. Take a reading near each of the three setscrews. If a wide variance in measurement is noted, loosen the setscrews and rotate the shaft several turns to again seat the rollers. Retighten the setscrews as described in Step 16d. Record the average measured depth as *initial aft driven gear dimension*.

f. Subtract the *front driven gear shims* calculation recorded in Step 13f from the *initial aft driven gear dimension* recorded in Step 16e. Record the results as *corrected aft driven gear dimension*.

g. Locate the correction number stamped or etched into the aft driven gear (**Figure 135**). Record the correction number in thousandths of an inch. For example, if +5 is etched into the gear, write it as +0.005 in. If -5 is etched into the gear, write it as -0.005 in. Add or subtract (as indicated by the plus or minus sign) the correction number to or from the *corrected aft driven gear dimension* recorded in Step 16f. Perform the calculation using the actual correction number and record the results as *aft driven gear shims*.

h. Remove the propeller shaft, gears and bearing races from the shim fixtures.

17. Install the driven gear shims (26, **Figure 108**) and bearing race as follows:

a. Select driven gear shims with a total thickness equal to the predetermined value recorded in Step 13f. Install the shims into the gearcase opening and rest them on the step in the bore.

b. Install the guide plate (OMC or Volvo Penta part No. 3855863-1) over the large bolt (OMC or Volvo Penta part No. 3850623-4). Make sure the larger diameter of the guide plate faces the bolt head as shown in **Figure 136**.

c. Thread the large nut (from Volvo Penta tool part No. 3855860-7) approximately halfway onto the bolt threads. Install the installation tool (OMC or Volvo Penta part No. 3855864-9) onto the bolt with the larger diameter facing the nut as shown in **Figure 136**.

d. Lubricate the bearing race and place it into the recess with the tapered side facing outward.

e. Guide the bearing installer into the bearing race while installing the tool assembly into the housing. Fit the alignment plate into the bearing carrier bore of the housing and seat it against the step in the bore.

f. If the housing is not mounted into a suitable holding fixture, rest the leading edge of the gearcase against a block of wood to prevent damage to the gearcase.

g. Drive against the bolt head until the race seats against the shim(s) and the step in the bore. Remove the tool and inspect the bearing for proper installation. Remove and replace the bearing if found to be damaged or tilted in the bore.

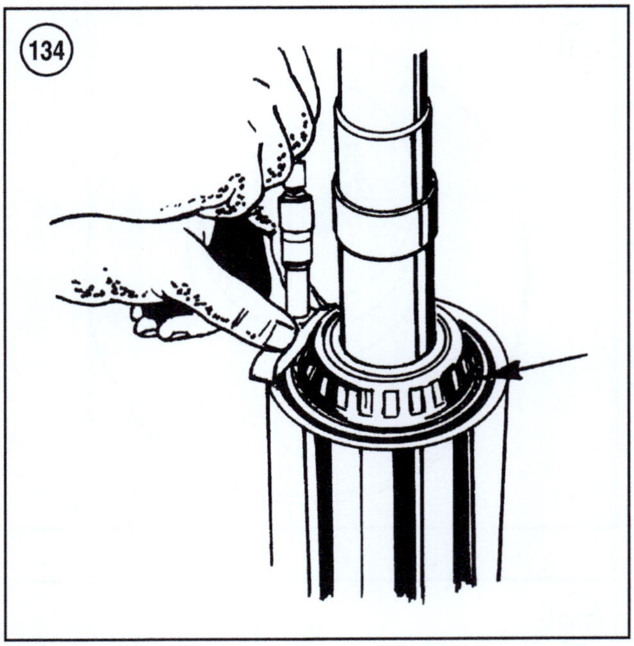

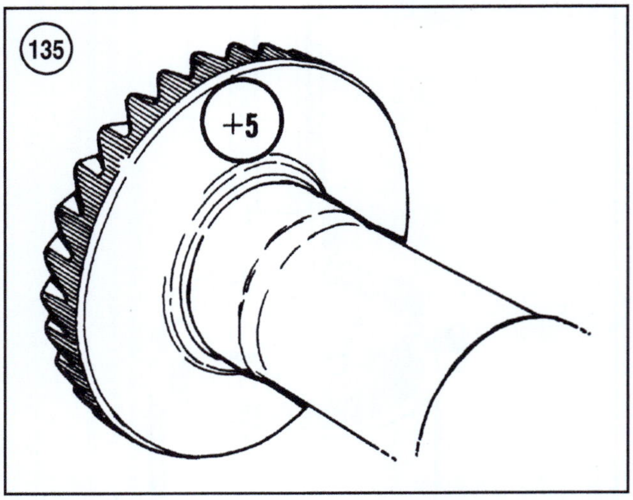

18. Lubricate the front driven gear and inner propeller shaft assembly. Guide the driven gear roller bearing into the bearing race while installing the assembly into the gearcase. Push on the inner propeller shaft to seat the driven gear roller bearing into the race at the front of the housing.

19. Mesh the gear teeth and hold the pinion gear in position directly aligned with the drive shaft bore. Carefully guide the lower drive shaft assembly into the bore. Work carefully to avoid dislodging the rollers in the lower drive shaft bearing. Rotate the drive shaft to align the shaft and pinion gear splines and then seat the smaller roller bearing against the race in the bore.

NOTE

Use the original pinion nut during assembly at this time to check the gear lash and gear

136

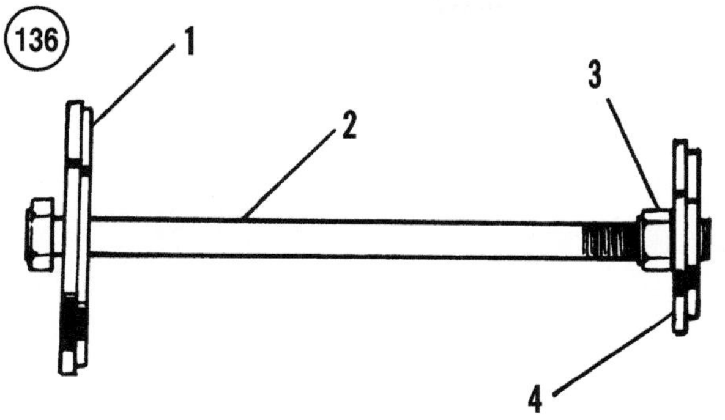

1. Guide plate
2. Large puller bolt
3. Nut
4. Bearing installer

137

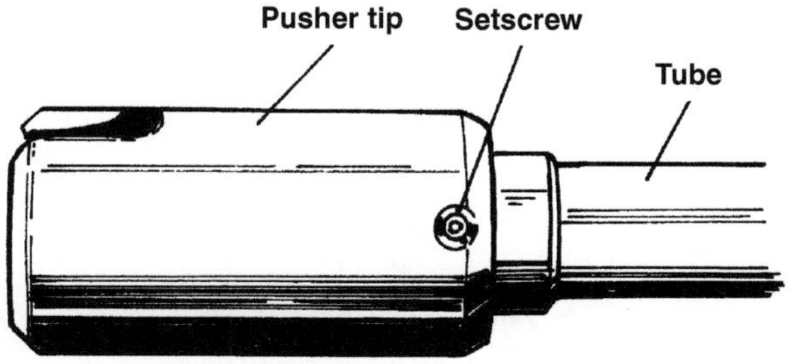

pattern. Install a new pinion nut during the final assembly.

20. Use mechanical fingers to install the *used* pinion nut (11, **Figure 108**) onto the lower drive shaft. Rotate the drive shaft to tighten the nut by hand only at this time.

21. Install the larger drive shaft roller bearing race (4, **Figure 108**) over the drive shaft with the tapered side facing the housing. Seat the bearing race onto the tapered roller bearing on the drive shaft.

22. Lubricate the drive shaft bearing retainer (3, **Figure 108**) and install it over the drive shaft with the threaded end facing the housing. Thread the retainer into the housing by hand only at this time.

23. Tighten the pinion nut as follows:

a. Thread the adapter (OMC or Volvo Penta part No. 3855919) onto the inner propeller shaft threads. Fit the pusher tip (OMC or Volvo Penta part No. 3855921-7) onto the tube (OMC or Volvo Penta part No. 3855922-5) as shown in **Figure 137**. Hold the tip onto the tube and securely tighten the pusher tip setscrew.

b. Fit the pusher tip and tube assembly over the inner propeller shaft. Rotate the drive shaft to engage the flats on the pinion nut into the slot in the pusher tip.

c. Thread the nut (OMC or Volvo Penta part No. 3855920-9) onto the adapter by hand. Align the splines and then install the splined adapter (OMC or Volvo Penta part No. 3850598-8) onto the lower drive shaft.

d. Attach a socket and an appropriate torque wrench to the adapter. Turn the adapter to tighten the pinion nut to the exact torque recorded in Step 11c.

24. Rotate the gearcase assembly to position the bearing carrier bore facing straight up. Adjust the drive shaft rolling torque as follows:

a. Attach a dial torque wrench capable of reading 2-4 in.-lb. (0.23-0.45 N•m) to the splined adapter installed in Step 23. Turn the drive shaft clockwise several turns to seat the bearing rollers.

b. Note the torque wrench reading while rotating the drive shaft at approximately 3 rpm.

c. If the reading is less than 2 in.-lb. (0.23 N•m), tighten the drive shaft retainer (3, **Figure 108**) *slightly* with the spanner wrench (**Figure 112**) and recheck the rolling torque.

d. If the reading exceeds 4 in.-lb. (0.45 N•m), loosen the drive shaft retainer (3, **Figure 108**) *slightly* with the spanner wrench (**Figure 112**) and recheck the rolling torque.

e. Do not proceed with assembly until the drive shaft rolling torque is adjusted to 2-4 in.-lb. (0.23-0.45 N•m). Remove the tools. Do not disturb the retainer position.

25. If removed, install the small tapered roller bearing (31, **Figure 108**) and shim (30) onto the propeller shaft as follows:

a. Select inner propeller shaft shim(s) (30, **Figure 108**) with the total thickness equal to the *aft driven gear shims* measurement recorded in Step 16g. Slide the shims over the propeller end of the inner propeller shaft and seat them against the shoulder.

b. Lubricate the roller bearing and slide it over the propeller shaft with the roller facing outward. Slide the bearing onto the propeller shaft as far as possible with hand pressure.

c. Thread the adapter (A, **Figure 111** [Volvo Penta part No. 3855919-1]) onto the inner propeller shaft threads.

d. Slide the tube (B, **Figure 111** [Volvo Penta part No. 3855922-5]) over the propeller shaft and seat against the inner bearing race.

e. Lubricate the nut (C, **Figure 111** [Volvo Penta part No. 385590-9]) and thread it on the adapter as far as possible by hand.

f. Hold the tube with an adjustable wrench and tighten the nut until the bearing fully seats against the shims and propeller shaft shoulder. Remove the tools from the propeller shaft.

g. Make sure the bearing seats fully against the shims and propeller shaft shoulder. Reinstall the tools and repeat the installation procedure as necessary.

26. Coat at least three teeth on the aft driven gear with GM Gear Marking Compound, available from any GM automotive dealership.

27. Guide the aft driven gear and outer propeller shaft assembly over the inner propeller shaft. Rotate the lower drive shaft to mesh the gear teeth. Then seat bearing race in the face of the aft driven gear against the small tapered roller bearing on the inner propeller shaft. Lubricate the tapered roller bearing on the rear of the aft driven gear.

28. Lubricate both flanges and the bearings in the propeller shaft bearing carrier (39, **Figure 108**). Guide the carrier over the outer propeller shaft and seat it into the gearcase bore. Make sure the lubricant drain/fill plug opening is on the lower side or toward the skeg.

29. Install the tab washer (45, **Figure 108**) over the propeller shaft. Align the tab with the opening at the top of the carrier bore and then seat the washer against the bearing carrier. Lubricate the retaining ring (46, **Figure 108**) and thread it into the housing as far as possible by hand.

30. Set the overall rolling torque for the gear lash and gear pattern check procedure as follows:

a. Install the splined adapter (OMC or Volvo Penta part No. 3850598-8) onto the drive shaft. Attach an accurate dial torque wrench capable of reading 10-40 in.-lb. (1-5 N•m) to the splined adapter with an appropriate socket. Rotate the drive shaft several revolutions to seat the bearing rollers.

b. Note the torque wrench reading while rotating the drive shaft at approximately 3 rpm.

c. If the reading is less than 18 in.-lb. (2 N•m) for 1.68:1 and 1.78:1 gear ratios, or 27 in.-lb. (3 N•m) for all other gear ratios, tighten the retaining ring *slightly* with the removal tool (Volvo Penta part No. 3855877-1) as shown in **Figure 109** and recheck the rolling torque.

d. If the reading exceeds 27 in.-lb. (3 N•m) for 1.68:1 and 1.78:1 gear ratios, or 35 in.-lb. (4 N•m) for all other gear ratios, loosen the retaining ring *slightly* with the removal tool (Volvo Penta part No. 3855877-1) as shown in **Figure 109** and recheck the rolling torque.

e. Do not proceed with assembly until the overall rolling torque is adjusted to 18-27 in.-lb. (2-3 N•m) for 1.68:1 and 1.78:1 gear ratios, or 27-35 in.-lb. (3-4 N•m) for all other gear ratios.

f. Record the actual rolling torque (preload) measurement for final assembly purposes. Remove the tools. Do not disturb the retaining ring position.

31. Install the drive shaft puller (OMC or Volvo Penta part No. 3855923-3) onto the drive shaft as shown in **Figure 113**. Tighten the clamp screws (A, **Figure 113**) to hold the tool securely onto the drive shaft. Rotate the drive shaft to position the bolts (B, **Figure 113**) over a flat surface on the gearcase. Tighten the bolts *lightly* to prevent drive shaft movement during the gear lash measurement.

138

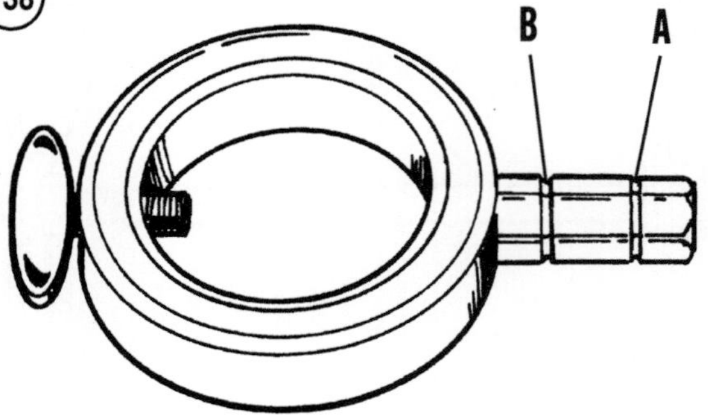

139

SHIM LOCATION AND FUNCTION (DP-S DUOPROP GEARCASE)

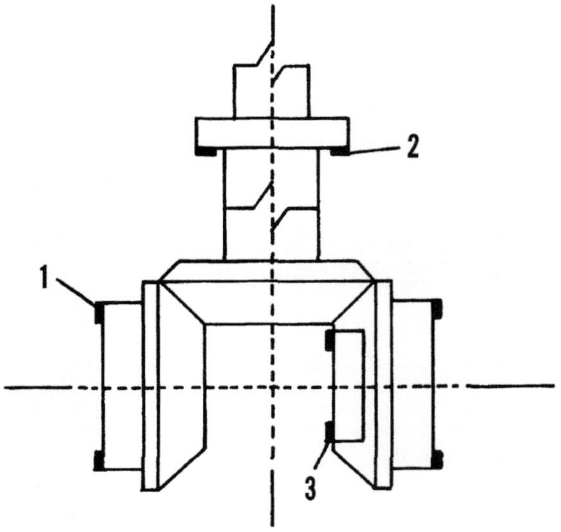

1. Front driven gear location
2. Pinion gear height
3. Aft driven gear location

32. Install the gear lash indicator tool (**Figure 138** [Volvo Penta part No. 3855873-0]) onto the smooth portion of the inner propeller shaft. Do not position the tool on the propeller shaft splines. Tighten the thumb screw to secure the tool onto the propeller shaft.

33. Securely attach a dial indicator to the gearcase housing. Position the dial indicator plunger aligned with the outer machined groove (A, **Figure 138**) on the gear lash tool and at a 90° angle to the tool arm. The plunger must extend perfectly horizontally to the tool arm. Zero the dial indicator.

NOTE

Gear lash is a measurement of the clearance between the pinion gear and the driven gear

140

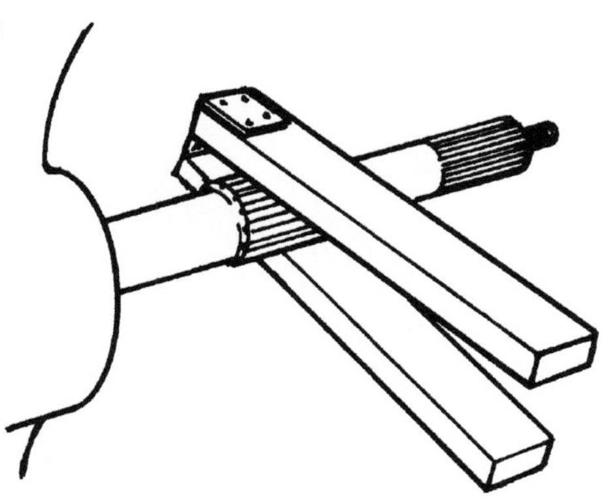

141

teeth. The pinion gear must not move during the measurement and must be clean. The gears should make a distinct clicking noise as the driven gear is rotated back and forth against the pinion gear.

34. Check the gear lash by *lightly* rotating the inner propeller shaft back and forth. Record the reading as *front driven gear lash*. Remove the indicator from the inner propeller shaft. Do not remove the dial indicator or drive shaft puller at this time.

35. Install the gear lash indicator tool (**Figure 138** [Volvo Penta part No. 3855873-0]) onto the smooth portion of the outer propeller shaft. Do not position the tool on the propeller shaft splines. Tighten the thumb screw to secure the tool to the propeller shaft.

36. Position the dial indicator plunger aligned with the inner machined groove (B, **Figure 138**) on the gear lash tool and at a 90° angle to the tool arm. The plunger must extend perfectly horizontally to the tool arm. Zero the dial indicator.

37. Check the gear lash by *lightly* rotating the outer propeller shaft back and forth. Record the reading as *aft driven gear lash*. Remove the indicator from the outer propeller shaft. Remove the dial indicator and drive shaft puller.

NOTE

If the gear lash requires correction, change approximately 0.001 in. (0.03 mm) in shim thickness for every 0.001 in. (0.03 mm) of correction required. Move both gears (pinion and driven) toward each other to decrease the gear lash. Move both gears away from each other to increase gear lash. Make one-half of the shim change at the pinion gear shim and the other half of the shim change at the driven gear shim. Always change equal amounts of shim thickness at both locations.

NOTE

If it is determined that a large change in shim thickness is required to correct gear lash, recheck the measurements and calculations for incorrect measurement, improper assembly or mathematical errors before changing the shims.

38. The *front driven gear lash* must be 0.006-0.012 in. (0.152-0.305 mm). The *aft driven gear lash* must be 0.006-0.014 in. (0.152-0.356 mm). Refer to the following:

a. If both gear lash measurements are within the specifications, proceed to Step 39.

b. If one or both gear lash measurements are not within the specification, disassemble the gearcase and make the shim changes specified in **Table 4**. Reassemble the gearcase and recheck the rolling torque and gear lash. Do not proceed with assembly until the rolling torque and gear lash are correct. **Figure 139** shows shim location and function.

10

39. Install the splined adapter (OMC or Volvo Penta part No. 3850598-8) onto the drive shaft. Attach a breaker bar and suitable socket to the adapter.

NOTE

The gears must be loaded to produce a contact pattern. A simple clamping brake can be fabricated from two 2 × 4 pieces of wood approximately 18 in. (457 mm) long. Hinge the boards together so they can be folded upon each other and pinch the propeller shaft between them.

40. Place an appropriate clamping brake (**Figure 140**) over the outer propeller shaft. Apply enough clamping load to resist rotation and load the aft driven gear. Rotate the drive shaft 6-8 revolutions in each direction.

41. Remove the propeller shaft bearing carrier and outer propeller shaft as described in the gearcase disassembly instructions (in this chapter). Compare the pattern on the front and aft driven gears with the following patterns.

a. **Figure 141** shows a satisfactory contact pattern slightly biased toward the root of the tooth and toward the outside diameter of the gear. Correction is not required.

b. **Figure 142** shows an unsatisfactory contact pattern too close to the root of the gear and too close to the outside diameter of the gear. To correct the pattern, move the pinion gear up and the driven gear(s) toward the pinion gear.

c. **Figure 143** shows an unsatisfactory contact pattern too close to the crest of the tooth and too close to the inner diameter of the gear. To correct the pattern, move the pinion gear down and driven gear(s) away from the pinion gear.

42. If the gear pattern is unsatisfactory, disassemble the gearcase and make the shim changes specified in **Table 6**. Make shim thickness changes in 0.004 in. (0.102 mm) increments. **Figure 139** shows the location and function of applicable shims. Reassemble the gearcase and first check and correct the rolling torque, gear lash and gear pattern. Do not proceed with assembly until the rolling torque, gear lash and gear pattern are correct.

43. Remove the propeller shaft, lower drive shaft, pinion gear and front driven gears from the gearcase while keeping all shims in the correct locations. Do not remove any pressed bearings from the housing, gears or shafts. Discard the pinion nut.

44. Clean all of the marking compound from the gear teeth and lubricate all bearings and gears with clean gearcase lubricant.

45. Install the front driven gear and inner propeller shaft, pinion gear and drive shaft as described in the assembly procedures. Install a new pinion nut (11, **Figure 108**). Tighten the pinion nut to the specification recorded in Step 11c or the specification in **Table 1**.

46. Install *new* seals (36 and 37, **Figure 108**) into the outer propeller shaft (32) as follows:

a. Place the double lip, single case seal (36, **Figure 108**) onto the wide-shouldered edge of the seal installer (OMC or Volvo Penta part No. 884975-4) with the steel edge (**Figure 144**) facing away from the tool shoulder.

b. Place the propeller shaft on a press with the seal bore facing upward. Using the tool, insert the double lip seal into the bore. Press against the tool until the tool contacts the propeller shaft.

c. Place the single lip seal (37, **Figure 108**) onto the narrow-shouldered side of the seal installer with the spring side facing away from the tool shoulder.

d. Using the tool, insert the single lip seal into the bore. Press against the tool until the tool contacts the propeller shaft. The outer seal should be approximately 0.039 in. (0.991 mm) recessed in the end of the shaft as shown in **Figure 144**.

e. Pack the seal lips and area between the seals with OMC Triple Guard Grease.

47. Insert the seal protector (OMC or Volvo Penta part No. 884976-2) into the outer propeller shaft seals. Carefully guide the outer propeller shaft over the inner propeller shaft. Rotate the drive shaft to mesh the gears while seating the outer propeller shaft against the bearing on the inner propeller shaft. Remove the seal protector.

(142) UNSATISFACTORY

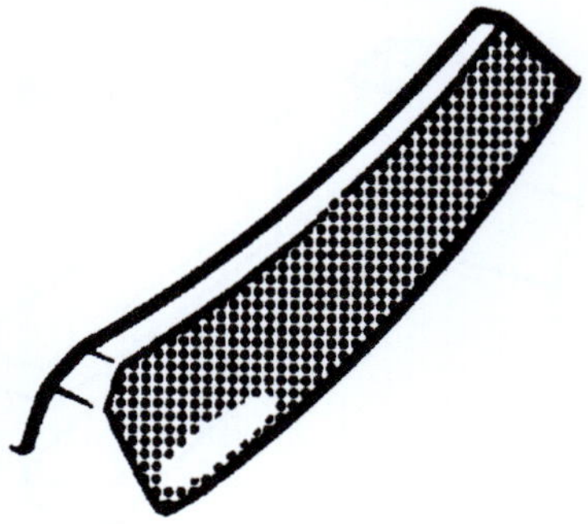

(143) UNSATISFACTORY

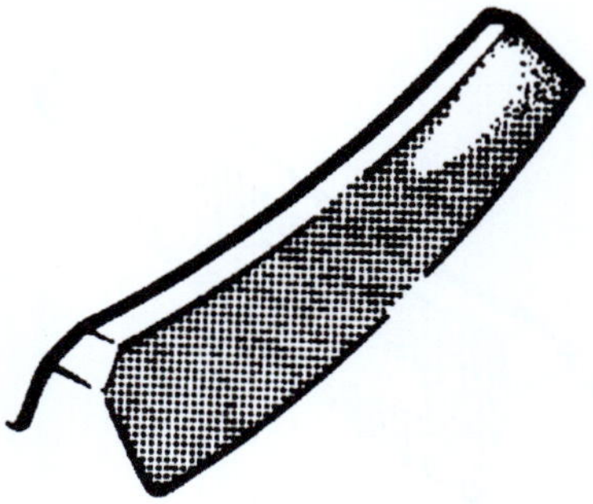

NOTE

Install the new seals into the propeller shaft bearing carrier with the seals ***back-to-back*** *with the spring side of each seal facing away from each other.*

48. Install *new* propeller shaft seals (43, **Figure 108**) into the bearing carrier as follows:

a. Install both seals onto the short end of the installation tool (OMC or Volvo Penta part No. 3855925-8). Make sure the seals are oriented *back-to-back*. The spring side of each seal must face away from each other.

b. Place the bearing carrier onto a press with the seal bore facing upward. Place a block of wood or plastic under the carrier to protect the surfaces.

c. Use the tool to insert the seals into the bore. Press against the tool until it seats against the bearing carrier.

d. Pack the seal lips and area between the seals with OMC Triple Guard Grease.

49. Coat the two *new* O-rings (38. **Figure 108**) with Volvo Penta White Sealing Compound (part No. 1141570-0). Install the O-rings onto the bearing carrier grooves. Coat both bearing carrier flanges with the sealing compound.

50. Install the seal protector (OMC or Volvo Penta part No. 884807-9) in the bearing carrier seals. Carefully guide the carrier over the outer propeller shaft and seat it into the gearcase housing. Make sure the drain/fill plug opening is

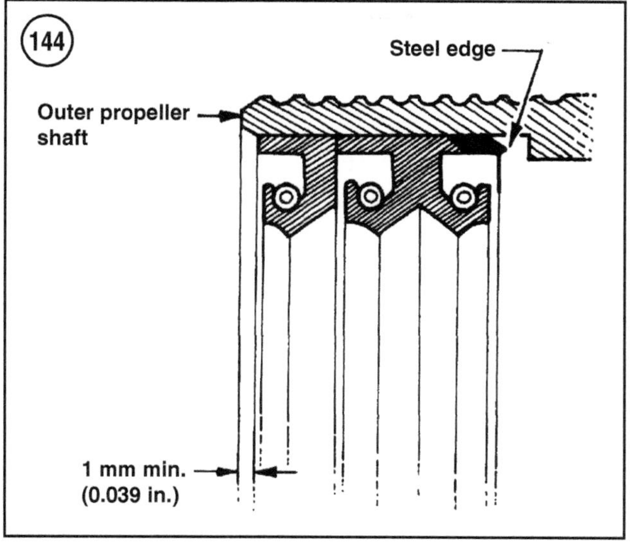

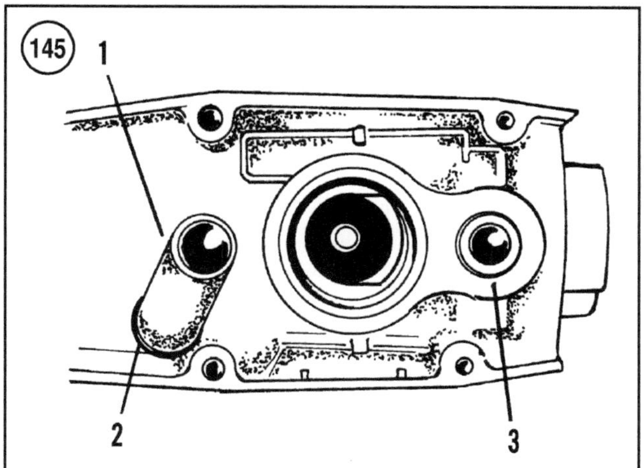

1. Copper water tube
2. O-ring
3. Lubricant screen

positioned toward the lower end (skeg) of the housing. Remove the seal protector.

51. Coat the *new* O-ring (44, **Figure 108**) with Volvo Penta White Sealing Compound (part No. 1141570-0) and install it into the thin groove in the bearing carrier flange.

52. Install the tab washer (45, **Figure 108**) over the propeller shaft. Align the tab with the opening at the top of the carrier bore and then seat the washer against the bearing carrier.

53. Coat the two *new* O-rings (47, **Figure 108**) with Volvo Penta White Sealing Compound (part No. 1141570-0). Install the thicker O-ring into the groove in the outside diameter of the retaining ring (46, **Figure 108**). Install the thinner O-ring into the groove on the carrier mating surface of the retainer ring.

54. Lubricate the retaining ring and thread it into the housing as far as possible by hand.

55. Set the *final* overall rolling torque as follows:

a. Install the splined adapter (OMC or Volvo Penta part No. 3850598-8) onto the drive shaft. Attach an accurate dial torque wrench capable of reading 10-40 in.-lb. (1-5 N•m) to the splined adapter with an appropriate socket. Rotate the drive shaft several revolutions to seat the bearing rollers.

b. Note the torque wrench reading while rotating the drive shaft at approximately 3 rpm. Tighten the retaining ring *slightly* with the removal tool (**Figure 109** [Volvo Penta part No. 3855877-1]) until the rolling torque just reaches the value recorded in Step 30f or the specification in **Table 3**.

56. Install the lubricant drain plug (41, **Figure 108**) with a *new* O-ring (40). If necessary, tighten the retaining ring *slightly* to align the plug with a recess in the retaining ring. Tighten the drain plug to 26 ft.-lb. (35 N•m).

57. Install the lower gearcase as described in this chapter.

58. Pressure and vacuum test the drive unit as described in this chapter.

59. If removed, install the drive unit as described in this chapter.

60. Install the propeller as described in this chapter.

61. Fill the drive unit with lubricant as described in Chapter Three.

Installation

1. Lubricate and then install a new O-ring into the groove surrounding the lubricant screen passage (3, **Figure 145**).

2. Lubricate and then install a new O-ring onto the drive shaft retainer. Seat the O-ring against the housing.

3. Insert a new screen into to the lubricant passage (3, **Figure 145**) with the cone end facing upward. Seat the screen flush with the mating surface.

4. Install the water tube grommet into the upper gearcase with the larger-diameter opening facing outward. Apply a light coat of grease to the inner surfaces to ease installation of the copper water tube.

5. Install the copper water tube into the upper water tube grommet. Rotate the lower end of the water tube to be aligned with the lubricant screen bore and the lower drive shaft.

6A. On earlier Cobra SX drives, verify that the plastic water tube guide (1, **Figure 146**) is secured to the gearcase housing. Apply a light coating of OMC Type M adhesive to the seal in the guide bore.

6B. On later Cobra SX and all DP-S DuoProp drives, insert the plastic guide (1, **Figure 147**) into the gearcase with the protruding edges facing downward. Apply a coat of OMC Gasket Sealing Compound (part No. 508235) to the outer surfaces of the lower grommet (2, **Figure 147**). Purchase the compound from an OMC dealership. Insert the lower grommet into the opening with the tapered end facing downward. Seat the grommet against the guide.

7. Install the intermediate drive shaft coupler onto the lower drive shaft with the grooved end facing upward.

8. To reduce corrosion damage, apply a coat of OMC or Volvo Penta White Sealing Compound (part No. 1141570-0) to the upper drive shaft housing and lower gearcase mating surfaces.
9. With the aid of an assistant, carefully guide the drive shaft into the opening in the upper housing. Be sure to align the copper water tube with the lower grommet while raising the gearcase onto the upper housing. Rotate the propeller shaft(s) to align the shaft splines. The housings will mate when the splines align.
10. Apply OMC Gasket Sealing Compound to the threads of the six fasteners. Install the fasteners for the upper drive shaft housing (**Figure 148**) and tighten them evenly to 22-34 ft.-lb. (30-46 N•m) for 3/8-16 thread fasteners and 32-40 ft.-lb. (43-54 N•m) for 7/16-14 thread fasteners.
11. Pressure and vacuum test the drive unit as described in this chapter.
12. Install the drive unit as described in this chapter.
13. Install the propeller(s) as described in this chapter.
14. Adjust the shift cable as described in Chapter Five.
15. Fill the drive unit with lubricant as described in Chapter Three.
16. Connect the negative battery cable

DRIVE UNIT PRESSURE AND VACUUM TESTS

Use the recommended pressure tester (OMC or Volvo Penta part No. 3810152-3) or a suitable equivalent (such as Stevens S-34 available from tool suppliers) to pressure test the drive unit. Use a vacuum tester (such as Stevens V-34) to vacuum test the drive unit.

Pressure Test

1. Disconnect the negative battery cable.
2. Remove the propeller(s) as described in this chapter.
3. Drain the drive unit lubricant as described in Chapter Three.
4. Make sure the lubricant drain/fill plug and level/check plugs are installed and securely tightened.
5. Remove the dipstick and match the dipstick threads to the pressure tester fitting. Use only the fitting that matches the dipstick threads. Thread the pressure tester fitting fully into the dipstick opening and tighten securely.
6. *Slowly* pump the tester until the pressure reaches 3-5 psi (21-35 kPa). Wait approximately 3 minutes and check the pressure reading. If a pressure drop is evident, spray a soap-and-water solution onto the gearcase mating and all external surfaces to help pinpoint the leak. Maintain pressure with the tester and look for bubbles forming at the point of leakage.
   a. If the drive unit is removed for service, observe the pressure gauge while rotating the universal drive shaft and propeller shaft(s) and shifting the unit out of gear. If a pressure drop is evident, spray a soap-and-water solution onto the universal drive shaft-, propeller shaft(s)- and shift arm-to-seal mating surfaces. Maintain pressure and look for bubbles forming to pinpoint the leak.
   b. If the drive unit is mounted onto the intermediate housing/transom shield, shift the drive unit into NEUTRAL. Then observe the pressure gauge while rotating the propeller shaft(s). If a pressure drop is evident, spray a soap-and-water solution onto the propeller shaft(s)- and shift arm-to-seal mating surfaces. Maintain pressure with the tester and look for bubbles forming to pinpoint the leak.

146 GEARCASE WATER TUBE ADAPTER (EARLIER COBRA SX-S DRIVE UNITS)

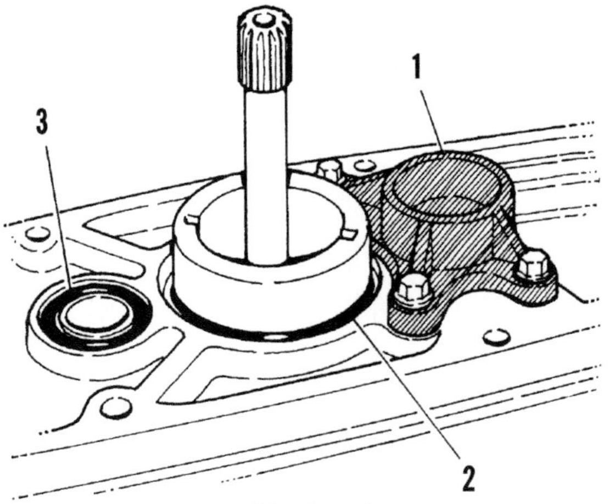

1. Plastic water tube guide
2. O-ring
3. O-ring

7. *Slowly* pump the tester until the pressure reaches 7 psi (48 kPa). Wait approximately 3 minutes and check the pressure reading. If pressure drop is evident, spray a soap-and-water solution onto the gearcase mating and all external surfaces to help pinpoint the leak. Maintain pressure with the tester and look for bubbles forming at the point of leakage.
   a. If the drive unit is removed for service, observe the pressure gauge while rotating the universal drive shaft and propeller shaft(s) and shifting the unit into and out of gear. If a pressure drop is evident, spray a soap-and-water solution onto the universal drive shaft-, propeller shaft(s)- and shift arm-to-seal mating surfaces. Maintain pressure and look for bubbles forming to pinpoint the leak.
   b. If the drive unit is mounted onto the intermediate housing/transom shield, shift the drive unit into NEUTRAL. Then observe the pressure gauge while rotating the propeller shaft(s). If a pressure drop is evident, spray a soap-and-water solution onto the

(147) **GEARCASE WATER TUBE AND GROMMETS (LATER COBRA SX AND DP-S DUOPROP DRIVE UNITS)**

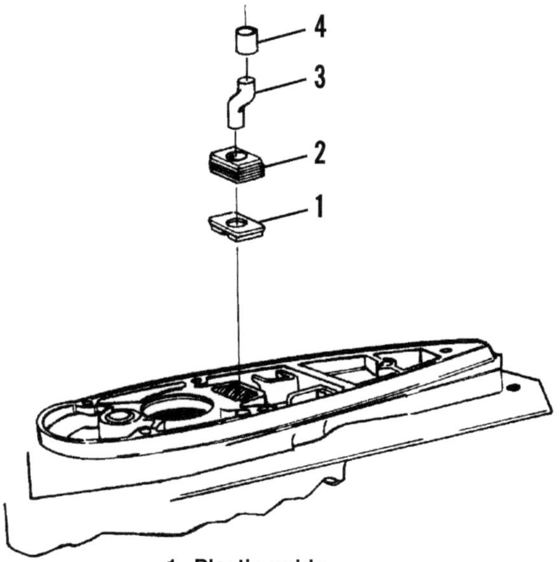

1. Plastic guide
2. Lower grommet
3. Copper water tube
4. Upper drive shaft housing grommet

(148)

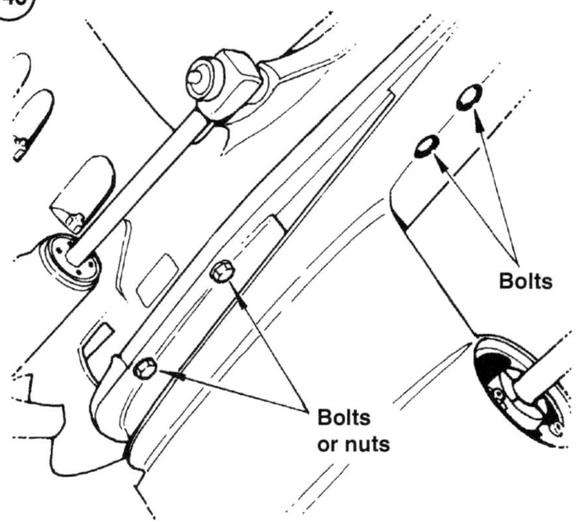

propeller shaft(s)- and shift arm-to-seal mating surfaces. Maintain pressure with the tester and look for bubbles forming to pinpoint the leak.

8. *Slowly* pump the tester until the pressure reaches 16-18 psi (110-124 kPa). Wait approximately 3 minutes and check the pressure reading. If pressure drops more than 1 psi (7 kPa) during the 3-minute period, spray a soap-and-water solution onto the gearcase mating and all external surfaces to help pinpoint the leak. Maintain pressure with the tester and look for bubbles forming at the point of leakage.
   a. If the drive unit is removed for service, observe the pressure gauge while rotating the universal drive shaft and propeller shaft(s) and shifting the unit into and out of gear. If a pressure drop is evident, spray a soap-and-water solution on the universal drive shaft-, propeller shaft(s)- and shift arm-to-seal mating surfaces. Maintain pressure and look for bubbles forming to pinpoint the leak.
   b. If the drive unit is mounted onto the intermediate housing/transom shield, shift the drive unit into NEUTRAL. Observe the pressure gauge while rotating the propeller shaft(s). If a pressure drop is evident, spray a soap-and-water solution onto the propeller shaft(s) and seal mating surfaces. Maintain pressure with the tester and look for bubbles forming to pinpoint the leak.
9. Remove the pressure tester.
10. If the point of leakage is verified, disassemble the drive unit and replace the faulty seal, gasket or O-ring to correct the leak.
11. If the point of leakage cannot be verified, disassemble the drive unit and replace all seals and O-rings as described in this chapter. Repeat the pressure test after the repair.
12. Refill the lubricant as described in Chapter Three. Install the drive unit and propellers as described in this chapter. Connect the battery cable. 10

Vacuum Test

1. Disconnect the negative battery cable.
2. Remove the propeller(s) as described in this chapter.
3. Drain the drive unit lubricant as described in Chapter Three.
4. Make sure the lubricant drain/fill plug and level/check plugs are installed and securely tightened.
5. Remove the dipstick and match the dipstick threads to the pressure tester fitting. Use only the fitting that matches the dipstick threads. Thread the vacuum tester fitting fully into the dipstick opening and tighten securely.
6. Pump the tester to achieve a vacuum of 3-5 in. Hg (10-17 kPa). Wait 3 minutes and check the vacuum reading. If a vacuum drop is evident, maintain the vacuum with the tester and apply gear lubricant to the propeller shaft(s)-to-seal mating surfaces and all external mating surfaces. The lubricant will be drawn in at the point of leakage.
   a. If the drive unit is removed for service, observe the vacuum gauge while rotating the universal drive shaft and propeller shaft(s) and shifting the unit into and out of gear. If a vacuum drop is evident, maintain the vacuum with the tester and apply gear lubricant to the propeller shaft-, shift arm- and universal joint-to-seal mating surfaces. The lubricant will be drawn in at the point of leakage.

b. If the drive unit is mounted onto intermediate housing/transom shield, shift the drive unit into NEUTRAL. Observe the vacuum gauge while rotating the propeller shaft(s) and shifting the unit into and out of gear. If a vacuum drop is evident, maintain vacuum with the tester and apply gear lubricant to propeller shaft(s)- and shift arm-to-seal mating surfaces. The lubricant will be drawn in at the point of leakage.

7. Pump the tester to achieve a vacuum of 14-16 in. Hg (47-54 kPa). Wait 3 minutes and check the vacuum reading. If the vacuum drops more than 1 in. Hg (3.4 kPa) during the 3-minute period, maintain the vacuum with the tester and apply gear lubricant to the propeller shaft(s)-to-seal mating surfaces and all external mating surfaces. The lubricant will be drawn in at the point of leakage.

a. If the drive unit is removed for service, observe the vacuum gauge while rotating the universal drive shaft, propeller shaft(s) and shifting the unit into and out of gear. If a vacuum drop is evident, maintain the vacuum with the tester and apply gear lubricant to the propeller shaft-, shift arm- and universal joint-to-seal mating surfaces. The lubricant will be drawn in at the point of leakage.

b. If the drive unit is mounted onto the intermediate housing/transom shield, shift the drive unit into NEUTRAL. Observe the vacuum gauge while rotating the propeller shaft(s) and shifting the unit into and out of gear. If a vacuum drop is evident, maintain vacuum with the tester and apply gear lubricant to the propeller shaft(s)- and shift arm-to-seal mating surfaces. The lubricant will be drawn in at the point of leakage.

8. Remove the vacuum tester.

9. If the point of leakage is verified, disassemble the drive unit and replace the faulty seal, gasket or O-ring to correct the leak.

10. If the point of leakage cannot be verified, disassemble the drive unit and replace all seals and O-rings as described in this chapter. Repeat the vacuum test after the repair.

11. Refill the lubricant as described in Chapter Three. Install the drive unit and propellers as described in this chapter. Connect the battery cable.

Table 1 DRIVE UNIT TORQUE SPECIFICATIONS

Fastener	in.-lb.	ft.-lb.	N•m
Drive unit to pivot housing locknuts			
First step	–	25	34
Final step	–	50	68
Lower gearcase housing			
Bearing carrier anode screw			
Cobra SX (single-propeller) drive	50-84	–	6-10
Bearing carrier anode bracket screws			
Cobra SX (single-propeller) drive	144-168	–	16-19
Bearing carrier setscrew			
Cobra SX (single-propeller) drive	42-60	–	5-7
Bearing carrier retainer block screw			
Cobra SX (single-propeller) drive	–	20-25	27-34
Gearcase drain plug			
DP-S DuoProp (twin-propeller)	–	26	35
Pinion gear nut			
Cobra SX (single-propeller)	–	150-160	203-217
DP-S DuoProp	–	72-87	98-118
Propeller shaft bearing carrier			
Cobra SX (single-propeller) drive	–	200-225	271-305
Trim tab bolt			
Cobra SX (single-propeller) drive	168-195	–	19-22
Upper drive shaft housing			
3/8–16 fasteners	–	22-34	30-46
7/16–14 fasteners	–	32-40	43-54
Water tube guide			
Cobra SX-S drives	120-144	–	14-16
Propeller nut			
Cobra SX drive unit	–	70-80	95-108
DP-S DuoProp drive unit			
Front propeller nut	–	44	60
DP-S DuoProp drive unit			
Rear propeller nut	–	52	71
Trim cylinder pivot rod nut	124-142	–	14-16

(continued)

Table 1 DRIVE UNIT TORQUE SPECIFICATIONS (continued)

Fastener	in.-lb.	ft.-lb.	N•m
Upper drive shaft housing			
Drive gear bearing carrier to shift housing bolts	142-168	–	16-19
Lower driven gear retainer nut	–	145-165	197-224
Oil level dipstick	48-72	–	5-8
Shift housing	60-84	–	7-10
Shift shoe stop bolt	124-142	–	14-16
Top cover bolts	195-212	–	22-24
Upper output shaft nut	–	96-110	130-149
Water cover	60-84	–	7-10

Table 2 BEARING PRELOAD (ROLLING TORQUE) SPECIFICATIONS

Model	Specification
Drive gear bearing carrier (tapered roller bearings)	
SX and DP-S drives	9-14 in.-lb. (1.0-1.6 N•m)
Drive shaft rolling torque	
SX and DP-S drives	2-4 in.-lb. (0.23-0.45 N•m)*
Gearcase overall rolling torque	
SX drives	
1.43:1, 1.51:1 and 1.60:1 gear ratio	14-23 in.-lb. (1.58-2.60 N•m)*
1.66:1, 1.85:1, 1.97:1 and 2.18:1 gear ratio	11-18 in.-lb. (1.24-2.03 N•m)*
DP-S models	
1.68:1 and 1:78:1 gear ratio	18-27 in.-lb. (2-3 N•m)*
1.95:1 and 2.30:1 gear ratio	27-35 in.-lb. (3-4 N•m)*

*Measured at the drive shaft while rotating at 3 rpm.

10

Table 3 GEAR LASH SPECIFICATIONS

Model	Specification
DP-S DuoProp (twin-propeller)	
Upper drive shaft housing output gears[3]	0.006-0.010 in. (0.152-0.254 mm)
Lower gearcase driven gears[1]	
Inner propeller shaft gear	0.006-0.012 in. (0.152-0.305 mm)
Outer propeller shaft gear	0.006-0.014 in. (0.152-0.356 mm)
Upper drive shaft housing	
Output shaft end play	0.002-0.010 in. (0.051-0.254 mm)
Thrust plate measurement	5.007-5.036 in. (12.712-12.791 cm)
SX (single-propeller)	
Upper drive shaft housing output gears[3]	0.006-0.010 in. (0.152-0.254 mm)
Lower gearcase driven gear[2]	0.005-0.012 in. (0.127-0.305 mm)
Upper drive shaft housing	
Output shaft end play	0.002-0.010 in. (0.051-0.254 mm)
Thrust plate measurement	5.007-5.036 in. (12.712-12.791 cm)

1. Measured at the propeller shaft(s).
2. Measured at the lower drive shaft.
3. Measured on the driven gear tooth.

Table 4 GEAR LASH AND PATTERN CORRECTION GUIDE (DP-S DUOPROP [TWIN-PROPELLER] DRIVES)

Pattern	Pinion gear	Front driven gear	Inner propeller shaft bearing or rear driven gear
Both gears lash below specification			
Changes required[1]:	add	subtract	add[2]

(continued)

Table 4 GEAR LASH AND PATTERN CORRECTION GUIDE (DP-S DUOPROP [TWIN-PROPELLER] DRIVES) (continued)

Pattern	Pinion gear	Front driven gear	Inner propeller shaft bearing or rear driven gear
Both gears lash above specification			
Changes required[1]:	subtract	add	subtract[2]
Front gear lash below specification			
Changes required[3]:	no change	subtract	add
Front gear lash above specification			
Changes required[3]:	no change	add	subtract
Rear gear lash below specification			
Changes required:	no change	no change	add
Rear gear lash above specification			
Changes required:	no change	no change	subtract
Pattern too close to outer diameter			
Changes required[3]:	add	add	subtract[2]
Pattern too close to inner diameter			
Changes required[3]:	subtract	subtract	add[2]

1. Changed one-half of the total desired change at the pinion gear. Change the other one-half by changing the same quantity (thickness) at the pinion gear at each driven gear.
2. Since the position of this bearing is dependent on the front driven gear, any change to the front driven gear requires the exact opposite change at this bearing to maintain position. Make any required shim changes in addition the shims needed to maintain position.
3. Change equal amounts (thickness) at each location, except where noted.

Table 5 DRIVE UNIT IDENTIFICATION AND AVAILABLE GEAR RATIOS

Drive	Drive unit description	Available gear ratios (overall drive ratio)
Cobra SX	Later design, single propeller	1.43:1
		1.51:1
		1.60:1
		1.66:1
		1.79:1
		1.89:1
		1.97:1
		2.18:1
DP-S DuoProp	Later design, twin propeller	1.68:1
		1.78:1
		1.95:1
		2.32:1

Chapter Eleven

Transom Bracket

Always inspect the universal joint bellows, exhaust bellows and water hose anytime the transom bracket requires service. Removal, inspection and installation of the bellows and hose are described in the appropriate removal and installation instructions in this chapter.

Inspect the anode (where applicable) as described in the removal instructions anytime the transom bracket assembly requires service.

Table 1 lists torque specifications for most intermediate housing and transom bracket fasteners. Use the general torque recommendations in Chapter One for fasteners not listed in **Table 1**.

Common repairs to the transom bracket, including the gimbal bearing, gimbal bearing seal, universal joint bellows and exhaust bellows, and water hose replacement are described under the appropriate sections. Replacement of the gimbal ring, tilt bushing, swivel bushing and steering arm are covered under *Gimbal Ring*. Removal and installation of the transom bracket are described under *Inner Transom Plate and Bracket Housing* in this chapter. Service for the trim hoses, trim cylinders and trim position senders are described in Chapter Twelve.

PIVOT HOUSING

The pivot housing can be quickly removed to allow easy access to the universal joint bellows, exhaust bellows, water hose, trim hoses, trim system manifold gimbal bearing grease tube and fitting. If desired, the gimbal bearing and gimbal bearing seal can be replaced without removing the pivot housing.

Removal and Inspection

Refer to **Figure 1** for this procedure.

1. Remove the drive unit as described in Chapter Ten.
2. Pull the shift cable out of the pivot housing.
3. Carefully lift the lip of the universal joint bellows out of the pivot housing bore. Push the bellows through the bore and free from the pivot housing.
4. Remove the snap ring (**Figure 2**) from the aft end of the exhaust bellows. Lift the lip of the bellows out of the groove in the pivot housing and then push the bellows through the bore until free from the pivot housing.

1

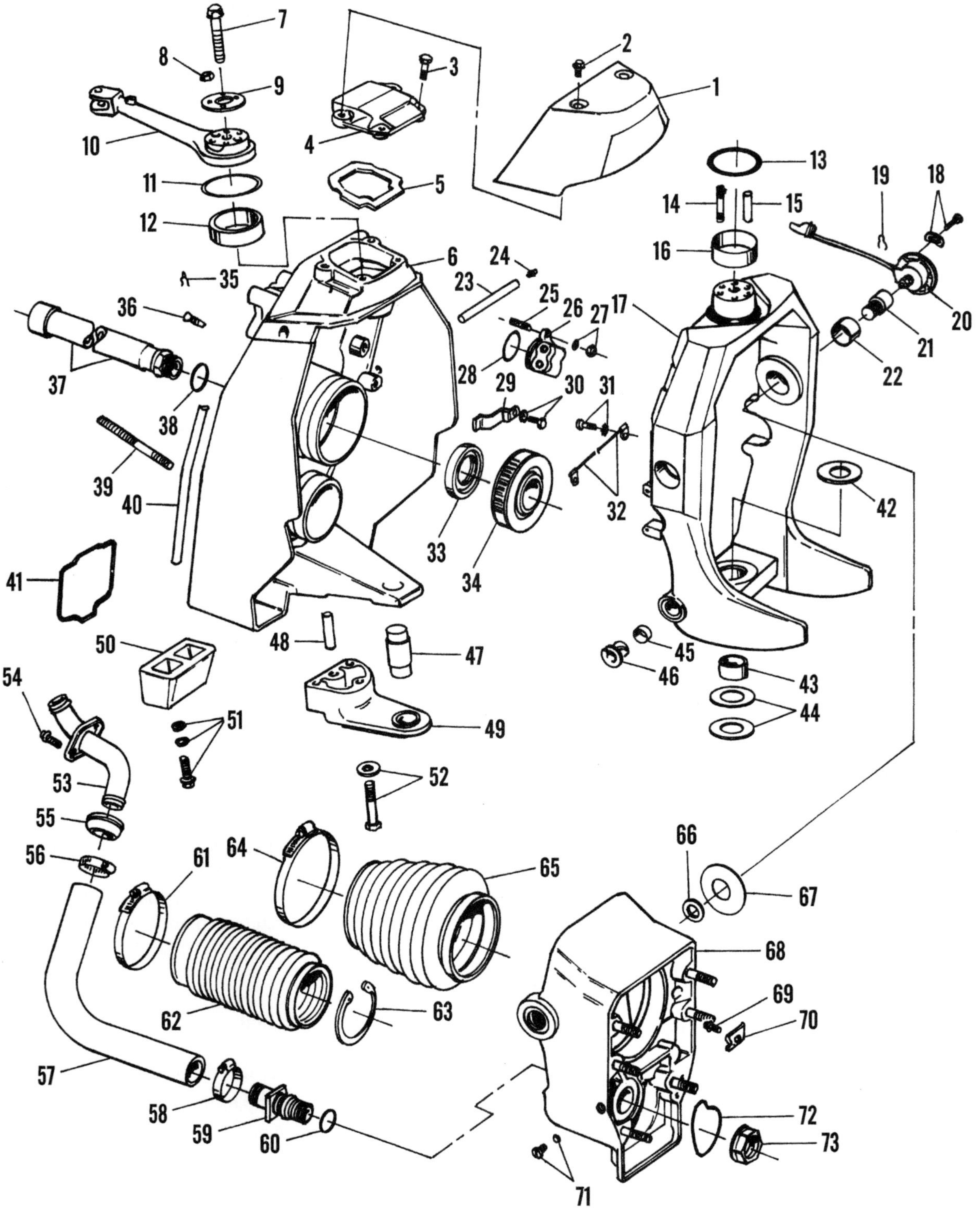
1
2
3
4
5
6
7
8
9
10
11
12
13
14
15
16
17
18
19
20
21
22
23
24
25
26
27
28
29
30
31
32
33
34
35
36
37
38
39
40
41
42
43
44
45
46
47
48
49
50
51
52
53
54
55
56
57
58
59
60
61
62
63
64
65
66
67
68
69
70
71
72
73

TRANSOM BRACKET COMPONENTS (COBRA SX AND DP-S MODELS)

1. Plastic cover
2. Screw
3. Screw
4. Inner cover (metal)
5. Foam gasket
6. Gimbal housing
7. Steering arm bolt
8. Elastic locknut
9. Washer plate
10. Steering arm
11. Thrust washer
12. Upper steering bushing
13. O-ring
14. Stud
15. Dowel pin
16. Bushing
17. Gimbal ring
18. Screw and washer*
19. Retaining clip*
20. Trim sending unit*
21. Hinge pin
22. Bushing
23. Grease tube
24. Grease fitting
25. Stud
26. Trim system manifold
27. Elastic locknut and washer
28. O-ring
29. Retainer
30. Screw and washer
31. Screw and washer
32. Ground strap
33. Seal
34. Gimbal bearing
35. Retaining clip
36. Plug
37. Shift cable sleeve
38. O-ring
39. Stud
40. Transom seal
41. Exhaust pipe seal
42. Thrust washer
43. Steering bushing
44. Shim washers
45. Rubber grommet
46. Plastic bushing
47. Lower steering pin
48. Dowel pin
49. Lower support bracket
50. Anode
51. Screw, lockwasher and flat washer
52. Bolt and washer
53. Water tube and retainer
54. Screw
55. Water tube grommet
56. Clamp
57. Water hose
58. Clamp
59. Water inlet nipple
60. O-ring
61. Clamp
62. Exhaust bellows
63. Snap ring
64. Clamp
65. Universal joint bellows
66. Thrust washer (metal)
67. Thrust washer (plastic)
68. Pivot housing
69. Screw
70. Retainer (shift cable)
71. Screw and O-ring*
72. Seal
73. Plastic nut (water nipple)

*Not used on all models.

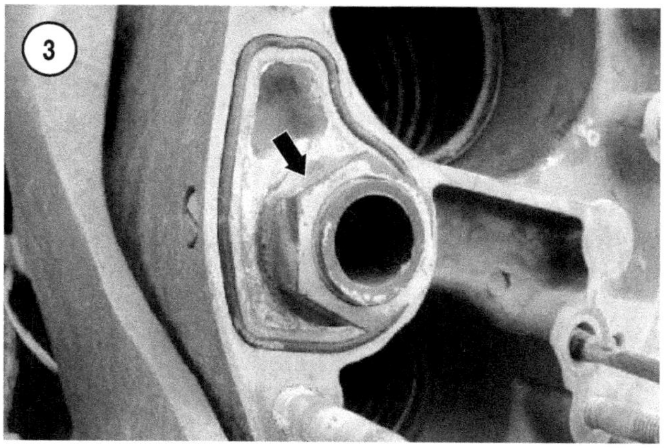

5. Unthread and remove the plastic nut (**Figure 3**) from the rear of the pivot housing. Push the nipple (59, **Figure 1**) through the bore until free from the pivot housing (68). Remove the O-ring (60, **Figure 1**) from the nipple or pivot housing. Discard the O-ring.

6. Remove the screw and washer (31, **Figure 1**) and disconnect the ground wire (32) from the upper boss on the port side of the gimbal ring (17).

7A. On earlier Cobra SX models (1994), remove the two screws and then pull the sender (**Figure 4**) off the starboard side of the gimbal ring. Carefully guide the sender lever away from the pivot housing and gimbal ring while remov-

11

ing the sender. Do not disconnect the sender leads for this procedure.

7B. On later Cobra SX models (1995-2000), remove the two screws and washers (18, **Figure 1**) and then remove the trim position sender (**Figure 5**) from the starboard side of the gimbal ring. Do not disconnect the sender leads for this procedure.

8. Insert a 1/2 in. hex socket into the starboard side hinge pin (**Figure 6**, typical). Place a block of wood between the lower end of the pivot housing and the gimbal ring to prevent contact while loosening the pin. Use a breaker bar to loosen the hinge pin. Do not remove the hinge pin at this time. Block the pivot housing with a wooden block and loosen the port side hinge pin using the same procedure.

9. Support the housing while removing both hinge pins (**Figure 7**, typical). Pull the pivot housing off the gimbal ring.

10. Remove the seal (**Figure 8**) from the pivot housing. Discard the seal. Remove all old adhesive from the seal groove.

11. Inspect the self-adhesive plastic thrust washers (67, **Figure 1**) on each side of the pivot housing for cracks, excessive wear, missing sections or other damage. Replace the plastic washers if these or other defects are evident or if they fit loosely onto the housing. To remove the washer, simply pry them loose with a suitable instrument. Inspect the smaller metal thrust washers (66, **Figure 1**) that are held into the bore by the plastic washers. Replace the washers if they are excessively worn or damaged.

12. Thoroughly clean the pivot housing with solvent and dry with compressed air.

13. Inspect the drive mounting studs on the pivot housing for damaged threads, bending or other damage. If damaged, replace the stud(s) as follows:

   a. Measure the distance from the end of the stud to the mating surface. Record the distance. The replacement stud must be installed to the original depth. Otherwise, the mounting nut might bottom out before actually securing the drive unit.
   b. Remove the defective stud with a suitable stud removal tool.
   c. Clean all old sealant from the threaded opening.
   d. Apply OMC Gasket Sealing Compound (OMC part No. 508235) to the threads and then thread the *new* stud into the pivot housing.
   e. Stopping frequently for measurement, thread the new stud into the housing until it reaches the depth recorded prior to removal.

14. Inspect the ground wire (**Figure 9**) for a loose fit to the clamp, broken wire sections or other defects and replace as needed.

15. Inspect the hinge pin threads in the pivot housing. The threads are usually Heli-Coil inserts but may be a threadlocking version of the Heli-Coil inserts. The threadlocking version is slightly distorted in an octagonal

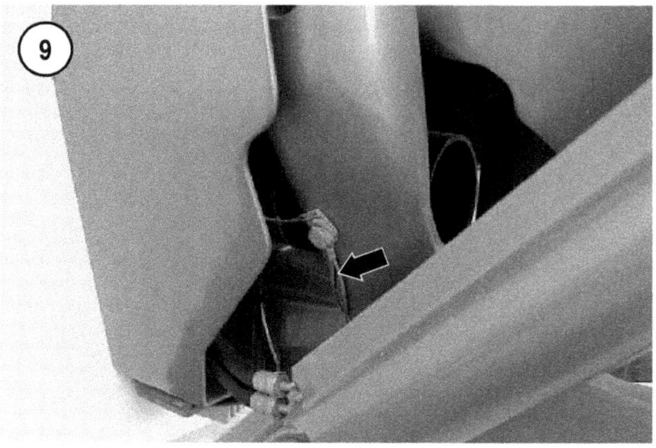

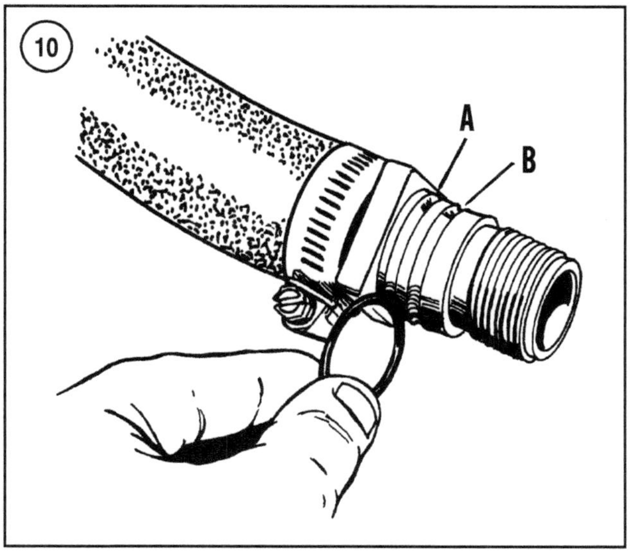

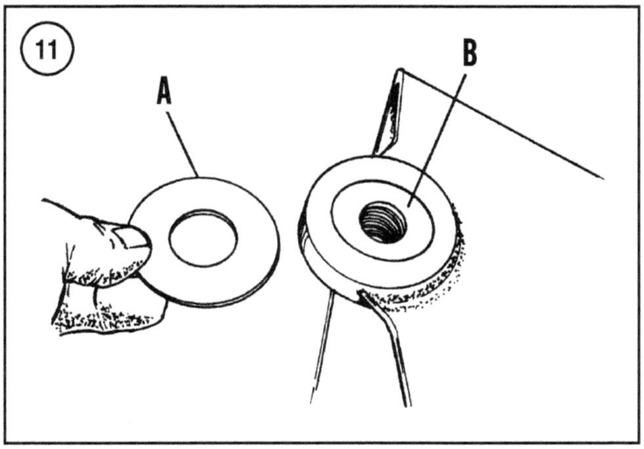

fashion. If either type of threads is damaged or loose, replace the insert as follows:

a. Remove the old insert by gripping the outermost coil with a pair of needlenose locking pliers. Turn the insert counterclockwise until free from the pivot housing.
b. Purchase the replacement insert from an OMC or Volvo Penta dealership. Purchase the installation kit from an OMC dealership.
c. Run the tap from the installation kit through the bore to clean and recondition the pivot housing threads.
d. Fit the replacement insert onto the installation tool. Coat the outer threads of the new insert with Loctite 271 threadlocking adhesive.
e. Using the installation tool, thread the insert into the pivot housing as described in the instructions included with the installation kit.

16. Inspect the hinge pin bushings in the gimbal ring for a missing bushing, broken bushings or excessive wear. If defective, replace the bushing(s) as described in this chapter. Refer to *Gimbal Ring, Removal and inspection*.

Installation

1. Install a *new* O-ring into the water nipple groove (A, **Figure 10**) located nearest the nipple shoulder. *Do not* install an O-ring into the water drain groove (B, **Figure 10**).
2. If removed, install the smaller metal thrust washer (B, **Figure 11**) into the recess in the pivot housing. Place the self-adhesive plastic thrust washer (A, **Figure 11**) over the metal washer with the adhesive side facing downward. Center the plastic washer over the threaded opening and then press the washer firmly in place.
3. Apply a *light* coat of OMC Type M adhesive (OMC part No. 318535) to the seal (72, **Figure 1**) and then install it into the pivot housing groove (**Figure 8**). Hold the seal in position until the adhesive cures.
4. Guide the pivot housing into the gimbal ring (17, **Figure 1**). Position the universal joint and exhaust bellows at the entrances to the pivot housing bores.
5. Coat the threads of the hinge pins (21, **Figure 1**) with an antiseize compound. Purchase the compound at an automotive parts store.
6. Align the openings and then insert the hinge pins into the starboard and then port side openings (**Figure 6**, typical). Insert a 1/2 in. hex socket into the starboard hinge pin (**Figure 6**, typical) and tighten by hand at this time. Using the same procedure, tighten the port side hinge pin by hand.
7. Move the pivot housing to the full up and down positions to check for binding and check for excessive side play. If binding or excessive side play is evident, remove the pivot housing and check for improperly installed or worn hinge pins, hinge pin bushings and thrust washers.
8. Block the pivot housing to prevent excessive upward movement and then use a torque wrench and the 1/2 in. hex socket to tighten the starboard side hinge pin to 105-120 ft.-lb. (142-163 N•m). Block the pivot housing to prevent excessive downward movement and then use the same procedure to tighten the port side hinge pin to the same specification.
9. Pull the water tube nipple (**Figure 10**) through the pivot housing and seat against the inner surface of the pivot housing. Hold the nipple in firm contact with the pivot housing while threading the plastic nut (**Figure 3**) onto the threaded end. Tighten the plastic nut to 97-124 in.-lb. (11-14 N•m).

If correctly seated, several threads of the nipple will be visible through the nut.

10. Carefully pull the aft end of the universal joint bellows through the pivot housing bore. Fit the lip on the bellows over the lip in the pivot housing bore (**Figure 12**). If necessary, tilt the pivot housing and reach up behind the housing to help feed the bellows through the bore. Inspect the bellows for flat spots on the inner diameter of the lip. Flat spots indicate the bellows lip is not fully engaging the pivot housing lip. Reposition the bellows as needed to make sure it fully seats in the opening.

11. Pull the aft end of the exhaust bellows through the pivot housing bore. Fit the groove in the bellows over the lip in the pivot housing. The inner diameter of the bellows must be perfectly round at the pivot housing bore. Otherwise, the bellows is not fully seated in the pivot housing. Reposition the bellows as needed.

12. Install the snap ring (**Figure 2**) into the bellows groove with the gap at the top of the opening.

13. Connect the pivot housing ground wire (32, **Figure 1**) to the boss at the upper port side of the gimbal ring (17).

14. Install the trim position sender (**Figure 4** or **Figure 5**) onto the starboard side of the gimbal ring as described in Chapter Twelve.

15. Install the drive unit as described in Chapter Ten.

16. Adjust the trim position sender as described in Chapter Five.

BELLOWS AND WATER HOSE

Inspect the universal joint bellows, exhaust bellows and water hose for kinking; tears; excessive wear; weathering; sponginess; and brittle, cracked or burned surfaces. Replace the bellows and/or water hose if these or other defects are evident. Otherwise, leave the bellows and hose intact unless the gimbal housing must be replaced.

There are no orientation marks or specified up or down installed position for the universal joint bellows; however, the exhaust bellows must be installed with the relief slots facing straight down. All clamps must be installed as specified to prevent interference with other clamps, bellows and the water hose.

Removal

1. Remove the pivot housing as described in this chapter.
2. Loosen the hose clamps (64 and 61, **Figure 1**) that secure the universal joint (65) and exhaust bellows (62) to the gimbal housing (6). Pull the bellows off the transom bracket flange. If a tool must be used to separate the bellows from the flange, work carefully to avoid damaging the flange or the bellows (if it must be reused).
3. If the water hose must be replaced, loosen the clamp (**Figure 13**) that secures the hose to the water tube. Carefully twist the hose off the tube. If necessary, split the end of the hose over the tube for removal. Do not cut the tube.

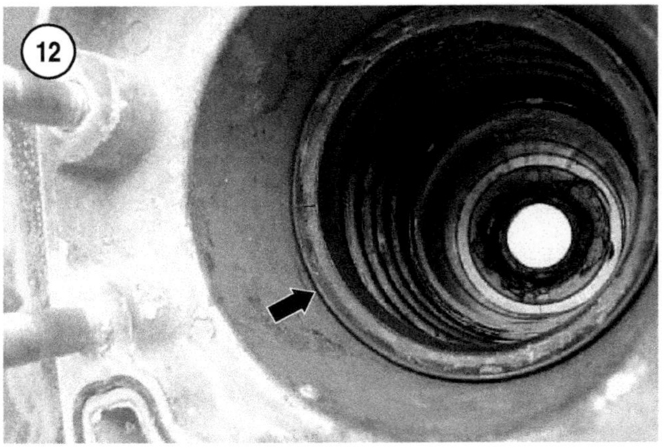

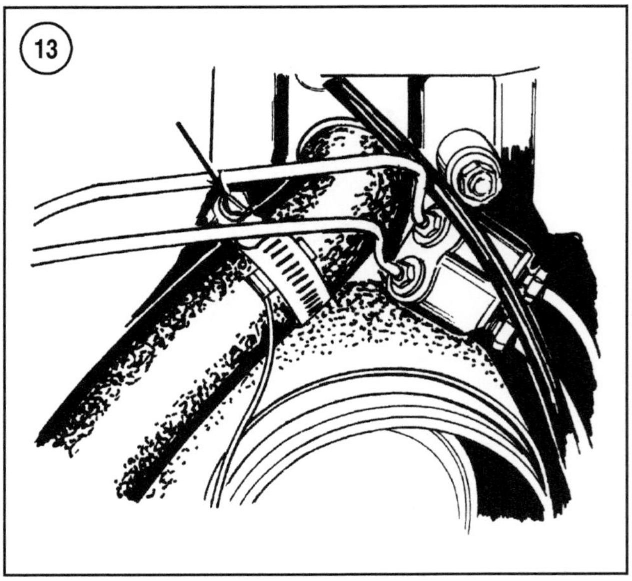

Loosen the clamp (58, **Figure 1**) and then twist the water nipple (59) off the water hose.

4. If the water tube (53, **Figure 1**) is damaged or must be removed to replace the gimbal housing (6), remove the tube as follows.

   a. Working from inside the boat, remove the two screws (54, **Figure 1**) from the water tube plate.

   b. Pull the water tube and plate out of the gimbal housing. The water tube is curved and will require a fair amount of manipulation for removal.

   c. Remove the grommet (55, **Figure 1**) from the water tube plate or gimbal housing opening. Discard the grommet.

   d. Thoroughly clean the water tube and grommet bore in the gimbal housing. Make sure to remove all remnants of the old grommet. Any material in the bore can prevent the grommet from seating properly and allow water to leak into the boat.

5. Clean all old sealant from the water tube opening (outside the boat) and both bellows mounting flanges.

14

15

GROUND WIRE ROUTING

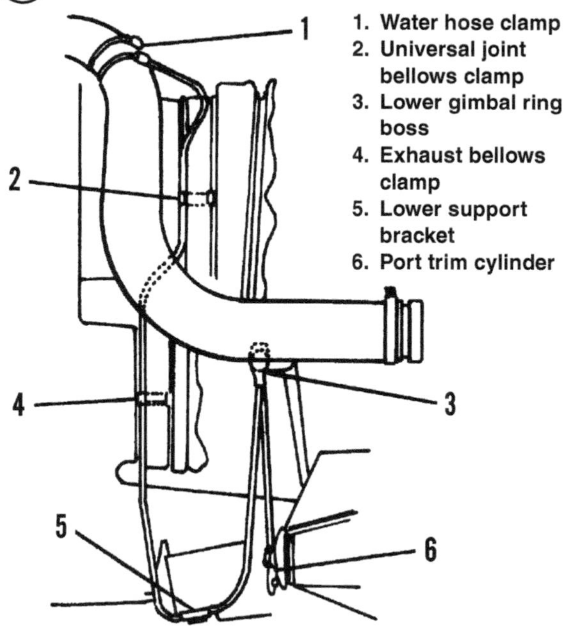

1. Water hose clamp
2. Universal joint bellows clamp
3. Lower gimbal ring boss
4. Exhaust bellows clamp
5. Lower support bracket
6. Port trim cylinder

Installation

1. If removed, install the water tube and *new* grommet as follows:

   a. Install the new grommet (55, **Figure 1**) over the hose end of the tube with the larger-diameter side facing the plate. Seat the grommet against the plate.

   b. Working from inside the boat, guide the water tube and grommet into the gimbal housing opening. Seat the plate and grommet against the housing.

   c. Rotate the plate to align the screw openings. Apply OMC Gasket Sealing Compound (OMC part No. 508235) to the threads and then install the two screws (54, **Figure 1**). Tighten the screws evenly to 124-142 in.-lb. (14-16 N•m).

2. If the universal joint bellows (65, **Figure 1**) was removed, install as follows:

   a. Apply a very light coat of OMC Gasket Sealing Compound to the bellows mounting flange on the gimbal housing.

   b. Install the universal joint bellows over the flange and engage the molded rib in the bellows onto the case groove in the flange.

   c. Install the clamp (64, **Figure 1**) over the bellows and fit it onto the clamping surface provided on the bellows. Rotate the clamp to position the clamp screw facing down and aligned as shown in A, **Figure 14**.

   d. Insert the ground wire tab under the clamp as shown in **Figure 15**.

   e. Hold the clamp in position and then securely tighten the clamp.

3. If the exhaust bellows (62, **Figure 1**) was removed, install as follows:

   a. Apply a light coat of OMC Gasket Sealing Compound to the bellows mounting flange on the gimbal housing.

   b. Install the exhaust bellows over the flange and engage the molded rib in the bellows onto the case groove in the flange. Rotate the bellows on the flange to position the exhaust relief slots facing downward.

   c. Install the clamp (61, **Figure 1**) over the bellows and fit it onto the clamping surface provided on the bellows. Rotate the clamp to position the clamp screw facing down and aligned as shown in B, **Figure 14**.

   d. Insert the ground wire tab under the clamp as shown in **Figure 15**.

   e. Hold the clamp in position and then securely tighten the clamp.

4. If the water hose was removed, install as follows:

   a. Install the clamp (56, **Figure 1**) over the gimbal housing end of the water hose (57).

   b. Push the water hose fully onto the water tube (53, **Figure 1**).

   c. Rotate the clamp to position the screw between the trim hose fittings as shown in **Figure 13**.

   d. Insert the ground wire tab under the clamp as shown in **Figure 15**.

   e. Hold the clamp in position and then securely tighten the clamp.

5. If removed, install the water inlet nipple (59, **Figure 1**) as follows:

   a. Fit the hose clamp (58, **Figure 1**) over the aft end of the water hose.

   b. Push the hose completely over the nipple.

   c. Rotate the hose clamp to position the screw on the bottom side of the nipple with the screw facing the port side (when installed).

11

d. Hold the clamp in position and then securely tighten the clamp.

6. Install the pivot housing as described in this chapter.

7. Have an assistant pivot the steering in the full port and starboard directions while checking for interference between the gimbal ring and universal joint bellows clamp screw. Any contact can cause the clamp to loosen during operation and cause a serious water leak. Reposition the clamp as needed to prevent any contact with the gimbal ring and clamp.

GIMBAL BEARING AND SEAL

Inspect the gimbal bearing anytime the drive unit is removed and if gimbal bearing failure is suspected of causing unusual noise during engine operation.

With the drive removed, rotate the inner portion of the bearing to check for binding or roughness. Check for corrosion of the exposed surfaces and other indications of water damage. Replace the bearing if binding or roughness and water damage is present. Water damage is usually the result of a hole in the universal joint bellows, improper drive unit installation or a high water level in the bilge compartment. Check for these conditions and correct the cause before installing the new gimbal bearing.

A suitable three-jaw slide hammer (such as OTC part No. OEM-4184) is required to remove the gimbal bearing. Use the installer available from an OMC dealership (OMC part No. 3850038) to install the gimbal bearing. The tool available from a Volvo Penta dealership (Volvo Penta part No. 3854357-5) can be used to install the bearing provided that it is used properly. If using the Volvo Penta tool, fit the bearing onto the tool before installation to check for adequate clearance between the inner portion of the bearing and the tool. One side of the bearing will provide enough clearance, whereas the other side might contact the tool. The bearing can be installed with either side facing out. Install the bearing with the side that provides clearance to the tool facing out.

CAUTION
Do not remove the gimbal bearing unless it or the gimbal housing must be replaced. The removal process damages the bearing.

Removal

1. Remove the drive unit as described in Chapter Ten.
2. Insert the three-jaw puller into the bore of the gimbal bearing (34, **Figure 1**). Place a rubber band over the jaws to hold them in the retracted position to allow easier installation.
3. Turn the wheel on the puller to expand the jaws enough to contact the inner bore of the gimbal bearing.
4. Use short hammer strokes to pull the bearing out of the gimbal housing.

16

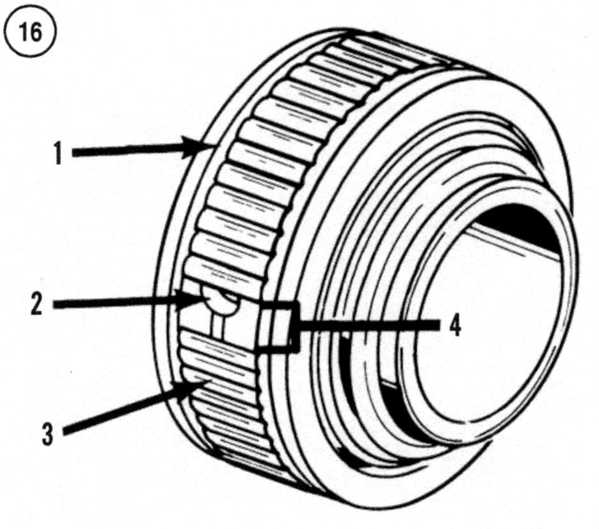

1. Carrier
2. Lubrication opening
3. Tolerance ring
4. Gap

5. Using the same method for gimbal bearing removal, remove the seal (33, **Figure 1**) with the three-jaw puller.
6. Discard the gimbal bearing and seal.
7. Wipe all sealant and grease from the bearing bore. Clean residual grease from the bore with aerosol carburetor cleaner and dry with compressed air.
8. Inspect the bore for nicks, burrs or other surface irregularities. Correct minor surface defects by polishing with emery cloth. The raised bars on the tolerance ring (on the outer diameter of the gimbal bearing) usually leave depressions around the bore. This is a normal occurrence and does not require correction. Inspect the bore for mechanical damage from a failed gimbal bearing rotating in the bore, deep corrosion pitting or cracks. Replace the gimbal housing if cracked, deeply pitted from corrosion or damaged from mechanical failure of the gimbal bearings.
9. Inspect the grease tube and fitting (23 and 24, **Figure 1**) for corrosion or mechanical damage. If damaged, pull the tube from the housing with a pair of pliers. Discard the tube and fitting.

Installation

1. If the grease tube was removed, install the new tube and fitting as follows:
   a. Apply Loctite 271 threadlocking adhesive to the tube bore in the gimbal housing.
   b. Carefully drive the new tube (23, **Figure 1**) into the bore until fully seated.
   c. Thread the new fitting (24, **Figure 1**) into the tube and securely tighten.
2. Install the *new* seal (33, **Figure 1**) as follows:

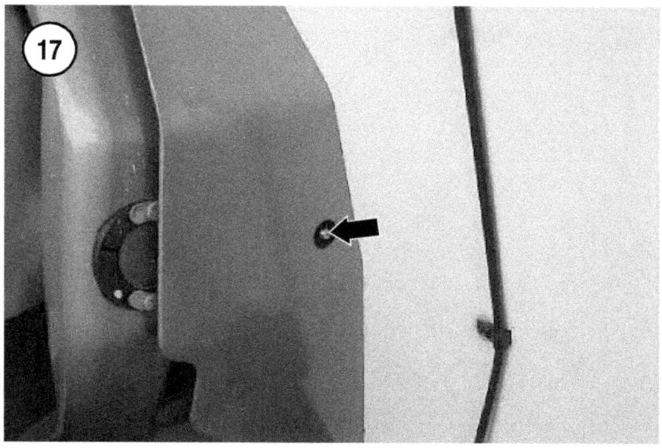

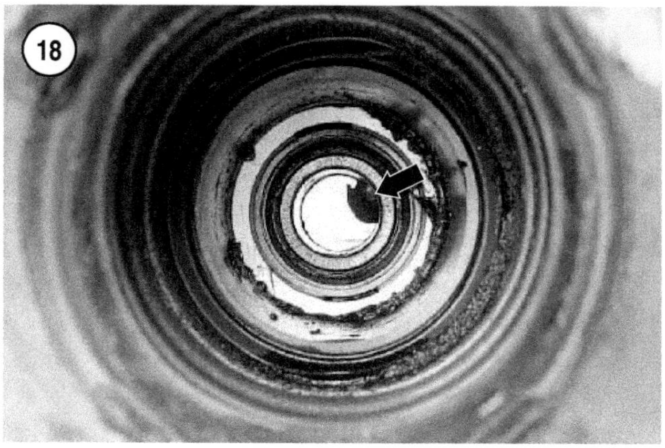

a. Thread the installer (OMC part No. 3850038 or Volvo Penta part No. 3854357-5) onto the drive handle (Volvo Penta part No. 3850609-3) with the recessed side facing the handle.
b. Apply OMC Gasket Sealing Compound (OMC part No. 508235) to the metal case of the seal. Fit the seal over the installation tool with the lip (spring) side facing the tool.
c. Guide the seal and tool into the bore. Drive the seal into the bore until the tool just seats against the housing.

3. Install the *new* gimbal bearing as follows:
a. Thread the installer (OMC part No. 3850038 or Volvo Penta part No. 3854357-5) onto the drive handle (Volvo Penta part No. 3850609-3) with the recessed side facing *away from* the handle.
b. Rotate the tolerance ring (corrugated band) of the gimbal bearing until the gap (4, **Figure 16**) aligns with the lubrication hole (2) in the carrier.
c. On the outward facing surface that aligns with the gap and lubrication hole, mark the carrier (1, **Figure 16**) with a felt-tip marker.
d. Install the gimbal bearing onto the installation tool. Make sure the installation tool *does not* contact the inner bearing race.
e. Coat the bore with OMC EP/Wheel Bearing Grease.
f. Guide the bearing and tool into the bore. Rotate the bearing to align the marking with the grease tube in the gimbal housing.
g. Drive the bearing into the bore until fully seated.
h. Pump OMC EP/Wheel bearing grease into the starboard side fitting (**Figure 17**) until grease starts flowing out of the gimbal bearing (**Figure 18**).

4. Align the engine as described in Chapter Six.
5. Install the drive unit as described in Chapter Ten.

GIMBAL RING

The gimbal ring must be removed to replace the upper and lower bearings/bushings and the O-ring (13, **Figure 1**) that prevents water entry. Shimming adjustments performed during installation eliminate up-and-down gimbal ring movement that often causes unusual noise during operation.

Removal and Inspection

1. Remove the pivot housing, universal joint bellows, exhaust bellows and water hose as described in this chapter.
2. Remove the two screws (2, **Figure 1**) and then lift the plastic cover (1) off the gimbal housing.
3. Remove the two screws (3, **Figure 1**) then lift the metal cover (4) off the gimbal housing. Remove the seal (5, **Figure 1**) from the cover or gimbal housing. Discard the seal.
4. Note the routing and connection points and then remove the grounding wire from the gimbal ring (17, **Figure 1**).
5. Remove the bolts and washers (52, **Figure 1**) that secure the support bracket (49) to the gimbal housing. 8.2L models have five bolts and washers. All other models have three bolts and washers.
6. Alternately tap on the port and then starboard sides of the support bracket to break it free from the gimbal housing. Pull the bracket from the housing. Remove the lower pin (47, **Figure 1**) from the bracket or gimbal housing bore. Remove the shim washers (44, **Figure 1**) from the lower edge of the support bracket or housing.
7. Remove the two plastic caps from the trim cylinder rod that passes through the gimbal ring. Remove the elastic locknuts from the ends of the rod. Slide each trim cylinder off the rod and allow them to hang temporarily from the hydraulic hoses. Remove the plastic bushings (46, **Figure 1**) and rubber grommets (45) from the pin bores.
8. Working outside the boat, remove the four elastic locknuts (8, **Figure 1**) and the center bolt (7) from the steering arm (10). Lift the washer plate (9, **Figure 1**) off the steering arm. Discard the elastic locknuts.
9. Affix the steering arm puller (**Figure 19** [OMC or Volvo Penta part No. 3854361-7]) to the top of the steering arm. Thread the puller into the bolt opening until it contacts the two dowels (15, **Figure 1**). Support the gimbal ring and tighten the puller bolt to push the gimbal ring out of the

11

steering arm. When it is free, pull the gimbal ring down and out of the upper steering bushing (16, **Figure 1**).

10. Remove the thrust washer (11, **Figure 1**) from the gimbal housing or steering arm. Remove the thrust washer (42, **Figure 1**) from the gimbal ring or the upper support portion of the gimbal housing.

11. Remove the O-ring (13, **Figure 1**) from the gimbal ring. Discard the O-ring.

12. Inspect the lower pin (47, **Figure 1**) for excessive wear, corrosion pitting or other damage. Replace the pin if these or other defects are evident.

13. Inspect the hinge pin bushings (22, **Figure 1**) in the gimbal ring for excessive wear, cracks, looseness or missing sections. Loose or missing bushings is a common occurrence. Replace the bushing(s) if these or other defects are evident. Remove the bushings as follows:

  a. If removed, place the gimbal ring with the side down on a suitable work surface. Otherwise, remove the bushing(s) with the gimbal ring installed onto the transom bracket.
  b. Insert the large diameter of the removal tool (OMC or Volvo Penta part No. 3854358-3) into the faulty bushing.
  c. Tap on the tool with a mallet until the bushing is free.
  d. Repeat the procedure for the other bushing if necessary.

14. Inspect the lower bushing (43, **Figure 1**) for excessive wear, cracks, looseness or missing sections. Replace the bushing if these or other defects are evident. Remove the bushing as follows:

  a. Insert the small end of the removal tool (OMC or Volvo Penta part No. 3854355-9) into the upper surface of the bushing.
  b. Use a mallet to drive the bushing (from the top side) until the bushing exits the bore at the bottom of the gimbal ring.

15. Inspect the upper bushing (12, **Figure 1**) for excessive wear, cracks, looseness or missing sections. Replace the bushing if these or other defects are evident. Remove the bushing as follows:

  a. Thread the removal tool (OMC or Volvo Penta part No. 3854360-9) onto the driver handle (OMC or Volvo Penta part No. 3854356-7) with the small diameter facing away from the handle.
  b. Insert the removal tool into the lower end of the bushing.
  c. Use a mallet to drive the bushing up and out of the gimbal housing.

CAUTION

If performed improperly, the procedure in Step 16 can easily damage the gimbal ring. The sleeve must be split without damaging the gimbal ring surfaces. Proceed with caution.

16. Inspect the bushing (16, **Figure 1**) for excessive wear, cracks, looseness or missing sections. Replace the sleeve if these or other defects are evident. Remove the sleeve as follows:

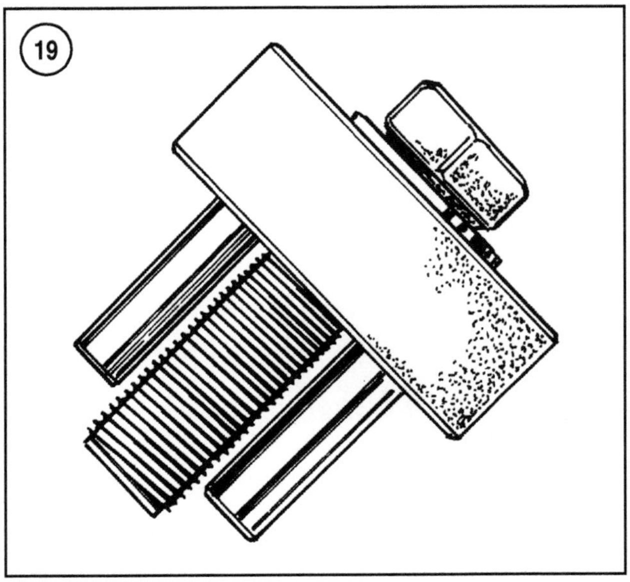

  a. Locate a sharp chisel with the same width as the vertical height of the sleeve.
  b. Position the chisel against the sleeve and *carefully* split the sleeve from top to bottom with controlled strikes from a mallet.
  c. Remove and discard the sleeve.
  d. Clean the gimbal ring steering post and remove any corrosion, nicks or minor surface imperfections with emery cloth.

Installation

1. If the upper bearing bushing (16, **Figure 1**) was removed, install the *new* sleeve as follows:

  a. Apply Volvo Penta Propeller Shaft Grease (Volvo Penta part No. 1141644-3) to the gimbal ring post and the inner diameter of the sleeve.
  b. Position the sleeve over the post.
  c. Install the steering arm (10, **Figure 1**) over the four steering post studs and rest it against the sleeve.
  d. Apply grease (Step 1a) to the thread of the bolt (7, **Figure 1**) and then thread it into the gimbal ring.
  e. Tighten the bolt while supporting the arm and gimbal ring until the sleeve fully seats on the gimbal ring.
  f. Remove the bolt. Using the procedure described in the removal procedures (Step 9), remove the steering arm from the gimbal ring.
  g. Inspect the sleeve for damage and proper seating. Seat or replace the sleeve as needed.

2. If the upper steering bushing (12, **Figure 1**) was removed, install the *new* bushing as follows:

  a. Place the installer (1, **Figure 20** [OMC or Volvo Penta part No. 3854359-1]) into the steering arm cavity with the recessed side facing downward. The tool

20

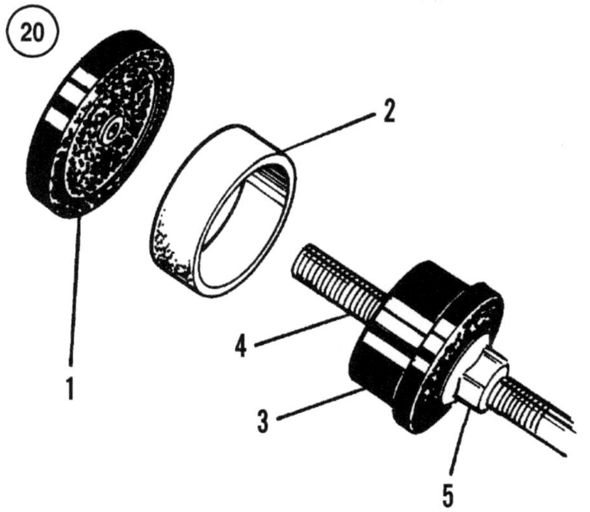

1. Installer
2. Bushing
3. Bearing installer
4. Threaded rod
5. Nut

functions as an alignment guide and stop for this procedure.

b. Lubricate the outer diameter of the bushing with gearcase lubricant. Place the bushing (2, **Figure 20**) into the bearing installer (3 [OMC or Volvo Penta part No. 3854360-9]). Slide the tool and bearing over the threaded rod (4, **Figure 20**) and seat against the flange nut (5).

c. Position the tool, bushing and threaded rod assembly under the bushing bore in the gimbal housing. Thread the threaded rod into the installer (Step 2a).

d. Engage a wrench on the threaded rod to prevent rotation and then tighten the nut (5, **Figure 20**) to push the bushing into the bore. Make sure the bushing enters the bore straight. Reposition the tools as needed. Continue tightening the nut until the installation tools mate and the bushing is fully seated in the bore.

e. Remove the tools. Inspect the bushing for damage and proper seating. Seat or replace the bushing as needed.

3. If the lower steering bushing (43, **Figure 1**) was removed, install the *new* bushing as follows:

a. Apply gearcase lubricant to the bushing.

b. Fit the bushing onto the larger-diameter (short) side of the installation tool (OMC or Volvo Penta part No. 3854358-3).

c. Fit the bushing and installation tool into the bottom of the bore in the gimbal housing support.

d. Use a mallet to tap the bushing into the bore until the tool seats against the lower surface of the support.

e. Remove the tool and inspect the bushing for damage and proper seating. Seat or replace the bushing as needed.

4. If removed, install the *new* hinge pin bushing(s) (22, **Figure 1**) as follows:

a. Apply gearcase lubricant to the outer diameter of the new bushing.

b. Fit the bushing onto the larger-diameter (short) side of the installation tool (OMC or Volvo Penta part No. 3854358-3).

c. Insert the bushing and tool into the gimbal ring bore from the outside.

d. Use a mallet to drive the bushing into the bore until flush to 0.010 in. (0.254 mm) below the inner surface of the gimbal ring.

e. Remove the tool and check the bushing for damage. Replace the bushing as needed.

f. Repeat the procedure for the other bushing as needed.

CAUTION

Do not apply any type of lubricant to pivoting surface of the hinge pin bushings, upper steering bushing or lower steering bushing. The bushings are designed to run dry. Lubricant will attract dust and result in binding and increased wear.

5. Install a *new* O-ring (13, **Figure 1**) over the gimbal ring post and seat around the base of the post. Then grease the O-ring with Volvo Penta Propeller Shaft Grease.

6. Install the thrust washer (11, **Figure 1**) around the upper steering bushing. Apply a small amount of grease to hold the washer in position.

7. Position the steering arm in the steering cavity and center over the bushing bore. Do not dislodge the thrust washer.

8. Carefully guide the gimbal ring post through the upper bushing and into the steering arm. Do not dislodge the thrust washer.

9. Hold the gimbal ring in position while installing the washer plate (9, **Figure 1**) and four *new* elastic locknuts onto the studs. Thread the bolt (7, **Figure 1**) into the steering arm and gimbal ring. Tighten the bolt and nuts by hand at this time.

10. Install the thrust washer onto the gimbal ring (**Figure 21**) and center over the lower pin bore. Insert the lower steering pin (47, **Figure 1**) through the gimbal ring, thrust washer and into the support arm. If necessary, have an assistant hold the pin in place during the procedure in Step 11.

11. While holding the gimbal ring in position, tighten the steering arm bolt (7, **Figure 1**) to 65-72 ft.-lb. (88-98 N•m). Then tighten the four elastic locknuts to 159-177 in.-lb. (18-20 N•m).

12. Pivot the gimbal ring back and forth several times while checking for free movement. If binding is evident, the upper bushing or sleeve was damaged during installation. Replace the bushing and sleeve before proceeding with the assembly.

13. Shim the vertical gimbal ring movement as follows.

a. Verify that only one thrust washer (**Figure 21**) is installed between the gimbal ring and support and that the lower steering pin remains in position. Correct as needed.
b. Temporarily install the lower support bracket (49, **Figure 1**) onto the gimbal housing. Use only one bolt to secure the bracket at this time.
c. Hold the bracket upward and pinch the single thrust washer between the lower support and the gimbal ring.
d. Using thrust washers as a gauge, measure the distance between the raised rib (around the steering pin) on the lower support bracket and the lower edge of the gimbal ring. Record the number of thrust washers that can be inserted with the gimbal ring held up as far as possible.
e. Remove the lower support bracket. Add one more thrust washer to the total recorded in Step 13d.
f. Remove the lower steering pin and install the number of thrust washers determined in Step 13e.
g. Install the lower steering pin without dislodging the thrust washers.
h. Install the lower support bracket (49, **Figure 1**) onto the gimbal housing. Make sure the lower steering pin enters the bore in the bracket. Hold the bracket in position while threading the bolts and washers (52, **Figure 1**) into the bracket and gimbal housing openings. Tighten the bolts evenly to 212-239 in.-lb. (24-27 N•m).
i. Pivot the gimbal ring several times to check for proper shimming. The ring should pivot with *slight* drag and no measurable vertical movement. Remove and install the bracket and steering pin to add or subtract thrust washers to achieve the desired results.

14. Coat the trim cylinder rod with Volvo Penta Propeller Shaft Grease. Then insert the rod through the gimbal ring openings. Place a rubber grommet (45, **Figure 1**) and plastic bushing (46) over each end of the rod and guide into the recess on each side of the gimbal ring.
15. Make sure the trim cylinder bushings and grommets are installed and then fit the cylinders onto the rod. Install a flat washer and *new* elastic locknut onto each end of the rod. Tighten the elastic locknuts to 124-142 in.-lb. (14-16 N•m).
16. Connect one end of the ground wire to the boss on the lower port side of the gimbal ring. Tighten the first ground wire screw to 60-84 in.-lb. (7-10 N•m).
17. Install a *new* gasket (5, **Figure 1**) onto the metal steering arm cover (4, **Figure 1**). Install the cover and secure with the two screws (3, **Figure 1**). Tighten the screws to 60-84 in.-lb. (7-10 N•m).
18. Install the plastic cover (1, **Figure 1**) and secure with the two screws. Tighten the screws to 60-84 in.-lb. (7-10 N•m).
19. Connect the other end of the ground wire to the steering arm. Tighten the last ground wire screw to 60-84 in.-lb. (7-10 N•m).

21

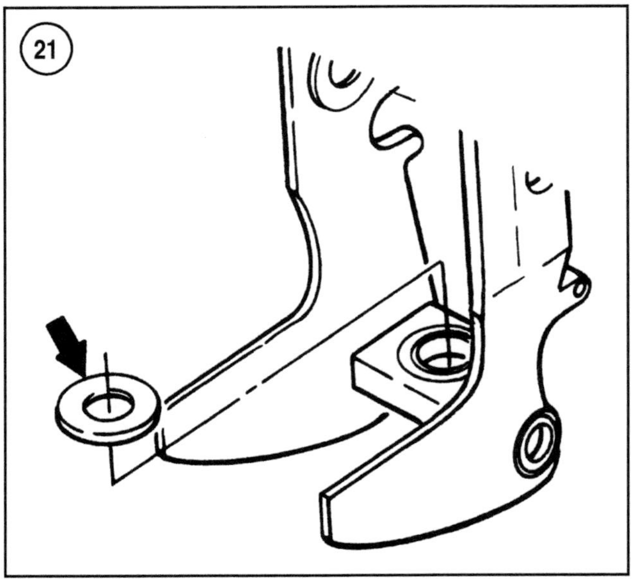

22

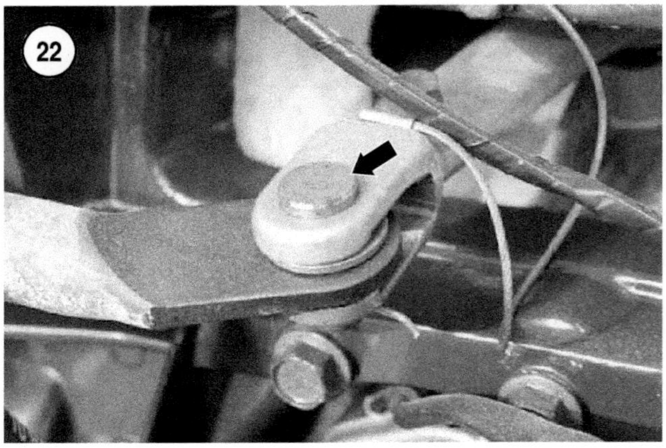

20. Install the pivot housing as described in this chapter.

INNER TRANSOM PLATE AND BRACKET HOUSING

This section describes the procedures for removing and installing the transom bracket as an assembly. The transom bracket can be removed and installed as an assembly or partially disassembled if necessary.

Removal and Inspection

1. Remove the drive unit as described in Chapter Ten.
2. Remove the engine as described in Chapter Six.
3. Pull the remote control shift cable from the sleeve (37, **Figure 1**) and rest in the bilge area.
4. Remove the four bolts and washers securing the exhaust pipe to the bracket. Pull the pipe away from the bracket. Remove the seal (41, **Figure 1**) from the exhaust pipe or transom bracket. Discard the seal.

5A. If the engine is equipped with manual steering, disconnect the steering system as follows:

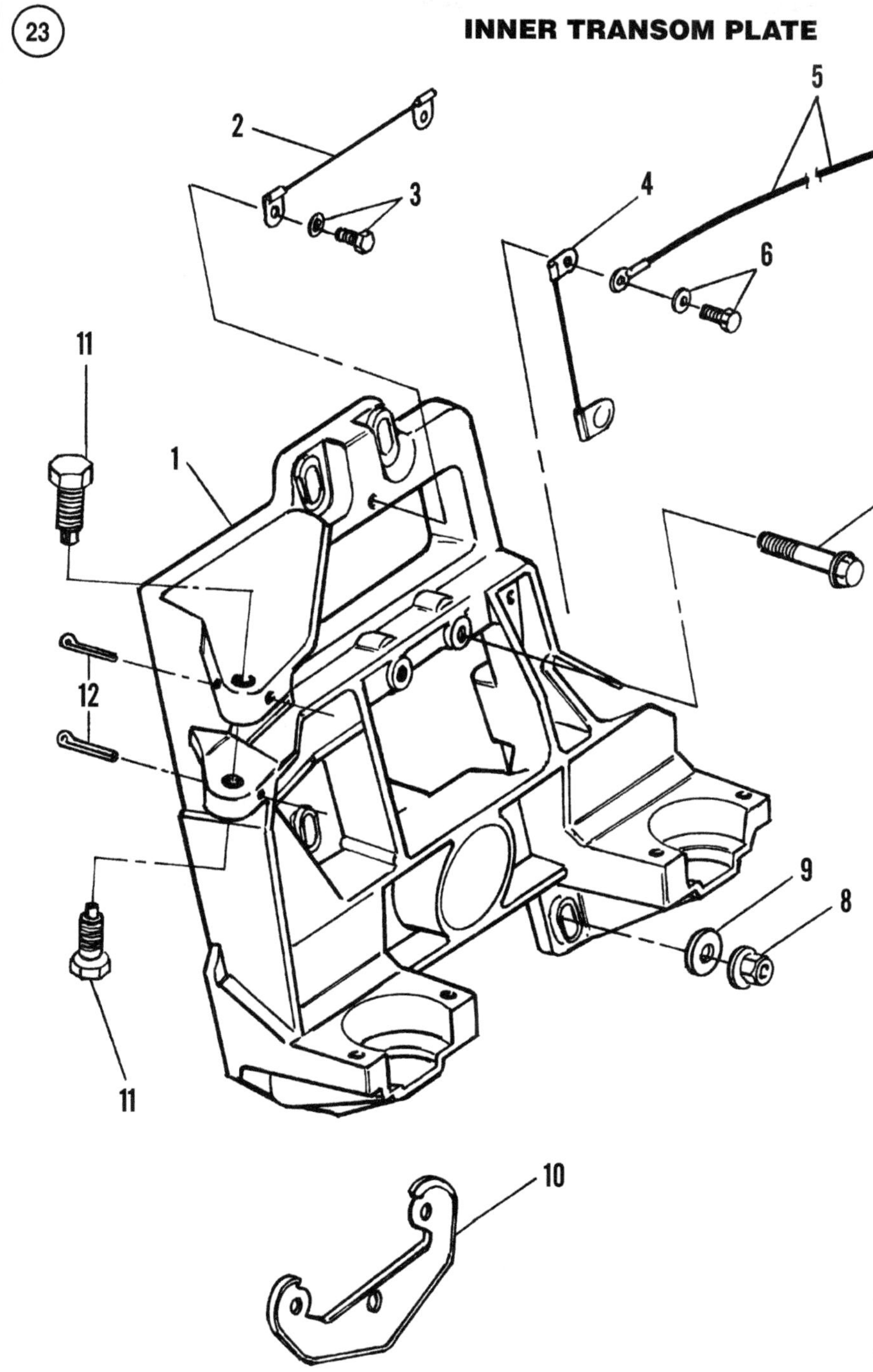

1. Inner transom plate
2. Ground wire (steering arm to inner plate)
3. Screw and washer
4. Ground wire (inner plate to mounting stud)
5. Ground wire (inner plate to flywheel housing stud)
6. Screw and washer
7. Bolt
8. Washer-faced elastic locknut
9. Flat washer
10. Washer plate
11. Steering pivot bolts
12. Cotter pin

a. Remove the cotter pin securing the clevis pin (**Figure 22**) to the steering arm. Discard the cotter pin. Lift the clevis pin out of the cable clevis end and steering arm openings.

b. Remove the cotter pins (12, **Figure 23**) securing the upper and lower pivot bolts (11) to the inner transom plate.

c. Remove the upper and lower pivot bolt to free the steering tube and anchor block from the inner transom plate. Move the cable, tube and block to the side to prevent interference during the removal and installation procedures.

5B. If the engine is equipped with power steering, disconnect the steering system as follows:

a. Remove the large cotter pin securing the longer clevis pin (**Figure 24**) to the steering cylinder and steering arm. Discard the cotter pin. Lift the clevis pin out of the cable clevis end and steering arm openings.

b. Remove the cotter pins (12, **Figure 23**) securing the upper and lower pivot bolts (11) to the inner transom plate.

c. Remove the upper and lower pivot bolts (11, **Figure 23**) securing the steering cylinder assembly to the inner transom plate. Move the cable and steering cylinder assembly to the side to prevent interference during the removal and installation procedures.

6. Disconnect the two-pin main power lead and three-pin trim switch harness quick connectors (**Figure 25**) from the trim pump or trim limit control module (if so equipped). Disconnect the trim position sender harness connector (**Figure 26**) from the engine wire harness or trim limit module harness (if so equipped). A two-wire sender harness is used on earlier models and a three-wire sender is used on later models. Remove any tie-wires from the trim pump harness to the transom. Remove the screws securing the trim pump to the transom or bilge compartment floor. Set the trim pump temporarily to the side.

7. Remove the screw and washer (3, **Figure 23**) and then disconnect the ground wire (2) from the steering arm.

8. Remove the two washer-faced bolts (1, **Figure 27**) from the openings at the top of the inner transom plate.

9. Remove the two elastic locknuts (2, **Figure 27**) from the lower studs. Pull the washer plate (10, **Figure 23**) off the boat transom.

10. While an assistant supports the transom bracket from outside the boat, remove the four elastic locknuts (3, **Figure 27**) and washers from the mounting studs. If necessary, tap on the inner surface of the transom bracket with a rubber bracket to break the bond. Discard the elastic locknuts.

11. With the aid of an assistant, pull the transom bracket away from the boat transom. Feed the trim pump and hydraulic cable through the cutout while removing the transom bracket.

12. Remove the transom seal (40, **Figure 1**) from the transom bracket. Thoroughly clean the seal groove in the bracket. Any remaining material may prevent the seal from seating and allow water to leak into the boat.

13. Inspect the transom bracket mounting studs for corrosion pitting and damaged threads. Replace the studs if damaged or corroded. Use a stud removal tool to remove the studs. Replace all of the studs if any of them are defective.

14. Inspect the shift cable sleeve (37, **Figure 1**) for burned, worn or otherwise damaged surfaces. Replace the sleeve if in questionable condition. Remove the sleeve by unthreading from the gimbal housing. Remove the O-ring (38, **Figure 1**) from the sleeve fitting. Discard the O-ring.

15. Inspect the anode(s) for deterioration or loose fasteners. Replace the anode if it has lost one-third or more of the material.

16. Thoroughly clean all old sealant from the boat transom. Paying particular attention to the cutout surfaces, inspect the transom. If the wood in the cutout is rotted, it must be replaced before installing the transom bracket. Inspect the surfaces surrounding the cutout for cracks or other damage. If damage is evident, have the transom wood replaced or fiberglass repaired at a reputable marine fiberglass repair shop. If the wood is wet but otherwise in good condition, allow the wood to dry completely before installing the transom bracket. Seal any bare wood or exposed surfaces with marine epoxy or several coats or marine varnish. Allow the epoxy or varnish to cure or dry completely before installing the transom bracket.

24

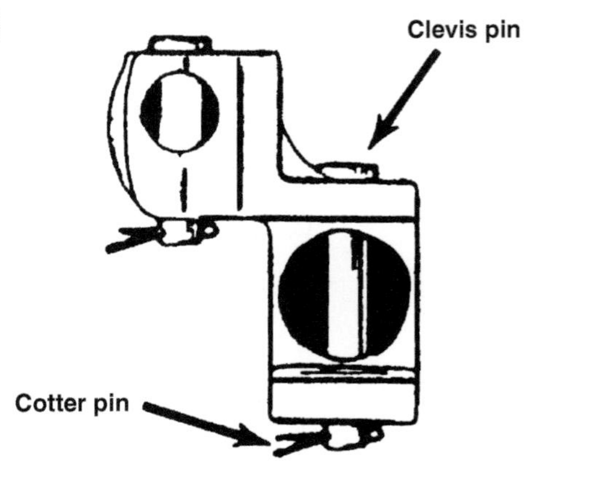

25

Installation

Coat the threads of all fasteners that are installed into the boat transom with 3M Marine Sealant 101 or equivalent. Purchase the sealant from an OMC dealership (OMC part No. 506852). This will help prevent water entry and subsequent wood rot in the transom.

1. If the studs were removed from the gimbal housing, coat the threads of the *new* studs with Volvo Penta White Sealing Compound (Volvo Penta part No. 1141570-0). Thread the studs into the openings by hand. Then use a common stud installation tool to securely tighten the studs in the openings.

2. Glue the transom seal (40, **Figure 1**) into the transom bracket groove with OMC Type M adhesive (OMC part No. 318535). Butt the ends of the seal together tightly at the top of the transom bracket groove and then fit the seal into the groove.

3. If the shift cable sleeve (37, **Figure 1**) and cable were removed, reinstall as follows:

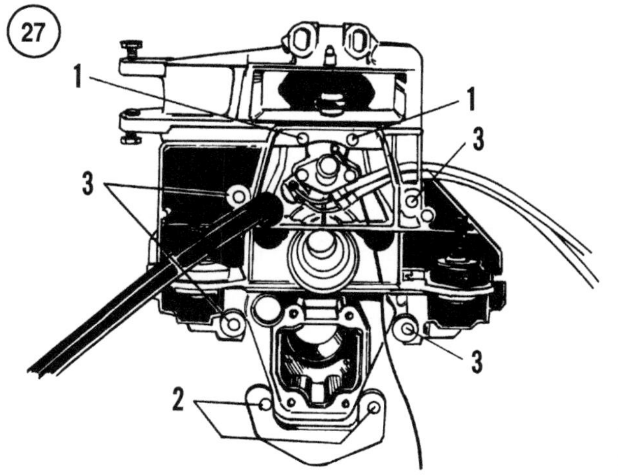

1. Washer faced bolts (2)
2. Elastic locknuts (2)
3. Elastic locknuts (4)

a. Apply Volvo Penta White Sealing Compound to the threads and then fit a *new* O-ring (38, **Figure 1**) over the end and seat against the fitting.

b. Thread the shift cable sleeve fitting into the gimbal housing and tighten to 46-52 ft.-lb. (62-71 N•m).

4. Coat the shank of the transom bracket studs with 3M Marine Sealant 101. With the aid of an assistant, feed the trim pump and hoses into the transom cutout. Align the studs with the openings and then seat the transom bracket against the boat transom.

5. Apply Volvo Penta Propeller Shaft Grease (Volvo Penta part No. 1141644-3) into the bores of the inner transom plate (1, **Figure 23**). Then position the plate over the mounting studs. Make sure the trim hoses are routed over the top port stud and into the relief at the top port corner of the plate.

6. Coat the threads of the two bolts (7, **Figure 23**) with Volvo Penta Propeller Shaft Grease and then thread the bolts into the inner transom plate and gimbal housing openings. Tighten the bolts by hand at this time.

7. Install the four flat washers and four *new* elastic locknuts (3, **Figure 27**) over the studs protruding through the inner transom plate. Make sure the ground wire (4, **Figure 23**) is installed over the stud before the nut. Tighten the nuts by hand at this time.

8. Fit the washer plate (10, **Figure 23**) over the lower studs and then thread the two *new* elastic locknuts onto the studs. Tighten the nuts by hand at this time.

9. Starting in the middle and working outward, tighten the six elastic locknuts (2 and 3, **Figure 27**) in a crossing pattern to 20-25 ft.-lb. (27-34 N•m). Wait approximately 10 minutes and then tighten the nuts in a crossing pattern a second time to the same specification. Wait approximately 10 minutes and then tighten the nuts in a crossing pattern a final time to the same specification.

10. Tighten the two transom bracket bolts (1, **Figure 27**) to 142-168 in.-lb. (16-19 N•m).

11. Position the trim pump onto the transom or bilge floor mounting surface with the mounting fastener openings aligned with the original openings. Apply 3M Marine Sealant 101 to the threads and then secure the trim pump with the original fasteners or suitable replacements. Securely tighten the trim pump fasteners.

12. Connect the two-pin main power lead and three-pin trim switch harness quick connectors (**Figure 25**) to the trim pump or trim limit control module harness (if so equipped). Connect the trim position sender harness connector (**Figure 26**) to the engine wire harness or trim limit module harness (if so equipped). A two-wire sender harness is used on earlier models, and a three-wire sender is used on later models. Route the wiring to prevent interference with the steering lever or other moving components. Secure the wiring with plastic locking clamps as needed.

13. Slide the remote control cable into and through the cable sleeve.

CAUTION

Failure to properly pilot the pivot bolts into the bushings in the anchor block or steering cylinder will result in a broken inner transom plate when the bolts are tightened. Tighten the bolts slowly and carefully while holding the anchor block or steering cylinder bushing in perfect alignment with the pivot bolts. The anchor block or steering cylinder must pivot freely after the bolts are tightened.

14A. If the engine is equipped with mechanical steering, connect the steering system to the transom bracket as follows:

a. Position the steering cable, tube and anchor block assembly between the inner transom plate mounting bosses.

b. Thread the upper and lower pivot bolts (11, **Figure 23**) in the bosses while guiding the bolts into the anchor block bushings.

c. Verify proper engagement of the pivot bolts into the anchor block bushings. Then tighten the bolts to 40-45 ft.-lb. (54-61 N•m). Make sure the anchor block pivots freely after tightening the bolts. Other-

11

wise, remove the bolts and repeat the installation procedure.

d. Insert *new* cotter pins (12, **Figure 23**) into the openings in the bosses to prevent the pivot bolts from loosening. Bend over both prongs on each cotter pin.

e. Move the steering arm to align the clevis pin opening in the arm and clevis pin end of the steering cable. Insert the clevis pin (**Figure 22**) into the opening with the flange on top. Make sure the clevis pin is fully seated against the arm and the cotter pin opening at the bottom extends past the bottom surface of the steering arm. Install a *new* cotter pin into the opening at the bottom of the clevis pin. Bend both prongs of the cotter pin over to secure the clevis pin.

14B. If the engine is equipped with power steering, connect the steering system to the transom bracket as follows:

a. Position the steering cable and cylinder between the inner transom plate mounting bosses.

b. Thread the upper and lower pivot bolts (11, **Figure 23**) into the bosses while guiding the bolts into the cylinder bushings.

c. Verify proper engagement of the pivot bolts in the cylinder bushings. Then tighten the bolts to 40-45 ft.-lb. (54-61 N•m). Make sure the steering cylinder pivots freely after tightening the bolts. Otherwise, remove the bolts and repeat the installation procedure.

d. Insert *new* cotter pins (12, **Figure 23**) into the openings in the bosses to prevent the pivot bolts from loosening. Bend over both prongs on each cotter pin.

e. Move the steering arm to align the clevis pin opening in the arm and clevis pin end of the steering cylinder. Insert the clevis pin into the opening with the flange on top (**Figure 22**). Make sure the clevis pin is fully seated against the arm and that the cotter pin opening at the bottom extends past the bottom surface of the steering arm. Install a *new* cotter pin into the opening at the bottom of the clevis pin. Bend both prongs of the cotter pin over to secure the clevis pin.

15. Apply a light coat of OMC Gasket Sealing Compound (OMC part No. 508235) to the surfaces and then fit the new O-seal (41, **Figure 1**) into the groove in the gimbal housing. Coat the threads of the four exhaust pipe mounting bolts with the same compound. Install the exhaust pipe onto the transom bracket. Make sure the seal was not inadvertently dislodged and then thread the four bolts and washers into the pipe and gimbal housing openings. Tighten the bolts in a crossing pattern to 20-25 ft.-lb. (27-34 N•m).

16. Reconnect the ground wire (2, **Figure 23**) to the steering arm. Secure with the screw and washer (6, **Figure 23**). Tighten the screw to 60-84 in.-lb. (7-10 N•m).

17. Install and align the engine as described in Chapter Six.

18. Install the drive unit as described in Chapter Ten.

Table 1 TRANSOM BRACKET TORQUE SPECIFICATIONS

Fastener	in.-lb.	ft.-lb.	N•m
Exhaust pipe to transom bracket	–	20-25	27-34
Grounding wire screws	60-84	–	7-10
Hinge pin (pivot housing)	–	105-120	142-163
Lower support bracket bolts	212-239	–	24-27
Shift cable sleeve	–	46-52	62-71
Steering arm bolt	–	65-72	88-98
Steering arm cover screws			
Metal	60-84	–	7-10
Plastic	60-84	–	7-10
Steering arm elastic nuts (4)	159-177	–	18-20
Steering cable anchor (pivot bolts)	–	40-45	54-61
Transom bracket to inner plate			
Elastic locknuts (6)			
First step	–	20-25	27-34
Second step	–	20-25	27-34
Final step	–	20-25	27-34
Bolts (2)	142-168	–	16-19
Trim cylinder anchor rod nuts	124-142	–	14-16
Water tube grommet plate screws	124-142	–	14-16
Water tube nipple (plastic nut)	97-124	–	11-14

Chapter Twelve

Power Trim System

Very few of the trim system components are individually serviceable. The trim cylinders can be resealed; however, the piston, ram and cylinder bore are not available separately. Each cylinder must be replaced as an assembly if it cannot be restored to normal operation by resealing.

The electric motor on the trim pump is not serviceable and must be replaced if faulty. The pressure relief valves, reservoir, inlet screen, shaft coupling and hydraulic pump assembly are available separately. Internal hydraulic pump components are not available separately. The hydraulic pump assembly must be replaced if the pump cannot be repaired by replacing the pressure relief valves, O-rings and seals. This chapter describes replacement and repair (when applicable) of the power trim system components.

Table 1 lists torque specifications for most power trim system fasteners. Use the general torque recommendations provided in Chapter One for fasteners not listed in **Table 1**.

ELECTRICAL TRIM SYSTEM COMPONENTS

This section describes replacement procedures for the trim relays, trim switch fuse, trim limit control module and trim motor. Replacement of the trim position sender is described later in this chapter.

Circuit Board Replacement

Early models use a circuit board (**Figure 1**) mounted onto the side of the trim pump to operate the trim pump motor. The up and down relays integrated into the circuit board are controlled by the remote control or dash-mounted trim switch. The relays reverse polarity and motor rotational direction to provide both up and down trim operation. On DP-S (twin-propeller) models, the trim switch wiring and circuit board wiring connect to the transom-mounted trim limit control module (**Figure 2**). Using the resistance reading from the trim position sensor, the module limits the trim down to 2° negative and the trim up to 5° positive by internally opening the trim switch-to-relay circuits. To trim up or down beyond the limits, the operator must depress the override switch on the control *and* the trim switch. The override switch provides current to a module terminal that disables the limit function. This allows the additional trim needed for shallow water operation or loading the boat onto a trailer or additional down trim as needed.

1. Disconnect the negative battery cable.

2. Make a sketch of the wire routing and terminal connection points for the trim pump and trim switch harness that connect to the circuit board.

3. Remove the terminal screw or nut and then disconnect each wire from the circuit board terminals.
4. Clean all corrosion from the wire terminals.
5. Remove the single screw that secures the circuit board to the trim pump reservoir.
6. Pivot the board to release the bottom tab from the pocket in the reservoir. Use compressed air to remove any loose material from the external reservoir surfaces. Wipe oily deposits from the surfaces with a clean shop towel.
7. Guide the tab on the bottom of the replacement circuit board into the pocket in the reservoir. Pivot the board in position. Align the openings and then thread the screw into the circuit board and reservoir openings. Securely tighten the screw.
8. Connect the trim switch and trim motor wiring to the corresponding terminals on the circuit board. Install the terminal screws or nuts and tighten securely. To prevent corrosion-related failure, thoroughly coat the terminals with Black Neoprene Dip. Purchase the neoprene dip (OMC part No. 909570) from an OMC dealership or automotive parts store.
9. Connect the battery cable. Check for proper trim system operation.

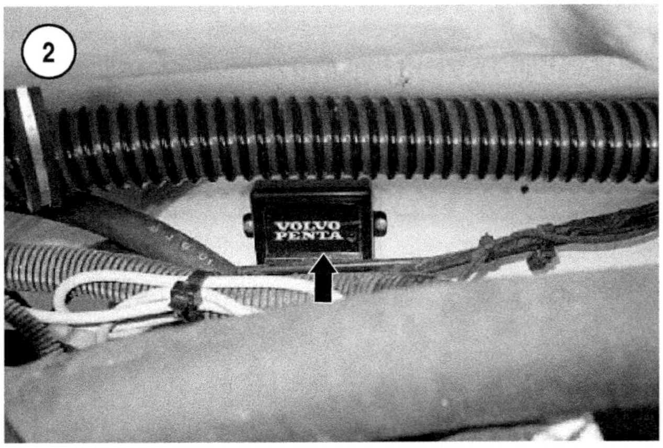

Trim Relay Replacement

Later models use two trim pump-mounted relays (A, **Figure 3**) to operate the trim motor. The relays are controlled by the remote control or dash-mounted trim switch. The relay reverses the polarity and motor rotational direction to provide both up and down trim operation. On DP-S (twin propeller) models, the trim switch wiring and relay wiring connect to the transom-mounted trim limit control module (**Figure 2**).

The relays have a basic five-terminal (**Figure 4**) plug-in design. The wire harness connector plugs onto the bottom of the relay and is held in place with a locking tab. A screw holds the relay onto the pump mounting bracket.

1. Disconnect the negative battery cable.
2. Lift the locking tab and then unplug the harness from the relay.
3. Remove the screw and then remove the relay from the bracket. Remove the grommet and sleeve from the relay opening. Replace the grommet if it is torn or deteriorated.
4. Use electrical contact cleaner to thoroughly clean the relay and wire harness or mounting bracket terminals. Corrosion of the terminals is common, and cleaning the terminals will often restore proper relay operation.
5. Use terminals 30 and 87A (**Figure 4**) for reference and carefully align the wire harness terminals with the relay terminals. *Do not* apply grease or sealant to the terminals. Grease or other foreign material will wick into the relay through the terminal openings, contaminate the electrical contacts and cause premature relay failure. Carefully plug the wire harness connector onto the relay. The locking tab on the harness connector must engage the tab on the relay housing.
6. Install the grommet and sleeve into the relay opening. Align the screw and sleeve openings and then thread the screw into the sleeve and bracket. Orient the relay with the harness pointing down and then securely tighten the screw.
7. Connect the negative battery cable.
8. Check for proper trim system operation.

Trim Switch Fuse Replacement

On later models using the trim pump with relays, the trim switch fuse (B, **Figure 3**) is located on the trim pump

4

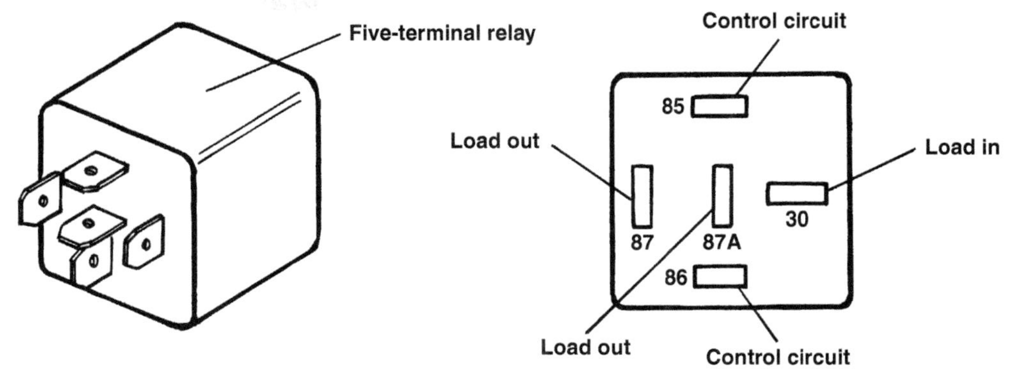

5

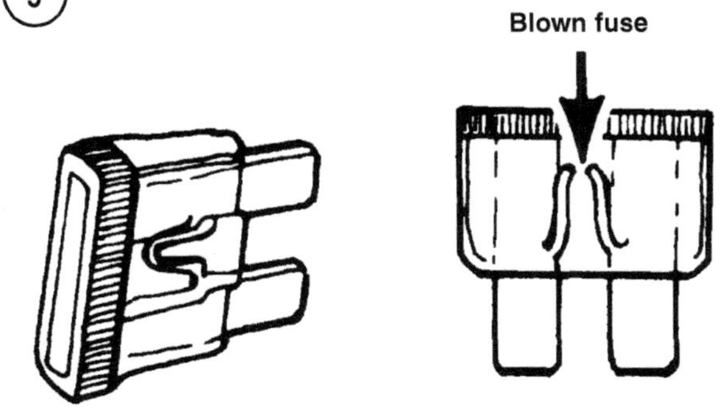

mounting bracket. Remove and inspect or replace the trim switch fuse if the trim pump does not operate in either direction. The fuse is designed to protect the trim switch circuit if a short to ground or an overload occurs. If the fuse is blown, inspect the trim switch and switch harness for damaged wiring before replacing the fuse. Unless the fault is corrected, the fuse will probably fail again. Never install a fuse with a higher capacity than the original 10-amp fuse. Anytime it is removed, test the fuse with an ohmmeter as described in this procedure. Never rely solely on a visual inspection to determine the fuse condition.

1. Disconnect the negative battery cable.
2. Pull up on the locking tab and then pull the fuse and harness out of the cap. Do not remove the cap from the bracket. Inspect the rubber seal on the harness connector for tears or deterioration and replace as needed.
3. Grip the fuse and pull it out of the harness plug. Clean corrosion or other contaminants from the fuse and harness terminals with electrical contact cleaner (available at automotive parts stores).
4. Inspect and test the fuse as follows:
   a. Visually inspect the fuse for a break in the element or a burned appearance indicating a blown fuse (**Figure 5**).
   b. Touch ohmmeter leads to the fuse terminals. The meter must show *continuity*. If otherwise, the fuse has failed open and must be replaced.
   c. If the fuse is blown, inspect the *entire* length of the wiring connecting the trim switch (at the controls) to the trim pump. Replace the fuse only after correcting any faults and verifying the wiring is in satisfactory condition.
5. Align the terminals and then plug the replacement fuse fully into the connectors. Align the locking tabs and then plug the harness onto the cap. Route the trim switch wiring to prevent interference with moving components (especially the steering cable and related components). Moving components can abrade the wire jacket and eventually short the wiring. This is a common cause of intermittent fuse failure.
6. Connect the battery cable.

Trim Limit Control Module Removal and Installation

On models using a DP-S (twin-propeller) drive unit, the trim limit control module (**Figure 2**) is mounted on the boat transom.

1. Disconnect the negative battery cable.
2. Lift the locking tabs and then disconnect the trim switch and trim pump harness from the module harnesses.
3. Remove the two screws and pull the module off the mounting surface. Wipe dust or debris off the mounting surfaces.
4. Use electrical contact cleaner to remove corrosion and other contaminants from the terminals in the module, trim switch and trim pump harness connectors. The up and down circuits pass through the module. Corroded terminals can cause the entire trim electrical system to malfunction. Corroded terminals can allow trim operation beyond the up limit or prevent operation of the tilt function.
5. Install the replacement module onto the mounting surface and secure it with two stainless steel screws.
6. Lift the locking tabs and then plug the trim switch and trim pump harnesses onto the module harnesses. Release the tabs and tug on the wiring to verify a secure connection.
7. Connect the battery cable. Adjust the trim limit module as described in Chapter Five.

Trim Pump

Two different styles of pumps are used on the models covered in this manual. Earlier models use the pump shown in **Figure 1**. The earlier pump has a top-mounted fluid reservoir and a circuit board attached to the side of the pump.

Later models use the pump shown in **Figure 3**. The later pump has a bottom-mounted fluid reservoir with relays and a fuse attached to the pump mounting bracket.

Removal and installation

WARNING
The hydraulic system contains fluid under very high pressure. Wear safety glasses and protective clothing when working with the hydraulic system.

1A. On models using the early-style pump (**Figure 1**), position and support the drive unit as follows:

a. If the pump is operational, trim the drive unit to the full up position. Provide adequate support to continually support the drive in the full up position. Use an overhead hoist securely attached to the drive or a suitable support positioned under the drive unit.
b. If the pump is not operational, attach an overhead lift to the drive unit. Place a suitable container under the hose fittings and cylinder openings to capture spilled fluid. Then, disconnect the trim hoses from the trim cylinder(s) as described in this chapter. Refer to *Trim Cylinder and Hoses*. *Slowly* lift the drive unit to the full up position. Provide enough lifting force to lift and continually support the weight of the drive unit. Use an overhead hoist securely attached to the drive or use a suitable support positioned under the drive unit.

1B. Operate the trim pump to place the drive unit in the full down position. If the pump is not operational, attach an overhead lift to the drive unit for support. Provide enough lifting force to support the weight of the drive unit. Place a suitable container under the hose fittings and cylinder openings to capture spilled fluid. Then disconnect the trim hoses from the trim cylinder(s) as described in this chapter. Refer to *Trim Cylinder and Hoses*. *Slowly* lower the drive unit to the full down position. Disconnect the drive from the overhead lift.

2. Disconnect the negative battery cable.

3A. On models using the early-style pump (**Figure 1**), remove the circuit board as described in this chapter.

3B. On models using the later-style pump (**Figure 3**), trace the trim pump power and trim switch harnesses to the connections (**Figure 6** and **Figure 7**) to the engine wire harness. Disconnect the power and switch harnesses.

4A. On models using the earlier-style pump (**Figure 1**), mark the trim hoses (23, **Figure 8** or 23, **Figure 9**) and the trim pump to identify the connection points. The trim system will operate opposite the desired direction if the hoses are installed in crossed locations.

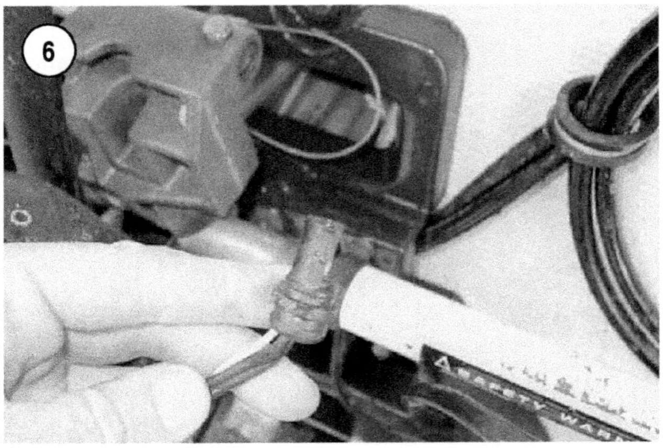

4B. On models using the later-style pump (**Figure 3**), mark the trim hoses and the trim pump to identify the connection points (A and B, **Figure 10**). The trim system will operate opposite the desired direction if the hoses are installed in crossed locations.

5. Place a suitable container under the trim hose fittings and pump and then *slowly* loosen the hose fittings. Allow residual pressure to dissipate and then unthread the hose fittings. Cover the hose fittings and pump openings to prevent contamination.

6A. On models using the earlier-style pump (**Figure 1**). Support the pump and then remove the two lag screws (9, **Figure 8** or 9, **Figure 9**) and remove the pump. Position the pump with the fill plug opening over the fluid container and then empty the reservoir. Install the fill plug to prevent contamination.

6B. On models using the later-style pump (**Figure 3**), remove the mounting screws (C, **Figure 10**) and then remove the assembly. The remaining mounting screws are accessible from the rear of the pump. Remove the pump from the mount. Remove the fill cap (**Figure 11**) and then position the fill cap opening over the fluid container. Then invert the pump and empty the reservoir.

7A. On models using the early-style pump (**Figure 1**), fit the replacement pump onto the mounting surface with the reser-

8 **EARLY-STYLE TRIM PUMP ASSEMBLY (1994-EARLY 1996 MODELS)**

1. Circuit board
2. Screw
3. Fluid reservoir
4. Screws (4)
5. Fill plug and O-ring
6. O-ring
7. Valve body
8. Pump housing
9. Lag screw
10 Screw
11. Trim down relief valve and spring
12. Trim down relief seat and O-ring
13. Trim up relief valve and spring
14. Trim up relief seat and O-ring
15. Pump control piston and check ball
16. Manual release check ball, retainer and O-ring
17. Inlet check ball
18. Check valve spring
19. Check valve and seat
20. O-ring
21. Pump gear set
22. O-ring (hose fitting)
23. Hydraulic hose
24. Ring filter
25. Coupler
26. O-ring
27. Plastic locking clamp
28. Electric trim motor
29. Screw
30. Snap ring
31. O-ring
32. Manual release valve
33. O-ring
34. Nylon ring
35. Filter

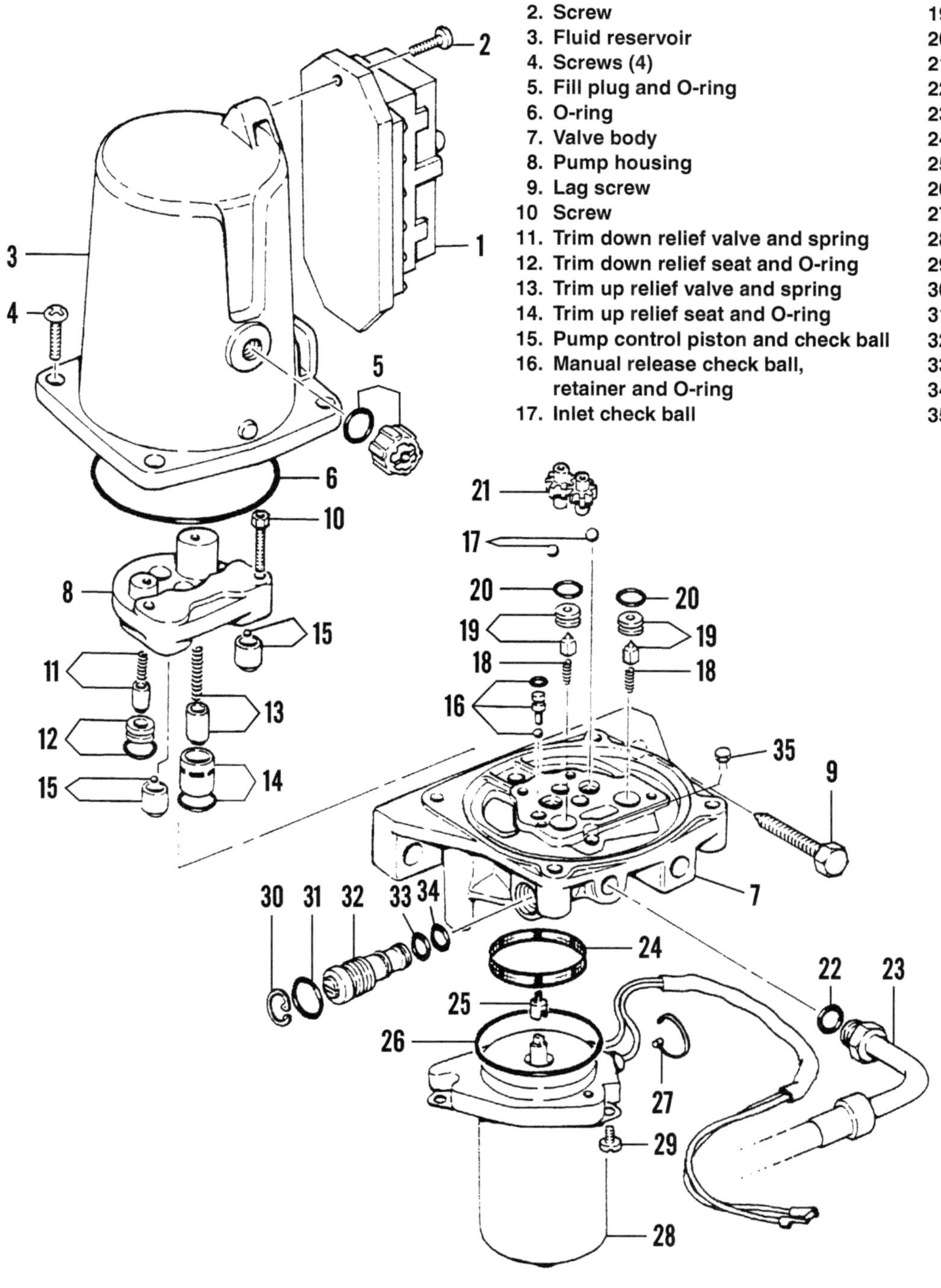

voir side facing up. Align the openings and then thread the two lag screws (9, **Figure 8** or 9, **Figure 9**) into the pump mount and mounting surface openings. Securely tighten the lag screws.

7B. On models using the later-style pump (**Figure 3**), secure the replacement trim pump assembly to the mount with the mounting screws (C, **Figure 10**). Use only good grade stainless steel screws and securely tighten them to ensure secure mounting.

8A. On models using the early-style pump (**Figure 1**), remove the plugs from the pump openings and caps from the hose fittings. Remove and replace the O-ring (22, **Figure 8** or 22, **Figure 9**) from the hoses. Thread the up and down pressure hose fittings (23, **Figure 8** or 23, **Figure 9**) into the corresponding trim pump openings. Grip the crimped section of the hose with pliers to prevent hose twisting and then use a flare nut wrench to securely tighten the fittings.

12

9

EARLY-STYLE TRIM PUMP ASSEMBLY (LATE 1996-1997 MODELS)

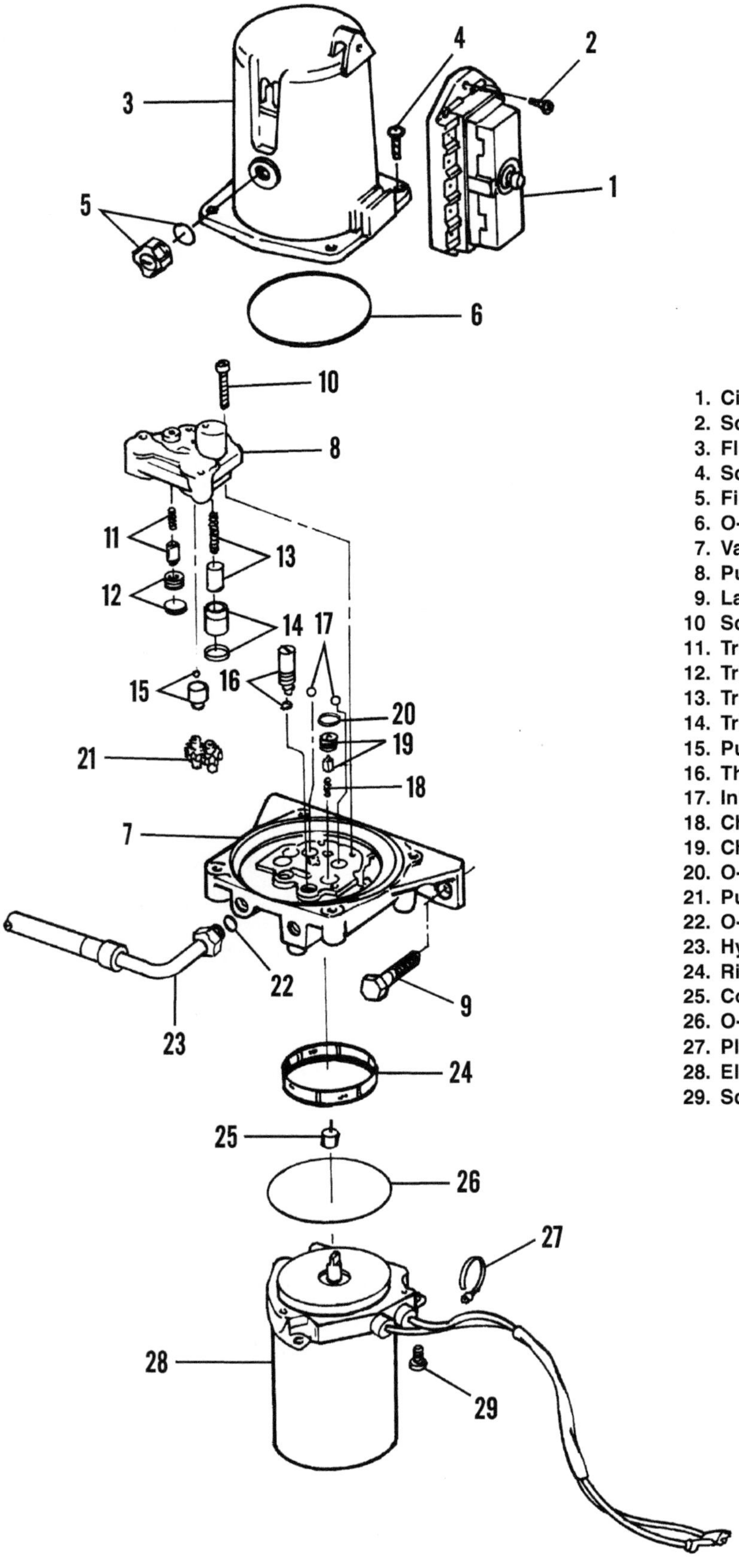

1. Circuit board
2. Screw
3. Fluid reservoir
4. Screws (4)
5. Fill plug and O-ring
6. O-ring
7. Valve body
8. Pump housing
9. Lag screw
10 Screw
11. Trim down relief valve and spring
12. Trim down relief seat and O-ring
13. Trim up relief valve and spring
14. Trim up relief seat and O-ring
15. Pump control piston and check ball
16. Thermal relief valve and O-ring
17. Inlet check ball
18. Check valve spring
19. Check valve and seat
20. O-ring
21. Pump gear set
22. O-ring (hose fitting)
23. Hydraulic hose
24. Ring filter
25. Coupler
26. O-ring
27. Plastic locking type clamp
28. Electric trim motor
29. Screw

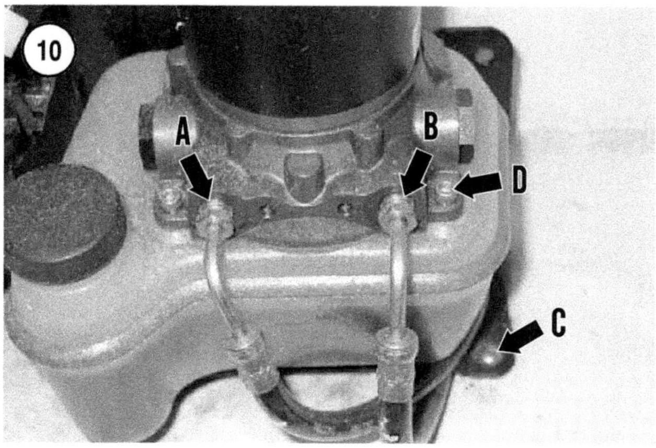

8B. Remove the plugs and caps and then thread the up pressure (A, **Figure 10**) and down pressure (B) hose fittings into the corresponding trim pump openings. Grip the crimped section of the hose with pliers to prevent hose twisting and then use a flare nut wrench to securely tighten the fittings.

9A. On models using the early-style pump, install the circuit board as described in this chapter.

9B. On models using the later-style pump (**Figure 3**), connect the trim switch (**Figure 6**) and trim power (**Figure 7**) harness connectors to the engine wire harness connectors. Route the wiring to prevent contact with moving components. Secure the wiring with plastic locking clamps as needed. Contact with moving components, especially the steering system components, is a common cause of trim electrical system failure.

10. Connect the battery cable.

11. Fill and bleed air from the trim system as described in this chapter.

Trim Motor Removal and Installation

Two different styles of pumps are used on the models covered in this manual. Earlier models use the pump shown in **Figure 1**. The earlier pump has a top-mounted fluid reservoir and a circuit board attached to the side of the pump.

Later models use the pump shown in **Figure 3**. The later pump has a bottom-mounted fluid reservoir with relays and a fuse attached to the pump mounting bracket.

WARNING
The hydraulic system contains fluid under very high pressure. Wear safety glasses and protective clothing when working with the hydraulic system.

Early-style pump (with top-mounted reservoir)

Two different versions of the early-style trim pump are used. The main difference is the components of the valve body. Refer to **Figure 6** for 1994-early 1996 models. Refer to **Figure 7** for late 1996-1998 models.

1. Remove the trim pump assembly as described in this chapter.
2. Wipe the outside of the pump with mild solvent, such as mineral spirits. Wipe the solvent off the pump with a lint-free shop towel.
3. Place the pump on a clean work surface with the reservoir facing down.
4. Remove the three screws (29, **Figure 8** or 29, **Figure 9**) and then carefully separate the trim motor from the valve body. Remove the O-ring (26, **Figure 8** or 26, **Figure 9**) from the trim motor or valve body. Discard the O-ring.
5. Remove the coupler (25, **Figure 8** or 25, **Figure 9**) from the valve body or trim pump shaft.
6. Carefully lift the ring filter (24, **Figure 8** or 24, **Figure 9**) from the valve body.
7. Use solvent to clean the filter and the exposed surfaces of the valve body. Dry the filter and valve body with compressed air.
8. Cover the valve body to prevent dust and dirt from contaminating the surfaces. Dust or dirt buildup will cause malfunction and/or rapid wear of the pump components.
9. Remove the cover just before starting the assembly. Fit the coupler (25, **Figure 8** or 25, **Figure 9**) onto the gear shaft protruding from the valve body. Install the ring filter (24, **Figure 8** or 24, **Figure 9**) onto the valve body with the foam side facing up or away from the valve body.
10. Fill the cavity within the ring filter with the recommended trim fluid (refer to Chapter Three). Fit the tip of a small (clean) screwdriver into the slot in the coupler. Use the screwdriver to rotate the coupler in both directions until a noticeable increase in effort is required to rotate the coupler. This indicates that the air has been purged from the pump passages.
11. Lubricate the surfaces with trim fluid and then install a *new* O-ring (26, **Figure 8** or 26, **Figure 9**) onto the trim motor and seat it against the step.
12. Position the pump motor over the valve body with the armature shaft aligned with the coupler.
13. Rotate the pump to align the armature shaft with the coupler. Then seat the pump onto the valve body.

12

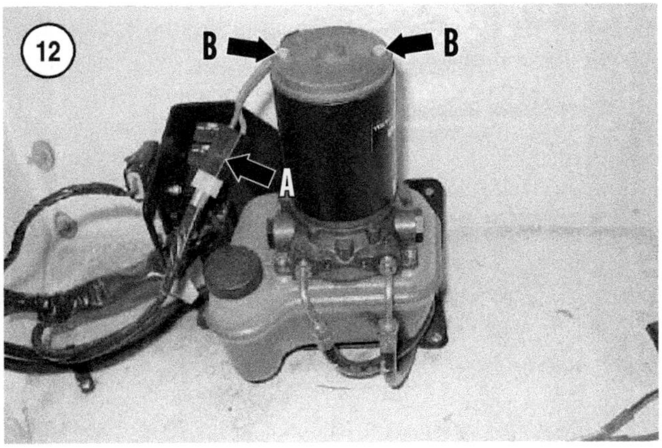

14. *Slowly* rotate the pump motor to align the screw openings. Verify that the pump motor is fully seated against the valve body. Then thread the three screws (29, **Figure 8** or 29, **Figure 9**) into the motor and valve body openings. Tighten the three screws evenly to 35-52 in.-lb. (4-6 N•m).
15. Install the trim pump assembly as described in this chapter.

Later-style trim pump (with bottom-mounted reservoir)

1. Disconnect the negative battery cable.
2. Use compressed air to blow debris and loose material off the trim pump.
3. Pull up on the locking tab and then unplug the relay harness connector (A, **Figure 12**) from the trim motor harness connector.
4. Fully loosen the two throughbolts (A, **Figure 12**). Do not remove the bolts.
5. Lift the trim motor (1, **Figure 13**) off the hydraulic pump and valve body (4).
6. Remove the O-ring (**Figure 14**) from the motor or pump. Replace the O-ring if torn, flattened or deteriorated.
7. Remove the shaft coupler from the hydraulic pump (**Figure 15**). Wipe hardened grease or debris off the coupler, motor shaft and hydraulic pump shaft.
8. Cover the hydraulic pump to prevent dust and dirt from contaminating the surfaces. Dust or dirt buildup will cause rapid wear of the shaft seal in the pump.
9. Apply a light coat of the recommended trim fluid (refer to Chapter Three) to the O-ring and then fit it into the recess in the hydraulic pump (**Figure 14**).
10. Apply OMC Triple Guard Grease or equivalent to the shaft coupling. Then fit the coupling onto the hydraulic pump shaft (**Figure 16**).
11. Install the trim motor onto the hydraulic pump and align the motor shaft with the slot in the shaft coupling. Rotate the motor to align the throughbolt openings and trim motor harness as shown in **Figure 12**.
12. When the shafts, coupling, throughbolts and threaded openings align, seat the motor onto the pump.

13

LATER-STYLE TRIM PUMP ASSEMBLY (1998-2000 MODELS)

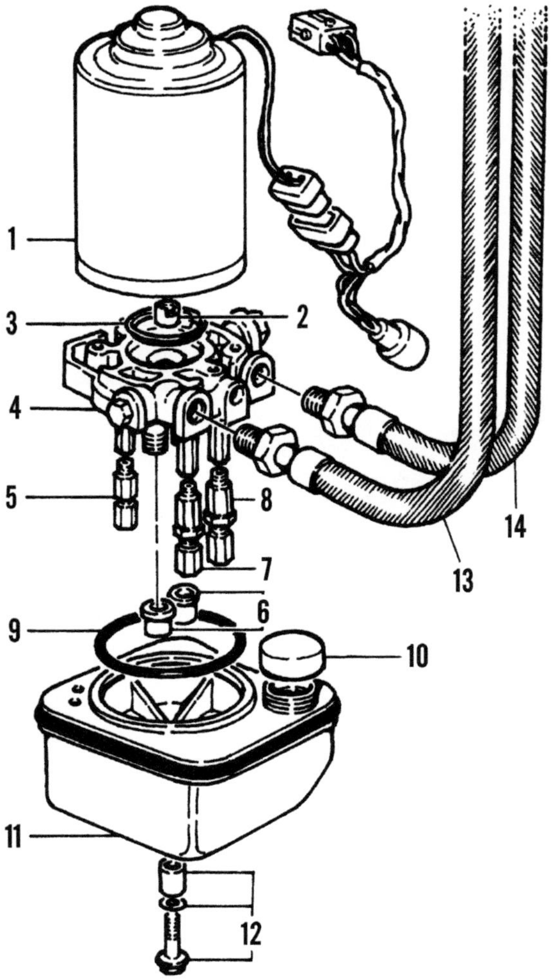

1. Trim motor
2. Shaft coupler
3. O-ring
4. Hydraulic pump and valve body
5. Thermal relief valve
6. Fluid screens
7. Up pressure relief valve
8. Down pressure relief valve
9. O-ring
10. Cap
11. Fluid reservoir
12. Screw, O-ring and sleeve*
13. Up pressure hose
14. Down pressure hose

*Not used on all models

14

15

16

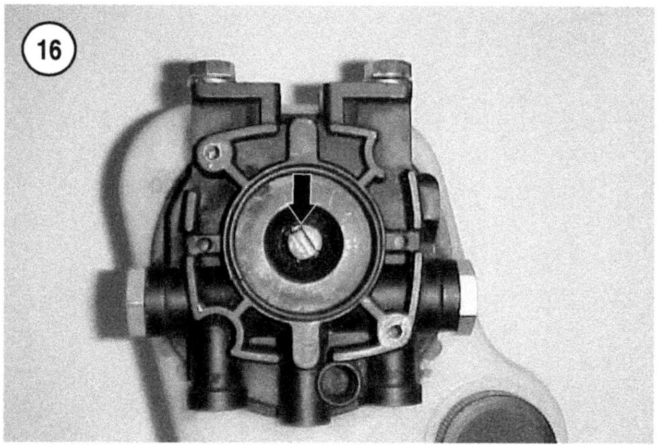

13. Thread the throughbolts by hand into the pump. Verify that the motor is seated on the pump and then securely tighten the screws.
14. Pull up on the locking tab and then plug the motor harness connector into the relay harness connector. Tug on the wiring to verify a secure connection.
15. Connect the battery cable.

HYDRAULIC TRIM SYSTEM COMPONENTS

This section describes removal and installation of the trim pump; fluid reservoir; pressure relief valves; fluid screens; trim cylinders; and related lines, hoses and manifolds.

Two different styles of pumps are used on the models covered in this manual. Earlier models use the pump shown in **Figure 1**. The earlier pump has a top-mounted fluid reservoir and a circuit board attached to the side of the pump.

Later models use the pump shown in **Figure 3**. The later pump has a bottom-mounted fluid reservoir with relays and a fuse attached to the pump mounting bracket.

WARNING
The hydraulic system contains fluid under very high pressure. Wear eye protection and protective clothing when working with the hydraulic system.

Fluid Reservoir Removal and Installation

Early-style pump (with top-mounted reservoir)

Two different versions of the early-style trim pump were used. The main difference is the components of the valve body. Refer to **Figure 8** for 1994-early 1996 models. Refer to **Figure 9** for late 1996-1997 models.

1. Remove the trim pump assembly as described in this chapter.
2. Wipe the outside of the pump with mild solvent, such as mineral spirits. Wipe the solvent off the pump with a lint-free shop towel.
3. Place the pump on a clean work surface with the reservoir facing up.
4. Remove the four screws (4, **Figure 8** or 4, **Figure 9**). Then separate the fluid reservoir (3, **Figure 8** or 3, **Figure 9**) from the valve body.
5. Remove the O-ring (6, **Figure 8** or 6, **Figure 9**) from the reservoir or valve body. Discard the O-ring.
6. Use solvent to clean the reservoir and the exposed surfaces of the valve body. Dry the reservoir and valve body with compressed air.
7. Cover the valve body to prevent dust and dirt from contaminating the surfaces. Dust or dirt buildup will cause malfunction and/or rapid wear of the pump components.
8. Remove the cover just before starting the assembly. Lubricate the surfaces with the recommended trim fluid (refer to Chapter Three) and then install a new O-ring (6, **Figure 8** or 6, **Figure 9**) onto the fluid reservoir. Make sure the O-ring fits within the recess in the reservoir.
9. Guide the reservoir onto the valve body. Align the screw openings and then seat the reservoir onto the valve body. Make sure the O-ring was not dislodged and then thread the four screws (4, **Figure 8** or 4, **Figure 9**) into the reservoir and valve body openings. Tighten the screws in a crossing pattern to 35-52 in.-lb. (4-6 N•m).
10. Install the trim pump as described in this chapter.

Later-style trim pump (with bottom-mounted reservoir)

Two different types of reservoirs are used on this style of pump. Some reservoirs are secured to the hydraulic pump with a screw, O-ring and sleeve (12, **Figure 13**). This type of reservoir must be drained before removing the hydraulic pump. The later-style reservoir is secured to the hydraulic pump with four screws (D, **Figure 10**) that thread into metal inserts embedded in the reservoir. On the later design, the pump can be removed without draining the reservoir. The reservoirs are not interchangeable and must be matched to a pump with the proper attaching bosses.

NOTE

*Reservoir failure is generally caused by impact with boat structure due to a poor mounting location or loose mounting fasteners. Another cause of damage is inadequate venting of the reservoir. Unless properly vented, the rising and falling fluid level in the reservoir that occurs during normal trim system operation will cause internal pressure and vacuum that stresses and eventually cracks the reservoir. Improper venting is generally caused by failure to remove the shipping plug (**Figure 17**) or from installing a nonvented cap onto the opening.*

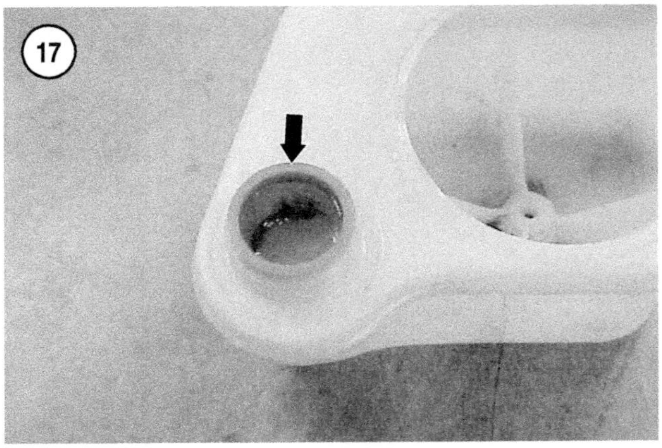

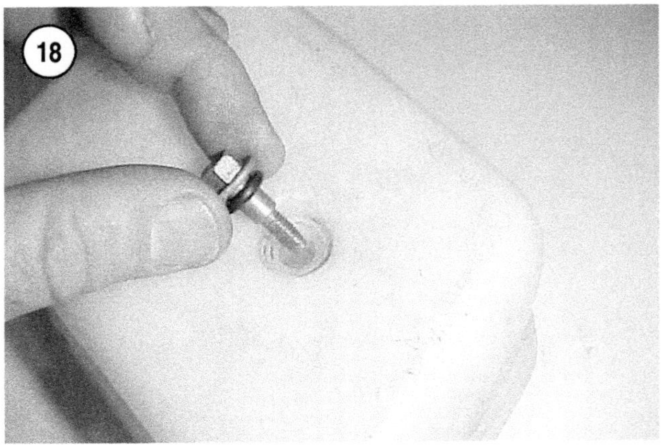

1. Disconnect the negative battery cable.
2. Remove the trim pump as described in this chapter.
3. On models using a screw, O-ring and sleeve (12, **Figure 13**), remove the fill cap (**Figure 11**). Invert the pump and pour the fluid into a suitable container. If present, remove and discard the shipping plug.

4A. On models using a screw, O-ring and sleeve (12, **Figure 13**), place the trim pump into a container with the motor side facing downward. Remove the screw, O-ring and sleeve (**Figure 18**) and then pull the reservoir off the trim pump. Discard the O-ring.

4B. On models with the four screws (D, **Figure 10**), place the trim pump on a clean work surface with the motor side facing upward. Remove the four screws (D, **Figure 10**) that secure the reservoir to the hydraulic pump. Lift the trim pump and motor assembly out of the reservoir.

5. Remove the O-ring (**Figure 19**) from the reservoir or hydraulic pump. Discard the O-ring.
6. To prevent contamination, wrap the hydraulic pump and threaded-in valves with clean lint-free shop towels.
7. Lubricate with the recommended trim fluid (refer to Chapter Three) and then place the *new* O-ring (**Figure 19**) onto the replacement reservoir. Make sure the O-ring fits into the step in the opening.
8. If the pump will be reused, thoroughly clean the reservoir with clean (mild) solvent and dry thoroughly with compressed air.
9. Carefully guide the hydraulic pump and motor assembly into the reservoir and seat against the O-ring.

10A. On models using a screw, O-ring and sleeve (12, **Figure 13**), install the reservoir onto the hydraulic pump as follows:

a. Rotate the pump in the opening to orient the fill cap opening to the right of the opening for the down hose fitting. Then rotate the pump and reservoir to position the reservoir locating boss into the notch in the pump (**Figure 20**).
b. Lubricate with trim fluid and then fit a *new* O-ring onto the screw and sleeve (12, **Figure 13**).
c. Hold the reservoir in firm contact with the hydraulic pump and then invert the pump and place the assembly on a clean work surface.
d. Guide the screw, sleeve and O-ring into the reservoir opening and thread into the mounting boss on the bottom of the pump. Verify proper alignment and seating of the reservoir and pump and then securely tighten the screw. Place the pump assembly onto the work surface with the motor side facing upward.

10B. On models with the four screws (D, **Figure 10**), install the pump into the reservoir as follows:

a. Rotate the pump in the opening to orient the fill cap opening to the left of the opening for the up hose fitting (A, **Figure 10**).
b. Rotate the pump and reservoir as needed to align the four mounting screw openings.

19

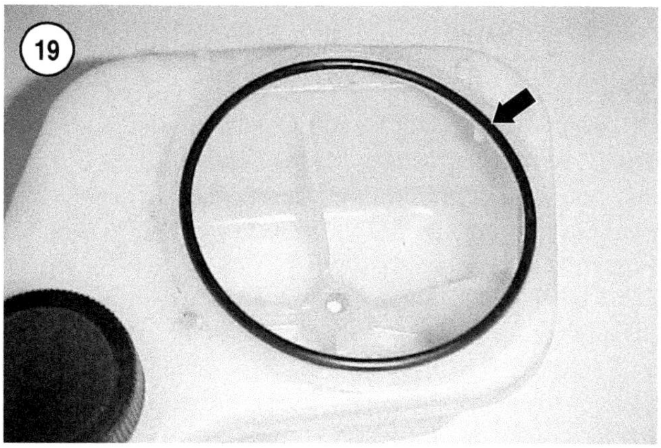

20

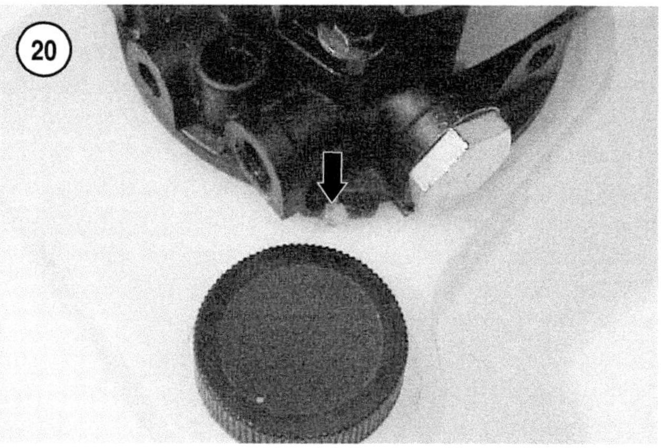

c. Thread the screws (C, **Figure 10**) into the four openings. Securely tighten the four screws in a crossing pattern.

11. *Slowly* pour trim fluid into the reservoir until the lever just reaches the *FULL* mark on the reservoir. Install the fill cap.
12. Install the trim pump as described in this chapter.

Fluid Screen Replacement

The fluid screens, used on the later-style pump with a bottom-mounted reservoir, can be replaced without removing the hydraulic pump or trim motor. Do not remove the screens unless they must be replaced. The removal process damages the screens.

1. Remove the fluid reservoir as described in this chapter.
2. Grip the metal ring portion of the screens (6, **Figure 13**) with pliers and then pull the screens straight off the hydraulic pump bosses. Do not twist the screens off the bosses. The twisting action can damage the bosses and produce metal particles that contaminate the system.
3. Use clean (mild) solvent to remove contaminants from the pump bosses.
4. Fit the replacement screen onto the pump boss. Use the open end of a 5/8 in. socket to push the screen onto the boss until the metal ring section of the screen fits fully over the boss and the boss is just visible in the screen when viewed from the side. Installing the screen too far over the boss will reduce the effective screen area. Repeat this step for the remaining screen.
5. Install the fluid reservoir as described in this chapter.
6. Fill and bleed air from the system as described in this chapter.

Hydraulic Pump Disassembly and Assembly

Two different styles of pumps are used on the models covered in this manual. Earlier models use the pump shown in **Figure 1**. The earlier pump has a top-mounted fluid reservoir and a circuit board attached to the side of the pump.

Later models use the pump shown in **Figure 3**. The later pump has a bottom-mounted fluid reservoir with relays and a fuse attached to the pump mounting bracket.

Early-style pump (with top-mounted reservoir)

Two different versions of the early-style trim pump were used. The main difference is the components of the valve body. Refer to **Figure 8** for 1994-early 1996 models. Refer to **Figure 9** for late 1996-1997 models.

1. Remove the trim pump assembly as described in this chapter.
2. Remove the trim motor and then the fluid reservoir as described in this chapter.
3. Remove the four socket head screws (10, **Figure 8** or 10, **Figure 9**) from the pump housing. To prevent pump housing or valve body damage, loosen the screws in one-turn increments.
4. Prepare to capture the two pump control pistons and check balls (15, **Figure 8** or 15, **Figure 9**). Then carefully lift the pump housing (8, **Figure 8** or 8, **Figure 9**) from the valve body. If they do not fall out, carefully remove the pistons and check balls from the housing. Work carefully to avoid damaging the piston bores.
5. Carefully remove the relief valve seats (12 and 14, **Figure 8** or 12 and 14, **Figure 9**) from the pump housing. Then remove the relief valves and springs (11 and 13, **Figure 8** or 11 and 13, **Figure 9**).
6. Mark the gears to indicate the top side and gear teeth that are currently meshed. The gears must be installed exactly as removed.
7. Remove the gears (21, **Figure 8** or 21, **Figure 9**) and inlet check balls (17, **Figure 8** or 17, **Figure 9**) from the valve body.
8. Completely fill the trim hose ports with needle bearing assembly grease. Purchase the grease from an OMC dealership. Thread a fine-thread 7/16 in. bolt into each of the trim hose ports. Withdraw each bolt and add more grease as needed until the hydraulic pressure exerted by the grease and bolt forces the check valve and seat (19, **Figure 8** or 19, **Figure 9**) out of the valve body. Remove the spring (18, **Figure 8** or 18, **Figure 9**) from the valve body. Remove the O-rings (20, **Figure 8** or 20, **Figure 9**) from the seats. Discard the O-rings.

12

9. On early models, insert a suitable tool into the opening in the trim motor side of the valve body and then push the manual release check ball, retainer and O-ring (16, **Figure 8**) out of the valve body. Discard the O-ring.

10. On early models, remove the snap ring (30, **Figure 8**) and then unthread the manual release valve (32, **Figure 8**) from the valve body. Remove the O-rings (31 and 33, **Figure 8**) and nylon ring (34) from the valve. Discard the O-rings and nylon ring. Remove the filter (35, **Figure 8**) from the valve body. Replace the filter if damaged.

11. On later models, remove the thermal relief valve (16, **Figure 9**) from the valve body. Remove the O-ring from the valve or valve body. Discard the O-ring.

12. Clean the pump housing, valve body, gears and seats with mild solvent and dry with compressed air.

13. Inspect the pump housing-to-valve body mating surfaces for scoring or other surface imperfections. The pump will not develop full pressure if these surfaces are damaged. If the surfaces are damaged, replace the pump assembly.

14. Inspect the pump control pistons and corresponding bores in the pump housing for wear, roughness or other surface imperfections. If damaged, replace the pump assembly.

15. Inspect all valve seats, springs, check valves and check balls for defects and replace as needed.

16. On early models, lubricate with trim fluid and then install the *new* O-rings (31 and 33, **Figure 8**) and nylon ring (34) onto the manual release valve. Thread the manual release valve into the valve body opening, then tighten the valve to 45-55 in.-lb. (5-6 N•m). Install the snap ring (30, **Figure 8**) into the groove within the valve opening.

17. On early models, install a *new* O-ring onto the retainer and then insert the check ball and then the retainer (16, **Figure 8**) into the valve body opening until the retainer is flush with the mating surface.

18. On early models, carefully push the filter (35, **Figure 8**) into the valve body opening.

19. On later models, lubricate with trim fluid and then install a *new* O-ring onto the thermal relief valve (16, **Figure 9**). Thread the valve into the valve body. Tighten the relief valve to 45-57 in.-lb. (5-6 N•m).

20. Lubricate with trim fluid and then install *new* O-rings (20, **Figure 8** or 20, **Figure 9**) onto the check valve seats (19, **Figure 8** or 19, **Figure 9**). Insert the check valve springs (18, **Figure 8** or 18, **Figure 9**) into the valve body. Install the check valve into the openings and rest on the springs with the pointed end facing away from the springs. Carefully press the check valve seats into the valve body until flush with the mating surface.

21. Lubricate the two inlet check balls (17, **Figure 8** or 17, **Figure 9**) and gears (21) with trim fluid. Insert the check balls into the valve body openings. Orient the gear to align the up side and gear teeth mesh and then carefully install the gears into the valve body. The shallow recess in the gears must face up or toward the fluid reservoir.

22. Lubricate with trim fluid and then install a *new* O-ring onto the trim up relief valve seat (14, **Figure 8** or 14, **Figure 9**). Install the spring and then the trim up relief valve (13, **Figure 8** or 13, **Figure 9**) into the pump housing. Then install the relief valve seat into the housing.

23. Lubricate with trim fluid and the install a *new* O-ring onto the trim down relief valve seat (12, **Figure 8** or 11, **Figure 9**). Install the spring and then the trim up relief valve (11, **Figure 8** or 11, **Figure 9**) into the pump housing. Then install the relief valve seat into the housing.

24. Lubricate with trim fluid and then install the two check balls and the pump control pistons (15, **Figure 8** or 15, **Figure 9**) into the pump housing.

25. Working carefully to avoid dislodging any components., position the pump housing over the valve body. Make sure all components are in place, then seat the housing onto the valve body. Rotate the housing to align the screw opening, then hand-thread the four socket head screws (10, **Figure 8** or 10, **Figure 9**) into the housing and valve body. Tighten the screws in one-turn increments, and in a crossing pattern, until the check valve springs compress and the pump housing seats against the valve body. Tighten the screws in a crossing pattern to 35 in.-lb. (4 N•m).

26. Insert a small (clean) screwdriver into the recess in one of the gear shafts. Use the screwdriver to rotate the gears while checking for binding or roughness. If binding or roughness is evident, loosen the four screws one-half turn and reposition the pump housing on the valve body. Retighten the screws and again check for binding or roughness. Continue this step until the gears turn freely.

27. Install the fluid reservoir and then the trim motor as described in this chapter.

28. Install the trim pump as described in this chapter.

Later-style trim pump (with bottom-mounted reservoir)

1. Remove the trim pump, trim motor and then the reservoir as described in this chapter.

2. Remove the two bolts that secure the hydraulic pump to the mounting bracket.

3. Place the trim pump on a clean work surface with the trim motor side facing downward.

4. To remove the thermal relief valve (C, **Figure 21**), provide continual downward force on the valve while unthreading it from the pump. The valve is spring-loaded. When completely unthreaded, *slowly* reduce the pressure and pull the valve, spring, ball and seat from the pump.

CAUTION

Do not remove the up and down pressure relief valves unless they must be replaced. The removal process can change the pressure relief settings. Operating the trim system with incorrect pressure will cause the system to

malfunction and can damage to the trim system.

NOTE

The replacement up and down pressure relief valves are of a different design than the factory originals and can be removed and installed without changing the pressure relief setting.

5. To remove the up pressure relief valve (A, **Figure 21**), loosen the jam nut at the base of the valve and then unthread the valve from the pump. Discard the valve upon removal.
6. To remove the down pressure relief valve (B, **Figure 21**), loosen the jam nut at the base of the valve (D) and then unthread the valve from the pump. Discard the valve upon removal.
7. Clean oil residue and other contaminants from the pump and thermal relief valve components (if removed) with clean (mild) solvent. Dry the pump with compressed air. Direct air into the passages to remove solvent residue.
8. Inspect the fluid screens for blockage or damage. Replace the screens as described in this chapter if these or other defects are evident.

CAUTION

Using only the shank portions of the valves, tighten the replacement relief valves. Tightening the valves by using the jam nut or threaded-in section might change the pressure setting. The new valve must be replaced if the jam nut or threaded-in section is disturbed.

9. If removed, thread the replacement up pressure relief valve (A, **Figure 21**), down pressure relief valve (B) and thermal relief valve (C) into the pump. Tighten the valves using only the shank portions of the valves. The shank portion threads into the hydraulic pump. Tighten the replacement valves to 71 in.-lb. (8 N•m).
10. Install the trim motor and fluid reservoir as described in this chapter.
11. Install the trim pump as described in this chapter.

TRIM CYLINDER AND HOSES

This section describes trim cylinder and hose removal, cylinder resealing and installation.

Cylinder

Removal

Refer to **Figure 22** for this procedure.

1. Disconnect the negative battery cable.
2. Remove the drive unit as described in Chapter Ten.
3. Remove the screw (3, **Figure 22**), grounding wire (4) and retainer (5) from the front of the cylinder. Then lift the plastic cover (2, **Figure 22**) off the trim cylinder (1).
4. Place a suitable container under the hose fittings to capture spilled fluid.
5. Remove the screw and washer (6, **Figure 22**) and then pull the line clamp (7) off the down pressure line and trim cylinder.
6. Wrap a clean shop towel around the up pressure hose (15, **Figure 22**) fitting. Then use a flare nut wrench to *loosen* the fitting. Allow residual fluid pressure to dissipate and then unthread the fitting from the cylinder. Remove the O-ring (17, **Figure 22**) from the hose fitting or threaded opening. Discard the O-ring. Cap the line fitting and plug the threaded cylinder opening to prevent contamination and minimize fluid loss.
7. Following the procedure described in Step 6, disconnect the down pressure hose (16, **Figure 22**) fitting.
8. Remove the plastic cap (8, **Figure 22**) from each end of the front cylinder rod (14). Remove the nut (9, **Figure 22**) and washer (10) from the cylinder. Pull the trim cylinder off the rod and place on a clean work surface.
9. Pull the plastic bushings (11, **Figure 22**) and grommets (12) out of the trim cylinder (1). Remove the plastic bushing (11, **Figure 22**) from the cylinder rod and rod bore in the gimbal ring (18).
10. If necessary, remove the opposing cylinder as described in Steps 3-10.
11. If both cylinders are removed, pull the rod (14, **Figure 22**) out of the gimbal ring bore and inspect the surfaces. Replace the rod if it is bent, is worn or has damaged threads that cannot be renewed with a die.
12. Inspect the bushings for excessive wear, cracks or distortion. Inspect the rubber grommet(s) for excessive wear, tears or deterioration. Replace the bushings and/or grommets unless in excellent condition.

Disassembly and inspection

Disassemble the cylinder if it is suspected of leaking internally. A reseal kit is available; however, major components (such as the cylinder rod, piston, rod, end cap or eyelet) are not available separately. The cylinder must be replaced as an assembly if any of the major components are

(22) **TRIM CYLINDER AND HOSES**

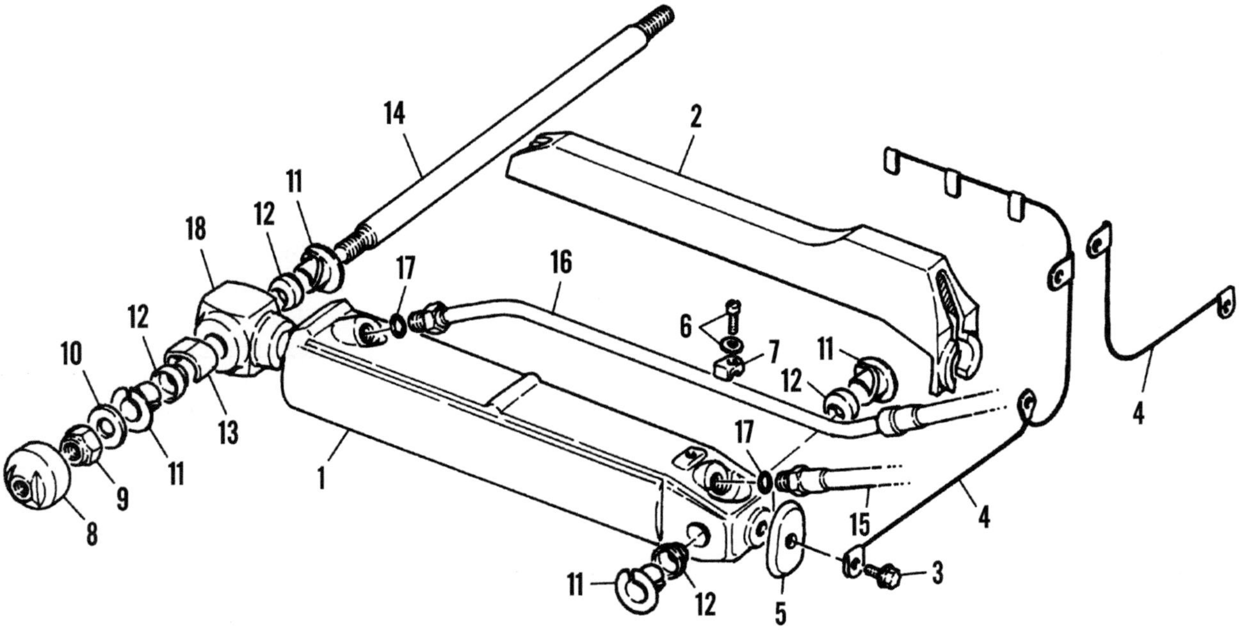

1. Trim cylinder
2. Plastic cover
3 Screw
4. Grounding wire
5. Retainer plate
6. Screw and washer
7. Plastic retainer
8. Plastic cap
9. Nut
10. Flat washer
11. Plastic bushing
12. Rubber grommet
13. Ground clip
14. Cylinder rod
15. Up pressure hose
16. Down pressure hose
17. O-ring
18. Gimbal ring

damaged. Bear in mind that the condition affecting one cylinder is probably affecting the other. It is good to reseal both cylinders.

1. Clamp the front pivot rod boss of the cylinder into a vise with the trim hose fittings facing down. Place suitable containers under the fittings to capture spilled fluid.

2. *Slowly* extend the ram to expel fluid from the down pressure fitting. Then *slowly* push the ram into the cylinder to expel fluid from the up pressure fitting. Repeat this several times to remove as much fluid as possible from the cylinder.

3. Clamp the front pivot rod boss into a vise with the ram facing upward. Use a spanner wrench (OMC part No. 3854365-8) or equivalent to loosen the end cap (**Figure 23**). If necessary, tap on the spanner wrench with a plastic mallet to loosen the cap. Unthread the cap until there is approximately one thread left engaged.

4. *Slowly* extend the cylinder until the attached piston (in the cylinder) just contacts the cap. Unthread the cap and pull the ram and cap assembly out of the cylinder.

5. Remove the eyelet from the ram as follows:

   a. Thoroughly clean all oil and grease from the cylinder ram surfaces.

(23)

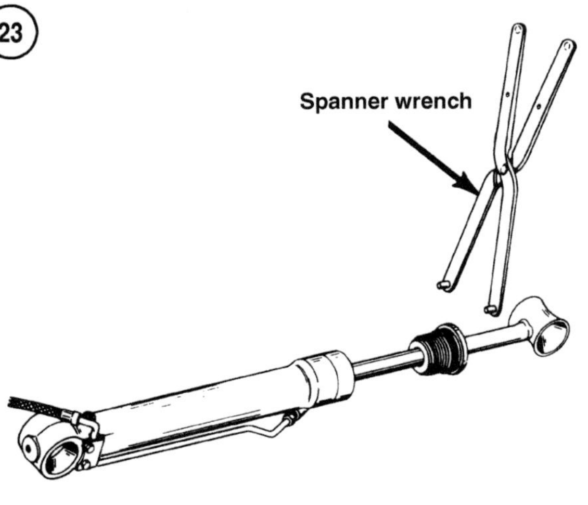

   b. Wrap paper around the ram and then fit the clamp blocks (OMC part No. 3854367-4) onto the paper and ram surfaces. Clamp the blocks and ram firmly into a vise as shown in **Figure 24**.

   c. Insert one of the cylinder anchor rods (removed with the cylinder or drive unit) into the eyelet opening.

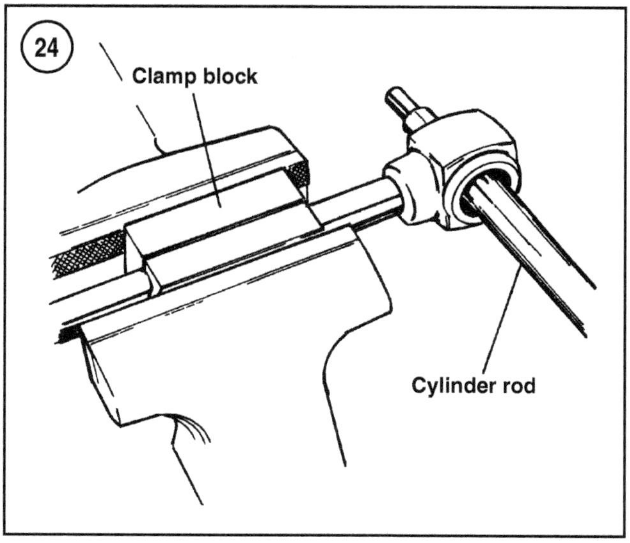

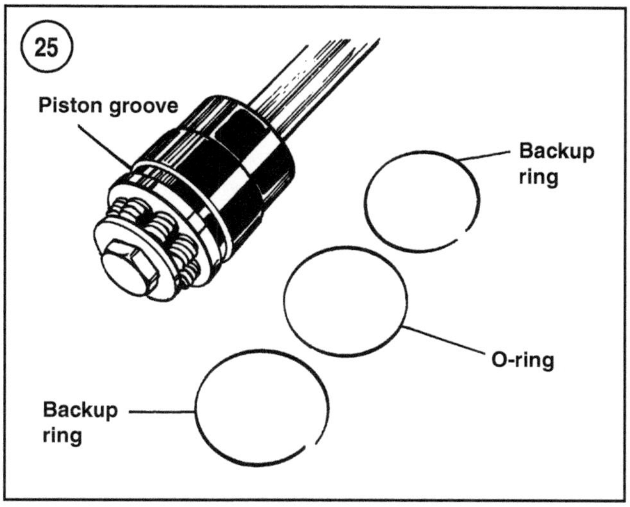

d. Use the rod to unthread the eyelet. If necessary, apply mild heat to the eyelet to soften the threadlocking compound. *Do not* allow the cylinder ram to rotate in the clamping block. Tighten the vise and apply more heat to the eyelet if necessary.

6. Slide the end cap off the cylinder ram. Use a dental pick or other suitable instrument to hook and remove the O-ring and backup rings from the piston (**Figure 25**) groove. *Do not* scratch the piston surfaces with the removal tool. Discard the O-ring and backup rings.

CAUTION
Work carefully when removing the O-rings and scraper from the cylinder end cap. Do not allow the removal tools to contact and damage any surfaces. The end cap is not available separately. If the cap is damaged, the cylinder assembly must be replaced.

7. Use a pick or other suitable instrument to hook and remove the internal and external O-rings from the end cap. Discard the O-rings.

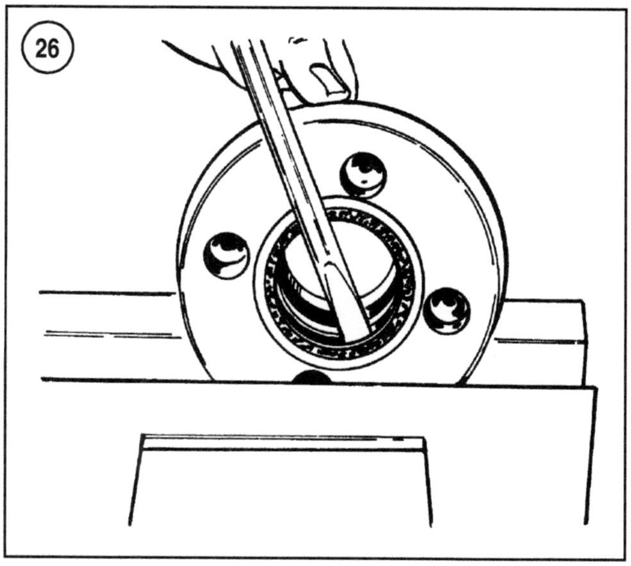

8. Clamp the end cap in a vise with protective jaws as shown in **Figure 26**. Do not clamp on the threaded surfaces. Carefully pry the scraper ring out of the end cap. *Do not* allow the removal tool to contact and damage the scraper bore in the end cap. The cap is not available separately. If damaged, the cylinder assembly must be replaced.

9. Wash all components with clean solvent and dry with compressed air.

10. Remove all threadlocking sealant from the ram and eyelet threads.

11. Inspect the internal cylinder walls and end cap threads for scoring, excessive wear, corrosion pitting or other damage. Replace the cylinder assembly if these or other defects are evident.

12. Inspect the cylinder end cap for excessive corrosion damage and mechanical damage to the threads, O-ring grooves or scraper ring bore. Replace the cylinder assembly if these or other defects are evident. The end cap is not available separately.

13. Inspect the relief valves in the cylinder piston for broken or damaged springs. Use a suitable drift to press against each check valve. The check valve must by held firmly in place and be movable with the drift. If not, replace the cylinder assembly.

14. Remove any debris from the check valve as follows:

a. Clamp the ram into a vise and clamp blocks as described in Step 5.

b. Loosen the large bolt head on the cylinder piston just enough to allow opening of the check valve balls with a drift.

c. Open each valve individually and flush the seat and ball with clean solvent.

d. After all valves are thoroughly flushed, securely tighten the large bolt head.

Assembly

Lubricate all components with the recommended trim fluid (refer to Chapter Three) during assembly. Do not make any dry assemblies.

1. Install the *new* scraper ring into the end cap as follows:
   a. Place the end cap on a clean work surface with the threaded end facing downward.
   b. Fit the scraper into the bore. Use the installation tool (OMC part No. 38554364-1) to press the scraper into the bore until flush with the end cap surface.
2. Lubricate the surfaces and then fit *new* O-rings into the internal and external end cap grooves.
3. Slide the seal protector (OMC part No. 3854363-3) over the cylinder ram. Then slide the end cap, threaded end first, over the protector and position on the cylinder ram. Remove the protector.
4. Install the cylinder eyelet as follows:
   a. Clean the threads and then apply a light coat of Loctite 242 threadlocking adhesive (OMC part No. 500418) to the ram and eyelet threads.
   b. Wrap paper around the ram and then fit the clamp blocks (OMC part No. 3854367-4) onto the paper and ram surfaces. Clamp the blocks and ram firmly into a vise as shown in **Figure 24**.
   c. Insert one of the cylinder anchor rods (removed with the cylinder or drive unit) into the eyelet opening. Use the rod to securely tighten the eyelet.
5. Install a *new* O-ring and backup rings into the piston groove. Make sure the O-ring fits between the backup rings and the backup rings fit securely on the piston. If necessary, remove the ring(s) and compress *slightly* to achieve a secure fit.
6. Clamp the front pivot rod boss of the cylinder into a vise with the end cap opening facing upward.
7. Lubricate the cylinder bore, ram and piston liberally, then carefully insert the piston into the cylinder bore. Be careful not to catch the backup rings between the piston and cylinder wall. Once the piston reaches the tapered bore, apply steady downward pressure to the eyelet until the piston fully seats in the bore.
8. *Slowly* pour trim fluid into the end cap opening until the level just reaches the taper. Make sure to allow room in the cylinder for the end cap.
9. Thread the end cap by hand into the cylinder. Use a spanner wrench (OMC part No. 3854365-8) or equivalent (**Figure 23**, typical) to tighten the end cap to 25-30 ft.-lb. (24-41 N•m).
10. Plug the up and down pressure ports in the cylinder to prevent contamination and fluid loss.
11. Install the cylinder(s) as described in this chapter.

Installation

1. If removed, install the cylinder rod as follows:
   a. Apply OMC Triple Guard or Volvo Penta Propeller Shaft Grease (Volvo Penta part No. 1141644-3) to the cylinder rod.
   b. Guide the cylinder rod (14, **Figure 22**) into the gimbal ring bore until centered. Slide plastic bushing (11, **Figure 22**) over each end of the rod with the flange side facing away from the gimbal ring. Guide the bushings over the rod and into the gimbal ring rod bores. Push the bushing into the bore until the flange seats against the gimbal ring.
2. Fit the two rubber grommets (12, **Figure 22**) over the smaller-diameter end of the bushings (11). Slide one of the bushing assemblies over the rod with the rubber grommet side facing away from the gimbal ring. Seat the assembly against the gimbal ring.
3. Position the rod opening at the front of the cylinder over the rod. Hold the opposing side of the rod and carefully guide the cylinder opening over the bushing assembly.
4. Support the rod and cylinder and then fit the second bushing assembly over the rod with the rubber grommet side facing the cylinder. Hold the opposing side of the rod and guide the bushing into the cylinder opening.
5. Install the washer (10, **Figure 22**) over the end and seat it against the plastic bushing. Thread the nut by hand onto the rod.
6. If removed, install the opposing cylinder as described in Steps 3-6.
7. Tighten the nut (9, **Figure 22**) on each side of the cylinder rod anchor pin to 124-142 in.-lb. (14-16 N•m). Thread the plastic cover(s) (8, **Figure 22**) onto the rod and tighten securely.
8. Remove the caps and plugs from the trim hose fittings and corresponding cylinder openings.
9. Apply a light coat of automatic transmission fluid and then install *new* O-rings (17, **Figure 22**) onto the hose fittings. Do not cut the O-rings while installing them over the fitting threads.
10. Thread the hose fittings by hand into the cylinder openings. Do not tighten the fittings at this time.
11. Fit the retainer (7, **Figure 22**) onto the down pressure line and cylinder and then thread the screw and washer (6) into the clamp and cylinder opening. Securely tighten the screw.
12. Tighten the hose fittings to 84-108 in.-lb. (10-12 N•m). Grip the crimped end of the hose with pliers while tightening to prevent the hose from twisting. Twisting will promote loosening of the fitting and cause premature hose failure.
13. Hook the plastic cover (2, **Figure 22**) over the rear of the cylinder(s) and rotate down over the front of the cylinder(s). Fit the retainer plate (5, **Figure 22**) onto the cover and cylinder and then thread the screws (3) with grounding wire attached into the cover and cylinder. Securely tighten the screw.
14. Install the drive unit as described in Chapter Ten.
15. Connect the battery cable.

27

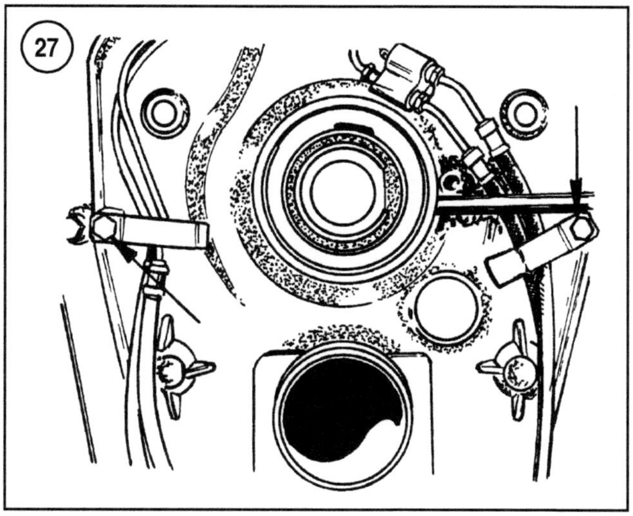

16. Fill and bleed air from the system as described in this chapter.

Trim Hoses and Manifold

This section describes replacement of the internal and external trim hoses and the trim hose manifold on the transom bracket.

External hoses (removal and installation)

Refer to **Figure 22** for this procedure.

1. Disconnect the negative battery cable.
2. Remove the drive unit as described in Chapter Ten.
3. Remove the pivot housing as described in Chapter Eleven.
4. Remove the screw (3, **Figure 22**), grounding wire (4) and retainer (5) from the front of the cylinder. Then lift the plastic cover (2, **Figure 22**) off the trim cylinder (1).
5. Place a suitable container under the hose fittings to capture spilled fluid.
6. Remove the screw and washer (6, **Figure 22**) and then pull the line retainer (7) off the down pressure line and trim cylinder.
7. Wrap a clean shop towel around the up pressure hose (15, **Figure 22**) fitting. Then use a flare nut wrench to *loosen* the fitting. Allow residual fluid pressure to dissipate and then unthread the fitting from the cylinder. Remove the O-ring (17, **Figure 22**) from the hose fitting or threaded opening. Discard the O-ring.
8. Following the procedure described in Step 6, disconnect the down pressure hose (16, **Figure 22**) fitting.
9. Repeat Steps 3-7 to disconnect the opposite trim cylinder hoses.
10. Loosen the screws (**Figure 27**) and then swing the clamps away from the hoses.
11. Clearly mark the hoses and connection points to the transom bracket manifold to ensure proper installation. Use a flare nut wrench to loosen and disconnect the hose fittings from the manifold.
12. Plug the hose fittings in the cylinders and transom manifold to prevent contamination and minimize fluid loss.
13. Remove the O-rings from the hoses or manifold openings. Discard the O-rings.
14. Lubricate with the recommended trim fluid (refer to Chapter Three) and then install *new* O-rings onto the hose fittings. Do not cut the O-rings while installing them over the fitting threads.
15. Remove the plugs and then thread the hose fittings by hand into the respective openings in the trim manifold and cylinders. Do not tighten the fittings at this time.
16. Fit the retainer (7, **Figure 22**) onto the down pressure line and cylinder and then thread the screw and washer (6) into the clamp and cylinder opening. Securely tighten the screw.
17. Tighten the cylinder hose fittings to 84-108 in.-lb. (10-12 N•m). Grip the crimped end of the hose with pliers while tightening to prevent the hose from twisting. Twisting will promote loosening of the fitting and cause premature hose failure.
18. Hook the plastic cover (2, **Figure 22**) over the rear of the cylinder(s) and rotate down over the front of the cylinder(s). Fit the retainer plate (5, **Figure 22**) onto the cover and cylinder, then thread the screws (3) with grounding wire attached into the cover and cylinder. Securely tighten the screw. Connect and tighten the hose fittings for the opposite cylinder as described in Steps 15-18.
19. Route the hoses in the transom bracket and then position the clamps over the hoses as shown in **Figure 27**. Securely tighten the clamp screws.
20. Use a flare nut wrench to tighten the hose fittings at the manifold to 84-108 in.-lb. (10-12 N•m). Grip the crimped end of the hose with pliers while tightening to prevent the hose from twisting. Twisting will promote loosening of the fitting and cause premature hose failure.
21. Install the pivot housing as described in Chapter Eleven.
22. Install the drive unit as described in Chapter Ten.
23. Connect the battery cable.
24. Fill and bleed the system as described in this chapter.

Internal hoses and manifold (removal and installation)

Two different styles of pumps are used on the models covered in this manual. Earlier models use the pump shown in **Figure 1**. The earlier pump has a top-mounted fluid reservoir and a circuit board attached to the side of the pump.

Later models use the pump shown in **Figure 3**. The later pump has a bottom-mounted fluid reservoir with relays and a fuse attached to the pump mounting bracket

1. Disconnect the negative battery cable.
2. Remove the drive unit as described in Chapter Ten.

3. Remove the pivot housing as described in Chapter Eleven.

4. Remove the engine as described in Chapter Six.

5. Remove the inner transom plate as described in Chapter Eleven.

6A. On models using the earlier-style pump (**Figure 1**), mark the trim hoses (23, **Figure 8** or 23, **Figure 9**) and the trim pump to identify the connection points. The trim system will operate opposite the desired direction if the hoses are installed in crossed locations.

6B. On models using the later-style pump (**Figure 3**), mark the trim hoses and the trim pump to identify the connection points (A and B, **Figure 10**). The trim system will operate opposite the desired direction if the hoses are installed in crossed locations.

7. Place a suitable container under the trim hose fittings and pump and then *slowly* loosen the hose fittings. Allow residual pressure to dissipate and then unthread the hose fittings. Clean up any spilled fluid. Route the hoses to a location higher than the drive unit to minimize fluid loss. Plug the hose fittings and corresponding openings in the trim pump to prevent contamination.

8. Clearly mark the internal hoses and connection points onto the manifold to ensure correct installation. The system will operate opposite the desired direction if the hoses are connected in crossed locations. Use a flare nut crowfoot adapter to loosen the hose fittings at the manifold (**Figure 28**). Disconnect the hoses. Cap the hoses. Plug the manifold openings and cap the hoses to prevent contamination and minimize fluid loss.

9. If only replacing the internal hoses, proceed to Step 19. If the manifold must be replaced, proceed to Step 10.

10. Clearly mark the external hoses and connection points to the transom bracket manifold to ensure proper installation. Use a flare nut wrench to loosen and disconnect the hose fittings from the manifold.

11. Plug the hose fittings and transom manifold openings to prevent contamination and minimize fluid loss.

12. Remove the O-rings from the hoses or manifold openings. Discard the O-rings.

13. Remove the elastic locknut and washer from the manifold mounting stud (**Figure 29**). Discard the locknut.

14. Using a suitable drift, drive the manifold out of the transom bracket. Remove the O-ring from the manifold. Discard the O-ring. Remove all corrosion, sealant or other material from the manifold and the opening in the transom bracket.

15. Install a *new* O-ring into the groove on the replacement manifold. Apply a coat of water-resistant grease to the O-ring and mating surfaces of the manifold. Working from outside the boat, guide the manifold into the transom bracket opening. Fit the manifold over the mounting stud.

16. Install the washer and then the *new* elastic locknut (**Figure 29**) over the mounting stud. Tighten the nut to 106-133 in.-lb. (12-15 N•m).

28

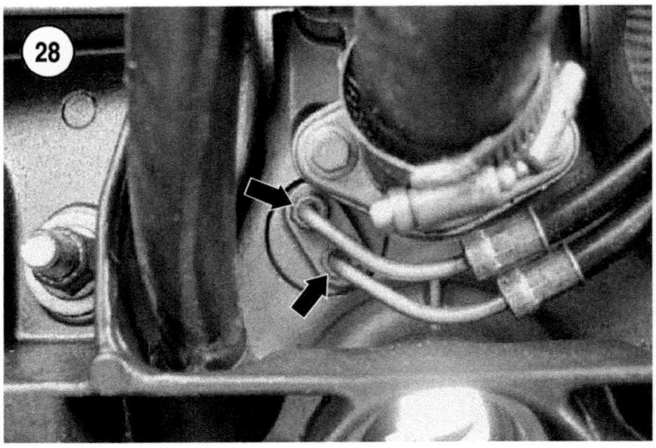

29

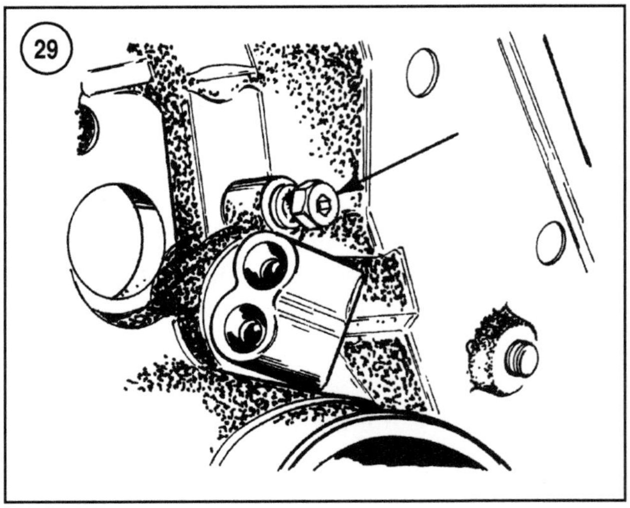

17. Lubricate with trim fluid and then install *new* O-rings onto the outer hose fittings. Do not cut the O-rings while installing them over the fitting threads.

18. Remove the caps and plugs and then thread the hose fittings by hand into the respective openings in the trim manifold and cylinders. Tighten the cylinder hose fittings to 84-108 in.-lb. (10-12 N•m). Grip the crimped end of the hose with pliers while tightening to prevent the hose from twisting. Twisting will promote loosening of the fitting and cause premature hose failure.

19. Apply trim fluid to them and then install *new* O-rings onto the internal hose fittings. Do not cut the O-rings while installing them over the fitting threads.

20. Remove the plugs and then thread the hose fittings by hand into the respective manifold openings. Route the hoses and position the fittings as shown in **Figure 28**. Tighten the manifold hose fittings to 84-108 in.-lb. (10-12 N•m). Grip the crimped end of the hose with pliers while tightening to prevent the hose from twisting. Twisting will promote loosening of the fitting and cause premature hose failure.

21. Install the inner transom plate as described in Chapter Eleven.

22. Install the engine as described in Chapter Six.

30

31

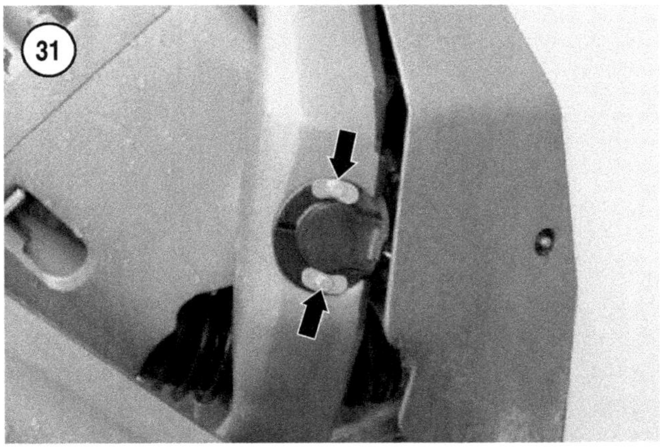

23. Install the pivot housing as described in Chapter Eleven.
24. Install the drive unit as described in Chapter Ten.
25. Connect the battery cable.
26. Fill and bleed air from the system as described in this chapter.

TRIM POSITION SENDER

This section describes removal and installation of the trim position sender. The trim position sender operates the dash-mounted trim gauge and the trim limit system used on DP-S (twin-propeller) models.

On 1994 Cobra SX models, two screws secure the trim position sender (**Figure 30**). The sender lever contacts the pivot housing and operates the sender as the drive moves up or down.

On 1995-2000 Cobra SX and DP-S models, two screws secure the trim position sender (**Figure 31**) to the gimbal. The rotor in the center of the sender fits into the recess in the hinge pin. The rotating hinge pin operates the sender as the drive moves up or down.

Removal and Installation

Do not remove the sending unit from the transom bracket unless it must be replaced. The removal process usually damages the sender leads as they are pulled from the bracket opening.

1. Disconnect the negative battery cable.
2. Remove the drive unit as described in Chapter Ten.
3. Remove the pivot housing as described in Chapter Eleven.
4. Disconnect the three-pin sender harness connector (**Figure 6**) from the engine harness (SX [single-propeller] models) or trim limit control module harness (DP-S [twin-propeller] models).
5. Remove the leads and terminals from the harness connector as follows:
   a. Note the position of each of the leads in the sender harness connector.
   b. Lubricate the pin removal tool (OMC part No. 3854350-0) with isopropyl (rubbing) alcohol.
   c. Insert the tool into the plug and push the terminals and leads out of the plug as shown in **Figure 32**.
6. Remove the spring locking clip from the sender grommet. The sender grommet protrudes from the inner transom plate. The locking clip is accessible from inside the engine compartment.

7A. On 1994 Cobra SX models, remove the two screws and then pull the sender (**Figure 30**) off the starboard side of the gimbal ring. Carefully guide the sender lever away from the pivot housing and gimbal ring while removing the sender.

7B. On 1995-2000 Cobra SX models, remove the two screws and washers (**Figure 31**) and then remove the trim position sender from the starboard side of the gimbal ring.

8. Steer the drive unit to full port. Reach behind the gimbal ring and cut the plastic locking clamp securing the sender leads to the trim hoses. *Do not* cut the trim hoses, bellows or water hose.
9. Pull the sender leads and grommet out of the transom bracket opening. Clean all corrosion and old sealant from the opening.
10. Apply a coat of water-resistant grease to the replacement sender lead grommet. Feed the sender leads into the transom bracket opening. Then seat the grommet into the opening. Have an assistant hold the grommet in the bracket. Then working inside the boat, install the spring locking clip into the groove in the grommet. Have the assistant tug *lightly* on the lead to verify a secure connection.
11. Install the leads into the harness connector plug as follows:
    a. Guide the wire terminals and leads into the respective openings in the plug.
    b. Lubricate the pin installation tool (OMC part No. 3854349-2) with isopropyl (rubbing) alcohol.
    c. Insert the tool into the plug and push the terminals and leads in position as shown in **Figure 32**.
12. Install a new plastic locking clamp over the external sender lead and *lightly* clamp it onto the trim hoses near the gimbal bearing grease tube.
13. Install the pivot housing as described in Chapter Eleven.

12

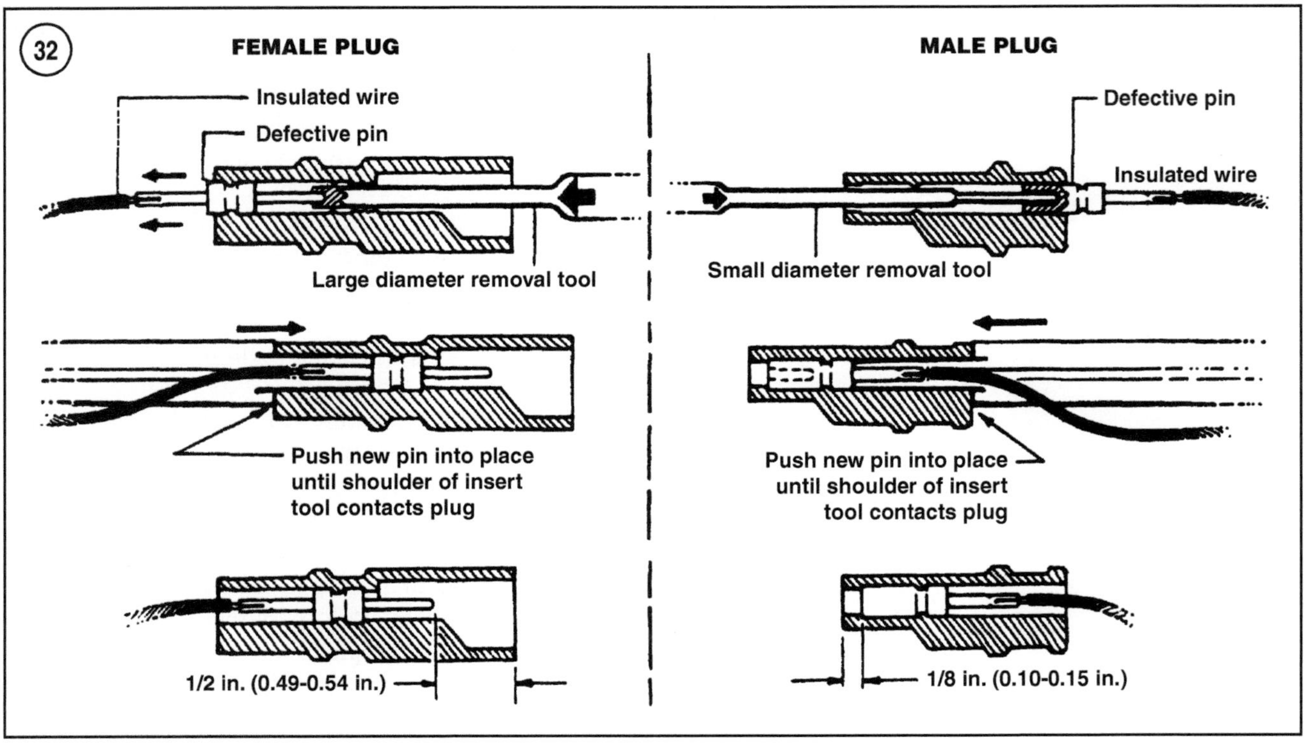

14A. On 1994 Cobra SX models, install the trim position sender as follows:

a. Guide the sender lever in position against the pivot housing.
b. Position the sender on the gimbal ring and align the screw openings.
c. Install the two screws and washers. Tighten the trim position sender screws to 18-24 in.-lb. (2-3 N•m).

14B. On 1995-2000 Cobra SX or DP-S models, perform the preliminary adjustment and install the trim sender onto the gimbal ring as follows:

a. Calibrate an ohmmeter onto the 20- or 40-ohm scale.
b. Connect the meter leads to pin A and pin C of the connector plug (installed in Step 11).
c. *Slowly* rotate the sender shaft until the ohmmeter reads approximately 11 ohms.
d. Move the pivot housing to the full down position. Hold the housing and position and carefully guide the sender shaft into the hinge pin. Rotate the sender body *slightly* if necessary to align the sender shaft with the hinge pin opening. *Do not* rotate the sender shaft to align the shaft and hinge pin.
e. Hold the sender and pivot housing in position and then thread the screws and washers into the sender and gimbal ring openings. Tighten the trim position sender screws to 18-24 in.-lb. (2-3 N•m).
f. Note the resistance on the meter. Remove the sender and readjust the sender if the meter does not indicate 10-12 ohms.

15. Connect the three-pin sender harness connector (**Figure 6**) to the engine harness (SX [single-propeller] models) or trim limit control module harness (DP-S [twin-propeller] models). Route the wiring to prevent interference with steering system components or other moving components. Secure the wiring with clamps as needed.

16. Install the drive unit as described in Chapter Ten.

17. Connect the battery cable.

18. Adjust the trim position sender as described in Chapter Five.

FILLING AND BLEEDING THE TRIM SYSTEM

Follow the filling procedure after replacing or repairing any trim system components except the electrical components.

Always prefill the trim cylinders and trim pump as described in the assembly procedures before installing them. Otherwise, the system will ventilate and foam the fluid and dramatically increase the time needed to fill and bleed the system.

Perform the bleeding procedure if air in the system is not removed during the filling procedure or if air entered the system due to a low fluid level in the reservoir, leaking hose or line, or other means. Air in the system will cause the drive unit to move in (down) under load in forward or move out when under load in reverse or during deceleration.

Two different styles of pumps are used on the models covered in this manual. Earlier models use the pump shown in **Figure 1**. The earlier pump has a top-mounted fluid reservoir and a circuit board attached to the side of the pump.

Later models use the pump shown in **Figure 3**. The later pump has a bottom-mounted fluid reservoir with relays and a fuse attached to the pump mounting bracket.

WARNING
The hydraulic system operates under very high pressure. Wear safety glasses and protective clothing when working with the hydraulic system.

Fluid Filling

1. Operate the trim pump to place the drive unit in the full DOWN position. If the pump ventilates or the drive does not move, attach an overhead lift to the drive unit for support. Provide enough lifting force to support the weight of the drive unit. Place a suitable container under the line fittings and cylinder openings to capture spilled fluid. Then disconnect the trim hoses from the trim cylinder(s) as described in this chapter. Refer to *Trim Cylinder and Hoses*. Slowly lower the drive unit to the full DOWN position. Disconnect the drive from the overhead lift. Reconnect the trim hoses before beginning the filling procedure.

2A. On models using the earlier-style pump (**Figure 1**), remove the fill plug (5, **Figure 8** or 5, **Figure 9**). Add the recommended trim fluid (refer to Chapter Three) until the fluid level is even with the bottom of the fill plug opening. Install the fill plug and securely tighten.

2B. On models with a later-style pump (**Figure 3**), remove the fill cap (**Figure 11**) from the fluid reservoir. Add trim fluid into the reservoir until the level just reaches the *FULL* mark on the reservoir. Reinstall the fill cap. *Do not* install the shipping plug into the opening. If present, remove and discard the shipping plug.

3. Operate the trim in the UP direction. Stop immediately if the trim motor suddenly increases in speed. This indicates a low fluid level is ventilating the pump. Using the hydraulic system, immediately lower the drive unit and correct the fluid level as described in Step 2. Continue until the drive reaches the full up and down positions without ventilating the pump.

4A. On models using the earlier-style pump (**Figure 1**), place the drive unit in the full DOWN position. Allow a few minutes for the foam on the fluid surface to dissipate and then correct the fluid level as described in Step 2.

4B. On models using the later-style pump (**Figure 3**), place the drive unit in the full DOWN position. Allow a few minutes for the foam on the fluid surface to dissipate and then correct the fluid level as described in Step 2.

5. If air in the system cannot be completely removed by the filling procedures, perform the air bleeding procedures as described in the following section.

Air Bleeding

1. Operate the trim pump to place the drive unit in the full DOWN position. If the pump ventilates or the drive does not move, attach an overhead lift to the drive unit for support. Provide enough lifting force to support the weight of the drive unit. Place a suitable container under the line fittings and cylinder openings to capture spilled fluid. Then disconnect the trim hoses from the trim cylinder(s) as described in this chapter. Refer to *Trim Cylinder and Hoses*. Slowly lower the drive unit to the full DOWN position. Disconnect the drive from the overhead lift. Reconnect the trim hoses before beginning the filling and bleeding procedures.

2A. On models using the earlier-style pump (**Figure 1**), remove the fill plug (5, **Figure 8** or 5, **Figure 9**). Add the recommended trim fluid (refer to Chapter Three) until the fluid level is even with the bottom of the fill plug opening. Install the fill plug and securely tighten.

2B. On models with a later-style pump (**Figure 3**), remove the fill cap (**Figure 11**) from the fluid reservoir. Add trim fluid into the reservoir until the level just reaches the *FULL* marking on the reservoir. Reinstall the fill cap. *Do not* install the shipping plug into the opening. If present, remove and discard the shipping plug.

3. Place a suitable container under the hose fitting to capture spilled fluid.

4. Disconnect the up pressure hose from the front fitting of the trim cylinder(s) as described in this chapter. Refer to *Trim Cylinder and Hoses*.

5. Direct the hose(s) into the container. Have an assistant operate the trim pump in the UP direction until there are no bubbles in the fluid stream exiting the hose(s). Reconnect the up pressure hose(s) as described in this chapter. Refer to *Trim Cylinder and Hoses*.

6. Correct the fluid reservoir as described in Step 2.

7. Disconnect the down pressure hose from the rear fitting of the trim cylinder(s) as described in this chapter. Refer to *Trim Cylinder and Hoses*. Cap the hose fitting(s) to prevent air introduction. Purchase suitable cap fittings from an automotive parts store. Thread used trim hose(s) or other suitable hose(s) into the cylinder fitting(s) farthest from the transom and direct the hoses into a suitable container to capture fluid exiting the cylinder(s). *Do not* plug the fitting opening in the cylinder(s).

8. Have an assistant operate the trim pump in the up direction until the drive reaches the full UP position. Stop immediately if the trim motor suddenly increases in speed. This indicates the pump is ventilating due to a low fluid level. Add a small amount of the recommended fluid to the reservoir. Continue until the drive reaches the full UP position.

9. Remove the caps from the down pressure hoses. Direct the hoses into the container with the hoses attached in Step 7. Have an assistant operate the trim pump in the DOWN direction until no bubbles are present in the fluid exiting the hoses.

10. Remove the temporary hoses connected to the cylinder(s) in Step 7. Remove the caps and then connect the down pressure hoses to the cylinder(s) as described in this chapter. Refer to *Trim Cylinder and Hoses*.

11. Fill the system and bleed residual air from the system as described in the *Fluid Filling* procedures in this section.

Table 1 appears on the following page.

Table 1 POWER TRIM SYSTEM TORQUE SPECIFICATIONS

Fastener location	in.-lb.	ft.-lb.	N•m
Cylinder rod (anchor pin) nut	124-142	–	14-16
Down pressure relief valve			
Later-style pump (bottom-mounted reservoir)	71	–	8
External manifold hose fittings	84-108	–	10-12
Fluid reservoir to valve body			
Early-style pump (top-mounted reservoir)	35-52	–	4-6
Manual release valve			
Early-style pump (top-mounted reservoir)	45-55	–	5-6
Internal manifold hose fittings	84-108	–	10-12
Pump housing to valve body			
Early-style pump (top-mounted reservoir)	35	–	4
Thermal relief valve			
Early-style pump (top-mounted reservoir)	45-57	–	5-6
Later-style pump (bottom-mounted reservoir)	71	–	8
Trim cylinder end cap	–	25-30	24-41
Trim cylinder hydraulic hose	84-108	–	10-12
Trim manifold locknut	106-133	–	12-15
Trim motor to valve body			
Early-style pump (top-mounted reservoir)	35-52	–	4-6
Trim position sender	18-24	–	2-3
Up pressure relief valve			
Later-style pump (bottom-mounted reservoir)	71	–	8

Chapter Thirteen

Steering System

OMC Cobra stern drive engines may be equipped with a mechanical or manual (not power-assisted) or power steering system. Proper operation of the stern drive steering system is essential for safe boating. The steering system should be rigged *only* by an experienced marine technician and serviced (whenever possible) by one who is equally qualified. This chapter covers steering safety precautions and repair to the steering system components. **Table 1** lists torque specifications for most steering system fasteners. Use the general torque recommendations provided in Chapter One for fasteners not listed in **Table 1**.

SAFETY PRECAUTIONS

The steering system connects the drive unit to the steering wheel (**Figure 1**).When properly installed and maintained, the steering system gives the boater control over the vessel. A steering system that hesitates or jams prevents avoiding obstacles, such as other boats. If the steering system is loose, the boat will weave regardless of any attempt to maintain a straight course. Steering system failure can cause complete loss of control, which can result in a serious accident and possible loss of life.

The most important safety precaution to observe is proper lubrication and maintenance of the steering system. It is especially important to check the steering system if the drive unit receives a severe blow from hitting a piling or other underwater object or while trailing the boat. Damaged or weak components might fail at a later time while the boat is on the water. Know what to look for and have any problems corrected before operating the boat.

If an experienced marine technician does not make these adjustments or repairs, then make them carefully and use only the fasteners supplied with the steering attachment kit or equivalent fasteners sold as replacement items by marine dealers. After completing adjustments or repairs, let an experienced marine technician make sure no safety hazards exist.

Whenever the engine is removed or other service that affects the steering system is performed, make sure that:

1. Cable movement is not restricted. See **Figure 2**. The cable must make a smooth bend (not exceeding 90°) to the steering cylinder assembly and not bind against the boat structure or other components. Cable restrictions can result in possible jamming of the system. On power steering models, a cable restriction can cause the power steering not to provide enough assistance or cause the drive unit to unexpectedly go into a full turn without turning the wheel.

1
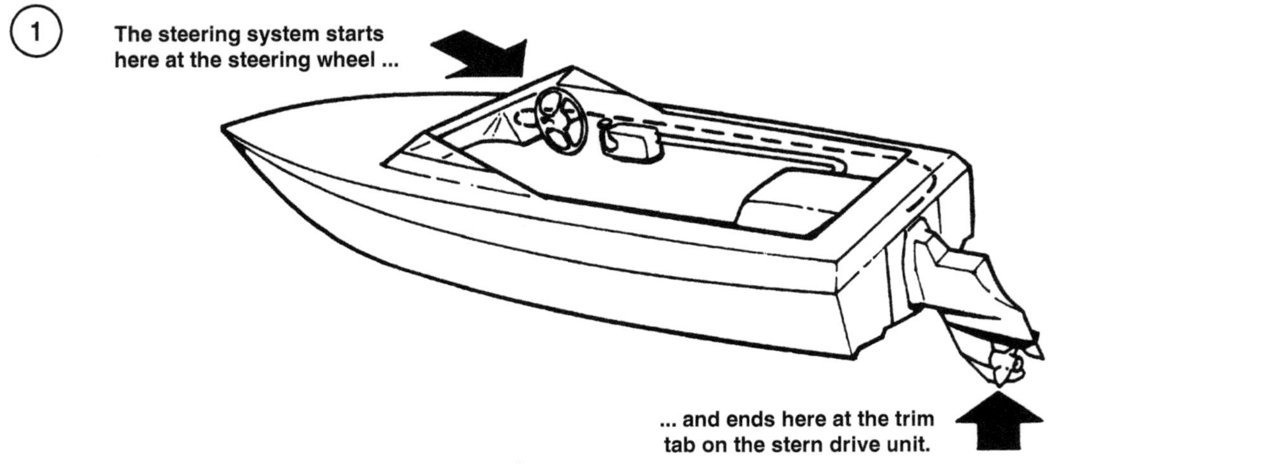

2
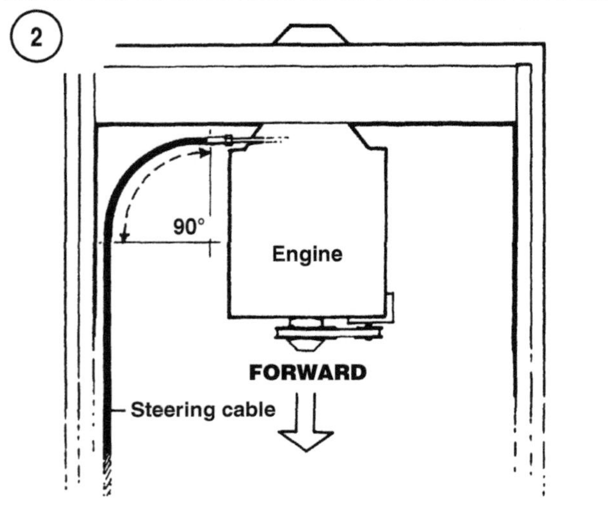

3
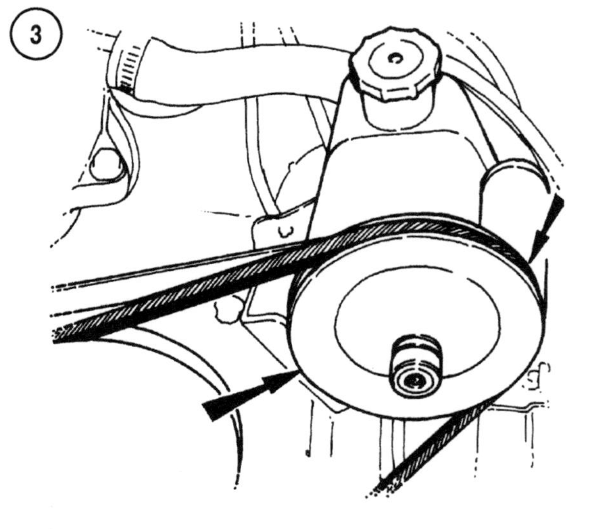

2. On models with power steering, boat structure or other components do not contact the power steering pump, pulley (**Figure 3**) or drive belt.
3. On models with power steering, the power steering cable moves freely and operates the steering cylinder assembly (**Figure 4**) only when turning the steering wheel.

MECHANICAL (MANUAL) STEERING SYSTEM

A mechanical steering system is used only on some 3.0L and 4.3L models. All other models are factory-equipped with power steering. A mechanical system (**Figure 5**, typical) consists of the helm-and-steering wheel assembly, the steering cable and hardware connecting the cable to the inner transom plate and steering lever.

WARNING
Failure of the steering system can cause a serious accident. Do not operate a boat with loose or binding steering.

4
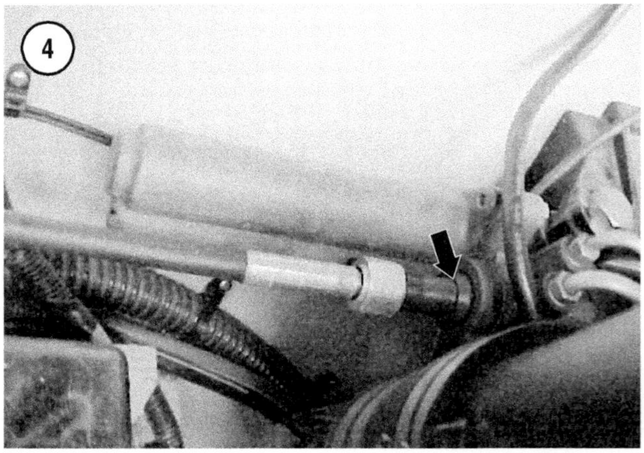

WARNING
*The steering cable must have a self-locking coupler nut (**Figure 6**) or coupler nut locking mechanism (**Figure** 7, typical) to prevent the cable nut from loosening. A loose cable nut results in excessive steering play and eventual lack of steering control. Do not operate*

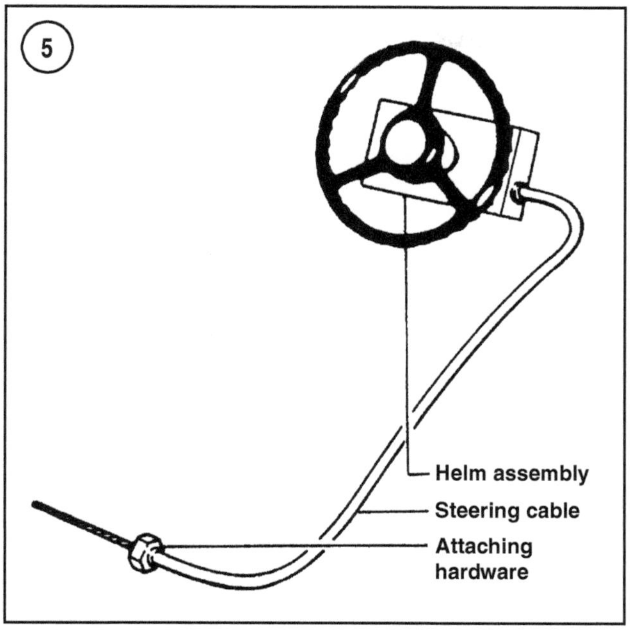

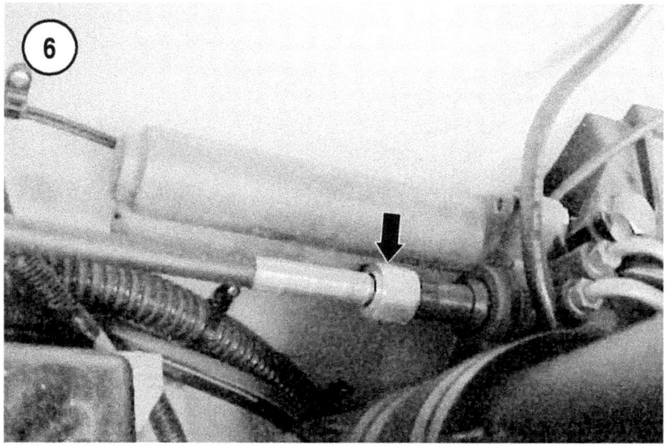

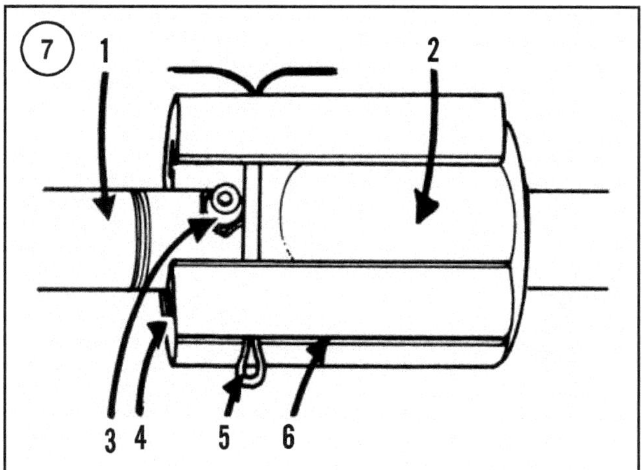

1. Steering cable
2. Steering cable coupler
3. Grease fitting
4. Bumper
5. Cotter pin
6. Locking sleeve

the boat if this important safety feature is not working properly.

Steering Cable and Helm

Replace the steering cable and helm as an assembly. Both components are subject to wear and the effects of corrosion, and failure of one usually precedes the other. Purchase the helm and cable from a marine dealership or marine supply business. Follow the manufacturer's instructions to install the helm into the boat. Removal and installation of the cable anchor block are described in Chapter Eleven. Refer to *Inner Transom Plate and Bracket*.

Removal

1. Remove the cotter pin and clevis pin (**Figure 8**) that connects the cable end to the steering arm.

2A. If the cable uses a self-locking cable coupler (**Figure 6**), loosen the cable coupler. Pull the cable from the guide tube and pivot block.

2B. If the cable uses a coupler locking mechanism (**Figure 7**), disconnect the cable as follows:

a. Turn the steering to the full port position.

b. Remove the cotter pin (5, **Figure 7**). Slide the locking sleeve (6, **Figure 7**) off the nut.

c. Fully loosen the coupler. Pull the cable from the guide tube and anchor block.

3. Remove the faulty steering cable and helm from the boat. Following the manufacturer's instructions, install the replacement cable and helm onto the boat.

4. Clean all grease, corrosion or other contaminants from the bore of the guide tube. Inspect the tube for cracked, pitted or damaged surfaces. Replace the tube and pivot block if they are damaged.

Installation

1. Apply a light coat of special steering cable lubricant (available from most marine suppliers) to the bore of the guide tube.

13

2. Rotate the steering wheel to the full starboard steering position. Guide the cable through the guide tube. Align the openings and then install the clevis pin into the cable and steering arm openings. The shouldered side of the pin must rest against the top of the steering lever. Install the *new* cotter pin and bend over both prongs.
3. Guide the steering cable onto the pivot block. The cable must not make any sharp bends or bind against the engine, boat structure or other objects.
4. Rotate the steering wheel one full turn in the port steering direction. Thread the steering cable coupler nut (**Figure 6**) or (2, **Figure 7**) fully onto the tube guide.
5A. If the cable uses a self-locking cable coupler nut, tighten the coupler nut to 35 ft.-lb. (47 N•m).
5B. If the cable uses a coupler locking mechanism, attach the cable as follows:
   a. Tighten the coupler (2, **Figure 7**) to 35 ft.-lb. (47 N•m).
   b. Fit the locking sleeve (6, **Figure 7**) over the cable jacket. Align the slot of the sleeve with the grease fitting (3, **Figure 7**) while sliding the sleeve over the coupler. Rotate the cable jacket as required to align the sleeve, grease fitting and cable coupler.
   c. Insert the cotter pin as shown in **Figure** 7. Bend both cotter pin prongs over to secure the sleeve.
6. Turn the steering wheel lock to lock. The drive unit must traverse fully in each direction without binding or excessive play. Check for a faulty cable or helm by disconnecting the cable and checking for free movement or excessive play with the drive out of the system. If the cable is ruled out as the cause, pivot the steering arm back and forth to check for binding or looseness.

POWER STEERING SYSTEM

Power steering is standard equipment on 4.3L electronic fuel injection (EFI) models and all V-8 models. Power steering is a factory option on 3.0L and carburetor-equipped 4.3L models.

The power steering system consists of the same components used on the mechanical (manual) steering system plus: the belt driven power steering pump (**Figure 9**), power steering fluid cooler (**Figure 10**) and steering cylinder assembly (**Figure 11**). The power steering pump is mounted onto the front of the engine and is belt-driven from the crankshaft pulley. The steering cylinder assembly mounts onto the inner transom plate and connects to the steering arm and steering cable ram.

Figure 12 shows the fluid flow diagram for the typical power steering system. The pump moves the fluid through the system under pressure. The relief valve functions to limit the maximum pressure developed by the pump while it is working and lowers the pressure when the system is static (no steering cable movement). Special passages in the relief valve hydraulically sense when the system is loaded, and the valve reacts accordingly. Lowering the pressure when the system is static reduces parasitic power loss, fluid temperature and wear on the power steering system components. The spool valve, in the control valve portion of the steering cylinder assembly, moves in response to movement of the steering cable. Although the spool valve movement is small (approximately 0.125 in. [3.175 mm]), it is enough to open the valve and direct pressurized fluid to either side of the piston. This provides the power assistance. The spring portion of the control valve pushes the spool valve to the static position as the cable movement ceases. If the cable movement is restricted by clamps, excessively tight bends or boat structure, the spool valve might not react properly to steering cable movement and might cause hard steering in one or both directions. Wire harness and other cables must not be attached to the steering cable.

9

10

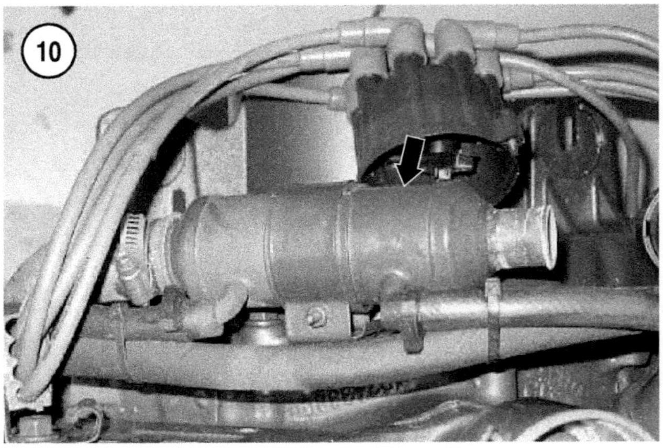

This section describes replacement of the steering cable and steering cylinder assembly, power steering pump removal, pump repair and pump installation. The steering cylinder assembly is nonserviceable and must be replaced if faulty. Power steering system troubleshooting is described in Chapter Two.

Steering Cable and Helm

Replace the steering cable and helm as an assembly. Both components are subject to wear and the effects of corrosion, and failure of one usually precedes the other. Purchase the

11

POWER STEERING CYLINDER ASSEMBLY

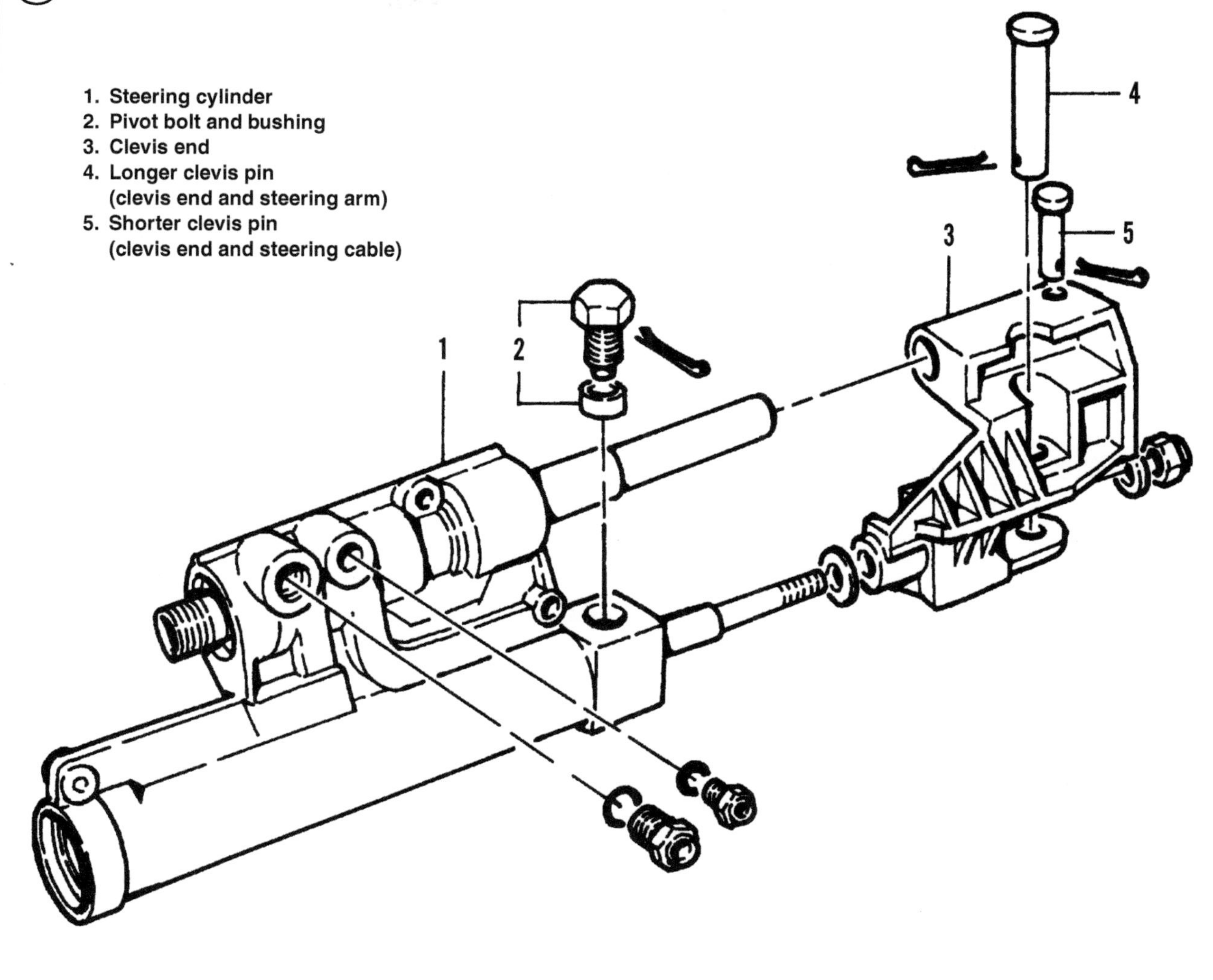

helm and cable from a marine dealership or marine supply business. Follow the manufacturer's instructions to install the helm into the boat. Disconnect the steering cylinder from the inner transom plate as described in Chapter Eleven. Refer to *Inner Transom Plate and Bracket*.

Removal

1. Remove the cotter pin and clevis pin (**Figure 8**) that connects the steering cylinder to the steering arm.

2A. If the cable uses a self-locking cable coupler nut (**Figure 6**), engage a wrench onto the flat surfaces of the tube guide (**Figure 13**). Then fully loosen the cable coupler (**Figure 6**). Pull the cable out of the steering cylinder guide tube.

2B. If the cable uses a coupler nut locking mechanism (**Figure 7**), disconnect the cable as follows:

a. Turn the steering to the full port position.
b. Remove the cotter pin (5, **Figure 7**). Slide the locking sleeve (6, **Figure 7**) off the nut.
c. Engage a wrench onto the flat surfaces of the tube guide (**Figure 13**) and then fully loosen the cable coupler (**Figure 6**). Pull the cable out of the steering cylinder guide tube.

3. Remove the faulty steering cable and helm from the boat. Following the manufacturer's instructions, install the replacement cable and helm onto the boat.

4. Clean all grease, corrosion or other contaminants from the bore of the guide tube. Inspect the tube for cracks, pits or damage. Replace the steering cylinder assembly if it is damaged.

Installation

1. Apply a light coat of special steering cable lubricant (available from most marine suppliers) to the bore of the guide tube.

2. Rotate the steering wheel to the full starboard steering position. Guide the cable through the guide tube. Align the openings and then install the clevis pin into the cable and

12

TYPICAL POWER STEERING SYSTEM

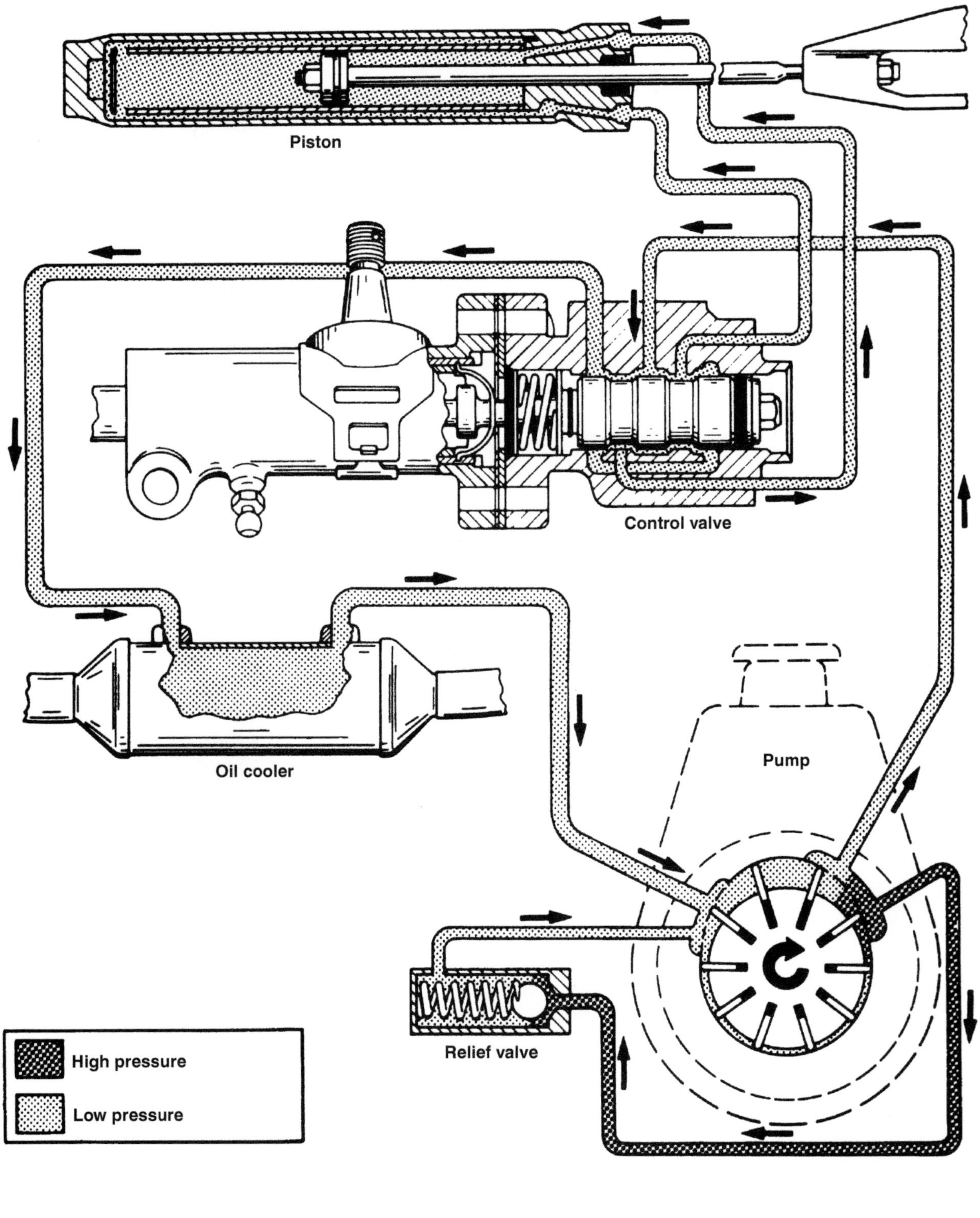

steering arm openings. The shouldered side of the pin must rest against the top of the steering lever. Install the *new* cotter pin and bend over both prongs.

3. Guide the steering cable onto the pivot block. The cable must not make any sharp bends or bind against the engine, boat structure or other objects.

4. Rotate the steering wheel one full turn in the port steering direction. Thread the steering cable coupler (**Figure 6** or 2, **Figure 7**) fully onto the tube guide.

5A. If the cable uses a self-locking cable coupler nut, engage a wrench onto the flat surfaces of the guide tube (**Figure 13**) and then tighten the coupler nut (**Figure 6**) to 35 ft.-lb. (47 N•m).

5B. If the cable uses a coupler locking mechanism, attach the cable as follows:

a. Tighten the coupler (2, **Figure 7**) to 35 ft.-lb. (47 N•m).

b. Fit the locking sleeve (6, **Figure 7**) over the cable jacket. Align the slot of the sleeve with the grease fitting (3, **Figure 7**) while sliding the sleeve over the coupler. Rotate the cable jacket as required to align the sleeve, grease fitting and cable coupler.

c. Insert the cotter pin as shown in **Figure 7**. Bend both cotter pin prongs over to secure the sleeve.

6. Turn the steering wheel lock to lock. The drive unit must traverse fully in each direction without binding or excessive play. Do not confuse normal control valve movement with excessive play. Check for a faulty cable or helm by disconnecting the cable and checking for free movement or excessive play with the drive out of the system. If the cable is ruled out as the cause, pivot the steering arm back and forth to check for binding or looseness.

Steering Cylinder

The steering cylinder is nonserviceable and must be replaced if faulty. Failure of the steering cylinder is extremely rare. The steering cylinder should be replaced to correct a steering system malfunction only if all other components of the system have been ruled out as the cause.

Removal

Refer to **Figure 11** for this procedure.

1. Disconnect the steering cable from the cylinder as described in this section. Refer to *Steering Cable and Helm*.

2. Place shop towels or a suitable container under the steering cylinder hose fittings (**Figure 14**, typical). Disconnect the hose fittings and drain residual fluid from the hoses. Inspect the O-rings on the fittings for cuts, flattening or deterioration. Replace the O-rings if they are damaged. Tie the hoses to a location higher than the power steering pump to minimize fluid loss. Cover the fittings with a towel to prevent contamination.

3. Disconnect the steering cylinder from the inner transom plate as described in Chapter Eleven. Refer to *Inner Transom Plate and Bracket*.

Installation

1. Apply Volvo Penta Propeller Shaft Grease (Volvo Penta part No. 1141644-3) or equivalent to the pivot bushings. If removed or replaced, insert them into the pivot pin bores of the steering cylinder assembly (1, **Figure 11**).

2. Apply a coat of automatic transmission fluid (ATF) to the O-rings on the fittings and then thread the hoses (**Figure 14**) into the steering cylinder by hand. Route the hoses to prevent kinking or interference with the steering arm or other moving components. Hold the hoses in position and then tighten the fittings to 177-204 in.-lb. (20-23 N•m).

3. Connect the steering cylinder to the inner transom plate as described in Chapter Eleven. Refer to *Inner Transom Plate and Bracket*.

4. Connect the steering cable to the cylinder as described in this section. Refer to *Steering Cable and Helm*.

5. Fill and bleed the power steering system as described in this section.

Power Steering Pump

This section describes power steering pump removal, pulley removal and installation, pump disassembly and assembly, and pump installation. Replacement and adjustment for the power steering belt is described in the pump removal and installation procedures. The power steering pump is pivoted in the mounting bracket to adjust the belt tension.

CAUTION
The power steering pump hose fittings and mounting fasteners have metric thread and bolt head sizes. To prevent rounding of fittings and fasteners, use only metric wrenches.

Removal

1. Disconnect the negative battery cable.
2. Remove the drive belt (**Figure 15**) as follows:
   a. Note the belt routing around all of the pulleys prior to removing the belt. Depending on the model, one or more of the other belts must also be removed. Refer to Chapter Nine for alternator belt removal procedures. Refer to Chapter Eight for seawater pump belt removal procedures. If necessary, make a sketch indicating the belt routing around *all* of the pulleys.
   b. Loosen but *do not* remove the pivot and bracket clamping bolts (**Figure 16**, typical). Carefully pivot the pump toward the crankshaft and recirculation pump pulleys. *Do not* push against the filler neck portion of the pump. The weld at the pump body might break and allow fluid to leak. If so equipped, use a breaker bar inserted into the 1/2 in. square opening in the bracket to pivot the pump.
   c. Slip the belt off the power steering pulley. Inspect the belt for worn, damaged or deteriorated surfaces and replace as needed.
3. Place a suitable container or shop towels under the pump hose fittings to capture spilled fluid (A and B, **Figure 17**). Loosen the clamp on the return hose. Push the hose off the fitting (B, **Figure 17**). Discard the clamp. Drain residual fluid from the hose. Use a metric flare nut wrench to loosen the pressure hose fitting (A, **Figure 17**). Do not loosen the larger fitting that threads into the pump body. Allow residual fluid to flow from the loosened fitting for a few minutes and then disconnect the hose. Drain residual fluid from the hose.
4. Remove the nut and washers from the stud at the rear and bottom of the pump. This nut retains the pump in the lower support bracket. Support the pump while removing the two bolts and washers that retain the pump in the front support bracket. The bolts are located behind the pulley, and a thin profile wrench might be required.
5. Lift the pump out of the mounting brackets. Position the pump fittings over a suitable container and drain all residual fluid from the pump.
6. Clean rust or corrosion from the pulley surfaces. Rust or corrosion pitting on the belt contact surfaces will allow slippage and quickly wear out a new belt. Clean, sand and repaint the pulley as necessary. Replace the pulley if a

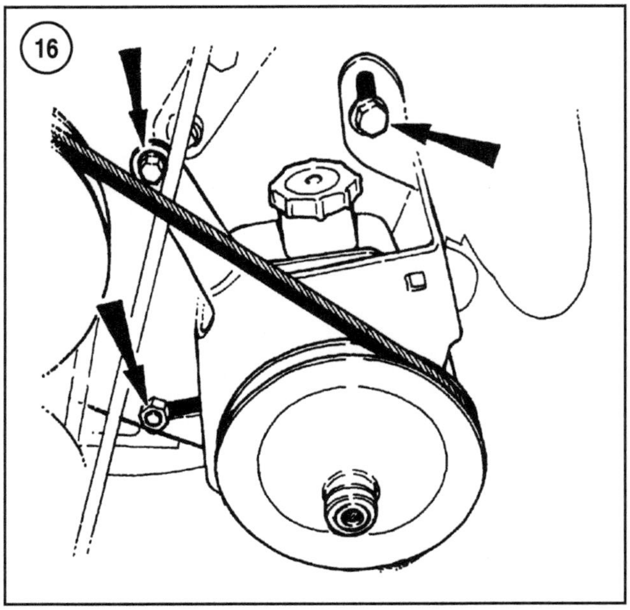

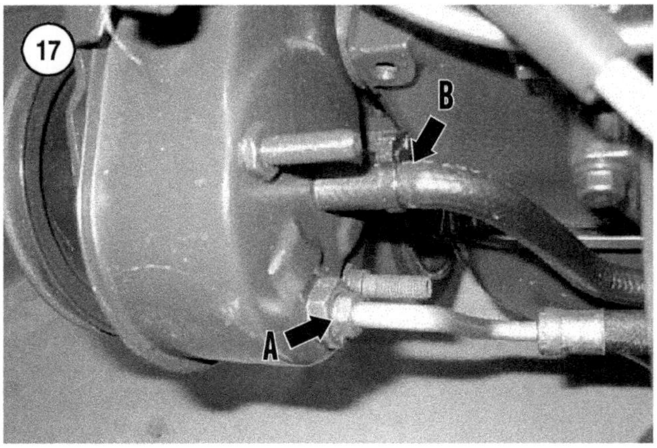

smooth belt surface cannot be achieved by sanding and painting.

Pulley removal

The power steering pump can be removed with the pump on or off the engine. Use a power steering removal tool

18

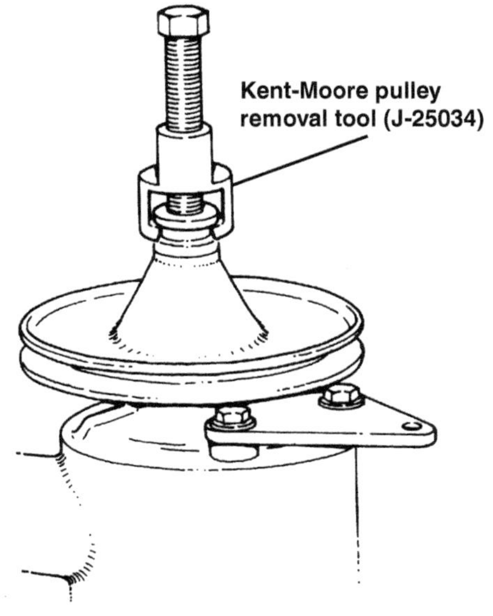

19

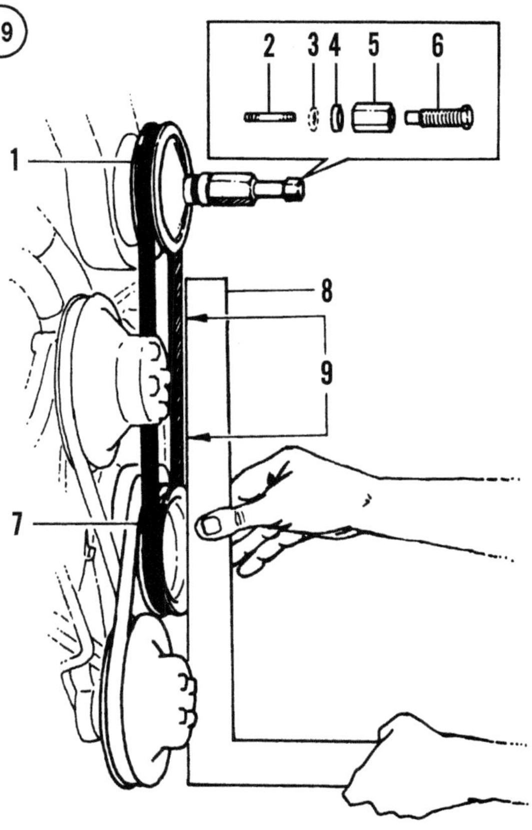

1. Power steering pump pulley
2. Stud
3. Spacer
4. Bearing
5. Nut
6. Bolt
7. Crankshaft pulley
8. Straightedge
9. Parallel surfaces

(**Figure 18**) to pull the pulley from the pump shaft. This tool is commonly used in automotive applications. Purchase the tool from a tool supplier or rent this tool from an automotive parts store. Use a pulley installation tool (OMC or Volvo Penta part No. 9995444-8) and a long straightedge to install the pulley. Install the pulley with the pump installed on the engine. This is necessary to properly align the belts and pulleys. Improper alignment results in noisy operation, increased wear on the pulley(s) and premature drive belt failure.

The pulley must fit tightly onto the pump shaft. If the fit is loose, the pulley bore or pump shaft or both are excessively worn. To correct a loose fit, first install a new pulley. If the new pulley fits loosely, replace the power steering pump. The pump shaft is not available separately.

1. Disconnect the negative battery cable.
2. If the pump is mounted on the engine, remove the drive belt from the pulley as described in the pump removal procedures in this section.
3. Slide the pulley removal tool onto the hub of the pulley as shown in **Figure 18**. Thread the puller bolt by hand into the puller until it contacts the pump shaft. Hold the body of the tool with a suitable wrench and tighten the bolt to remove the pulley.
4. Clean grease from the pulley with warm soapy water. Sand surface corrosion from steel pulleys with Carborundum paper. Thoroughly clean all abrasive material from the pulley. Paint the pulley to help prevent corrosion. Inspect the pulley and pump shaft for cracks or wear. Replace as needed.

Pulley installation

The pump must be mounted onto the engine to install the pulley.

Refer to **Figure 19** for this procedure.

1. Position the pulley on the pump shaft. Fit the drive belt around the pulleys. Do not adjust the belt tension or fit the belt onto the tensioner at this time.
2. Thread the stud (2, **Figure 19**) from the pulley installation tool (OMC or Volvo Penta part No. 9995444-8) as far as possible into the pump shaft.
3. Position the bearing from the tool assembly over the stud. Do not use the spacer (3, **Figure 19**) from the tool kit.
4. Thread the nut (5, **Figure 19**) onto the shaft (6). Thread the shaft and nut as far as possible into the stud.
5. Position the straightedge against the crankshaft and power steering pulleys as shown in **Figure 19**. Tighten the nut (5, **Figure 19**) until the drive belt surfaces are parallel to the straightedge.
6. Remove the straightedge and installation tool components.
7. Install the belt and adjust the belt tension as described in the pump installation procedures in this section. Connect the battery cables.

13

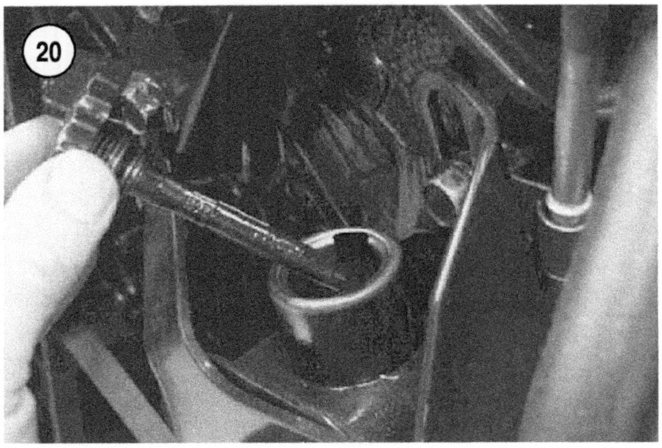

Pump disassembly

The power steering pump can be disassembled, but internal components (except for seals and O-rings) are not available separately. Replace the pump if an inspection reveals faulty components. This procedure is primarily of value if the engine has been submerged and water has entered the pump. If completed promptly, this procedure can prevent corrosion of the internal components and the need to replace the entire pump.

Replace all seals and O-rings if they are removed from the pump. Purchase a seal kit from a local GM dealership (part No. 5688044).

1. Remove the power steering pump as described in this chapter.
2. Remove the front pump mounting bracket from the engine. Clamp the bracket into a vise and then temporarily attach the pump to the bracket with the original fasteners.
3. Remove the power steering pump pulley as described in this chapter. Remove the pump from the bracket and reinstall the bracket onto the engine. Securely tighten the fastener bracket.
4. Remove the fill cap (**Figure 20**). Invert the pump and pour any remaining fluid into a suitable container.
5. Remove the fitting (1, **Figure 21**). Pull the control valve (3, **Figure 21**) and spring (4) from the pump. Remove and discard the O-rings (2 and 5, **Figure 21**).
6. Remove the stud(s) (11, **Figure 21**) from the reservoir and pump body. Carefully tap the reservoir (6, **Figure 21**) to remove it from the pump body (9). Remove and discard the O-rings (7, 8, and 10, **Figure 21**).
7. Place the pump body (shaft side down) onto a vise with open jaws. Do not clamp the body with the jaws.
8. Insert an awl into the opening on the side of the body (**Figure 22**) and push the ring from the groove. Then pry the ring from the body. Lift the end plate and spring (1, **Figure 23**) from the pump.
9. Lift the pressure plate (2, **Figure 23**) and pump ring (3) from the pump. Push on the exposed end of the pump shaft to remove the rotor and shaft assembly (5, **Figure 23**). Remove the two O-rings and dowels (6, **Figure 23**). Discard the O-rings.

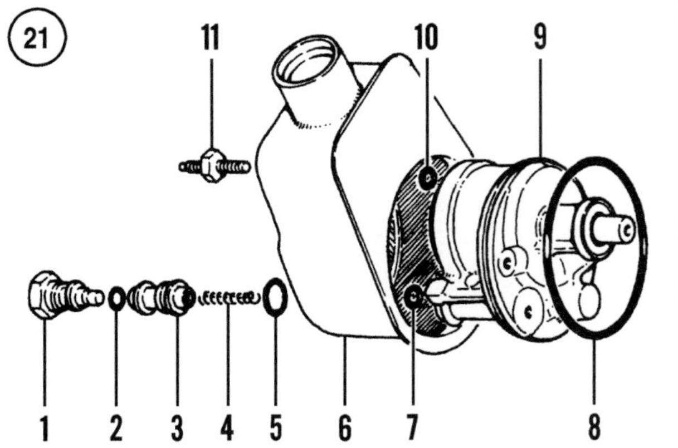

1. Hose fitting
2. O-ring
3. Control valve
4. Spring
5. O-ring
6. Reservoir
7. O-ring
8. O-ring
9. Pump assembly
10. O-ring
11 Stud bolt

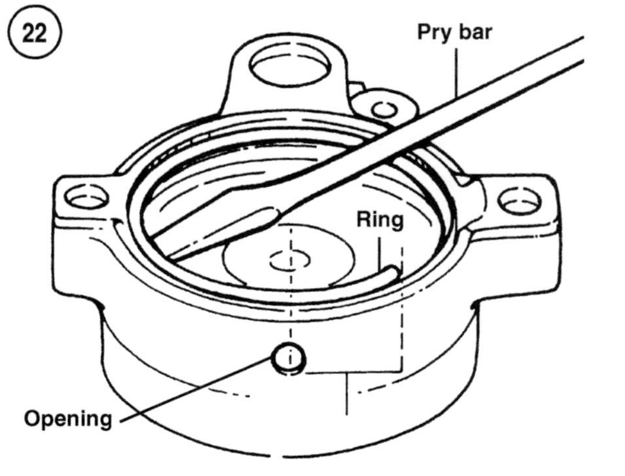

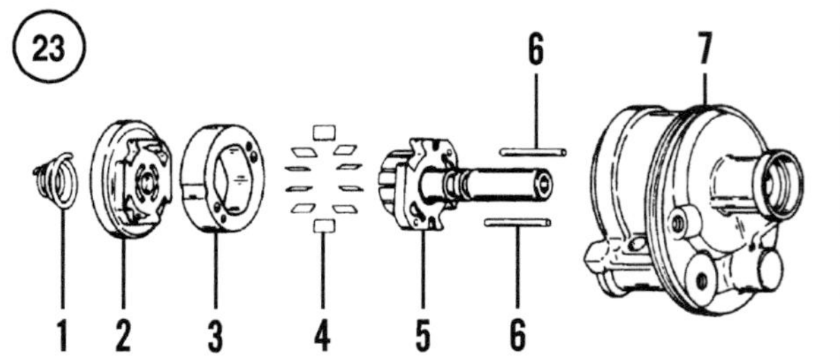

1. Spring
2. Pressure plate
3. Pump ring
4. Pump vanes
5. Pump shaft and rotor assembly
6. Dowel pins
7. Pump body

24

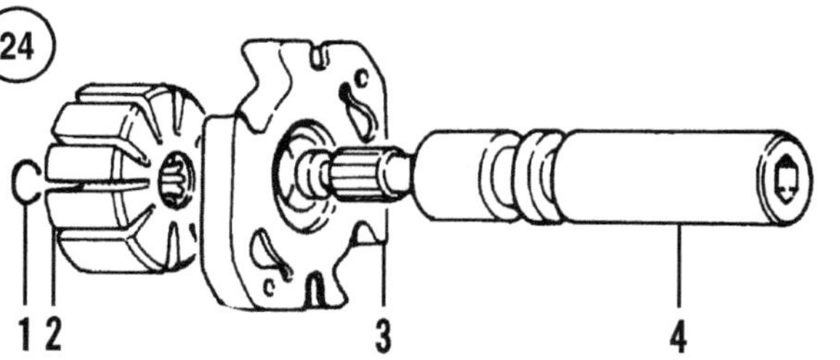

1. Retaining ring
2. Rotor
3. Thrust plate
4. Pump shaft

25

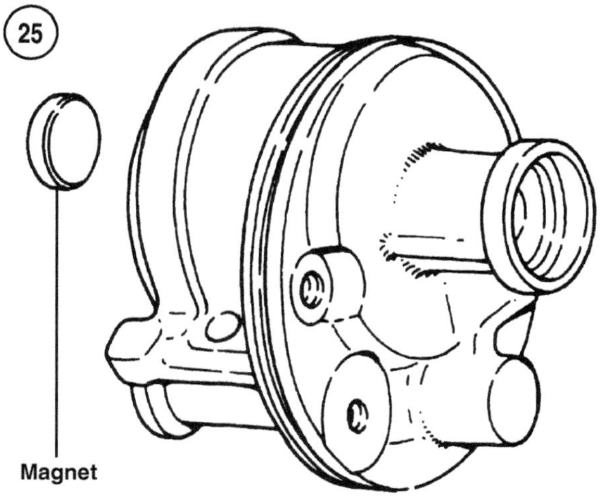

26

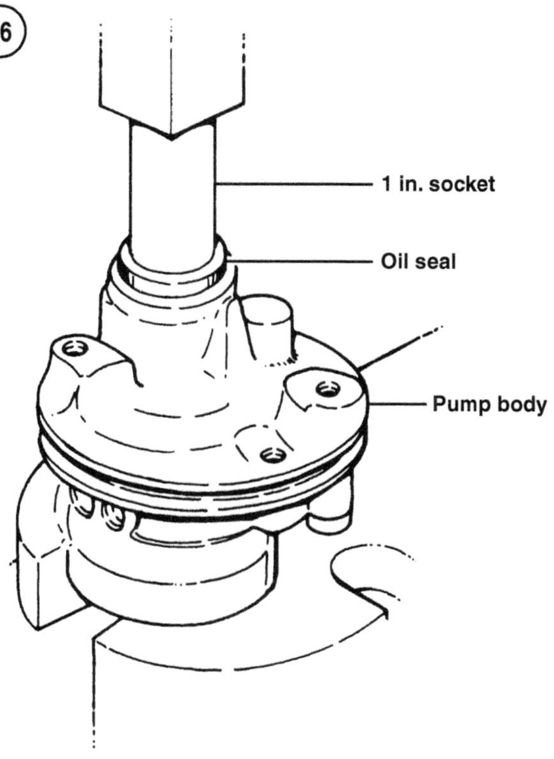

10. Slide the pump vanes (4, **Figure 23**) from the rotor. Remove the retaining ring (1, **Figure 24**) and then slide the rotor and thrust plate (2 and 3) from the pump shaft (4).
11. Remove the magnet from the pump body (**Figure 25**). Carefully pry the oil seal from the pump shaft bore. Work carefully to avoid damaging the seal bore.
12. Clean all components with clean solvent and dry them with compressed air.
13. Inspect the pump body for cracked or damaged surfaces.
14. Slide each vane (rounded side out) into each of the slots in the rotor. The vanes must slide freely into and out of the slots.
15. Inspect the pump shaft, pressure plates and rotor for wear, discoloration or roughness.
16. Replace the pump assembly if the pump body is cracked; if the pump shaft, pressure plate or rotor is worn, damaged or discolored; or if the vanes bind or stick in the rotor slots. Internal components are not available separately.
17. Use crocus cloth to polish burrs or minor imperfections from the control valve surfaces. The valve must slide freely within the pump bore. Replace the pump if the valve surfaces are damaged or the spring is corroded because the parts are not available separately.

Pump assembly

Replace all seals and O-rings if they are removed from the pump. Purchase a seal kit from a local GM dealership (part No. 5688044). Prior to assembly, lubricate all O-rings with Dexron III automatic transmission fluid.

1. Install the pump shaft seal as follows:
   a. Clean the seal bore in the pump body with a quick-drying solvent such as carburetor cleaner.
   b. Place the pump body on the base of a press as shown in **Figure 26**. The body must be well-supported. Otherwise, the body might distort during seal installation.
   c. Place the oil seal (lip side down) into the bore opening. Use a 1 in. socket to press the seal fully into the bore.
2. Lubricate all metal components and the seal lip with Dexron III automatic transmission fluid. Install a new pressure plate O-ring into the third groove of the housing. Install the two dowel pins (**Figure 27**).
3. Slide the thrust plate and rotor (2 and 3, **Figure 24**) onto the pump shaft (4). Install the retaining ring into the shaft groove. Slide the rotor and shaft assembly into the pump body (**Figure 28**).
4. Fit the two smaller holes over the dowel pins while installing the pump ring into the body (**Figure 29**).
5. Insert the rotor vanes into the rotor slots with the rounded edges facing the ring (**Figure 30**).
6. Install the pressure plate (2, **Figure 23**) with the spring groove facing outward. Install a new end plate O-ring into the second groove in the pump body (**Figure 31**). Install the spring (1, **Figure 23**) and align with the groove in the pressure plate. Place the end plate onto the spring (**Figure 32**).

13

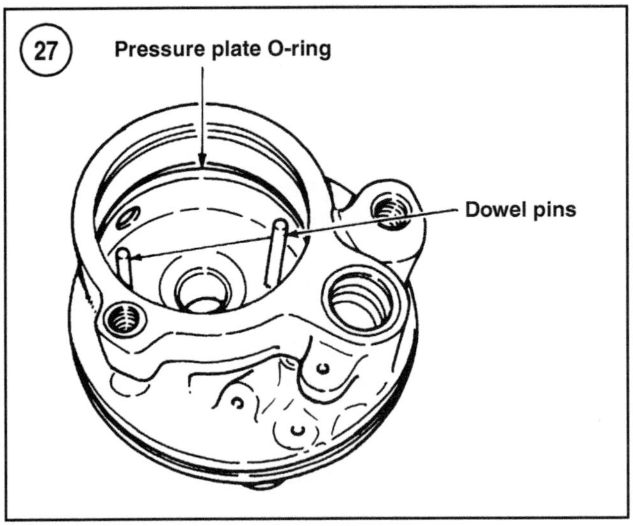

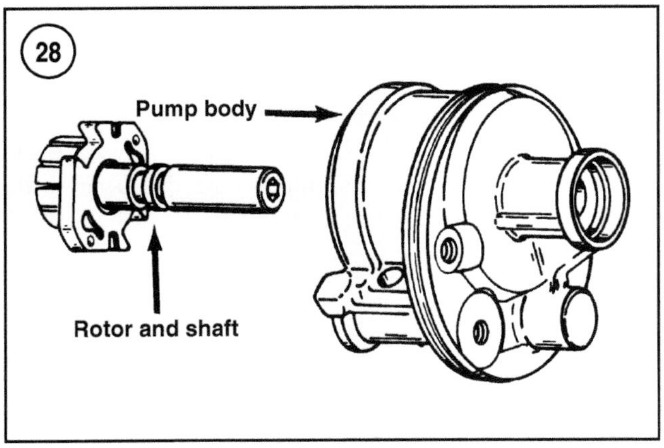

7. Press down on the end plate just enough to expose the groove for the ring (**Figure 22**). Carefully work the ring into the groove. The ring must fit fully into the groove around the entire perimeter of the body. Do not align the ring gap with the opening.
8. Install new O-rings onto the pump body (**Figure 33**). Slide the body into the reservoir. Insert the stud bolt (11, **Figure 21**) through the reservoir and thread them by hand into the pump body. Tighten the stud bolt to 25 ft.-lb. (34 N•m).
9. Slide the spring and control valve into the pump.
10. Lubricate them with automatic transmission fluid and then place the O-rings (2 and 5, **Figure 21**) onto the fitting. Thread the control valve fitting by hand into the pump. Tighten the fitting to 15-26 ft.-lb. (20-35 N•m).
11. Install the power steering pump as described in this chapter.
12. Install the power steering pulley as described in this section.

Pump installation

1. Place the pump into the mounting bracket. Install the mounting fasteners (**Figure 16**, typical). Do not tighten the fasteners at this time.

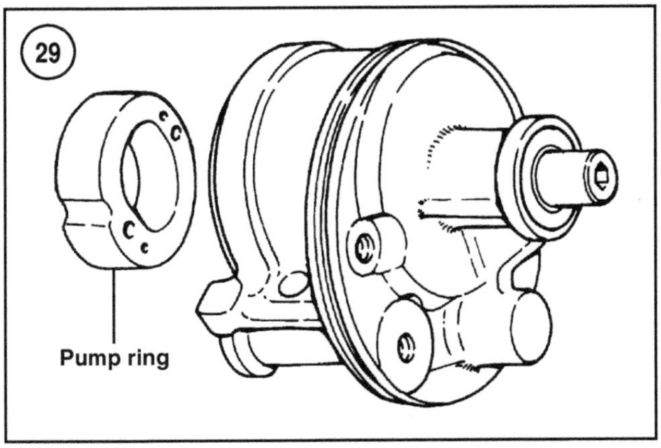

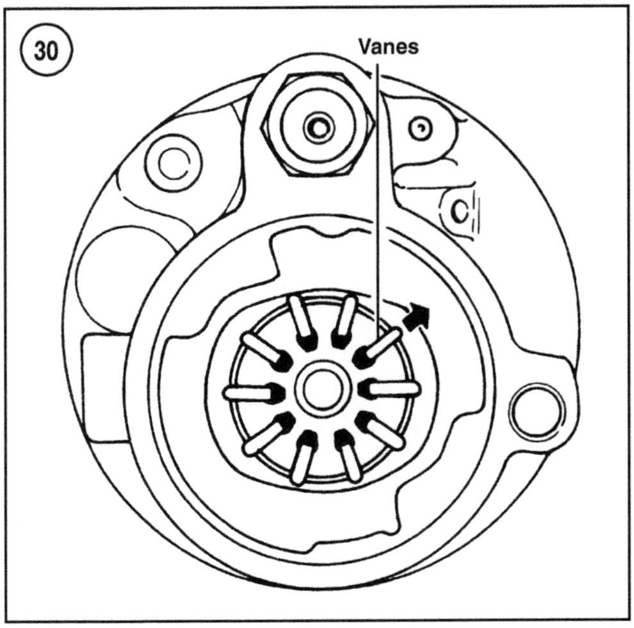

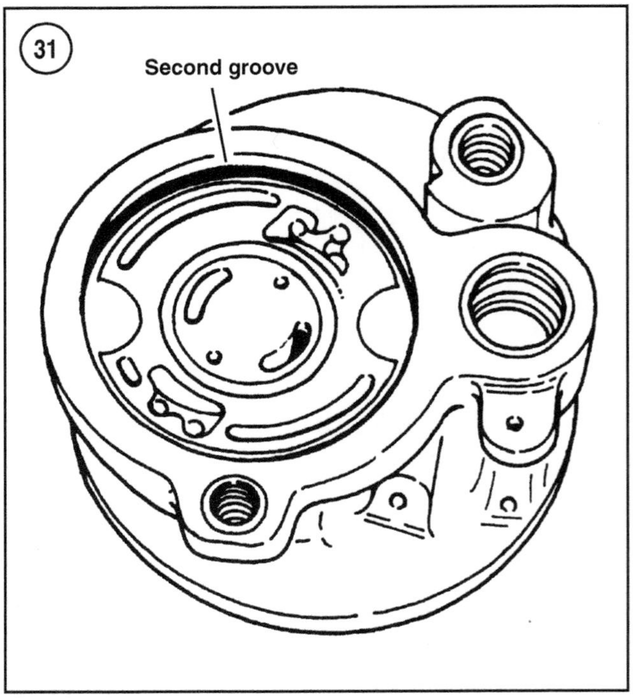

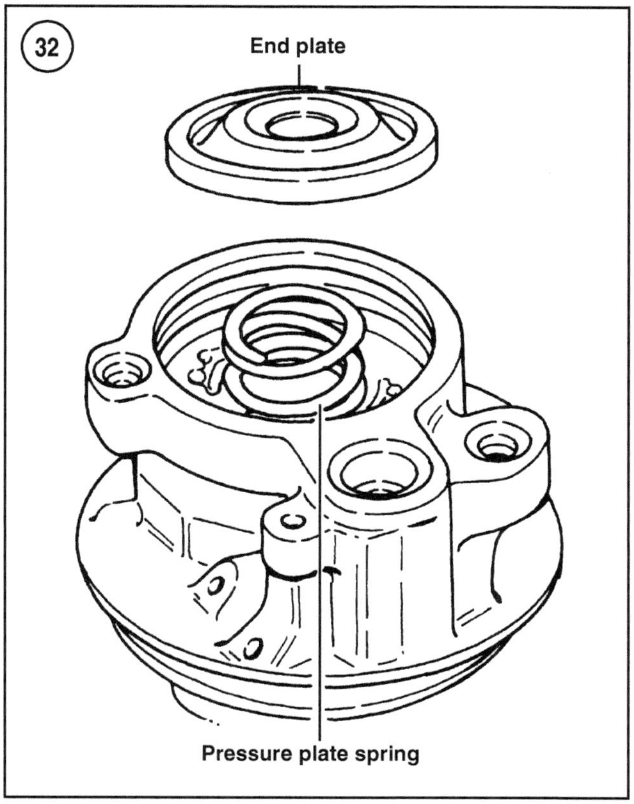

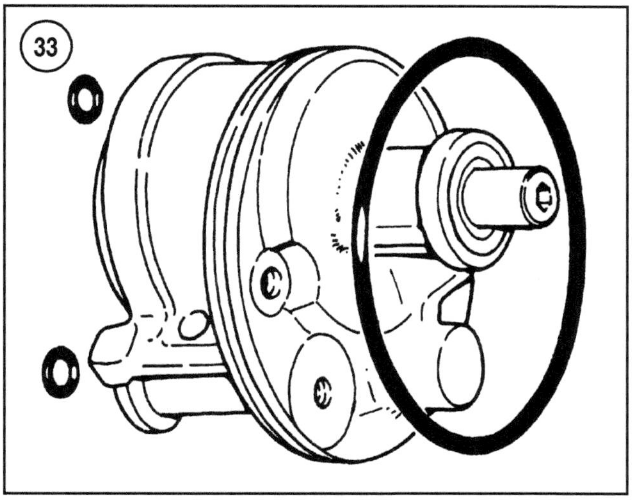

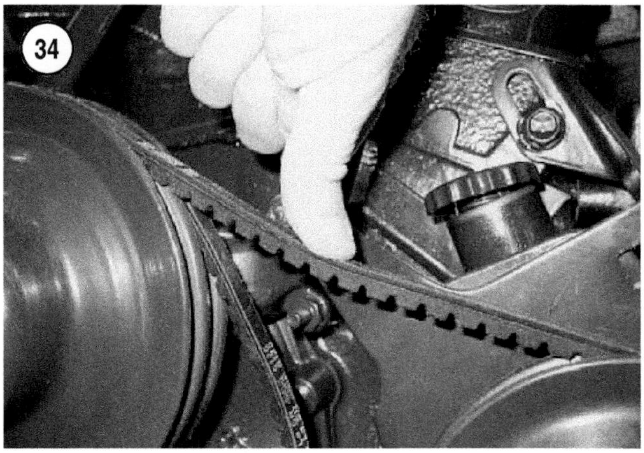

2. Thread the pressure hose fitting (A, **Figure 17**) into the control valve fitting. Rotate the fitting to prevent contact with the pump mounting bracket and other components. Hold the position and then tighten the fitting to 177-204 in.-lb. (20-23 N•m).
3. Fit a *new* clamp over the return hose. Install the return hose fully into the pump fitting (B, **Figure 17**). Securely tighten the hose clamp.
4. If removed, install the pulley as described in this chapter.
5. Install and adjust the drive belt as follows:
   a. Pivot the pump toward the crankshaft pulley and then slip the drive belt onto pulleys. Make sure the belt is not twisted and fits into the groove in each pulley.
   b. Pivot the pump away from the crankshaft pulley to apply tension to the drive belt. If so equipped, use a breaker bar inserted into the 1/2 in. square opening in the bracket to pivot the pump. When properly tensioned, the belt will deflect approximately 1/4-1/2 in. (6-13 mm) with thumb pressure applied at a point midway between the pulleys (**Figure 34**).
   c. Hold the pump in position to maintain belt tension while tightening the mounting bolts to 25 ft.-lb. (34 N•m).
6. Fill and bleed the system as described in this section.
7. Connect the battery. Start the engine and immediately check for fluid leakage. Have an assistant rotate the steering through several lock-to-lock cycles while checking for fluid leakage or unusual noises. Correct fluid leakage before proceeding.
8. Correct the fluid level as described in this chapter.
9. Seat the new belt and correct the tension as follows:
   a. Start the engine and allow it to run at idle speed for approximately 10 minutes.
   b. Stop the engine and disconnect the negative battery cable.
   c. Apply thumb pressure at a point midway between the pulleys (**Figure 34**). The belt must deflect approximately 1/4-1/2 in. (6-13 mm). If necessary, loosen the mounting bolts (**Figure 16**, typical) and pivot the pump to correct the tension. Hold the pump in position and tighten the mounting bolts to 25 ft.-lb. (34 N•m).
   d. Connect the battery cable.

Filling and Bleeding the Steering System

Fill and bleed the system if it was operated with a low fluid level and after repairing or replacing any steering system component. Use Dexron III automatic transmission fluid or one of the power trim/tilt or power steering fluids (such as Volvo Penta part No. 3851039 or Mercruiser part No. 92-90199A12) in the power steering system.

CAUTION

Never operate the power steering system with a low or high fluid level. Operating the system while low on fluid causes foaming of the fluid, increased steering effort and damage to the hydraulic pump components. Operating the system with a high fluid causes fluid overflow.

1. Place the drive unit in the straight-ahead position. Disconnect the negative battery cable. Remove the fill cap (**Figure 20**).
2. Wipe the dipstick portion of the fill cap with a shop towel and reinstall it into the pump reservoir.
3. Remove the cap and note the fluid level on the dipstick (**Figure 35**). Correct the fluid level as follows:
   a. If the engine is warm, add fluid into the cap opening until the fluid level just reaches the upper end of the range (**Figure 35**).
   b. If the engine is cold, add fluid into the cap opening until the fluid level just reaches the groove below the *ADD* marking.
4. Bleed air from the system as follows:
   a. Without starting the engine, rotate the steering wheel in the port and starboard turning directions while stopping frequently and checking the fluid level and for bubbles in the reservoir. Immediately stop rotating the wheel and add fluid if the level drops below the tip of the dipstick. Otherwise foam might appear in the fluid.
   b. Continue until the wheel reaches a minimum of five turns lock to lock and bubbles stop appearing in the reservoir.

35

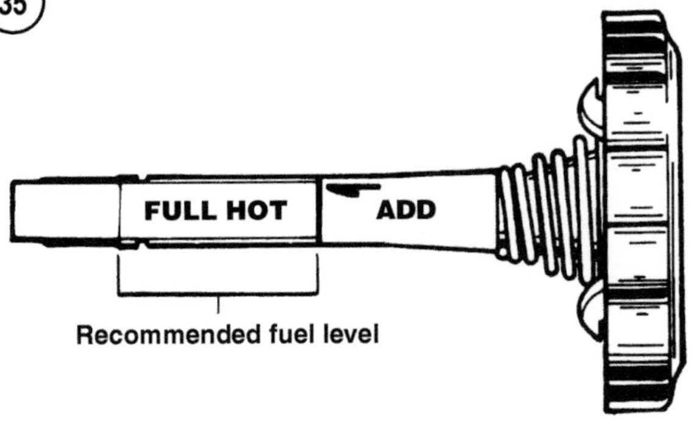

   c. Position the steering wheel in the straight-ahead position. Correct the fluid level as described in step 3. Connect the battery cables.

5. Start the engine and cycle the steering wheel through several turns lock to lock while checking the fluid level and for foam or bubbles in the reservoir. Immediately stop the engine if foam appears in the reservoir. Allow a few minutes for the foam to dissipate and then repeat Step 4. The air bleeding procedure is complete when neither bubbles nor foam appear in the reservoir.

6. Install the fill cap.

Table 1 STEERING SYSTEM TORQUE SPECIFICATIONS

Fastener location	in.-lb.	ft.-lb.	N•m
Hose fittings	177-204	–	20-23
Control valve fitting (at power steering pump)	–	15-26	20-35
Power steering pump mounting bolts	–	25	34
Steering cable coupler nut	–	35	47
Stud bolt (power steering pump)	–	25	34

Index

A

Adjustments
- engine operating speed
  - specifications . . . 189
- gauge settings for trim limit control
  - module . . . 191
- ignition timing . . . 177-183
  - and fuel requirement . . . 189-191
- shift cable . . . 183-185
- throttle cable . . . 185
- torque specifications . . . 188
- trim limit control module . . . 185-188
- trim position sender . . . 185-188

Alternator . . . 366-368

B

Balance shaft . . . 237-239
Battery . . . 360-366
- cable recommendations . . . 413
- capacity (hours) . . . 413
- state of charge . . . 413-414

Bearing preload (rolling torque)
- specifications . . . 487

Bellows and water hose . . . 494-496
Bracket housing . . . 500-504
Break-in procedure . . . 261-262

C

Camshaft
- and lifters . . . 239-245
- lobe lift . . . 205-209
- service specifications . . . 265-266

Carburetor . . . 283-299
- adjustment . . . 174-177
- four-barrel . . . 291-299
- fuel inlet filter replacement . . 283-284
- two-barrel . . . 284-291

Charging system
- troubleshooting . . . 46-50, 130-130

Connecting rod
- and piston . . . 245-254
- service specifications . . . 268-270

Conversion formulas . . . 33-34
Coolant temperature
- sender . . . 401-402
- sensor . . . 402-403

Cooling system
- cooler replacement . . . 340
- hose replacement . . . 332-333
- recirculation pump
  - replacement . . . 333-335
- seawater pump . . . 335-340
- thermostat . . . 329-332
- torque specifications . . . 359
- troubleshooting . . . 110-114, 133-134

Crankshaft
- and rear main seal . . . 254-261
- service specifications . . . 268-270

Cylinder
- bore specifications . . . 271
- head . . . 209-214

D

Diagnostic trouble codes
- GM EFI models . . . 135-136
- troubleshooting . . . 93-97

Drive coupler . . . 221-223
Drive system troubleshooting . . . 125-126
Drive unit . . . 422-444
- bearing preload (rolling torque)
  - specifications . . . 487
- gear lash
  - and pattern correction guide (DP-S DuoProp [twin-propeller] drives) . . . 487-488
  - specifications . . . 487
- gear ratios . . . 417-418, 488
- high altitude operation . . . 418
- identification . . . 415-416, 488
- installation . . . 424-425
- lower gearcase . . . 449-484
  - DP-S DuoProp (twin-propeller)
    - drives . . . 465-483
  - installation . . . 483-484
  - removal . . . 449-465
- lubricant capacities . . . 162
- operation . . . 416-417
- outdrive service precautions . . . 418
- pressure and vacuum tests . . . 484-486
- propeller . . . 420-422
  - rotation and shifting . . . 419-420
- removal . . . 422-424
- sacrificial anodes . . . 418-419
- shift housing service . . . 446-449
- torque specifications . . . 486-487
- universal joint service . . . 444-446
- upper drive shaft housing . . . 425-444

E

Electrical system
- alternator . . . 366-368
- battery . . . 360-366
  - cable recommendations . . . 413
  - capacity (hours) . . . 413
  - state of charge . . . 413-414
- coolant temperature
  - sender . . . 401-402
  - sensor . . . 402-403
- electronic fuel injection
  - (EFI) relay . . . 399-400
- engine
  - control
    - assembly (ECA) . . . 412
    - unit (ECU) . . . 411
  - firing order . . . 414

Electrical system
engine (continued)
overheat switch 403-404
fuel pump relay 398-399
fundamentals 25-27
idle air control
motor (IAC) 407-408
idle speed control
solenoid (ISC) 408-409
ignition system 386-397
intake air temperature
(IAT) sensor 405-406
knock sensor (KS) 409-410
amplifier 411
module 410-411
manifold air pressure
sensor 404-405
oil pressure
sender 400-401
switch 401
starter
and starter solenoid 368-386
installation 369-371
removal 369
component cleaning and
inspection 384-386
disassembly and assembly . 373-384
relay 398
slave solenoid 397-398
solenoid removal and installation
. . . . 371-372
throttle position sensor
(TPS) 406-407
torque specifications 412-413
wiring diagrams 545-560
Electronic fuel injection (EFI) . . 299-301
relay 399-400
resistance specifications 135
troubleshooting 75-93
EFI system relay test 81-83
fuel delivery system 75-78
fuel injection control system . 78-81
fuel injector test 91-93
fuel pressure test 87-91
fuel pump relay test 83-87
Engine
alignment 199-200
balance shaft 237-239
break-in procedure 261-262
camshaft
and lifters 239-245
lobe lift 205-209
service specifications 265-266
connecting rod 245-254
service specifications 268-270
control
assembly (ECA) 412
unit (ECU) 411
coolant 154-156
crankshaft 254-261
service specifications 268-270
cylinder
bore specifications 271
head 209-214
drive coupler 221-223
firing order 414
flywheel and flywheel
housing 221-223
harmonic balancer 231-233
identification 36
installation 198-199
lifter cover 205
lifters and camshaft 239-245
oil
capacities 162
pan 223-225
pump 225-231
service specifications . . . 268-270
seal 233-236
operating speed
specifications 136, 189
operation 4, 192-193
overheat switch 403-404
piston 245-254
service specifications 270-271
pushrod 214-221
rear main seal 254-261
removal 194-198
replacement parts 193
rocker arm 214-221
cover 200-201
special service precautions . . . 193-194
specifications
camshaft service 265-266
crankshaft/oil pump and connecting
rod service 268-270
cylinder bore 271
piston service 270-271
torque 262-265
valve service 267-268
timing
chain and sprockets 236-237
gear or chain cover 233-236
tools 193
top dead center (TDC) firing
position 204
torque specifications 262-265
troubleshooting 117-125
valves
adjustment 201-204
and valve seats 214
service specifications 267-268
Exhaust system
bellows 349-354
elbow 341-348
exhaust tube 349-354
flapper valve 349-354
manifold 354-359
torque specifications 359

F

Fasteners 4-8
Fitting out 173
Flywheel and flywheel
housing 221-223
Fuel system
carburetor 283-299
four-barrel 291-299
fuel inlet filter replacement 283-284
two-barrel 284-291
electronic fuel injection
(EFI) 299-301
relay 399-400
flame arrestor and cover 281-283
fuel requirements 189-191
injectors 309-316
intake
manifold 323-327
plenum 309-316
maintenance 151-154
pressure
regulator 316-320
specifications 135
pump
high-pressure electric 320-321
low-pressure electric 279-281
mechanical low-pressure . . 274-279
relay 398-399
rail 309-316
reservoir 321-323
service 272-274
throttle body 301-308
torque specifications 327-328
troubleshooting 65-75, 128-130

G

Galvanic corrosion 10-12
Gear lash
and pattern correction guide (DP-S
DuoProp [twin-propeller]
drives) 487-488
specifications 487
Gear ratios 417-418, 488
Gimbal
bearing and seal 496-497
ring 497-500

H

Harmonic balancer 231-233
High altitude operation 418
Hose replacement 332-333

I

Idle air control motor (IAC) 407-408
Idle speed control solenoid
(ISC) 408-409
Ignition system 386-397
resistance specifications 135
troubleshooting 50-65, 127-128
timing
adjustment 177-183
specifications 189-191
Inner transom plate 500-504
Intake air temperature
(IAT) sensor 405-406

Intake manifold 323-327

K

Knock sensor (KS) 409-410
  amplifier 411
  module 410-411

L

Lay-up 165-168
Lifter cover 205
Lower gearcase 449-484
  DP-S DuoProp (twin-propeller)
    drives 465-483
  installation 483-484
  removal 449-465
Lubrication. 146-151
  capacities. 162
  recommended lubricants 163

M

Maintenance
  after each use 144-146
  before each use. 137-143
  engine coolant 154-156
  fuel system. 151-154
  intervals 161-162
  torque specifications 161
Manifold air pressure sensor. . . . 404-405
Model identification. 3-4

O

Oil
  cooler (replacement) 340
  pan 223-225
  pressure
    sender. 400-401
    switch 401
  pump 225-231
    service specifications 268-270
  seal. 233-236

P

Piston
  and connecting rod. 245-254
  service specifications 270-271
Pivot housing 489-494
Power steering system 530-540
  fluid cooler (replacement). 340
  troubleshooting 98-99, 131
Power trim system
  electrical components. 505-513
  filling and bleeding 524-525
  hoses 517-523
  hydraulic components 513-517
  torque specifications 526
  trim
    cylinder and hoses 517-523
    position sender. 523-524
  troubleshooting 99-110, 131-133

Propellers 12-16, 420-422
  rotation. 419-420
Pushrod 214-221

R

Rear main seal and crankshaft. . . 254-261
Recirculation pump
  replacement 333-335
Replacement parts 193
Rocker arm. 214-221
  cover 200-201

S

Sacrificial anodes 418-419
Seawater pump. 335-340
Service precautions
  drive unit
    outdrive 418
  engine. 193-194
Shift cable adjustment 183-185
Shift housing service 446-449
Shifting 419-420
Shop supplies 8-10
Spark plug specifications 163-164
Specifications
  adjustments 188
  battery
    cable recommendations. 413
    capacity (hours). 413
    state of charge 413-414
  bearing preload (rolling torque) . . 487
  camshaft service. 265-266
  charging system 130-130
  connecting rod service 268-270
  conversion formulas. 33-34
  cooling system 133-134, 359
  crankshaft service. 268-270
  cylinder bore 271
  diagnostic trouble codes
    (all GM EFI models) 135-136
  drive unit
    available gear ratios. 488
    identification 488
    lubricant capacities 162
    torque. 486-487
  electrical system. 412-413
  electronic fuel injection
    (EFI) resistance 135
  engine
    firing order 414
    identification 36
    oil capacities 162
    operating speed 136, 189
    torque. 262-265
  exhaust system 359
  fuel
    pressure 135
    requirement 189-191
    system 128-130
  gauge settings for trim limit
    control module 191
  gear lash 487
    and pattern correction guide (DP-S
      DuoProp [twin-propeller]
      drives) 487-488
  ignition system. 127-128
    resistance. 135
    timing. 189-191
  lubricants. 163
  maintenance
    intervals 161-162
    torque 161
  oil pump service. 268-270
  piston service 270-271
  power steering system. 131
  power trim system 131-133
    torque 526
  spark plug. 163-164
  starting system 126-127
  steering system 540
  tap and drill sizes
    metric 35
    U.S. standard 34-35
  technical abbreviations. 32-33
  torque
    adjustments 188
    cooling system. 359
    drive unit 486-487
    electrical system. 412-413
    engine. 262-265
    exhaust system 359
    fuel system. 327-328
    maintenance specifications 161
    power trim system. 526
    recommendations. 35-36
    steering system 540
    transom bracket. 504
  transom bracket. 504
  valve service 267-268
  warning system 134
Starter
  and starter solenoid 368-386
    installation 369-371
    removal 369
  component cleaning and inspection
    384-386
  disassembly and assembly . . . 373-384
  relay 398
  slave solenoid. 397-398
  solenoid removal and
    installation 371-372
Starting system
  troubleshooting 39-46, 126-127
Steering system
  mechanical (manual) 528-530
  power. 530-540
  safety precautions 527-528
  torque 540
Submersion 160-161

T

Tap and drill sizes
  metric 35

14

Tap and drill sizes (continued)
- U.S. standard 34-35

Technical abbreviations 32-33
Thermostat. 329-332
Throttle
- body 301-308
- cable adjustment 185
- position sensor (TPS). 406-407

Timing chain and sprockets 236-237
Timing gear or chain cover 233-236
Tools
- basic. 16-19
- precision measuring 19-25
- special 19, 193

Top dead center (TDC) 204
Torque specifications
- adjustments 188
- cooling system. 359
- drive unit 486-487
- electrical system. 412-413
- engine. 262-265
- exhaust system 359
- fuel system 327-328
- maintenance 161
- power trim system. 526
- recommendations. 35-36
- steering system 540
- transom bracket. 504

Transom bracket
- bellows. 494-496
- bracket housing 500-504
- gimbal
  - bearing and seal 496-497
  - ring. 497-500
- inner transom plate and
  - bracket housing 500-504
- pivot housing 489-494
- torque specifications 504
- water hose 494-496

Trim
- cylinder 517-523
- electrical components. 505-513
- filling and bleeding
  - the system 524-525
- hoses 517-523
- hydraulic components 513-517
- limit control module
  - adjustment 185-188
  - gauge settings 191
- position sender 185-188, 523-524

Troubleshooting
- charging system 46-50, 130-130
- cooling system 110-114, 133-134
- diagnostic trouble codes. 93-97
  - GM EFI models 135-136
- drive system 125-126
- electronic fuel injection
  - (EFI). 75-93
  - fuel
    - delivery system 75-78
    - injection control system . . . 78-81
    - injector test. 91-93
    - pressure test 87-91
    - pump relay test. 83-87
  - relay test. 81-83
  - resistance specifications 135
- engine. 117-125
  - operating speed 136
- fuel system 65-75, 128-130
  - pressure 135
- ignition system 50-65, 127-128
  - resistance specifications 135
- power steering system 98-99, 131
- power trim system. . . 99-110, 131-133
- preliminary inspection 37-38
- starting system 126-127
  - electric 39-46
- warning system. 114-117, 134

Tune-up 156-160
- spark plug specifications 163-164

U

Universal joint service 444-446

V

Valves
- adjustment 201-204
- and valve seats 214
- service specifications 267-268

W

Warning system. 114-117, 134
Water hose and bellows 494-496
Winterizing 168-173
Wiring diagrams. 545-560

GM 3.0GL AND 3.0GS MODELS (WITH EARLY-STYLE TRIM PUMP AND SLAVE SOLENOID)

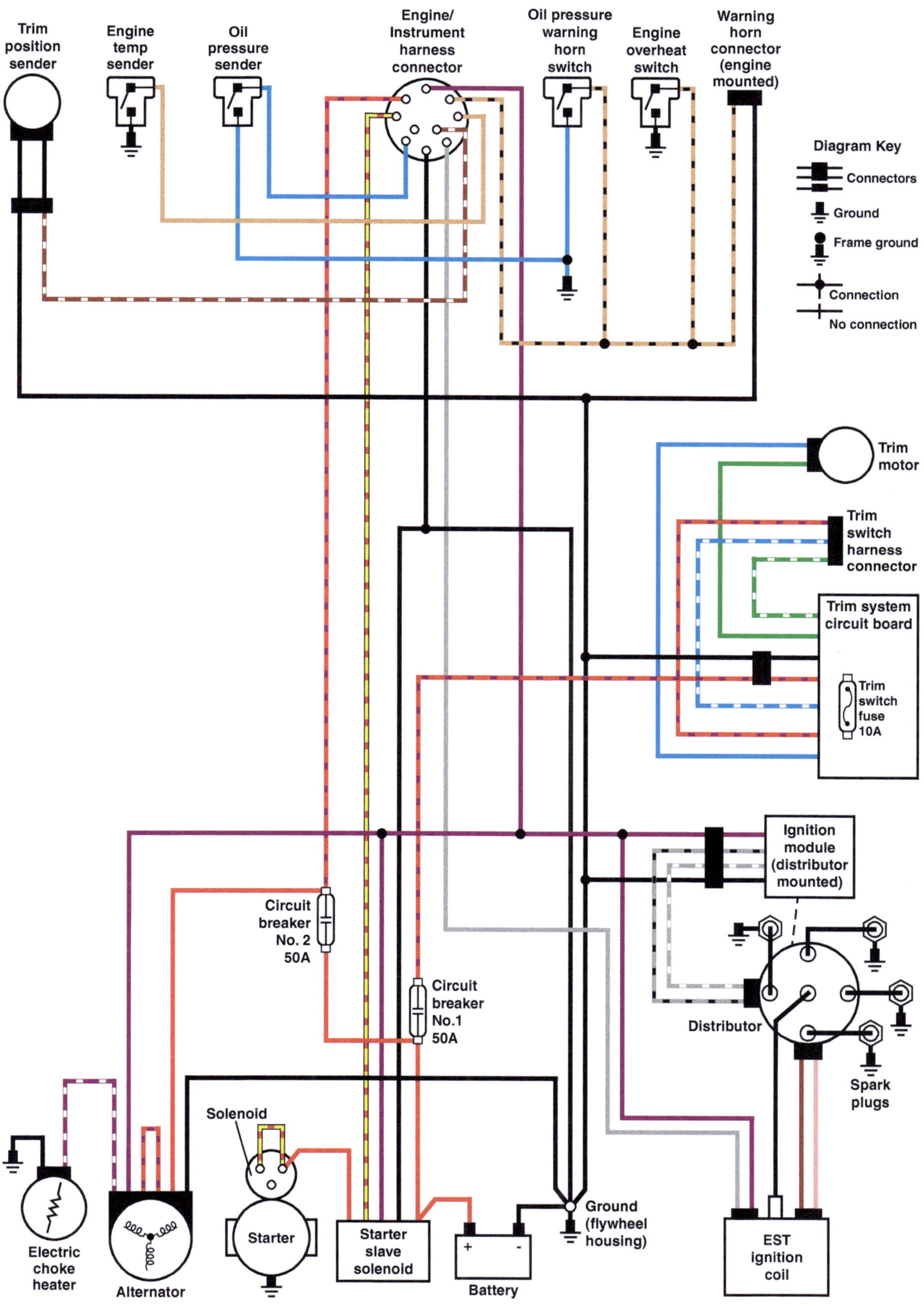

GM 3.0GL AND 3.0GS MODELS (WITH LATER-STYLE TRIM PUMP AND STARTER RELAY)

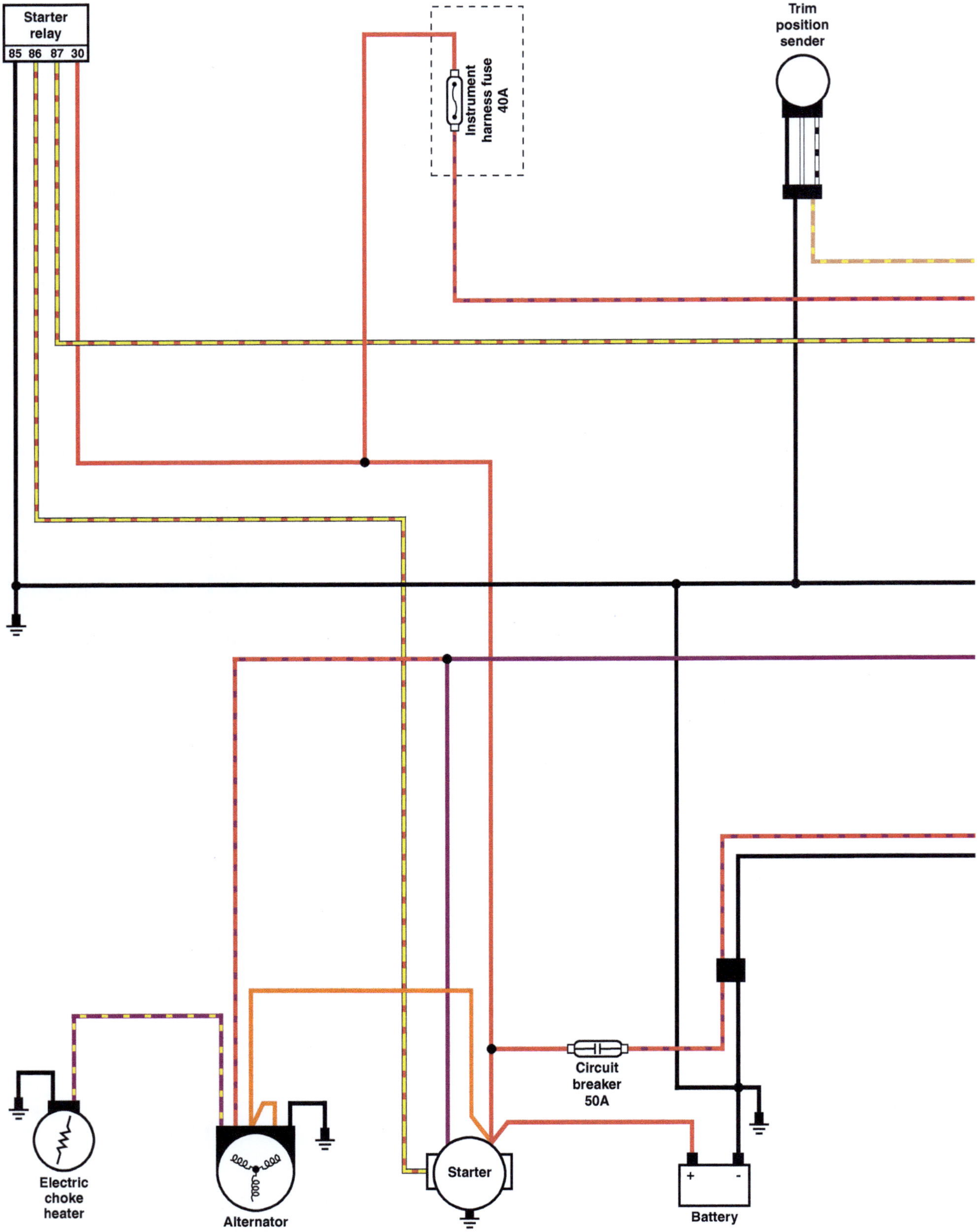

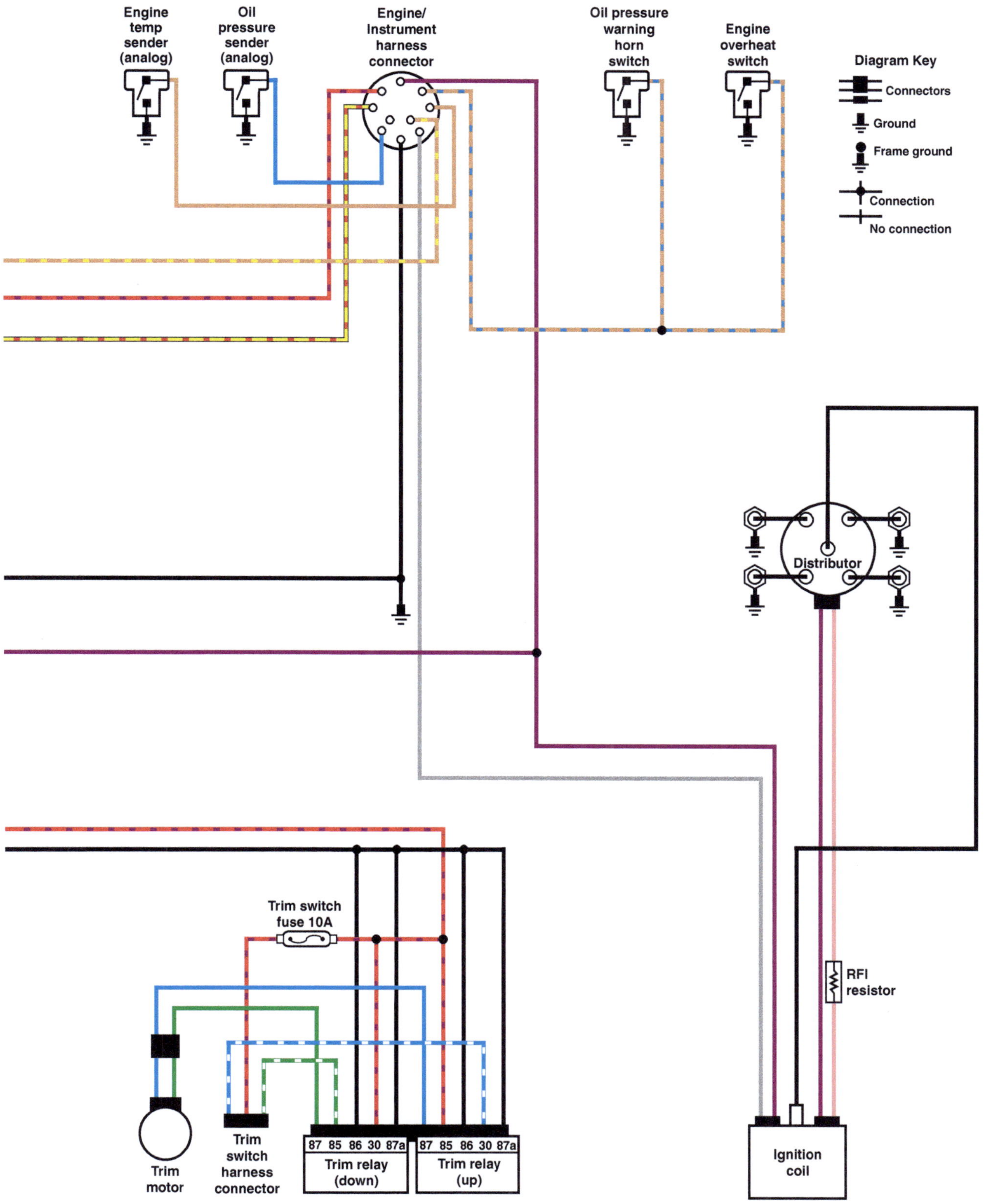
Engine temp sender (analog)
Oil pressure sender (analog)
Engine/ Instrument harness connector
Oil pressure warning horn switch
Engine overheat switch
Diagram Key
Connectors
Ground
Frame ground
Connection
No connection
Distributor
Trim switch fuse 10A
RFI resistor
Trim motor
Trim switch harness connector
87 85 86 30 87a
Trim relay (down)
87 85 86 30 87a
Trim relay (up)
Ignition coil

GM 4.3GL, 4.3GS, 5.0GL, 5.7GL AND 5.7GS MODELS (WITH DELCO EST IGNITION SYSTEM)

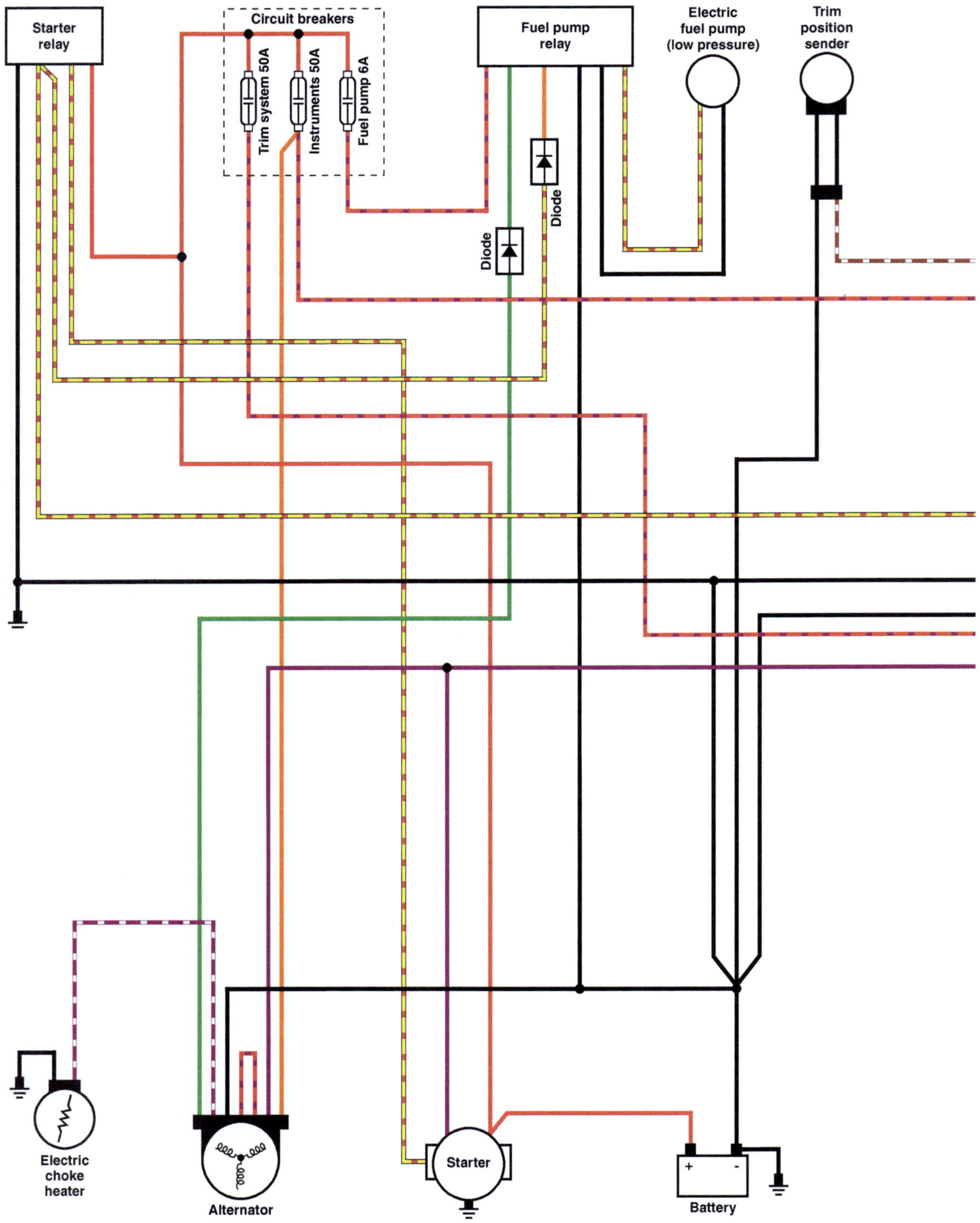

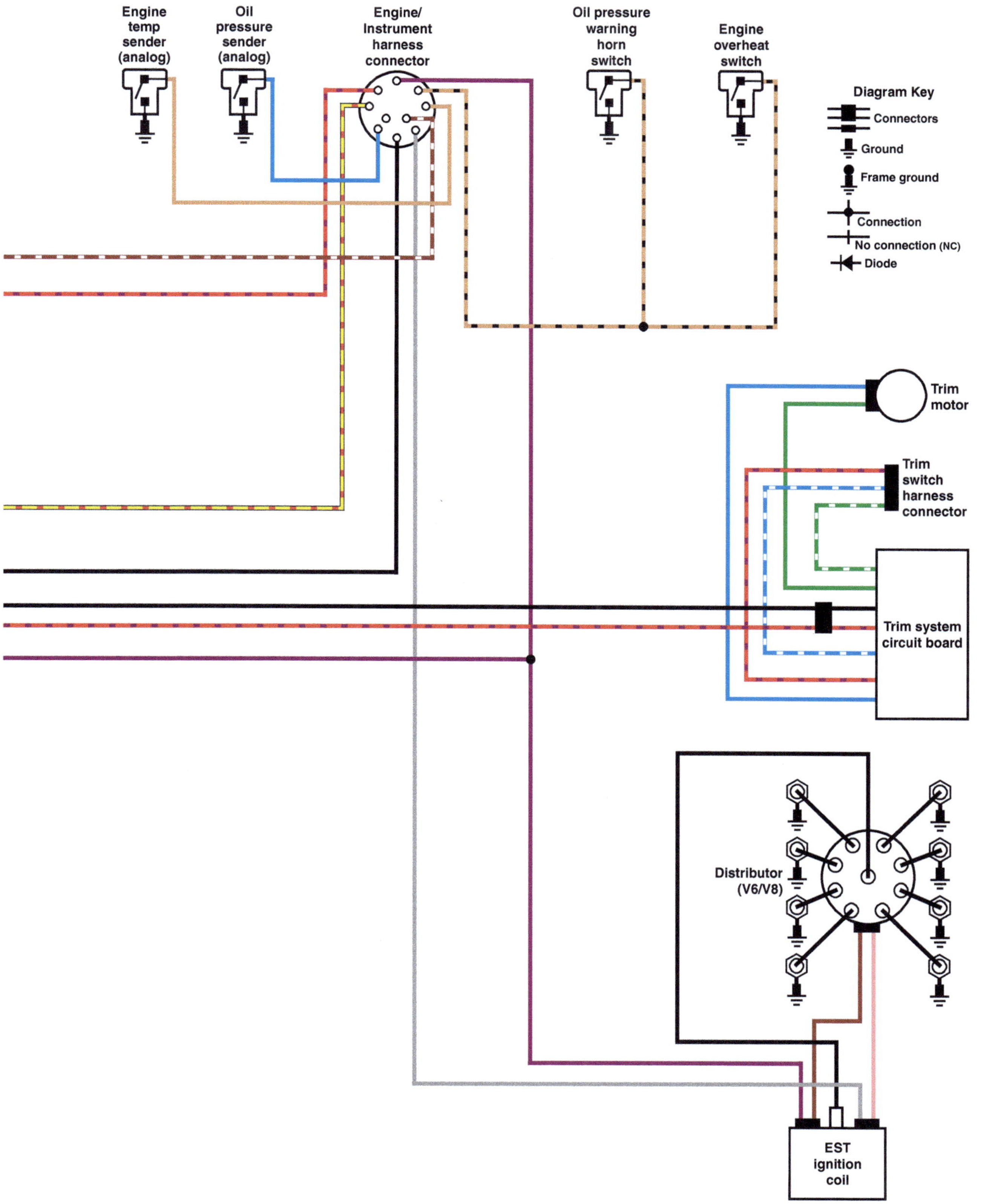

Engine temp sender (analog)
Oil pressure sender (analog)
Engine/ Instrument harness connector
Oil pressure warning horn switch
Engine overheat switch
Diagram Key
Connectors
Ground
Frame ground
Connection
No connection (NC)
Diode
Trim motor
Trim switch harness connector
Trim system circuit board
Distributor (V6/V8)
EST ignition coil

GM 4.3GI, 5.0GI AND 5.7GI MODELS (WITH THROTTLE BODY ELECTRONIC FUEL INJECTION)

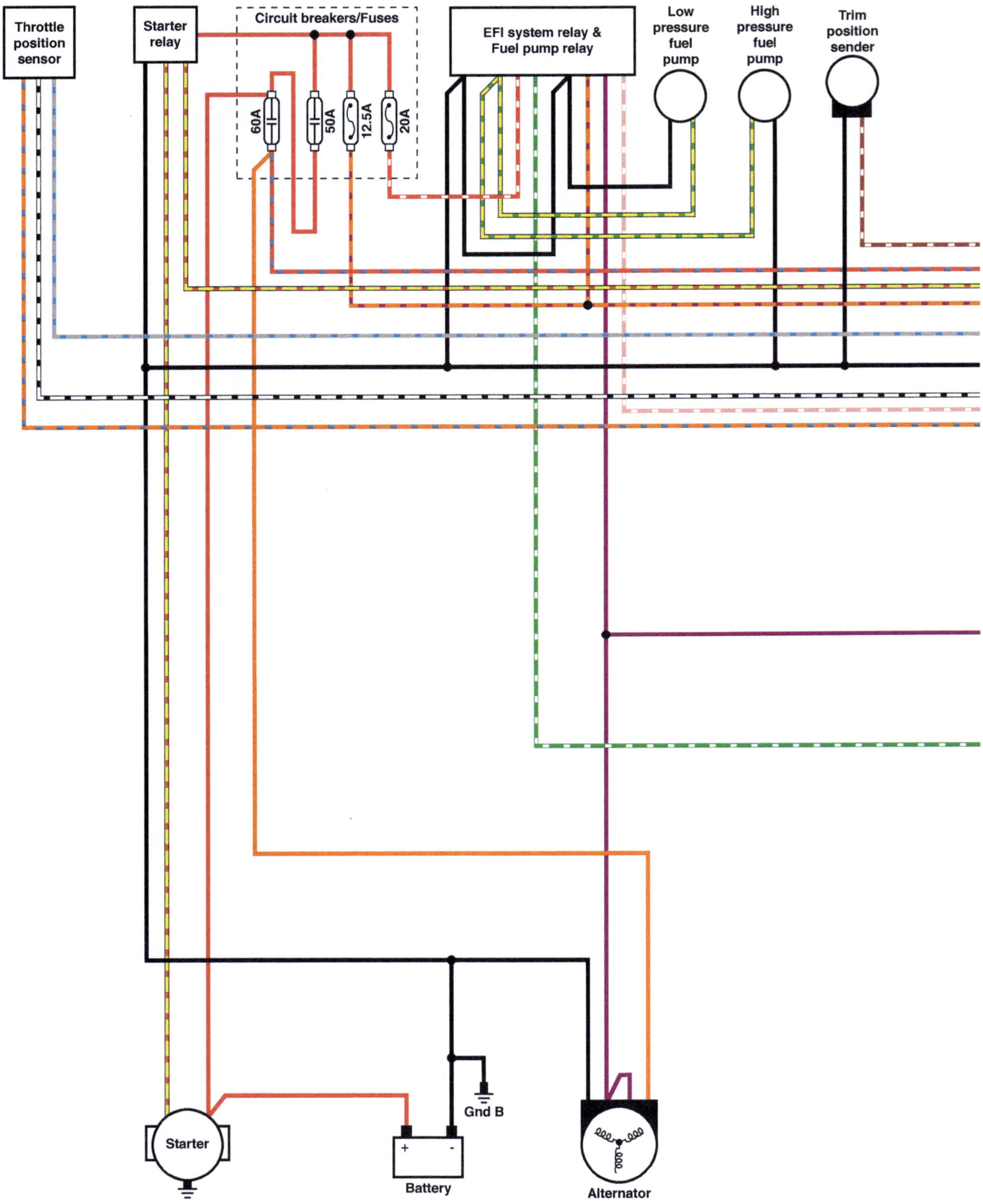

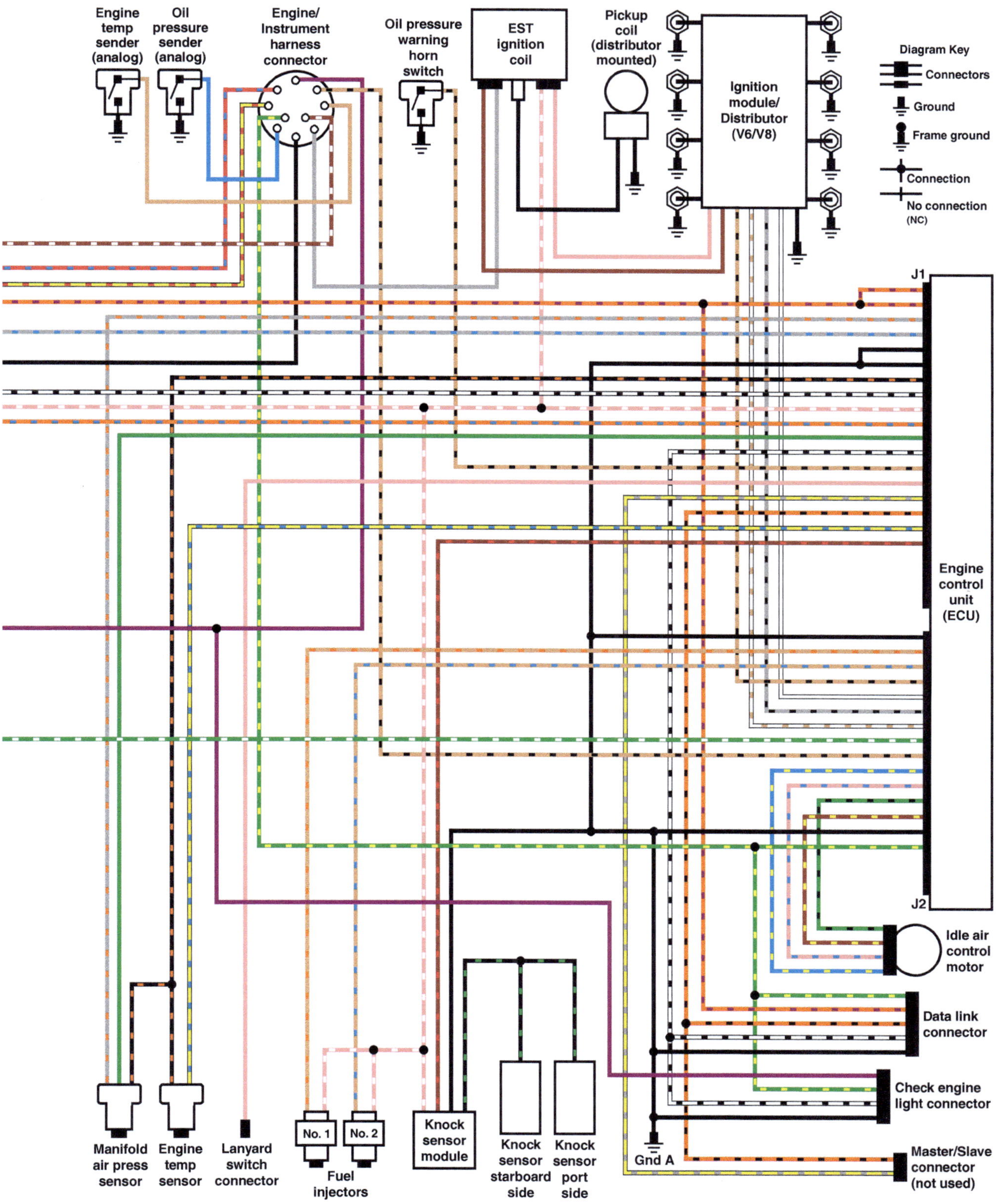

Engine temp sender (analog)
Oil pressure sender (analog)
Engine/ Instrument harness connector
Oil pressure warning horn switch
EST ignition coil
Pickup coil (distributor mounted)
Ignition module/ Distributor (V6/V8)
Diagram Key
Connectors
Ground
Frame ground
Connection
No connection (NC)
J1
Engine control unit (ECU)
J2
Idle air control motor
Data link connector
Check engine light connector
Master/Slave connector (not used)
Gnd A
Manifold air press sensor
Engine temp sensor
Lanyard switch connector
No. 1
No. 2
Fuel injectors
Knock sensor module
Knock sensor starboard side
Knock sensor port side

FORD 5.0FI, 5.8FI AND 5.8FSI MODELS (WITH MULTIPORT ELECTRONIC FUEL INJECTION)

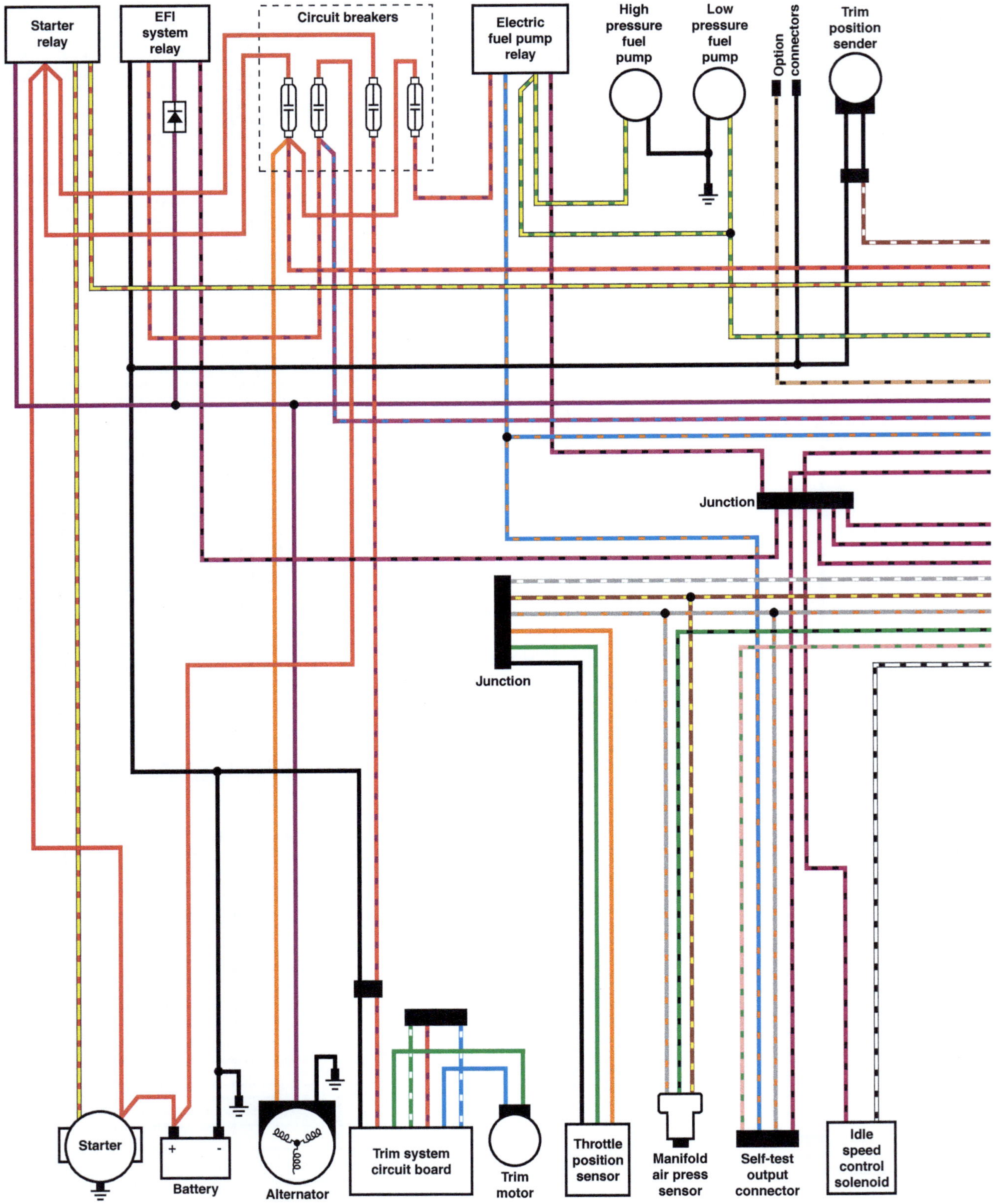

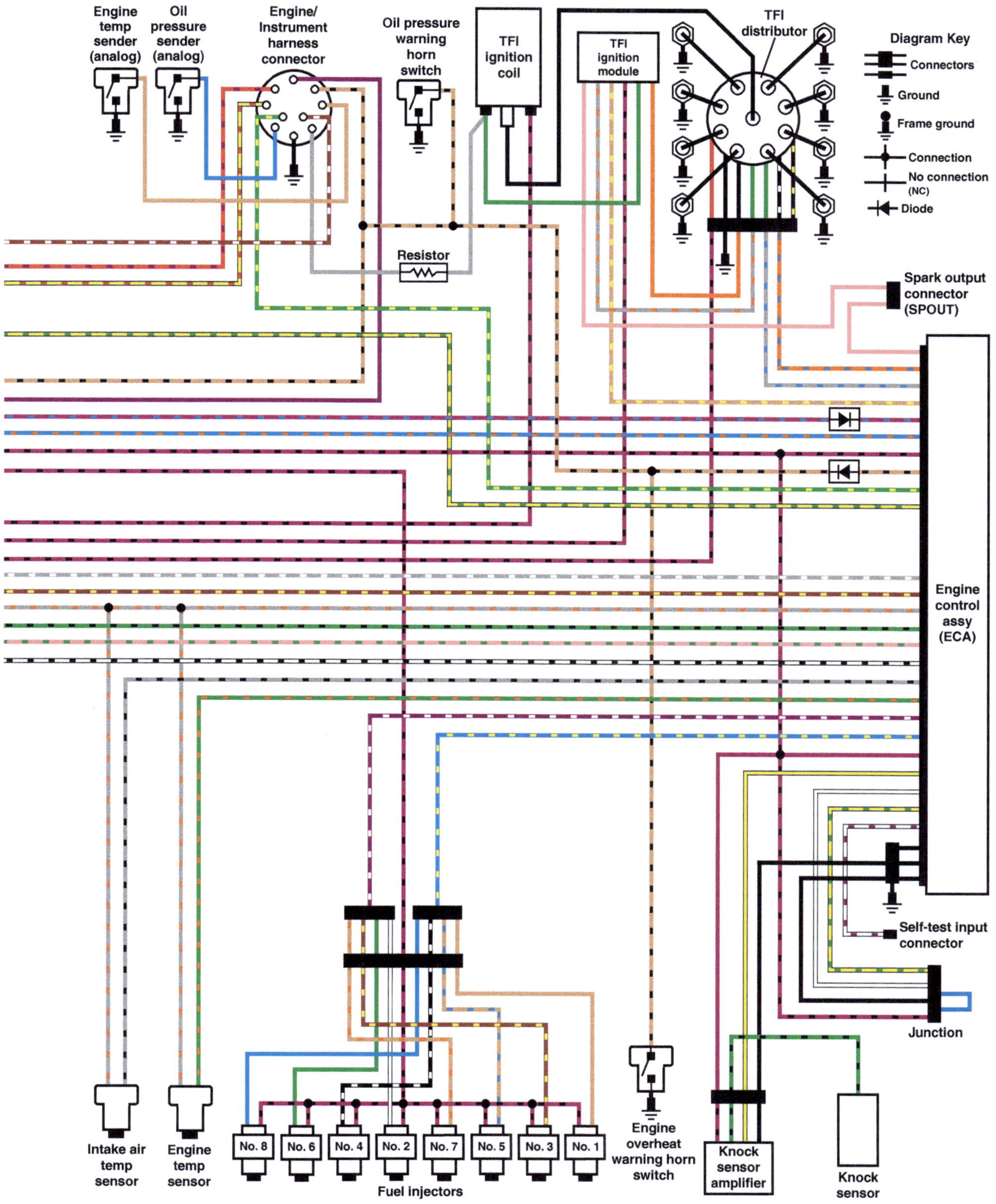
Engine temp sender (analog)
Oil pressure sender (analog)
Engine/ Instrument harness connector
Oil pressure warning horn switch
TFI ignition coil
TFI ignition module
TFI distributor
Diagram Key
Connectors
Ground
Frame ground
Connection
No connection (NC)
Diode
Resistor
Spark output connector (SPOUT)
Engine control assy (ECA)
Self-test input connector
Junction
Intake air temp sensor
Engine temp sensor
No. 8
No. 6
No. 4
No. 2
No. 7
No. 5
No. 3
No. 1
Fuel injectors
Engine overheat warning horn switch
Knock sensor amplifier
Knock sensor

GM 5.0GL, 5.7GL, 5.7GS, 7.4GL, 7.4GS, 8.2GL AND 8.2GS MODELS (WITH PRESTOLITE BID IGNITION SYSTEM)

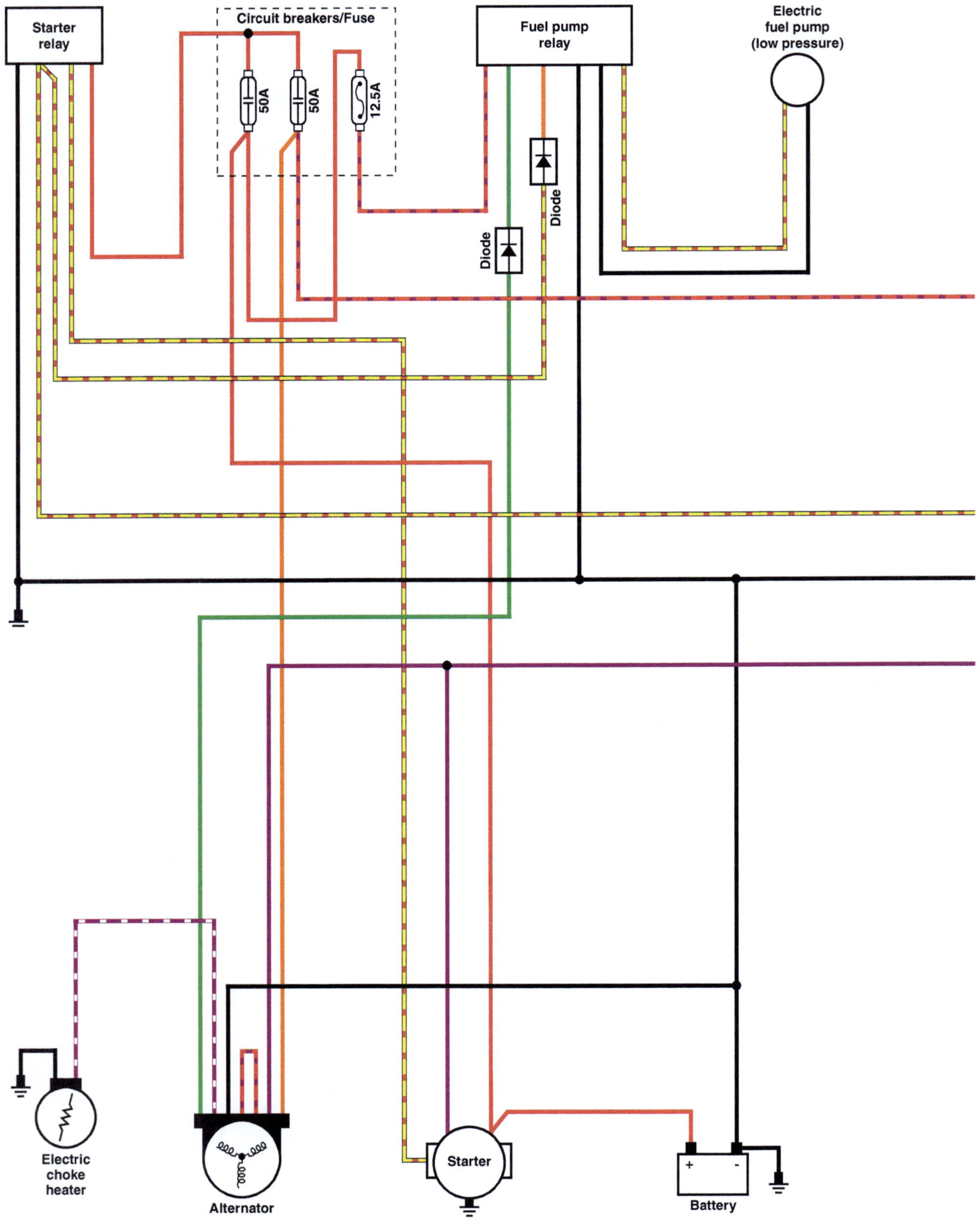

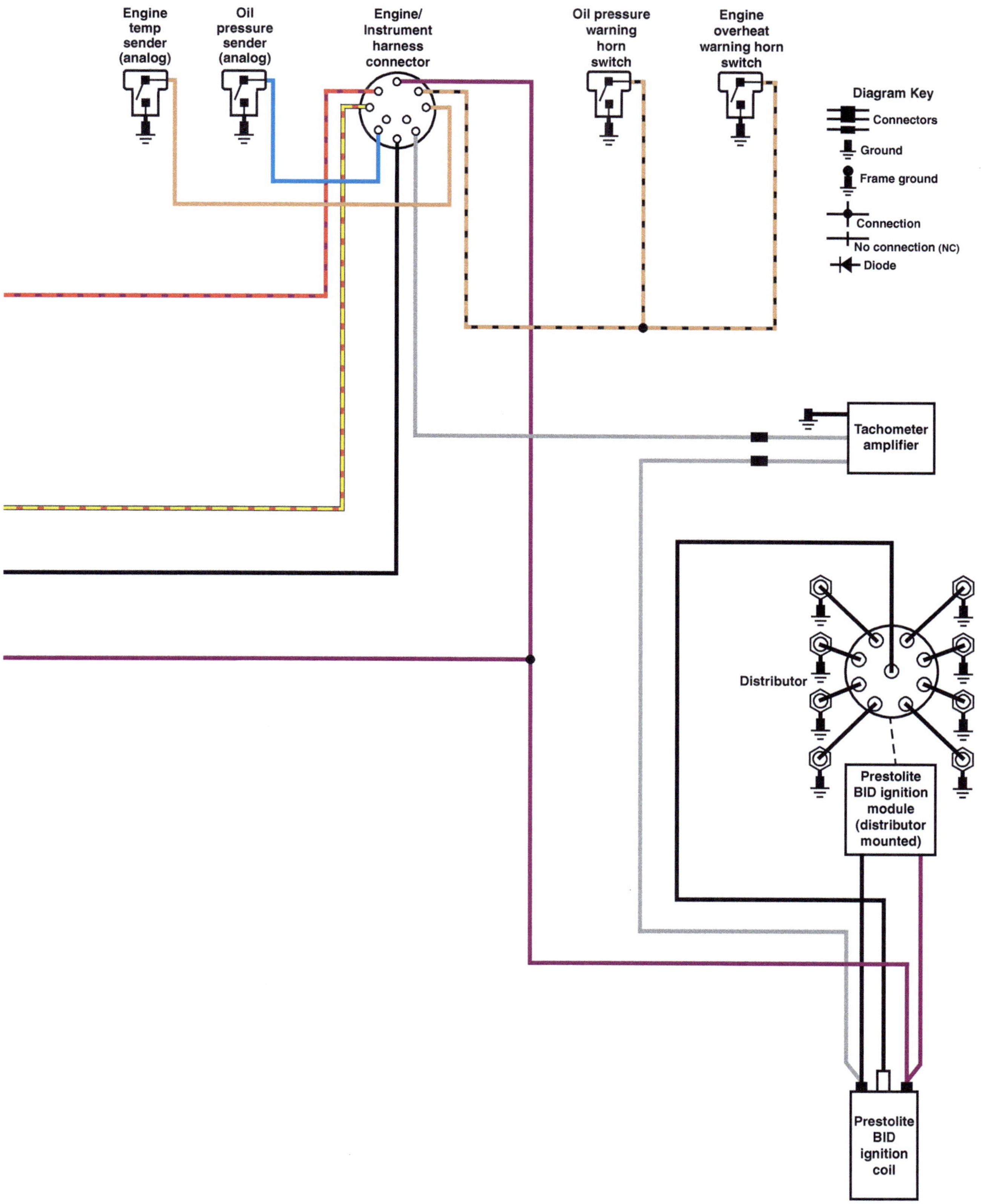
Engine temp sender (analog)
Oil pressure sender (analog)
Engine/ Instrument harness connector
Oil pressure warning horn switch
Engine overheat warning horn switch
Diagram Key
Connectors
Ground
Frame ground
Connection
No connection (NC)
Diode
Tachometer amplifier
Distributor
Prestolite BID ignition module (distributor mounted)
Prestolite BID ignition coil

GM 7.4GI, 7.4GSI AND 8.2GSI MODELS (WITH MULTIPORT ELECTRONIC FUEL INJECTION SYSTEM)

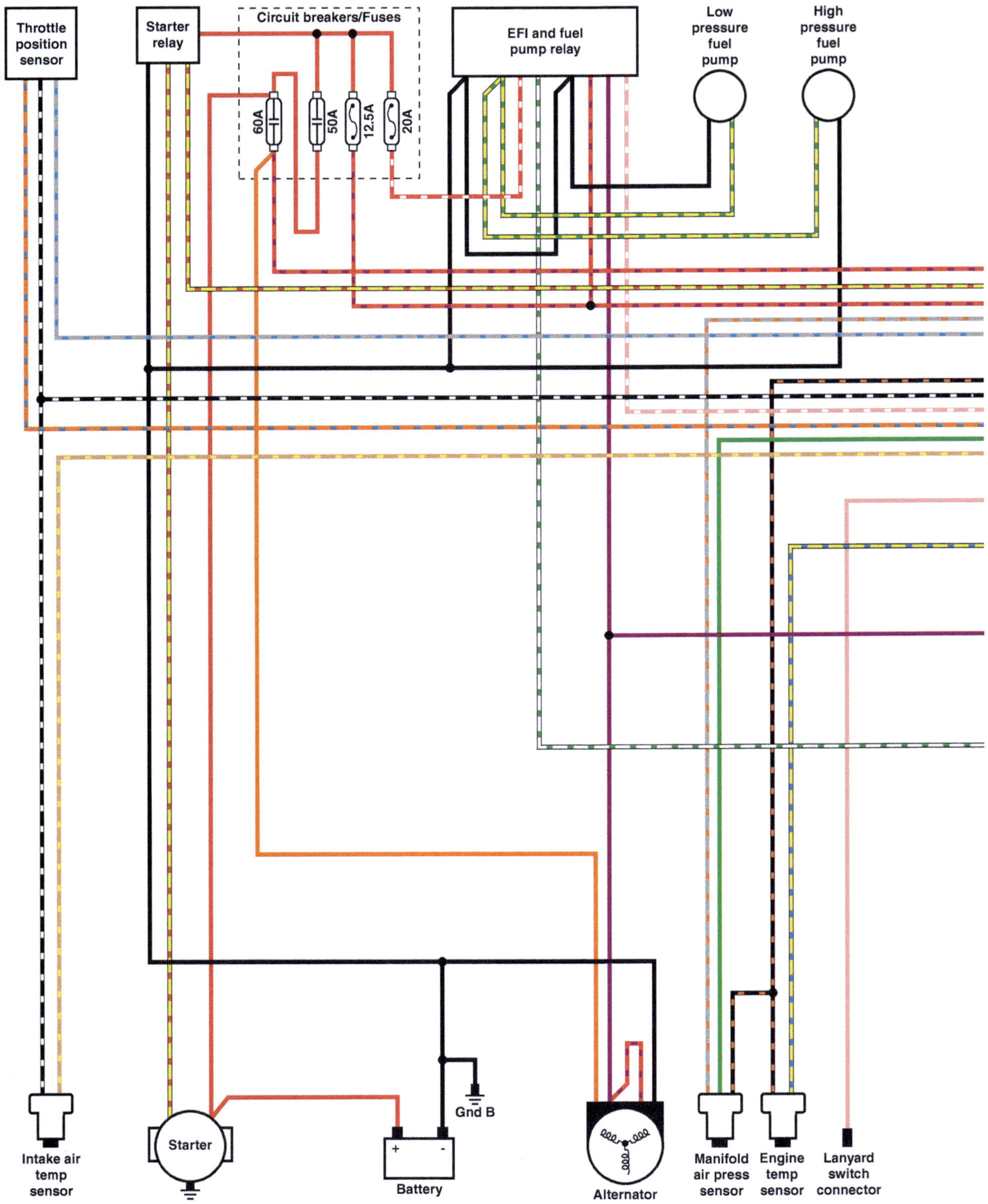

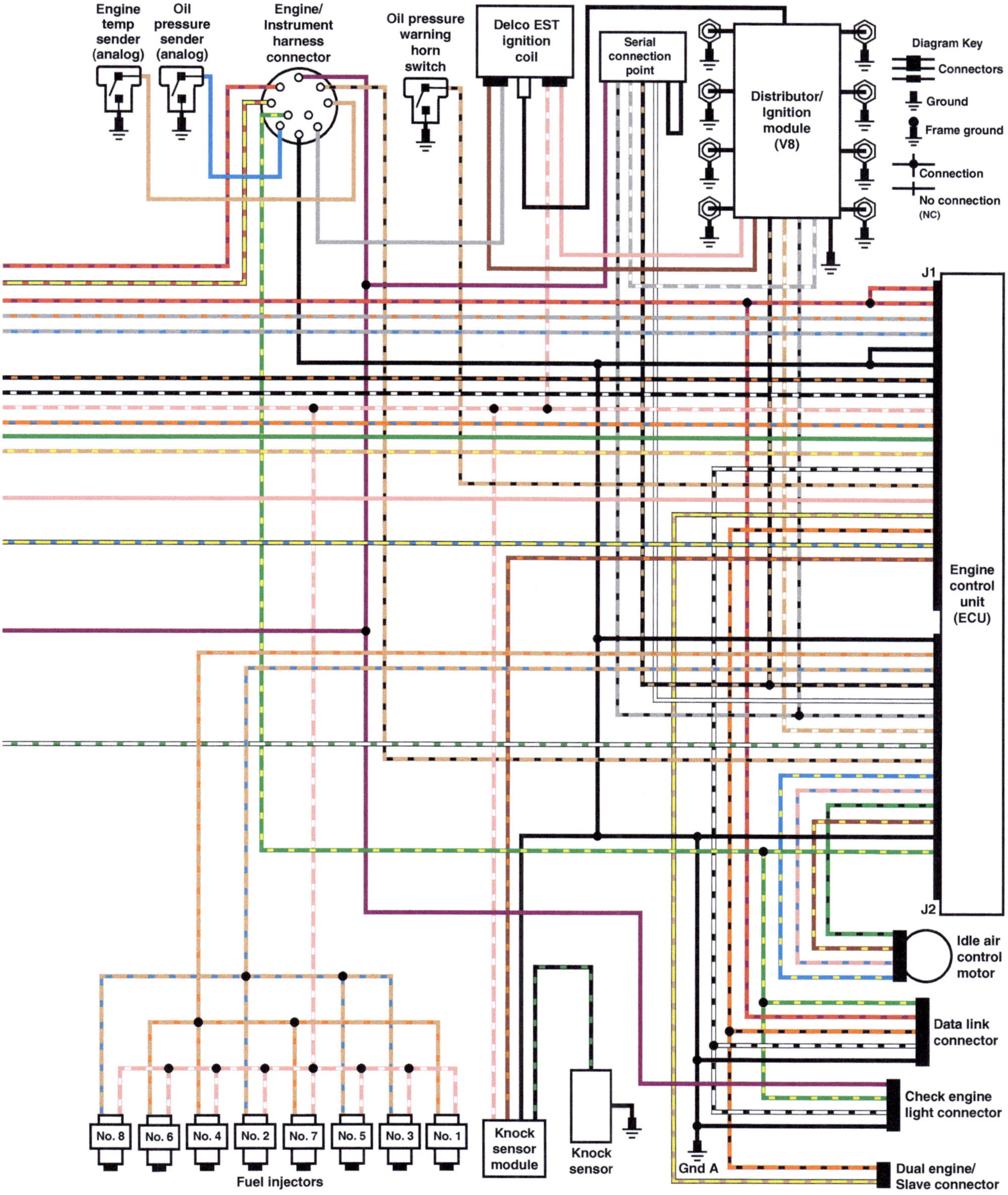
Engine temp sender (analog)
Oil pressure sender (analog)
Engine/ Instrument harness connector
Oil pressure warning horn switch
Delco EST ignition coil
Serial connection point
Distributor/ Ignition module (V8)
Diagram Key
Connectors
Ground
Frame ground
Connection
No connection (NC)
J1
Engine control unit (ECU)
J2
Idle air control motor
Data link connector
Check engine light connector
Dual engine/ Slave connector
Gnd A
Knock sensor
Knock sensor module
No. 8
No. 6
No. 4
No. 2
No. 7
No. 5
No. 3
No. 1
Fuel injectors

15

FORD 5.0FL AND 5.8FL MODELS

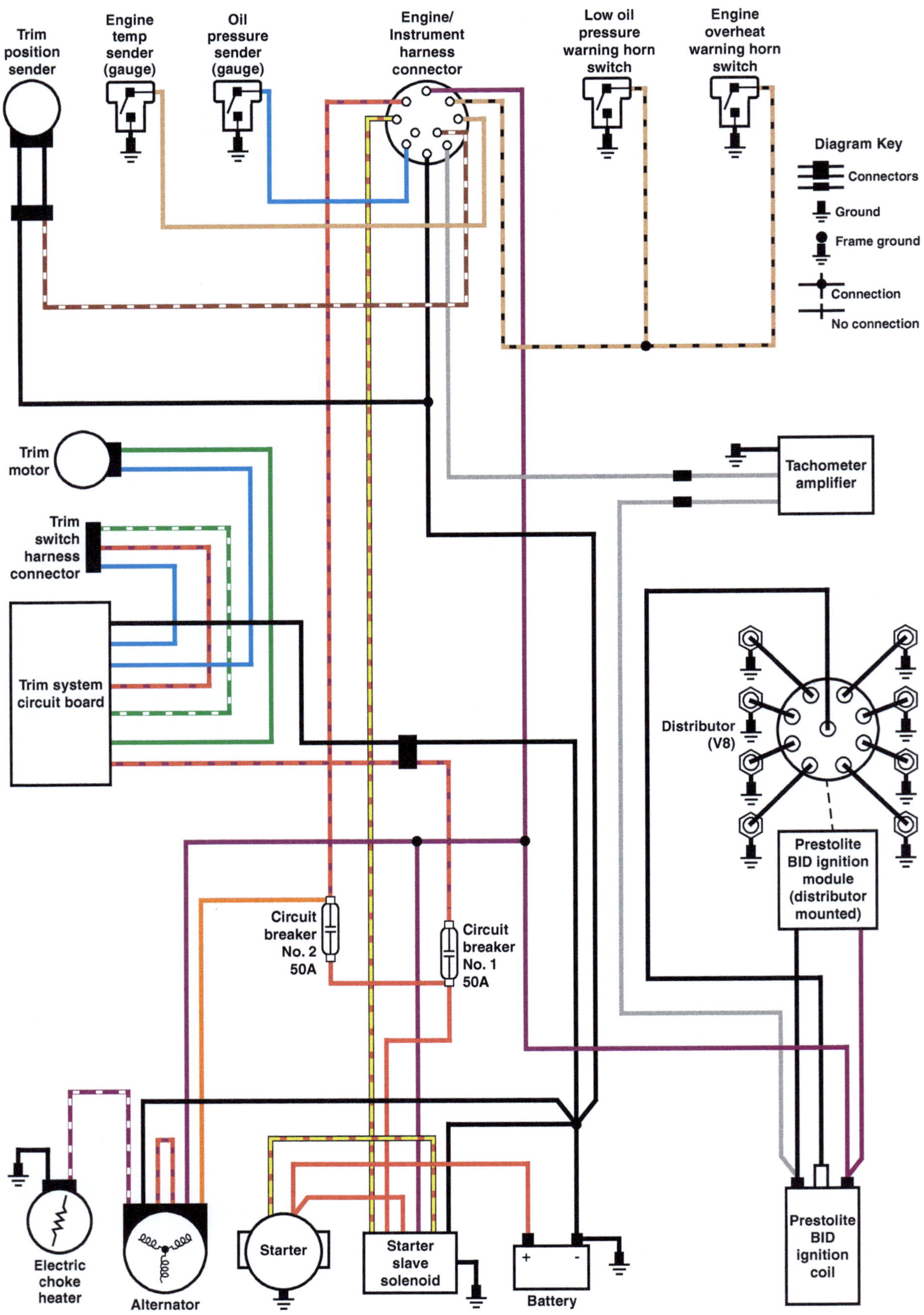

HYDRAULIC TRIM ELECTRICAL SYSTEM (SX DRIVE UNIT)

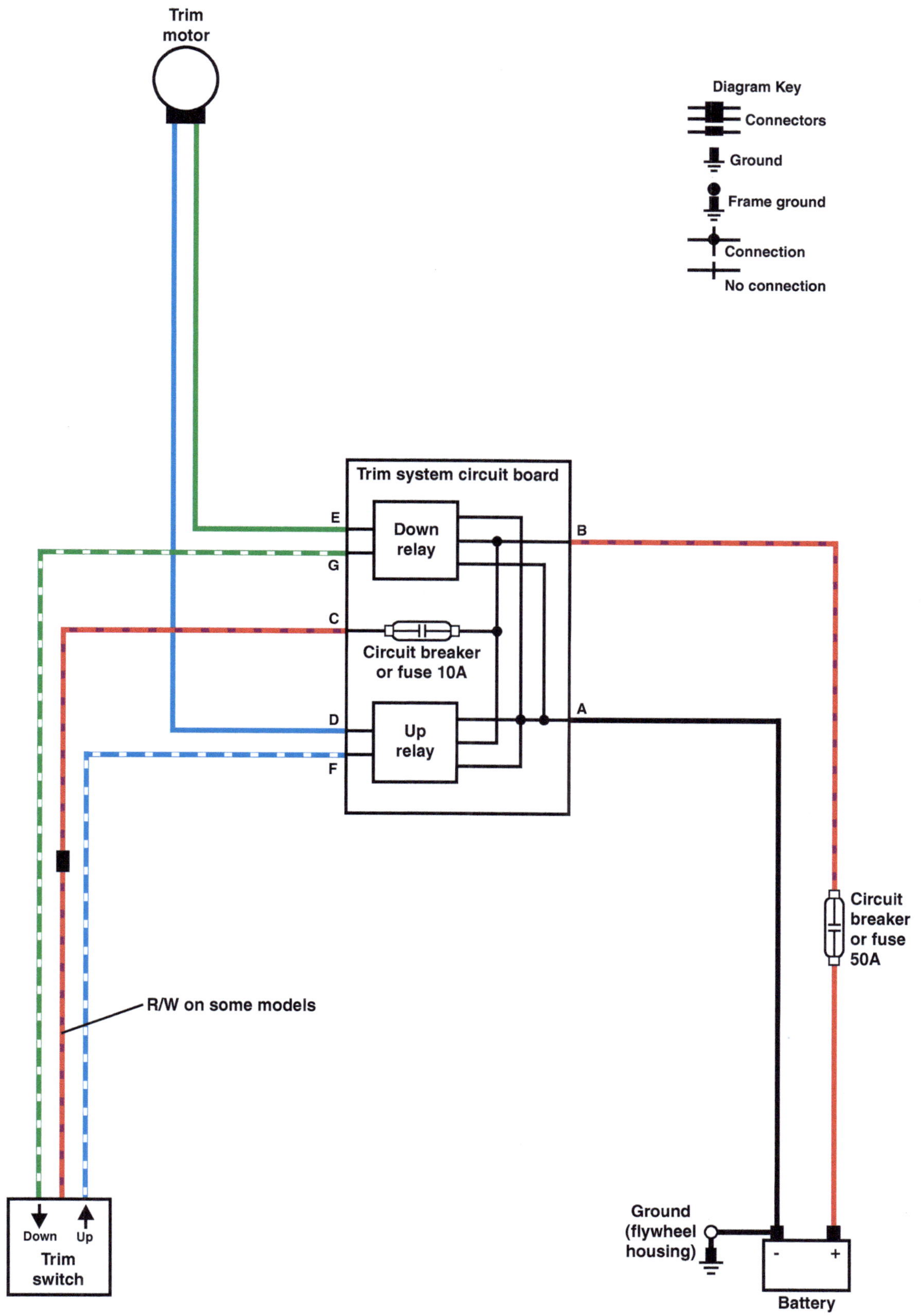

15

HYDRAULIC TRIM ELECTRICAL SYSTEM (DP-S DRIVE UNIT)

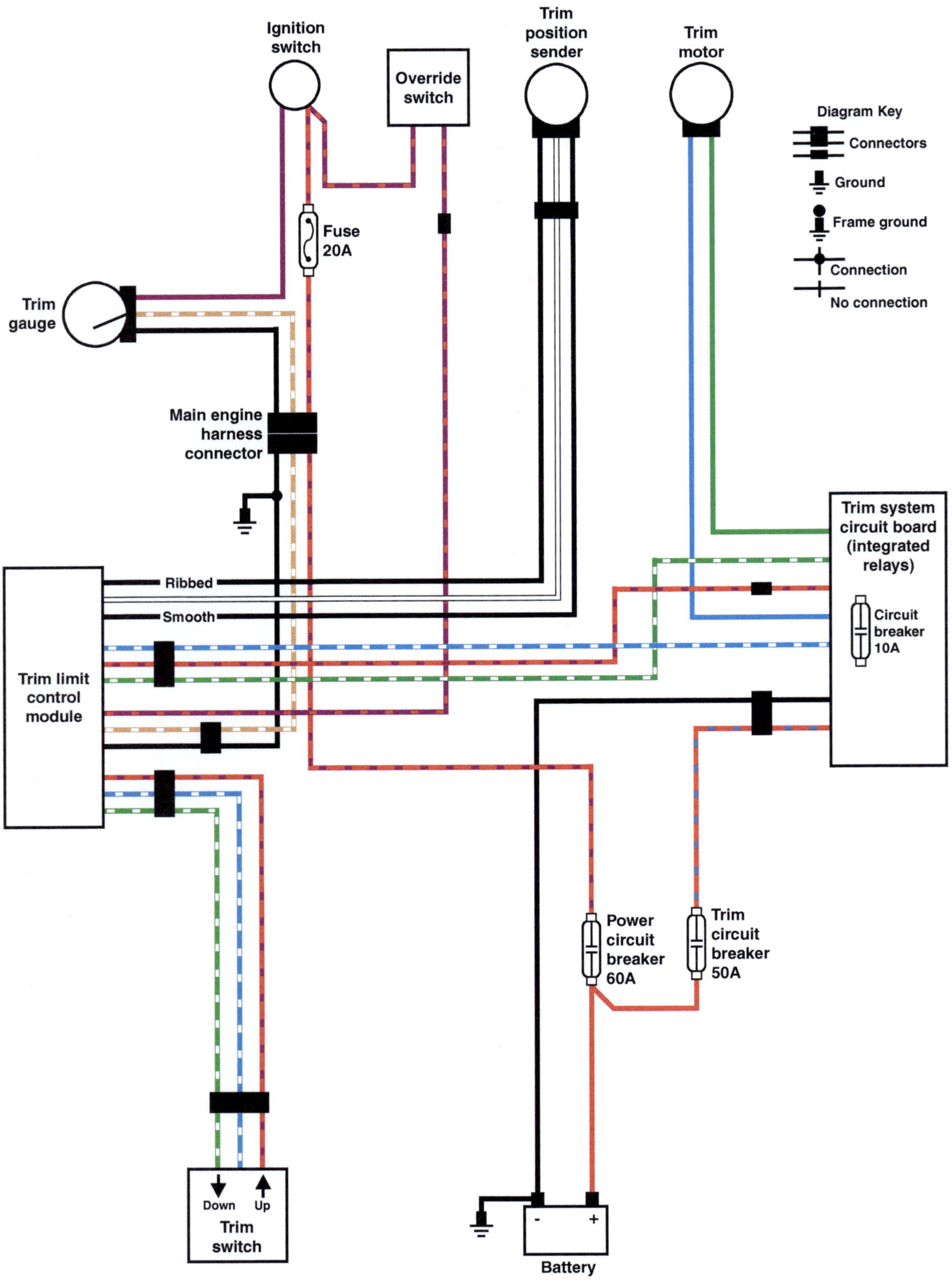